# GEOGRAPHY: REGIONS AND CONCEPTS

**To The Student:**
A Study Guide for the textbook is available through your college bookstore under the title *Study Guide to Accompany Geography: Regions and Concepts*, 4th edition, by Harm J. de Blij and Peter O. Muller. The Study Guide can help you with course material by acting as a tutorial, review, and study aid. If the Study Guide is not in stock, ask the bookstore manager to order a copy for you.

FOURTH EDITION

# GEOGRAPHY: REGIONS AND CONCEPTS





HARM J. DE BLIJ
University of Miami

PETER O. MULLER
University of Miami

JOHN WILEY & SONS
New York
Chichester
Brisbane
Toronto
Singapore

Text and Cover Designer: Rafael H. Hernandez
Production Manager: Rose Mary Scarano
Map Coordinator: Lisa Heard
Photo Researcher: Maura Grant



***Library of Congress Cataloging in Publication Data:***

de Blij, Harm J.
Geography, regions and concepts.

Includes indexes.
1. Geography—Text-books—1945– I. Muller,
Peter O. II. Title.
G128.D42 1985 910 84-13010
ISBN 0–471–88596–7

Printed in the United States of America

10 9 8 7 6 5 4 3

To our fathers
Hendrik de Blij and Hans Muller,
whose early lessons in geography we remember with affection

# PREFACE

The fourth edition of *Geography: Regions and Concepts* retains the overall structure of its predecessors, but it introduces substantial revisions and additions in the text and cartography. Among the major changes are the following: (1) a complete reorganization and rewriting of Chapter 3 on North America, (2) the addition of a Systematic Essay for each regional chapter to provide expanded coverage of topical human and physical geography, (3) the introduction of five Model Boxes to explore more fully the use of some familiar models in geography, (4) an expanded emphasis on urbanization throughout, beginning with a new section in the opening chapter, and (5) the appearance of several boxes to cover new concerns in geography (e.g., acid rain, remote sensing technology, and employment opportunities) as well as places that have become increasingly important on the world scene in the 1980s (e.g., Central America's spreading conflicts, the Falkland Islands issue, and the territorialization of Antarctica).

The general updating of the entire book, which preserves the innovations introduced in the third edition, has taken several forms. Stephen S. Birdsall of the University of North Carolina revised Chapter 7 on African Worlds, expanding certain topics and incorporating the Southern Africa section (a separate vignette in the previous edition) into the main body of the text. The glossary has been revised to reflect the updated and expanded coverage in each chapter. An extended box on map reading and interpretation has been added to the end of the introductory chapter. The world population table is now in the Appendix and the bibliographical listings at the end of each chapter are more inclusive than in earlier editions. To facilitate a clear understanding, the more basic concepts are introduced in boldface type; they are defined again in the glossary, which also includes brief definitions of other concepts as well as certain regional terminology. With the exception of two historic news photographs, this fourth edition has become an all-color volume.

Every effort was made to collect the newest population data. For national populations, the chief source was *World Population 1983: Recent Demographic Estimates for the Countries and Regions of the World*, a publication of the U.S. Census Bureau's Center for International Research issued less than one month before we went to press; 1985 projections are used in the text, and 1984 data are included in the complete table in the Appendix. The source of urban population figures, which involve a far greater problem in reliability and comparability, was *Patterns of Urban and Rural Population Growth*, a 1980 publication (No. 68) of the United Nations' Department of International Economic and Social Affairs; 1985 estimates are used throughout the text, and they represent *metropolitan-area totals* unless otherwise indicated. The demographic maps employed in the text could not always be updated to display the latest available population data, but they are quite reasonably consistent with the most current data compiled in the Appendix table.

Besides the aims that have characterized this book since its initial publication in 1971, we have been guided by an additional purpose during the preparation of this edition: increasing the international awareness of U.S. college students. In recent years, a great deal has been said and written about the inadequate international knowledge of the American people. We feel that the study of world regional geography is an ideal way to combat that ignorance.

To the student reader about to embark on the exploration of world geography, we leave you with the following exhortations offered by James Michener in his 1970 article in *Social Education*:

> *The more I work in the social-studies field the more convinced I become that geography is the foundation of all. . . . When I begin work on a new area—something I have been called upon to do rather frequently in my adult life—I invariably start with the best geography I can find. This takes precedence over everything else, even history, because I need to ground myself in the fundamentals which have governed and in a sense limited human development. . . . If I were a young man with any talent for expressing myself, and if I wanted to make myself indispensable to my society, I would devote eight or ten years to the real mastery of one of the earth's major regions. I would learn languages, the religions, the customs, the value systems, the history, the nationalisms, and* above all the geography [emphasis added], *and when that was completed I would be in a position to write about that region, and I would be invaluable to my nation, for I would be the bridge of understanding to the alien culture. We have seen how crucial such bridges can be* (pp. 764, 765–766).

# ACKNOWLEDGMENTS

In the course of this latest revision we were fortunate to receive advice and assistance from many individuals.

One of the rewards associated with the publication of a book of this kind is the correspondence that it generates. Over the years we have heard from colleagues, students, and lay readers. Geographers, political scientists, economists, and others have written us, almost always with helpful suggestions, often with fascinating enclosures. We have responded personally to every such letter, and our editor has communicated with many of our correspondents. We have, in addition, considered every suggestion made—and many who wrote or transmitted their reactions through other channels will see their recommendations in print in this edition. The list that follows is merely representative of a group of colleagues and students to whom we owe a special debt:

Randall Anderson (Emporia State University)
Jeane Bagwell and staff (Auburn University)
Marvin Baker (University of Oklahoma)
Melvin C. Barber (Memphis State University)
Bill Bishop (Georgian Court College)
Dennis Bozyk (Madonna College)
J. William Bradfield (Ohio State University at Mansfield)
Robert H. Brown (University of Wyoming)
Ken Calbeck (Mesa Community College)
William Clark (James Madison University)
G. Cosway (Howard Community College)
John Cross (University of Wisconsin-Oshkosh)
Claud Davidson (Texas Tech University)
Fred Day (Ohio State University)
Harold M. Elliott (Weber State College)
Roger E. Ervin (California State University, Fresno)
L. E. Estaville (University of Wisconsin-Oshkosh)
David Firman (Towson State College)
Fred Fryman (University of Northern Iowa)
John S. Haupert (SUNY Buffalo)
David Hendrickson (Fresno City College)
Vera L. Herman (Ohio State University)
Sam Hilliard (Louisiana State University)
George W. Hoffman (University of Texas at Austin)
James C. Hughes (Slippery Rock University)
Kenneth Israel (Florida Junior College, South Campus)
Herbert L. Jacobson (Escazú, Costa Rica)
Robert B. Johnson (California State University, Dominguez Hills)
John Kimura (California State University, Long Beach)
Max Kirkeberg (San Francisco State University)

Robert D. Klingensmith (Ohio State University at Newark)
John Ladd (Colorado School of Mines)
R. S. Langley (University of Connecticut)
William Leonard (Austin Community College)
Gordon Levine (University of Minnesota, Duluth)
Gordon Lewthwaite (California State University, Northridge)
Jonathan Lu (University of Northern Iowa)
Bernard McGonigle (Community College of Philadelphia)
Jesse O. McKee (University of Southern Mississippi)
Robert Monaghan (Meramec Community College)
James L. Mulvihill (California State University, San Bernardino)
Douglas C. Munski (University of North Dakota)
George N. Nasse (California State University, Fresno)
Midori Nishi (California State University, Los Angeles)
Gordon Oosterman (Calvin College)
Paul E. Phillips (Fort Hays State University)
Milton Rafferty (Southwest Missouri State University)
Roger Reede (Southwest [Minnesota] State University)
Jimmy Rogers (Marshall University)
Gregory Rose (Ohio State University at Marion)
K. M. A. Rouf (Jagannath College, Bangladesh)
Howard Salisbury (Northern Arizona University)
Robert A. and Rose M. Sauder (University of New Orleans)
Cyril Sawyer (University of Florida)
James T. Scofield (College of the Sequoias)
Lucy Spencer (Texarkana College)
David C. Streich (Carrboro, N.C.)
Malcomb Thomas (Laredo Junior College)
H. L. Throckmorton (San Diego Mesa College)
Jay Vanderford (Western Oregon State College)
Canute Vander Meer (University of Vermont)
Bob J. Walter (Ohio University)
Stuart White (University of New Mexico)

We also wish to single out two readers for special mention: Laurence Wolf of the University of Cincinnati for his meticulous observations on our maps and Gene Wilken of the Department of Economics at Colorado State University who, together with several of his students, identified many places where the written text could be improved.

A number of commissioned reviewers provided valuable and perceptive comments that were helpful in revising the regional chapters, and they will recognize many of their suggestions in print. They are William James Acker (Arizona State University), Thomas Anderson (Bowling Green State University), Patricia Humbertson (Youngstown State University), Roland Fuchs (University of Hawaii), Allen D. Bushong (University of South Carolina), Clifton Pannell (University of Georgia), John C. Lewis (Northeast Louisiana University), and Brian J. Murton (University of Hawaii). Besides these reviewers—whom we collectively thank here—the list also includes Stephen S. Birdsall (University of North Carolina), Roland Chardon (Louisiana State University), Truman Hartshorn (Georgia State University), Donald Janelle (University of Western Ontario), Henry Michael (the University Museum, University of Pennsylvania), and Charles Sargent (Arizona State University). The errors that remain are, of course, ours alone.

At the University of Miami, our colleagues Thomas Boswell, Donald Capone, Ira Sheskin, and Felipe Prestamo (Planning) helped us resolve a number of questions in the rewriting of certain chapters. A variety of important supporting tasks were cheerfully and tirelessly performed by our office staff, Esther Nedelman (coordinator), Alina Cruz, and Wendy Lewengrub—as well as Lisa Burrus in the office of the Department of Anthropology. The manuscript was typed by Kathy Kelleher of the College of Arts and Sciences' Word Processing Center, and we also thank Patty Wilson who supervises the Center, Judy Thompson of the Dean's Office, and Richard G. Banks—our special friend and honorary geographer who retired as Assistant Dean in 1983. Two graduate students, Donald Rallis and Alicia Goicoechea, provided observations on their native cities of Johannesburg and Caracas, respectively, and Donald supplied the slide that became Figure 7–34.

At the Colorado School of Mines, where the work on this new edition began, Mrs. Helen Mogensen laid the groundwork for the project, and we acknowledge her work with gratitude. Back in Miami, Ralph Clem, a geographer on the faculty of Florida International University, kindly provided publications on the Soviet Union. And Alan C. G. Best of Boston University very graciously supplied us with some last-minute slides.

In the quest for the latest population data and maps, we are most grateful for the materials and excellent service we received from Betty Adamek, Sylvia Quick, and Les Solomon of the U.S. Bureau of the Census. In our enthusiasm to find the best collection of photographs, we got personally involved in the search process and wish to thank Barbara Shattuck of the National Geographic Society as well as several commercial photographers and their agencies who patiently dealt directly with us.

From start to finish, the preparation of this fourth edition has benefited from the outstanding professionalism of the John Wiley team. The entire project was expertly guided by Geography Editor Katie Vignery, who was constantly willing to go to extra lengths to achieve the best possible results while making us feel as though we were the only authors with whom she was working. The production process was again directed by Rosie Scarano, with an efficiency and maintenance of quality standards that were remarkable, given the abundant details and complexities that accompany the assembling of this now full-color volume in barely six months. Rafael Hernandez was responsible for the design of this handsome final product. We are also greatly indebted to Map Coordinator Lisa Heard for her many efforts on behalf of the cartographic additions and revisions. Maura Grant smoothly handled the multiple tasks associated with the photography program. Andrew Yockers capably edited the manuscript, giving careful attention to every detail; Kenneth Kimmel expeditiously managed the editing, and good-naturedly tolerated the fussiness of the authors he encountered. Others on the staff of John Wiley & Sons who supported us in various ways include Tom Gay, Ray O'Connell, and especially Beth Meder, who so diligently assisted the Geography Editor on our behalf; we also salute Jim Dewberry, who represents the publisher in South Florida, for his many courtesies.

Finally, we owe an enormous measure of gratitude to our wives—Bonnie and Nancy—for their encouragement and editorial advice.

**Harm J. de Blij**
**Peter O. Muller**

*Coral Gables, Florida*
*April 17, 1984*

# CONTENTS


## INTRODUCTION
## REGIONAL GEOGRAPHY OF THE WORLD 1

Regional Concepts and Classifications
Geographic Scale
Concepts of Culture and Landscape
Changing Natural Environments
  Pleistocene Influences
  Water Cycles
  Climatic Regions
  MODEL BOX 1 THE HYPOTHETICAL CONTINENT
  Vegetation Regimes
  Soil Distribution
World Population Patterns
Politics and Geography
Geography of Economic Development
Geographic Realms of the World
Regional Studies in Contemporary Geography
Map Reading and Interpretation

## PART ONE
## DEVELOPED REGIONS 53

### Chapter 1
### THE MOSAIC OF EUROPE 55

SYSTEMATIC ESSAY 1: Population Geography
Landscape and Rivalries
Heritage of Order
Empires, Rebirths, and Revolutions
Modern Geographic Dimensions
MODEL BOX 2 THE VON THÜNEN MODEL
Problems of Economic Decline and Political Fragmentation
Regions of Europe
  The British Isles
  Western Europe
  Nordic Europe
  Mediterranean Europe
  Eastern Europe
European Supranationalism and Unification

**AUSTRALIA: A European Outpost**

Migration and Transfer
Economic Activities and Urbanization
Population Policies
Politico-Geographical Structures
New Zealand

### Chapter 2
### THE SOVIET UNION: REGION AND REALM 139

SYSTEMATIC ESSAY 2: Climatology
A World Superpower

The European Heritage
Physiography
The Modern Soviet State and Its Centrally Controlled Economy
Regions of the Soviet Realm
Soviet Heartlands and Boundaries
The Changing Geography of the 1980s

Chapter 3
NORTH AMERICA: THE POSTINDUSTRIAL TRANSITION 173

SYSTEMATIC ESSAY 3: Urban Geography
Two Highly Advanced Nation-States
Physical Geography and Human Environmental Impacts
Population in Time and Space
Settling Rural America
Industrial Urbanization
MODEL BOX 3 CENTRAL PLACE THEORY
Contemporary Postindustrial America
Cultural Geography
Environmental Perception and Spatial Behavior
Changing Geography of Economic Activity
The Postindustrial Revolution
Regions of the North American Realm
The Emerging "Nine Nations" of North America

**PRODIGIOUS JAPAN: The Aftermath of Empire**

Imperial Japan
Modernization in the Context of Limited Assets
Japan's Spatial Organization
Japan in the Postindustrial Era

PART TWO
UNDERDEVELOPED REGIONS 245

Chapter 4
MIDDLE AMERICA: COLLISION OF CULTURES 247

SYSTEMATIC ESSAY 4: Historical Geography
Legacy of Mesoamerica
Collision of Cultures
Mainland and Rimland
Political Differentiation
Caribbean Patterns
Tourism: The Irritant Industry
The African Heritage
Troubled Mexico
Central America's Besieged Republics

Chapter 5
SOUTH AMERICA AT THE CROSSROADS 279

SYSTEMATIC ESSAY 5: Economic Geography
The Human Sequence
Culture Areas
Urbanization
The Republics: Regional Geography
The Caribbean North
The Andean West
The Mid-latitude South

**EMERGING BRAZIL**

Regions
Population Dynamics
Development Problems

Chapter 6
NORTH AFRICA AND SOUTHWEST ASIA 323

SYSTEMATIC ESSAY 6: Cultural Geography
A Greatness Past
Decline and Rebirth
MODEL BOX 4 SPATIAL DIFFUSION PRINCIPLES
Boundaries and Barriers
Arabian Oil Bonanza
Regions and States
Egypt and the Nile Basin
The Western Maghreb
The Middle East
The Arabian Peninsula
The Non-Arab Northern Tier
The African Transition Zone

Chapter 7
AFRICAN WORLDS 375

SYSTEMATIC ESSAY 7: Medical Geography
The Environmental Base
Continental Drift
Environmental Hazards and Diseases
Agricultural Predominance
Africa's Past
The Colonial Legacy
MODEL BOX 5 COLONIAL SEQUENCE OF TRANSPORT DEVELOPMENT
Contemporary African Regions
West Africa
East Africa
Equatorial Africa
Southern Africa

Chapter 8
INDIA AND THE INDIAN PERIMETER 423

SYSTEMATIC ESSAY 8: Geomorphology
Physiographic Regions
The Human Sequence
Federal India
India's Economic Geography and Development
Crisis of Numbers
Bangladesh
Pakistan
Sri Lanka

Chapter 9
CHINA OF THE FOUR MODERNIZATIONS 459

SYSTEMATIC ESSAY 9: The Geography of Development
China in Today's World
Evolution of the State
A Century of Convulsion
Regions of China
Physiographic Regions
Human Regional Geography
The New China
Taiwan
Korea

Chapter 10
SOUTHEAST ASIA: BETWEEN THE GIANTS 505

SYSTEMATIC ESSAY 10: Political Geography
Population Patterns
Indochina
European Colonial Frameworks
Territorial Morphology
Fragmented Malaysia, Indonesia, and the Philippines
The Insurgent Vietnamese State
The Domino Theory and Thailand
Land and Sea

**PACIFIC REGIONS**

Melanesia
Micronesia
Polynesia

APPENDIX 541
GLOSSARY 547
PHOTO CREDITS 561
MAP INDEX 565
INDEX 569

# INTRODUCTION

# REGIONAL GEOGRAPHY OF THE WORLD

IDEAS AND CONCEPTS

**Regional Concepts**
**Scale**
**Culture**
**Cultural Landscape**
**Natural Environments**
**Pleistocene Cycles**
**Climate Regions**
**Models in Geography**
**Population Concentrations**
**Urbanization**
**Development**
**World Geographic Realms**
**Regional and Systematic Geography**
**Map Reading and Interpretation**

This is a book about the world's great realms, discussed in geographical perspective. Each of the major geographic realms of the human world (such as China, North America, or Southeast Asia) possesses a special combination of cultural, physical, historical, economic, and organizational qualities. These characteristic properties are imprinted on the landscape, giving each region its own flavor and social environment. Geographers take a particular interest in the way people have decided to arrange and order their living space. The street pattern of a traditional Arab town differs markedly from the layout of an Indian place of similar size. The fields and farms of sub-Saharan Africa look quite unlike those of Europe. Thus, the study of world realms also provides the opportunity to examine the concepts and ideas that form the basis of the modern field of geography. These are our twin objectives.

**Figure I–1**
Regional boundary: the edge of the Nile Valley in central Egypt.

## Concepts of Regions

Modern scientific concepts can be complicated mathematical constructions, but we use others, almost without realizing it, in our everyday conversation. Among the most fundamental concepts of geography are those involving the identification, classification, and analysis of regions. When we refer to some part of our country (the Midwest, for example), or to a distant area of the world (such as the Middle East), or even to a section of the metropolitan area in which we may live, we employ a regional concept. We reveal our perception of local or distant space, our mental image of the region to which reference is made.

Everyone has some idea of what the word **region** means, and we use the regional concept frequently in its broadest sense as a frame of reference. But regional concepts are anything but simple. Take just one implication of a regional name just used: the Midwest. If the Midwest is indeed a region of the United States, then it must have limits. Those limits, however, are open to debate. In his book, *North America*, John Paterson states that the Midwest includes the states of Ohio, Michigan, Indiana, Illinois, Wisconsin, Minnesota, Iowa, and Missouri, "but in the cultural sense it can also be said to include much of the area of heavy industry in Pennsylvania and West Virginia." Compare this definition to that of Otis Starkey, J. Lewis Robinson, and Crane Miller, who define the Midwest in their book *The Anglo-American Realm* as consisting of "most of the west North Central states [North and South Dakota, Nebraska, Kansas, Minnesota, Iowa, and Missouri] to which have been added the western parts of Wisconsin and Illinois." These two perceptions of the Midwest as an American region obviously differ. Does this invalidate the whole idea of an American Midwest? Not necessarily: the apparent conflict arises from the use of different *criteria* to give specific meaning to a regional term that has long been a part of American cultural life. Your own personal impression of the Midwest as a region is based on certain properties you have reason to consider important. When you add to your information base, you may modify your definition. Regionalization is the geographer's means of classification or taxonomy, and regions, like all classes, have their basis in established criteria. Classification schemes are open to change as new knowledge emerges, and so are regional definitions.

Regions obviously have **location**. Various means can be employed to identify a region's position on the globe, as the authors quoted above did when they enumerated the states that form part of their conception of the American Midwest. Often a region's name reveals much about its location. During the Vietnam War the name *Indochina* became familiar to us; it is a regional appellation identifying an area that has received cultural infusions from India and human migrations from China. Sometimes we have a particular landscape in mind when we designate a region—for example, the Amazon Basin or the Rocky Mountains. It would be possible, of course, to denote a region's location by reference to the earth's grid system and to record its latitude and longitude. That would give us the extent of its **absolute location**, but such a numerical index would not have much practical value. Location attains relevance only when it relates to other locations. Hence, many regional names give reference to other regions (*Middle* America, *Eastern* Europe, *Equatorial* Africa). This indicates a region's **relative location**, a much more meaningful and practical criterion.

Regions also have *area*. Again, this appears to be so obvious that it hardly requires emphasis, but some difficult problems are involved here. For example, certain regions are identified as the (San Francisco) Bay Area, the Greater Chicago Area, or the New York Metropolitan Area. Everyone would probably agree that each of these areas is focused on a few internal urban concentrations, but what are the limits or boundaries of such urban-centered regions? In quite another context, we use such terms as the *Corn Belt*—an agricultural region in the central United States—and the *Sunbelt*—a broad zone across the southern United States that attracts a growing number of migrants, including numerous persons of retirement age who seek to escape the rigors of northern winters. The geographical or **spatial** extent of a region, whether Bay Area or Corn Belt, Midwest or Middle East, cannot be established and defined without reference to its specific areal contents.

An overriding characteristic of a region's contents may be its *homogeneity* or sameness. Sometimes the landscape leaves no doubt where one region ends and another begins: in Egypt, the break between the green, irrigated, cultivated lands adjacent to the Nile River and the desert beyond is razor-sharp and all-pervading (Fig. I–1). On the map, the line representing that break is without question a regional boundary. Everything changes beyond that line—population density, vegetation, soil quality. But regional distinctions are not always so clear. The example of the U.S. Corn Belt (mapped in Fig. 3–25) provides a good contrast. Traveling northward from Kentucky into Illinois or Indiana, you would undoubtedly be struck by the increasing number of cornfields. Since not all farmland is under corn, even within the Corn Belt, the difference between what you saw in Kentucky and Illinois is a matter of degree. Therefore, in order to define a Corn Belt and represent it on a map, it would be necessary to establish a criterion; for instance, 50 percent or more of all the cultivated land must be devoted to growing corn. The line so drawn would delimit an agricultural region, but on the landscape it would be far less evident than the border enclosing Egypt's Nile Valley farmlands.

It is possible, of course, to increase the number of criteria, so that more than one condition must be satisfied before the region is delimited. To define a particular cultural region, such items as the use of a certain language, adherence to a specific religion, perhaps even the spatial variations of architectural, artistic, and other traditions might be employed. Maps of the cultural geography of Canada, including those showing religious affiliation, dominant language used, land division, and settlement patterns, reveal the reality of Quebec as a discrete region within the greater Canadian framework.

A region can also be conceptualized as a **system**. Certain regions are marked not by internal uniformity but by a particular activity, or perhaps a set of integrated activities, that interconnects its various parts. This is how it is possible to perceive the Bay Area and similar metropolitan entities as true regions. A large city-suburban complex has a substantial surrounding area for which it supplies goods and services, from which it buys farm products, and with which it interacts in numerous ways. The metropolitan city's manufacturers distribute their products wholesale to regional subsidiaries. Its newspapers sell in the nearby smaller towns. Maps showing the orientation of road traffic, the sources and destinations of telephone calls, the readership of newspapers, the audiences of television stations, and other activities confirm the close relationship between the metropolis and its tributary region or *hinterland.* Here again, we have a region; this time it is not characterized by homogeneity but, instead, by a city-centered system of interaction that produces a nodal or *functional region*.

## Classifying Regions

Given the various qualities and properties of regions, their differences in dimensions and complexity, is it possible to establish a hierarchy, a ranking of regions based on some combination of their characteristics? Some geographers have proposed systems of classification, and although none of the results has gained general acceptance, it is interesting to see how they confronted this often frustrating problem. In a book entitled *Introduction to World Geography*, Robert Fuson

## REGIONAL TERMINOLOGY

The internal uniformity of a homogeneous region can be expressed by human (cultural) as well as natural (physical) criteria. A country constitutes such a political region, for within its boundaries certain conditions of nationality, law, and political tradition prevail throughout. Similarly, a natural region such as the Rocky Mountains or the Mississippi Delta is expressed by the dominance of a particular physical landscape. Quebec and the Corn Belt are uniform cultural and agricultural regions, respectively. Regions marked by this internal homogeneity are classified as **formal regions**.

Regions conceptualized as *spatial systems*—such as those centered on an urban core, a node, or a focus of regional interaction—are identified collectively as **functional regions**. Thus, the formal region might be viewed as static, uniform, and immobile; the functional region is seen to be dynamic, active, and continuously shaped by forces that modify it.

This distinction between formal and functional regions is still debated among geographers. A formal region's evenness, some argue, is also the result of the operation of shaping forces; perhaps formal regions are less affected by change, more durable, and, therefore, more visible, but they may not be fundamentally different from functional regions. And on the landscape itself, both regional types may often be recognized simultaneously: in Fig. I–2—which shows a small Iowa town within the Corn Belt—we can observe both the functional ties (roads, storage elevators) between the town and its surrounding farms as well as the formal-region homogeneity of the agricultural spatial pattern (evenly dispersed farmsteads, repetition of field size and land uses) beyond the edge of the built-up area in the foreground. Some contributions to the continuing discussion of regional terminology in the current geographical literature are cited in the bibliography at the end of this introductory chapter.

**Figure I–2**
Regional ties around Garden City, Iowa.

suggests a comprehensive, seven-level regional hierarchy that would divide the earth into the European and non-European world and, hence, into realms, landscapes, superregions, regions, districts, and subregions. This system is complicated and quite difficult to apply consistently, but it underscores the elusive nature of the problem.

Cultural geographers often employ a three-tier system. The *culture realm* (sometimes *culture world*) identifies the largest and most complex area that can be described as being unified by common cultural traditions. For example, North America (Canada and the United States) constitutes such a culture realm; Middle and South America form an additional pair of cultural realms that comprise "Latin" America. Each of these culture realms, in turn, consists of an assemblage of *culture regions*. Within the Latin American realms, Mexico is such a culture region, the Central American republics comprise a second region, and Brazil a third. Within the European culture realm, Mediterranean Europe and Eastern Europe rank as separate culture regions. These regions, in turn, consist of *subregions*. Canada is a region within the North American realm; French-speaking Quebec is a subregion within Canada. Similarly, the Balkan states form a subregion within Eastern Europe.

Note that the regional and subregional areas tend to be identified as countries or groups of countries, which can become misleading. The country names are used because of their convenient familiarity, but their boundaries do not necessarily coincide with those of the regions or subregions in question. Mexico's regional properties spill over into the southwestern margins of the North American realm. The Sahel—a distinct subregion of West Africa—extends across parts of several of that region's countries. Thus, in a world context, realms and regions are not bounded by sharply defined limits—instead, they are separated by transition zones that are, in places, quite broad. But if we must mark regional boundaries on maps, we should take advantage of the existing politico-geographical "grid" of countries. This has the advantage of comparative simplicity, as we will see later in this chapter (p. 38) when we discuss the geographic realms framework to be used in this book.

# Concepts of Scale

Regions can be conceptualized in various forms and at different levels of generalization or *scale*, which is defined as the ratio of map distance to actual ground distance.

Consider the four maps in Fig. I–3. On the first map (upper left) most of the North American realm is shown, but very little spatial information can be provided, although the political boundary between Canada and the United States is shown. On the second map (upper right), East and Central Canada are depicted in sufficient detail to permit display of the provinces, several cities, and some physical features (Manitoba's major lakes) not shown on the first map. The third map (lower left) shows the main surface communications of Quebec and immediate surroundings, the relative location of Montreal, and the St. Lawrence and Hudson/James Bay drainage systems. The fourth map (lower right) reveals the metropolitan layout of Montreal and environs in considerable detail.

Each of the four maps has a scale designation, which can be shown as a *bar graph* (in kilometers and miles in this case) and as a fraction—1:103,000,000 on the first map. The fraction is a ratio indicating that one unit of distance on the map (one inch or one centimeter) represents 103 million such units on the ground. The smaller the fraction (i.e., the larger the number in the denominator), the smaller the scale of the map. Obviously, this *representative fraction* on the first map (1:103,000,000) is the smallest of the four, and that of the fourth map (1:1,000,000) is the largest. Comparing maps nos. 1 and 3, we find that on the *linear* scale, no. 3 has a representative fraction that is more than four times larger than no. 1. When it comes to *areal* representation, however, 1:24,000,000 is more than 16 times larger than 1:103,000,000 because the linear difference prevails in both dimensions (the length *and* breadth of the map).

In a book that surveys world realms, it is obviously necessary to operate at relatively smaller scales. When studying regions or subregions in greater detail, our ability to specify criteria and to "filter" the factors we employ increases as we work at larger scales. On occasion that method will be used, for example, when urban centers are the topic of concern (as suggested by Montreal in Fig. I–3). But most of the time our view will be the more macroscopic and general—the small-scale view of the world's geographic realms.

Besides scale, maps exhibit a number of other basic properties. Because familiarity with them simplifies the task of reading and interpreting the maps in this book, these properties are discussed in the box at the end of this chapter (p. 49).

# Concepts of Culture

The realms and regions to be discussed in the chapters that follow

**EFFECT OF SCALE**

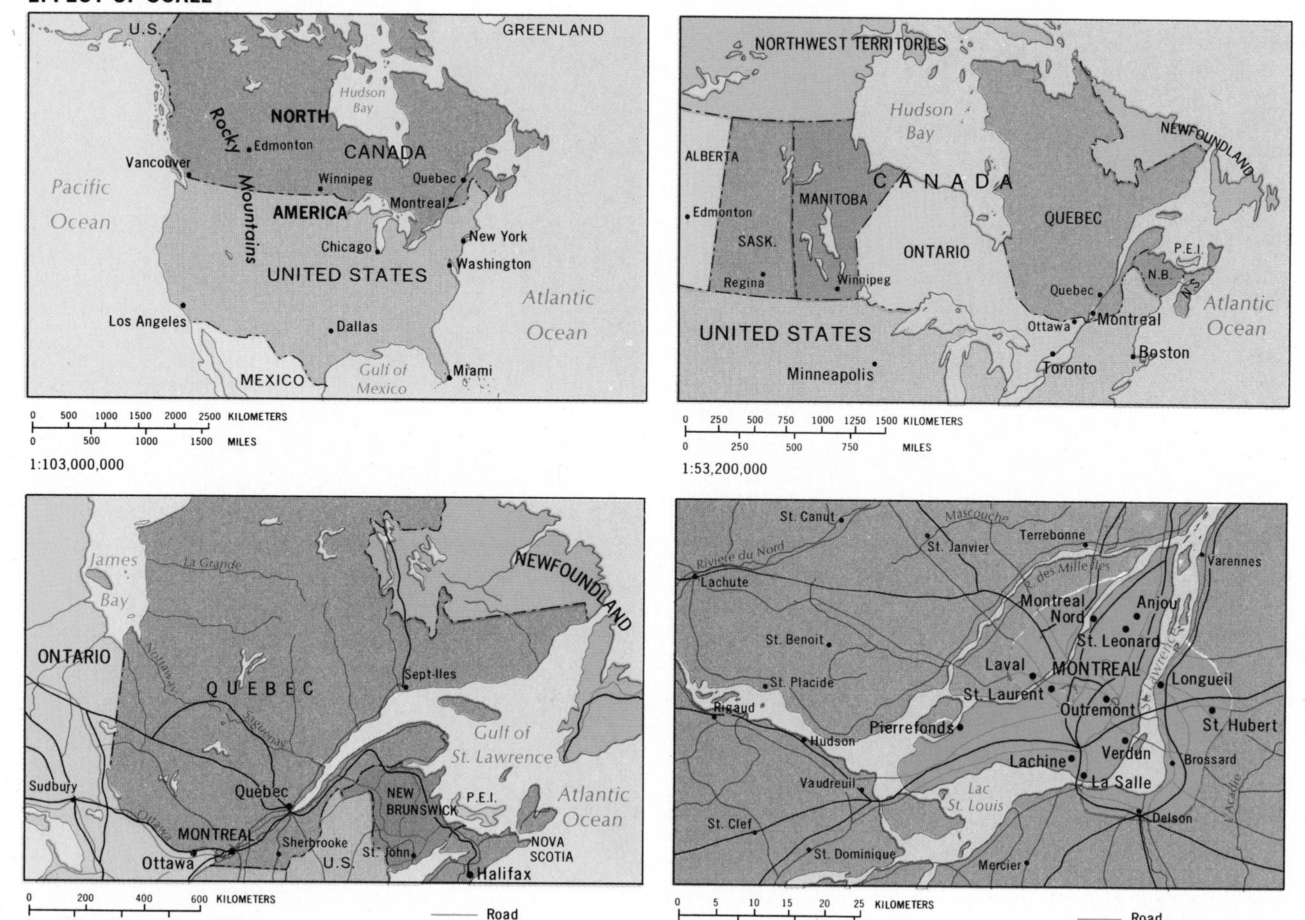


**Figure I–3**

are, in part, defined by humanity's *cultures*. Geographers approach the study of culture from several vantage points, and one of these, the analysis of *cultural landscape*, is central to our regional interests. Therefore, we should look rather closely at the concept of culture. The word **culture** is not always used consistently in the English language, which can lead to some difficulties in establishing its scientific meaning. When we speak of a "cultured" individual we tend to mean someone with refined tastes in music and the arts, a highly educated, well-read person who knows and appreciates the "best" attributes of his or her society. But as a scientific term, culture refers not only to the music, literature, and arts of a society, but also to all the other features of its way of life: prevailing modes of dress, routine living habits, food preferences, the architecture of houses as well as public buildings, the layout of fields and farms, and systems of education, government, and law. Thus, culture is an all-encompassing term that identifies not only the mosaic of life-styles of a people, but also their prevailing values and beliefs.

This is not to suggest that anthropologists and other social scientists haven't had problems treating the concept of culture. If you read some of the basic literature in anthropology, you will find that anthropologists have had as much difficulty with definitions of the culture concept as geographers have had with the regional concept. A culture may be the total way of life of a people—but is it their *actual* way of life ("the way the game is played") or the standards by which they give evidence of *wanting* to live through their statements of beliefs and values ("the rules of the game")? There are strong differences of opinion on this, and as a result the various definitions have become quite complicated.

Anthropologist E. Adamson Hoebel says in his *Anthropology: The Study of Man* that culture is

*the integrated system of learned behavior patterns which are characteristic of the members of a society and which are not the result of bio-*

> *logical inheritance . . . culture is not genetically predetermined; it is noninstinctive . . . [culture] is wholly the result of social invention and is transmitted and maintained solely through communication and learning.*

This definition raises still another question: how is culture carried over from one generation to the next? Is this entirely a matter of learning, as Hoebel insists, or are certain aspects of a culture indeed instinctive and, in fact, a matter of genetics? This larger question is the concern of sociobiologists and not cultural geographers, although some of its side issues such as *territoriality* (an allegedly human instinct for territorial possessiveness) and *proxemics* (individual and collective preferences for nearness or distance in different societies) clearly have important spatial dimensions.

But even without these theoretical additions, the culture concept remains difficult to define satisfactorily. In 1952, anthropologists Alfred Kroeber and Clyde Kluckhohn published a paper that identified no fewer than 160 definitions—all of them different—and from these they distilled their own:

> *Culture consists of patterns, explicit and implicit, of and for behavior and transmitted by symbols, constituting the distinctive achievements of human groups, including their embodiments in artifacts . . . the essential core of culture consists of traditional (that is, historically derived and selected) ideas and especially their attached values; culture systems may, on the one hand, be considered products of action, and on the other as conditioning elements of further action.*

Some of the definitions from which this one was synthesized, together with a few more recent ones, appear in the box on culture concepts.

# THE CULTURE CONCEPT

Below are several definitions of the concept of culture as developed by some prominent social scientists:

*That complex whole which includes knowledge, belief, art, morals, law, custom, and any other capabilities and habits acquired by man as a member of society.*
**E. B. Tylor (1871)**

*The sum total of the knowledge, attitudes, and habitual behavior patterns shared and transmitted by the members of a particular society.*
**R. Linton (1940)**

*The mass of learned and transmitted motor reactions, habits, techniques, ideas, and values—and the behavior they induce.*
**A. L. Kroeber (1948)**

*The man-made part of the environment.*
**M. J. Herskovits (1955)**

*A way of life which members of a group learn, live by, and pass on to future generations.*
**A. E. Larimore et al. (1963)**

*The learned patterns of thought and behavior characteristics of a population or society.*
**M. Harris (1971)**

*The acquired knowledge that people use to interpret experience and to generate social behavior.*
**J. P. Spradley and D. W. McCurdy (1975)**

*The sum of the morally forceful understandings acquired by learning and shared with the members of the group to which the learner belongs.*
**M. J. Swartz and D. K. Jordan (1976)**

For our purposes it is sufficient to stipulate that culture consists of a people's *beliefs* (religious, political), *institutions* (legal, governmental), and *technology* (skills, equipment). This notion is a good deal broader than that adopted by many modern anthropologists, who now prefer to restrict the concept to the interpretation of human experience and behavior as products of systems of symbolic meaning. It is also important to keep in mind that definitions of this kind are never final and absolute; rather, they are arbitrary and designed for a particular theoretical purpose. The culture concept is defined to facilitate the explanation of human behavior. Anthropologists today tend to focus on what people know, on codes and values, on the "rules of the game." This was not always the case, as the listing of definitions shows. Sociologists, political scientists, psychologists, and historians have different requirements and would construct contrasting "operational" definitions. The same is true of cultural geographers. Geographers would be particularly attracted to the Herskovits definition, because they have a particular interest in the way the members of a society perceive and exploit their resources, the way they maximize the opportunities and adapt to the limitations of their environment, and the way they organize that portion of the earth that is theirs.

This last aspect, the way human societies organize that part of the earth that is theirs, goes to the heart of our approach in this book. Human works carve long-lasting if not permanent imprints into the earth: Roman structures still mark some European countrysides and many Roman routes of travel have now evolved into Europe's major highways. Over time, regions take on certain dominant qualities that together create a regional character, a personality, a distinct atmosphere. This, in large part, is the basis for our division of the human world into geographic realms.

## The Cultural Landscape

Culture is expressed in many ways as it gives visible character to a region. Aesthetics play an important role in all cultures, and often a single scene in a photograph or a picture can reveal to us, in general terms, in what part of the world it was made. The architecture, the clothing of the people, the means of transportation, and perhaps even the goods being carried reveal enough to permit a good guess. This is because the people of any culture, when they occupy their part of the earth's available space, transform the land by building structures on it, creating lines of transport and communication, parceling out the fields, and tilling the soil. There are but few exceptions; nomadic people may leave a minimum of permanent evidence, and some people living in desert margins (such as the Bushmen) and in tropical forests (the Pygmies, for example) alter their natural environment very little. But most of the time humans are agents of change, creating asphalt roadways, irrigation canals, terraced hillslopes, fences and hedges, urban settlements, and a myriad of other visible artifacts.

This composite of human imprints on the surface of the earth is called the **cultural landscape**, a term that came into general use in geography in the 1920s. Carl O. Sauer, for several decades professor of geography at the University of California, Berkeley, developed a school of cultural geography that was focused around the concept of cultural landscape. In a paper written in 1927 entitled "Recent Developments in Cultural Geography," Sauer proposed his most straightforward definition of the cultural landscape: *the forms superimposed on the physical landscape by the activities of man*. He stressed that such forms result from the operation of cultural **processes**—causal forces that shape cultural patterns—which unfold over a long time period and involve the cumulative influences of successive occupants.

Sometimes these successive groups are not of the same culture. Farm settlements and villages built by European colonizers are now occupied by Africans. Minarets of Islam rise above the buildings of Eastern European cities, evincing an earlier period of hegemony by the Moslem Ottoman Empire. In 1929, Derwent Whittlesey introduced the term *sequent occupance* to categorize these successive stages in the evolution of a region's cultural landscape, a concept explored further in Chapter 7.

A cultural landscape consists of buildings and roads and fields and more; but it also has an intangible quality, an "atmosphere" or flavor, a sense of place that is often easy to perceive and yet difficult to define. The smells and sights and sounds of a traditional African market are unmistakable, but try recording those qualities on maps or in some other objective way for comparative study! Geographers have long grappled with this problem of recording the less tangible characteristics of the cultural landscape, which are often so significant in producing the regional personality.

Jean Gottmann, a Franco-British geographer, put it this way in his *A Geography of Europe*:

> *To be distinct from its surroundings, a region needs much more than a mountain or a valley, a given language or certain skills; it needs essentially a strong belief based on some religious creed, some social viewpoint, or some pattern of political memories, and often a combination of all three. Thus regionalism has what might be called* iconography *as its foundation: each community has found for itself or was given an icon, a symbol slightly different from those cherished by its neighbors. For centuries the icon was cared for, adorned with whatever riches and jewels the community could supply.*

Gottmann is trying here to define some of the abstract, intangible qualities that go into the makeup of a region's cultural landscape. The more concrete properties are a bit easier to observe and record. Take, for instance, the urban "townscape" (a prominent element of the overall cultural landscape), and compare a major U.S. city with, say, a leading Japanese city. Visual representations of these two metropolitan scenes would reveal the differences quickly, of course, but so would maps of the two urban places. The American city, with its rectangular layout of the central business district (*CBD*) and its far-flung sprawling suburbs contrasts sharply with the clustered, space-conserving Japanese city. Using a rural example, the spatially lavish subdivision and ownership patterns of American farmland (Fig. I–2) looks unmistakably different from that of the traditional African countryside, with its irregular, often tiny patches of land surrounding a village (Fig. I–4). Still, the whole of a cultural landscape can never be represented on a photograph or

Figure I–4
Africa's rural landscape in central Kenya.

map, because the personality of a region involves far more than its prevailing spatial organization: one must also include its visual appearance, its noises and odors, the shared experiences of its inhabitants and even their pace of life.

## Changing Natural Environments

Carl Sauer defined the cultural landscape as the forms superimposed on the natural (physical) landscape by human activity, and Derwent Whittlesey introduced the idea of cultural landscape evolution as sequent occupance. Both concepts contained the element of change—changing cultural traditions, changing societies, changing regional contents and personalities. The earth itself, the physical landscape, was presumed to be a comparatively static stage on which the cultural scenes were played out. Changes were brought to this natural landscape largely through humanity's works: building dams in river valleys, terracing hillsides, substituting cultivated crops for natural vegetation. The physical world was viewed as passive and stationary; the human world produced the dynamics of activity and change.

In recent decades, this assumption has been challenged. Of course, the earth has always been more receptive to human occupance in certain areas than in others, a variability that is reflected on every map of world population distribution. However, the physical world varies not only in space, but also in time. Today, archaeologists excavate ancient cities located in deserts, but when those cities were built no arid climate prevailed there. In Roman times, farmlands near North Africa's Mediterranean coast produced large quantities of farm produce from fields amply watered by rain and irrigation, but today the Roman aqueducts lie in ruins and the fields are abandoned. Icy tundra conditions prevailed where the heart of the Soviet Union lies today.

The earth is, therefore, a variable stage, too, both in time as well as in space. This is true even of the surface itself, with its component plains, plateaus, hills, and mountains (Fig. I–5). The planet may be nearly 5 billion years old, and its solid crust began to form between 3 and 4 billion years ago. Eventually the earth differentiated into a number of concentric shells, with the surface crust underlain by a thicker *mantle*. Physical geographers have known for more than a century that the surface of the crust has undergone momentous changes: mountain ranges have arisen only to be eroded away again, continental regions (such as the U.S. Great Plains) were invaded by the ocean and lay inundated for millions of years as sediments accumulated. Then, in 1915, Alfred Wegener published a book in which he presented evidence that the continents themselves are mobile and were once united in a gigantic landmass he called Pangaea. The breakup of this supercontinent, he reasoned, has taken place during the last 80 to 100 million years, and the continents are still moving.

Quite recently it was discovered that the earth's outer shell, the crust, and the uppermost layer of the mantle (together called the *lithosphere*) consist of a set of rigid geologic *plates*. The exact number and location of these dozen or so tectonic plates are still not certain, but it is known that they average some 100 kilometers (60 miles) in thickness and that the largest plates are of continental dimensions (Fig. I–6). The plates are in motion, however, emerging along great fissure zones in the ocean basin and colliding elsewhere. Where they meet, one plate tends to descend under the other, and there is great deformation and crumpling of the crust that often produces seismic (earthquake) and volcanic activity. Human communities located in such zones are accustomed to sudden and sometimes violent changes in their physical environment. What thousands of years of quiet weathering and erosion cannot accomplish, an earthquake or volcanic eruption can achieve in seconds as the U.S. Pacific Northwest discovered after the 1980 explosion of Washington's Mt. St. Helens (Fig. I–7).

A comparison of Figs. I–5 and I–6 helps to explain the spatial distribution of the earth's present landscapes. Great mountain ranges, complete with volcanic zones and earthquake-prone belts, extend from Western South America through Middle and North America to East Asia and the Pacific islands, including New Zealand east of Australia. Here the Pacific Plate and its neighbors, the Philippine, Cocos, and Nazca plates, collide against the Australian, Eurasian, and American plates; no wonder this contact zone is frequently called "The Pacific Ring of Fire." Africa and Australia, on the other hand, lie at the heart of the African and Australian plates, respectively, and are much less subject to such deformation. Other relationships between Figs. I–5 and I–6 are readily apparent, but it is important to remember that the map of the world tectonic plates is being modified as new evidence and interpretations come to light. For example, some physical geographers believe that the African Plate—shown here as an unbroken cohesive unit of the lithosphere—in fact consists of several segments.

Mobile tectonic plates carry the earth's landmasses along as they move, and Wegener's hypothesis of continental drift now appears substantially correct. But it is important to view the process of continental drift in the perspective of human evolution on this planet. Current research in East Africa indicates that the human family began to emerge between 12 and 14 million years ago and that the use of stone tools developed between 2 and 3 million years ago. A fossil site in Tanzania has yielded evidence of a stone structure, probably the foundation of a simple hut, dated as 1.8 million years old. But the emergence of larger human communities, made possible by the domestication of plants and animals and attended by the development of urban centers, is the product of the past 10,000 years—a tiny fraction of time in the context of earth (and even evolutionary human) history. Thus, recorded time has not been nearly long enough to observe all the momentous changes we know the earth's landscapes have undergone.

## Pleistocene Cycles

The earth's lifetime has been divided by geologists into four major stages or *eras* on the basis of information derived from the study of rocks and fossils. Obviously, the least is known about the oldest of these eras, the Precambrian, covering the period from the planet's beginnings until about 600 million years ago. Next comes the Paleozoic era, from 600 to 225 million years ago, followed by the Mesozoic, which lasted until about 70 million years before the present (B.P.). The latest era, the one during which we live, is the Cenozoic. Each of these lengthy eras is subdivided into *epochs*, and the Cenozoic is marked by five of these. The two most recent Cenozoic epochs, the Pleistocene (ca. 3 million to 10,000 years ago) and the Recent or Holocene (10,000 years B.P. to the present), witnessed the emergence of humanity and the eventual attainment of civilization.

Geologic eras and epochs are identified through the study of rock layering, fossil assemblages, mountain building, and other evidence yielded by the earth's crust. Many an epoch begins with the deposition of huge thicknesses of sedimentary rocks and ends when those rocks are bent and broken by tectonic activity (internal earth forces that deform the crust). But the Pleistocene is no ordinary epoch. Its beginning is marked by the development of large icesheets that eventually covered all of Canada, most of the north-central and northeast United States, and a substantial part of Europe (Fig. I–8). The earth's climatic and vegetative environments changed drastically as the glaciers of the Pleistocene first began to advance toward the mid-latitudes, probably between 2 and 3 million years ago. And the first outbreak of the icesheets was not the last. Throughout the nearly 3 million years of the Pleistocene, these icesheets expanded, then contracted toward the poles, only to push outward again. Many physical geographers believe that there were 4 major glacial advances and 3 periods of withdrawal (interglacials), but there is evidence to suggest that numerous additional advances and retreats occurred, perhaps as many as 14. The Pleistocene epoch, therefore, is often called the Ice Age.

Most geologic time charts suggest that the Pleistocene is at an end and that we now exist during a new epoch called the Holocene (or sometimes simply "Recent"). It is also still likely, however, that the human world of the past 10,000 years has evolved during yet another Pleistocene interglacial, whose mildness has enormously expanded the earth's living space—temporarily. In this view, the possibility is strong that the continental icesheets will once again overspread those large areas previously affected by Pleistocene glaciation (Fig. I–8). If this should occur, the renewed ice age will have an impact far beyond those regions directly affected. World climates everywhere will change; moist areas will dry up; areas now dry will begin to receive ample moisture; temperatures, even in tropical zones, will decline.

Whatever the future, there can be no doubt that the modern world's human cultures emerged and evolved in the wake of the last of the Pleistocene glaciers' withdrawal. Certainly there were active human communities during earlier phases, but the great transformations—plant and animal domestication, urbanization and civilization, mass migration, agricultural and industrial revolutions—occurred during the past 10,000 years. We cannot know as yet why these momentous changes occurred when they did, but they have been accompanied, especially during the past century, by an unprecedented increase in human numbers. This population explosion (about which more will be said in Chapters 1 and 8) is taking place now, while the earth's available living space is at a maximum. The implications of a return of the icesheets and their

**Figure I–5**

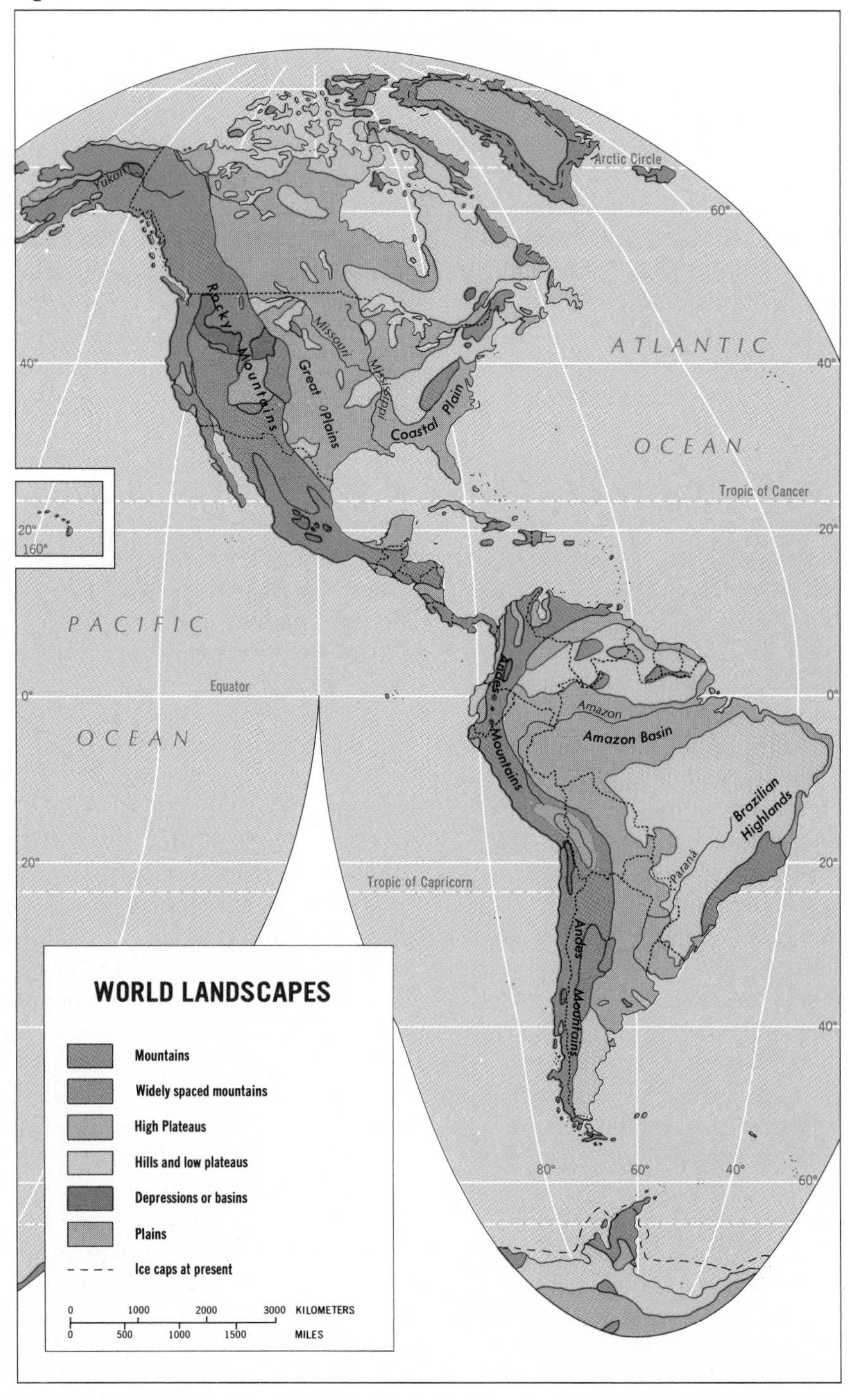

impact on the earth's climates and habitable space stagger the imagination. Nevertheless, many environmental scientists warn that such a sequence of events may well lie ahead.

## Water—Essence of Life

The French geographer Jean Brunhes, writing in his book *Human Geography,* remarked that "every state, and indeed, every human establishment, is an amalgam made up of a little humanity, a little soil, and a little water." He might have added that without water there would be neither humanity nor soil. When the United States in 1976 sent two space probes to Mars, scientists on earth waited for the crucial information to be relayed back from the landed vehi-

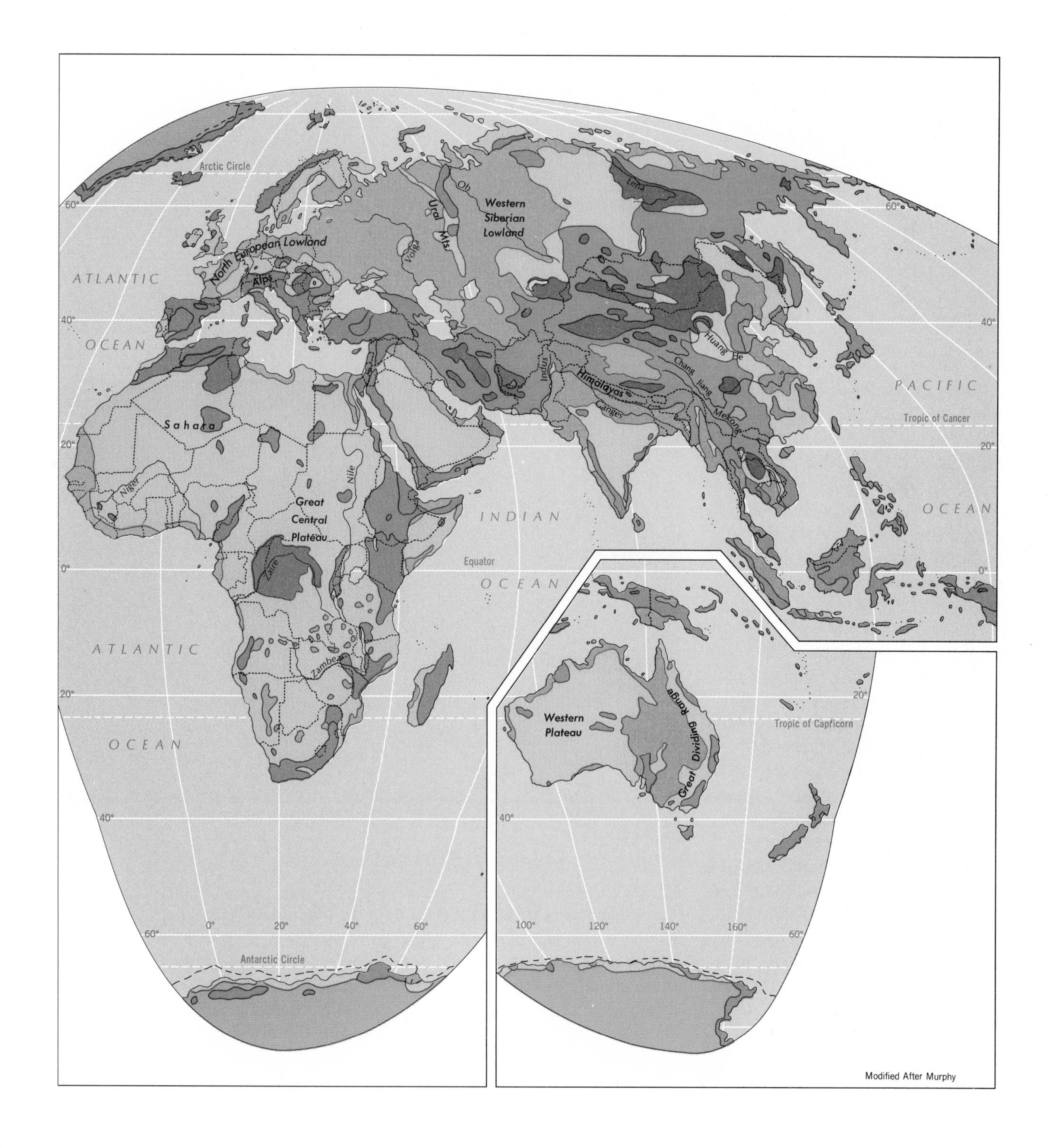

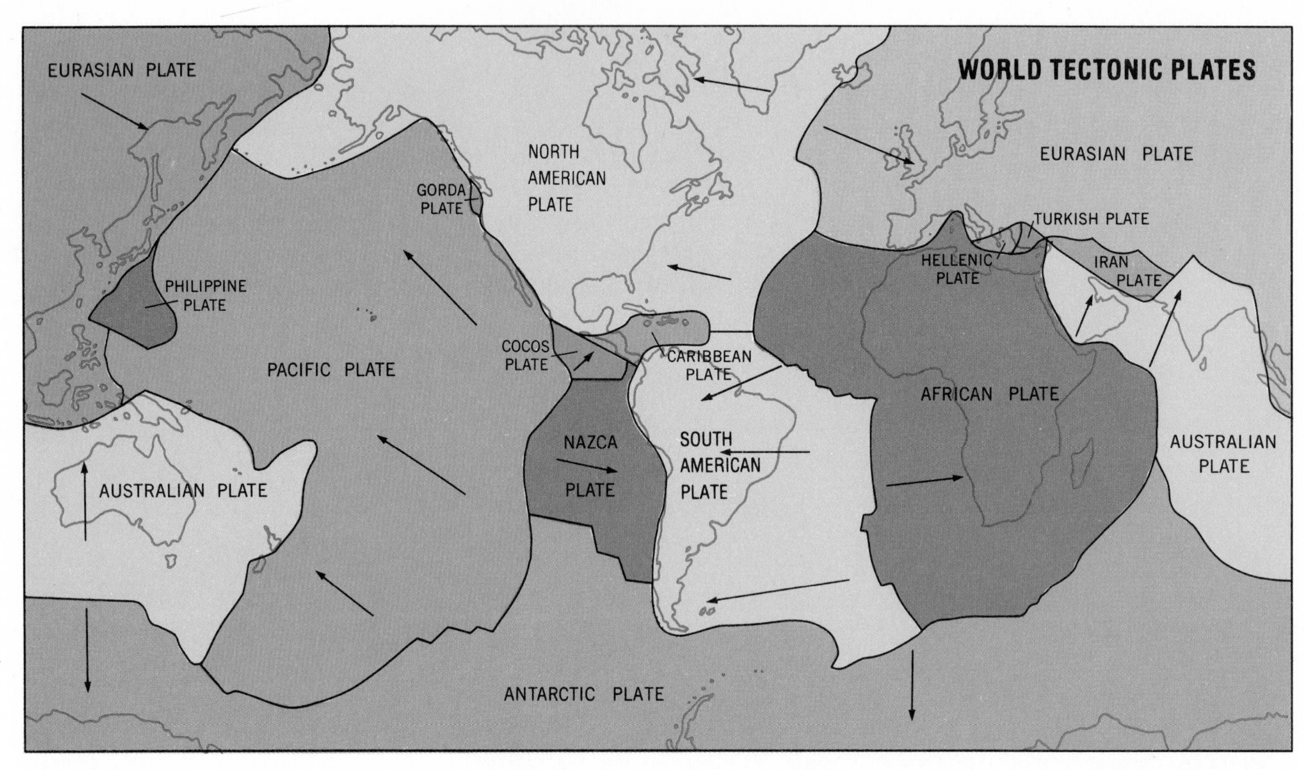


**Figure I–6**

cles: was any moisture present on our neighboring planet's surface? Moisture would be the key to life on Mars as here on earth; but the Martian surface proved to be as barren as the moon's. Alone among the planets of our solar system, the earth possesses a *hydrosphere*—a cover of water in the form of a vast ocean, frozen polar icesheets, and a moisture-laden atmosphere. Technically all of the earth's water, even in lakes and streams, is part of the hydrosphere, but the great world ocean constitutes about 97 percent of it all by volume. The world ocean covers just over 70 percent of the earth's surface, but it would do little good if a mechanism did not exist whereby moisture from the ocean is brought to the land. This mechanism, the *hydrologic cycle*, functions as a circulation system. Moisture evaporates into the air from the ocean's surface, and this humid mass of air then moves over land where, by various atmospheric processes, condensation occurs and precipitation falls. Much of it returns to the ocean as runoff via streams, and this continuous cycle again repeats itself.

**Figure I–7**
Mt. St. Helens following the 1980 eruption.

The earth's landmasses do not share equally in this provision of moisture, and much of the historical geography of humanity involves the search and competition for well-watered areas. Global patterns of precipitation are mapped in Fig. I–9, which should be viewed as the latest frame from a piece of Pleistocene film, a still picture of continually changing conditions. It represents the distribution of moisture conditions as they prevail today, but it differs from the world pre-

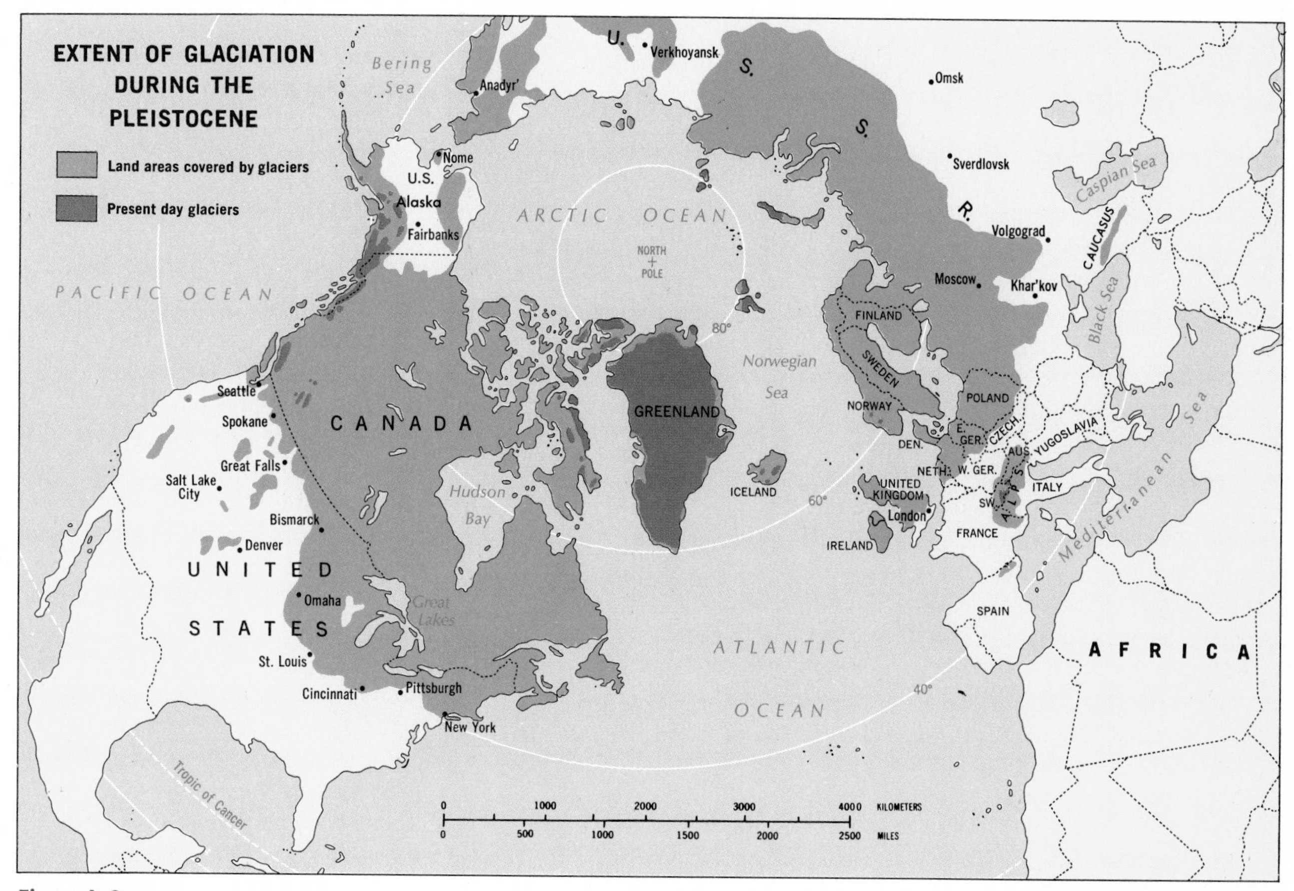


Figure I–8

cipitation map as it would have looked when the Middle East's Fertile Crescent witnessed the domestication of crops or when North Africa, several thousand years later, was a granary of the Roman Empire. Moreover, the map will look quite different a thousand years from now. Today, the map reveals an equatorial zone of heavy rainfall, where annual totals exceed 200 centimeters (80 inches), extending from Middle America through the Amazon Basin, across smaller areas of West and Equatorial Africa, and into South and Southeast Asia. This equatorial zone of high precipitation gives way to dry conditions in both poleward directions. In equator-straddling Africa, for example, the Sahara lies to the north of the low-latitude humid zone and the Kalahari Desert lies to the south. Interior Asia and central Australia also are very dry, as is southwestern North America.

The general pattern of equatorial wetness and adjacent dryness is broken along the coasts of all the continents, and it is possible to discern a certain consistency in this spatial distribution of precipitation. Observe that eastern coasts of continents and islands in tropical as well as mid-latitude locations receive comparatively heavy rainfall, as in the southeastern United States, eastern Brazil, eastern Australia, and southeastern China. Furthermore, a narrow zone of higher precipitation exists at higher latitudes on the western margins of the continents, including the Pacific Northwest coast of the United States and Canada, the coast of Chile, the southwestern tip of Africa, the southwestern corner of Australia, and, most importantly, the western exposure of the great Eurasian landmass—Western Europe.

The distribution of world precipitation, as reflected in Fig. I–9, results from an intricate combination of global systems of atmospheric and oceanic circulation as well as heat and moisture transfer. Whereas the analysis of these systems is a part of the subject of physical geography, we should remind ourselves that even a slight change in one of them can have a major impact on a region's habitability. Figure I–9 also displays the *average* annual precipitation on the continents, but no place on earth has a firm guarantee that it will receive, in any given year, precisely its average rainfall. In general, the variability of precipitation increases as the recorded average total decreases, which means that rainfall is least dependable just where reliability is needed most—in the drier

# EVAPOTRANSPIRATION

The map of world precipitation distribution (Fig. I–9) should be viewed in the context of temperature distribution. From the map it might be concluded that all the areas that receive over 100 centimeters (40 inches) of rainfall are thereby equally amply supplied with moisture. But there are equatorial areas where temperatures average over 25°C (77°F) that receive 100 centimeters and much cooler places—for example, parts of New Zealand and Western Europe—where 100 centimeters are also recorded. Obviously, evaporation from the ground goes on much more rapidly in the tropical areas than in the mid-latitude zones.

Similarly, evaporation from vegetation also speeds up in equatorial regions. This evaporation from leaf surfaces is actually a three-stage process. Roots of plants absorb water in the soil. This water is then transmitted through the organism and reaches the leafy parts, which transpire in warm weather much as we perspire. From the surface of the leaves, the moisture evaporates. Thus, a plant acts like a pump, and the process of evaporation from vegetation is actually a process of transpiration plus evaporation, or *evapotranspiration*.

Thus, 100 centimeters of rainfall in a tropical area may very well be inadequate, and if the amount lost by evaporation and evapotranspiration is calculated, it could exceed 100 centimeters—which means that the plants would use more moisture if it were available. Some of those "moist" tropical areas, even those with over 150 centimeters (60 inches) of rainfall annually, can be shown to be moisture deficient. In other areas, the seasonality of precipitation is so pronounced that there is deficiency during part of the year. On the other hand, in cooler parts of the world, just 75 centimeters (30 inches) of rainfall may be enough to keep the soil moist and the vegetation adequately supplied. A map such as Fig. I–9 is of necessity a generalization, and it is important to know what it fails to reveal.

portions of the inhabited world. A prominent region of low precipitation that has recently suffered greatly from variable rainfall is West Africa's Sahel, the semiarid southern margin of the Sahara Desert, where a seven-year drought intervened between moist years during the decade of 1967–1977 and caused hundreds of thousands to die of starvation.

Even heavy, year-round precipitation is no guarantee that an area can sustain large and dense populations. In equatorial areas the heavy precipitation, combined with high temperatures, leads to the faster destruction of fallen leaves and branches by bacteria and fungi; this severely retards the formation of *humus*, the dark-colored upper soil layer consisting of nutrient-rich decaying organic matter that is vital to soil fertility. The drenched soil is also subjected to the *leaching* of its best nutrients—the dissolving and downward transport by percolating water—so that only oxides of iron and aluminum remain to give the tropical soil its characteristic reddish color. Such tropical soils (called latosols) support the dense rainforest, but they cannot carry crops without massive fertilization. The rainforest thrives on its own decaying vegetative matter, but when the land is cleared, the leached soil proves to be quite infertile. Not surprisingly, the Amazon and Congo (Zaïre) basins are not among the world's most populous regions.

## Climatic Regions

It is not difficult to discern the significance of precipitation distribution on the map of world climates (Fig. I–10). The regionalization of climates has always presented problems for geographers. In the first place, climatic records are still scarce, short-term, or otherwise inadequate in many parts of the world. Second, weather and climate tend to change gradually from place to place, but the transitions must be represented as lines on the map. In addition, there is always room for argument concerning the criteria to be used and how these criteria should be weighed. Vegetation, for example, is a response to prevailing climatic conditions. Should boundary lines between climatic regions therefore be based on vegetative changes observed in the landscape, no matter what precipitation and temperature records show? This debate still goes on.

Figure I–10 displays a regionalization system devised by Wladimir Köppen and modified by Rudolf Geiger, which we later generalize further in our discussion of Fig. I–11. This scheme has the advantage of comparative simplicity and is based on a triad of letter symbols. The first (capital) letter is the critical one; the *A* climates are humid and tropical, the *B* climates are dominated by dryness, the *C* climates are still humid and comparatively mild, the *D* climates reflect increasing coldness, and the *E* climates mark the frigid polar and near-polar areas.

The *humid equatorial* climates (*A*), also referred to as the humid tropical climates, are marked by high temperatures all year and by heavy precipitation. In the *Af* subtype, the rainfall comes in substantial amounts every month; but in the *Am* areas, there is a sudden enormous increase due to the arrival of the annual wet *monsoon* (the Arabic word for season). The *Af* subtype is named after the vegetation association that develops there—the rainforest. The *Am* subtype, prevailing in part of peninsular India, in a coastal area of West Africa, and in sections of Southeast Asia, is appropriately referred to as the monsoon climate. A third tropical climate, the savanna (*Aw*), has a wider daily and annual temperature range and a more strongly seasonal distribution of rainfall. As Fig. I–9 indicates, savanna rainfall totals tend to be lower than those in the rainforest zone, and the associated seasonality often is expressed in a "double maximum." This means that each year produces two periods of increased rainfall separated by pronounced dry spells. In many savanna zones the people refer to the "long rains" and the "short rains" to identify these seasons, and a persistent problem in these regions in the unpredictability of the rain's arrival. Savanna soils are not among the most fertile, and when the rains fail the specter of hunger arises on these humid grasslands. Savanna regions are far more densely peopled than rainforest areas, and millions of residents of the savanna subsist on what they manage to cultivate. Rainfall variability under the savanna regime is their principal environmental problem.

The *dry* climates (*B*) occur in low as well as higher latitudes. The difference between the *BW* (true desert) and *BS* (semiarid steppe) varies, but may be taken to lie at about 25 centimeters (just over 10 inches). Parts of the central Sahara receive less than 10 centimeters (4 inches) of rainfall and, as Fig. I–9 shows, much of West Africa's Sahel is steppe country, the most tenuous of environments for farmers or herders. A pervasive characteristic of the world's dry areas is the enormous daily temperature range that may exceed 35°C (60°F) in low-latitude deserts. Recorded instances exist where the maximum daytime shade temperature was over 49°C (120°F) followed by a nighttime low of 9°C (48°F).

The *humid temperate* climates (*C*) are also referred to as the mid-latitude climates; as the map shows, almost all the areas of *C* climate lie just beyond the Tropics of Cancer and Capricorn. This is the prevailing clime in the southeastern United States from Kentucky to central Florida, on North America's west coast, in Western Europe and the Mediterranean, in southern Brazil and northern Argentina, in coastal South Africa and Australia, and in eastern China and southern Japan. None of these areas is associated with climatic extremes or severity, but the winters here can be fairly cold, especially away from the coasts. These areas lie about midway between the winterless equatorial climates and the summerless polar zones.

The humid temperate climates range from quite moist, as along the densely forested coasts of Oregon, Washington, and British Columbia, to relatively dry, as in the so-called Mediterranean (dry-summer) areas that include not only coastal southern Europe and northwest Africa, but also the southwestern tips of Australia and Africa, middle Chile, and southern California. In these Mediterranean areas the scrubby, moisture-preserving vegetation creates a natural landscape very different from that of richly green Western Europe.

The *humid cold* climates (*D*), or snow climates as they are also known, may be called the continental climates, for they seem to develop within the interior of large landmasses, as in the heart of Eurasia and North America. No equivalent land areas, at similar latitudes, exist in the Southern Hemisphere, and no *D* climate occurs there at all.

Great annual temperature ranges mark these humid continental climates, and very cold winters and relatively cool summers are the rule. In a *Dfa* climate, for example, the warmest summer month (July) may average as high as 21°C (70°F), but the coldest month (January) only −11°C (12°F). Total precipitation, a substantial part of which comes in the form of snow, is not very high, ranging from over 70 centimeters (30 inches) to a steppelike 25 centimeters (10 inches). Compensating for this paucity of precipitation are cool temperatures, which inhibit the loss of moisture via evaporation and evapotranspiration.

Some of the world's most productive soils lie in areas under humid cold climates, including the U.S. Midwest, parts of the Soviet Union's Ukraine, and North China. The period of winter dormancy, when all water is frozen, and the accumulation of plant debris during the fall combine to balance the soil-forming and enriching processes. The soil differentiates into well-defined nutrient-rich layers, and a substantial store of organic humus accumulates. And even where the annual precipitation is light, this environment sustains extensive coniferous forests.

The *cold polar* climates (*E*) are differentiated into true icecap conditions, where permanent ice and snow exist and no vegetation can gain a foothold, and the "tundra," where up to four months of the year average temperatures may be above freezing. Like *rainforest*, *savanna*, and *steppe*, the term *tundra* is a vegetative as well as a climatic appellation, and the boundary between *D* and *E* climates on Fig. I–10 can be seen to correspond quite closely to that between the needleleaf forests and arctic tundra on Fig. I–12. Finally, the *H* climates—undifferentiated highlands—resemble the *E* climes in a number of ways: their high elevations and complex topography, associated with major mountain systems, often produce near-arcticlike climates above the tree line, even in the lower latitudes (such as the Andes in equatorial South America).

We have already pointed out that the mapping of world climates in Fig. I–10 (and global precipitation in the preceding map) involves a good deal of generalizing with respect to the much more complicated and detailed environmental patterns that actually occur across the surface of the earth. Köppen and Geiger—who performed their work in the early decades of this century—realized the value of generalization, which allowed them to concentrate on the big picture unencumbered by less essential local complexities. This kind of methodology persists in contemporary geography, which in the last 30 years has greatly expanded the search for general principles and process-response relationships through the use of laboratorylike abstractions called **spatial models** (see *Model Box 1*, pp. 20–21).

## Vegetation Patterns

The map of world vegetation patterns (Fig. I–12) shows the closeness of the spatial relationship between climatic regions and plant associations. This map depicts the global distribution of natural vegetation, but much of that vegetation has been destroyed or modified by the human population. Hence, the regions shown on Fig. I–12 represent the natural vegetation that exists or would exist as a result of long-term plant succession and adaptation to prevailing climatic conditions. When climate over an area remains the same for several thousand years, a plant association develops that is in equilibrium with this environment. Such a *climax* vegetation may be a rainforest, as in the Amazon Basin, or it may be a scrub-and-bush association that just manages to take hold in the driest steppe.

We should not lose sight of the changeable character of Pleistocene environments, however; like the maps of precipitation and climates, this map of world vegetation is a still from a motion picture, not a permanent end product. Places exist where one plant association can

**Figure I–9**

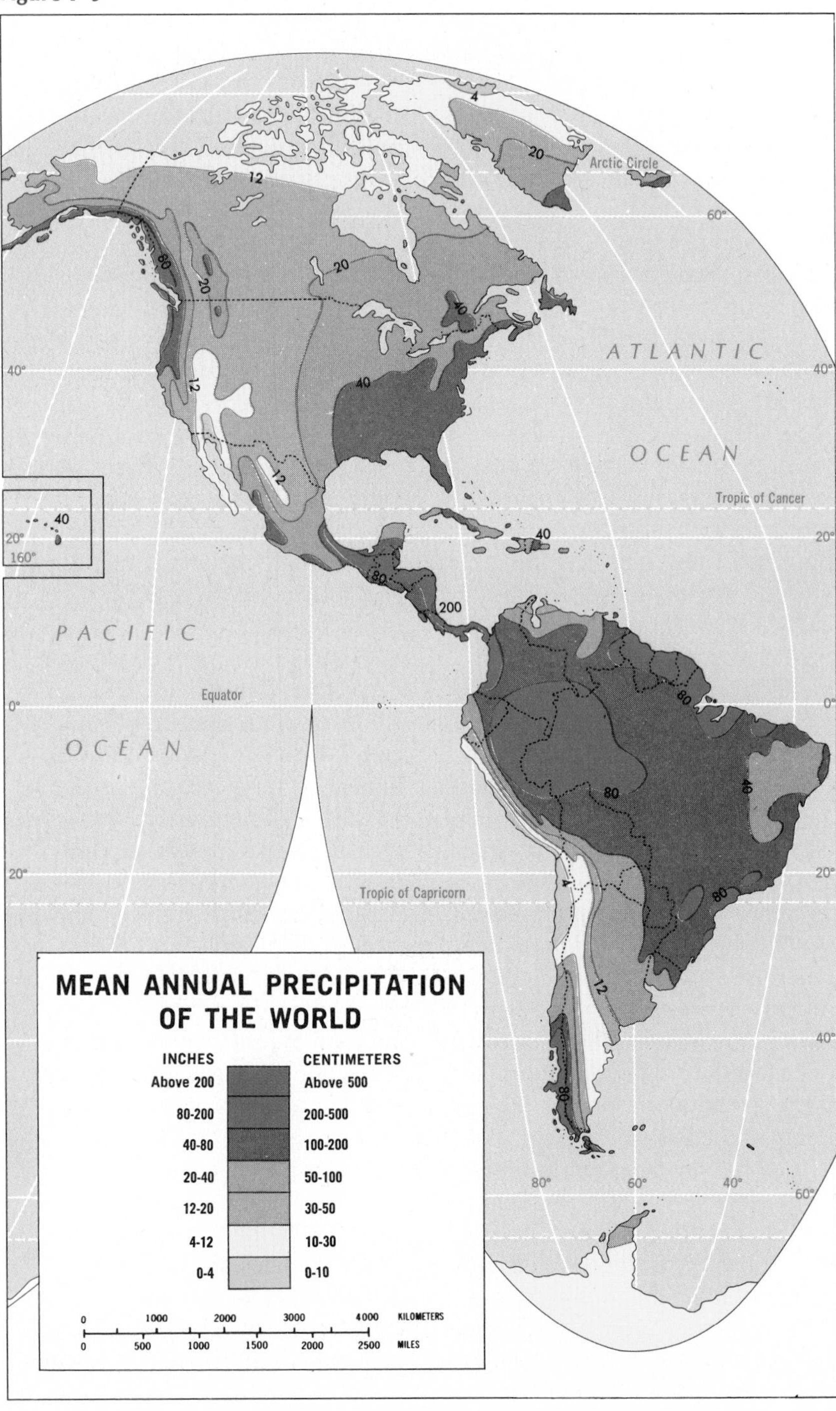

be observed to be gaining on another, signaling change. In West Africa, the steppe is encroaching southward on the savanna, and the Sahara, in turn, is gaining on the steppe. Inexorable climatic change may be involved in this, although human communities in the region have played a major role in modifying the natural environment and thereby hastening "desertification." In any case, large parts of the regions shown on Fig. I–12 do not actually support the climax vegetation communities shown in the map's legend.

The vegetative cover of the earth consists of trees, shrubs, grasses, mosses, and an enormous variety of other plants. Plant geographers group this mass of vegetation into the five broad vegetative regions (*biomes*) shown in the map legend: forest, savanna, grassland, desert,

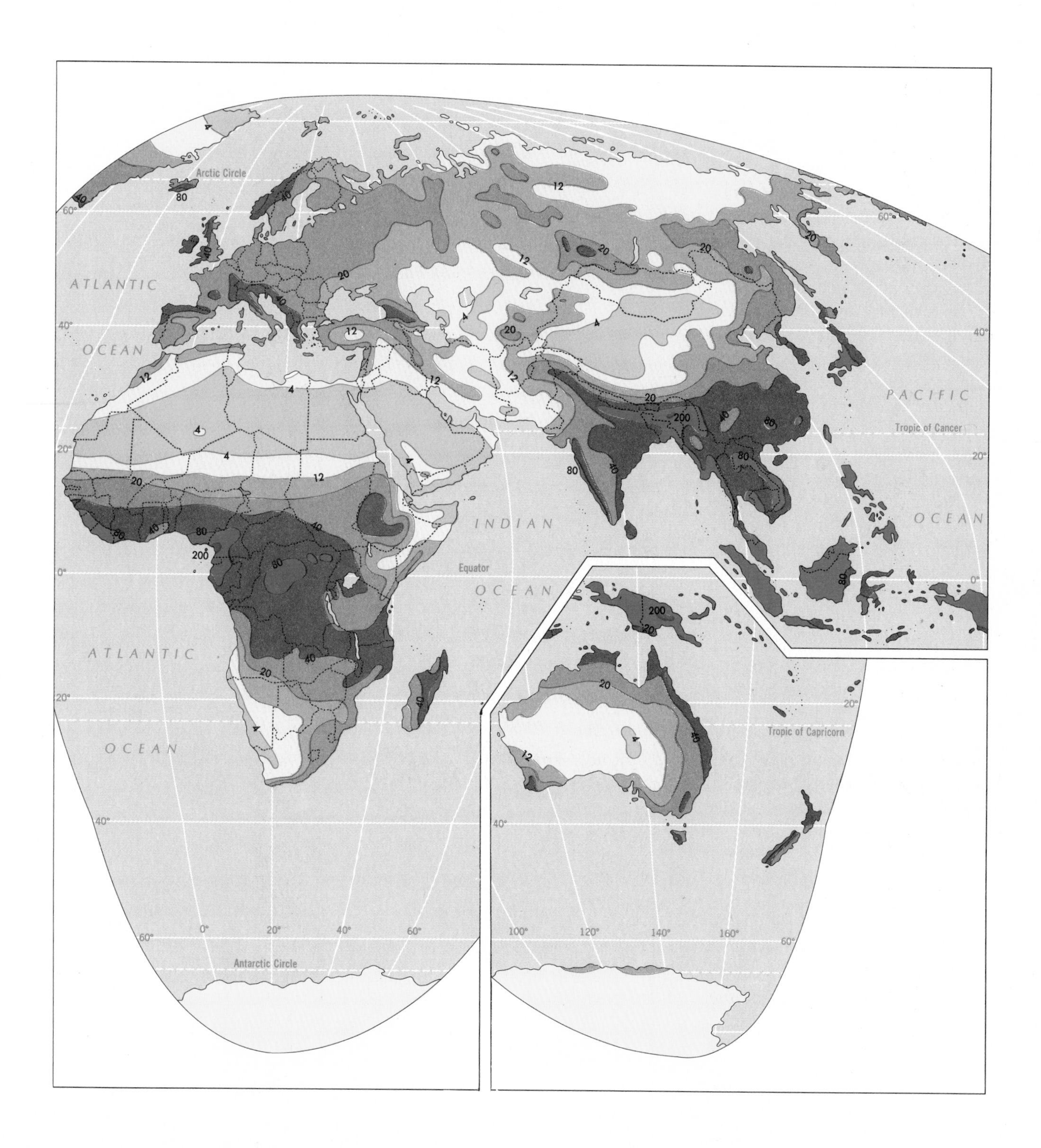

and tundra. Note that these biomes are not confined to particular latitudes; there is forest in the equatorial zone, in the subtropics poleward of the Tropics of Cancer and Capricorn (23½° North and South latitude, respectively), in mid-latitude areas such as the eastern United States, and in higher latitudes on the margins of the tundra. The adaptation of the forest species is what differs. Equatorial and temperate forests have leafy evergreen trees, whereas cold-climate, high-latitude forests have coniferous trees with thin needles. In the mid-latitudes, trees are deciduous and shed their leaves each autumn.

Another prominent impression gained from Fig. I–12 relates to the vastness of the savanna grasslands. The bulk of Africa south of the Sahara is savanna country, as is most of eastern Brazil and India. The savanna also prevails in interior Southeast Asia and in northern Australia; in later chapters there will be frequent occasion to refer to the vagaries of the savanna environment, with which hundreds of millions of the world's farmers must cope.

## Soil Distribution

We conclude this overview of the human habitat with an examination of another vital ingredient for the sustenance of life on this planet: the soil. The earth's soils have proved even more difficult to classify and regionalize than climate and vegetation, and Fig. I–13 is only one possible alternative among several. Research into the processes of soil formation continues to produce new data, and as this new evidence becomes available the regionalization schemas change.

Because the parent material (the rocks beneath), the temperature, the moisture conditions, the vegetation, and the terrain (degree of slope steepness) all vary from place to place across the globe, there is enormous diversity of soils as well. Some soils are infertile and cannot sustain crops; others can carry two or even three crops annually, and do so year after year. And still today, liberating technologies notwithstanding, the great majority of the world's people depend directly on the local soil for their food. Any map of global population distribution to a considerable degree reflects the productiveness of the soils of certain particular areas—and their infertility elsewhere.

Figure I–13 once again displays

Figure I–10

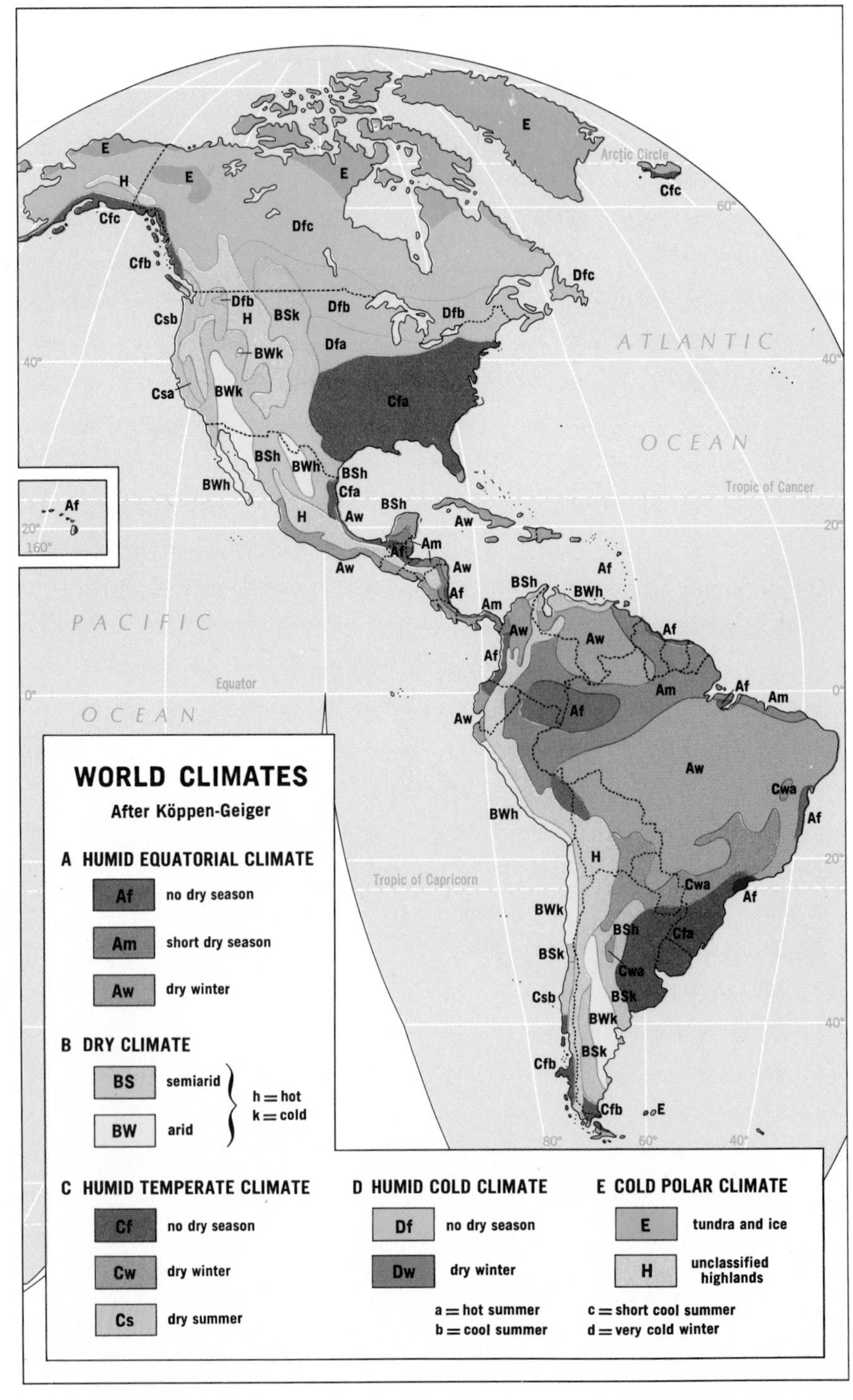

areas of correspondence with world distributions of climatic elements and vegetation, reminding us that the soil is a responsive element of the total environmental complex. It is important, again, to remember that this map of the world distribution of soils represents a generalization of the actual situation. After many years of experimentation, pedologists (soil scientists) produced an all-encompassing classification system known as the Comprehensive Soil Classification System (*CSCS*); because so many less successful attempts preceded it, they refer to it as the "Seventh Approximation." According to this classification, the world's soils are grouped into 10 *Orders*, which, in turn, are divided into 47 Suborders, 185 Great Groups, about 1000 Subgroups, 5000 Families, and 10,000 Series. No small-scale map

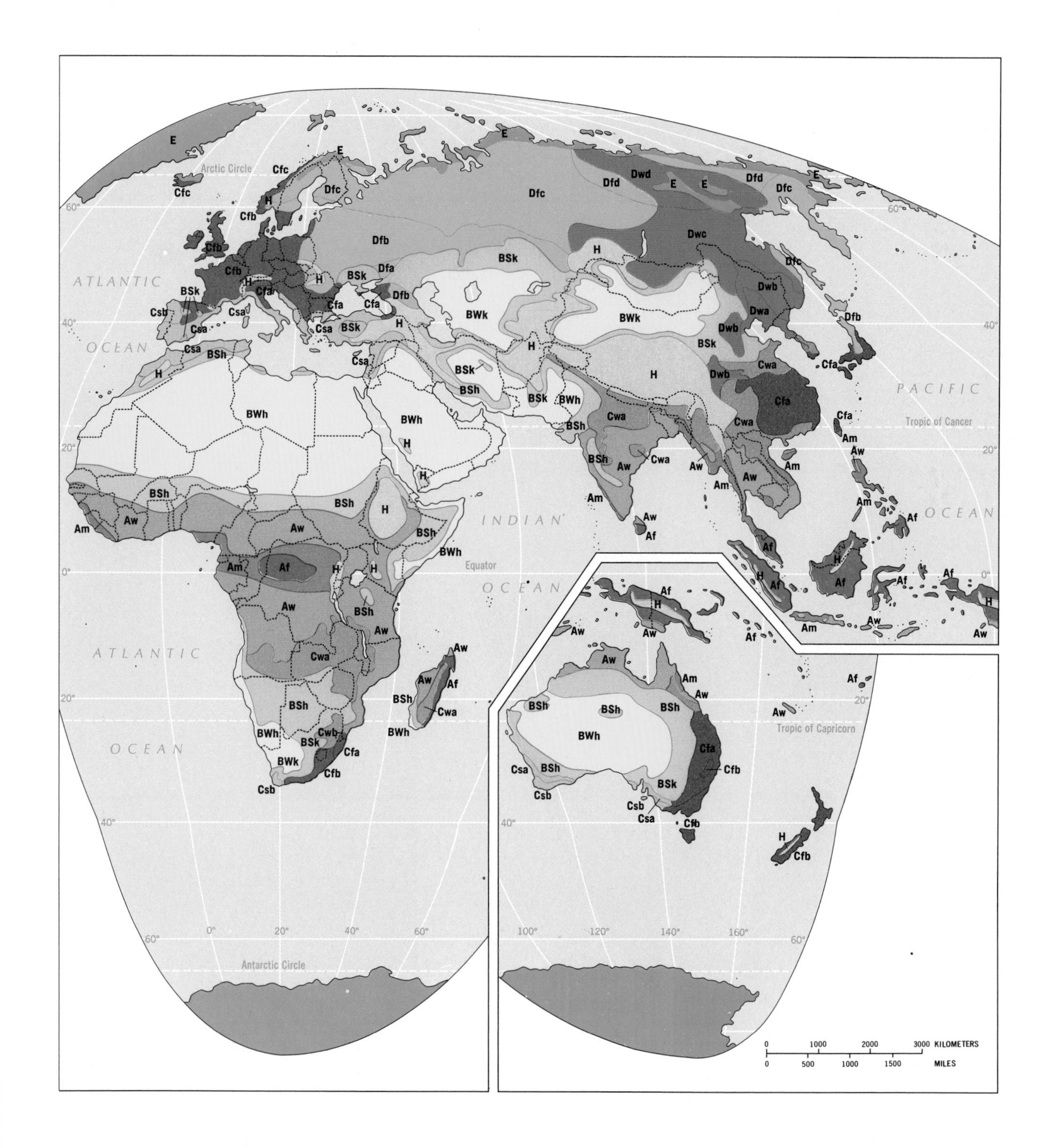

## MODEL BOX 1

# THE HYPOTHETICAL CONTINENT AND MODELS IN GEOGRAPHY

A modern approach to generalization in both physical and human regional geography is through the development of **models**. Peter Haggett, in his 1965 book *Locational Analysis in Human Geography*, offers an especially lucid definition: *in model-building we create an idealized representation of reality in order to demonstrate its most important properties.* He further points out that using models is made necessary by the complexity of reality—in order to understand how things work we must first filter out the main processes and responses from the myriad details with which they are embedded in a highly complicated world. In other words, models provide a simplified picture of reality, conveying perhaps not the entire truth but a useful and essential part of it.

The effectiveness of such modeling of reality is demonstrated in Fig. I–11, which presents the generalized distribution of climates on a *Hypothetical Continent* that tapers southward from the upper latitudes of the Northern Hemisphere to the mid-latitudes of the Southern Hemisphere. Note that the shape of this idealized land surface is based on a representation of the amount of land that is observed at each latitude on the world map: Eurasia and Canada are widest in their extent at about 50° North, whereas at 35° South only the relatively narrow southern extremities of South America, Africa,

Figure I–11

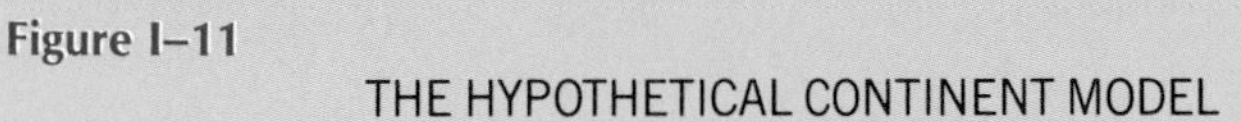


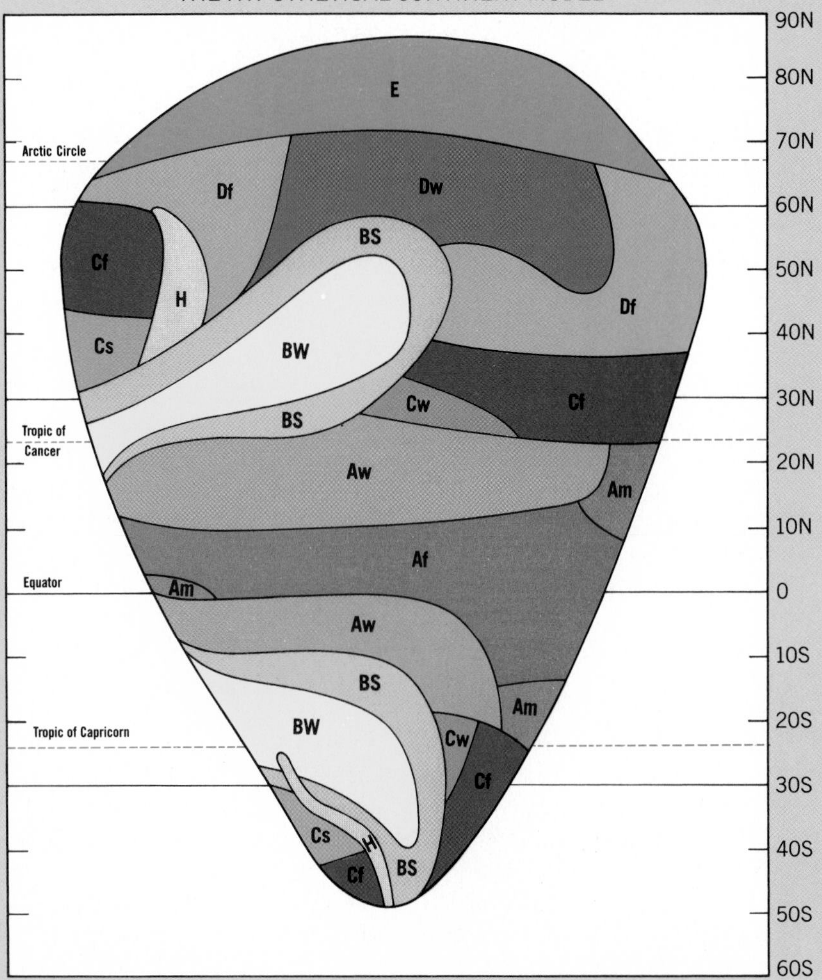

and Australia interrupt the vast Southern Ocean that girdles the globe. Within the Hypothetical Continent, the distribution of climate regions—utilizing the identical color scheme of Fig. I–10—represents a considerable smoothing out of the regionalization system of Köppen and Geiger. This simplification makes the overall geographical pattern of climate types much easier to comprehend and think about, and, with the exception of place references, is an equally effective spatial expression of the text discussion of climate regions. Once the generalities are understood, it is a simple matter to shift one's attention back and forth between the idealized and *empirical* (actual) maps—Figs. I–11 and I–10—to see how and why certain distortions occur in reality.

A similar Hypothetical Continent could be constructed to model the regionalization (and interrelationships) of precipitation, or vegetation, or soils, or even population—all of which are geographically linked to the world distribution of climates in this part of the chapter. Models play an increasingly vital role in understanding the concepts of geography and the regional spatial structure of the late twentieth-century world. Therefore, we will occasionally use *Model Boxes* in future chapters to elaborate on some of the most important of these geographical abstractions.

such as Fig. I–13 could even begin to communicate all this detail.

Some of the new soil names are self-explanatory and easy to understand. *Oxisols*, for example, are the excessively leached soils of the tropics, the familiar reddish-colored soils marked by high concentrations of the oxides of iron and aluminum. These soils, formerly called laterites and latosols, are what the subsistence farmers of the savannas and equatorial areas of Africa and South America must confront and cultivate. The oxisols are often thick and deeply weathered, but the needed nutrients have, to a large extent, been leached or washed downward. Tropical vegetation that grows in this soil sustains itself by absorbing nutrients directly from fallen leaf matter. When the farmer clears the land, the oxisols may carry a crop for a year or two, but then they are exhausted and fail.

Another soil order with an appropriate name comprises the *aridisols*. These are salt-rich, infertile soils of rainfall-deficient areas and, as Fig. I–13 shows, they are widely distributed across the earth, occurring in the southwestern United States, in western South America, and in vast areas of Africa, Asia, and Australia.

In the high latitudes, where soil-forming conditions are inhibited by periods of extreme cold, limited warmth from the sun, the thin covering of tundra mosses and lichens, and poor drainage, the incompletely developed soils are called *inceptisols* (which also occur on coastal and river floodplains in lower latitudes). This is another name with an obvious derivation, but the neighboring *spodosols* are not. The spodosols, which extend across northern Canada and much of high-latitude Eurasia, are better developed and support great stands of pine and spruce (the Needleleaf Forests on Fig. I–12). They owe their name to the peculiar, ashlike, bleached appearance that results from the constant and intense removal of matter from the upper soil layer; Russian farmers refer to this as *podzol* (ashy soil).

Between the oxisols (with the related ultisols and vertisols) and the aridisols of lower latitudes and the inceptisols and spodosols of higher latitudes lie the *mollisols* and the *alfisols*. As Fig. I–13 shows, these two soil orders extend across vast reaches of interior North America, Eurasia, and, to a lesser extent, South America. The mollisols (*mollis* is Latin for "soft") possess a dark, humus-rich upper layer, and they never become hard as the aridisols do, even under comparatively dry conditions. As the map indicates, mollisols are located in intermediate positions between dry and humid climates (see Fig. I–9). These are the grassland soils that support large livestock herds or, when farmed, sustain vast expanses of grain crops. Alfisols occur in a broad zone in Canada, in the U.S. Midwest, and in Europe and the Soviet Union. These soils evolve under a wide range of environmental conditions and are generally quite fertile, often supporting highly intensive agriculture.

Figure I–13 shows limited areas of *entisols* (rec*ent*) in North America and Eurasia (but larger zones in Africa and Australia). Entisols, like inceptisols, have not existed long enough to develop maturity. They may lie on sand accumulations, on recently deposited river alluvium (silt), or in areas subject to strong erosion where surface material is constantly removed. Except for intrinsically fertile alluvium, entisols are not usually good soils for farming. *Histosols* (from *histos*, Greek for "tissue") are the soils of bogs and moors, and they primarily consist of organic rather than mineral matter. Frequently waterlogged, histosols may develop as peat accumulations. The largest areas of histosol exposure lie in Canada, south of Hudson Bay, and in the far northwest, just below the permafrost limit.

Finally, Fig. I–13 shows the distribution of the world's mountain soils, the soils of high-relief terrain. A comparison with Fig. I–5, however, proves that not all mountainous or high-elevation areas must necessarily contain thin, poorly developed or stony soils. The highlands of Ethiopia, for example, form an area of comparatively fertile alfisols.

It is also interesting to compare Fig. I–13 with Fig. I–14 (showing the world distribution of population). The valley and delta of the lower Nile River in North Africa, the basin of the Ganges River in South Asia, and the plain of the Huang He in China all contain alluvial inceptisols, and through their nearly legendary fertility they sustain many millions of people. Today, fully 95 percent of Egypt's 48 million people live within 20 kilometers (12 miles) of the Nile River. Hundreds of millions of Indians and Chinese depend directly on alluvial soils in the river basins of the Ganges and Huang He (Yellow), where crops are grown that range from corn to cotton, wheat to jute, rice to soybeans. These are only the most prominent examples of such alluvium-based agglomerations: in Pakistan, Bangladesh, Vietnam, and many other countries, the alluvial soils of river valleys provide in abundance what the generally infertile upland soils do not.

These riverine population clusters also provide a link with humanity's past. In the fertile valleys of the Middle East's rivers, the art and science of irrigation may first have been learned. And two of the world's oldest continuous cultures —Egypt and China—still retain heartlands near their ancient geographical hearths of thousands of years ago.

# Space and Population

Less than 30 percent of the earth's surface is land area; but, as we observed in the course of our brief survey of global environments, large parts of that land area cannot support any substantial number of inhabitants. Arid deserts, rugged mountain ranges, and frigid tundras constitute just a few of the less habitable parts of the world, but together they cover nearly half the landmasses—and we have not even begun to account for places where unproductive soils, persistent diseases, frequent droughts, and other more localized conditions keep population numbers comparatively low.

As a result, the map of world

Figure I–12

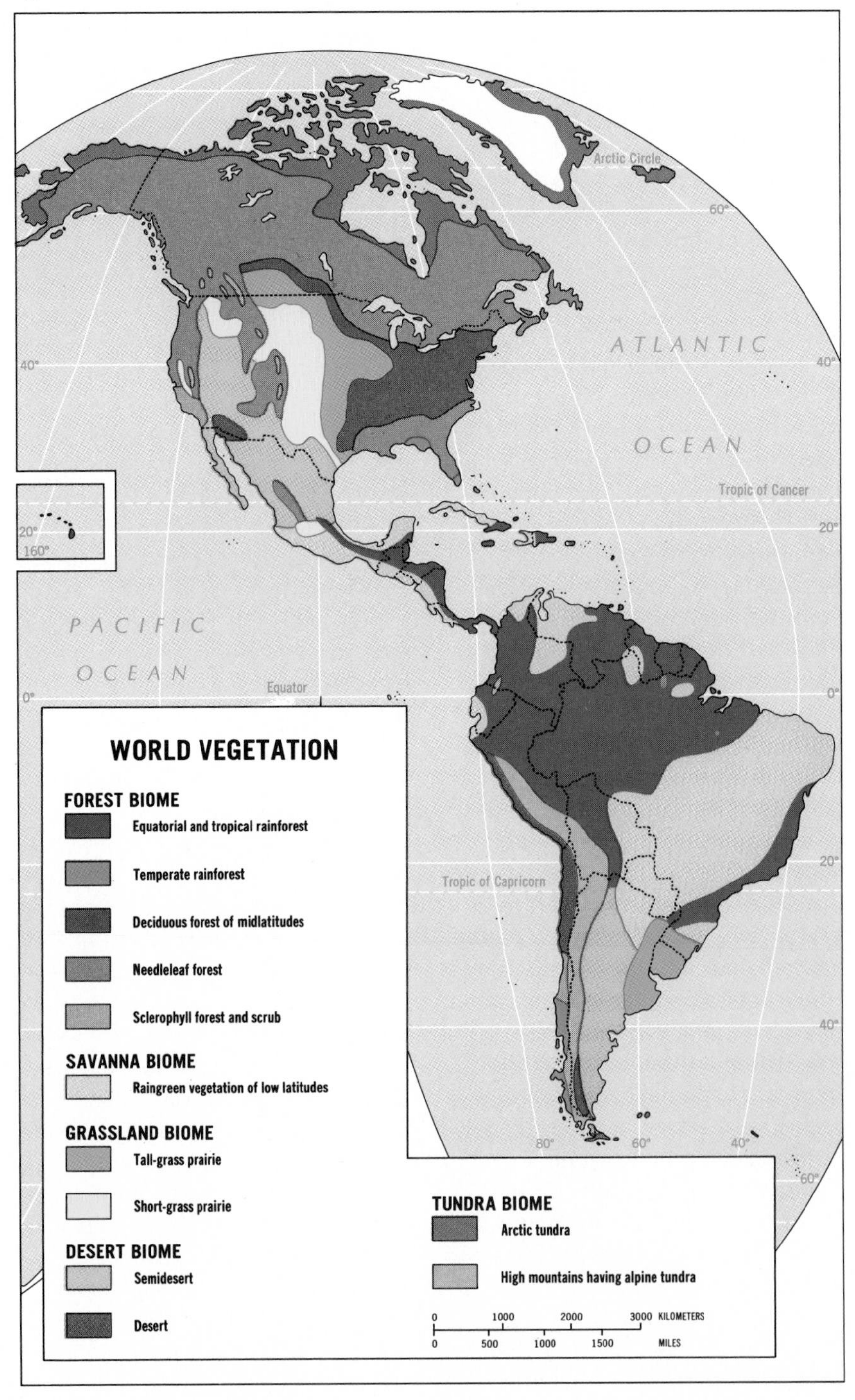

population distribution (Fig. I–14) prominently displays the clustering of huge numbers of people in areas of fertile and productive soils. For thousands of years, following the beginning of crop and animal domestication, communities remained dependent on soils and pastures in their immediate vicinity; the large-scale, worldwide food transportation of today is essentially a phenomenon of the past 200 years, a product of industrial and technological revolutions. True, the ancient Romans imported grains from North Africa, and colonial powers even centuries ago brought shiploads of spices and sugar across the oceans. But the overwhelming majority of the world's peoples continued to subsist on what they could cultivate locally. When groups migrated, they did so in search of new lands to be opened up, new

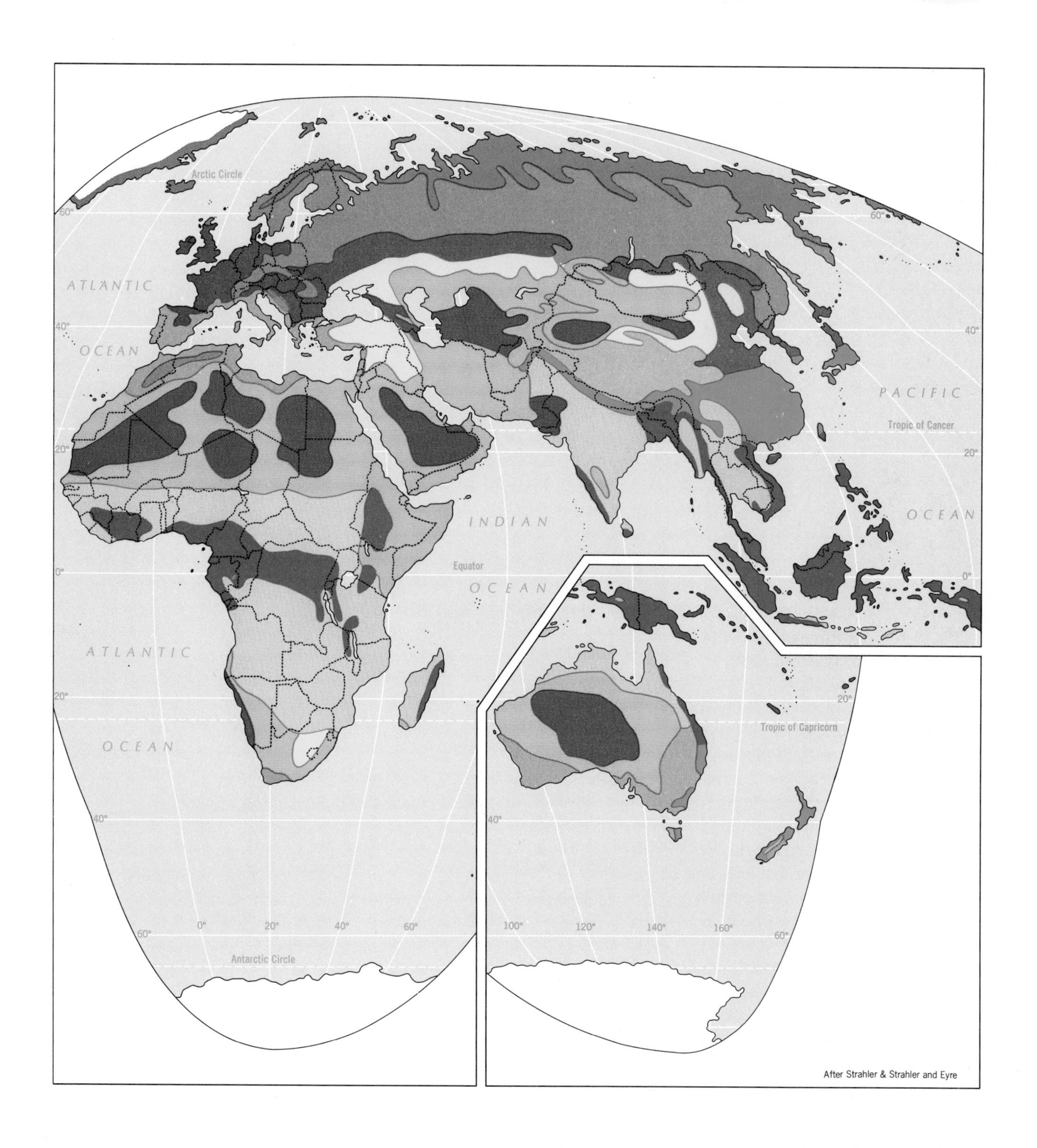

pastures to be exploited. If they succeeded, as the Chinese of North China did when they moved southward, they thrived. When they failed, they often starved. Today's map of world population is another one of those stills from that Pleistocene motion picture, for it shows the current stage of humanity's expansion and dispersal across the habitable space on the earth's surface. And the picture is changing rapidly. After thousands of years of relatively slow growth, world population during the past two centuries has been expanding at an increasing rate. It took about 17 centuries from the time of the birth of Christ for the world to add 250 million people. We will add that same number in less than the next 3 years. This is the subject of discussions in Chapters 1 and 8, but it is important to realize that this explosive modern growth is *not* simply a matter of filling in the remaining available living space on the globe. Rather, it is the already-crowded areas that are becoming even more so; thus, the food produced in the ricefields of Asia (plentiful though it is) must be shared by ever more and hungrier people.

At present we confine ourselves to a view of the world population as it is distributed today. The 1984 world population was estimated to be some 4.8 billion, and it is broken down by region and country in the Appendix (pp. 541–545). More than one-fifth of this total resides in one country, China, and the next most populous single state, India, has more than 750 million inhabitants. Yet, both India and China still contain extensive areas that are nearly devoid of permanent population, as Fig. I–14 reveals. China's habitable, agriculturally productive environments lie concentrated in that country's east, and the map leaves no doubt about the association between fertile land and the clustering of population.

Figure I–14 shows that the earth presently contains four major population agglomerations. The three largest are all located on a single landmass—Eurasia—and include East Asia, South Asia and Europe; the fourth of these world-scale clusters is found in eastern North America. The spatial prominence of these population concentrations on the world map of humankind is even more apparent in Fig. I–15. This specially transformed map or *cartogram*, instead of being based on the traditional scale representation of distance, is based on area data so that countries containing large numbers of people are "blown up" in population-space,

Figure I–13

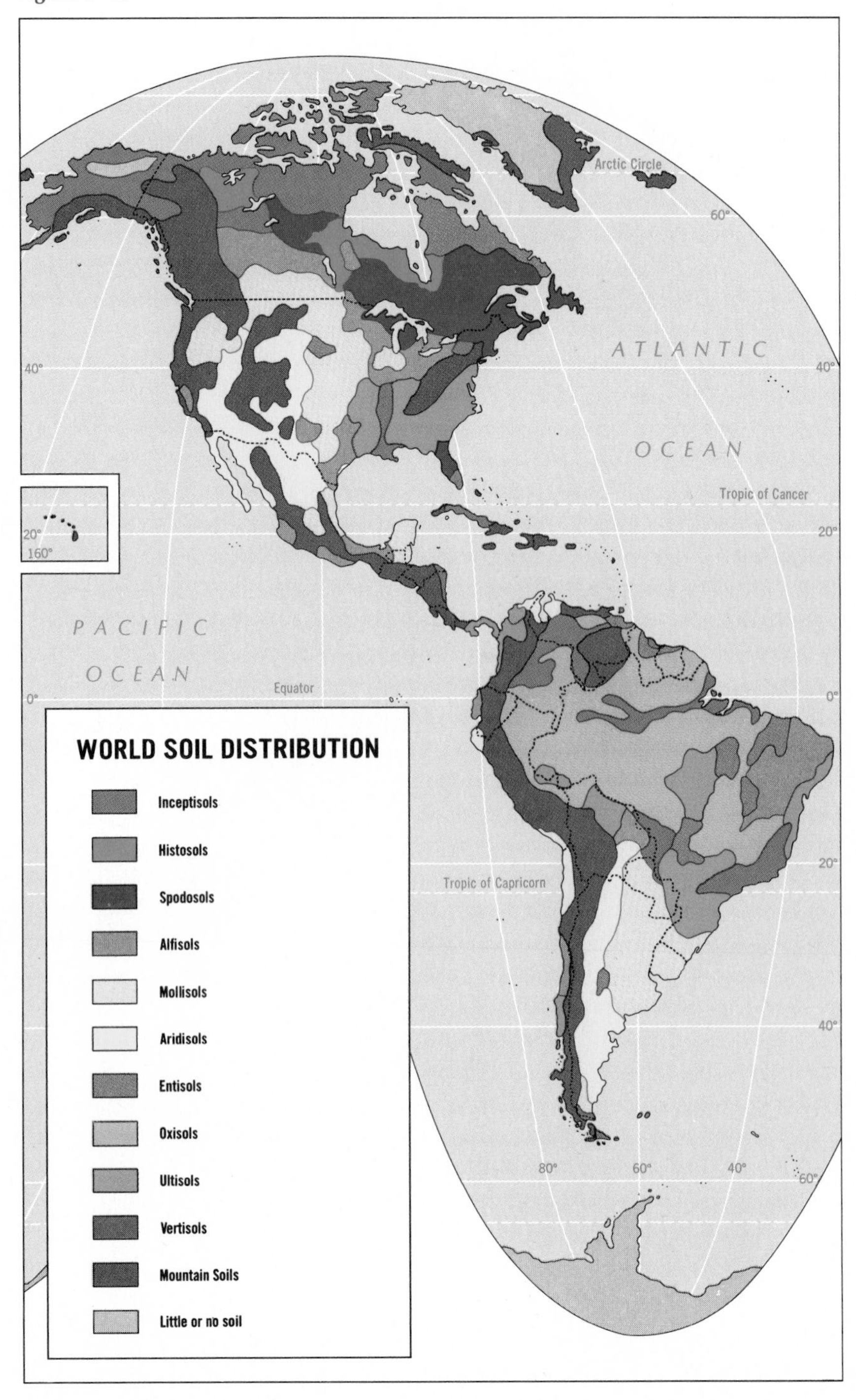

whereas those containing smaller numbers are shrunk in size accordingly.

The greatest of the four population agglomerations is the *East Asia* cluster, adjoining the Pacific Ocean from Korea to Vietnam and centering, of course, on China itself. The map indicates that the number of people per unit area tends to decline from the coastal zone toward the interior, but several ribbonlike extensions can be seen to penetrate the deeper interior (*A* and *B* on Fig. I–14). Reference to the map of world landscapes (Fig. I–5) proves that these extensions represent populations concentrated in the valleys of China's major rivers, especially the Huang He (Yellow) and the Chang Jiang (Yangtze). This serves to remind us that the great majority of the people of East Asia are farmers, not city dwellers. True, there are great cities in China, and some of them, such as Shanghai and Beijing (Peking), rank among

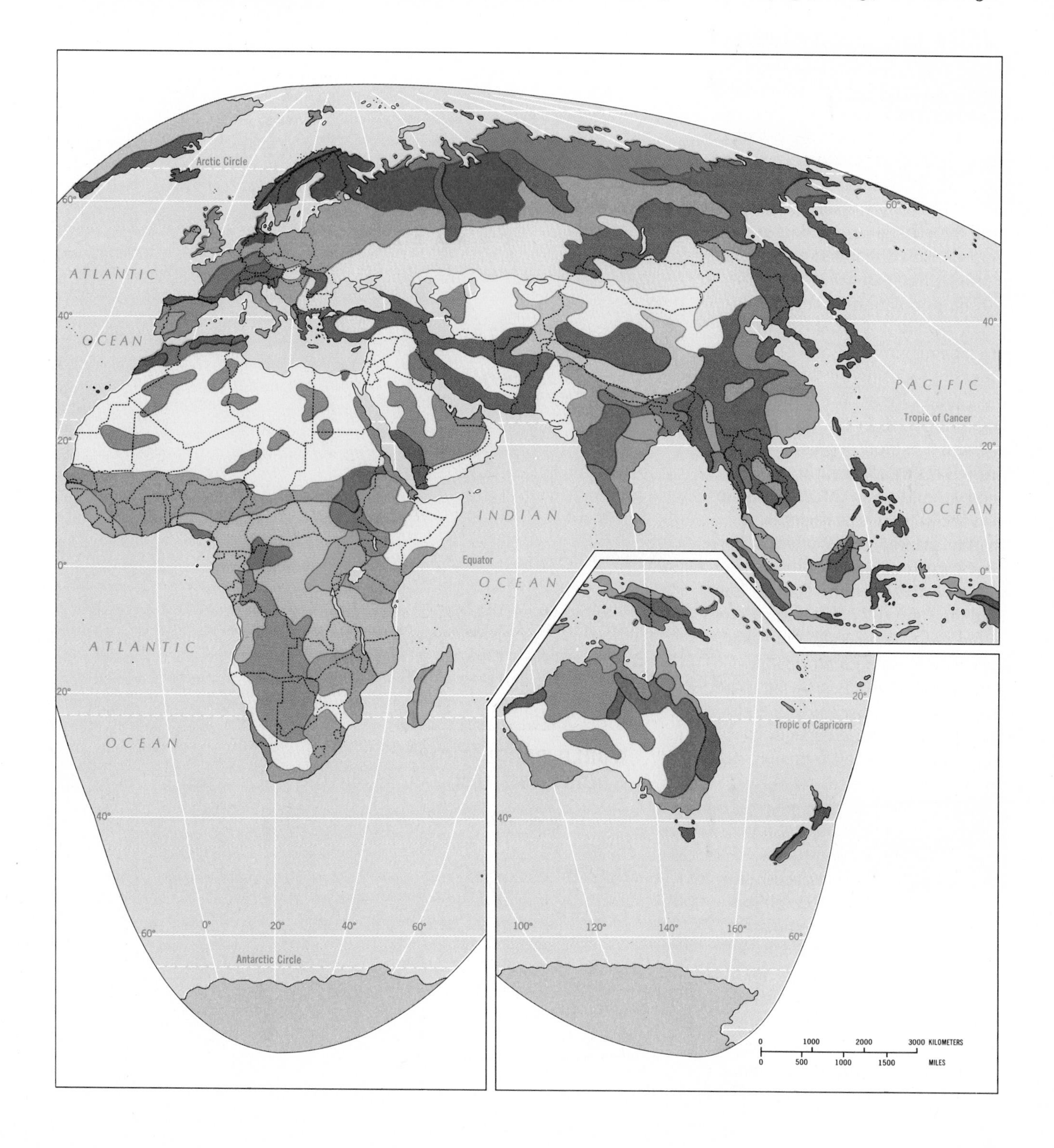

the largest in the world. But the total population of these and the other cities is far outnumbered by the farmers—those who need the river valleys' soils, the life-giving rains, and the moderate temperatures to produce crops of wheat and rice to feed not only themselves, but also those residing in the cities and towns.

The second major concentration of world population also lies in Asia and displays many similarities to that of East Asia. At the heart of this *South Asia* cluster lies India, but it extends also into neighboring Pakistan, Bangladesh, and island Sri Lanka (formerly Ceylon). Again, note the coastal orientation of the most densely inhabited zones, and the fingerlike extension of high-density population in northern India (C on Fig. I–14). This is one of the great agglomerations of people on earth, focusing on the valley of the Ganges River. The South Asia population cluster in 1985 numbers an even 1 billion people, and at present rates of growth it will exceed 1.3 billion in 2000. Our map also shows how this region is sharply marked off by physical barriers: the Himalaya Mountains rise to the north above the Ganges lowland, and the desert takes over west of the Indus River Valley in Pakistan. This is a confined region, whose population is growing more rapidly than almost anywhere else on earth and whose capacity to support it has, by all estimates, already been exceeded. As in East Asia, the overwhelming majority of the people are farmers, but here in South Asia the pressure on the land is even greater. In Bangladesh (formerly East Pakistan), 103 million people are crowded into an area about the size of Iowa. Nearly all of these people are farmers, and over large parts of Bangladesh the rural population density is more than 1000 per square kilometer. In comparison, the 1980 population of Iowa was about 2.9 million, with a rural density of just 8 per square kilometer.

Further inspection of Fig. I–14 reveals that the third-ranking population cluster—*Europe*—also lies on the biggest landmass, but at the opposite end of Eurasia from China. An axis of very dense population extends from the British Isles into Soviet Russia. It includes large parts of West and East Germany, Poland, and the western Soviet Union; it also incorporates the Netherlands and Belgium, parts of France, and northern Italy. This European cluster (including the contiguous

Figure I–14

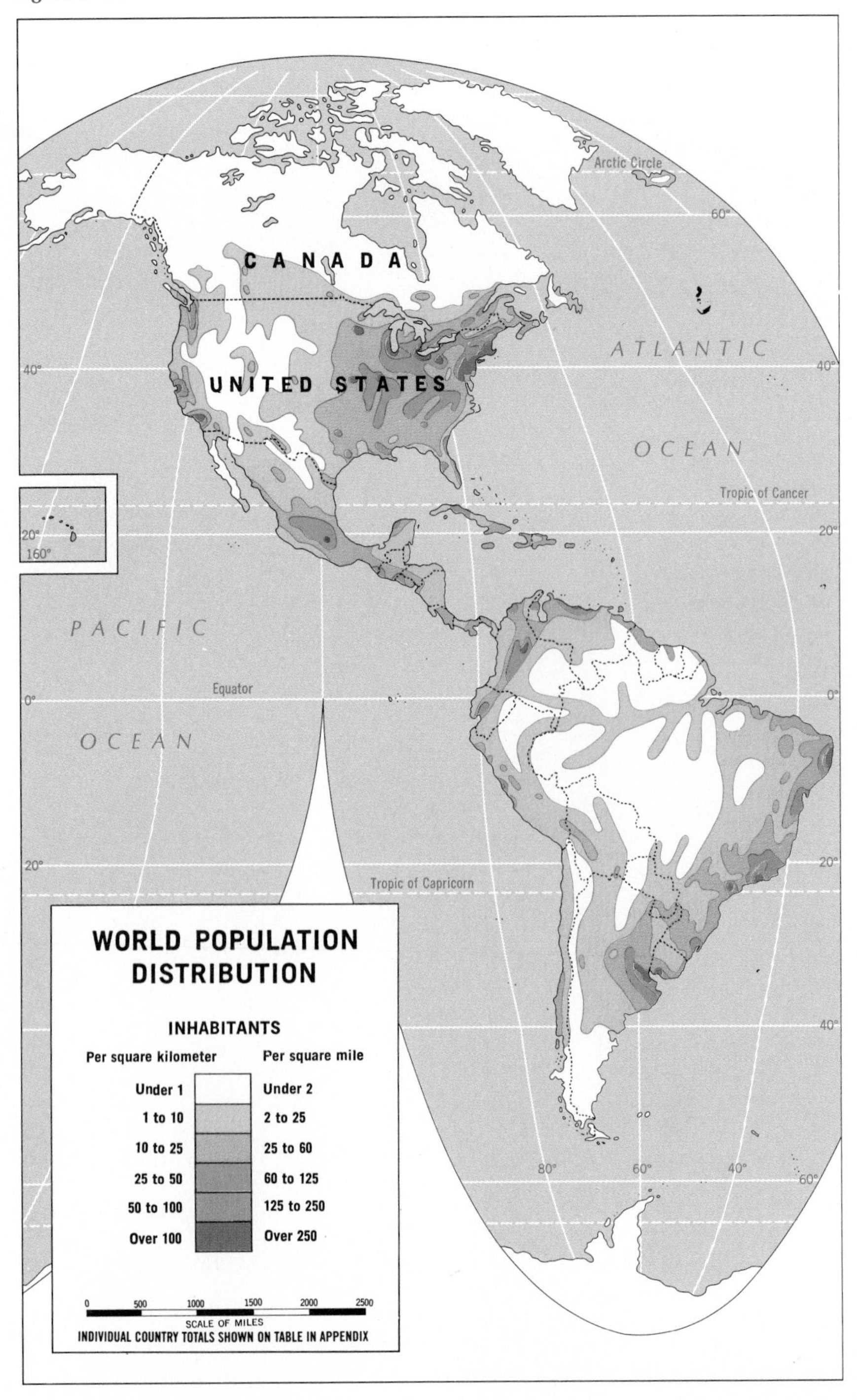

U.S.S.R.) counts over 700 million inhabitants, which puts it in a class with the South Asia concentration—but there the similarity ends. A comparison of the population and physical geography maps indicates that in Europe, terrain and environment appear to have less to do with population distribution than in the two Asian cases. See, for example, that lengthy extension marked *D* on Fig. I–14, which protrudes far into the Soviet Union. Unlike the Asian extensions, which reflect fertile river valleys, the European population axis is associated with the orientation of Europe's coalfields, the power resources that fired the industrial revolution. If you look more closely at the landform map (Fig. I–5), you will note that comparatively dense population occurs even in rather mountainous, rugged country—for example, along the

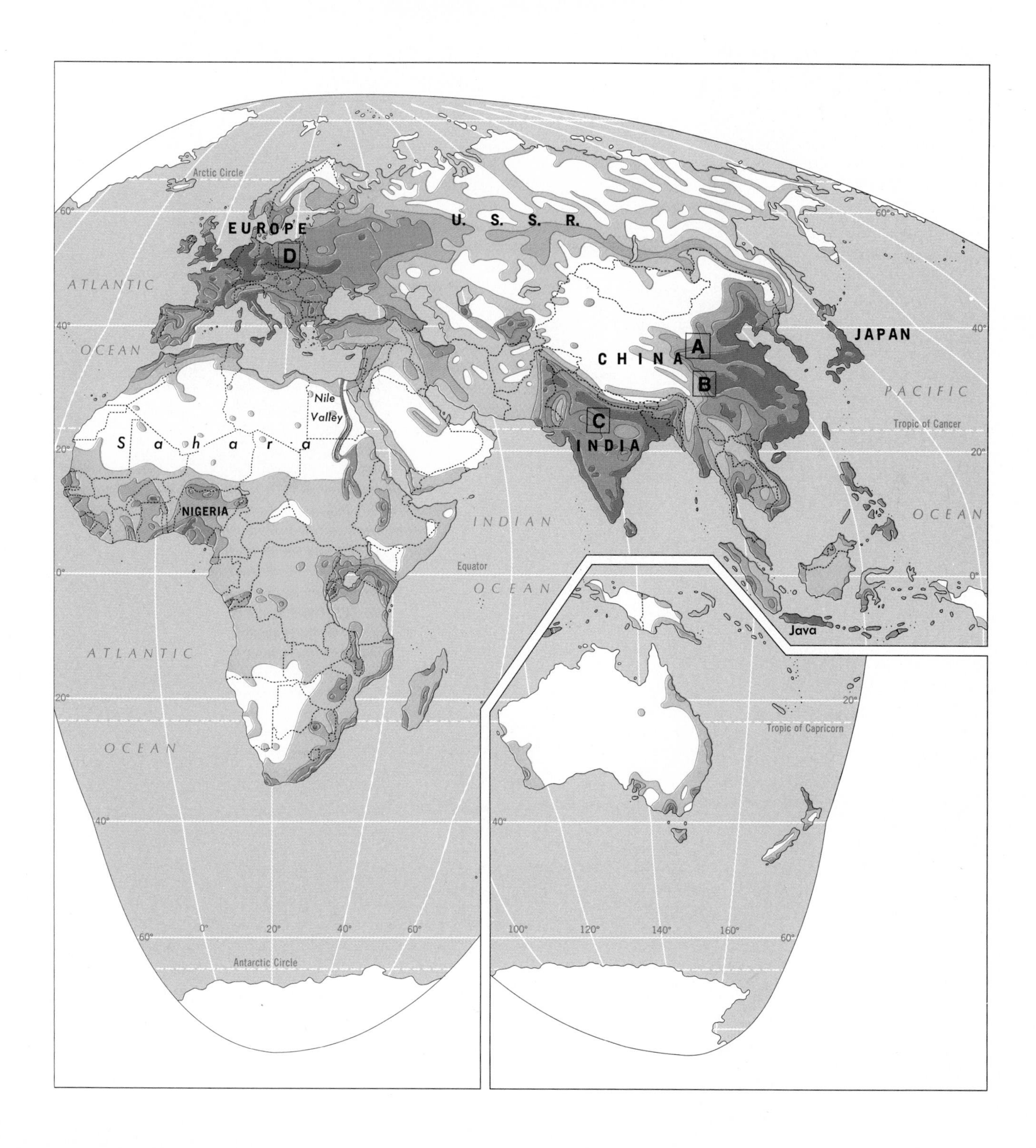

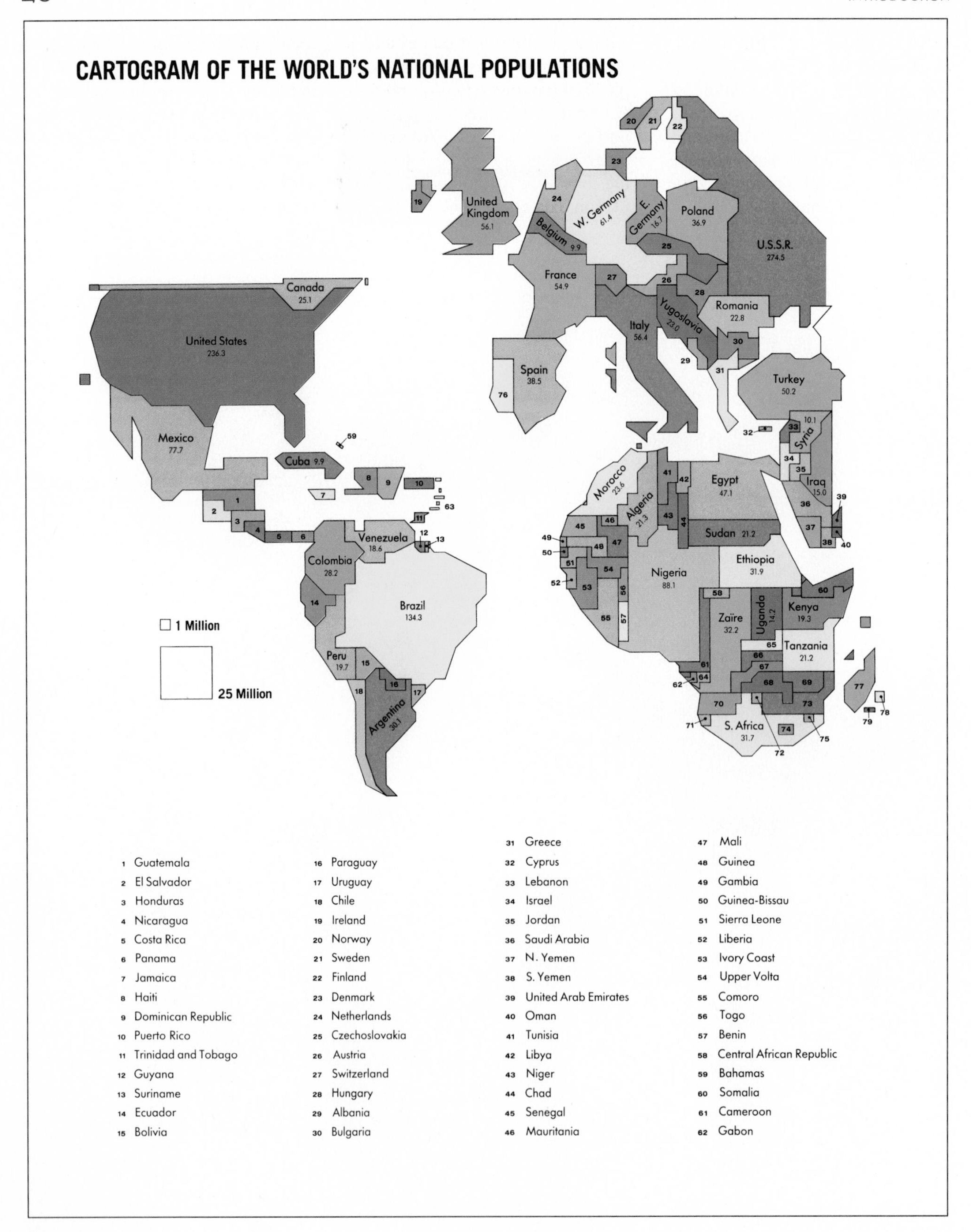
CARTOGRAM OF THE WORLD'S NATIONAL POPULATIONS
United Kingdom 56.1
W. Germany 61.4
E. Germany 16.7
Poland 36.9
U.S.S.R. 274.5
Belgium 9.9
France 54.9
Romania 22.8
Yugoslavia 23.0
Italy 56.4
Spain 38.5
Turkey 50.2
Canada 25.1
United States 236.3
Mexico 77.7
Cuba 9.9
Syria 10.1
Iraq 15.0
Morocco 23.6
Algeria 21.3
Egypt 47.1
Sudan 21.2
Ethiopia 31.9
Nigeria 88.1
Venezuela 18.6
Colombia 28.2
Brazil 134.3
Peru 19.7
Argentina 30.1
Zaïre 32.2
Uganda 14.2
Kenya 19.3
Tanzania 21.2
S. Africa 31.7
1 Million
25 Million
1 Guatemala
2 El Salvador
3 Honduras
4 Nicaragua
5 Costa Rica
6 Panama
7 Jamaica
8 Haiti
9 Dominican Republic
10 Puerto Rico
11 Trinidad and Tobago
12 Guyana
13 Suriname
14 Ecuador
15 Bolivia
16 Paraguay
17 Uruguay
18 Chile
19 Ireland
20 Norway
21 Sweden
22 Finland
23 Denmark
24 Netherlands
25 Czechoslovakia
26 Austria
27 Switzerland
28 Hungary
29 Albania
30 Bulgaria
31 Greece
32 Cyprus
33 Lebanon
34 Israel
35 Jordan
36 Saudi Arabia
37 N. Yemen
38 S. Yemen
39 United Arab Emirates
40 Oman
41 Tunisia
42 Libya
43 Niger
44 Chad
45 Senegal
46 Mauritania
47 Mali
48 Guinea
49 Gambia
50 Guinea-Bissau
51 Sierra Leone
52 Liberia
53 Ivory Coast
54 Upper Volta
55 Comoro
56 Togo
57 Benin
58 Central African Republic
59 Bahamas
60 Somalia
61 Cameroon
62 Gabon

**Figure I–15**

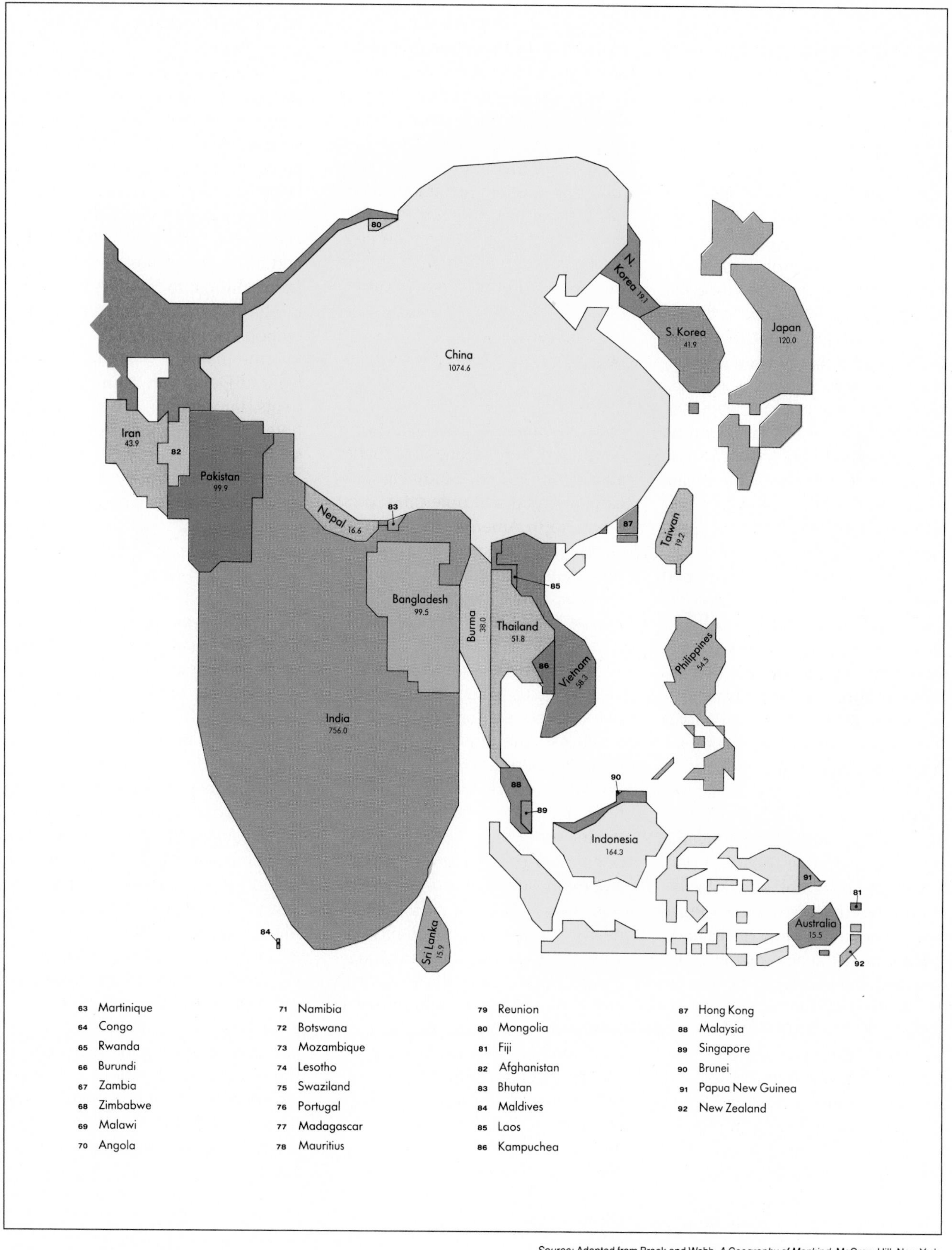


*Source:* Adapted from Broek and Webb, *A Geography of Mankind*, McGraw-Hill, New York, 3rd rev. ed., 1978.

boundary zone between Czechoslovakia and Poland. In Asia, there is much more correspondence between coastal and river lowlands and high population density than there generally is within Europe. Another contrast lies in the number of Europeans who live in cities and towns. Far more than in Asia, the Europe population agglomeration is composed of numerous cities and towns, many of them products of the industrial revolution. In the United Kingdom, about 91 percent of the people live in such urban places; in West Germany, nearly 86; in Sweden about 89. With so many people concentrated in urban areas, the rural countryside is more open and sparsely populated than in East and South Asia, where fewer than 40 percent of the people reside in cities and towns.

The three world population concentrations just discussed (East Asia, South Asia, and Europe) account for about 3.3 of the world's more than 4.8 billion people. Nowhere else on the globe is there any population cluster with a total even half of any of these. Look at the dimensions of the landmasses on Fig. I–14 and consider that the populations of South America, Black Africa, and Australia together equal *less* than that of India alone. In fact, the next-ranking cluster is *Eastern North America*, comprising the east-central United States and southeastern Canada; however, it is only about one-quarter the size of the smallest of the Eurasian concentrations. As Fig. I–14 clearly shows, this region does not possess the large, contiguous high-density zones of Europe or East and South Asia. The North American population cluster displays European characteristics, and it even outdoes Europe in some respects. Like the European region, much of the population is concentrated in several major metropolitan centers, whereas the rural areas remain relatively sparsely populated. The leading focus of the North American cluster lies in the urban complex along the U.S. northeastern seaboard, from Boston to Washington, which includes New York, Philadelphia, and Baltimore. This great urban agglomeration is called *Megalopolis* by urban geographers who predict that it is only a matter of time before the whole region coalesces into one enormous megacity. But there are other large urban concentrations within this North American cluster as well; in fact, two other megalopolislike regions can be identified, one linking Chicago-Detroit-Cleveland-Pittsburgh, the other Montreal-Toronto-Windsor in the Canadian provinces of Quebec and Ontario. If you study Fig. I–14 carefully, you will note other prominent North American metropolises standing out as small areas of high-density population, among them St. Louis, Minneapolis-St. Paul, Kansas City, San Francisco, and Los Angeles.

Still further examination of Fig. I–14 leads us to recognize substantial local population clusters in Southeast Asia. It is appropriate to describe these as discrete (discontinuous) clusters, for the map confirms that they are actually a set of nuclei rather than a contiguous population concentration. Largest among these nuclei is the Indonesian island of Djawa (Java), with nearly 100 million inhabitants. Elsewhere in the region, populations cluster in the lowlands of major rivers, some of which came to our attention frequently during the Vietnam conflict—for example, the Mekong Delta. Neither these river valleys nor the rural surroundings of the cities have population concentrations comparable to those of either China to the north or India to the west, and under normal circumstances Southeast Asia is able to export rice to its hungrier neighbors. However, decades of strife have disrupted the region (and continues today) to such a degree that its productive potential has not been attained.

The remaining continents do not sustain population concentrations comparable to those we have considered. Africa's 548 million inhabitants cluster in above-average densities in West Africa (where Nigeria has a population larger than its official census, probably 110 million) and in a zone in the east extending from Ethiopia to South Africa. Only in North Africa is there an agglomeration comparable to the crowded riverine plains of Asia: the Nile Valley and Delta with 48 million residents. Importantly, it is the pattern here—not the dimensions—that resembles Asia. As in East and South Asia, the Nile's Valley and Delta teem with farmers who cultivate every foot of the rich and fertile alluvial (riverine) soil. But the Nile's gift (see Fig. I–1) is a miniature compared to its Asian equivalents; the lowlands of the Ganges, Chang Jiang and Huang He contain many times the number of inhabitants who manage to eke out a living along the Nile.

The large light-shaded spaces in South America and Australia and the peripheral distribution of the modest populations of these continents suggest that here we might find considerable space for the world's huge numbers. And indeed, South America could probably sustain more than its present 270 million, as Australia can undoubtedly accommodate more than 16 million. But the population growth rate of South American countries is among the highest in the world and Australia's severe environmental limitations hardly qualify the smallest continent as a safety valve for Asia's farming millions.

## Urbanization

*Urbanization* increasingly characterizes the current stage of humanity's expansion throughout the world, although not everywhere with the same intensity. Hope Tisdale has defined urbanization as

> *. . . a process of population concentration. It proceeds in two ways: the multiplication of the points of concentration and the increasing in size of individual concentrations. . . . Urbanization is a process of becoming,* [implying movement] *from a state of less concentration to a state of more concentration.*

Over the past three decades, the population clustered in cities worldwide has more than doubled to a total of just over 2 billion, thereby making urbanites of more than 4 out of every 10 people on earth. The significance of this latest increase in global urbanization can be seen in Fig. I–16. Throughout this century urban population has been increasing at an accelerating rate, particularly since 1950; in fact, if projections hold, the urbanization pace of the second half of the twentieth century will just about double all world urban growth that occurred before 1950. Table I–1 breaks down this recent urbanization trend by major world region for each decade since 1950. Observe that urban population growth has steadily increased in every region over the last three decades. The latest evidence for the 1980s shows that urbanization may be slowing down in the more developed regions of the world, and some are now using the term "counter-urbanization" to describe an apparent new surge of growth in certain nonmetropolitan areas of North America and Europe. On the other hand, the urban growth rate in the less developed parts of the world may be accelerating. It is worth keeping in mind that although the overall percentage of urban population is lower in these regions, the much greater *absolute* number of people makes even a 1 percent cityward shift a major

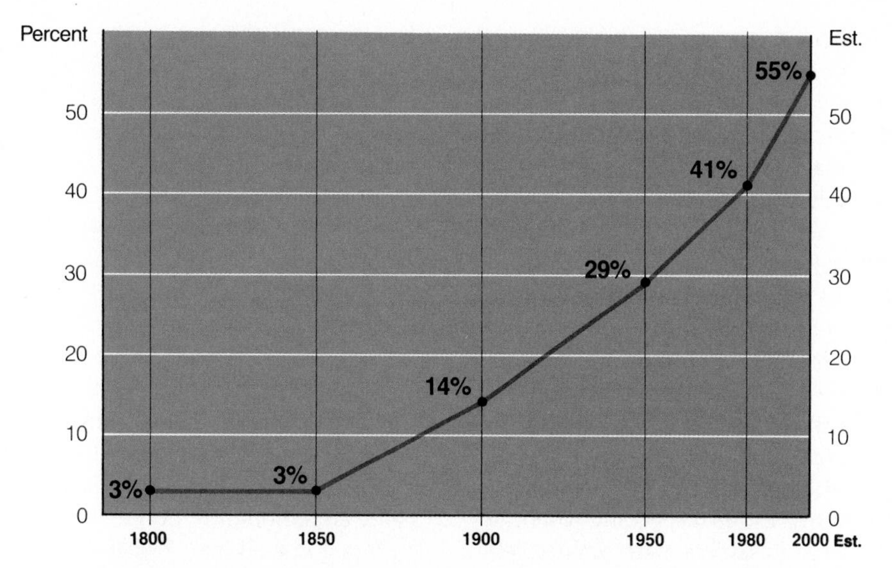


Figure I–16

movement; moreover, the impact of such migration is felt even more deeply because these cities are invariably unable to house or employ the newcomers, as we will shortly see in the section on underdeveloped countries.

Urbanization as a force of change is now so pervasive in contemporary world regional geography that we will be considering the various manifestations of this process and its responses throughout the book. As a prelude, we offer here some basic ideas about the nature of cities—the spatial focal points containing sizable concentrations of people and activities produced by the operation of world urbanization processes. Truman Hartshorn defines cities as "centers of power—economic, political, and social . . . they are where the action is in terms of innovations and control." He further points out that cities are agglomerations of people "with a distinctive way of life, in terms of employment patterns and organization," and that they contain "a high degree of specialized and segregated land uses and a wide variety of social, economic, and political institutions that coordinate the use of [urban] facilities and resources. . . ." We are also reminded that cities today have expanded into vast *metropolitan* complexes, which are composed of the older core or central city and a surrounding outer subur-

Table I–1 **Urban Population as a Percentage of Total Population by Major World Region, 1950–1980**

| Region | 1980 | 1970 | 1960 | 1950 |
|---|---|---|---|---|
| North America | 74 | 70 | 67 | 64 |
| Europe | 69 | 64 | 58 | 54 |
| Soviet Union | 65 | 57 | 49 | 39 |
| Latin America | 65 | 57 | 49 | 41 |
| Asia | 29 | 25 | 22 | 17 |
| Africa | 28 | 23 | 18 | 15 |
| World Total | 41 | 38 | 34 | 29 |

*Source: Patterns of Urban and Rural Population Growth* (New York: United Nations, Dept. of Intl. Economic and Social Affairs, Population Studies, No. 68, 1980).

ban city (at least in the developed world) that is rapidly enlarging in size and urban function. Although this definition by an urban geographer captures a wide range of functions and properties, we should also be aware of additional uses of the term on the world scene. Viewed specifically, as Jack Williams et al. tell us, "the term *city* is essentially a political designation, referring to a place governed by some kind of administrative body or organization." Seen more broadly, on the other hand, the largest cities (especially when they serve as capitals) are nothing less than the foci—indeed complete microcosms—of their national cultures. In succeeding chapters, we will introduce additional urban terminology and concepts, including a Model Box and a capsule survey of urban geography itself in Chapter 3.

# Political Geography

We return in later chapters to additional issues involving world population: problems of density, growth patterns, migration, well-being; and at the outset of the next chapter a brief overview of the subfield of population geography is presented. Our survey of world realms must, however, also be preceded by a brief introduction to the politico-geographical complexities of the world.

The limited land area of the globe is presently divided among nearly 160 national states and about 50 dependent territories. The states range in population from China's 1,088 million to Liechtenstein's 27,000. In terms of territory, the largest is the Soviet Union with 22,275,000 square kilometers (8,600,400 square miles), and the smallest is Monaco, containing less than 2 square kilometers. But even when the so-called *microstates* are left out of such comparisons (as is done in the table displaying national data in the Appendix), the range remains enormous. In fact, some full-fledged members of the United Nations such as Iceland and Qatar have populations of under 300,000.

Inequalities among the world's national states are not confined to size and population. Large territorial size is no guarantee of resource wealth: some of the world's larger states are among the poorest. Zaïre (in Equatorial Africa), Sudan (much of which is Sahara), India (with its huge population), and Brazil (better off but not rich despite its huge area) are among countries whose size is not matched by comparable shares of the earth's natural resources. Neither do large numbers of people ensure national strength. Indonesia has more than 165 million inhabitants; Nigeria contains nearly twice as many people as West Germany. Some of the states with small populations (Sweden, Switzerland) are among those enjoying high standards of living, and many populous countries are poor.

The familiar world political map (Fig. I–17) is a recent product of national territorial competition and adjustment. Just one century ago large areas of Africa and Asia remained beyond the jurisdiction of the encroaching imperialist powers. Finally, when colonial spheres of influence collided, boundaries were often drawn to assign dependencies and to facilitate European administration. The map of Africa was largely created or confirmed at the Berlin Conference held in the mid-1880s. At that time, most probably, no one realized that those hastily defined colonial boundaries would within a century constitute the borders of independent states.

The world's boundary framework is always under pressure in certain areas, and the past decade has witnessed several changes. The boundary between North and South Vietnam has been eliminated. Similarly, the border between Portuguese and Indonesian Timur has been erased. The boundary between Morocco and former Spanish Sahara was also erased, and the territory of the latter divided between Morocco and Mauritania along a new border. And following a major conflict on the island of Cyprus, a *de facto* boundary between Greeks and Turks appeared and functioned in some respects as a recognized international border. In Lebanon, too, demarcated boundaries appeared between Christian and Moslem zones during the conflicts of the 1970s, but they remain in doubt following the recent Syrian and Israeli invasions and the continuing upheaval of that country in the mid-1980s. In fact, the entire question of Israel's boundaries with its neighbors continues to plague Middle East political affairs.

These actual and prospective changes involve comparatively minor modifications of a boundary framework whose major outlines have proved quite durable. When African and Asian colonies attained independence, some observers predicted that the "boundaries of imperialism" would soon be abandoned in favor of more realistic dividing lines. But despite some significant steps in this direction (the India-Pakistan border, for example, created when the British colonial era came to an end in 1947), the chief elements of the preindependence boundary framework have remained intact—as they are today.

The concept of the linear political boundary is quite old. Romans and Chinese built walls for this purpose 2000 years ago, and relicts still remain on the cultural landscape; an outstanding example is Hadrian's Wall (Fig. I–18), built by the Romans to demarcate the edge of their empire along a line near the border between modern-day England and Scotland. In post-Roman Europe, rivers often served as trespass lines, and territorial competition and boundary delimitation were part of European political gestation. Europe's colonial expansion during the seventeenth and eighteenth centuries, and the new power arising from the industrial revolution, hastened a process that had begun early but had progressed slowly. European concepts of territorial acquisition and delimitation, even at sea, now were imposed on much of the rest of the world. Within a century, what remained of open frontiers, even on ice-covered Antarctica, disappeared. We are today witnessing the final stage of this territorialization process as states are about to claim and absorb the bulk (and perhaps all) of the seas and oceans and what lies beneath them. These and other spatial political realities will be discussed in several of the regional chapters that follow; and at the outset of Chapter 10, we review the subfield of political geography.

# The Geography of Economic Development

As a final prelude to our regional survey of the world's geographic realms, we need to consider two important economic dimensions of humankind: livelihood patterns and the relative wealth of national groups on a worldwide scale. The first of these dimensions involves *economic geography*; the second, the *geography of development and modernization.*

## Economic Geography

This subfield of geography is concerned with the various ways in which people earn a living and how the goods and services they produce in order to earn that income are spatially expressed and organized. Geographers tend to group various occupations broadly into such major aggregates as "manufacturing" and "services." Many concepts and principles of economic geography will be introduced throughout the following chapters. At this point, we introduce the subject by briefly contrasting some of the practices used in different world regions by the workers comprising humanity's leading occupation—*agriculture.*

The great majority of the earth's peoples, industrial and technological progress notwithstanding, still farm the soil for a living. This activity of agriculture may be defined as the deliberate tending of crops and livestock in order to produce food (or fiber). But farming in the tropical rainforest of Africa is something very different from growing rice in Asia—and Asia's paddyfields look nothing like the vast wheat farms of the North American Great Plains.

Agricultural practices and systems, therefore, vary widely. In the tropics, it is often necessary to cut down and burn the original forest vegetation to clear a patch of land that will support a crop (probably a root crop) for one year, perhaps two. It is all done by hand, and such *slash-and-burn agriculture* is energy efficient; machines mean very little in areas where cleared land must soon be abandoned—owing to the leaching of soil nutrients—in favor of a new patch. But neither can such shifting cultivation support a very dense population, even on the forest margins where the rainfall decreases a bit and the soils are somewhat less leached. On Fig. I–14, huge areas of lowland Equatorial Africa, South America, and Southeast Asia show a population density under 10 per square kilometer (25 per square mile). Extensive (nonintensive) subsistence farming is the rule here, even where the rainfall declines to 100 centimeters (40 inches) annually (Fig. I–9), and a dry season permits the harvesting of corn and other hardy grains.

The ricefields of South, East, and Southeast Asia sustain subsistence farmers, too, but cultivation here is highly intensive and population densities are great. Once again, the Asian rice culture is highly efficient in terms of energy inputs: most of the paddies are still prepared by ox-drawn plow, and the rice is planted by hand—hundreds of millions of hands. But much of the Asian riceland would be just another tropical zone of meager subsistence agriculture were it not for the alluvial soils in the great river basins. Highly fertile and replenished by rains that bring not only needed moisture, but also new coatings of silt, these soils are sometimes capable of sustaining two or even three crops in a single year—one *after* the other. The great masses of South and East Asian population depend on such soils, and these great human clusters have grown on the strength of their productivity.

Commercial agriculture in the wheatlands of the American Great Plains presents quite another picture. Vast, almost unbroken fields of grain cover the flat countryside; the soil was fertilized by machine, the sowing was done by machine, and so was the harvesting. This type of modern commercial agriculture is quite labor efficient, requiring few hands, but it is less efficient in terms of energy requirements.

Between these extremes lie other agricultural systems: the plantation, a specialized commercial enterprise in tropical and subtropical environments capable of supporting a population of medium density; and the complex, integrated exchange type of farming in the hinterlands of the large urban areas of Western Europe, the eastern United States, and Japan—where space is at a premium, distance to markets is often crucial, and the soil is heavily fertilized and tended so that it will produce as much as possible.

## Developed and Underdeveloped Countries

The world geographic realms discussed in this book are grouped under two main headings: *developed* regions and *underdeveloped* (or *developing*) regions. These terms are used because they have become commonplace, as have such adjectives as rich and poor, "haves" and "have-nots," and the like. But perhaps the best designation for the developed and underdeveloped countries would be *advantaged* and *disadvantaged.* In U.N. parlance, the disadvantaged countries are primarily the countries of the Third World, the capitalist and socialist systems dominating the first two. Generally speaking, Third World countries find themselves in a disadvantageous economic position in relation to the developed, economically powerful states of the world—whether capitalist or socialist.

Grouping the advantaged countries together, they include most of the states of Europe, the Soviet Union, Canada and the United States, Australia and New Zealand, Japan, Singapore, Israel, and, by some definitions, South Africa. Against this comparatively short list of countries (which contain barely 25 percent of the world population total) are the rest of the countries at various levels of underdevelopment: those of Middle and South America, North Africa, and South, Southeast, and East Asia. Not all of these less developed countries are at the same level of underdevelopment (just as some developed countries are ahead of others). Few would argue that Haiti, Bolivia, Ethiopia, Bangladesh, and Indonesia are not underdeveloped countries (*UDCs*). But to place Argentina, Uruguay, Chile, Venezuela, Mexico, Brazil, Turkey, Taiwan, and South Korea in the same category creates an issue. Some economic geographers suggest that it is appropriate to recognize, between the developed coun-

**Figure I–17**

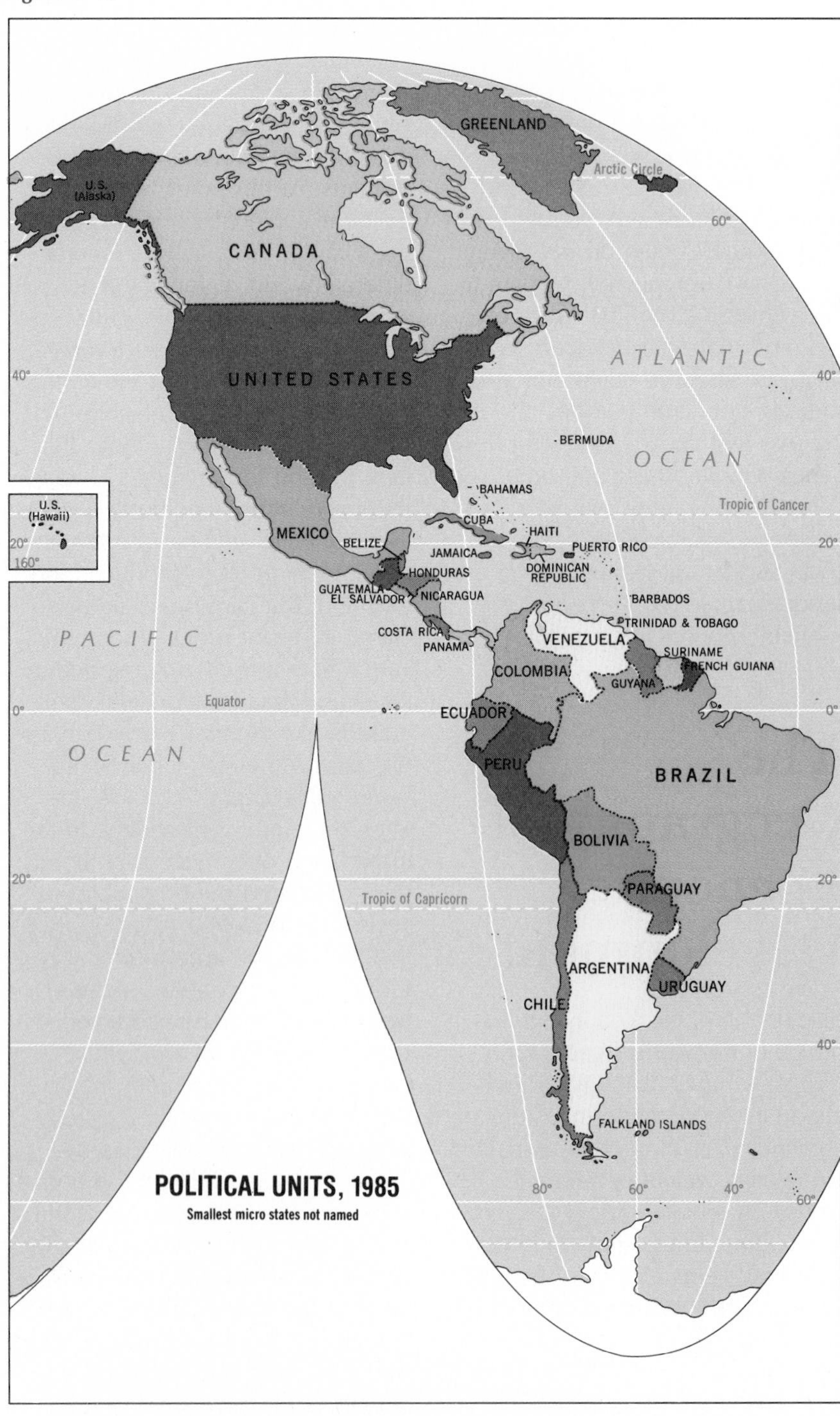

tries (*DCs*) and the UDCs, a middle tier of countries that, although still possessing many characteristics of UDCs, are now "emerging" or at a "takeoff" stage.

Although we routinely (and arbitrarily) divide our world into developed and underdeveloped portions, no universally applicable criteria exist to measure development accurately. Leading indexes are listed under seven headings (see Box, p. 37), but they remain arbitrary and subject to debate. And what these numerical indexes fail to convey is the time factor. The DCs are the advantaged countries—but why cannot the UDCs and "emerging" countries catch up? It is not, as is sometimes suggested, simply a matter of environment, resource distribution, or cultural heritage (a resistance to innovation, for example). The sequence of events that led to

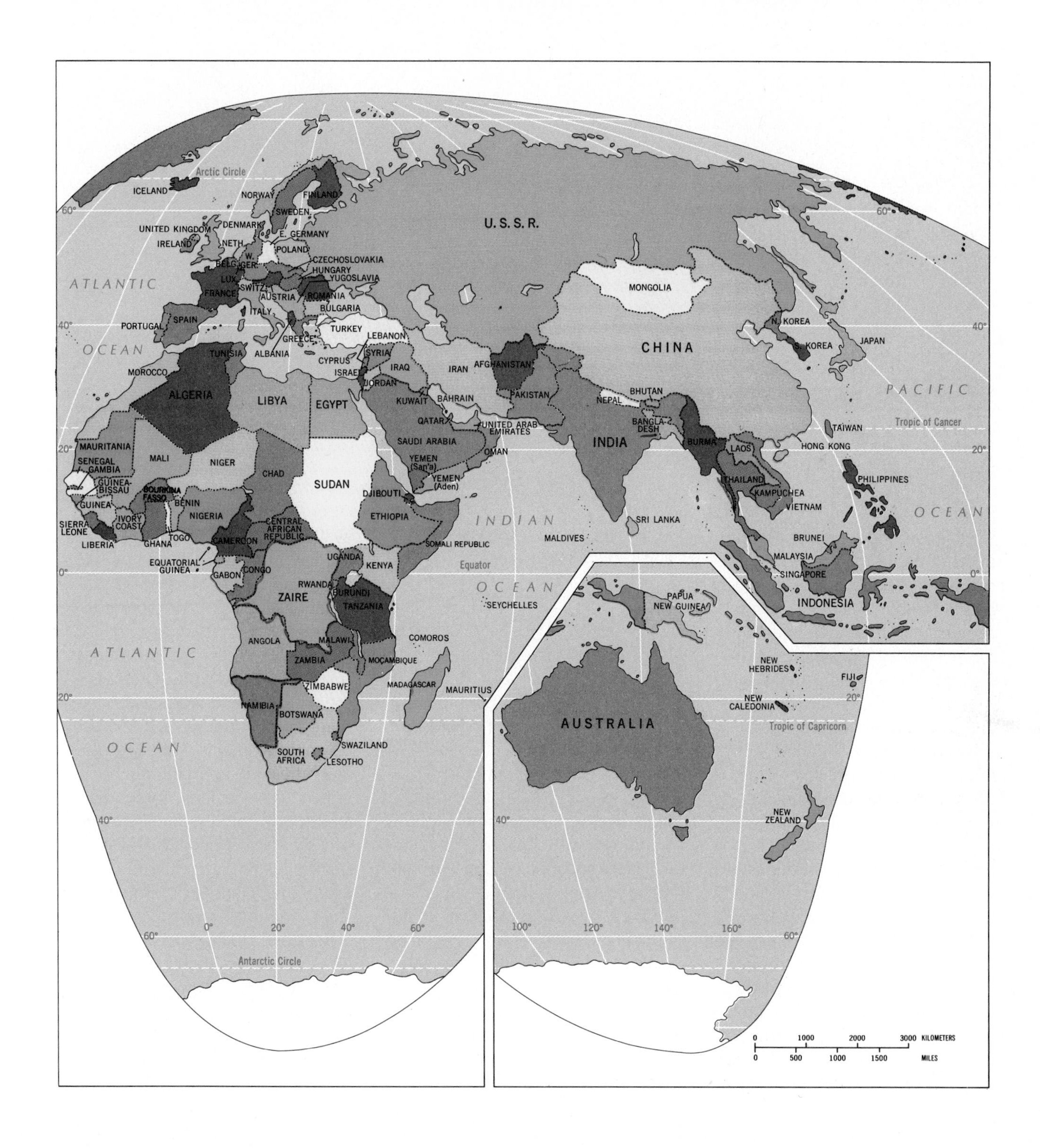

Figure I–18
Hadrian's Wall: a relict linear boundary between England and Scotland.

the present division of our world began long before the industrial revolution occurred. Europe, even by the middle of the eighteenth century, had laid the foundations for its colonial expansion; the industrial revolution magnified Europe's demands for raw materials and its products increased the efficiency of its imperial control. Although Western countries gained an enormous head start, colonial dependencies remained suppliers of resources and consumers of the products of the Western industries. Thus was born a system of international exchange and capital flow that really changed very little when the colonial period came to an end. Underdeveloped countries, well aware of their predicament, accused the developed world of perpetuating its advantage through *neocolonialism*—the entrenchment of the old system under a new guise.

## Symptoms of Underdevelopment

The disadvantaged countries suffer from numerous demographic, economic, and social ills. Their populations tend to exhibit high birth rates and moderate to high death rates; life expectancy at birth is comparatively low (see Chapter 8 and the Appendix). A large percentage of the population (as much as half) is 15 years or younger. Infant mortality is high. Nutrition is inadequate, and diets are not well balanced. Protein deficiency is a common problem. The incidence of disease is high; health care facilities are inadequate. There is an excessively high number of persons per available doctor; hospital beds are too few in number. Sanitation is poor. Substantial numbers of school-age children do not go to school; illiteracy rates are high.

Rural areas are overcrowded and suffer from poor surface communications. Men and women do not share fairly in the work that must be done; women's workload is much heavier, and children are pressed into the labor force at a tender age. Landholdings are often excessively fragmented, and the small plots are farmed with outdated, inefficient tools and equipment. The main crops tend to be cereals and roots; protein output is low, because its demand on the available land is higher. There is little production for the local market, because distribution systems are poorly organized and demand is weak. On the farms, yields per unit

# MEASURES OF DEVELOPMENT

What distinguishes a developed economy from an underdeveloped one? Obviously it is necessary to compare countries on the basis of certain measures; the question cannot be answered simply by subjective judgment. No country is totally developed; no economy is completely underdeveloped. We are comparing *degrees* of development when we identify DCs and UDCs. Our division into developed and underdeveloped economies is arbitrary, and the dividing line is always a topic of debate. There is also the problem of data. Statistics for many countries are inadequate, unreliable, incompatible with those of others, or simply unavailable.

The following list of measures is normally used to gauge levels of economic development:

1. *National product per person.* This is determined by taking the sum of all incomes achieved in a year by a country's citizens and dividing it by the total population. Figures for all countries are then converted to a single index for purposes of comparison. In DCs the index can exceed $2000; in some UDCs it is as low as $100.
2. *Occupational structure of the labor force.* This is given as the percentages of workers employed in various sectors of the economy. A high percentage of laborers engaged in the production of food staples, for instance, signals a low overall development level.
3. *Productivity per worker.* This is the sum of production over the period of a year divided by the total number of persons comprising the labor force.
4. *Consumption of energy per person.* The greater the use of electricity and other forms of power, the higher the level of national development. These data, however, must be viewed to some extent in the context of climate.
5. *Transportation and communication facilities per person.* This measure reduces railway, road, airline connections, telephone, radio, television, and so forth, to a per capita index. The higher the index, the higher the development level.
6. *Consumption of manufactured metals per person.* A strong indicator of development levels is the quantity of iron and steel, copper, aluminum, and other metals utilized by a population during a given year.
7. *Rates.* A number of additional measures are employed, including literacy rates, caloric intakes per person, percentage of family income spent on food, and amount of savings per capita.

area are low, subsistence modes of life prevail, and the specter of debt hangs constantly over the peasant family. Such circumstances preclude investment in such luxuries as fertilizers and soil conservation techniques. As a result, soil erosion and land denudation scar the rural landscapes of many UDCs. Where areas of larger-scale modernized agriculture have developed, these produce for foreign markets with little resultant impact on the improvement of domestic conditions.

In the urban areas appalling overcrowding, poor housing, inadequate sanitation, and a general lack of services prevail. Employment opportunities are insufficient, and unemployment is always high. Per capita income is low, savings per person are minimal, and credit facilities are poor. Families spend a very large proportion of their income on food and basic necessities. The middle class remains small; not infrequently, a substantial segment of that middle-income population consists of foreign immigrants (see Chapter 7).

These are some of the criteria that signal underdevelopment, and the list is not complete. For example, one of the geographic properties that mark UDCs is the problem of internal regional imbalance. Even in UDCs, there are local exceptions to the general economic situation. The capital city may appear as a skyscrapered model of urban modernization, with thriving farms in the immediate surroundings, factories on the outskirts; road and rail may lead to a bustling port where luxury automobiles are unloaded for use by the privileged elite. Here in the country's core area the rush of "progress" is evident—but travel a few kilometers into the countryside (or even to the squatter "shacktowns" at the edge of the city), and you may find that almost nothing has changed. And just as the rich countries become richer and leave the poorer ones farther behind, so the gap between progressing and stagnant regions within DCs grows larger. It is a problem of global dimensions.

There can be no doubt that the world economic system works to the disadvantage of the UDCs, but sadly it is not the only obstacle the less advantaged countries face. Political instability, corruptible leaderships and elites, misdirected priorities, misuses of aid, and traditionalism are among circumstances that commonly inhibit development. External interference by interests representing powerful DCs have also had negative impacts on the economic as well as the political progress of UDCs. Underdeveloped countries even get caught in the squeeze when other UDCs try to assert their limited strength; when the Organization of Petroleum Exporting Countries (OPEC), mostly underdeveloped themselves, raise the price of oil, energy and fertilizers slip still further from the reach of the poorer UDCs not fortunate enough to belong to this favored group. As the DCs get stronger and wealthier, they leave the underdeveloped world ever farther behind: the gap is still widening, and the prospects for the UDCs are not bright. The theme of development and modernization runs throughout the chapters (4–10) on Underdeveloped Regions; an overview of the role of geography in this interdisciplinary inquiry is presented at the outset of Chapter 9.

# Geographic Realms of the World

Earlier in the chapter, we discussed the classification of world regions and introduced the three-tier hierarchy of culture realm, region, and subregion. The world regional system to be used in this book, however, is based not on culture realms but on broader *geographic realms*. Thus, our global spatial framework (Fig. I–19) represents more than a regionalization of cultural-geographical phenomena and cultural landscapes. It also reflects aspects of economic, urban, political, physical, and historical geography, and it is a synthesis of human geography as a whole, not cultural geography alone. Cultural geographers, for example, might combine Middle and South America under the

**Figure I–19**

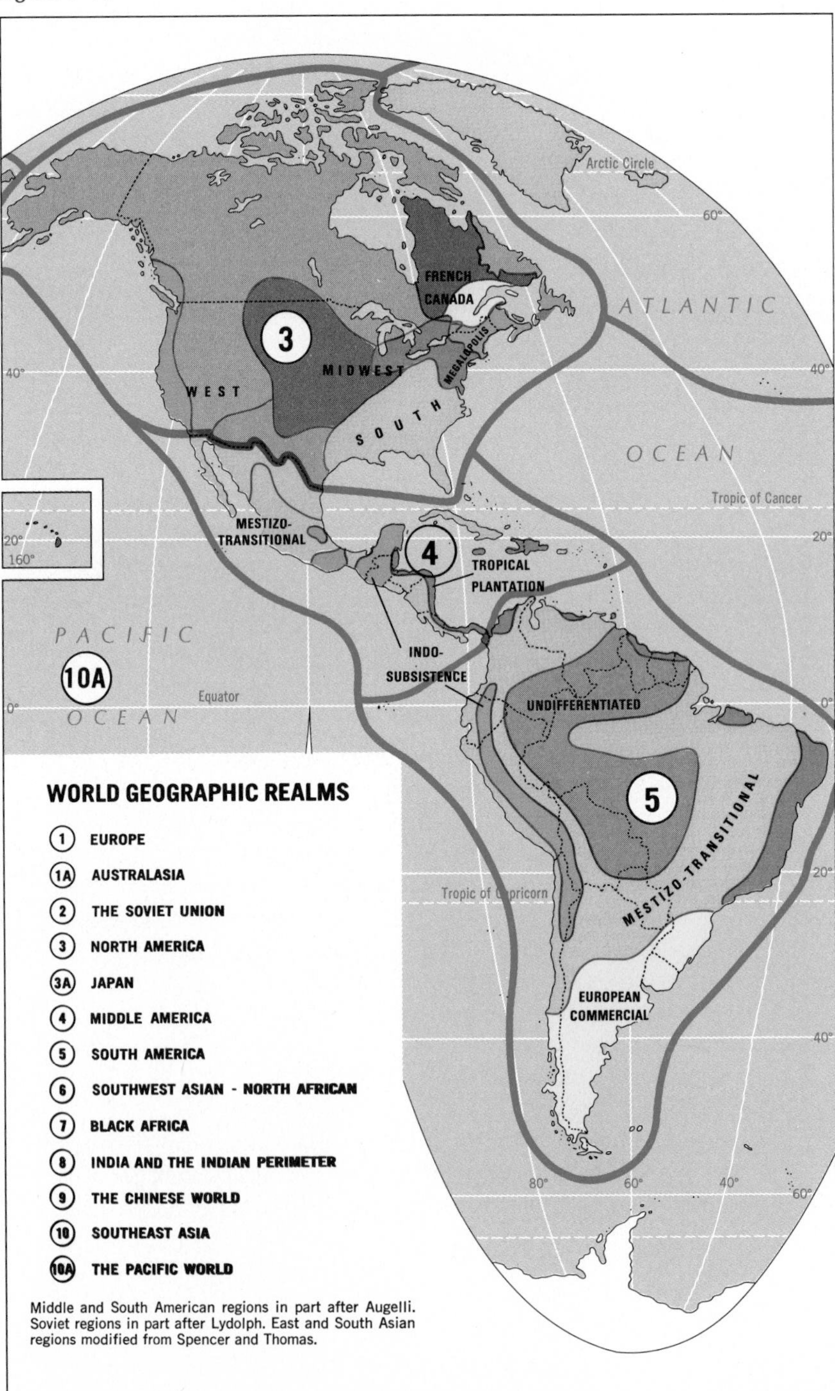

single rubric of "Latin" America because of the strength of the Latin cultural imprint. But that imprint is far stronger in South America than in Middle America, as will be noted later. The contrast is sufficiently strong to separate Middle America as a *geographic* realm from South America. Differences in physical, political, and economic geography confirm the boundary between realms (4) and (5) on the map.

Because Fig. I–19 defines geographic realms, cultural connotations have been avoided in all but one instance (the exception is Black Africa). This has been done because the cultural identification sometimes assigned to world realms can be misleading and even objectionable. The name "Anglo" America to denote the realm delimited by the United States and Can-

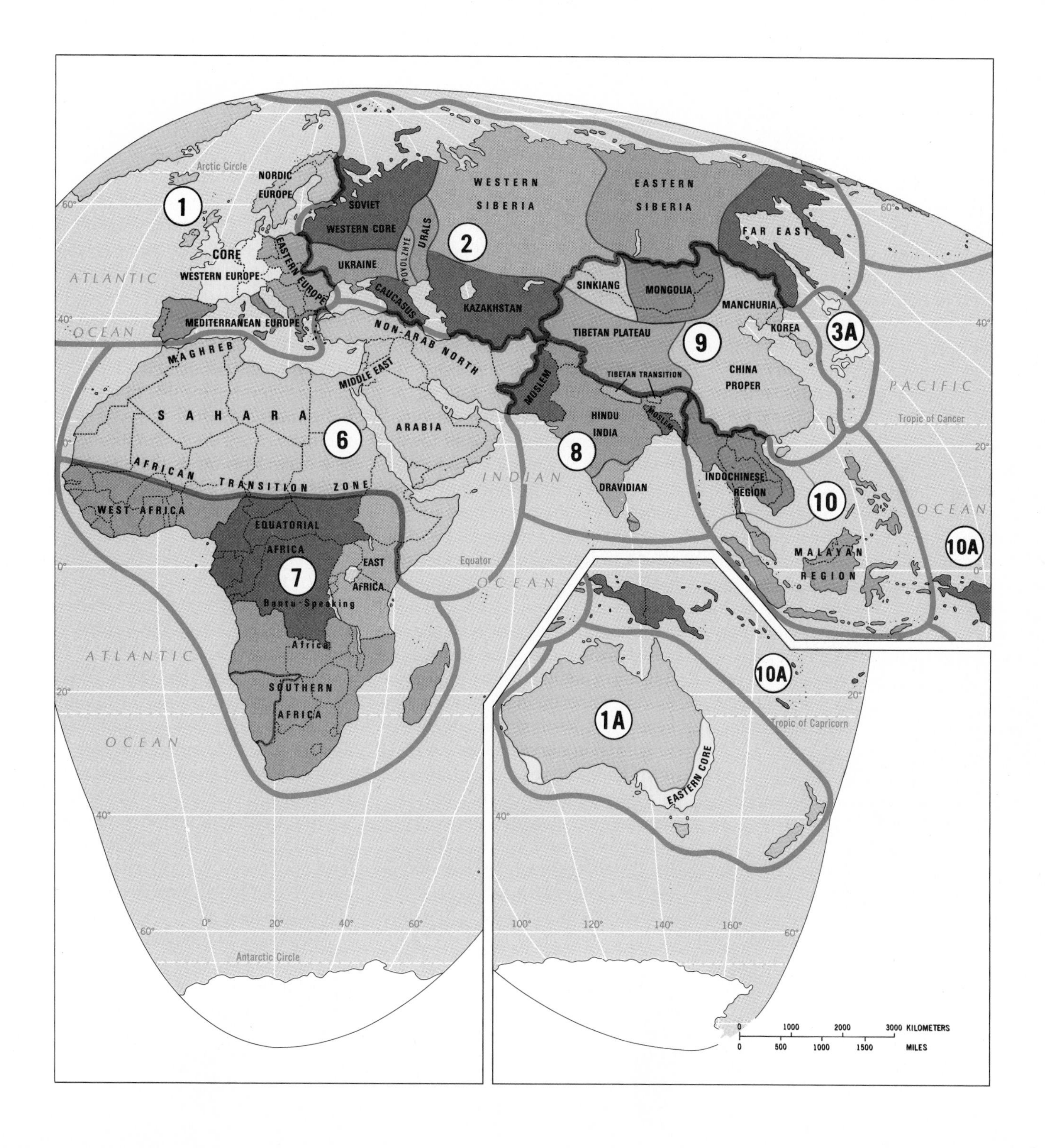

ada is used quite frequently, because Anglo- (English-) derived culture does dominate. But it masks the very pluralism that is one of this realm's essential properties, and, in any case, millions of North Americans of French, African, Indian, and Hispanic descent and heritage are not Anglos. The appropriate geographic appellation, therefore, is North America. The single exception on our map—Black Africa—does refer to a regional quality other than relative location. "Africa South of the Sahara" may be a better name, but it creates other problems. As Fig. I–19 shows, the "African Transition Zone" is one of the few areas where the geographic-realm boundary does not approximately conform to political geography. The Sahara's southern margin is not really the place where Africa's geographic transition is sharpest. Hence, the employment of a descriptive term for Africa's complex regional geography.

Thirteen geographic realms form the structural framework for our regional survey. Five of them comprise the developed world (Europe, Australia, the Soviet Union, North America, Japan), and the remaining eight consist of assemblages of underdeveloped and emerging countries (Fig. I–19). We begin with a series of regional vignettes (to be read in conjunction with this map) that form the introductions to the coming 10 chapters.

## Europe (1)

Europe merits designation as a world realm despite the fact that it occupies a small proportion of the total area of the Eurasian landmass—a fraction that, moreover, is largely made up of the continent's western peninsular extremities. Certainly Europe's size is no measure of its world significance; probably no other part of the world is or ever has been so packed full of the products of human achievement. Innovations and revolutions that transformed the world originated in Europe. Over centuries of modern times, the evolution of world interaction focused on European states and European capitals. Time and again, despite internal wars, despite the loss of colonial empires, despite the impact of external competition, Europe has proved to contain the human and natural resources needed for rebounding and renewed progress.

Among Europe's greatest assets is its internal natural and human diversity. From the warm shores of the Mediterranean to the frigid Scandinavian Arctic and from the flat coastlands of the North Sea to the grandeur of the Alps, Europe presents an almost infinite range of natural environments. An insular and peninsular west contrasts against a more continental east. A resource-laden backbone extends across Europe from England toward the east. Excellent soils produce harvests of enormous quantity and variety. And the population includes people of many different stocks, peoples grouped under such familiar names as Latin, Germanic, and Slavic. Europe has its cultural minorities as well—for example, the Hungarians and the Finns. Immigrants continue to stream into Europe, contributing further to a diversity that has been an advantage to Europe in uncountable ways. Today's Europe, especially in its west, is a realm dominated by great cities, intensive transport networks and mobility, enormous productivity, dynamic growth, a large and often very dense population, and an extremely sophisticated technology.

## Australia (1A)

Just as Europe merits recognition as a continental realm, although it is merely a peninsula of Eurasia, so has Australia achieved identity as the island realm of Australasia. Australia and New Zealand are European outposts in an Asian-Pacific world, as unlike Indonesia as Britain is unlike India. Although Australia was spawned by Europe and its people and economy are Western in every way, Australia as a continental realm is a far cry from the crowded European world. The image of Australia is one of impressive, large, open cities, wide expanses, huge herds of livestock (perhaps also the pests, rabbits and kangaroos), deserts, and beautiful coastal scenery. There is much truth in such a picture: 10.7 of Australia's nearly 16 million people are concentrated in the country's seven largest cities. In this as well as other respects, Australia is more like the United States than other realms. With its British heritage, its homogeneous population, single language (except for that of a small indigenous minority), and type of economy, Australia's identification as a realm rests on its remoteness and spatial isolation.

## The Soviet Union (2)

The Soviet Union—the world's largest country in areal extent—constitutes a geographic realm not just because of its size. To its south lie the Japanese, Chinese, Indian, and Southwest Asian realms, all clearly different culturally from the realm that Russia forged. The Soviet realm's western boundary is always subject to debate, but neither Finland nor Eastern Europe have been areas of permanent Russian or Soviet domination.

The events of the twentieth century have greatly strengthened the bases for the recognition of the Soviet Union as a geographic realm. Czarist Russian expansionism was halted, the revolution came, and a whole new order was created, transforming the old Russia and its empire into a strongly centralized socialist state whose hallmark was economic planning. The new political system awarded the status of Soviet Socialist Republic to areas incorporated into the Russian Empire, including those inhabited by Moslems and other minorities within the eastern U.S.S.R. In the economic sphere, the state took control of all industry and production, and agriculture was collectivized.

Nothwithstanding the disastrous dislocation of World War II, the past half century has seen much progress in the Soviet Union. The state has risen from a backward, divided, near-feudal country to the position of a world superpower with a strong individualistic (and also centrally directed) culture, a highly advanced technology, and a set of economic and social policies that have attracted world attention and, in some instances, emulation.

## North America (3)

The North American geographic realm consists of two of the most heavily urbanized and industrialized countries in the world. In the United States in 1981 there were 318 metropolitan areas with populations in excess of 65,000, within which lived fully 75 percent of all the people in the country; a substantial proportion of the remainder, moreover, lived in urban places larger than 10,000.

The North American realm is characterized by its large-scale sophisticated technology, its enormous consumption of the world's resources and commodities, and its unprecedented mobility and fast-paced life-styles. Suburbs grow toward each other as metropolises coalesce, surface and air transport networks intensify, and skylines change; the landscape of Los Angeles exquisitely captures these dynamics (Fig. I–20). Bustling shopping centers, traffic jams, waiting lines, noise, pollution of air and water—these are some of the attributes of North American technocracy.

Both the United States and Canada are pluralistic societies, with Canada's cultural sources lying in Britain and France and those of the United States in Europe, Africa, and Latin America. In both Canada and the United States, minorities remain separate from the dominant culture, thereby giving rise to a number of persistent social problems. Quebec is Canada's French province, and in the United States patterns of racial segregation endure with black Americans overwhelmingly concentrated in particular urban areas.

**Figure I–20**
Skyscrapers, suburbs, and freeways in Los Angeles.

## Japan (3A)

On Fig. I–19 Japan hardly seems to qualify as a geographic realm. Can a group of comparatively small islands qualify as a separate realm? Indeed it can—when 120 million people inhabit those islands and when they produce there an industrial giant, a modern urban society, and a vigorous national power. Japan is unlike any other non-Western country, and it exceeds many Western-developed countries in many ways. Like its European rivals, Japan built and later lost a colonial empire in Asia. During its tenure as a colonial power Japan learned to import raw materials and export finished products. Even the calamity of World War II, when Japan became the only country ever to suffer atomic attack, failed to extinguish Japan's forward push. In the postwar era, Japan's overall economic growth rate has been the highest in the world. Raw materials no longer come from colonies, but are purchased all around the world; products are sold practically everywhere. Almost no city in the world is without Japanese cars in its streets, few photography stores lack Japanese cameras, and many scientific laboratories use Japanese optical equipment. From cassette players to television sets, from oceangoing ships to electronic games, Japanese manufactures flood the world's markets. The Japanese seem to combine the quality precision skills of the Swiss with the massive industrial power of prewar Germany, the forward-looking designs of the Swedes with the innovativeness of the Americans.

Japan, therefore, constitutes a geographic realm by virtue of its role in the world economy today, the transformation—in a single century—of its national life and its industrial power. Yet Japan has not turned its back on its old culture; modernization is not all that matters to the Japanese. Ancient customs are still adhered to and honored. Japan is still the land of *Kabuki* drama and *sumo* wrestling, Buddhist temples and Shinto shrines, tea ceremonies and fortune-tellers, arranged marriages, and traditional song and dance. There is still a reverence for older people, and despite all the factories, huge cities, fast trains, and jet aircraft, Japan is still a land of shopkeepers, handicraft industries, home workshops, carefully tended fields, and space-conserving villages. Only by holding on to the culture's old traditions could the Japanese have tolerated the onrush of their new age.

## Middle America (4)

Middle and South America combined are often called "Latin" America because of the Iberian (Spanish and Portuguese) imprint placed on them during the European expansion when the ancient Indian civilizations of the region were submerged and destroyed. The Latin imprint is far more pervasive in South America than in Middle or Central America, however, for although South America contains only three small non-Iberian countries, Middle America has a much larger number of territories with British, Dutch, and French legacies.

Middle America consists of Mexico, the Central American republics from Guatemala to Panama, and the numerous islands of the Caribbean Sea to the east. In pre-European times this was a significant hearth of human development, where great cities arose a thousand years ago, crops and livestock were domesticated, progress was made in the sciences and the arts, and empires were forged. This realm, too, was the scene of the first European arrivals in the Americas nearly 500 years ago, and for some time it remained the New World core from which the white invaders' influences radiated outward. The early importance of the slave trade is still reflected in the high percentages of black people among Middle American island (and certain mainland) populations.

## South America (5)

The continent of South America is also a geographic realm—a region shared by Portuguese-influenced Brazil and a Spanish colonial domain now divided among nine separate countries. As in mainland Middle America, the Roman Catholic religion dominates life and culture; systems of land ownership, tenancy, taxation, and tribute were transferred from the Iberian Old World to the New. Today, South America still carries the impress of its source region in the architecture of its cities, in the visual arts, and in its music.

South America remains one of the more sparsely peopled realms of the world, and much of the continent's interior is virtually empty terrain (Fig. I–14). Population is not only coastally oriented, but is also strongly clustered, and to a considerable degree those clusters exist in isolation from one another. Understandably, internal communication and interaction have not been leading qualities in South America, whose individual countries often have stronger ties with Europe than they do with their neighbors.

Whereas South America does not yet suffer from the population pressure that bedevils other parts of the underdeveloped and emerging world (including several countries in Middle America to the north), the region nevertheless confronts some serious liabilities. These include the nature of land ownership, control of the means of production, and the resistance of those in authority against the pressures of change. Under such circumstances,

South American experiments in democratic government have failed, and authoritarian rule repeatedly arouses violent response.

## Southwest Asia-North Africa (6)

The Southwest Asian-North African realm, as we noted earlier, is known by several names, none of them satisfactory: the Islamic realm, the Arab world, the dry world. Undoubtedly, the all-pervading influence of the Islamic (Moslem) religion is this realm's overriding cultural quality, for Islam is more than a faith; it is a way of life, one that is also vividly expressed in the landscape (Fig. I–21). The contrasts within the realm are underscored when we note its contents, extending as it does from Morocco and Mauritania in the west to Iran and Afghanistan in the east and from Turkey in the north to Ethiopia in the south. Huge desert areas separate highly clustered populations whose isolation perpetuates cultural discreteness, and we can distinguish regional contrasts within the realm quite clearly. The term *Middle East* refers to one of these regions, countries of the eastern Mediterranean (Fig. I–19); the *Maghreb* is a region constituted by the population clusters in the countries of northwest Africa and centered on Algeria. Still another region extends along the sub-Saharan transition zone to Black Africa, where Islamic traditions give way to African lifestyles. In the realm's northern region, Turkey and Iran dominate as non-Arab countries.

Rural poverty, strongly conservative traditionalism, political instability, and conflict have marked the realm in recent times, but this is also the source area of several of the world's great religions and the site of ancient culture hearths and early urban societies. Had we drawn Fig. I–19 several centuries ago, we would have shown Arab-Turkish penetrations of Iberian and Balkan Europe and streams of trade and contact reaching to Southeast Africa and East Asia. So vigorously was Islam propagated beyond the realm that, to this day, there are more adherents to the faith *outside* the Arab world than within it.

## Black Africa (7)

Between the southern margins of the Sahara Desert and South Africa's southernmost Cape Province

**Figure I–21**
The unity of faith and culture in the Islamic townscape.

lies the Black Africa geographic realm. As we noted, its boundary with the realm of Islam is generally a zone of transition, and this is a case where coincidence with political boundaries cannot be employed. In Sudan, for example, the north is part of the Arab world, but the south is quite distinctly African. So strong are the regional contrasts within Sudan that it would be inappropriate to include the entire country in either the Islamic realm or the Black African realm. Therefore, Sudan (as well as Ethiopia, Chad, Nigeria, and Mali) is bisected.

The African realm is defined by its mosaic of hundreds of languages belonging to specific African language families and by its huge variety of traditional local religions that, unlike world religions such as Islam or Christianity, remain essentially "community" religions. With some exceptions, Africa is also a realm of farmers. A few people remain dependent on hunting and gathering, and some communities depend primarily on fishing. But the principal livelihood is farming. The subsistence way of life was changed very little by the European colonialists. Tens of thousands of villages all across Black Africa were never brought into the economic orbits of the Europeans (Fig. I–4). Root crops and grains are grown in ancient, time-honored ways with the hand hoe and the digging stick. It is a backbreaking low-yield proposition, but the African agriculturalist does not have much choice.

## India and the Indian Perimeter (8)

The familiar triangular shape of India outlines a subcontinent in itself—a clearly discernible physical region bounded by mountain range and ocean, inhabited by a population that constitutes one of the greatest human concentrations on earth. The scene of one of the oldest civilizations, it became the cornerstone of the vast British colonial empire. Out of the colonial period and its aftermath emerged four states: India, Pakistan, Bangladesh, and Sri Lanka.

Europe was a recipient of Middle Eastern achievements in many spheres; so was India. From Arabia and the Persian Gulf came traders across the sea. The wave of Islam came over land, across the Indus, through the Punjab, and into the Ganges Valley. Along this western route had also come the realm's modern inhabitants, who drove the older Dravidians southward toward the tip of the peninsula.

Long before Islam reached India, major faiths had already arisen here, religions that still shape countless lives and attitudes. Hinduism and Buddhism emerged before Christianity and Islam, and the postulates of the Hindu religion have dominated life in India for several thousands of years. They include beliefs in reincarnation, in the goodness of holy men and their rejection of material things, and in the inevitability of a hierarchical structure in life and afterlife. This last quality of the Hindu faith had expression in India's *caste system*, the castes being rigid social strata and steps of the universal ladder. It had the effect of locking people into social classes, the lowest of which—the untouchables—suffered a miserable existence from which there was no escape. Buddhism was partly a reaction against this system, and the invasion of Islam was facilitated by the alternatives it provided. Eventually the conflicts between Hinduism (India's primary religion today) and Islam led to the partition of the realm into India and Pakistan, an Islamic republic that subdivided again in 1971.

If this realm is one of contrasts and diversity, there are overriding qualities nevertheless. It is a region of intense adherence to the various faiths; a realm of thousands of villages and several teeming, overpopulated large cities; and an area of poverty and frequent hunger, where the difficulties of life are viewed with an acquiescence that comes from a belief in a better new life—later.

## The Chinese World (9)

China is a nation-state as well as a geographic realm. The Chinese world may be the oldest of the continuous civilizations. It was born in the upper basin of the Huang He (Yellow River) and now extends over an area of over 9.5 million square kilometers with a population of about 1.1 billion inhabitants. Alone among the ancient culture hearths we discussed earlier, China's spawned a major, modern state of world stature, with the strands to the source still intact. Chinese people still refer to themselves as the "People of Han," the great dynasty (202 B.C. to A.D. 220) that was a formative period in the country's evolution. But the cultural individuality and continuity of China were already established 2000 years before that time, perhaps even earlier.

In the lengthy process of its evolution as a regionally distinct culture and a great nation-state, China has had an ally in its isolation. Mountains, deserts, and sheer distance protected China's "Middle Kingdom" and afforded the luxury of stability and comparative homogeneity. Not surprisingly, Chinese self-images were those of superiority and security; the growing Chinese realm was not about to be overrun. There were invasions from the steppes of inner Asia, but invaders were repulsed or absorbed. There would always be a China. The European impact of the nineteenth century finally ended China's era of invincibility, but it was held off far longer than in India, it lasted a much shorter time, and it had less permanent effect.

Unlike India, China's belief system was always concerned more with the here and now, the state and authority, than with the hereafter and reincarnations. The century of convulsion that ended with the Communist victory in 1949 witnessed a breakdown in Chinese life and traditions, but there is much in the present system of government that resembles China under its dynastic rulers. China is being transformed into a truly Communist society, but Chinese attitudes toward authority, the primacy of the state, and the demands of regimentation and organization are not new.

China is moving toward world-power status once again, and there are imposing new industries and growing cities. But for all its modernization, China still remains a realm of crowded farmlands, carefully diked floodplains, intricately terraced hillsides, and cluttered small villages. Crops of rice and wheat are still meticulously cultivated and harvested; the majority of the people still bend to the soil.

## Southeast Asia (10)

Southeast Asia is nowhere nearly as well defined a geographic realm as either India or China. It is a mosaic of ethnic and linguistic groups, and the region has been the scene of countless contests for power and primacy. Spatially, the realm's discontinuity is quite obvious: it consists of a peninsular mainland, where population tends to be clustered in river basins, and thousands of islands forming the archipelagoes (island chains) of the Philippines and Indonesia.

During the colonial period, the term *Indochina* came into use to denote part of mainland Southeast Asia. The term is a good one, for it reflects the major sources of cultural influence that have affected the entire realm. The great majority of Southeast Asia's inhabitants have ethnic affinities with the people of China, but it was from India that the realm received its first strong, widely disseminated cultural imprints: Hindu and Buddhist faiths, architecture, and aspects of social structure. The Moslem faith also arrived via India. From China came not only ethnic ties, but also elements of culture: Chinese modes of dress, plastic arts, boat types, and other qualities were widely adopted in Southeast Asia. In recent times, a major migration from China to the cities of Southeast Asia further strengthened the impact of Chinese culture on this realm.

## Pacific Regions (10A)

Between Australia and the Americas lies the vast Pacific Ocean, larger than all the land areas of the world combined. In this great ocean lie tens of thousands of islands, large and small. This fragmented, culturally complex realm defies effective generalization. Population contrasts are reflected to some extent in the regional diversification of the realm. Thus, the islands from New Guinea eastward to the Fiji group are called *Melanesia* (*mela* means black). The people here are black or very dark brown and have black hair and dark eyes. North of Melanesia—that is, east of the Philippines—lies the island region known as *Micronesia* (*micro* means small), and the people here evince a mixture between Melanesians and Southeast Asians.

In the vast central Pacific, east of Melanesia and Micronesia and extending from the Hawaiian Islands to the latitude of New Zealand, is *Polynesia* (*poly* means many). Polynesians are widely known for their good physique; they have somewhat lighter skin than other Pacific peoples, wavier hair, and rather high noses. Their ancestry is complex, including Indian, Melanesian, and other elements. Anthropologists also recognize a second Polynesian group, the neo-Hawaiians, a blend of Polynesian, European, and Asian ancestries. Yet, despite the complexity of human occupance within the Pacific realm, the Polynesians, in their songs and dances, their philosophies regarding the nature of the world, their religious concepts and practices, their distinctive building styles, their work in stone and cloth, and in numerous other ways have built a culture of strong identity and distinction.

We have defined the 13 realms depicted in Fig. I–19 on the basis of a set of criteria that include not only cultural elements but also political and economic circumstances, relative location, and modern developments that appear to have lasting qualities. Undoubtedly our scheme possesses areas that are open to debate, but such a debate can by itself be quite instructive. We have indicated some locations where doubts may exist; there are more, as further reading and comparisons of our framework to others would underscore.

# The Place of Regional Studies in Contemporary Geography

Geography, which is often called the "parent" of all the sciences, is one of the very oldest fields of learning, and regional study has been a focus of the discipline throughout its evolution over the last 3000 years. It was during the third century B.C. that the ancient Greek scholar Eratosthenes first began to use the term "geography" (*geo*, earth + *graphy*, writing). By Roman times, the search for meaningful physical and human regions had become well established, and maps frequently incorporated such concepts as climatic zones. This valuable tool of regionalization greatly helped geographers to interpret the ever-expanding mass of environmental and cultural information they were accumulating. This approach also assisted them in developing a broad *spatial perspective* of the world and its contents, which was being assembled as regional study was integrated with three other maturing geographical traditions: earth science (physical geography), culture-environmental interaction (human ecology), and spatial organization (the locational structure of spatial distributions). In combination, these four basic traditions established geography as nothing less than a formal and comprehensive system of thinking that could be applied to any aspect of human affairs involving space, a *holistic* (integrating) discipline in the same sense that history is with respect to answering questions involving a time element. Besides its obvious academic applications, contemporary geography is also showing surprising strength in a more practical way: increasing numbers of geographers are being employed these days in the private and public sectors to help solve spatial problems of every variety (see Box, at right).

In modern practice, the four traditions have usually taken turns in overshadowing each other rather than always operating as coequals. For example, American geography opened this century with the earth-science tradition at the forefront, but by 1925 the culture-environment approach was dominant; by the eve of World War II in 1940, regionalism began a period of ascendancy that lasted until the early 1960s when the spatial-organization approach quickly eclipsed the position of area studies. The picture in the 1980s is somewhat more complex: although the spatial organization tradition still holds center stage, there are signs that the pendulum is again swinging back toward regional studies. This trend may well strengthen in the immediate future as the United States begins to cope with its citizens' inadequate understanding of the rest of the world at a time when the nation is entering a new era dominated by international relationships (see Box, p. 47).

As contemporary geography continues to evolve as both a social science tradition at the forefront, but by 1925 the culture-environment approach was dominant; by two. Reaffirming the role of regional study today, the 1982 President of the Association of American Geographers, Richard Morrill, has said that this approach is one of the two very justifications for having a distinct discipline of geography:

> . . . [its major contribution is] *its capacity to describe and analyze how diverse physical and human processes* [play out and] *interact to produce particular regional landscapes, cultures and places.* . . . [Regions] *are the manifest taxonomy of geography, the empirical, interrelated composites of phenomena which geographers, as both scientists and humanists, strive to explain and understand.*

## WHAT DO GEOGRAPHERS DO?

A systematic spatial perspective and an interest in regional study are the unifying themes and enthusiasms of geography. Geography's practitioners include physical geographers, whose principal interests are in the study of geomorphology (land surfaces), in research on climate and weather, in vegetation and soils, and in the management of water resources. There are also geographers who concentrate their research and teaching on the ecological interrelationships between the physical and human worlds; they study the impact of humankind on our globe's natural environments and the influences of the environment (including such artificial contents as air and water pollution) on human individuals and societies. Other geographers are regional specialists, concentrating their work for governments, planning agencies, and multinational corporations on a particular region of the world. Still other geographers—who now comprise the largest group of practitioners—are devoted to certain systematic subfields such as urban geography, economic geography, cultural geography, and many others (see Fig. I–22); they perform numerous tasks associated with the identification and resolution (through policymaking and planning) of spatial problems in their specialized areas. And, as in the past, there are still geographers who combine their fascination for spatial questions with technical know-how. Cartography, computer graphics, remote sensing, and even environmental engineering are among specializations listed by the 10,000+ professional geographers of North America.

## GEOGRAPHY AND INTERNATIONAL UNDERSTANDING

A number of governmental and private-sector reports in the early 1980s have criticized the quality of education in the United States. One of the greatest deficiencies identified was the appalling lack of international knowledge possessed by American citizens at precisely the time when their nation was entering a new era dominated by global economic and social interrelationships.

Responding to this problem of world ignorance within the United States, a special committee convened by the Association of American Geographers demonstrated that geography contributes heavily to international understanding and the development of global perspectives by emphasizing:

1. The relationships of societies, cultures, and economies around the world to specific combinations of natural resources and of the physical and biological environment.
2. The importance of location of places with respect to one another, as depicted on appropriate maps.
3. The diversity of the regions of the world.
4. The significance of ties of one country with another through the flow of commodities, capital, ideas, and political influence.
5. The world context of individual countries, regions, and problems.

As was pointed out in the Preface, the need for greater international awareness has become an added purpose in this latest edition of the book. Accordingly, all five emphases listed are stressed throughout the 10 regional chapters that follow.

Morrill's other justification is geography's concern with developing "[theories and] principles of human-environmental interaction and spatial organization [that] hold across broad classes of physical and human processes." Most often, this involves specializing in a topical or *systematic subfield* of geography. A political geographer, for instance, would study politico-spatial concepts and principles as they occurred across all the world's regions; he or she might be searching for a useful classification of boundary types, which can be developed by studying boundary evolution and perception in each of the geographic realms. Of course, by studying every major systematic branch of geography across the entire set of realms, the same total world picture obtained from a complete regional survey would emerge. This close linkage between regional and systematic geography (which in the final analysis are complementary means toward the same end) has been conceptualized by a number of geographers. A most appropriate model was developed by Nevin Fenneman in 1919; he saw regional geography as a unifying theme at the core, integrating the various systematic subfields that were distributed around the discipline's circumference, where they also spilled across geography's limits to connect with their own neighboring parent disciplines (e.g., political geography's link to political science). Because systematic geography is so closely tied to the regional geography of the 1980s, we will highlight one of these topical subfields in an essay at the beginning of each of the succeeding chapters. This program is illustrated in Fig. I–22, where the 10 subfields we have chosen to discuss are presented in the framework of the Fenneman scheme.

It is now time to commence our survey of the world's major regions.

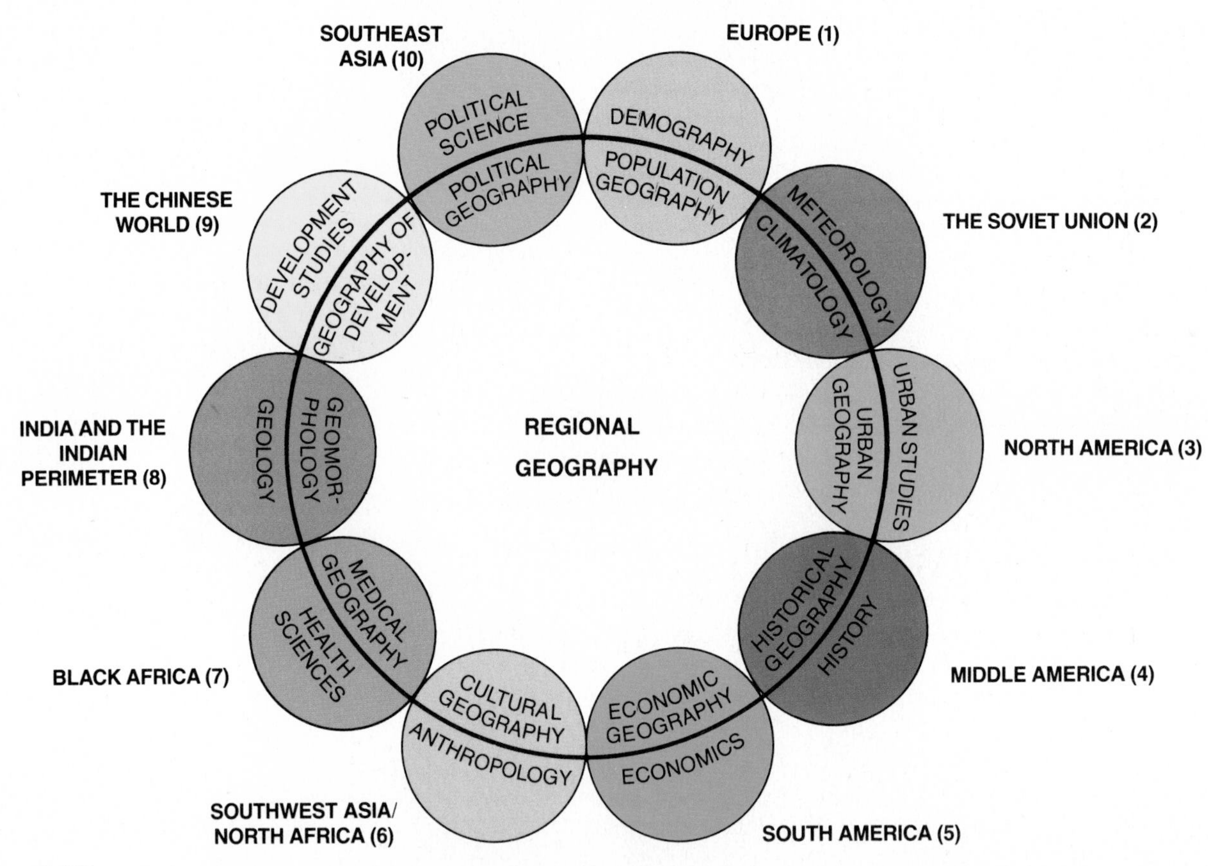


Figure I–22

## References and Further Readings


Abler, R. et al., eds. *Human Geography in a Shrinking World* (North Scituate, Mass.: Duxbury Press, 1975).

Abler, R. et al. *Spatial Organization: The Geographer's View of the World* (Englewood Cliffs, N.J.: Prentice-Hall, 1971).

Amedeo, D. & Golledge, R. *An Introduction to Scientific Reasoning in Geography* (New York: John Wiley & Sons, 1975).

Berry, B. J. L. *Comparative Urbanization: Divergent Paths in the Twentieth Century* (New York: St. Martin's Press, 2 rev. ed., 1981).

Broek, J. et al. *The Study and Teaching of Geography* (Columbus, Ohio: Charles E. Merrill, 2 rev. ed., 1980).

Brownell, J. "The Cultural Midwest," *Journal of Geography*, 59 (1960), 81–85.

Brunhes, J. *Human Geography* (London: George Harrap, trans. E. Row, 1952). Quotation taken from p. 40.

Brunn, S. & Williams, J., eds. *Cities of the World: World Regional Urban Development* (New York: Harper & Row, 1983).

Chorley, R. & Haggett, P., eds. *Models in Geography* (London: Methuen, 1967).

de Blij, H. J. *Human Geography: Culture, Society, and Space* (New York: John Wiley & Sons, 2 rev. ed., 1982).

de Souza, A. & Porter, P. *The Underdevelopment and Modernization of the Third World* (Washington: Association of American Geographers, Commission on College Geography Resource Paper No. 28, 1974).

Detwyler, T., ed. *Man's Impact on Environment* (New York: McGraw-Hill, 1971).

Dickinson, R. *The Regional Concept* (London: Routledge & Kegan Paul, 1976).

Durrenberger, R., ed. *Dictionary of the Environmental Sciences* (Palo Alto, Cal.: National Press Books, 1973).

"The Dynamic Earth," *Scientific American*, Special Issue (September 1983).

Ehrlich, P. & Ehrlich A. *Population, Resources, Environment* (San Francisco: W. H. Freeman, 1975).

Fenneman, N. "The Circumference of Geography," *Annals of the Association of American Geographers*, 9 (1919), 3–11.

Flint, R. F. *The Earth and Its History: An Introduction to Physical and Historical Geology* (New York: W. W. Norton, 1973).

Frejka, T. *The Future of Population Growth* (New York: Wiley-Interscience, 1973).

Fuson, R. *Introduction to World Geography: Regions and Cultures* (Dubuque, Iowa: Kendall/Hunt, 1977). Regional ranking system on pp. 2–5.

*Geography and International Knowledge* (Report) (Washington: A.A.G., Committee on Geography and International Studies, 1982).

# MAP READING AND INTERPRETATION

As we have seen throughout this chapter, maps are basic tools that are used to gain an understanding of patterns in geographic space. In fact, they comprise an important visual or *graphic communications medium* whereby encoded spatial messages are transmitted from the cartographer (mapmaker) to the map reader. This shorthand is necessary because the real world is so complex that a great deal of geographical information must be compressed into the small confines of maps that can fit onto the pages of this book. At the same time, cartographers must carefully choose which information to include; these decisions, thus, force them to omit many things in order to prevent cluttering a map with less relevant information. For example, Fig. I–23B shows several city blocks in central London, but avoids mapping individual buildings because they would interfere with the main information being presented—the spatial distribution of cholera deaths.

Deciphering the coded messages contained in the maps of this book is not difficult, and it becomes quite easy with a little experience in utilizing this "language" of geography. The need to miniaturize portions of the world onto small maps has already been discussed in the section on *map scale* (pp. 4–5), and two additional contrasting examples are provided in Figs. I–23A and I–23B below. *Orientation* or direction on maps can usually be discerned by refer-

**Figure I–23A**

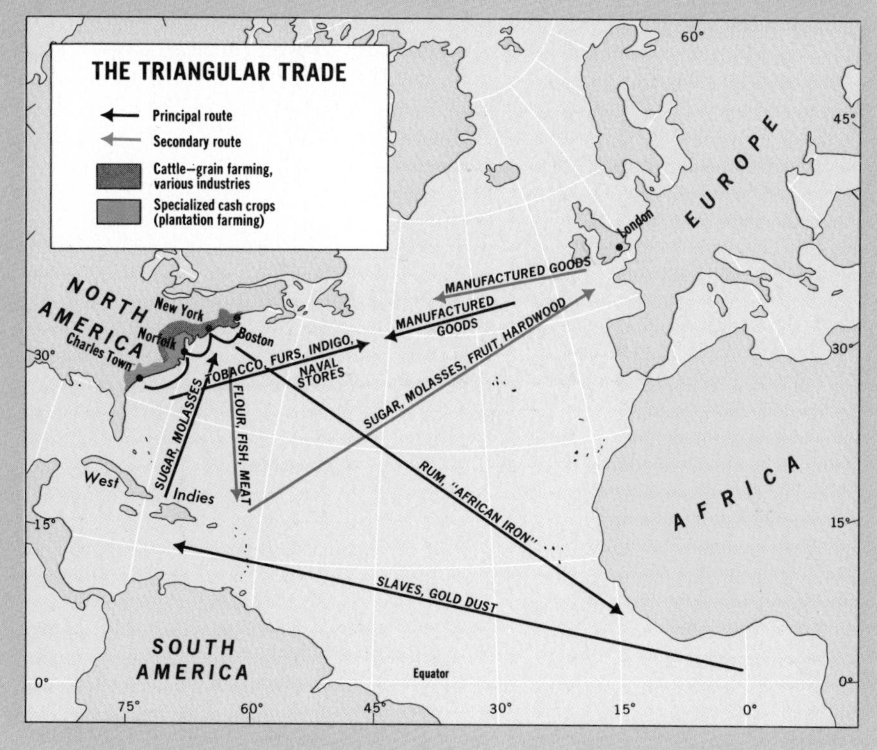

Glassner, M. & de Blij, H. J. *Systematic Political Geography* (New York: John Wiley & Sons, 3 rev. ed., 1980).

*Goode's World Atlas* (Chicago: Rand McNally, 16 rev. ed., 1982).

Gottmann, J. *A Geography of Europe* (New York: Holt, Rinehart & Winston, 4 rev. ed., 1969). Quotation taken from p. 76.

Haggett, P. *Locational Analysis in Human Geography* (London: Edward Arnold, 1965). Definition taken from p. 19.

Hare, F. K. *The Restless Atmosphere: An Introduction to Climatology* (New York: Harper Torchbooks, 3 rev. ed., 1963).

Hart, J. F. "The Highest Form of the Geographer's Art," *Annals of the Association of American Geographers*, 72 (1982), 1–29.

Hartshorn, T. *Interpreting the City: An Urban Geography* (New York: John Wiley & Sons, 2 rev. ed., 1986). Quotation taken from pp. 1–2 of the 1st ed.

Hoebel, E. A. *Anthropology: The Study of Man* (New York: McGraw-Hill, 4 rev. ed., 1972).

James, P. E. & Martin, G. *All Possible Worlds: A History of Geographical Ideas* (New York: John Wiley & Sons, 2 rev. ed., 1981).

Kroeber, A. & Kluckhohn, C. "Culture: A Critical Review of Concepts and Definitions," *Papers of the Peabody Museum of American Archaeology and Ethnology*, 47 (1952), entire issue. Definition taken from p. 181.

Lanegran, D. & Palm, R., eds. *An Invitation to Geography* (New York: McGraw-Hill, 2 rev. ed., 1978).

Larkin, R. & Peters, G., eds. *Dictionary of Concepts in Human Geography* (Westport, Conn.: Greenwood Press, 1983).

Leighly, J., ed. *Land and Life: A Selection from the Writings of Carl Ortwin Sauer* (Berkeley and Los Angeles: University of California Press, 1967).

Ley, D. & Samuels, M., eds. *Humanistic Geography: Prospects and Problems* (Chicago: Maaroufa Press, 1978).

Manners, I. & Mikesell, M., eds. *Perspectives on Environment* (Washington: Association of American Geographers, Panel on Environmental Education, General Publication No. 13, 1974).

"Mapping Out the Way We Live Now," *USA Today*, April 28, 1983, 1D–2D.

McDonald, J. *A Geography of Regions*


ence to the geographic grid of latitude and longitude. *Latitude* is measured from 0° to 90° degrees north and south of the equator (parallels of latitude always run in an east-west direction), with the equator being 0° and the North and South Poles being 90°N and 90°S, respectively. Meridians of *longitude* (which always run north-south) are measured 180° east and west of the *prime meridian* (0°), which passes through the Greenwich Observatory in London, England; the 180th meridian, for the most part, forms the International Date Line that runs down the middle of the Pacific Ocean. Inspection of Fig. I–23A shows that north is not automatically at the top of a map; instead, the direction of north curves along every meridian, with all such lines of longitude converging at the North Pole. Incidentally, the many minor directional distortions in this map are unavoidable—it is geometrically impossible to transfer the grid of a three-dimensional sphere (globe) onto a two-dimensional flat map. Therefore, compromises in the form of *map projections* must be devised in which properties such as areal size and distance are preserved, but directional constancy is sacrificed.

Once the background mechanics of scale and orientation are understood, the main task of decoding the map's content can proceed. Most of the maps in this book have their content sharply organized within the framework of point, line, and area symbols, which are made even clearer through the use of color. These symbols are usually identified in the map's *legend*, as in Fig. I–23A. Occasionally, the map designer omits the legend, but must tell the reader verbally in a caption or within the text what the map is about; Fig. I–23B, for instance, is a map of cholera deaths in the London neighborhood of Soho during the outbreak of 1854, with each Ⓟ symbol representing a municipal water pump and each red dot the location of a cholera fatality. *Point symbols* are shown as dots on the map and can tell us two things: the location of each phenomenon and sometimes its quantity. The cities of New York and London on Fig. I–23A and

**Figure I–23 B**

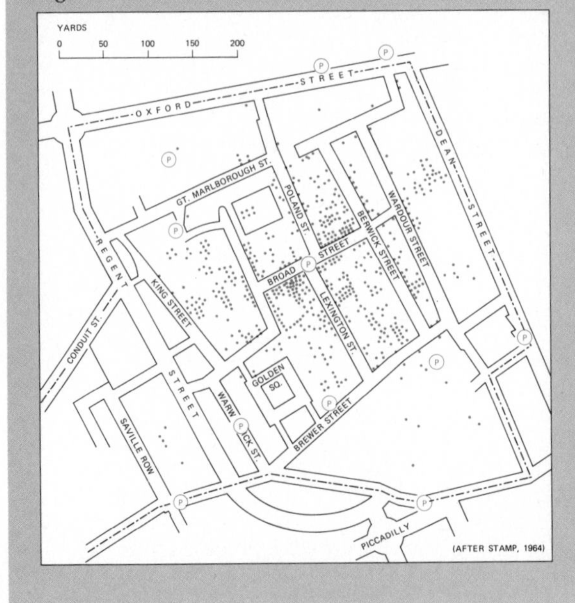

the dot pattern of cholera fatalities on Fig. I–23B (with each red dot symbolizing one death) are examples. *Line symbols* connect places between which some sort of movement or flow is occurring. The "triangular trade" among Britain and its seventeenth-century Atlantic colonies (Fig. I–23A) is a good example—each leg of these trading routes is clearly mapped, the goods moving along are identified, and the more heavily traveled principal routes are differentiated from the secondary ones. *Area symbols* are used to classify two-dimensional spaces and thus provide the cartographic basis for regionalization schemes (as we saw in Figs. I–9 through I–14). Such classifications can be developed at many levels of generalization. In Fig. I–23A, blue and beige areas broadly offset land from ocean surfaces; more specifically, red and green area symbols along the eastern North American coast delimit a pair of regions specializing in different types of commercial agriculture. Area symbols may also be utilized to communicate quantitative information: for example, the gray zones in Fig. I–9 delimit semiarid regions within which annual precipitation averages from 30–50 centimeters/12–20 inches.

*Map interpretation*—the explanation of cartographic patterns—is one of the geographer's most important tasks. Although that task is performed for you throughout this book, readers should be aware that today's researchers are using many sophisticated techniques and machines to analyze vast quantities of areal data. However, these modern methods notwithstanding, geographic inquiry still focuses on the search for meaningful *spatial relationships*. This long-standing concern of the discipline is nicely demonstrated in Fig. I–23B. By showing on his map that cholera fatalities clustered around municipal water pumps, Dr. John Snow was able to persuade city authorities to shut them off; almost immediately, the number of new disease victims dwindled to zero, thereby confirming Snow's theory that contaminated drinking water was crucial in the spread of cholera.

(Dubuque, Iowa: W. C. Brown, 1972).
Meadows, D. et al. *The Limits to Growth* (New York: Universe Books, 1972).
Meinig, D. W., ed. *The Interpretation of Ordinary Landscapes: Geographical Essays* (New York: Oxford University Press, 1979).
Miller, G. T., Jr. *Living in the Environment* (Belmont, Cal.: Wadsworth, 3 rev. ed., 1982).
Minshull, R. *Regional Geography: Theory and Practice* (London: Hutchinson, 1967).
Morrill, R. "The Nature, Unity and Value of Geography," *Professional Geographer*, 35 (1983). Quotations taken from pp. 5–6.
Natoli, S. *Careers in Geography* (Washington: Association of American Geographers, 3 rev. ed., 1983).
Newland, K. *City Limits: Emerging Constraints on Urban Growth* (Washington: Worldwatch Institute, Paper No. 38, 1980).
Paterson, J. *North America* (New York: Oxford University Press, 7 rev. ed., 1984). Quotation taken from p. 181 of the 5th ed.
Pattison, W. "The Four Traditions of Geography," *Journal of Geography*, 63 (1964), 211–216.
Paxton, J., ed. *The Statesman's Year-Book* (New York: St. Martin's Press, annual).
Robinson, A. et al. *Elements of Cartography* (New York: John Wiley & Sons, 5 rev. ed., 1984).
Robinson, J. "A New Look at the Four Traditions of Geography," *Journal of Geography*, 75 (1976), 520–530.
Russell, R. et al. *Culture Worlds* (New York: Macmillan, 3 rev. ed., 1969).
Salter, C., ed. *The Cultural Landscape* (Belmont, Cal.: Wadsworth, 1971).
Sauer, C. O. "Cultural Geography," *Encyclopedia of the Social Sciences* (New York: Macmillan, Vol. 6, 1931), pp. 621–623.
Shabad, T. "Americans Are Graded as Poor in Geography," *New York Times*, May 27, 1982, p. 7.
Starkey, O. et al. *The Anglo-American Realm* (New York: McGraw-Hill, 2 rev. ed., 1975). Quotation taken from p. 138.
Strahler, A. N. & Strahler, A. H. *Modern Physical Geography* (New York: John Wiley & Sons, 2 rev. ed., 1983).
Swartz, M. & Jordan, D. *Anthropology: Perspective on Humanity* (New York: John Wiley & Sons, 1976).
Symanski, R. & Newman, J. "Formal, Functional, and Nodal Regions: Three Fallacies," *Professional Geographer*, 25 (1973), 350–352.
Thomas, W., ed. *Man's Role in Changing the Face of the Earth* (Chicago: University of Chicago Press, 1956).
Tisdale, H. "The Process of Urbanization," *Social Forces*, 20 (1942), 311–316. As quoted in Berry, 1981, p. 27.
Vance, J. E., Jr. *This Scene of Man: The Role and Structure of the City in the Geography of Western Civilization* (New York: Harper's College Press, 1977).
Wegener, A. *The Origin of Continents and Oceans* (New York: Dover Publications, trans. J. Biram, 1966).
Wheeler, J. & Muller, P. *Economic Geography* (New York: John Wiley & Sons, 2 rev. ed., 1986).
Whittlesey, D. "Sequent Occupance," *Annals of the Association of American Geographers*, 19 (1929), 162–165.
Whittlesey, D. et al. "The Regional Concept and the Regional Method," in James, P. E. & Jones, C. F., eds., *American Geography: Inventory and Prospect* (Syracuse, N.Y.: Syracuse University Press, 1954), pp. 19–68.
Williams, J. et al. "World Urban Development," in Brunn, S. & Williams, J., eds., *Cities of the World: World Regional Urban Development* (New York: Harper & Row, 1983). Quotation taken from p. 6.
Zelinsky, W., guest ed. "Human Geography: Coming of Age," *American Behavioral Scientist*, 22 (1978), 3–167.

# PART ONE

# DEVELOPED REGIONS

BfG
COMMERZBANK

# CHAPTER 1

# THE MOSAIC OF EUROPE

IDEAS AND CONCEPTS

**Population Geography**
**Relative Location**
**Functional Specialization**
**Nation-State**
**Von Thünen's Isolated State**
**Industrial Location**
**Organic Theory of Political Evolution**
**Spatial Interaction Principles**
- **Complementarity**
- **Intervening Opportunity**
- **Transferability**

**Primate City**
**European Urbanization Trends**
**Conurbation**
**Site and Situation**
**Acid Rain**
**Balkanization**
**Irredentism**
**Supranationalism**

For centuries, Europe has been the heart of the world. European empires spanned the globe and transformed societies far and near. European capitals were the focal points of trade networks that controlled distant resources. Millions of Europeans migrated from their homelands to the New World as well as the Old, to create new societies from Canada to Australia. While Europe's colonial powers competed abroad, they fought each other at home. In agriculture, in industry, and in politics Europe went through revolutions—revolutions that were promptly exported throughout the world and served to consolidate the European advantage. Europe's decline (it may be temporary) began during the second of two disastrous twentieth-century wars. In the aftermath of World War II (1939–1945), Europe's weakened powers lost the colonial possessions that had so long provided wealth and influence. But it is a measure of the region's fundamental strength that European states, even without their empires, continued to thrive.

As noted in the introductory chapter, Europe constitutes a discrete geographic realm, despite its comparatively small size on the peninsular margin of Western Eurasia. For more than 2000 years Europe has been a focus of human achievement, a hearth of innovation and invention. And Europe's human resources were matched by its large and varied raw material base. When the opportunity or the need arose, Europe's physical geography proved to contain what was required. Most recently, the sea floor adjacent to Europe's North Sea coasts has yielded enough oil to make several surrounding countries *exporters* of energy in the 1980s.

For so limited an area (slightly more than half the size of the United States), Europe's internal natural diversity is probably unmatched. From the warm shores of the Mediterranean to the frigid

**Figure 1–1**
Juxtaposition of tradition and ultramodernity: the cityscape of West Germany's dynamic Frankfurt.

# SYSTEMATIC ESSAY

## Population Geography

**Population geography** is concerned with the *distribution, composition, growth* and *movement* of people as related to spatial variations in living conditions on the earth's surface. The subject differs from *demography*—the interdisciplinary study of population characteristics—in that it focuses on the geographic expression through which population data are linked to specific areas.

*Population distribution* is perhaps the most essential of all geographical expressions, because the way in which people have arranged themselves in space at any given time represents the sum total of the adjustments they have made to their overall environment. Worldwide, as we saw in the previous chapter (Fig. I–14), humanity has sorted itself out to the extent that we can identify four primary concentrations of population: East Asia, South Asia, Europe, and eastern North America. Since this chapter treats the geography of the third cluster, let us examine Europe in some detail.

The current distribution of Europe's population is shown on the map in Fig. 1–2 and reflects a cumulative response to the broad forces that have progressively shaped this continent's *ecumene* (habitable areas) for several centuries. Physical extremes of terrain and climate retarded settlement in rugged mountains and much of the Arctic margin to the north, whereas favorable environmental conditions encouraged the massive peopling of fertile valleys and coastal lowlands. Variable economic conditions related to perceived modern opportunities spawned several migrations toward prosperous areas, with allowances for deep-rooted ethnic affiliations that have preserved much of Europe's cultural mosaic since the Middle Ages. Descriptively, we may analyze the map according to three spatial criteria: pattern, dispersion, and density. By *pattern* is meant the geometric arrangement of a distribution; Europe's pattern is quite complex, but certain features do stand out such as overall unevenness and linearity of settlement within leading river valleys and other major transportation corridors. *Dispersion* describes the extent of spread of a distribution; although people are widely dispersed across the region as a whole, the clustering or localization of population in many major metropolitan centers is a dominant feature of the map. *Density* measures the frequency of occurrence of a phenomenon within a given area. It has already been pointed out that Europe's population concentration of 700+ million is one of the most densely settled parts of the world: continentwide, the figure is about 100 persons per square kilometer (260 per square mile). In certain countries, however, the measure is much higher; in the crowded Netherlands, this *arithmetic density* figure is 350 people per square kilometer (900 per square mile), but the *physiologic density*—the number of persons per unit area of arable land—jumps to 1200 per square kilometer.

*Population composition* is usually grouped by age and sex as shown in Fig. 1–3, which displays the population "profile" of Sweden. Since growth is a function of the birth rate, the relatively modest proportion of females in the childbearing years (ages 15–44)—a situation heightened through the widespread use of birth-control techniques—indicates this nation has a low growth rate. At the same time, an outstanding health-care system promotes longevity and fairly low death rates. Thus, the overall population is an aging one, which is typical of European and other advanced industrial (and postindustrial) countries in the late twentieth century.

In general, the degree of demographic change affecting a country can be represented in this simple formula:

$$\text{Total Population} = \text{Starting Population} + \text{Births} - \text{Deaths} + \text{Immigration} - \text{Emigration}$$

The population of the world as a whole is now growing older: by 2025, there will be more people over 60 than under 25. Nonetheless, Third World countries are not expected to share this characteristic to any great degree: younger people will continue to dominate age-sex structures there for many decades to come, and these nations will continue to account for about 90 percent of the planet's population growth until 2000, when 6.2 billion persons (vs. about 4.9 billion in 1985) are expected in the world.

*Population growth*, therefore, continues to be a major world crisis. As famine, disease, and other so-called growth "checks and balances" have been reduced in many parts of the world, population totals have accelerated; from 1650 to 1985 alone, the global total has surged tenfold, from 500 million to almost 5 billion. In a period of such rapid growth, population doubling times become compressed: whereas 170 years were required for a population of 500 million to double, only about 40 years are needed to double a population of 5 billion. Counteracting this discouraging outlook are the population

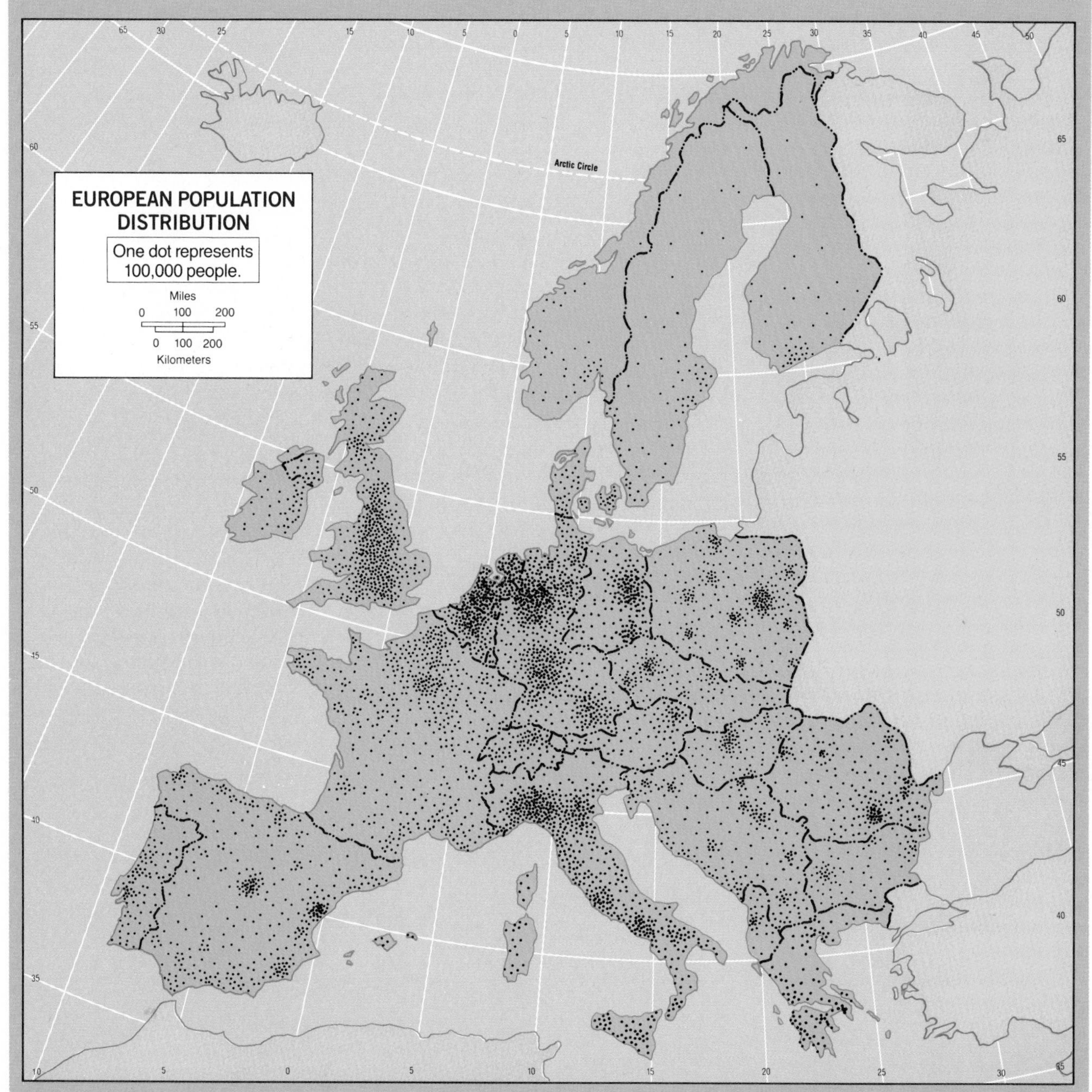


Figure 1–2

growth changes experienced by countries that have undergone modern economic development in the form of the industrial revolution. These may be summarized in the three-stage, **Demographic Transition Model**, which is a generalization derived from the historic case of the United Kingdom (Fig. 1–4). Stage One is the preindustrial era before 1800 (shown in green), in which a rural agricultural population predominated and exhibited rather high birth and death rates but an overall low growth rate. Following the onset of industrialization and the huge cityward migration it spawned, things began to change. Better health standards and new medical advances began to eradicate disease, and the effect on lowering death rates was dramatically demonstrated during the nineteenth century in Stage Two (dark orange). At the same time, however, birth rates stayed high, and the steady drop in mortality—particularly infant mortality—resulted in an enormous net population increase. Unchanged attitudes toward family formation supported persistently high birth rates, because the former rural residents of the new industrial cities had always viewed children as additional laborers who could augment a family's income. Eventually, overcrowded living conditions and the enactment of child-labor laws altered the social and

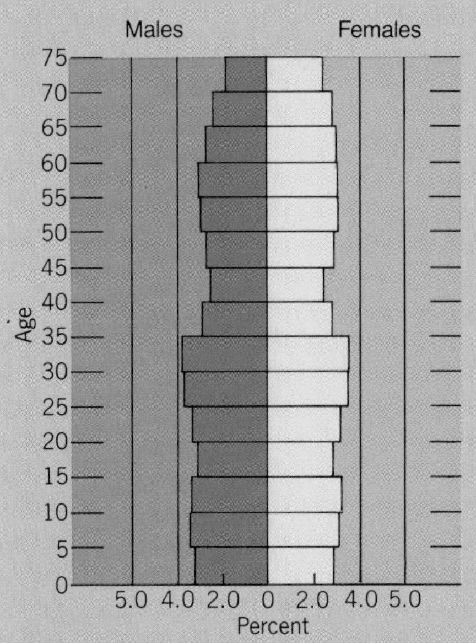


**Figure 1–3**

economic values that led to the breeding of large families; by 1900, birth rates began to drop as family size limitations finally came to prevail in industrial society, aided widely by the acceptance of birth-control methods. By the middle of this century, Stage Three (light orange) was achieved, marked again by a convergence of birth and death rates but at a far lower overall level; that condition persists today as Zero Population Growth has almost been accomplished, and the industrial countries now enjoy the lowest net natural increase rates in their history.

The British lesson seems to offer some hope for the future as developing nations begin to enter the industrialization phase. Yet, many demographers are uncertain that a further population explosion can be averted. Thus, although there may be some vague grounds for optimism, there is presently no effective machinery to defuse the population "bomb." Many of the fastest growing areas are associated with the world's expanding cities, a topic that will be treated in this and every subsequent chapter.

*Population movements* are another major concern of this subdiscipline, because the redistribution of people constantly changes the world's patterns of resource utilization. Indeed, **migration**—the purposeful relocation of one's permanent residence—is so pervasive a topic in contemporary world regional geography that it can only be introduced here. In nearly every realm this multifaceted subject will be raised anew in order to explore the reshaping of regional spatial structures; important generalizations can be found in the vignette on Australia (p. 124).

Economic causes often propel the world's major population flows: the search for a higher-paying job and a better life is universal, and perceived economic opportunity—the belief that the grass is greener in another location—annually convinces tens of millions to relocate their homes and families. Other broad forces also enter the picture and act as catalysts of migration, among them cultural pressures, changing political conditions, and the need to escape from hazardous natural environments. *Voluntary* and *involuntary* migrations occur side by side: although most movements are voluntary, the experiences of the Palestinians and Cambodians in the last few years is ample testimony that forced migrations are still very much a part of life in today's world. *External* and *internal* migrations, with respect to national borders, will also be frequently differentiated in this book.

**Figure 1–4**
**THE DEMOGRAPHIC TRANSITION**

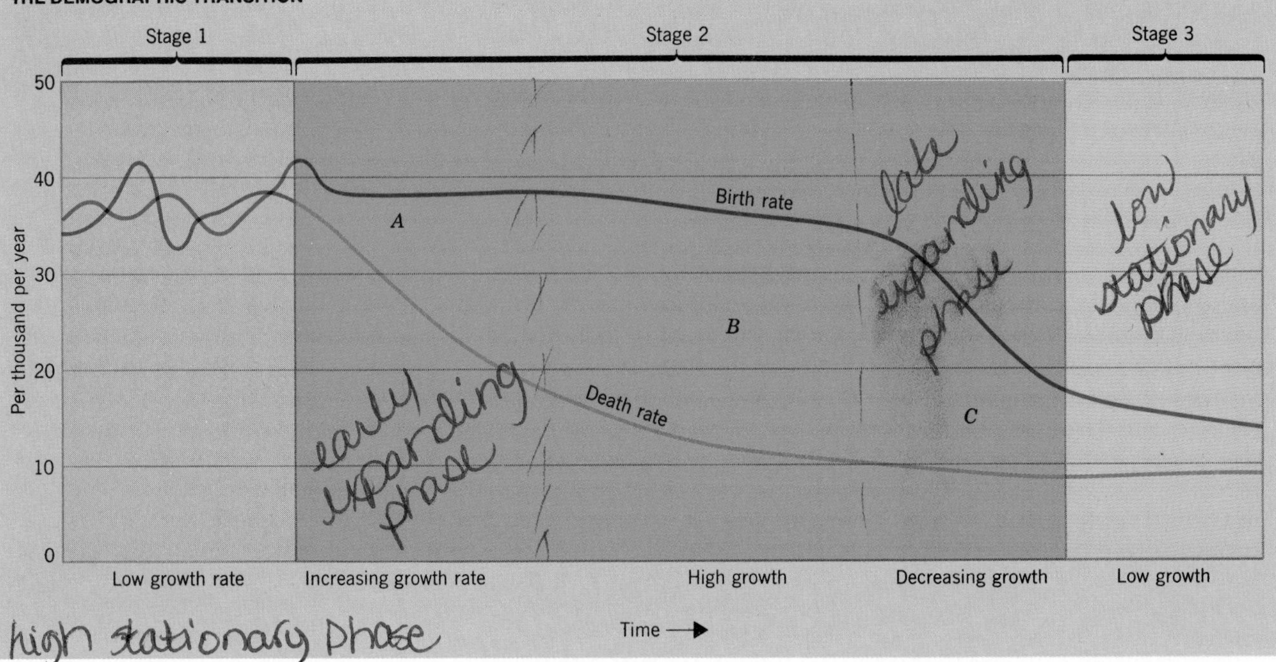

Scandinavian Arctic and from the flat coastlands of the North Sea to the grandeur of the Alps, Europe presents an almost infinite range of natural environments. The insular and peninsular west contrasts strongly against a more interior, continental east. A resource-laden backbone extends across Europe from England eastward to Poland, yielding coal, iron ore, and other valuable minerals; the North Sea floor contains oil and natural gas. And this diversity is not confined to the physical makeup of the continent. The European realm includes peoples of many different stocks. Not only does Europe contain Latins, Germans, and Slavs, but it also has numerous minorities, including Finns and Hungarians.

This diversity of physical and human content alone, of course, does not constitute such a great asset to Europe. If differences in habitat and race or culture automatically led to rapid human progress, Europe would have had many more competitors for its world position than it did. But in Europe, there were virtually unmatched advantages of scale and proximity. Globally, Europe's relative location is one of maximum efficiency for contact with the rest of the world (Fig. 1–5). Regionally, Europe is not just a western extremity of the Eurasian landmass: almost nowhere is Europe far from that essential ingredient of European growth—the sea—and, the water interdigitates with the land here as it does nowhere else. Southern and Western Europe are practically made of peninsulas and islands, from Greece, Italy, France, and the Iberian Peninsula (Spain and Portugal), to the British Isles, Denmark, Norway, and Sweden. Southern Europe faces the Mediterranean, and Western Europe virtually surrounds the North Sea as it looks out over the Atlantic Ocean. Beyond the Mediterranean lies Africa, and across the Atlantic are the Americas. Europe has long been a place of contact between peoples and cultures, of circulation of goods and ideas. The hundreds of kilometers of navigable waterways; the bays, straits, and channels between numerous islands and peninsulas and the mainland; the Mediterranean and North Seas; and later the oceans provided the avenues for these exchanges.

**RELATIVE LOCATION: EUROPE IN THE LAND HEMISPHERE**

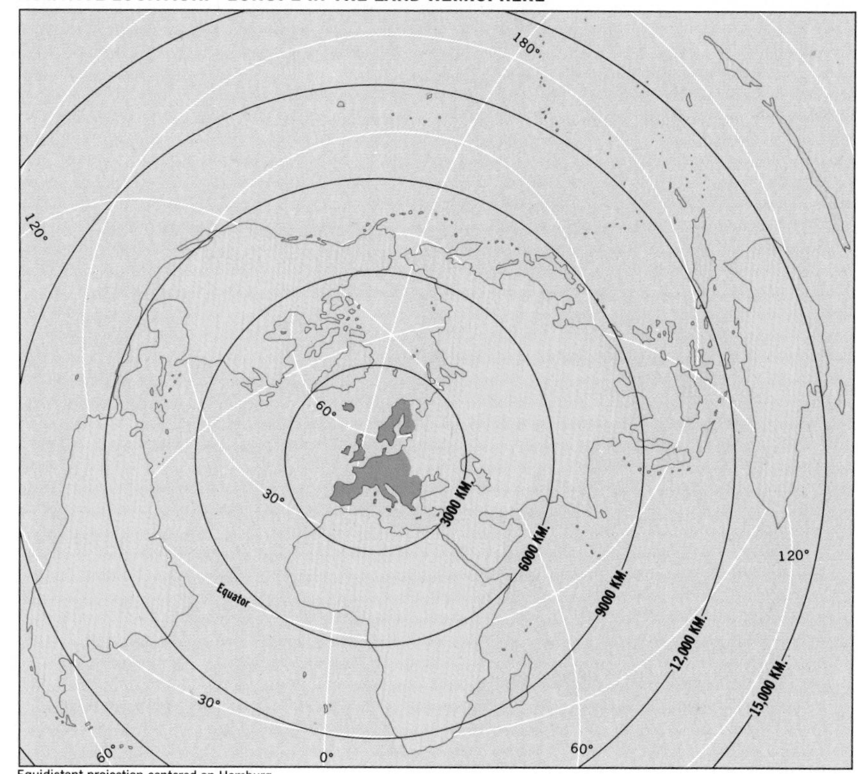


Equidistant projection centered on Hamburg

Figure 1–5

Small wonder, then, that navigation skills were quickly adopted, learned, and developed in Europe. Again, the advantage of scale presents itself—the advantage of moderate distances. With the exception of Iceland, the islands of Europe are all quite near the mainland. There is a remarkably low variance in the widths of such water bodies as the Aegean Sea, the Adriatic, the English Channel, the Skagerrak-Kattegat entrance to the Baltic Sea, and the Baltic itself and its Gulf of Bothnia. Again, the coasts of the North Sea are about as far from each other as are those of the western Mediterranean, the Tyrrhenian, and the Ionian seas. Europe's offshore waters, therefore, are not the almost endless expanses that surround the Americas, Africa, South Asia, and Australia. They invited exploration and utilization; what lay on the other side was not some dark unknown but, more often than not, a visible coast of land of known qualities.

This applies on the mainland as well. Europe's Alps seem to form a transcontinental divider, but what they divide still lies in close juxtaposition (and in any case, the Alps provide several corridors for contact). But consider Rome and Paris: the distance between these longtime headquarters of Mediterranean and northwestern Europe is less than that between New York and Chicago or Miami and Atlanta. No place in Europe is very far from anyplace else on the continent, and nearby places are often sharply different from each other in terms of economy and outlook. Short distances and large differences make

## TEN MAJOR GEOGRAPHIC QUALITIES OF EUROPE

1. The European realm is constituted by the western extremity of the Eurasian landmass, a location of maximum efficiency for contact with the rest of the developed world.
2. Europe's lingering and resurgent world influence results largely from advantages accrued over centuries of global political and economic domination.
3. Europe's nation-states emerged from durable power cores that formed the headquarters of world colonial empires.
4. Europe is marked by strong regional differentiation (cultural as well as physical), has a high degree of functional specialization, and provides multiple exchange opportunities.
5. The European natural environment displays a wide range of topographic, climatic, vegetative, and soil conditions.
6. European economies are dominantly industrial, and levels of productivity are high. Levels of development, in general, decline from west to east.
7. Europe's population is generally healthy, well fed, has low birth and low death rates, enjoys long life expectancies, and constitutes one of the three great population clusters on earth.
8. Europe's population is highly urbanized, highly skilled, and well educated.
9. Europe is served by efficient transport and communications networks, which promote extensive trade and other forms of international spatial interaction.
10. European agriculture achieves high levels of productivity and is mainly market oriented.

for much interaction—and that has marked the geography of Europe for many centuries.

# Landscapes and Rivalries

Europe may be small (it is among the three smallest world regions discussed in this book), but its landscapes are varied and complex. It would be possible to establish a large number of physiographic regions, but in doing so we might lose sight of the overall regional pattern—a pattern that has much to do with the way the human drama developed.

Europe's landforms can be grouped regionally into four units (Fig. 1–6): the Central Uplands, the Alpine Mountains in the south, the Western Uplands, and the great North European Plain. The very heart of Europe is occupied by an area of hills and small plateaus, with forest-clad slopes and fertile valleys. These *Central Uplands* also contain the majority of Europe's productive coalfields, and when the region emerged from its long medieval quiescence and stirred with the stimuli of the industrial revolution, towns on the uplands' flanks grew into cities, and farms gave way to mines and factories.

The Central Uplands are flanked on the south by the much higher *Alpine Mountains* and to the west and north by the North European Lowland. The Alpine Mountains include not only the famous Alps themselves, but also other ranges that belong to this great mountain system. The Pyrenees between Spain and France (one of Europe's few true barriers), Italy's Appennines, Yugoslavia's Dinaric Ranges, and the Carpathians of Eastern Europe all are part of this system that extends even into North Africa (the Atlas Mountains) and eastward into Turkey and beyond. Although these Alpine mountains are rugged and imposing, they have not been a serious obstacle to contact and communication. Europe's highest mountain, Mont Blanc (4813 meters; 15,781 feet), towers over valleys that have for centuries served as pass routes for traders and warriors.

Europe's western margins also are quite rugged, but the *Western Uplands* of Scandinavia, Scotland, Ireland, Brittany (France), Portugal, and Spain are not part of the Alpine system. Maximum elevations are lower than those in the Alps, ranging to over 2500 meters (8200 feet) in several locations in Spain and approaching such heights in a few locales in Scandinavia. This western arc of highlands represents an older segment of Europe's geomorphology compared to the relatively young, still active, earthquake-prone Alpine mountains. Scandinavia's uplands form part of an ancient shield zone, underlain by old crystalline rocks now bearing the marks of the Pleistocene glaciation. Spain's plateau or *Meseta* is also supported by comparatively old rocks, worn down now to a tableland.

The last of Europe's landscape regions is also its most densely populated. The *North European Lowland* (other names have also been given to this region) extends from southwestern France through

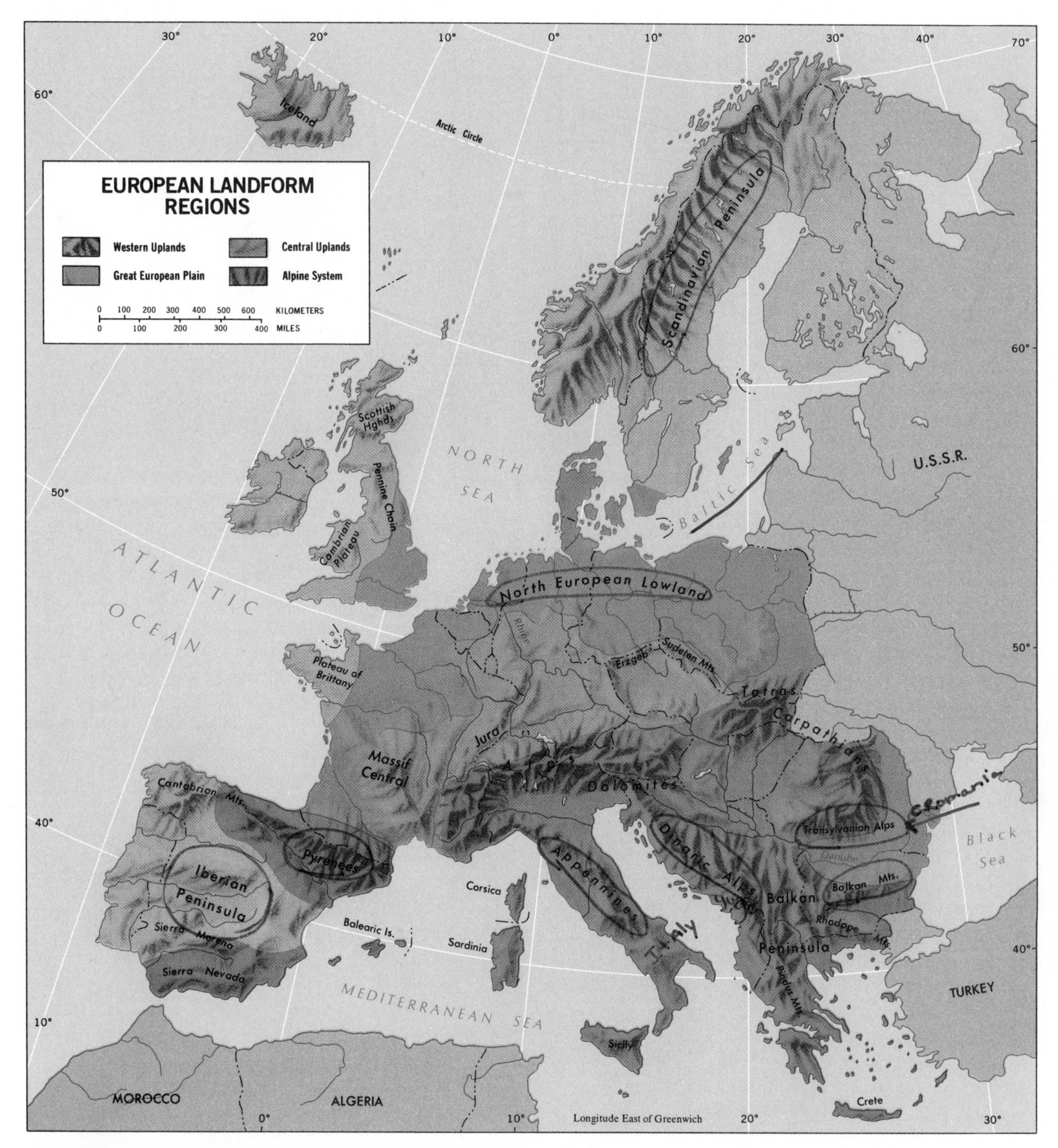


Figure 1–6

the Paris Basin and the Low Countries across northern Germany and eastward through Poland. Southeastern England and the southern tip of Sweden also are part of this region, which forms a continuous belt on the mainland from southern France to the Gulf of Finland (Fig. 1–6). Most of the North European Lowland lies below 150 meters (500 feet) in elevation, and local relief rarely exceeds 30 meters (100 feet). But make no mistake; this region may be topographically low lying and flat or gently rolling, but there its uniformity ends. Beyond this single topographic factor there is much to differentiate it internally. In France, it includes the basins of three major rivers, the Garonne, Loire, and Seine. In the Netherlands, for a good part, it is made up of land reclaimed from the sea, enclosed by dikes and lying below sea level. In southeast England, the higher areas of the Netherlands, northern Germany, southern Sweden, and farther eastward, it bears the marks of the Pleistocene glaciation that withdrew only a few thousand years ago (see Fig. I–8).

Each of these particular areas affords its own opportunities for human endeavor, as soils and climates vary. Along with wheat and corn, there are areas of viticulture in the Garonne lowland. There are fine pastures in the Netherlands and England, where dairying is an important industry. Fields of rye cover the North German countryside. Toward the northeast the glaciated country is clothed by stands of mixed forest (including areas of conifers familiar to anyone who has seen the woodlands of Canada and northern Michigan), with pastures and cropland in intervening belts.

**Figure 1–7**
Village and farmlands on the North European Lowland.

The North European Lowland has been one of Europe's major avenues of human contact. Entire peoples migrated across it; armies marched through it. At an early stage, there were trade ties with southern parts of the continent. As stock raising gave way to agriculture, the forests that once covered most of this region were cut down. As time went on, new techniques of cultivation and new crops led to ever-greater diversification. Land use came to be dominated by intensive farming organized around a myriad of small village settlements from which farmers commuted to their nearby fields (Fig. 1–7). It is hardly possible to speak of a "wheat belt" or "ranching area" in Europe, such as exist in North America. Even where one particular activity or crop dominates the farming scene, some other pursuit takes place only a few hundred yards away. A short train ride almost anywhere will lead past fields of carefully demarcated pasture, then suddenly along fields of well-tended potatoes, back to pasture, perhaps through fields of bulb flowers, a stretch of rye, and then—we must be nearing a city—neat rows of vegetable gardens. Trimmed hedges, narrow canals, or fences limit the various fields; every bit of available land seems somehow in productive use. Small wonder that trade in agricultural produce long was Europe's mainstay.

It is not just the low relief that has favored contact and communication within Europe's lowland. The region has another crucial advantage: its multitude of navigable rivers, emerging from higher adjacent areas and wending their way to the sea. In addition to the rivers of France already mentioned, the Rhine-Maas (Meuse) river system serves one of the world's most productive industrial and agricultural areas and reaches the sea in the Netherlands; the Weser, Elbe, and Oder penetrate northern Germany; the Vistula traverses Poland. In Eastern Europe, the Danube rivals the Rhine in regularity of flow and navigability. Thus, Europe's major rivers create a radial pattern outward from the region's higher central areas, although the Mediterranean south is less well endowed in this respect. France's Rhône-Saône system flows from Alpine flanks into the Mediterranean Sea; Italy's Po River, also fed from Alpine slopes, enters the Adriatic.

In this way, the natural waterways as well as the land surface of the North European Lowland favor traffic and trade; over the centuries, the Europeans have improved the situation still more. Navigable stretches of rivers were connected by artificial canals, and through these and a system of locks a network of water transport routes evolved. Waterways, roads, and later railroads combined to bring tens of thousands of localities into contact with each other. New techniques and innovations spread rapidly; trade connections and activities intensified continuously. No region in Europe provided greater opportunities for all this than the Lowland.

# Heritage of Order

Modern Europe was peopled in the wake of the Pleistocene's most recent glacial retreat—a lingering withdrawal that saw cold tundra turn into deciduous forest and ice-filled valleys into grassy vales. It was on Mediterranean shores that Europe witnessed the rise of its first

## EUROPE: THE EASTERN BOUNDARY

The European realm is bounded on the west, north, and south by Atlantic, Arctic, and Mediterranean waters, respectively. But Europe's eastern boundary has always been a matter for debate. Some scholars place this boundary at the Ural Mountains, thereby recognizing a "European" Soviet Union and, presumably, a non-European one as well. Others argue that, because there is transition from west to east (a transition that continues into the Soviet Union), there is no point in trying to define a boundary.

Still, the boundary used in this chapter (marking Europe's eastern boundary as the border with the Soviet Union) has geographic justifications. Here, diversity changes into uniformity, variety into sameness. Eastern Europe shares with Western Europe its fragmentation into several states with distinct cultural geographies—a condition that sets it apart from the giant to the east. Historical and cultural contrasts mark large segments of this boundary, notably in Finland, Hungary, and Romania (see Fig. 1–13). The Soviet Union's postwar ascendancy in Eastern Europe has not wiped out certain ideological and iconographic differences between the two regions. The Soviet Union remains the great monolith. Eastern Europe retains its discrete nationalisms, latent conflicts, pressures, and vicissitudes, qualities vividly demonstrated in this decade by the Polish people.

great civilizations—on the islands and peninsulas of Greece and later in Italy. Greece lay exposed to the influences radiating from the advanced civilizations of the Middle East (see Chapter 6), and the eastern Mediterranean was crisscrossed by maritime trade routes.

As the ancient Greeks forged their city-states and intercity leagues, they made intellectual achievements as well. Their political philosophy and political science became important products of Greek culture—and, indeed, here was a culture of world significance. Plato (428–347 B.C.) and Aristotle (384–322 B.C.) are among the most famous of the Greek philosophers of this period, but they stand out among a host of other contributors to the lasting greatness of ancient Greece. The writings they left behind have influenced politics and government ever since, and constitutional concepts that emerged at that time are still being applied in modified form today. But there was more to Greece than politics. In such fields as architecture, sculpture, literature, and education, the Greeks displayed their unequaled abilities as well. Because of the fragmentation of their habitat, there was local experimentation and success—followed by exchanges of ideas and innovations.

Individualism and localism were elements the Greeks turned to great advantage, but internal discord was always present, actually or potentially. Eventually, it got the better of them, especially because of the endless struggle between the two major cities, Athens and Sparta. After the fourth century B.C., there was a period of ups and downs, and finally the Romans defeated the last sovereign Greek intercity league, the Achaean, in 147 B.C. But what the Greeks had accomplished was not undone. True, they had borrowed from their Middle Eastern neighbors in such fields as astronomy and mathematics, but they had by their own energy transformed the eastern Mediterranean into one of the culture cores of the world.

Greek culture was a major component of Roman civilization; but the Romans made their own essential contributions. The Greeks never achieved politico-territorial organization on the scale accomplished by Rome. In such fields as land communications, military organization, law, and public administration, the Romans made unprecedented progress. Comparative stability and peace marked the vast realm under their domination, and for centuries these conditions promoted social and economic advances. The Roman Empire during its greatest expansion (which came during the second century A.D.) extended from Britain to the Persian Gulf and from the Black Sea to Egypt (Fig. 1–8). In North Africa, the desert and the mountains formed its boundaries; to the west it was the sea; and to the north tribal forest peoples lived beyond the Roman domain. Only in Asia was there contact with a state of any significance. Thus, the Empire could organize internally without interference. But its internal diversity was such that isolation of this kind did not lead to stagnation. The Roman Empire was the first truly interregional political unit in Europe. Apart from the variety of cultures that had been brought under its control and the consequent diffusion of ideas and innovations, there were a multitude of possibilities for economic exchange.

This process of economic development (for such it really was) made a profound impact on the whole structure of southern and western Europe. Areas that had hitherto supported only subsistence modes of life were drawn into the greater economic framework of the

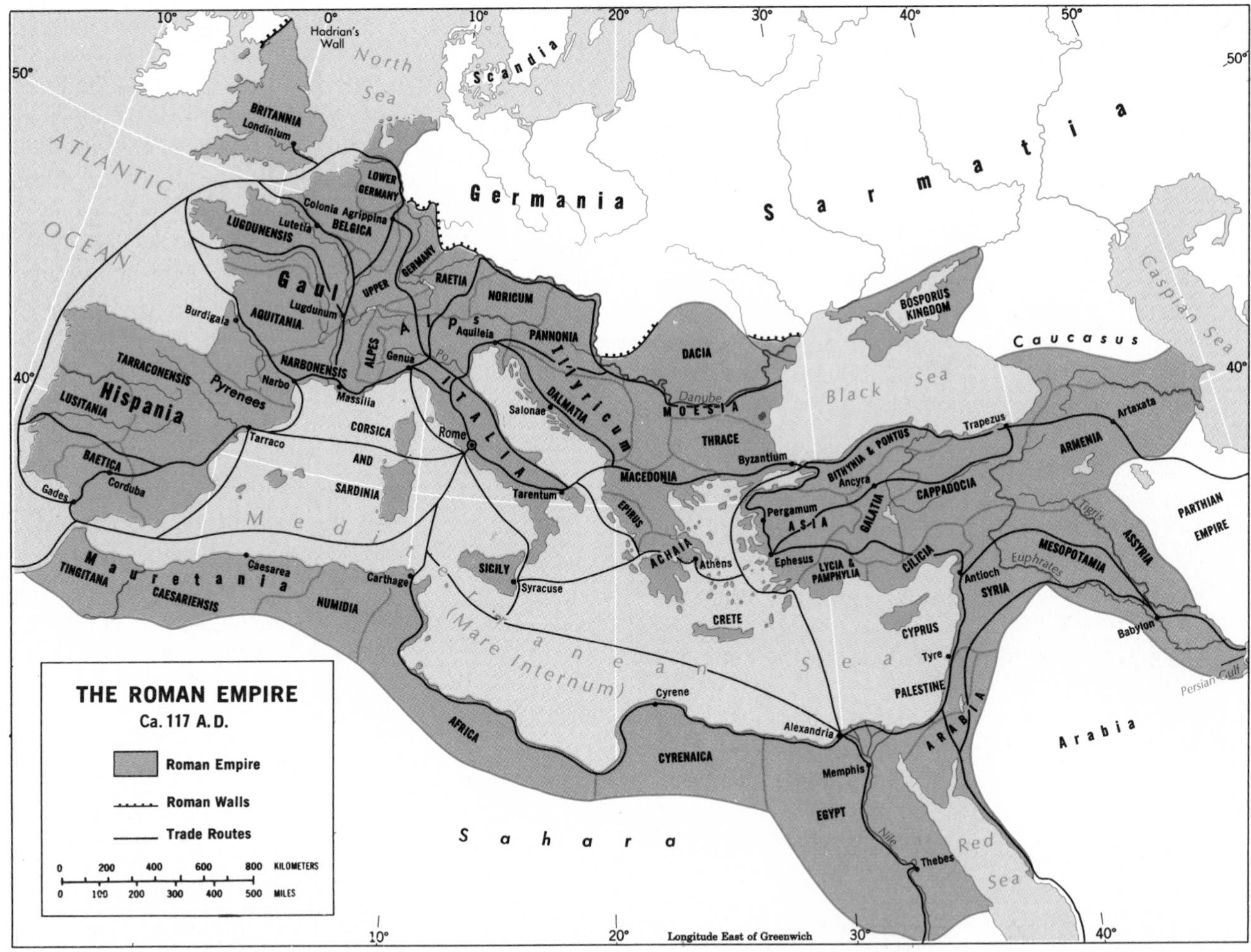


Figure 1–8

state, and suddenly there were distant markets for products that had never found even local markets before. In turn, these areas received the farming know-how of the heart of the Roman state, so that they could increase their yields and benefit even further. Foodstuffs came to Rome from across the Mediterranean as well as from southeastern and southwestern Europe; with an ever more complex population of perhaps a quarter of a million, the city itself was the greatest individual market of the empire and the first real metropolitan urban center of Europe. In addition to these agricultural advances, the Romans undertook a sociopolitical reorganization of their dominions that served economic growth—and was to have a lasting impact on Europe in general. This was their decision to support the development of a number of secondary Roman capital cities throughout the empire, cities that would have authority over surrounding rural areas and that would be the chief organizing centers for such areas. It is not difficult to see the Greek city-state concept in this, modified to serve Roman needs. But unlike the Greek cities, the Roman towns were connected by that unparalleled network of highways referred to earlier. European integration was proceeding at an unprecedented pace. On current maps, Roman towns (and, here and there, Roman routes as well) are still represented: Lugdunum (Lyons in France), Singidunum (Belgrade), Eboracum (York), and Colonia Agrippensis (Cologne), among others.

Roman culture was, above all else, an urban culture, and many urban centers founded or developed by the Romans continue to function and grow today. Whereas the economic base of many of these places has changed, others, notably the ports, still perform as they did in Roman times. Roman artifacts proved quite durable and frequently survive in the cultural landscape, such as the famous aqueduct that runs through the city of Segovia in central Spain (Fig. 1–9). Although the original surface of the old Roman roads has mostly disappeared, some modern highways in Europe follow exactly the routes laid out by Roman engineers. But more than anything else, the Roman Empire left Europe a legacy of ideas, of concepts that long lay dormant but eventually played their part when Europe again found the

**Figure 1–9**
The Roman cultural imprint: Segovia, Spain.

path of progress. In terms of political and military organization, effective administration, and long-term stability, the Roman Empire was centuries ahead of its time. Concepts of federal and confederal relationships as applied by the Romans had great currency more than a thousand years later. At no time was a larger part of Europe unified by acquiescence than was the case under the Romans. Never did Europe come closer to obtaining a *lingua franca* (common language) than it did during Roman times.

Europe's transformation under Roman hegemony also involved the geographic principle of **areal functional specialization**. Before the Romans brought order and connectivity to their vast empire, much of Europe was inhabited by tribal peoples whose livelihood was on a subsistence level; many of these groups lived in virtual isolation, traded little, and fought over territory when encroachment occurred. Peoples under Rome's sway, however, were brought into Roman economic as well as political spheres, and farmlands, irrigation systems, mines, and workshops appeared. Thus, Roman areas of Europe began to take on a characteristic that has marked Europe ever since: *particular peoples and particular places produced specific goods.* Parts of North Africa became granaries for urbanizing (European) Rome. Elba, a Mediterranean island, produced iron ore that was shipped to Puteoli, near present-day Naples. Inland from today's Cartagena in Spain lay important silver and lead mines whose yields were exported through Cartagena's Roman predecessor, the port of Nova Carthago. Many other locales in the Roman Empire specialized in the production of certain farm products, manufactured goods, or minerals. The Romans knew how to exploit their natural resources; they also used the talents of their subjects.

# Heirs to the Empire

Today, 2000 years later, the contributions of ancient Rome still are imprinted in the cultural landscape of Europe (Fig. 1–9). The eventual breakdown and collapse of the empire could not undo what the Romans had forged in the diffusion of their language, in the dissemination of Christianity (in some ways the sole strand of permanence through the ensuing Dark Ages), in education, in the arts, and in countless other spheres. But Rome's collapse was attended by a momentous stirring of European peoples as Germanic and Slavic populations moved to their present positions on the European stage. The Anglo-Saxons invaded Britain from Danish shores, the Franks moved into France, the Allemanni traversed the North European Lowland and settled in Germany. Capitalizing on the wane of Roman power, numerous kings, dukes, barons, and counts established themselves as local rulers. Europe was in turmoil, and in its weakness it was invaded from Africa and the Middle East. In Spain and Portugal the Arab-Berber Moors conquered a large region; in Eastern Europe the Ottoman Turks created a large Islamic empire. The townscapes of Iberia and the Balkans still carry the cultural imprint of the Moslem invasions.

Europe's politico-geographic map still shows the marks of feudal fragmentation during the Dark Ages. Counties, duchies, and marches (*marks*), although now incorporated into modern nation-states, still evince by their boundaries the domains of the old feudal rulers. And it was not until the second half of the eleventh century that reform began. In England the Norman invasion (1066) destroyed the Anglo-Saxon nobility and replaced it with one drawn from the new immigrants. A great Germanic state evolved as an eastern fragment of Charlemagne's empire (it was named the Holy Roman Empire to confirm its political, religious, and cultural inheritance derived from the conquest and absorption of Rome itself). By the end of the twelfth century, elements of the modern map of Europe were already evident (Fig. 1–10).

# The Rebirth of Europe

The emergence of modern Europe dates from the second half of the fifteenth century—some put this rebirth at 1492, the year of Columbus's first arrival in the New World. At home, the growing strength of monarchies forged the beginnings of nation-states; feudal lords and landed aristocracies began to lose their power and privilege. Abroad, Western Europe's emerging states were on the threshold of discovery—the discovery of continents and riches across the oceans. Europe's developing nations were fired by a new national consciousness and pride, and there was renewed interest in Greek and Roman achievements in science and government. Appropriately, the period is referred to as Europe's *Renaissance*.

The new age of progress and rising prosperity was centered in Western Europe, and this was not a matter of chance. Western Europe's countries lay open to the new pathways to wealth—the oceans. While Columbus ventured to the Americas, Eastern Europe was under attack from the Ottoman Turks. Having captured Constantinople, capital of the eastern remnant of the Roman Empire, the Ottomans pushed into the Balkans. They took Hungary and penetrated as far as present-day Austria. Had the Ottomans not been preoccupied with conflicts in Asia and the Near East as well, they might have succeeded, as the Romans had done, in making the Mediterranean Sea an interior lake to their empire. In that case, Western Europe would have been concerned with self-defense instead of with a scramble for the wealth of the distant lands of which it had become newly aware.

Thus protected, the competing monarchies of Western Europe could afford to engage in economic rivalry without interference from the east. Indeed, to the political nationalism that focused on the monarchies we must now add economic nationalism in the form of *mercantilism*. The objectives of the policy of mercantilism were the accumulation of as large a quantity of gold and silver as possible, and the use of foreign trade and colonial acquisition to achieve this. Mercantilism was a policy promoted and sustained by the state; it was recognized to be in the interests of the people of the state in general. Wealth lay in precious metals, and precious metals could be obtained either directly by the conquest of peoples in possession of them (as Spain did in Middle and South America) or indirectly by achieving a favorable balance of international trade. Thus, there was stimulus not only to seek new lands where such metals might lie, but also to produce goods at home that could be sold profitably abroad. The government of the state sought to protect home industries by imposing high duties on imported goods and by entering into favorable commercial treaties. The network of nation-states in Western Europe was taking shape more and more rapidly; Europe had entered the spiral that was to lead to great empires and temporary world domination.

# The Revolutions

Europe's march to world domination did not proceed without strife and dislocation. Much of what was achieved during the Renaissance was destroyed again as monarchies struggled for primacy, religious

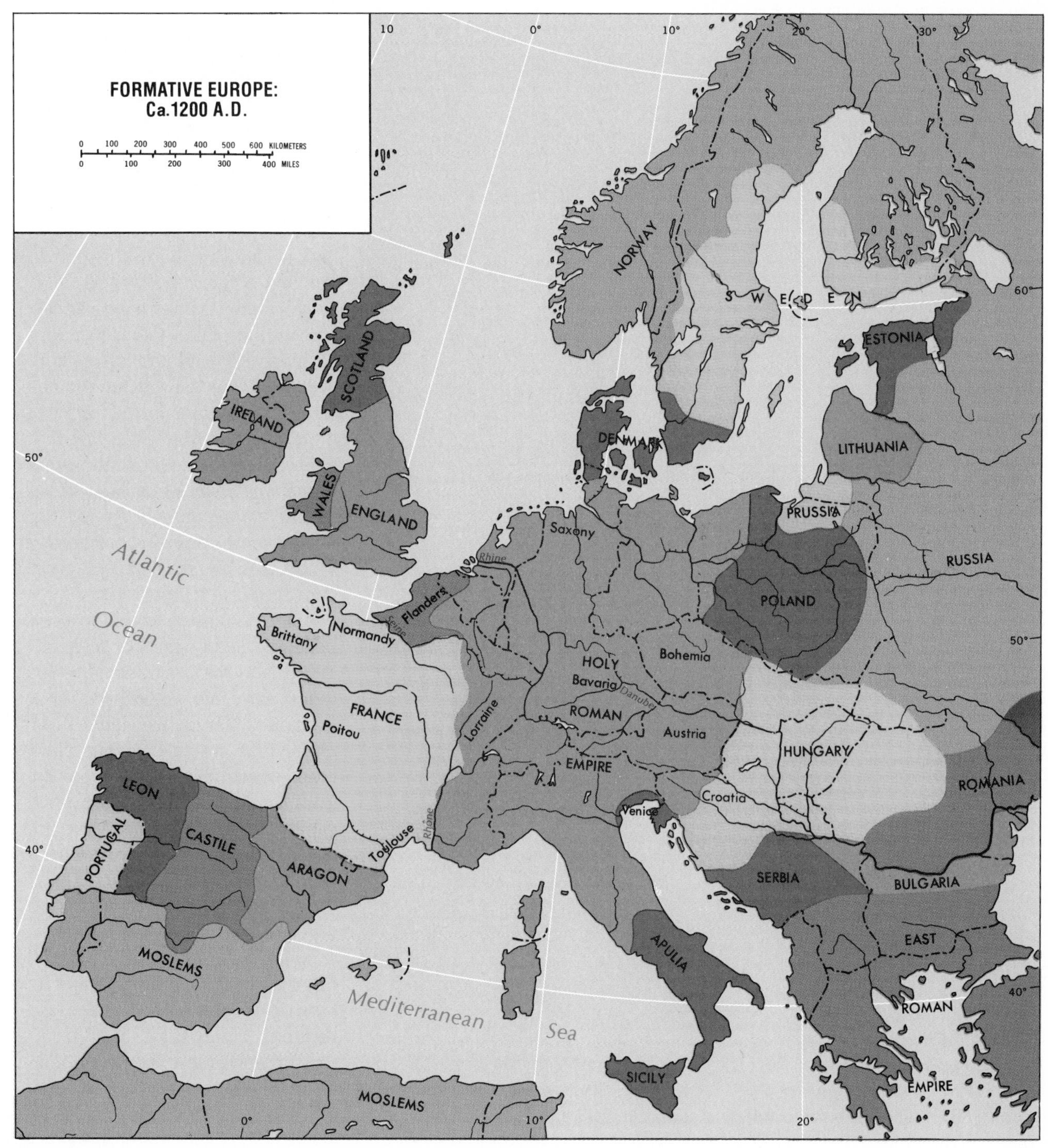


Figure 1–10

conflicts dealt death and misery, and the beginnings of parliamentary government fell under new despotisms. Once again, the landowning nobility rose to power and privileged status. France's King Louis XIV (1638–1715) personified the absolutism that had returned to Western Europe. Builder of Versailles and patron of the arts, Louis XIV also mercilessly persecuted the Huguenots and financed wars to achieve his personal ambitions.

Revolutions, however, were in the making—and in several spheres. Economic developments in Western Europe ultimately proved to be the undoing of monarchical absolutism and its systems of patronage. The urban-based merchant gained wealth and prestige, and the traditional measure of affluence—land—began to lose its relevance in the changing situation. The merchants

## NATION-STATE AS CONCEPT

As Europe went through its periods of rebirth and revolutionary change, the politico-geographic map was transformed. Smaller entities were absorbed into larger units, conflicts resolved (by force as well as negotiation), boundaries defined, and internal divisions reorganized. European **nation-states** were in the making.

But what is a nation-state and what is not? The question centers in part on the definition of the term *nation*. The definition usually involves measures of homogeneity: a nation should comprise a group of tightly knit people who speak a single language, have a common long-range history, share the same racial-cultural background, and are united by common political institutions. Accepted definitions of the term suggest that many states are not nation-states because their populations are divided in one or more important ways.

But cultural homogeneity may not be as important as a more intangible "national spirit" or emotional commitment to the state and what it stands for. One of Europe's oldest states, Switzerland, has a population that is divided along linguistic, religious, and historical lines, but Switzerland has proved a durable nation-state nonetheless. A nation-state may, therefore, be defined as a political unit comprising a clearly defined territory and inhabited by a substantial population, sufficiently well organized to possess a certain measure of power, with the people considering themselves to be a nation having certain emotional and other ties that are expressed in their most tangible form in the state's legal institutions, political system, and ideological strength.

This definition essentially identifies the European model that emerged in the course of the region's long period of evolution and revolutionary change. France is often cited as the best example among Europe's nation-states, but Italy, the United Kingdom (Great Britain), Germany (before World War I), Spain, Poland, Hungary, and Sweden are also among countries that satisfy the terms of the definition to a great extent. European states that cannot at present be designated as true nation-states include Yugoslavia and Belgium.

and businessmen of Europe demanded political recognition on grounds the noblemen could not match. Urban industries were thriving; Europe's population, more or less stable at about 100 million since the mid-sixteenth century, was on the increase.

Already, an *agrarian revolution* was in progress. In our changing world we sometimes tend to look on the industrial revolution as the beginning of a new era, thereby losing sight of a less dramatic but, nevertheless, significant and revolutionary transformation in European farming. This began even earlier than the industrial revolution, and it helped make possible the sustained population increase in Europe during the late seventeenth and eighteenth centuries.

The agrarian revolution was focused in the Netherlands and northern Belgium as well as northern Italy. These areas experienced population growth and increased urbanization because of their success in commerce and manufacture, and the stimulus provided by the expanding markets led to improved organization of land ownership and cultivation. Some of the new practices spread to England and France, where traditional communal land ownership began to give way to individual landholding by small farmers. Land parcels were marked off by fences and hedges, and the owners readily adopted new innovations to improve yields and increase profits. Methods of soil preparation, crop rotation and care, and harvesting improved. For the first time in centuries, innovations in the form of more effective farm equipment were adopted on European farms. There was better planning and experimentation; storage and distribution systems became more efficient. In the growing cities and towns, farm products fetched higher prices. The pastoral industry also benefited. New breeding techniques and the provision of winter fodder for cattle, sheep, and other livestock enlarged the herds and increased the animals' weight. New crops were introduced, especially from the Americas; the potato became a European staple. More and more European farmers were drawn from subsistence into marketing economies. Later, the products of the industrial revolution further stimulated the transformation of agriculture in Europe. But on the farms, a revolution had already begun.

Europe was also not without industries before the *industrial revolution* began. In Flanders and England, specialization had been achieved in the manufacture of woolen and linen textiles. In several parts of present-day Germany (especially in Thüringen, now in southwestern East Germany), iron was mined, and smelters refined the ores. European manufacturers produced a wide range of goods for local markets. But the quality of those products could not match the superior textiles and other wares from India and China. So good were the Indian textiles, for example, that British textile makers staged a riot in 1721, demanding legislative protection against these imports. Thus, the European manufacturers had much incentive to refine and mass produce their products. Raw materials could be shipped home in virtually unlimited quantity. If they could find ways to mass produce these raw materials into finished products, they could bury the Indian and Chinese industries under growing volumes and declining prices. The search for improved machines was on, especially improved spinning and weaving equipment. The first steps in the industrial revolution were not especially revolutionary, for the larger spinning and weaving machines that were built were still driven by the old source of power: water running downslope. But then Watt and others who were trying to devise a steam-driven engine succeeded during the period 1765–1788, and soon this new invention was adapted for various uses. At about the same time, it was realized that coal could be converted into more-carbon-rich coke and that coke was a greatly superior substitute for charcoal in the smelting of iron.

These momentous innovations had rapid effect. The power loom revolutionized the weaving industry. Iron smelters, long dependent on Europe's dwindling forests for fuel, could now be concentrated near coalfields. Engines could move locomotives as well as power looms. Ocean shipping entered a new age. England had an enormous advantage, for the industrial revolution occurred when British influence was worldwide and the significant innovations were made in Britain itself. The British controlled the flow of raw materials, they held a monopoly over products that were in world demand, and they alone possessed the skills necessary to make the machines that manufactured the products. Soon the fruits of the industrial revolution were being exported, and British influence around the world reached a peak.

Meanwhile, the spatial pattern of modern, industrial Europe began to take shape. In Britain, industrial regions, densely populated and heavily urbanized, developed in the "Black Belt"—near the coalfields. The largest complex, called the Midlands, is positioned in central England. Secondary coal-industrial areas developed near Newcastle in northeast England, in southern Wales, and along the Clyde River in Scotland.

In mainland Europe, a belt of major coalfields extends from west to east, roughly along the southern margins of the North European Lowland—due eastward from southern England across northern France and southern Belgium, the Netherlands, Germany (the Ruhr), western Bohemia in Czechoslovakia, and Silesia in Poland. Iron ore is found in a broadly similar belt, and the industrial map of Europe reflects the resulting concentrations of economic activity (Fig. 1–11). But nowhere were the coasts, coalfields, and iron ores located in such close proximity as in Britain.

Another set of industrial regions developed in and near the growing urban centers of Europe, as Fig. 1–11 demonstrates. London already was Europe's greatest urban focus, and Britain's largest and richest domestic market. Many local industries had been established here, taking advantage of the large supply of labor, the ready availability of capital, and the proximity of so great a number of potential buyers. Although the industrial revolution thrust other places into prominence, London's primacy was sustained, and industries in and around London multiplied. Similar developments occurred in the Paris region and near other major urban concentrations in Europe. As the new methods of manufacturing were adopted in mainland Europe, old industrial areas were rejuvenated, and regional growth was stimulated in France and Italy (Fig. 1–11).

Europe had had long experience with experiments in democratic government, but the *political revolution* that swept the realm from the 1780s brought transformation on an unprecedented scale. Overshadowing these events was the French Revolution (1789–1795), but Europe's political catharsis lasted into the twentieth century and eventually affected every monarchy on the continent.

In France, the popular revolution plunged the country into years of chaos and destruction. There had been dissatisfaction everywhere: among the nobility and clergy critical of the rule of Louis XVI; among the middle classes, where a constitutional, parliamentary monarchy along English lines was wanted; and among the peasants, who faced rising prices, declining living standards, and harsh exploitation by landowners. Ultimately, the most extreme elements came to control the movement, and in 1792, a republic was proclaimed amid terrible excesses. The king was executed in 1793, and revolutionary armies began to advance into neighboring countries, calling on all peoples to dispose of their rulers and to join in a war of liberation. Europe's growing nationalism proved to be the downfall of many dynastic monarchies.

Even while the revolutionary forces swept into Europe, disorder continued in France itself. Only when Napoleon took control (1799) was stability restored. Napoleon personified the new French republic; he reorganized France so completely that he laid the foundations of the modern nation-state. His empire extended from Prussia and Austria to Spain and from the

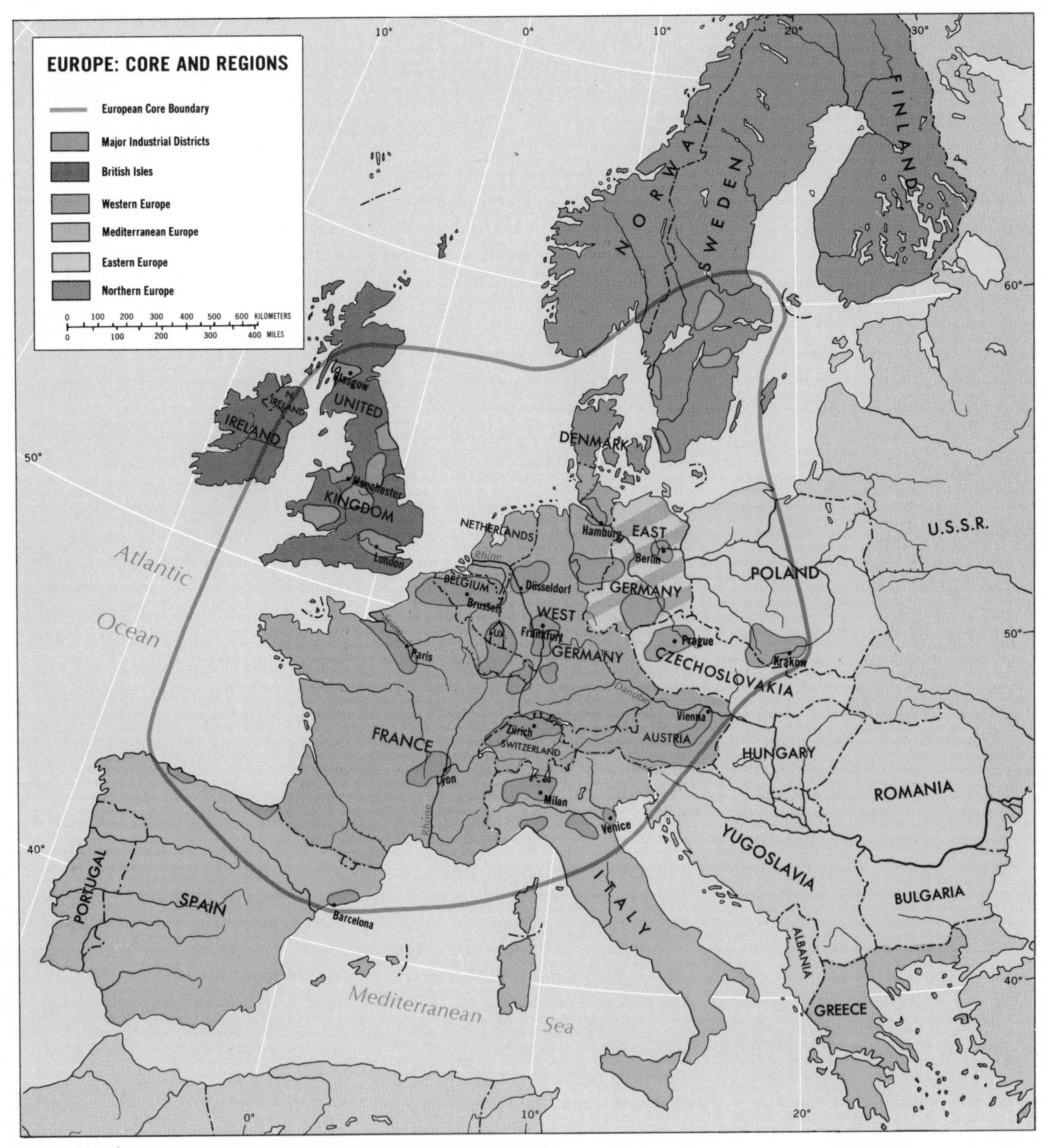


Figure 1–11

Netherlands to Italy. Although his armies were repelled by the British in Portugal and more seriously by the Russians two years later (1812), Napoleon had forever changed the politico-geographical map of Europe. His French forces had been joined by revolutionaries from all over the continent; one monarchy after another had been destroyed. Even after his defeat at Waterloo, there were popular uprisings in Spain, Portugal, Italy, and Greece. In France itself, the monarchy was temporarily reestablished through outside intervention—but France was to be a republic, as it still is today. If the monarchy survived elsewhere, its powers were limited or short lived. Europe had its taste of democracy and nationalist power, and it would not revert to its old ways.

# Geographic Dimensions

Modern Europe has emerged from an age of revolutionary change—in livelihoods and commerce, industrialization and technology, revolution and war. During the nineteenth century, some geographers tried to interpret the forces and processes that were shaping the new Europe, and some of their work still holds interest today.

In particular, the agrarian revolution that accompanied the industrial revolution was observed attentively by an economist-farmer named Johann Heinrich Von Thünen (1783–1850), who in 1826 fashioned one of the world's first geographical models. Von Thünen owned a large farming estate not far from the German town of Rostock. For four decades he kept meticulous records of his estate's transactions, and he became interested in a subject that still excites economic geographers today—the effects of distance and transportation costs on the location of productive activity. Using all the data he gathered, Von Thünen began to write about the spatial structure of agriculture. His studies were published under the title *Der Isolierte Staat* (*The Isolated State*), and his methods in many ways constitute the foundations of location theory. In fact, Von Thünen's conclusions are still being discussed and debated in the modern literature, and his main ideas are presented in *Model Box 2* (pp. 72–74).

## Industrial and Urban Intensification

It is not surprising that the industrialization of Europe (or, rather, its industrial *intensification* following the industrial revolution) also became the topic of geographic research. What influences affected the location of industries? How was Europe's industrialization channeled? Again, the first important studies were done by German geographers and economists, but well after Von Thünen's time—during the second half of the nineteenth century. Much of this work was incorporated in a volume by Alfred Weber (1868–1958), published in 1909, entitled *Über den Standort der Industrien* (*Concerning the Location of Industries*). Like Von Thünen, Weber began with a set of limiting assumptions in order to minimize the complexities of the real Europe. But unlike Von Thünen, Weber dealt with activities that take place at particular *points* rather than across large areas. Manufacturing plants, mines, and markets are located at specific places, and so Weber created a model region marked by sets of points where these activities would occur. He eliminated labor mobility and varying wage rates and thereby could calculate the "pulls" exerted on each point in his theoretical region.

In the process, Weber discerned various factors that affect industrial location. He defined these in various ways. For example, he recognized what he called "general" factors that would affect all industries, such as transport costs for raw materials and finished products; and "special" factors such as perishability of foods. He also differentiated between "regional" factors (transport and labor costs) and "local" factors. Local factors, Weber argued, involve *agglomerative* (concentrating) and *deglomerative* (deconcentrating) forces. Take the case of London we discussed previously: industries located there, in large part, because of the advantages of locating together. The availability of specialized equipment, a technologically sophisticated labor force, and a large-scale market made London (and Paris, and other cities not positioned on rich resources) an attractive site for many manufacturing plants that could benefit from agglomeration. On the other hand, such concentration may eventually create strong disadvantages, chiefly in terms of competition for space, rising land rents, and environmental pollution. Eventually, an industry might move away and deglomerative forces set in. This subject arises again in the context of North American urbanization in Chapter 3.

Weber singled out transport costs as the critical determinant of regional industrial location and suggested that the least-transport-cost location is the place where it would be least expensive to bring raw materials to the point of production and finished products to the consumers. But economic geographers following Weber have concluded that some of Weber's assumptions seriously weakened the usefulness of his conclusions, especially in the area of markets and consumer demand. Consumption does not take place at a single location, but over a wide (in some cases a worldwide) area. But practically all modern study of industrial location has a direct relationship to Weber's work, and he was a pioneering model builder in a class with Von Thünen.

Europe's industrialization also speeded the growth of many of its cities and towns. In Britain in the year 1800, only about 9 percent of the population lived in urban areas, but by 1900 some 62 percent lived in cities and towns (today the figure approaches 92 percent); all this was happening as the total population skyrocketed as well (see Fig. 1–4). As industrial modernization came to Belgium and Germany, to France and the Netherlands, and to other areas of Western Europe, the whole urban pattern changed. The nature of this process—the growth and

## MODEL BOX 2

# THE VON THÜNEN MODEL

Von Thünen called his model *The Isolated State* because he wanted to establish, for purposes of analysis, a self-contained country, devoid of outside influences that would disturb the internal workings of the economy. Thus, he created a sort of regional laboratory within which he could identify the factors that influence the locational distribution of farms around a central urban area. In order to do this, he made a number of limiting assumptions. First, he stipulated that the soil and climate would be uniform throughout the region. Second, there would be no river valleys or mountains to interrupt a completely flat land surface. Third, there would be a single, centrally positioned city in the Isolated State, which was surrounded by an empty, unoccupied wilderness. Fourth, Von Thünen postulated that the farmers in the Isolated State would transport their own products to market (no transport companies here), and that they would do so by oxcart, directly overland, straight to the central city. This, as you can imagine, is the same as assuming a system of radially converging roads of equal and constant quality; with such a system, transport costs would be directly proportional to distance.

Von Thünen integrated these assumptions with what he had learned from the actual data collected while running his estate, and he now asked himself: What would be the best spatial arrangement of agricultural activities within his Isolated State? He concluded that farm products would be grown in a series of concentric zones outward from the central market city. Nearest to the city would be grown those crops that perished easily and/or yielded the highest returns (such as vegetables), because this readily accessible farmland was in great demand and, therefore, quite expensive; dairying would also be carried on in this innermost zone. Farther away would be potatoes and grains. And eventually, since transport costs to the city increased with distance, there would come a line beyond which it would be uneconomical to produce crops. There the wilderness would begin.

The Isolated State

Figure 1–12A

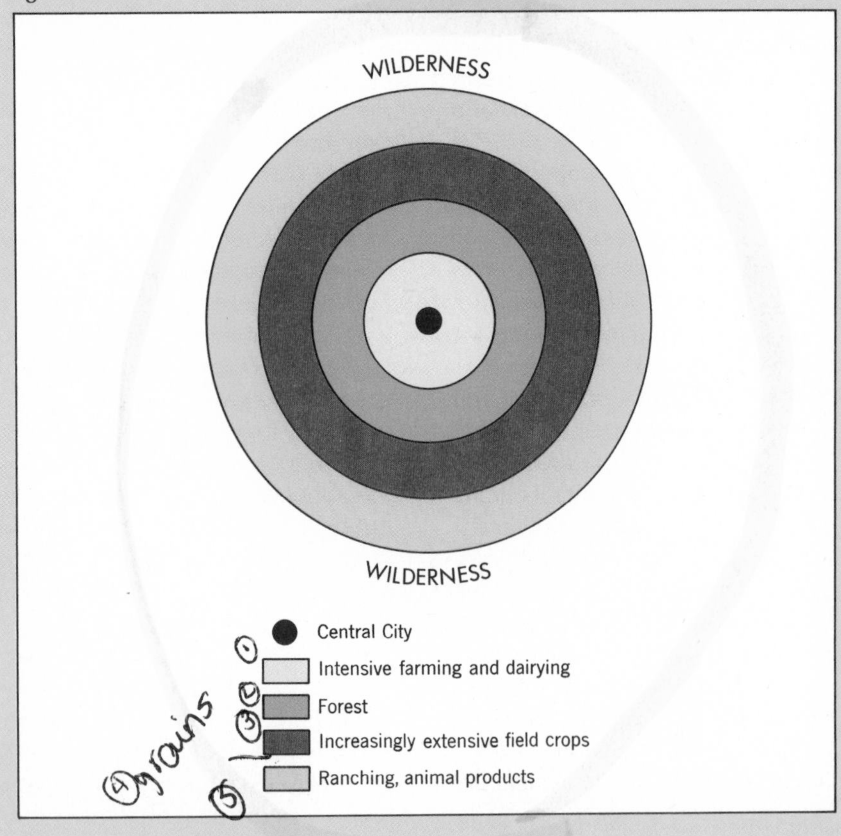

Von Thünen's model, then, incorporated four zones surrounding the market center (Fig. 1–12A). The first and innermost belt would be a zone of intensive farming and dairying. The second zone, Von Thünen said, would be an area of forest, used for timber and firewood. Next, there would be a third belt of increasingly extensive field crops. The fourth and outermost zone would be occupied by ranching and animal products, beyond which began the wilderness that isolated the region from the rest of the world.

If the location of the second zone, the forest, seems inappropriate, it should be noted that the forest was still of great importance during Von Thünen's time as a source of building materials and fuel. All that was about to change with the onslaught of the industrial revolution, and modern applications of the Von Thünen model to the developed world no longer contain a forestry ring. But there are lots of towns and cities left in the world that are still essentially preindustrial. When Ronald Horvath made a study of Addis Ababa, the capital

**Figure 1–12B**

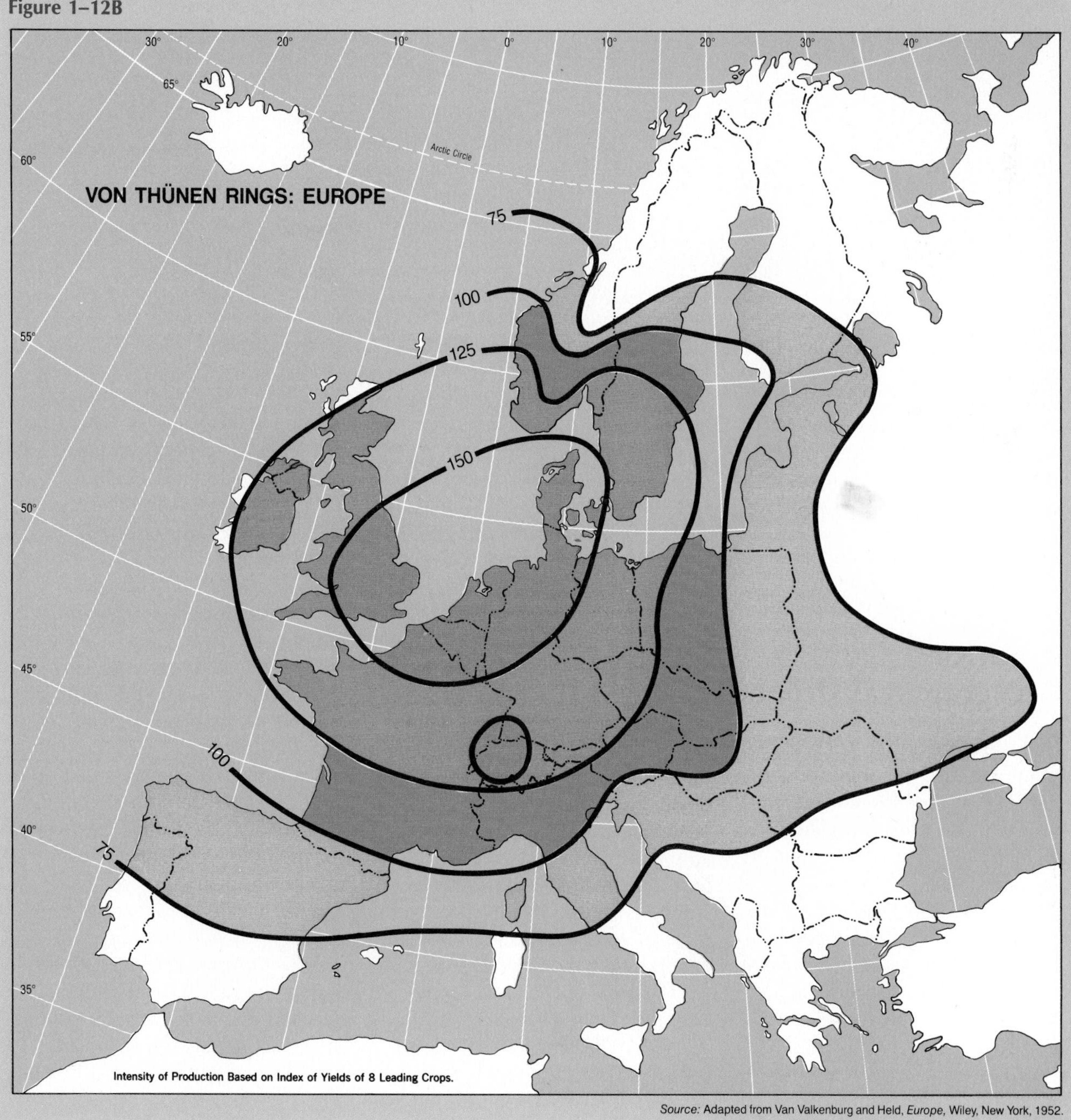


*Source:* Adapted from Van Valkenburg and Held, *Europe*, Wiley, New York, 1952.

of Ethiopia, in this context (see Fig. 6–30), he found a wide and continuous belt of eucalyptus forest surrounding that city, positioned more or less where Von Thünen would have predicted it to be, and serving functions similar to those attributed to the forest belt of the Isolated State.

Von Thünen knew, of course, that the real Europe (or the world) did not present idealized situations exactly as he had postulated them. Transport routes serve certain areas more efficiently than others. Physical barriers can impede the most modern of surface communications. External economic influences invade every area. But Von Thünen wanted to eliminate these disruptive conditions in order to discern the fundamental processes at work in shaping the spatial layout of the agricultural economy. Later, the distorting factors could be introduced one by one and their influence measured. But first he developed his model in theoretical isolation and based it on total regional uniformity.

It is a great tribute to Von Thünen that his work still retains the attention of geographers. The economic-geographic landscape of Europe has changed enormously since his time, but geographers still compare present-day patterns of economic activity to the Thünian model. Such a comparison was made by Samuel van Valkenburg and Colbert Held (Fig. 1–12B), whose map of Europe's agricultural intensity reveals a striking ringlike concentricity. The major spatial change since Von Thünen's time is the improvement in transportation technology, which permitted the Isolated State to expand from *micro-* to *macro-scale*. Thus, the model is no longer centered on a single city, but on the vast urbanized area lining the southern coasts of the North Sea, which now commands a continentwide Von Thünen agricultural system.

The role of distance (to markets, from raw materials, and so on) in the development of the spatial pattern of the economy remains the subject of many other modern studies, and geographers acknowledge that it was Von Thünen who first formulated the crucial questions.

strengthening of towns and cities—and related questions also became topics of geographic study, just as agriculture and industrialization had. The German geographer Walter Christaller (1893–1969) addressed himself to these topics in formulating a theory of location for cities, which we will review in Chapter 3.

## Politico-Geographical Order

Europe's convulsive political revolution was also the object of geographic study, and thereby was born the field of modern political geography. Still another German geographer, Friedrich Ratzel (1844–1904), observed the fortunes and failings of European states and sought to identify underlying spatial principles and processes. Ratzel suggested that political states are like biological organisms, passing through stages of growth and decay. Just as an organism requires food, Ratzel reasoned, so the state needs space. Any state, to retain its vigor and to continue to thrive, must have access to more space, more territory—the essential, life-giving force. Only by acquiring territory and through the infusion of newly absorbed cultures could a state sustain its strength. The acquisition of frontier areas with their resources and populations would renew the state's energies, Ratzel argued, but boundaries were dangerous straightjackets.

Ratzel believed that a nation, being constituted by an aggregate of organisms (human beings), would itself behave as an organism capable of birth, growth, maturity, and death. From his study of European historical geography, he proposed an *organic theory* of state evolution, and he linked the struggles among European nation-states to biological competition. The German nation-state was forged during Ratzel's lifetime (1871), and, in the 1880s, the German *Reich* acquired a colonial empire overseas. In his book *Political Geography*, Ratzel detailed the rise of empires, the fortunes of alliances, and the subjugation of weaker countries. Invariably, states in ascendancy were gaining territory; those in decline were losing it. But what might determine whether a state would rise or fall? Like Von Thünen and other geographers, Ratzel sought to identify the fundamental processes that affected the course of events he could observe. In an article entitled "Law of the Spatial Growth of States," published in 1896, he attempted to discern laws that govern the growth of states. Essentially, each of these seven laws forms an element of the organic theory: growth is vitality.

Certainly, Europe appeared to provide ample confirmation of Ratzel's theory. The British Empire spanned the world, and British power was nearing its zenith. France, having lost its European empire, expanded its sphere of influence in Africa and Asia. Other European states thrived on their overseas domains. Ratzel's writings, of course, appeared in scholarly journals and were couched in theoretical language. But some of his readers were less theoretical and sophisticated than he, and ultimately there arose in Germany a school of *Geopolitik* (geopolitics). This group gave practical expression to the *deterministic* character of Ratzel's writings, and they laid the intellectual foundations for German military expansionism—not in distant colonies, but in Europe itself. Twice during the twentieth century, Germany plunged Europe and the world into war; the recovery from the second of these conflicts (1939–1945) has been the formative period of contemporary Europe.

# The European Realm Today

The nation-states of Europe are among the world's oldest, and the colonial empires of European powers have been among the most durable. Wars and revolutions notwithstanding, European nations survived, and out of their long-term stability they forged a confidence that is a hallmark of European culture. Centuries of exploitation of overseas domains amassed fortunes at home and established a foreign influence that continued when the colonial era ended. The impressive merchants' homes that line the canals of Amsterdam; the architectural splendor of Paris, London, Lisbon; and those castles and country villas—much of all this was built by the slave trade, by the domination of colonial economies, and by policies that no longer prevail in the modern world. But all of these advantages were nearly destroyed forever in the horror of World War II.

However, like the proverbial phoenix rising from its ashes, Europe has emerged from the devastation of that war to enter a new era of reconstruction, realignment, and resurgence. Reconstruction was aided by the $12 billion made available through the Marshall Plan (1948–1952) by the United States. Realignment came in the form of the *Iron Curtain* separating the Soviet-dominated East from Western Europe; it also involved the emergence of several international "blocs," consisting of states seeking to promote multinational cooperation. The resurgence of Europe continued, despite the loss of colonial empires, recurrent political crises, energy shortages, labor problems, and other obstacles. Europe's momentum has carried the day once again.

Although Europe quite clearly constitutes a geographic realm, it has little of the overall homogeneity that such regional identity might imply. It is sometimes postulated that Europe may be viewed as a regional unit, because its peoples share Indo-European languages (Fig. 1–13), Christian religious traditions (Fig. 6–3), and common European racial ancestry (Fig. 7–3). But these human cultural and physical traits extend well beyond Europe; in any case, they are not strong unifying elements in the European mosaic. As Fig. 1–13 shows, Hungarians and Finns are among European groups who do not speak Indo-European languages; for so small a region Europe is a veritable Tower of Babel. As for common religious traditions, Europe has a history of intense and destructive conflict over religious issues. Shared Christian principles have done little to bring unity to Northern Ireland, for example, and the recurrent conflict there is only the most recent manifestation of the depth of latent religious division that affects certain parts of Europe. And again, Europe's common racial ancestry (the term *European race* has now replaced *Caucasian*) masks strong internal differences between Spaniard and Swede, Scot and Sicilian.

We now consider several of the important geographical properties of contemporary Europe.

## Opportunities for Exchange

Perhaps to a greater degree than any world area of similar dimensions, Europe's environments and resources present opportunities for exchange and interaction. The concept of **spatial interaction** is best organized around a triad of principles developed by the American geographer, Edward Ullman (1912–1976): complementarity, transferability, and intervening opportunity. **Complementarity** occurs when two places or areas are in a position that the first area has a *surplus* of an item demanded by the second area. The mere existence of a resource in a locality is no guarantee that trade will develop—that resource must specifically be needed elsewhere; thus, complementarity arises from regional variations in both the supply and demand of human and natural resources. **Transferability** refers to the ease with which a commodity may be transported between two places. Sheer distance, in terms of both the cost and time of movement, may be the major obstacle to the transferability of a good; therefore, even though complementarity may exist between a pair of areas, the problems of economically overcoming the distance separating them may be so great that trade cannot begin. The third interaction concept, **intervening opportunity**, maintains that potential trade between two places—even if they satisfy the necessary conditions of complementarity and transferability—will *only* develop in the absence of a closer, intervening source of supply.

European economic-geographic development has always been stimulated by internal complementarities. A specific example involves Italy. Among Mediterranean countries other than France, Italy is foremost in terms of economic development. But Italy is a coal-poor

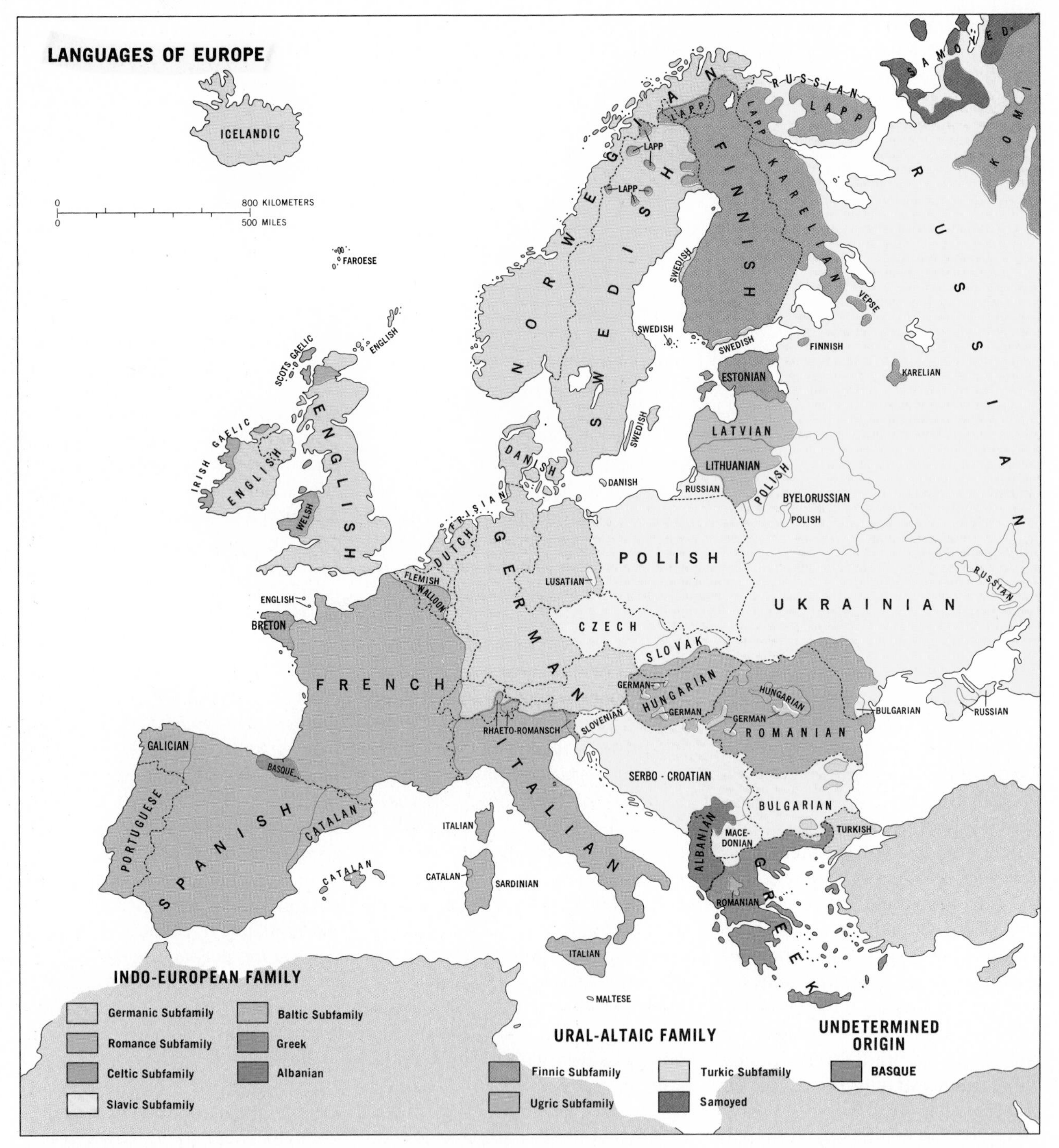


Figure 1–13

country, and for a long time its industries depended on coal supplies from the rich resources of Western Europe. At the same time, Italian farmers grow crops that cannot be grown in the north: citrus fruits, olives, grapes, early vegetables. Such products are wanted on Western Europe's markets. Hence, Italy needs Northern Europe's coal; Northern Europe imports Italian fruits and wines. This case of double complementarity is not counteracted by transferability restrictions: the physical barrier of the Alps has long been breached by rail and highway routes, and the two-way trade flow is most attractive to shippers, because freight-carrying vehicles will not have to return home empty. Moreover, because Northern Europe and Italy are the closest sources of coal and fruits, respectively, there are no intervening opportunities to disrupt spatial interaction, and a robust mutual trading of these surplus commodities is encouraged.

Historically, there are thousands of examples of this kind within Europe alone. Thus, if we possessed all trade information for the past several hundred years and drew a series of maps to represent these interregional transactions, we would see an expanding web of ever-increasing complexity and sophistication. Governed by the three interactance principles, improved transport technology spawned trade flows involving more and more places, eventually overflowing the European continent and reaching out across the oceans to encompass most of the other parts of the world.

## Problems of Political Fragmentation

Unlike the United States or the Soviet Union, Europe's stronger competitors in terms of world power and influence, the European realm is fragmented politically and ideologically—and economically as well—although efforts are being made to overcome this problem in Western Europe. Ever since the unifying period of the Roman Empire, Europe has been divided into numerous political entities. Not counting the smallest of these (Monaco, Andorra, Liechtenstein, etc.) there are 26 states in 1985, as the table in the Appendix indicates. Many of these states are true nation-states, with strong identities and traditions.

World War II left Eastern Europe under Soviet domination, Germany fractured, and Western Europe in chaos. Even before the end of the war, three countries (the Netherlands, Belgium, and Luxembourg) had laid plans to create a cooperative organization to be called *Benelux*. Later, the terms of the Marshall Plan required greater international cooperation, and 16 countries of Western Europe formed the Organization for European Economic Cooperation (OEEC). Ever since the OEEC began operations (1948), the states of Western Europe have pursued the prospect of European unification. Today, the greatest success is the 10-member European Economic Community (EEC), incorporating most of Europe's core area. But Germany and Berlin remain divided, and Eastern and Western Europe continue to be separated along an externally imposed ideological barrier.

## Benefits of Well-Being

The European population, in general, is well housed, well educated, and well fed. European cities do have their slums, but nothing to compare to the miserable shantytowns that surround so many cities of the underdeveloped world. Levels of education are high, and educational achievement is held in esteem throughout Europe. Measures of such attainment include literacy rates (over 95 percent everywhere, except Spain, Portugal, Italy, and the Balkan countries), newspapers and magazines printed (among the largest number per capita in the world), book titles published (over 30,000 annually in Britain alone) and, less tangibly, the premium placed on high school and university graduation.

Europeans also eat well-balanced and adequate meals, and they enjoy good health. Several countries rank ahead of the United States in such areas as longevity and infant survival, and medical services are generally very good. As we noted in the Systematic Essay at the beginning of this chapter, European countries no longer confront the explosive population growth that marked the turbulent nineteenth century (and still beleaguers underdeveloped countries today). Under such circumstances, progress and improved living standards are attainable.

## Industrial Stagnation

Europe today no longer enjoys the industrial monopolies it held during the early stages of the industrial revolution. It is true that European industries still constitute one of the four great manufacturing complexes in the world (the other three being in the United States, the Soviet Union, and Japan) and that many European specialized products still command world markets. But European industry now faces some serious problems.

These are perhaps most evident at the very source of the industrial revolution—in the United Kingdom. British industry, once the world's leader, now is comparatively inefficient. Output per worker is low and declining, and overall productivity lags far behind other industrial countries, including West Germany, Japan, and the United States. British manufacturing plants, once among the world's most modern, are now outmoded; the investments needed to keep them current and competitive have not been made. As the British economy has suffered, particularly in the global recession of the early 1980s, government representatives and labor leaders have struggled for control over the course of events. Will the latest British experience diffuse into mainland Europe as the industrial revolution itself did? Are the political problems of Italy and the labor unrest (and alarming industrial decline) in France signs of a similar situation in the making? The tenuous fabric of Europe's Common Market may not be able to survive such an economic crisis.

## Commercial Agriculture

When Von Thünen calculated his farming estate's transactions, most Europeans were still farmers, and enough food was produced in Europe itself to feed city dwellers and farmers alike. But in the aftermath of two centuries of modern industrialization and urbanization, only one European in six still farms, and foodstuffs must be imported to supplement what the local farmers can produce. With very few exceptions (and those mostly in southeastern Europe), non-Communist Europe's farmers grow their crops for the market, and for profit.

European agriculture is not as large scale nor as mechanized as it is in the United States, but European farmers are very productive, especially in the Netherlands and Britain. In Eastern Europe, agriculture is being reorganized on the Soviet pattern, but, despite the advantages (greater availability of mechanized equipment on the collectives), production per person and per hectare still remain far below those of the West.

## Achievements in Transport and Communication Systems

Circulation and interaction have characterized Europe from Roman times onward, and Europeans have always been among the most mobile peoples in the world. Today, the barge traffic on rivers and canals is still substantial, but Europe is connected even more efficiently by excellent railroad and highway networks, air routes, and pipelines. Again, these systems are best developed in Western Europe, but even in Eastern Europe few areas remain remote from some means of modern communication.

The evolution of Europe's transport networks was not a coordinated process. Individual countries built systems designed to serve their internal needs; in fact, a century ago each country deliberately possessed its own unique railway gauge, an international incompatibility that can still be found today where tracks cross the Iron Curtain. Indeed, the full regional integration of the various networks only became a concern of the postwar planners of a unified Europe. Especially in the Common Market countries of the European core area, considerable progress has been made not only in the integration of the transport routes, but also, as will be noted later, in the reduction of the barrier effects of international boundaries.

## Contemporary European Urbanization

Europe is among the most highly urbanized regions in the world (see Table I–1), though more so in its west than in the east. In West Germany in 1985, more than 85 percent of the people lived in cities and towns; in Britain and France the figures were 91 and 80 percent, respectively. Even in Eastern Europe, where the percentage of urbanization is lower (55–65 percent average), the towns and cities contain a much larger share of the population than in Asian or African countries.

As we saw in the previous chapter, cities—especially the largest metropolitan centers—are the crucibles of their nation's culture. In his 1939 study of the pivotal role of great cities in the development of national cultures, Mark Jefferson postulated the **"law" of the primate city**, which states that "a country's leading city is always disproportionately large and exceptionally expressive of national capacity and feeling." Although rather imprecise, this "law" can readily be demonstrated using European examples. Certainly Paris personifies France in countless ways, and there is nothing in England to rival London, where the culture and history of a nation and empire are indelibly etched in the urban landscape. For similar reasons, Amsterdam is a microcosm of the Netherlands, Warsaw is the heart of Poland, Stockholm is Sweden, Athens is Greece, and Vienna is Austria. Ironically, perhaps no city reflects its nation as faithfully as divided Berlin represents fragmented Germany.

Europe's great urban population clusters also contain concentrations of one of the world's most technologically advanced labor forces. Specializations and skills that have been learned over many years are widely available to urban-based industries. And Europe's cities constitute large and wealthy markets, because personal incomes in Europe (although not quite as high as in the United States and Canada) are still among the world's highest.

Although the trend of the past three decades has slowly been toward U.S.-style suburbanization, the overall appearance of the cityscape in Western Europe is still quite different from the North American urban scene. Lawrence Sommers has concisely captured this contrast:

> *Age is a principal factor, but ethnic and environmental differences also play major roles in the appearance of the European city. Politics, war, fire, religion, culture, and economics also have played a role. Land is expensive due to its scarcity, and capital for private enterprise development has been insufficient, so government-built housing is quite common. Land ownership has been fragmented over the years due to inheritance systems that often split land among sons. Prices for real estate and rent have been government*

*controlled in many countries. Planning and zoning codes as well as the development of utilities are determined by government policies. These are characteristics of a region with a long history, dense population, scarce land, and strong government control of urban land development.*

The internal spatial structure of the European *metropolis*—which consists of both the central city and its suburban ring—is very much typified by the urban pattern of the London region (Fig. 1–16). The whole metropolitan area is still focused on the large city at its center, especially the downtown *central business district* (CBD), which is the oldest part of the urban agglomeration and contains the region's main concentration of business, government, and shopping facilities as well as its wealthiest and most prestigious residences; although the CBD continues to be revered as the political, economic and cultural core, a growing number of skyscrapers increasingly transform its richly historical landscape (Fig. 1–1). Broad residential sectors radiate outward from the CBD across the rest of the central city, each one being home to a particular income group and thereby perpetuating geographically the strong class system in European society; lower-income groups are usually located in crowded neighborhoods adjacent to, and downwind of, the inner-city industrial district, whereas middle-income residents congregate at lower densities well away from the factories and their effluents. Beyond the central city is a sizable suburban ring, but residential densities here are much higher than in the United States, because the European tradition is one of living in apartments rather than detached single-family houses and because there is a far greater reliance on public transportation, which necessitates a compact development pattern. Although automobile-based suburbanization has increased with the proliferation of high-speed freeways during the past two decades—the Germans, in fact, pioneered these superhighways with the earliest *autobahns* in the 1930s—it has not yet produced a significant amount of low-density residential sprawl (quite possibly because home heating oil and gasoline are priced up to three times current American levels). Therefore, the typical European suburb is still a high-density satellite town or village surrounded by open countryside that is heavily utilized for recreational purposes.

The preservation of open space and avoidance of sprawl development in the metropolitan periphery, reinforced by strong governmental control over urban growth—as Sommers pointed out—is the legacy of an *urban and regional planning movement* that has been a popular tradition throughout Europe for much of this century. Modern planning originated in Britain shortly before 1900—as a reaction to the horrors of the industrial urbanization of the nineteenth century—and soon spread across the rest of the continent. The centerpiece of these efforts was the New Towns Movement, which involved the controlled dispersal of certain people and economic activities (including light manufacturing) to new self-sufficient lower-density towns located in the outer suburbs within an hour's train ride of the CBD. Although the most vigorous period of new-town construction has passed (the early postwar 1940s and 1950s), it is still expected that more than 5 percent of the population of England and Scotland will reside in more than 30 of these settlements by the year 2000. Postwar continental Europe has adopted this innovation even more widely, and such showcase new towns as Tapiola (Finland), Vallingby (Sweden), and Lelystad (the Netherlands) rank among the most advanced state-of-the-art urban communities in the world. Another major planning feature of the contemporary European metropolis is the encircling band of open space or "greenbelt" that separates most large cities from their suburbs (see Figs. 1–16 and 1–19); this represents a reaction to the rapid urbanization of the pre-World War II period, which threatened to overrun the entire urban countryside, and its successful implementation guaranteed most city dwellers the same access to nearby open space enjoyed by their suburban counterparts.

As this century draws to a close, urban Europe continues to change. The latest development is the persistent and now-accelerating suburbanization of population, which is today quite pronounced in Britain, France, West Germany, Switzerland, and the Low Countries. But is the ring road around Paris (*Boulevard Péripherique*) really about to become the "Main Street" of that metropolis in the manner that the Capital Beltway has transformed the urban geography of the Washington, D.C. area? The answer for now is almost surely no, given the apparently enduring commitment of most Europeans toward preserving their rich urban heritage; but we must remind you that we began this overview of the spatial nature of Europe's metropolitan areas by observing that the slow but steady trend since mid-century has been toward Americanization, and that might ultimately be a force that is simply too powerful to resist.

# Regions of Europe

Earlier in this chapter a European core area was defined (see Fig. 1–11). This is Europe's heartland, a functional region of dynamic growth, dense population, high degree of urbanization, enormous mobility, and great productivity. We now turn to a more traditional view of Europe's regions as we attempt to group sets of countries together, so that the regional groups reflect their proximity, historical-geographical association, cultural similarities, common habitats, and shared economic pursuits. On these bases it is possible to identify five regions of Europe: (1) the British Isles, (2) Western Europe, (3) Nordic or Northern Europe, (4) Mediterranean Europe, and (5) Eastern Europe.

## The British Isles

Off the coast of mainland Western Europe lie two islands called the British Isles (Fig. 1–14). The largest of the two, also lying nearest to the mainland (a mere 34 kilometers or 21 miles at the closest point) is the island of Britain, and the smaller island is known as Ireland. Surrounding Britain and Ireland are several thousand small islands and islets, none of great significance.

Although Britain and Ireland continue to be known as the British Isles, the British government no longer rules over the state that occupies most of Ireland, the Republic of Ireland or *Eire*. After a very unhappy period of domination, the London government gave up its control over that part of the island, which was overwhelmingly Roman Catholic; in 1921, it became the Irish Free State. The northeastern corner of Ireland, where English and Scottish Protestants had settled, was retained by the British and called *Northern Ireland*.

Although all of Britain is part of the United Kingdom, political divisions exist here too. *England* is the largest of these units, the center of power from where the rest of the Isles were originally brought under unified control. The English conquered *Wales* in the Middle Ages, and *Scotland* was first tied to England in the seventeenth century when a Scottish king ascended the English throne. Thus, England, Wales, Scotland, and Northern Ireland became the *United Kingdom* (*U.K.*). In recent years, however, there has been a revival of nationalism in both Wales and Scotland, and demands for greater local autonomy are being heard.

**Highland and Lowland Britain** Britain's *physical geography* (i.e., the island's topography and relief, climate and drainage, soils and vegetation—the total natural landscape) is varied, and the relative location of the different areas matters much. The most simple, and at the same time the most meaningful, division of Britain is into a *lowland* region and a *highland* region. Most of England, with the exception of the Pennine Mountains and the peninsular southwest, consists of lowland Britain. Highland Britain consists mainly of Scotland and Wales. Lowland Britain, therefore, lies opposite the western periphery of the North European Lowland on the mainland, and it is essentially a geological continuation of it. In this part of England, the relief is low, soils are generally good, rainfall is ample, and agricultural productivity is relatively high.

Throughout the evolution of British society, the Thames River (on which London lies) and its basin was an avenue of penetration into England. Several good natural harbors and places of easy access also rendered Britain open to foreign influences. From the east, northeast, and southeast they came: the prehistoric builders of the great Stonehenge structures; the Celtic-speaking peoples, whose linguistic imprints still remain on the map as Irish and Scottish Gaelic (Fig. 1–13); the Romans, who made Dover their port of entry and who established routes that remain to this day among England's major highways; the Angles and the Saxons, Germanic peoples who settled on England's eastern bulge and in the Thames Basin; Danish and Norwegian Vikings, who penetrated the interior and settled on many surrounding islands; and the Normans, those Scandinavian Vikings who had been acculturated through generations of residence in French Normandy and whose invasion under William the Conqueror in 1066 closed one era in British historical geography and started another.

Through all this, Lowland Britain held center stage in the British Isles. True, Christianity—which had reached Britain during Roman times—proved more durable in remote Ireland than in exposed, frequently strife-torn England. But the British Isles's destiny was determined in the English countryside, where the invaders struggled for power and where, after the eleventh century, the British nation and empire were born. In its political development, Britain benefited from its insular separation from Europe, and England became the center of regional power and political as well as economic integration and organization. Centered on London, England steadily evolved into the core area of the British Isles. Ultimately, the British achieved a system of parliamentary government that had no peer in the Western world, and England rose from a national core area into the headquarters of a global empire.

The economic developments that to a considerable extent precipitated this political progress came in stages that could be accommodated without wrecking the whole society. Britain's original subsistence economy gave way to the commercial era of mercantilism, and the industrial revolution was foreshadowed by the development of manufacturing based on

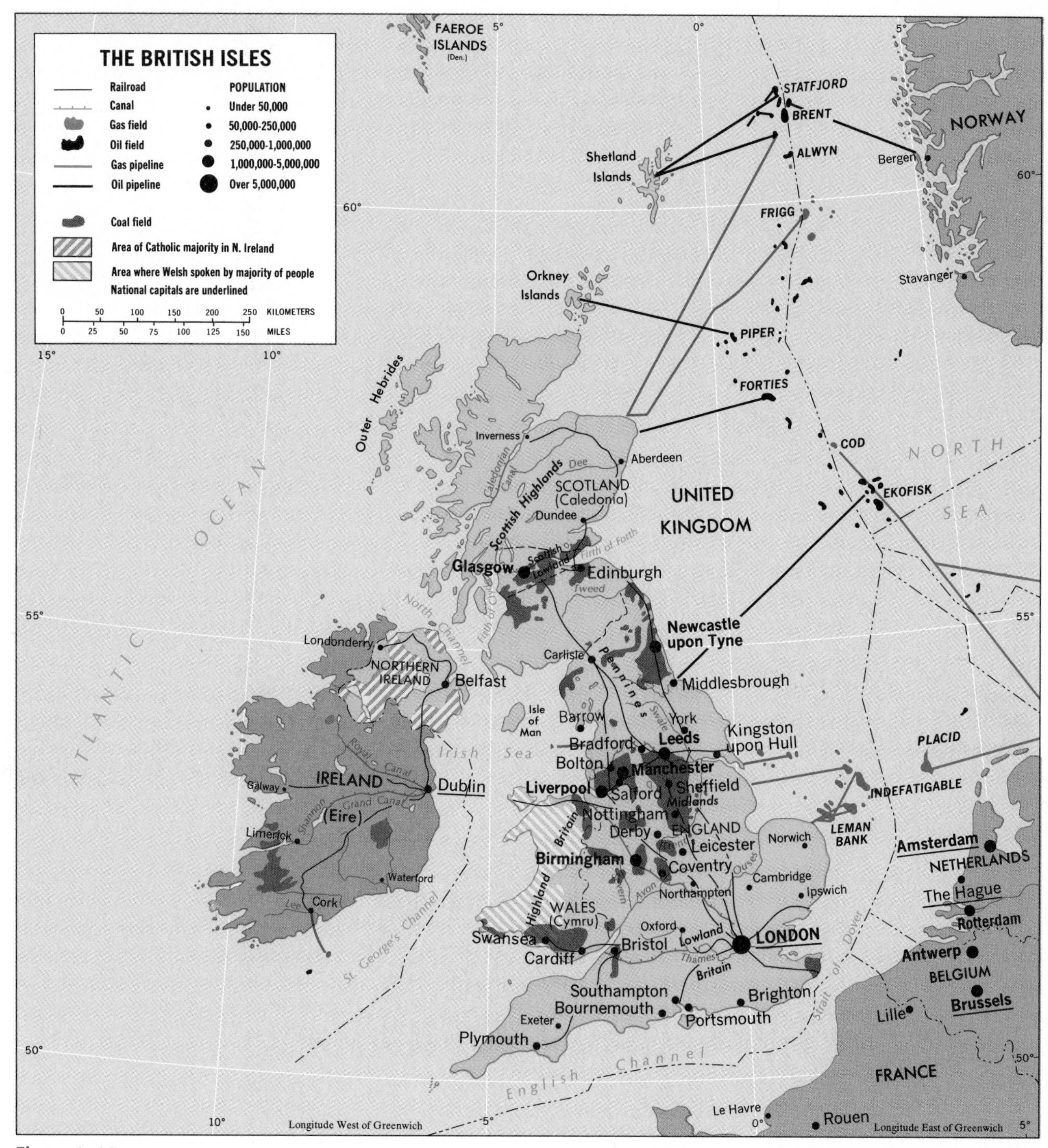


Figure 1–14

the energy provided by streams flowing off the Pennines, Lowland Britain's mountain backbone. Then, when the industrial revolution came, coalfields in the Midlands, just south of the Pennines, provided the needed power source. The eastern Midland deposits supported the industries of Sheffield, Derby, and Nottingham; those to the west sustained Manchester, Salford, and Bolton. Smaller coalfields were south of the Pennines, near Birmingham, and to the northeast, near Newcastle.

The impact of the industrial revolution was felt most strongly in the industrial regions of England near the Midlands and northeast coalfields—London was not directly involved. Working conditions in the burgeoning areas were bad and living conditions were often worse, but population totals soared. Labor organized and became a potent force in British politics. Meanwhile, Britain prospered. Nowhere is the principle of functional specialization better illustrated than here. Woolen-producing cities lie east of the Pennines, centered on Leeds and Bradford. The cotton textile industries of Manchester are on the other side of the range. Birmingham, now a city of 2.9 million people, and its nearby cities concentrate on the production of steel and metal products, including automobiles, motorcycles and bicycles, and airplanes. The Nottingham area specializes in hosiery, as does Leicester; boots and shoes are made in Leicester and Northampton. In northeast England, shipbuilding and the manufacture of chemicals are the major large-scale industries, along with iron and steel production and, of course, coal mining. Coal became an important British export, and with its seaside locations in northeast England, south Wales, and central Scotland, it could be shipped directly to almost any part of the world—more cheaply than it could be produced from domestic sources in those faraway countries.

Today, the industrial specialization of British cities continues, but the resource situation has changed drastically as local raw materials became exhausted. A one-time exporter of large quantities of coal, England has in recent years been forced to import even this commodity as domestic supplies dwindle. Iron ores have also approached depletion, and two-thirds of the ore required by British industries is now bought elsewhere. Of course, Britain's industries always did use raw materials from other countries, such as wool from Australia, New Zealand, and South Africa; and cotton from the United States, India, Egypt, and the Sudan. But with the breakdown of the British Empire and greater competition from the rising industrial powers, Britain is now hard pressed to maintain its position as a leading Western industrial power. West Germany far outdistances Britain in the production of steel, and Britain has lost the lead in the manufacture of other products as well. These developments have induced some industrial enterprises to flee the declining cities of the Midlands and relocate near the old primate city, London. This reflects not only the Midlands's decline, but also industries' declining dependence on coal (use of nuclear energy is rising), the decision to start afresh with new modern machinery, and the attraction of London not only as a huge and still wealthy domestic market, but also as a convenient port and a source of specialized labor.

The U.K.'s energy situation, on the other hand, has improved markedly since 1970, and Britain is now self-sufficient in oil and natural gas thanks to recent discoveries beneath the North Sea. European nation-states facing the North Sea have allocated sectors of sea floor to themselves (see Fig. 1–14); the British sector has proven to contain rich fuel reserves, and the United Kingdom has also purchased concessions in Norway's productive Ekofisk field, from which a pipeline was opened in 1974. Several offshore oilfields are now operating (Fig. 1–15), particularly benefiting Britain's northeastern coast. British holdings on the North Sea floor continue to yield new discoveries of petroleum and natural gas reserves, and it is estimated that today's self-sufficiency should last well into the twenty-first century.

Lowland Britain, in addition to its industrial development, also has the vast majority of Britain's good agricultural land. Actually, the land is not nearly enough to feed the British people (56 million of them) at present standards of calorie intake. The good arable land is concentrated largely in the southeastern part of England, although there is a sizable belt of good soils behind the port of Liverpool in the west. Where the soils are adequate there is very intensive cultivation, and grain crops (wheat, oats, barley) are grown on tightly packed, irregularly shaped fields that cover virtually the entire countryside. Potatoes and sugar beets are also grown on a large scale. But much of Lowland Britain is suitable only for pasture and cannot—for reasons of soil quality, coolness, excessive moisture, or other factors—sustain field crops. Hence, Britain requires a large quantity of food imports every year; the country cannot adequately feed even half its people, perhaps not even one-third. There are various ways to estimate this, but in all probability Britain imports nearly two-thirds of the food consumed by its population. This makes Britain the greatest importer of food in the world, and, for those countries that have food to sell, one of the world's top markets; for meat, grains, dairy products, sugar, cocoa, tea, and fruits, Britain is either the leading or one of the world's leading importers.

In a country where some 91 percent of the population live in urban centers of one kind or another and where only 1 of every 40 gainfully employed persons makes a living out of farming, none of this is surprising. Most Britishers work in commerce, manufacturing, transportation, or other urban-oriented economic activities; this is the most highly urbanized society in the world. The dependence on food importing developed during the nineteenth century, with Britain's population explosion and the competition from outside (especially Australia) of cheap farm produce—cheaper wheat came from America and Australia, for instance, than British farmers could profitably grow at home. Hence, they turned in ever-larger numbers to raising cattle and sheep—the main domestic livestock—and to converting

Figure 1–15
An offshore rig pumping North Sea oil.

more acreage to pasture and meadow. With their milk and meat, they would have a better chance to beat foreign competitors to the local market than with grain crops. In recent decades, the British government has attempted to turn this tide and has subsidized farmers to encourage them to return to cultivating wheat and potatoes, among other crops. But it has been an expensive and not very fruitful exercise. Only a crisis involving a cutoff of overseas supply lines is likely to stimulate British agriculture to its full calorie-producing capacity.

**Regions of Britain** The British Isles constitute a region of the European realm, and both Britain and Ireland can be divided into subregions. In the case of Britain, nine such units can be identified: (1) the Southeast, including London; (2) the Southwest; (3) the Midlands, industrial heartland; (4) East Anglia, historic settlement zone; (5) the Northeast, between the Pennines and the North Sea; (6) the Northwest, or Lancashire; (7) Northern England with the Pennines and the famous Lake Country; (8) Scotland; and (9) Wales.

***The Southeast, including London*** England's *Southeast* is essentially London and its immediate hinterland, a region than contains nearly 20 million of the U.K.'s 56 million people. London itself is a city of more than 8 million people; it also centers one of the world's great **conurbations**, the general term used by geographers to describe vast metropolitan complexes (such as "megalopolis," which was discussed on p. 30) that formed from the coalescence of two or more major urban areas (Fig. 1–16). It is the country's historic focus, the seat of government, the headquarters of numerous industrial and commercial enterprises, the major port, the most concentrated and richest market: in short, the primate city (Fig. 1–17). London lies at the center of Britain's transport networks, and the metropolitan area is served by two major international airports (Heathrow and Gatwick).

Figure 1–16 reveals the dimensions of the London metropolitan area and its classic European-city form. To stem and channel the city's vast urban sprawl, a "Green Belt" was designated in a development plan for Greater London published in 1944. This plan recommended that a zone surrounding built-up London would be permanently set aside for nonresidential uses, including recreation and farming. The original proposal was modified under the pressures of London's continued growth, so that today there are not only segments of Green Belt, but also "green wedges" between residential areas. Notwithstanding these modifications, the Green Belt proposal had the effect of preserving open space in and around a city badly in need of breathing room. Thus, the most recent episode of suburbanization has occurred beyond this zone, more than 40 kilometers (25 miles) from Central London; in addition to residential development, much of

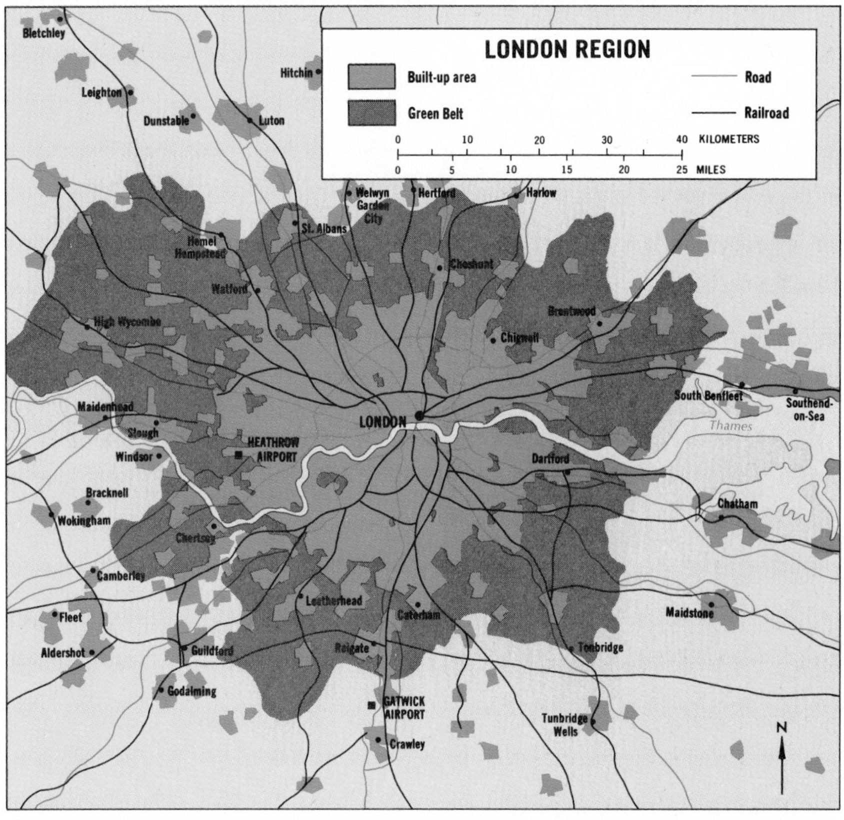


Figure 1–16

this growth increasingly involves offices, shopping facilities, and other major economic activities.

***The Southwest*** The *Southwest* is a world apart from the crowded, populous Southeast. This is a portion of Highland Britain, including Cornwall, a declining stronghold of Celtic Britain. This area has long been in economic stagnation, but its mild climate, spectacular coastlines, and impressive national parks (e.g., Dartmoor) have generated a growing tourist industry. Eastward, the counties of Hampshire, Wiltshire, and Gloucestershire display, through new industries and revived economies, their proximity to the burgeoning London area.

***The Midlands, industrial heartland*** The *Midlands* are sometimes divided into the East Midlands, centered on Nottingham and Sheffield (with its world-famous steel), and the West Midlands, whose focus is Birmingham. The Midlands are appropriately named, for they do occupy the center of England and many of the region's cities have prospered because of their fortuitous location. Industry thrived in the Midlands from the time of the industrial revolution and, as noted earlier, sophisticated functional specialization occurred here. The cities and towns of the Midlands are not England's most attractive (much of this "Black Country" is, indeed, afflicted by grime and soot from factories and mines), but the labor force is large, the unions strong, and wages are higher than in other manufacturing areas.

***East Anglia, historic settlement zone*** *East Anglia* is England's historic landfall, a region of early settlement by the Angles, invaders from Germany. Long before the industrial revolution, this area exported wool to the textile makers of Flanders, and its outward orientation continued afterward when fishing grew in importance. Without coalfields, East Anglia could not compete with other regions in Britain; therefore, early on it stressed fishing and mechanized agriculture. Today, East Anglia looks outward once more because its location is opposite the newly productive North Sea oil and gas reserves; in the late 1970s, pipelines were brought to its coast from the gasfields at Leman Bank (Fig. 1–14).

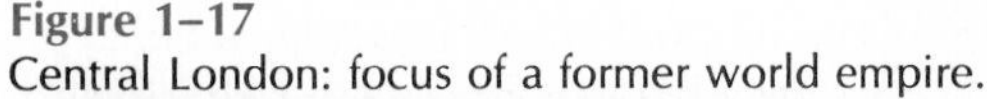
Figure 1–17
Central London: focus of a former world empire.

***The Northeast, between the Pennines and the North Sea*** The *Northeast* is not only the region of Newcastle and its heavy manufacturing and of Leeds and Bradford and their textiles, but also of the farms of Yorkshire, the deep-sea fishing fleets of Hull, the tourist hotels along a scenic North Sea coast, the impressive mansions of a verdant countryside, and the magnificent churches and cathedrals in towns and cities. The Northeast, too, has shared in the development of the North Sea's energy reserves: an oil pipeline connects Middlesbrough to the Ekofisk oilfield. Diversification is the hallmark of the productive and varied Northeast, which, like New England in the United States, is undergoing economic revival following decades of stagnation.

***The Northwest (Lancashire)*** The *Northwest* is centered on the Manchester-Liverpool conurbation. This is the region of Lancashire, one of England's leading areas when the industrial revolution gained momentum. With water power and available coal, soft water (needed in the dyeing process of cotton textiles), and moist air (which facilitated spinning before air conditioning became possible), this region had all the advantages—including its port of Liverpool, where raw cotton from America and Egypt was unloaded. Liverpool grew from a fishing village into England's second port; Manchester rose to become the country's textile capital. But today, the Northwest shows the effect of modern competition, the obsolescence of unreplaced equipment, overspecialization, and inadequate planning. The grimy factory towns of Lancashire, set against gray sandstone ridges defoliated by two centuries of smoke and chemical pollution, symbolize the current problems of a region that once was the pulse of industrializing Britain.

***Northern England (the Pennines and the Lake Country)*** *Northern England* consists of the island's mountain backbone, the Pennines, as well as the scenically beautiful Lake District near its northern end. This is a segment of what we previously identified as Highland Britain, and there is much rugged topography with pastureland and high-relief coastlines studded with fishing villages. Fewer opportunities for industrialization presented themselves here, and the major city in the region is Barrow with some steel mills and shipyards, which are based, in part, on high-grade iron ores scattered through the mountains and valleys of the Lake District.

***Scotland and Wales*** *Scotland* and *Wales* are distinct regions of Britain because they have political as well as cultural, economic, and physiographic identity. Their relationships with long-dominant England, furthermore, are changing. By European standards of scale, these are by no means small provinces; Scotland covers nearly 80,000 square kilometers (30,000 square miles) and has just over 5.5 million inhabitants; Wales, with 2.9 million people, extends over 21,000 square kilometers (8000 square miles). These are Britain's most rugged, remote, highland territories, where ancient Celtic peoples found refuge from encroaching, alien invaders. Celtic languages still survive here (Fig. 1–13), even after centuries of domination by the English and their strong culture. The same hills and mountains that provided sanctuary and long repelled the power of London also tended to keep the Scots and Welsh divided into relatively isolated groups or clans. But communications are better now, and Scottish and Welsh nationalisms are reviving.

Wales and Scotland shared in Britain's industrial revolution, but with mixed results. Initially, the coalfields of southern Wales generated valuable export revenues and attracted industries, but when the most accessible coal seams were exhausted and production costs went up, competition from fields in England proved too strong. A creeping depression struck Cardiff, Swansea, and the other cities that had attracted Welsh labor from the hills. The industrial revolution left Wales's countryside ravaged by strip mines, its people dislocated, its cities slum ridden. Many Welsh citizens left their country for other parts of the world and new opportunities.

Scotland was more fortunate. Along the narrow "waist" of the Clyde and Firth of Forth, Scotland possessed an extensive coalfield, a nearby iron ore deposit, and (near Glasgow) an excellent port. This narrow lowland corridor between Scotland's mountains has become the region's core area, with industrial Glasgow at one end and the cultural focus of Edinburgh, the primate city, at the other. The Scottish coalfields were not simply a supply area for other parts of the world, they formed the basis for major manufacturing development, notably in shipbuilding and textiles. A variety of healthy and competitive industries managed to stave off the fate that had befallen Wales, and today, the Scottish economy still is in better condition than that of the Welsh, although it shares with England the decline caused by obsolescence. Scotland now finds itself fronting a North Sea that produces not only the old commodity, fish, but also the new ones, oil and natural gas. A new age may be dawning—and not only in politics.

Wales and Scotland share a resurgent nationalism that has modern economic-geographic as well as historical roots. The government in London (where Scottish and Welsh representatives are far outnumbered by their English counterparts) has always pursued economic policies that favor England, often at the clear disadvantage of Scotland and Wales. The 1970s brought a lingering recession that left unemployment high. Few major industrial or financial firms are in Scottish or Welsh hands, and no political or economic decisions affecting the two territories are made by the Scots or Welsh themselves. The centralization of authority in London and the disadvantageous position of Wales and Scotland sometimes lead local newspapers to refer to their "colonial" status.

Strong demands by Scottish and Welsh nationalists for regional autonomy continue to be heard. These demands have a varied background. For several decades, local leaders had been encouraging the revival of Celtic languages not only in Scotland and Wales, but even in the Southwest, where the *Mebyon Kernow* (Sons of Cornwall) led this effort. Furthermore, Scotland's eastern cities are experiencing an economic resurgence as a result of the success of the new energy industries of the North Sea. And the crisis in Northern Ireland is a constant warning against any failure to adjust at an early stage to regional pressures. Thus, the British Parliament in 1976 began formal discussions under the theme of *devolution* (see box)—the redistribution of authority in the United Kingdom and the restructuring of the country's political framework. Although separatist proponents had only limited successes in Scotland and Wales during recent elections, the issue is likely to remain current in British politics for decades to come.

## DEVOLUTION—GLOBAL SYMPTOM?

The possibility that regional self-government may come to Wales and Scotland is unthinkable to many Britons; numerous Canadians view the prospect of an independent Quebec as inconceivable; and Nigeria went to war in the late 1960s to prevent the secession of Biafra. Has the evolution of the state system come to an end? Will a reverse process—devolution—become commonplace?

The indications are that many states will confront the specter of fragmentation. Yugoslavia's fragile federal structure, especially since the death of Marshal Tito, is under pressure from Croatian separatists. Spain faces a secessionist effort in the Basque country and strengthening regionalism in the northeast. Corsican nationalists want independence from France. Eritrean partisans are fighting for separation from Ethiopia. Pakistan has for years combated a separatist movement in Pathanistan, its northwestern zone. Kurds in Iran, Moslems in the Philippines, Saharans in Morocco, Turks in Cyprus—the number of regional secessionist movements appears to be growing, and the list of potential problem areas expands as well. Such movements are vulnerable to involvement in the greater ideological struggle in the modern world as the major powers try to exploit them to their own advantage.

## Fractured Ireland

**Northern Ireland** The smallest of the U.K.'s four units continues to be London's most serious domestic problem in the 1980s. Northern Ireland (14,000 square kilometers/5450 square miles) occupies the northeastern one-sixth of the island of Ireland, a legacy of England's colonial control. With a population of 1.7 million, Northern Ireland also is the U.K.'s poorest area in terms of resources for primary industries. Still, a diversified economy has emerged here as a result of infusions of British capital and raw materials. Belfast's shipbuilding industry grew to significant proportions, as did Londonderry's textile (mostly linen) manufacturing. But unemployment has been an erosive reality, and Northern Ireland has not fared well since 1970—economically or socially.

Northern Ireland, which was severed from the Republic of Ireland when the latter became independent from London in 1921, incorporates six of the nine counties of the historic Irish province of Ulster. What makes Northern Ireland different from the rest of Ireland is its substantial Protestant population. About two-thirds of the people trace their ancestry to Scotland or England, from where their predecessors emigrated to settle across the Irish Sea. Only about 35 percent of Northern Ireland's people are Catholics, in contrast to the Republic of Ireland, which is overwhelmingly Roman Catholic. No spatial separation marks the religious geography of Northern Ireland: Protestants and Catholics live in clusters throughout the area. The two eastern counties are most strongly, but by no means exclusively, Protestant (see Fig. 1–14).

For the past several years, Northern Ireland has remained on the brink of civil war as Protestants and Catholics attacked each other in terrorist campaigns. Catholics have long protested what they perceive to be discrimination by the Protestant-dominated local administration, especially in housing and employment; Protestants accuse Catholics of seeking union with the Republic. The deterioration of the economy during the early 1970s plunged the region into a deepening crisis, and in 1972, the British Parliament suspended local administration and began to rule the territory directly from London. Neither British administration nor a British armed force sent to maintain order has been able to end the sectarian violence, which is occasionally even carried to London itself in the form of random bombings. Nevertheless, there continues to be hope that discussions on devolution might produce a solution for Northern Ireland as well as Scotland and Wales, but intermittent destruction and violence continues.

**Republic of Ireland** Ireland (Eire) is one of Europe's youngest independent states, and the country from which it fought itself free was its neighbor, Britain. In fact, it is more correct to say that the struggle was one between the Irish and English, for certainly in Wales and Scotland there are pro-Irish sympathies. But to those who think that colonialism and its oppressive policies and practices, especially where land ownership is concerned, is something that can only be perpetrated by European peoples on non-European people, Ireland's history is a good object lesson.

Without protective mountains and without the resources that might have swept their country ahead in the great forward rush of the industrial revolution, the Irish—a nation of peasants—faced their English adversaries across a narrow sea; and a sea was slim protection against England, whatever its width. Wales had given up the struggle against English authority centuries earlier (shortly before 1300); Scotland had been incorporated soon after 1600. But the Irish continued to resist, and their resistance led to devastating retaliation. During the seventeenth century the English conquered the Irish, expropriated their good agricultural land, and turned it over to (mostly absentee) landlords; those peasants who remained faced exorbitant rents and excessive taxes. Many moved westward to less fertile farmlands along the Atlantic margins, there to seek subsistence as far away from the British sphere as possible. The English placed restrictions on every facet of life—especially in Catholic southern Ireland, though rather less in the Protestant north. Irish opportunity and initiative were put down.

The island on which Eire is located is shaped like a saucer with a wide rim, a rim that is broken toward the east where lowlands open to the sea. Although less prominent or rugged than their Scottish and Welsh counterparts, the hills that form the margins of Ireland are made of the same kinds of rocks, and, in places, elevations exceed 900 meters (3000 feet). A large lowland area is thus enclosed, and, certainly, Ireland would seem to have better agricultural opportunities than Scotland and Wales. But, in fact, the situation is not much better at all. As Fig. I–9 shows, Ireland is even wetter than England, and excessive moisture is the great inhibiting element in agriculture here. Practically all of Ireland receives between 100 and 200 centimeters (40 and 80 inches) of rain annually. Hence, pastoralism is, once again, the dominant agricultural pursuit, and a good deal of potential cropland is turned over to fodder. The country's major exports are livestock and dairy products, and most goes to Britain.

How can 150 centimeters of rainfall be too much for cultivation when there are parts of the world where 150 centimeters is barely enough? Here, the *evapotranspiration* factor comes into play (p. 14), and this is a good illustration of the deceptive nature of such climatological maps as Fig. I–9—deceptive unless more is known than the map tells us. Ireland's rain comes in almost endless, soft showers that drench the ground and keep the air cool and damp—in Scotland, they are called "Scottish mists." Rarely is the atmosphere really dry enough to permit the ripening of grains in the fields. Without such a dry period the crops are damaged or destroyed, and so there are severe limitations on what a farmer can plant without too great a risk. In Ireland, the potato quickly took the place of other crops as a staple when it was introduced from America in the 1600s; it does particularly well in such an environment. But even the potato could not withstand the effects of such wetness as prevailed in several successive years during the 1840s. Having long been the nutritive basis for Ireland's population growth, which by 1830 had reached 8 million, the potato crop, ravaged by blight and soaking, failed repeatedly; the potatoes simple rotted in the ground. Now the Irish faced famine in addition to British rule. Over a million people died, and nearly twice that number left the country within 10 years of this new calamity. It set a pattern that has only recently been broken, as emigration offset and even exceeded natural increase. Ireland and Northern Ireland combined had a 1985 population of 5.3 million (3.6 million in Ireland proper), barely 60 percent of what it was when the famine of the 1840s struck. Not many places in the world can point to a decline in population since the days of the industrial revolution.

The Irish have not forgotten their colonial experiences at the hands of the British. Independence in 1921 was followed by neutrality during World War II and withdrawal from the Commonwealth in 1949. The continued partition of Ireland is a source of friction, although Dublin has officially remained quite aloof from Northern Ireland's civil strife. Nevertheless, other signs point toward improving relations. Britain is Ireland's leading trade partner, and more Irish emigrants leave for the United Kingdom than any other destination. Still, Ireland seeks to foster domestic nationalism by encouraging the use of Gaelic and through vigorous support of Irish cultural institutions.

A determined effort is being made to diversify Ireland's economy and to speed the county's development. Dublin, the capital, has become the center for a growing number of light industries. For many years, the great majority of Ireland's exports were meats, live animals, dairy items, and other farm-derived products, but the balance is swinging toward manufactures, now accounting for about half the total sales annually. Ireland's decision to join the EEC in 1973 is further evidence of the country's desire to modernize. But in the face of its resource and environmental limitations, the task is difficult. In many ways, Ireland is still more reminiscent of agrarian Eastern Europe than the European heartland it has recently joined.

## Western Europe

The essential criteria on the basis of which a Western European region can be recognized are those that give Western Europe the characteristics of a regional core: this is the Europe of industry and commerce, of great cities and enormous mobility, of functional specialization and areal interdependence. This is dynamic Europe, whose countries founded empires while forging democratic parliamentary governments and whose economies staged an astounding recovery from the devastation of World War II four decades ago.

But if it is possible to recognize some semblance of historical and cultural unity in British, Scandinavian, and Mediterranean Europe, that quality is absent here in the melting pot of Western Europe. Europe's western coreland may be a contiguous area, but in every other way, it is as divided as any part of the continent is or ever was. Unlike the British Isles, there is no *lingua franca* here. Unlike Scandinavia, there is considerable religious fragmentation, even within individual countries. Unlike Mediterranean Europe, there is not a common cultural heritage. Indeed, the core region we think we can recognize today based on economic realities could be shattered at almost any time by political developments. It has happened before.

**France and West Germany** The leading states of Western Europe, of course, are West Germany and France. In addition, there are the Low Countries of Belgium, the Netherlands, and Luxembourg (also known collectively as *Benelux*), and the Alpine states of Switzerland and Austria. It will be recalled that northern Italy, by virtue of its industrialization and its interconnections with Western Europe, was identified as part of the European regional core; for the purposes of the present regionalization of Europe, however, all of Italy will be considered as part of Mediterranean Europe. Thus, the region comprises nearly 1 million square kilometers (414,000 square miles), not including East Germany's 108,000 square kilometers (41,800 square miles).

The partition of Germany into West and East poses a regional problem, for East Germany has stronger cultural-geographical and historical ties with Western rather than Eastern Europe. But East Germany, following World War II, was incorporated into the Soviet sphere of influence and has been reorganized along socialist lines, in common with Eastern European countries. East Germany will, therefore, be mentioned both in the overall German context in the present section and again under Eastern Europe. Reference to *Germany* is intended to involve all of the divided German state, West and East.

Territorially, Western Europe is dominated by France; economically, West Germany is paramount. In many other ways, the two mainland powers present interesting and sometimes enlightening contrasts. France is an old state, by most measures the oldest in Western Europe; Paris has for centuries been its primate city and cultural focus. Germany is a young country, born in 1871 after a loose association of German-speaking states had fought a successful war against . . . France! Berlin became the national focus only during the nineteenth century (although it was a prominent Prussian center earlier), having reached 100,000 inhabitants about 1750. Modern France bears the imprint of Napoleon, who died just a few years after the political architect of Germany, Bismarck, was born.

From the map of Europe it would seem that Germany, smaller than France in area, also has a disadvantageous position on the continent in comparison with its rival (Fig. 1–18). France has a window on the Mediterranean, where Marseilles—at the end of the cross-continental route, of which the Rhône-Saône Valley is the southernmost link—is its major port. In addition, France has coasts on the North Atlantic Ocean, the English Channel, and, at Calais, even a corner on the North Sea. Germany, on the other hand, has short coastlines only on the North Sea and the icy Baltic, and the rest of its territory seems heavily landlocked by the Netherlands and Belgium to the west, by the Alps to the south, and by Poland, to which Germany lost important territory after World War II, in the east. But such appearances can be deceptive. In effect, France is at a disadvantage when it comes to

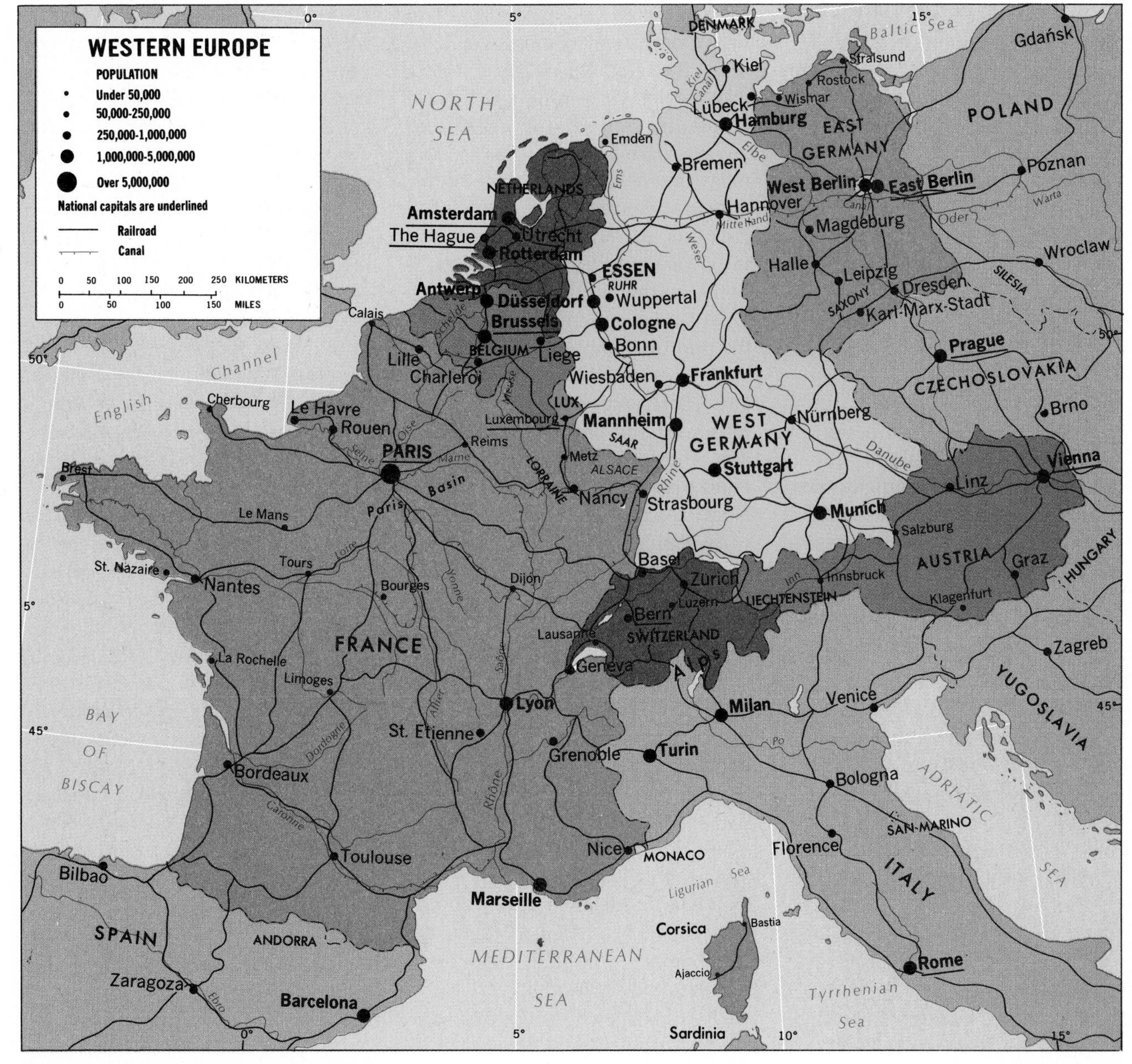


Figure 1–18

foreign trade, not Germany. None of France's natural harbors, including that of Marseilles, is particularly good. Very few of France's many miles of rivers and inland waterways are navigable by modern, large oceangoing vessels. Indeed, France in an earlier day was better served by water transport than it is at present; larger ships with deeper drafts have caused a decline in the usefulness of France's waterways. Among the Atlantic ports that France does have, several possess major disadvantages. Le Havre and Rouen serve as outlets for Paris and the Paris Basin; yet, large ships can navigate up the River Seine only as far as Rouen, now France's second port. Brest and Cherbourg lie at the end of peninsulas, far from the major centers of production and population. Nantes, La Rochelle, St. Nazaire, and Bordeaux have only local significance. Therefore, a good deal of trade destined for France goes not through French, but through other European ports such as London, Rotterdam, and Antwerp.

Germany is much better off. Although the mouth of the great Rhine waterway lies in the Netherlands, Germany contains most of its course—a course that runs past Europe's leading industrial complex, the *Ruhr*. For the western part of Germany, the Rhine is almost as effective a connection with the North Sea and the Atlantic as a domestic coast and harbor would be; Rotterdam, the world's largest port in terms of tonnage handled, is a more effective outlet for Germany than any French port is for France. But Germany has its own ports as well: Hamburg, 100 kilometers (60 miles) inland on the Elbe River, is the major **break-of-bulk point**, where cargoes from oceangoing vessels are transferred to barges, and vice versa. Hamburg has suffered from the consequences of World War II; through a series of canals its hinterland had extended as far across the North European Lowland as Czechoslovakia, but the partition of Germany has all but eliminated its trade volume with most of this area. To partially compensate, a North-South Canal has been constructed to connect Hamburg with the Mittelland Canal within West Germany, a major tributary to the Rhine network. Not far to the west, on the Weser River, is the port of Bremen, a smaller city located in a less productive area, but of historic and, more recently, of military importance, as it became the major American military port after World War II. Prior to the war, Hamburg and Bremen had benefited from the completion of the Kiel Canal across the German "neck" of Denmark; they had thereby attained new significance as gateways to the Baltic. And on the Baltic side, Germany (including East Germany) has several ports that are easily in a class with the French entries of Nantes, La Rochelle, and Bordeaux: such places as Kiel, Lübeck, and Wismar, each with well over a quarter-of-a-million population and some industrial as well as port development.

In the aggregate, then, Germany's outlets are far superior to those of France. The same is true for the inland waterways that interconnect each country's productive areas. In France, although canals link the Garonne and Rhône rivers and the Loire and Seine, the important waterways that carry the heavy traffic lie in the northeast and north, where the productive areas of Lorraine, the Paris Basin, and the Belgian borderlands are opened to the Meuse (Maas) and Rhine, and, thus, to Rotterdam and Antwerp. In Germany, the Mittelland Canal, with an east-west orientation, literally connects one end of the country to the other. In turn, it intersects each of the northward-flowing rivers: the Oder-Neisse system, the Elbe (to the port of Hamburg), the Weser (to Bremen), and the Dortmund-Ems Canal, which links with the Ems River and with the port of Emden. The Dortmund-Ems Canal provides the Ruhr with a German North Sea outlet (and inlet—most of Sweden's Kiruna iron ores arrive through the port of Emden) and, by also linking the North German canal system to the Rhine and Rotterdam, completes a countrywide network of great effectiveness.

Another strong contrast between France and Germany lies in the nature and degree of urbanization in the two countries. True, Paris, is without rival in France and in mainland Europe, but there is a huge gap in France between it (population: over 7 million in the city proper, over 10.5 million counting adjacent urbanized areas) and the next ranking city, Lille, near the coalfields on the Belgian border (population: about 1.2 million). Two questions arise, one no less interesting than the other: Why should Paris, without major raw materials in its immediate vicinity, be so large; and why should Lille, Lyons, and Marseilles (all about 1.2 million in population) be no larger than they are?

***Paris*** Paris owes its origins to advantages of **site** and its later development to a fortuitous **situation** (Fig. 1–19). Whenever an urban center is considered, these are two very important locational aspects to take into account. A city's **site** refers to the actual, physical attributes of the place it occupies—whether the land is flat or hilly, whether it lies on a river or lake, whether the port (if any) has shallow or deep water, whether there are any obstacles to future expansion such as ridges or marshes. By **situation** is meant the geographical position of the city with reference to surrounding or nearby areas of productive capacity, the size of its hinterland, the location of competing towns—in other words, the greater regional framework within which the city finds itself. Paris was founded on an island in the Seine River, a place of easy defense and in all probability also a place where the Seine was often crossed. Exactly when settlement on this site, the *Île de la Cité*, actually began is not known (probably it was in pre-Roman times), but it functioned as a Roman outpost some 2000 years ago. For many centuries, this defense aspect of the site continued to be of great importance, for the authority that existed here was by no means always strong. Of course, the island soon proved to be too small, and the city began to expand outward along the banks of the river (Fig. 1–20), but that did not diminish the importance of the security it continued to provide to the government.

The advantages of the situation of Paris were soon revealed. The city lies near the center of a large and prosperous agricultural area, and as a growing market its *focality* increased continuously. In addition, the Seine River, itself navigable by river traffic, is joined near Paris by several tributaries, all of them navigable as well, leading to various sections of the Paris Basin. Via these rivers (the Marne, Oise, Yonne) converging from the northeast, east, and southeast, and the canals that later extended them even farther, Paris can be reached from the Loire, the Rhône-Saône, the Lorraine industrial area, and from the Franco-Belgian border

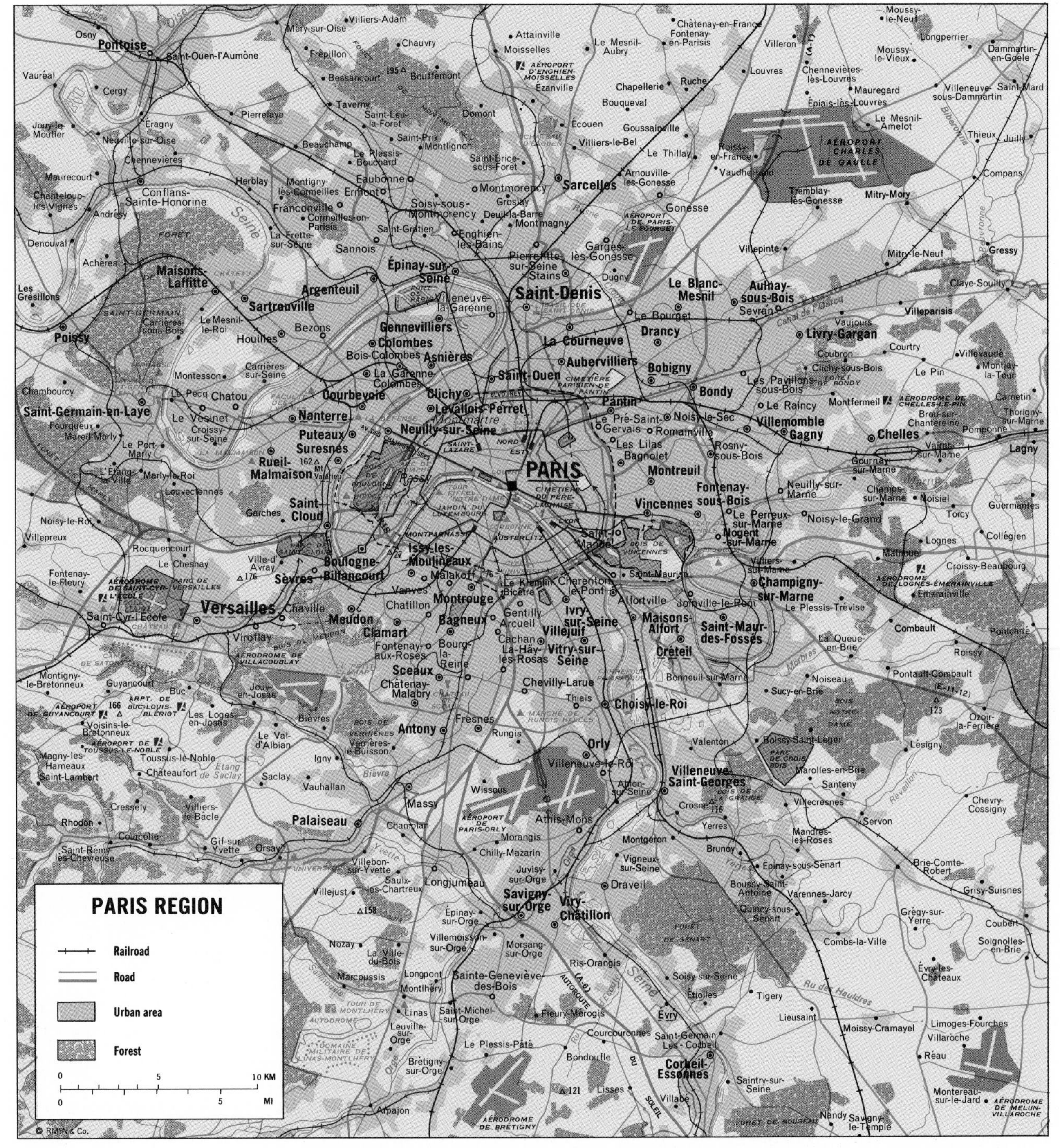


Figure 1–19

zone of coal-based manufacturing. Of course, there are land connections as well. The political reorganization brought to France by Napoleon involved not only a reconstruction of the internal administrative divisions of the country, but a radial system of roads that, like Roman roads focused on Rome, all led directly to Paris.

If it is not difficult to account for the greatness of Paris, it is quite another matter to account for the relatively limited development of French industrial centers elsewhere. In France, there is no Birmingham, no Glasgow—certainly, no Ruhr. And yet, there is coal, as we have seen earlier, and there is good quality iron ore. There is also a long history of manufacturing: the linen industry of Lille existed in the Middle Ages and the silk industry of Lyons also is no newcomer to

**Figure 1–20**
Central Paris: *Île de la Cité* in the Seine River.

the manufacturing scene. What is lacking? For one thing, large amounts of readily accessible high-quality coal; for another, the juxtaposition of such coal with cheap transport facilities and large existing population concentrations. So French manufacturers have done what Europeans have done almost everywhere—they have specialized. French industry produces precision equipment of many kinds, high-quality china, luxury textiles, automobiles, and, of course, wines and cheese. Unable to compete in volume, they compete in quality and specialty.

France achieved greater strength in agriculture with a much larger area of arable land than Britain or Germany and with the benefits of a temperate climate free of extremes in moisture or temperature. Apart from the Paris Basin and adjacent sections of the North European Lowland in France, there are several other major agricultural areas, including the valleys of the Rhône-Saône, Loire, and Garonne. Elsewhere, as in the sandy and marshy parts of the southwestern Aquitaine, in the higher parts of the Massif Central, and in the Jura and Vosges mountains to the east, the soil is not suitable or the slopes are too steep for farming. Between these extremes there is a wide variety of conditions, each with its special opportunities and limitations; wheat is grown on the best soils, and it is France's leading crop. Oats will grow under slightly less favorable conditions, and barley will tolerate even greater disadvantage. Southern and Mediterranean France (the *Midi*), of course, produce grapes and the usual association of fruits. French agriculture is marked by an enormous diversity of production, and apart from the predominance of wheat it is hardly possible to isolate regions of exclusive land use—except in the vineyards of the Rhône-Saône, Garonne, and Loire valleys. Much crop rotation is practiced; animal manure and commercial fertilizer are widely used (France is Europe's top beef and milk producer), and, despite a rather low level of farming efficiency, especially in the south, France's annual agricultural output is second to none in Europe.

Germany—West or East—has no Paris, and West Germany is far from self-sufficient in farm production. But, despite the absence of a city with the primacy and centrality of Paris, West Germany today is a more highly urbanized country than France (by official definitions the comparison is over 85 against 80 percent; for East Germany, 78 percent). West Germany has nearly twice the number of urban centers containing over a half-million population: 12 in 1985 (France has 7). And as might be expected, West Germany also is more strongly industrialized than its southwestern neighbor.

The majority of Germany's cities lie in or near the zone of contact between the two major physiographic regions, the plain of the north (part of the North European Lowland) and the uplands of the south (Fig. 1–6). This, as we know, is also the chief zone of coal deposits, and, of course, the development of several of these cities is bound up with the availability of this energy-resource base. Prewar Germany counted three major areas of industrial growth in this zone, extending from west to east: *the Ruhr*, based on the Westphalian coalfield, *the Saxony area* near the Czechoslovakian boundary (now in East Germany), and *Silesia* (now in Poland). In the south, near the French border, there are coalfields in *the Saar*, and minor coal deposits lie scattered in many other parts of Germany. The best fields, however, are those serving the Ruhr and Silesia; Germany has lost Silesia, but the Ruhr has become the greatest industrial complex of Europe.

Whereas the Ruhr specializes in heavy industries, Germany's second industrial complex, Saxony, is more oriented to skill and quality. Today, this region, which includes such large cities as Leipzig (printing and publishing), Dresden (ceramics), and Karl-Marx-Stadt (textiles), lies in East Germany. Again, Saxony was a region of considerable urban and manufacturing development long before the industrial revolution occurred in the early nineteenth century, and nearby coal supplies, though of lower quality than those of the Ruhr, have maintained industries and steady growth in this century.

West and East Germany together have a population of nearly 80 million, and a great deal of good agricultural land would be required to feed so large a population. Farming is limited by extensive areas of hilly country in the south, and there are problems with sandy and otherwise difficult soils in the north. Nevertheless, with staples, including the ever-present potato, rye, and wheat, Germany manages to produce about three-quarters of the annual caloric intake of its population. This is not quite as good as France, but it is a great deal better than Britain manages to do; Germany also has greater industrial productivity than France. In view of the size of the population (even West Germany alone has substantially more people than the United Kingdom or France), the continuing German achievement in agriculture is remarkable.

# DIVIDED BERLIN

Berlin was the centrally positioned capital of the German state when Germany still included East Prussia and other areas of what is today Poland. As the German headquarters, the city grew rapidly; it was endowed with magnificent, wide avenues and impressive public architecture. In the 1930s, its population exceeded 4 million, and Berlin was a major industrial center in Germany, although (like Paris and London) the city does not lie near major sources of raw materials.

Berlin was devastated during World War II. The Allied forces in 1944, even before the war's end, divided defeated Germany into occupation zones and carved Berlin (although it lay in the Soviet zone) into four sectors: a Soviet East Berlin, and a West Berlin divided among the United States, the United Kingdom, and France. In effect, the city was split in half, with 403 square kilometers (156 square miles) and about 1 million people in the Soviet sector and 480 square kilometers (185 square miles) and just over 2 million in the

**Figure 1–21**
The multiple barriers separating East and West Berlin.

combined West. Soon, West Germany's recovery and free economy began to contrast sharply with East Germany's more austere existence, a contrast that was especially vivid in the divided city. When the Allies unified their occupation zones in 1948 and included West Berlin in a new currency structure, the Soviets blockaded the western sectors, but the blockade was quickly broken by a massive airlift.

In 1961, the East Germans—with Soviet approval—built a wall nearly 3 meters (10 feet) high and topped with barbed wire to stop the flow of refugees from Communist-dominated East Germany who used the West Berlin "window" as an escape route. Numerous people have been shot attempting to scale that wall, which has since been reinforced by a "death strip" of electrified fences and vehicle traps (Fig. 1–21). Thus, Berlin today is two cities reflecting the two very different systems of modern Germany, interacting only slightly. West Berlin is an exclave of the German Federal Republic, accessible only with East German and Soviet acquiescence; East Berlin is the chief stronghold of the German Democratic Republic.

**Benelux** Three political entities are crowded into the northwestern corner of Western Europe: Belgium, the Netherlands, and tiny Luxembourg, together referred to by their first letters (*BeNeLux*). In combination, these nations also are often called the Low Countries, which is a very appropriate name: most of the land is extremely flat and lies near (and, in the Netherlands, even below) sea level. Only toward the southeast, in Luxembourg and eastern Belgium's Ardennes, is there a hill-and-plateau landscape with elevations in excess of 1000 feet.

As with France and Germany, there are strong contrasts between Belgium and the Netherlands. Indeed, there are such contrasts that the two countries find themselves in a position of complementarity. Belgium, with its coal base along an east-west axis through Charleroi and Liège, where there are heavy industries, and with its considerable manufacturing of lighter and varied kinds along a zone extending north from Charleroi through Brussels to Antwerp, produces a surplus of industrial products, including metals, textiles, chemicals, and specialties such as pianos, soaps, and cutlery. The Netherlands, on the other hand, has a large agricultural base (along with its vitally important transport functions); it can export dairy products, meats, vegetables, and other foods. Hence, there was mutual advantage to the Benelux economic union (initiated in 1948), for it facilitated the reciprocal flow of needed imports to both countries and it doubled the domestic market.

The Benelux countries are among the most densely populated in the world, and space is at a premium. Some 25 million people inhabit an area about the size of West Virginia. For centuries, the Dutch have been expanding their living space—not at the expense of their neighbors but by wresting it from the sea. The greatest project so far is the draining of almost the entire Zuider Zee, which should be completed by 2000; in the southwest, the islands of Zeeland are being connected by dikes and the water pumped out, creating additional *polders* (reclaimed lands). Another future project involves the islands that curve around northern Holland, which may be connected to dry and reclaim the intervening Wadden Sea.

The three cities of the Netherlands' triangular core area each contain about 1 million inhabitants. Amsterdam, the constitutional capital, has canal connections to the North Sea as well as the Rhine, but it obviously does not have the situational advantages of Rotterdam. Nonetheless, Amsterdam remains very much the focus of the Netherlands, with a busy port, a bustling commercial center, and a variety of light manufactures. Rotterdam—the world's busiest port (Fig. 1–22)—is the leading gateway to Western Europe, commanding the entries to the Rhine and Maas (Meuse) Rivers, and its development, especially during the past century, mirrors that of the German Ruhr-Rhineland. With its major shipbuilding industry, Rotterdam has importance also in fields other than transportation. The third city in the triangle, The Hague, lies near the North Sea coast but is without a port. It is the seat of the Dutch government, the home of the U.N.'s World Court and, in addition to its administrative and legal functions, it benefits from the tourist trade of the beaches of suburban Scheveningen. Collectively, the three cities of the core triangle form a conurbation called *Randstad*, with their coalescence creating a ring-shaped multimetropolitan complex that surrounds a still rural center (thus, "ring-city"—the literal translation of *rand* is edge or margin). A more precise labeling of the conurbation would be Randstad-Holland, named for the country's western provinces in which it is located; "Holland" is also used interchangeably with "Netherlands" by many of the Dutch (see Box, p. 92).

Whereas these cities of Randstad-Holland stand out among a dozen other urban centers with populations in excess of 100,000, the Netherlands is still less heavily urbanized than Belgium. Hence, the population densities of the rural areas are very high, and practically every square foot of available soil

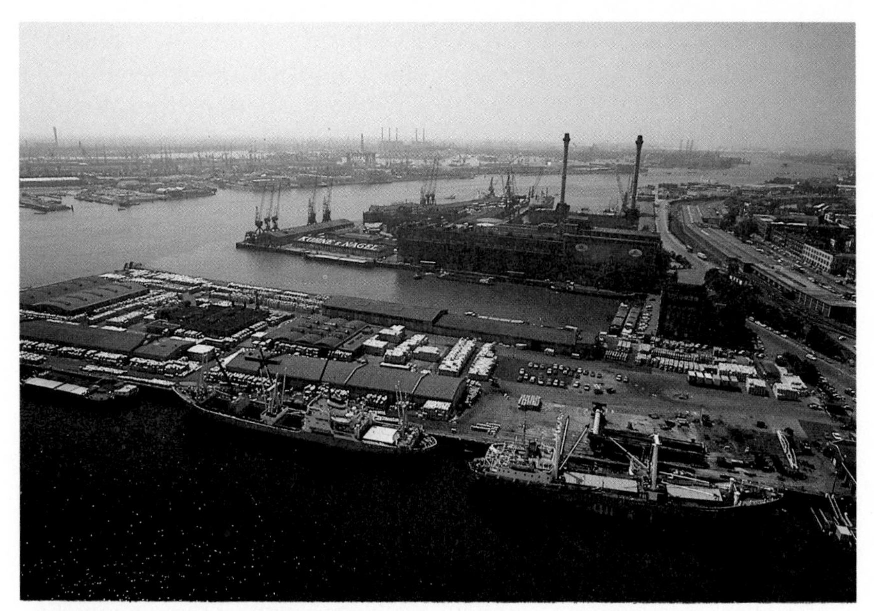

**Figure 1–22**
A panorama of central Rotterdam's port facilities.

is in some form of productive use, even the medians of highways and the embankments of railroad lines. In contrast to Belgium, the local Dutch resource base has always been largely agricultural. Only in the southeastern Dutch province of Limburg, wedged between Belgium and Germany, does the Netherlands share the Campine belt of good coal deposits that extends across northern Belgium. But other than ensuring self-sufficiency in coal for heat and fuel—and some minor industrial development within the province itself—this coal has not been as important to the Netherlands as have adjacent deposits to its neighbors. Not long ago, newly discovered natural gas reserves were opened up in the northeast, and future development of this resource is promising. But all in all, what the Dutch need most is farmland—and this they are adding by their own hands through the continuing reclamation of polders, a practice dating from the Middle Ages.

With the existing limitations of space and raw materials, the Netherlands especially, but Belgium as well, has turned to managing international trade as a leading economic activity. It is hard to think of any other countries that could be better positioned for this. Not only do the Rhine and Maas form primary arteries that begin in the distant interior and terminate in the Low Countries, but Benelux itself is also surrounded by the three most productive countries in Europe.

This is not to suggest that the Low Countries share equally in this great locational advantage. Holland had the better position, and during centuries of competition between Rotterdam and Antwerp, the latter has fared badly. Antwerp can only be reached via the Schelde River, which cuts through the Dutch province of Zeeland. After the Eighty-Year War (1568–1648), the Dutch closed the Schelde, cut off Antwerp, and did not permit the city full access to the oceans until nearly a century and a half later. The Netherlands subsequently turned down Belgian requests for a canal to link Antwerp with the Maas (in part this canal was to be routed through Dutch territory). And repeatedly, the maintenance of the Zeeland portion of the Schelde has been an issue of contention.

## THE NETHERLANDS OR HOLLAND?

The Netherlands is among countries that have two names (the United Kingdom—England; the Soviet Union—Russia) of which only one is technically correct. The Kingdom of the Netherlands has 11 provinces, of which 2 are named North Holland and South Holland. These 2 provinces also are the country's most important, containing the three leading cities, the largest population cluster, most of the heavy industry, and much of the agricultural capacity. North and South Holland form the historic heartland of the Netherlands, and many citizens simply call themselves "Hollanders."

Both names reflect the country's dominant physical property. *Nederland* (the Dutch spelling) means "low" country; *Holland* means "hollow" country—which, given the fact that more than one-third of it lies below sea level, is not an inappropriate name.

But not all Netherlanders approve of the other designation. Even in this small country, there is regionalism. Frieslanders or Groningers of the two northernmost provinces are not likely to call themselves Hollanders. Thus, the correct geographic name, the Netherlands, should be employed when reference is made to the country as a whole; Holland may be used to identify the Dutch heartland that faces the North Sea.

Thus, Antwerp, after an early period of greatness and promise, became mostly a Belgian outlet; Rotterdam forged ahead to a lead it has never yielded. In this respect, Antwerp resembles Amsterdam; Amsterdam, too, has canal and rail connections with the interior, but they are simply not efficient and economical enough to affect Rotterdam's primacy. Hence, Amsterdam, after its early period of greatness as the chief base for Dutch fleets and the center of a growing colonial empire, also gradually yielded its economic position to Rotterdam. Today, Amsterdam's port handles mainly Netherland's trade, whereas Rotterdam specializes in the transit of West German and other international trade.

Belgium's capital city, Brussels, lies in Antwerp's hinterland, connected by river and canal. But Brussels, an agglomeration of 19 municipalities with a combined population of over 1 million, is not a port city of consequence. Instead, this historic royal headquarters, positioned awkwardly astride the Flemish-French (Walloon) dividing line across Belgium (but with dominantly French-speaking residents in the capital city), has become an international administrative center. Hundreds of international corporations with European interests have their main offices here, enhancing the city's role as a financial center and commercial-industrial complex. Moreover, Brussels has become the administrative headquarters for international economic and political associations such as the Common Market (EEC) and its many subsidiaries and the North Atlantic Treaty Organization (NATO).

### Switzerland and Austria

Switzerland and Austria share a landlocked, increasingly peripheral location and the mountainous topography of the Alps—and little else. On the face of it, Austria would seem to have the advantage over its western neighbor; it is twice as large in area and has a substantially larger population than Switzerland. A sizable portion of the upper Danube Valley lies in Austria, and the Danube is to Eastern Europe what the Rhine is to Western Europe and the Volga to the Soviet Union. In addition, no city in Switzerland can boast of a population even half as large as that of the famous Austrian capital, Vienna (1.6 million). Austria also has considerably more land that is relatively flat and cultivable—a prize possession in this part of Europe—and, as Fig. I–12 suggests, more of the remainder of it is under forest, a valuable resource. That is not all. From what is known of the raw materials buried within the Alpine topography, it appears that Austria is again the winner; in the east, where the Alps drop to the lower elevations, iron ores have been found, and elsewhere there are deposits of coal, bauxite, graphite, magnesite, and even some petroleum. Switzerland, for all practical purposes, does not have any exploitable mineral deposits.

From all this, we might infer that Austria should be the leading country in this part of Europe, and that impression would probably be strengthened by a look at some cultural geography. Take, for example, the map of European languages (Fig. 1–13). Austria is a unilingual state in which only one language (German) is spoken throughout the country. In Switzerland, on the other hand, no less than four languages are in use. German is spoken in the largest (northern) part; French over the western quarter; in the extreme southeast, there is an area where Italian dominates; and in the mountains of the central southeast lies a small remnant of Romansch usage. This is hardly a picture of unity, and when we examine a map of religious preferences, we note that somewhat more than half of Switzerland's people are of Protestant orientation and most of the remainder are Catholic—a division similar to that prevailing in the Netherlands—whereas Austria is 90 percent Roman Catholic.

Yet, in the final analysis, it is the Swiss people who have forged for themselves a superior standard of living, and it is the Swiss state, not Austria, that has achieved greater stability, security, and progress. Although poor in raw materials, Switzerland is an industrial country and exports on an average about four times as high a value of manufactured products annually as Austria. Its limited resources amount to the hydroelectric power supplied by mountain streams and the specialized skills of its population; yet over half of all Swiss employment is in industry. Whereas both Austria and Switzerland are clearly Western European countries, a good case can even be made for the conclusion that Switzerland is part and parcel of the European core, whereas Austria is not.

The world map sometimes seems to suggest that mountainous countries share a set of limitations on development that preclude them from joining the "developed" nations. It is very tempting to generalize about the impact of mountainous terrain (and the frequent corollary, *landlocked location*) as preventing productive agriculture, obstructing the flow of raw materials and the discovery of resources, and hampering the dissemination of new ideas and the diffusion of innovations. Tibet, Afghanistan, Ethiopia, Lesotho, and the Andean portions of South American states seem to prove the point. That is why Switzerland is such an important lesson in human geography: all the tangible evidence suggests that here, at last, is a European area that will be stagnant economically and lack internal cohesion—but the real situation is exactly the opposite. The Swiss, through their skills and abilities, have overcome a seemingly restrictive environment; they have made it into an asset that has permitted them to keep pace with industrializing Europe. First, they took advantage of their Alpine passes to act as "middlemen" in interregional trade, then they used the water cascading from those mountains to reduce their dependence on power from imported coal by local hydroelectric means, and, finally, they learned to accommodate those who came to visit the mountain country—the tourists—with professional excellence.

Farmers as well as manufacturers in Switzerland are skilled at getting the most (in terms of value) out of their efforts. The majority of the country's population is concentrated in the central plateau, where the land is at lower elevations. Here lie three of the major cities, Bern (the capital), Zurich (the largest city), and Geneva (the city of international conference headquarters). Here, too, despite the fact that the land on this "plateau" is far from flat, lie most of the farms. The specialization is dairying, for several reasons: (1) the industry produces items that can command a high price, (2) little of the central plateau is suitable for the cultivation of grain crops, and (3) the industry affords an opportunity to use a mountain resource—namely, the Alpine pastures that spring up at higher elevations when the winter snows melt. In the summer, much of the dairy herd is driven up the slopes to these high pastures. Their herders—and sometimes the whole farm family—take up living in cottages built specifically for this purpose near the snow line. With the arrival of autumn the cattle, goats, and their keepers abandon pasture and cottage to descend to the protection of the plateau or intermontane valleys. Swiss farming is famous for this seasonal practice, known as *transhumance*, but it is by no means unique to Switzerland.

Thus, Swiss farmers are engaged in substantially the same activity as their British and Dutch counterparts. Dairy products, cheeses, and chocolate are the chief products, and the last two are exported to a worldwide market. In manufacturing, too, the situation in Switzerland is quite like that in other countries of the European core—except that no other country has to import virtually *all* its raw materials. But again, specialization is the rule and high-value exports the result. Swiss industries attempt to import as few bulky raw materials as possible, although the prices of manufactured items are determined more by the skills that have gone into them than the materials used in them. Precision machinery, instruments, tools, fine watches, and luxury textiles are among the major exports, and the prestigious reputation of Swiss manufactures guarantees them a major place on the world market.

Nevertheless, Switzerland must import over one-third of its food requirements in addition to its industrial raw materials, and without further sources of income, the unfavorable trade balance would have the country in trouble. Instead, there are no problems—thanks to the thriving tourist industry as well as the country's role as an international banker and insurer. The tourist industry, of course, capitalizes on Switzerland's spectacular scenery (Fig. 1–23), but the Swiss have made tourism another field of specialization. In their hotels, lodges, rest homes, and other temporary abodes for visitors, they have set a standard of excellence that is enough to draw millions of tourists each year. And the banking-insurance industry is founded on centuries of confidence inspired by Switzerland's stability, sovereignty, and neutrality.

Austria is a much younger state than Switzerland, and its history is far less stable. Austria at one time lay at the center of the Austro-Hungarian Empire, one of the great concentrations of power in Europe and a major casualty of World War I. Modern Austria, a remnant of the empire, suffered through convulsions reminiscent of Eastern Europe rather than Switzerland's peaceful Alpine isolation. Thus, although Switzerland kept abreast of economic changes in industrializing Europe, Austria was constantly involved in costly power struggles. When modern Austria emerged as a separate entity in 1919, it faced not a time of prosperity but a period of reconstruction and national economic reorganization. And before the country had the opportunity to recover from its misfortunes of the 1920s and 1930s, Austria lost its independence to Nazi Germany, which forced incorporation on it in 1938. More recently, Austria has regained its independence in stages, first in 1945, with the end of World War II (but under Allied occupation), then in 1955, when Soviet and Allied forces withdrew under condition of continued Austrian neutrality.

Since 1945, Austria's most difficult problem has been its reorientation to Western Europe. Even Austria's physical geography seems to demand that the country look eastward: it is at its widest, lowest, and most productive in the east; the Danube flows eastward; and even Vienna lies nearer the eastern perimeter. But the days of domination over Eastern European countries are gone, and the markets there are no longer available. With the interruptions and setbacks of the twentieth century, Austria is not well equipped to catch up with competitive Western Europe.

## Nordic Europe (Norden)

In three directions from its core area, Europe changes quite drasti-

**Figure 1–23**
The Alpine topography of Switzerland.

cally. To the south, Mediterranean Europe is dominated by Greek and Latin influences and by the special habitat produced from the combination of Alpine topography and a "Mediterranean" climatic regime; to the east lies "continental" Europe, which is less industrialized, urbanized, and migratory than the core; and to the north lies the Nordic Europe of Scandinavia, Finland, and Iceland, almost all of it separated by water from what we have here defined as the European core (Fig. 1–24).

Despite its peripheral location, Nordic Europe is not "underdeveloped" Europe. Quite the contrary; in general, a great deal has been achieved in many places without much help from nature. But in terms of resources and in a European context, the northern areas of Europe are not particularly rich, although the discovery and exploitation of the North Sea oilfields have lately added an important dimension to the Norwegian economy.

Scandinavia and Finland (as well as Iceland) possess Europe's most difficult environments. Cold climates, poorly developed soils, steep slopes, and sparse mineral resources mark much of this region, although small Denmark presents better opportunities with its more fertile soils and lower relief. Northern Europe's conditions are reflected by its comparatively small population: its five countries contain a total population (1985) of just 22.7 million, which is less than that of Benelux. The total area, on the other hand, is some 1,257,000 square kilometers (486,000 square miles)—larger than the entire European core, which includes Britain, France, Germany, Benelux, Switzerland, and northern Italy. People go where there is a living to be made; but the living in most of Scandinavia is not easy.

Several aspects of Nordic Europe's location have much to do with this. First, this is the world's northernmost group of states; although the Soviet Union, Canada, and the United States possess lands in similar latitudes, each of these much larger countries has its national core area in a more southerly position. The North Europeans themselves call their region *Norden*, an appropriate term indeed. Second, Norden, as viewed from Western Europe, is on the way to nowhere. How different would the relative location of the Norwegian coast be if important world steamship routes rounded the North Cape and paralleled the shoreline on their way to and from the European core? Norway's ports of Trondheim and Bergen, and the capital of Oslo as well, would be far different places today. Third, all of Norden, except Denmark, is separated by water from the rest of Europe. As we know, water has often been an ally rather than an enemy in the development of Europe—but mostly where it could be used for the *interchange of goods*. Norden lies separated from Europe *and* relatively isolated in its northwestern corner. Denmark and southern Sweden are really extensions of the

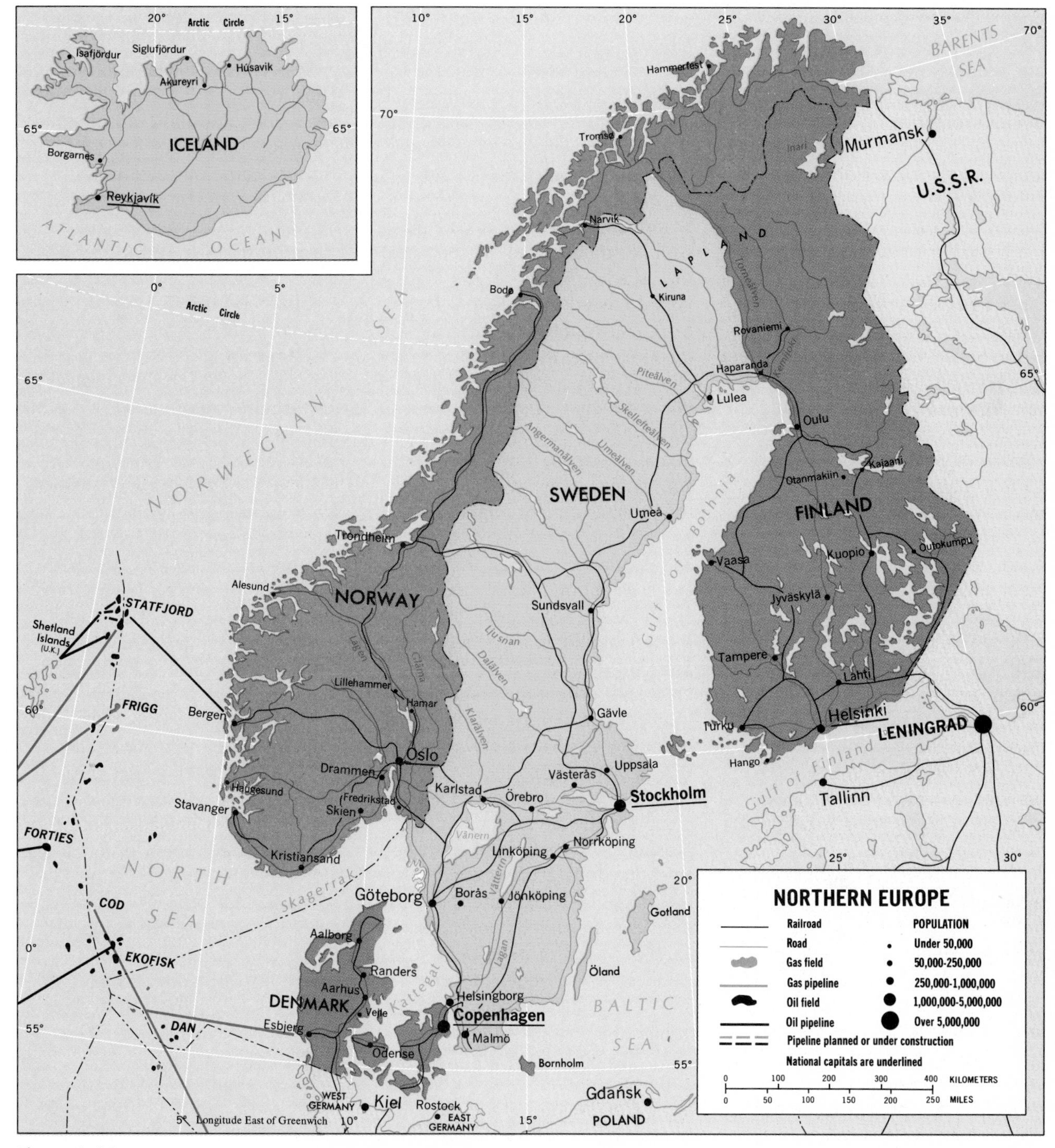


Figure 1–24

North European Lowland and an exception to the bleak Scandinavian rule. Denmark, the most populous Scandinavian country after Sweden, is also the smallest state in the region; but its average population density is over six times that of Sweden. On Denmark's flat and partially reclaimed terrain, an intensive dairy industry has developed—so intensive, in fact, that cattle feed has to be imported in exactly the same way an industrial country imports iron ore. The dairy products produced are largely sold to Europe's greatest food importer, Britain, and across the border in Germany; for these transactions, Denmark is very favorably situated (Fig. 1–24).

Nordic Europe's relative isolation did have some positive consequences as well. Its countries have a great deal in common. They were not repeatedly overrun or invaded by different European groups, as was so much of the rest of the continent. The three major languages—Danish, Swedish, and Norwegian—are mutually intelligible, allowing people to converse with each other without requiring an interpreter. Icelandic belongs to the same language family; only Finnish is of totally different origins—but the long period of contact with Sweden has left a sizable Swedish-speaking resident population in Finland, where this language has achieved official recognition. Furthermore, in each of the Scandinavian countries, there is overwhelming adherence to the same Lutheran church, and in each it is the recognized state religion. Finally, there is considerable similarity in the political evolution of the Scandinavian states and in their socioeconomic policies. Democratic, representative parliaments emerged early, and individual rights have long been carefully protected. This progress was possible, as it was in England, because of the lack of an immediate outside threat to Scandinavia through most of its modern history. The manner in which the common cultural heritage that marks Scandinavia evolved in many ways parallels that of the British Isles.

The five Nordic countries share more than cultural ties. In their northerly location, they also—within a certain range of variation, of course—face common conditions of climate and habitat. The three largest countries (Sweden, Finland, and Norway) all have major concentrations of population in the southern parts of their land area. Although the warm waters of the North Atlantic Ocean help temper the Arctic cold and keep Norway's Atlantic ports open, they become less of an amelioration both northward and landward. Northward, the frigid polar conditions reduce the temperature of the water and thereby its effect on overlying air masses. But most important is Scandinavia's high mountain backbone, which everywhere stands in the way of air moving eastward across Norway and into Sweden. Not only does this high upland limit the milder maritime belt to a narrow strip along the Norwegian coast, but by its elevation it also allows Arctic conditions to penetrate southward into the heart of the Scandinavian Peninsula. This is well illustrated in Fig. I–10, as is the "shadow" effect of the highland on Sweden. Note that Denmark and southern Sweden lie in the temperate, marine-influenced C climatic region and that the remainder of Sweden, despite its peninsular position, has a climate of continental character. One cannot help speculating on what Western Europe's environment might have been if that Scandinavian backbone had continued through the Netherlands, Belgium, France, and into Spain! At the same time, the lack of a physiographic barrier between Scandinavia and Europe has allowed airborne industrial pollution to become a major environmental hazard in southern Norway (see Box, p. 101–102).

As we have seen, only Denmark combines the advantages of temperate climate with land sufficiently level and soils good enough to sustain intensive agriculture. Norway, with its long Atlantic coast, consists almost entirely of mountains whose valley soils have been stripped away by glaciation (Fig. 1–26); only in the southeast (around Oslo), the southwest (south of Bergen), and on the west coast (near Trondheim) are there limited areas of agricultural land and reasonably good soils (apart from tiny patches of bottomland in fjorded valleys near the coast). Less than 4 percent of Norway's area can be cultivated, and even in the small areas that are arable, conditions are far from ideal. Norway is by far the wettest part of Norden and, as we saw in Ireland, with cool temperatures the moisture is soon excessive. So it is in Norway that potatoes and barley generally replace the crops farmers would rather grow—namely, wheat and rye. Thus, much of the farmland lies under fodder, and pastoralism is the chief agricultural activity. Certainly, Norway came off second best in the division of the Scandinavian Peninsula, and it turned to the sea to make up for this. And in the seas, the Norwegians have found considerable profit; the Norwegian fishing industry is one of the world's largest, and fishing fleets from Norway's coasts ply all the oceans of the world. One of the most productive fishing grounds of all happens to lie very close to Norway, partially in its own territorial waters. Here, in the North Atlantic, Norwegian fishermen take large catches of herring, cod, mackerel, and haddock. But the oceans have served the Norwegians in yet another manner. Over the years, Norway has developed a merchant marine that, in tonnage, is the sixth largest in the world, competing with such giants as the British, American, and Japanese fleets. This fleet carries little in the way of Norwegian products; it handles those of other countries, performing transfer functions. Norway, through its location, was otherwise denied. Very recently, Norway's share of the North Sea began to yield undreamed-of wealth in petroleum and natural gas (see Box, p. 103), so that the seas are favoring Norway once again.

Sweden, Norway's eastern neighbor on the Scandinavian Peninsula, is more favored in almost every respect; it may not have Norway's access to the Atlantic, but it needs the Atlantic less. Sweden has two agricultural zones, the leading one lies at the southernmost end of the country, just across the Kattegat Straits from Denmark. In fact, this area resembles Denmark in many ways, including its agricultural development, except that more grain crops, especially wheat, are grown here. Malmö, the area's chief urban center, is Sweden's third largest city with 1 million people; its main function is that of an agricultural service center. Sweden's other agricultural zone lies astride a line drawn southwest from the capital, Stockholm, to Göteborg. Here dairying is the main activity, but farming is overshadowed by industry. Swedish manufacturing, unlike that of some of the Western European countries we have considered, is scattered through dozens of small and medium-sized towns. However, unlike Denmark and Norway, there are resources in Sweden to sustain such industries. For a long time Sweden served in large part as an exporter of raw or semifinished materials to industrializing countries, but increasingly the Swedes are making finished products themselves, specializing much the way the Swiss have done. Already, the list of famous products is quite long; it includes Swedish safety matches, furniture, stainless steel and ball bearings (based on steel produced through electric refining processes), and automobiles. And there is a great deal more, much of it based on relatively small local ores; apart from iron, of which there is a great deal in the area around Kiruna north of the Arctic Circle, there are also copper, lead, zinc, manganese, and even some silver and gold. There are small metallurgical industries at one end of the scale—and the huge shipbuilding works at Göteborg at the other. In electronics, engineering, glassware, and textiles, Sweden shows, once again, that the skill and expertise of the people is as important as the resource base itself.

All three of the larger Nordic countries possess extensive forest resources and all three have exported large quantities of pulp and paper. At first the mills used direct water power, but with the advent of steam power, they relocated at

# ACID RAIN

One of the most discussed environmental perils of recent years is *acid rain*. This is caused by the considerable quantities of sulfur dioxide and nitrogen oxides released into the atmosphere as fossil fuels (coal, oil, and natural gas) are burned. These pollutants combine with water vapor contained in the air to form dilute solutions of sulfuric and nitric acids, which are subsequently washed out of the atmosphere by rain or other types of precipitation such as fog and snow.

Although acid rain usually consists of relatively mild acids, they are sufficiently caustic to do great harm over time to certain natural *ecosystems* (the mutual interactions between groups of plant or animal organisms and their environment). Already, there is much evidence that this deposition of acid is causing lakes and streams to acidify (with resultant fish kills), forests to become stunted in their growth, and acid-sensitive crops to die in affected areas; in cities the corrosion of buildings and monuments is both exacerbated and accelerated. To some extent, acid rain has always been present in certain humid environments, originating from such natural events as volcanic eruptions, forest fires, and even the bacterial decomposition of dead organisms. However, during the past century as the worldwide industrial revolution has spread ever more widely, the destructive capabilities of natural acid rain have been greatly enhanced by human actions.

The geography of acid rain occurrence is most closely associated with patterns of industrial concentration and middle- to long-distance wind flows. The highest densities of coal- and oil-burning are associ-

Figure 1–25

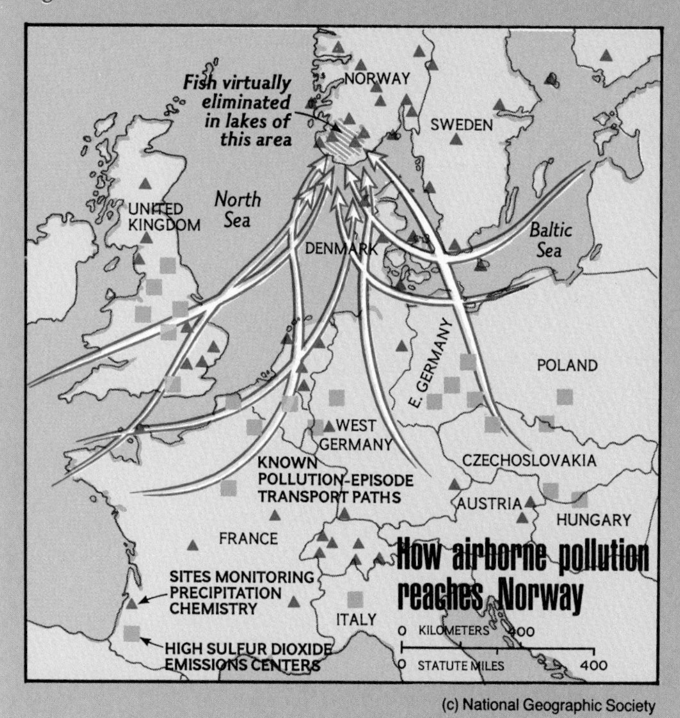

ated with large concentrations of heavy manufacturing such as those already discussed for Britain and Western Europe. As these industrial areas began to experience increasingly severe air pollution problems in the second half of the twentieth century, many nations (including the United States in 1970) enacted environmental legislation to establish minimal clean-air standards for the first time. For industry, the easiest solution has often been the construction of very high smokestacks (1000 feet [300 meters] or higher is now quite common) in order to disperse pollutants away from source areas through the use of higher-level winds. These longer-distance winds have been effective as transporters, with the result, of course, that more distant areas have become the dumping grounds for sulfur and nitrogen oxide wastes. Regional wind flows all too frequently steer these acid rain ingredients to wilderness areas, where livelihoods depend heavily on tourism, agriculture, fishing, and forestry; where international borders are crossed by such airborne pollution, political problems develop (notably between the United States and Canada) and can be expected to intensify in the future.

The spatial distribution of acid rain within Europe offers a classical demonstration of all of this. As Fig. 1–25 shows, high-sulfur-emission sources are located in the major manufacturing complexes of England, France, Belgium, the Germanys, Poland, and Czechoslovakia. The map also indicates that, most unfortunately, prevailing wind flows from these areas all converge northward to Scandinavia, with a particularly severe acid rain crisis occurring in southern Norway. Lake acidity there is already in the moderately caustic 4-to-5 range on the pH scale of 0 to 14 (7 is neutral), and most fish species, the phytoplankton they feed on, and numerous aquatic plants have been decimated.

**Figure 1–26**
A glacial trough on Norway's fjorded coastline.

river mouths and used, in part, their own waste as fuel. No Nordic country depends to a greater extent on its forest resources than does Finland, whose wood and wood-product exports normally account for between two-thirds and three-quarters of all annual export revenues. In this respect, the area long resembled the colonial dependencies of Western European powers, underscoring their "underdeveloped" position compared to raw material-consuming Western Europe. But even logs and timber can be transformed into something specialized before export: the Norwegians make much high-quality paper; the Swedes, matches and furniture; and the Finns plywood, multiplex veneers, and even prefabricated cottages. Not all the forest resources are thus transformed, of course, and Finland, like Sweden and Norway, continues to export much pulp and low-grade paper. In Finland, however, the alternatives are fewer. Most of the country is too cold and its glacial soils too thin to sustain permanent agriculture. Where farming is possible, mostly along the warmer coasts of the south and southwest, the objective is self-sufficiency rather than export. Known mineral deposits are few; some copper lies at Outokumpu and iron ore at Otanmakiin. Nevertheless, the Finns have succeeded in translating their limited opportunities into a healthy economic situation, in which, once again, the skills of the population play a major role. The country is nearly self-sufficient in its farm products, and the domestic market sustains a textile industry (centered at Tampere) and metal industries for locally needed machinery and implements at Turku and the capital, Helsinki. For these and for the shipyards at Helsinki and Turku, raw materials must, of course, be imported.

The westernmost of Norden's countries—and the most westerly

## EKOFISK! STATFJORD!

The electrifying news came in 1970: exploration by the United States Phillips Petroleum Company had proved the existence of a large oil and natural gas field in the continental shelf under the North Sea—and in the sector allocated to Norway. By the late 1960s, Norway had been consuming some 8 million tons of oil annually, but the new Ekofisk source promised to produce at least 15 million tons per year. Norway could become Europe's first petroleum-exporting country.

Recent years have proved that initial estimates had been low, although technical difficulties and exceptionally bad weather in the North Sea delayed full production. An oil pipeline from Ekofisk to Teeside, England, was completed in 1974. A gas pipeline to Emden in West Germany was laid in 1975, and in 1976 Norway became, for the first time in its history, an exporter of energy resources. Norway's own petrochemical industry thrived. And then, in the past few years, the full dimensions of the Statfjord field have become clear: reserves here are larger even than those at Ekofisk (Fig. 1–24). In the Barents Sea, north of the Scandinavian Peninsula, Norway has a maritime boundary with the Soviet Union—and here, it now appears, lies still another major reserve. But the Norway-Soviet boundary in the Barents Sea is under dispute, and the political difficulties must be resolved before exploitation can begin. Europe's historical marginal seas are now becoming a lifeline of a different sort.

state of Europe—is Iceland, a hunk of basaltic rock that emerges above the surface of the frigid waters of the North Atlantic just south of the Arctic Circle. Three-quarters of Iceland's 103,000 square kilometers (40,000 square miles) are barren and treeless; one-eighth of the island's rough and mountainous terrain is covered by imposing glaciers. Below, the earth still rumbles with earthquakes and hot springs are numerous. Volcanic eruptions have left their mark on the country's past and have claimed thousands of lives.

Iceland's population has the dimensions of a microstate (241,000), and nearly half this population is concentrated in the capital, Reykjavik. Iceland shares with Scandinavia its difficulties of terrain and climate (even more severe here), and it also has ethnic affinities with continental Norden. The majority of the earliest settlers came from Norway during the ninth century, and the Icelanders claim that theirs was Europe's first constitutional democracy; in 930 A.D., a parliament was elected. Later, Iceland fell under the sway of Norway and then Denmark; it was not until 1918 that it regained its autonomy. In 1944, all remaining ties with the Danish crown were renounced, and the republic was finally resurrected.

The possibilities for agriculture are severely restricted in Iceland, and so the country has turned to the sea for its survival. More than 80 percent of Iceland's exports today consist of fish products, but Iceland has had its problems protecting its nearby fishing grounds. To avoid overfishing, Iceland in 1972 announced that it would claim exclusive rights in a zone 93 kilometers (50 nautical miles) wide around the island. This led to confrontations at sea between Icelandic fishing boats and British fishing boats protected by naval vessels; eventually, a treaty was signed that permitted British trawlers inside the 93-kilometer zone in certain areas. Then, in 1975, Iceland extended its claim to 370 kilometers (200 nautical miles) from its coasts, and the agreement with the United Kingdom expired. Again, the North Atlantic was the scene of angry challenges. The Icelanders point to Britain's diversified economy and argue that they have no choice: the preservation of their fishing grounds is a matter of national survival. To Britain and Germany, those fish catches are merely another minor element in a much wider economy.

Denmark is by far the smallest of the countries of Northern Europe, but its population of 5.1 million is the second largest. Consisting of the Jutland Peninsula and numerous adjacent islands between Scandinavia and Western Europe, Denmark has a comparatively mild, moist climate; more than 70 percent of its area is devoted to agriculture. Denmark exports dairy products, meats, chickens, and eggs, mainly to its chief trading partners, the United Kingdom and West Germany. It has been estimated that Denmark's farm production could feed some 17 million people annually.

Denmark's capital, Copenhagen (1.4 million) is also Northern Europe's largest urban center. It has long been the place where large quantities of goods are collected, stored, transhipped, and dispatched, because it lies at the point where many large oceangoing vessels must stop and unload; they cannot enter the shallow Baltic Sea. Conversely, ships with smaller tonnages ply the Baltic and take their cargoes to Copenhagen's collecting stations. Thus, Copenhagen is an *entrepôt*, and its break-of-bulk functions have long maintained the city's position as the lower Baltic's leading port.

Like Copenhagen, Stockholm (1.1 million) and Oslo (0.7 million) are ancient cities, centers of old national states, and the primate cities of their countries today. It is a measure of Copenhagen's primacy that Denmark's second city, Aarhus, has fewer than 250,000 inhabitants. Stockholm, which has lost its position as Sweden's first port to Göteborg, still is three times as large as its competitor.

Nordic Europe has overcome its resource limitations through ingenuity and the pursuit of diversification. Long before oil was found beneath the North Sea, Scandinavia had secured energy from its hydroelectric power sources. Iceland made its mid-Atlantic location an asset and turned strategic position into revenues. The sea became Norway's avenue to prosperity. But problems persist: Denmark's productive economy is unbalanced and Finland's dependence on its forest resources is excessive. Still, a great deal of actual and potential complementarity and interdependence exists among the Nordic countries. Norway, with its merchant marine, buys tankers built in the shipyards of Göteborg. Sweden and Norway import Danish meat and dairy products. Denmark needs timber, paper, pulp, and associated products and can import them from Sweden, Finland, and Norway. Norway, best suited of all to cheaply produce hydroelectric power, could supply this vital commodity to Denmark (which needs it most) and Sweden. Norway, in turn, can use equipment and machinery built in Denmark, Sweden, and Finland. The opportunities are many; regional associations are already strong, and they will undoubtedly grow stronger.

## Mediterranean Europe

From Northern Europe, we turn to the four countries of the south: Greece, Italy, Spain, and Portugal (Fig. 1–27). From near-polar Europe, we now look into near-tropical Europe; it is reasonable to expect strong contrasts, and there are many. But there are also similarities, and some very telling ones. Once again, we are dealing with peninsulas—three of them. This time, two are occupied singly by Greece and Italy, and one jointly by Spain and Portugal (the Iberian Peninsula). And once again, there is effective separation from the Western European core. Greece lies at the southern end of Eastern Europe and has the sea and the Balkans between it and the west; Iberia lies separated from France by the Pyrenees, which through history has proved to be a major barrier; and then there is Italy. Southern Italy lies far removed from Western Europe, but the north is situated very close to it and presses against France, Switzerland, and Austria. For many centuries, northern Italy was in close contact with Western Europe and developed less as a Mediterranean area than as a part of the core area of Western Europe. In a very general way, northern Italy is as much an exception to Mediterranean Europe as Denmark is to Scandinavian Europe, and these two areas happen to lie opposite one another across Germany and Switzerland in physical as well as functional contact with the European core.

The Scandinavian countries share a common cultural heritage, and so do the countries of Mediterranean Europe. Firm interconnections were established early by the Greeks and Romans. Under the Greeks, ports as far away as Marseilles and Alexandria formed part of an integrated trading system; under the Romans, virtually the whole Mediterranean region was endowed with similar cultural attributes. This unity, as we know, did not last. New political arrangements re-

Figure 1–27

placed the old, and the Romanic language differentiated into Portuguese, Spanish, and Italian. In Greece, the Roman tide was resisted to a surprising extent, but the underlying shared legacy remains strong to this day.

Like Nordic Europe, Mediterranean Europe lies largely within a single climatic region, and from one end to the other, the opportunities and problems created by this feature of the environment are similar. The opportunities lie in the warmth of the near-tropical location, and the problems are largely related to the moisture supply, especially its quantity and the seasonality of its arrival. In topography and relief, too, Mediterranean countries share similar conditions—conditions that would look quite familiar to a Norwegian or Swede. Much of Mediterranean Europe is mountainous or upland country with excessively steep slopes and poor, thin, rocky soils. For its agricultural productivity, the region largely depends on river basins and valleys and coastal lowlands. As with practically every rule, there are exceptions, as in northern Italy and northern and interior Spain, but generally the typical Mediterranean environment prevails.

Neither is Mediterranean Europe much better endowed with mineral resources than Scandinavia. Both Greece and Italy are deficient in coal and iron ore; for many years, Italy, industrializing despite this shortcoming, has been one of the world's leading coal importers. Recently, Italy's fuel position has improved somewhat through the exploitation of minor oilfields in Sicily and by the use of natural gas found beneath the Po Valley, but these are far from enough to satisfy domestic needs. Only in Spain, in the Cantabrian Mountains of the north, are there really sizable fields of coal and iron ore positioned close to each other. But Spain, best endowed with raw materials of all Mediterranean countries, has not responded with manufacturing industries on the scale of Western Europe. On the contrary, a good deal of the highest-grade iron ore and coking coal have been exported as though Spain were destined to be an underdeveloped, raw material-supplying country rather than an industrial power. In this respect, Spain only mirrors the whole Mediterranean region; there are scattered mineral deposits, some of them valuable and capable of sustaining local, skill-dependent metallurgical industries, chemical industries, and other enterprises—instead, almost all these raw materials are exported to Europe's core region. Scandinavia and Mediterranean Europe may share resource poverty, but Scandinavia gets the most out of what it has and Mediterranean Europe often the least.

This is, to a lesser extent, the case with both Scandinavian and

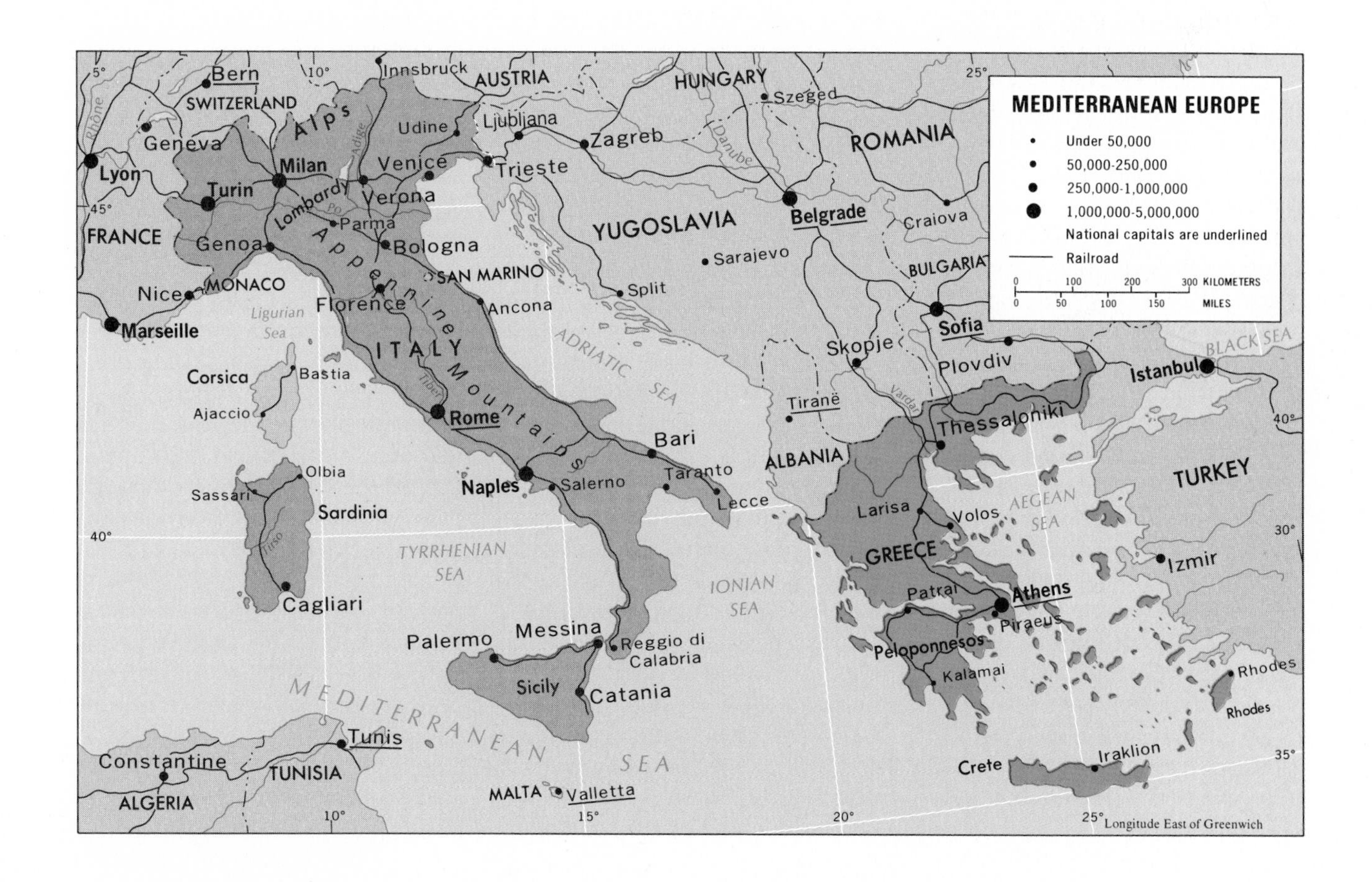

Mediterranean Europe's answer to the fuel shortage: hydroelectric power. Like Norway, Italy is one of the world's leading states in the development of its hydroelectric-power potential; in the 45 years since 1939, the country has nearly quintupled its output, further reducing its dependence on imported sources of power. But Italy is most favorably endowed in this respect (much more so than either Greece or Iberia): its hydroelectric production is about double that of its three Mediterranean partners combined. The Mediterranean environment, for all its mountains and uplands, is not all that advantageous for hydroelectric-power development; with its seasonal and rather low precipitation, there are frequent water shortages when streams fall dry and water levels behind dams go down. In northern Italy, where the largest market is located, conditions are better: rainfall distribution is much less seasonal and water supply from the nearby Alpine ranges is much more dependable.

Some centuries ago, Scandinavians and southern Europeans would have recognized another area of similarity between their environments: the prevalence of an extensive forest cover, a valuable resource in preindustrial as well as modern times. The Scandinavians still have it, and their export revenues include a sizable percentage derived from these forests—but Mediterranean Europe stands largely bare and denuded. Trees were cut down for centuries of housebuilding, shipbuilding, fuel utilization, and to make way for agriculture. In the period of Spain's greatness, the forests of the Meseta Plateau constituted one of Iberia's major assets. Today, this area stands windswept and treeless.

Mediterranean Europe's four countries in 1985 had a combined population of 115 million, notwithstanding enormous emigration, especially from Spain and Portugal, during the colonial period. From what we know of Mediterranean dependence on agriculture, the historical geography of trade routes and urban growth, and the general topography of Mediterranean Europe, we can fairly well predict the distribution of population. In Italy, nearly 25 million of the country's 57 million people are concentrated in and near the basin of the Po River, and elsewhere heavy concentrations also exist in coastal lowlands and riverine basins all the way from Genoa to Sicily. Thus, the least heavily peopled part of the peninsula is the Appennine mountain chain (together with the Alpine rim in the far north), but the eastern and western flanks of the Appennines are heavily populated. In Greece, today as in ancient times, densely settled coastal lowlands are separated from each other by relatively empty and sometimes barren highlands. Especially dense concentrations occur in the lowland dominated by Athens and on the western side of the Peloponnesus Peninsula. Both Spain and Portugal also contain heavy settlement on coastal lowlands, although the interior Meseta is somewhat more hospitable than Greece's rocky uplands. The Mediterranean shorelands around Barcelona and Valencia have attracted a high population total, and Barcelona stands at the center of Spain's leading industrial area (Catalonia). The Basque provinces from Bilbao to the Pyrenees in north-central Spain are a major manufacturing area, especially for metal and machine industries. Spain's other population agglomerations have developed in the northwest near the mining areas of the Cantabrian Mountains; in the south, where the broad lowland of the Guadalquivir River opens into the Atlantic and where the Huelva-Cadiz-Córdoba triangle incorporates a sizable urban and rural population; and in the center of the country, in and near—especially southwest of—the capital of Madrid. In Portugal, which possesses Atlantic but no Mediterranean coasts, the majority of the population is nevertheless located on the coastal lowlands rather than on the Iberian Plateau; Lisbon and Porto form the leading centers for these coastal concentrations.

Thus, Mediterranean population distribution is marked by a dominant peripheral location, by a heavy clustering of high concentrations and great densities in productive areas, usually coastal and riverine lowlands, and by a varying degree of isolation on the part of these clusters. Spatial interaction between central and southern Greece, between the east and west coasts of Italy, and between Atlantic and Mediterranean Spain is, therefore, not always effective. Terrain and distance are the obstacles. Although it is difficult to say exactly what constitutes overpopulation, there obviously is excessive population pressure on the land and resources in many parts of the Mediterranean Basin. Other than Ireland, no country in Europe has sent a greater share of its people to overseas realms than has Portugal; standards of living here, in Greece, and to a lesser extent in Spain and Italy are quite low.

Perhaps the sharpest contrast between Scandinavian and Mediterranean Europe lies in the living standards of the people. Whereas economic specialization and limited population growth, along with generally enlightened government policies and attempts to ensure a fairly equitable distribution of wealth, have produced standards of living in most of Scandinavia that are comparable and even superior to those prevailing in the European core, much of Mediterranean Europe seems to lag well behind. Greece is perhaps the least favorably endowed of the four countries: less than a third of its area is presently capable of supporting some form of cultivation, and on this land the average density of population is around 300 persons per square kilometer (800 persons per square mile)—less than an acre per person. Thus, many of Greece's farmers are engaged in near-subsistence agriculture, their income is low, and their ability to buy improved farm equipment or fertilizer is minimal. Water supply is an ever-present problem, and the capital required to enhance it through irrigation is very scarce. Yet, agriculture is Greece's mainstay, for industrial opportunities are few; in fact, Greek workers by the thousands leave each year to seek work in the industrial centers of Western Europe. Greek farmers, where conditions permit, raise crops such as wheat and corn for the home market and tobacco, cotton, and typical Mediterranean produce such as olives, grapes, citrus fruits, and figs for exporting. Greece ranks third among the world's leading exporters of olive oil (after Spain and Italy), and in some years nearly half its export revenues come from tobacco.

If this list of Greek agricultural products has an unusual ring to it, it should be remembered that only the Peloponnesus really constitutes Greece's Mediterranean zone from which the Mediterranean crops are derived, whereas northward Greece takes on more continental characteristics—including more field and less garden-type agriculture. Eastward, the two areas meet in a point at Athens, whose metropolitan area (including the port of Piraeus) counts some 3.5 million inhabitants. Athens is the administrative, commercial, financial, cultural, and, indeed, the historic focus of Greece, and, despite its limited industry, it has grown to a size far beyond what would seem reasonable for such a relatively poor and agrarian country. With Piraeus it stands at the head of the Aegean Sea, and Athens also has a large and busy international airport. With its heritage of ancient structures still a direct reminder of past glory, Athens has become one of the Mediterranean's major tourist attractions, and, thus, a source of much-needed revenues; unfortunately, the Greek capital experiences a serious air pollution problem, which now poses a grave threat to the fragile monuments of the ancient civilization. Another source of income is one resembling that developed by Norway—a large, worldwide merchant-marine fleet competing for cargoes wherever and whenever they need to be hauled.

At the other end of the Mediterranean, the Iberian Peninsula is less restrictive in the opportunities it presents for development. Iberia is much larger than Greece, and proportionately less of it is as barren and rocky as the Greek land. Also, raw materials are in far more plentiful supply, but these opportunities have hardly been put to maximum use. The countryside is overpopulated, and one price Spain has paid for its slow industrial development was that its population "explosion" had to be accommodated largely in the rural areas, where pressures already were high. Land was divided and subdivided, farms grew smaller and smaller and less and less efficient; poorer soils were utilized for farming, even though they were marginal, and their productivity was bound to be low. Most of northern Spain is fragmented into these tiny parcels. The situation is worst in Galicia, the country's northwest corner, where farms have been subdivided beyond the minimum level of reasonable economic efficiency.

These, then, are some of the reasons why Southern Europe's per hectare (1 hectare = 2.47 acres) yields are always so much lower than those of Western Europe—60 percent lower, on the average—and why so many farmers are caught up in a cycle of poverty. Another reason lies in the division of land, something that is less of a problem in Greece, where most of the land is already held in small private holdings. In both Spain and Portugal, problems of land ownership have impeded agricultural development, because huge estates known as *latifundias* remain in the hands of (often absentee) landowners, but are farmed by tenants. Land reform has been resisted by the conservative estate owners, although a program of reallocation was begun in Portugal following the political upheavals of the 1970s. In Spain, there has been political reform since the end of the long Franco regime in 1975, and agrarian reform is beginning.

Spain's major industrial area, as we saw, is located in the northeast in Catalonia, not on the coast of the Bay of Biscay along the mineral-rich Cantabrian ranges. Thus, Cantabrian coal would have to be shipped all the way around the Iberian Peninsula to provide power for Catalonian industries; that being the case, it might as well be imported from outside Spain, indeed, much coal is introduced from abroad. No, it is not the favorable location or the rich local resource base that has stimulated industrialization in Catalonia, rather, it seems in the first instance to be the different attitude and outlook of the Catalans that has produced this development. Vigorous and progressive, these people have forged ahead of the rest of Spain, aided by a strong regional identity in the form of a distinct language, Catalan, which differs from Castilian Spanish, the language used in most of Spain. Moreover, Barcelona is a major urban-cultural focus that always competes aggressively with Madrid and nurtures a certain local Catalonian nationalism.

Catalonia does what many industrializing although resource-poor areas do—it imports most of its raw materials and depends on a few local assets for success. Among the assets are the hydroelectric power available from the streams flowing off the Pyrenees, the local labor force and its skills, and the local market, comparatively poor as it may be. Although it has had considerable success, it cannot be counted among Europe's leading industrialized areas. Unlike Italy's Po Valley, Catalonia never stimulated effective trading ties across the mountains to the north. Compared to the Midlands of England, there is less diversification here, although the last decade has seen considerable expansion; most of the industrial establishments still produce either textiles (mostly cotton goods) or chemicals. In large measure, of course, this reflects the very limited capacity of the Spanish market, with its millions of poor families and its underpaid labor. But by going against the trend in reluctant Spain, the Catalonian achievement is a major one, and it reminds us how people through their determination and skills can transform the economic map.

Very slowly, Spain has recently been altering its course—all the while falling further behind the accelerating development of the European core. In the northwest, local iron and steel production is increasing, as is coal production. The hydroelectric-power output has multiplied. But the general situation has changed little; there is a characteristically underdeveloped economy—exporting a number of untreated raw materials and importing a wide variety of foods and consumer goods. The exports reveal Spain's varied resources: in addition to iron ore and coal, they include copper, zinc, lead, potash, mercury (Spain is the world's leading producer of this element); the agricultural exports sound more familiar, including olive oil from Andalusia in the south, citrus fruits from the eastern coastal zone around Valencia, and wines from the Ebro Valley. At the center of it all, at the foot of the Guadarrama Range, stands Madrid, capital since the sixteenth century and still by far the dominant city of the entire Iberian Plateau. Chosen because of its position of centrality on the Iberian Peninsula, Madrid (5.1 million) mirrors the current problems of Spain; a facade of splendor hides large areas of severe urban blight, just as the tourist-admired beauty of Iberia conceals a great need for further social reform.

Much of what has been said concerning Mediterranean habitats and economies applies to southern Italy and the islands of Sicily and Sardinia. But, in a way, Italy is not one country—it is two. Whereas the north has had the opportunities and advantages (including those of proximity to the European core) to sustain development in the style of Western Europe, the south has for centuries been a laggard and stagnant region. Although the north has metropolitanized and now counts several cities with over a half-million people and many more with over 100,000, the south contains only one major city—Naples (4.1 million), undoubtedly the poorest of Italy's large cities, with staggering urban blight. Together, north and south count over 57 million inhabitants (more than Spain, Portugal, and Greece combined), bound by Rome, situated, fortuitously from this point of view, in the transition zone between the two contrasting regions. In every way, Italy is Mediterranean Europe's leading state—in the permanence of its contributions to Western culture, in the productivity of its agriculture and industries, in the percentage of its people engaged in manufacturing, and in living standards. The full richness of the Mediterranean cultural landscape is most evident here too, captured quintessentially in the red-tiled roofs and narrow streets of the historic city of Florence (Fig. 1–28).

Today, the focus of Italy has shifted from where it was during Roman times: Latium and Rome no longer form the peninsula's center of gravity. True, government and church are still headquartered in the historic capital and its adjunct, Vatican City; but in terms of population totals, Milan, the northern Po Valley industrial rival, has established a clear lead. However, Rome (3.8 million) is not likely to lose its special position in Italy and the world: it was chosen for symbolic reasons to be the new Italy's capital (in 1870) and, no doubt, would be chosen today if the choice had to be made again. Nonetheless, Italy's core area, certainly in economic terms, has moved north into the area called Lombardy, centering on the valley of the Po River.

Northern Italy has a number of advantages. The Appennines, which form the backbone of the Italian Peninsula, bend westward, leaving the largest contiguous low-lying area in the Mediterranean between it and the Alps. This area, narrow in the west, where the Alps and Appennines meet near the city of Turin, opens eastward to a wide and poorly drained coastal plain on the Adriatic Sea. Here lies one of the great centers of medieval Europe, Venice, Italy's third largest port and a city that still carries the imprint of the splendor brought by that early age. As the climatic map (Fig. I–10) shows, this area has an almost wholly non-Mediterranean regime, with a much more even rainfall distribution throughout the year. Certainly, the Po Valley has great agricultural advantages, but what marks the region today is the greatest development of manufacturing in Mediterranean Europe. It is all a legacy of the early period of contact with Flanders and the development of transalpine routes; when the stimulus of the industrial revolution came, the old exchange was vigorously renewed. As we have seen, hydroelectric power

**Figure 1–28**
The townscape of Florence, Italy.

from Alpine and Appennine slopes is the only local resource, other than a large and skilled labor force. But northern Italy imports large quantities of iron ore and coal and today ranks as Europe's fourth largest steel producer, after France—although its production is less than half that of France. The iron and steel are put to a variety of uses. Competing for the first position among the industries are metals and textiles, the latter enjoying the benefit of a much longer history, dating back to the days of glory during the Middle Ages. The metal industries are led by the manufacturing of automobiles, for which Turin is the chief center. Italy is famous for high-quality automobiles; other metal products such as typewriters, sewing machines, and bicycles are also produced. Italian industry seeks to create precision equipment, in which a minimum of metal and a maximum of skill produce the desired products and revenues. Italy also has an impressive shipbuilding industry at Genoa on the Ligurian coast; with 1.2 million people, Genoa is Italy's leading port, located as it is on the side of the peninsula closest to the Atlantic at the head of the Gulf of Genoa.

The principal city in the northern region is Milan (7 million), located in the heart of Lombardy and the leading industrial center in Mediterranean Europe. This is Italy's financial and manufacturing headquarters; although it has seen unprecedented growth in recent years, it, too, has its roots in an earlier age of greatness. This period is still visible in the urban landscape, with its impressive public buildings, palaces, churches, cathedrals and the La Scala Opera House; but now looming above these are the modern multistoried office buildings that house the offices of Italian industrial firms. No city in Italy rivals the range of industries based here—from farm equipment to television sets, from fine silk (Milan competes with Lyons in this field) to medicines, from chinaware to shoes. The Milan-Turin-Genoa triangle is Italy's industrial heart, and it is the center of a larger integrated region that forms part of the greater Western European core area (see Fig. 1–11).

## Eastern Europe

Between the might of the Soviet Union and the wealth of industrialized Western Europe, lies a region of transition and fragmentation—Eastern Europe (Fig. 1–29). Its position with reference to the major cultural influences in this part of the world at once explains a great deal: to the west lie Germanic and Latin cultures, politically repre-

Figure 1–29

sented by Germany and Italy, and to the east looms the Slavic culture realm that is Russia's. And to the south lie Greece and Turkey, whose impact has also been felt strongly in Eastern Europe; the Byzantine and Ottoman Empires took in sizable portions of this ever-unstable region.

As here defined, then, Eastern Europe consists of seven states: Poland, the largest in every way; Czechoslovakia and Hungary, both landlocked; Romania and Bulgaria, facing the Black Sea; and Yugoslavia and Albania, political mavericks that possess Adriatic coastlines. These countries form the fifth and easternmost regional unit of Europe, and their eastern boundaries also form the eastern limit of Europe itself. This is Europe at its most continental, most agrarian, and most static. Overall, it was not as well endowed as Western Europe or the Soviet Union with the essentials for industrialization, and it was the most remote of all parts of Europe from the sources of those innovations that brought about the industrial revolution. Since World War II, Eastern Europe has looked eastward rather than westward for directions in its political and economic development. The Soviet Union gained hegemony here in the mid 1940s—thanks to the presence of the Red Army as well as the cooperation of local Communist parties—and Communist forms of totalitarian control, resource utilization, and political organization were introduced. Although the countries of Eastern Europe were not remade into Soviet Socialist Republics (as were the former Baltic States of Estonia, Latvia, and Lithuania), they did become satellites of the Soviet Union and were drawn completely into the Soviet economic and military sphere.

But nothing has ever succeeded in unifying Eastern Europe, and it is doubtful that even Soviet power can do it. As early as 1948, one of the satellites, Yugoslavia, began to move away from the Russian course. An uprising was suppressed by invading Soviet armed forces in Hungary in 1956, and similar force brutally put down Czechoslovakia's "springtime of freedom" reforms in 1968. Worker disturbances in Poland's industrial regions wrested certain political reforms from the authorities in the 1960s and 1970s, and the impact of the dramatic *Solidarity* union movement in 1980–1981 and subsequent imposition of martial law continues to be felt there (Fig. 1–30). Even Albania, Eastern Europe's smallest state, has deviated and for a time aligned itself with Maoist China. The pendulum of power has swung across Eastern Europe many times, and it is likely to do so again.

The boundary framework of Eastern Europe today has evolved from the 1919 Paris Peace Conference, where a new Eastern European map was drawn following World War I. But the boundaries with which Eastern Europe was endowed in 1919 did not eliminate the internal ethnic problems of the region. In fact, it really was impossible to arrive at any set of boundaries that would totally satisfy all the peoples involved; so intricate is the ethnic patchwork that different social groups simply had to be joined together. Hence, people who had affinities with each other were also separated by the new international borders, and every country in Eastern Europe as constituted in 1919 found itself with sizable minorities to govern. Frequently, these minorities were located near boundaries, and adjacent states began to call for a transfer of their authority, on ethnic, historical, or some other grounds. For instance, Transylvania, the eastern part of the Hungarian Basin, was severed from Hungary

## BALKANIZATION

The southern half of what has been defined here as Eastern Europe is sometimes referred to as the Balkans, or the Balkan Peninsula (Fig. 1–29). This refers to the triangular landmass whose points are at the tip of the Greek Peloponnesus, the head of the Adriatic Sea, and the northwestern corner of the Black Sea. The name comes from a mountain range in the southeasterly part of this triangle, but it has become more than merely a name for this area—it has become a concept, born of the reputation for division and fragmentation of this whole Eastern European region. Any good dictionary of the English language will carry a definition for the terms *balkanize* and *balkanization*—for example, "to break up (as a region) into smaller and often hostile units," to quote Webster. Certainly division and hostility have been outstanding characteristics of Eastern Europe. Its peoples—Poles, Czechs, Slovaks, Magyars, Bulgars, Romanians, Slovenes, Croats, Serbs, and Albanians—have often fought each other; at other times, they have been in conflict with forces from outside the region. Various empires have tried to incorporate all or parts of Eastern Europe in their domains; none was completely successful, and the most recent effort, by the Soviet Union, is not likely to succeed either. Countless boundary shifts have taken place: time and again local minorities rose in revolt against their rulers. Culturally, Eastern Europe is divided by strong regional differences, expressed in language and religion, and sustained by intense nationalisms. Among these, none has emerged with sufficient strength to impose unification on the region; the nearest thing to an "indigenous" empire was that of Austria-Hungary (1867–1918).

Balkanized Eastern Europe is a region of endless contrast and division, with Slavic and non-Slavic peoples, Roman Catholic, Eastern Orthodox, and even Moslem religions, and with mutually unintelligible languages derived from Slavic, Romanic, and Asian roots. History has often seen peoples with common ties separated by political boundaries, and peoples with few similarities thrown together in a single state. However, in a world in which boundary disputes and struggles for territory have long been commonplace, the case of Eastern Europe has been sufficiently outstanding to have given its name to this phenomenon in political geography: **Balkanization**.

**Figure 1–30**
A 1980 *Solidarity* demonstration in Warsaw.

and attached to Romania—but Hungary openly laid claim to it. Macedonia had been divided between Yugoslavia and Greece, but both Bulgaria and Albania wanted parts of it too. Between Yugoslavia and Italy, the Trieste area known as the Julian March became the scene of territorial competition. These are just a few instances, and there were several more—not to mention German claims on Poland and Czechoslovakia and Soviet claims on Romania.

When a certain state, through appeals to a regionally concentrated minority in an adjacent state, seeks to acquire the people and territory on the other side of its boundary, its action is termed *irredentism*. We shall discuss irredentist policies in greater detail in Chapter 7, but the term comes from the name of a political pressure group in northernmost Italy whose principal objective was the incorporation of *Italia Irredenta* (Unredeemed Italy), a neighboring part of the South Tyrol in Austria's Alps.

Eastern Europe has been plagued by such irredentist problems. Only after the end of World War II were some of them eliminated by further boundary revision (Fig. 1–29). Poland was literally shoved westward for well over 100 miles: the Soviet Union annexed much of the eastern half of that country and compensated now-communist Poland by attaching to it a large chunk of defeated Germany in the west. The Soviet Union also took for itself the eastern portions of Czechoslovakia and Romania, and Bulgaria and Romania settled their dispute over the southern Dobruja, which was returned to Bulgaria. The problem of the Julian March was also settled, though not immediately: through U.N. mediation, the port of Trieste came under Italian administration, whereas most of the disputed surrounding territory went to Yugoslavia.

Another major factor in the simplification of the ethnic situation in Eastern Europe has been the hegemony of the Soviet Union, not only over this region, but over East Germany as well. East Germany in a very real sense has now become an Eastern European state, certainly as far as economic and political reorganization are concerned; but the new order has also involved a migration, voluntary as well as forced, of Germans from several East European countries—where they formed significant prewar minorities—to this part of their homeland. Thus, the German minorities of Poland and Czechoslovakia have been drastically reduced (and, in the former, practically eliminated), and long-troublesome East Prussia no longer exists. Elsewhere, migrations have carried Hungarians from Czechoslovakia and Romania to Hungary, Ukrainians and Belorussians from Poland to the Soviet Union, and Bulgars from Romania to Bulgaria. This is not to suggest that Eastern Europe's minority problems are solved. Many Hungarians still live in Romania, and Hungary has not forgotten its interests in Transylvania. Yugoslavia still has a large Albanian minority in the south, a Hungarian minority in the north, and some Romanian clusters in the east. When the umbrella of Soviet dominance is removed, as it surely will be some day, Eastern Europe may still not be free of its irredentist problems.

**Poland and Czechoslovakia** Several geographic bases exist to separate Eastern Europe's two northernmost countries from the others. Poland, facing the Baltic Sea, is the region's largest state territorially (313,000 square kilometers; 121,000 square miles) and demographically (37.2 million in 1985); it ranks sixth in Europe in both categories. Czechoslovakia, mountainous and surrounded, is less than half as large—but it is ahead in economic development. Poland's strongly agrarian economy began to yield to industrialization only after World War II, when the innovation of centralized national planning made possible the exploitation of the country's industrial opportunities. Czechoslovakia, on the other hand, has been more directly exposed to influences from Western Europe. It lay more directly in the paths of commercial exchange and industrial development along Europe's east-west axis, and the Czechs had long-standing ties with the West. One of these ties is the Elbe River, which originates in the basin that is Bohemia, cuts through the Erzgebirge (Ore Mountains), and flows through central Germany to the North Sea port of Hamburg. Before the Iron Curtain was lowered in the mid-1940s, this waterway was a major factor in Czechoslovakia's westward orientation.

Poland lies largely in the Northern European Lowland, so that there are few internal barriers to effective communication. In the south, along the Czech border, Poland shares the foothills of the Sudeten and Carpathian Mountains, where the raw materials necessary for heavy industry are located. Warsaw, the capital (2.1 million), lies near the center of the country in a productive agricultural area near the head of navigation on the Vistula River and has made itself the focus of a radiating network of transportation routes that reaches all parts of the country. Warsaw is Poland's primate city, the economic, historical, cultural, and political center of Polish life. The leading manufacturing complex is found in Silesia in southern and southwestern Poland, anchored by the Krakow-Czestochowa-Wroclaw triangle, which has become the country's industrial core area. Polish planners also want to stimulate development in the country's heartland, which includes Warsaw and the textile center of Lódz (with a population of 1 million, Poland's second city), the "Polish Manchester." But, in almost every way, Poland's best opportunities lie in the southern half of the country. This includes good quality farmland: southern Poland has a black soil belt—broadening eastward into the Ukraine—that sustains intensive farming, with wheat as the major crop. In central Poland, thin and rocky glacial soils support only rye and potato crops; the Baltic rim possesses such poor soils that cultivation must give way to pastureland and moors.

Poland has not accepted Soviet economic innovations and political dominance without reservations, which have at times been expressed in violent resistance. However, it cannot be doubted that Soviet "cooperation" has boosted Poland's industrial and urban progress. On the other hand, a measure of the failure of communism here can be gained from a review of the attempts that were made to collectivize agriculture. In Poland, as in certain other countries of Eastern Europe, this program had to be slowed down or temporarily shelved because of peasant resistance, and, today, less than a quarter of Polish farmland is under collective management. The Soviet pattern of great investment and support for industry, even at the expense of badly needed agricultural progress, is evident in Eastern Europe as well.

Czechoslovakia shares with Poland the Silesian industrial region, of which the Krakow-Czestochowa-Wroclaw triangle is the leading part. Local sources of raw materials lie astride the gap between the Sudeten Mountains and the Tatra extension of the Carpathians. This area is referred to as Moravia, and the gap as the Moravian Gate because of its vital importance as a low-lying passageway between the Danube Valley and the North European Plain. Lying midway between Bohemia in the west and Slovakia in the east, this growing industrial area is of increasing importance to Czechoslovakia; with its coal supply and its heavy industry, Ostrava (345,000) lies at the focus of a manufacturing complex that is dominated by chemicals and metals fabrication.

The western sector of Czechoslovakia, focused on the mountain-enclosed Bohemian Basin, has always been an important core area in Eastern Europe, cosmopolitan in character and Western in its exposure and development. With its Elbe River outlet, this westward orientation of Bohemia was maintained for centuries. But, in 1919, the Slovaks were attached to Bohemia. The Slovak eastern part of Czechoslovakia, mountainous and rugged, lies in the drainage basin of Eastern Europe's greatest river, the Danube. However, the Danube flows not westward, but south and then eastward to the Black Sea. Slovakia is, therefore, much more representative of Eastern Europe than is the Czech part of the state. It is mostly rural and neither as industrialized nor as urbanized as the Bohemian-Moravian west. In the days of the Austro-Hungarian Empire, this was a peripheral frontier of comparatively little importance, whereas Bohemia was significant as a manufacturing area even then. During the interwar decades of the Czechoslovakian state, the region and its inhabitants took second place to the more advanced west. This situation has continued under communism: a steady stream of manpower leaves Slovakia every year in search of work in the factories of Bohemia and Moravia. One result is that labor shortages occur on the farms at every harvest time.

Czechoslovakia's center of gravity, then, lies in the west, in Bohemia-Moravia. Unlike Warsaw, Prague (1.2 million) is not a centrally positioned capital; it is located in the west, and it is very clearly a Bohemian city. It is three times as large as the country's next urban center (Moravia's Brno with 382,000 people). Slovakia's capital, Bratislava (on the Danube), is somewhat smaller still. In every way, Prague is Czechoslovakia's primate city. It is the political and cultural focus of the country and constitutes a major industrial region. Founded at a place where the Vltava River can be easily crossed, the city lies near the upper Elbe River and at the middle of the country's greatest concentration of wealth in the Bohemian Basin. The surrounding mountains contain a variety of ores, and, in many of the valleys, stand small manufacturing towns that specialize, Swiss-style, in certain kinds of manufacture—for example, pencils are produced from local graphite at Budejovice, glass and crystal at Teplice-Sanov and Yablonec, and so on. In Eastern Europe, the Czechs have always led in the field of technology and engineering skills. Prague became a major manufacturing center, and nearby Pilsen is the site of the famous Skoda steelworks and (perhaps even more famous) the breweries that developed Pilsner lager beer. Czech products, from automobiles to textiles, find their way to capitalist as well as Communist markets.

**Hungary and the Balkan Peninsula** The five countries that make up the southern tier of Eastern Europe (excluding Greece) seem almost to have been laid out at random, with little regard for the potential unifying features of this part of Eastern Europe. This apparent randomness is the result of centuries of territorial give and take, of migrations and invasion. In fact, the consolidation and shattering of states and empires here has been so pronounced throughout modern history, that the area is often called a *shatter belt*.

The comparative orderliness of Poland and Czechoslovakia, one a land of uniform plains and the other a rather well-defined mountain state, is lost here, especially in the Balkans, where imposing mountain ranges and masses abound, large and small basins are sharply differentiated and separated, and the ethnic situation is even more confused than elsewhere in Eastern Europe. At least the Slovaks had this in common with the Czechs: they were neither Poles nor Hungarians and their attachment to the Czech state seemed a reasonable solution. But, in the Balkans, such solutions have been harder to come by.

The great Balkan unifier that might have been is the Danube River, which originates in southern Germany, traverses northern Austria, and then crosses central Eastern Europe, forming first the Czech-Hungary boundary, then the Yugoslavia-Romania boundary, and, finally, the Romania-Bulgaria border. It is, indeed, anomalous that such a great transport route, which could form the focus for a large region, is, instead, a constant dividing line. After it emerges from the Austrian Alps, the Danube crosses the Hungarian Basin, which, although largely occupied by Hungary, is shared also by Yugoslavia, Romania, and Czechoslovakia. Then, it flows through the Transylvanian Alps at the Iron Gate (near Orsova), and into the basin that forms its lower course, which is shared by Romania and Bulgaria. No other river in the world touches so many countries, but the Danube has not been a regional bond. Only two Eastern European capitals—Budapest (Hungary) and Belgrade (Yugoslavia)—lie directly on the river. Only Hungary is a truly Danubian state, as the river turns southward just north of Budapest and crosses the entire country along with its tributary, the Tisza.

***Hungary*** In a very general way, it can be argued that progress and development in Eastern Europe decline from west to east and also from north to south. In both Poland and Czechoslovakia, the western parts of the country are the most productive. Hungary contains the largest and most productive share of the Hungarian Basin and is better off than Yugoslavia immediately to the south. In Yugoslavia itself, the core of the country lies in the Danubian lowland, and the southern mountainous areas lag by comparison. As a whole, however, Yugoslavia is far ahead of its small southern neighbor, Albania. On the eastern side of the Balkan peninsula, Romania has the advantage over Bulgaria to the south. Thus, Hungary is southeastern Europe's leading state in several respects; given its pivotal position in Eastern Europe and its former role in the Austro-Hungarian Empire, this is not surprising.

The Hungarians (or Magyars) themselves form a minority in the Balkans since they are neither of Slavic nor of Germanic stock. They are a people of Asian origins (distantly related to the Finns) who arrived in Hungary in the ninth century A.D. Ever since, they have held on to their fertile lowland, retaining their cultural identity (including their distinctive language), although at times losing their political sovereignty. The capital, Budapest, was a Turkish stronghold for more than a century during the heyday of the Ottoman Empire. The city's recent growth (to over 2.3 million) was achieved during the period of the Austro-Hungarian Empire (1867–1918) and following the creation of the all-Magyar state of Hungary after World War I. Today, Budapest is about 10 times as large as the next-ranking Hungarian city, a reflection of the rural character of the country. With its Danube port and its extensive industrial development, its cultural distinctiveness and its nodal location within the state, Budapest epitomizes the general situation in Eastern Europe, where urbanization has been slow and where the capital city is normally the only urban center of any magnitude.

Hungary's rural economy has not been without problems. When the country was delimited in 1919, there was a great need for agrarian reform. Large estates were carved into small holdings and productivity rose. Then, World War II and its attendant destruction, especially of livestock, set the rural economy back; after the war, the Soviets coveted fertile Hungary as a potential breadbasket for Eastern Europe. When farm production failed to rise, collectivization was encouraged, but the peasants who had become small landholders resisted the effort. But Hungary had been on the German side during World War II, and the Soviets were conquerers here, not liberators; hence, they pushed their reform program vigorously (as they also did in East Germany). Today, three-quarters of Hungary's land is in collective operation, but farm yields are not expanding nearly as fast as the country's planners would like. Certainly the country is agriculturally self-sufficient, with its harvests of wheat, corn, barley, oats, and rye, but there is less surplus for export than desired.

Industrially, also, Hungary has not yet been able to take full advantage of its potential. There is some coal, notably near Pecs, not far from the Danube in the southern part of the country, and iron ore can be brought in via the Danube for steel production. This has been done in quantity only recently; for a long time, Hungary, like so many countries in the developing world, was exporting millions of tons of raw materials to other producing areas. This is especially true of the one mineral Hungary has in major quantities—bauxite—mined near Gant but refined for manufacture in Czechoslovakia and the Soviet Union rather than in Hungary itself.

***Yugoslavia*** Yugoslavia, south of Hungary, shares a part of the Danube Basin, and this area has become Yugoslavia's heartland. Yugoslavia was created after World War I, when Serbs, Croats, Slovenes, Macedonians, Montenegrons, and lesser ethnic groups were combined in an uneasy union under the royal house of Serbia. The monarchy collapsed in the chaos of World War II, but the country was resurrected as a federated socialist republic under the leadership of Marshal Tito in 1945. Tito, who had led the fight against the Germans, took Yugoslavia on a Communist course. But when the Soviets tried to impose their directives on him, he asserted his country's independence from Moscow. Since 1948, the relationship between the Soviet Union and Yugoslavia has been difficult. Seeking a balanced approach to its economic problems, Yugoslavia went slowly in its collectivization efforts (less than one-sixth of the farmland is under collectivized production today), whereas industry was guided by national planning. Before the war, Yugoslavia had been an exporter of some agricultural products, some livestock, and a few ores. After the war, with effective political control and economic centralization, its government sought to change this—not at the expense of agriculture, as was the case in so many Communist-influenced countries, but in addition to it.

Yugoslavia today is a federal state consisting of the Socialist Republics of Slovenia, Croatia, Bosnia-Hercegovina, Montenegro, Macedonia, and Serbia (Serbia is divided into two "autonomous regions," Vojvodina and Kosovo). This complex federal structure dates from 1963, when a new constitution was approved. But the republics, although created along ethnic lines, contain many minorities: Yugoslavia has Hungarians, Albanians, Slovaks, Bulgarians, Romanians, Italians, and even Turks within its borders. About 43 percent of Yugoslavia's 23 million people (1985) are Serbian, and perhaps 24 percent are Croat (or Croatian). Less than 10 percent of the population is Slovene. Yugoslavia's politico-geographical problems can be discerned from Fig. 1–29—the complicated mixture of peoples,

## YUGOSLAVIA AND CARINTHIA

In southern Austria, on the border with Yugoslavia and Italy, lies the province of Carinthia (Fig. 1–29), home of an estimated 40,000 Slovenes. However, Slovenia, the Slovenes's ancient homeland, is today Yugoslavia's northernmost "socialist republic."

A minority census proposed for 1976 by the Austrian government produced oppositon by the Slovenes (and also by the Croats, another Yugoslavian minority living in Burgenland, Austria's easternmost area). The Slovenes expressed fear of intimidation by Austrian nationalist groups. They also worred that too many Slovenes would count themselves as members of a third borderland minority group, the Windisch, whose language and culture contains both Germanic and Slavic elements; if this should occur, the numerical strength of the Carinthian Slovenes would be diminished, and the Windisch—who apparently prefer more contact with Austria—would be given a stronger local identity. In the Yugoslavian capital, Belgrade, the newspapers began to support the Austrian Slovenes's position; when their leaders visited the late President Tito, the Yugoslav government proposed to take the issue to the United Nations. As the quarrel intensified, there were calls for the "incorporation" by Yugoslavia of areas of Austrian Carinthia where Slovenes are in the majority. It was, for Eastern Europe, a familiar sequence of events; the issue remains unresolved as another of the region's numerous latent territorial disputes.

their unequal access (through location) to the country's productive opportunities, and regional inequalities in progress and development. Add to these residual nationalism and latent hostilities, and it is something of a miracle that Yugoslavia has survived so long with stability. Internal dissent, however, has been put down quite ruthlessly, and nationalist elements representing several Yugoslavian groups are in exile. It is a situation that could be exploited now that Tito's personal unifying force no longer prevails.

***The lesser Balkan states*** If our generalization concerning declining development and southerly location in the Balkans is to hold true, then Albania, Yugoslavia's Adriatic neighbor, should be less developed than even mountain-and-plateau Yugoslavia. And so it is. With less than 3 million people and under 30,000 square kilometers (11,000 square miles), Albania ranks last in Europe (excluding Luxembourg and Iceland) in both territory and population. Its percentage of urbanized population (about 39 percent) is also among Europe's lowest. Most Albanians eke a subsistence out of livestock herding and farming the one-seventh or so of this mountainous country that can be cultivated at all; the largest town is the capital, Tirane, which has about 320,000 inhabitants and a few factories.

Perhaps because of its abject poverty and the limited opportunities for progress (consisting of some petroleum exports, tobacco cultivation, and chrome ore extraction), Albania turned to Maoist China for ideological as well as material support. Unlike Yugoslavia, which moved away from the Soviet orbit in the direction of moderation, Albania committed itself to the stricter Maoist version, thus choosing the opposite way. Certainly, Albania was worth more to China in this context than it was to the Soviet Union, which already dominated much of Eastern Europe; Albania clearly preferred a prominent place in China's priorities than a lowly one in the Soviet Union's. This country has little to bargain with except a somewhat strategic position in Eastern Europe, on the Mediterranean at the entry to the Adriatic Sea. Most recently, however, China's new ideological directions are depriving Albania of its political alternative.

The region's two Black Sea states, Romania and Bulgaria, also confirm the southward lag in progress and development. Romania is both richer and larger—twice as large as its southern neighbor in terms of territory as well as population. Both countries' boundaries epitomize those of the Balkans in general. Areas of considerable homogeneity are divided (e.g., the Danube lowland shared by these two countries); areas that would seem to fit better with other countries are incorporated (such as the Transylvanian part of the Hungarian Basin, now part of Romania); and national aspirations are denied (such as Bulgaria's long-time desire for a window on the Aegean Sea). But on the whole, Romania, despite some major territorial losses to the Soviet Union as a result of World War II, remains quite well endowed with raw materials.

The potential advantages of Romania's compact shape are to some extent negated by the giant arch of the Carpathian and Transylvanian Mountains. To the south and east of this arch lie the Danubian plains and the hills and valleys of the Siret River (Moldavia), and to the west lies the Romanian share of the Hungarian Basin—with a sizable Hungarian population forming one of Eastern Europe's ubiquitous minorities here. As we have noted on previous occasions, a mountain-plain association often provides mineral wealth, and Romania is no exception. In the foothills of the Carpathians, near the westward turn of the Transylvanian Alps, lies Europe's major continental oilfield. From Ploesti, the urban center of this area, pipelines lead to the Black Sea port of Constanta, to Odessa in the Soviet Union, to nearby Bucharest, and to Giurgiu on the Danube. Although the Soviet Union is, of course, Romania's chief customer, both oil and natural gas are sold in Eastern Europe as well; the gas is piped not only to Bucharest, but to Budapest also. Pipelines for oil are now being laid into several Eastern European countries.

After World War II, the Romanian state found itself with some boundary revisions costing it over 50,000 square kilometers (20,000 square miles), and 3 million people. It also found itself under Soviet domination, and, thus, began the familiar sequence of centralized planning, an emphasis on industrialization, further agrarian reform involving collectivization, and accelerated urbanization. But progress has not been spectacular—not as impressive, even, as in some of the other Soviet-dominated Eastern European countries. Agricultural mechanization and the provision of adequate fertilizers have been delayed, and farm yields never have reached expected levels. Despite Romania's considerable domestic resources, industrial development has largely been geared to local markets. Bucharest at 1.9 million people is still several times as large as the next-ranking town, the interior industrial center of Cluj with chemical, leather, and wood manufacturers (480,000). The causes of all this are difficult to pinpoint, but there seems to have been a lack of national commitment to the cause of planned development, which may be a short-term factor; in the longer sense, Romania has suffered from a chronic ailment of Eastern Europe—political instability.

The Romanians are a people of complex roots, whose distant Roman heritage is embodied in a distinctive language, but whose subsequent adventures have brought a strong Slavic influence, especially to the people in the rural areas, as well as an admixture with Hungarian, German, and Turkish elements.

The Bulgars, too, have been affected by Slavic as well as other influences. These people, numbering 9 million today, also have a history of Turkish domination, and a Bulgarian state did not appear until 1878. Bulgaria, in a way, was the westernmost buffer state in the long zone that emerged in Eurasia between Russian and English spheres of influence; the Russians helped push the Turks from Bulgaria, and British fears of a Russian penetration to the Aegean led to the Treaty of Berlin, at which time Bulgaria's boundaries were delimited.

Bulgaria is a mountain state, except in the Danube lowland, which in this country is narrower than in Romania, and in the plains of the Maritsa River, which to the south forms the boundary between Greece and Turkey. Even the capital, Sofia, is located in the far interior, away from the plains, the rivers, and the Black Sea; the mountains were used as protection for the headquarters of the weak embryo state, and it has remained here. In any case, Bulgaria is not a country of towns—it would be more appropriate to see it as a country of peasants. Over one-third of the population is still classed as rural, and after Sofia with 1.1 million people, the next-ranking city (Varna) contains 420,000. The east-west mountain range that forms the country's backbone and separates the Danubian lowland area from the southern valleys carries the region's name—the Balkan Mountains. Much of the rest of Bulgaria, especially the west and southwest, is mountainous as well, and people live in clusters in basins and valleys and are separated by rough terrain. The Turkish period destroyed the aristocracy and eliminated the wealthy landowners, and, for many decades, Bulgarian farms have been small gardens, carefully tended and quite productive. These garden plots survive today, although collectivization has been carried further here than in any other Eastern European country, except perhaps Albania. On the bigger collective farms, the domestic staples of wheat, corn, barley, and rye are grown, but the smaller gardens produce vegetables and fruits; plums, grapes, and olives remind us that we are approaching the Mediterranean. From this point of view, Bulgaria is a valuable economic partner to the Soviet Union, where fruits, vegetables, and tobacco are always needed and where the conditions favoring the cultivation of this particular crop assemblage are very limited.

At the time of transition to Soviet domination, Bulgaria was in a different position from both of its Eastern European neighbors, Romania and Yugoslavia. It was poorer than both, and it had still less in terms of industrial and agricultural resources. But there was a lengthy history of association of one kind or another with the Russians and Soviets; the Bulgars are also more Slavicized than the Romanians and have proven to be the U.S.S.R.'s most loyal Eastern European satellite.

Thus, Eastern Europe is going through a phase of eastward orientation and Communist political and economic organization—an organization that is in most cases interrelated with that of the Soviet Union itself. But, viewing the history and historical geography of Eastern Europe, it appears likely that this phase will eventually lead to something else. Stability has never been a quality of Eastern European life, and the superimposed stability of the postwar period has already been disturbed in some instances. And despite the repatriation of large numbers of people who formed minorities in Eastern Europe's countries, old national goals have not been forgotten. Bulgaria still considers the Macedonian question to be unsettled—and through Macedonia, of course, this country could obtain a way to the open sea. Hungary is very well aware of the Magyar minority in Romania's Transylvania. And there are signs of dissatisfaction with the status quo as it relates to the Soviet Union—in Czechoslovakia, in Romania, in Hungary and, above all, in Poland. Eastern European nationalism is a potent force, and national pride and Soviet planning suggestions are not always compatible.

## SUPRANATIONALISM IN EUROPE

| | |
|---|---|
| 1944, September | Benelux Agreement signed. |
| 1947, June | Marshall Plan proposed. |
| 1948, January | Organization for European Economic Cooperation (OEEC) established. |
| 1949, January | Council for Mutual Economic Assistance (Comecon) formed. |
| 1949, May | Council of Europe created. |
| 1951, April | European Coal and Steel Community (ECSC) Agreement signed (effective July 1952). |
| 1957, March | European Economic Community (EEC) Treaty signed (effective January 1958). European Atomic Energy Commission (Euratom) Treaty signed (effective January 1958). |
| 1959, November | European Free Trade Association (EFTA) Treaty signed (effective May 1960). |
| 1965, April | EEC-ECSC-Euratom Merger Treaty signed (effective July 1967). |
| 1972, January | Accession Treaty for entry to EEC by United Kingdom, Denmark, Ireland, and Norway signed. |
| 1972, September | Norway's voters defeat EEC membership proposal. |
| 1973, January | United Kingdom, Denmark, Ireland become members of EEC, creating "The Nine." |
| 1976, November | Greece makes application for EEC membership. |
| 1977, March | Portugal makes application for EEC membership. |
| 1977, December | Spain makes application for EEC membership. |
| 1978, April | The nine member states of the European Community (EC) (EEC has been shortened) commit themselves to admit Greece, Portugal, and Spain when formalities are concluded. |
| 1978, November | Liechtenstein is admitted to the Council of Europe as its twenty-first member. |
| 1979, March | European Monetary Agreement (EMA) linking the currencies of EC members goes into effect. |
| 1979, June | First general elections to a European Parliament are held. First session of the 410-member legislature is held the following month and Simone Weil presides. |
| 1981, January | Greece formally enters membership in the EC, creating "The Ten." |

# European Unification

Individually, no European country constitutes a power of world stature comparable to the United States or the Soviet Union. But in combination, Europe west of the Iron Curtain would indeed have world power status. Europeans have long recognized the potential advantages of unification; as long ago as the 1500s the philosopher Erasmus called for it, and many after him have also done so, including Rousseau, Kant, and Victor Hugo. But, until the twentieth century, only the Romans had succeeded in unifying Italy and Iberia, Greece and Britain.

In the aftermath of World War I, the nations of Europe held a congress in Vienna to consider the possibility of Pan-European unification, but their aspirations soon were dashed by the events of the 1930s and World War II. Since 1945, however, Europe has made unprecedented progress toward economic integration and political unification (see Box). There is, as yet, no United States of Europe—but critical first steps have been taken, and the push toward still stronger coordination continues.

It all began with Benelux, as we noted earlier, and this first move was followed by the infusion of Marshall Plan aid (1948–1952) and the creation of the Organization for European Economic Cooperation (OEEC). An important principle of *supranationalism* (international cooperation involving the voluntary participation of three or more nations in an economic, political, or cultural association) is that political integration, if it is ever to occur, follows rather than precedes economic cooperation. In this way the infusion of Marshall Plan funds, by generating the need for an international European economic-administrative structure, also promoted coordination in the political sphere. As early as 1949, the year after the Marshall Plan began, the *Council of Europe* convened for the first time; it met in a city that was proposed as a future capital for a united Europe, Strasbourg, on the French bank of the Rhine River. For 30 years, the Council was little more than a forum for the exchange of ideas and opinions; it discussed such issues as cultural cooperation, legal coordination, human rights, crime problems, and migration and resettlement. The representatives of the Council's 17 member states had no executive authority, but their views did have an impact on national governments.

The year 1979 brought a momentous development when, for the first time, representatives to a European Parliament were elected rather than appointed. European political parties played a role in the formation of this regional legislature, and its composition began to reflect more accurately the political mainstreams of Europe west of the Soviet sphere. It was another step in the direction of a united Europe, and the Council of Europe may yet come to be recognized as the beginning of a true European government.

Achievements in economic cooperation have been more difficult to come by—but so far, they have been more consequential. Soon after the OEEC was established, the (then) foreign minister of France, Robert Schuman, proposed the creation of the *European Coal and Steel Community* (ECSC), with the principal objective of lifting the restrictions and obstacles in the way of the flow of coal, iron ore, and steel among the mainland's six prime producers: France, West Germany, Italy, and the three Benelux countries. The mutual advantages of this arrangement are obvious, even from a map showing simply the distribution of coal and iron ore and the position of the political boundaries of Western Europe; but the six participants did not stop here. Gradually, through negotiation and agreement, they enlarged their sphere of cooperation to include reductions and eliminations of tariffs, and a free flow of labor, capital, and nonsteel commodities; ultimately, in 1958, they launched the "Common Market," the *European Economic Community* (EEC). This organization linked virtually all of the European core area of the mainland, and its total assets in terms of resources, skilled labor, and markets were enormous. Moreover, its jurisdiction was strengthened by various commissions and by legislative and judicial authorities.

One very significant development related to the creation of the Common Market was the decision of the United Kingdom not to join it. This was itself a move based on supranational considerations. There was fear in Britain that participation would damage evolving relationships with its Commonwealth countries; for many of these countries, Britain was the chief trade partner and there were pressures on the United Kingdom not to endanger these ties—pressures both within the country and from the far-flung Commonwealth. Thus, Britain stayed out of the Common Market, but it made its own effort to create closer economic bonds in Europe; in 1959, it took the lead in establishing the so-called *European Free Trade Association* (EFTA), comprising, in addition to the United Kingdom, three Scandinavian countries (Sweden, Norway, and Denmark), the two Alpine states (Switzerland and Austria), and Portugal; Finland, ever wary of Soviet disapproval of any real change in the neutrality status of its neighbor, also joined, but carefully continues to limit its participation to that of an associate member. This scattered group of countries, with their relatively small populations, generally limited resources, and restricted purchasing power, added up to something much less than the Common Market; they soon became known as the "Outer Seven," whereas the contiguous states of the core came to be known as the "Inner Six."

Within a few years of the creation of the Outer Seven, the United Kingdom changed its position on EEC membership and decided to seek entry. Now, however, the political attitude on the continent had changed; France took an inflexible position under de Gaulle and obstructed British participation, despite a desire on the part of other Common Market members to ratify admission. But early in the 1970s, the path was cleared for Britain's entry, and the British officially became members of the Common Market in 1973; at the same time, Denmark and Ireland also joined the EEC (Fig. 1–31A). Although the nine member states of the Common Market in 1978 voted in principle to view favorably the future admission of Greece, Spain, and Portugal, only Greece has been admitted (in 1981) to date. Meanwhile, EFTA

EUROPEAN SUPRANATIONALISM

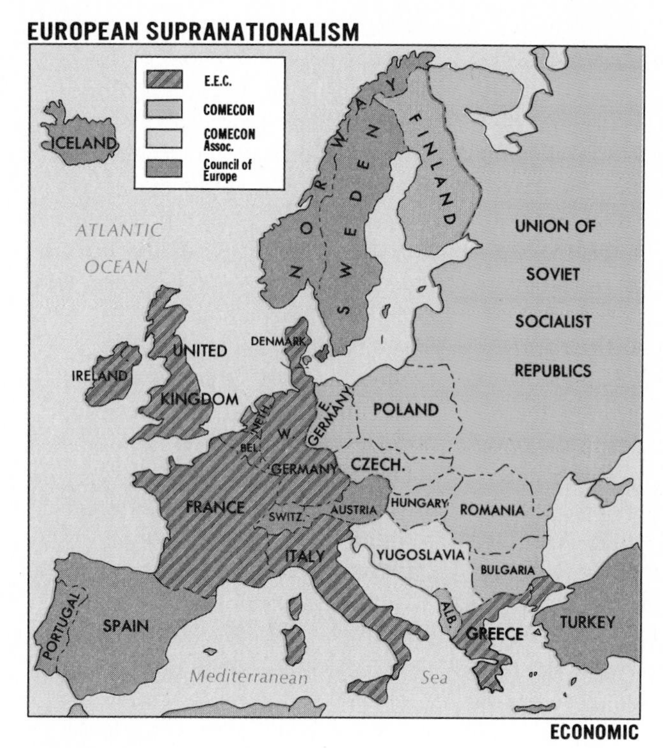


Figure 1–31A

Figure 1–31B

persists in a weakened state: Britain and Denmark departed in 1972 to join EEC and were replaced only by Iceland. Thus, pending additional changes, the lineup in the mid-1980s consists of an "Inner Ten" and an "Outer Six."

Even before this expansion of the EEC occurred, still another step had been taken toward greater unity on the OEEC model, but with the United States and Canada as full members in a nonmilitary Atlantic association. The objective was to reduce the divisive nature of the EEC-EFTA split and, at the same time, to broaden the basis for Western economic cooperation. Ratified in 1961, the *Organization for Economic Cooperation and Development* (OECD) counted 20 members, including all those who signed the OEEC papers in 1948. The OECD, however, unlike the Common Market, has little power and no binding authority, and some participants argued that it represents nothing very constructive in the European drive toward unification. But it may yet play its role in finding a way to overcome some of the remaining obstacles in the path to this objective.

Behind the highly visible Common Market, a large number of other European associations function to further European unity. The European Atomic Energy Community (Euratom) aims at the development of peaceful uses of atomic energy and the sharing of research and information; the formation of the European Space Research Organization (ESRO) promotes coordinated space research. The Western European Union (WEU) coordinates defense in Western Europe. And, of course, many of the states of Europe west of the Iron Curtain are members of the North Atlantic Treaty Organization (NATO), adding military cooperation to coordination in other spheres (Fig. 1–31B).

The existence of the Iron Curtain and the regional ideological division of Europe prevents still wider European integration. In 1949, the Soviet Union took the lead in establishing the Council for Mutual Economic Assistance (COMECON). This union was designed to promote industrial specialization in Eastern European countries, to finance investment projects planned by two or more countries, and generally to promote closer economic integration.

World economic problems since the mid-1970s have slowed Europe's drive toward unification. Nor were all countries determined to participate: for example, Norway continued to disdain membership in the Common Market. Nonetheless, the new Europe provides strong evidence that its old divisions are fading. Mobility is greater than ever before; Italians work in Amsterdam, French workers labor in German factories, Belgian capital invests in Italy. International boundaries function less to divide than ever before, and license plates bearing the exhortation "Europa!" can be seen on automobiles from Rotterdam to Rome.

Europe in the past has transformed the world—today, Europe vigorously continues to restructure itself.

## References and Further Readings


Beckinsale, M. & Beckinsale, R. *Southern Europe* (London: University of London Press, 1975).

Boal, F. & Douglas, J., eds. *Integration and Division: Geographical Perspectives on the Northern Ireland Problem* (New York and London: Academic Press, 1982).

Burtenshaw, D. et al. *The City in West Europe* (New York: John Wiley & Sons, 1981).

Chapman, K. *North Sea Oil and Gas* (London: David and Charles, 1976).

Chisholm, M. *Rural Settlement and Land Use: An Essay in Location* (London: Hutchinson University Library, 3 rev. ed., 1979).

Clayton, K. & Kormoss, I., eds. *Oxford Regional Economic Atlas of Western Europe* (London: Oxford University Press, 1971).

Clout, H., ed. *Regional Development in Western Europe* (New York: John Wiley & Sons, 2 rev. ed., 1981).

Demko, G. et al., eds. *Population Geography: A Reader* (New York: McGraw-Hill, 1970).

Diem, A. *Western Europe: A Geographical Analysis* (New York: John Wiley & Sons, 1979).

Dury, G. H. *The British Isles* (New York: Barnes & Noble, 5 rev. ed., 1973).

East, W. G. *An Historical Geography of Europe* (London: Methuen, 1962).

Evans, E. E. *The Personality of Ireland: Habitat, Heritage and History* (London: Cambridge University Press, 1973).

French, R. & Hamilton, F. E. I., eds. *The Socialist City: Spatial Structure and Urban Policy* (New York/Chichester, Eng.: John Wiley & Sons, 1979).

Fullerton, B. & Williams, A. *Scandinavia: An Introductory Geography* (New York: Praeger, 1972).

Gottmann, J. *A Geography of Europe* (New York: Holt, Rinehart & Winston, 4 rev. ed., 1969).

Hoffman, G., ed. *A Geography of Europe: Problems and Prospects* (New York: John Wiley & Sons, 5 rev. ed., 1983).

Hoffman, G., ed. *Eastern Europe: Essays in Geographical Problems* (London and New York: Methuen and Praeger, 1971).

Houston, J. *A Social Geography of Europe* (London: Gerald Duckworth, 1963).

Houston, J. *The Western Mediterranean World: An Introduction to Its Landscapes* (London: Longmans, 1964).

Ilbery, B. *Western Europe: A Systematic Human Geography* (New York: Oxford University Press, 1981).

Jefferson, M. "The Law of the Primate City," *Geographical Review*, 29 (1939). Quotation taken from p. 226.

Johnston, R. & Doornkamp, J., eds. *The Changing Geography of the United Kingdom* (London: Methuen, 1983).

Jones, H. *A Population Geography* (New York: Harper & Row, 1981).

Jordan, T. *The European Culture Area* (New York: Harper & Row, 1973).

Kasperson, R. & Minghi, J. *The Structure of Political Geography* (Chicago: Aldine, 1969).

Kosinski, L. *The Population of Europe: A Geographical Perspective* (London: Longmans, 1970).

Lichtenberger, E. "The Changing Nature of European Urbanization," in Berry, B., ed., *Urbanization and Counter-Urbanization* (Beverly Hills, Cal.: Sage Publications, 1976), pp. 81–107.

Likens, G. et al. "Acid Rain," *Scientific American*, 241 (October 1979), 43–51.

Malmstrom, V. *Geography of Europe* (Englewood Cliffs, N.J.: Prentice-Hall, 1971).

Manners, I. *North Sea Oil and Environmental Planning: The United Kingdom Experience* (Austin: University of Texas Press, 1982).

Mead, W. *An Economic Geography of the Scandinavian States and Finland* (London: University of London Press, 1964).

Mellor, R. *Eastern Europe* (New York: Columbia University Press, 1975).

Mellor, R. *The Two Germanies: A Modern Geography* (New York: Barnes & Noble, 1978).

Monkhouse, F. *A Regional Geography of Western Europe* (London: Longmans, 4 rev. ed., 1974).

Newman, J. & Matzke, G. *Population: Patterns, Dynamics, and Prospects* (Englewood Cliffs, N.J.: Prentice-Hall, 1984).

Nystrom, J. W. & Hoffman, G. *The Common Market* (New York: D. Van Nostrand, 2 rev. ed., 1976).

Orme, A. *Ireland* (Chicago: Aldine, 1970).

Parker, G. *An Economic Geography of the Common Market* (New York: Praeger, 1969).

Riley, R. & Ashworth, G. *Benelux: An Economic Geography of Belgium, the Netherlands and Luxembourg* (New York: Holmes & Meier, 1975).

Sommers, L. "Cities of Western Europe," in Brunn, S. & Williams, J., eds., *Cities of the World: World Regional Urban Development* (New York: Harper & Row, 1983), pp. 84–121. Quotation taken from p. 97.

Sullivan, S. "The Decline of Europe," *Newsweek*, April 9, 1984, 44–56.

Thompson, I. *Modern France: A Social and Economic Geography* (Totowa, N.J.: Rowman & Littlefield, 1970).

Wagret, P. *Polderlands* (New York: Barnes & Noble, 1972).

Walker, D. *Mediterranean Lands* (New York: John Wiley & Sons, 1962).

Wheeler, J. & Muller, P. *Economic Geography* (New York: John Wiley & Sons, 1981), Chapter 13.

White, P. *The West European City: A Social Geography* (New York: Longman, 1984).

Wild, T. *West Germany: A Geography of Its People* (New York: Barnes & Noble, 1980).

Zelinsky, W. *A Prologue to Population Geography* (Englewood Cliffs, N.J.: Prentice-Hall, 1966).


## Bibliographical Footnote

Individual countries have also been the subject of many geographies (or groups of countries such as the Low Countries and the Alpine states), both for Europe and the remaining 12 world realms. A good beginning is a continuing series of sketches in *Focus*, published five times a year by the American Geographical Society in New York City. The Van Nostrand *Searchlight Series* of paperbacks, many dating back to the 1960s, includes discussions of many of the world's nations and areas; unfortunately, these are out of print in the 1980s (and are therefore not cited in this book), but they can still be found in many libraries. Aldine's *World Landscapes Series* contains many incisive works. Also refer to Praeger's *Country Profiles Series* (many now published by its branch, Westview Press, Boulder, Colo.) and Methuen's *Advanced Geographies Series*.

# AUSTRALIA: A European Outpost

IDEAS AND CONCEPTS

**Migration**
**Intervening Opportunity (2)**
**Remoteness**
**Federalism**

Australia evokes images of faraway isolation, enormous expanses and uninhabited spaces, modern cities, vast livestock ranches, beautiful coastlines. Australia is a continent and a nation-state—a discrete realm of the world, because its population and culture are European. If Australia were populated by peoples of Malayan stock with ways and standards of living resembling those of Indonesia and mainland Southeast Asia (and had a comparable history), then, it is quite possible that it would be viewed today only as an exceptionally large island sector of the Asian continent. But, just as Europe merits recognition as a continental realm—despite the fact that it is merely a peninsula of "Eurasia"—so Australia has achieved identity as the island realm of "Australasia." Australia and New Zealand are European outposts in an Asian Pacific world, as unlike Indonesia as Britain and America are unlike India.

Although Australia was spawned by Europe, and its people and economy are Western in every way, Australia as a geographic realm is a far cry from the crowded, productive, complex European world. Australia's entire population (nearly 16 million in 1985) is only slightly larger than that of the Netherlands, but Australia is about 200 times as large territorially. This gives an average population density of under 2 persons per square kilometer (4.6 per square mile) and suggests that Australia is a virtual population vacuum on the very edge of overpopulated Asia. But so much of Australia

**Figure A–1**
Sydney: Australia's largest city and port.

is arid or semiarid (Fig. I–10) that only about 8 percent of its total area is agriculturally productive, and much of this moister part of the continent is too rugged for farming. By some calculations, only 1 percent of Australia's total area of more than 7.6 million square kilometers (3 million square miles) is prime land for intensive cultivation. Certainly, Australia could support many more people than it does today, but it is no feasible outlet for Asia's millions. Australia's *total* population is less than India's annual *increase*.

## Migration and Transfer

Australia's indigenous peoples probably arrived on the island continent from Southeast Asia by way of New Guinea across a land bridge that existed when enlarged Pleistocene glaciers lowered global sea levels. The original Australians were black peoples with ancestral roots in East Asia, and they never numbered more than 300,000 to 350,000; they were hunters and gatherers who lived in small groups. Cultural anthropologists have identified over 500 tribes and clans using different and mutually unintelligible languages, and cultural geographers have determined that they were more numerous in eastern than in western Australia. The first Australians skillfully adapted themselves to a difficult natural environment and developed a strong sense of territoriality. Almost every narrative about Australia's indigenous peoples remarks on their detailed knowledge of the terrain, its opportunities and limitations.

Today, the first Australians are a nearly forgotten people. They could not withstand the impact of the European invasion, which was delayed because of Australia's remoteness and its apparent unattractiveness. Portuguese, Spanish, and Dutch explorers saw and landed on Australian coasts during the sixteenth, seventeenth, and eighteenth centuries (Abel Tasman, the Dutch seaman, reported on Tasmania's coast in 1642), but it was not until the journeys of James Cook, the British sea captain, that Australia finally entered the European orbit. When Cook visited the east coast in 1770, he was the first European to see this part of Australia. At the behest of the British government, he returned in 1772 and again in 1776, and, by that time, the British were determined that *terra australis incognita* should become a British settlement. In 1778—just over two centuries ago—white settlement in Australia began, and with it the demise of the aboriginal societies.

The Europeanization of Australia involved the movement of hundreds of thousands of emigrants, most of them British, from Western Europe to the shores of a little-known landmass in another hemisphere. This type of external migration has affected human communities from the earliest times, but it reached an unprecedented level during the nineteenth century. Before the 1830s, fewer than 3 million Europeans had left their homelands to settle in the colonies. But, between 1835 and 1935, perhaps as many as 75 million departed for other lands—for the Americas, for Africa, and for Australia (Fig. A–2).

If so many people had not left Europe during the great population expansion that accompanied the industrial revolution, Europe's population problems would have been

## TEN MAJOR GEOGRAPHIC QUALITIES OF AUSTRALIA

1. Australia lies remote from the places with which it has the strongest cultural and economic ties.
2. Australia and New Zealand constitute a geographic realm by virtue of territorial dimensions, relative location, and cultural distinctiveness, not population size.
3. Australia has the lowest average elevation and the lowest overall relief of all the continental landmasses.
4. Australia is marked by a vast arid and semiarid interior, extensive, open plainlands, and marginal moister zones.
5. Australia has a large and diverse natural resource base and contains substantial untapped potential.
6. Australia's population has a very low arithmetic and a low physiologic density.
7. A very high percentage of Australia's population is concentrated in a small number of major urban centers.
8. Australia's population distribution is strongly peripheral as well as highly clustered.
9. Australia's indigenous (black) population was almost completely submerged under the European invasion, remains numerically small, and participates only slightly in the modern society.
10. Australian agriculture is highly mechanized and produces large surpluses for sale on foreign markets. Huge livestock herds feed on vast pasturelands.

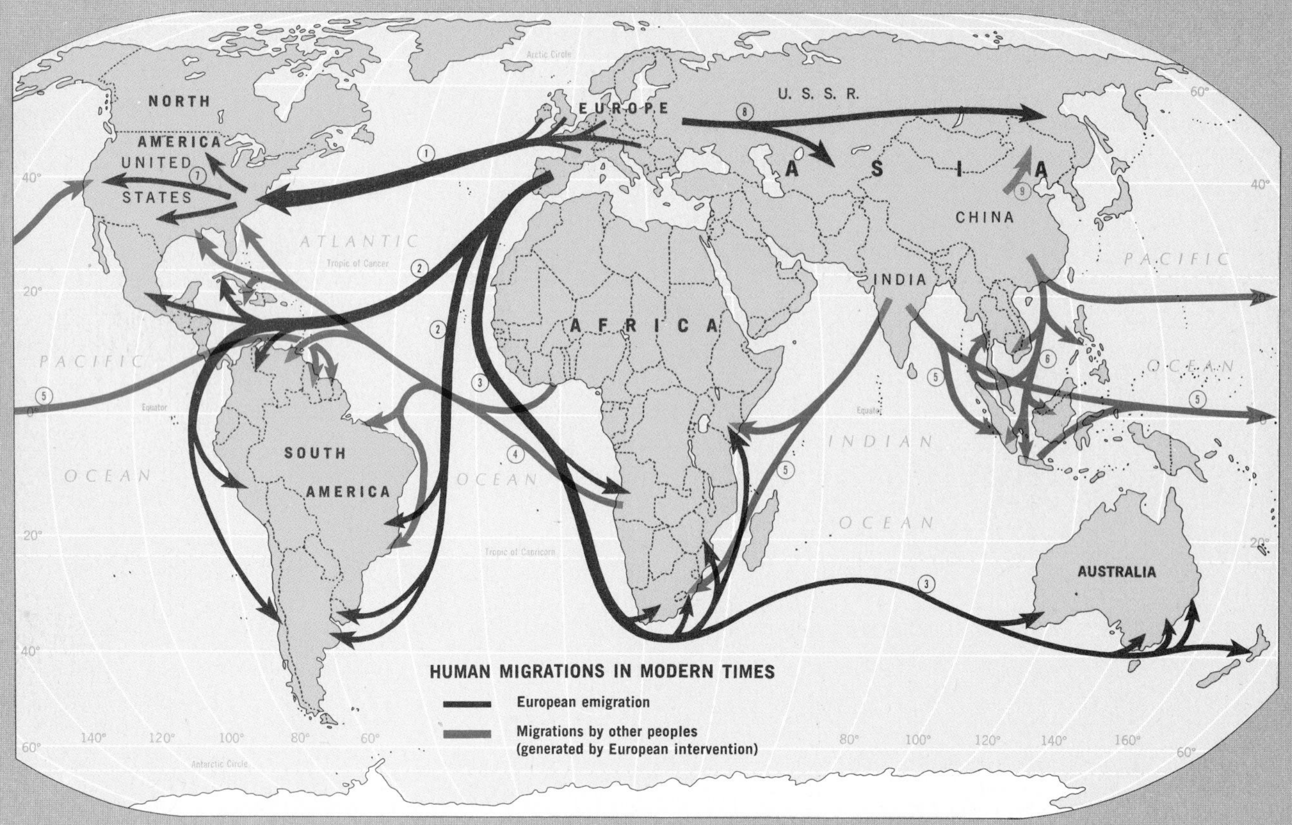


Figure A–2

far more serious than they were. Even so, Europe during the nineteenth century was, for many people, an unpleasant place to be. Famine swept Ireland; war and oppression overtook parts of the mainland; and the cities of industrializing Britain, Belgium, and Germany contained some of the worst living conditions ever. Many Europeans abandoned their homes to seek a better life across the ocean, even though they were uncertain of their fate in their new environments.

Many more Europeans came to the Americas than went to Africa or Australia. What led to their decision to select the New World? Studies focusing on the migration decision indicate that the intensity of a migration flow varies with such factors as: (1) the perceived degree of difference between one's home and the planned destination, (2) the effectiveness of the information flow—the news sent back by those who migrated to those who stayed behind awaiting details, and (3) distance. A century ago, a British social scientist, E. G. Ravenstein, studied migration in England and concluded that an inverse relationship exists between the incidence of migration and the distance between source and destination. Subsequent studies have modified Ravenstein's conclusions. The concept of intervening opportunity, for example, holds that people's perception of a faraway destination's comparative advantages is changed when there are closer opportunities. In the days when Britishers emigrated to lands other than North America, many chose South Africa rather than cross yet another ocean to Australia.

Studies of migration also refer to *push* factors (conditions that tend to motivate people to move away) and *pull* factors (circumstances that attract people to a new destination). Australia was a new frontier, a place where one might acquire a piece of land, a herd of livestock, or where a piece of property might prove to hold valuable minerals; the skies (it was said) were clear, the air fresh, the climate much better than that of England.

Many of the first Europeans to arrive in Australia, however, were not free, but in bondage. We have just described voluntary migration, in which movers made their own decisions to relocate; this is not the way thousands of the first European Australians reached their new homeland. European countries used their overseas domains as dumping grounds for people who had been convicted of some offense and sentenced to deportation—or "transportation," as the practice was called. This is a form of forced migration, in which voluntary push-pull factors play no role. "Transportation" had long been bringing British convicts to American shores, but, in the late 1770s, this traffic was impeded by the Revolutionary War. British jails soon were overcrowded, and judges continued to sentence violators to be deported; this was a time when Britain was undergoing its own economic and social revolution and offenders were numerous. An alternative to America for purposes of "transportation" simply had to be found, and it was not long before Australia was suggested. From the British viewpoint, the far side of Australia, the east coast, was an ideal place for a penal colony. It lay several thousand miles away from the nearest British colony and would hardly be a threat; its environment appeared to be such that the convicted deportees might be able to farm and hunt. In 1786, an order was signed making the southeast corner of Australia, known as New South Wales, such a penal colony. Within two years, the first party of convicts arrived at what is today the harbor of Sydney and began to try to make a living from scratch.

For those of us who have learned of the horrible treatment of black slaves transported to America, it is revealing to see that European prisoners sentenced to deportation were not any better off, despite the fact that the offense of many of them was simply their inability to pay their debts. The story of the second group of deportees to Australia gives an idea: more than 1000 prisoners were crammed aboard a small boat; 270 died on the way and were thrown overboard. Of those who arrived at Sydney alive, nearly 500 were sick; another 50 died within a few days. What the colony needed was equipment and healthy workers; what it got was hardly any tools and ill and weakened people. The first white Australians hardly seemed the vanguard of a strong and prosperous nation. But, by the middle of the nineteenth century, as many as 165,000 deported offenders had reached Australian shores, and London was spending a great deal of money on the penal stations. Meanwhile, the numbers of free colonists were increasing, convicts were entering free society after serving their sentences, and new settlements were founded. Australia's image in England began to change. As its penal image faded, the continent's spacious beauty and its economic opportunities beckoned. The last convict ship arrived in 1849. The flow of voluntary immigrants grew, and along Australia's coastline several thriving colonies emerged whose tentacles soon reached into the interior.

# Livelihoods and Urbanization

Australia's physiographic regions are identified on Fig. A–3, but another regionalization of Australia is more meaningful. Australia is a coastal rimland with cities, towns, farms, and forested slopes giving way to a dry, often desertlike interior that Australians themselves call the "outback" or the "inland" (Fig. A–4). The habitable coastal rimland is not continuous, for both in the south and in the northwest the desert reaches the coast (Fig. I–10). In the east, lies the Great Dividing Range, extending from Cape York in the north to the island of Tasmania in the south; between these highlands and the sea lie the fertile, well-watered foothills where Europeans in Australia had their start. Across the Great Dividing Range lie the extensive grassland pastures that catapulted Australia into its first commercial age—and on which still range one of the greatest sheep herds in the world (nearly 135 million sheep in 1979 produced 28 percent of all the wool sold in the world that year).

But Australia is not just a nation of stockbreeders. The steppe and near-desert country where sheep are raised and the moister northern and eastern regions where cattle by the millions are ranched—these are Australia's sparsely inhabited areas (Fig. I–14) where sheep stations lie separated by dozens of kilometers of parched countryside, where towns are small and dusty, where a true frontier still exists. The great majority of modern Australians—about 90 percent by the latest census—live in the cities and towns on or near the coast. In this respect, Australia's situation is quite similar to that of Europe itself, for in Britain, too, 9 out of 10 people live in urban areas.

Australia's cities began as penal colonies (as was the case at Sydney Cove and Brisbane), as strategic settlements to protect trade and confirm spheres of influence (Perth in the southwest), or as centers of local commerce and regional markets (Adelaide, Melbourne). They became the foci of self-governing colonies; for a long time, Australia and Tasmania were seven ocean ports with separate hinterlands. These political regions were delimited by Australia's now-familiar pattern of straight-line boundaries, which were completed by 1861 (Fig. A–5). Notwithstanding their shared cultural heritage, the Australian colonies found themselves at odds not only with London over colonial policies, but also with each other over economic and political issues. National integration was a slow and difficult process, even on the ground. Each of the Australian colonies laid down its own railroad lines, not with a view toward continentwide transport, but with local objectives; in the process, three different railroad gauges came into use. The reorientation and unification of these separate systems was costly and difficult; not until 1970 was it possible to go by railroad from Sydney to Perth without changing trains.

Australians share with Europeans their increasing mobility. Australia

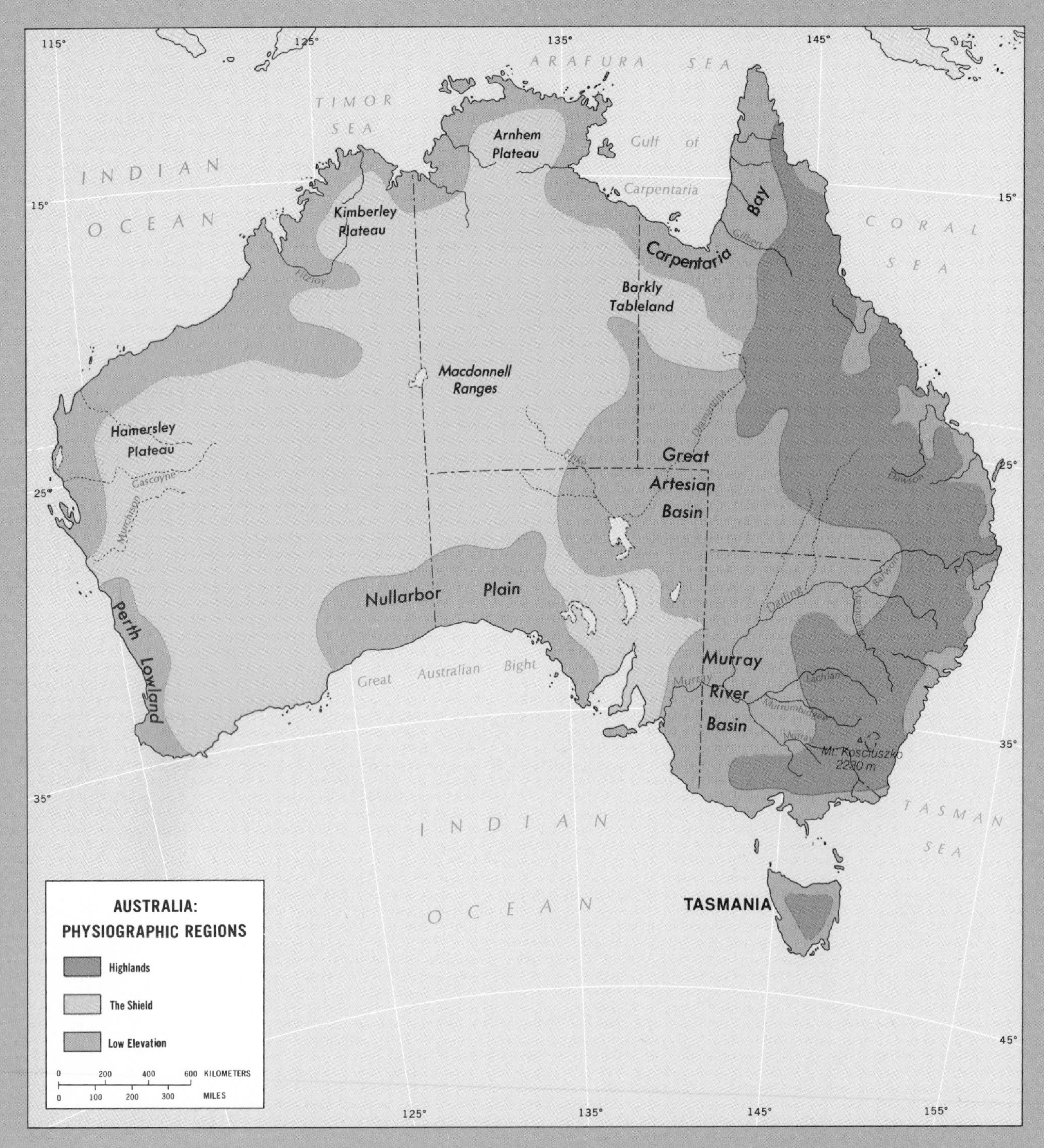


Figure A–3

now counts more automobiles per 100 persons than even the United States, and Australians, like North Americans, are quite accustomed to long-distance travel. Australia is nearly as large as the contiguous United States, but its population of about 15 million is what the United States white population was in 1825. In Australia, neighbors are often far removed. Perth in Western Australia is as far from Sydney (New South Wales) as Los Angeles is from Atlanta, and Brisbane and Adelaide are as far apart as Washington, D.C., and Miami. Not surprisingly, there are today more road kilometers per person in Australia than in the United States.

Australia's cities retain some of the competitive atmosphere of the colonial period, although they are as modern, spacious, generally well planned, and suburbanized as many cities in the Western world are. Sydney (3.5 million), capital of New South Wales and Australia's

Figure A-4
The Olga Mountains in the barren Outback.

largest city and leading port, has a long history of rivalry with Melbourne, which at one time overtook Sydney in size and was Australia's temporary capital. Sydney lies on a magnificent site, its skyscrapered central city overlooking the famed Sydney Harbour Bridge (Fig. A–1); but apart from the new Opera House, its architecture is no match for less spectacular, but historically more interesting, Melbourne, which is still the capital of Victoria. Between them, Sydney and Melbourne contain nearly 7 of Australia's 15.7 million inhabitants, and Brisbane (Queensland) and Adelaide (South Australia) each have a population of approximately 1 million. Perth, Western Australia's capital, has more than 1.1 million residents, whereas Hobart, Tasmania, has under 165,000. By far the smallest of the regional capitals is Darwin (Northern Territory), devastated by the Christmas Day tropical cyclone of 1974, when it had a population of about 37,000; thousands have never returned to the scene of Australia's worst natural disaster. Australia's federal capital, the planned city of Canberra, still has a population of less than 500,000.

# Farming

Sheepraising thrust Australia into the commercial age, and the technology of refrigeration brought European markets within reach of Australian beef producers (who also comprise an important branch of the national economy)—but there is more to Australian rural land use than herding livestock. Commercial crop farming concentrates on the production of wheat; Australia is one of the world's leading exporters of wheat, which brings in nearly half the amount of income derived from wool.

Wheat production is concentrated in a broad belt that extends from the vicinity of Adelaide into Victoria and New South Wales (along the rim of the Murray Basin) and even into Queensland, where a separate area is shown in Fig. A–6 in the hinterland of Brisbane. It also exists in an area behind Perth in Western Australia. Although these zones are mapped as "commercial grain farming," a unique rotation system exists here, whereby sheep and wheat share the land. Under this mixed crop and livestock farming, the sheep use the cultivated pasture for several years, and it is then plowed for wheat sowing; after the wheat harvest, the soil is rested again. This system plus the innovations in mechanized equip-

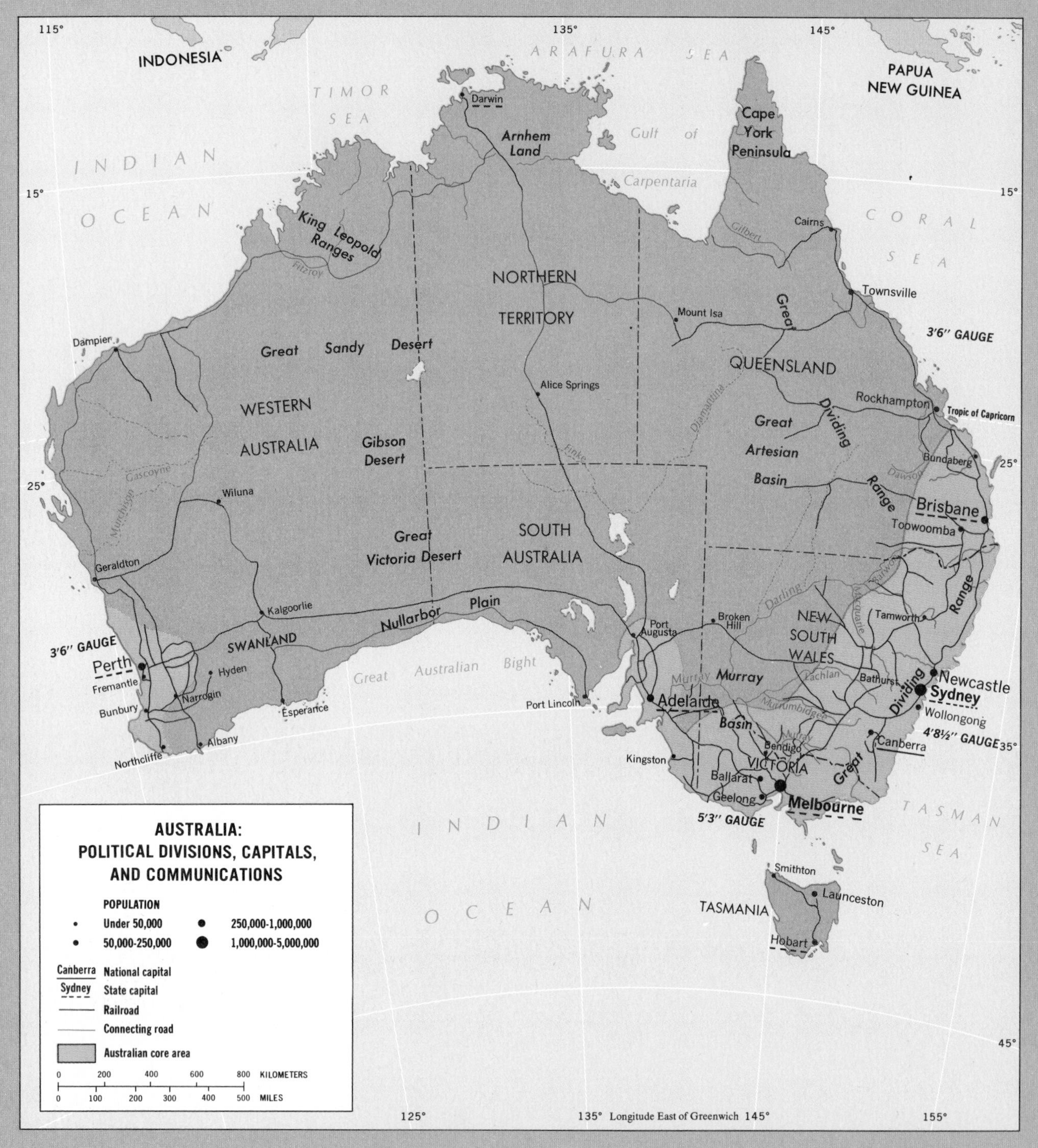


Figure A–5

ment the Australians themselves have made has created a highly lucrative industry. Australian wheat yields per hectare are still low by American standards, but the output per worker is about twice that of the United States.

As Fig. A–6 shows, dairying, with cultivated pastures, takes place in what Fig. I–9 shows to be areas of high precipitation. The location of the dairy zones also suggests their development in response to Australia's comparatively large urban markets. Even these humid areas of Australia are occasionally afflicted by damaging droughts, but the industry is normally capable of providing more than the local market demands. Dairying is, in fact, the most important rural industry as measured by the number of people who find work in it.

It is natural that irrigation should be attempted in a country as dry as Australia. Unfortunately, the opportunities for irrigation-assisted agri-

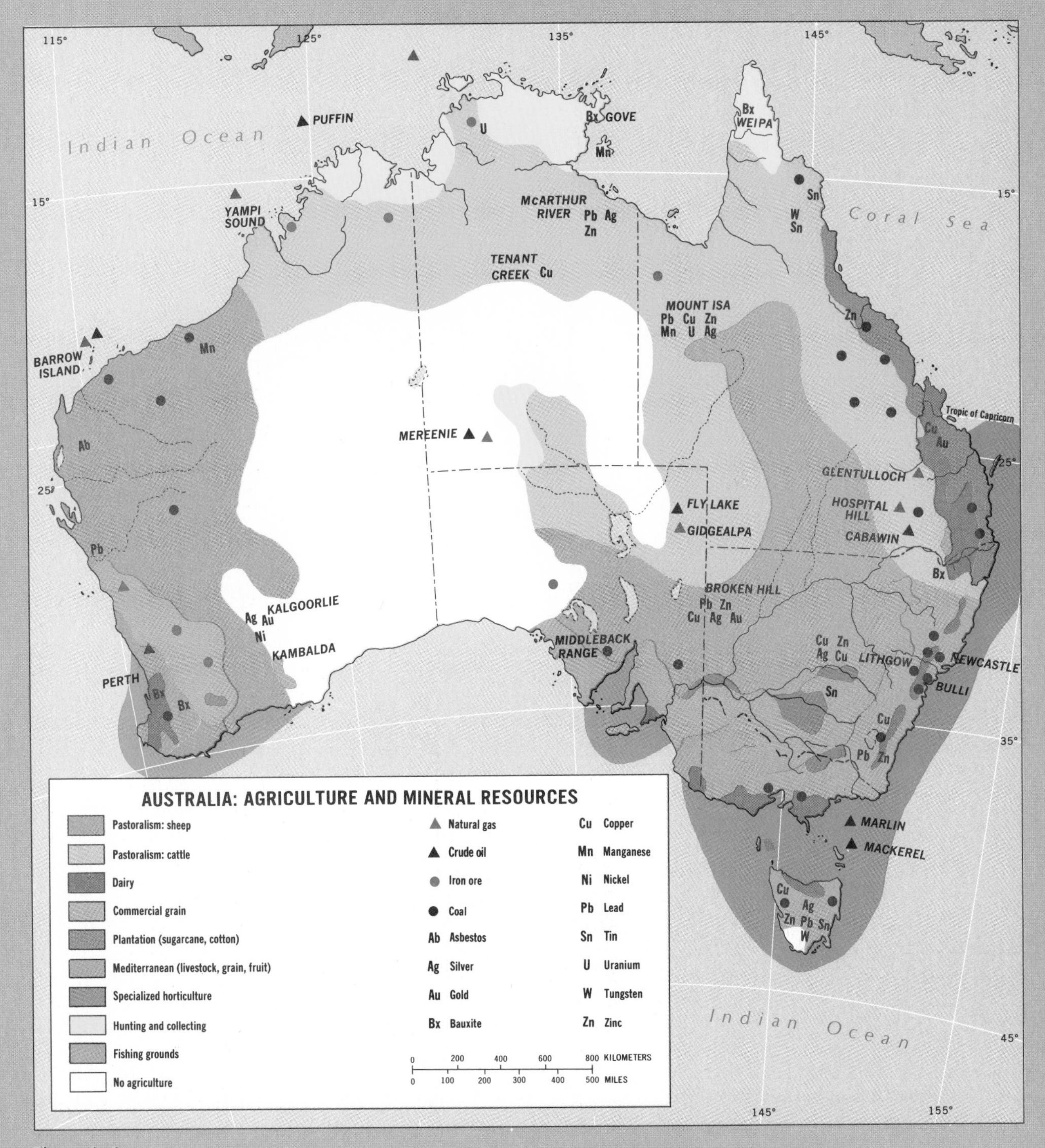


Figure A–6

culture are quite limited: the major potential lies in the basin of the Murray River, shared by the states of Victoria (which leads in irrigated acreage) and New South Wales. In addition, rice (produced at one of the highest per hectare yields in the world) in the Wakool and Murrumbidgee valleys and grapes and citrus fruits are among the irrigated crops in the Murray River Basin. On the east coast, in Queensland and northern New South Wales, the cultivation of sugarcane is partly under irrigation.

# Mineral Resources

Nothing shook the Australian economy as much as the discovery of gold in 1851 in the territories of New South Wales and Victoria. In no way—politically, socially, or materially—was Australia prepared for what followed the news of the gold finds. True, gold had been known to exist prior to 1851, but only when some Australian diggers returned from California's fields to prove the lucrative character of the Australian ores did the rush begin. During the decade from 1851 to 1861, the population of the colony of Victoria increased more than sevenfold, from just over 75,000 to well over a half million; yet, Melbourne, the colony's leading town, was just 16 years old when the rush began and had none of the amenities needed to serve so large a number of people. New South Wales, which in those days still included all of what is now Queensland and the Northern Territory, saw its population rise to over 350,000, and its pastoral economy was rudely disturbed by the rush of diggers to Bathurst. Overall, Australia's population nearly tripled in the 1850s; gone was the need for subsidized immigration in all areas but the west and northeast.

The new wealth of Australia brought problems of accommodation for the new settlers, and inevitably it brought on some of the disorder and political conflict that also marked the gold rushes in North America and Africa. But it also brought great advantages. A new prosperity came to every Australian colony; in the 1850s, the continent produced 40 percent of the world's gold. It brought successful searches for other minerals (Tasmania, with its tin and copper, was an early beneficiary). Discoveries are still being made. Long believed poor in petroleum and natural gas (oil always ranks high on the import list for this urbanized, industrialized society), Australia seems to have sizable supplies of both fuels after all, if recent finds on the margins of the Great Artesian Basin are large enough to warrant exploitation. Additional reserves have been located on the continental shelf between the mainland and Tasmania, and these, in the Bass Strait, now produce half of Australia's crude oil requirements. Recently, a deposit of nickel was found near the center of Western Australia's gold mining area, Kalgoorlie. For nearly 80 years, people have been mining and searching in this area, yet, the nickel discovered recently at Kambalda may constitute the world's largest known reserve of this important ferroalloy. Even today, the Australian frontier still holds its secrets and surprises.

As Fig. A–6 suggests, Australia's mineral deposits are scattered and quite varied. The country is fortunate in being so well endowed with coal deposits when its waterpower prospects are minimal (except in Tasmania and the southeast) and its petroleum discoveries so long delayed (the Bass Strait fields were not found until 1964). The chief coalfields, as the map indicates, lie in the east, notably around Sydney—north at Newcastle, west at Lithgow, and south at Bulli. In the hinterland of Brisbane and about 200 kilometers (75 miles) inland from Rockhampton (Queensland), bituminous coal is mined. From the New South Wales field, this energy resource is sent by coastal shipping to areas that are deficient in coal. But coal is widely distributed—even Tasmania and Western Australia have some production and can, thus, keep their import necessities down. There is no doubt that coal is Australia's more important mineral asset: it is used for the production of 90 percent of the electricity the country consumes, for railroads, factories, manufactured gas production, and a host of other purposes.

The most famous Australian mining district, undoubtedly, is Broken Hill, which has neither gold, coal, nor iron. Discovered in the aftermath of the gold rush, Broken Hill became one of the world's leading lead- and zinc-producing areas (in the beginning it was important also for silver), and the enormous income derived from the export sales of these minerals provided much of the capital for Australia's industrialization. Japan has become one of the principal buyers, and Australia's "Japanese connection" has proved very lucrative. In the early 1980s, Japan bought about 30 percent of all Australian exports. In Queensland, the Mount Isa ore body yields a similar mineral association, and, in both areas, important uranium deposits have also been mined. But these products are only a small part of the total Australian inventory. Tasmania's copper, South Australia's large bauxite (aluminum ore) deposit in the York Peninsula, Queensland's tungsten, and Western Australia's asbestos represent the vast range of the continent's minerals. There are only a few nonmetallic resources in which the country is truly deficient, and the real search has just begun. In terms of mineral resources, Australia is, indeed, in a fortunate position.

# Manufacturing

For a long time, Australia was to Britain what colonies always were, a supplier of needed resources and a ready market for British-manufactured products. But World War I, at a crucial stage, cut these trade connections—and Australia for the first time was largely on its own. Now it had to find ways to make some of the consumer goods it had been getting from Britain. By the time the war was over, Australian industries had made a great deal of progress. Wartime also pushed the development of the food-processing industries, and, today, Australian manufacturing is varied, producing not only machinery and equipment made of locally manufactured steel, but also textiles and clothing, chemicals, foods, tobacco, wines, and paper—among many other items.

Australia's industries, naturally, are situated where the facilities for manufacturing are best (and where markets exist)—the large cities. Of a labor force numbering over 6.5 million in 1985, nearly one-quarter is employed in manufacturing, although agriculture (which employs only 5 percent of the labor force) and mining remain the mainstays of the economy. Australia is not, as yet, a major exporter of finished products; manufacturing of steel products, automobiles, textiles, chemicals, and electrical equipment are oriented to the domestic market. Although this market is small numerically, it is not a minor factor: per capita income in Australia in 1983 was almost $8,000. Although Australia in 1980 had only 15 million inhabitants, it had over 6.5 million vehicles on its roads, nearly 5.5 million of them passenger cars. (The United Kingdom, with a population of 56 million, had fewer than three times as many.)

The success of domestic industries is a matter of prime concern for the Australian government. Australian manufactures have not taken their place on world markets, and foreign products pose a challenge. During the early 1980s, Australian industries experienced a serious downturn accompanied by unemployment problems; at the same time, Japan pressed Australia to lower tariff barriers against Japanese imports, notably automobiles. Thus, plentiful raw materials notwithstanding, Australian manufacturing faces high production costs and long distances to foreign markets, and it cannot expect to compete overseas. Indeed, Australia's list of imports still includes machinery and equipment, chemicals, foods and beverages that could, in part at least, be produced at home. In terms of its manufacturing industries, Australia still exhibits some symptoms of a developing country.

# Population Policies

Not surprisingly, the state capitals of Australia became the country's major industrial centers. These urban clusters, at the foci of the railroad networks of the individual states, were also the main ports for overseas as well as coastwise shipping; they had the best amenities for industry—concentrated labor force, adequate power supply, water, access to government—and they were, at the same time, the major markets. The process of agglomerative self-perpetuation, to which reference was made in the context of some of Europe's larger cities, is well illustrated here. In addition, the cities and manufacturing industries have absorbed the vast majority of Australia's recent European immigrants (e.g., Melbourne now contains more Greeks than any other city outside Greece).

The Australian government pursues a vigorous policy of encouraging selective immigration; the majority of the new Australians come from Britain, but numerous Italians, Hollanders, Germans, Turks, and Greeks also seek new homes in Australia. About half of Australia's population today is of English origin; about 20 percent has Irish roots; 10 percent is of Scottish ancestry—in all, 20 percent has non-British (mostly mainland European) roots. In recent years, about 110,000 immigrants have annually come to Australia, and most of them have settled in the cities and found work in industry. The advantages of immigration to Australia are obvious: a strengthening of the country's still small population, an expansion of the skills and numbers of the labor force, an increase in the size of the domestic market, and a growing consolidation of the Australian presence on the large island-continent (from a strategic viewpoint).

Australia has long been known for its selective immigration policies, virtually excluding persons who are not white from the opportunity to settle there. Since 1901, successive Australian governments adhered to an unofficial "White Australia" policy; after World War II, international criticism mounted, so, in the 1950s, the practice began to soften. First, the government approved the admission of family members of Australians who were not white; in 1966, it became possible for Asians to apply for Australian residence—but only those with needed skills are admitted even today. The rate of immigration by people who are not white is still quite small but has approached 10,000 annually in recent years.

In the meantime, the indigenous Australian population, which numbered as many as 350,000 two centuries ago, has been reduced to a mere remnant. It is the familiar story of contact between indigenous peoples living a life of subsistence in small groups and the onslaught of an advanced society: the black Australian communities were driven away, broken up, often destroyed. Today, fewer than 50,000 Australians still have a totally aboriginal ancestry; another 100,000 have mixed origins. Only a few communities still live the original life of hunting and gathering, mainly in the Northern Territory, Western Australia, and northern Queensland. The remainder are scattered over the continent, working on the cattle stations, living near Christian missions, or subsisting on reserves set aside for them by the government; many perform menial jobs in the towns. In recent years, the Australian conscience has been aroused, and an effort has begun to undo centuries of persecution and neglect—these attempts are met, understandably, with suspicion and distrust. In affluent Australia, the original Australians remain victims of poverty, disease, insufficient education, and even malnutrition.

Most recently, Australia has faced a new immigration problem: the arrival of boatloads of Asians (70,000 between 1976 and 1983), mostly from Vietnam and Kampuchea (Cambodia), whose desperation drove them out to sea in the hope of a landfall in a hospitable place (see Fig. 10–14, p. 524). The phenomenon of the "boat people" is not solely a Southeast Asian one. Overcrowded boats with Haitians aboard have been arriving on the shores of Caribbean islands and the United States for years; their accommodation has presented governments in the Americas with problems as well.

In Australia, immigration policies and procedures were modified significantly in 1979, raising immigration quotas and liberalizing admissions requirements. These changes were not adopted without opposition; labor leaders charged that the immigration of a large number of workers without guaranteed jobs would create a docile work force and sustain unemployment, which would be advantageous to employers but not to workers and their unions. Certainly, the 1970s had been difficult for Australia as the energy crisis of 1973 and its aftermath reduced the economy's ability to absorb newly arrived workers. In 1978, unemployment among new immigrants was reported to be nearly 20 percent.

Against this background came the issue of the Vietnamese (and other) refugees. They came by way of Malaysia and Indonesia and arrived, mostly, at or near the northern port of Darwin. Soon, the inevitable problems of adjustment mounted, and a national debate arose. The Australian labor movement always was the focus for the "White Australia" immigration policy and traditionally opposed Asian immigration; furthermore, the Australian Labor Party (the opposition party) had opposed the Vietnam War. Thus, refugees from the new order in Vietnam were not welcome on several grounds—in addition to their circumvention of established immigration policy. Australian government leaders were dispatched to Thailand and Malaysia to seek assistance in restraining would-be "boat people," but they reported that many in Southeast Asia, officials and refugees alike, believed Australia to be a vast open land of promise and opportunity.

Quite probably, Vietnam's "boat people" are but the vanguard of what is likely to be a growing stream of irregular immigrants. Just as U.S. soil beckons the oppressed and the poor in Middle America, so is Australia perceived as a desirable objective by Asian emigrants—and not just the Vietnamese "boat people." Political and economic crises in South, Southeast, and East Asia as well as the Pacific will generate new waves of emigrants. Indonesia's stepping stones, its corridor of islands (or *archipelago*) stretching toward Australia, lead toward a threshold unlike any other in the entire Asian-Pacific hemisphere. Therefore, Australia's relative location may, during the final decade of the twentieth century, become its greatest challenge.

# Australian Federalism

Australia may be a European offshoot, but in at least one respect it differs significantly from most European countries. Australia is a federal state; all but a few European countries are unitary states.

The term *unitary* derives from the Latin *unitas* (unity), which in turn comes from *unus* (one). A **unitary state**, then, is unified and centralized. It is not surprising that many European states, long ruled with absolute authority from a royal headquarters, developed into unitary states. The distant colonies had no such tradition, whether in Canada, the United States of America, or Australia. In Australia, six colonies for most of the nineteenth century went their own way, competing for hinterlands and arguing over trade policies. Eventually, however, the idea that they might combine in some greater national framework began to take hold. But this would be no unitary state: each colony wanted to retain certain rights. Thus, Australia became a **federal state**.

The federal concept also has ancient Greek and Roman roots, and the term originates from the Latin *foederis*, meaning league or association. In practice, the federal system is one of alliance and coexistence and the recognition of differences and diversity. Of course, there must be an underlying foundation for a federal union: in Australia, this was, in the first instance, the colonies' common cultural heritage. But there were other bases as well. The colonists worried about a foreign invasion from Asia or some mainland European power; there was a particular fear of Germany, for, in the 1880s, the Germans embarked on a vigorous colonization campaign in Africa and the Pacific. In 1883, the colonial government of Queensland took control of the eastern part of New Guinea because it feared a German takeover there. Also, the colonies shared the objectives of selective immigration. A coordinated White Australia policy would be more easily administered on a national basis. The colonies realized that their internal trade competition was damaging and precluded a coordinated commercial policy on world markets.

With such positive prospects, it would seem that federation would have been accomplished quickly and without difficulties, but this was not the case. Victoria's manufacturers wanted protective tariffs; public opinion in New South Wales favored freer trade. There was wrangling over the site of the future federal capital, and no colony was willing to yield to another capital city the advantage of becoming the federal headquarters. But in 1900, lengthy conferences succeeded in creating a federal constitution approved by the voters in the colonies and confirmed by London. From January 1, 1901, the colonies became the *Commonwealth of Australia* consisting of six states and the federal territories of the future capital as well as the Northern Territory, the sparsely peopled rectangular area that extends from Arnhem Land to the heart of the Australian Desert (Fig. A–5).

## Canberra: The Federal Capital

More than one federal government has found the selection of a federal capital a difficult task requiring compromise and consent. Australia experienced this problem, arising mainly from the rivalry between the adjacent southeastern states of Victoria and New South Wales. Victoria's capital, Melbourne, was the chief competitor of Sydney, headquarters of New South Wales. Shortly before the implementation of federation, the premier of New South Wales managed to secure an agreement that the federal capital would lie in that state—in return for two concessions. First, the new capital would *not* be Sydney, but a totally new city to be built at a site to be jointly selected. Second, Melbourne would function as the temporary capital until the federal headquarters was ready for occupation.

**Figure A–7**
Canberra, the planned capital city.

The site of Canberra was chosen in 1908, and the buildings necessary for the government functions were ready by 1927. Canberra, unlike other major Australian cities, lies away from the coast, at the foot of Mount Ainslie in the Great Dividing Range (Fig. A–5). Once the site was selected, a worldwide architectural design competition was held in order to acquire the best city plan, for this would be a totally new city. The U.S. architect Walter Burley Griffin was chosen for the task; today, Canberra is a spacious, modern, widely dispersed urban area with some 475,000 residents (Fig. A–7). Its nearly 2500 square kilometers (940 square miles) comprise the Australian Capital Territory, a federal district separated from the state of New South Wales.

Canberra shares with other comparatively new federal capital cities a sense of remoteness and isolation. In 1980, the highway to Sydney, just under 300 kilometers (200 miles) long, still was interrupted by lengthy stretches of overcrowded two-lane roads. The city itself, endowed with impressive public architecture, ceremonial places, and parks, still retains an atmosphere of coldness and distance. It lacks Sydney's drive and bustle, Melbourne's style and comfort, and Adelaide's sense of history. The presence of the government and the Australian National University notwithstanding, Canberra is a far cry from the pulse of Australia.

## A Youthful State

It states do, indeed, experience life cycles, Australia exhibits the energy and vigor of youth. Many observers have linked Australia today to the United States of a century and a half ago: still discovering major resources, still in the early stages of the penetration of national territory, still essentially peripheral, the interior still a frontier—albeit a modern frontier with airstrips rather than stagecoaches.

Australia's cities, where the great majority of Australians live, reflect the country's youthfulness in several ways. The largest, Sydney, contains well over 20 percent of the entire Australian population. Sydney anchors a metropolitan area of over 12,000 square kilometers/ 4600 square miles (larger than New York or London), but internal circulation is hampered by the absence of adequate highways, as available facilities have been overwhelmed by the still mushrooming population and its automobiles. The city itself is also severely congested: traffic jams on the approaches to picturesque but inadequate Harbour Bridge (opened in 1932), linking Sydney's north and south (Fig. A–1), are among the world's worst. Major arteries, optimistically called the Great Western Highway and the Pacific Highway, begin, at best, as wide city streets whose traffic is slowed by numerous stoplights, lane inconsistencies, and inadequate posted directions.

Australia never has had a space problem, and the cities reflect this: suburbs sprawl far and wide. Downtown Sydney lies at the heart of a conurbation consisting of five individual cities and more than 30 municipalities, a built-up region that extends more than 80 kilometers (50 miles) from north to south. Melbourne's dimensions are only slightly less. Few remnants in the urban landscape reveal the cities' ages: modern towering skyscrapers dominate the central business districts. Melbourne, capital of Victoria, retains more architectural interest and is a more cultured city than Sydney; it is also better served by major traffic arteries. Perth, the capital of enormous Western Australia, is the youngest of the state capitals, a thriving urban center with a population now exceeding 1 million, and still the most frontierlike among Australia's large cities.

The Australian economy also reflects the country's youth. Notwithstanding its urbanized and industrialized character, Australia still imports large quantities of consumer goods—many of which could be manufactured at home. Certainly, Australians can afford to pay for these imports; they enjoy one of the world's highest living standards. But the situation reveals the limitations of a small domestic market and the liabilities of distance to world markets as well as the absence of any urgent need to compete in faraway marketplaces. Unlike the United Kingdom, where manufactures must be sold, so that food may be bought, Australia has a small population and secure world markets for its wool, beef, and wheat. Explanations for the condition of Australia's industries sometimes refer to the Australian preference for the leisurely life, and, indeed, Australians are a great sports people, excelling in competition from tennis to cricket and rugby to lawn bowling. But it is more likely to be a question of opportunity rather than preference: Australia does not (yet) face the need to maximize efficiency. In their remote "Down Under," Australians have advantages of which many Europeans can only dream.

# New Zealand

Like Australia, New Zealand is a product of European expansion. In another era, New Zealand, located about 1700 kilometers (over 1000 miles) southeast of Australia, would have been included in the Pacific realm, for before the Europeans' arrival, it was occupied by the Maoris, a people with Polynesian roots. Today New Zealand's population of 3.2 million is more than 90 percent European, and the Maoris form a minority of about 300,000—with many of mixed ancestry.

New Zealand consists of two large mountainous islands and numerous scattered smaller islands. The two large islands, with South Island somewhat larger than North Island, together cover just over 268,000 square kilometers (104,000 square miles); although they look diminutive in the great Pacific Ocean, they are, nevertheless, larger than Britain (Fig. A–8). In sharp contrast to Australia, the islands are mainly mountainous or hilly, with several mountains rising far higher than any on the Australian landmass. South Island has a spectacular, snowcapped range appropriately called the Southern Alps, with peaks reaching beyond 3500 meters (11,700 feet). Smaller

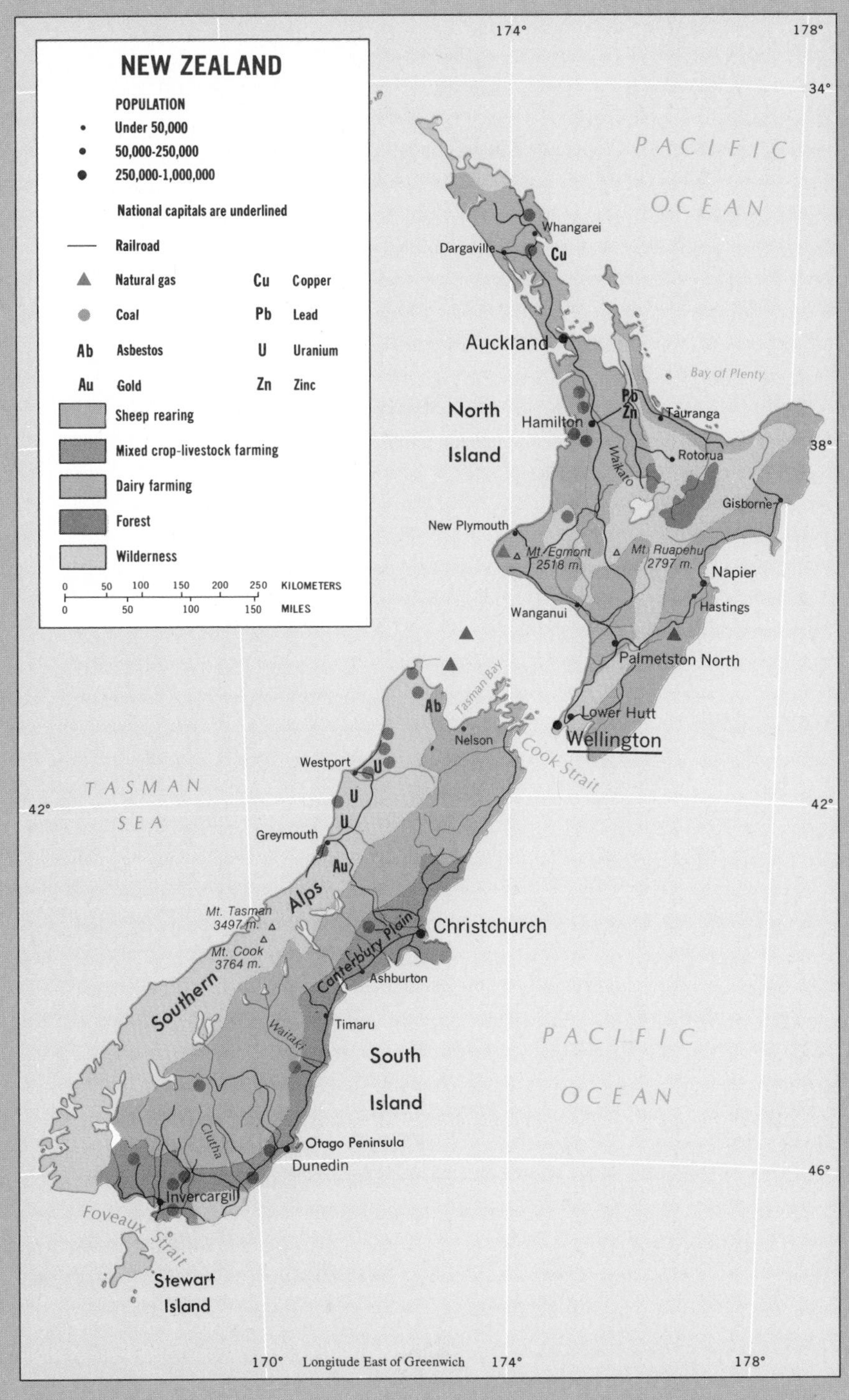


Figure A–8

North Island has proportionately more land under low relief, but it also has an area of central ranges and high mountains on whose lower slopes lie the pastures of New Zealand's chief dairying district. Hence, whereas Australia's land lies relatively low in elevation and has much low relief, New Zealand's is on the average quite high and has mostly rugged relief. Thus, the most promising areas must be the lower-lying slopes and lowland fringes on both islands. On North Island, the largest urban area, Auckland (just over 900,000) occupies a comparatively low-lying peninsula; on South Island, the largest lowland is the agricultural Canterbury Plain, centered on Christchurch (340,000). What makes these lower areas so attractive, apart from their availability as cropland, is their magnificent pastures. Such is the range of soils and pasture plants that both summer and winter grazing can be carried on; and a wide variety of vegetables, cereals, and fruits can be produced in the Canterbury Plain (Fig. A–9), the chief farming region. About half of all New Zealand is pastureland, and much of the farming is done to supplement the pastoral industry in order to provide fodder when needed. In the drier areas of South Island, wheat is grown; truck farming exists around the cities, and fruit orchards of apples, pears, and grapes are widely distributed from Otago in the south to Nelson in the north.

Thus, New Zealand's pastoral economy is based on sheep and cattle—wool, meat, and dairy products. As in Australia, sheep greatly outnumber cattle (68 million against about 8 million in 1980); in modern times, New Zealand has been among the world's leaders in terms of per capita trade; also, its standard of living is very high. Meat (mutton and beef) combine with wool to provide nearly 75 percent of the islands' export revenues.

Despite their contrasts in size, shape, physiography, and history, New Zealand and Australia have a great deal in common. Apart from their joint British heritage, they share a substantially pastoral economy, a small local market, the problem of great distances to world markets, and a desire to stimulate and develop domestic manufacturing.

The fairly high degree of urbanization in New Zealand indicates another similarity to Australia: sizable employment in city-based industries. About 85 percent of all New Zealanders live in cities and

**Figure A–9**
New Zealand's agricultural Canterbury Plain.

towns, where the industries are still mainly those that process and package the products of the pastoral-agricultural economy. But, although New Zealand is behind Australia in its industrial development, the country's manufactures are becoming more diverse all the time. Textiles and clothing, wood products, fertilizers, cement and building materials, metal products, and some machinery are now made in New Zealand. Among important developments in the industrial sphere are the recent opening of the first steel mill near Auckland and the construction of an aluminum smelter (converting Australian bauxite) at the port of Bluff on South Island.

Spatially, New Zealand shares with Australia its pattern of peripheral development, imposed not by desert but by high rugged mountains. The country's major cities, Auckland and the capital of Wellington (together with its satellite of Hutt, 310,000) on North Island, and Christchurch and Dunedin (125,000) on South Island, are all located on the coast, and the entire railway and road system (Fig. A–8) is peripheral in its orientation. This is more pronounced on South Island than in the north, for the Southern Alps form New Zealand's most formidable barrier to surface contact.

Compared to brash, progressive, and modernizing Australia, New Zealand seems quiet. A slight regional contrast might be discerned perhaps between more forward-looking North Island and conservative South Island, but the distinction fades in the light of Australia's historic internal urban and regional rivalries. Nothing in New Zealand compares to Australia's urban architecture; a sameness pervades New Zealand that is quite unlike the variety of Australia. This may be partly why, in the early 1980s, New Zealand's population was actually dwindling slightly, its natural increase and small immigration exceeded by a high emigration rate, mainly to Australia.

If New Zealand fails to satisfy many of its younger citizens, it is not because of a lack of personal security. The government has developed an elaborate cradle-to-grave system of welfare programs that are affordable because of the country's high incomes and standards of living (taxes also are high). Whether this has suppressed entrepreneurship and initiative is a matter for debate, but New Zealand is a country with few excesses and much stability.

Nevertheless, significant change can be discerned. Auckland not only is New Zealand's largest city, it also may be the largest Polynesian city of all, with nearly 90,000 Maoris, Samoans, Cook Islanders,

Tongans, and others making up 10 percent of the total population. The twentieth century has witnessed a revival of Maori culture, and a still slow but quickening pace of Maori integration into New Zealand society (nearly all the Maoris reside on North Island, where nearly three-quarters of all New Zealanders live). This has enlivened New Zealand's visual arts and music. But Australia and New Zealand prove a point: there is a difference between remoteness and isolation.

## References and Further Readings


Bolton, G. *Spoils and Spoilers: Australians Make Their Environment, 1788–1980* (Winchester, Mass.: Allen & Unwin, 1981).

Cameron, R. *Australia: History and Horizons* (New York: Columbia University Press, 1971).

Cumberland, K. & Whitelaw, J. *New Zealand* (Chicago: Aldine, 1970).

Jeans, D., ed. *Australia: A Geography* (New York: St. Martin's Press, 1978).

Johnston, R. J. *Urbanisation in New Zealand* (Wellington: Reed, 1973).

Linge, G. & Frazer, R. *Atlas of New Zealand Geography* (Wellington: Reed, 1966).

McKnight, T. L. *Australia's Corner of the World* (Englewood Cliffs, N.J.: Prentice-Hall, 1970).

McKnight, T. L. *The Camel in Australia* (Melbourne: Melbourne University Press, 1969).

Moran, W. & Taylor, M., eds. *Auckland and the Central North Island* (Auckland: Longman Paul, 1979).

Neutze, G. M. *Urban Development in Australia* (Sydney: Allen & Unwin, 1977).

Powell, J., ed. *Urban and Industrial Australia* (Melbourne: Sorrett, 1974).

Rapoport, A. *Australia as Human Setting* (Sydney: Angus and Robertson, 1972).

Robinson, K. *Australia, New Zealand, and the Southwest Pacific* (London: University of London Press, 3 rev. ed., 1974).

Rose, A. J. "Cities of Oceania," in Brunn, S., & Williams, J., eds. *Cities of the World: World Regional Urban Development* (New York: Harper & Row, 1983), pp. 162–197.

Rose, A. J. *Patterns of Cities* (Melbourne: Nelson, 1968).

Spate, O. *Australia* (New York: Praeger, 1968).

White, R. *Inventing Australia: Images and Identity, 1688–1980* (London: Allen & Unwin, 1981).

# CHAPTER 2

# THE SOVIET UNION: REGION AND REALM

IDEAS AND CONCEPTS

**Climatology**
**Imperialism**
**Centrally Planned Economy**
**Socialist-Republic Federation**
**Acculturation**
***Kolkhoz* and *Sovkhoz***
**Analogue Area Analysis**
**Core Area**
**Heartland Theory**
**Frontier**
**Buffer State**
**Migration (2)**

From the Baltic Sea in Europe to the Asian shore of the Pacific Ocean and from the Arctic coast in the north to the borders of Iran in the south lies the Soviet Union—political region, culture area, historic empire, ideological hearth, and world superpower. Its 22.3 million square kilometers (8.6 million square miles) constitute nearly one-sixth of the earth's land surface and half the vast landmass of Eurasia; territorially, the Soviet Union is about 2.5 times as large as the United States or China. A country of massive continental proportions, the Soviet Union stretches across 11 time zones. When people awake in the morning in Vladivostok on the Sea of Japan, it is still early the previous evening in Leningrad on the Gulf of Finland. The ground is permanently frozen across the northern tundra that borders the Arctic Ocean, but subtropical crops grow on the balmy Georgian shore of the Black Sea.

**Figure 2–1**
Leadership in transition: confronting Soviet challenges and opportunities in the post-Brezhnev period.

The Soviet Union does not, however, have a population to match its huge territory. Its 277 million people (1985) are far outnumbered by 1.1 billion Chinese and 760 million Indians, neighbors on the Asian continent. And, as Fig. I–14 reveals, the Soviet population is still strongly concentrated in what is sometimes called "European" Russia—the Soviet Union west of the Ural Mountains. Most of the remainder, Siberia and the Soviet Far East, still remain virtually empty country, an unchanged, little-known frontier. In those vast trans-Ural expanses, the Soviets may discover additional mineral resources—but the cost of building transport systems to carry them to the industrial centers of the western U.S.S.R. may be prohibitive for a long time to come. As the map of Soviet railroads (Fig. 2–14) shows, there is nothing east of the Urals to rival the network that has developed in the west. Surface contact between the Soviet Far East and the populous west is still tenuous.

SYSTEMATIC ESSAY

# Climatology

**Climatology**, the branch of physical geography that studies climates, is concerned with the *average weather* conditions that prevail over the various portions of the earth's surface, with *weather* being the state of the atmosphere at any given moment. Climate, therefore, is a phenomenon that develops over an extended period of time (at least 30 years) and is a synthesis of the succession of weather events we have learned to expect in any particular location. Because there is a degree of organization in the heat and water exchanges at the earth's surface and in the general circulation of the atmosphere, there is a broadly simple pattern of climates across the globe, which is mapped in Fig. I–10 on pages 18–19. The climate regions that characterize this pattern may be viewed as geographic-environmental aggregates within which conditions of heat, moisture, and air movements are generally alike. These atmospheric conditions comprise the *climatic elements* of temperature, precipitation, pressure, winds, and their regional groupings are shaped by a number of *climatic controls* or processes.

Eight major climatic controls may be identified: (1) latitude, (2) semipermanent highs and lows, (3) wind belts, (4) jet streams, (5) ocean currents, (6) altitude, (7) mountain barriers, and (8) land-and-water heating differentials. *Latitude* determines the amount of solar energy (the triggering force in all atmospheric activity) received at a given location, particularly the angle of its intensity and the duration of daylight which depends on the season of the year. As surface air is heated, it tends to rise, decreasing its pressure, and as air in the upper atmosphere cools, it tends to sink, thereby increasing its pressure (*pressure* is the weight of an overlying column of air). This creates pressure differentials—with "lows" associated with warm rising air and "highs" with cool subsiding air—that, on a global scale, produce a system of *semipermanent highs and lows* (or pressure cells) that occur in areas where vertical air movements are more or less constant throughout the year. Since wind is the horizontal movement of air from places of higher pressure to those of lower pressure (everything in nature seeks equilibrium), regular wind flows develop among the semipermanent pressure cells. These highs and lows are aligned latitudinally, and the air currents they generate produce a series of six lateral *wind belts* whose internal motions are further shaped by the rotational effects of the planet: the Northeast and Southeast Tradewinds (whose average extent is 0° to 30°N and 0° to 30°S, respectively), the Prevailing Westerlies (approximately between 30° to 60° latitude in both the Northern and Southern hemispheres), and the Polar Easterlies (between 60° latitude and each pole). High-speed, upper-atmosphere winds, which are channeled into "rivers" of air called *jet streams*, snake irregularly along the poleward and subtropical boundaries of the Westerlies wind belt and often steer outbreaks of extreme weather into the mid-latitudes as well as guiding storms and fair-weather air masses through this zone. *Ocean currents* closely follow the wind belts and can greatly moderate the climates of nearby coastal areas: a good example is Western and Northern Europe, as we saw in Chapter 1, where the warm North Atlantic Drift current ameliorates what would otherwise be a much colder and drier climate on the subarctic margins of the mid-latitudes. Increasing *altitude* in mountain and highland-plateau areas influences climate in exactly the same way as increasing latitude—a progression of steadily colder climatic zones until arctic-like conditions are encountered at the snow line, even in the tropics (see discussion on page 269). *Mountain barriers* modify climates by blocking and redirecting air flows, as seen in the cases of the Alps and Scandinavia in the previous chapter; mountain ranges can also block moisture flows and thereby produce arid and semiarid environments, as will be seen in the discussion of the "rain shadow effect" in the western United States (page 182). Finally, the *land/water heating differential* or "continentality" can markedly affect inland climates, a control so important for central Eurasia that it is elaborated separately below.

The Soviet Union's climatic regions are displayed in Fig. 2–2, a blowup of the U.S.S.R. portion of the world map of climate types (Fig. I–10). The larger-scale Fig. 2–2, of course, reveals many additional details concerning the regionalization of Soviet climates, especially where several dissimilar regions are closely bunched—as in the Caucasus area between the Black and Caspian Seas. The U.S.S.R. is certainly a good laboratory for studying the geography of climates, because it contains nearly every type except the tropical (*A*) climates. The overall distribution shows a predominance of the *B* and *D* climates (the Köppen classification was covered on pages 14–16), indicating the widespread occurrence of dry and cold environments. The semiarid and arid (*B*) climates are concentrated in Soviet Central Asia, but are followed by

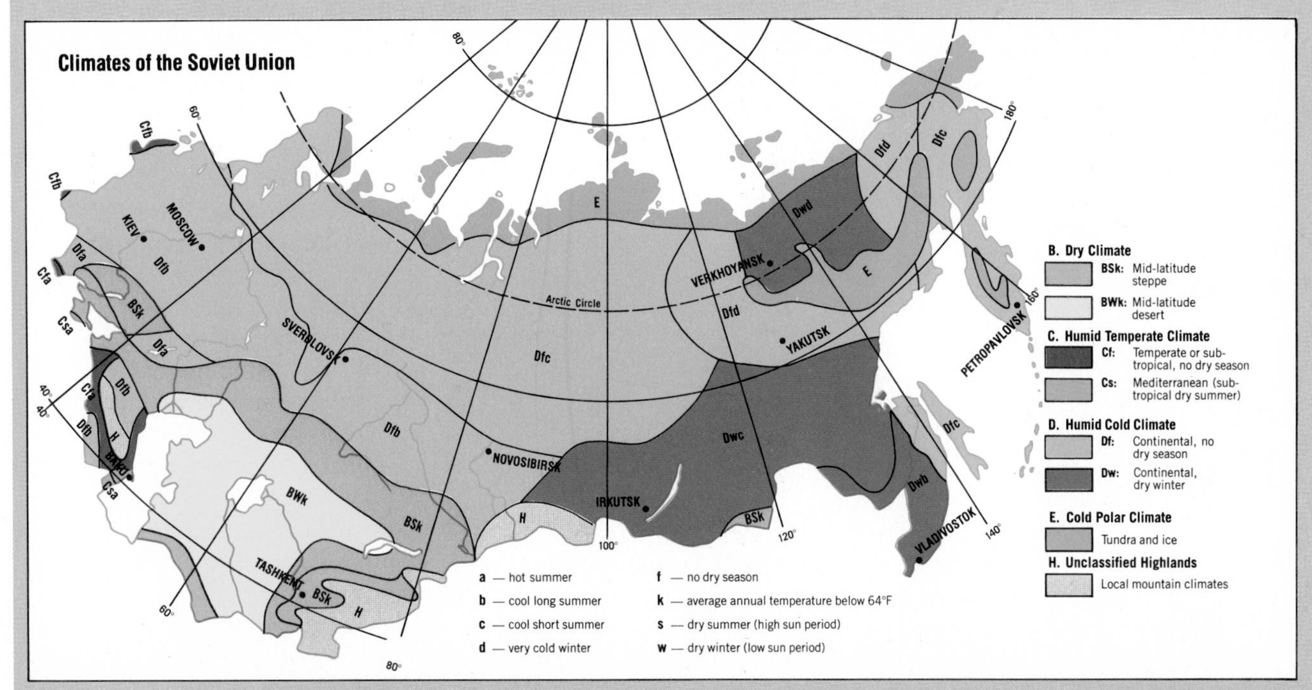


Figure 2–2

the suffix *k*, which means they are "cold" drylands exhibiting average annual temperatures below 18°C (65°F). The humid temperate (C) climates are clustered along the western and southwestern fringes of the U.S.S.R., near the moderating influences of the two inland seas and in places where the easternmost remnants of the North Atlantic maritime effect are felt. The humid cold (*D*) climate types cover most of the Soviet Union poleward of 50°N—about two-thirds of the entire country—and are interrupted only by the arctic-fringe belt of cold polar (*E*) climates and scattered local highland climates (*H*) along the Chinese border and in the north of the Far Eastern area. It is also worth noting that the *D* climates are dominated by *Dw* rather than *Df* in the interior areas of the eastern half of the U.S.S.R., an indication of dryness there, because *w* signals a dry winter, whereas *f* means the absence of a dry season.

Each of the world's climate regions is a response to the climatic controls that shape its climatic elements; to understand the regime of any given one, we need to know the relative importance of each of the climatic controls that affect that portion of the planetary surface. For almost all of the Soviet Union, one control is so pervasive that it completely overshadows the remaining seven—*continentality*, the effects of heating and cooling of the world's largest landmass. Any solid surface will heat up and cool off rapidly in response to the temperature of the air with which it comes in contact, and the bigger the surface, the more important this property becomes. In climatology, this phenomenon is most noteworthy when the moderating influences of warm water bodies are absent; nowhere on earth is this situation more pronounced than in the Soviet Union's interior, which is too far east of the Atlantic to feel its warming effects, is blocked by high mountain barriers from the Indian Ocean to the south, is protected by prevailing west-to-east winds from the inland penetration of Pacific air, and is open only to the harsh dry winds that blow across the frigid ice-covered Arctic Ocean. In winter, the land cools off sharply and develops a high-pressure cell from which winds blow eastward to the warmer (and, therefore, lower-pressure) air overlying the northwestern Pacific. In summer, the process is reversed as the land heats up markedly, generates a low-pressure cell that sucks in moisture-laden air from over the cool and now relatively higher-pressure Pacific, and brings most of the area's yearly precipitation. Further south on the Asian continent, this same mechanism helps to create the annual reversal of onshore and offshore wind flows known as *monsoons* (an Arabic word meaning *season*): the *dry monsoon* is the offshore airflow during the winter season, and the *wet monsoon* is the onshore flow of rain-bearing winds during the hot summer months. Incidentally, continentality in the interior eastern U.S.S.R. also produces the greatest annual range of average monthly temperatures on earth: for example at Verkhoyansk (Fig. 2–2) in far northeastern Siberia, the yearly range is 66°C (118° F)—from −50°C (−58°F) in January to 16°C (60°F) in July.

# TEN MAJOR GEOGRAPHIC QUALITIES OF THE SOVIET UNION

1. The Soviet Union is by far the largest territorial state in the world; its area is more than twice as large as the next ranking country (Canada).
2. The Soviet Union's enormous area lies mainly at high latitudes; much of it is very cold or very dry. Large, rugged mountain zones accentuate harsh environments.
3. The Soviet Union has more neighbors than any other state in the world. Boundary conflicts and territorial disputes involve several of these neighboring states, ranging from capitalist Japan and revisionist China to theocratic Iran, satellite Romania, and nonaligned Finland.
4. For so large an area, the Soviet Union's population of 275-odd million is comparatively small; it is, moreover, overwhelmingly concentrated in the westernmost one-quarter of the country.
5. Development in the Soviet Union remains focused in "European" Russia west of the Ural Mountains; here lie the major cities, industrial regions, densest transport networks, and most productive farming areas.
6. The Soviet Union is the heart of the socialist world (as opposed to the capitalist world and the "Third World").
7. The Soviet Union constitutes the world's largest-scale experiment in national economic planning.
8. National integration and economic development east of the Ural Mountains in Soviet Asia extend mainly along a narrow corridor that connects Omsk, Novosibirsk, Irkutsk, Khabarovsk, and Vladivostok.
9. Its large territorial size notwithstanding, the Soviet Union suffers from land encirclement in Eurasia; it has few good and suitably located ports.
10. The Soviet Union is a contiguous, multinational empire dominated by Russians. It is the product of Russian colonialism and imperial subjugation of nations, ranging from Estonia, Latvia, and Lithuania in the northwest to Moslem khanates in the south and Siberian natives in the east.

# A World Superpower

The Soviet Union today is a state whose military power and political influence in the world rival only that of the United States of America. Some observers suggest that the Soviet Union is in the process of surpassing the United States in terms of military strength, endangering the approximate balance between the two giants that has so far forestalled a nuclear holocaust. But the competition between the two countries is not only a matter of military buildup. The world has become an arena of ideological and political as well as economic rivalry and conflict between the United States and its scattered allies and the Soviet Union and its Eurasian and third-world clientele. At issue is the social system that will prevail in the world of the future: Will capitalist enterprise survive or will communism ultimately triumph? The cost to the world of this twentieth-century power struggle has been incalculable. Apart from the billions of rubles and dollars that have been poured into Soviet and American military-industrial complexes, the greater conflict has magnified minor discords among nations, damaged smaller economies, intensified national politics, destroyed democracies, and fueled terrorism. When states quarrel, the great powers are quick to exploit the dispute, and soon leaders who might otherwise have been identified as representing national viewpoints are reported to be "pro Western" or "leftist leaning." Then, the flow of money and weapons starts. The civil strife of Southeast Asia, the political convulsions of Chile, the decolonization of Angola, the widening instabilities of the Middle East and Central America—the list of countries and regions caught up in the cold-war struggle is lengthening.

Although a champion of peoples trying to throw off repressive colonial regimes, the Soviet Union must be seen as a colonial empire itself. Just as the British and French built colonial realms in nearby (Ireland, Algeria) and distant (India, Vietnam) lands, the Soviet Union is the ultimate product of Russian imperialism—an imperialism whose acquisitions were not liberated when the Bolshevik Revolution changed the social order in 1917. Nor has Moscow been tolerant of national self-determination in its Eastern European sphere of influence, acquired and consolidated when

Western European colonial powers were beginning to lose their dependencies. The Soviets brutally deployed tanks and other heavy arms to suppress assertions of independence in Hungary (1956) and Czechoslovakia (1968); their Polish puppets harshly clamped down on the Solidarity labor movement in the early 1980s, just managing to avoid military violence against the civilian population. Soviet fear and paranoia vis-à-vis the free world has also prompted vicious acts against other nations, the most outrageous being the shooting down of an unarmed Korean passenger jet in 1983 for an accidental intrusion into Soviet airspace within its Pacific boundary. This cowardly crime against humanity, knowingly committed in full defiance of international aviation treaties, resulted in the massacre of 269 civilians and produced a worldwide revulsion that will color external perceptions of the U.S.S.R. for years to come. Finally, at home, such behavior is directed against ideological dissidents, who are subjected to an unending reign of terror.

## A European Heritage

For many centuries, this geographic realm's main area of human interaction and spatial organization was in what is today western Russia. Although our knowledge of Russia before the Middle Ages is but fragmentary, it is clear that peoples moved in great migratory waves across the plains on which the modern state eventually was to emerge. The dominant direction seems to have been from east to west; many groups came from central Asia and left their marks in the makeup of the population. Scythians, Sarmatians, Goths, and Huns—they came, settled, fought, and were absorbed or driven off. Eventually, the Slavs emerged as the dominant people in what is today the Ukraine; they were peasants, farming the good soils of the plains north of the Black Sea. Their first leadership came not from their own midst, but from a Scandinavian people to the northwest known as the Varangians, who had for some time played an important role in the fortified trading towns (*gorods*) of the area.

The first Slavic state (ninth century A.D.) came about for reasons that will be familiar to anyone who remembers the importance of the transalpine routes across Western Europe. The objective was to render stable and secure an eastern crossing of the continent, from the Baltic Sea and Scandinavia in the northwest to Byzantine Europe and Constantinople in the southeast. This route followed the Volkhov River to Novgorod (positioned on Lake Ilmen), then led to Kiev and southward to the Black Sea along the Dnieper River. Novgorod, so near Scandinavia and close to the shores of the Baltic, was the "European" center with its cosmopolitan population and its Hanseatic League trade connections. Kiev, on the other hand, lay in the heart of the land of the Slavs near an important confluence on the Dnieper and not far from the zone of contact between the forests of middle Russia and the grassland steppes of the south. Kiev had centrality; it served as a meeting place of Scandinavian and Mediterranean Europe; for a time during the eleventh and twelfth centuries, it was a truly European urban center as well as the capital of Kievan Russia. Kiev and Milan had something in common: both were positioned near major breaks in the natural landscape, Milan at the gateway to the difficult Alpine crossing, Kiev at the edge of the dense and dangerously perceived forests that covered central Russia.

About the middle of the thirteenth century, Kievan Russia fell to yet another invasion from the east. Now the Mongol empire deep inside Asia, which had been building under Genghis Khan, sent its Tatar hordes into Kiev (led, incidentally, by the grandson of Genghis, Batu), and the city as well as the state fell. Many Russians sought protection by fleeing into the forests that lay north of Kiev, because the horsemen of the steppes were not very effective in that terrain. What remained of Russia, then, lay in the forest between the Baltic and the steppes; and there, for a time, a number of weak feudal states arose, many of them ruled by princes who paid tribute to the Tatars in order to retain their positions. From among these feudal states, the one centered on Moscow emerged as supreme. In the fifteenth century, its ruler conquered Novgorod; then in the sixteenth, Ivan IV took over authority and assumed the title of tsar (another Mediterranean contribution—czar or *caesar*). Ivan the Terrible, as this tsar came to be known, made an empire out of the Moscow-centered state of Muscovy. He soon established control over the entire basin of the Volga River, succeeding in his campaigns against the Tatars, and he eventually extended Moscow's authority into western Siberia.

This eastward expansion of Russia was carried out by a relatively small group of seminomadic peoples who came to be known as Cossacks. Opportunists and pioneers, they sought the riches of the east, chiefly fur-bearing animals, as early as the sixteenth century. By the middle of the seventeenth century, they reached the Pacific Ocean, defeating Tatars in their path and consolidating their gains by constructing *ostrogs*, strategic fortified way stations along river courses. Before the eastward expansion halted in 1812 (Fig. 2–3), the Russians had moved across the Bering Strait to Alaska and down

# SOVIET UNION

The Soviet Union is the product of the Revolution of 1917, when more than a decade of rebellion against the rule of Nicholas II led to the tsar's abdication. Russian revolutionary groups were called *soviets* (meaning "councils"), and these soviets had been engaged in revolutionary activity since the first Russian workers' revolution in 1905. In that year, thousands of workers marched on the palace in protest, and the tsar's soldiers opened fire on the marchers, killing and wounding hundreds. Throughout the years that followed, Russia was engaged in costly armed conflicts, government became more corrupt and less effective, and social disarray prevailed. In 1917, a coalition of military and professional men forced the tsar to abdicate, and Russia was ruled by a Provisional Government until November (by the Gregorian calendar), when the country experienced its first and only truly democratic election. In the meantime, however, the Provisional Government had allowed the return to Russia of exiled Bolshevik activists: Lenin from Switzerland, Trotsky from New York, and Stalin from Siberia. In the political struggle that ensued, the Bolsheviks gained control over the revolutionary soviets, and this ushered in the Communist era. Since 1924, the country has officially been known as the Union of Soviet Socialist Republics (U.S.S.R.), for which Soviet Union is shorthand. Sometimes the country is still referred to by its prerevolutionary name, *Russia*. But Russia is only one (though by far the largest and dominant) of the 15 republics in the Union.

the western coast of North America into what is now northern California (see Box).

By the time Peter the Great took over the leadership of Russia (he reigned from 1682 to 1725), Moscow lay at the center of a great empire—great, at least, in terms of the territories under its hegemony. As such, emergent Russia had many enemies. The Mongols were finished as a threat, but the Swedes, no longer allies in trade, threatened from the northwest, as did the Lithuanians. To the west there was a continuing conflict with the Germans and the Poles. And to the southwest, the Ottoman Turks were heirs to the Byzantine Empire and they, too, posed a threat. Peter consolidated Russia's gains and did much to make a modern European-style state out of the loosely knit country. He wanted to reorient the empire to the Baltic, to give it a window on the sea and to make it a maritime as well as a land power. In 1703, following his orders, the building of Petersburg (later St. Petersburg) began; it was built by Italian architects at the tsar's behest, and they designed numerous ornate buildings that were arranged around the grandiose city's many waterways in a manner reminiscent of Venice (Fig. 2–4). St. Petersburg, known today as Leningrad, is positioned at the head of the Gulf of Finland, which opens into the Baltic Sea. Not only did it provide Russia with an important maritime outlet, but the city was also designed to function as a forward capital: it lay on the doorstep of Finland, which at that time was a Swedish possession. In 1709, Peter's armies defeated the Swedes, confirming Russian power on the Baltic coast; in 1713, Peter took the momentous step of moving the Russian capital from Moscow to the new Baltic headquarters, where it remained until 1918.

Peter was an extraordinary leader, in many ways the founder of modern Russia. In his desire to remake Russia, to pull it from the forests of the interior to the western coast, to open it to outside influences, to end its comparative isolation, he left no stone unturned. Not only did he move the capital, but he himself, aware that the future of Russia as a major force lay in strength at sea as well as power on land, went to Holland to work as a laborer in the famed Dutch shipyards to learn how ships were most efficiently built. Peter wanted a European Russia, a maritime Russia, a cosmopolitan Russia. He developed Petersburg into a leading seat of power as well as one of the most magnificent cities in the world, and it remains to this day a European-style city apart from most others in the Soviet Union.

During the eighteenth century, the tsarina Catherine the Great, who ruled from 1762 to 1796, continued to build Russian power, but on another coast and in another area: the Black Sea in the south. Here the Russians confronted the Turks, who had taken the initiative from the Greeks; the Byzantine Empire had been succeeded by the Turkish Ottoman Empire. But the Turks were no match for the Russians. The Crimean Peninsula soon fell, as did the old and important trading city of Odessa; before long, the whole northern coast of the Black Sea was in Russian hands. Soon afterward, the Russians penetrated the area of the Caucasus, and, early in the nineteenth century, they took Tbilisi, Baku, and Yerevan. But as they pushed farther into the corridor between the Black and the Caspian seas, they faced

## RUSSIANS IN NORTH AMERICA

The first white settlers in Alaska were Russians, not Western Europeans, and they came across Siberia and the Bering Strait, not across the Atlantic and overland. Russian hunters of the sea otter, valued for its high-priced pelt, established their first Alaskan settlement at Kodiak Island in 1784. Moving southward along the North American coast, the Russians founded additional villages and forts to protect their tenuous holdings, until they reached as far as the area just north of San Francisco Bay, where they built Fort Ross in 1812. But the Russian settlements were isolated and vulnerable. European fur traders began to put pressure on their Russian competitors, and Moscow found the distant settlements a burden and a risk. In any case, American, British, and Canadian hunters were decimating the sea otter population, so that profits declined. When U.S. Secretary of State William Seward offered to purchase Russia's holdings in 1867, Moscow quickly agreed—for $7.2 million. Thus, Alaska, including its lengthy southward coastal extension, became American territory; although Seward was ridiculed for his decision—Alaska was called "Seward's Folly" and "Seward's Icebox"—his reputation was redeemed when gold was discovered there in the 1890s. The twentieth century has proved Seward's action one of great wisdom, strategically as well as economically. At Prudhoe Bay, off Alaska's northern Arctic slope, large oil reserves are being developed; and like Siberia, Alaska probably contains other, yet unknown, riches.

growing opposition from the British—who held sway in Persia (now Iran)—as well as the Turks, and their advance was halted short of its probable ultimate goal: controlling a coast on the Indian Ocean.

But Russian expansionism was not yet satisfied. While extending the empire southward, the Russians also took on the Poles, old enemies to the west, and succeeded in taking most of what is today the Polish state, including the capital of Warsaw; to the north, Russia took over Finland from the Swedes (1809). During most of the nineteenth century, however, Russian preoccupation was with Asia, where Tashkent and Samarkand came under St. Petersburg's control. The Russians were still bothered by raids of nomadic Mongol horsemen, and they, therefore, sought to establish their authority over the central Asian steppe country to the edges of the high mountains that lay to the south. Thus, Russia gained a considerable number of Moslem subjects, for this was Islamic Asia they were penetrating, but, under tsarist rule, these people acquired a sort of ill-defined protectorate status and retained some autonomy. Much farther to the east, a combination of Japanese expansionism and a decline of Chinese influence led Russia to annex from China several provinces along the Amur River. Soon afterward in 1860, the port of Vladivostok was founded.

Now began the course of events that was to lead, after five centuries of almost uninterrupted expansion and consolidation, to the first setback to the Russian drive for territory. In 1892, the Russians began building the Trans-Siberian Railway, in an effort to connect the distant frontier more effectively to the western core. As the map shows, the most direct route to Vladivostok was across Chinese Manchuria. The Russians wanted China to permit the construction of the last link of the railway across Manchurian territory, but the Chinese resisted. Taking advantage of the 1900 Boxer Rebellion in China, Russia responded by annexing Manchuria and occupying it. This brought on the Russo-Japanese War of 1905, in which the Russians were disastrously defeated; Japan took possession of southern Sakhalin Island (which they called Karafuto). For the first time in nearly five centuries, Russia sustained a setback that resulted in a territorial loss.

Thus, Russia—recipient of British and European innovations in common with Germany, France, and Italy—was just as much a colonizer too. But where other European powers traveled by sea, Russian influence traveled overland into central Asia, China, and the Pacific coastlands of northeastern Asia. What emerged was not the greatest empire, but the largest *territorially contiguous* empire in the world; it is tempting to speculate what would have happened to this sprawling realm had European Russia (for such it still was) developed politically and economically in the manner of other European power cores. At the time of the Japanese war, the Russian tsar ruled over more than 22 million square kilometers (8.5 million square miles), just a fraction less than the area of the Soviet Union today. Thus, the modern Communist empire, to a very large extent, is a legacy of St. Petersburg and European Russia—not the product of Moscow and the Socialist Revolution.

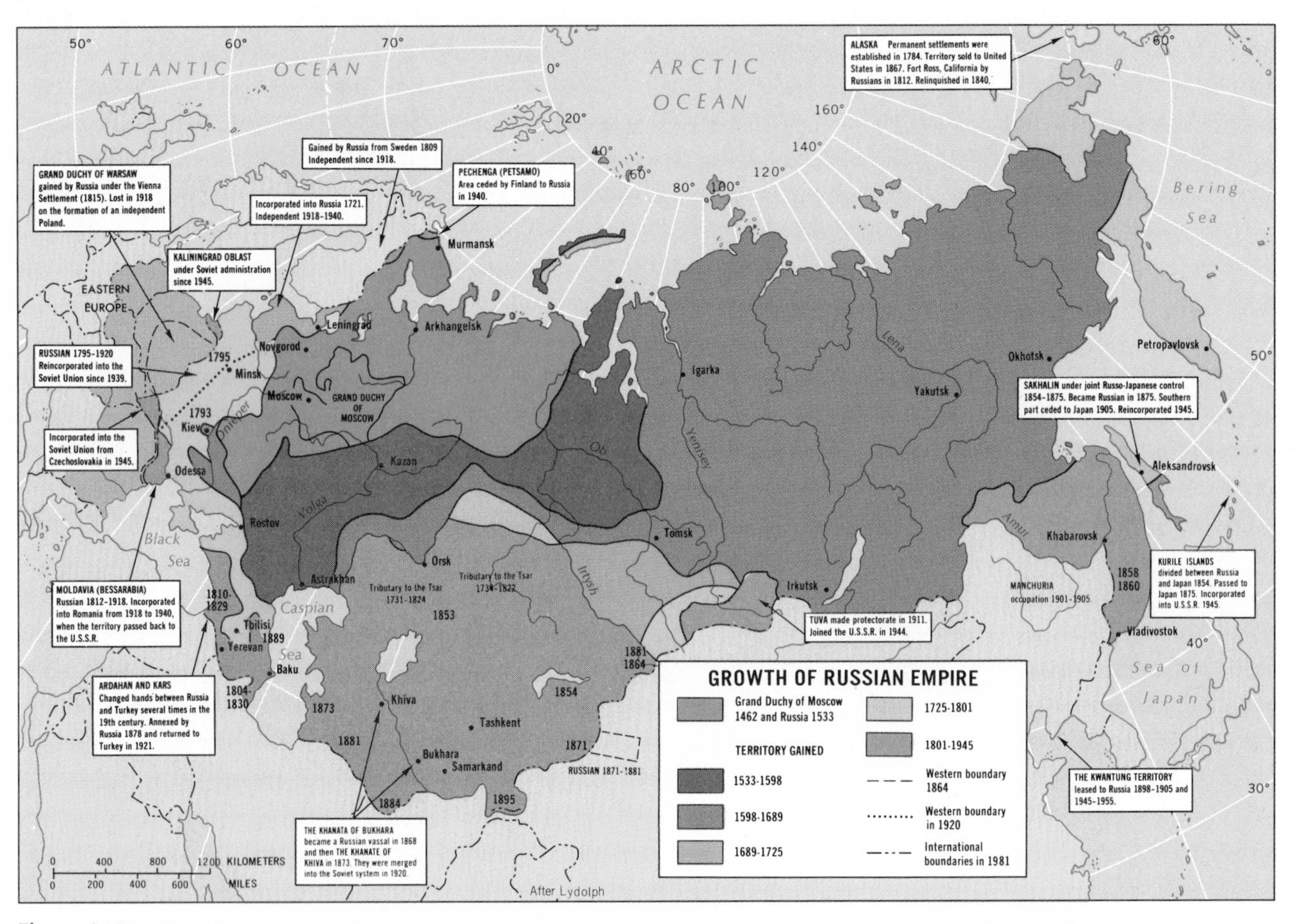


Figure 2–3

# Physiographic Regions

Before returning to the Soviet Union's human geography, we briefly examine the physiographic regions of this vast realm. As Figs. I–10 and 2–2 indicate, the Soviet Union's natural environments are harsh, even bleak. More than 80 percent of the total area of the country lies farther to the north than the American Great Lakes. It, therefore, lies open to the frigid air masses of the Arctic, which strike here in full force, because the U.S.S.R. is shielded by distance from the moderating influences of Atlantic waters to the west and by mountain barriers from the warmer climates of the south. The prevalence of *D* climates (Fig. I–10) underscores the prevailing coldness, whose severity increases eastward toward the heart of the Eurasian landmass. The map of precipitation (Fig. I–9) emphasizes that low and variable precipitation is another environmental problem in the Soviet Union. And Fig. I–13 reveals that the country's best soils lie in a belt extending from Eastern Europe into the center of the Soviet Union—but a large part of this belt lies in a zone of low and variable rainfall. For an insight, locate the comparable soil zones in North America and the Soviet Union and relate them to the precipitation map (Fig. I–9) in both regions.

Its great size nothwithstanding, the Soviet Union is practically a landlocked country whose massive continentality is accentuated by coldness, drought, short growing seasons, high winds, and other environmental hazards that make the productive areas of the southwest and south mere islands in a forbidding, severe expanse of tundra and forest, mountain and desert (Fig. I–12). Neither can one discern physiographic regions as distinct as those of North America: the Soviet land surface, for all its size, divides much less clearly. A map of Soviet terrain shows that the east and south tend to be high and rugged, whereas the north and west are lower and more level. Only the Ural Mountains, trending north-south right through the middle of the western lowlands, constitute a major exception to this broadest division of the Soviet Union's physiography.

The simplest subdivision of the Soviet Union's physical geography produces nine regions (Fig. 2–5). The *Russian Plain (1)* is the east-

**Figure 2–4**
Leningrad, often called the Venice of the Baltic.

ward continuation of the North European Plain—the theater within which arose the Russian state. At its heart lies the Moscow Basin; to the north the countryside is covered by needleleaf forests similar to those of Canada, whereas to the south lie the grain fields of the grassland Ukraine. The Russian Plain is bounded to the east by the *Ural Mountains (2)*, not a high range but prominent because it separates two extensive plains. The Ural range forms no barrier to transportation, and its southern end is quite densely populated; the region has yielded a variety of minerals, including oil.

South of the Urals lies the *Caspian-Aral Basin (3)*, a region of windswept plains of steppe and desert. Its northern section is the Kirghiz Steppe; in the south, the Soviets are trying to conserve what little water there is here through irrigation schemes along rivers draining into the Aral Sea. East of the Urals lies Siberia. The *West Siberian Plain (4)* has been described as the world's largest true unbroken lowland; it is the basin of the Ob and Irtysh Rivers. Over the last 1600 kilometers (1000 miles) of its course, the Ob falls only 90 meters (less than 300 feet). The northern area of the West Siberian Plain is permafrost ridden, and the central area marshy; but the south carries significant settlement, including the cities of Omsk and Novosibirsk, within the corridor of the Trans-Siberian Railroad (Fig. 2–6).

East of the West Siberian Plain the country begins to rise, first into the *Central Siberian Plateau (5)*, a sparsely settled, remote, permafrost-afflicted region. Here winters are long and extremely cold, and the summers short. Some mineral finds have been made—including nickel, copper, and platinum around Norilsk in the northwest—but the area remains barely touched by human activity. Beyond the *Yakutsk Basin (6)*, the terrain becomes mountainous and the relief high. The *Eastern Highlands (7)* are a jumbled mass of ranges and ridges, precipitous valleys, and volcanic mountains. Lake Baykal lies in a trough that is over 1500-meters (5000-feet) deep; on the Kamchatka Peninsula, volcanic Mount Klyuchevskaya reaches nearly 4750 meters (15,600 feet). The northern area is the Soviet Union's most inhospitable zone, but southward, along the Pacific coast, the climate is less severe. Nonetheless, this is a

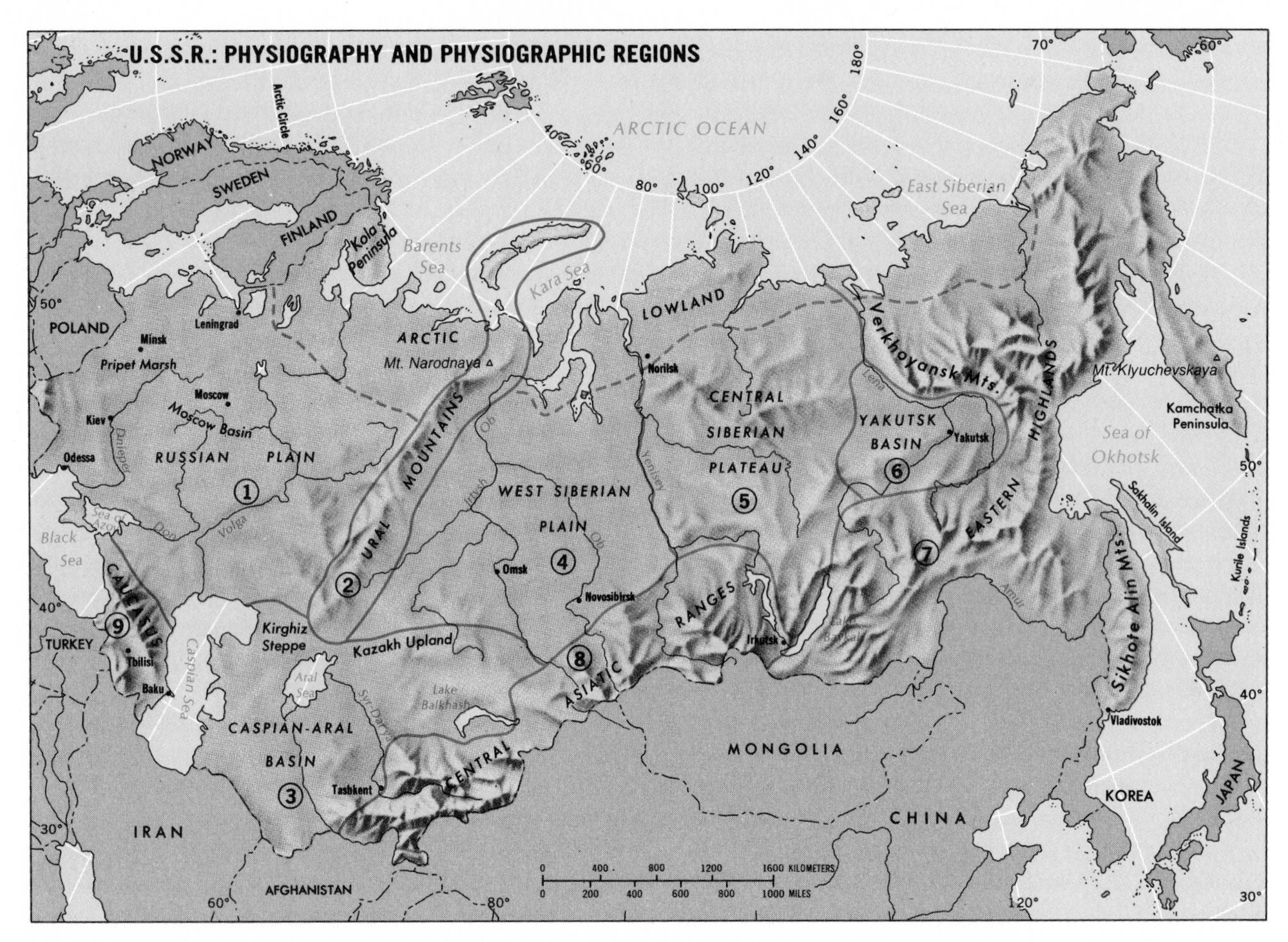


**Figure 2–5**

true frontier region. The forests provide opportunities for lumbering, a fur trade exists, and gold and diamonds are found by hardy miners. The newest resource to be discovered here is solid gas, a frozen fossil fuel said to be present in such huge quantities that if complex problems of extraction could be solved, enough energy is available to supply total world needs for thousands of years. To help develop this promising area, the Soviets hope to complete construction of the new Baykal-Amur Mainline (BAM) railroad in the mid-1980s, a 3540-kilometer-(2200-mile)-long route that would roughly parallel the aging Trans-Siberian line.

The southern margins of the Soviet Union are also marked by mountains: the *Central Asiatic Ranges (8)* from the vicinity of Tashkent in the west to Lake Baykal in the east, and the *Caucasus (9)* between the Black and Caspian seas. The Central Asiatic Ranges rise above the snow line and contain extensive mountain glaciers. The annual melting brings down alluvium to enrich the soils on the lower slopes and water to irrigate them. The Caucasus form an extension of Europe's Alpine Mountains and exhibit a similar topography, but they do not provide convenient passes. The trip by road from Tbilisi to Ordzhonikidze (via the Georgian Military Highway) is still a circuitous adventure in surface travel.

# The Soviet Union in the Twentieth Century

Russia's enormous territorial expansion produced an empire, but nineteenth-century Russia was infamous for the abject serfdom of its landless peasants, the exploitation of its workers, the excesses of its nobility, and the palaces and riches of its rulers. The arrival of the industrial revolution had introduced a new age of misery for those laboring in factories. There were ugly strikes in the cities, and, from time to time, the peasants rebelled against the nobility. But retribution was always severe, and the tsars re-

**Figure 2–6**
The Trans-Siberian Railroad in south-central Siberia.

mained in tight control. And then, in 1904, Russian forces fighting in the Far East against the Japanese faced defeat. It was a milestone in the destruction of the old order, because this loss was followed by ever-increasing opposition throughout Russia against royal authority. Local revolutions took place, upheaval was the order of the day. Certainly Russia was not ready for its own defense when World War I erupted, and again Russian forces were disastrously defeated.

Finally, the 1917 Revolution succeeded where earlier rebellions had failed. The revolution was no more a unified uprising than its predecessors had been; while the Bolsheviks' "Red" armies battled the "Whites," both fought against the forces of the tsar. Ultimately, the Bolsheviks prevailed and the capital, Petrograd (as St. Petersburg had been renamed in 1914 to rid it of its German nomenclature), witnessed the end of tsarist rule. Even before the Bolshevik victory was complete, the signs of things to come were already evident. In 1918, the headquarters of the government was moved from Petrograd to Moscow, the focus of the old state of Muscovy, deep in the interior, not even on a major navigable waterway, and amid the remnants of the same forests that centuries earlier had afforded the Russians protection from their enemies. It was a symbolic move, ushering in a new period in Russian history, and an expression of distrust toward a Europe that had contributed war and misery to Russia, but whose promises—of maritime power, of industrial modernization, of political advancement—had never been fulfilled. The new Russia looked inward; it sought a means whereby it could achieve with its own resources and labor force the goals that had for so long eluded it. The chief political and economic architect in this effort was also a revolutionary leader—V. I. Lenin. Among his solutions to the country's problems are the present-day political framework of the Soviet Union and its centrally planned economy.

## The Political Framework

Russia's great expansion had brought a large number of nationalities under tsarist control, and now it was the turn of the revolutionary government to seek the best method of organizing this heterogeneous social mosaic into a smoothly functioning state. The tsars had conquered, but they had done little to bring Russian culture to the peoples they ruled; the Georgians, Armenians, Tatars, and residents of the Islamic Khanates of Central Asia were among dozens of individual cultural, linguistic, and religious groups that had not been "Russified." The Russians themselves in 1917, however, constituted only about one-half of the population of the entire country (the proportion today remains about the same). Thus, it was impossible to instantly establish a Russian state over the whole of this vast political region, and these diverse national groups had to be accommodated.

The question of the nationalities became a major issue in the young Soviet state after 1917. Lenin, who brought the philosophies of Marx to Russia, talked from the begin-

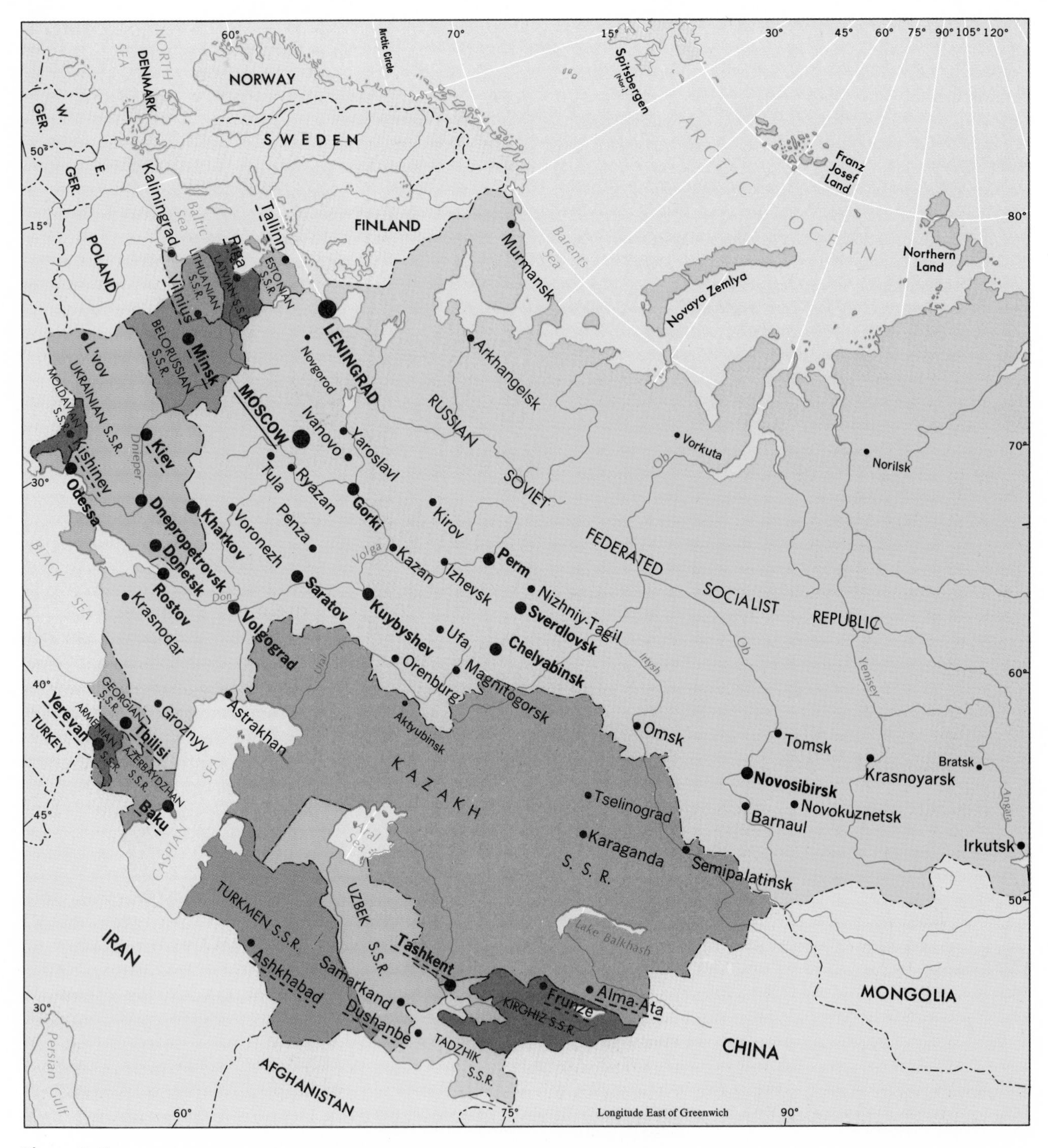


Figure 2–7

ning about the "right of self-determination for the nationalities." The first response by many of Russia's subject peoples was to proclaim independent republics, as was done in the Ukraine, Georgia, Armenia, Azerbaydzhan, and even in Central Asia. But Lenin had no intention of permitting any breakup of the state; in 1923, when his blueprint for the new Soviet Union went into effect, the last of these briefly independent units was fully absorbed into the sphere of the Moscow government. Nevertheless, the whole framework of the Soviet Union remains essentially based on the cultural identities of its incorporated peoples. Although there have been modifications since the days of Lenin, the major elements of the original system are still there. The country is divided into 15 Soviet Socialist Republics or S.S.R.s (Fig. 2–7), each

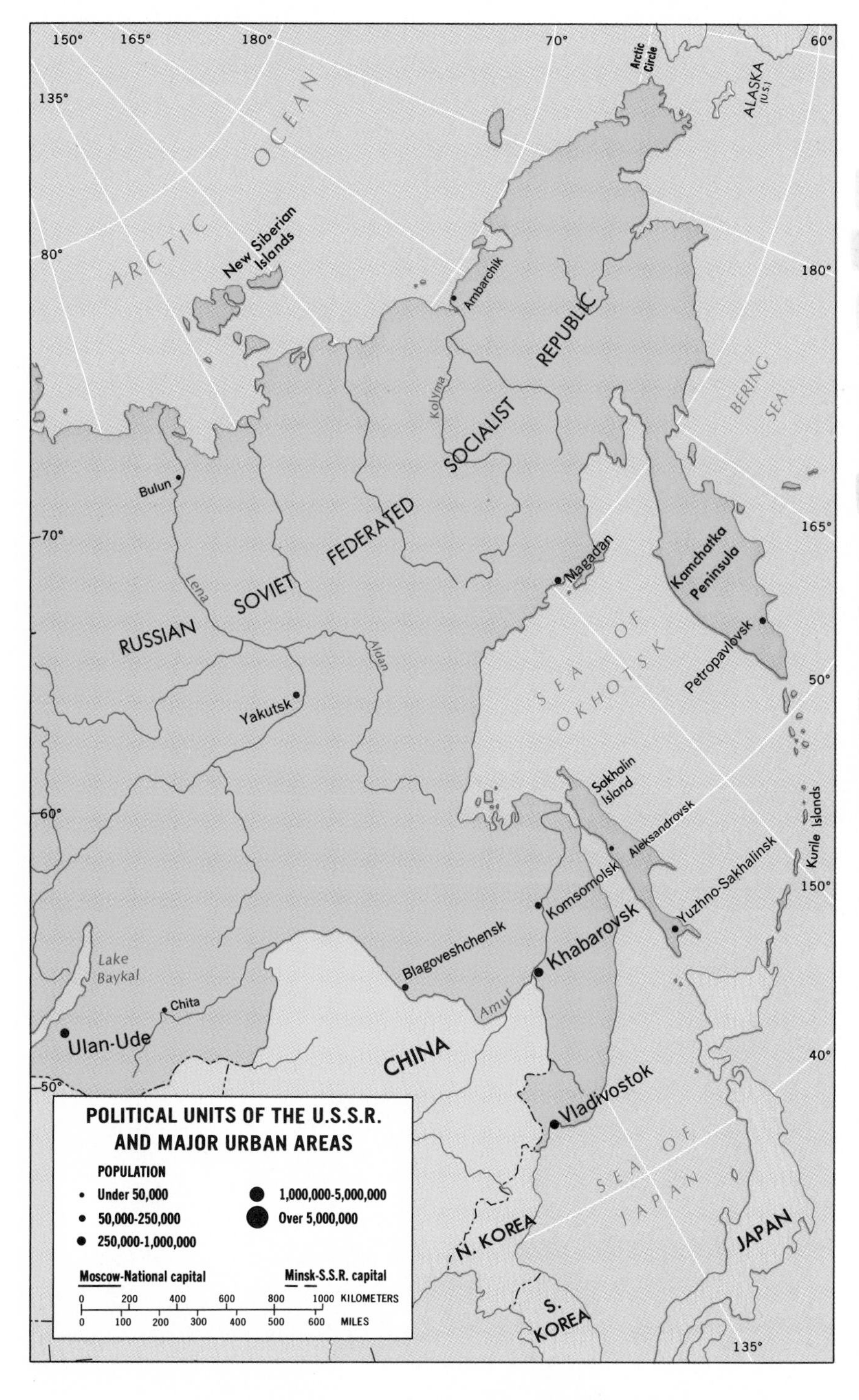


of which broadly corresponds to one of the major nationalities. As the map shows, the largest of these S.S.R.s by far is the Russian Soviet Federated Socialist Republic (R.S.F.S.R.), which extends from west of Moscow halfway around the globe to the apex of the Pacific Ocean in the east. The remaining 14 republics lie in a belt that extends southeast from the Baltic Sea (where Estonia, Latvia, and Lithuania retain their identities as S.S.R.s) to the Black Sea (north of which lies the Ukrainian and Moldavian S.S.R.s), and then eastward through the Georgian and Armenian Republics to the Caspian Sea, beyond which lie the five additional Asian S.S.R.s.

Theoretically, these 15 republics possess equal standing in the Soviet Union; in practice, the Russian Republic leads and the others look upward. With half the country's population, the capital, most of its major cities, and over three-quarters of its territory, Russia remains the nucleus of the Soviet Union. The Russian language is taught in the schools of all the other republics, but none of the languages of these other areas are required to be taught within the R.S.F.S.R. Although, again theoretically, the country's constitution would permit each republic to carry on its own foreign policy, to issue its own money, and even to secede from the Union, nothing even slightly suggestive of this has ever occurred in practice. On the other hand, even though it can be argued that the Soviet Union is highly centralized and a federal state only in theory, no republic has made use of wartime conditions to break away. The Germans, during their invasion of the Soviet Union in World War II, hoped that an anti-Russian nationalism would provide them with support in the Belorussian S.S.R. and the Ukrainian S.S.R., but they found only isolated instances of cooperation and no attempts were made to proclaim sovereign states outside Moscow's sphere. Undoubtedly, there is a great deal of acquiescence to the *status quo* among Soviet peoples; in view of the racial, cultural, linguistic, and religious complexity of the population (Fig. 2–8), this is no small achievement. Yet, surprising local variations from dominant national themes are tolerated in certain areas. For example, residents of peripheral Soviet Georgia are widely allowed to grow food on private farmsteads, own real-estate property, and even publicly demonstrate

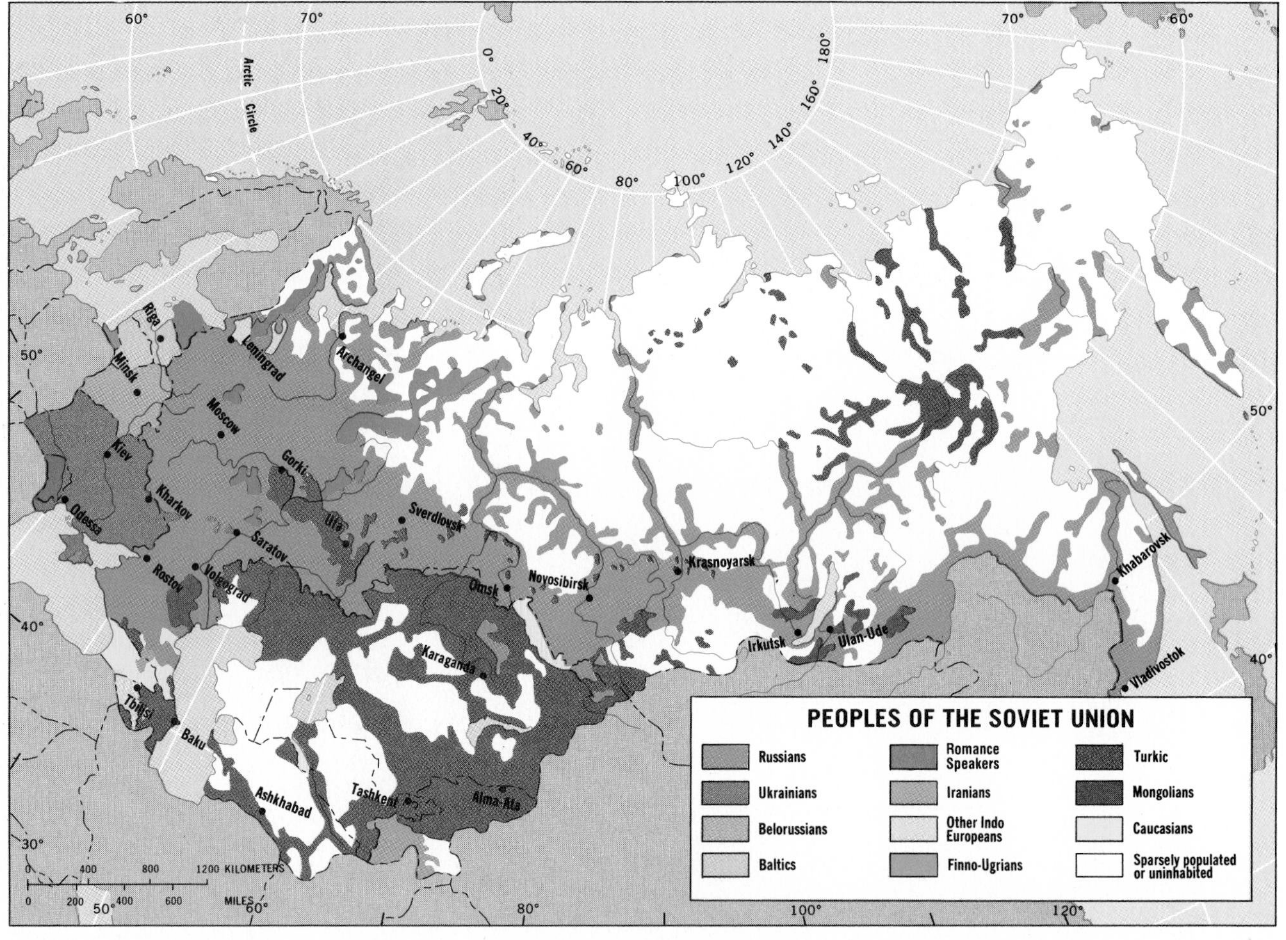


Figure 2–8

to keep Georgian the main local language—all without reprisal from the central authorities. Perhaps their most outrageous behavior is their open adherence to the cult of Stalin—a native Georgian whose brutal dictatorship (1924–1953) is officially frowned on—who has become the symbol of Georgian nationalism to the extent that his humble birthplace in the village of Gori is now a grotesque shrine (Fig. 2–9).

If the system developed after 1917 proved to have some of the liabilities of the pre-Communist era, it also had some hitherto unknown assets. For the first time, the entire country was brought under effective national control. For the first time, the Russians, recipients of European innovations, now began to transmit these innovations to non-

Figure 2–9
Stalin's birthplace, Gori, Georgian S.S.R.

Russian and non-Slavic parts of the realm (Fig. 2–10). For the first time, there were elements of compromise in the relationships between Russians, other Slavic peoples, and the non-Slavic population of the country. Certainly, there were conflicts, and many individual groups at various times have held separatist feelings, but, in general, there has been an awareness of progress brought by the Soviet administration. After the terror of the Russian conquest itself, the Soviet contributions to development in such areas as Armenia and Central Asia are proof positive that membership in the Soviet Union does have its advantages. Standards of living have been raised, educational opportunities have improved, housing projects have taken the place of city slums, and the economy has accelerated.

If these accomplishments resemble those boasted of by colonial powers defending the benevolence of their rule, we are reminded that the Soviet Union is a Russian empire. The landscape of the town of Khiva (Fig. 2–11), located in the Uzbek S.S.R. near the Aral Sea, illustrates this point clearly: any Western power controlling an urban center that exhibited so distinct an Islamic cultural landscape might be accused of colonialism—but, in the case of the U.S.S.R., this scene merely represents another Soviet Republic. Although local culture has been sustained (with organized religion discouraged wherever possible), Russification is relentlessly pursued; this kind of pressure exerted by powerful outsiders is known as *acculturation*, a process we will encounter in stronger perspective in Chapter 4. The Soviets, of course, like to claim that their seven decades of control have brought unanimous approval of their system throughout the country. Although this is not so, their treatment of minority nationalities does appear to have achieved greater domestic approval than that of some colonial powers and, indeed, of some governments of sovereign states.

## The Centrally Planned Economy

The sweeping changes that came to Russian politics after the success

Figure 2–10

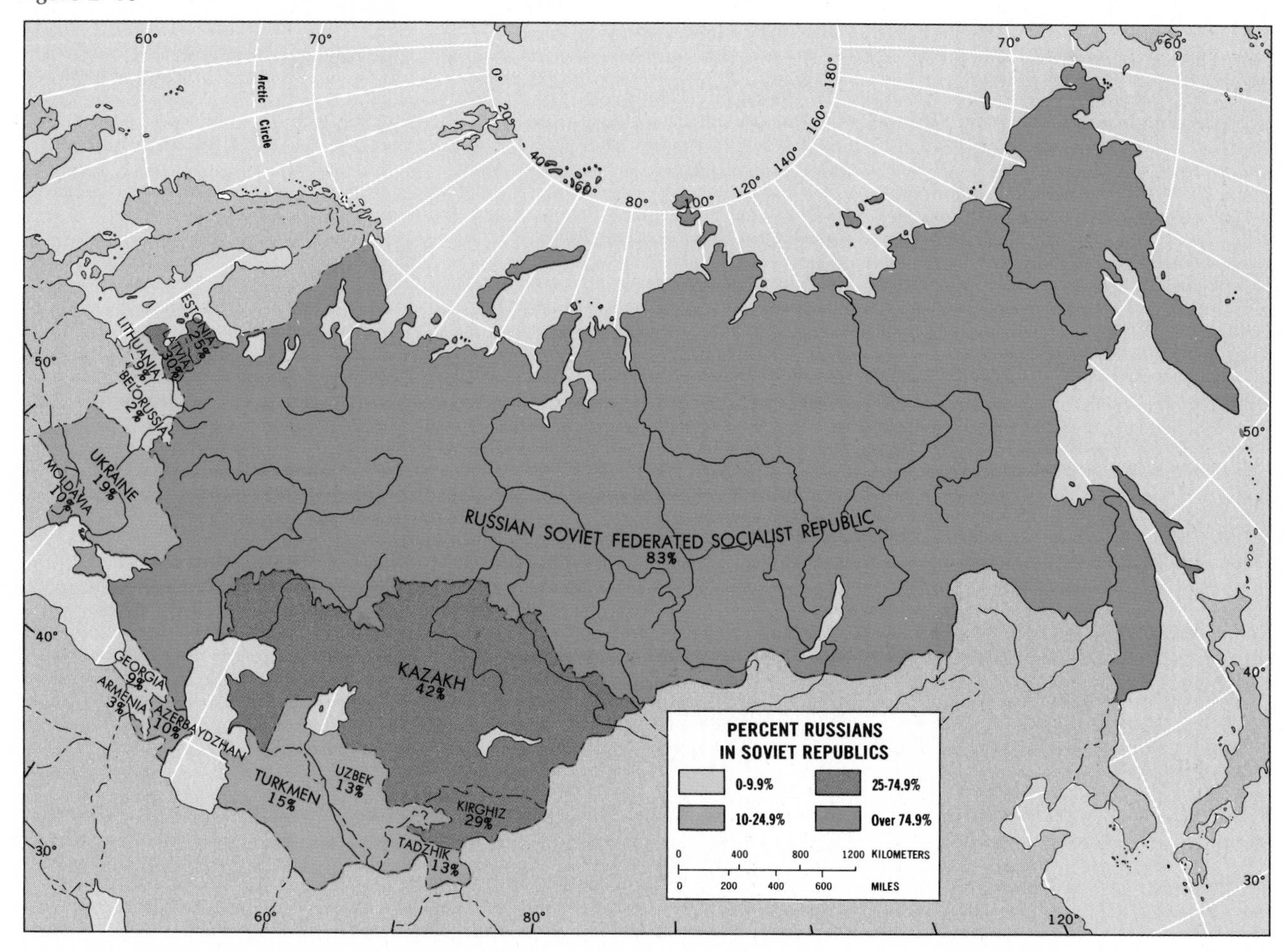

**Figure 2–11**
Moslem-dominated Khiva in Uzbekistan.

of the 1917 Revolution are more than matched by the great Soviet economic experiment. After an early transitional period that saw some private enterprise continue while Soviet economists charted the socialist course, the revolutionary leaders' economic aspirations were embodied in the first of a series of *five-year plans*. The objectives of the Soviet economic planners were: (1) to speed the industrialization of the country and (2) to collectivize Soviet agriculture. In order to accomplish this, the entire country was mobilized, with a State Planning Commission (*Gosplan*) at the helm. For the first time ever on such a scale and in accordance with Marxist principles, a whole state and its peoples worked in concert toward national goals that were prescribed by a central government. The five-year plans serve first as guidelines and aims for the economy during the particular period (although the guidelines are often changed and the aims almost never achieved) and second as exhortations for Soviet workers to labor as hard as they can toward the goals put before them by the state, goals that are invariably tied to (usually false) promises of better days ahead.

The first Five-Year Plan was designed to accomplish the elusive collectivization of agriculture, delayed by then for fully a decade after the revolution. A major objective of this program was to secure more grain for the urban-industrial population through the extension of government control over the farmers. The farmers, dissatisfied with the terms of trade under the new order, had been withholding grain from the markets. The 1928–1933 collectivation program led to the consolidation of nearly 70 percent of all individual peasant holdings into larger units operated jointly and under government supervision. There was much opposition to the program, from the slaughter of livestock to avoid their being surrendered to the larger collective herds to outright agrarian revolt. On the other hand, the poorest peasants without land or livestock voluntarily and gladly participated. But the initial impact on grain supplies was negligible; in 1929, it was agreed that the process would have to be accelerated. The Communist ideal was the huge state farm or *sovkhoz*, literally a grain and meat factory in which agricultural efficiency through mechanization and minimum labor requirements would be at its peak. But such a huge enterprise is not easily established, and it can often result in overmechanization and inefficiency. Therefore, by way of an intermediate step, the smaller local collective farm or *kolkhoz* was recognized as a more easily attained substitute. By the end of the second Five-Year Plan, in the late 1930s, well over 90 percent of all peasant farms were collectivized.

**Soviet Agriculture** A generalized map of agricultural regions within the Soviet Union (Fig. 2–12) underscores the limitations of the Soviet environment. The leading agricultural zone—region (1)—lies south of Moscow, extending eastward from the northern shores of the Black Sea to the valley of the Irtysh River in south-central Siberia. This region is named *Large-scale Diversified* because here lie the country's most extensive, highly mechanized collective farms where the bulk of the country's wheat is harvested and crop variety is enormous. From press reports of recent Soviet crop failures, we can conclude that this is also a region of unreliable rainfall and that disastrous droughts frequently affect certain areas toward the drier east. Besides winter and spring wheat, this region can produce a wide range of food and industrial crops, including sugar beets, potatoes, fruits and vegetables; it also supports large livestock herds of cattle, sheep, and pigs. Although this landscape frequently reminds one of the North American Great Plains, the house-and-barn units that mark the U.S. countryside are completely lacking. Instead, the old Russian villages of the tsarist era have been converted into Soviet collectives (Fig. 2–13), at whose focal points lie the newer administration buildings—including

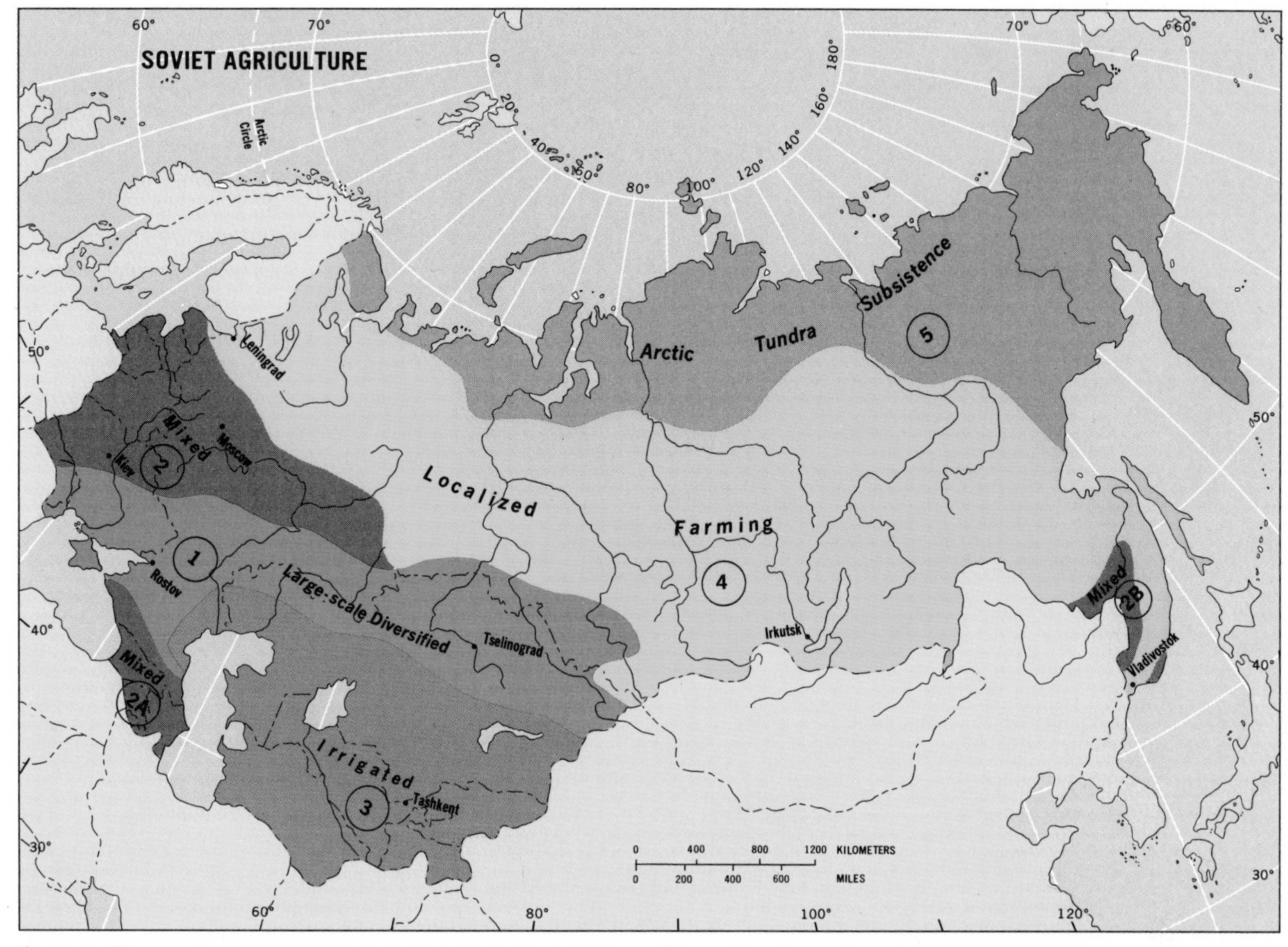


Figure 2–12

the medical clinic, school, recreation facilities, and meeting hall. Large patriotic billboards still exhort the Soviet worker to labor hard for the country. But walk to some of the small, often old and outmoded wooden houses of the collective's inhabitants and you may find that each has a small plot of land where the farmer can still grow some crops for his own use (and even for sale). In those meticulously cultivated gardens, the Soviet farmer clearly retains some individuality.

In order of importance, the discontinuous region identified as *Mixed* (2) follows. Actually, three subregions fall into this category: the large triangular area that contains Moscow, the mountains and foothills of the Caucasus (2A), and the small but significant zone in the Soviet Far East that fronts the Chinese border and the Pacific Ocean (2B). Farming here occurs at a rather smaller scale than in region (1), but the variety of crops and livestock is great. Rye, barley, potatoes, fodder crops, and dairy cattle dominate the farm scene, which in many ways looks like the landscape adjacent to the North American Great Lakes. The zone centered on the Caucasus has a different mixture of crops, because the lower-lying areas here are located in a subtropical environment (citrus groves and tea plantations mark these gentler slopes). On the eastern side of the Caspian Sea, the climate is also warm and the growing season long—but rainfall is so low that crops must be *Irrigated* (3). Sheep and hardy cattle support themselves on scattered pastures, but the crops—cotton, tobacco, fruits, vegetables, and some grains, including rice—must be artificially watered.

These three regions are clustered in the southwestern portion of the Soviet Union, and the two remaining areas cover the rest of this vast realm. Region (4) is characterized by *Localized Farming*. In the difficult terrain, which would be familiar to Canadians, farming tends to concentrate near the scattered, often isolated settlements that serve as local markets for such products as potatoes, rye, barley, oats, some hardy vegetables, and milk, butter, cheese, eggs, and meats. Although growing seasons are short in this region and farming difficult, things get even worse in region (5), where *Arctic–Tundra Subsistence* prevails. A few root crops manage to

Figure 2–13
Tsarist-era village in the age of collectivization.

mature in the brief but intense summer, however, this is mainly hunting country. As in arctic environments elsewhere, reindeer help sustain human communities.

Despite the Soviet Union's overall economic development in recent years, agriculture remains a national disaster. Poor management and constantly recurring weather problems have thwarted most efforts aimed at making meaningful gains, and the U.S.S.R. is forced to keep importing much of its annual grain supply. The opening of the post-Brezhnev era in 1983, at least, is being accompanied by a refreshing frankness as the new leaders confront the food-production fiasco. The realization seems to be sinking in at last that agriculture is the hardest (not easiest) sector to run and that the government must now begin applying its most talented people to resolve the worsening mess. By the mid-1980s, agriculture is supposed to become a leading priority, ranking only behind the military, and greater pressures than ever will be unleashed from Moscow to increase farm-sector productivity. One new method already in use involves the replacement of piecework remuneration by "collective contracts" between the state and *kolkhoz* work teams; by rewarding these worker brigades for the results of the entire collective-farm operation, instead of for specific tasks, it is hoped that labor will respond to its new autonomy and incentive by improving its morale and overall output.

**Soviet Industry** Since the 1917 Revolution, the Soviet Union has been recognized less for its achievements in agriculture than for its massive industrialization. From the beginning, the goals of the first Five-Year Plan, even in the field of agriculture, were in considerable measure related to the high-priority objectives of industrial development. The collectivization and mechanization of agriculture would free thousands upon thousands of peasants for factory work in the cities, and, with a larger and more dependable volume of farm production, the growing urban population would be better fed. Major investments were made in transportation, electrification, and other heavy-industry-oriented facilities, and, certainly, the results were spectacular (Fig. 2–14). During the first Five-Year Plan, while the rest of the world experienced the effects of the Great Depression, Soviet industrial production climbed continuously and sharply. In terms of steel production, for example, the Soviet Union in 1932 was ahead of all countries but the United States; needless to say, it trumpeted this achievement as proof of the effectiveness of the plan, and it would be difficult to argue otherwise. From then on, Soviet manufacturing has been in the forefront of the world.

In a way, then, the Soviet Union operates as a giant corporation, with the government and its agencies acting as a board of directors that determines the levels of productivity, the resources that will be exploited or opened up, the kinds of goods that will be produced and made available, the wages paid, and so forth. The government's economic planners set the country on a certain course; although there are year-to-year modifications in all such programs, the major features of the plans normally remain visible. But it would be a mistake to view the Soviet Union's economy (and that of other Communist states) as the only economy in the world that involves "planning." Economic changes in the United States, too, are the result of planning activities, but here the plans are laid at different levels, from the giant corporations to small businesses. Certainly, the government takes a hand in shaping the progress of the U.S. economy, but it does not wield control in the manner of the Soviet government. On the other hand, it is easy to exaggerate the effectiveness of Soviet national planning in bringing about the desired goals. The measures used to gauge the degree of success in meeting production schedules have varied, and the psychological impact that is so important a part of each plan has not always had the desired result. Agriculture always

## ANALOGUE AREA ANALYSIS

The small private plots that Soviet farmers still are permitted to farm produce a huge harvest—over 40 percent of all the eggs and more than 30 percent of all the vegetables and milk consumed. If Soviet economic planners had not collectivized the farmlands, would production of wheat and other crops be higher than it is on the *kolkhozes* and *sovkhozes*?

To try and answer that question, Stephen Birdsall employed the method of analogue area analysis to compare Soviet and U.S. agriculture. He chose two areas with essentially the same physical base (climate, soils, relief, drainage): the Minsk Oblast in the central part of Belorussia—region (2) in Fig. 2–12—and six farming counties in central Wisconsin. Birdsall reported: the two areas "appear nearly identical in terms of their respective physical bases for agriculture . . . areas so widely separated on the earth's surface could hardly be more similar physically." Having established the areas as approximately analogous, Birdsall next compares their productivity: in grain yields, the Wisconsin area produces 1.5 to 2.0 times as much as Minsk. In pastoral output, too, the Wisconsin counties are far ahead: in milk yield per cow (1.8 to 3.6) and egg production (2.0 to 3.0). Only in the production of root crops do farmers of the Minsk Oblast match American yields, exceeding in one—potatoes. But, overall, Soviet production lags far behind. Birdsall leaves no doubt about his conclusion: "The lower agricultural productivity of [the Minsk Oblast] cannot be said to be caused by factors beyond the control of those ultimately in charge of farming operations . . . [this area] could ultimately have agricultural productivity at least 150 percent of its present level. This could be achieved by adopting the managerial methods (including decisions on technology utilization) used in the selected counties in Wisconsin."

seems to have been the problem sector; the long haul toward adequate farm production was probably unnecessarily long, because of the low prices paid, the lack of incentives, and the enduring backseat position vis-à-vis manufacturing. But the Soviets sought the fastest route to world power, and that road, they quickly realized, lay via heavy industry. In this field, there really has been revolutionary development, so much so that two generations after the 1917 Revolution, the Soviet Union had achieved the status of a superpower. Agriculture still presents problems—its production insufficient to release the country from foreign imports, an enormous labor force still tied to the land, farming at comparatively low efficiency while industrial labor shortages persist. Still, the situation has improved as investment in agriculture has increased, yields have risen, and lasting improvement may finally be becoming a leading planning goal in the mid-1980s. Consumer goods, of which the Soviet citizen has long been virtually deprived, are increasing in volume on the Soviet market. Accompanying this are some signs of political and economic relaxation; the profit motive is helping to solve agricultural problems on collectives where farmers may be rewarded for individual output as well as time.

It is tempting, naturally, to speculate whether the Soviet economy could have developed so spectacularly without state control over all productive capacity and without the centralized planning for which it has become known. Many Western economists, pointing to the rapid industrialization of Russia during the 1880s and 1890s, argue that if the trends set at that time had been continued, the country would be further ahead industrially than it is today. But the revolution and ensuing civil wars largely obliterated this early progress, and what the Soviets remember (and are constantly reminded of) is the rise of their country out of a divided, agrarian, poor, unproductive, inefficient, and stagnant past. Undoubtedly, the vast majority of the country's citizens are true believers: hundreds of thousands have witnessed this emergence of the Soviet Union from poverty and obscurity to greatness, and millions have seen promises made by economic planners come true.

# Regions of the Soviet Realm

The Soviet Union today is a country of large impressive cities, efficient communications, modern industries, and mechanized farms. It is also a country of vast empty spaces, remote settlements, inadequate transport facilities, and badly outmoded rural dwellings. Against the confidence and accomplishments of the Soviet planners, the treatment of many Jewish citizens and dissident authors, scientists, and artists stands in stark contrast. The Soviet Union has powerful armed forces, great universities,

Figure 2–14

magnificent orchestras, spectacular space programs—but it still deploys the sinister State Security Committee (KGB) to spy on its citizens, sends nonconformists to "mental hospitals," and nurtures the intimidating specter of Siberian exile. Soviet world influence works to liberate faraway colonies from foreign domination, but Moscow finds itself totally unable to accommodate expressions of nationalism in the satellite states of Czechoslovakia, Hungary, or Poland. In short, the modern Soviet Union is a complex of contradictions. Some of those contradictions can be identified in the country's regional geography. Earlier in this chapter we examined its physical geography; we now turn to the human regions that give the Soviet Union its varied character.

## 1. The Soviet Core

Geographers who study the properties of states call the heartland of a state its *core* or *core area*. In the core area, the state is likely to display its national iconographies most strongly; here lie its largest cities (normally including the capital), its most productive industries, its greatest population cluster. In the core area, surface communications are more efficient, networks denser than anywhere else. States that have grown from ancient core areas have their lengthiest history there, and cultural imprints remain strongest.

The Soviet Union's core area centers on Moscow and includes the major cities of the western U.S.S.R. As we saw, Moscow was heir to Kiev as the early center of Russian consolidation and expansion. Although located in rather meager forest country, Moscow possessed centrality—a situation that proved highly favorable during the periods of conquest and growth. On the map, it is the river system west of the Urals that really emphasizes Moscow's unrivaled position, with the Volga and its Oka tributary draining toward the southeast, the Northern Dvina north to the Arctic, the Western Dvina west to the Baltic, and the Dnieper south to the Black Sea (Fig. 2–15). These were the routes of expansion; in the lands between the rivers, the tsars' control was secured through land grants to loyal citizens. Southward, the frontier of expansion penetrated the better soils of the Ukrainian black earth belt, and a Russian breadbasket de-

Figure 2–15

veloped there. *Ostrogs* on the Volga developed into towns, and Moscow began to feel some urban competition: as early as the seventeenth century, these Volga sites were manufacturing products from locally grown flax and hemp, from wood, and from the fur-bearing animals trapped in the forest margins.

Moscow lies at the heart of what is commonly called the *Central Industrial Region*. The precise definition of this region varies, as all regional definitions are subject to debate. Some geographers prefer to use the name "Moscow Region," thereby emphasizing that for over 400 kilometers (250 miles) in all directions from the capital, regional orientations are pulled toward the historic focus of the state. One advantage in this last name is that it avoids the suggestion that this region is the only industrial center of the Soviet Union or that it leads in all sectors of the industrial economy. The fact is that the Moscow region has several competitors; the focus of heavy industry, for example, has long been to the south in the Ukraine, and certainly, Leningrad to the northwest has been an important center—there was a time when it seemed destined to take over the lead permanently from Moscow itself. Therefore, other geographers prefer a wider definition of the Central Industrial Region, one that contains not only the Moscow-Gorky industrial concentration, but also most of southern Russia to the Ukrainian border, including such cities as Kursk, Voronezh, and Borisoglebsk.

The Moscow-Gorky area is the leading urban-industrial cluster within the Central Industrial Region (Fig. 2–15). Moscow (8.1 million) still remains exceptionally large and exceptionally expressive of national culture and ideology, traits that helped Moscow hold its strength in the Russian state during the rise of St. Petersburg. Given the present distribution of population, Moscow has maintained its decisive centrality: roads and railroads radiate in all directions to the Ukraine in the south; to Minsk, Belorussia, and Europe to the west; to Leningrad and the Baltic coast in the northwest; to Gorky and the Urals in the east; and to the cities and waterways of the Volga in the southeast (a canal links the city to this, the Soviet Union's most important navigable river).

**Figure 2–16**
Moscow: new apartment buildings are reshaping the urban scene.

Moscow itself is a transforming city as high-rise apartment complexes increasingly dominate the residential townscape (Fig. 2–16). Although this helps to alleviate the Soviet capital's housing shortage, nearly all Russian cities are horribly overcrowded, with most people forced to accept unreasonably cramped personal living spaces while they endlessly wait to move up on long waiting lists for roomier quarters. Moscow is also the focus of an area that includes some 50 million inhabitants (about one-sixth of the country's total population), many of them concentrated in such major cities as Gorky (1.5 million)—the automobile-producing "Soviet Detroit"; Yaroslavl (0.7 million)—the tire-producing center; Ivanovo—the "Soviet Manchester," with its textile industries; and Tula—the mining and metallurgical center where lignite (brown coal) deposits are worked.

Leningrad, as the Soviets renamed Petrograd in 1924 after Lenin's death, remains the Soviet Union's second city, with 4.7 million people. Leningrad has none of Moscow's locational advantages, at least not with reference to the domestic market. It lies well outside the Central Industrial Region and, in effect, in the northwestern corner of the country, 650 kilometers (400 miles) from Moscow. Neither is it better off than Moscow in terms of resources; fuels, metals, and foodstuffs all must be brought in, mostly from far away. The Soviet emphasis on self-sufficiency has even reduced Leningrad's asset of coastal location, because some raw materials might be imported much more cheaply across the Baltic from foreign sources than from domestic sites in distant Central Asia. Only bauxite deposits lie nearby, at Tikhvin. But Leningrad was at the vanguard of the industrial revolution in Russia, and its specialization and skills have remained. Today, the city and its immediate environs contribute about 5 percent of the country's manufacturing, much of it through high-quality machine building. In addition to the usual association of industries (metals, chemicals, textiles, and food processing), Leningrad has major shipbuilding plants and, of course, its port and the nearby naval station of Kronstadt. This productive complex, although not enough to maintain its temporary advantage over Moscow, has kept the city in the forefront of modern Soviet development.

## 2. *Povolzhye*: The Volga Region

*Povolzhye* is the Russian name for an area that extends along the middle and lower Volga River. It would be appropriate to call this the Volga Region, for this greatest of Soviet rivers is its lifeline, and most of the cities that lie in the *Povolzhye* are situated on its banks. In the 1950s, a canal was completed to link the lower Volga with the lower Don River, extending this region's waterway system still further.

The Volga River was an important historic route in old Russia, but, for a long time, neighboring regions overshadowed it. The Moscow area and the Ukraine were far ahead in industry and agriculture; the industrial revolution that came late in the nineteenth century to the Moscow area left the *Povolzhye* little affected. Its major function remained the transit of foodstuffs and raw materials to and from other regions.

This transport function is still important, but things have changed in the *Povolzhye*. First, World War II brought a time of furious development, because the Volga region, located east of the Ukraine, was protected by distance from the invading German armies. Second, at the time, the Volga region proved to be the greatest source of petroleum and natural gas in the entire Soviet Union. From near Volgograd (formerly Stalingrad) in the southwest to Perm on the Urals flank in the northeast, an enormous oilfield exists. Once this was believed to be the Soviet Union's largest reserve, but recent discoveries of oil and gas in Western Siberia indicate that even more extensive fields lie beyond the Urals. Still, because of their location and size, the Volga region's fossil fuel reserves retain their importance. Third, the transport system has been greatly expanded. The Volga-Don Canal directly connects the Volga waterway to the Black Sea; the Moscow Canal extends the northern navigability of the system into the very heart of the Central Industrial Region; and the Mariinsk Canals provide a link to the Baltic Sea. Today, the Volga region's population approaches 21 million, and the cities of Kuybyshev, Volgograd, Kazan, and Saratov are all in the 1.0–1.3 million range.

## 3. The Ukraine

Soviet strength has been built on a mineral wealth of great variety and volume. With its immense area, the country possesses an almost limitless range of raw materials for industry, although some deposits are located in remote areas and require heavy investments in transportation. In the Ukraine, the Soviet Union has one of those regions where major deposits of industrial resources lie in relatively close proximity. The Ukraine began to emerge as a major region of heavy industry toward the end of the nineteenth century, and one major reason for this was the Donets Basin, one of the world's greatest coalfields. This area, known as *Donbas* for short, lies north of the city of Rostov; in the early decades, this *Donbas* field produced over 90 percent of all the coal mined in the country. Most of the *Donbas* coal is high grade. Today, the Donets Basin alone still accounts for between one-quarter and one-third of the total Soviet output, which is still about double that of the second-ranking producing area.

What makes the Ukraine unique in the Soviet Union is the location, less than 320 kilometers (200 miles) from all this *Donbas* coal, of the Krivoy Rog iron ores. Again, the quality of the deposits is high, although the better ores are being worked out, and the industry is now turning to the concentrating of poorer-grade deposits. Major metallurgical industries arose on both the Donets coal and the Krivoy Rog iron: Donetsk and its satellite Makeyevka dominate the *Donbas* group (these constitute what might be called a "Soviet Pittsburgh"), whereas Dnepropetrovsk is the chief center of the Krivoy Rog cluster (the city of Krivoy Rog remains the largest iron and steel producer). One way or another, all the major cities located nearby have benefited from the fortuitous juxtaposition of minerals in the southern Ukraine: Rostov, Volgograd, and Kharkov near the cluster of *Donbas* cities, the leading Black Sea port of Odessa, and even Kiev not far from the Krivoy Rog agglomeration. And like West Germany's Ruhr, the Ukraine industrial region lies in an area of dense population (and, hence, available labor), good agricultural productivity, and adequate transportation systems; it also lies near large markets. Moreover, it has provided alternatives when exhaustion of the better ores began to threaten; not only are extensive lower-grade deposits capable of sustaining production through the foreseeable future, but additional iron reserves are located near Kerch, on the eastern point of the Crimean Peninsula, quite close enough for use in the established plants of the Ukraine. However, the biggest development in recent decades is the opening up of the Kursk Magnetic Anomaly south of Moscow; this area may technically lie outside the Ukraine (by 50 or so miles/80 kilometers), but its iron ores are increasingly critical to keeping the heavy industrial complexes of the Ukraine going.

And this is not all. The Ukraine and areas immediately adjacent to it yield several other essential raw materials for heavy industry. Chief among these is manganese, a ferroalloy needed in the manufacture of steel. About 12 pounds (5.4 kilograms) of manganese ore, on the average, go into the making of a ton of steel. The Soviet Union has ample supplies of this vital commodity. Deposits just south of Krivoy Rog at Nikopol and between the greater and lesser ranges of the Caucasus at Chiatura are the two leading deposits in the world.

Still there is more. To the southeast, along the margins of the Caucasus in the Russian Soviet Federated Socialist Republic (R.S.F.S.R.) and on the southwestern shores of the Caspian Sea in the Azerbaydzhan S.S.R., there are significant oil deposits—certainly not far from the Ukrainian manufacturing complexes by Soviet standards of distance. A pipeline connection runs from the Caspian coast along the Caucasus foothills to Rostov and Donetsk. The Western Siberia and Volga-Urals fields have now overtaken these southern oil fields, which center on the old city of Baku, but production continues here.

The Ukraine provided the Soviets with their opportunity to gain rapidly in strength and power; when World War II broke out, this was the center of heavy industry in the country. But favorable as its concentration of resources and manufacturing was, it also contributed to Soviet vulnerability. The industries of the Ukraine and the Caucasian oilfields were prime objectives of Germany's invading armies. As the Soviets were forced to withdraw, they dismantled and even destroyed more than 1000 manufacturing plants, so that they might not fall into the hands of the invaders. Thus began a series of developments that has pulled the Soviet center of economic gravity steadily eastward; Soviet planners were impressed with the need for greater regional dispersal of industrial production, not just because the Ukrainian and other western mines might eventually become exhausted, but also for strategic reasons.

The Ukraine (loosely translated, the name means "frontier"), for all the Soviets' encouragement of the eastward march of population and economic development, remains one of the cornerstones of the country. With over 60 million people, it contains about one-fifth of the entire Soviet population (though less than 3 percent of the land), and its industrial and agricultural production is enormous. With all that spectacular industrial growth, it is easy to forget that much of the Ukraine was pioneer country because of its agricultural possibilities, not its known minerals. Even today, half of its people live on the land rather than in those industrial cities; in the central and western parts of the republic, rural densities exceed 150 per square kilometer (400 per square mile). Wheat and sugar beets cover the landscape where the soils are best—in the heart of the Ukraine—and where moisture conditions are optimal.

## 4. The Urals

The Ural Mountains, seen by some as the eastern boundary of Europe, form a north-south break in the vast Russian-Siberian plainland. In the north the Urals are rather narrow, but, to the south, the range broadens considerably; nowhere are the Urals particularly high, nor do they form any real obstacle to east-west transportation. Roads, railroads, and pipelines cross the range in many places. An enormous array of abundant metallic mineral resources, located in and near the Urals, makes this area a natural for substantial industrial development.

The Urals regions, more or less on a par with the Ukraine in terms of total manufacturing output, rose to prominence during World War II and its immediate aftermath. The whole Urals industrial region is now so well developed, so specialized, and the product of such heavy investment that it is likely that Soviet planners will see to it that the flow of raw materials into the Urals is not interrupted as local resources are exploited ever more heavily.

It would perhaps be reasonable to speak of a Urals-Volga industrial region, for the problem of energy supply in the coal-poor Urals area has been considerably relieved by the oilfields discovered to lie in the area between the Urals and Volga Regions (Fig. 2–15). This fuel is pipelined into various parts of the Soviet core area, to Eastern Europe, and even as far eastward into Soviet Asia as Irkutsk near Lake Baykal. Natural gas, too, is piped cheaply from distant sources to the Urals, for instance, from Bukhara in the Uzbek S.S.R. Thus, the Urals industrial area, the Volga area (centered on Kuybyshev), and the Moscow-Gorky area are growing toward each other, and the core region of the Soviet Union is being consolidated. But to the east, as well, there is growth; the core region by any measurement has long ago crossed the Urals and is expanding deeper into the Asian interior.

## 5. The Karaganda Area

More than 1000 kilometers (600 miles) east-southeast of the Urals manufacturing center of Magnitogorsk, lies one of those "islands" of mining and industrial development that form part of the ribbonlike eastward extension of the Soviet heartland. This is the Karaganda-Tselinograd area, positioned in the northeastern part of the Kazakh S.S.R. (Fig. 2–7). Although this region is much smaller than the Ukraine and the Urals, it is no longer merely a mining outpost as it was before World War II when it sent coal to the Urals. Today, it possesses a variety of industries, including iron and steel, chemicals, and related manufactures; and Karaganda, no longer a frontier town, has a population greater than three-quarters of a million. Rather than a supplier of raw materials, the Karaganda area now exchanges commodities with other productive areas; in exchange for the coal it continues to send to the Urals, it receives iron ores from one of the Urals own source areas at Kustanay.

Among Karaganda's advantages is its position in the path of the Soviet economy's eastward march. There are rail connections and cost-reducing trade with the Urals, but opportunities for such exchanges also exist to the east; moreover, further mineral discoveries have been made locally, including a sizable deposit of medium-grade iron ore. Less than 320 kilometers (200 miles) to the northwest and on the railroad to the Urals, Tselinograd has emerged as the urban focus for the troubled Virgin and Idle Lands agricultural project, launched in the 1950s to expand domestic wheat production into semiarid Central Asia. The problem of water supply, a serious one here in the dry steppe of Kazakhstan, has been somewhat relieved by the construction of a 500-kilometer (310-mile) canal from the Irtysh River to the north.

Nevertheless, Karaganda and its environs cannot compare to the Ukraine or the Urals, with their cluster of resources, urban centers, transport networks, agricultural productivity, and dense populations. The Karaganda-Tselinograd region remains far from the present-day hub of the Soviet Union and is still a marginally located, subsidiary developing area; distances are great, markets are far, and costs of production are high. But the future is likely to see the ever-greater integration of the Karaganda area into the spreading Soviet core. By bringing vast eastern agricultural areas into production and by their willingness to have the state bear the costs of major industrial expansion here—again in large measure for strategic reasons born of World War II—Soviet planners have demonstrated their commitment to this region's continuing development.

## 6. The Kuznetsk Basin (*Kuzbas*)

Fully 2000 kilometers (1200 miles) east of the Urals lies the third-ranking region of heavy manufacturing in the Soviet Union—the Kuznetsk Basin, or *Kuzbas*. In the 1930s, this area was opened up as a supplier of raw materials (especially coal) to the Urals, but that function has steadily diminished in importance as local industrialization has accelerated. The original plan was to move coal from the *Kuzbas* to the Urals and to let the returning trains carry iron ore to the coalfields, but good-quality iron ore deposits were subsequently discovered in the area of the Kuznetsk Basin itself. As the new resource-based *Kuzbas* industries grew, so did its urban centers; the leading city of Novosibirsk (with over 1.5 million inhabitants) stands at the intersection of the Trans-Siberian Railroad and the Ob River as the very symbol of Soviet enterprise in the vast eastern interior. To the northeast lies Tomsk, one of the oldest Russian towns in all of Siberia, founded three centuries before the Bolshevik uprising and now caught up in the modern development of the *Kuzbas* area. Southeast of Novosibirsk lies Novokuznetsk, a city of over a half million people, specializing in the fabrication of such heavy engineering products as rolling stock for the railroads; aluminum products using Urals bauxite are also made here.

Impressive as the concentration of coal, iron, and other resources may be in the Kuznetsk Basin, the industrial and urban development that has taken place here must in large measure be attributed, once again, to the ability of the state and its planners to promote this kind of expansion, notwithstanding what capitalists would see as excessive investments. In return, they were able to push the country vigorously ahead on the road toward industrialization, with the hope that certain areas would successfully reach "takeoff" levels, after which they would require progressively fewer direct investments as growth would to an ever-greater extent become self-perpetuating. The *Kuzbas*, for instance, was expected to grow into one of the Soviet Union's major industrial agglomerations, with its own important market and with a location, if not favorable to the Urals and points west, then at least fortuitous with reference to the developing markets of the Soviet Far East.

Between the *Kuzbas* and Lake Baykal lies one of those areas that has not yet reached the takeoff point. This area has impressive resources, including coal, but Soviet planners have been uncertain exactly where to apply the necessary investments to make this a significant industrial district. In addition to hydroelectric power plants already operating at Bratsk and elsewhere along the Angara River, there are further possibilities of harnessing water power from the Yenisey River. Timber is in plentiful supply, brown coal is used for electricity production, and oil is piped all the way from the Volga-Urals fields to be refined at Irkutsk. Thus, Krasnoyarsk and Irkutsk, both cities in the half-million-plus class, lie several hundred miles farther again from the established core of the country; although the distant future may look bright—especially if Soviet plans for its Pacific margins become reality—the situation today still presents difficulties.

## 7. The Pacific Margin

The Soviet Union has about 8000 kilometers (5000 miles) of Pacific coastline—more than the United States (including Alaska). But most of its coastline lies north of the latitude of the state of Washington, and from the point of view of coldness, the Soviet coasts lie on the "wrong" side of the Pacific. The port of Vladivostok (population: 700,000), the eastern terminus of the Trans-Siberian Railway, lies at a latitude about midway between San Francisco and Seattle, but must be kept open throughout the winter by icebreakers—something unheard of in these American ports. The climate, to say the least, is harsh. Winters are long and bitterly cold; summers are cool. Nevertheless, the Soviets are determined to develop their Far Eastern Region as intensively as possible, and their resolve has recently been spurred by the growing ideological conflict between the Soviet Union and China. Farther to the west, high mountains and empty deserts mark the Chinese-Soviet boundary; south of the Kuznetsk Basin and Lake Baykal, the Republic of Mongolia functions as a sort of buffer between Chinese and Soviet interests. But in the Pacific area, the Soviet Far Eastern Region and Chinese Manchuria confront each other across an uneasy river boundary (along the Amur and its tributary, the Ussuri). Hence, the Soviets want to consolidate their distant outpost, and are offering inducements to western residents to "go east."

Although it has a severe climate and difficult terrain, the Far Eastern Region is not totally without assets. Although most of it still consists of very sparsely populated wilderness, the endless expanses of forest (see Fig. I–12) have slowly begun to be exploited by a timber industry whose major problems are the small size of the local market, the bulkiness of the product, and—as always in Soviet Asia—the enormous distances to major markets. Lumbering centers and fishing villages break the emptiness of the countryside; fish is still the region's major product; from Kamchatka to Vladivostok, Soviet fishing fleets sail the Pacific in search of salmon, herring, cod, and mackerel, to be frozen, canned, and shipped via rail to the western markets.

In terms of minerals, too, the Far Eastern Region has possibilities. A deposit of coking-quality coal lies in the Bureya River Valley (a northern tributary of the Amur), and a lignite (brown coal) field near Vladivostok is being exploited; on Sakhalin Island, there are minor quantities of bituminous coal and oil. Not far from Komsomolsk (350,000) lies a body of inferior iron ore, and this city has become the first steel producer in the region. In addition to lead and zinc, the tin deposits north of Komsomolsk are important because they constitute the U.S.S.R.'s major source.

Thus, the major axis of industrial and urban development within the Pacific margin has emerged along the Ussuri-lower Amur River system, from Vladivostok in the south—with its naval installations, shipbuilding, and fish-processing plants—to Komsomolsk, the iron and steel center, in the north. Near the confluence of the Amur and Ussuri Rivers lies the city with the greatest advantages of centrality, Khabarovsk. Here metal manufacturing plants process the iron and steel of Komsomolsk, chemical industries benefit from Sakhalin oil, and timber is used for the burgeoning furniture industry. Khabarovsk, a city of well over a half million people, has become the region's leading urban center. Railroads lead south toward Vladivostok, north to Komsomolsk and beyond, and west to the Bureya coalfields, the *Kuzbas*, and eventually on to Moscow.

The farther eastward the Soviet Union's developing areas lie, the greater is their isolation and the larger their need for self-sufficiency in view of the ever-increasing cost of transportation. Therefore, the Pacific Region exchanges far fewer commodities with points westward than does the *Kuzbas*, although some raw materials from the Pacific area do travel to the west in return for relatively small tonnages of *Kuzbas* resources. In terms of food supply, the Pacific area cannot come close to feeding itself, despite its rather small population (something over 7 million), and the farmlands of the Ussuri and lower Amur valleys. But Soviet planners have not been sure just how to direct industrialization here. Large investments are needed in energy supply to take advantage of some magnificent opportunities for hydroelectric development in the Amur and tributary valleys. But the choice is difficult: to spend here—in an area of long-term isolation, modest resources, labor shortage, and harsh weather—or to spend money where conditions are less bleak. Future relationships with China may have much to do with the pace of Soviet development in the Far East.

# Soviet Heartlands

In the late nineteenth century, there was little indication that Russia would within a half century emerge as a world power. Still an agrarian country, remote from the major sources of industrial innovation, vast but poorly organized, subject to unrest and resistant to change, Russia seemed to most scholars to have little prospect for rapid development. When Japan's forces defeated the Russians in 1904 and the country plunged into more than a decade of chaos, its fundamental weakness appeared to be confirmed.

But some remarkable scholarship was produced by those nineteenth-century European geographers we identified in Chapter 1 (Von Thünen, Ratzel, and their contemporaries), and one of them, Sir Halford Mackinder (1861–1947), did foresee the rise of Russian power. Mackinder studied Britain's imperial expansion along the world's maritime routes as well as Russia's overland growth. He noted that areas subject to penetration along river routes were vulnerable to a strong maritime power, for most of Europe's colonial acquisitions were first accomplished along the corridors provided by rivers. Thus, he was led to speculate on the comparative strengths of land and sea power. Scanning the world map, Mackinder concluded that a large area of Eurasia, not penetrated by major navigable rivers, would be safe as a fortress and, thus, able to develop strength in its secure isolation. This area, he argued in an article published in 1904, was located in western Russia, stretching from the Moscow Basin across the Volga Valley and the Urals into central Siberia (Fig. 2–17). Mackinder predicted that this region—which he called the *pivot area*—would be of crucial significance in the political geography of the twentieth century; moreover, he flatly asserted that any power based in the pivot area could gain such strength as eventually to dominate the world.

Mackinder's article aroused a heated debate, and he was an enthusiastic participant, convinced to the end that he had been correct. He was even willing to adjust his concept of the pivot area in 1919 when it was pointed out that much of adjacent Eastern Europe possessed similar qualities plus greater known resources. But long before the Soviets became masters of Eastern Europe, Mackinder proposed his heartland theory:

> *Who rules East Europe commands the Heartland*
> *Who rules the Heartland commands the World Island*
> *Who rules the World Island commands the World*

In Mackinder's hypothesis, the *heartland* was the new name for the pivot area with its Eastern European appendage, and *world island* meant the entire land cluster of Eurasia and Africa. Soon, other geographers sought to evaluate the resource content and strategic assets of the Soviet-East European heart-

Figure 2–17
**EURASIAN HEARTLANDS**

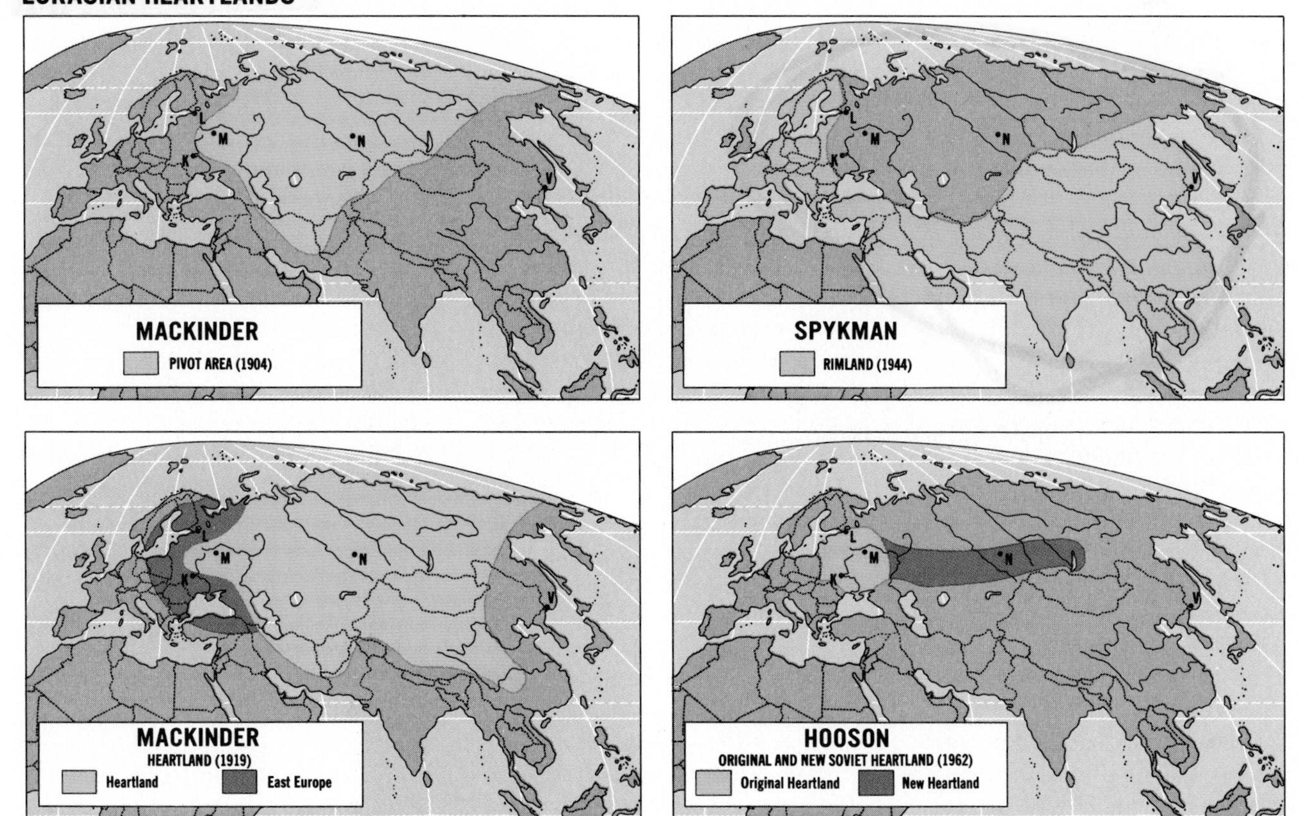

land, and there were those who concluded that Mackinder had overestimated the capacities of the region. Nicholas Spykman (1894–1943), concluding that the countries on the outer edges (*rimland*) of Eurasia contained greater power potential than the heartland, responded to Mackinder in a parody of his "theory":

> *Who controls the Rimland rules Eurasia*
> *Who rules Eurasia controls the destinies of the World*

But the rimland, of course, always consisted of a large number of countries, and at no time were all these countries under a single authority. The heartland, on the other hand, fell to the Soviet Union when, after World War II, Eastern Europe came under Moscow's domination. When Mackinder wrote his famous essay, Germany was dominant in Central Europe and parts of Eastern Europe, and Russia hardly seemed likely to challenge German power there. Mackinder could not foresee the political consolidation of the heartland—but the fact is that the heartland today sustains one of the world's two superpowers.

The Soviet Union has made major investments in its drive to develop its Siberian and Far Eastern regions. Among the reasons for this eastward push is the awareness that Soviet productive capacity and strategic strength are heavily concentrated in a core area that, by virtue of its clustering, is vulnerable to attack by modern nuclear weapons of mass destruction. In addition, Soviet planners remember the shattering impact of the German invasion during World War II, when cities, factories, and farms were laid waste in western Russia, but the east lay protected by distance. Hence, the urge to develop the eastern interior is not only a matter of economic policy, it is also a question of strategic advantage; in fact, the publicity following the 1983 Soviet downing of the Korean passenger jet above Sakhalin Island clearly revealed the key role the Sea of Okhotsk area plays in the Russian deployment of nuclear missiles aimed at the United States. And as David Hooson has pointed out in a further contribution to the "heartland" discussion, this process is unmistakably creating a new central Siberian Soviet heartland that is an eastward extension of the original Soviet core.

# Eastern Boundaries

A stronger confirmation of the Soviet presence in its Far Eastern Region is important to Moscow for other reasons as well. The Soviet Union shares boundaries with as many as 12 countries, from Norway and Finland in the northwest to Mongolia, China, and North Korea in the Asian east. The European boundaries have undergone several shifts (Poland and Romania lost territory to the U.S.S.R. following World War II), but the Central Asian boundaries lie in uncontested, remote, high mountains. In the Far Eastern Region, however, the Soviet Union faces China across a boundary that may yet be contested (Fig. 2–18).

When states were still expanding into unclaimed areas, as Russia did during its era of territorial growth under the tsars, boundaries were often temporary trespass lines, obliterated with the next forward push. In those times, the core areas of national states were separated as much by *frontiers* as by boundaries. A frontier was a zone of separation, perhaps one of contention; it might be an area of dense forest, high mountains, swampland, or desert. Or a frontier might exist simply by virtue of distance. But eventually, the effects of Europe's technological and political revolutions, the colonial scramble, explosive population growth, and the rush for resources and empires combined to produce competition for even the most remote frontier areas (see Box on Mongolia). Before long the last frontiers disappeared, divided and parceled out among states. Boundaries replaced them—borders that were often established without much planning or concern for the peoples whose lives they would affect. With the creation of such boundaries in Asia, Africa, and America were sown the seeds of modern conflict.

So it was in East Asia. To the Chinese and Japanese, the Russians had as much business on Pacific shores as the British had in East Africa, but, for a long time, the Russians had the power, China was weak, and Japan was yet to become a major force in regional affairs. Russia was able to acquire huge areas of East Asia and to secure its territorial position through treaties imposed on the Chinese between 1858 and 1864. But the terms of those treaties are under dispute; at various times, China has announced its intentitons to reclaim lands lost during the Russian imperial expansion (even though Chinese control in most of those areas was marginal and the Chinese presence minimal).

Although the boundaries between the Soviet Union and China extend over some 6000 kilometers (nearly 4000 miles), problems have so far arisen in two areas (Fig. 2–18). One of these lies in Central Asia between Chinese Xinjiang (Sinkiang) and the Soviet Kazakh, Kirghiz, and Tadzhik Republics. At various times, people under Soviet rule have sought refuge in China (e.g., during Stalin's excesses); at

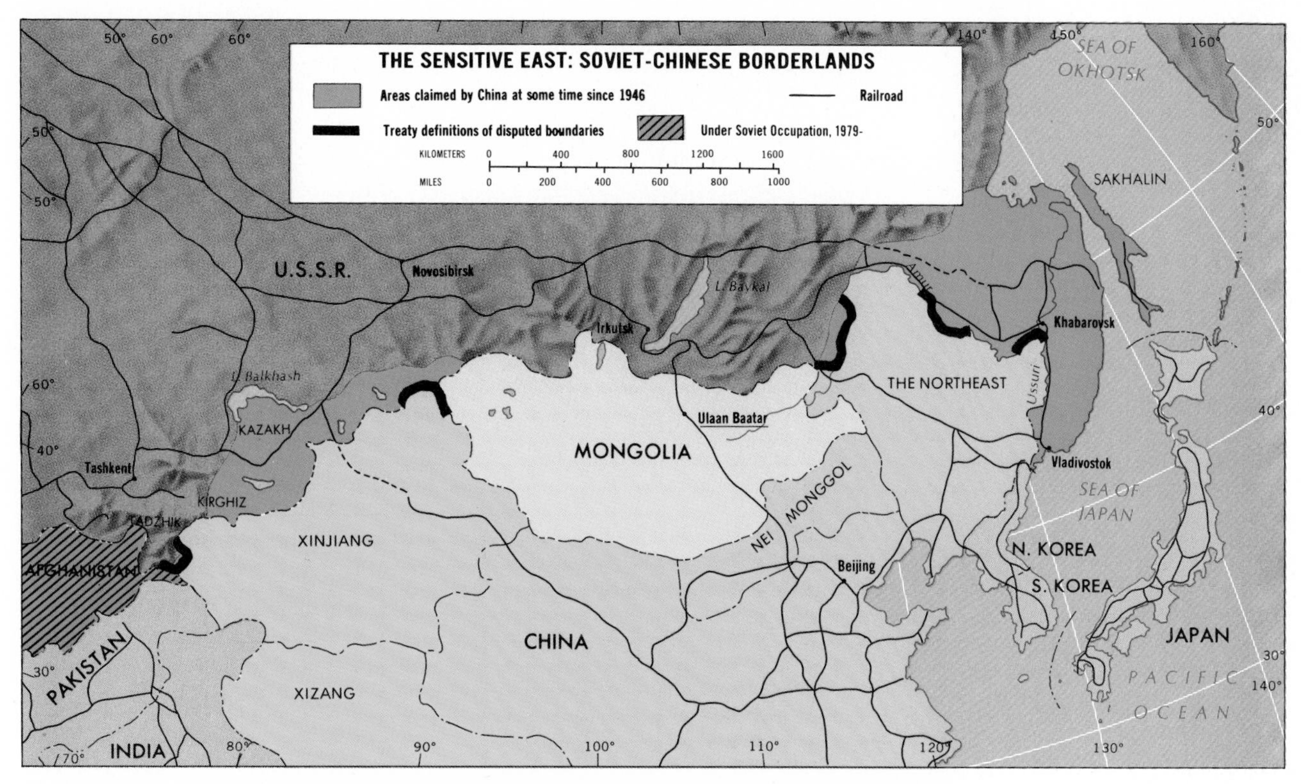


Figure 2–18

other times, Chinese minorities living in Xinjiang have crossed the border to seek safety from Beijing's (Peking's) harsh rule. Inevitably, the border region became a zone of stress and frequent violence, worsened since 1950 by the growing ideological split between the two countries. In recent years, serious conflicts have involved the lengthy Sino-Soviet boundary along the Amur and Ussuri Rivers in the Far East. In 1969, there was fighting between Soviet and Chinese troops at Chen Pao (Damansky) Island in the Ussuri River, and a dispute over navigation rights on the border rivers had to be settled anew. Soon thereafter, the Central Asian boundary conflict flared up again, with major armed forces skirmishing and a substantial number of casualties on both sides. Although both conflicts have quieted in the early 1980s, the Soviet Union may yet find itself in difficulties quite familiar to colonial powers trying to hold remote, vulnerable territories.

# Population Movements

In discussing Australia, we took note of the migration process, specifically in the context of voluntary and forced population movements (pp. 124–125). These migration streams that carried people from Europe to Australia (and to the Americas and Africa) can be classified as *intercontinental* migrations. But it is also possible to identify *regional* migrations—migration within a geographic realm but across political boundaries (e.g., the movement of people from Mediterranean European countries into the states of the European core area in search of work)—and *internal* migrations—a substantial population shift that occurs within a single country.

The United States of America has experienced large-scale internal migration flows, notably from the country's eastern heartland toward the West and now the Sunbelt (see Chapter 3). Another internal population movement of large dimensions involves the Soviet eastward drive (Fig. A–2, p. 124, route 8). As Fig. I–14 (p. 27) reveals, this movement has proceeded along a fairly narrow, well-defined corridor toward the Soviet Far East, but it has not even begun to bring to the eastern side of the Urals a population comparable in size to that of the western Soviet Union. For so vast a country, the Soviet population remains quite small.

Little more than a century ago, Russia had twice as many inhabitants as the United States. Even as late as 1900, the Russian population was about 125 million, whereas that of the United States was not much more than half this figure. Yet today the Soviet population is not much larger than that of the United States (277 versus about 239 million in 1985). What suppressed Soviet population growth? The unhappy answer is that the

## MONGOLIA: FRONTIER ZONE TO BUFFER STATE

Although unclaimed frontier areas have virtually disappeared from the world political map, territories still exist whose origins lie in the period of great-power competition for land. Landlocked Afghanistan, Nepal, Sikkim, and Bhutan have such a past, interposed between the British colonial sphere in South Asia, on one hand, and Russian and Chinese interests, on the other. Mongolia is another such *buffer* state—as these countries are appropriately called—lying between the inner Asian spheres of the Soviet Union and the People's Republic of China.

Mongolia, a vast country of more than 1.5 million square kilometers (600,000 square miles) but less than 2 million people, was under Chinese control from the end of the seventeenth century until 1911 when it declared independence amid the chaos of the Chinese revolution against the Manchu rulers. Bypassed by the major thrust of Russian imperialism, Mongolia, nevertheless, was a pawn in the Russian-Chinese settlement and treaty of 1915 when it virtually became a Chinese colony. Chinese attempts to confirm their hegemony, however, were aborted by the 1917 Russian Revolution, and fighting between "red" and "white" Russian armies spilled over into Mongolia. Mongolians saw the Bolshevik victory as a chance for help against the covetous Chinese, and, in 1921, Mongolia became a People's Republic, although outside the Soviet Union.

Today, with its vast deserts, grassy plains, forest-clad mountains, and fish-filled lakes, Mongolia lies uneasily between the Soviet Union and China, neither of whom can permit its incorporation into the other. Despite its historic associations, ethnic affinities, and cultural involvement with China, Mongolia's development in the past 60 years has been guided by the Soviets. In the 1980s, Soviet army divisions are based in Mongolia, and the country is a member of Soviet-dominated COMECON, The capital, Ulan Bator (Ulaan Baatar), with the largest population concentration, lies near the Soviet boundary; effectively, China lies far from here, across the nearly endless Gobi Desert and Altai Mountains. Although the cultural landscape remains East Asian, the Soviet presence has transformed Mongolia in many ways: apartment buildings now crowd the urban scene in the capital, a railroad traverses the country, and trucks increasingly replace camels on the roads. Mongolia survives as a true buffer state by the acquiescence of powerful adversaries. Will it escape Afghanistan's fate?

twentieth century brought repeated disaster to the Soviet people, in spite of the fact that their country rose to the position of world superpower during that period. By comparison, the population of the United States, though suffering casualties in war as well, sustained far fewer losses. Part of the Soviet Union's losses were self-inflicted; for example, during the collectivization period of the early 1930s, famines and state-directed killings probably cost almost 5 million lives. But the major destruction, staggering in its magnitude, was caused by the two world wars.

World War I, its aftermath of civil war, and the attendant famines resulted in some 17 million deaths and about 8 million deficit births (i.e., births that would have taken place had the population not been reduced by such large numbers). World War II cost the Soviet Union approximately 27 million deaths and 13 million deficit births. Thus, during the twentieth century, the Soviet population lost over 70 million in destroyed lives and unborn children, whereas American losses in all its twentieth-century wars stand well below 1 million. Moreover, the Soviet Union was affected by substantial emigration, whereas the United States received millions of immigrants. Hence, the American population grew rapidly, whereas the Soviet increase was severely limited: in the half century from the beginning of World War I to 1964, the United States gained some 90 million people (from a base of 100 million), whereas the Soviet Union gained about 65 million. Thus, during that period, the U.S. population nearly doubled while the Soviet population grew by only one-third.

Naturally, the results of these recent calamities have concerned Soviet governments and planners. The deficit of births is today reflected in a shortage of younger people in the labor force; at higher ages, women still outnumber men to a greater extent than perhaps in any other country in the world. For a period after World War II, the government pursued policies whereby families were encouraged to have a large number of children, and mothers who had 10 children received the Order of Mother Heroine from the Presidium of the Supreme Soviet. This vigorous pursuit of expansionist population policies has now been slackened, and Soviet planners have other objectives: to accommodate the strong movement from rural areas to the cities and, especially, to lure labor to population centers in the more remote trans-Ural areas of the country. Still, concern over low birth rates has emerged again in recent years, and incentives for larger families may be reinstituted.

Just 60 years ago, less than one-fifth of the Soviet population was urbanized, a reflection of the country's state of development shortly after the takeover of the Communist administration. In 1940, the urban proportion reached one-third of the population; by 1960, just under half of the population (then 210 million) resided in cities and towns. Today, about two-thirds live in urbanized areas, a figure that still lags behind Western European standards. The continuing migration toward the cities has produced housing problems and serious overcrowding in the many miles of medium-sized apartment buildings that line the streets of Soviet cities from Leningrad to Vladivostok. The desire to open the more remote eastern areas of the country for development has led Soviet governments to offer incentives to families willing to move to such places as Komsomolsk and Khabarovsk. These incentives include better living space and higher wages.

# The Changing Geography of the 1980s

The first half of the 1980s has been a period of transition for the Soviet Union following the death of Leonid Brezhnev in 1982, the subsequent regime of the late Yuri Andropov that lasted only 15 months, and the emergence of the Chernenko government in 1984 (Fig. 2–1). It is by no means certain that Konstantin Chernenko's stewardship will endure for any length of time, and several observers expect a new longer-term leadership group to emerge in the foreseeable future, dominated by some of the younger politicians in the highest ranks of the Communist Party. On the international scene, besides a worsening relationship with the United States, Chernenko inherited the Afghanistan problem, involving a stalemated Soviet military occupation that some have compared to the frustrating American experience in Vietnam during the 1960s and 1970s (it is believed that the U.S.S.R. is quietly searching for a way to pull out its forces). On the domestic scene, major regional development projects dot the Soviet realm in the mid-1980s, with two being of particular importance. The first is the (3540-kilometer/2200-mile-long) Baykal-Amur Mainline railroad (Fig. 2–14), one of the world's largest construction projects (costing about $15 billion in U.S. currency), which is scheduled to open in 1986 or 1987; the BAM line will not only provide a higher-speed rail link across the Far Eastern Region, but it will also open up a corridor rich in a number of mineral resources. The other new facility is the natural gas pipeline that will transport this fossil fuel from Siberia to Western Europe (Fig. 2–19), beginning in 1984 if pumping problems can be minimized; the route of this 4500-kilometer (2800-mile), 1-meter (36-inch) pipeline will extend from Urengoi, near the Arctic Circle just east of the northernmost Urals, across the entire western U.S.S.R. and then through Czechoslovakia into West Germany, France, and Italy.

Of all the Soviet Union's developing regions, Siberia continues to receive the greatest encouragement by economic planners. Clearly, the more that is learned about this vast area, the richer its resource base becomes. The latest estimates of coal and natural gas alone are staggering: at least 50 percent of the earth's coal reserves and sufficient gas (though most of it is locked in a solid, frozen state) to supply the *entire world* with fuel for thousands of years to come at today's consumption rates. These treasures notwithstanding, the costs of developing resources in such a remote and inhospitable place are enormous—perhaps four times greater than in the European U.S.S.R.; realizing this economic reality, the Soviets are courting the Japanese and Europeans to help them jointly develop those resources, and it is not inconceivable that the United States might be approached to form a similar partnership when the world political climate is favorable. Within Siberia, the eastern region of Yakutia (approximately covering region ⑥ in Fig. 2–5) is one of the most actively developing areas. Fifty years ago, only 25 percent of the Soviet population resided east of the Urals; today, that proportion exceeds 40 percent. Yakutia's population has doubled to 1 million since the mid-1960s; at least another 300,000 people are expected to be residents by 2000. Besides the plethora of coal and natural gas already mentioned, this region contains the country's leading deposits of gold and diamonds plus substantial reserves of copper, tin, and a dozen other strategic minerals.

A final pair of public-works projects are the most grandiose of all: reversing the flows of major rivers to transport water to the parched flatlands of Central Asia. Russian planners have long been aware that a hydrographical paradox dominates the physical geography of their realm, wherein 80 percent of the U.S.S.R.'s river water courses northward into the remote Arctic Ocean, whereas expanding agricultural lands to the south and east are subject to chronic drought. Certain Russian scientists, including geographers, have argued for years that

**Figure 2–19**
The new Siberian gas pipeline tunnels beneath the mighty Volga.

the government should seriously consider diverting some of these unneeded water flows from north to south. The first and more modest of the two diversion schemes is located about 400-kilometers (250-miles) north of Moscow in the vicinity of the city of Vologda (Fig. 2–15) and could be completed by 2000. A 40-kilometer (25-mile) artificial channel would be dug to divert water from a tributary of the Northern Dvina River south into the headwaters of the Volga, greatly enhancing the capability of that mighty river to carry water to its semiarid lower reaches near the Caspian Sea. The second project, a far more ambitious and longer-term venture, involves the Irtysh River in western Siberia, which would be diverted southward via a (2400-kilometer/1500-mile-long) canal across Kazakhstan to the Aral Sea area. Although it is not possible to predict an accurate timetable, the Soviet Union's record of single-minded doggedness in pursuit of its developmental goals is such that these colossal potential achievements should not be scoffed at, particularly if the government puts all of its efforts (and the necessary investments) behind the two diversion schemes. If they should become reality in the next century, Soviet climates and hydrography could be altered to the extent that the projects would amount to the largest-scale impacts on the natural environment ever achieved by humans.

## References and Further Readings

Allworth, E., ed. *Ethnic Russia in the U.S.S.R.: The Dilemma of Dominance* (Elmsford, N.Y.: Pergamon Press, 1980).

Armstrong, T. et al. *The Circumpolar North: A Political and Economic Geography of the Arctic and Sub-Arctic* (London and New York: Methuen, 1978).

Bandera, V. & Melnyk, Z., eds. *The Soviet Economy in Regional Perspective* (New York: Praeger, 1973).

Birdsall, S. "The Effect of Management on Crop Yields in Soviet Agriculture," *Journal of Geography*, 67 (1968), 95–103. Quotations taken from p. 103.

Brown, A. et al., eds. *Cambridge Encyclopedia of Russia and the Soviet Union* (Cambridge: Cambridge University Press, 1982).

Chew, A. *Atlas of Russian History: Eleven Centuries of Changing Borders* (New Haven, Conn.: Yale University Press, 1967).

Clem, R. "Regional Patterns of Population Change in the Soviet Union, 1959–1979," *Geographical Review*, 70 (1980), 137–156.

Clem, R. "Russians and Others: Ethnic Tensions in the Soviet Union," *Focus*, September–October 1980.

*Climate and Man: 1941 Yearbook of Agriculture* (Washington: U.S. Government Printing Office, 1941).

Cole, J. *Geography of the Soviet Union* (London: Butterworths, 1984).

Critchfield, H. *General Climatology* (Englewood Cliffs, N.J.: Prentice-Hall, 4 rev. ed., 1983).

Demko, G. & Fuchs, R., eds. *Geographical Perspectives in the Soviet Union* (Columbus: Ohio State University Press, 1974).

Dewdney, J. *A Geography of the Soviet Union* (Elmsford, N.Y.: Pergamon Press, 3 rev. ed., 1979).

Dewdney, J. *U.S.S.R. in Maps* (New York: Holmes & Meier, 1982).

Dewdney, J. *The U.S.S.R.: Studies in Industrial Geography* (Boulder, Colo.: Westview Press, 1976).

Dienes, L. & Shabad, T. *The Soviet Energy System* (New York: Halsted Press/V. H. Winston, 1979).

Florinsky, M. *Russia: A History and an Interpretation* (New York: Macmillan, 2 vols., 1953).

French, R. & Hamilton, F. E. I., eds. *The Socialist City: Spatial Structure and Urban Policy* (New York: John Wiley & Sons, 1979).

Gedzelman, S. *The Science and Wonders of the Atmosphere* (New York: John Wiley & Sons, 1980).

Gibson, J. *Imperial Russia in Frontier America* (New York: Oxford University Press, 1976).

Giese, E. "Transformation of Islamic Cities in Soviet Middle Asia into Socialist Cities," in French, R. & Hamilton, F. E. I., eds., *The Socialist City: Spatial Structure and Urban Policy* (New York: John Wiley & Sons, 1979), pp. 145–165.

Glassner, M. & de Blij, H. J. *Systematic Political Geography* (New York: John Wiley & Sons, 3 rev. ed., 1980).

Goldhagen, E. *Ethnic Minorities in the Soviet Union* (New York: Praeger, 1968).

Greenland, D. & de Blij, H. J. *The Earth in Profile: A Physical Geography* (San Francisco: Canfield Press/Harper & Row, 1977).

Gregory, S. *Russian Land, Soviet Peo-*

*ple: A Geographical Approach* (London: George Harrap, 1968).

Hamilton, F. E. I. *The Moscow City Region* (Oxford: Oxford University Press, 1976).

Hare, F. K. *The Restless Atmosphere: An Introduction to Climatology* (New York: Harper Torchbooks, 3 rev. ed., 1963).

Harris, C. D. *Cities of the Soviet Union* (Chicago: Rand McNally, 1970).

Hooson, D. "A New Soviet Heartland?," *Geographical Journal*, 128 (1962), 19–29.

Hooson, D. *The Soviet Union: Peoples and Regions* (Belmont, Cal.: Wadsworth, 1966).

Howe, G., ed. *The Soviet Union: A Geographical Study* (Plymouth Eng.: MacDonald & Evans, 3 rev. ed., 1983).

Hunter, H. *The Soviet Transport Experience: Its Lesson for Other Countries* (Washington: Brookings Institute, 1968).

"Inside the U.S.S.R.," *Time*, Special Issue, June 23, 1980.

Jensen, R. et al., eds. *Soviet Natural Resources in the World Economy* (Chicago: University of Chicago Press, 1983).

Laird, R. *Soviet Agriculture: The Permanent Crisis* (New York: Praeger, 1965).

Lydolph, P. *The Climate of the Earth* (Totowa, N.J.: Rowman & Allanheld, 1984).

Lydolph, P. *Geography of the U.S.S.R.* (New York: John Wiley & Sons, 3 rev. ed., 1977).

Mackinder, H. J. *Democratic Ideals and Reality* (New York: Holt, 1919).

Mackinder, H. J. "The Geographical Pivot of History," *Geographical Journal*, 23 (1904), 421–444. Reprinted in Glassner & de Blij, pp. 281–297.

Mather, J. *Climatology: Fundamentals and Applications* (New York: McGraw-Hill, 1974).

Mellor, R. *The Soviet Union and Its Geographical Problems* (London: Macmillan, 1982).

Nove, A. *The Soviet Economic System* (Winchester, Mass.: Allen & Unwin, 1977).

Osleeb, J. & ZumBrunnen, C. *The Soviet Iron and Steel Industry* (Totowa, N.J.: Rowman & Allanheld, 1984).

Pipes, R. *The Formation of the Soviet Union* (New York: Atheneum, 2 rev. ed., 1968).

Scherer, J., ed. *U.S.S.R.: Facts and Figures Annual* (Gulf Breeze, Fla.: Academic International Press, annual).

Shabad, T. *Basic Industrial Resources of the U.S.S.R.* (New York: Columbia University Press, 1969).

Shabad, T., ed. *Soviet Geography: Review and Translation*. (Published quarterly by the American Geographical Society.)

Shabad, T. & Mote, V. *Gateway to Siberian Resources* (New York: Scripta, 1977).

Spykman, N. *The Geography of the Peace* (New York: Harcourt, 1944).

Suslov, S. *Physical Geography of Asiatic Russia* (San Francisco: W. H. Freeman, 1961).

Symons, L. et al. *The Soviet Union: A Systematic Geography* (New York: Barnes & Noble, 1982).

Symons, L. & White, C., eds. *Russian Transport: An Historical and Geographical Survey* (London: Bell, 1975).

# CHAPTER 3

# NORTH AMERICA: THE POSTINDUSTRIAL TRANSITION

IDEAS AND CONCEPTS

**Urban Geography**
**Plural Society**
**Time-Space Convergence**
**Rain Shadow Effect**
**Population Mobility**
**Culture Hearth**
**Epochs of Metropolitan Evolution**
**Megalopolis**
**Location Theory**
**Central Place Theory**
**Urban Hierarchies**
**Eras of Intraurban Structural Evolution**
**Hypothesis of the Galactic Metropolis**
**Urban Realms Model**
**Mental Maps**
**Economies of Scale**
**Postindustrial Revolution**
**Nine-Nations Hypothesis**

The North American realm consists of two countries that share numerous characteristics. In the United States as well as Canada, European cultural imprints dominate. Indeed, the realm is often called Anglo-America: English is the official language of the United States and shares equal status with French in Canada. The overwhelming majority of churchgoers adhere to Christian faiths. Most (but not all) of the people trace their ancestries to various European countries. In the arts, architecture, and other spheres of cultural expression, European norms prevail. North American society is the most highly urbanized in the world, and nothing symbolizes the New World quite as strongly as the skyscraper panorama of New York, Toronto, Chicago, or San Francisco. North Americans are also hypermobile, with networks of superhighways, railroads, and air routes efficiently interconnecting the realm's far-flung cities and regions. Commuters stream into and out of American cities and suburbs by the millions each working day; a typical family relocates, on the average, once every six years.

In the mid-1980s, North America finds itself at the threshold of an entirely new age, the third since the discovery of the New World by Columbus nearly 500 years ago. The first four centuries were dominated by agriculture and rural life; the second age—industrial urbanization—has endured over the past century but is now showing definite signs of coming to an end. In its place, the United States and Canada are now experiencing the swift emergence of a *postindustrial* society and economy, which is dominated by the production and manipulation of information, skilled services and "high-tech" manufactures, and operating within an increasingly global-scale framework of business interactions. As blue-collar rapidly yields to white-collar employment and as the factory gives way to the automated office,

**Figure 3–1**
The burgeoning outer suburban city: the Virginia side of the Potomac as seen from downtown Washington, D.C.

# SYSTEMATIC ESSAY

## Urban Geography

**Urban geography** is concerned with the spatial interpretation of those parts of the earth's surface that contain city-centered population concentrations exhibiting a high-density, continuously built-up settlement landscape, and the impact of such agglomerations on their surrounding countrysides. The significance of contemporary worldwide urbanization has already been discussed (pp. 30–32), including definitions of such key terms as *urbanization, city, suburb,* and *metropolitan complex* (*metropolis*). Table I–1 revealed that North America is the most heavily urbanized of the world's realms—and, as we will presently see, is a dimension of vital importance to a geographic understanding of the United States and Canada in the 1980s. Therefore, this is a particularly appropriate place to present a capsule survey of this very active subfield of human geography.

In his monumental survey of the spatial evolution of urbanism in western civilization, *This Scene of Man*, the historical-urban geographer James Vance points out that not only is the city humanity's largest and most durable artifact, but it is also the very basis of modern life. This fundamental observation is echoed in the organization of contemporary urban geography into four major components. The first is an appreciation for the *historical evolution* of urban society. The second and third involve the study of contemporary spatial patterns at two distinct levels of generalization: macroscale or *interurban geography*, which treats cities as a system of interacting points that serve large surrounding areas, and microscale or *intraurban geography*, which focuses on the internal structuring and functioning of individual metropolitan complexes. The fourth deals with *planning and policy-making*, applications of urban-geographical knowledge to help resolve spatial problems experienced by certain localities.

*Urban evolution* is concerned with the historical geography of cities, emphasizing the forces that produced urban growth during each of the major periods of human development since the city emerged in the Mesopotamian area of the Middle East about 7000 years ago. The *raison d'être* of cities from the beginning was to provide their *hinterlands* with goods and services, the most successful being those urban centers that occupied locations maximizing accessibility to surrounding populations. As urbanization spread out from its Middle Eastern hearth, it was incorporated into the cultures of various realms, as we will see in each of the coming chapters. In Chapter 3, we trace the origin and growth stages of the American metropolis, treating both the macroscale and microscale expressions of what has become the world's most advanced urban tradition.

*Interurban geography* deals with the broad scope of national-scale urban systems, viewing individual cities as points in a network that interact with each other and serve hinterlands whose territories are commensurate with the size and functional diversity of each city. The study of city types, functions, and spheres of influence within urban systems has led to the development of many useful classifications and models; examined over time, these spatial generalizations have shed light on the processes of city growth. Many of these notions were linked together in the development of *central place theory* (covered in this chapter's *Model Box 3*) by the German geographer Walter Christaller more than 50 years ago. This theory, one of the discipline's most significant scientific achievements, attempts to describe some of the basic rules that govern the locations and relationships among places within urban systems.

*Intraurban geography* is involved with the internal spatial organization of the metropolis, emphasizing structural form as well as the detailed distribution of people and activities. The first attempts at modeling urban structure date back to the 1920s work of the sociologist Ernest Burgess, who viewed the city in terms of *concentric zones* of declining land-use intensity and increasingly affluent residents as distance from the city center was increased (Fig. 3–2A); a decade later, economist Homer Hoyt proposed an alternative model based on his research into residential rent patterns, which showed that the differential influence of radial transport routes was so important that it shaped an intracity structure based on contrasting *sectors* (Fig. 3–2B); this was followed in 1945 by yet another model—devised by geographers Chauncy Harris and Edward Ullman—in which it was argued that automobile-based intraurban dispersal was creating a *multiple*

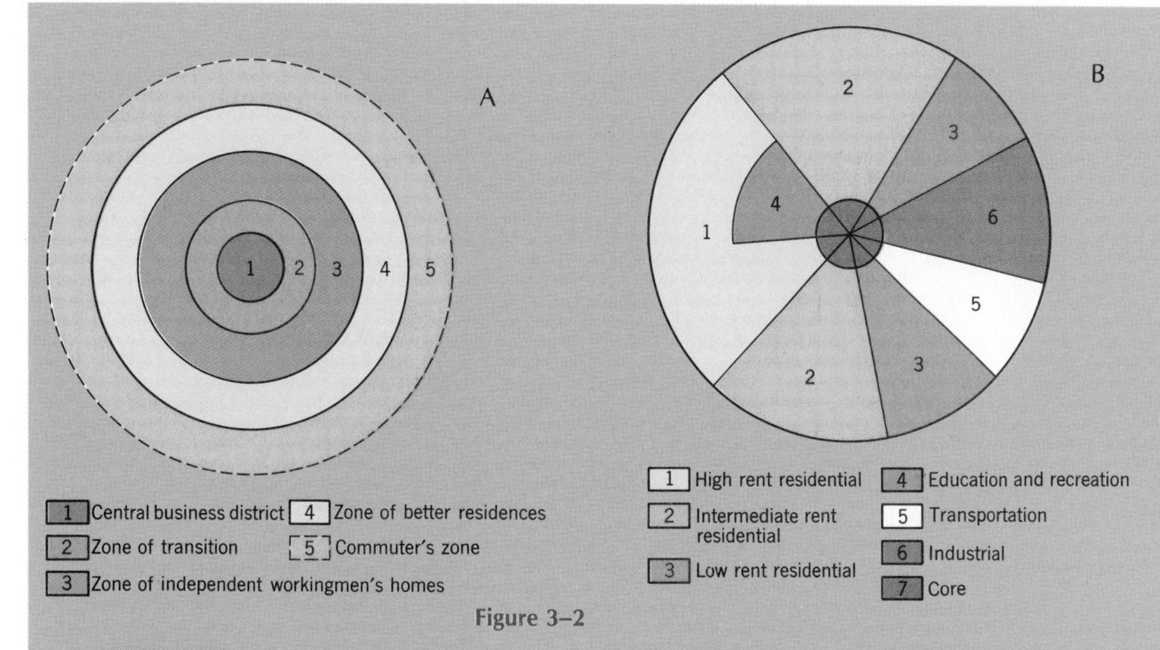


Figure 3–2

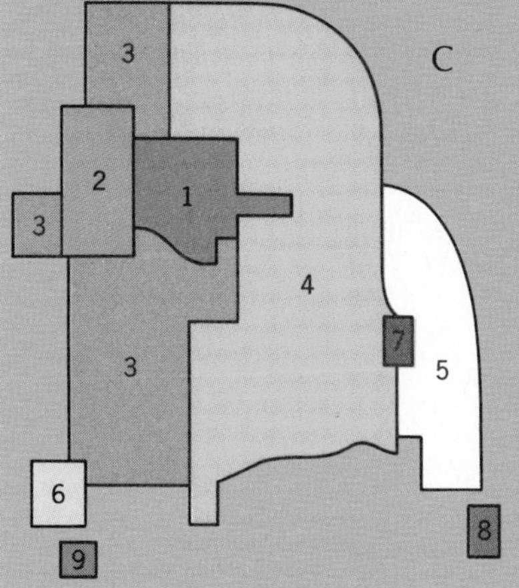


*nuclei* structure (Fig. 3–2C) that was reducing the pull of the downtown central business district (CBD). Collectively, concentric zones, sectors, and nucleations still provide a good interpretation of the land-use organization of today's central cities. But the contemporary metropolis has spilled out of its central-city confines in the second half of this century, and these models are no longer capable of accommodating a new urban reality in which the suburbs increasingly become the essence of the American city. Curiously, the last 40 years have not produced a comprehensive structural model to supersede these three "classical" generalizations—nonetheless, the elements of an updated model are available, and they are presented in this chapter's section on Urbanization in Postindustrial America Since 1970.

The spatial arrangement of people within the metropolis is the domain of *urban social geography*, one of the subdiscipline's newest specializations. The residential territorial mosaic is the focus of attention and is studied from a number of perspectives, including ethnicity, race, socioeconomic status, community formation, housing-market operations, and the dynamics of intraurban migration. These studies are steadily cross-fertilizing with the work of urban sociologists—much to the credit of geographers—particularly in the understanding of contemporary neighborhood change. The distribution of intrametropolitan nonresidential activities is the concern of *urban economic geography*, the more traditional branch of the subdiscipline that dates back more than 50 years. Yet, here too, much change is evident. Twenty years ago, the first U.S. textbook in urban geography dealt heavily with commercial, manufacturing, and transportation land uses within the central city; by contrast, one of the newest contains chapters on metropolitan retailing, industry, wholesaling, hotels, and office complexes that perceive a new coequalness between economically struggling central cities and the economically robust suburbia they spawned (but which is now approaching outright functional independence).

The discipline's growing concern with the contemporary human condition is expressed in the deepening involvement of professional geographers with *planning and policy making* approaches, which provide the opportunity for utilizing spatial concepts and methods to help solve current social, economic, and environmental problems. Urban geographers have been in the vanguard of such efforts, building on ties with the urban-planning community that extend back to the pre-World War II

**Figure 3–3**
The aging inner city: South Philadelphia.

period. Planning is defined by Harold Mayer as a "process of decisionmaking [involving] the conscious intervention of people, individually or collectively, into the evolutionary process in order to facilitate progress toward predetermined goals." Since this process is fundamentally a spatial one, geographers have much to contribute to the development and implementation of plans that seek to improve urban locational patterns and environmental conditions. Spatial data gathering and analysis are particularly important in the devising of plans today, increasingly calling on the application of such advanced geographical skills as computerized cartography, interpretation of remotely sensed satellite imagery, data-base management, and the statistical modeling of spatial structures. Geographers also participate in policy-making by demonstrating the likely outcomes of various policy options, and many practitioners have gone on to argue that the achievement of social justice means the elimination of spatial inequalities that continue to fragment the metropolitan population mosaic.

In the 1980s, these geographical disparities may well intensify as new forces continue to transform the metropolis. The aging city, its once-dominant industrial role all but ended, is trying to carve out a new economic role for its downtown office complex, although its vast inner residential areas increasingly house the least fortunate people in urban America (Fig. 3–3). This is but one of the many paradoxes that mark the metropolitan scene as the nation enters a new postindustrial era, which raises a set of planning challenges that will require the services of urban geographers for many decades to come.

fundamental dislocations are felt throughout North America. Already, the aging Manufacturing Belt of the northeastern quadrant of the United States is being called the "Rust Bowl," whereas the glamour and prestige attached to such Sunbelt locales as California's "Silicon Valley" seem to grow with every newscast. Not surprisingly, the human geography of the United States and Canada is starting to undergo a parallel transformation as dynamic new locational forces surface; in fact, the realm's durable internal re-

## TEN MAJOR GEOGRAPHIC QUALITIES OF NORTH AMERICA

1. North America comprises two of the world's biggest states territorially (Canada is the second largest in size in the world, the United States is fourth).
2. The North American realm is marked by clearly defined physiographic regions.
3. Both Canada and the United States are federal states, but their systems differ. Canada's is adapted from the British parliamentary system, and that Commonwealth is divided into 10 Provinces and 2 Territories. The United States separates its executive and legislative branches of government, and it consists of 50 states, the Commonwealth of Puerto Rico, and a number of island territories under U.S. jurisdiction in the Caribbean Sea and the Pacific Ocean.
4. Both Canada and the United States are plural societies. Although ethnicity is of increasing importance, Canada's pluralism is most strongly expressed in regional bilingualism. In the United States, major divisions occur along racial/ethnic lines.
5. Despite North America's internal social cleavages and regional economic inequalities, the realm is unified by the prevalence of European cultural norms.
6. By world standards, North America is a rich realm where high incomes and high rates of consumption prevail. Raw materials are consumed prodigiously.
7. North America's population, not large by international standards, is the most highly urbanized and mobile in the world.
8. The North American realm is the world's largest and most productive manufacturing complex. The region's industrialization generated its unparalleled urban growth, but a new postindustrial society and economy are rapidly maturing in both countries.
9. Agriculture in North America employs less than 5 percent of the labor force; it is overwhelmingly commercial, mechanized, and specialized; and it produces a huge annual surplus for sale in overseas markets.
10. North America possesses a highly diversified resource base, but rates of depletion are sizable and energy prospects remain uncertain.

gional structure is being reshaped to the extent that a new mosaic of *nine nations* can and will be discussed at the conclusion of this chapter. And Americans appear to be quite interested in their changing spatial milieu: the architect of the "nations" scheme—*Washington Post* editor Joel Garreau—has become a best-selling author and popular talk-show guest. Whatever the outcome, the winners and losers in the current scramble to adapt to the changing infrastructure will determine the geography of these two leading countries in the maturing postindustrial age, which is certain to last well into the twenty-first century.

# Two Highly Advanced Nation-States

Although it is true that Canada and the United States share a number of historical, cultural, and economic qualities, the two countries do differ in important ways. The differences, in fact, have solid geographical dimensions. The United States, somewhat smaller territorially than Canada, occupies the heart of the North American continent and, as a result, encompasses a greater environmental range. Whereas, in the United States, we tend to think of an "East" and "West" as a major geographical breakdown, in Canada it is appropriate to identify a *Southern* Canada and a Canadian *North* (Fig. 3–4). The overwhelming majority of Canadians reside in Southern Canada, mostly within a 160-kilometer-wide (100-mile-wide) zone that adjoins the U.S. boundary from Maine west to Washington State. The United States also includes North America's northwestern extension, Alaska (offshore Hawaii belongs to the Pacific Realm); thus, unlike Canada, the United States is a *fragmented* state, a discontinuous country whose territory consists of more than one major areal political unit (see Chapter 10).

Differences also become apparent when population numbers and composition are examined. In 1985, the population of the United States was estimated to be 238.6 million; that of Canada was believed to be 25.4 million, just over

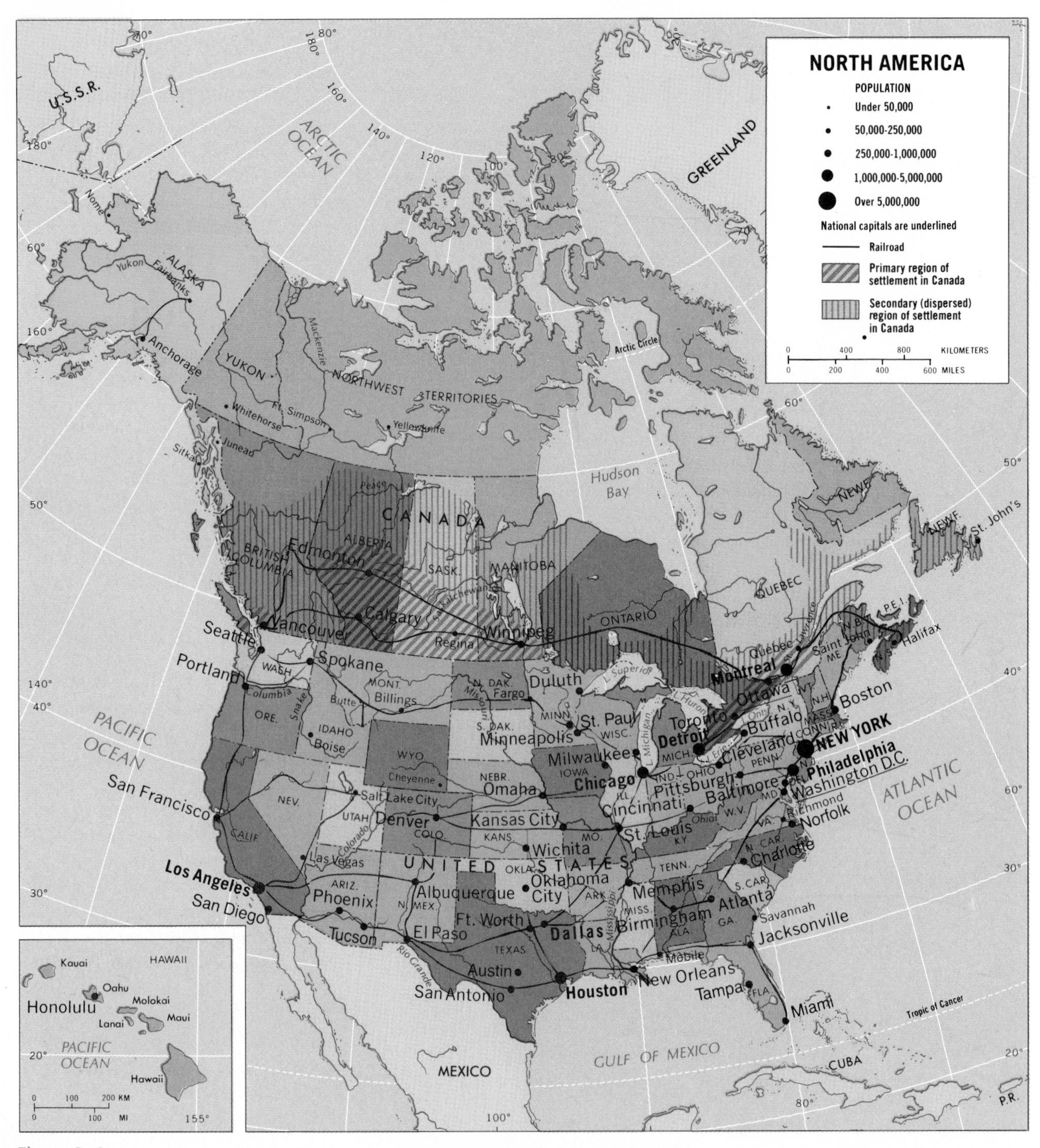


Figure 3–4

one-tenth as large. Although comparatively small, Canada's population is divided by culture and tradition, and this division has a pronounced regional expression. According to the recent census, 40 percent of Canada's citizens trace their ancestries to Britain, 27 percent have French origins, and about 25 percent come from other European stock. Today, English is the home language of 60 percent of Canadians, French is spoken by 27 percent, and other languages are used by 13 percent of the population (Indians and Eskimos make up less than 1.5 percent of the total). Certainly such a multilingual situation is not unique in the world; peoples of countries from Guatemala to Nigeria to Switzerland to India speak several different languages.

Canada's problem, however, is compounded by the strong spatial concentration of French speakers in one of the country's 10 provinces,

Quebec. With 6.6 million people, Quebec ranks second in population (after Ontario), and the province contains one of Canada's two largest metropolises, Montreal (3.2 million). Nearly 85 percent of Quebec's population is French Canadian, and Quebec is the historic, traditional, and emotional focus of French culture in Canada. Quebec straddles the St. Lawrence River, one of the access routes into North America for the early French settlers. In recent decades, a strong nationalist movement has arisen in Quebec, at times demanding nothing less than outright separation from the rest of Canada, an issue to be discussed later in the regional survey under French Canada.

Internal regionalism also affects Canada west of Francophone Quebec. The country's core area lies mainly in Ontario, Canada's most populous province, centered on metropolitan Toronto (3.6 million). Ontario, which faces the United States across the Great Lakes, has a population of almost 9 million, with a French-speaking minority of 6 percent, mostly clustered in the zone near the Quebec border. Farther west lie the interior prairie provinces of Manitoba, Saskatchewan, and Alberta, where other minorities contribute to Canada's complex cultural mosaic: Germans (6 percent in Saskatchewan, 3 percent in Alberta, 7 percent in Manitoba), Ukrainians (6 percent in Manitoba, 4 percent in Saskatchewan and Alberta), and others, including French—who account for about 4 percent in each prairie province. These western people of central and eastern European stock have been unwilling to accede to the separatist demands of Francophone Canadians, given their own minority status and language preferences; in 1977, this attitude led to formal opposition against any future economic association with a future independent Quebec. This position is even more popular in Canada's westernmost province, British Columbia, which is third in population (3.0 million) and focuses on Greater Vancouver (1.5 million). Here 82 percent of the population claims English ancestry, and only 1.7 percent are ethnic French. But British Columbia's main opposition to French Canadian nationalist aspirations is based on economic geography: this most distant Pacific coast province owes its continuing prosperity to favorable long-distance trading accessibility, both transcontinental and transoceanic, and to an advantageous relative location that would be permanently weakened by the secession of Quebec.

No problem comparable to the Quebec question affects the unity of the North American realm's other federation, but pluralism of another kind prevails south of the border. The population of the United States is also a divided one, but not by language or ethnicity—the so-called "melting pot" has consistently absorbed the vast upwardly mobile majority. More persistent in the U.S. cultural mosaic is the division between peoples of European descent (about 85 percent of the population) and those of African origins (12 percent). Despite the significant progress of the postwar "civil rights" movement, which decidedly weakened *de jure* racial segregation in public life, whites still refuse to share their immediate living space with blacks; thus, *de facto* residential segregation is all but universal, as dualistic housing-market mechanisms overwhelmingly concentrate blacks into separate nonwhite neighborhoods. Although separatist regional thinking on the provincial/state scale of Quebec does not exist, a strong case can be made that persisting local racial segregation has, in effect, produced two societies—one white and one black—that are both separate and unequal.

The United States and Canada rank among the most highly advanced countries of the world by every measure of national development. Among the myriad assets and amenities that its citizens enjoy are superior housing, clothing, electrical appliances, outstanding food and nutrition, easy long-distance transportation and communication, health and education, and a plethora of leisure options. These are the end products of the highest living standards on earth, but they are not shared equally by all of North America's residents. Deprivation is surprisingly widespread, with notable spatial concentrations in the United States inside inner-ring slums of big cities and on rural "reservations" containing Indians, Eskimos, and other "native" Americans; in the worst of these poverty pockets malnutrition is commonplace, although its severity pales by comparison to the daily misery experienced by the Third World poor.

North America's highly developed societies have clearly achieved a global leadership role, which arose from a combination of history and geography. Presented with a rich abundance of natural and human resources over the past 200 years, Americans and Canadians have brilliantly converted these productive opportunities into continentwide affluence and worldwide influence as their booming industrial revolution surpassed even Europe's by the early decades of this century. Perhaps the greatest triumph accomplished by human ingenuity in the North American realm was the hard-won victory over a difficult, large-scale environment: by persistently improving transport and communication, the United States and Canada were finally able to spatially organize their entire countries across great distances (exceeding 4000 kilometers/2500 miles) in the needed east-west direction across a terrain in which the grain of the land—particularly mountain barriers—is consistently oriented north-south. As breakthroughs in railroad, highway, and air transportation technology have succeeded each other during the past 150 years, geographic "space shortening" took place, in that the overcoming of distance was steadily made easier. Such *time-space convergence* allowed far-off places to become ever nearer to each other and simultaneously enabled people and activities to disperse ever more widely; this process of spatial reorganization has been studied by Donald Janelle, who discovered that larger demands for better accessibilty in bigger population centers decidedly steered the time-distance convergence factor in favor of cities, which quickly became the control points in the maturing long-distance spatial interaction networks.

Although they continue to have their share of differences, the United States and Canada maintain close cordial relations, and the border running between them is by far the longest open international boundary on earth. The present balance of closeness and tension between these two outspoken allies was summed up by former Canadian Prime Minister Trudeau, who, on the occasion of the first major-league baseball all-star game in Canada (Montreal) in 1982, quipped that Canada exports hockey players and cold fronts to the United States, while the U.S. exports baseball players and acid rain to Canada. Yet, putting such transitory disagreements aside, the two countries are firmly locked together in a mission of free-world leadership. What is now happening in the North American realm—and why—matters enormously to every other country. The United States and Canada constitute the most advanced region in the world; wherever it goes in the dawning postindustrial age, many of the rest will eventually follow.

# North America's Physical Geography

Before we proceed to examine North America's human geography more closely, it is useful to consider briefly the physical setting in which this geographic realm is rooted. The North American continent extends from Alaska to Panama, but we will confine ourselves here to the region north of Mexico—a region that still extends from the near-tropical latitudes of southern Florida and Texas to the Arctic lands of Alaska and Canada's Northwest Territories. The remainder of the North American continent, which constitutes a separate realm, will be designated *Middle America* and discussed in Chapter 4.

## Physiography

North America's physiography is characterized by its clear and well-defined division into physically homogenous regions called *physiographic provinces*. Such a region is marked by a certain degree of uniformity in relief, climate, vegetation, soils, and other environmental conditions, resulting in a scenic sameness that comes readily to mind. For example, we identify such regions when we refer to the Rocky Mountains, the Great Plains, the Appalachian Highlands. Not all the physiographic provinces of North America are so easily delineated however.

A complete picture of the continent's physiography is seen in Fig. 3–5. The most obvious aspect of this map of North America's physiographic provinces is the north-south alignment of the continent's great mountain backbone, the Rocky Mountains, whose rugged, often snow-covered topography dominates the western segment of the continent from Alaska to New Mexico. The eastern part of North America is dominated by another, much lower set of mountain ranges called the Appalachian Highlands. These eastern mountains also trend approximately north-south and extend from Canada's Atlantic Maritime Provinces to Alabama. The orientation of the Rockies and Appalachians is important, because they do not form a topographic barrier to polar or tropical air masses flowing southward or northward across the continent's interior. In Europe, the generally east-west trending of the Alps does form such a barrier, protecting the Mediterranean region from arctic cold.

Between the Rocky Mountains and the Appalachian Highlands lie North America's vast interior plains that extend from the shores of Hudson Bay to the coast of the Gulf of Mexico. These flatlands can be subdivided into several provinces. The major regions are: (1) the great Canadian Shield, which is the geologic core area containing North America's oldest rocks; (2) the Interior Plains or Lowlands, covered mainly by glacial debris laid down by meltwater and wind during the Pleistocene glaciation; and (3) the Great Plains, the extensive sedimentary surface that leads up to the Rocky Mountains. Along the southern margin, these interior plainlands merge into the Gulf-Atlantic Coastal Plain, which extends along the seaward margin of the Appalachian Highlands and the neighboring Piedmont until it ends at New York's Long Island.

On the western side of the Rocky Mountains lies the zone of Intermontane Basins and Plateaus. This physiographic province includes the Colorado Plateau in the south, with its thick sediments and spectacular Grand Canyon; the lava-covered Columbia Plateau in

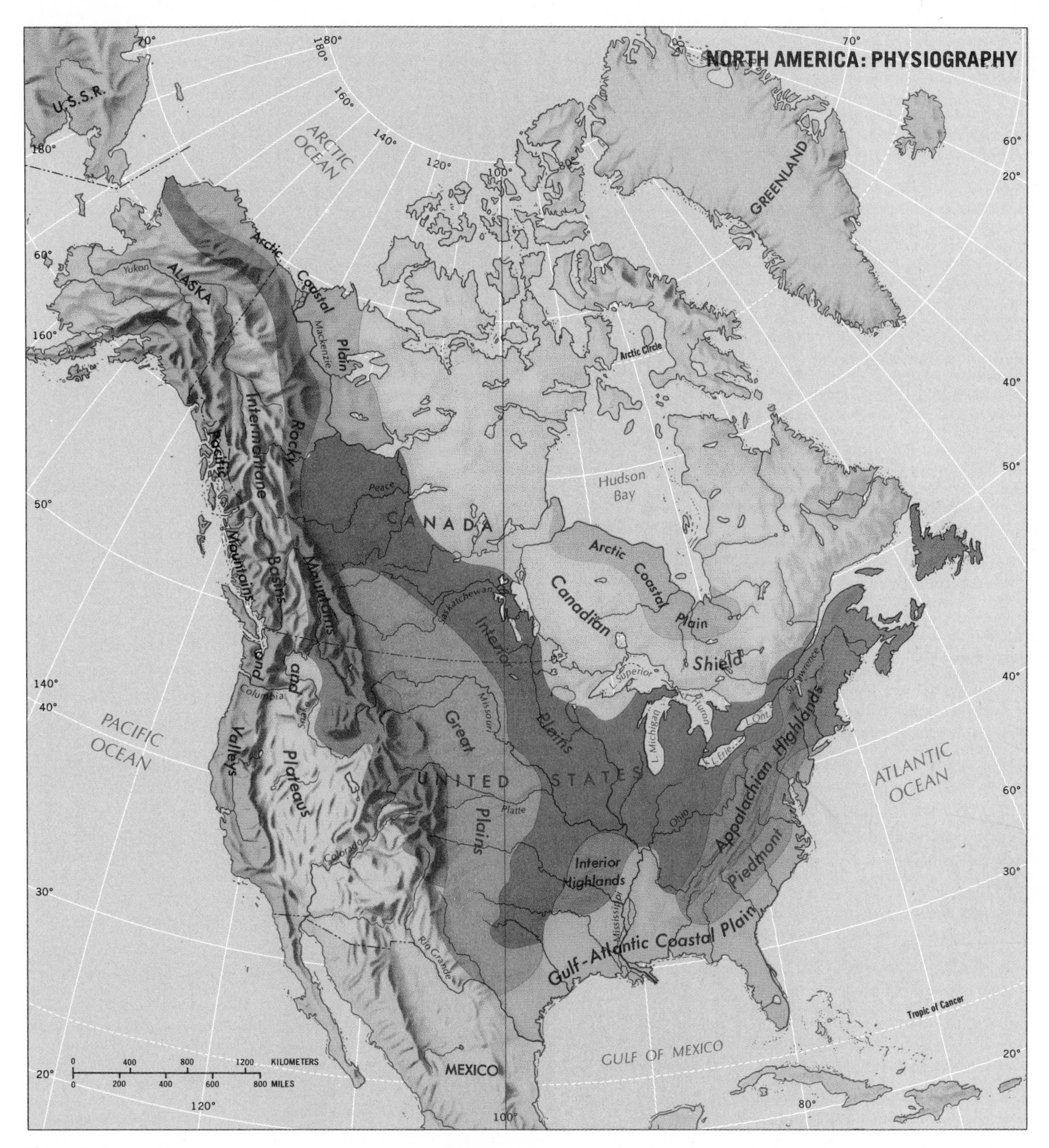


Figure 3–5

the north, which forms the watershed of the Columbia River; and the central Basin-and-Range country (Great Basin) of Nevada and Utah, which contains several extinct lakes from the glacial period as well as the surviving Great Salt Lake. Although the zone of intermontane topography continues into Canada across the Yukon and eventually central Alaska, the landscape here is much more complex, and distinct subregions such as those just described cannot be identified.

The reason this province is called *intermontane* has to do with its position between the Rocky Mountains to the east and the Pacific Coast Mountain System to the west. From Alaska to California, the west coast of North America is dominated by an unbroken corridor of high mountain ranges whose origins stem from the contact between the American and Pacific plates (Fig. I–6). The major components of this coastal mountain belt include California's Sierra Nevada, the Cascades of Oregon and Washington, and the long chain of highland massifs that line the British Columbian and southern Alaska coasts. Two broad valleys in the U.S. portion of the province, which contain dense populations, are the only noteworthy interruptions: California's Central (San Joaquin-Sacramento) Valley, and the Puget Lowland of Washington State that extends southward into western Oregon's Willamette Valley.

## Climate

The various climate regimes and regions of North America are clearly depicted on the world climate map (Fig. I–10) and model (Fig. I–11). In general, temperature varies latitudinally: the further north one goes, the cooler it gets. Local land and water heating differentials, however, distort this broad maxim. Because solid land surfaces heat and cool far more rapidly than water bodies, yearly temperature ranges are much larger where "continentality" (lack of nearby marine influences) is greatest: for example, the coldest and warmest average monthly temperatures in interior Minneapolis (−12°C/11°F and 23°C/73°F, respectively) are far more extreme than in maritime Portland, Oregon (5°C/40°F and 19°C/66°F) at the same latitude.

Precipitation generally tends to decline toward the west—except for the Pacific coastal strip itself—as a result of the "rain shadow effect," whereby most Pacific Ocean moisture is effectively screened from the continental interior (see *Box*). This broad division into Arid and Humid America, however, is marked by a very fuzzy boundary that is best viewed as a wide transitional zone. Although the separating criterion of 50 cm. (20 inches) of annual precipitation is easily mapped,

## THE RAIN SHADOW EFFECT

The uniform dryness of most of western North America is a result of the relationship between climate and physiography. At these latitudes, the prevailing wind direction is from west to east, but the moisture-laden air moving onshore from the Pacific is unable to penetrate the continent because of the blocking effect of the high west coast mountains that stand in the way. When eastward-moving humid Pacific air reaches the foot of the Sierra Nevada and other north-south ranges, it is forced to rise up the western (windward) slope in order to surmount the topographic barrier. As the air rises toward the 8000- to 12,000-foot (2400- to 3600-meter) summit level, it steadily cools. Since cooler air is less able to hold moisture, falling temperatures produce substantial rainfall (and snowfall), in effect "squeezing out" the Pacific moisture as if the air were a sponge filled with water. This altitude-induced precipitation is called *orographic* (mountain) rainfall; the two other major types of precipitation are associated with *fronts* (boundaries along which air masses of considerably different temperature come in contact, sometimes spawning spiraling storm systems called cyclones) and with *convection* (the localized rising and cooling of warm moist air).

Although precipitation totals are quite high on the unpopulated upper windward slopes of the Sierra Nevada-Cascade Mountain chain, the major impact of the orographic phenomenon is felt to the east. Robbed of most of its moisture content by the time it is pushed across the summit ridge, the eastward-moving air now rushes down the leeward (downwind) slopes of the mountain barrier. As this air warms, its capacity to hold moisture greatly increases, and the result is a warm dry wind known as the *chinook*, which can blow strongly for hundreds of kilometers inland. This widespread existence of semiarid (and in places even truly arid) environmental conditions is known as the *rain shadow effect*—with mountains quite literally creating a leeward "shadow" of dryness. In fact, as air masses move west to east across the Intermontane and Rocky Mountain provinces, they may again be subjected to orographic uplift (and even further drying out) as they pass over additional highland barriers. This reinforces the spreading of dryness, in most years, well east of the Rockies. Here in the Great Plains, Pacific moisture sources are usually blocked, and the area must depend on fickle south-to-north winds from the Gulf of Mexico for much of its precipitation; no wonder this region is susceptible to recurring droughts and was the scene of "dust bowl" environmental disasters in the 1930s and 1950s. In all, the vast dry area extending from the west coast mountains halfway across the continent to the eastern margin of the Great Plains (in Southern Canada as well as the United States) comprises what may be called *Arid America*; alternatively, the eastern half of the realm and the narrow strip between the Pacific shore and its nearby parallel ranges—where annual precipitation exceeds 50 centimeters (20 inches)—can be viewed as *Humid America*.

Elsewhere in the world, many additional examples of rain shadows can be observed downwind from where mountains block prevailing moist wind flows. Most major deserts are, at least in part, enlarged by the phenomenon. Even Europe exhibits three such areas of relative dryness: just north of the Alps, Sweden inland from Norway's coastal range, and east of the Pennines in central Britain.

that generally north-south *isohyet* (the line connecting all places receiving exactly 50 centimeters per year) can and does swing widely across the drought-prone Great Plains from year to year, as highly variable rains from the Gulf of Mexico come and go in unpredictable fashion. On the other hand, in Humid America, precipitation is far more regular. The prevailing westerly winds (blowing from west to east—winds are always designated by the direction *from which they come*) that normally come up dry for the large zone west of the 100th meridian (see Fig. 3–5), pick up considerable moisture over the Interior Lowlands and distribute it throughout eastern North America. A large number of storms develop here on the highly active weather front between tropical gulf air to the south and polar air to the north; unlike much of Europe, outbursts of polar air are not impeded by any topographic barrier and can flow directly southward from the Arctic Ocean. Even if major storms do not materialize, local weather disturbances created by sharply contrasting temperature differences are always a danger—it is no accident that there are more tornadoes (nature's most violent weather) in the central United States each year than anywhere else on earth. And, in winter, the northern half of this region receives large amounts of snow, especially just downwind from the Great Lakes (such as "Snow Alley" in upstate New York south and east of Lake Ontario), whose moisture often reinforces the precipitative potential of moving storms.

Fig. I–10 reveals the absence of humid temperate climates from Canada (except along the narrow Pacific coastal zone) and the prevalence of coldness in Canadian natural environments. East of the Rocky Mountains, Canada's most *moderate* climes correspond to U.S. *coldest* climes. Nevertheless, Southern Canada does share the environmental conditions that mark the Upper Midwest and Great Lakes areas of the United States, so that agricultural productivity in the prairie provinces and in Ontario is substantial. Canada is a leading food exporter as is the United States (chiefly wheat), in spite of its comparatively short growing seasons.

## Soils and Vegetation

As the world soil distribution map (Fig. I–13) shows, the wide variety of climates in North America has helped spawn an even more complex pattern of soil regions. In general, the realm's soils also reflect the broad environmental partitioning into Humid and Arid America. Where precipitation annually exceeds 50 centimeters (20 inches), humidland soils tend to be leached and *acidic* in chemical content. Since crops do best in soils that are neither acidic nor *alkalinic* (higher in salt content), fertilization is necessary to achieve the desired level of neutrality between the two. Arid America's soils are typically alkalinic and must be fertilized back toward neutrality by adding acid compounds. Although many of these dryland soils, particularly in the Great Plains, are quite fertile, settlers learned over a century ago that water is the main missing ingredient in achieving their agricultural potential. Whereas many watering schemes were attempted, it was not until the 1970s that the *center-pivot irrigation* method (Fig. 3–6) was perfected and provided the opportunity to expand more intensive farming west from the Corn Belt into the drier center and western portions of the Plains. Glaciation has also enhanced the rich legacy of fertile soils in the central United States, both from the deposition of mineral-rich glacial debris left by meltwater and from thick layers of fine windblown glacial material (*loess*) in and around the central Mississippi Valley.

Vegetation patterns are displayed in the world map (Fig. I–12), but the enormous human modification of the North American environment in modern times (the greatest of any geographic realm) has all but reduced this regionalization scheme to the level of the hypothetical. Nonetheless, a few useful observations should be made. The Humid/Arid America dichotomy is again a valid generalization: the natural vegetation of areas receiving more than 50 centimeters (20 inches) of water yearly is *forest*, whereas the drier climes give rise to a *grassland* cover. The forests of North America east of 100°W longitude—plus those of the Pacific rim and the higher elevations of inland mountain ranges—tend to make a broad transition by latitude. In the Canadian North needleleaf forests dominate, but these coniferous trees become mixed with broadleaf deciduous trees as one crosses the border into the U.S. Northeast. As one proceeds toward the Southeast, broadleaf vegetation becomes dominant, except for large stands of pine forests along the drier sandy soils of the Gulf-Atlantic Coastal Plain; in southern Florida, the realm's only tropical climate begets

**Figure 3–6**
Circular fields, the hallmark of center-pivot irrigation.

a savannalike grassland in combination with such large coastal swamps as the Everglades. The only noteworthy departure in the limited western forest zone is the brushy tree-and-shrub landscape (*chaparral*) of Mediterranean southern California. Arid America mostly consists of steppes or short-grass prairies. The only area of true desert (less than 25 centimeters [10 inches] of annual rainfall) is found in southeastern California and adjoining southwestern Arizona, with cactus and other leathery shrubs predominating; however, overgrazing of dry steppes in certain places like southwestern Texas has created new scrublands that arguably represent an extension of the desert environment. Astride the transition zone between Humid and Arid America one finds the tall-grass prairie, which appears to be a contradiction in such places as Iowa and Illinois where abundant precipitation should have spawned forests; although unproven, some scholars insist this is no mystery at all but simply a case of prehistoric Indian groups repeatedly burning the original forest until fire-resistant grass replaced the trees, an hypothesis that could well apply to the similarly paradoxical savanna grasslands of the humid tropics.

## Hydrography (Water)

Surface water patterns in North America are dominated by the two major drainage systems that lie between the Rockies and the Appalachians: (1) the five Great Lakes that drain into the St. Lawrence River and (2) the mighty Mississippi-Missouri river network, supported by such major tributaries as the Ohio, Tennessee, Platte, and Arkansas Rivers. Both products of the last episode of Pleistocene glaciation, together they amount to nothing less than the best natural inland waterway system in the world; human intervention has further enhanced this network of navigability, mainly through the building of canals that link the two systems as well as the ambitious St. Lawrence Seaway (opened in 1959), which can accommodate ships up to depths of 8 meters (27 feet). Artificial flood control, especially along the lower course of the Mississippi River, is another element of this particular human-environment interaction. For much of this century, the U.S. Army Corps of Engineers has battled successfully to keep riverbanks (levees) high enough to prevent overspill into surrounding lower-lying populated floodplains in Louisiana. A parallel struggle in the 1980s is the containment of the Mississippi in its traditional channel near Baton Rouge until a new riverwall is built, because the river is seeking a shorter outlet to the Gulf via the Atchafalaya Basin, which would inundate thousands of farms and cut off the port of New Orleans from the main route to its huge interior hinterland. Elsewhere, the northern east coast of the continent is well served by a number of short rivers leading inland from the Atlantic; in fact, most of the big northeastern seaboard cities are located at the waterfalls that marked the limit to tidewater navigation. Rivers in the Southeast and west of the Rockies at first offered little practical value owing to orientation and navigability problems. In the Far West, however, the Colorado and Columbia Rivers have now become critically important as suppliers of drinking and irrigation water and hydroelectric power as the west coast continues its regional development.

Groundwater is also a crucial resource in many parts of the United States. *Aquifers*, or underground reservoirs contained within porous water-bearing rock layers, underlie many states and have become essential to the maintenance of people and activities on the surface above. In dry areas, especially, they are the cornerstones of development—such as in the central Great Plains where the Ogallala Aquifer is the only steady source of water for the mainly agricultural economy. Problems quickly arise, however, when increasing usage begins to exceed the leisurely rate of natural recharge—California's heavily agricultural Central Valley is a good case in point. Even in wet areas such groundwater can be vital: for instance, metropolitan Miami obtains its drinking water from the Biscayne Aquifer during the summer rainy season. And an even more critical problem may be looming in the Biscayne Aquifer, because the subterranean dumping of industrial wastes and sewage has reached a point where contaminants are beginning to seep into the aquifer. Such issues raise a whole new dimension in North American physical geography, and merit a closer look.

## Human Environmental Impacts

More than a century of advanced industrial technology has taken its toll on the natural environment of North America. For decades, the growing problems of air and water pollution were ignored until a public outcry in the United States during the late 1960s forced the federal government to establish the Environmental Protection Agency (EPA) and take a leading role in enforcing new pro-environmental legislation. Although substantial progress was made during the 1970s, the Reagan administration was widely criticized for weakening the federal enforcement apparatus in the early 1980s. At the same time, many new environmental hazards were coming to light: Love Canal in Niagara Falls, N.Y., Pennsylvania's Three Mile Island, and Times Beach, Missouri, are three places (there are many more) that have become synonymous with recent incidents involving threats to residents by highly dangerous chemical and nuclear substances. It was also during the 1970s that cancer surpassed heart disease as the leading killer in the United States; since many cancers—especially lung cancer—are environmentally related, medical geographers have increased their study of the subject (this subdiscipline is reviewed in the Systematic Essay in Chapter 7). The spatial distribution of cancer mortality is mapped in Fig. 3–7, which reveals that respiratory-system cancer coincides with a number of major manufacturing and refining centers, including the large northeastern seaboard cities, Buffalo, Detroit, Chicago-northwest Indiana, the San Francisco Bay Area, and the lower Mississippi region.

One of the most severe *air pollution* problems of large metropolitan complexes is smog. St. Louis (Fig. 3–8) is just one of dozens of major cities that experience this annoying hazard—Los Angeles and Denver being the worst offenders on the continent. Smog is usually created by a "temperature inversion" in which a warm, dry layer of air hundreds of meters above the ground prevents cooler underlying air from rising; this causes the surface air to become stagnant, thus trapping automobile and industrial emissions that intensify and rapidly turn into chemical smog (the word is a combination of "smoke" and "fog"). The realm's most serious *water pollution problem*, alluded to in Prime Minister Trudeau's earlier remark, is once again acid rain (see Box, p. 101). Although at least 75 percent of North America's sulfur and nitrogen emissions emanate from the U.S. Manufacturing Belt (particularly Illinois, Indiana, Ohio, and Michigan), the lion's share of the resultant acid rain appears to fall in southeastern Canada as prevailing winds blow from the Midwest toward eastern Ontario and southern Quebec. As environmental damage similar to Scandinavia's intensifies, the issue could harm relations between the two countries in the future: the American attitude is that reducing factory and power-plant emissions would threaten thousands of additional blue-collar jobs in the already declining industrial Midwest, whereas the equally insistent Canadian position is that its citizens can no longer tolerate living at the end of a gigantic geographical exhaust pipe.

# Population in Time and Space

The current distribution of North America's population is shown in Fig. 3–9. This map is once again the latest still in a motion picture, one that has been unreeling for the nearly four centuries that have elapsed since the founding of the first permanent European settlements on the northeastern coast of the United States. Slowly at first,

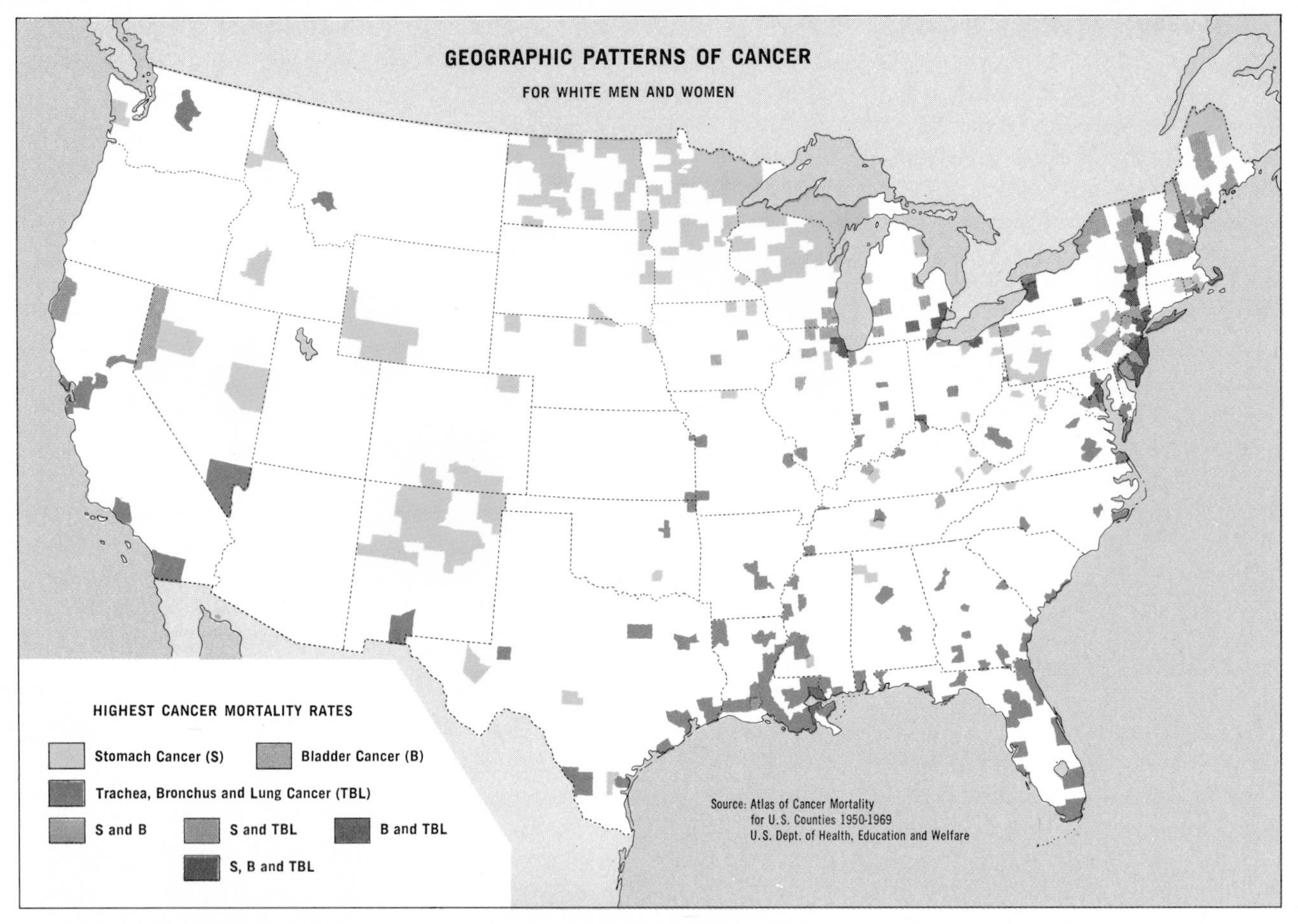


**Figure 3–7**

then with accelerating speed after 1800 as one major technological breakthrough followed another, Americans and Canadians took charge of their remarkable continent and pushed the settlement frontier westward to the Pacific. Undoubtedly, the population distribution of North America has undergone the swiftest and most dramatic changes of any realm over the past 150 years. Moreover, its inhabitants are still the most mobile in the world by far (see *Box*), a trait deeply rooted in the American national character and nativist culture. Migrations in the 1980s continue to reshape the United States, perhaps the most important being the persistent interregional shift of people and livelihoods toward the south and west (the so-called "Sunbelt"), away from the north and east; one significant measure is the 1980 census finding that, for the first time, the geographic center of the American population had moved west of the Mississippi, culminating a 200-year movement in which a westward shift was reported in every decennial census.

In order to have a good understanding of the contemporary popu-

**Figure 3–8**
Smog bathes the midcontinent city of St. Louis.

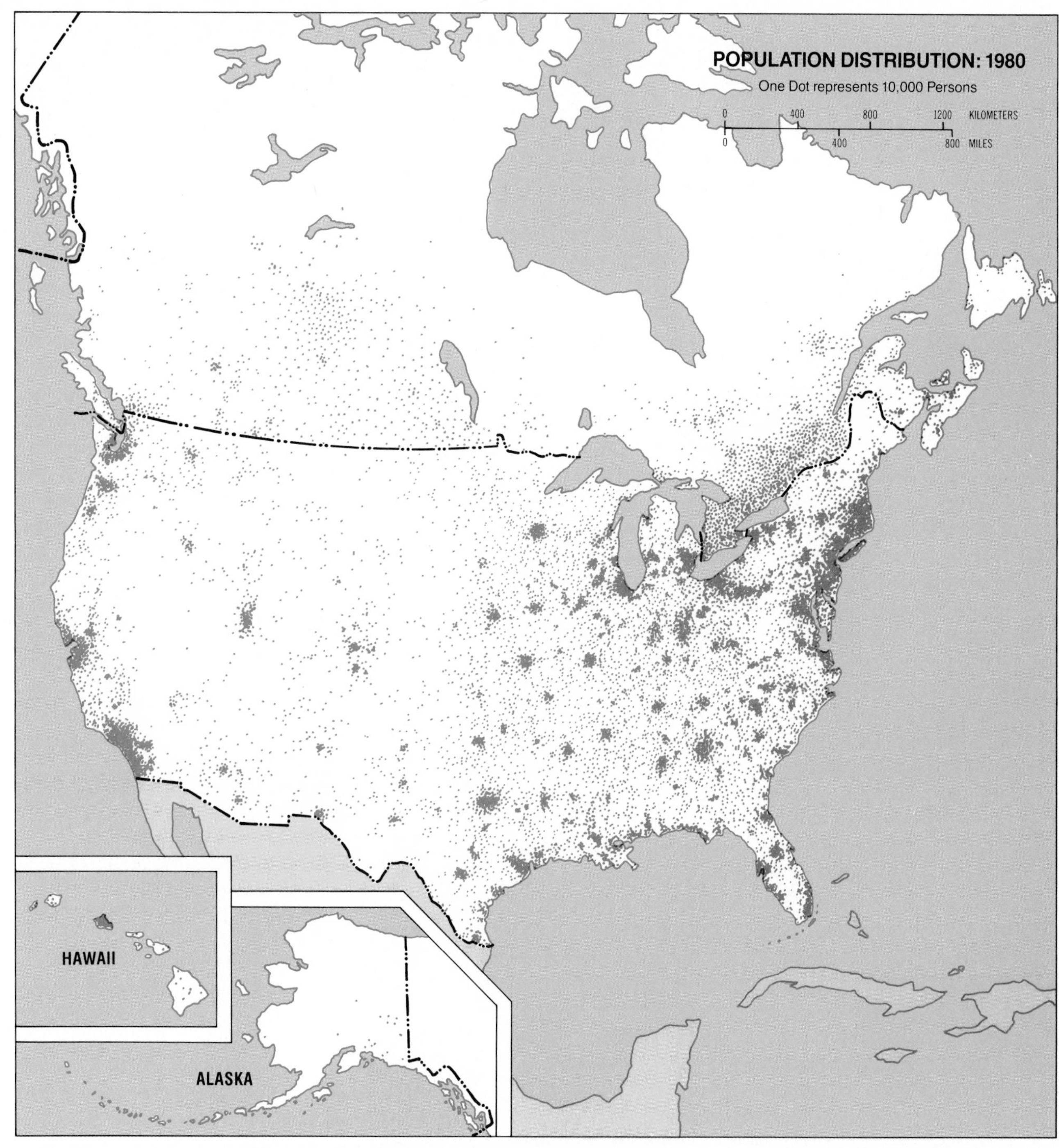


Figure 3–9

*Source:* Adapted from U.S. Census Bureau and Census Canada.

lation map, we need to review the more important forces that have shaped (and continue to shape) the distribution of North Americans and their activities. Practically from the moment of its independence, the United States has been perceived as the world's premier "land of opportunity," thereby attracting a steady influx of immigrants who were rapidly assimilated into the societal mainstream (despite tight quotas, that immigration continues with well over 1 million new arrivals annually—most of them illegal aliens). Within the realm, people have sorted themselves out to maximize their proximity to existing economic opportunities, and they have shown little resistance to relocating as the nation's changing economic geography successively favored different sets of places over time.

During the past century, these transformations have spawned a number of major migrations. Although the American frontier closed almost exactly 100 years ago, the westward shift of popula-

# THE WORLD'S MOST MOBILE POPULATION

The latest census has revealed that nearly half of the U.S. population (about 100 million) had moved to another house or apartment at least once between 1975 and 1980. On an annual basis, 18 percent of all Americans changed their addresses each year, a rate that actually reflected a slight decrease compared to the preceding 25 years during which 20 percent relocated annually. Demographers are now seeking to understand why this change occurred. Are short-term causes at work—such as economic recession and high mortgage interest rates—that may soon yield to a renewed higher mobility rate? Or are more permanent forces coming to prevail, among them a heightened desire to put down roots, persistent housing-price inflation, higher energy costs, a preference for smaller families, and greater resistance to corporate transfers when (more commonly today) two working spouses are involved? The inquiry goes on as the reduced mobility rate carried over into the early 1980s.

The geography of this hypermobility is quite interesting. Most relocations involve short distances—fully 80 percent of all movers stay within the same local area, with 60 percent remaining in the same county. This clearly reflects the close linkage between spatial and upward social mobility (another prevalent American culture trait): people are always eager to move to a nearby "better neighborhood" when a promotion or salary increase permits access to the turf of the next higher-status level in local society. Taking a broader view of aggregate trends, population gainers and losers are sharply contrasted. From 1970 to 1980, the nation's central cities lost 375,000, whereas their surrounding suburbs grew by 14.5 million; at the same time, the remaining nonmetropolitan areas grew by 8.0 million, a reversal of the trend of preceding decades as we noted. At the regional level, the most mobile populations are found in the burgeoning "wide-open-spaces" states of the west (e.g., Alaska and Colorado), whereas the lowest mobility rates are recorded in the crowded industrial states of the northeast (e.g., Pennsylvania and Massachusetts).

Population mobility in the rest of the world ranks far behind the rates exhibited for North America, although recent technological development has been accompanied by increasing mobility throughout the developed realms. A decade ago, although the average American could expect to move 14 times in a lifetime, the British relocated only 8 times, the French 7 times, and the Japanese 5 times. Today, there is evidence that the gap is closing, particularly in Europe. This entails not only growing internal mobility, but also a sharp increase in migration across Europe's international boundaries. The search for employment, as we saw in Chapter 1, is attracting large numbers of Europeans to move from the rural south into the industrialized northwest. Expanding ghettoes of foreign nationals are now commonplace in western Europe's major manufacturing centers; in fact, West Germany's Munich even has a daily Turkish-language newspaper. In the underdeveloped realms, very low mobility rates remain unchanged; however, a number of population movements are stirring there and will be noted in the following chapters.

tion—as indicated above—has continued strongly, although it now is decidedly deflecting to the south. The explosive growth of cities, triggered by the industrial revolution in the last three decades of the nineteenth century, launched a rural-to-urban migratory stream that lasted into the 1960s; to some extent, that trend may now be reversing as nonmetropolitan growth is occurring for the first time in several decades, but metropolitan suburbs strongly continue to be the mainstay of recent U.S. population increases. The middle decades of this century also witnessed significant migration from south to north—especially by blacks—but that, too, has ended and may even be reversing as many middle-income blacks are returning south. This could also be part of the much larger north-to-south movement into the Sunbelt mentioned above, a migratory stream that may still be gathering strength for the following reasons: (1) the U.S. economy and its jobs are clearly moving in that direction, (2) the historically dominant westward push is increasingly inseparable from the shift toward the southern tier, (3) the large postwar retirement migration of the affluent elderly to such states as Florida and Arizona shows no sign of abating, and (4) the new tide of Latin American immigration is overwhelmingly directed at the zone adjacent to the southern border. Let us now look more closely at the historical geography of these changing population patterns, viewing first the inital rural influence and, then, the decisive impact of industrial urbanization.

## Populating Rural America Before 1900

The geographic distribution of North America's population is rooted in the colonial era of the seventeenth and eighteenth centuries. That era began just after 1600 with a number of discoveries and land claims made by such European powers as England, France, Spain, and the Netherlands. The French were concerned with penetrating the continental interior, propelled by their desire to establish a lucrative fur-trading network and aided by the St. Lawrence/Great Lakes waterways that lay at the heart of their colonial territory. The other major (and larger) colonizing force, the English, who also held interior territories, concentrated their settlement efforts along the coast of what is today the northeastern U.S. seaboard. These British colonies quickly became differentiated in their local economies, a diversity that was to endure and eventually shape American cultural geography. The northern colony of New England (Massachusetts Bay and environs) specialized in commerce; the southern Chesapeake Bay colony (Tidewater Virginia and Maryland) emphasized the large-scale plantation farming of tobacco; the Middle Atlantic area in between (southeastern New York, New Jersey, eastern Pennsylvania) was home to a number of smaller independent-farmer colonies. As these three adjacent colonial regions thrived and yearned to expand after 1750, the British government kept the inland frontier closed and exerted ever-tighter economic controls, thereby unifying the three colonies in their growing dislike for the now heavy-handed mother country. By 1776, escalating tensions produced open rebellion on the American side of the Atlantic and the onset of the seven-year-long Revolutionary War, which resulted in English defeat and independence for the newly formed United States of America. In the aftermath of this upheaval, the British Empire Loyalists fled north to a safe refuge in Ontario and the Maritime Provinces, where they later (in 1867) formed the Canadian confederation together with the descendants of the early French settlers who had come to occupy the lower St. Lawrence Valley of Quebec and with an older Scottish maritime colony (Nova Scotia) located on a peninsula extending into the Atlantic east of Maine. Meanwhile, in the fledgling nation to the south, the western frontier swung open and the old British Northwest Territory (Ohio-Michigan-Indiana-Illinois-Wisconsin) was settled rapidly, as much of its soil proved to be highly favorable for agriculture—a remarkable improvement over the relatively infertile seaboard soils. This added yet another element to the increasingly varied American environment; but diversity had by then been converted to a positive force through the establishment of widespread trading ties—including trans-Appalachian commodity flows—that were based on interregional complementarities and the emergence of an economy whose spatial organization was assuming national-scale proportions.

By the time the westward-moving frontier swept across the Mississippi Valley in the 1820s (Fig. 3–10), it was clear that the three former seaboard colonies had become separate **culture hearths**—source areas and innovation centers from which emigrants carried cultural traditions (*A*, *B*, and *C* on Fig. 3–10). The New England region (*A*) influenced the southern and western margins of the Great Lakes, with settlers here creating a cultural landscape that reflected New England's house types and village patterns. Immediately to the south in the trans-Appalachian West lay a much larger area that focused on the Ohio Valley but stretched as far south as southern Tennessee, which similarly resembled Middle-Atlantic Pennsylvania (*B*); and along the Atlantic and Gulf Coastal Plain, from Delaware to newly purchased Louisiana, lay an emerging culture area dominated by the tobacco (and now cotton) plantation tradition that had originated in the Tidewater Maryland/Virginia hearth (*C*). The northern half of this vast interior space—with cultural and trading ties to both New England and the Middle Atlantic region—soon became well unified as transport linkages improved; by 1860, the railroad had replaced earlier plank roads and canals, providing a highly efficient network that offered fast and cheap long-distance transportation, thereby effecting significant time-space convergence among once-distant regions within the northeast quadrant of the United States. The American South, however, did not wish to integrate itself economically with the north to any great degree, preferring, instead, to pursue cotton and tobacco exporting in the overseas marketplace. This divergent regionalism, together with its insistence on preserving slavery, soon led the South into secession and the ruinous Civil War (1861–1865); in the aftermath, the South took decades slowly to rebuild, and it was not until a full century later that this region, once again, became a full participant and beneficiary vis-à-vis the prosperous national economy.

The second half of the nineteenth century also saw the frontier cross the western United States, although it was a "hollow" frontier, inasmuch as parts of California and Oregon were settled before the drier Great Plains region (Fig. 3–10). In fact, the frontier had stalled at the eastern edge of the Plains prior to mid-century as the pioneers first encountered Arid America and propagated the myth that these steppelands constituted a "Great American Desert." As news of lush farm country in Oregon—and especially the discovery of gold in California in 1848—filtered back east, the frontier quickly leapfrogged to the Pacific as thousands of migrants annually streamed

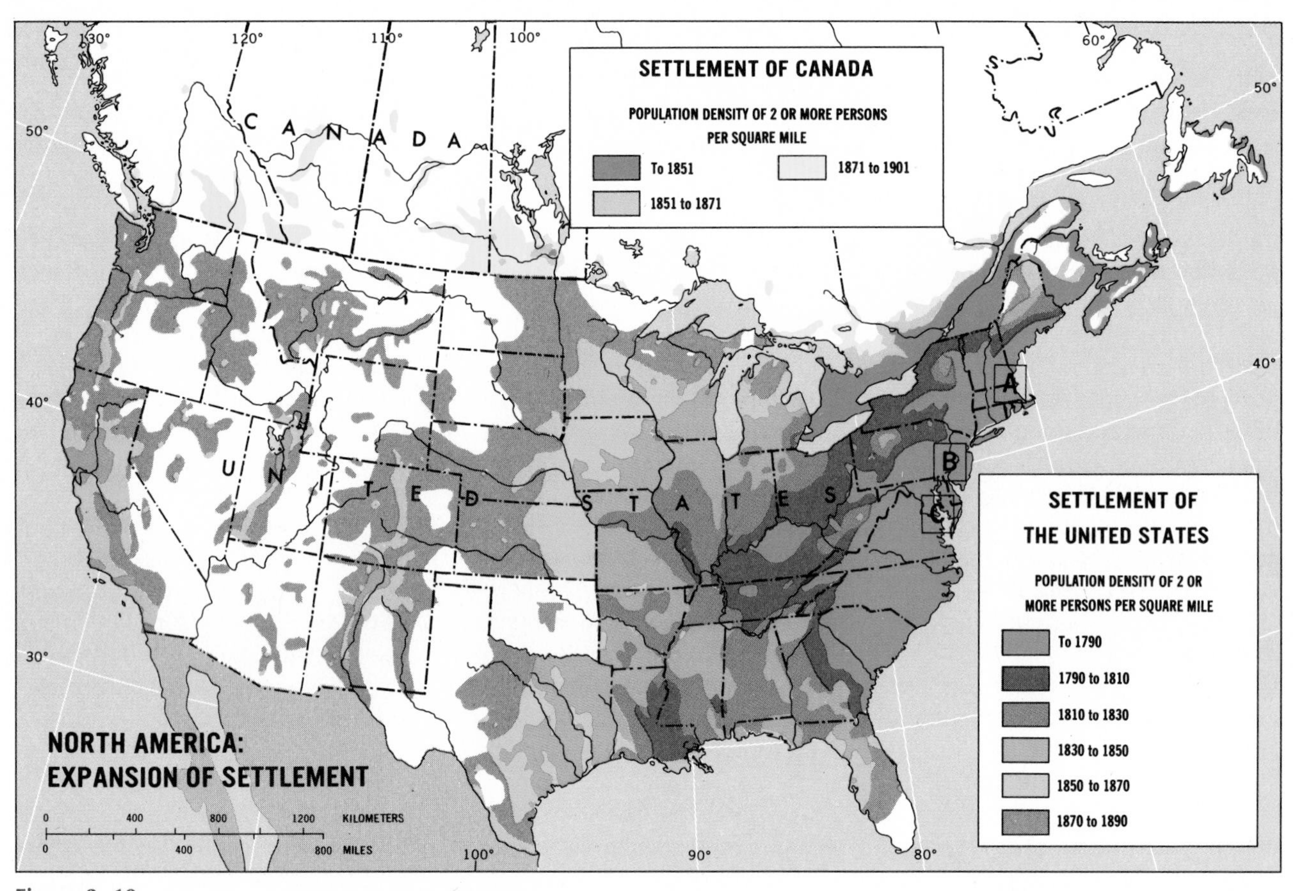


Figure 3–10

across the difficult physiographic provinces of the west. By 1869, agriculturally booming Calfornia was linked to the rest of the nation by transcontinental railroad, and these same steel tracks also opened up the Great Plains. The semiarid Plains, it turned out, were quite fertile after all, particularly for raising wheat on a large scale; the area was deficient only in water supplies, which could be made to last longer through careful farming methods or enhanced by irrigation near river and aquifer sources. The arrival of the railroads, automated farm machinery for large wheatfields, and barbed wire fences—which protected croplands from grazing cattle, another major agricultural activity in Arid America—further supported the expansion of dryland grain farming and the development of the Plains's economy.

The Canadian experience in westward regional development, although it involved a far smaller national population, was a parallel one. A strong bicultural division prevailed in the east here as well (British vs. French settlers), although these differences resulted in the 1867 British-led confederation agreement rather than civil war. The initial push to the west also encountered physical obstacles, not the Appalachians but the barren Canadian Shield—a heavily glaciated and swampy flatland totally devoid of productive soils, which stretched from just north of Lake Ontario all the way to the 100th meridian (Fig. 3–5). The Great Plains on the western side of the Shield—known here as the prairie provinces—offered an environment for dryland wheat farming identical to the one south of the U.S. border (except for a shorter growing season) and also provided Canada's only opportunity to develop a large-scale agricultural region. To the west of the Plains, there again arose the barrier of the Rockies-intermontane plateaus-Pacific highlands, narrower in Canada but lacking in the convenient mountain passes that were available in the U.S. portions of these physiographic provinces. Nevertheless, a transcontinental railway was heroically pushed across to Vancouver in the 1880s, providing a similar growth stimulus to the interior wheat region, which was largely settled at that time (Fig. 3–10). Thus, despite some formidable obstacles, Canada, too, had connected its Atlantic and Pacific coasts by rail, often urged forward in the more difficult moments by the fear that U.S. westward expansion would surge north across the 49th parallel (the U.S.-Canada boundary between the Pacific and the Lake Superior area) if Canadians did not utilize and populate their westernmost lands.

By the time the American frontier closed in the 1890s, today's rural settlement pattern was firmly in place, anchored to a set of enduring national agricultural regions (Fig. 3–25), which will be discussed later in this chapter. At the local level, except for the thirteen original states and Texas, which used other surveying systems, rural population was distributed within the square geometry of the *Township-and-Range* land-division system: this nationwide checkerboardlike scheme—an imprint clearly evident on the Iowa farming landscape shown in Fig. I–2—was designed by Thomas Jefferson and his associates before 1800 to survey properties easily and disperse settlers evenly across newly opened farmlands. To this day, this system still represents the largest area of planned rural settlement anywhere on earth.

The closure of the frontier also coincided with an accelerating shift in the location of the U.S. population: the census of 1900 revealed that more people lived in newly developed metropolitan areas than in the countryside. An urban revolution was following in the wake of the industrial revolution, which had taken hold in North America after the Civil War a generation earlier. To be sure, cities had been vitally important since colonial times, and they commanded the expanding farming regions of the nineteenth century through investments in transportation technology and by functioning as markets for the goods produced by their now greatly enlarged agricultural hinterlands. But the degree and extent of urbanization by 1900 was approaching massive proportions, and it was clear, as the twentieth century opened, that the tilt from rural to urban America was just beginning. Over the next seven decades that persistent tilt intensified and transformed the United States and Canada into the world's most urbanized societies, dominated by the huge concentrated cities of the industrial age; these events also set the stage for the further transformation of urban America in the postindustrial age that emerged after 1970.

## North American Industrial Urbanization, 1900–1970

We have already reviewed in Chapter 1 the economic and social transformation wrought by Europe's industrial revolution, as people concentrated at high densities in urban manufacturing centers. In the United States, this revolution occurred almost a century later, but when it finally did cross the North Atlantic in the 1870s, it took hold so successfully and advanced so robustly that only 50 years later America was surpassing Europe as the world's mightiest industrial power. Thus, the far-reaching societal changes and demographic transition (Fig. 1–4) experienced in Europe's industrializing countries were greatly accelerated in the United States, fueled further by the arrival of more than 25 million European immigrants—who overwhelmingly headed for jobs in the major manufacturing centers—between 1865 and 1914. The impact of industrial urbanization occurred simultaneously at two levels of generalization. At the national scale or *macroscale*, a network or system of new cities rapidly emerged, specializing in the collection, processing, and distribution of raw materials and manufactured goods, linked together by an ever more efficient web of long-distance and local railroad lines. Within that urban system, at the *microscale*, individual cities prospered in their new roles as manufacturing centers, generating a wholly new internal structure that still forms the spatial framework of most of the central cities of America's large metropolitan areas. We now examine the urban trend at both of these scales.

### Macroscale urbanization

The rise of the national urban system in the late nineteenth century was based on the traditional external role of cities: providing goods and services for their hinterlands in return for raw materials. This function was present since colonial times, with preindustrial North American cities in the early nineteenth century (which were quite small in size and spatial extent) located at the most accessible points on the existing transportation network, so that movement costs could be minimized. Since handicrafts and commercial activities were already concentrated in cities, the emerging industrialization movement naturally gravitated toward them. These urban centers, of course, also contained concentrations of labor and capital, provided a large market for finished goods, and possessed favorable transport and communication connections. Furthermore, they could readily absorb the horde of newcomers who would cluster by the thousands around new factories built within rail corridors in and just outside these compact cities. Their constantly growing incomes, in turn, permitted the newly industrialized cities to invest in a bigger local infrastructure of private and public services as well as housing, and thereby convert each round of industrial expansion into a new stage of urban development. Moreover, this whole process unfolded so quickly that planning was impossible. Quite literally, America awoke one morning near the turn of the twentieth century to discover it had built a number of large cities, even though (as we will see) they ran counter to some fundamental values of American culture.

The rise of the national urban system, unintended though it may have been, was a necessary by-

## MODEL BOX 3

# CENTRAL PLACE THEORY

The urbanization of population that accompanied industrialization was a distinct departure in scale, composition, and geography from urban patterns associated with the preindustrial city. The process of growth and the acquisition of functional specializations that shaped cities and towns of the industrial age particularly intrigued the German geographer Walter Christaller (1893–1969), who devised a location theory for such urban centers, which was published in his classic 1933 book, *Central Places in Southern Germany*. A **location theory** is a logical explanation of the distribution of an economic activity and the way in which its component regions are spatially interrelated; it is usually based on formal mathematical expressions, but also includes a body of empirical (real-world) evidence and applications. Christaller's **central place theory**, a continuing subject of research today, entails all of these dimensions as geography's foremost location theory. Our focus here is on a description of the basic model.

Christaller was concerned with the spacing of urban centers of different sizes and the forces that govern their distribution, focusing on *central places*—towns and cities centrally located within a surrounding area whose population they served. He began with a set of assumptions formulated to isolate the most critical factors affecting urban distributions. Environmental differences were "smoothed" by assuming an even dispersion of resources. Demographic and income variations were eliminated by assuming a uniform population density outside the urban places and an equal income distribution among this consuming population. Distance was made important by presuming each person could travel in every direction toward any center, with the travel cost proportional only to the distance to the center. Finally, it was assumed that each person desiring something in a town would seek to minimize the expense involved by traveling to the closest center supplying the good or service sought.

Within this set of assumptions, each town's activities drew on the surrounding population for its set of customers. Nearby populations would be certain to shop in the closest center. As the radial distance from the urban place increased, a population would eventually be reached that was about the same distance from several towns. To avoid the difficulty of a set of consumers who were indifferent to alternative central places, Christaller argued that competition between equivalent centers for the surrounding population would "fill" the landscape with a series of hexagonal service areas (Fig. 3–11). If the central places are too far apart to serve the rural population adequately, an intervening center will develop, so, the spatial distribution of urban places should be predominantly even.

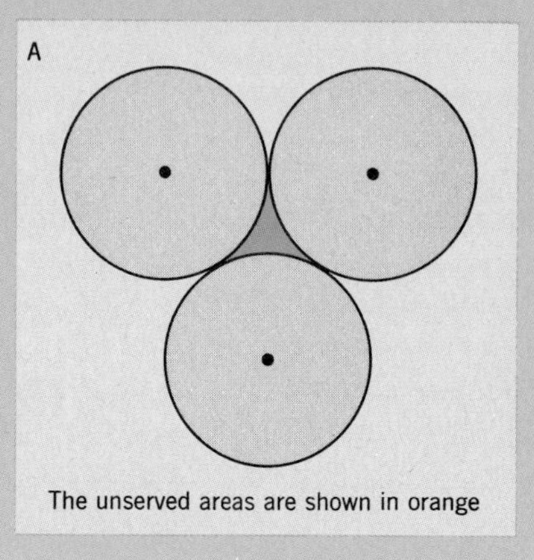


The unserved areas are shown in orange

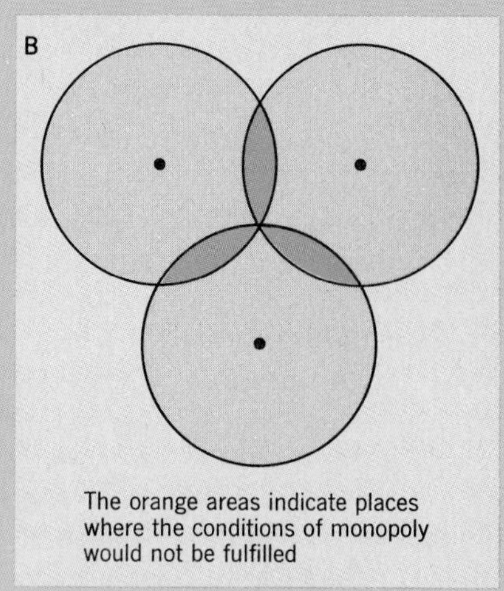


The orange areas indicate places where the conditions of monopoly would not be fulfilled

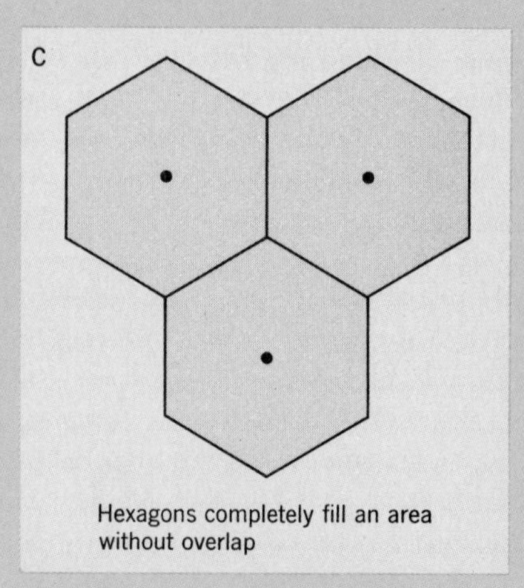


Hexagons completely fill an area without overlap

Figure 3–11

Any casual inspection of an actual landscape, however, demonstrates the purely theoretical nature of Christaller's proposed hexagonal distribution of central places. Some regions appear to approximate this pattern better than others, but there are several critical explanations for realistic departures from theory. In part, towns are not evenly distributed because resources (environmental and human) are not evenly distributed. Also, social scientists have recently realized that many people do not behave in such a way that their costs are minimized. Finally, as Christaller himself observed, urban places are not all the same size. Most regions, therefore, contain a varied array of small, medium, and large urban places in complex interrelation with each other.

The basic relationship between urban size and urban spacing can be viewed as an outgrowth of Christaller's initial assumptions and arguments. The simple hexagonal distribution is ideal if the assumptions are met *and* the urban places are about the same size. Some central place functions—activities that take place in large urban centers—will not be used as often or by as high a proportion of the total population as will other functions. A grocery store, for example, is likely to be used often and by almost everyone nearby, whereas a shop that specializes in the sale of high-quality imported glassware will be visited less frequently by the average consumer. Some central place functions can, therefore, be found in even the smallest central places, whereas other functions are located only in large towns and cities. Small central places, then, contain relatively few activities—called "lower-order functions"—and large centers contain a greater variety of activities, some very specialized in addition to the lower-order functions. Because higher-order functions must "reach" outward farther to include enough potential customers for the function's economic survival (i.e., the "range of the good" is greater), these functions cannot be located too near each other. And because higher-order functions are found in larger central places, fewer such large urban centers can be located in a given area; they will, therefore, tend to be farther apart than small places.

Christaller and other researchers who followed him extrapolated these conclusions to identify an overall *urban hierarchy*. Because higher-order functions could not survive unless they had access to populations exceeding some "threshold" and because central place theory based the spacing of urban places on the smallest hexagonal lattice of the lowest-order centers, it seemed apparent that the higher-order places would occur in discrete size categories, just as they appeared on the basic lattice in discrete spacing patterns.

At first glance, this concept of an urban hierarchy fits with Mark Jefferson's idea of a primate city (page 78) and with the observed pattern of cities in North America. At the highest level, there is only one New York City. Both Chicago and Los Angeles might be placed in the next category; both contain about half New York's population, fewer higher-order functions, and are located at great distances from each other and from New York. The third level in the hierarchy contains even more cities such as Philadelphia, Montreal, Toronto, Houston, and others—and so on down the "hierarchy" to the very smallest centers containing urban functions. Some researchers, however, have found it very difficult to identify discrete central place categories and the distinct functional thresholds that would create such categories. They argue that the variety of functions occurring in medium- and large-size towns creates a continuous system of central places—without "breaks" between various size-levels of cities and towns—too complex for simple hierarchical categorization.

Regardless of the validity of these alternative positions, many aspects of central place theory are worth considering. Despite the confounding complexities of reality (few places can meet Christaller's simplifying assumptions), the theory does permit identification and testing of the fundamental relationships that govern the growth and distribution of systems of cities. That the general basis of the theory is correct is borne out on the landscape. There *are* fewer and more widely spaced large cities than small ones. Large cities *are* more numerous and closer to each other in the economically intense continental core area (Fig. 3–12) than outside that region. Specialized urban activities tend *not* to be found in smaller, closely spaced towns and villages unless the surrounding population is unusually wealthy.

product of industrialization, without which rapid U.S. economic development could not have taken place. This far-flung hierarchy of cities and towns now blanketed North America, and came to serve its local populations with all of the conveniences of modern life. The rules governing the spatial distribution of these urban service centers—organized within the framework of the *central place model*—are presented in *Model Box* 3, and constitute some of the most significant theoretical work performed by geographers to date.

Even though it first emerged during the industrial revolution of the 1870–1910 period, the American urban system was in the process of formation for several decades preceding the Civil War. The evolutionary framework of the system from 1790 to 1970 is best summarized within the four-stage model developed by the geographer John Borchert, which he based on key changes in transportation technology and industrial energy. The preindustrial *Sail-Wagon Epoch* (1790–1830) was the first stage, marked by slow and primitive overland and waterway movements. The leading cities were such northeastern ports as Boston, New York, and Philadelphia—none yet emerged as the primate city—which were at least as heavily oriented to the European overseas trade as they were to their still rather inaccessible western hinterlands (although the Erie Canal was opened as the epoch came to a close). Next came the *Iron Horse Epoch* (1830–1870), dominated by the arrival and spreading of the steam-powered railroad, which steadily expanded its network from east to west until the transcontinental line was completed as the epoch ended. Accordingly, a nationwide transport system had been forged, coal-mining centers boomed (to keep locomotives running), and—aided by the easier and cheaper movement of raw materials—small-scale urban manufacturing began to spread outward from its New England hearth, in which the factory-system innovation had been transplanted from Britain in the early 1800s. And the national urban system started to take shape as New York advanced to become the primate city by 1850, and the next level in the hierarchy was increasingly occupied by such burgeoning new industrial centers as Pittsburgh, Detroit, and Chicago. This economic/urban development process crystallized during the third stage—the *Steel-Rail Epoch* (1870–1920)—which coincided with the American industrial revolution. Among the massive forces now shaping the growth and full establishment of the national metropolitan system were the rise and swift dominance of the all-important steel industry along the Chicago-Detroit-Pittsburgh axis (as well as its coal and iron ore supply areas in the northern Appalachians and Lake Superior district, respectively), the increasing scale of manufacturing that necessitated greater agglomeration in the most favored raw material and market locations for industry, and the steel-related improvements of the railroads—much more durable tracks of steel (that replaced iron), more powerful steam locomotives, heavier and larger (also refrigerated) freight-cars—that permitted significantly higher speeds, longer hauls of bulk commodities, and the time-space convergence of hitherto distant rail nodes. The *Auto-Air-Amenity Epoch* (1920–1970) comprised the final stage of North American industrial urbanization and maturation of the national urban hierarchy. The key innovation was the gasoline-powered internal combustion engine, which underwrote ever greater automobile- and truck-based regional and metropolitan dispersal. And as technological advances in manufacturing spawned the increasing automation of blue-collar jobs, the U.S. labor force steadily shifted toward a new emphasis on white-collar personal and professional services to manage the industrial economy—a productive activity that responded less to traditional cost- and distance-based location forces and ever more strongly to the amenities (pleasant environments) available in suburbia as well as the outlying Sunbelt states in a nation now fully interconnected by jet travel and long-distance communication networks.

The growth of the American urban system and the national integration of its industrial-based economy produced dramatic spatial changes as population relocated to keep abreast of employment opportunities. The most notable regional transformation was the early twentieth century emergence of the continental *core area* or American Manufacturing Belt, which contained the lion's share of industrial activity in both the United States and Canada. As Fig. 3–12 shows, the geographic form of the core region—which includes Canada's southern Ontario—was a great rectangle whose four corners were Boston, Milwaukee, St. Louis, and Washington, D.C. However, because manufacturing is such a spatially concentrated activity, the core area should not be thought of as a factory-dominated landscape; in fact, less than 1 percent of the territory of the Manufacturing Belt is actually devoted to industrial land use, with most of its mills and foundries clustered tightly into a dozen districts centering on metropolitan Boston, Hartford-New Haven, New York-northern New Jersey, Philadelphia, Baltimore, Buffalo, Toronto-Hamilton-Windsor, Pittsburgh-Cleveland, Detroit-Toledo, Chicago-Milwaukee, Dayton-Cincinnati-Louisville, and St. Louis.

At the subregional scale, as transportation breakthroughs permitted progressive urban decentralization, the expanding suburbs of major cities soon coalesced to form a number of conurbations (see pp. 83–84). The most important of these by far was the *Atlantic Seaboard Megalopolis* (Fig. 3–13), the 975-kilometer-long (600-mile-long) urbanized northeastern coastal strip stretching from southern Maine to Virginia, which contains metropolitan Boston, New York, Philadelphia, Baltimore, and Washington; this is the economic heartland of the realm's core region, the seat of American politics, business, and culture, as well as the trading

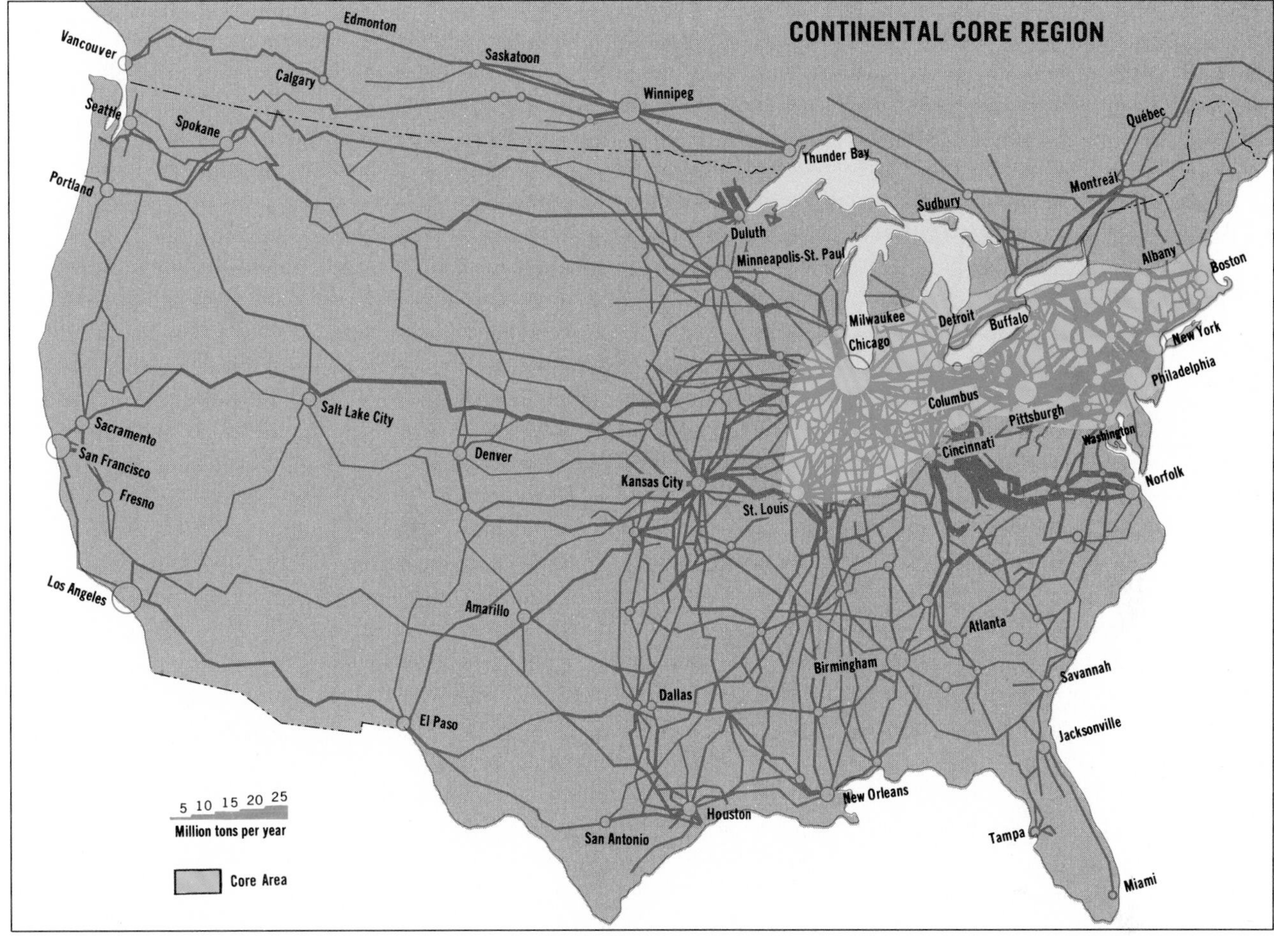


Figure 3–12

"hinge" between the continent and the rest of the world. Three other primary American conurbations also emerged—Lower Great Lakes (Chicago-Detroit-Pittsburgh), California (San Diego-Los Angeles-San Francisco), and Peninsular Florida (Jacksonville-Orlando-Miami)—and are expected to continue growing along with a number of secondary megalopolitan concentrations (Fig. 3–13). And north of the border, the same forces also created a nationally predominant conurbation linking Quebec City-Montreal-Toronto-Windsor (which the geographer Maurice Yeates has christened *Main Street*), capping a parallel episode of rapid industrial urbanization in Canada between 1885 and 1940.

## Microscale urbanization

The internal structure of the American city (metropolis after 1900) closely reflected the mixture of forces that shaped the continental urban system. As such, the North American industrial city represented a departure from its European parentage: whereas Europe's major cities were historically centers of political and military power—onto which industrialization was grafted almost as an afterthought—most U.S. cities came into being as *economic machines* to manufacture the goods and services required to sustain a burgeoning industrial revolution. Thus, right from the beginning, the performances of America's modern, unintended cities were judged mainly in terms of their profit-making abilities, the less successful ones callously discarded in a society that has yet to develop an enduring urban tradition. And, in retrospect, the chief social function of the city was to receive and process foreign (as well as domestic rural) immigrants for assimilation into the mainstream American nativist culture—the so-called "melting pot"—which increasingly concentrated in the middle- and upper-income neighborhoods of the rapidly growing suburbs after 1920.

At the microscale or intraurban scale, too, transportation technology was a decisive force in the shaping of geographic patterns. The influence of rails—in this instance, lighter street rail lines—once again governed spatial structure, as horse-drawn trolleys were succeeded by electric streetcars in the late nineteenth century. The coming of the automobile after World War I changed all that, and America began to turn from building compact single-core cities to the widely dispersed multicentered metropolises of the post-World War II highway era. By 1970, the intraurban expressway network was completed,

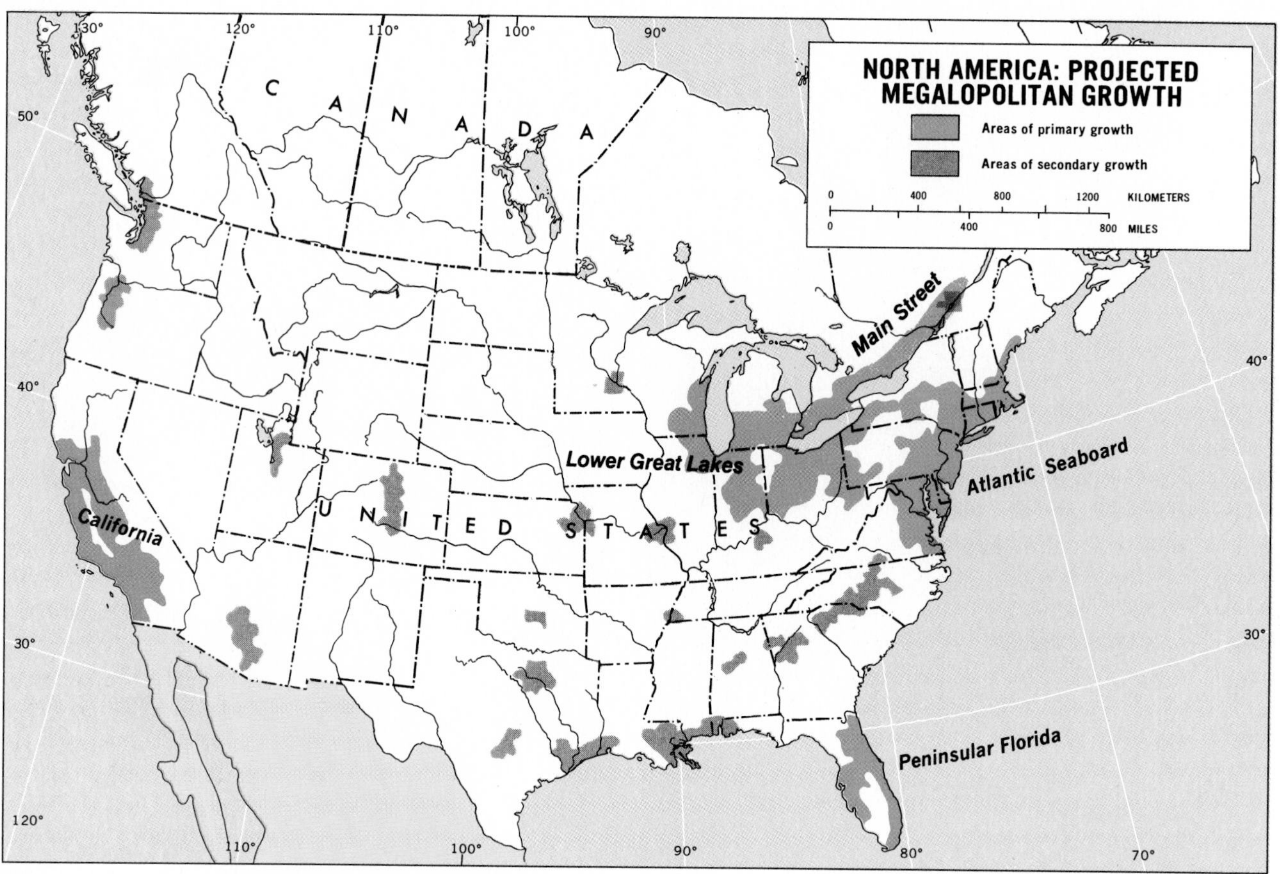


Figure 3–13

equalizing location costs throughout the metropolis, and the stage was set for suburbia to transform itself swiftly from a residential preserve into a complete outer city whose amenities and prestige proved attractive even to the highest-order central functions. As the newly urbanized suburbs started to capture major economic activities and thereby gain a surprising degree of functional independence, many large cities saw their status diminish to that of coequal, their once thriving CBDs reduced to serving the increasingly less affluent population that now inhabited the central city's residual rings, sectors, and nucleations (Figs. 3–2 and 3–3).

Intrametropolitan spatial evolution may also be organized within a four-stage, transport-based model—this one developed by the urban geographer John Adams (Fig. 3–14). The initial stage (I), the *Walking-Horsecar Era* (prior to 1888), was basically a pedestrian city in which people had to get around on foot, although after 1850 some could afford to travel by the new but not-much-faster horse-drawn trolley; urban structure was dominated by compactness—everything and everybody had to be within a 30-minute's walk—and relatively little land-use specialization could occur. The invention of the electric traction motor in 1888, a device that could easily be attached to horsecars, launched the *Electric Streetcar Era* (1888–1920); speeds of up to 32 kilometers per hour (20 miles per hour) enabled the 30-minute travel radius and the urbanized area to be expanded considerably along outlying trolley corridors (II in Fig. 3–14), spawning several streetcar suburbs and helping to differentiate space within the older core city—whose CBD, industrial, transport, and residential land uses emerged in their modern form (see Fig. 3–2). Stage III was the *Recreational Automobile Era* (1920–1945), during which the initial impact of cars and highways constantly improved the accessibility of the outer metropolitan ring, thereby launching a wave of mass suburbanization throughout the ring, which significantly extended the urban frontier (and the half-hour time-distance radial) of the once-again circular metropolis; meanwhile, the still-dominant central city experienced its economic peak and the full partitioning of residential space into neighborhoods sharply defined by income level, ethnicity, and race. The final stage (IV), the *Freeway Era* (1945 to the present), saw the full impact of automobiles as high-speed expressways pushed the metropolitan development and time-distance limits in certain sectors to 50 kilometers (30 miles) or more from downtown, thereby spawning round after round of massive suburbanization; structurally, the growing distinction between central city and booming suburbia (hardly *sub* to the *urb* anymore) mirrored a steady deconcentration of the metropolis, which, by the 1970s, resulted in the emergence of an increasingly independent outer suburban city.

The social geography of the evolving industrial metropolis over the past 150 years has been marked by the development of a residential mosaic that exhibited the congregating of ever-more-specialized groups. Before the arrival of the electric streetcar around 1890, which finally introduced "mass" transit affordable by urbanites of every social class, the heterogeneous city population had been unable to sort itself into homogeneous neighborhoods. Immigrants pouring into the industrializing cities were forced to reside within walking distance of their workplaces in heavily crowded tenements and row houses, whose tiny apartments were literally filled in the order that their tenants arrived in town; thus, ethnically diverse communities arose but were constantly beset by the stresses and conflicts caused by misunderstanding and intolerance, as a jumble of cultures were thrown together under the worst of conditions. Once the inner-city neighborhood pattern was able to form in the 1890s, the familiar residential rings, sectors and nucleations quickly materialized (see Fig. 3–2). When the United States all but closed its doors to foreign immigration in the 1920s, industrial managers had to search elsewhere for an alternative source of cheap unskilled labor. They soon discovered the large black population of the rural Deep South—increasingly unemployed there as cotton-related agriculture declined—and began to recruit them by the thousands to work in the factories of the Manufacturing Belt cities; this new migration had an immediate impact on the social geography of the industrial city because whites refused to share their living space with the racially unalike newcomers. The result was the involuntary segregation of these newest migrants, who were steered to geographically separated all-black areas that by 1945 became large expanding *ghettoes*, speeding the departure of many white central-city communities in the postwar era and helping to create the racially divided society described at the outset of this chapter. The suburban component of the residential mosaic, although characterized by numerous clusters of segregated lower-income as well as black populations, is generally home to the more affluent residents of the metropolis. The social makeup of these congregations is largely determined by income—the residential turf of affluent suburbia is marked by a plethora of minor class- and status-related variables, perfectly suited to a society in which frequent upward social and spatial mobility go hand in hand.

The economic geography of industrial urbanization has already been outlined. Manufacturing was originally attracted to cities because of their preexisting locational advantages as transport nodes, labor and product markets, and centers of investment capital. As industry thrived, cities grew and lured many additional mills, factories, and supporting activities, until huge manufacturing complexes were established. But, at the same time,

Figure 3–14

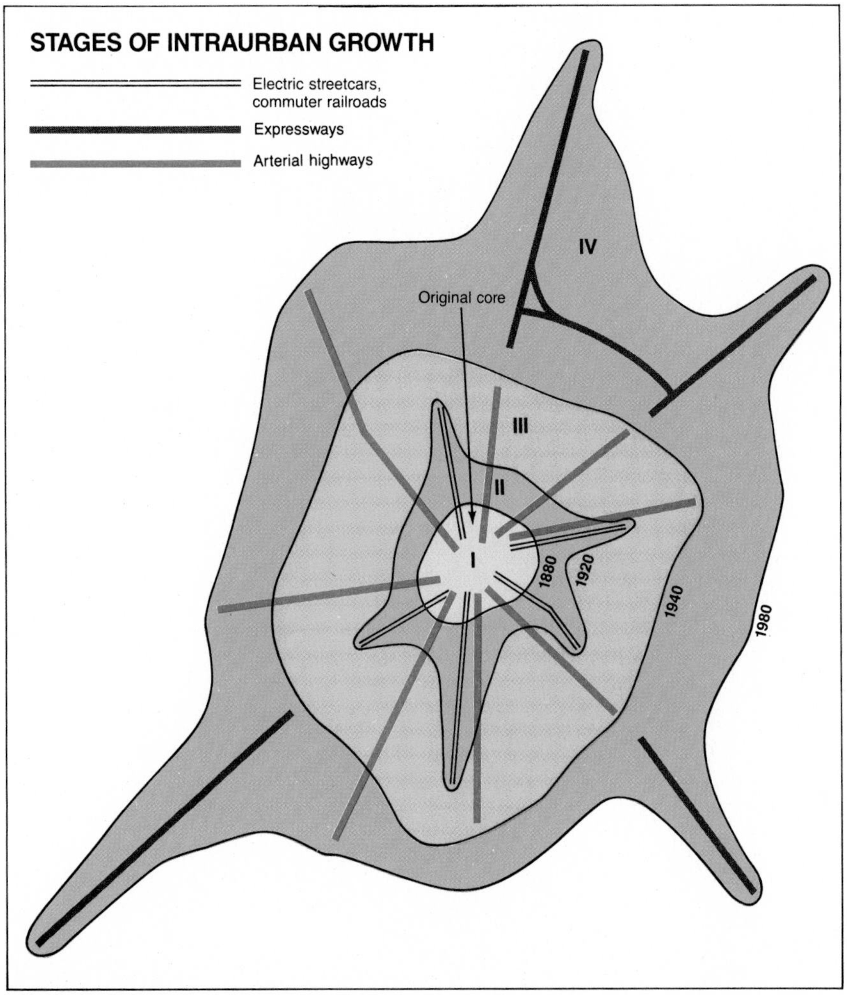

Reproduced by permission from the *Annals of the Association of American Geographers*, Volume 60, 1970, p. 56, fig. 15.

large-scale agglomeration produced negative side effects such as congestion, pollution, and the like; these problems were accepted for decades as part of the cost of doing business, but when cities began to lose their concentrative advantages after 1960, many manufacturers simply abandoned them for suburbia and beyond without a second thought. This decentralization trend was also reinforced by the declining position of heavy industry in the American economy, which in the last 30 years has been tilting ever more strongly toward the services and information-producing sectors. By 1980, this shift had become so pervasive that an entirely new postindustrial society was emerging; although all of the implications of the passage across this historic watershed are not yet clear, the geography of the realm's overwhelmingly urban population is already undergoing further change that merits our attention.

## Urbanization in Postindustrial America Since 1970

By 1970, the long arc of industrial urbanization had been traversed: in 1800, only 5 percent of the agriculture-dominated U.S. population lived in cities—today less than 2.5 percent live on the farms and well over 90 percent reside within a two-hours' drive of a metropolitan center. The rise of the postindustrial economy (see *Box* p. 214) has coincided with a number of recent urban spatial changes. At the macroscale, metropolitan growth has leveled off nationally, although many Sunbelt cities as well as suburbs everywhere continue to expand; at the same time, many nonmetropolitan places are experiencing new growth. At the microscale, a new intraurban spatial structuring is evident: a quarter-century of deconcentration has now practically turned the central city inside out (Fig. 3–1), producing a metropolis so widely dispersed that large portions of it are now splitting up into self-contained functional areas. We now examine each of these levels of generalization.

### Macroscale urbanization

The current national- and regional-level changes have been grouped by Phillips and Brunn as the fifth stage of Borchert's model of urban-system evolution—the *Slow-Growth Epoch* (1970 to the present). The key trends at work are slower natural population increases, the opening of a new age of expensive and uncertain energy supplies (always cheap and abundant before 1970), the "renaissance" of rural areas, the substitution of communication for transportation, and the growth of the Sunbelt economy at the direct expense of the now-declining Manufacturing Belt. However, in the mid-1980s, not all of these forces are so clearly apparent. Energy costs and supplies have, at least temporarily, ceased to be major worries, and the birth rate is higher than 10 years ago; moreover, the 1980 census showed a much higher growth rate for suburbs than for nonmetropolitan areas (although central-city losses do result in an overall near-zero metropolitan growth equation), and the substitutability of communication for transport has so far had a greater impact on the suburbanization of business at the intraurban level. Only the regional shift toward the Sunbelt persists, perhaps propelled more by the alarming decline of the Manufacturing Belt than the continued boom of the southern tier—Houston and several other Sunbelt cities were surprisingly hard hit by the Great Recession of the early 1980s. This transformation signals the arrival of interregional equalization in location costs, especially for the white-collar office-based industries; noneconomic factors, therefore, are becoming supreme in locational decision making—geographical prestige, local amenities, and proximity to recreational activities head the list. Perhaps no other place better symbolizes the postindustrial age in the Sunbelt than *Silicon Valley* (Fig. 3–15): this is the headquarters of the ultrasophisticated microprocessor industry (which produces vital components for computers and electronic appliances), located near Stanford University in the San Francisco Bay Area's high-amenity Santa Clara Valley. Nonetheless, spectacular success has produced overcrowding, exorbitant housing prices, and a diminishing quality of life since 1975, and the industry—to keep attracting outstanding young scientific talent—is dispersing to other glamorous Sunbelt/suburb locations in the outer cities of San Diego, Dallas, Miami-Ft. Lauderdale, Raleigh-Durham (N.C.), Denver—and even back to the Boston area, where the electronics industry grew up in the 1940s.

Macrospatial changes in the new epoch have still not emerged with any great clarity. The national urban system still functions as it did a decade ago, although certain Manufacturing Belt cities will likely yield high positions in the hierarchy in the foreseeable future to such up-and-coming metropolises as San Antonio, Phoenix, and San Diego. The search also continues for new geographical perspectives on large-scale changes in America's urbanization patterns, the latest being the provocative hypothesis of the "galactic metropolis" (see *Box*).

### Microscale urbanization

The most dramatic changes in postindustrial urban geography to date have occurred at the scale of the metropolis. As completion of the radial-and-circumferential urban freeway network resulted in near-equal levels of time-space convergence for most locations within the

Figure 3–15
California's Silicon Valley, innovation center for the computer industry.

metropolitan area, people and high-order activities dispersed across the ever-widening outer suburban city. In effect, the expressway system destroyed the regionwide centrality advantage of the core city's CBD, making most places on the freeway network just as accessible to the rest of the metropolis as only downtown had been before the 1970s. Industrial and commercial employers quickly realized that most of the advantages of being located in the CBD were now eliminated; several companies in the office-based services sector chose to remain and even enlarge their presence downtown, but many others, together with myriad firms in every other sector, chose to respond to the new economic-geographic reality by voting with their feet—or rubber tires—and headed for outlying sites. As early as 1973, the suburbs surpassed the central cities in total employment; by the early 1980s, even some major metropolises in the Sunbelt were experiencing the suburbanization of a critical mass of jobs (greater than 50 percent of the urban-area total).

As the outer city grew rapidly, the volume and level of interaction with the central city began to lessen and suburban self-sufficiency increased. This shift toward functional independence was heightened as new suburban nuclei sprang up (to serve the new local economies), the major ones locating near key freeway intersections. These multipurpose activity nodes developed around big regional shopping centers, whose prestigious images attracted scores of industrial parks, office companies, hotels, restaurants, entertainment facilities, and even major-league sports stadiums, which together formed burgeoning new *minicities* that were an automobile-age version of the CBD (Fig. 3–16). As minicities flourished, they attracted tens of thousands of nearby suburbanites to organize their lives around them—offering workplaces, shopping, leisure activities, and all of the other elements of a complete urban environment—thereby not only further loosening ties to the central city, but also to other portions of suburbia as well. These spatial elements of the contemporary metropolis are assembled in the model displayed in Fig. 3–17, which should be regarded as an updating and extension of the classical models of intraurban structure (Fig. 3–2). The rise of the outer city has now produced a *multicentered* metropolis consisting of the traditional CBD and a set of increasingly coequal suburban minicities, with each "downtown" serving a discrete and self-sufficient surrounding area. The urban geographer James Vance has called these new tributary areas *urban realms*, recognizing in his studies of the San Francisco Bay Area that each such realm maintains a separate and distinct economic, social, and political significance and strength.

The position of the central city within the emerging multinodal metropolis of realms is an eroding one. No longer the dominant metropolitanwide center for urban goods and services, the CBD is being reduced to serving the less affluent residents of the innermost realm and those working there. As core-area manufacturing employment declined precipitously, many large cities adapted successfully by

## THE HYPOTHESIS OF THE GALACTIC METROPOLIS

The rapid outward spreading of urban America since World War II has been studied intently by geographers and planners. Although slowed by the two energy "crises" of 1974 and 1979—in which temporary disruptions of normal gasoline supplies gave millions of Americans pause about the wisdom of relying so heavily on their automobiles—urbanization continues to spill out of its traditional metropolitan confines. Much of this deconcentration, which involves industrial and commercial activities as well as people, is directed toward *exurbia*—the semirural zone just beyond the suburban frontier that contains millions who commute to the nearby metropolis. But in the 1970s, a substantial proportion of this migration appears to have also been aimed at small cities and towns in genuinely rural areas located well away from major urban complexes.

The geographer Peirce Lewis (1983a) views this phenomenon as part of a vast continuum of urbanization that now stretches from coast to coast. He argues that, in most of the settled portions of the United States today, one rarely has far to travel to reach an interstate freeway, which almost always spawns clusters of houses and industrial or commercial structures around its interchanges no matter how far out in the "boondocks" they might be. To describe his contemporary national vision of huge urban concentrations interspersed with loose separated clusters of people and buildings, Lewis coined the term *galactic metropolis*. In his own words, "residential subdivisions, the shopping centers and industrial parks—all seemed to float in space, and seen together they resembled a galaxy of stars and planets, held together by mutual gravitational attraction, but with large emtpy spaces between the clusters."

This newest macroscale model of the American urban tissue will require additional research and field testing to both discern its dimensions and learn how it operates. It is also a fine example of a significant new interpretation of the North American landscape, based on careful spatial observation, that demonstrates how geographers can make valuable contributions to our understanding of ongoing cultural change.

promoting a shift toward the growing service industries. Accompanying this switch is downtown commercial revitalization, which has been widespread since 1970, but, in many cities, for each shining new skyscraper that goes up, several old commercial buildings are abandoned; in many younger cities such as Denver and Los Angeles, a whole forest of new office towers contains suburban commuters whose only contact with the city below is the short drive between the freeway exit and their building's parking garage. Residential revitalization in and around the CBD also occurred in many cities during the 1970s, but the numbers of people involved were not as significant as first believed; since most reinvestment was undertaken by those already residing in the central city, a "return-to-the-city" movement by suburbanites clearly did not take place. Such downtown-area neighborhood redevelopment involves *gentrification*—the upgrading of residential areas by new higher-income settlers—but, in order to succeed, this usually requires the displacement of established lower-income residents, an emotional issue that has sparked many conflicts. Beyond the CBD zone, the vast inner city remains the problem-ridden domain of low- and moderate-income people, most forced to reside in ghettoes; financially ailing big-city governments are unable to fund adequate schools, crime-prevention programs, public housing, and sufficient social services, and the downward spiral, including adandonment in the old industrial cities, continues unabated in the mid-1980s (Fig. 3–18). Moreover, no new planning solutions are on the horizon, particularly in a decade when federal aid to the beleaguered cities is deliberately being cut back. Peirce Lewis (1983b) depressingly reminds us the throwaway attitude has prevailed throughout the urban past, warning that "it is not so easy to recycle obsolete cities. They are too big to haul away—and besides, people live in them."

In sum, the splitting asunder of the functions of most large American cities in the postindustrial age has restructured the organization of metropolitan space and lends additional support to Lewis's hypothesis of the galactic city. This truly historic transformation of the metropolis is fraught with consequences for the future population geography of the continent, not the least of which is the possibility of achieving an urban civilization lacking traditional cities.

Figure 3–16
The suburban answer to downtown: Newport Beach outside Los Angeles.

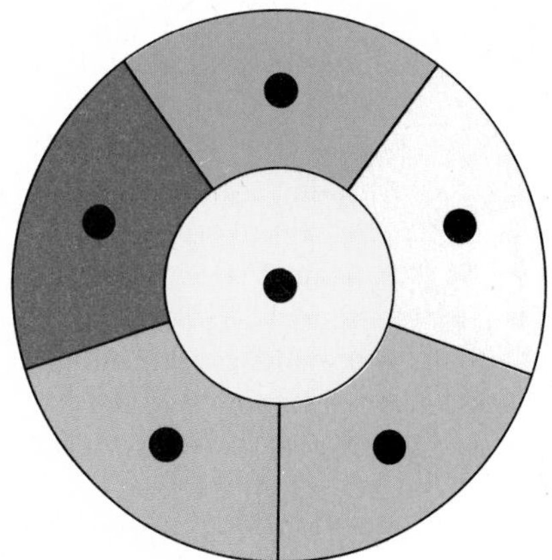

Figure 3–17

# Cultural Geography

North America constitutes one of the world's youngest geographic realms, yet the contributions of a wide spectrum of immigrant groups over the past two centuries have shaped a rich and varied cultural mosaic. Linguistic bifurcation between English speakers and French speakers has inhibited the formation of a single overriding culture in Canada, although other unifying societal elements continue to bind that country together; it is also worth pointing out that millions of Europeans not of British or Gallic descent have been absorbed into Canadian society in this century. In the United States, however, newcomers were far more willing to set their original cultural baggage aside in favor of assimilation into the emerging culture of their adopted homeland, which in itself was a hybrid nurtured by constant infusions of new influences. For most upwardly mobile immigrants, this plunge into the "melting pot" provided a ticket for acceptance into mainstream American society. But for millions of other would-be mainstreamers, the melting pot has proven to be a lumpy stew; whether by choice or not, they continue to stick together in ethnic communities (especially within the inner cities of the Manufacturing Belt), with a large proportion of these neighborhoods inhabited by racial minorities who do not fully participate in the national culture.

Figure 3–18
Inner-city devastation: New York's South Bronx.

## American Cultural Bases

As the American nativist culture matured, it came to develop a set of powerful values and beliefs: (1) love of newness, (2) a desire to be near nature, (3) freedom to move, (4) individualism, (5) societal acceptance, (6) aggressive pursuit of goals, and (7) a firm sense of destiny. The geographer Brian Berry has discerned these cultural traits in the behavior of people throughout the evolution of urban America. A "rural ideal," based on the philosophy of Thomas Jefferson and others, has prevailed throughout U.S. history and is still expressed in a strong bias against residing in cities; when industrialization made urban living unavoidable, those able to afford it soon moved to the emerging suburbs (*newness*) where a form of country life (*close to nature*) was possible in a semiurban setting. The fragmented metropolitan residential mosaic, composed of a myriad of slightly different neighborhoods, encouraged frequent *unencumbered mobility* as middle-class life revolved around the *individual* nuclear family's *aggressive pursuit* of its aspirations for *acceptance into the next higher stratum of society*. These accomplishments confirmed to most Americans that their goals could be attained through hard work and perseverance—perhaps adding as an eighth trait the "work ethic"—and that they possessed the ability to realize their *destiny* by achieving the "American Dream" of home ownership, affluence, and total satisfaction.

## Language

Although linguistic variations play a far more important role in Canada, 11 percent of the American population spoke a primary language other than English in 1980 (mainly Spanish). U.S. linguistic differences are also evident at the subnational level, where regional variations (*dialects*) are still widespread, despite the recent trend toward a truly national society—the Deep South and New England immediately come to mind as areas that possess distinctive accents. The geography of word usage is also a varied one; if a person were to drive from New York to Chicago and avoid national-chain restaurants, the observer would recognize a steady change in the local name of the big-bun sandwich—"submarine" in New York, "hoagie" in Pennsylvania, "Big Boy" in northern Ohio, and "Poor Boy" in Chicago. An even closer connection between language and landscape is established through *toponymy*, the naming of places. U.S. place-name geography provides important clues to the past movements of cultural influences and national groups: for example, the preponderance of such Welsh place-names as Cynwyd, Bryn Mawr, and Uwchlan to the west of Philadelphia is the final vestige of an erstwhile colony of Celtic-speaking settlers from Wales.

## Religion

North America's Christian-dominated kaleidoscope of religious faiths contains important spatial variations. Many major Protestant denominations are clustered in particular regions, with Baptists localized in the southeastern quadrant of the United States including Texas, Lutherans in the Upper Midwest and northern Great Plains, and Mormons in Utah and southern Idaho. Roman Catholics are most visibly concentrated in two locations: the Manufacturing Belt metropolises as well as nearby New England, which received huge infusions of Catholic Europeans over the past century, and the entire southwestern borderland zone that is home to a burgeoning Mexican-American population. Judaism is the most highly agglomerated major religious group on the continent, its largest congregations located in the cities and suburbs of Megalopolis, southern California, South Florida, the Midwest, and Canada's Main Street conurbation.

## Ethnicity

Ancestry is another key element in the cultural diversity of this realm. As we have seen, *ethnicity* (meaning "nationality") was a decisive influence in the shaping of American culture, which, in turn, also reshaped the cultural traditions of newcomers who were assimilated into mainstream society. The current U.S. ethnic tapestry is, as ever, characterized by an increasingly complex mosaic (see *Box*). The spatial distribution of ethnic minorities is mapped in Fig. 3–19, which also includes the native American or Indian population. Because the latter largely occupy tribal lands on reservations ceded by the federal government, they are undergoing very little distributional change. The Hispanic population, however, is growing rapidly through in-place natural increases and in-migration from Latin America (much of it illegal); spatially, Hispanics are both increasing in density along the southwestern border and are fanning out toward the north and east, particularly in larger metropolitan areas. Blacks, also exhibiting birth rates higher than the national average, are not as mobile as Hispanics, undoubtedly constrained by the racially dualistic housing market that offers fewer opportunities to blacks than to nonblacks of similar income levels; nonetheless, upwardly-mobile blacks have increased their presence in both suburbia and the more affluent rural areas of the South since 1970, although settling in new and existing communities that remain highly segregated. The newest element in the ethnic tapestry is the arrival of Asian-Americans, whose tight clustering in Los Angeles and other west coast metropolitan areas is not yet sizable enough to show up on Fig. 3–19.

American cultural geography continues to evolve. What is now

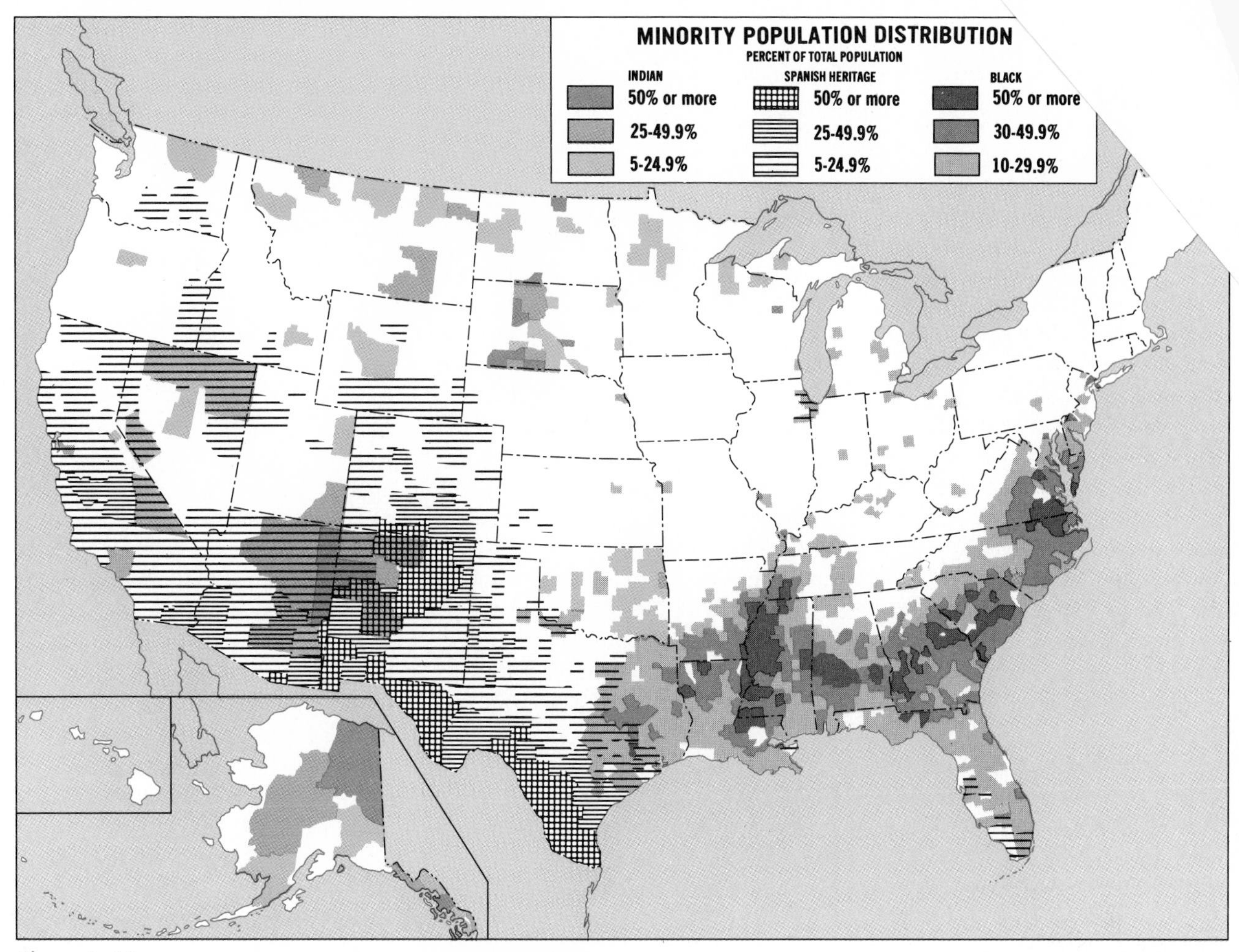


**Figure 3–19**

taking place is a new fragmentation into a *mosaic culture*, an increasingly heterogeneous complex comprising a myriad of separate, uniform tiles that cater to more specialized groups than ever before. No longer based solely on such larger divisions as income, race, and ethnicity, today's residential communities are also forming along the dimensions of age, occupational status, and especially lifestyle. Their success and rapid proliferation reflects an obvious satisfaction by a majority of Americans. But such balkanization—which is fueled by people choosing to interact only with others exactly like themselves—bears watching, because it could threaten the survival of some important democratic values that have prevailed throughout the evolution of U.S. society.

## Environmental Perception and Spatial Behavior

The cultural geographer Jan Broek has written:

> *. . . each society in each era perceives and interprets its physical surroundings and its relations to other [places] through the prism of its own way of life. . . . It is always and only in relation to culture that the parts of the earth receive specific meaning.*

Recent research on cultural appraisals of the environment has focused on the way in which a person's perceptions shape his or her spatial decision making and behavior. This work has revealed that the impressions and images people develop about various places shape their attitude toward them and that this information is structured within personal **mental maps** that all of us carry around in our minds. Because American society is both affluent and predisposed toward frequent mobility, interpreting these mental maps represents much more than a fascinating academic exercise—it can be a useful tool in the current understanding and forecasting of migration flows. There are two separate but closely interrelated di-

## THE AMERICAN ETHNIC TAPESTRY IN 1980

The diversity and complexity of America's ancestral makeup was reported by the latest census, which showed that over 83 percent of the U.S. population—which totaled 226.5 million on April 1, 1980—identifies with one of 134 different national backgrounds. There are today more Americans of full or partial English descent (50 million) than the total population of Great Britain, more than one-half as many Americans of German stock (49 million) as currently reside in the two Germanys, and the Irish-Americans outnumber the population of the Republic of Ireland by a ratio of 12 to 1. Moreover, 21 million Americans listed themselves as Afro-Americans—a number exceeding the populations of all but four Black African states. Other significant ethnic ancestries include French (13 million), Italian (12 million), Scottish (10 million), Polish and Mexican (each 8 million); those of native American background numbered 7 million. The degree of interethnic mixing is also considerable—52 percent of those born in the United States of American-born parents reported multiple ancestry, further confirming the notion of the melting pot society.

Geographically, the dominant European ancestries—French, Irish, and those originating in the United Kingdom—were dispersed throughout the United States. Most of the others showed an affinity for a particular region: Italians, Portuguese, and Russians clustered in the northeast, whereas Scandinavians and Czechs were localized in the north-central states. California, the most populous state, best exhibited the nation's diverse ethnicity, with more Americans of English, German, Irish, French, Scottish, Dutch, Swedish, and Danish origin concentrated there than in any other state.

Although the Afro-American population remained fairly stable in relative size from 1970 to 1980 (about 11 percent of the U.S. total), two other minority groups substantially increased their presence in the national ethnic tapestry—Asians and Hispanics. During the decade, the number of Asians and Pacific Islanders more than doubled from 1.5 to 3.5 million, remaining largely concentrated in the far western United States. The biggest single Asian group today is the Chinese; they supplanted Japanese-Americans from that position to third during the 1970s, the Japanese also being surpassed by the rapidly expanding Filipinos; at the same time, the Koreans also gained quickly, quintupling their numbers from 70,000 to 350,000. The Hispanic population, which may become America's largest minority by the end of the century, grew by over 60 percent from 1970 to 1980, only trailing the black population 14 to 26 million. The fast-growing Mexican component (up 93 percent in the 1970s) accounts for three-fifths of this population, Puerto Ricans 14 percent, and Cubans 6 percent. However, whereas Mexican-Americans and Puerto Ricans are now dispersing from their main centers of concentration—the U.S. Southwest and New York City, respectively—the Cuban population is agglomerating ever more tightly in South Florida's Miami area.

mensions of mental mapping.

The first dimension involves *designative* mental mapping, in which spatial information is received and objectively placed on the "map" in an individual's mind. Places often mentioned in the news are easily recognized, and most American citizens have a good idea where New York City, California, and Florida are located. Sometimes an obscure place comes to our attention and remains fixed in mind—Three Mile Island and Mt. St. Helens are two recent examples. People also learn a great deal more about nearby places that they encounter frequently; for example, most Chicago-area residents know where Woodfield Mall is located, but very few people living elsewhere have heard of this shopping center in suburban Schaumburg, Illinois (which happens to be the biggest one in the United States). At the national level, the geographer Wilbur Zelinsky has carefully mapped the "perceptual regions" of North America (Fig. 3–20), which represent a summarization of individual perceptions as to where such traditional regions as "South," "Midwest" and "West" are situated. However, these vernacular or popularly perceived regions—which were derived by studying local telephone directories and interviewing people residing throughout the nation—do not readily coincide with the formal regionalization scheme that is developed later in this chapter (Fig. 3–28). This indicates a knowledge lag between the regional realities of the 1980s and their comprehension by Americans, further evidence of the inadequacy of geographical education in the United States (*Box* p. 47).

The other dimension involves *appraisive* mental maps, in which spatial information is subjectively processed according to an individual's personal biases. A good example is the souvenir postcard that can be purchased throughout Texas

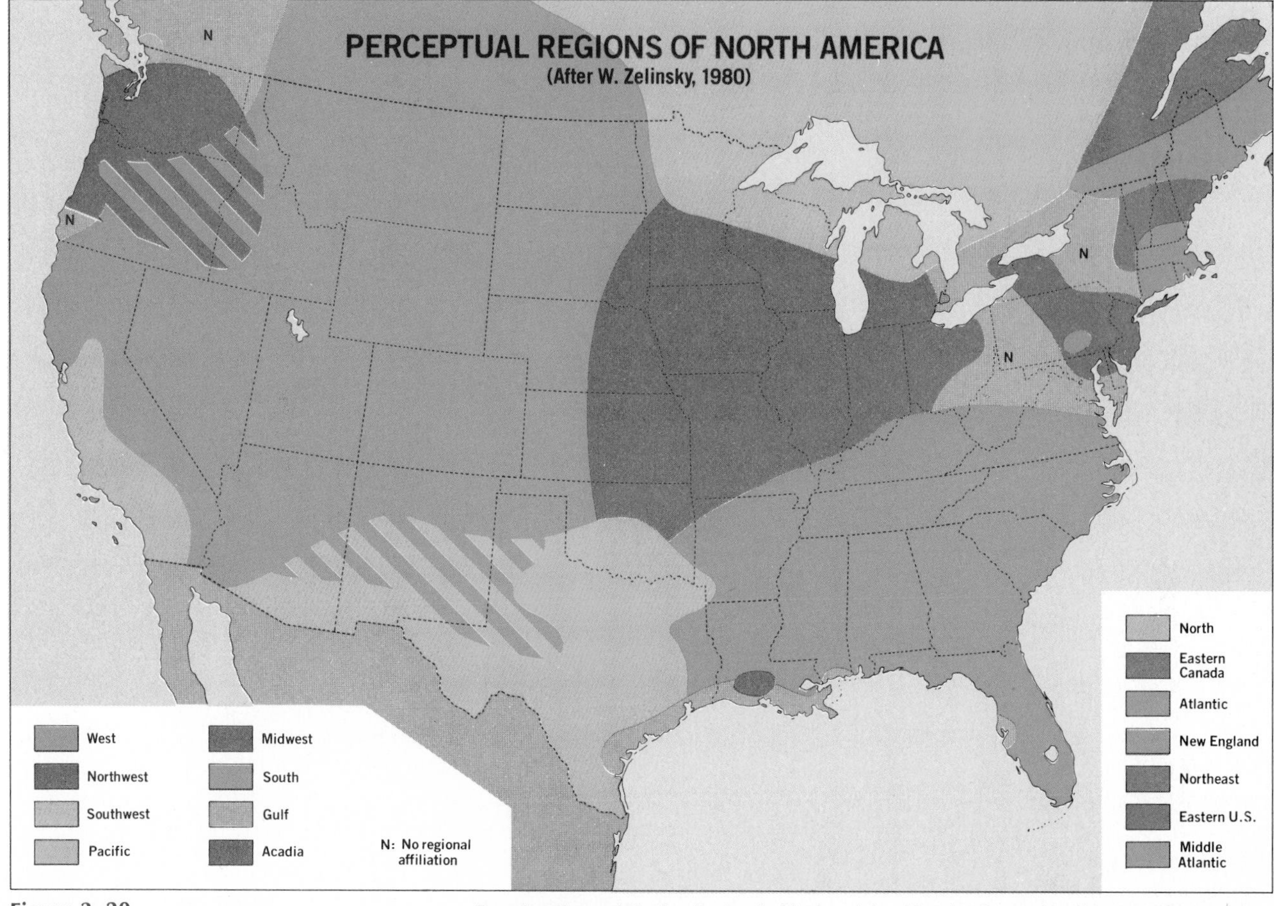


**Figure 3–20**

Reproduced by permission from the *Annals of the Association of American Geographers*, Volume 70, 1980, p. 14, fig. 5.

displaying a humorous map entitled "A Texan's View of the United States," which shows the state of Texas occupying about three-quarters of the nation. In addition to this positive bias concerning the prominence of the Lone Star State, the "map" contains negative biases toward all other states. Besides diminishing them in size, the remaining states are renamed in jest; the most unflattering names are reserved for those regions that have historically opposed the South or competed against Texas—for example, New England is "Damned Yankee Land," Illinois is "Ill Noise," Washington is "Snafu, D.C.," Florida is "Swampland," and California is simply "Uninhabitable." Although everybody possesses place biases, the process leading to their formation is not well understood. Why should Cleveland and Philadelphia suffer persistently negative images, whereas Boston and Denver do not? And how did Dallas, the scene of a presidential assassination in 1963, manage to reverse its image from bad to excellent only a few years later? These questions cannot yet be answered with much certainty, though it is likely that an explanation rests with some combination of the following location variables: employment opportunities, weather patterns, political climate, social setting, and recreational possibilities.

A well-developed application of appraisive mental mapping has been the preparation of maps that reveal the collective future residential preferences of college students. Various student groups around the United States were asked to rank the 50 states in answer to the question, "Where would you like to live?" One such map produced in 1981 showed the relative attractiveness of the states as perceived by students at the University of Miami in Coral Gables, Florida, an institution that enrolls a sizable out-of-state student population: the 12 most popular and unpopular states are of greatest interest, and they are shown in Fig. 3–21. Based on follow-up classroom discussions, the majority of students revealed the following: (1) Florida, North Carolina, Texas, and the west coast states were picked for their employment opportunities, mild climate, and recreational offerings; (2) Colorado, Wyoming, Vermont, and New Hampshire were chosen for their wide-open spaces and mountain scenery; (3) Virginia and New York represented the job opportunities of the Washingtion and New York City metropolitan areas; (4) the 5 Deep South states were rejected for their social and political climes; (5) Ohio, Michigan, and

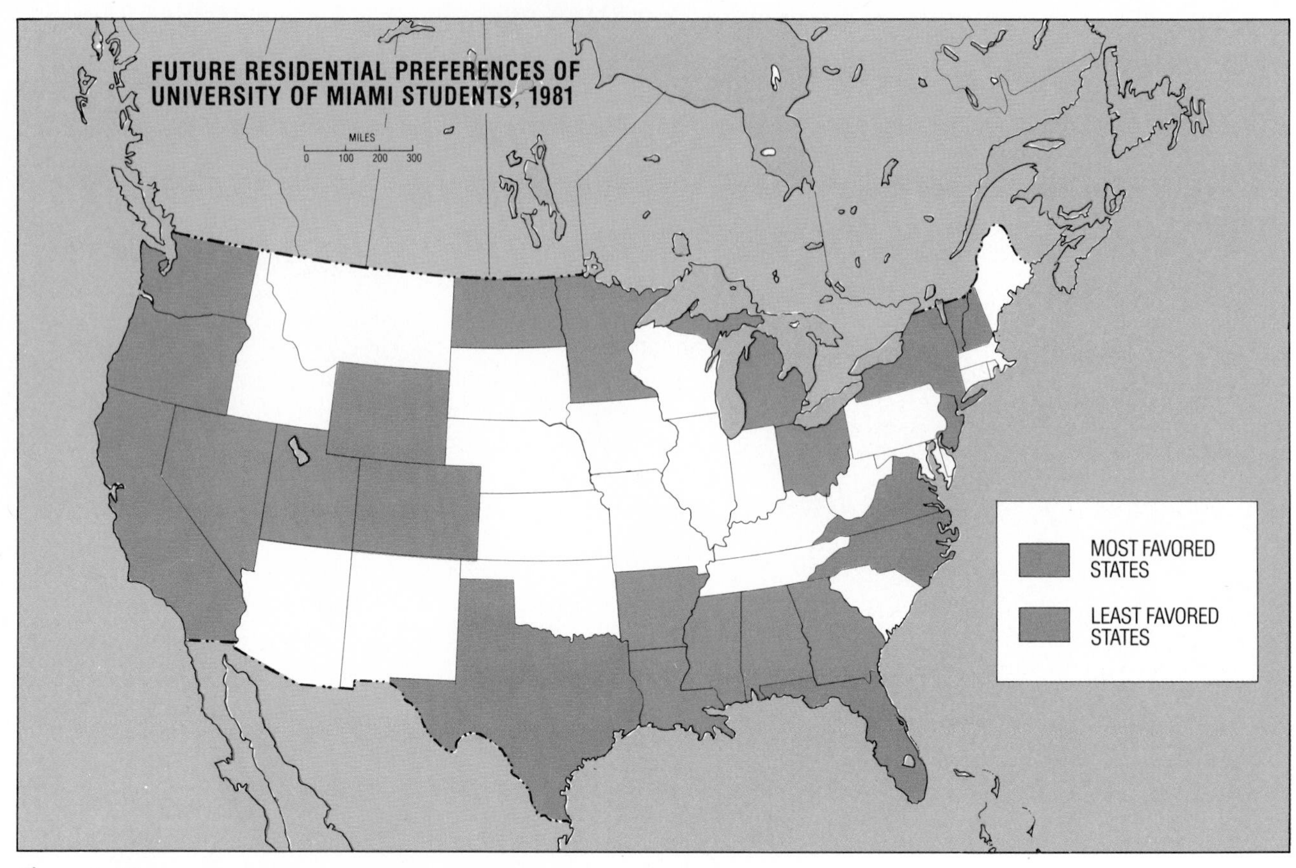


Figure 3–21

New Jersey were seen as undesirable overindustrialized places with chronic economic and employment problems, and (6) Minnesota, North Dakota, Utah, and Nevada were regarded as "dull" places, possessing harsh weather for at least part of the year. In follow-up studies of similar groups at other universities, these kinds of maps provided valuable clues to the future migration behavior of the hypermobile college-student population—many actually did later convert perceptions into migration decisions, with generally satisfactory results.

# The Changing Geography of Economic Activity

The economic geography of North America in the late twentieth century is the product of all the foregoing, as bountiful environmental and human resources have been cumulatively blended together to create the most advanced economy and society in world history. Perhaps the greatest triumph was the overcoming of the "tyranny" of distance—a geographic quality that the Soviets can only view with the deepest of envy—as people and activities were organized into a truly continental spatial economy that took maximal advantage of agricultural, industrial, and urban-development opportunities. Yet, despite these past achievements, American economic geography today is, once again, in upheaval as the nation experiences the swift transition from industrial to postindustrial society.

As we have seen, a new set of locational forces has been unleashed wherein the distribution of key economic activities is now being shaped by noneconomic variables. Thus, business-location decisions are increasingly dominated by the same concerns that would govern a person's choice of residence if he or she had vast financial resources to draw on—finding a site and building that maximizes geographical prestige, amenities, commuting convenience, and access to recreational activities, and that also exemplifies the newest in fashionable architectural trends. Judging by the recent spatial behavior of U.S. companies, these prized locations are most frequently found in the suburban outer city nationwide as well as in selected central cities of the Sunbelt. As economic activities become more spatially footloose, the major casualty in this regional and intraurban transformation is the industrial city, especially those within the Manufacturing Belt. With the disappearance of the concentrative forces that created the high-density industrial city, many such urban centers have been scrambling recently in order to adapt to a new era in which economic-spatial deconcentration is the overriding trend. The most successful cities have revitalized their CBDs by encouraging repeated waves of skyscraper construction since 1960, hoping to attract sufficient white-collar office jobs to replace rapidly declining employment in manufacturing industry; the less successful industrial cities have only managed to put up a few new cosmetic symbols in their downtowns, unable to stem the tide of employment decentralization. This is the background against which the new postindustrial spatial economy of the 1980s and 1990s is materializing, with the metropolitan connection crucial as ever to its structure and functioning.

## Major Components of the Spatial Economy

Economic geography is heavily (although not exclusively) concerned with the locational analysis of productive activities. Four major sets may be identified:

**Primary Activity** The extractive sector of the economy, in which workers and the environment come into direct contact—especially *mining* and *agriculture*.

**Secondary Activity** The *manufacturing* sector, in which raw materials are transformed into finished industrial products.

**Tertiary Activity** The *services* sector, including a wide range of activities from retailing to finance to education to routine office-based jobs.

**Quaternary Activity** The fast-growing sector involving the collection, processing, and manipulation of *information*; a subset, sometimes referred to as **quinary activity**, is managerial or control-function activity associated with *decision making* in large organizations.

Historically, each of these activities has successively dominated the American labor force for a time over the past 200 years (with the quaternary sector now poised to dominate the near-future economy). Agriculture was dominant until late in the nineteenth century, giving way to manufacturing by 1900; the steady growth of services after 1920 finally surpassed manufacturing industry in the 1950s but now shares a dwindling portion of the limelight with the fast-rising quaternary sector. The approximate breakdown by major sector of employment in the U.S. labor force today is agriculture—3 percent; manufacturing—17 percent; services—20 percent; and quaternary—60 percent (with about 10 percent in the quinary sector). We now treat these major productive components of the spatial economy in the following coverage of resource utilization, agriculture, manufacturing, and the postindustrial revolution.

## Resource Utilization

The North American continent was blessed with abundant deposits of industrial and energy resources. Fortunately, these were usually concentrated in large enough quantities to make long-term extraction an economically feasible proposition, and most of the richest raw material sites are still the scene of major drilling or mining operations—the Gulf Coast oilfields, the soft (bituminous) coal beds of Appalachia, and the iron and nickel ores of the southern Canadian Shield. Moreover, the continental and offshore mineral storehouse may yet contain outstanding resources that will attract future exploitation: substantial oil and gas supplies in the North American Arctic have barely begun to be tapped (including Alberta's tar sands), and enormous deposits of oil shale and other energy resources in the central Rockies await the perfection of new extractive technologies (as well as environmental protection strategies).

**Industrial mineral resources** North America's mineral deposits are localized in three zones: the Canadian Shield north of the Great Lakes, the Appalachian Highlands, and scattered areas throughout the western mountain ranges. The Shield's most noteworthy minerals are iron ore (Minnesota's Mesabi Range just west of Lake Superior and the eastern Shield in Quebec and Labrador), nickel (around Sudbury, Ontario, near Lake Huron's north shore), and gold, uranium, and copper (from upper Saskatchewan north to the Arctic coast). Besides vast deposits of soft coal, the Appalachian region also contains hard (anthracite) coal in northeastern Pennsylvania and iron ore in central Alabama. The U.S.-Canadian western mountain zone contains significant deposits of coal, copper, lead, zinc, molybdenum, uranium, silver, and gold (which spawned "rushes" to both California and the Alaska-Yukon area in the last century). The mineral wealth of North America is one of the most varied of any realm—the only high-quality raw materials lacking here are tin, bauxite (aluminum ore), and manganese, which along with a few additional minor minerals must be imported.

**Fossil fuel energy resources** The most strategically important resources of North America are its coal, petroleum (oil), and natural gas supplies—the *fossil fuels*, so named because they were formed by the geologic compression and transformation of plant and tiny animal organisms that lived hundreds of millions of years ago. These energy supplies are mapped in Fig. 3–22 and reveal abundant deposits and distribution networks. The realm's *coal* reserves are among the greatest anywhere on earth, the U.S. portion alone comprising at least a 300-year supply. Three coal regions are evident: (1) Appalachia, which produces about half of all U.S. coal, is still the largest region but is declining because its high-

Figure 3–22

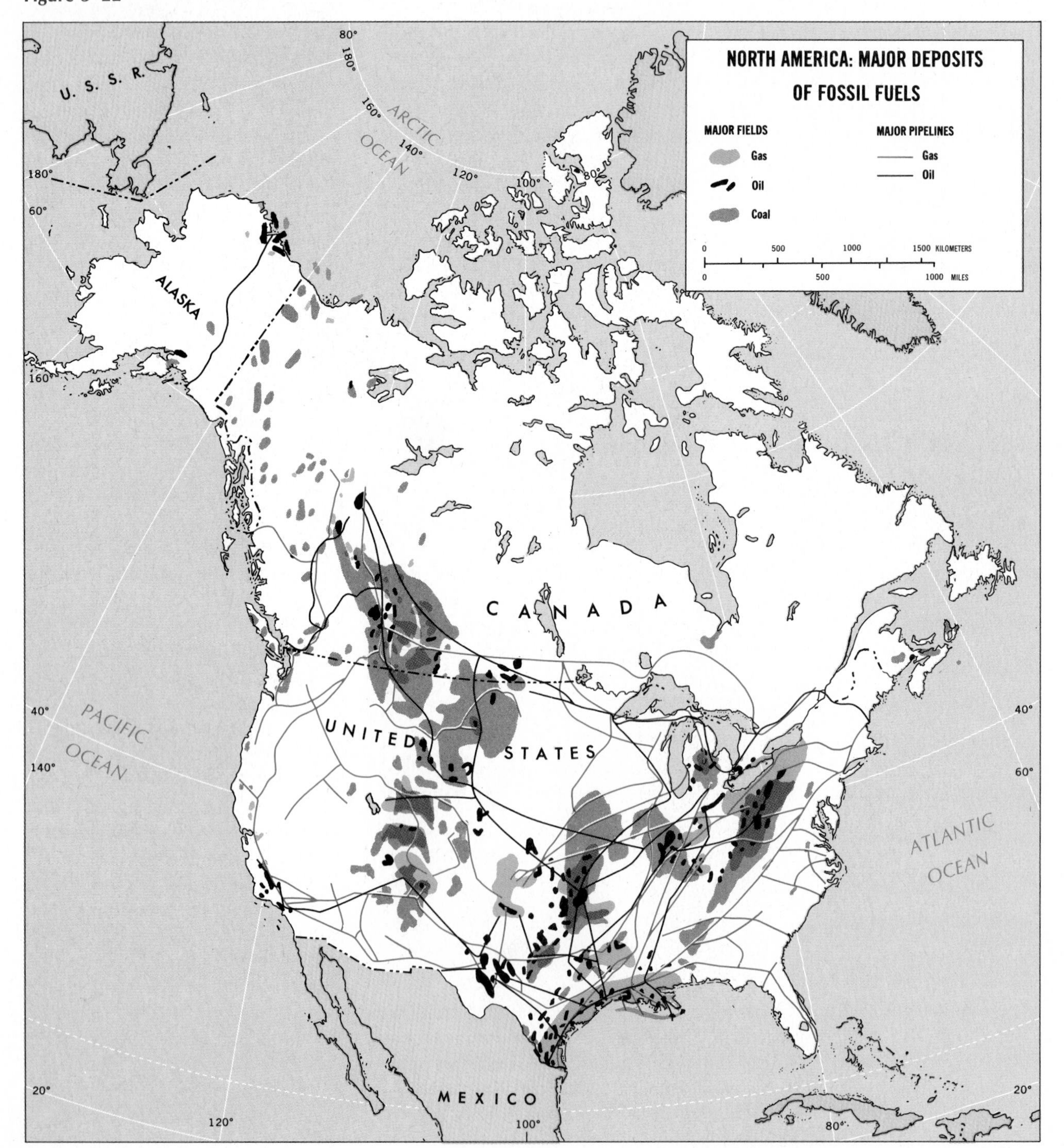

sulfur coal must be expensively filtered to meet the standards set by federal clean-air statutes; (2) the Western coal region, centered on the Great Plains within 500 miles north and south of the U.S.-Canada boundary, is expanding vigorously, thanks to the deposition of its vast low-sulfur supplies in thick near-surface seams—ideal for lucrative strip-mining—and has even prompted the construction of the first long-distance railroad line in the United States in over 50 years (into Wyoming's Powder River Basin); and (3) the Midcontinent coalfields, a large arc of high-sulfur, strippable deposits centered on southern Illinois and western Kentucky, which is also declining as the West surges ahead. North America's major *oil*-production areas are located along and offshore from the Texas-Louisiana Gulf Coast, in the Midcontinent district extending through western Texas, Oklahoma, and eastern Kansas, and along the front of the Canadian Rockies in central and northern Alberta; lesser oilfields are also found in southern California, west-central Appalachia, the northern Great Plains, and the southern Rockies. Both the United States and Canada have recently turned their attention to the Arctic, where ample supplies exist along Alaska's North Slope and adjacent territory in northwestern Canada. In 1977, the Trans-Alaska Pipeline was opened (Fig. 3–23)—a triumph over the harsh Arctic environment—enabling the pumping of North Slope petroleum south 1300 kilometers (800 miles) to the warm-water port of Valdez for transhipment by tanker to the lower 48 states; both countries have also agreed to jointly build another pipeline to carry North Slope natural gas all the way south overland through western Canada. The distribution of *natural gas* deposits resembles the geography of oilfields because both petroleum and gas are usually found in similar geologic formations (the floors of ancient shallow seas). Accordingly, major gasfields are located in the Gulf, Midcontinent and Appalachian districts. However, when subsequent geologic pressures are exerted on underground oil, the liquid is converted into natural gas; this has happened frequently in mountainous zones, so that western gas deposits in and around the physiographic province of the Rockies tend to stand apart from oilfields.

**Figure 3–23**
The Trans-Alaska Pipeline, carrying Arctic Slope oil south.

Their vast petroleum supplies notwithstanding, North Americans experienced a pair of serious oil-supply disruptions during the 1970s. These "crises," however, were not a simple matter of inadequate domestic supply and excessive local demand: they were crises of convenience. In order to keep petroleum prices low, Americans began significantly to increase oil imports after World War II rather than begin the more costly exploitation of adequate but less accessible oil deposits at home. By 1973, this dependency had become so great that when political events in the Middle East temporarily halted U.S.-bound oil shipments (as they did again in 1979), the nation quickly experienced widespread shortages followed by substantial price increases. The United States has since begun to develop its less readily available supplies, weaning itself slowly away from depending heavily on imports; in the process, it has ended (at least for now) the age of cheap energy. The search also continues for nonfossil fuel alternatives: with solar power and nuclear fission proving to be overly expensive and controversial, nuclear fusion appears to be the most promising avenue for the future, with harnessing of this inexpensive and virtually unlimited energy source possible by 2000.

**Figure 3–24**
Ultramechanized wheat farming in Washington State's Palouse country.

## Agriculture

Despite the twentieth century emphasis on urbanization and the development of the non-primary sectors of the spatial economy, agriculture remains an important element in the realm's human geography. Because it is the most *extensive* (space-consuming) economic activity, vast expanses of the American and Canadian landscape are clothed by fields of grain; moreover, great herds of livestock are sustained by pastures and fodder crops—because this wealthy realm can afford the luxury of feeding animals from its farmlands and has come to demand vast quantities of red meat in its diet. North Americans consume more meat per capita than the people of any other world realm, and large grain and dairy-product surpluses permit the exporting of food to less fortunate countries.

The increasing application of the latest technology to farming—particularly mechanization (Fig. 3–24)—has steadily increased both the volume and value of total agricultural production: a single farmer's output in 1979 fed 60 people, a tremendous gain over the 28 that could be fed in 1965. This remarkable recent gain in productivity, through automation and more efficient larger-scale farming, has been accompanied by a sharp reduction in the number of those actively engaged in agriculture (to only about 3 percent of the U.S. work force today). Small family farms have borne the brunt of this change: those unable or unwilling to modernize and enlarge their operations increasingly find it impossible to survive. Many critics have castigated big food-processing companies as the force behind this displacement: actually, most agricultural corporations (or "agribusinesses") are now family-scale operations, managed by shrewd farmers who realized that continued success depended on adaptation to rapidly changing times.

The regionalization of U.S. agricultural production, which has been well established for over a century, is shown in Fig. 3–25. Its overall spatial organization developed largely within the framework of the Von Thünen model (pp. 72–74); as in Europe (Fig. 1–12B), the early nineteenth century original-scale model of town and hinterland—in a classical demonstration of time-space convergence—expanded outward, with constantly improving transportation technology, from a locally "isolated state" to encompass the entire continent by 1900. As the macrogeographical structure formed, the greatly enlarged original Thünian production zones (Fig. 1–12A) were modified: (1) the first ring now differentiated into an inner fruit-vegetable zone and a surrounding band of dairying; (2) the forestry ring was displaced to the outermost limits of the regional system, because railroads could now transport wood quite cheaply; (3) the "field crops" ring subdivided into an inner mixed crop-and-livestock ring to produce meat (the *Corn Belt*, as it came to be called) and an outer zone that specialized in the mass production of wheat grains; and (4) the ranching area remained in its outermost position, a grazing zone that supplied young animals to be fattened in the meat-producing Corn Belt as well as supporting an indigenous sheep-raising industry. The "supercity" anchoring this macro-Thünian regional system was the northeastern Megalopolis, well on its way toward coalescence and already the dominant food market and transport focus of the whole country.

Although the circular rings of the

model are not apparent, many spatial regularities can be observed (remember that Von Thünen, too, applied his model to reality and thereby introduced several distortions of the theoretically ideal pattern). Most important is the sequence of regions as distance from the national market increased, especially westward from Megalopolis toward central California, which was the main directional thrust of the historic inland penetration of the United States. The Atlantic Fruit and Vegetable Belt, Dairy Belt, Corn Belt, Wheat Belts, and Grazing Zone are indeed consistent with the model's logical structure, each region successively farther inland astride the main transcontinental routeway. Deviations from the scheme may be attributed to irregularities in the environment and unique conditions: (1) central Appalachia and the dry western mountains cannot support anything other than isolated-valley ("General") farming or "oasis" cropping; (2) the near year-round growing seasons of California and the Gulf Coast-Florida region permit those distant areas (with the help of efficient refrigerated transport) to produce fruits and vegetables in competition with the innermost zone; (3) the bifurcation into two wheat regions results from hilly and sandy country in western Nebraska, with the milder-climate winter wheat zone concentrated to the south and with spring wheat deflected to the much colder northern Great Plains, where it extends deep into the Canadian prairie provinces (Fig. 3–31); (4) the small farming regions of the Pacific Northwest serve local populations, the nation with specialty crops, and raise wheat for export to Pacific Basin customers; and (5) the Old Cotton Belt is a remnant of the bygone era when the South was not well integrated into the national economy, which specializes today in the production of beef, poultry, soybeans, and increasingly lucrative timber.

## Manufacturing

The geography of North America's industrial production has long been dominated by the Manufacturing Belt, the continental core region bounded by the great rectangle that connects Boston-Milwaukee-St. Louis-Washington-Boston (Fig. 3–12). As we saw earlier, the internal structure of the Belt is built around a dozen urban-industrial districts, linked together by a dense transportation network, which also interconnects the manufacturing centers with such major resource deposits as the Appalachian coalfields and the Lake Superior-area iron mines. This highly agglomerated spatial pattern developed be-

**Figure 3–25**

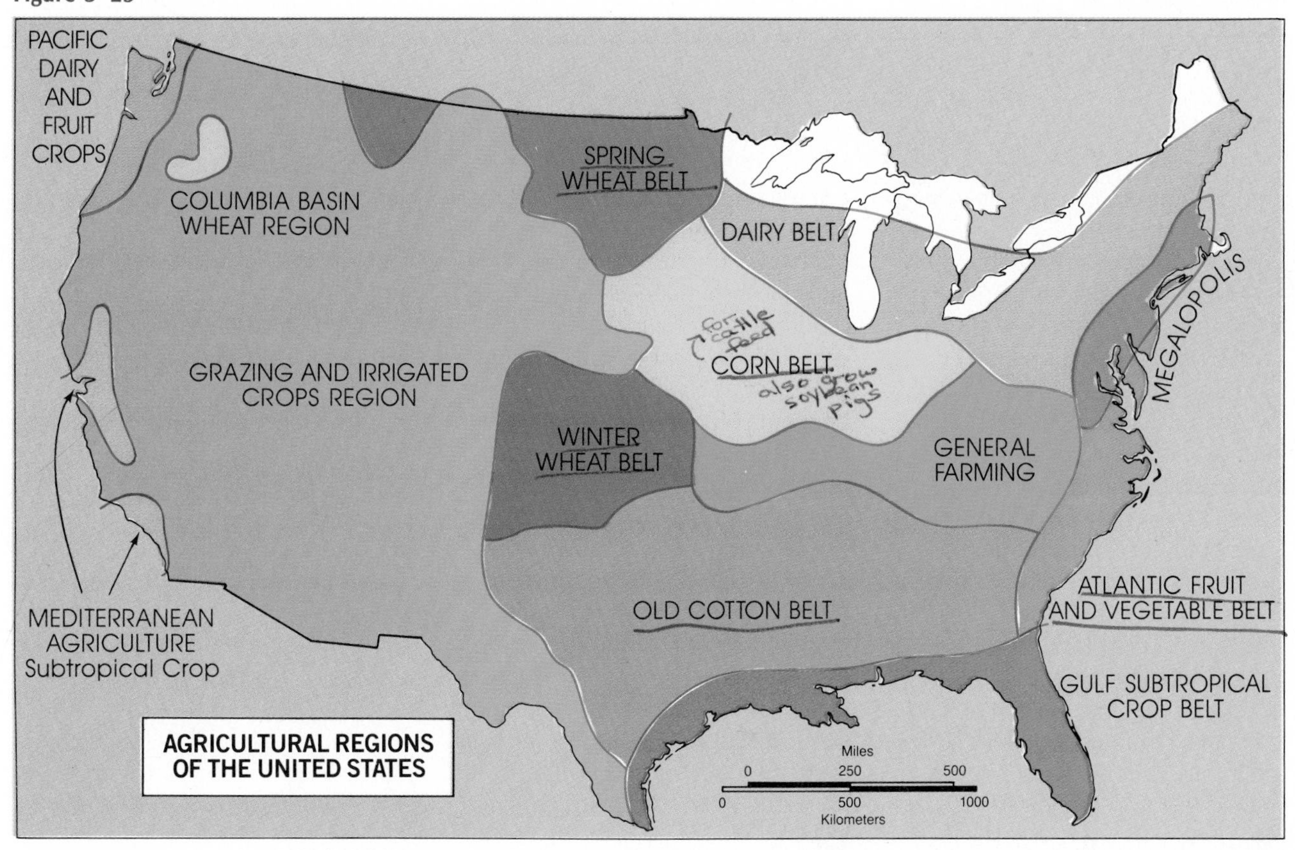

cause of the economic advantages of clustering industries in cities, which contained the largest concentrations of labor and investment capital, constituted major markets for finished goods, and were the most accessible nodes on the national transport network for assembling raw materials and distributing manufactures. As these industrial centers expanded rapidly during the late nineteenth century, they also achieved **economies of scale**, savings accruing from large-scale production in which the cost of manufacturing a single item was further reduced as individual companies mechanized assembly lines, specialized their work forces, and purchased raw materials in huge quantities (a situation analogous to the ability of a supermarket to translate its lower overhead costs into a lower sales price for an item than would be charged in a small corner store).

This efficient production pattern served the nation well throughout the remainder of the industrial age. Because of the effects of *historical inertia*—the need to continue using hugely expensive manufacturing facilities for their full multiple-decade lifetimes to cover initial long-term investments—the Belt is not about to disappear for a long time to come. However, as aging factories are retired, a process already under way in certain parts of the Manufacturing Belt, the distribution of American industry will change. As transportation costs even out among U.S. regions, as energy costs increasingly favor the south-central oil- and gas-producing states, as high-technology manufacturing advances reduce the need for lesser-skilled labor, and as locational decision making intensifies its attachment to noneconomic factors, industrial management is signaling its willingness to relocate to more desirably perceived regions in the south and west. The latest economic census (1978) revealed that the Belt states' share of total U.S. manufacturing had dropped to less than 48 percent; for most of the preceding century, this region had contained over 60 percent of American industry, reporting a figure of 56 percent as recently as 1967. On the other hand, from 1967 to 1978, the four Sunbelt states of California, Texas, North Carolina, and Georgia advanced their collective share of national manufacturing employment from 17 to 21 percent; interestingly, these states gained 856,000 blue-collar jobs over the 11-year span, whereas the Manufacturing Belt was losing an almost identical 873,000 jobs.

The decline of the Manufacturing Belt and most of its heavy industries continues unabated during the 1980s, and one congressional study forecasts that smokestack-industry employment will drop from 20 percent of the U.S. work force in 1980 to 8 percent by the mid-1990s. Undoubtedly, this sectoral shift is being hastened by the rise of the postindustrial economy; its dislocations are also heightened by chronic unemployment which was ratcheted upward by the Great Recession of 1982–1983, the worst economic downturn in the United States since the Depression of the 1930s. Within the Belt today, problems of obsolescence and reduced productivity are widespread, being especially evident in such leading industries as steel, automaking, rubber, and shipbuilding (which all operate well below their full capacities). Very little is being done about the region's deteriorating physical plant, with major factory closures all too regularly reported in the evening newscasts; on the landscape of the inner industrial city, the abandoned manufacturing complex is already a depressingly familiar sight (Fig. 3–26). Industrial management is sometimes criticized for not being aggressive enough in holding onto traditional markets, let alone finding new ones at home and overseas; even the U.S. Steel Corporation was publicly questioned about its choosing to diversify through purchasing the Marathon Oil Company in 1981, instead of investing those same billions of dollars in the modernization of its aging steel-making facilities. As industrial employment opportunities shrink, it will become necessary to retrain large numbers of blue-collar laborers for alternative careers in the services sector—a looming crisis that government has all but totally ignored. And perhaps the biggest problem of all faced by the North American manufacturing sector is the growing competition from abroad, with high-quality industrial imports increasing their share of the U.S. market from 9 percent in 1970 to 19 percent in 1982. Once based exclusively on cheap labor, America's chief foreign competitors in Europe and non-Communist Asia now possess superior technologies and productive efficiencies in a growing list of industries—the Japanese, for example, utilize five times as many robots as the Americans in running some of the world's most sophisticated factories.

The weakened position of the United States is particularly well demonstrated in the case of steel, the primary raw material for the nation's basic heavy industries and the very foundation on which the Manufacturing Belt was built. The world leader in domestic and export production as recently as the late 1960s, American steel-making has fared so poorly in the competition with foreign producers since 1975 that more than 20 percent of the steel consumed in the United States today is derived from imports; even more ominously, sales to the U.S. automobile industry—domestic steel's biggest customer—were halved between 1976 and 1983 owing to depressed retail sales of domestic cars. The main shortcoming is the failure of the American steel industry to modern-

**Figure 3–26**
Manufacturing Belt decline: an abandoned plant in Lowell, Massachusetts.

ize its increasingly inefficient foundries, rolling mills, and fabrication plants (in spite of import restrictions that were imposed to allow the industry to catch up with its more aggressive overseas competitors). Japan, as well as several European countries, has in recent years led the way in applying the latest high-technology methods to steel-making and, as a result, can now produce this metal more cheaply and efficiently than the United States, despite inferior resources and greater distances to major markets. And another measure of how far the domestic industry has slipped in the last two decades was the announcement in 1983 that the U.S. Steel Corporation was considering shutting down a large part of its gigantic Fairless Works near Philadelphia (a state-of-the-art complex as recently as the early 1970s), eliminating 7500 jobs and sharply curtailing operations to the finishing of slabs manufactured by the British Steel Corporation; in the end the plan was abandoned, but this kind of corporate thinking has now become well established.

## The Postindustrial Revolution

The signs of postindustrialism are visible throughout the North American realm in the 1980s, and they are popularly grouped under such appellations as "the information society" and "the electronic era." The term *postindustrial* by itself, of course, tells us mainly what the theme of the American economy is no longer; yet the term is also used by many scholars to refer to a specific set of societal and business-organizational changes that comprise an historic break with the recent past (see *Box* p. 214). Many of the urban-spatial expressions of the currently transforming society have already been highlighted. The discussion here focuses on the broader reshaping of U.S. economic geography as this century draws to a close.

High-technology (*high-tech*) activities are the leading growth industries of the postindustrial economy, and they are highly prized by local area-development agencies. The epitome of such development is still the San Francisco Bay Area's Silicon Valley (Fig. 3–15), which is rapidly spawning competitors in suburban Dallas ("Silicon Prairie"), South Florida's Boca Raton ("Silicon Coast"), and several other metropolises possessing the right blend of pace-setting locational qualities. A study presented to the Urban Land Institute in 1983 outlined the "10 golden rules" that must be satisfied in order to attract a critical mass of high-tech companies to a given locality:

1. A nearby major university that offers an excellent graduate engineering program.
2. Close proximity to a cosmopolitan urban center.
3. A large local pool of skilled and semiskilled labor.
4. Three hundred days of sunshine a year.
5. Recreational water within an hour's drive.
6. Affordable nearby housing.
7. Even closer prestigious luxury housing for top executives.
8. Start-up capital worth $1 billion to lure new high-tech firms.
9. Lower-than-normal risk for establishment of profitable high-tech businesses.
10. Cooperative spirit among landowners, lenders, government, and business.

Despite their glamour, there is an emerging problem as high-tech industries proliferate: on the whole, such economic activity is bound to eliminate more jobs than it creates. With growing automation, high-tech companies in the future will mainly employ highly trained scientists and fewer lesser-skilled production workers. Indeed, some policymakers now fear that thousands of blue-collar and routine-skill service workers—and even middle-level managers—may soon join an expanding population of the chronically unemployable unless major

## THE LINEAMENTS OF POSTINDUSTRIAL SOCIETY

Sociologist Daniel Bell, in a landmark 1973 book entitled *The Coming of Postindustrial Society*, sketched the distinguishing features of the emerging economy and its impact on American life. However, the subsequent pace of change has been unrelenting and many of his then future speculations have already become reality.

The new society and economy is marked by a fundamental change in the character of technology use—from fabricating to *processing*—in which telecommunications and computers are vital for the exchange of knowledge. Information becomes the basic product, the key to wealth and power, and is generated and manipulated by an intellectual rather than a machine technology. Yet, postindustrial does not displace industrial society, just as manufacturing did not obliterate agriculture. Instead, in Bell's words, "the new developments overlie the previous layers, erasing some features and thickening the texture of society as a whole."

Several hallmarks of postindustrialism can be identified. Knowledge is central to the functioning of the economy, which is led by science-based or *high-tech* industry. The technical/professional or "knowledge" class increasingly dominates the work force: quaternary and quinary activity already employed 40 percent of the U.S. labor force in 1975, employs about 60 percent today, and is expected to employ more than 70 percent by the mid-1990s. The nature of work in the information society focuses on person-to-person interaction rather than person-product or person-environment contacts. With the eclipsing of manufacturing, which largely was man's work alone, more and more women participate in today's labor force. A meritocracy now prevails in which individual advancement is based on education and acquired job skills.

Among the many spatial implications of this socioeconomic transformation, Bell observed that postindustrial occupations would gravitate toward five major types of "situses" or locations: economic enterprises, government, universities, social-service complexes, and the military. Central North Carolina's famous *Research Triangle Park*, located near the center of the "triangle" circumscribed by the cities of Raleigh, Durham, and Chapel Hill, is the quintessential example of a prestigious Sunbelt high-tech manufacturing and research complex. Even a small sampling of the list of tenants fully supports Bell's hypothesis: (1) among its business enterprises are such corporate giants as IBM, TRW, and the headquarters of Burroughs Wellcome Pharmaceuticals; (2) government is represented by the U.S. EPA, the Southern Growth Policies Board, and North Carolina's Board of Science and Technology; (3) three universities are prime movers in the Park's research operations—Duke, North Carolina, and North Carolina State—which have also attracted the National Humanities Center; (4) social services are the business of the National Center for Health Statistics Laboratory and the International Fertility Research Program; and (5) the military presence is embodied by the U.S. Army Research Office.

retraining programs are initiated to better match people with jobs in the new economy.

The geographic impact of the postindustrial revolution through the early 1980s can be gauged from Fig. 3–27, which maps state per capita income changes since 1975. Benefits have been rather unevenly distributed: only 20 of the 50 states exceeded the average national increase. The biggest gainers were the energy-rich states of Texas, Oklahoma, and Louisiana as well as Connecticut with its expanding affluent suburbs, exurbs, and corporate-headquarter complexes (spun off by deconcentration away from nearby Manhattan). States offering high-amenity environments fared quite well, especially Florida, California, Colorado, Wyoming, Virginia, and northern New England; not surprisingly, the Manufacturing Belt states as a group performed weakly. This map also destroys the notion that the Sunbelt is a uniformly booming area (let alone a valid geographical region): even at the state level, significant internal variations are obvious. Moreover, within each Sunbelt state, stark contrasts are common as the burgeoning economy of one county frequently leaves a neighbor completely untouched and mired in the preindustrial past. It is worth pointing out, as well, that even prosperous corners of the Sunbelt are not immune to changing economic currents—the huge glut of unused new office space in downtown Houston and Denver during the early 1980s was a testament to the belief that energy prices would never stop rising. However spotty the overall economic development pattern, it is still extremely likely that certain southern-tier metropolises will continue to capture an expanding share of major U.S. business activity. A 1982 survey of corporate managers rated Florida as the leading state in business climate, followed closely by Texas,

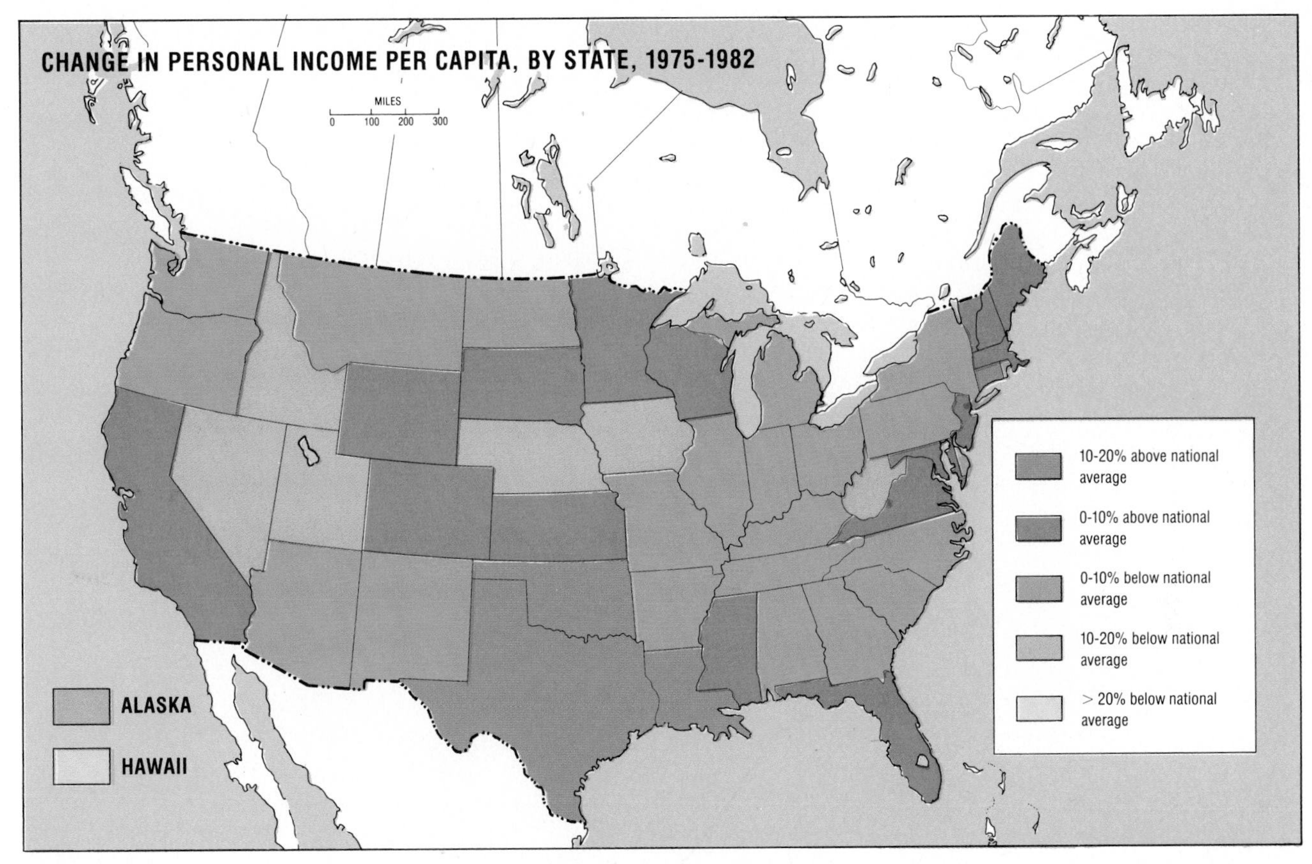


Figure 3–27

California, Colorado, Louisiana, and Mississippi (which all ranked above average in Fig. 3–27)—with the first three of these states predicted to contain 20 percent of all of America's jobs by 2000.

# Regions of the North American Realm

The ongoing transformation of North America's human geography is also reflected in its internal regional organization. As we pointed out earlier (see Fig. 3–20), traditional popularly perceived regions are increasingly irrelevant to understanding the new spatial realities of postindustrial America. Whereas many established regions still exhibit strong ties, those bonds may be assuming some new dimensions (New England and the American South are good examples). Elsewhere, long-dormant areas are now vigorously asserting new regional identities (such as the U.S. Southwest and the interior west in both countries). And in other cases—particularly the Anglo-American Core—old regions are struggling to find new answers to problems that threaten to intensify as the postindustrial revolution favors other more dynamic areas.

Quite clearly, the decade of the 1980s catches North America's regions at a time of deep-seated change as new forces redistribute people and activities. Although a number of old locational rules no longer apply, the varied character of the continent's physical, cultural, and economic landscapes does assure that meaningful regional differences will persist. Accordingly, the current areal arrangement of the United States and Canada is presented within a framework of eight broad regions (Fig. 3–28). Each will be briefly treated, reviewing in the process much of the material that has been discussed in this chapter. We will then conclude Chapter 3 with an overview of a provocative new regionalization scheme based on the hypothesis that the realm is now composed of a mosaic of nine diversifying "nations."

## 1. The Anglo-American Core

This continental core area (Fig. 3–12)—synonymous with the American Manufacturing Belt—has been discussed earlier. Serving as the historic workshop for the closely linked spatial economies of the United States and Canada, this region was the unquestioned leader and centerpiece during the century

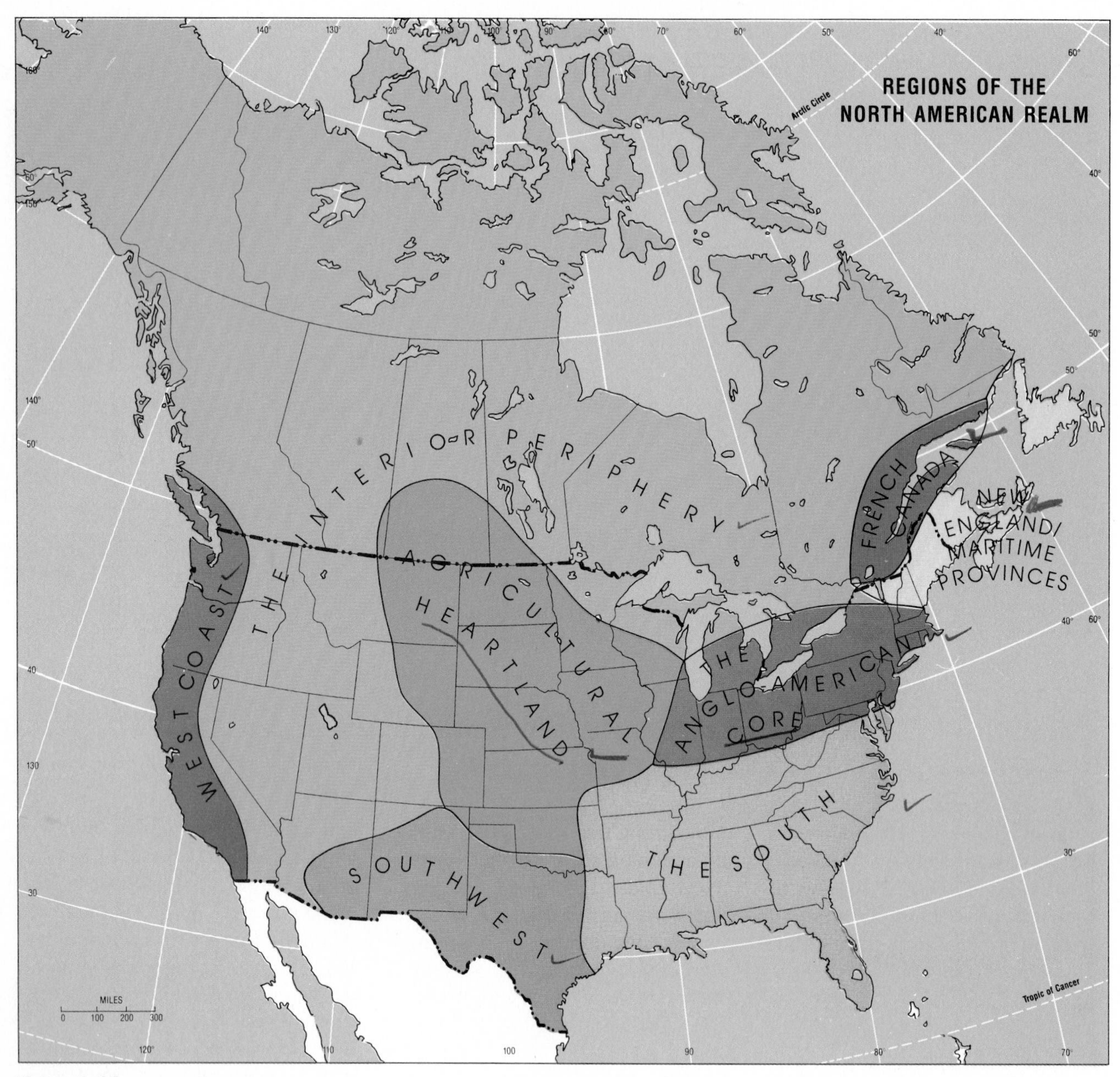


Figure 3–28

between the Civil War and the close of the industrial age (1865–1970). With the coming of the post-industrial revolution, however, that linchpin regional role is unraveling today as the Anglo-American Core is forced to share a growing number of major functions with fast-rising areas to the west and south. Unfortunately for the Manufacturing Belt, the sharp decline in blue-collar employment is not being accompanied by the creation of sufficient replacement jobs in other sectors. Nor is any real effort being made to retrain the thousands of displaced factory workers who will probably never find another job in heavy industry, and the region now faces the distinct possibility of becoming home to an ever-larger group of the permanently unemployed; such problems are most apparent in Chicago (8.1 million), Detroit (4.7 million), Philadelphia (5.8 million), Pittsburgh (2.4 million), Cleveland (2.8 million), and Baltimore (2.3 million). Manufacturing will remain a highly important activity within the American economy, but the productivity and obsolescence problems that sharply elevate production costs in the core area present staggering obstacles in an era when its traditional industries are increasingly able to respond to more footloose locational forces that operate in favor of other regions. To remain competitive, the key factor for the Manufacturing Belt in the immediate future is how much new investment capital it can attract. That largely depends on the region forging a new (and diminished) role for itself in a wholly new age, something it has been reluctant to do because its leaders still view the Belt as the permanent headquarters of the continent and the exodus of talented people and economic resources as a temporary aberration.

Irreversible manufacturing decline is widespread and has already produced a lasting impact, but there are many bright spots left within the Core. The major metropolitan complexes of Megalopolis—already the scene of much quaternary and quinary economic activity—are adjusting well. New York City (7.0 million) remains the national leader in finance and advertising, and it houses the broadcast media; yet, even here, the city increasingly shares its once-exclusive decision making leadership with other places—notably its own outer suburban city (10.7 million)—which, in 1978, surpassed Manhattan in the total number of Fortune-1000 corporate headquarters' facilities. In fact, these outer cities easily constitute the healthiest category of economic subareas throughout the Anglo-American Core, containing thriving minicities, office campuses, and industrial park complexes. But, as these new economic centers succeed within the framework of the multinodal metropolis of realms (Fig. 3–17), they provide an increasingly stark contrast to nearby dying industrial areas, which are littered with closed factories, bankrupt businesses, and struggling blue-collar neighborhoods. The latter are overwhelmingly concentrated in the inner portions of big central cities, the gritty smaller cities that line the freight-rail tracks between major metropolises, and certain aging inner railroad suburbs that were always a social world apart from their affluent commuter-suburb neighbors.

Perhaps the core-area metropolis that has gained the most from the emergence of postindustrialism is Washington (3.5 million). As the information and control-function sectors have blossomed, and as the U.S. federal government extends its connections ever deeper into America's business operations, the District of Columbia (620,000), together with its surrounding outer city of hyperaffluent Maryland-and-Virginia suburbs (2.9 million), has amassed an enormous complex of office, research, trade organization, lobbying, and consultant firms (Fig. 3–1). Suburbanization of facilities has been heightened by a lack of space in the District's center, avoidance of its adjacent sprawling black ghettoes, and particularly the lure of the Capital Beltway—a 106-kilometer (66-mile), eight-lane freeway that encircles Washington and connects the most prestigious suburbs. The federal government has been very active in the suburbs as well, decentralizing so many of its operations that it now employs more workers in the D.C. suburbs than inside the capital city.

## 2. New England/ Maritime Canada

New England, one of the continent's historic culture hearths, has retained a powerful regional identity for well over 300 years. Although the urbanized southern half of New England has been the northeastern anchor of the Anglo-American Core since the mid-nineteenth century—where it must rightfully be classified—its six states (Maine, New Hampshire, Vermont, Rhode Island, Massachusetts, and Connecticut) still share many common characteristics. Besides this overlap with the Manufacturing Belt (and Megalopolis) in its south, the New England region also extends northeastward across the Canadian border to encompass the three Maritime Provinces of New Brunswick, Nova Scotia, and Prince Edward Island and the outer island of Newfoundland.

A long association based on economic and cultural similarities has tied upper New England to Maritime Canada. Both are rural in character, possess rather difficult environments in which land resources are limited, and were historically bypassed in tavor of more fertile inland areas. Thus, economic growth here has always lagged behind the rest of the realm, and the region remains relatively disadvantaged. Development has centered on primary activities, mainly fishing the rich offshore banks of the nearby North Atlantic, forestry in the uplands, and farming in the few fertile valleys that have materialized. Recreation and tourism have boosted the regional economy in recent times, with the scenic beauty of rocky coasts and low mountains attracting millions from the neighboring core region; the heightened popularity of skiing has particularly helped, extending the tourist season through the harsh winter months. A rich sense of history and tradition permeates the close cultural affinity of New England and the Maritimes. Both possess homogeneous English-based cultures that have long withstood persistent Francophone incursions from adjacent Quebec to the northwest, and both have given rise to a staunchly self-sufficient, pragmatic, and conservative population. Village settlement is overwhelmingly preferred—a sharp contrast to the dispersed rural population of the continental interior that clings to its individual farmsteads. In fact, the fiercely independent small-town governments of New England are still regarded as the true citadel of American-style democracy, and the charming village landscape itself (Fig. 3–29) as an idealized residential community.

Northern New England, after two centuries of struggling, is fi-

**Figure 3–29**
New England's townscape: Kennebunk in southern Maine.

nally showing signs of significant development. Long an area of out-migration, population has increased substantially here since 1960, as newly dominant noneconomic locational forces have permitted more and more employers to move their firms to the high-amenity vicinities of the Green and White Mountains. Current growth is also spurred by the improved accessibility of the region via its all-weather interstate freeways, exurban spillover (especially in southern New Hampshire, where local taxes are much lower than in neighboring "Taxachusetts"), and the proliferation of second homes in the country within a half-day's drive of metropolitan Boston (4.0 million).

## 3. French Canada

Francophone Canada comprises the effectively settled (southern) portion of the province of Quebec, straddling the lower St. Lawrence Valley from where the river crosses the Ontario-Quebec boundary just upstream from Montreal to its mouth beyond the Gaspé Peninsula in the Gulf of St. Lawrence; also included are sizable concentrations of French speakers who reside just across the provincial border in New Brunswick and the U.S. border in the potato-farming district of northernmost Maine. Significantly, this is the only North American region that is defined by culture alone, but the reasons are compelling. More than 80 percent of French Canada's population speaks French as a mother tongue and over 85 percent adhere to the Roman Catholic faith. About three-quarters of the English speakers (many of whom use French as a second language) are clustered in the Montreal metropolitan area (3.2 million), which is one of the strictest bilingual cities on earth (Fig. 3–30). The old-world charm of Quebec's cities is matched by an equally unique rural cultural landscape—narrow rectangular farms known as "long lots" are laid out in sequence perpendicular to the St. Lawrence and other rivers (allowing each farm access to the waterway), a settlement system the French also introduced to colonial Louisiana, where it still survives along the lower Mississippi.

The economy of French Canada is no longer a rural one (although dairying remains a leading agricultural pursuit), exhibiting urbanization rates equivalent to the rest of the country. Industrialization is widespread, and the region produces much of Canada's clothing, cigarettes, pulp and paper, aircraft, and particularly aluminum, whose refining requires a vast amount of hydroelectric energy, which is generated cheaply at huge dams in northern Quebec. Tertiary and postindustrial commercial activities are centered in Montreal, but the recent surge of separatist activity has intensified the migration of firms to Toronto—the city 485 kilometers (300) miles to the southwest, which shares the national primacy position. Tourism and recreation are also important, annually attracting millions from the nearby continental core area and abroad.

As was discussed near the beginning of this chapter, heightened nationalism in Quebec during the 1970s has raised the specter of that province's secession from the confederation, which would be a disaster for Canada. Campaigning on a separatist platform, the *Parti Québécois* was elected in 1976 to run the provincial government. Following enactment of a series of new laws that strengthened the French language and culture within the province, the secession issue was placed before the electorate in a special referendum on May 20, 1980. Separatism was soundly defeated by a 60-to-40-percent vote, defusing the issue for the time be-

**Figure 3–30**
Bilingualism rigorously observed in Montreal.

ing. Although Québécois nationalism could arise again in the future, the feeling in the mid-1980s is that separatism is an increasingly outdated method for protesting an English domination that no longer threatens to submerge French Canadian culture on its home turf.

## 4. The Agricultural Heartland

The miracle of American food production was described earlier, and it is here in the heart of the continent that agriculture becomes the predominant feature of the landscape. Whereas the innermost fruit-and-vegetable and dairying belts in the macro-Thünian regional system (Fig. 3–25) are in competition with other economic activities in the Anglo-American Core, by the time one reaches the Mississippi Valley, meat and grain production prevail for almost 1000 miles west to the base of the Rocky Mountains. As will be recalled, this portion of North America possesses abundant supplies of some of the best soils on this planet. Because the eastern half of the Agricultural Heartland lies in Humid America and closer to the national food market on the northeastern seaboard, mixed crop-and-livestock farming wins out over less competitive wheat raising, which is relegated to the fertile but semiarid environment of the central and western Great Plains on the dry side of the 100th meridian. The latter area also contains Canada's agricultural heartland north of the 49th parallel border; however, although these fertile prairie provinces are somewhat less subject to serious dry spells than the U.S. high plains to the south, they are situated at a higher latitude, which makes for a shorter growing season.

The distribution of North American corn and wheat production is shown in Fig. 3–31, which rather neatly defines the boundaries of the Heartland region. The Corn Belt to the east is focused on Illinois and Iowa, with extensions into neighboring states (see Fig. 3–25). The area of north-central Illinois, which displays a major cluster of corn farming, is a classic example of transition along a regional boundary: the Agricultural Heartland/Anglo-American Core dividing line connecting Milwaukee (1.6 million) and St. Louis (2.4 million) (Fig. 3–28) passes right through this zone, which contains a number of sizable urban-industrial concentrations interspersed by some of the most productive corn-producing counties in the nation. On its western margins, the Corn Belt quickly gives way to the dryland Wheat Belts, whose bifurcated presence has already been explained. The growth of foreign wheat sales in recent years has prompted the intensification of production in the Winter Wheat Belt through the use of center-pivot irrigation (Fig. 3–6), a costly but so far profitable means of overcoming the moisture deficit as long as groundwater supplies hold up.

Throughout the Heartland region, nearly everything is oriented to agriculture. Its major cities—Kansas City (1.5 million), Minneapolis (2.3 million), Winnipeg (560,000), Omaha (735,000), and even the Denver area (1.9 million)—are main processing and marketing points for beef packing, flour milling, and pork production. People in this region are generally of northern European ancestry, and conservative. Yet, rapid technological advances require that farmers keep abreast of increasingly scientific agricultural techniques and business methods if they are to survive in a hypercompetitive atmosphere. The challenges and opportunities in today's ever-more-sophisticated farming may also be persuading more of the region's young people to stay here—traditionally high outmigration rates have leveled off since the 1960s.

## 5. The South

The American South is by far the most rapidly changing of the realm's eight regions. Choosing to

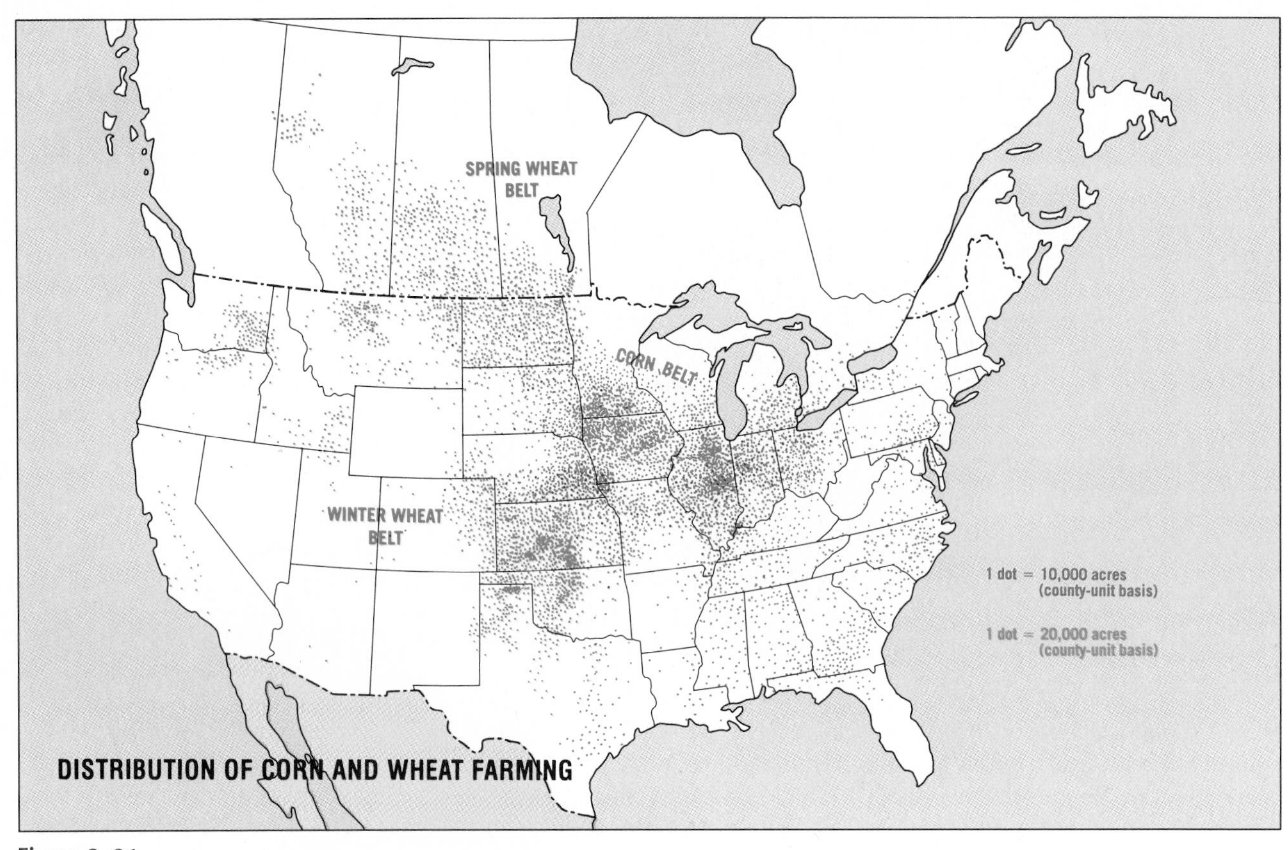


Figure 3–31

pursue its own sectional interests practically from the advent of nationhood, this region's economic and cultural isolation from the rest of the United States deepened during the long and bitter aftermath of the ruinous Civil War. For over a century, the South languished in economic stagnation, until the 1970s witnessed so stunning a reversal in the nation's perception of the region that a Georgian was elected President in 1976. This reassessment was accompanied by a wave of growth and change unparalleled in the region's history that persists into the 1980s. Propelled by the forces that created the Sunbelt phenomenon, people and activities began to stream into the South and launched widespread urbanization; cities such as Atlanta (2.4 million), Houston (4 million), Miami (1.9 million), and New Orleans (1.4 million) spawned booming metropolises practically overnight, and conurbations swiftly formed in such places as southern and central Florida, the Carolina Piedmont, and along the Gulf coast from Houston to Mobile, Alabama. This "bulldozer revolution" was matched in the more favored rural areas by an agrarian renaissance: replacing the defunct Cotton Belt was a new agricultural complex that stressed the production of such high-value commodities as beef, soybeans, poultry, and lumber. And on the social front, institutionalized racial segregation was dismantled; although certain problems in minority relations persist, inequalities in the South today are generally no worse than in other parts of the country (in fact, the mid-1980s saw black mayors in Atlanta, Birmingham [910,000], and New Orleans; a woman in Houston; and a Hispanic in Miami).

Yet, for all the growth that has taken place since the 1960s, the South remains a region beset by many economic problems because the geography of its development has been quite uneven. Although certain urban and farming areas have benefited, many others containing large populations have not; even counties adjacent to boom areas have often been left unaffected—as spillover effects have been disappointing overall—and the juxtaposition of progress and backwardness is frequently encountered in the Southern landscape (Fig. 3–32). The geographic selectivity of nonagricultural growth is noteworthy in another regard, because—with the exception of the centrally located Atlanta metropolis—the most important development has occurred near the region's periphery—in the Washington suburbs of northern Virginia, the North Carolina Piedmont, the Houston area, and Florida's central and southern-coast postindustrial frontiers (which include urban Miami's Latin-American-gateway function). As the checkerboard pattern of development intensifies, even places within boom areas may be falling behind: this is especially true of large Southern central cities, whose prospects appear to be swiftly converging with their northern counterparts as burgeoning outer suburban cities around Atlanta, Memphis (950,000), Houston, Miami, Orlando (875,000), and New Orleans capture an increasing majority of metropolitanwide employment activity. Moreover, outside of the few high-tech areas in North Carolina, Florida, and Texas, the regionwide emphasis on attracting traditional manufacturing industry has resulted in a slowdown of the 1970s growth spurt; Southern cost-of-living advantages are being leveled, narrowed regional freight-rate differences enable the shrinking pool of mobile industries to relocate to other parts of the nation, and the resources that lure postindustrial facilities are not being developed fast enough.

In retrospect, Southern development over the past two decades has produced much beneficial change and has helped to remove the region's long-standing inferiority complex. But one must also keep in mind that the South had a long way to go, and much of it still lags behind the times. One of its foremost students, the historian C. Vann Woodward, offers this evaluation of the "New South":

> *The old Southern distinction of being a people of poverty among a people of plenty lingers on. There is little prospect of closing the gap overnight.*

**Figure 3–32**
Old and new often contrast sharply in today's South: central Atlanta.

## 6. The Southwest

As recently as the 1960s, North America geography textbooks did not identify a distinct southwestern region, classifying the Mexican borderland zone into separate southward extensions of the Great Plains, Rocky Mountains, and Intermontane Plateaus. Today, however, this area must be recognized as a major regional entity, albeit the youngest of this realm (and quite possibly of the world). The emerging Southwest is also unique in the United States, because it is a bicultural regional complex peopled by in-migrating Anglo-Americans atop the crest of the Sunbelt wave as well as by the quickly expanding Mexican-American population anchored to the sizable long-time resident Hispanic group, which traces its origins to the Spanish colonial territory that once extended from the Texas Gulf coast to San Francisco. In fact, if one counts the large native American population—which is overwhelmingly concentrated on isolated rural Indian reservations—then, the Southwest is actually a *tricultural* region. Recent rapid development in Texas, New

Mexico, and Arizona is essentially built on a three-pronged foundation: (1) availability of huge amounts of generated *electricity* to power air conditioners through the long, brutally hot summers of this semiarid and arid climate; (2) sufficient *water* to supply everything from swimming pools to irrigated crops, so that large numbers may reside in this dry environment; and (3) the *automobile*, so that affluent newcomers may spread themselves out at much-desired low densities, yet overcome the distances engendered by such a settlement pattern (the aerial photograph of the Phoenix metropolis [population: 1.8 million] in Fig. 3–33 demonstrates the triumph of this way of life over the surrounding desert). The first and third of the foundation prongs have been rather easily attained, since the eastern flank of the Southwest is abundantly endowed with oil and natural gas (Fig. 3–22). The future of water supplies, however, is a far more worrisome dilemma, particularly if present population-growth rates were to continue beyond the 1980s.

So far, the generation of new wealth in this most rapidly growing North American region makes it an undeniable success. However, since much of this achievement is linked to the fortunes of the oil business, the Southwest may well be entering a future beset by unpredictable economic ups and downs if one uses as benchmarks the sudden swings of the energy industry since 1973. The postindustrial revolution is also prominently represented in this region—with a specialized activity complex located in every major metropolis and a world-class concentration of electronic and space-technology facilities in the eastern Texas triangle formed by connecting Houston, San Antonio (1.3 million), and Dallas-Fort Worth (3.4 million); indeed, the state capital of Austin (450,000) near its center is becoming one of the leading international high-technology research complexes in a partnership between private manufacturers and the main campus of the University of Texas.

## 7. The Interior Periphery

Despite its apparent contradiction, the Interior Periphery is an appropriate name for North America's largest region by far—it covers most of Canada, all of Alaska, the northern salients of Minnesota-Wisconsin-Michigan, New York's Adirondacks, and the inland American west between (and including) the Sierra Nevada-Cascades and the Rockies. There can be no doubt about the interior position of this vast slice of the North American realm; its peripheral nature stems from its isolation and rugged environment, which have attracted only the sparsest of populations relative to the other seven regions—in the U.S. portion, even after counting notable recent growth, population density is only 5 per square kilometer (12 per square mile) in contrast to 25/64 overall. Yet, its disadvantages notwithstanding, the Interior Periphery contains great riches, because it is one of the earth's major storehouses of mineral and energy resources.

Accordingly, the region's history has been a frustrating one of boom-and-bust cycles, determined by technological developments, economic fluctuations, and corporate and government decisions made in the Anglo-American Core. The events of the 1975–1985 period exquisitely demonstrate this bittersweet development process. The decade opened with a headlong rush in the search for additional petroleum and natural gas to avert future energy crises, with efforts concentrating on the Alaskan North Slope (Fig. 3–23) and southwestern Wyoming's Overthrust Belt; the latter area and neighboring Colorado were also invaded by energy companies seeking to perfect methods for obtaining gasoline from huge local oil shale deposits. With the inauguration of the Trans-Alaska Pipeline in 1977, optimism reached an all-time high, and western Wyoming and Colorado were overrun by tens of thousands of in-migrants who taxed local community facili-

**Figure 3–33**
Arizona's booming Phoenix: spread city in the dry Southwest.

ties to the breaking point. But, as has happened so many times before, the boom here fizzled out with a bewildering suddenness. Everything from oilfield home building to the wave of skyscraper construction in nearby Denver was predicated on a continuing rise in the price of oil. Therefore, when that price unexpectedly leveled off in the early 1980s, the repercussions were swift and painful. Oil companies halted much of their exploration activity—most notably Exxon's abandonment of its large Colony Oil Shale project in 1982—and the prospering newcomers found themselves faced with unemployment in a part of the country that offered few other opportunities. Elsewhere in the region, the simultaneously occurring Great Recession reinforced the negative impact on the mining industry: nearly all of northeastern Minnesota's Iron Range closed down (perhaps permanently) and once-productive copper mines in Montana, Utah, and Arizona faced the widest slowdown in this century. These transitory difficulties aside, the future still beckons brightly. Rocky Mountain coal- and uranium-extraction operations were relatively untouched by the economic misfortunes of the early 1980s, and the mining of silver, lead, zinc, and nickel persisted at scattered sites throughout western North America and on the Canadian Shield to the east. And, surely, with the steady decline of fossil fuels worldwide, oil- and gas-drillers will return in force to the central Rockies in the not-too-distant future.

The recent sharp swings of the mining economy also mask a steadier influx of population and nonprimary activities to other parts of the Interior Periphery. In fact, from 1970 to 1983, the U.S. segment grew from 8.3 to 12.4 million, thereby advancing its relative position from 4 to 6 percent of the national population. The guiding force, once again, was the search for high-amenity locations, with the clean, stress-free wide-open spaces of certain parts of the Rockies and Intermontane Plateaus a particular attraction. Accordingly, Nevada, Alaska, Arizona, Utah, Wyoming, Colorado, and Idaho all ranked among the 10 fastest-growing states in the 1970s, and that trend carries over into the 1980s. Not all of this growth has been welcomed, by the way, since many long-time residents decry the urbanization, pollution, and new pressures on limited water resources that are changing forever the character of a conservative region that relishes its detachment from the hectic pace of life elsewhere in the realm.

## 8. The West Coast

The Pacific Coast of the conterminous United States and southwestern Canada has been a powerful lure to migrants since the Oregon Trail was pioneered 150 years ago. Unlike the remainder of the North American west, the narrow strip of land between the Sierra/Cascade mountain wall and the sea receives adequate moisture; it also possesses a far more hospitable environment with generally delightful weather south of San Francisco, highly productive farmlands in California's Central Valley, and such scenic glories as the Big Sur coast, Washington's Olympic Peninsula, and the spectacular waters surrounding San Francisco (4.2 million), Seattle (2.4 million), and Vancouver (1.5 million). Most major development here took place during the post-World War II era, accommodating enormous population and economic growth, and the West Coast is just now beginning to face the less pleasant consequences of regional maturity.

Forty years of unremitting growth have especially taken their toll in California, because the massive development of America's most populated state has been overwhelmingly concentrated in the "San-San" conurbation—the teeming metropolitan corridor extending south from San Francisco through San Jose (2.2 million), the San Joaquin Valley, the Los Angeles Basin, and the southwestern coast into San Diego (2.1 million) on the Mexican border. Environmental hazards threaten this entire strip, including inland droughts, coastal flooding, mudslides, brush fires, and earthquakes—with the ominous San Andreas fault practically the axis of megalopolitan coalescence. To all this, humans have added their own abuses of the fragile natural habitat, from overuse of water supplies (requiring vast aqueduct systems to import water from hundreds of miles away) to the incredible air pollution of Los Angeles caused by the emissions of 4 million automobiles, which are vital to movement within this widely dispersed metropolis (population: 12.5 million). Lost somewhere in the shuffle is yesterday's glamorous image of Southern California—communicated so effectively in the motion pictures of the postwar period—that is based on relaxed outdoor living in luxurious horticultural suburbs amid one of the most agreeable climates to be found anywhere on earth. Undeniably, economic development has brought California new prosperity, and the state continues its leadership as a national innovator—most recently in the technological marvels emanating from Silicon Valley beside San Francisco Bay (Fig. 3–15). But much of this affluence has also been subject to sudden shifts, because the Southern California economy is overly tied to the aerospace industry, which continues to experience sharp cyclical fluctuations. Clearly, the Golden State today is a changed place from what it was in the recent past: the years of innocence and unbridled optimism are over and have been replaced by a new problem-ridden era, in which the magic has worn off and the future appears increasingly as a struggle to maintain the state's still-considerable advantages in the face of growing competition from would-be new Californias elsewhere on the continent.

The upper portion of the West Coast region is the Pacific Northwest, focused on Oregon's Willamette Valley, the Cowlitz-Puget Sound Lowland of western Washington, and the British Columbia coast of southwesternmost Canada. Originally built on timber and fishing—primary activities that still thrive here—the impetus for industrialization came from the massive Columbia River dam projects of the 1930s and 1950s that created cheap hydroelectricity. This, in turn, attracted aluminum and aircraft manufacturers, and the huge Boeing aerospace complex around Seattle makes that metropolis one of the world's biggest company towns. Unique environmental amenities—zealously safeguarded here—have lured hundreds of growth companies, and the Pacific Northwest should have little trouble accommodating to the postindustrial economy. Perhaps one of its greatest advantages, which is shared with urban California to the south, is as a gateway to the emerging Pacific Basin. As North America is increasingly enmeshed in the global-scale economy, that part of the world will be a critically important marketplace, and Seattle (as well as Vancouver in Canada) is bound to benefit, because its ultramodern airport is the closest to Tokyo of any in the conterminous United States.

# The Emerging "Nine Nations" of North America

At the outset of our regional survey, we pointed out that the American-Canadian regional system was in the throes of change, a dynamism highlighted in the profile of each region. An important overview of the direction and significance of all this change is contained in *Washington Post* editor Joel Garreau's 1981 book, *The Nine Nations of North America*, which (as mentioned earlier) has caught the fancy of the American public. His central thesis is that the realm is rapidly reorganizing into nine separate "nations":

> *Each has its capital and its distinctive web of power and influence. A few are allies, but many are adversaries. . . . Some are close to being raw frontiers; others have four centuries of history. Each has a peculiar economy; each commands a certain emotional allegiance from its citizens. These nations look different, feel different, and sound different from each other, and few of their boundaries match the political lines drawn on current maps. . . . Each nation has its own list of desires. Each nation knows how it plans to get what it needs from whoever's got it. Most important, each nation has a distinct prism through which it views the world.*

Garreau's regionalization scheme is mapped in Fig. 3–34, and it covers not only the United States and Canada, but Mexico and the Caribbean Basin as well. Briefly reviewing the essence of each "nation," *The Foundry* is dominated by a declining industrial infrastructure of aging gritty cities and faces a bleak future as its human and economic resources relocate to areas possessing lower production costs. *New England*, which has survived innumerable economic crises, is turning local amenities, outstanding educational resources, and a much-admired regional tradition into a haven for new growth activities. *Quebec* is asserting its cultural identity as never before and gaining new leverage within Canada. *Dixie* is now permeated with change, but the "Old South" is not being replaced by a clear regional personality and sense of direction. *The Islands* is the newest nation, built on growing economic ties between Miami and nearby Latin America, which are at least in part based on illicit drug flows. *The Breadbasket*, the "nation that works best," is characterized by a prospering agriculture and a conservative population at peace with itself. *The Empty Quarter* is endowed with an abundance of energy resources that will lead to much albeit scattered development. *Ecotopia* is the preserve of environmentalists who guard the *status quo* but, ironically, it is also pioneering the new technologies of the postindustrial future. And *MexAmerica* is swiftly being shaped into a distinctive region by its burgeoning Hispanic populations, who are helping to generate new wealth in the border zone from South Texas clear across to central California.

The geographic pattern of this regionalization scheme invites comparison with ours (Fig. 3–28). Certainly there are many elements in common, and it is quite possible that our eight regions are evolving into something resembling Garreau's spatial classification of the continent. In fact, the only "nation" in our scheme that is "missing" is *The Islands*, which could well be in a formative stage (as residents of South Florida, we acknowledge that this area is becoming ever more different from the rest of the South and is already the

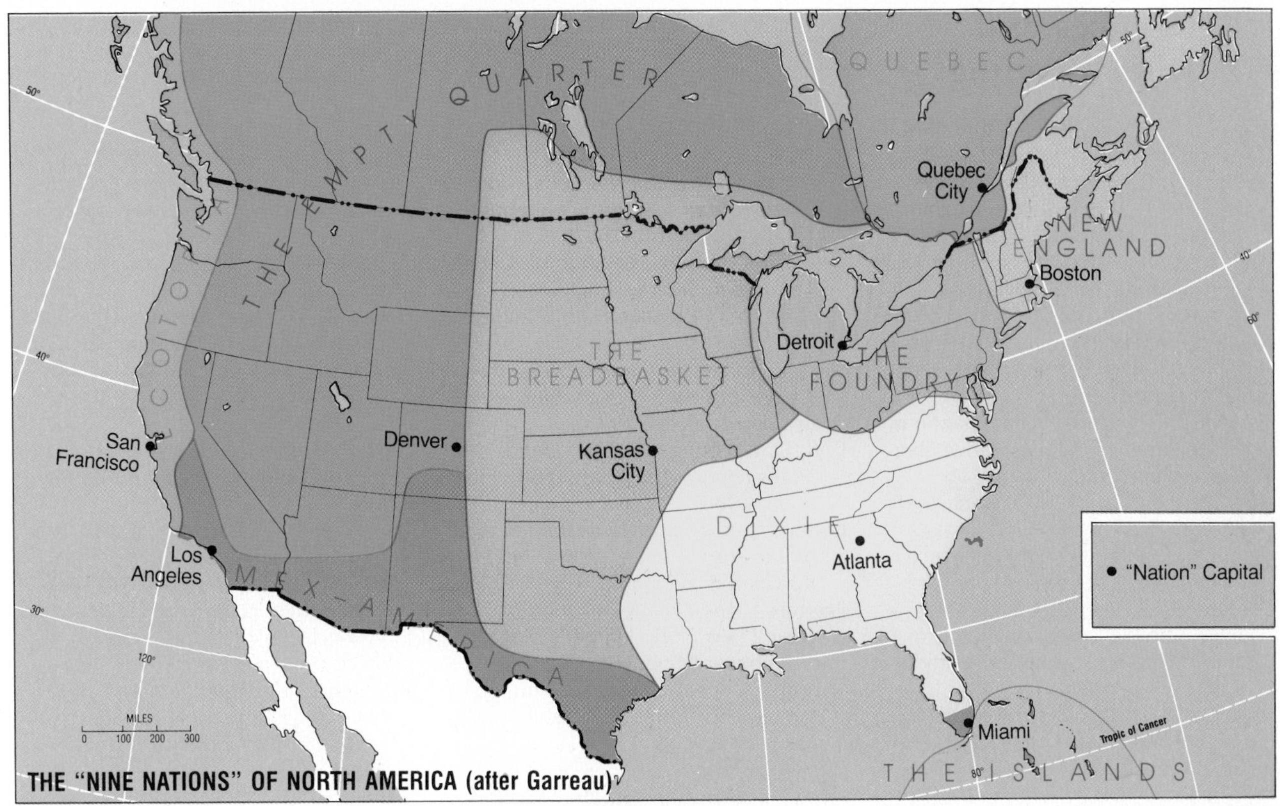


Figure 3–34



financial capital of the Caribbean). Proceeding across the map, the Anglo-American Core is somewhat more rectangular than the square-shaped Foundry, incorporating the important southern Illinois coalfields and St. Louis manufacturing complex and taking in all of Megalopolis in the northeast (Garreau does not specifically treat urbanization in his analysis and thereby misses some meaningful subnational trends). Our New England/Maritimes region is, therefore, somewhat smaller, as is French Canada, which Garreau insists on treating as the political entity of Quebec (which contains vast empty spaces to the north and misses significant Francophone minorities in Maine and New Brunswick). The South and Dixie concur closely, differing only slightly around the edges and on the matter of the incipient Islands "nation." Agricultural Heartland and Breadbasket also agree about regional cores, with variations again observed at the fringes: actually, Garreau may well have a stronger case vis-à-vis the Breadbasket/Empty Quarter boundary in Wyoming and Montana—coal strip-mining is quickly becoming the leading activity just to the west of that dividing line, which offsets places where "hydrocarbons become more important than carbohydrates." The Interior Periphery and Empty Quarter are similarly in accord, varying only in the minor matters of Quebec, the upper Great Lakes area, and the southern and western margins. The West Coast region does not separate Ecotopia from Mex-America, but it does recognize internal California differences, which could, indeed, lead to a full regional split in the near future. Finally, the ripening Southwest was viewed as an increasingly robust region containing a decidedly Hispanic flavor, but we have reservations about projecting it west to the Pacific, because the Southern California international-banking and high-tech boom—particularly in burgeoning suburban Orange County (2.3 million) between Los Angeles and San Diego—is more an expression of West Coast economic geography than the assertiveness of an up-and-coming ethnic group.

As we already know from previous chapters, no two regionalization schemes will ever be exactly the same. The differences here are fairly minor ones, based on superficial form rather than supporting logic. Only one thing is certain: North America's dynamic regions will continue to change, in some cases quite rapidly. The *nine nations* hypothesis is a valuable and thought-provoking contribution to the fascinating business of monitoring the geographical restructuring of this remarkable realm, and could well prove to be correct in its central arguments. Time will tell.

## References and Further Readings

Adams, J. "Residential Structure of Midwestern Cities," *Annals of the Association of American Geographers*, 60 (1970), 37–62. Model diagram adapted from p. 56.

Adams, J., ed. *Contemporary Metropolitan America* (Cambridge: Ballinger, 4 vols., 1976).

Atwood, W. *The Physiographic Provinces of North America* (New York: Ginn, 1940).

Baerwald, T. "Urban Transportation in North America," *Focus*, November-December, 1983.

Bell, D. *The Coming of Postindustrial Society* (New York: Basic Books, 1973). Quotation taken from p. xvi of the 1976 [paperback] Foreword.

Bernard, R. & Rice, B., eds. *Sunbelt Cities: Politics and Growth Since World War II* (Austin: University of Texas Press, 1983).

Berry, B. J. L. "The Decline of the Aging Metropolis: Cultural Bases and Social Process," in Sternlieb, G. & Hughes, J., eds., *Post-Industrial America: Metropolitan Decline and Inter-Regional Job Shifts* (New Brunswick, N.J.: Center for Urban Policy Research, Rutgers University, 1975), pp. 175–185.

Birdsall, S. & Florin, J. *Regional Landscapes of the United States and Canada* (New York: John Wiley & Sons, 3 rev. ed., 1985).

Borchert, J. "American Metropolitan Evolution," *Geographical Review*, 57 (1967), 301–332.

Bourne, L., ed. *Internal Structure of the City: Readings on Urban Form, Growth, and Policy* (New York: Oxford University Press, 2 rev. ed., 1982).

Broek, J. *Geography: Its Scope and Spirit* (Columbus: Charles E. Merrill, 1965). Quotation taken from p. 72.

Brown, R. *Historical Geography of the United States* (New York: Harcourt, Brace & World, 1948).

Browning, C., ed. *Population and Urbanized Area Growth in Megalopolis, 1950–1970* (Chapel Hill: University of North Carolina, Studies in Geography No. 7, 1974).

Brunn, S. *Geography and Politics in America* (New York: Harper & Row, 1974).

Brunn, S. & Wheeler, J., eds. *The American Metropolitan System: Present and Future* (New York: Halsted Press/V. H. Winston, 1980).

Christaller, W. *Central Places in Southern Germany*, [Original title: *Die Zentralen Orte in Süddeutschland*, 1933.] (Englewood Cliffs, N.J.: Prentice-Hall, trans. C. Baskin, 1966).

Christian, C. & Harper, R., eds. *Modern Metropolitan Systems* (Columbus: Charles E. Merrill, 1982).

Chudacoff, H. *The Evolution of American Urban Society* (Englewood Cliffs, N.J.: Prentice-Hall, 2 rev. ed., 1981).

Clay, G. *Close-Up: How to Read the American City* (Chicago: University of Chicago Press, 1973).

Cuff, D. & Young, W. *The United States Energy Atlas* (New York: Free Press, 1980).

Cutter, S., et al. *The Geography of Natural Resource Use* (Totowa, N.J.: Rowman & Allanheld, 1984).

Egerton, J. *The Americanization of Dixie* (New York: Harper's Magazine Press, 1974).

"Energy: A Special Report in the Public Interest," *National Geographic*, February 1981, 1–114.

Garreau, J. *The Nine Nations of North America* (Boston: Houghton Mifflin, 1981). Quotation taken from pp. 1–2.

Gastil, R. *Cultural Regions of the United States* (Seattle: University of Washington Press, 1975).

Gottmann, J. *Megalopolis: The Urbanized Northeastern Seaboard of the United States* (New York: Twentieth Century Fund, 1961).

Gould, P. & White, R. *Mental Maps* (Baltimore: Penguin, 1974).

Harris, C. D. & Ullman, E. L. "The Nature of Cities," *Annals of the American Academy of Political and Social Science*, 242 (1945), 7–17.

Hart, J. F. *The Look of the Land* (Englewood Cliffs, N.J.: Prentice-Hall, 1975).

Hart, J. F., ed. *Regions of the United States* (New York: Harper & Row, 1972). Reprint of the June 1972 issue of the *Annals of the Association of American Geographers.*

Hartshorn, T. *Interpreting the City: An Urban Geography* (New York: John Wiley & Sons, 2 rev. ed., 1986).

Hunt, C. *Natural Regions of the United States and Canada* (San Francisco: W. H. Freeman, 2 rev. ed., 1974).

Janelle, D. "Spatial Reorganization: A Model and Concept," *Annals of the Association of American Geographers*, 59 (1969), 348–364.

Lewis, P. "The Galactic Metropolis," in Platt, R. & Macinko, G., eds., *Beyond the Urban Fringe: Land-Use Issues of Nonmetropolitan America* (Minneapolis: University of Minnesota Press, 1983a), pp. 23–49. Quotation taken from p. 34.

Lewis, P. "The Geographic Roots of America's Urban Troubles," *Earth and Mineral Sciences Newsletter* (Pennsylvania State University), Spring 1983b. Quotation taken from p. 27.

Lewis, P. "Learning from Looking: Geographic and Other Writing About the American Cultural Landscape," *American Quarterly*, 35 (1983c), 242–261.

Louv, R. *America II* (Los Angeles: Jeremy Tarcher/Houghton Mifflin, 1983).

Mayer, H. M. "Geography in City and Regional Planning," in Frazier, J., ed., *Applied Geography: Selected Perspectives* (Englewood Cliffs, N.J.: Prentice-Hall, 1982). Quotation taken from p. 27.

McCann, L., ed. *Heartland and Hinterland: A Geography of Canada* (Scarborough, Ont.: Prentice-Hall Canada, 1982).

Meinig, D. "Symbolic Landscapes: Some Idealizations of American Communities," in Meinig, D., ed., *The Interpretation of Ordinary Landscapes: Geographical Essays* (New York: Oxford University Press, 1979), pp. 164–192.

Muller, P. *Contemporary Suburban America* (Englewood Cliffs, N. J.: Prentice-Hall, 1981).

*National Atlas of the United States of America* (Washington: U.S. Geological Survey, 1970).

Noyelle, T. & Stanback, T., *The Economic Transformation of American Cities* (Totowa, N.J.: Rowman & Allanheld, 1984).

*Oxford Regional Economic Atlas of the United States and Canada* (New York: Oxford University Press, 2 rev. ed., 1975).

Palm, R. *The Geography of American Cities* (New York: Oxford University Press, 1981).

Paterson, J. *North America: A Geography of Canada and the United States* (New York: Oxford University Press, 7 rev. ed., 1984).

Phillips, P. & Brunn, S. "Slow Growth: A New Epoch of American Metropolitan Evolution," *Geographical Review*, 68 (1978), 274–292.

Putnam, D. & Putnam, R. *Canada: A Regional Analysis* (Toronto: Dent, 2 rev. ed., 1979).

Rooney, J., et al., eds. *This Remarkable*

*Continent: An Atlas of United States and Canadian Society and Cultures* (College Station: Texas A&M University Press, 1982).

Saxinen, A. "The Urban Contradictions of Silicon Valley," in Sowers, L. & Tabb, W., eds., *Sunbelt/Snowbelt: Urban Development and Regional Restructuring* (New York: Oxford University Press, 1984), pp. 163–197.

Schertz, L. et al. *Another Revolution in U.S. Farming?* (Washington: U.S. Department of Agriculture, Agricultural Economic Report 441, 1979).

Stewart, G. *U.S. 40: Cross Section of the United States of America* (Boston: Houghton Mifflin, 1953). See also Vale, T. & Vale, G. *U.S. 40 Today: Thirty Years of Landscape Change in America* (Madison: University of Wisconsin Press, 1983).

Thernstrom, S., ed. *Harvard Encyclopedia of American Ethnic Groups* (Cambridge: Belknap-Harvard University Press, 1980).

Vance, J. E., Jr. *This Scene of Man: The Role and Structure of the City in the Geography of Western Civilization* (New York: Harper's College Press, 1977). Urban realms model discussed on pp. 411–416.

Ward, D. *Cities and Immigrants: A Geography of Change in Nineteenth Century America* (New York:, Oxford University Press, 1971).

Ward, D., ed. *Geographic Perspectives on America's Past* (New York: Oxford University Press, 1979).

Watson, J. W. *Social Geography of the United States* (New York: Longman, 1979).

Watson, J. W. & O'Riordan, T., eds. *The American Environment: Perceptions and Policies* (New York: John Wiley & Sons, 1976).

Wheeler, J. & Muller, P. *Economic Geography* (New York: John Wiley & Sons, 2 rev. ed., 1986). Map adapted from p. 315 of the 1st ed.

White, C., et al. *Regional Geography of Anglo-America* (Englewood Cliffs, N.J.: Prentice-Hall, 6 rev. ed., 1985).

Woodward, C. V. "The South Tomorrow," *Time*, The South Today—Special Issue, September 27, 1976. Quotation taken from p. 99.

Yeates, M. *Main Street: Windsor to Quebec City* (Toronto: Macmillan, 1975).

Yeates, M. *North American Urban Patterns* (New York: Halsted Press/ V. H. Winston, 1980).

Zelinsky, W. *The Cultural Geography of the United States* (Englewood Cliffs, N. J · Prentice-Hall, 1973).

Zelinsky, W. "North America's Vernacular Regions," *Annals of the Association of American Geographers*, 70 (1980), 1–16. Map adapted from p. 14.

# PRODIGIOUS JAPAN: The Aftermath of Empire

IDEAS AND CONCEPTS

**Modernization**
**Relative Location (2)**
**Exchange Economy**
**Industrial Location (2)**
**Areal Functional Organization**
**Physiologic Density**
**Postindustrial Revolution (2)**

In the non-Western, non-European world, there is no country quite like Japan. None of the metaphors of underdevelopment applies here: Japan is an industrial giant, a fully urbanized society, a political power, a vigorous nation. Probably no city in the world today is without Japanese cars in its streets; few photography stores lack Japanese cameras; many laboratories use Japanese optical equipment. From tape recorders to television sets, from oceangoing ships to children's electronic games, Japanese manufactures flood the world's markets.

The Japanese seem to combine the precision skills of the Swiss with the massive industrial power of prewar Germany, the forward-looking designs of the Swedes with the innovations of the Americans. How have the Japanese done it? Does Japan have the kind of resources and raw materials that helped boost Britain into its early, revolutionary industrial leadership? Are there locational advantages in Japan's position off the eastern coast of the Eurasian landmass, just as Britain benefited from its situation off the western coast? Could Japan's rise to power and prominence have been predicted, say, a century ago?

Japan consists of four large, mostly mountainous islands (plus numerous small islands and islets) whose total area amounts to a mere 377,000 square kilometers (146,000 square miles), a territory smaller than California. Japan's population of 121 million (1985) is mostly crowded onto the limited coastal plains, valleys, and lower slopes of these islands, principally on the largest (Honshu) and southernmost (Kyushu and Shikoku) (Fig. J–2). The northernmost island of Hokkaido still possesses frontier characteristics. And had it not been for Russian imperialism, Japan might also control Sakhalin Island, the fifth largest island of the Japanese archipelago.

Is Japan a discrete geographic realm or should it be viewed as an insular extension of the Chinese world? Japan was peopled from Asia mostly via the Korean Peninsula, and, over many centuries, it received numerous cultural infusions from the Asian mainland. By the sixteenth century a highly individualistic culture already had evolved. But for more than a century Japan has taken new directions—technological, cultural, and political—so unlike those prevailing in mainland Asia that ample justification exists for its designation as a separate realm. By the turn of the twentieth century, Japan was an in-

**Figure J–1**
Mimicry overrides innovation as a cultural force: Japan's Disneyland just outside Tokyo, shortly after its 1983 opening.

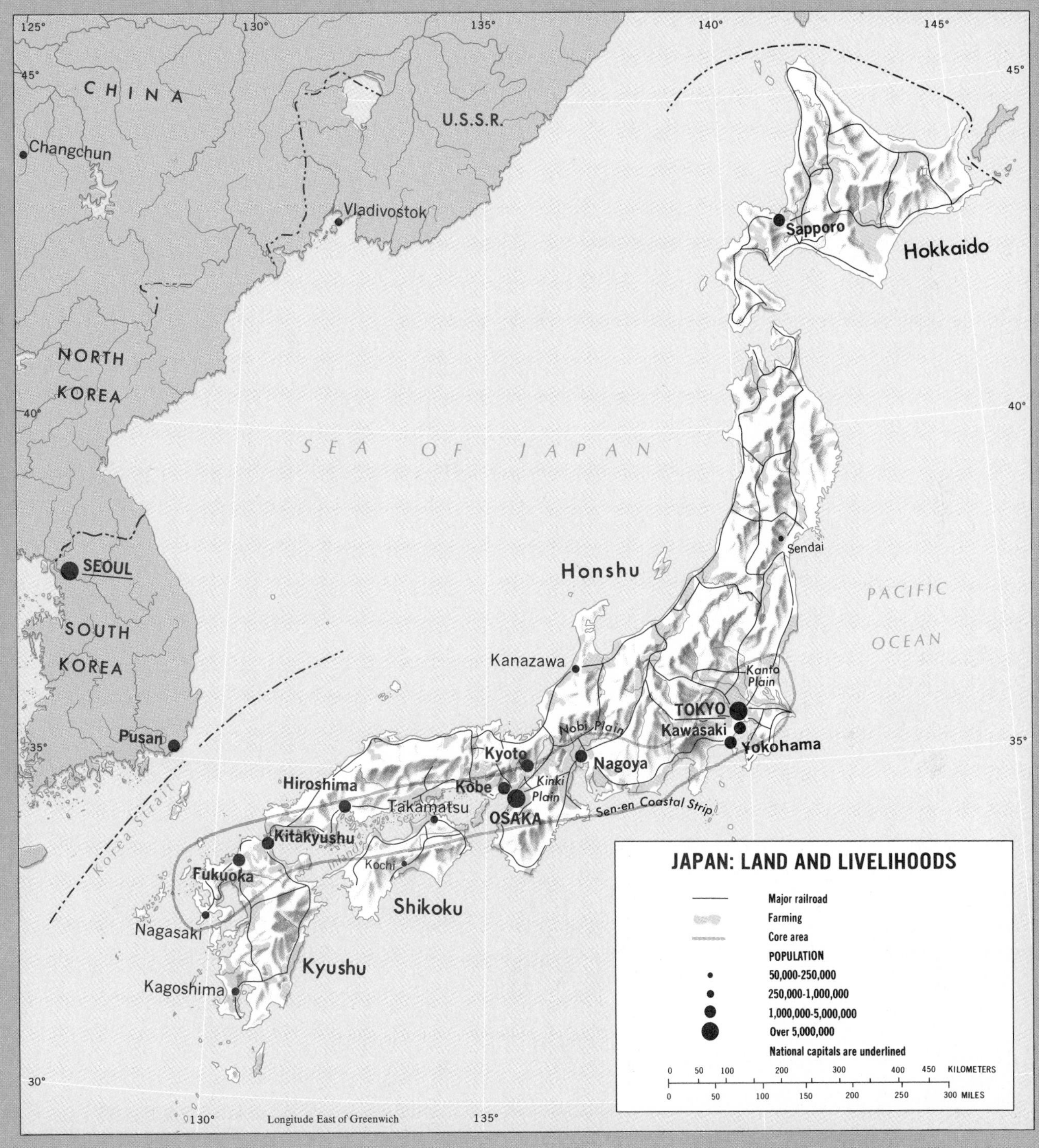


Figure J–2

dustrial force, a colonial power. Through a unique blend of tradition and modernization, borrowing and innovation, discipline and ambition, and with enormous energy and organizational ability, the Japanese transformed their country, overcame defeat and disaster, and created a society unlike any other in Asia.

The Japanese realm is a reality, despite its comparatively small territorial size and relatively modest population numbers (although Japan has about six times as many inhabitants as Austrialia and New Zealand combined). Not the numbers, but the remarkable homogeneity of that population constitutes one of the main reasons why Japan is a separate realm. Centuries of insulation and isolation, coupled with government policies designed to perpetuate the condition, have

## TEN MAJOR GEOGRAPHIC QUALITIES OF JAPAN

1. Japan is an archipelago of mountainous islands off the east coast of Eurasia.
2. Japan buys raw materials and sells finished products the world over; it lies remote from most of the locations with which it has close economic contact.
3. Japan is the prime example in the world of modernization in a non-Western society.
4. Japan was the non-Western world's major modern colonial power; during the colonial era it strongly influenced the development of the Western Pacific.
5. Although recent interaction has been limited, Japan lies poised for a major role in development in East Asia, including Eastern Siberia as well as China.
6. Japan's modern industrialization depends on the acquisition of raw materials (including energy sources) from distant locales; Japan's own mineral resource base is limited.
7. Modern and traditional society exist intertwined in Japan.
8. Japan has the highest physiologic population densities in the world.
9. Japan's agriculture, given conditions of climate, soil, and slope, is the most efficient and productive in Asia.
10. Japan's areal functional organization, based on regional specialization, is fully developed.

given Japan the most homogeneous population of its dimensions in the world. A recent example of the government's concern in this area involves the outflow of hundreds of thousands of "boat people" escaping from Communist-dominated Vietnam in the mid-1970s. Although other countries accepted nearly 1 million of these refugees, Japan agreed to permit fewer than 200 to enter, leaving no doubt about the reason. "Japan is a country of Japanese," said a government representative—"We are not a nation of minorities and subgroups."

Closely linked with this ethnic homogeneity is a cultural uniformity that confirms and emphasizes Japan's geographical identity. From Hokkaido to Kyushu, and from one end of Honshu to the other, Japan exhibits a sameness that is pervasive and even baffling. Japan has its huge, modern cities and its small, traditional villages—in that sense, there is contrast. But so do most other societies. In Japan, uniformity is assured by history and custom, by inherited pattern and controlled behavior. From the universal language to the Shinto national belief system, from the strength of family ties to the role of education and the schools (everyone in Japan wears the approved school uniform), from diets to diversions, the Japanese exhibit a cultural unity unmatched in modern times.

A further confirmation of Japan's status as a separate realm lies in its unmatched technological development and, hence, its economic geography. Using raw materials from virtually all corners of the planet, the Japanese have created a manufacturing complex whose production far outstrips that of giant China, the two Koreas, and Taiwan combined. On this basis alone (more will be said about it later) Japan is no mere corner of an East Asian geographic realm but merits its own discrete status.

In the mid-nineteenth century, Japan hardly seemed destined to become Asia's leading power. For 300 years, the country had been closed to outside influences; Japanese society was stagnant and tradition bound. Very early during the colonial period, European merchants and missionaries were tolerated and even welcomed, but, as the European presence on the Pacific shores grew greater, the Japanese began to shut their doors. By the end of the sixteenth century, Japan's overlords, fearful of Europe's imperialism, decided to expel all foreign traders and missionaries. Christianity (especially Catholicism) had gained a considerable foothold in Japan, but it was now viewed as a prelude to colonial conquest; in the first part of the seventeenth century, it was practically stamped out in a massive, bloody crusade. Determined not to share the fate of the Philippines, which had by then fallen to Spain, the Japanese permitted only the most minimal contact with Europeans. Thus, a few Dutch traders, confined to a little island near the city of Nagasaki, were for many decades the sole representatives of the European realm. So Japan retreated into a long period of isolation, which lasted past the middle of the nineteenth century.

Japan could maintain its aloofness when other areas were falling victim to the colonial tide because of a timely strengthening of its central authority, because of its insular character, and because of its distant position along East Asia's difficult northern coast. There were other factors as well—Japan's isolation was far less splendid than that of

China, whose silk and tea and skillfully-made wares always attracted hopeful traders. But, in the eighteenth and nineteenth centuries, the modernizing influences brought by Europe to the rest of the world passed by Japan as they did China. When Japan finally came face to face with the steel "black ships" and the firepower of the United States as well as Britain and France, it was no match for its enemies, old or new.

In the 1850s, the United States first showed the Japanese to what extent the balance of power had shifted. American naval units sailed into Japanese harbors—beginning with Commodore Perry's flotilla in 1853—and extracted trade agreements through a show of strength. Soon the British, French, and Dutch were also on the scene seeking similar treaties. When there was resistance in local areas to these new associations, the Europeans and Americans quickly demonstrated their superiority by shelling parts of the Japanese coast. By the late 1860s, no doubt remained that Japan's lengthy isolation had come to an end.

# Imperial Japan

In the century following 1868, Japan emerged from its near-colonial status to become one of the world's major powers, finally overtaking a number of its old European adversaries in the process. The year 1868 is important in Japanese history, for it marks the overthrow of the old rulers that brought to power a group of reformers whose objective was the introduction of Western ways in Japan. The supreme authority officially rested with the emperor, whose role during the previous militaristic period had been pushed into the background. Thus, the 1868 rebellion came to be known as the Meiji Restoration—the return of enlightened rule, centered around Emperor Meiji. But, despite the divine character ascribed to the royal imperial family, Japan was, in fact, ruled by the revolutionary leadership, a small number of powerful men whose chief objective was to modernize the country as rapidly as possible in order to make Japan a competitor, not a colonial prize.

The Japanese success is written on the map; even before the turn of the twentieth century a Japanese empire was in the making, and the country was ready to defeat encroaching Russia. Japan claimed the Ryukyus (1879), took Taiwan from China (1895), occupied Korea (1910), and established a sphere of influence in Manchuria. Various archipelagoes in the Pacific Ocean were acquired by annexation, conquest, or mandate (Fig. J–3). In the early 1930s, Manchuria was finally conquered, and the Tokyo government began calling for a Greater East Asia Co-Prosperity Sphere, which would combine—under Japanese leadership, of course—all of China, Southeast Asia, and numerous Pacific island territories. By the late 1930s, deep penetrations had been made into China, but Japan got its big chance in the early years of World War II. Unable to defend their distant Asian possessions while engaged by Germany at home, the European colonial powers offered little resistance to Japan's takeover in Indochina, Burma, Malaya, and Indonesia; neither was the United States able to protect the Philippines. Japan's surprise 1941 attack on Pearl Harbor, a severe blow to the U.S. military installations there, was a major success—and is still admired in Japanese literature and folklore. The Japanese war machine was evidence of just how far Japan's projected modernization had gone: airplanes, tanks, warships, guns, and ammunition were all produced by Japanese industries, which were now a match for their Western counterparts. Japan, as we know, was ultimately defeated by superior American power (Hiroshima and Nagasaki were annihilated by nuclear bombs in 1945), but this defeat, coupled with the loss of its prewar as well as wartime empire, still failed to destroy the country's progress. Postwar Japan rebounded with such vigor that its overall economic development now ranks among the most advanced in the world.

With the Meiji Restoration a little over a century ago, Japan began its phenomenal growth, which brought it an empire, involved it in a disastrous war, and eventually saw the country achieve an industrial capacity unmatched not only in the non-Western world, but in much of the Western world as well. As if to symbolize their rejection of the old, isolationist position of the country, the new Meiji rulers moved the capital from Kyoto to Tokyo (or *Eastern Capital*), the new name for the populous city of Edo in the Kanto Plain. Kyoto had been an interior capital; Tokyo lay on the Pacific Coast, alongside an estuary that was soon to become one of the busiest harbors on earth. The new Japan looked to the sea as an avenue to power and empire, and it looked to industrial Europe for the lessons that would help it achieve those ends. Although China suffered as conservatives and modernizers argued the merits of an open-door policy to Westernization, Japan's leaders committed themselves—with spectacular results.

# Modernization

What has happened in Japan during the past century is a process that is far-reaching, transforming society from "traditional" to "modern"—but the process is complex,

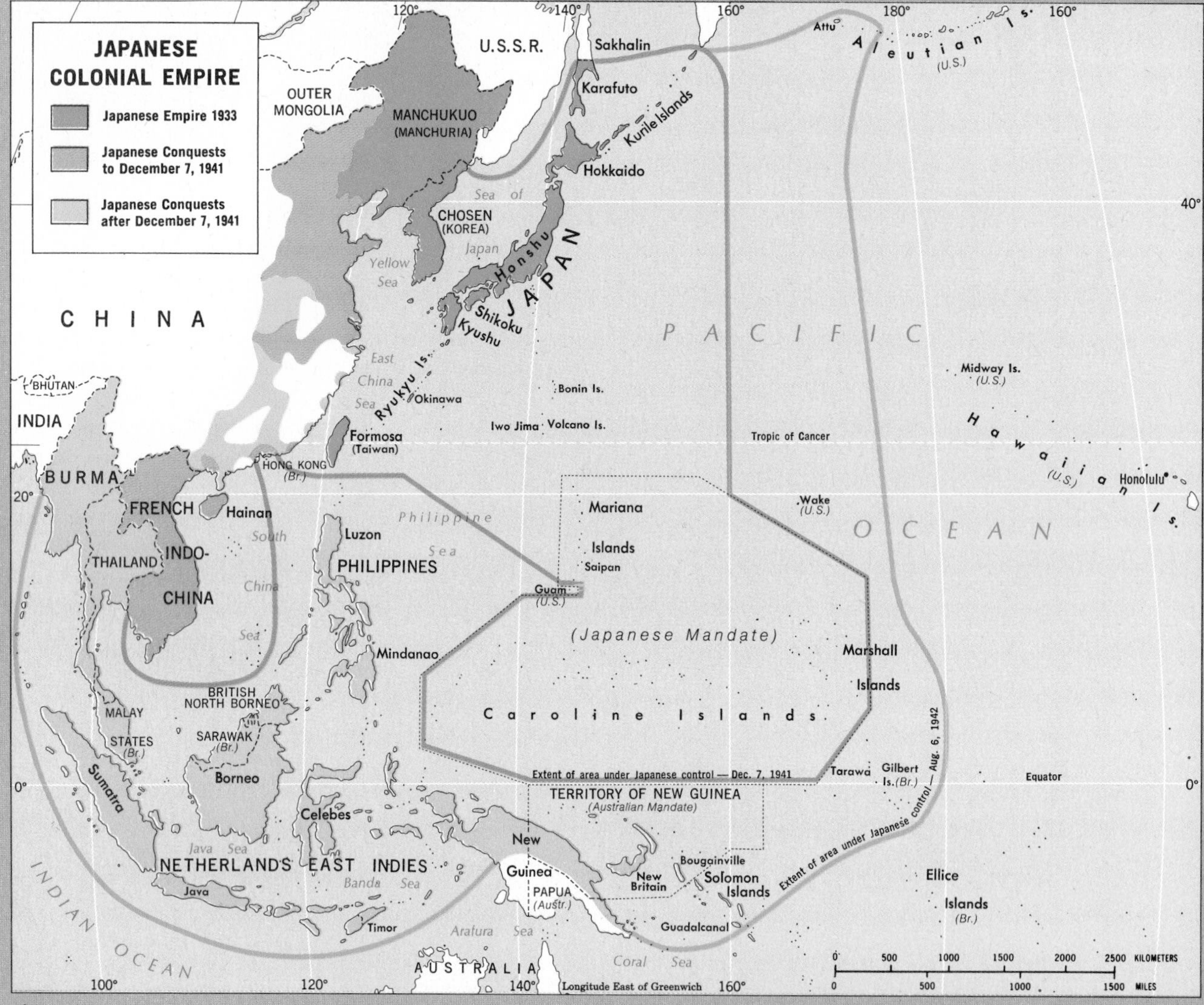


Figure J–3

difficult to define, and subject to much debate among social scientists. We in the Western world tend to view modernization as identical to Westernization—urbanization, the spread of transport and communications facilities, the establishment of the market (money) economy, the breakdown of local (tribal) communities, the diffusion of formal schooling, and the acceptance and adoption of foreign innovations. In the non-Western world, modernization (described perhaps with another word or phrase) is often seen as an outgrowth of colonial imperialism, the perpetuation of a system of wealth-accumulation introduced by alien invaders who were driven by greed. The local elites who have replaced the colonizers in the newly independent states, in that view, only carry on the disruption of traditional societies, not their modernization. The capital cities of Africa and Asia, with their crowded avenues and impressive skyscrapers, continue to grow—but they do not signal *development*. Traditional societies can be modernized without being Westernized.

Japan's modernization, in this context, is unique in many ways. Having long resisted foreign intrusion, the Japanese did not achieve the transformation of their society by importing a Trojan horse; it was done by Japanese planners, based on the existing Japanese infrastructure, and it fulfilled Japanese objectives. Certainly, Japan imported foreign technologies and adopted innovations from outside, but the Japan that was built, a unique combination of modern and traditional, was basically an indigenous achievement. Like the European powers they often emulated, the Japanese created a large colonial empire to supply the homeland with raw materials. The ripple effect of the Japanese period still continues in the economies of such former dependencies as Taiwan and Korea—and there the fundamental similarities between imperial Japan and imperialist Europe can still be discerned. But modernization in Japan itself defies the generalizations that apply to the process in the underdeveloped world.

# Limited Assets

Japan's success might lead to the assumption that the country's raw material base was sufficiently rich and diverse to support the same kind of industrialization that had characterized England. Britian's industrial rise, after all, was accomplished largely because the country possessed a combination of resources that could sustain an industrial revolution. But Japan is not nearly so well off. Although Japan still has coal deposits in Kyushu and Hokkaido, there is little good coking coal for steel production. The most accessible coal seams were soon exhausted. Before long, the cost of coal mining was rising steadily. There was a time when Japan was a net coal exporter, when coal was shipped to the coastal cities of China, but today Japan must import this commodity. Nevertheless, Japan did have enough coal, fortuitously located near its coasts, to provide a vital stimulus for initial industrialization during the late nineteenth century. Both in northern Kyushu and in Hokkaido, the coal deposits lie so near the coast that transport is not a problem, and, by the nature of Japan's landscape, the major cities and industrial areas also lie on or near the coast. Thus, coal could be provided quite cheaply and efficiently to every developing manufacturing center, and the fringes of the Inland Sea were among the first areas to benefit from this spatial relationship.

As for the other raw material of major industry—iron ore—Japan has only a tiny domestic supply, and one of small scattered deposits of variable but generally low quality. In this respect, there is simply nothing to compare to what Britain had at its immediate disposal, and neither is Japan rich in the various ferroalloy minerals that are vital to the steel-making process. The Japanese extract what they can, including ores that are so expensive to mine and refine that it is hardly worth the effort—in view of import alternatives—and they carefully collect and reprocess all available and imported scrap iron. But iron ore still ranks high on each year's import list, and Japan buys iron all over the world, even as far away as Swaziland in southern Africa.

Japan also possesses negligible petroleum reserves and was forced to import 99.8 percent of its oil in the early 1980s. A strict energy conservation program was instituted following the sharp oil price increases of the 1970s, but Japan still pays about twice as much for its annual petroleum purchases as it does for its iron ores. The few barrels of oil that are produced at home are hardly worth mentioning; and the hydroelectric power that is derived from the country's many favorable sites is not enough to fill its needs. This in itself is a measure of the amount of electricity Japan consumes; with its high mountain backbone, its ample rainfall, deep valleys, and urban markets clustered on the mountains' flanks, all conditions are right for efficient hydroelectric-power production and use. But less than one-fifth of Japan's total energy needs come from these dams, the demand for hydroelectricity rises faster than does production, and the day seems far off when petroleum and coal imports can be reduced. Even a strong commitment to nuclear power will not alter this fossil fuel dependency during the 1980s—Japanese facilities are encountering the same cost and safety problems that currently beset U.S. atomic-fission power plants.

This, then, answers one of the questions we posed in the first paragraph of this essay. Japan really has nothing to compare to what Britain had during its industrial transformation. In fact, after what has just been said, it would seem that we are dealing with just another of those many countries whose underdevelopment can be attributed largely to a paucity of resources and whose future is, therefore, deemed to depend on agriculture. Certainly, it would have been difficult in 1868 to predict that the enthusiastic leaders of the Meiji Restoration would really have much success with their plan for the Westernization and strengthening of their country. What, then, gave Japan its opportunity? Obviously, foreign trade and technology transfer played a great part in it. Was Japan's location the key to its fortune?

# Domestic Foundations

It is tempting to turn immediately to Japan's external connections and to call on those factors of location to account for Japan's great industrial growth. After all, the Japanese built an empire, much of it right across the Korea Strait, and this empire contained many of the resources Japan lacked at home. But before we rush into such a deduction, let us consider what Japan's domestic economy was like in the mid-nineteenth century, for, as we will see, what Japan achieved was primarily based on its internal human—and to a large extent natural—resources.

In the 1860s, manufacturing—light manufacturing of the handicraft type—was already widespread in Japan. In cottage industries and community workshops, the Japanese produced silk and cotton-textile manufactures, porcelain, wood products, and metal goods. At that time, the small metalliferous deposits of the country served their purpose: they were sufficient to supply these local industries. Power came from human arms and legs and from wheels driven by water; the chief source of fuel was charcoal. Thus, there was industry and an experienced labor force that possessed appropriate manufacturing skills.

The planners who took over the country's guidance after 1868 realized that this was an inadequate base for industrial modernization but that it might, nevertheless, generate some of the capital needed for this process. The community and home workshops were integrated into larger units, thermal and hydroelectric power began to be made available to replace more primitive sources of energy, and, for the first time, Japanese goods—still of the light manufacture variety of course—began to compete with Western products on the world's markets. Meanwhile, the Japanese continued to resist any infusion of Western capital, which might have accelerated the industrialization process but would have cost Japan its economic autonomy. Instead, the farmers were more heavily taxed, and the money thus earned by the state poured into the industrialization effort.

Now Japan's layout and topography contributed to the modernization process. Coal could be mined and—in the relatively small quantities then needed—transported almost wherever it was required. Throughout the country, a hydroelectric site was usually nearby, and electric power became available anywhere—in the cities as well as the populated countryside. Japan was still managing on what it possessed at home, and the factories and workshops multiplied. Soon the beginnings of a chemical industry emerged, and government subsidies led to the establishment of heavy industries. The period after 1890 was a time of great progress, stimulated by the wars against China (1894–1895) and Russia (1904–1905). Nothing served to emphasize the need for industrial diversification as much as war, and Japan's victories vindicated the ruthlessness with which industrial objectives were sometimes pursued.

Japan's war effort contributed to its improved world economic position as the imperial period ended. When the Meiji Restoration took place, Britain lay at the center of a global empire and the Europeanization of the world was in full swing; the United States still was a developing country. The wars of the twentieth century saw the Japanese defeat British and French forces in East Asia; the United States and its allies crushed Germany and Italy. Although Japan also came out of World War II a defeated and devastated country, the global situation had changed dramatically. The United States, Japan's trans-Pacific neighbor, had become the world's most powerful and wealthiest state, whereas distant Britain was fading, with its empire on the verge of disintegration. Japan's *relative location* had changed—its location relative to the economic and political foci of the world—and therein lay much of Japan's new postwar opportunity.

# Japan's Spatial Organization

Japan is not a very large country, but it was already quite populous when its modern economic revolution began: about 30 million people were crowded into its confined living space. Since then, despite a habitable area limited to a meager 18 percent of its national territory, Japan's population has quadrupled and its productive complex has grown enormously. With 1 million people, Tokyo (then still named Edo) may already have been the world's largest city in the 1600s, but today the capital is about 10 times as large and is part of an even bigger urbanized area, which includes Yokohama as well. This is the world's largest urban agglomeration, now containing a population of nearly 23 million, which exceeds the size of the second largest conurbation (Mexico City) by almost 4 million and third-place metropolitan New York by 5 million.

Given such growth, space in Japan has become increasingly scarce, and urban and regional planners have shared with farmers the problem of how to make every square kilometer count. Before the onrush of urbanization, it was often said that the most densely populated rural areas in Japan were more crowded than any other such areas in the world. Now the cities, towns, and villages must also vie for the country's limited livable terrain. When Japan's industries were still of the light handicraft variety, the myriad workshop-type establishments were located in the cities as well as in villages; the small-scale smelting of iron could be carried on in many places (charcoal was widely available), as could the manufacturing of textiles. Then, the age of the factory arrived, and the modern rules of industrial location went into effect. We discussed some of these rules earlier, notably the advantages of processing heavy or bulky raw materials at the spot where they are mined to save movement costs and the transport of more easily moved lighter raw materials to areas containing heavier less-transferable resources. In Britain during the industrial revolution, this led to the rapid growth of certain towns into industrial cities, the decline of others, and the founding of still other manufacturing centers. In Japan, on the other hand, urban-industrial growth was straitjacketed. Apart from northern Kyushu's coal, no great reserves of raw materials were found, and it soon was clear that these would have to be imported from elsewhere. Internally, the distribution of coal was mainly by water; externally, the iron ore and other required commodities, of course, came by sea. Hence, the coastal cities, where the labor forces were also concentrated, became the major centers of manufacturing activity.

The modern industrial growth of Japan was shaped by the economic development that was already taking place. Although the country did not have the resources to merit any major internal reorientation, some cities did possess advantages over others in terms of their existing facilities and their situation vis-à-vis the source areas of raw materials. Accordingly, there now began the kind of differentiation that has marked regional organization everywhere in the world, governed by the principle that Allen Philbrick has called *areal functional organization*. Five interrelated ideas are involved. First, human activity has spatial focus, in that it is concentrated in some locale—a farm or factory or store. Second, such "focal" activity is carried on in certain particular places. Obviously, no two establishments can occupy exactly the same spot on the earth's surface, so, every one of them has a finite or absolute location; but what is more relevant, every establishment has a location *relative* to other establishments and activities. Since no human activity is carried on in complete isolation, the third idea is that interconnections develop among the various establishments. Farmers send crops to markets and buy equipment at service centers. Mining companies buy gasoline from oil companies, lumber from sawmills, and send ores to refineries. Thus, a system of interconnections emerges and grows more complex as human capacities and demands expand, and they are expressed spatially as units of area organization. Philbrick's fourth idea is that the evolution of these units of areal organization—or regions—is the product of human "creative imagination" as people apply their total cultural experience as well as technological know-how when they decide how to organize and rearrange their living space. Finally, it is possible to recognize *levels* of development in areal organization, a ranking or hierarchy based on type, extent, and intensity of exchange. We detect here a relationship to Christaller's attempt to discern order in the geographic arrangement of urban centers possessing different levels of centrality. To quote Philbrick, "The progression of area units from individual establishments to world regions and the world as a whole are formed into a hierarchy of regions of human organization."

In the broadest sense, regions of human organization can be categorized under subsistence, transitional, and exchange types, with Japan's areal organization reflecting the last of these. But within each type of organization, especially within the complex exchange type of unit, individual places can also be ordered or ranked on the basis of the number and kinds of activities and interconnections they generate. A map of Japan showing its resources, urban settlements, and surface communications can tell us a great deal about the kind of economy the country has. It looks just like maps of other parts of the world where an exchange type of areal organization has developed—a hierarchy of urban centers ranging from the largest cities to the tiniest hamlets, a dense network of railroads and roads connecting these places, and productive agricultural areas near and between the urban centers. In Japan's case, Fig. J–4 shows us something else: it reflects the country's external orientation, its dependence on foreign trade. All primary and secondary regions lie on the coast; of all the cities in the million-size class, only Kyoto (1.5 million) lies in the interior, located at the head of the Kinki Plain. If we were to deduce that Kyoto does not match Tokyo, Yokohama, Nagoya (2.2 million), or Kobe-Osaka (10.1 million) in terms of industrial development, that would be correct—the old capital remains a center of small-scale light manufacturing. Actually, Kyoto's ancient character has been preserved deliberately, and large-scale industries have been discouraged. With its old temples and shrines, its magnificent gardens, and its many workshop and cottage industries, Kyoto remains a link with Japan's pre-modern past (Fig. J–5).

As Fig. J–4 shows, Japan's dominant region of urbanization and industry, along with very productive agriculture, is the Kanto Plain, which is focused on the Tokyo-Yokohama metropolitan area (22.6 million). This giant cluster of cities and suburbs, interspersed with intensively cultivated farmlands, forms the eastern end of the country's elongated, fragmented core area (Fig. J–2). Besides its flatness, the Kanto Plain possesses other advantages: its fine natural harbor at Yokohama, its relatively mild and moist climate, and its central location with reference to the country as a whole. It has also benefited enormously from Tokyo's designation as the modern capital, which coincided with Japan's embarkation on its planned economic growth. Many industries and businesses chose Tokyo as their headquarters in view of the advantages of proximity to the government's decision makers (Fig. J–6).

The Tokyo-Yokohama conurbation has become Japan's leading manufacturing center, producing

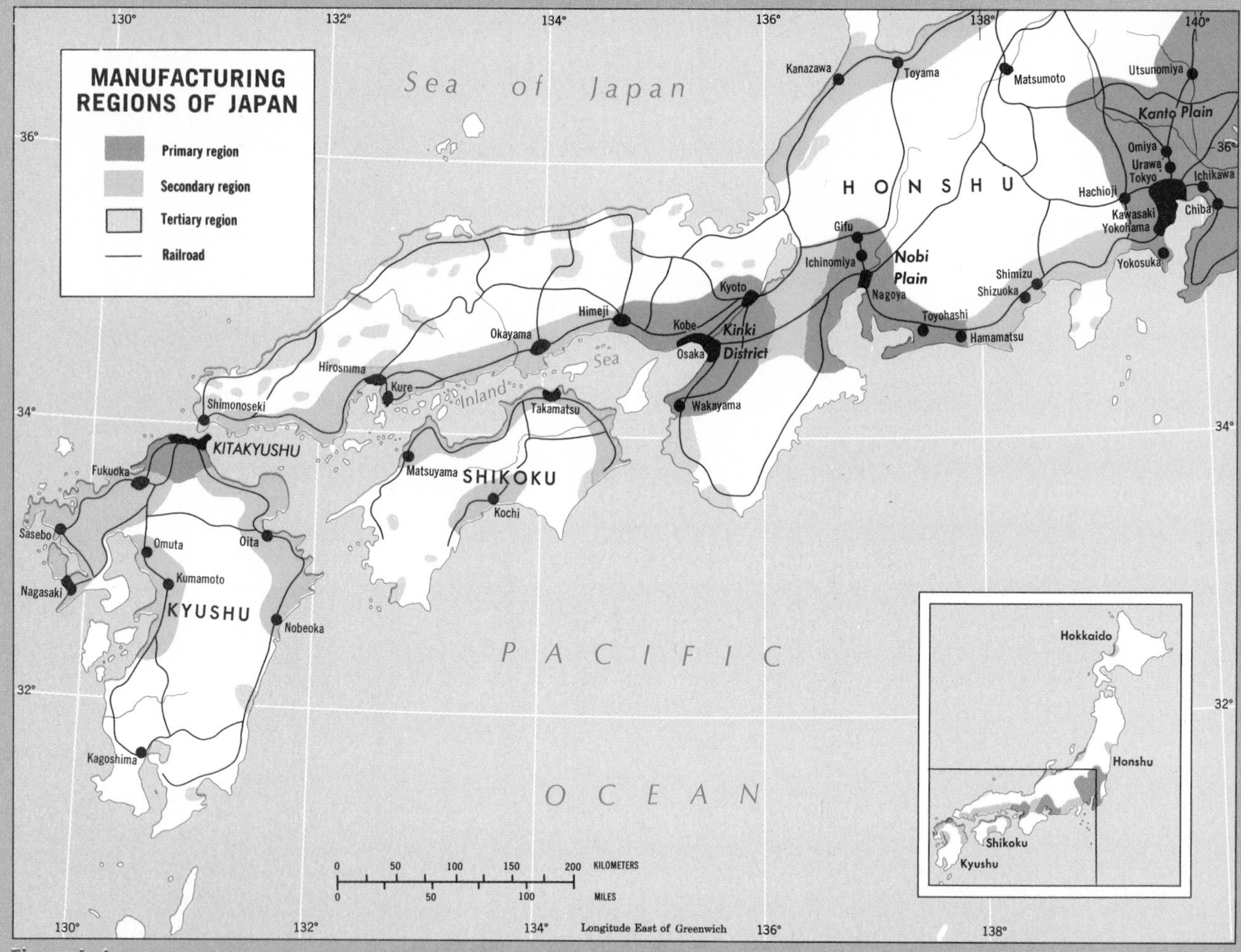


Figure J–4

Figure J–5
The preserved landscape of Kyoto, the pre-modern capital.

between one-fifth and one-quarter of the country's annual output. But the raw materials for all this industry come from far away; for example, the Tokyo district is among the chief steel producers in Japan, using iron ores from the Philippines, Malaya, Australia, India, and even Africa. Although electric power comes from the mountain valleys to the northwest and some coal is shipped from Hokkaido, most coal is imported from Australia and America and petroleum from Southwest Asia and Indonesia. As for the supply of food to its huge population, the Kanto Plain cannot produce nearly enough—imports come from Canada, the United States, and Australia as well as from some other areas in Japan itself. Thus, Tokyo depends completely on its external trading ties for food, raw

Figure J–6
The heart of Tokyo, focus of the world's largest urban agglomeration.

materials, and markets for its wide variety of products, which run the gamut from children's toys to high-precision optical equipment to massive oceangoing ships.

The second-ranking economic region in Japan's core area is the Kobe-Osaka-Kyoto triangle, at the eastern end of the Seto-Naikai (Inland Sea). The position of the Kobe-Osaka conurbation, with respect to the old Manchurian empire created by Japan, was advantageous. Situated at the head of the Inland Sea, Osaka was the major Japanese base for the China trade and the exploitation of Manchuria, but it suffered when the empire was destroyed, and its trade connections with China were not regained after World War II. Kobe (like Yokohama, Japan's chief shipbuilding center) has remained one of the country's busiest ports, handling the traffic of the Inland Sea as well as its extensive overseas connections. Among these linkages, for many decades, was the importing of cotton for Osaka's textile factories. With its large skilled labor force and its long history of productive settlement, Osaka, until recently, was Japan's first-ranked textile producer: today, textiles (mainly synthetics) form only part of this metropolitan area's huge and varied industrial base. Kyoto, as we noted, remains much as it was before Japan's great leap forward: with 1.5 million people, it nonetheless remains a city of small workshop industries. Similar to the Kanto Plain, the Kobe-Osaka-Kyoto region—also known as the Kinki District—is an important farming area as well. Rice, of course, is the most intensively grown crop in the warm moist lowlands, but agriculture here is less widespread than on the Kanto Plain. Thus, we observe another cluster of people that requires a large infusion of foodstuffs every year.

The Kanto Plain and Kinki District, as Fig. J–4 suggests, are the two leading primary regions within Japan's core. But between them lies the Nagoya or Nobi Plain, focused on the industrial center of Nagoya, the city that recently ousted Osaka from first place among Japan's textile producers. The map indicates some of the Nagoya area's advantages and liabilities. The Nobi Plain is larger than that of the Kinki District, thus, its agricultural productiveness is greater, although not as great as that of the Kanto Plain. But Nagoya has neither Tokyo's centrality nor Osaka's position on the Inland Sea: its connections to Tokyo are by the tenuous Sen-en Coastal Strip, along which runs the famous Tokaido "bullet train" (Fig. J–7). Its westward connections are somewhat better, and there are signs that the Nagoya area is coalescing with the Kobe-Osaka region. Still another disadvantage of Nagoya lies in the quality of its port, which is not nearly as good as Tokyo's Yokohama or Osaka's Kobe and has been plagued by silting problems.

Westward from the three regions just discussed extends the Inland Sea, along whose shores the remainder of Japan's core area is continuing to develop. The most impressive growth has occurred around the western entry to the sea, where Kitakyushu (a conurbation of five northern Kyushu Island cities) constitutes the fourth Japanese manufacturing complex. Honshu and Kyushu are connected by road and railway tunnels, but the northern Kyushu area does not have any equivalent on the Honshu side of the Strait of Shimonoseki. The Kitakyushu conurbation (2 million) includes Yawata, site of northwest Kyushu's large coal mines, and it is on the basis of this coal that the first steel plant in Japan was built there—a facility that for many years was Japan's largest. The advantages of transportation here at the western end of the Inland Sea are obvious. No place in Japan is better situated to do business with Korea and China; as relations with China expand, this area will reap many of the benefits of that trade. Elsewhere on the Inland Sea coast, the Hiroshima-Kure urban area (800,000) has a manufacturing base that includes heavy industry. And, on the coast of the Korea Strait, Fukuoka and Nagasaki (350,000) are the principal centers, the former an

**Figure J–7**
The high-speed Tokaido "bullet train," which links Honshu's 3 primary manufacturing regions.

industrial town, the latter a center of large shipyards.

The map of Japan's areal organization shows the country's core area to consist largely of four individual primary regions, each of them primary because each duplicates to some degree the contents of the others. Each contains iron and steel plants, each is served by one major port, and each lies in or near a large and productive farming area. What the map does not show is that each has its own external connections for the overseas acquisition of raw materials and the sale of finished products. These linkages (Fig. J–8) may even be stronger than those among the four internal regions of the core area; only in the case of Kyushu and its coal have domestic raw materials played much of a role in shaping the nature and location of heavy manufacturing. In the structuring of the country's areal functional organization, therefore, more than just the contents of Japan itself is involved. In this respect, Japan is not unique: all countries that have exchange-type organizations must to some degree adjust their spatial forms and functions to the external interconnections required for progress. But it would be difficult to find a country in which this is truer than in Japan.

# Food Production

The rapid modernization of the Japanese economy over the past half-century is reflected in its changing occupational structure. A recent study by Chauncy Harris has shown that in 1920, prior to the urban-industrial transformation, 55 percent of Japan's workers were employed in the primary sector, 21 percent in manufacturing, and 24 percent in the tertiary sector; his findings for 1980 revealed a complete reversal, with only 11 percent remaining as primary workers, 34 percent engaged in industry, and 55 percent employed in the postindustrial tertiary-quaternary-quinary sectors. The most remarkable changes have occurred in agricultural employment: whereas 51 percent of the labor force were farmers in 1920, the proportion that makes possible today's far more productive domestic agriculture is only 9 percent—with two-thirds of them part-timers who derive most of their income from nonagricultural sources. Let us take a closer look.

Japan's economic modernization so occupies the center stage that it is easy to forget that considerable achievements have been made in the field of agriculture. Japan's planners, no less interested today in closing the food gap than in expanding industries, have created extensive networks of experiment stations to promote mechanization, proper seed selection and fertilizer use, and information services to distribute to farmers as rapidly as possible information that is useful for enhancing crop yields. Although this program has been very successful, Japan faces the unalterable reality of its stubborn topography: there simply is not the necessary land to farm. Less than one-fifth of the country's total area is in cultivation; although there still may be some land that can be brought into production, it is so mountainous or cold that its contribution to the total harvest would be minimal anyway. Japanese agriculture may resemble Asian agriculture in general, but nowhere else do so many people depend on so little land. Japan's population density overall is 318 per square kilometer (824 per square mile) to begin with; when the *arable* land is measured it turns out that more than 1800 people depend on the average square kilometer of farmland (4600 per square mile). This measure, persons per unit of cultivable land, is the *physiologic density*, and Japan's is one of the highest in the world. Even Egypt and Bangladesh, notoriously crowded in confined agricultural areas, exhibit a lower physiologic density than Japan; and the physiologic density of India is less than *one-fifth* as high as Japan's!

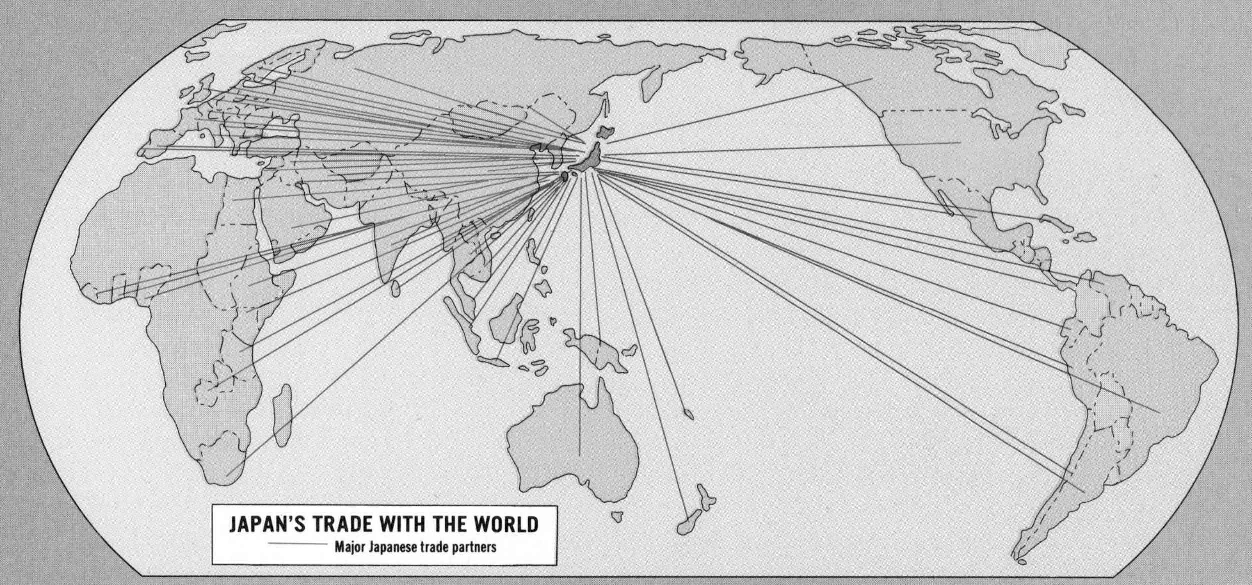


Figure J–8

Japanese agriculture stands apart from farming elsewhere in rice-growing Asia. It is by far the most efficient in all the Asian continent, given the existing conditions of slope, soil, and climate. Rice yields in Japan are among the highest in the world: about 60 quintals per hectare (110 bushels per acre), whereas China has under 35, the United States 51, and South and Southeast Asia under 20. Moreover, Japanese researchers constantly find new ways to improve this productivity—in the early 1980s, for instance, they innovated better hybrid varieties of rice, which mature so quickly that it is now possible to raise two crops every year, even in the colder climates.

Japan's population is now 121 million, having tripled over the past century; although the annual growth rate of 0.6 percent is much lower than that of other Asian countries, Japan no longer enjoys its one-time luxury of self-sufficiency in food. But it is remarkable that the country is somehow able to produce approximately two-thirds of its annual needs—enough to feed 80 million people. This is achieved through a number of parallel methods: turning over more than 90 percent of all farmland to food crops (with 53 percent of this rice), through diligent irrigation (more than half of Japan's agriculture is irrigated), through painstaking terracing of steep slopes, multiple cropping (now greatly enhanced by the newest hybrid rice), intercropping, intensive fertilization, and transplanting—that is, the use of seedbeds to raise the young plants of the next crop while its predecessor matures in the field.

So much agricultural space is devoted to rice that Japan's second and third crops, usually wheat and barley, occupy less than one-quarter of the farmland. Wheat is used as a winter crop in rotation with rice in the warmer parts of the country, and barley is grown in the north as a summer crop. Sweet and white potatoes, as the major root crop, occupy less than 10 percent of the farmland. Still less space (much of it on hilly slopes) is given over to cash crops such as tea, grown for local consumption, and mulberry, the plant whose leaves are used to feed the silkworms—the basis of Japan's once-rich silk industry. There are also vegetable gardens that cling to the outskirts of the large cities. But Japan's major agricultural effort goes into producing its staple crop—rice. Much of the remainder must be imported, and the Japanese annually bring in from abroad substantial quantities of wheat, corn, sugar, and soybeans.

With their national diet so rich in such starchy foods as rice, wheat, barley, and potatoes, the Japanese need protein to balance it. Happily, this can be secured in sufficient quantities, not through foreign purchase, but by harvesting it from rich fishing grounds near the Japanese islands. With customary thoroughness, the Japanese have developed a fishing industry that is larger than that of the United States or any of the long-time fishing nations of northwestern Europe, and it now supplies the domestic market with a second staple after rice.

Although mention of the Japanese fishing industry brings to mind fleets of ships and travelers scouring the oceans and seas far from Japan, the fact is that most of this huge catch (more than 25 percent of the world's annual total) comes from waters within a few dozen miles of Japan itself. Where the warm Kuroshio and Tsushima currents meet colder water off Japan's coasts, a rich fishing ground exists that contains sardines, herring, tuna, and mackerel in the warmer waters, and cod, halibut, and salmon in the seas to the north. Along Japan's coasts there are about 4000 fishing villages, and tens of thousands of small boats ply the waters offshore to bring home catches that are distributed both locally and in city markets. The Japanese also practice *aquaculture*—the raising of freshwater fish in artificial ponds and flooded paddy fields, seaweeds in home aquariums, and oysters, prawns, and shrimp in shallow bays; they are even experimenting with the cultivation of algae for their food potential.

# Japan in the Post-industrial Era

In the mid-1980s, Japan has become one of the world's most advanced nations. Its population of 121 million is now as heavily urbanized as that of Western Europe and North America—81 percent in 1985, and still climbing—a complete reversal since 1920 when only 18 percent resided in cities and towns. Although Western-style prosperity was a bit late in coming, the past decade has witnessed a significant narrowing of the affluence gap between Japan and the United States; since 1970, the proportion of Japanese families owning automobiles has risen from 17 percent to nearly 65 percent, those owning air conditioners from 6 to over 40 percent, and those owning color televisions from 25 to more than 99 percent.

As Japan's advancing economy shifts ever more steadily away from primary and secondary activities, there is little doubt that the country is fast approaching a full-scale North American-style postindustrial revolution. The rise of an information-based economy is being led by major research into the development of so-called "fifth-generation" computers, an ultrasophisticated technology which will bring such breakthroughs as problem solving by machines using humanlike reasoning, automatic language-translation machinery, and the repairing of complex electronic equipment. The new thrust into scientific research and development is best symbolized by the recent construction of a new town in Tokyo's outer suburbs called Tsukuba Science City (Fig. J–9). This 28,000-hectare (70,000-acre) complex, which is the Japanese version of North Carolina's Research Triangle Park (*Box* p. 214), is built around a leading national university campus and more than 50 scientific institutes and government research agencies that have relocated from central Tokyo 65 kilometers (40 miles) away; more than 7500 scientists, engineers, and technicians work here—performing experiments in every field from high-energy physics to earthquake simulation—and Tsukuba is well on its way to becoming a self-sufficient urban center of 100,000 residents.

If past performance is any guide to the future, Japan can be expected to share (and even aggressively promote) the new products of these technological breakthroughs with the rest of the world. Yet, despite its vigorous marketing and exporting efforts, Japan will probably not substantially increase its trading activities. In fact, the overall foreign relations of the Japanese are often exaggerated by outside observers, because the country's economy is not all that heavily dependent on sales in the international marketplace. For example, Japan in 1982, exported just 13 percent of its gross national product (*GNP*), with only the United States exhibiting a lower proportion (7 percent) among the most developed nations; included among the far more active exporters were Britain (21 percent), West Germany (27 percent) and Canada (29 percent). What is most remarkable about Japan's foreign sales activity is its success, which is largely based on a superior reputation for product quality and modest price, which are a result of survival in a hypercompetitive domestic market that is one of the toughest on earth.

The selective new trading ties the Japanese will encourage in the future are the ones that will most enhance their economic and political stability. With its unavoidable reliance on imported oil, Japan can be expected to maintain cordial relations with the suppliers of the Arab world; for instance, the Japanese kept buying petroleum from Iran during most of the 1979–1981 U.S.-hostage crisis and, in 1983, resumed joint construction of a huge petrochemical facility at Bandar Khomeini near the head of the Persian Gulf, despite the persistence of the nearby hostilities between the Iranian and Iraqi armies. Future trade with the United States (by far Japan's leading trading partner) will intensify as well, but much more heavily on the importing side: not only do the Americans supply such critical raw materials as cotton and scrap metal, but as Japanese consumerism continues to grow in an increasingly wealthy domestic market only America has the range and surplus quantity of luxury goods to supplement local production if these widening demands are to be satisfied without long delays. Closer ties with the Communist world may also be in the offing. The immense market of nearby China does offer an enormous opportunity, but the normalization of diplomatic and trade relations since 1972 have progressed slowly. The Soviet Union has also approached Japan since the mid-1970s in order to jointly develop projects in Eastern Siberia; although some energy and transportation ventures were initiated, the barbaric 1983 Russian

Figure J–9
Tsukuba, a new city devoted to scientific research at the edge of the Tokyo metropolis.

downing of a Korean passenger jet (in which 269 persons were killed) for inadvertent U.S.S.R. airspace violations near Japanese waters, was a grim reminder of the politico-geographical situation wherein Japan continues to oppose Soviet control over the Kurile Islands.

Perhaps the biggest adjustment required of the Japanese in the dawning postindustrial age will be in the way they view themselves. Despite its enormous postwar advances, Japan retains an understandable inferiority complex—after all, this was until quite recently a tradition-bound nation of farmers (many living at the edge of poverty) that became a prosperous urban society practically overnight. Thus, the Japanese are bewildered that certain outsiders should see them as an economic threat to other nations, because their self-perception is one of vulnerability based on their poor-quality land, meager natural-resource base, and a powerful dependency on foreign countries for vital food and fuel supplies. Japan's cultural traditions, which also reflect these shortcomings, reinforce this view: the crowding engendered for centuries by their unkind habitat produced a culture that has coped with the problem of numbers by emphasizing rigid adaptation, consensus, and miniaturization. Therefore, as contacts with the West were constantly strengthened after 1850, the Japanese have overwhelmingly preferred to copy new ideas rather than undertake their innovation; the opening of their own Disneyland in the eastern Tokyo suburb of Urayasu in 1983 (Fig. J–1)—which not only recreates the "Magic Kingdom" theme park, but also replicates a little piece of America itself by banning Japanese signs and refreshments—is ample evidence that mimicry remains a cultural force to be reckoned with. But postindustrial Japan must soon confront a wholly new reality—this transforming nation has reached its pinnacle of success and affluence by copying and improving the inventions of others, and it has now achieved the unintended position of world leader in many endeavors. Further success will depend on the seizing of this mantle of leadership in the near future; thus far, most Japanese have demonstrated a decided reluctance to do so.

## References and Further Readings


Ackerman, E. *Japan's Natural Resources and Their Relation to Japan's Economic Future* (Chicago: University of Chicago Press, 1953).

Christopher, R. *The Japanese Mind: The Goliath Explained* (New York: Linden Press/Simon & Schuster, 1983).

Dempster, P. *Japan Advances: A Geographical Study* (New York: Barnes & Noble, 1967).

Eyre, J. *Nagoya: The Changing Geography of a Japanese Regional Metropolis* (Chapel Hill, N.C.: University of North Carolina, Studies in Geography No. 17, 1982).

Hall, P. "Tokyo," in *The World Cities* (New York: McGraw-Hill, 2 rev. ed., 1977), pp. 219–239.

Harris, C. D. "The Urban and Industrial Transformation of Japan," *Geographical Review*, 72 (1982), 50–89.

Huddle, N. et al. *Islands of Dreams: Environmental Crisis in Japan* (New York: Autumn Press, 1975).

Ito, T. & Nagashima, C. "Tokaido—Megalopolis of Japan," *GeoJournal*, 4 (1980), 231–246.

"Japan: A Nation in Search of Itself," *Time*, Special Issue, August 1, 1983.

Kornhauser, D. "A Selected List of Writings on Japan Pertinent to Geography in Western Languages, with Emphasis on the Work of Japan Specialists," University of Hiroshima, 1979.

Kornhauser, D. *Japan: Geographical Background to Urban-Industrial Development* (London and New York: Longman, 2. rev. ed., 1982).

Morton, W. *The Japanese Way: How They Live and Work* (New York: Praeger, 1973).

Murata, K. & Ota, I., eds. *An Industrial Geography of Japan* (New York: St. Martin's Press, 1980).

Patrick, H., ed. *Japanese Industrialization and Its Social Consequences* (Berkeley: University of California Press, 1976).

Philbrick, A. "Principles of Areal Functional Organization in Regional Human Geography," *Economic Geography*, 33 (1957), 299–336.

Philbrick, A. *This Human World* (New York: John Wiley & Sons, 1963).

Pitts, F. *Japan* (Grand Rapids, Mich.: Fideler, 1974).

Reischauer, E. *The Japanese* (Cambridge: Belknap-Harvard University Press, 1977).

Reischauer, E. *The United States and Japan* (New York: Viking, 1965).

Scheiner, E. *Modern Japan: An Interpretive Anthology* (New York: Macmillan, 1974).

Taeuber, I. *Population of Japan* (Princeton N.J.: Princeton University Press, 1958).

Takahashi, N. "A New Concept in Building: Tsukuba Academic New Town," *Ekistics*, 48 (1981), 302–306.

Tiedemann, A. *Introduction to Japanese Civilization* (New York: Columbia University Press, 1974).

Trewartha, G. *Japan: A Geography* (Madison: University of Wisconsin Press, 1965).

# PART TWO

# UNDERDEVELOPED REGIONS

ISSSTE

# CHAPTER 4

# MIDDLE AMERICA: COLLISION OF CULTURES

IDEAS AND CONCEPTS

**Historical Geography**
**Land Bridge**
**Culture Hearth (2)**
**Environmental Determinism**
**Cultural Landscape (2)**
**Transculturation**
**Mainland-Rimland Concept**
**Plural Society (2)**
**Altitudinal Zonation**
**Third World City Structure**

Middle America is a realm of vivid contrasts, matchless variety, turbulent history, current political turmoil, and an uncertain future. Physiographically, Middle America extends from the southern borders of the United States to the northern limits of the South American continent, thereby including Mexico and the smaller countries from Guatemala to Panama as well as the large and small islands of the Caribbean Sea to the east (Fig. 4–2). Culturally, crowded Middle America spills over into its neighbors—across the Rio Grande into the United States, where Spanish-speaking communities prevail in a large region (see Fig. 3–19), and onto coastal South America, where peoples and ways of life in places resemble Caribbean rather than prevailing South American norms (see Fig. 5–8, p. 291).

Middle America's northern boundary quite obviously represents a significant cultural transition, but some scholars argue that a differentiation between Middle and South America is unnecessary since both are part of a larger "Latin" American realm; others suggest a more appropriate regional delineation lies between the Brazilian-Caribbean area, with its strong (if varied) European and African plantation heritage, and mainland Spanish America, with its notable highland Indian and Castilian imprints. Certainly, there are numerous countries in Middle America that share their dominant cultural imprints with South America, notably the Spanish language and Roman Catholic religion. But in Middle America, to a far greater degree than in South America, there is variety. Large population groups have African and Asian as well as European ancestries. Nowhere in South America has Indian culture contributed to modern civilization as strongly as it has in Mexico. The Caribbean area is a patchwork of independent states, territories in political transition, and dependencies.

**Figure 4–1**
Cultural contact in explosive expression: old and new in the landscape of Mexico City.

The Dominican Republic speaks Spanish, adjacent Haiti uses French; Dutch is spoken in Curaçao and its neighbors, whereas English is spoken in Jamaica. Middle America, thus, gives vivid definition to concepts of cultural-geographical pluralism.

Middle America's cultural diversity is matched by its highly varied physiography. The numerous Caribbean islands range from nearly flat, coral-fringed platforms to mountainous, volcanic landscapes; some have an area of less than 1 hectare (2.5 acres), whereas others extend over thousands of square kilometers (the largest, Cuba, has 115,000 square kilometers [45,000 square miles], slightly larger than Tennessee). Most of these islands represent the tops and crests of mountain ranges that rise from the floor of the Caribbean Sea and protrude above the water. Like the mountains of the Middle American mainland, these ranges are susceptible to earthquakes and volcanic eruptions. Because hurricanes also affect the Caribbean islands and mainland coasts, the Middle American environment is unusually hazardous.

The Middle American mainland is funnel-shaped, widest in northern Mexico, then narrowing irregularly to a slim 65 kilometers (40 miles) in Panama. This 6000-kilometer (3750-mile) connection between North and South America is defined by physical geographers as a *land bridge*. Because sea level varies over time, such land bridges come and go. When the level of the world ocean was lower, Australia and New Guinea were connected, as were the Soviet Far East and Alaska across what is now the Bering Strait. Geologic processes combine with sea-level changes (which are related to global glaciation trends) to create and destroy land bridges, but these overland links have, undoubtedly, played crucial roles in the distribution of animals and peoples in various parts of the world. Many scholars believe the Bering land bridge was the gateway for the first humans entering the Americas; the Middle American land bridge still rises between Atlantic and Pacific waters, and it, too, has been a factor in the dispersal of Indian peoples across the Americas. We return to this theme in Chapter 5, but it is important to note here that the Middle American mainland affords no easy

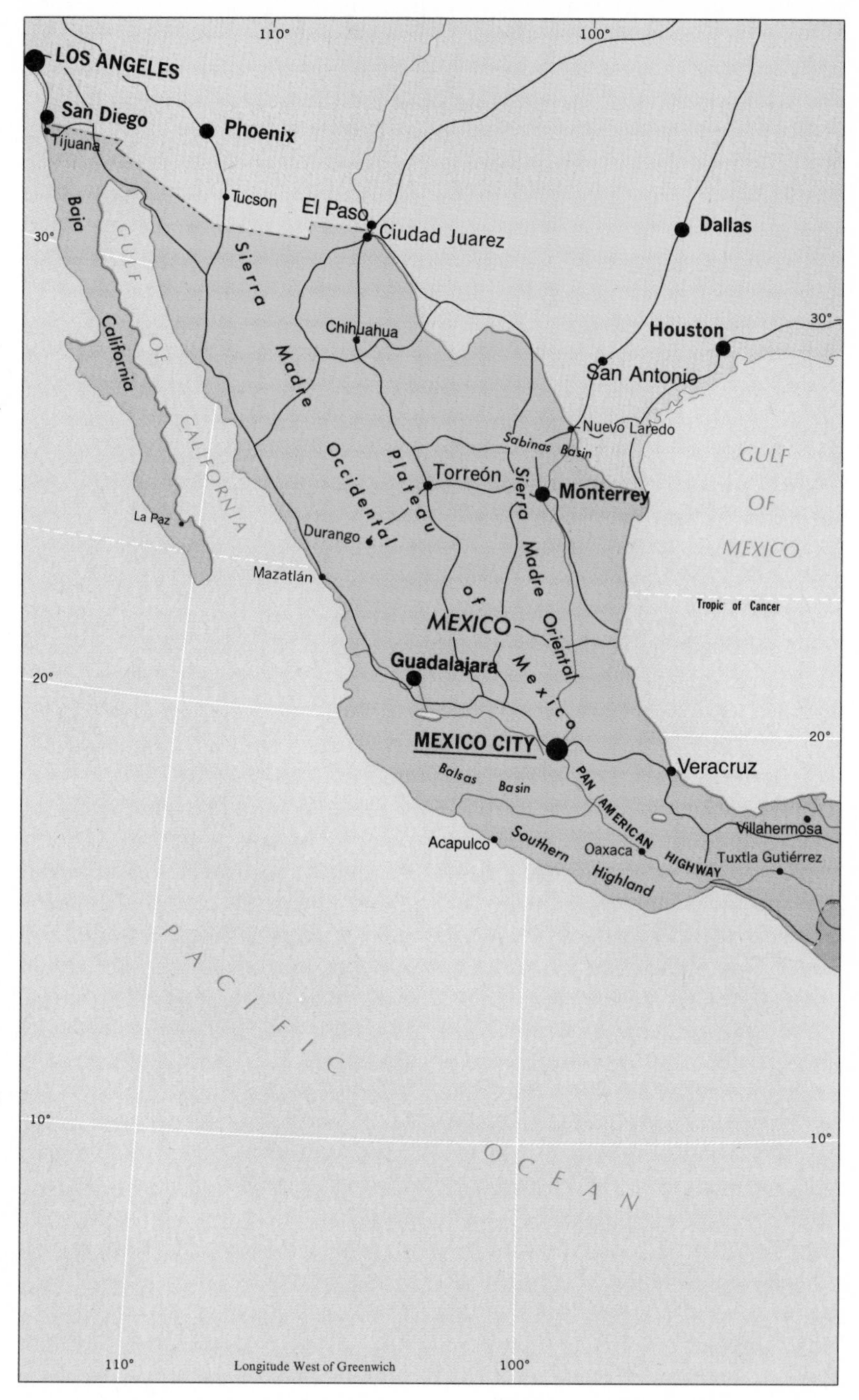


Figure 4–2

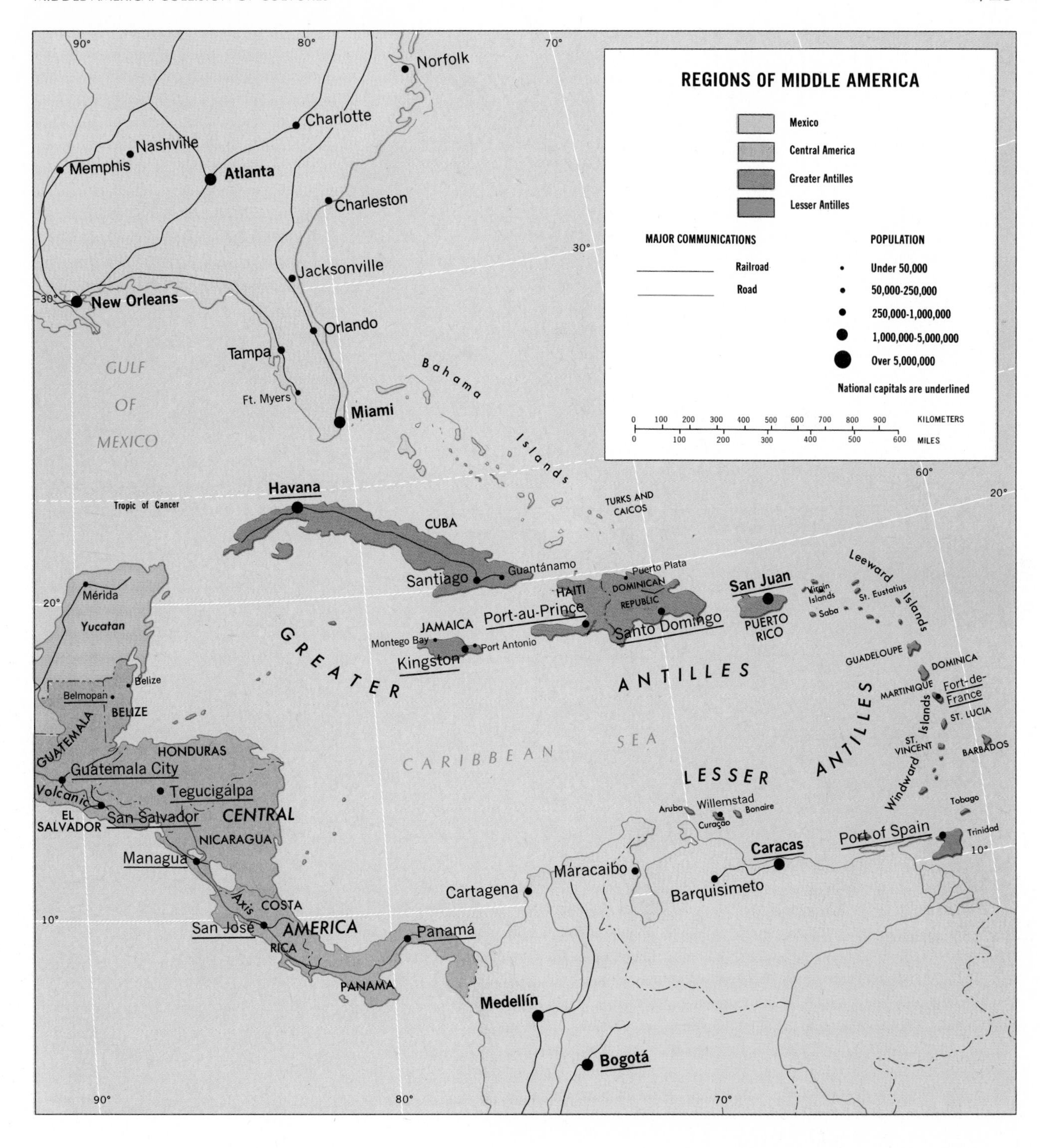


passage. In the north, the extensive Mexican Plateau is flanked by high mountain ranges, and Central America (see *Box*) is dominated by a mountainous, volcanic, earthquake-prone, lake-studded backbone. Coastlands are usually swampy, and dense and nearly impenetrable forest covers interior lowlands. Forested eastern Panama has long delayed the completion of the intercontinental Pan-American Highway across what is still the Darien gap. And the construction of the Panama Canal across the 65-kilometer-wide (40-mile-wide) *isthmus* in the early years of this century was a truly herculean effort, which would still pose a formidable challenge to the earth-moving technology we possess today.

SYSTEMATIC ESSAY

# Historical Geography

Even though our survey of world realms and spatial concepts is only one-third complete, it is already quite evident that the geographical present is a product of the past; a glance at the opening image in this chapter (Fig. 4–1) shows that observation to be just as true of underdeveloped world regions as it was in the developed realms we have already covered. The study of how regions evolve and continue to change is the domain of **historical geography**. Because it treats the all-encompassing dimension of time, the concepts and methods of historical geography—like those of regional geography—are applicable to every branch of the discipline. Historical geography, however, is clearly distinct from the field of history—it is dominated by the work of geographers whose main task is the interpretation of spatial changes on the earth's surface.

The rich historical dimension of regional geography was amply demonstrated in Chapters 1 through 3 and, indeed, is indispensable to an understanding of contemporary landscapes and societies; for example, the currently transforming distribution of economic activities within U.S. metropolitan areas, as we saw in Chapter 3, is best understood as a response to changing geographic forces as the ending industrial age gives way to its postindustrial successor. These dynamics remind us again of the central role of *spatial processes*, the interplay of causal forces that act and unfold over time to shape the spatial distributions we observe today. In turn, these geographical patterns—as cumulative responses to the various processes that shaped them—may be regarded as "stills" or individual frames in a relentlessly advancing film. Thus, spatial organization at any given moment of time is a point in a long continuum and is clearly related to what went before (as well as to what will develop in the future). Appreciating this central concern, historical geographers today approach their broad and diverse subdiscipline from a number of different (although not mutually exclusive) perspectives.

The first of these might be called the study of *the geographic past*, in which earlier "typical" spatial patterns are carefully reconstructed and compared for certain time periods that were crucial in the formation of the present regional structure. The raw materials for assembling these snapshots of the past come from a variety of sources: historical maps, books, censuses, public and private archives, and the interviewing of long-time residents of the study area. Besides the valuable portraits of individual regions that emerge from such scholarly investigations, this approach can also be used to build evolutionary models of spatial organization: the four-stage model of U.S. intraurban growth (Fig. 3–14, p. 197) is a classic example, with the structural pattern of each era shaped by the dominant transportation technology. Another fruitful avenue is that of *simulation modeling*, in which processes are regarded as "rules" that govern the actions of "players" in a "game" that deliberately resembles geographical reality; as each "round" of the game is played, a new spatial pattern results. The object is to fine-tune the model, so that the simulation of process and structure most closely approximates real-world outcomes—thereby replicating the forces that have actually shaped the region's organization (and providing important clues as to the directions of future change).

The second perspective focuses on *landscape evolution*, which entails the historical analysis and interpretation of current cultural landscapes. This search for "the past in the present" takes a number of forms. The study of individual relics such as Hadrian's Wall (Fig. I–18) is one approach. Another is comparative regional analysis of such landscape artifacts as house types, which facilitates the mapping of cultural change. Some researchers even perform simulation experiments; for instance, anthropologist Thor Heyerdahl recently attempted (with partial success) to sail his crude papyrus boat *Ra* from North Africa to Mexico in order to demonstrate that the ancient Egyptians possessed seafaring technology sufficient to transmit such innovations as the pyramid across the Atlantic.

A third approach to historical geography can be called *perception of the past*, the application of modern behavioral perspectives to recreate and study the attitudes of previous generations toward their environments. Migration decision making is of particular interest where it involved the acquisition of accurate spatial information on the "mental maps" of potential migrants; for instance, before the U.S. Civil War, the Great Plains were widely perceived as part of a "Great American Desert," a myth that helped delay the settling of this region until the late nineteenth century. Abstractions are often valuable here as well; a classic was Ralph Brown's (now out-of-print) book, *Mirror for Americans*, a highly creative reproduction of the geography of the U.S. eastern seaboard in 1810 as seen through the eyes and mental images of a fictitious Philadelphian named Thomas Keystone. A more recent example of successful generalizing is Donald Meinig's analysis of symbolic historical landscapes, from which he derives three idealized American communities that still exert a powerful hold on residential preferences throughout the United States—the New England village, the small-town Main Street, and the California horticultural suburb. More formal modeling approaches are now appearing as well, a promising one being the *time-geography* perspective developed by the Swedish geographer Torsten Hägerstrand, in which the allocations of time to perform various tasks in space is measured and regionally analyzed.

One of the most active areas of historical-geographical research—which has effectively integrated the various approaches to the subject—is the study of *settlement geography*, the facilities people build while occupying a region; because these facilities are usually durable and, therefore, likely to survive beyond the era of their original functions, they often provide some of the clearest expressions of the past in the contemporary landscape. The colonial landscapes of the New World, because of their sharp contrasts and relative recency, have received considerable attention, especially the layout of towns that represented the focal points in the evolving human settlement fabric. We now examine the Spanish colonial town of Middle America—a vital cultural interface between Old and New World.

The layout of this urban-settlement type is displayed in Fig. 4–3, and its historical-geographical significance is described by Charles Sargent:

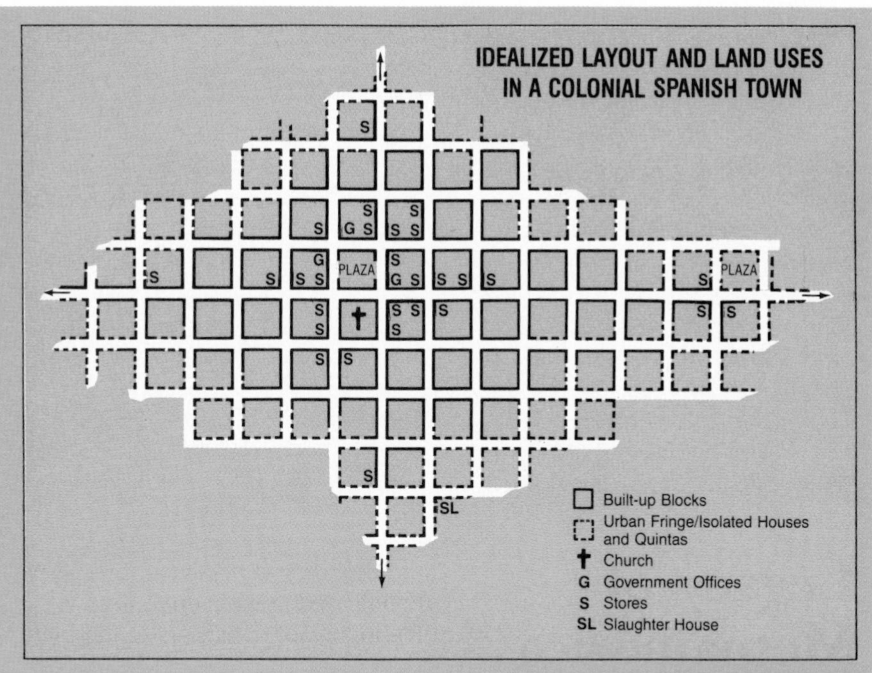




Figure 4–3

> *The morphology, or form, of any city reflects its past, and the clearest reflections are found in the street pattern, the size of city blocks, the dimensions of urban lots, and the surviving colonial architecture. In the typical Spanish American city, this colonial past is today on view in what is now the city core, commonly laid out in a gridiron of square or rectangular blocks subdivided into long, narrow lots along equally narrow streets.*

This particular spatial pattern hardly came about by accident—it was decreed by royal regulations down to the smallest details of street, building, and central-plaza construction; by the late sixteenth century, these rules became codified in the *Laws of the Indies* and assured regularity in the morphology of towns and cities throughout Hispanic America (including former Spanish-controlled areas of the southwestern United States such as New Mexico). Since most Spaniards were urban dwellers, it was hardly surprising that they chose exclusively to build towns in their colonial territories. But why this particular town form? The answer rests with that other major quality of an urban settlement—*function*.

The Spanish colonial town possessed several functions that, taken collectively, were best suited to the compact gridiron-street-pattern layout. Since the main function of the settlement was administrative (a point developed later in this chapter), town sites were chosen to maximize accessibility to regional trade routes and sources of tribute from local Indians. This control function also extended to the internal structure of the town: everything was tightly focused on the central plaza or market square, under the watchful eye of government authorities in adjacent buildings (G symbols in Fig. 4–3). It is often said that the threefold aims of the rulers of New Spain were "God, glory, and gold." The last two are embodied in the trade- and tribute-related functions of towns; the first is another leading town function, expressed in the central role of the church, which always faced the plaza as well (Fig. 4–3). The main activity of the church was to convert as many Indians to Christianity as possible. The easiest way to do this was forcibly to resettle the dispersed aboriginal population in

Spanish towns, where the collection of tribute, the recruitment of mine workers, and the farming of land surrounding the town were also facilitated. The congregation of large numbers of non-Spaniards within the settlement may also have had much to do with the choice of the gridiron street plan (which was not a feature of urban centers in Spain): with so many potentially hostile people living in the town, insurrections could readily be contained by having a small superior military force seal off the affected blocks and root out the troublemakers. The Greeks and Romans learned this lesson when they established far-flung empires, and it is very likely that the grid-plan tradition was passed down to their Mediterranean European successors. The idea even has modern applications: the battles for Algiers and Saigon (nongridiron cities) in the 1950s and 1960s favored guerillas who could move at will through twisting streets and alleyways, whereas the urban riots of 1967 in gridded Newark and Detroit were squelched in a matter of days by the U.S. Army, which systematically surrounded and pacified block after block.

Historical geography will continue to be a prominent theme in each subsequent chapter. Two additional important concepts that involve the meshing of time and space will also be introduced: *spatial diffusion* in Chapter 6 (pp. 334–338) and *sequent occupance* in Chapter 7 (p. 410).

# Legacy of Mesoamerica

Mainland Middle America was the scene of the emergence of a major ancient Indian civilization. Here lay one of the world's true culture hearths (see Fig. 6–4, p. 330), an area where population could increase and make significant advancements. Agricultural specialization developed, urbanization occurred, transport networks matured, and writing, science, art, architecture, religion, and other spheres of achievement witnessed major progress. Anthropologists refer to the Middle American culture hearth as *Mesoamerica*, and what is especially remarkable about this development is that it happened in very different geographic environments. In the low-lying, tropical, hot and humid plains of Honduras, Guatemala, Belize, and Mexico's Yucatan Peninsula, the Mayan civilization arose. On the higher plateau of present-day Mexico, the Aztecs founded their well-organized civilization. In the process, the Maya and the Aztecs overcame some serious environmental obstacles. Mayan Yucatan may not have been as hot and humid as it is today, but the integration of so large an area was a huge accomplishment; the Aztecs solved problems of distance and managed to unify people over a wide area, despite the topographic barriers of the interior upland (Fig. 4–4).

In the 1910s and 1920s, a school of thought developed in American geography that favored the view that human cultural, polit-

## TEN MAJOR GEOGRAPHIC QUALITIES OF MIDDLE AMERICA

1. Middle America is a fragmented region that consists of all mainland countries from Mexico to Panama and all the islands of the Caribbean Sea.
2. Middle America's mainland constitutes a crucial barrier between Atlantic and Pacific waters. In physiographic terms, this is an intercontinental *land bridge*.
3. Middle America's tropical location and climates are in important places ameliorated by altitude and its resultant vertical zonation of natural environments.
4. Middle America is a realm of intense cultural and political fragmentation. The political geography defies unification efforts, and instability is at an all-time high in the 1980s.
5. Middle America's cultural geography is complex. African influences dominate in the Caribbean, whereas Indian traditions survive on the mainland.
6. Middle America's historical geography is replete with involvement by its powerful neighbor, the United States.
7. Underdevelopment is endemic in Middle America; the realm contains the Americas' least-developed territories.
8. In terms of area, population, and economic strength, Mexico dominates the realm.
9. Out-migration of population for economic and political reasons prevails.
10. The realm (notably Mexico) contains major actual and potential reserves of mineral fuels.

## MIDDLE AMERICA AND CENTRAL AMERICA

*Middle America,* as we define it, includes all the mainland and island countries and territories that lie between the United States of America and the continent of South America. Sometimes the term *Central America* is used to identify the same realm, but Central America actually is a region within Middle America. Central America comprises the republics that occupy the strip of mainland between Mexico and Panama: Guatemala, Honduras, El Salvador, Nicaragua, and Costa Rica. The self-governing British colony of Belize (formerly British Honduras), adjacent to Guatemala, is also part of Central America. Panama itself is regarded here as belonging to Central America as well; however, it should be noted that many Central Americans do not consider Panama to be part of their realm, because that country was, for the most of its history, a part of South America's Colombia.

ical, and economic progress can only occur under particular environmental circumstances. Pointing to the present concentration of wealth and power in the middle latitudes, especially in the Northern Hemisphere, these geographers postulated that the tropics were simply not conducive to human productivity. Mostly, they related their views to one aspect of the environment—climate. Tropical climates, with their monotonous heat and humidity, were supposed to retard human progress; mid-latitude climates, with their variable weather, were purported to stimulate human achievement. This school of *environmental determinism* (or simply, *environmentalism*) held that the natural environment, principally climate, to a large extent dictates the course of civilization. Its leading proponent and popularizer was Ellsworth Huntington (1876–1947), who believed that human progress rested on three bases: climate, heredity, and culture. This, of course, is a controversial area of research, but Huntington did not engage in idle speculation. He wrote no fewer than 28 books and contributed to 30 others (in addition to some 240 articles) in his attempt to measure the influence of climate on civilization. Nevertheless, his conclusions came under severe criticism as ill founded and supportive of "master race" philosophies. Although he

**Figure 4–4**

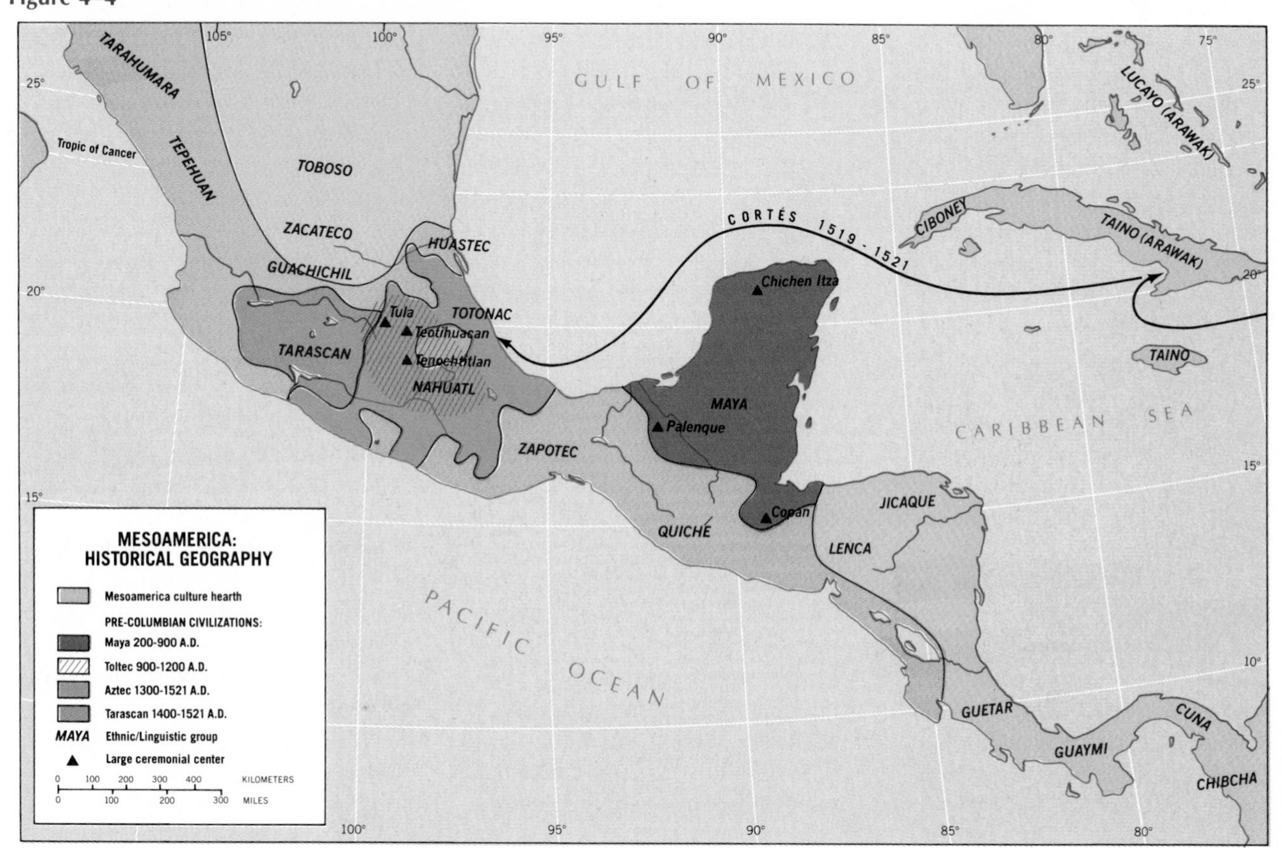

may have generalized carelessly and spoken injudiciously, Huntington was posing crucial questions still unanswered today—and now the sociobiologists, going at them from another angle, face similar obstacles. Human societies and natural environments interact, but how? When does a combination of particular environmental circumstances, inherited capacities, and cultural transmissions stimulate a new cultural explosion?

Certainly, few environmentalists could easily account for the rise of Mayan civilization in tropical Middle America. Some cultural geographers have suggested that cultural stimuli from ancient Egypt actually reached Middle American shores and that the pyramidlike stone structures of Mayan cities represent imitations and variations of Egyptian achievements. More likely, Mayan civilization arose as a spontaneous and independent accomplishment. It experienced successive periods of glory and decline, reaching its zenith—in present-day Guatemala—from the fourth to the tenth centuries A.D.

**Figure 4–5**
The ruins at Tikal, Guatemala testify to the scale of the Mayan achievement.

The Mayan civilization unified an area larger than any of the modern Middle American states except Mexico. Its population was probably somewhere between 2 and 3 million; the Mayan language—the local *lingua franca*—and some of its related languages remain in use in the area to this day. This was a theocratic state with a complex religious hierarchy, and the great cities that today lie in ruins, overgrown by tropical vegetation, served in the first instance as ceremonial centers. Their structures testify to the architectural capabilities of the Maya and include huge pyramids and magnificent palaces (Fig. 4–5). Intricate stone carvings, impressive murals, and detailed ideograms hewn into palace walls tell us that Mayan culture had its artists and writers. We know also that they were excellent mathematicians, had accumulated much knowledge of astronomy and calendrics, and were well ahead of the world's other advanced peoples in these endeavors. No doubt the Mayan civilization also had its poets and philosophers, but it had a practical side too. In agriculture and trade, these people accomplished a great deal. They grew cotton, had a rudimentary textile industry, and even exported finished cotton cloth by seagoing canoes to other parts of Middle America in return for, among other things, the cacao so highly prized by the Maya. This involvement in foreign trade is yet another reflection of the degree of organization that was achieved by this empire.

Thus, tropical lowland Middle America was the scene of impressive cultural achievements—but no more so than the uplands. At about the same time that the Mayan civilization emerged in the forests, the Mexican highland, too, witnessed the rise of a great Indian culture, similarly focused on ceremonial centers and also marked by major developments in agriculture as well as in architecture, religion, and the arts. Certainly, these achievements were comparable to those of the Maya, but they were to be overshadowed by what followed.

One of the important successors to the early highland civilization were the Toltecs, who moved into this area from the north, conquered and absorbed the local Indian peoples, and formed a powerful state centered on one of the first true cities of Middle America, Tula. The Toltecs' period of hegemony in the region was relatively brief, lasting for less than three centuries after their rise to power around A.D. 900. But during this period, they conquered parts of the Mayan domain, absorbed many of the Mayas' innovations and customs, and introduced them on the plateau. When their state was, in turn, penetrated by new elements from the north, it was already in decay. Still, Toltec technology was not lost: it was readily adopted and developed by their successors, the Aztecs.

The Aztec state, the pinnacle of organization and power in Middle America, is thought to have originated in the early fourteenth century when a community of Nahuatl- (or Mexicano-) speaking Indians founded a settlement on an island in one of the many lakes that lie in the Valley of Mexico. This village and ceremonial center, named Tenochtitlan, was soon to become the greatest city in the Americas and the capital of a large and powerful state. Through a series of alliances with neighboring peoples, the early Aztecs gained control over the whole Valley of Mexico, the pivotal geographic feature of Middle America and today still the heart of the modern state of Mexico. This 49-by-65 kilometer (30-by-40 mile) region is, in fact, a mountain-encircled basin positioned nearly 2500 meters (about 8000 feet) above sea level. Elevation and interior location both affect its climate; for a tropical area, it is quite dry and very cool. A number of lakes lie scattered through the region, and these water bodies formed valuable means of internal communication for the Aztec state. The Indians of Middle America never developed the wheel; so, they relied heavily on porterage and, where possible, the canoe for the water transportation of goods and people. The Aztecs connected several of the Mexican lakes by canals and maintained a busy canoe traffic on their waterways. Tens of thousands of canoes carried agricultural produce to the cities as well as tribute paid by its many subjects to the headquarters of the rulers and nobility.

Throughout the fourteenth century, the Aztec state strengthened its position, developing a strong military force and organizing its territory into provinces ruled by their respective governors and district commissioners. By the early fifteenth century, the Aztecs were ready to launch their conquest of neighboring peoples.

The Aztec drive to expand the empire was directed primarily eastward and southward. To the north, the rugged and unproductive land quickly became drier, inhabited only by a sparse nomadic population. To the west lay a powerful competing state, that of the Tarascans, with whom the Aztecs sought no quarrel, given the openings to the east and south. Here, then, they penetrated and conquered almost at will. The Aztec objective was not the acquisition of territory, but the subjugation of peoples and towns for the purpose of extracting taxes and tribute. They did not introduce their concepts of religion nor their language; they were not "colonists" in the European sense. On the other hand, they carried off thousands of people for purposes of human sacrifice in the ceremonial centers of the Valley of Mexico, a practice that would have made colonization rather difficult and "pacification" a self-defeating aim. The Aztecs needed a constant state of enmity with weaker people in order to take their human prizes.

As Aztec influence spread throughout Middle America, the volume of goods streaming back to the Valley of Mexico increased. Gold, drawn largely from the Balsas Basin, cacao beans, mostly from coastal areas, cotton and cotton cloth, the feathers of tropical birds, and the skins of wild animals were among the items that were carried back to Tenochtitlan and its surroundings. The state grew ever richer, its population mushroomed, and its cities expanded. Tenochtitlan probably had over 100,000 inhabitants at its peak; some put the estimate as high as a quarter of a million. These cities were not just ceremonial centers, but true cities, with a variety of economic and political functions and large populations among which there were labor forces possessing particular skills and specializations.

Aztec civilization produced an enormous range of impressive accomplishments, although the Aztecs seem to have been better borrowers and refiners than they were innovators. They practiced irrigation by diverting water from streams via canals to farmlands, and they built elaborate walls to terrace hillslopes where soil erosion threatened. Indeed, when it comes to measuring the legacy of the Mesoamerican Indians to their successors and, indeed, to the world, then the greatest contributions surely come from the field of agriculture. Corn (maize), various kinds of beans, the sweet potato, a variety of manioc, the tomato, squash, the cacao tree, tobacco—these are just a few of the crops that grew in Mesoamerica when the Europeans first made contact.

# Collision of Cultures

We in the Western world all too often are under the impression that history began when the Europeans arrived in some area of the world and that the Europeans brought such superior power to the other continents that whatever existed there previously had little significance. Middle America bears out this perception: the great, feared Aztec state fell before a relatively small band of Spanish invaders in an incredibly short time. But let us not lose sight of a few facts. At first, the Spanish were considered to be "White Gods," whose arrival was predicted by Aztec prophecy; having entered Aztec territory, the earliest Spanish visitors were able to determine that great wealth had been amassed in the Aztec cities. And Hernán Cortés, for all his 508 soldiers, did not single-handedly overthrow the Aztec authority. What Cortés brought on was a revolt, a rebellion by peoples who had fallen under Aztec domination and who had seen their relatives carried off for human sacrifice to Aztec gods. Led by Cortés with his horses and artillery, these Indian peoples rose against their Aztec oppressors and followed the Spanish band of men toward Tenochtitlan, where thousands of them died in combat against the Aztec warriors. They fed and guarded the Spanish soldiers, maintained connections for them with the coast, carried supplies from the shores of the Gulf of Mexico to the point Cortés had reached, and they secured and held captured territory while the white men moved on. Cortés started a civil war; he got all the credit for the results. But it is reasonable to say that Tenochtitlan would not have fallen so easily to the Spaniards without the sacrifice of many thousands of Indian lives.

Actually, in the Americas as well as in Africa, the Spanish, Portuguese, British, and other European visitors considered many of the peoples they confronted to be equals—equals to be invaded, attacked, and, if possible, defeated—but equals nonetheless. The cities and farms of Middle America, the urban centers of West Africa, the great Inca roads of South America, all reminded the Europeans that technologically they were, if anything, a mere step ahead of their new contacts. If a gap developed between the European powers and the indigenous peoples of many other parts of the world, it only emerged clearly when the industrial revolution came to Europe—centuries after Vasco da Gama, Columbus, and Cortés.

In Middle America, the confrontation between Hispanic and Indian cultures spelled disaster for the Indians in every conceivable way. The quick defeat of the Aztec state (as well as that of the Tarascans) was followed by a catastrophic decline in population. Whether there were 15 or 25 million Indians in Middle America when the Spaniards arrived (estimates vary), after only one century, just 2.5 million survived. The Spanish were ruthless colonizers, but not much more so than other European powers that subjugated other cultures. True, the Spanish first enslaved the Indians and were determined to destroy the strength of Indian society. But biology accomplished what ruthlessness could not have achieved in so short a time. As in the Caribbean islands, the Indians were not immune to the diseases the Spaniards introduced by their presence: smallpox, typhoid fever, measles, influenza, and mumps. Neither did they have any protection against diseases introduced by white people through their African slaves—such as malaria and yellow fever, which took enormous tolls in the hotter and more humid lowland areas of Middle America.

Middle America's cultural landscape, its great cities, its terraced fields, and the dispersed villages of the Indians were drastically modified. The Indian cities ceased to function as they had before, and the Spanish brought to Middle America new traditions and innovations in urbanization, agriculture, religion, and other pursuits. Having destroyed Tenochtitlan, the Spaniards nevertheless recognized the attributes of its site and situation and chose to rebuild it as their mainland headquarters. Whereas the Indians had used stone almost exclusively as their building material, the Spaniards employed great quantities of wood and utilized charcoal for purposes of metal smelting, heating, and cooking. Thus, the onslaught on the forests was such that great rings of deforestation quickly formed around the major Spanish towns. And not only around the Spanish towns; the Indians also adopted Spanish methods of house construction and charcoal use and they, too, contributed to forest depletion. Soon, the scars of erosion began to replace the stands of tall trees.

The Indians had been planters but had no domestic livestock that would make demands of the original vegetative cover. Only the turkey, the dog, and the bee (for honey and wax) had been domesticated in Mesoamerica. The Spaniards, on the other hand, brought with them cattle and sheep—in numbers that multiplied rapidly and made increasing demands not only on the existing grasslands, but on the cultivated crops as well. Again, the Indians soon adopted the practice of keeping livestock, putting further pressure on the land. The net effect on food availability was not favorable. Cattle and sheep became avenues to wealth, and the owners of the herds benefited. But the livestock competed with the people for the available food—requiring the opening up of large new areas of marginal land in higher and drier locations—contributing to a gross disruption of the food-production balance that existed in the region. Therefore, hunger quickly became a major problem in Middle America during the sixteenth century and, no doubt, heightened the susceptibility of the Indian population to many diseases.

The Spaniards also introduced their own crops, notably wheat, and their own farming equipment, of which the plow was the most important. Thus, large fields of wheat began to make their appearance alongside the small plots of corn of the Indian cultivators. Inevitably, the wheatfields encroached on the Indians' lands, reducing them further. The wheat was grown by and for the Spaniards, so that what the Indians lost in farmlands was not made up in available food. And neither were their irrigation systems spared. The Spaniards needed water for their fields and power for their mills, and they had the technological know-how to take over and modify the regional drainage and irrigation systems. This they did, leaving the Indian fields either waterless or insufficiently watered, thereby diminishing even further the Indians' chances for an adequate supply of food.

The most far-reaching changes in the cultural landscape brought by the Spaniards had to do with their traditions as town dwellers. To facilitate control, the decision was made to bring the Indians from their land into nucleated villages and towns established and laid out by the Spaniards. In these settlements, the kind of government and administration to which the Spanish were accustomed could be exercised. The focus of each town was the Catholic church; indeed, until 1565 the resettlement of Indians was mainly the responsibility of missionary orders. The location of each of these towns was chosen to lie near what was thought to be good agricultural land, so that the Indians could go out each day and work in the fields. Unfortunately, the selection was not always a good one, and a number of villages lay amid land that was not suitable for Indian farm practices. Here food shortages and even famine resulted—but once created, the village must try to survive. Only rarely was a whole village abandoned in favor of a better situation.

## Acculturation

In the towns and villages, the Indians came face to face with Spanish culture (Fig. 4–6). Here they learned the white invaders' religion, paid their taxes and tribute to a new master, or found themselves in prison or in a labor gang according to European laws and rules. Packed tightly in a concentrated settlement, they were rendered even more susceptible than they were in their own villages to the diseases that regularly ravaged the people. Despite all this, the nucleated Indian village survived. Its administration was taken over by the civil government and later by the independent governments of the Middle American states; today, it is a noteworthy feature of Indian areas of the Mexican and Guatemalan landscape. Anyone who wants to see remnants of the dispersed Indian dwellings and hamlets must travel into the most isolated parts of Middle America, where Indian languages still prevail (Fig. 4–7).

The Spanish towns and cities were administrative centers, located in the interior to function as centers of control over trade, tax collection, labor recruitment, and the like. At times the Spaniards, recognizing an especially favorable location chosen by their Indian predecessors, would build on the same site; other sites were chosen be-

**Figure 4–6**
Small town in Guatemala: a typical settlement founded by the Spanish colonizers.

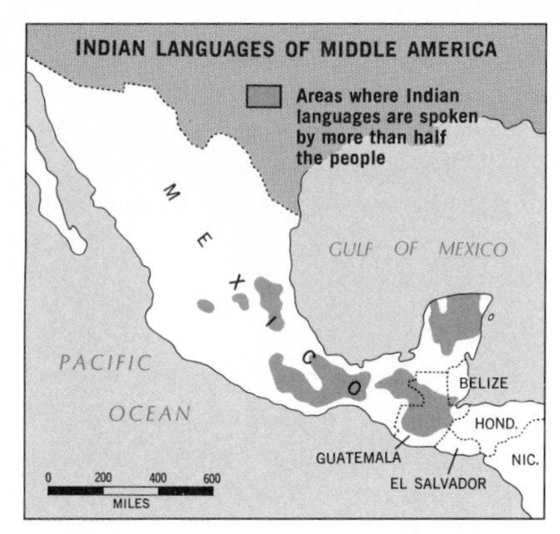


Figure 4–7

cause they served Spaniards better than preexisting Indian sites could. Within a half century of the Aztec defeat, the Spanish conquerors founded more than 50 new towns, some of which have grown into major modern cities. Along the Caribbean and Pacific coasts, cities emerged that were part of the globe-girdling chain of communications within the rising Spanish empire: Veracruz on the Gulf of Mexico rose to prominence in this way, as did Acapulco on the Pacific.

## Mining

The Spanish in Middle America had not come simply to ransack the riches in gold, silver, precious stones, works of art, and other valuables that had been accumulated by Indian kings, priests, and nobles; that aftermath of conquest was soon over. Next, the Spaniards wanted to organize the sources of this wealth for their own benefit. Mining, commercial agriculture, livestock ranching, and profitable trade were the avenues to affluence, and among these, mining held the greatest promise. The initial mining phase, during the first half of the sixteenth century, involved *placering* or the "washing" of gold from small streams carrying gold dust and nuggets. This is how the Indians had found their gold, but it was (and still remains) a small-scale operation. The Spaniards simply used the Indian placer workers for their own profit, something that was especially easy during the period of Indian slavery.

As the placers produced less and less gold and Indian slavery was abolished, Spanish prospectors began searching for other valuable minerals. They were quickly successful, finding lucrative silver and copper deposits. Between 1530 and 1570, a number of mining towns were founded that have since developed into modern urban centers, and a new phase of mining began. The initial focus lay to the southwest and west of the Valley of Mexico, but the most significant finds were made somewhat later and farther to the north, along the eastern foothills of the Sierra Madre Occidental, from Guanajuato and Zacatecas northward to the Chihuahua area (see Fig. 4–14).

The mining industry set in motion a host of changes in this portion of Middle America. Mining towns drew laborers by the thousands; the mines required equipment, timber, mules; the people in the towns needed food. Since many of the mining towns, especially those of the north, were located in dry country, irrigated fields were laid out wherever possible. Mule trains and two-wheeled carts connected farm-supply areas to the mining towns, and these towns to the coasts. More than a network of administrative towns could have achieved, the mining towns integrated and organized the Spanish domain in Middle America. And more than government and church could have done, the mining towns brought effective and permanent Spanish control to some very far-flung parts of the New Spain. Mining, indeed, was the mainstay of colonial Middle America.

# Mainland and Rimland

Outside Mesoamerica, only Panama, with its twin attractions of interoceanic transit and gold supply, became an early focus of Spanish activity. The Spaniards founded the city of Panama in 1519 and, apart from their use of the area of the modern Canal Zone as an Atlantic-Pacific link, their main interest lay on the Pacific side of the isthmus. From here, Spanish influence began to extend northwestward into Central America. Indian slaves were taken in large numbers from the densely peopled Pacific lowlands of Nicaragua and shipped to South America via Panama. The highlands, too, fell into the Hispanic sphere; before the middle of the sixteenth century, Spanish exploration parties based in Panama met those moving southeastward from Mesoamerica.

But the leading center of Spanish activity was in what is today central and southern Mexico, and the major arena of international competition in Middle America lay not on the Pacific side, but on the islands and coasts of the Caribbean. Here the Spaniards faced the English, French, and Dutch, all interested in the lucrative sugar trade, all searching for instant wealth, and all seeking to expand their empires (Fig. 4–8). Only the English gained a real foothold on the mainland, which otherwise remained an exclusively Spanish colonial domain. Belize (formerly British Honduras) is all that remains of a coastal sphere of English influence that once extended from the Yucatan Peninsula to the Nicaragua-Costa Rica boundary. For understandable reasons, the Spaniards were more interested in securing their highland holdings than in amassing territory in the rainy, disease-ridden, forested, swampy coastal areas, and

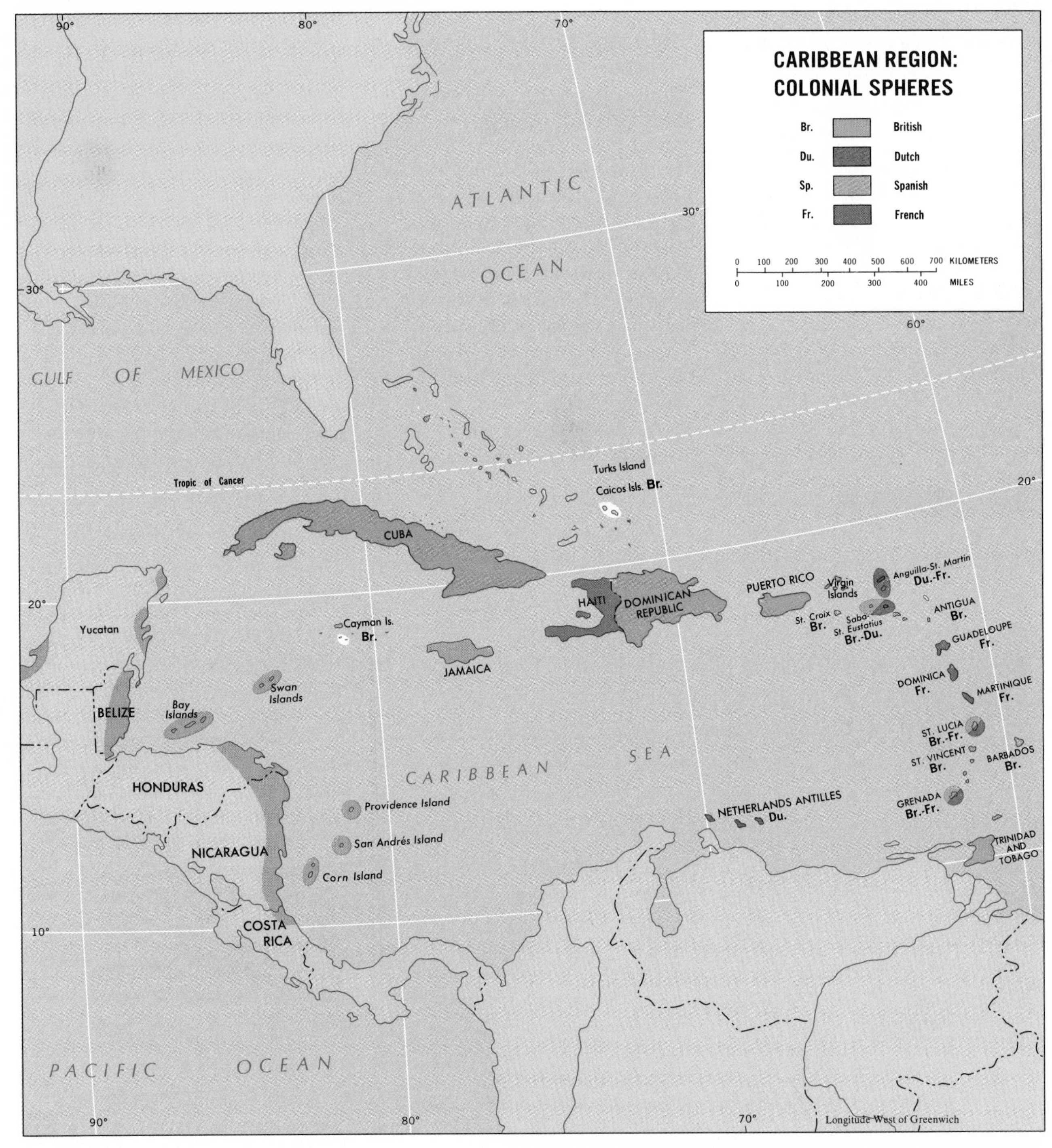


Figure 4–8

they formally recognized lowland British interests in 1670. Later, after centuries of European colonial rivalry in the Caribbean, the United States entered the picture and made its influence felt in the coastal areas of the mainland—not through colonial conquest, but by the introduction of large-scale and widespread plantation agriculture. The effects were as far-reaching as the colonial impact on the Caribbean islands. The economic geography of the Caribbean coastal areas was transformed, as hitherto unused alluvial soils in the many river lowlands were planted with thousands of acres of banana trees. Since the diseases the Europeans had brought to the New World had been most rampant in these hot and humid areas, the Indian population that survived was small and provided an insufficient labor force; the Indians of the Caribbean islands faced an even harsher fate (see *Box*). Tens of thousands of black laborers were brought to the coast from Jamaica and other islands, completely altering the demographic situation. In many physical ways, the coastal belt already resembled the islands more than the Middle American plateau, and now the economic and cultural geography of the islands were extended to it.

These contrasts between the Middle American highlands, on the one hand, and the coastal areas and Caribbean islands, on the other hand, were conceptualized by John Augelli into a *Mainland-Rimland* framework. Augelli recognized (1) a Euro-Indian *Mainland*, consisting of mainland Middle America from Mexico to Panama, with the exception of the Caribbean coast from Yucatan southeastward, and (2) a Euro-African *Rimland*, comprising this coastal zone and the islands of the Caribbean (Fig. 4–9). The terms *Euro-Indian* and *Euro-African* underscore the cultural heritage of each region: on the Mainland, European (Spanish) and Indian influences are paramount; in the Rimland, the heritage is European and African. As the map shows, the Mainland is subdivided into several areas on the basis of the strength of the Indian legacy; in southern Mexico and Guatemala, Indian influences are prominent, whereas, in northern Mexico and parts of Costa Rica, Indian influences are limited. Between these areas lie sectors with moderate Indian influence. The Rimland, too, is subdivided: the most obvious division is into the mainland-coastal plantation zone, on the one hand, and the islands, on the other. But the islands themselves can be classified according to their cultural heritage. Thus, there is a group of islands with Spanish influence (Cuba, Puerto Rico, and the Dominican Republic on old Hispaniola) and an additional group with other European influences, including the former British West Indies, the French islands, and the Netherlands Antilles.

The contrasts of human habitat described are supplemented by regional differences in outlook and orientation. The Rimland was an area of sugar and banana plantations, of high accessibility, of seaward exposure, and of maximum cultural contact and mixture. The Mainland, on the other hand, was farther removed from these contacts; it has been an area of greater isolation, of greater distance from all these forces. The Rimland was the region of the great plantation, and, thus, its commercial economy was susceptible to fluctuating world markets and tied to overseas investment capital; the Mainland was the region of the hacienda, more self-sufficient, and, therefore, less dependent on outside markets.

In fact, this contrast between the plantation and the hacienda in itself constitutes strong evidence for a Rimland-Mainland division. The hacienda was a Spanish institution, but the modern plantation, Augelli argued, was the concept of Europeans of more northerly origin (later it also became an Anglo-American one). In the hacienda, Spanish landowners possessed a domain whose productivity they might never push to its limits: the very possession of such a vast estate brought with it the social prestige

## THE ISLAND INDIANS

Mesoamerican Indian cultures have in some measure survived the European invasion. Indian communities still remain, and Indian languages continue to be spoken by about 3 million people in southern Mexico and Yucatan and another million in Guatemala (Fig. 4–7). But in the islands of the Caribbean, the Indian communities were smaller and more vulnerable. At about the time of Columbus's arrival, there probably were about 1.5 million Indian residents in the Caribbean, a majority of them in Cuba, Jamaica, Puerto Rico, and Hispaniola (the island containing modern-day Haiti and the Dominican Republic). These larger islands (the *Greater Antilles*) were peopled by the Arawaks, whose farming communities raised root crops, tobacco, and cotton. In the eastern Caribbean, the smaller islands of the *Lesser Antilles* (Guadeloupe, Martinique, Dominica, and several others) had more recently been peopled by the adventuresome Caribs, who traversed the waters of the Caribbean in huge canoes carrying several dozen persons. When the European sailing ships began to arrive, the Caribs were in the process of challenging the Arawaks for their land, just as the Arawaks centuries earlier had ousted the Ciboney.

The Europeans quickly laid out their sugar plantations and forced the Indians into service. But Arawaks and Caribs alike failed to adapt to the extreme rigors of labor to which they were subjected, and they perished by the thousands. Some fled to smaller islands where the European arrival was somewhat delayed. After just a half century, only a few hundred survived in the dense interior forests of some of the islands. There they were soon joined by runaway African slaves (brought to the Caribbean to replace the dwindling Indian labor force), and, with their mixture, the last pure Caribbean Indian strain disappeared forever.

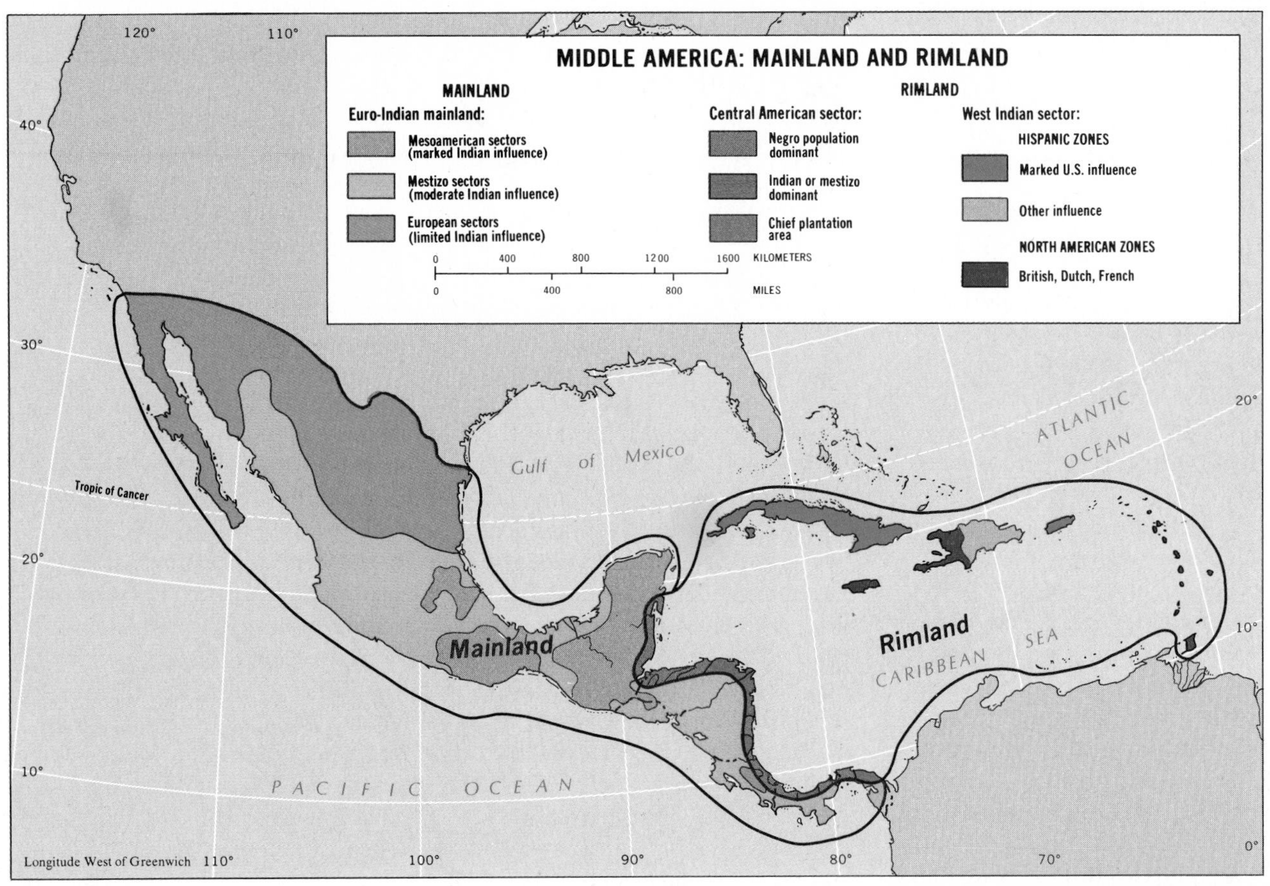


**Figure 4–9**

*Source:* Adapted from J. P. Augelli.

and comfortable life-style they sought. The workers lived on the land—even though it may once have been *their* land—and had plots where they could grow their own subsistence needs. Traditions survived: in the Indian villages incorporated into the early haciendas, in the methods of farming that were used, and in the means of transporting produce to the markets. All this is written as though it is mostly in the past, but the legacy of the hacienda, with its inefficient use of land and labor, is still visible throughout mainland Middle America.

The plantation, on the other hand, was conceived as something very different. In their book *Middle America: Its Lands and Peoples*, Robert West and John Augelli listed five characteristics of Middle American plantations that always apply. These quickly illustrate the differences between hacienda and plantation: (1) plantations are located in the humid tropical coastal lowlands of the region; (2) they produce for export almost exclusively—usually a single crop; (3) capital and skills are often imported, so that foreign ownership and an outflow of profits occur; (4) labor is seasonal—needed in large numbers during the harvest period but often idle at other times—and such labor has been imported because of the scarcity of Indian workers; and (5) with its "factory in the field" operation, the plantation is more efficient in its use of land and labor than the hacienda. The objective was not self-sufficiency but profit, and wealth rather than social prestige was a leading motive in the plantation's establishment.

During the past century, both the hacienda and plantation have changed a great deal. The vast U.S. investment in the Caribbean coastal zones of Guatemala, Honduras, Nicaragua, Costa Rica, and Panama transformed that area and brought a whole new concept of plantation agriculture to the region. On the Mainland, the hacienda has been under pressure from national governments that view them as economic, political, and social liabilities. Indeed, some haciendas have been parceled out to small landholders, whereas others have been pressed into greater specialization and productivity. Still other land has been placed in *ejidos*, and is communally owned by groups of families (to be discussed later with Mexico). Both institutions for centuries contributed to the different social and economic directions that have given the Mainland and Rimland their respective regional personalities.

# Political Differentiation

Mainland Middle America today is fragmented into eight different countries, all but one of which (Belize, the former British Honduras) have Hispanic origins. Largest of them all—in fact, the giant of Middle America—is the United Mexican States (Mexico), whose 1,972,500 square kilometers (762,000 square miles) constitute over 70 percent of the entire land area of Middle America (the islands included) and whose 80 million people outnumber those of all the other countries and islands of Middle America combined (Fig. 4–2). The cultural variety in Caribbean Middle America is much greater. Here Cuba dominates: its area is larger than that of all the other islands together, and its population of just over 10 million is well ahead of the next-ranking country (Haiti, with about 6 million). But, as we have seen, the Caribbean is hardly an area of exclusive Spanish influence: whereas Cuba has an Iberian heritage, its southern neighbor, Jamaica (population 2.4 million, mostly black), has a legacy of British involvement, and Haiti's strongest imprint has been French. The crowded island of Hispaniola is shared between Haiti and the Dominican Republic, where Spanish influence survives in the latter; it predominates also in nearby Puerto Rico where, however, it has been modified by the impact of the United States in this century.

In the Lesser Antilles, too, the cultural diversity is great. There are the (once Danish) American Virgin Islands; French Guadeloupe and Martinique; a group of British-influenced islands, including Dominica, Barbados, St. Lucia, and St. Vincent; and Dutch St. Maarten (shared with the French), Saba, St. Eustatius, and the A-B-C islands—Aruba, Bonaire, Curaçao—off the northwestern Venezuelan coast. Standing apart from the Antillean arc of islands is Trinidad, another former British dependency that, with its smaller neighbor of Tobago, became a sovereign state in 1962.

Although the mainland countries (except Belize, which became independent in 1981) ended their colonial status at a relatively early stage, the Caribbean islands remained colonized for a much longer time. No doubt this was due in large measure to their smaller size and insularity. Spain, which had to yield to demands and struggles for independence on the mainland, held Cuba and Puerto Rico, where there was much agitation for reform as well, until the Spanish-American War at the end of the nineteenth century. Spain ceded western Hispaniola (now Haiti) to France in 1697, Frenchmen having developed sugar plantations there, and, in 1795, gave up the rest of the island to the French. One of the earliest independent republics to emerge in the Caribbean was Haiti, whose black (95 percent) and mulatto (5 percent) population fought a successful slave revolt. In 1804, the Republic of Haiti was established, and the Haitians, having defeated the French—in the process destroying much of the economic structure of their country—then turned on their Dominican neighbors. From 1822 to 1844, Haiti ruled all of Hispaniola; then, after the Dominicans had fought themselves free, Spain briefly reestablished a colony there from 1861 to 1865. The most recent foreign intervention in Hispaniola came during the twentieth century, when the United States occupied both Haiti (1915–1934) and the Dominican Republic (1916–1924).

In the Greater Antilles, then, Spanish influence was strong. By the late nineteenth century, Cuba and Puerto Rico resembled the mainland republics in the composition of their population and in their cultural imprint. But Spain's colonial archrival, Britain, also had a share of the Greater Antilles in Jamaica, and it gained a large number of footholds in the Lesser Antilles to the southeast. The sugar boom and the strategic character of the Caribbean Sea brought the European competitors to the West Indies; in the twentieth century, the United States of America made its presence felt as well—although the plantation crop was now the banana and the strategic interest even more immediate than that of the European powers. In parts of Caribbean Middle America, the period of colonial control is only just ending. Jamaica attained full independence from Britain in 1962, as did Trinidad and Tobago. An attempt by the British to organize a Caribbeanwide West Indies Federation failed, but other long-time British dependencies, including Barbados, St. Lucia, Dominica, and Grenada, were steered toward a precarious independence nevertheless (with disastrous results in Grenada and a U.S. invasion in 1983). France, on the other hand, has made no moves to end the status of Martinique and Guadeloupe as Overseas Departments of the French national state; in 1985, the Dutch A-B-C islands were still within the Netherlands empire. The European imprint in the cultural landscape is, therefore, a strong continuing one, with Willemstad—Dutch Curaçao's capital—a particularly striking example (Fig. 4–10).

# Caribbean Patterns

Caribbean America today is a land crowded with so many people that, as a region, it is the most densely populated part of the Americas. It is

**Figure 4–10**
The Dutch colonial impress, vividly etched in the townscape of Willemstad, Curaçao.

a place of poverty and, in all too many locales, much misery and little chance for escape. In many respects, U.S.-affiliated Puerto Rico (see *Box*) and Communist Cuba constitute exceptions to any such generalizations made about Caribbean America. But, on most of the other islands, life for the average person is difficult, often hopeless, and tragically short.

All this is in almost incredible contrast to the early period of riches based on the sugar trade. But that initial wealth was gained while an entire ethnic group (the Indians) was being wiped off the Caribbean map and while another (the Africans) was being imported in bondage; the sugar revenues, of course, always went to the planters not the laborers. Then, the economy faced the rising competition of other tropical sugar-producing areas, and it soon lost its monopoly of the European market. The cultivation of sugar beets in Europe and America also cut into the sales of tropical sugar, and difficult times prevailed. Meanwhile, the Europeans helped stimulate the rapid growth of the island population, just as they did in other parts of the world. Death rates were lowered, but birth rates remained high, and explosive population increases occurred. With the declining sugar trade, millions of people were pushed into a life of subsistence, malnutrition, and hunger. Many sought work elsewhere; tens of thousands of Jamaican laborers went to the mainland coast when the Panama Canal was constructed (just prior to World War I) and when the United Fruit Company's banana plantations began to be developed there. Large numbers of British West Indians went to England in search of a better life, whereas Puerto Ricans have for decades been coming to the United States. But this outflow has failed to stem the tide of population growth; today, there are well over 30 million people on the Caribbean islands.

The Caribbean islanders just have not had many alternatives in their search for betterment. Their habitat is fragmented by water and mountains; even on the smaller islands, the amount of good, flat, cultivable land may only be a small fraction of the whole area. Although there has been some economic diversification, agriculture remains the area's mainstay; sugar is still the leading product, and it heads the export lists of Cuba, the Dominican Republic, and Jamaica. In Haiti, coffee has become the chief export (sugar now ranks second), whereas Trinidad, just off the northeastern Venezuelan coast, is fortunate in possessing sizable oil fields (petroleum accounts now for nearly 80 percent of that country's exports). In the Lesser Antilles, sugar has retained a somewhat less prominent position, having been replaced by such crops as bananas, sea-island cotton, limes, and nutmegs. Even here, in Barbados, St. Kitts, St. Lucia, and Antigua, sugar remains the leading revenue producer.

All of the crops grown in the Caribbean—Haiti's coffee, Jamaica's bananas, the Dominican Republic's cacao, the Lesser Antilles's citrus fruits as well as the sugar industry—face severe competition from other parts of the world and have not become established on a scale that could begin to have a real effect on standards of living. Those minerals that do exist in this area—Jamaica's bauxite (aluminum ore), Cuba's iron and chromium, Trinidad's oil—do not support any significant industrialization within Caribbean America itself. As in other parts of the developing world, these resources are exported for use elsewhere.

So, the vast majority of the people in this area continue to eke out a precarious living from a small plot of ground, mired in poverty and threatened by disease. The availability of food is not always adequate on Caribbean islands, and some countries—most notably Haiti—are chronically food deficient; where foodstuffs are disseminated, they tend to be sold in small markets, whose produce is dominated by a few local staples that hardly offer a balanced diet (Fig.

## PUERTO RICO: CLOUDED FUTURE?

The largest and most populous U.S. domain in Middle America is Puerto Rico, easternmost and smallest of the Greater Antilles chain. The 9000-square-kilometer (3500-square-mile) island, now with about 3.4 million inhabitants, fell to the United States during the Spanish-American War and was formally ceded to the United States as part of the resulting Treaty of Paris (1898), wherein the United States also acquired the Philippines and Guam.

Puerto Rico's struggle for independence from Spain was, thus, terminated (Cuba, on the other hand, achieved sovereignty under the same Treaty of Paris), and U.S. administration was at first difficult. Not until 1948 were Puerto Ricans allowed to elect their own governor, but, in 1952, following a referendum, the island became the autonomous Commonwealth of Puerto Rico. San Juan and Washington share governmental responsibilities in a complicated arrangement in which Puerto Ricans have U.S. citizenship but pay no federal taxes on local incomes. The political situation fails to satisfy everyone: there still is a small pro-independence segment, a fairly substantial number of voters who favor statehood, and (by most recent count) a majority who want to continue the Commonwealth status, but with certain modifications.

Puerto Rico stands in sharp contrast to Hispaniola and Jamaica. Long dependent on a single-crop economy (sugar), Puerto Rico during the postwar period industrialized rapidly as a result of tax advantages for corporations, comparatively cheap labor, governmental incentives of various kinds, political stability, and special access to the U.S. market. Today textiles—not bananas or sugar—rank as the leading export. Puerto Rico does not have substantial mineral resources, and it lies far from the U.S. core area: San Juan is about 2600 kilometers (1600 miles) from New York. These disadvantages have been overcome to a great extent by the development program started during World War II. Nonetheless, the Puerto Rican economy responds to, and reflects, the mainland economic picture; times of recession and low employment generate an exodus of islanders to the mainland, where they can qualify for welfare support, even if unable to find work. In the decade of the 1960s, the Puerto Rican population of New York City alone grew by some 450,000, accounting for the comparatively low population growth rate of the island. Association with the United States has undoubtedly speeded Puerto Rico's development. The question now is whether the island's social order and political system will be able to withstand the pressures that lie ahead.

4–11). Farm tools are still primitive, and cultivation methods have undergone little change over the generations. Inheritance customs have divided and redivided peasant families' plots until they have become so small that the owner must sharecrop some other land or seek work on a plantation or estate in order to supplement the harvest. Bad years of drought, cold weather, or hurricanes can spell disaster for the peasant family. Soil erosion constantly threatens; much of the countryside of Haiti and Jamaica is scarred by gulleys and ravines. Where soils are not eroded, their nutrients are depleted, and only the barest of yields are extracted. The good land lies under cash crops for the export trade, not food crops for local consumption. Those expanses of sugarcane and banana trees symbolize the disadvantageous position of the Caribbean countries, dependent on uncertain markets for their revenues and trapped in an international economic order they cannot change.

With such problems, it would be unlikely for Caribbean America to have many large cities; after all, there is little basis for major industry, little capital, little local purchasing power. Indeed, the figures reflect this situation, with less than one-half of the population classified as urban. Only Cuba and Puerto Rico have as much as 60 percent of their populations in urban areas; the first and third largest cities, Cuba's Havana (2.4 million) and Puerto Rico's San Juan (1.4 million), owe a great deal of their development to earlier U.S. influences. The second largest city is the Dominican Republic's Santo Domingo (2.2 million); Jamaica's capital, Kingston, and Port-au-Prince (Haiti) each contain about 800,000 inhabitants. Cities on these less fortunate islands, however, often exhibit even more miserable environments than the poorest rural areas. In Port-au-Prince, for example, slum residents are forced to do their laundry in a trickle of polluted water (Fig. 4–12); no wonder that such abysmal conditions drive thousands of Haitians away as desperate "boat people" in search of a better life elsewhere.

Figure 4–11
Local staples are disseminated through small-market trade: Bridgetown, Barbados.

## Tourism: The Irritant Industry

The cities of Caribbean America constitute a potential source of income as tourist attractions and places of call for cruise ships. The tourist industry (based in Miami) is growing again after a period of stagnation during the recession of the mid-1970s, and places such as Port Antonio (Jamaica) and Puerto Plata (Dominican Republic) have been added to cruise itineraries, which already include San Juan, Port-au-Prince, and Jamaica's Ocho Rios and Montego Bay. The Caribbean has long been known for its magnificent beaches and spectacular island landscapes, but visitors also are attracted by the night life and gambling of San Juan, the cuisine and shopping of Martinique's Fort-de-France, and the picturesque architecture of Curaçao's Willemstad (Fig. 4–10).

Certainly, tourism is a prospective money-earner for many Caribbean islands; it already ranks at or near the top in such places as the U.S. Virgin Islands, Martinique, Curaçao, and others. But tourism has serious drawbacks. The invasion of overtly poor communities by wealthier visitors at times leads to hostility, even actual antagonism among the hosts. For some island residents, tourists have a "demonstration effect," which leads locals to behave in ways that may please or interest the visitors but is disapproved by the larger community. Free-spending, sometimes raucous tourists contribute to a rising sense of local anger and resentment. Moreover, tourism has the effect of debasing local culture, which is adapted to suit the visitors' tastes. Anyone who has witnessed hotel-staged "culture" shows has seen this process. And many workers say that employment in the tourist industry is especially dehumanizing; expatriate hotel and restaurant managers demand displays of friendliness and servitude that locals find difficult to sustain. Nevertheless, the Caribbean's tourist trade has stimulated several local industries to produce for the visitors, as evidenced by Bahamian Nassau's straw market and similar activities elsewhere. One handicraft "industry" that has mushroomed with the tourist business is Haiti's visual arts: when cruise ships are in port, dockside areas are bedecked with countless paintings and carvings by local artists (Fig. 4–13).

Tourism does generate income in the Caribbean where alternatives are few, but the flood of North

Figure 4–12
Laundry day for the poorer residents of Port-au-Prince, capital of Haiti.

**Figure 4–13**
Local artwork is for sale at tourist sites throughout the Caribbean.

American tourists cannot be said to have a beneficial effect on the great majority of Caribbean residents. In the popular tourist areas, the intervention of multinational corporations and governments has removed opportunities from local entrepreneurs in favor of large operators and major resorts; tourists are channeled on prearranged trips in isolation from the local society. There are some cultural advantages to this, since tourists do not enhance international understanding when they invade Caribbean communities, but, on the other hand, it deprives local small establishments of potential income.

Tourism, then, is a mixed blessing to underdeveloped Caribbean territories. Given the region's limited options, it provides revenues and jobs where none would otherwise accrue. But there is a negative cumulative effect that intensifies contrasts and disparities: gleaming hotels tower over modest housing, luxury liners glide past poverty-stricken villages, opulent meals are served where, down the street, children suffer from malnutrition. The tourist industry contributes positively to island economies, but it strains the fabric of the local communities involved.

## African Heritage

The Caribbean region is a legacy of Africa, and there are places where cultural landscapes strongly resemble those of West and Equatorial Africa. In the construction of village dwellings, the operation of rural markets, the role of women in rural life, the preparation of certain kinds of food, methods of cultivation, the nature of the family, artistic expression, and in a host of other traditions, the African heritage can be read throughout the Caribbean-American scene.

Nonetheless, in general terms, it is still possible to argue that the European or white person is in the best position in this area, politically and economically; the *mulatto* (mixed white-black) ranks next; and the black person ranks lowest. In Haiti, for example, where 95 percent of the population is "pure" black and 5 percent mulatto, it is this mulatto minority that holds the reigns of power. In Jamaica, the British have only recently given up control—and the 17 percent "mixed" sector of the population plays a role of prominence in island politics far out of proportion to its numbers. In the Dominican Republic, the pyramid of power puts the 25-percent-white sector at the top, the 60-percent-mixed group next, and the 15-percent-black population at the bottom; here is the country that clings tenaciously to its Spanish-European legacy in the face of a century and a half of hostility from neighboring, Afro-Caribbean Haiti. In Puerto Rico, likewise, Spanish values are adhered to in the face of American cultural involvement and a nonwhite sector accounting for one-fifth of the island's 3.4 million people. In Cuba, too, the 15 percent of the population that is black has found itself less favored than the 15 percent *mestizo* (mixed white-Indian) sector and the 70 percent white Cuban population.

In very general terms, it is still possible to see on the map what were the original functions of Caribbean America's black peoples: there is still a correlation between the distribution of sugar-producing areas and black people on the islands. There was also a positive correlation between northwestern European control and the lack of Spanish control in the seventeenth century. The result was a "plantation" Caribbean with, today, blacks dominating the population of areas formerly controlled by northwest European countries (such as Haiti, Jamaica, and the Lesser Antilles), as contrasted with today's Spanish Caribbean, which exhibits a rather different population mix (e.g., Cuba and the Dominican Republic).

The composition of the population of the islands is further complicated by the presence of Asians from both China and India. During the nineteenth century, the emancipation of slaves and the ensuing localized labor shortages brought some far-reaching solutions. Accordingly, during the third quarter of the nineteenth century, some 100,000 Chinese emigrated to Cuba as indentured laborers (their numbers have dwindled considerably since then), and Jamaica, Guadeloupe-Martinique, and especially Trinidad saw nearly a quarter of a million East Indians arrive for similar purposes. To the Afro-modified forms of English and French heard in the Caribbean, therefore, can be added several Asian languages; Hindi is particularly strong in Trinidad, whose overall population is also now about one-third East Asian. The ethnic and cultural variety of Caribbean America is indeed endless.

# Mexico: Land of Troubled Revolution

Mexico is a land of distinction. This is the giant of Middle America, with a 1985 population of 80 million—exceeding that of all the other countries and islands of the realm combined—and a territory more than twice as large (Fig. 4–14). In all of Spanish-influenced "Latin" America, no country has even half as large a population as Mexico (Argentina and Colombia are next), and only a few can match Mexico's growth rate. In fact, Mexico grew so rapidly during the 1970s that its 1970 population will *double* before the end of the century.

The United Mexican States—the country's official name—consists of 31 states and the Federal District of Mexico City, the capital. About 50 percent of the Mexican population resides in towns and cities over 2500, with more than 19 million in the conurbation centered on Mexico City alone. Like all Mexico, the capital is growing at a staggering rate (about 800,000 per year); the country's second city, Guadalajara, has about 3.6 million inhabitants.

The Indian imprint on Mexican culture is extraordinarily strong. Today, about 56 percent of Mexicans are mestizos, nearly 30 percent are Indians. Only some 14 percent are Europeans, and there is a Mexican

Figure 4–14

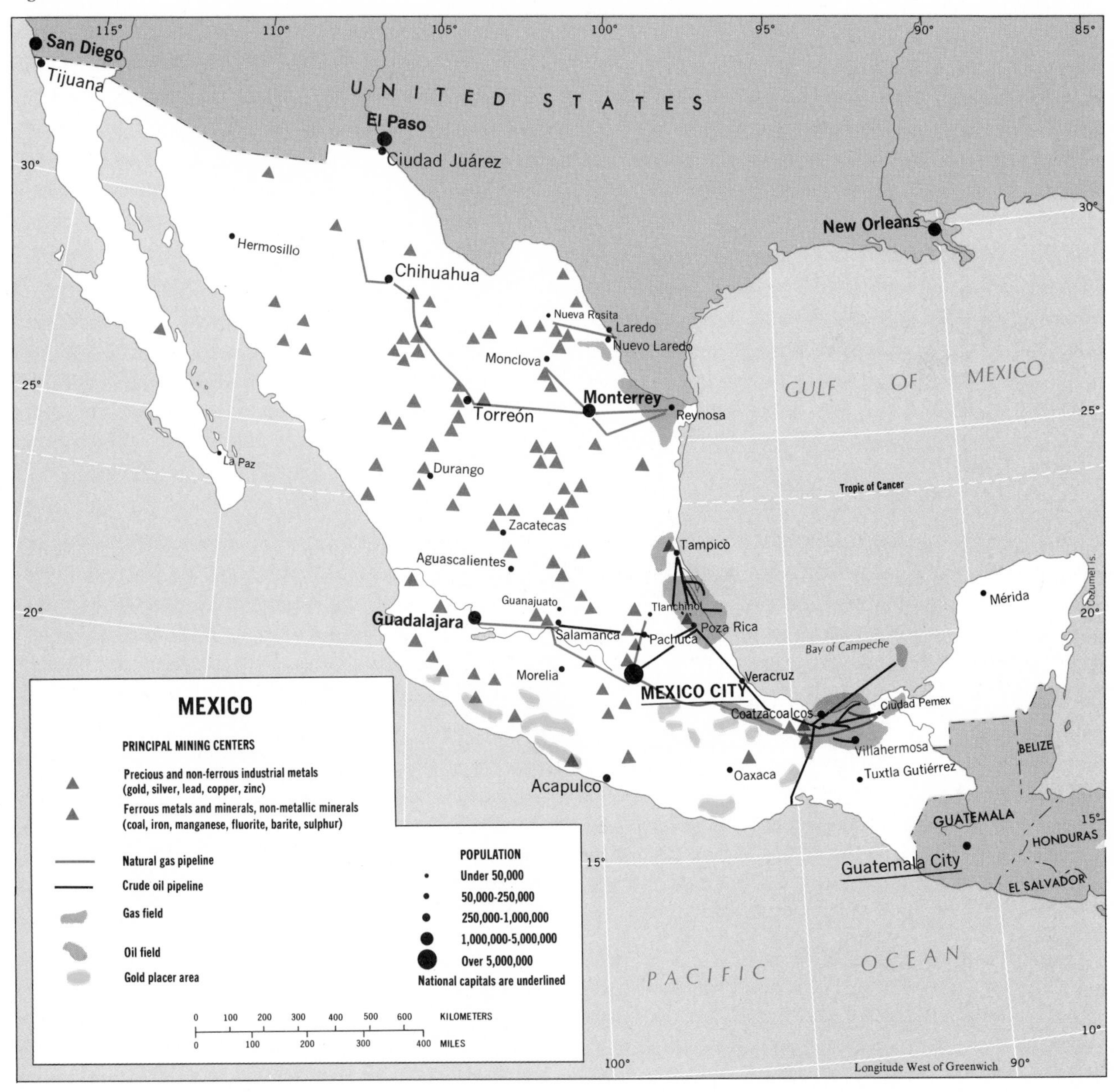

saying that Mexicans who do not have Indian blood in their veins, nevertheless, have the Indian spirit in their minds. Certainly, the Mexican Indian has been Europeanized, but the Indianization of Mexican society is so powerful that it would be inappropriate here to speak of acculturation; what happened in Mexico is *transculturation*, the two-way exchange of culture traits between societies in close contact. More than 1 million Mexicans still speak only an Indian language, and as many as 5 million still use Indian languages in everyday conversation, although they also speak Mexican Spanish (Fig. 4–7). Mexican Spanish has been strongly influenced by Indian languages, but this is only one aspect of Mexican culture that has received an Indian impress. Distinctive Mexican modes of dress, architectural styles, sculpting and painting, folkways, and foods also reflect the Indian contribution vividly.

Mexico is a large country, but its population is substantially concentrated in a zone that centers on Mexico City and extends across the southern "waist" from Veracruz in the east through Guadalajara in the west. This zone contains over half the Mexican people, as is shown on the world population distribution map (Fig. I–14). Half of Mexico's population lives in the rural areas associated with the country's better farmlands. Although some impressive developments have recently taken place in agriculture, there are still enormous difficulties to be overcome. After achieving independence from Spain in the early 1800s, Mexico failed for nearly a century to come to grips with the problems of land distribution, which were a legacy of the colonial period. In fact, the situation got worse rather than better; during the 31 years of rule by strongman Porfirio Díaz (1876–1880; 1884–1911), matters came to a head. In the first years of the twentieth century, the situation was such that 8245 haciendas covered about 40 percent of Mexico's entire area; approximately 96 percent of all rural families owned no land whatsoever, and these landless people worked as *peones* on the haciendas. There was deprivation and hunger, and the few remaining Indian lands and the scattered *ranchos* (smaller private holdings owned by mestizo or white families) could not produce enough food to satisfy the country's needs. Meanwhile, thousands of acres of cultivable land lay idle on the haciendas, which occupied just about all the good farmland of the country.

## Revolution

The revolution began in 1910 and set in motion a sequence of events that is still going on today. One of its major objectives was the redistribution of Mexico's land, and a program of expropriation and parceling out of the haciendas was made law by the Constitution of 1917. Since then, about half the cultivated land of Mexico has been redistributed, mostly to peasant communities consisting of 20 families or more. Such communally owned lands are called *ejidos*, and they are an adaptation of the old Indian-village farm community: the land is owned by the group, and parcels are assigned to each member of the group for cultivation. Not surprisingly, most of the *ejidos* carved out of haciendas lie in central and southern Mexico, where Indian traditions of land ownership and cultivation survived and where the adjustments were most successfully made. A few *ejidos* are true collectives, in that the members of the group do not farm assigned parcels but work the whole area in return for a share of the profits. These are located mainly in the oases of the north, where cotton and wheat are grown, and in some of the former plantations of the Yucatan.

With such a far-reaching program, it is not surprising that productivity temporarily declined; the miracle is that it has been carried off without a major death toll and that the power of the wealthy landowning aristocracy could be broken without ruin to the state. Mexico alone among the countries with large Indian populations has made major strides toward solving the land question, although there is still much malnutrition and poverty in the country. But the revolution that began in 1910 did more than that. It resurrected the Indian contribution to Mexican life and blended Spanish and Indian heritage in the country's social and cultural spheres. It brought to Mexico the distinctiveness that it alone possesses in "Latin" America.

The revolution could change the distribution of land, but it could not change the land itself nor the methods by which it was cultivated. Corn, beans, and squash continue to form the main subsistence food of most Mexicans, with corn the chief staple and still occupying over half the cultivated land. It used to be more than that, and even this proportion suggests that some crop diversification is taking place. But corn is grown all too often where the conditions are not right for it, so that yields are low; wheat might do better, but the people's preference determines the crop, not soil suitability. And if the people's preference has not changed a great deal, neither have farming methods over much of the country, in a vertical as well as horizontal geographic setting (see *Box*).

## Economic Geography

Commercial agriculture in Mexico has made major strides in recent decades with respect to both

## ALTITUDINAL ZONATION

Middle America and western South America are areas of high relief and strong local contrasts. People live in clusters in hot tropical lowlands, in temperate intermontane valleys, and even just below the snow line in the Andes. In these several zones, distinct local climates, soils, crops, domestic animals, and modes of life prevail. As a result, the zones are known by specific names as if they were regions with distinct properties.

The lowest zone, from sea level to about 750 meters (2500 feet), is known as the *tierra caliente*, the "hot land" of the coastal plains and low interior basins where tropical agriculture (including banana plantations) dominates. Above this lowest zone lie the tropical highlands containing Latin America's largest population clusters, the *tierra templada* of temperate land reaching up to about 1700 meters (somewhat over 5500 feet). Temperatures are cooler; prominent among the commercial crops is coffee, and corn (maize) and wheat are the staple grains. Still higher, from about 1700 meters to 3500 meters (nearly 12,000 feet) is the *tierra fria*, the cold country of the higher Andes where hardy crops such as potatoes and barley are the people's mainstays. Only small parts of the Middle America highlands reach into the *fria* zone; in South America, this environment is much more extensive in the Andes. The highest zone of all is the *tierra helada*, or "frozen land," also called the *paramós*, a zone of barren exposed country above the upper limit of tree growth that reaches to the icy peaks of Andean mountains.

the home market and for export. The greatest productivity is still in the hands of private cultivators, although much of the land has been subdivided into *ejidos*. The central plateau is geared mainly to the domestic production of food crops, but, in the north, large irrigation projects have been built on the streams coming off the interior highlands. Cotton is cultivated for the domestic market as well as for export (Mexico leads Middle and South America in this product); for the home market, wheat and winter vegetables are raised. The unlikely boom of large-scale farming in Mexico's arid north is due in large measure to the adoption of U.S. irrigation and mechanized-agriculture technology.

The importance of manufacturing in the Mexican economy is also rising. Mexico possesses a wide range of raw materials, many of which are located in the north and northeast. Quite early, an iron and steel industry was built in Monterrey, using iron ore located near Durango and coking coal from the Sabinas Basin in northern Coahuila; a second plant was built later at Monclova near the source of coking coal. The majority of Mexico's industrial production takes place in the cities, and that means that there is a heavy concentration of manufacturing in and around Mexico City. Indeed, by some measures, nearly two-thirds of all Mexican industry is located here.

Mexico's metal mining industries are less important today than they once were. The country still exports about one-quarter of the world's silver, and other important commodities include lead, zinc, copper and gold. The mines lie scattered throughout northern and north-central Mexico (Fig. 4–14), but many of the mines that were important in the colonial period—and near which urban centers of some size developed—have been worked out.

On the other hand, a spectacular recent success has been the petroleum industry, centered on the southern Gulf Coast's Bay of Campeche around the city of Villahermosa (Fig. 4–14). A series of discoveries of oil and natural gas reserves in this area—the largest found in the world since 1970—have made Mexico self-sufficient in these fossil fuels, an energy situation reinforced by major additional production and reserves located in oilfields lying along the Gulf coast between Veracruz and Tampico. Additional discoveries in the last few years—in the area south of Villahermosa and northeastward into the Yucatan—now rank Mexico with Saudi Arabia as one of the world's leading storehouses of petroleum and natural gas (the country was the fourth largest oil producer in 1981). The latest explorations suggest that the Bay of Campeche oilfields may be linked together in one gigantic crescent stretching from Tampico to Mérida (Fig. 4–14); if this is confirmed, Mexico's reserves would probably climb even higher than the 20 percent of the world total the country was estimated to possess in the mid-1980s. In any event, these huge Mexican oil supplies are certain to affect global distribution for decades to come. At home, Mexico's economic geography will be transformed as the government-controlled energy industry's enormous revenues are invested in industrialization and overall national development. However, this economic growth will have to be deferred until the late 1980s at the earliest, because the worldwide recession and oil glut of the early 1980s brought Mexico perilously close to bankruptcy. Expecting huge revenues from oil and gas exports, Mexico spent and borrowed abroad to the extent that it accumulated a foreign debt in excess of $85 billion; when the price of petroleum plunged, the debt crisis arose and domestic inflation in 1982 surpassed the 100 percent level. The newly elected government of President de la Madrid responded to these economic jolts by introducing a deep austerity program and elevating international loan repayment to the highest national priority, and recovery appeared to be under way in 1984. Although the longer-term prospect is a bright one, Mexico still faces tough decisions in formulating a development policy that matches the expansion of its petroleum production with a steady flow of domestic benefits that can effect real improvements in the lives of its growing, underemployed, and mostly poor population.

Mexico has begun to industrialize; it is addressing itself with new determination to agrarian reform; it also seeks to integrate all sectors of the population into a true Mexican nation. After a century of struggle and oppression, this country has taken long strides toward representative government. But its progress is threatened by the explosive growth of its population, for no amount of reform ultimately can keep pace with a growth rate that will, if unchecked, produce a Mexican population of about 115 million by the turn of the century. Increasingly, this growth is being steered toward the country's burgeoning urban areas (see *Box*). In the countryside, there is already local strife over land allocations; and tens of thousands of Mexicans move illegally across the U.S. border each year in search of a means of survival. Small areas are made unsafe by outlaw bands that prey on highway travelers, and they may foreshadow serious regional insurgencies. No political or economic

# THE LATIN AMERICAN CITY

As Table I–1 showed, Latin America is by far the most rapidly urbanizing of the underdeveloped realms, its population advancing from 41 to 65 percent urban between 1950 and 1980. Although the urban experience has been a varied one in the Middle and South American realms—a function of diverse historical, cultural, and economic influences—there are many common threads that have prompted geographers to search for meaningful generalizations. One of the more successful is the model of the intraurban spatial structure of the Latin American city proposed by Ernst Griffin and Larry Ford (Fig. 4–15), that may well have even wider applications to cities throughout the geographic realms of the Third World.

The basic spatial framework of city structure—which blends traditional elements of Latin American culture with modernization forces now reshaping the urban scene—is a composite of radial sectors and concentric zones. Anchoring the model is the thriving CBD, which, like its European counterpart, remains the primary business, employment, and entertainment focus of the surrounding metropolitan agglomeration; efficient public transit systems and nearby residential concentrations of the affluent assure the dominance of the CBD, whose landscape increasingly exhibits modern high-rise buildings (Fig. 5–11). Emanating outward from the urban core along the city's most prestigious axis is the commercial *Spine*, which is surrounded by the *elite residential sector*; this widening corridor is essentially an extension of the CBD, featuring offices, shopping, high-quality housing for the upper and upper-middle classes, restaurants, theaters, and such amenities as parks, zoos, and golf courses that give way to wealthy suburbs, which carry the elite sector beyond the city limits. The three remaining concentric zones are home to the less fortunate residents of the city (who comprise the great majority of the urban population), with socioeconomic levels and housing quality decreasing markedly as distance from the city center increases. The *zone of maturity* in the inner city contains the best housing outside of the Spine sector, attracting the middle classes, who invest sufficiently to keep their solidly built but aging dwellings from deteriorating. The adjacent *zone of in situ accretion* is one of much more modest housing, interspersed with unkempt areas, which represent a transition from inner-ring affluence to outer-ring poverty; the residential density of this zone is usually quite high, reflecting the uneven assimilation of its occupants into the social and economic fabric of the city. The outermost *zone of peripheral squatter settlements* is home to the impoverished and unskilled hordes that have recently migrated to the city from rural areas; although housing in this ring mainly consists of teeming, high-density shantytowns (see Fig. 5–10), residents here are surprisingly optimistic about finding work and eventually bettering their living conditions—a realistic aspiration documented by researchers, who confirm a process of gradual upgrading as squatter communities mature. A final structural element of many Latin American cities is the *disamenity sector* that contains relatively unchanging slums, known as *favelas;* the worst of these poverty-stricken areas often include sizable numbers of people who are

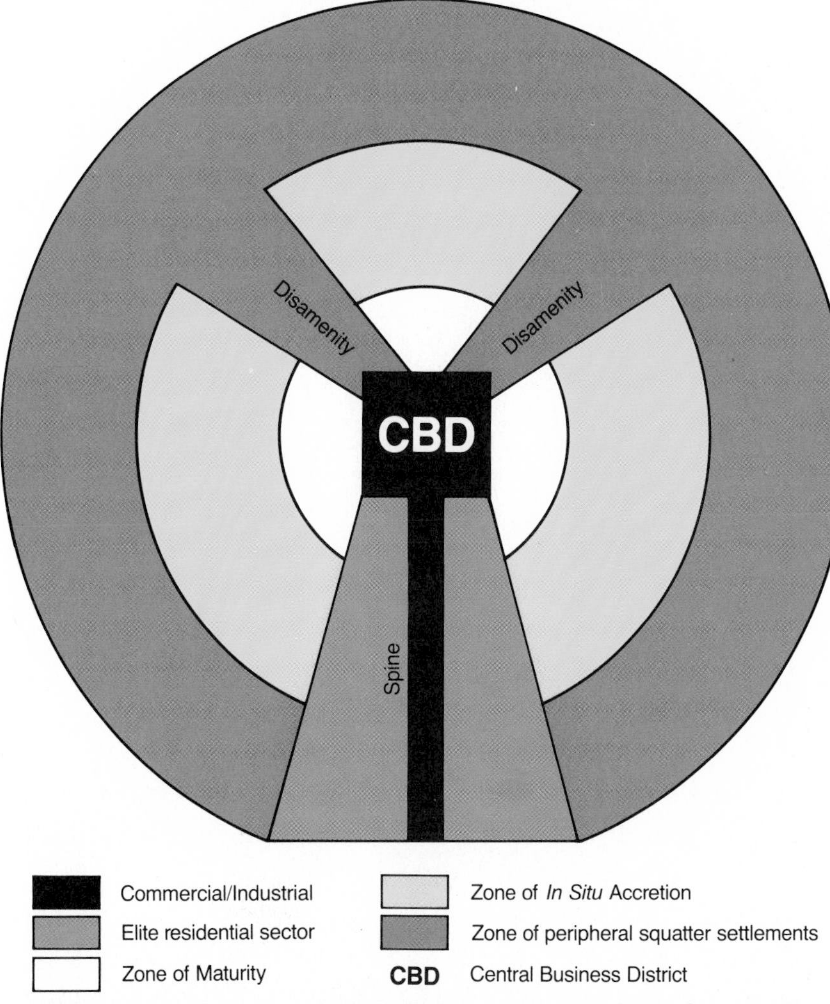


Griffin and Ford, *Geographical Review*, Vol. 70, 1980, p. 406. With the permission of the American Georgraphical Society.

Figure 4–15

so poor that they are forced literally to live in the streets.

As with all models, this particular construct can be criticized for certain shortcomings; among criticisms directed at this generalization are the need for a sharper sectoring pattern, a stronger time element involving a multiple-stage approach, and the observation that the form perhaps too closely resembles the U.S. city. Alternatively, others view this work as not only an especially useful abstraction of the Latin American city but of the Third World city in general. Later on in Chapter 10 (p. 517), we will explore this topic further when we consider a similar model developed for the Southeast Asian city, which incorporates the lasting spatial imprints of colonialism on contemporary urban structure.

system in the underdeveloped world could long withstand the impact of population change faced in Mexico, and the nation's latest accomplishments remain under the cloud of this threat.

# Central American Republics

Crowded onto the narrow portion of the Middle American isthmus between Mexico and the South American continent are seven countries collectively known as the Central American Republics (Fig. 4–16). Territorially, they are all quite small: only one, Nicaragua, is larger than the Caribbean island of Cuba. Populations range from Guatemala's 8.2 million down to Panama's 2.1 million in the six Hispanic republics, whereas the sole former British territory, Belize (which used to be called British Honduras), has barely 160,000 inhabitants. As elsewhere in Middle America, the ethnic composition of the population is varied, with Indian and white minorities and a mestizo majority. The exceptions are in Guatemala, where approximately 54 percent of the population remains relatively "pure" Indian (Maya and Quiché; see Fig. 4–7) with another 42 percent of strongly Indian mestizo ancestry, and in Belize, where 60 percent of the population is black or mulatto, a situation resembling the social geography of the Caribbean. Demographic complexity is at its least in Costa Rica, where there is a large white majority of Spanish and relatively recent European immigrants; in a population of 2.8 million, fewer than 15,000 Indians are counted, and the black population constitutes only 2 percent of the total.

The narrow land bridge on which these republics are situated

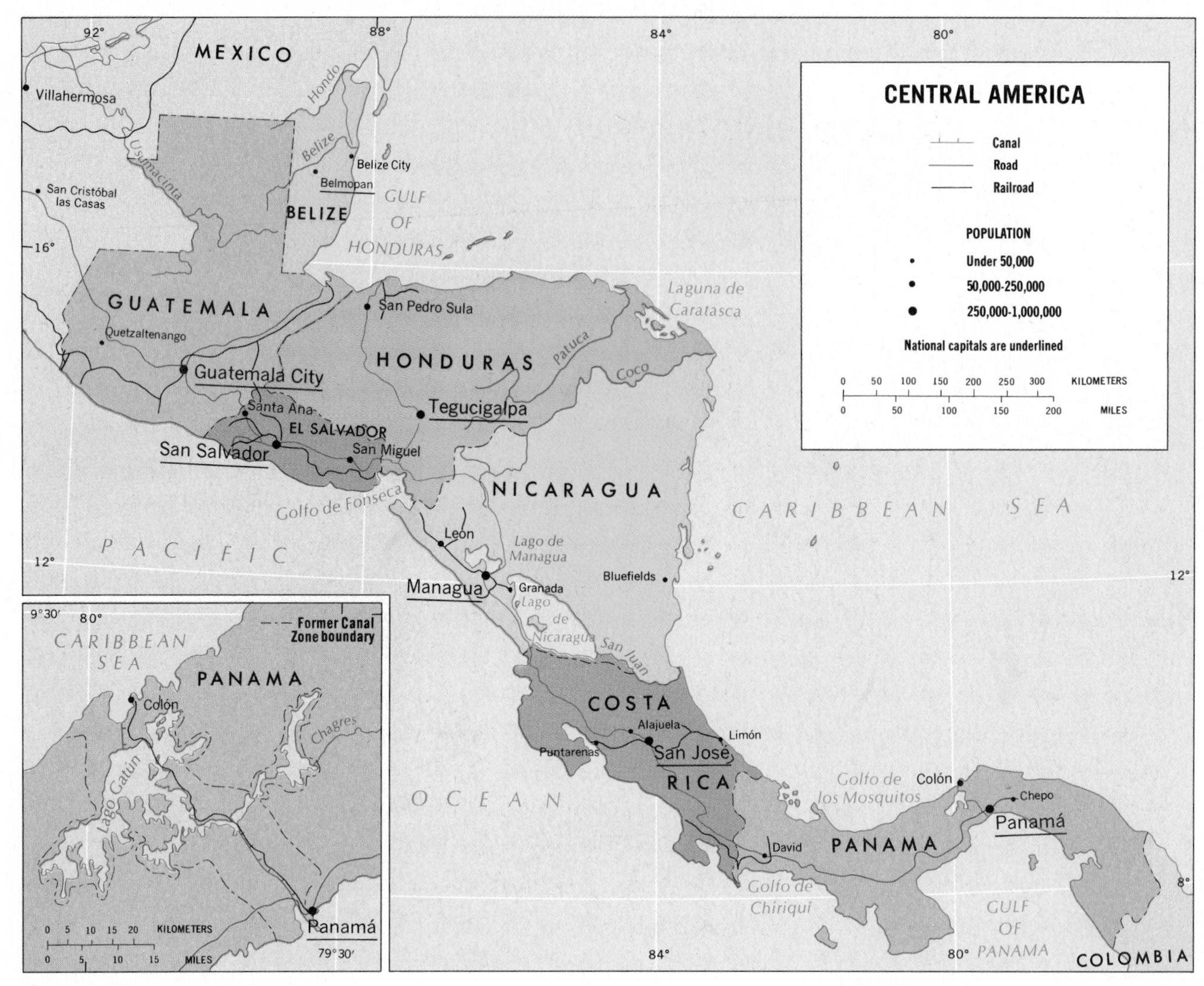


**Figure 4–16**

consists of a highland belt flanked by coastal lowlands on both the Caribbean and Pacific sides; from the earliest times, the people have been concentrated in the *templada* (temperate) zone. Here tropical temperatures are moderated by elevation, and rainfall is adequate for the cultivation of a variety of crops. As noted earlier, the Middle American highlands are studded with volcanoes—Costa Rica's active Mt. Arenal is shown in Fig. 4–17—and local areas of fertile volcanic soils are scattered throughout the region. The old Indian agglomerations were located in these more fertile parts of the uplands, and this human distribution persisted during the Spanish period. Today, the capitals of Guatemala (Guatemala City, 1.2 million), Belize (Belmopan, 4000), Honduras (Tegucigalpa, 575,000), El Salvador (San Salvador, 525,000), Nicaragua (Managua, devastated by the 1972 earthquake, now recovered to 850,000), and Costa Rica (San José, 770,000), all lie in the interior, most of them at 1000 meters (3300 feet) or more in elevation; in all of mainland Middle America, Panama City (850,000) is the only coastal capital (Fig. 4–16). The size of those cities, in countries whose populations average about 4 million, is a reflection of their primacy, identical to the dominance of Mexico City over the rest of Mexico; on an average, the next ranking town is only one-fifth as large as the capital city.

The distribution of population within Central America, besides its concentration in the region's higher sections, also exhibits greater densities toward the Pacific than toward the Caribbean coastlands (Fig. I–14, p. 26). El Salvador, with Belize and to a lesser degree Panama, is an exception to the rule that people in mainland Middle America are concentrated in the *templada* zone; most of El Salvador is *tierra caliente*, and the majority of its 5 million people are crowded, hundreds to the square kilometer, in the intermontane plains lying less than 750 meters (2500 feet) above sea level. In Nicaragua, too, the Pacific areas are the most densely popu-

**Figure 4–17**
Costa Rica's Arenal, one of Middle America's numerous active volcanoes.

lated; the early Indian centers lay near Lake Managua, Lake Nicaragua, and in the adjacent highlands. The frequent activity of volcanoes in this Pacific area is accompanied by the emission of volcanic ash, which settles over the countryside and quickly weathers into fertile soils (see Fig. 4–17). By contrast, the Caribbean coastal lowlands—hot, wet, and awash in leached soils—support comparatively few people. In the most populous republic, Guatemala, the heartland has also long been in the southern highlands; although the large majority of Costa Rica's population is concentrated in the central uplands around San José, the Pacific lowlands have been the scene of major in-migration since banana plantations were first established there. Even in Panama, there is a strong Pacific orientation—more than half of all Panamanians (and that means over 70 percent of the rural people) live in the southwestern lowlands and adjoining mountain slopes; another 25 percent live and work in the rapidly changing Panama Canal corridor (see *Box* pp. 274–275); of the remainder, a majority, many of them descendants of black immigrants from the Caribbean, live on the Atlantic side of the isthmus.

With the single exception of Costa Rica, Middle America's mainland republics face the same problems as Mexico, only more so; they also share many of the difficulties confronting the Caribbean islands. Although efforts are being made to improve the land situation, the colonial legacy of hacienda and peon hangs heavily over the area. Earlier, we referred to the brevity of life for the peoples of the Caribbean: demographic statistics indicate that life expectancy in Guatemala, where nearly one-third of the mainland's people outside Mexico live, is only 58 years. It is 71 years in comparatively well-off Costa Rica, and 63 in El Salvador.

Generations of people have seen little change in their way of life and have had even less opportunity to effect any real improvement; dwellings are still made of mud and straw, sanitary facilities have not reached them, schools are overcrowded, and hospital facilities are inadequate. As for their livelihood, each year brings a renewed struggle to extract a subsistence, with no hope that the next year will bring something better. One of the staggering contrasts in Middle America is that between the attractive capital city and its immediate surroundings, and the desolation of outlying rural areas; another is that between the splendor, style, and refined culture of the families who own the coffee *fincas* and the destitute, rag-wearing peasant. In the 1980s, driven by the leftist Sandinista revolution in Nicaragua and the intensifying civil war in El Salvador, these disparities have finally exploded into both guerilla and formal warfare throughout much of Central America (see *Box* pp. 276–277); with armed intervention by the United States remaining a possibility as of 1984, violent turmoil seems certain greatly to aggravate this region's serious problems for the foreseeable future.

Coffee—not bananas—is the most important export crop for all but three of the republics (in Honduras and Panama bananas lead; in Belize, lumber, naval stores, and chicle). Although introduced as early as the 1780s, major coffee plantings were first established in the 1850s, mostly in Guatemala. This is a *templada* crop that does best on the mountain slopes; it spread quickly to other countries, especially El Salvador and Costa Rica. During the second half of the nineteenth century, the banana plantation began to make its appearance along the Caribbean coast; with it, came an involvement of American big business in the affairs of Middle America. Black laborers came from Jamaica and elsewhere to take the newly available jobs (as they did when the Panama Canal was dug), and a flow of people toward the Caribbean coastal plain was stimulated on the mainland.

The corporations that introduced banana-plantation agriculture in Middle America—among them the United Fruit (now United Brands), Standard Fruit and Steamship, and Boston Fruit companies—were not always benevolent contributors toward a better life in Middle America. There was much interference in domestic affairs; thus, considerable resentment was aroused in what many Americans unflatteringly referred to as the "banana republics." As time went on, more positive contributions were made—although the industry has sustained major setbacks. Crop disease caused a large-scale decline in the Caribbean lowlands, and the focus later shifted to the Pacific coastal areas, which, in contrast to the always-wet Caribbean shores, are seasonally dry. In recent years, however, new disease-resistant varieties of fruit have been developed and the fruit has been reintroduced in the Caribbean zone (which had not lost its locational advantages vis-à-vis the primary eastern U.S. market), and there are signs that the old pattern of Caribbean dominance in the banana industry of Middle America is reviving.

## PANAMA AND THE PANAMA CANAL

The Republic of Panama owes its birth to an idea: the construction of an artificial waterway to connect Atlantic and Pacific waters and thereby avoid the lengthy circumnavigation of South America. In the 1880s, when Panama was still an extension of Colombia, a French company directed by Ferdinand de Lesseps (1805–1894)—builder of the Suez Canal in the 1860s—tried and failed to build such a canal; thousands of workers died of yellow fever, malaria, and other tropical diseases, and the company went bankrupt. At the turn of the century, U.S. interest in a Panama canal (which would shorten the sailing distance between the East and West coasts by 8000 nautical miles) rose sharply, and the United States in 1903 proposed a treaty that would permit a renewed effort at construction across Colombia's Panamanian isthmus. When the Colombian Senate refused to approve, Panamanians rebelled and the United States supported this uprising by preventing Colombian forces from intervening. The Panamanians, at the behest of the United States, declared their independence from Colombia and the new republic immediately granted the United States rights to the Canal Zone averaging about 16 kilometers (10 miles) in width and just over 80 kilometers (50 miles) in length.

Soon canal construction commenced, an epic struggle documented by historian David McCullough in his classic book, *The Path Between the Seas*; this time, the project succeeded as American engineering and technology triumphed over a formidable set of obstacles. The Panama Canal was opened in 1914, a symbol of U.S. power and influence in the Caribbean and Middle America. The Canal Zone was held by the United States under a treaty that granted it "all the rights, powers, and authority" in the area "as if it were the sovereign of the territory."

Such language might suggest that the United States held rights over the Canal Zone in perpetuity, but the treaty nowhere stated specifically that Panama permanently yielded its own sovereignty in that zone. In the 1970s, as the Canal was transferring 20,000 ships per year (Fig. 4–18) and generating hundreds of millions of dollars in tolls, Panama sought to terminate U.S. control in the Canal Zone. Delicate negotiations began; in 1977, an agreement was reached on a staged withdrawal by the United States from the territory, first from the Canal Zone and, by 2000, from the Panama Canal itself. This agreement took the form of two treaties, and following the signing by President Carter and President Torrijos, they were ratified in spite of stubborn opposition in the U.S. Congress. Long fragmented, Panama is finally on the way to becoming a territorially coherent state (a concept covered in Chapter 10).

**Figure 4–18**
One of the busy locks of the nearly 75-year-old Panama Canal.

## References and Further Readings

Augelli, J. P. "The Rimland-Mainland Concept of Culture Areas in Middle America," *Annals of the Association of American Geographers*, 52 (1962), 119–129.

Blakemore, H. & Smith, C., eds. *Latin America: Geographical Perspectives* (London and New York: Methuen, 2 rev. ed., 1983).

Blouet, B. & Blouet, O., eds. *Latin America: An Introductory Survey* (New York: John Wiley & Sons, 1982).

Brown, P. & Shue, H., eds. *The Border That Joins: Mexican Migrants and U.S. Responsibility* (Totowa, N.J.: Rowman & Littlefield, 1983).

Browning, D. *El Salvador: Landscape and Society* (Oxford, Eng.: Clarendon Press, 1971).

Carlstein, T. et al., eds. *Timing Space and Spacing Time* (New York: John Wiley/Halsted, 3 vols., 1978).

"Central America in Turmoil: An NBC News Guide to Central America" (New York: National Broadcasting Corporation—NBC News, 1983).

Clark, A. H. "Historical Geography," in James, P. E. & Jones, C. F., eds., *American Geography: Inventory and Prospect* (Syracuse, N.Y.: Syracuse University Press, 1954), pp. 70–105.

Crowley, W. & Griffin, E. "Political Upheaval in Central America," *Focus*, September–October 1983.

Dickenson, J. et. al. *A Geography of the Third World* (London and New York: Methuen, 1984).

Driver, H., ed. *The Americas on the Eve of Discovery* (Englewood Cliffs, N.J.: Prentice-Hall, 1964).

Floyd, B. *Jamaica: An Island Microcosm* (New York: St. Martin's Press, 1979).

Gilbert, A. & Gugler, J. *Cities, Poverty, and Development: Urbanization in the Third World* (New York: Oxford University Press, 1982).

Gilbert, A. et al., eds. *Urbanization in Contemporary Latin America: Critical Approaches to the Analysis of Urban Issues* (Chichester, Eng.: John Wiley & Sons, 1982).

Gourou, P. *The Tropical World: Its Social and Economic Conditions and Its Future Status* (London and New York: Longman, 5 rev. ed., trans. S. Beaver, 1980).

Griffin, E. & Ford, L. "Cities of Latin America," in Brunn, S. & Williams, J., eds., *Cities of the World: World Regional Urban Development* (New York: Harper & Row, 1983), pp. 198–240.

Griffin, E. & Ford, L. "A Model of Latin American City Structure," *Geographical Review*, 70 (1980) 397–422. Model diagram adapted from p. 406.

Helms, M. *Middle America: A Culture History of Heartland and Frontiers* (Englewood Cliffs, N.J.: Prentice-Hall, 1975).

Hoy, D. & Macfie, S. "Central America: A Bibliography of Economic, Political, and Cultural Conditions" (Athens: University of Georgia, Department of Geography, 1982).

Huntington, E. *Mainsprings of Civilization* (New York: John Wiley & Sons, 1945).

Jakle, J. "Time, Space, and the Geographic Past: A Prospectus for Historical Geography," *American Historical Review*, 76 (1971), 1084–1103.

"Jamaica—Special Issue," *Journal of Geography*, 82 (September–October, 1983), 189–252.

Jones, R., ed. *Spatial Patterns of Undocumented Migration: Mexico and the United States* (Totowa, N.J.:

Rawman & Allanheld, 1984).
Lewis, P. "Axioms for Reading the Landscape: Some Guides to the American Scene," in Meinig, D., ed., *The Interpretation of Ordinary Landscapes: Geographical Essays* (New York: Oxford University Press, 1979), pp. 11–32.
Lowenthal, D. *West Indian Societies* (New York: Oxford University Press, 1972).
Macpherson, J. *Caribbean Lands* (New York: Longman, 4 rev. ed., 1980).
Martin, G. *Ellsworth Huntington: His Life and Thought* (Hamden, Conn.: Archon Books, 1973).
McCullough, D. *The Path Between the Seas: The Creation of the Panama Canal, 1870–1914* (New York: Simon & Schuster, 1977).
Meinig, D. "Symbolic Landscapes: Some Idealizations of American Communities," in Meinig, D., ed., *The Interpretation of Ordinary Landscapes: Geographical Essays* (New York: Oxford University Press, 1979), pp. 164–192.
Picó, R. *The Geography of Puerto Rico* (Chicago: Aldine, 1974).
Pred, A. "The Choreography of Existence: Comments on Hägerstrand's Time-Geography and Its Usefulness," *Economic Geography*, 53 (1977), 207–221.
Richardson, B., ed. *Caribbean Migrants* (Knoxville: University of Tennessee Press, 1983).
Sargent, C. "The Latin American City," in Blouet, B. & Blouet, O., eds., *Latin America: An Introductory Survey* (New York: John Wiley & Sons, 1982), pp. 201–249. Quotation taken from p. 221; diagram adapted from p. 223.
Sauer, C. O. "Foreword to Historical Geography," *Annals of the Association of American Geographers*, 31 (1941), 1–24.
Smith, C. "Historical Geography: Current Trends and Prospects, " in Chorley, R. & Haggett, P., eds., *Frontiers in Geographical Teaching* (London: Methuen, 1965), pp. 118–143.
Stanislawski, D. "Early Spanish Town Planning in the New World," *Geographical Review*, 37 (1947), 94–105.
Tata, R. & Lee, D. "Puerto Rico: Dilemmas of Growth," *Focus*, November–December 1977.
Wagley, C. & Harris, M. *Minorities in the New World: Six Case Studies* (New York: Columbia University Press, 1958).

# CENTRAL AMERICA IN UPHEAVAL

With startling suddenness, Central America in the mid-1980s has become one of the world's leading crisis areas. Long-standing internal inequities and episodic local violence spawned civil wars during the late 1970s in El Salvador (Fig. 4–19) and Nicaragua, in both cases pitting rightist governments against strong Marxist-led guerilla movements. The United States—which in the past has intervened actively (including the sending of American troops) in the region to protect its self-interests—has increasingly aided the various military and para-military groups resisting the advance of communism in the affected republics. The successful 1979 Sandinista revolution in Nicaragua, for the first time, established a Marxist base on the mainland of Middle America, which aroused the immediate interest and active support of Cuba—and implicitly the Soviet Union. With the two global superpowers facing a possible confrontation here, American foreign policy debates are now focusing on this region as a whole rather than on its still separate individual conflicts. The eventual resolution of its many arguments and issues will undoubtedly further transform the cultural, economic, and political geography of Central America. We offer below a brief summary of developments within each of the seven republics as of early 1984, proceeding from Mexico toward South America.

Figure 4–19 Guerilla action in the El Salvador conflict.

*Guatemala:* A series of U.S.-backed extreme rightist military governments has ruled for the past 30 years. The latest *coup* (1983) installed a junta vowing a return to civilian government as soon as elusive political stability is achieved. A growing leftist insurgency movement, centered in the western highlands, is gathering strength. Brutal government response—assassination by death squad (including moderates as well as leftists), secret executions, and the like—produced an abysmal human rights record. Nonetheless, the U.S. government is convinced that conditions are improving, and it restored economic aid as well as sales of military weapons in 1983.

| | |
|---|---|
| *Belize:* | Former British colony is now ruled by a parliamentary democratic government. Although no insurgency problems have arisen, the country is acutely concerned about the spillover of hostilities from neighboring Guatemala, which has occasionally claimed Belize's territory as its own. |
| *El Salvador:* | Faces situation similar to polarized Guatemala, but with a much stronger Marxist-led insurgency that has exploded into all-out civil war in the 1980s. Especially violent rightist movement, according to the U.S. government, has curbed its excesses since the 1982 national election. Growing U.S. economic and military support has not weakened the position of the insurgents (whose aims are backed by the Catholic Church) and has raised comparisons with the Vietnam War in U.S. domestic politics. |
| *Honduras:* | The poorest of the republics is ruled by a civilian government, but the military is widely regarded as the real controlling force. Accelerating U.S.-backed military buildup is not aimed at Honduran opposition (still small and unorganized, however growing steadily) but to create a fortress from which to support anti-Communist movements in neighboring El Salvador and Nicaragua. |
| *Nicaragua:* | Marxist Sandinista junta has ruled the country since the ousting of the rightist Somoza government in 1979. The United States has led the effort to isolate Nicaragua in the hemisphere; meanwhile, the Soviet-supported Cuban presence grows. The civilian population is not enamored of communism, but the anti-Somoza revolution and its reduction of domestic inequalities is quite popular today; with the United States aiding the rightist guerillas (or "contras"), support for the well-entrenched Sandinista government widens on nationalist grounds. |
| *Costa Rica:* | The only truly democratic republic (Latin America's oldest), it has managed to endure without an army since 1948. A rapidly developing nation, Costa Rica sincerely prefers to remain neutral in the current regional struggle (although it openly favored the Nicaraguan revolution). This position is certain to be increasingly challenged by its long-time ally—the United States. Will this country be able to maintain its precarious role as the Switzerland of Central America? |
| *Panama:* | Growing canal-generated revenues buttress an expanding economy. Elections scheduled for 1984 are the first to be held under a really democratic process, but the military presence is a strong one and could return to power at any time. Insurgency movements are nonexistent, but serious social problems persist. Panama also has the advantage of being separated by stable Costa Rica from the region's most strife-torn territory. |


Wauchope, R., ed. *Handbook of Middle American Indians* (Austin: University of Texas Press, 1964).

Weaver, M. *The Aztecs, Maya, and Their Predecessors: Archaeology of Mesoamerica* (New York and London: Academic Press, 2 rev. ed., 1981).

West, R. & Augelli, J. *Middle America: Its Lands and Peoples* (Englewood Cliffs, N.J.: Prentice-Hall, 2 rev. ed., 1976).

Wolf, E. *Sons of the Shaking Earth* (Chicago: University of Chicago Press, 1959).

# CHAPTER 5

# SOUTH AMERICA AT THE CROSSROADS

IDEAS AND CONCEPTS

**Economic Geography**
**Agricultural Systems**
**Land Alienation**
**Isolation**
**South American Culture Spheres**
**Capital Cities Classification**
**Latin American Urbanization**
**Functional Classification of Cities**
**Rural-to-Urban Migration**
**Boundaries**

No continent has as familiar a shape as South America, that giant triangle that is connected by mainland Middle America's tenuous land bridge to its sister continent in the north. What is less often realized about South America is that it not only lies south, but mostly east of its northern counterpart as well. Lima, the capital of Peru—one of the continent's westernmost cities—lies farther east than Miami, Florida. Thus, South America juts out much more deeply and prominently into the Atlantic Ocean than does North America, and South American coasts lie much closer to Africa and even to southern Europe than do the coasts of Middle and North America. Lying so far eastward, South America on its western flank faces a much wider Pacific Ocean than does North America. From its west coast to Australia is nearly twice as far as from San Francisco to Japan, and South America has virtually no interaction with the Pacific world of Australasia—not just because of vast distances, but also because both lie in the insular and less populous southern hemisphere, whereas North America faces Japan and the crowded East Asian mainland.

**Figure 5–1**
Itaipu Dam on the Paraná River astride the Brazil-Paraguay border: mighty symbol of the new development thrust into the continental interior.

As if to reaffirm South America's northward and eastward orientation, the western margins of the continent are rimmed by one of the world's longest and highest mountain ranges, the Andes, a gigantic wall that extends from Tierra del Fuego near the southern tip of the triangle to Venezuela in the north (Fig. 5–2). Every map of world physical geography clearly reflects the existence of this mountain chain—in the alignment of isohyets (lines connecting places of equal precipitation totals) (Fig. I–9, p. 16), in the elongated zone of highland climate (Fig. I–10, p. 18), and in the regional distributions of vegetation and soils (Figs. I–12 and I–13, pp. 22 and 24, respectively). More-

# TEN MAJOR GEOGRAPHIC QUALITIES OF SOUTH AMERICA

1. South America's physiography is dominated by the Andean Mountains in the west and the Amazon Basin in the north; much of the remainder is plateau country.
2. Half the realm's area and almost half of its population are concentrated in one country, Brazil.
3. South America's population remains concentrated in peripheral zones; much of the interior is sparsely peopled.
4. Comparatively low population densities prevail in South America, but growth rates are among the world's highest.
5. Regional economic contrasts and disparities, both in the realm as a whole and within individual countries, are strong; in general, the south is the most developed, the northeast the least.
6. Interconnections among the states of the realm remain comparatively weak. External ties are frequently stronger.
7. Strong cultural pluralism exists in the majority of the realm's countries, and this pluralism frequently is expressed regionally.
8. With the exception of three small countries in the north, the realm's modern cultural sources lie in a single subregion of Europe, the Iberian Peninsula. Spanish and Portuguese (in Brazil) are the *linguae franca*.
9. Lingering politico-geographical problems beset the realm; boundary disputes and territorial conflicts persist.
10. The Catholic Church dominates life throughout the realm, and constitutes one of its unifying elements.

over, as Fig. I–14 (p. 26) reveals, South America's biggest population clusters are located along the eastern and northern coasts, overshadowing those of the Andean west.

# The Human Sequence

Although modern South America's largest populations are situated in the east and north, there was a time during the height of the Inca Empire when the Andes Mountains contained the most densely peopled and best organized state on the continent. Whereas the origins of Inca civilization are still shrouded in mystery, it has become generally accepted that the Incas were descendants of ancient peoples who came to South America via the Middle American land bridge (possibly after earlier migrations from Asia via the Bering land bridge). But even this is not totally beyond doubt; some anthropologists believe that the first settlers in this part of the western hemisphere may have reached the Chilean and Peruvian coasts from distant Pacific islands. In any case, for thousands of years before the Europeans arrived on the scene in the sixteenth century, Indian communities and societies had been developing. About 1000 years ago, a number of regional cultures thrived in the valleys within the Andean Mountains and those along the Pacific Coast. The llama had been domesticated as a beast of burden, a source of meat, and a producer of wool. Religions flourished and stimulated architecture as well as the construction of temples and shrines. Sculpture, painting, and other art forms were practiced. It was over these cultures that the Incas extended their authority from their headquarters at Cuzco in the Cuzco Basin of Peru, beginning late in the twelfth century, to forge the greatest empire in the Americas prior to the coming of the Europeans.

Nothing to compare with the cultural achievements of this central Andean area existed anywhere else in South America. In addition to the Andean civilizations, anthropologists recognize three other major groupings of Indian peoples: those of the Caribbean fringe, those of the tropical forest of the Amazon Basin and other lowlands, and those called "marginal," whose habitat lay in the Brazilian Highlands, the headwaters of the Amazon River, and far southern South America (Fig. 5–4). It has been estimated that the Caribbean, forest, and marginal peoples together constituted only about one-quarter of the continent's total native (indigenous) population.

## The Inca Empire

When the Inca civilization is compared to that of ancient Mesopotamia, Egypt, the old Asian civilizations, and the Aztecs' Mexica Empire, it quickly becomes clear that this was an unusual achievement. Everywhere else, rivers and waterways provided avenues for interaction and the circulation of

Figure 5–2

goods and ideas. Here, however, an empire was forged out of a series of elongated basins in the high Andes, basins called *altiplanos,* which were created when mountain valleys between parallel and converging ranges filled with erosional materials from surrounding uplands. The altiplanos are often separated from each other by some of the world's most rugged terrain, with high often-snowcapped mountains alternating with precipitous canyons. Individual altiplanos accommodated regional cultures; the Incas themselves were first established in the intermontane basin of Cuzco (Fig. 5–4). From that hearth, they proceeded, by military conquest, to extend their authority over the peoples of coastal Peru and other altiplanos. Their first thrust was southward, and it seems to have occurred toward the close of

# SYSTEMATIC ESSAY

## Economic Geography

In the preceding chapters and vignettes we have discussed a number of concepts, principles, and examples of **economic geography**. In the introductory chapter, this subdiscipline was defined (p. 33) as being concerned with the diverse ways in which people earn a living and how the goods and services they produce are expressed and organized spatially. In the North America chapter (p. 207), we identified four sets of productive economic activities as the major components of the *spatial economy*: (1) primary activities (agriculture, mining, and other extractive industries), (2) secondary activities (manufacturing), (3) tertiary activities (services), and (4) quaternary activities (information and decision making). Since most of the world's secondary, tertiary, and quaternary activities are located within the developed realms—and were treated in Chapters 1 through 3—we focus here on *agriculture*, which overwhelmingly employs the hundreds of millions of workers who inhabit the remaining underdeveloped realms.

The global distribution of agricultural activities is displayed in Fig. 5–3. The spatial organization of agriculture in the advanced commercial economies of Europe and the United States has already been explained in the context of the Von Thünen model expanded, thanks to modern transportation technology, to the continental scale. To some extent, the macro-Thünian framework can even be applied to the world as a whole—the "global city" would be the European and American edges of the North Atlantic Basin, and many of the colonially-generated farming systems would fit a sequence of concentric and increasingly distant agricultural zones (such as Middle American fruit and sugar, Argentinian beef, Australian wheat, and New Zealand wool). Although this worldwide Von Thünen structuring has eroded with the ending of the colonial era, many of that era's features

Figure 5–3

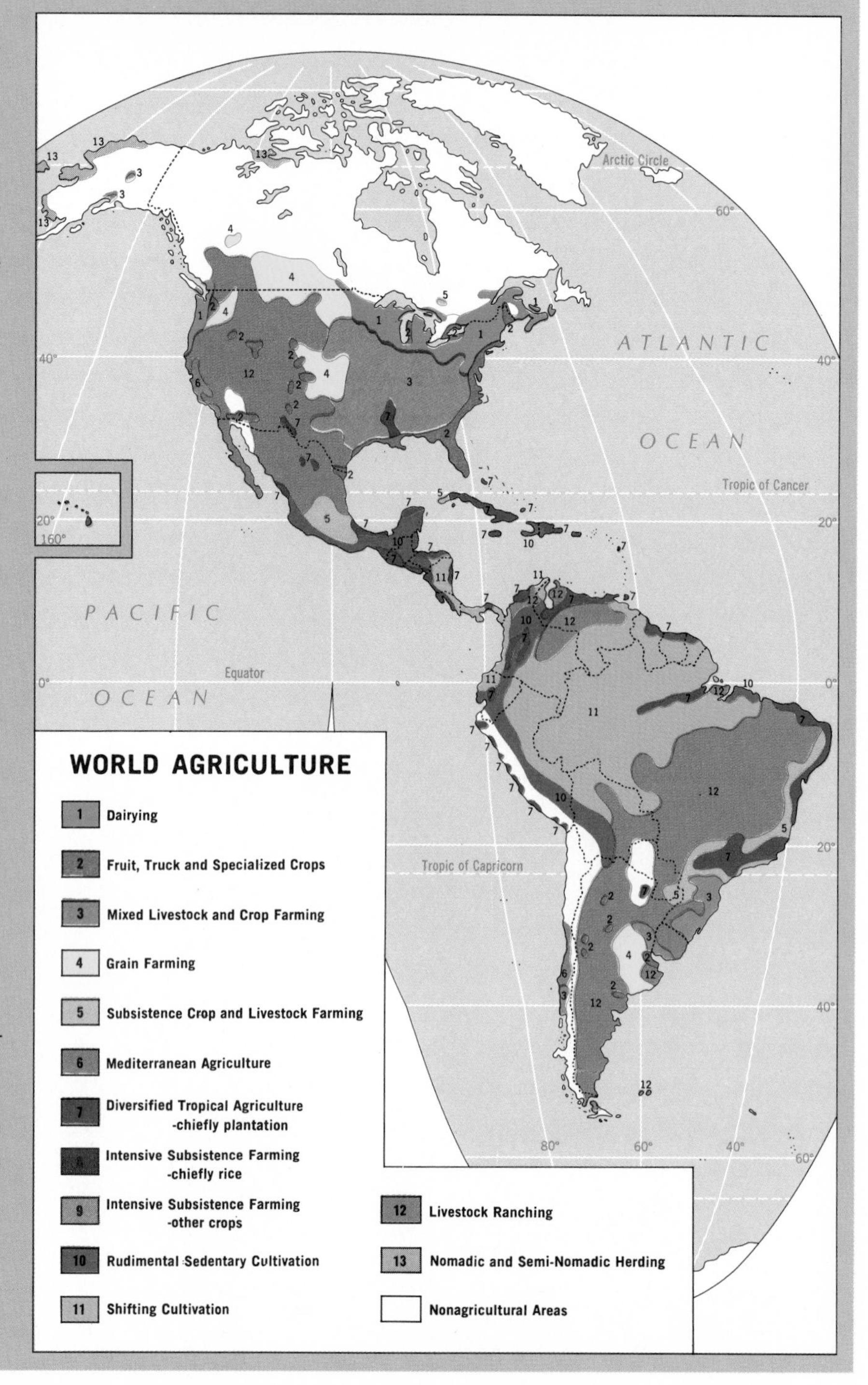

are still apparent (e.g., the plantation survives as a widespread institution), reminding us that most nonsubsistence Third World agriculture remains firmly oriented to the markets of the Western powers that first reorganized primary production in countries of the underdeveloped realms. The spatial pattern of those commercial farming activities can be observed in the distribution of Categories 1 through 4, 6 through 7, and 12 in Fig. 5–3.

The six remaining categories are all associated with the various types of subsistence agriculture. Shifting cultivation and nomadic/seminomadic herding (Categories 11 and 13) are the least intensive of these activities, relegated to marginal environments that—together with vast (unnumbered/white) nonagricultural areas—occupy more than two-thirds of the planet's land surface. Categories 5 and 8 through 10 involve more intensive forms of subsistence agriculture. As we saw

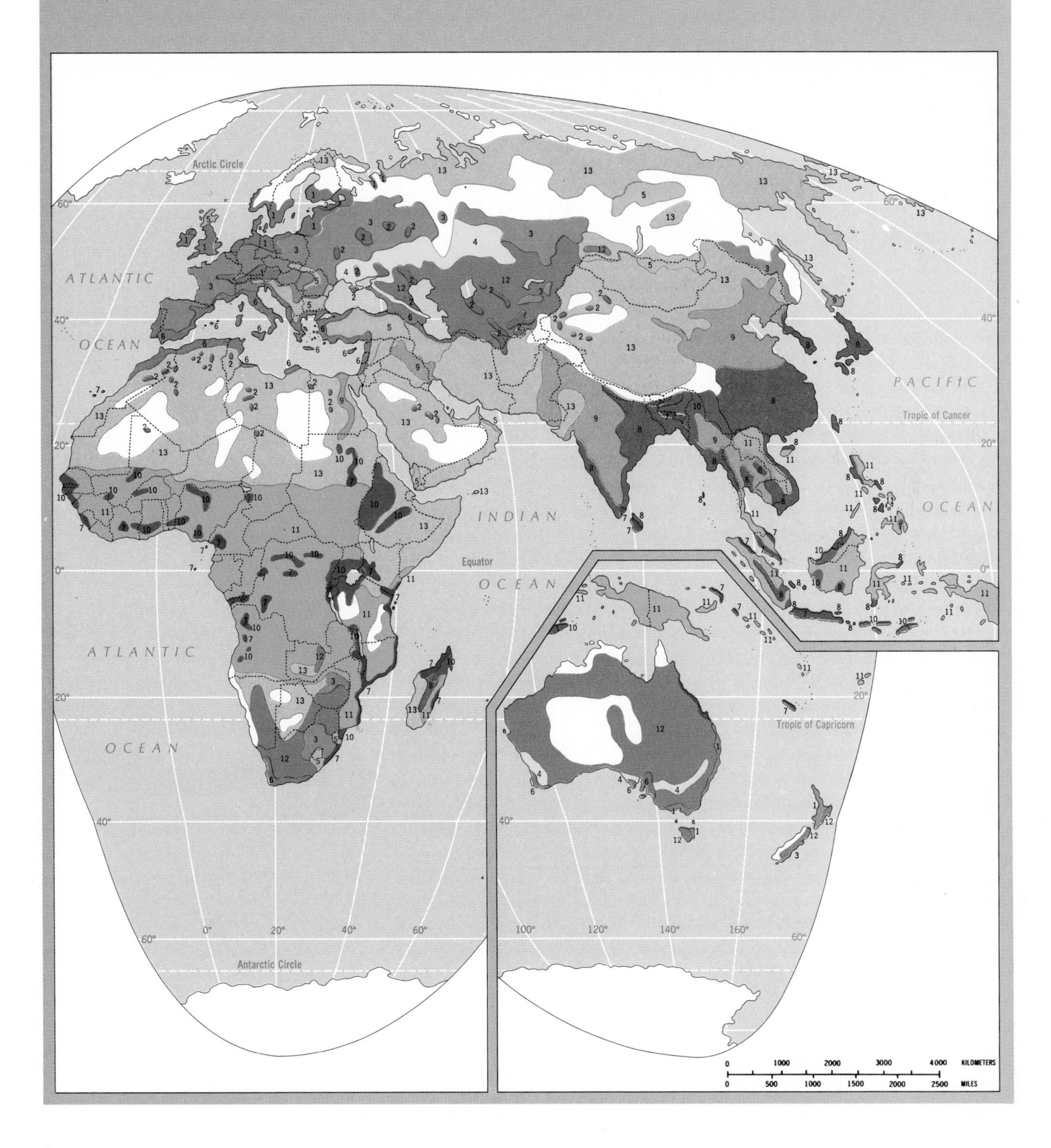

in the introductory chapter, the intensive rice farming of Category 8 supports the earth's largest and densest nonurban population clusters on the alluvial soils of Asia's great river valleys and coastal plains; the intensive subsistence cultivation of other crops (Category 9)—especially such grains as wheat—can also support huge population concentrations (northern China and northwestern India are two outstanding examples). Categories 5 and 10 are occasionally associated with sizable populations as well (although of much lesser magnitude than Categories 8 and 9), but these farming systems are far less productive and often represent a transition between intensive and extensive agriculture (the latter embodied in Categories 11 and 13).

Looking at the South American portion of Fig. 5–3, commercial and subsistence farming exist side by side here to a greater degree than in any other realm (where one of the two always geographically dominates the other); this, of course, does not represent a planned "balance" between the two, instead, it reflects the continent's deep internal cultural and economic divisions, which will be elaborated in this chapter. The commercial agricultural side of South America is expressed in a huge near-coastal cattle-ranching zone (Category 12) that stretches southwest from northeastern Brazil to Patagonia; Argentina's wheat-raising Pampa (Category 4), which is comparable to the U.S. Great Plains; a Corn Belt-type crop and livestock zone (Category 3) in Uruguay, southern Brazil, and south-central Chile; a number of seaboard tropical plantation strips (Category 7) located in Brazil, the Guianas, Venezuela, Colombia and Peru; and a Mediterranean agricultural zone (Category 6) in Middle Chile. In stark contrast to the foregoing food-production systems, subsistence farming covers the rest of the realm's arable land—primitive shifting cultivation (Category 11) occurs in the rainforested Amazon Basin; rudimental sedentary cultivation (Category 10) dominates the Andean plateau country from Colombia in the north to the Bolivian Altiplano in the south; and a wide strip of mixed subsistence farming (Category 5) courses through most of eastern Brazil between the coastal plantation and interior grazing zones.

the fourteenth century. But more impressive than the Incas' military victories was their subsequent capacity to integrate the peoples and regions of the Andean realm into a stable and smoothly functioning state. All the odds would seem to have been against them; as they progressed farther southward, into northern and central Chile and northwestern Argentina, their domain became ever-more elongated, making effective control more and more difficult. The Incas, however, were expert road and bridge builders, colonizers, and administrators; in an incredibly short time, they consolidated their southern territory. Shortly before the Spanish arrival (1531), they even conquered Ecuador and a part of southern Colombia.

The early sixteenth century was a critical period in the Inca Empire, because the conquest of Ecuador and nearby areas for the first time placed stress on the existing administrative framework. Until that time, the undisputed center of the state had been Cuzco, but now it was decided that the empire should be divided into two units—a southern one to be ruled from Cuzco and a northern sector to be centered on Quito. This decision was related to the continuing problem of control over the dissident and rebellious north and the possibilities of further expansion into central Colombia. Now the empire was beset by a number of internal difficulties, notably an uncertain frontier in the north and rising tensions between Cuzco and Quito. And just as the Aztec Empire had been ripe for internal revolt when Cortés and his party entered Mexico, so the Spanish arrival in western South America happened to coincide with a period of stress within the Inca Empire.

When it was at its zenith, the Inca Empire may have counted as many as 25 million subjects. Of course, the Incas themselves were always in a minority in this huge state, and their position eventually became one of a ruling elite in a rigidly class-structured society. The Incas, representatives of the emperor in Cuzco, formed a caste of administrative officials who implemented the decisions of their monarch by organizing all aspects of life in the conquered territories. They saw to it that all harvests were divided between the church, the community, and the individual family; they maintained the public granaries; they recognized and reported the need for investments to improve and maintain roads, terrace hillsides, and expand irrigation works. The life of the subjects of the far-flung Inca Empire was strictly controlled by this omnipresent bureaucracy of Inca administrators, and there was little personal freedom. Farm yields were predetermined, and there was no true market economy; the produce, like the soil on which it was grown, belonged to the state. Marriages were officially arranged, and families could live only where the Inca supervisors would permit. Indeed, the family (as a productive entity within the community) was considered to be the basic unit of administration, not the individual. Inca rule was amazingly effective, and obedience was the only course for its subjects. So highly centralized was the state and so complete the subservience of its efficiently controlled population that a takeover at the top was enough to gain power over the entire empire—as the Spaniards proved in the 1530s.

The Inca Empire, which had risen to greatness so rapidly (some argue that it may have taken less than a century and not, as others believe, several centuries) disintegrated abruptly with the impact of the Spanish invaders. Perhaps it was the swiftness of its development that contributed to its fatal weakness. At any rate, besides spectacular ruins (such as those at Peru's Machu Picchu—Fig. 5–5), the empire left behind numerous social values that have remained a part of Indian life in the Andes to this day—and that continue to contribute to fundamental divisions between the Iberian and Indian populations in this part of South America. The Inca state language—Quechua—was so firmly rooted that it is still spoken today by more than 3 million Indians living in Peru, Ecuador, and Bolivia. An even more basic cultural tradition relates to the ownership of land and other property. Even before the advent of the Inca Empire, the Indians of the various regional cultures practiced communal land ownership—if not by the state, then by villages or groups of villages. The Inca period confirmed and intensified this outlook through rigid state control of all land and resources. Personal wealth simply could not be achieved through the acquisition of land or the control of resources such as minerals or water supplies; not only did the system not permit it, the concept itself was all but unknown. Therefore, it is not surprising that land ownership conferred little or no social prestige. The Spaniards, as we know, possessed almost precisely opposite values and traditions. Today, nearly five centuries after the fall of the Inca Empire, these conflicting outlooks still divide Iberian and Indian South America.

**Figure 5–4**

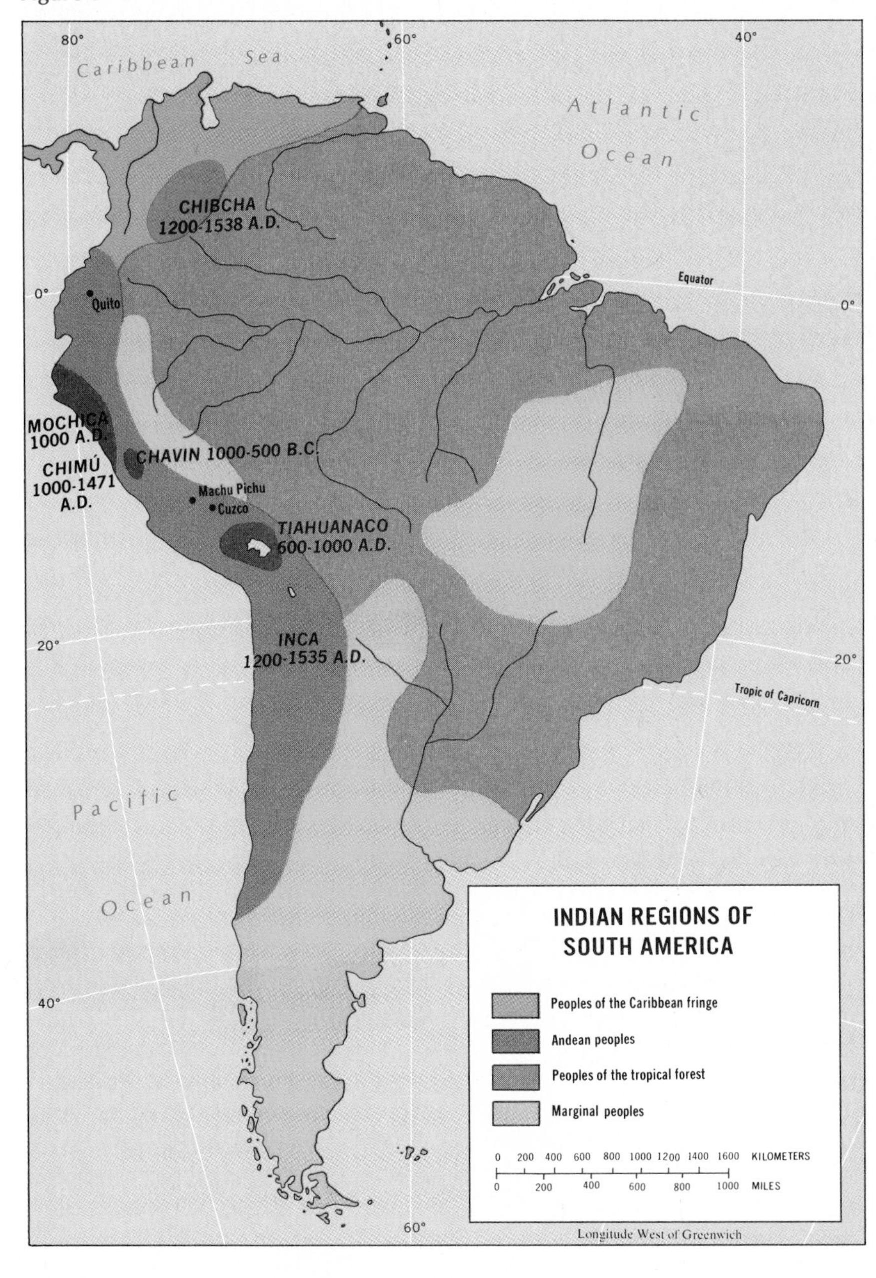


## The Iberian Invaders

In South America as in Middle America, the location of Indian peoples determined, to a considerable extent, the direction of the thrusts of European invasion. The Incas, like the Maya and the Aztec people, had accumulated gold and silver in their headquarters, they possessed productive farmlands, and they constituted a ready labor force. Not long after the 1521 defeat of the Mexica Empire of the Aztecs, the Spanish conquerors crossed the Panamanian isthmus and sailed down the continent's west coast. Francisco Pizarro on his first journey heard of the existence

**Figure 5–5**
Machu Picchu in the high Peruvian Andes, scene of the famous Inca ruins.

of the Inca Empire, and after a landfall in 1527 at Tumbes, located on the northern coast of Peru very near the Ecuador boundary, he returned to Spain to organize the penetration of the Incan domain. He returned to Tumbes with 183 men and a couple of dozen horses in 1531, a time when the Incas were preoccupied with problems of royal succession and strife in the northern provinces. The events that followed are well known; less than three years later, the party rode, victorious, into Cuzco. Initially, the Spaniards kept the structure of the empire intact by permitting the crowning of an emperor who was, in fact, in their power, but soon the land- and gold-hungry invaders were fighting among themselves, and the breakdown of the old order began.

The new order that eventually emerged in western and southern South America placed the Indian peoples in serfdom to the Spaniards. Land was alienated into great haciendas, taxes were instituted, and a forced-labor system was introduced to maximize the profits of exploitation. As in Middle America, the Spanish invaders were mostly people who had little status in Spain's feudal society, but they brought with them the values that prevailed in Iberia—land meant power and prestige, gold and silver meant wealth. Lima, the west coast headquarters of the Spanish conquerors, was founded by Pizarro in 1535, approximately 600 kilometers (375 miles) northwest of the Andean center of Cuzco (Fig. 5–6); before long it was one of the richest cities in the world, reflecting the amount of wealth being yielded by the ravaged Inca Empire. Soon, Lima became the capital of the viceroyalty of Peru, as the authorities in Spain began to integrate the new possession into their colonial empire. Subsequently, when Colombia and Venezuela became Spanish controlled and later on when Spanish settlement began to expand in the coastlands of the Plata Estuary in present-day Argentina and Uruguay, two additional viceroyalties were established: New Granada in the north and Rio de la Plata in the south.

Meanwhile, another vanguard of the Iberian invasion was penetrating the east-central part of the continent—the coastlands of present-day Brazil. This area had become a Portuguese sphere of influence almost by default. It was visited by

Spanish vessels early in 1500; later that year, the Portuguese navigator Pedro Cabral saw the coast on his way to the Far East. The Spaniards did not follow up their contact, but the Portuguese in 1501 sent Amerigo Vespucci with a small fleet to investigate further. After this journey, the area remained virtually neglected by the Portuguese for nearly 30 years; they were too absorbed in the riches of the East Indies. And the Spaniards, we know, had other geographic interests in the Americas. Moreover, the two countries had agreed in 1494, under the Treaty of Tordesillas, to recognize a north-south line drawn 370 leagues west of the Cape Verde Islands as the boundary between their New World spheres of influence. This border ran approximately along the meridian of 50° west longitude, thereby cutting off a sizable triangle of eastern South America for Portugal's exploitation (Fig. 5–6).

A brief look at the political map of South America (Fig. 5–7) shows that the Treaty of Tordesillas did not succeed in limiting Portuguese colonial territory to the east of the agreed-on meridian. True, the boundaries between Brazil and its northern and southern coastal neighbors (French Guiana and Uruguay) both reach the ocean near the 50° line, but, then, the Brazilian boundaries bend far inland to include almost the entire Amazon Basin as well as a good part of the Paraná-Paraguay Basin. Portugal's push into the interior was reflected in the Treaty of Madrid (1750), whose terms included a westward shift of the 1494 Papal Line. Brazil, with over 8.5 million square kilometers (3.3 million square miles), came to be only slightly smaller than all the other South American countries combined (9.6 million square kilometers/3.7 million square miles). In population, too, Brazil now has about half the total of all of South America; history and geography, therefore, dealt the Portuguese sphere a fairer hand than any treaties did. This enormous westward thrust was the work of many Brazilian elements—missionaries in search of converts, explorers in search of quick wealth—but no group did more to achieve this penetration than the so-called *Paulistas*, the settlers of São Paulo. From early in its colonial history, São Paulo had been a successful settlement, with thriving plantations and an ever-growing need for labor. The *Paulistas* organized huge expeditions into the interior, seeking Indian slaves, gold and precious stones, and, incidentally, were also intent on reducing the influence of Jesuit missionaries over the Indian population there.

Figure 5–6

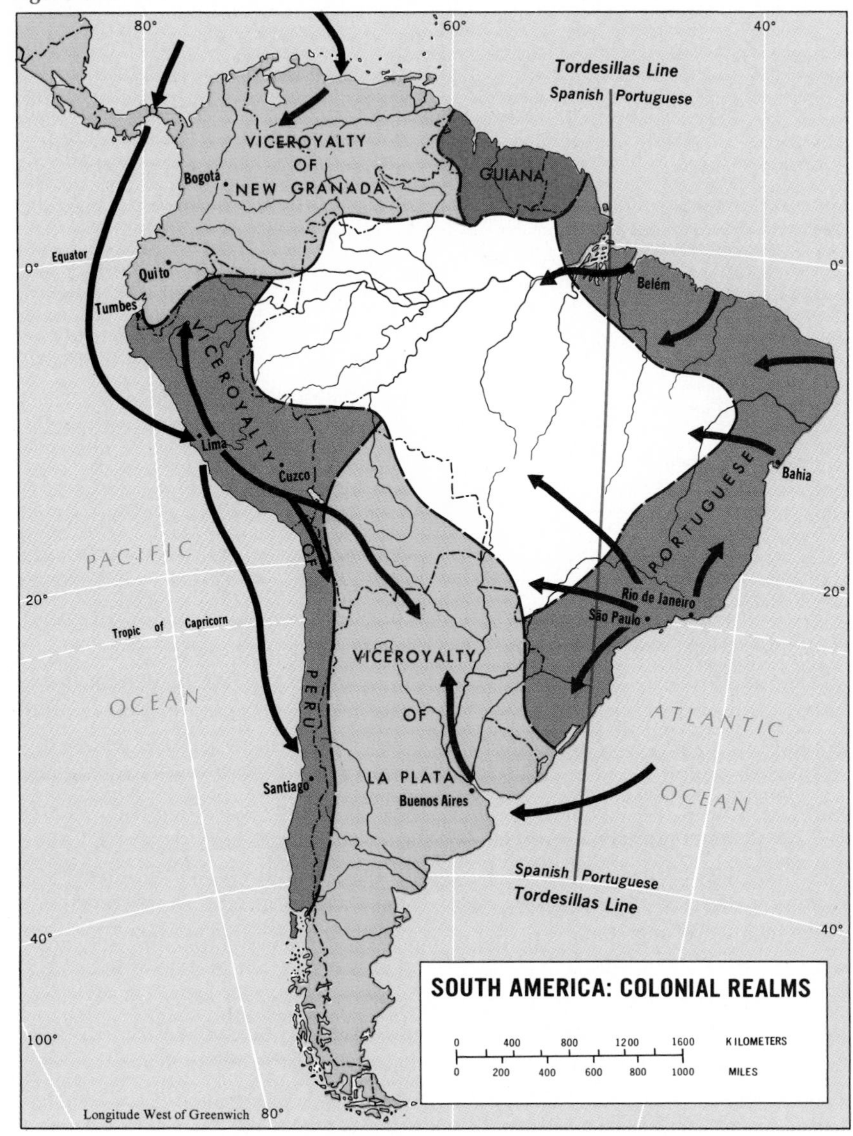


## The Africans

As Fig. 5–6 shows, the Spaniards initially got very much the better of the territorial division of South America—not just quantitatively, but, initially at least, qualitatively as

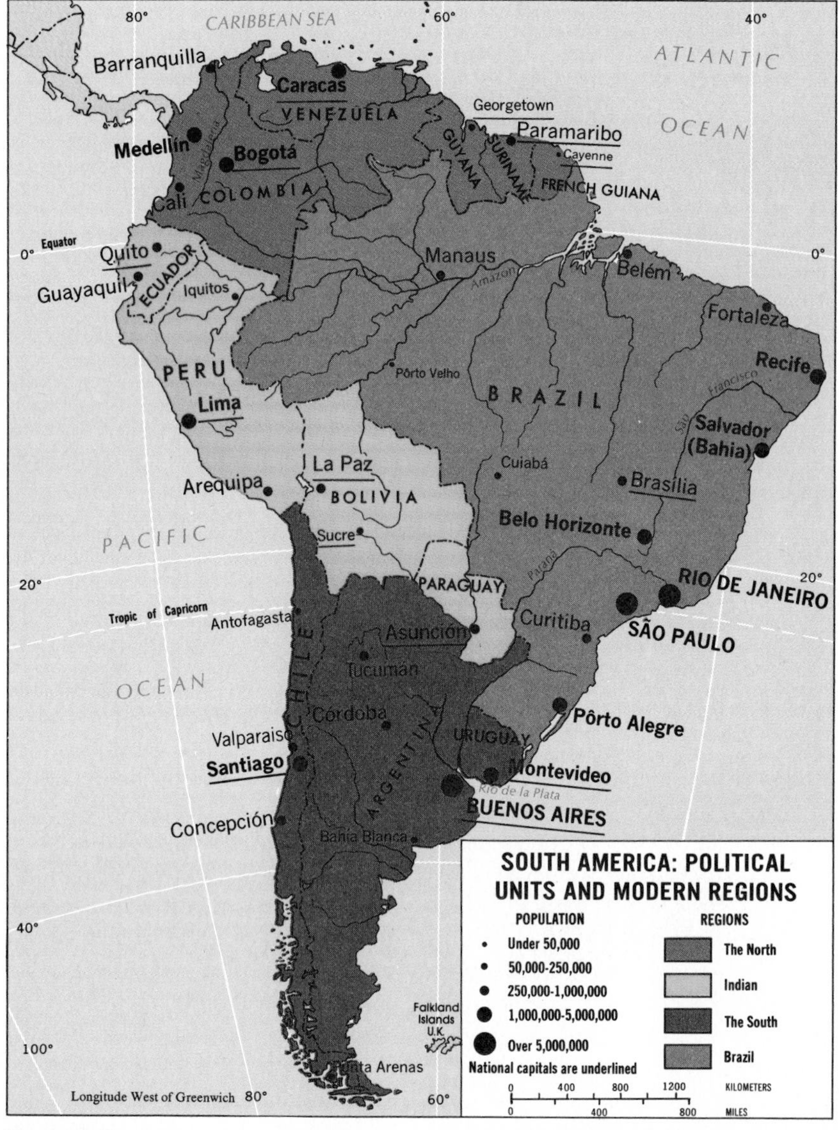


Figure 5–7

well. There were no rich Indian states to be conquered and looted in the east, and no productive agricultural land was under cultivation. The eastern Indians were comparatively few in number and constituted no usable labor force. It has been estimated that the whole area of present-day Brazil may have been inhabited by only about 1 million aboriginal people. When the Portuguese finally began to look with renewed interest to their New World sphere of influence, they turned to the same lucrative activity pursued by their Spanish rivals in the Caribbean—the plantation cultivation of sugar for the European market. And they found their labor force in the same region too, as millions of Africans were brought in slavery to the northern and northeastern Brazilian coast. Again, estimates of the total number of Africans comprising this forced migration flow vary; over the centuries, the figure probably exceeded 6 million. Not surprisingly, Brazil now has South America's largest black population, which is still heavily concentrated in the country's poverty-stricken northeastern states; today, with the overall population of Brazil exceeding 137 million, more than 11 percent of the people are black and another one-third are of mixed African, white, and Indian ancestry. Africans, then, definitely constitute the third major immigration of foreign peoples into South America (see Fig. A–2).

## Persistent Isolation

Despite their common cultural heritage (at least insofar as their European-mestizo population is concerned), their adjacent location on the same continent, their common language, and their shared national problems, the countries that arose out of South America's Spanish viceroyalties have existed in a considerable degree of isolation from each other. Distance, physiographic barriers, and other factors have reinforced this separation. To this day, the major population agglomerations of South America adhere to the coast, mainly the eastern and northern coasts. Of all the continents, only Australia has a population distribution that is as markedly peripheral, but there are only some 16 million people in Australia against 270-million-plus in South America.

Compared to other world realms, South America may be described as underpopulated, not just in terms of its low total for a continental area of its size, but also in view of the resources available or awaiting development. The continent never drew as large an immigrant European population as did North America. The Iberian Peninsula could not provide the numbers of people that western and northwestern Europe did, and colonial policy, especially Spanish policy, had a restrictive effect on the European inflow. The New World viceroyalties existed primarily for the purpose of the extraction of riches and the filling of Spanish coffers; in Iberia, there was little interest in developing the American lands for their own sake. Only after those who had made Hispanic and Portuguese America their permanent home and who had a stake there rebelled against Iberian authority did things begin to change, and then very slowly. South America was saddled with the values, economic outlook, and social attitudes of seventeenth- and eighteenth-century Iberia—not the best tradition from which to begin the task of forging modern nation-states.

## Independence

Some isolating factors had their effect even during the wars for independence. Spanish military strength was always concentrated at Lima, and those territories that lay farthest from this center of power—Argentina and Chile—were the first to establish their independence from Spain in 1816 and 1818, respectively. While the Argentine general José de San Martin led the combined Argentinian and Chilean armed forces to the coast of Peru, Simón Bolivar commanded the north, New Granada, in its fight for independence. Eventually, Bolivar organized an assault on the remaining Spanish forces, still fortified in the Andean Mountains; in 1824, two decisive battles finally terminated Spanish power in South America. Thus, in little more than a decade, the Spanish countries had fought themselves free, and the significance of their cooperation in this effort can hardly be overstated. But the joint struggle did not produce unity: no fewer than nine countries emerged out of the three former viceroyalties, including Bolivia, hitherto known as Upper Peru and named after Bolivar when it was declared independent in 1825. Bolivar's Colombian Confederacy, which had achieved independence in 1819, broke up in 1831 into Venezuela, Ecuador, and New Granada, which in 1861 was renamed Colombia. Uruguay was temporarily welded to Brazil, but it, too, attained separate political identity by 1828. Paraguay, once a part of Argentina, also appeared on the map as a sovereign state before 1830 (Fig. 5–7).

It is not difficult to understand why this fragmentation should have taken place; with the Andes intervening between Argentina and Chile, and the Atacama Desert between Chile and Peru, overland distances seem even greater than they really are, and these obstacles to contact have proven very effective. Thus, the countries of South America began to grow apart, a separation process heightened sometimes by uneasy frontiers. In fact, friction and even wars over Middle and South American borders have been frequent, and a number of boundary disputes remain unsettled to this day. Bolivia, for example, at one time had a direct outlet to the sea via northern Chile, but it lost this access corridor in a series of conflicts involving Chile, Peru, and indirectly, Argentina. Chile and Argentina themselves were long locked in a dispute over their Andean boundary (which still continues in the Tierra del Fuego area); Peru and Ecuador both laid claim to the upper Amazon Basin and the Peruvian town of Iquitos.

Brazil attained independence from Portugal at about the same time the Spanish settlements in South America were struggling to end overseas domination, although the sequence of events was quite different. In Brazil, too, there had been revolts against Portuguese control—the first as early as 1789; but the early 1800s, instead of witnessing a steady decline in Portuguese authority, actually brought the Portuguese government (Prince Regent Dom João and a huge entourage) from Lisbon to Rio de Janeiro. Thus, Brazil in 1808 was suddenly elevated from colonial status to the seat of empire, and it owed its new position to Napoleon's threat to overrun Portugal, which was allied with the British.

At first it seemed that the new era would bring progress and development to Brazil. Although there was some agitation against the regime based in Rio de Janeiro, notably in the Brazilian northeast, where the city of Pernambuco (now named Recife) was the center of a revolt in 1817, the real causes of Brazilian independence lay in Portugal, not Brazil. Dom João did not return to Lisbon immediately after the departure of the French, and by the time he did, it was in response to a revolution there; the Napoleonic period had left behind a great deal of dissatisfaction with the *status quo*. Worse, the regime in Lisbon wanted to end the new status of equality for Brazil and once again make it a colony. Accordingly, Dom João appointed his son, Dom Pedro, as regent and in 1821 set sail for Portugal. It was to no avail: the national assembly of Portugal was determined to undo Dom João's administrative innovations, and Dom Pedro was soon ordered to return to Lisbon as well. This he refused to do; in 1822, he proclaimed Brazil's independence and was crowned emperor. He had overwhelming support from the Brazilian people in this decision, and loyalist Portuguese forces still in the country were ignominiously forced to return to Lisbon.

The postindependence relationships of Brazil to its Spanish-influenced neighbors have been similar to the relationships among the individual Spanish republics themselves. Distance, physical barriers, and cultural contrasts serve to inhibit contact and interaction of a positive kind. Brazil's orientation toward Europe, like that of the other republics, remained stronger than its involvement with the countries on its own continent.

# Culture Areas

When we speak of the "orientation" or "interaction" of South American countries, it is very important to keep in mind just who does the orienting and interacting, for there is a tendency to generalize the complexities of these countries away. The fragmentation of colonial South America into 10 individual republics, and the nature of their subsequent relationships, was the work of a small minority of the people in each country. The black people in Brazil at the time of independence had little or no voice in the course of events; the Indians in Peru, numerically a vast majority, could only watch as their European conquerors struggled with each other for supremacy. It would not even be true to say that the European minorities *in toto* governed and made policy: it was the wealthy, landholding, upper-class elite who determined the posture of the state. These were—and in some cases still are—the people who made the quarrels with their neighbors, who turned their backs on wider American unity, and who kept strong the ties with Madrid and Paris (Paris had long been a cultural focus for "Latin" America's well-to-do, and their children were usually sent to French rather than Spanish schools).

So complex and heterogeneous are the societies and cultures of Middle and South America that practically every generalization has to be qualified. Take the one used generally—the term *Latin* America. Apart from the obvious exceptions that can be read from the map, such as Jamaica, Guyana, and Suriname, which are clearly not "Latin" countries, it may be improper even to identify some of the Spanish-influenced republics as "Latin" in their cultural milieu. Certainly the white, wealthy upper classes are of Latin European stock, and they have the most influence at home and are most visible abroad; they are the politicians and the businesspeople, the writers and the artists. Their cultural environment is made up of the Spanish (and Portuguese) language, the Catholic church, and the picturesque Mediterranean architecture of Middle and South America's cities and towns; these things provide them with a common bond and with strong ties to Iberian Europe. But, in the mountains and villages of Ecuador, Peru, and Bolivia there live millions of people to whom the Spanish language is still alien, to whom the white people's religion is another unpopular element of acculturation, and to whom decorous Spanish styles of architecture are meaningless when a decent roof and a solid floor are still unattainable luxuries.

South America, then, is a continent of pluralistic societies, where Indians of different cultures, Europeans from Iberia and elsewhere, Africans from the west coast and other parts of tropical Africa, and Asians from India, Java, and Japan have produced a cultural and economic kaleidoscope of almost endless variety. Certainly to call this human spatial mosaic "Latin" America is not very useful. But is there a more meaningful approach to a regional generalization that would better represent and differentiate the continent's cultural and economic spheres? One such attempt was made by John Augelli, who also developed the *Rimland-Mainland* concept for Middle America. His map (Fig. 5–8) shows five culture spheres—internal cultural regions—within the South American realm. This scheme is quite useful, but it should be kept in mind that the realm is undergoing rapid economic change in certain areas today and that these culture spheres are generalized and subject to further change as well.

The first of these, the *Tropical plantation* region, in many ways resembles the Middle American Rimland. It consists of several separated areas, of which the largest lies along the northeastern Brazilian coast, with four others lying along the Atlantic and Caribbean northern coast of South America. Location, soils, and tropical climates favored plantation crops, especially sugar; the small indigenous population led to the introduction of millions of African slave laborers, whose descendants today continue to dominate the racial makeup and strongly influence the cultural expression of these areas. The plantation economy later failed, soils were exhausted, the slavery system was terminated, and the people were largely reduced to poverty and subsistence—socioeconomic conditions that now dominate much of the region mapped as tropical plantation.

The second region on Augelli's map, identified as *European-commercial,* is perhaps the most truly "Latin" part of South America. Argentina and Uruguay, each with a population that is between 80 and 90 percent "pure" European and with a strong Hispanic cultural imprint, constitute the bulk of the European-commercial region. Two other areas also lie within it: most of Brazil's core area, and the central Chile core of that country. Southern Brazil shares the temperate grasslands of the Pampa and Uruguay (see Fig. I–12), and this area is important as a zone of livestock raising as well as corn production; Brazil fostered European settlement here at an early stage for strategic reasons. Middle Chile is an old Spanish settlement, with the rest of that country containing a much larger mestizo population than either Argentina or Uruguay. The one-quarter of the Chilean people who claim pure Spanish ancestry is concentrated in the valleys between the Andes and the coastal ranges (like 90 percent of the total

Figure 5–8

national population), and between the Atacama Desert of the north and the mountainous, forested, sea-indented south. Here, in an area of Mediterranean climate (Fig. I–10), cattle and sheep pastoralism plus mixed farming are practiced. In general, then, the European-commercial region is economically more advanced than most of the rest of the continent. A commercial economy prevails rather than subsistence modes of life, living standards are better, literacy rates are higher, transportation networks are superior and, as Augelli pointed out in his article, the overall development of this region is well ahead of that of several parts of Europe itself.

The third region is identified as *Indo-subsistence*, and it forms an elongated area running the length of the central Andes from southern Colombia to northern Chile and Argentina, an area that coincides approximately with the old Indian empires. The feudal socioeconomic structure that was established here by the Spanish conquerors still survives. The Indian population forms a large, landless peonage living by subsistence or by working on the haciendas, far removed from the Spanish culture that forms the primary force in the national life of their country. This region includes some of South America's poorest areas, and what commercial activity there is tends to be in the hands of the white or mestizo. The Indian heirs to the Incan Empire live, often precariously, at high elevations (as much as 3500 meters/11,000 feet) in the Andes. Poor soils, uncertain water supply, high winds, and bitter cold make farming an always difficult proposition; what little exchange exists takes place at remote upland Indian markets (Fig. 5–9).

The fourth region, *Mestizo-transitional*, surrounds the Indo-subsistence region, covering coastal and interior Peru and Ecuador, much of Colombia and Venezuela, most of Paraguay, and large parts of Argentina, Chile, and Brazil (including the Amazon and its tributary river valleys). This is the zone of mixture between European and Indian (or African in Brazil, Venezuela, and Colombia). The map, thus, reminds us that countries like Bolivia, Peru, and Ecuador are dominantly Indian and mestizo; in Ecuador, for example, these two groups make up nearly 90 percent of the total population, of which a mere 10 percent can be classified as white. The term *transitional* has an economic connotation too, because, as Augelli puts it, this region "tends to be less commercial than the European sphere but less subsistent in orientation than dominantly Indian areas."

The fifth region on the map is marked as *Undifferentiated*, because it is difficult to classify its characteristics. Some of the Indian peoples in the interior of the Amazon Basin have remained almost completely isolated from the momentous changes in South America since the days of Columbus, and isolation and lack of change are still two of the dominant aspects of this region. The Amazon backlands (plus the Chilean and Argentinian southwest) are also sparsely populated and exhibit only very limited

**Figure 5–9**
An Indian market in the highlands of Peru.

economic development; poor transportation and difficult location have contributed to the unchanging nature of this region. Trans-Amazonian highways, built in the last few decades, are the first incursions into the more isolated parts of this region.

The framework of culture spheres described is necessarily a generalization of a rather complex spatial reality, but, even in its simplicity it underscores the diversity of South American peoples, cultures, and economies. This heterogeneity is further underscored in the continent's agricultural geography, which was reviewed in the chapter-opening essay and map (Fig. 5–3).

# Urbanization

As in other parts of the world, people in South America are leaving the land and moving to the cities. This urbanization process intensified sharply after the Second World War; today, three of the realm's cities (Buenos Aires; Rio de Janeiro and São Paulo in Brazil) are among the world's 10 largest metropolitan agglomerations. A measure of the pace of urbanization is provided by the following indexes: in 1925, about one-third of South America's people lived in cities and towns and, as recently as 1950, the percentage was just over 40; but, in 1975, the continentwide figure had exceeded 60 percent and, by the mid-1980s, three out of four persons resided in urban South America. Of course, these percentages mask the actual numbers. Between 1925 and 1950, the continent's towns and cities grew by about 40 million residents as the urbanized percentage rose from 33 to 40; but between 1950 and 1975 more than 125 million people crowded into the teeming metropolitan areas—more than *three times* the total for the previous quarter-century—and an additional 60 million since 1975 swelled the continental total to 200 million by 1985.

South America's population of over 270 million (about 15 percent higher than that of the United States in the mid-1980s) has a high growth rate, but nowhere are the numbers increasing more rapidly than in the towns and cities. We usually assume that the populations of rural areas grow more rapidly than urban areas, because farm families traditionally have more children than city dwellers. But overall, the urban population of South America has grown by nearly 5 percent per year since 1950, whereas the rural areas increased by only about 1.5 percent. These figures reveal the dimensions of the migration stream from the countryside toward the towns and cities—still another kind of migration process, which affects modernizing societies throughout the underdeveloped realms.

In South America—as in Middle America, Africa, and Asia—people are attracted to the cities and driven from the poverty of the rural areas; both *pull* and *push* factors are at work. Rural land reform has been slow in coming, and many farmers simply give up and leave, seeing little or no change, year after year. The cities attract because they seem to provide opportunity—the chance to earn a regular wage. Visions of education for their children, better medical care, and the excitement of life in a big city draw millions to places such as Rio de Janeiro and Bogotá. Road and rail connections continue to improve, so that access is easier and exploratory visits can be made. City-based radio stations beckon the listener to the place where the action is.

But the actual move can be traumatic. As we saw in the box on the Latin American City in Chapter 4 (Fig. 4–15), South America's cities are surrounded and sometimes invaded by slums (*favelas* or *barrios*) that are often among the world's worst, and this is where the uncertain urban immigrant usually finds a first (and frequently permanent) abode, in a makeshift shack without even the most basic ameni-

**Figure 5–10**
Extreme social contrasts often characterize the realm's teeming cityscapes: Caracas, Venezuela.

ties. Many move in with relatives who have already made the transition, but whose dwelling can hardly absorb another family. And unemployment is high, sometimes as high as 25 percent of the available labor force. Jobs for unskilled workers are hard to find and they pay minimal wages. But the people still come, the crowding in the *barrios* worsens, and the threat of epidemic-scale disease rises. It has been estimated that in-migration accounts for over 50 percent of urban growth in some developing countries—in South America, urban populations have unusually high natural growth rates as well.

Hence, South American cities present almost inconceivable contrasts between wealth and poverty, splendor and squalor; nowhere is this harsh juxtaposition more evident than in the urban landscape of Venezuela's capital city, Caracas (Fig. 5–10). Largest of all is the city that represents more than any other the incredible mass agglomeration, crowding, pluralism, and energy of the realm's urban clusters—São Paulo (Fig. 5–11). Historically, Brazil's second city (before it surpassed Rio de Janeiro in the 1950s to become the national leader), São Paulo is now growing at a prodigious rate: the 1970 census reported just over 8 million inhabitants, but 1985 estimates reveal that the population today has at least doubled to more than 16.5 million. If São Paulo and its surroundings continue to expand at this pace, the metropolis will surpass 25 million residents by 2000. Already the city proper is described as having the world's worst traffic jams, greatest air pollution, highest noise levels, and perhaps the lowest quality of life. None of this, of course, deters the annual flood of new arrivals. South America's cities truly are crucibles in which the survival habits of a whole realm's population are being transformed.

## Capital Cities

South America's exploding cities, with few exceptions, would fit Mark Jefferson's concept of the primate city. In all but 2 of the 12 countries (and in the 13th territory, French Guiana, as well), the largest urban center overshadows all other towns, functions as the capital, and is the historic, cultural, political, and economic focus of the state. The exceptions are Brazil, where

**Figure 5–11**
São Paulo, at nearly 17 million (1985) one of world's fastest growing metropolises.

Rio de Janeiro, long the capital and national center, was replaced as governmental headquarters by newly built Brasilia in 1960, and Ecuador, where Quito (925,000), the capital in the interior Andes, is a smaller city than coastal Guayaquil (1.3 million), the country's port. But, in the other countries the capital is the national heart in every way.

Capital cities hold special interest for geographers (particularly political geographers), because they embody national qualities and aspirations, reflect cultural preferences, and represent the essence of the state. When Japan entered its modern age, its leaders boldly moved the national capital from venerable, revered Kyoto to coastal Tokyo. It was a symbolic as well as practical move, and one of many instances where governments have used their own headquarters to signal a new era, or to underscore a new orientation. Brasilia, 650 kilometers (400 miles) inland from the Rio de Janeiro coast (Fig. B–2), was designed to symbolize Brazil's age of development and its—hoped for—new westward orientation, and it has, indeed, become a growth pole in the country's vast inner frontier zone.

Capital cities, therefore, can be classified according to several criteria. Geographically, the most obvious is location relative to the state's territory. Santiago (4.5 million), capital of Chile, lies in the heart of its country's core area and near the geographical center of the state. In Colombia, Bogotá (7.2 million) also lies in the interior, nearer the center of the country's population than any coastal capital could be; Ecuador's Quito (925,000) is similarly an interior capital. On the other hand, Montevideo (1.5 million), capital of Uruguay, lies at the very edge of the national territory, as does Buenos Aires, Argentina.

The extent to which a capital's relative location can matter to a state is again underscored by Brazil. The government's desire to accelerate interior development led to Brasilia's construction in the late 1950s; by the mid-1980s, the new capital city—which has already attracted sizable squatter shacktowns on its peripheries—anchored a population agglomeration of nearly 2.5 million. Brasilia represents what political geographers call a *forward capital.* There are times when a state will relocate its capital to a sensitive area, perhaps near a zone under dispute with an unfriendly neighbor, partly to confirm its determination to sustain its position in the contested zone. A recent example is the decision by Pakistan to move its capital from coastal Karachi to northern Islamabad, near disputed Kashmir. These are called "forward" capitals because of their position in an area that would be first to be engulfed by conflict in case of strife with a neighbor. At one time, Berlin was a forward capital, a German headquarters near the margins of Slavic Eastern Europe. Brasilia, of course, does not lie in or near a contested area—but Brazil's interior has been a kind of internal frontier, one to be conquered by a developing nation. In that drive, the new capital has a forward position.

Another way to classify capital cities involves their origins and durability. Thus, Quito (Ecuador), Bogotá (Colombia), and La Paz (Bolivia) have Indian antecedents—Quito is the oldest capital city in South America. Bogotá (7.2 million) began as a Chibcha Indian settlement. As a Spanish center, it witnessed the rise and fall of the Viceroyalty of New Granada (Fig. 5–6) and the Confederation of Gran Colombia; since 1835, Bogotá has been the capital of the Republic of Colombia. In another category, Lima (5.8 million), capital of Peru, was founded by the Spanish colonizers in 1535 in preference to Cuzco, the interior Incan headquarters. In Venezuela, Caracas (3.7 million) began as a hamlet built on a colonist's ranch just 11 kilometers (7 miles) from the Caribbean coast, but on the landward side of an Andean mountain spur.

In still another category of capital cities, the administrative functions are divided among more than one headquarters. This particular situation has occurred in Libya (to resolve regional quarrels), in Laos (where royal and public capitals emerged), in the Netherlands (where Amsterdam and The Hague share capital roles), and in South Africa (where legislative and administrative functions are handled in Cape Town and Pretoria, respectively). South America, too, has a situation of this kind in Bolivia, in which La Paz is the seat of national government but Sucre is the seat of the supreme court and the legal capital. La Paz, the world's highest capital city at 3570 meters (11,700 feet) above sea level, has a population approaching 1 million, more than ten times as large as its historic rival.

## Urban Structure

In Chapter 4, we introduced the model of Latin American city structure (pp. 270–271), a generalization that applies to most of the large cities in the Middle and South American realms. One of the assumptions underlying this model is that this "ideal" city form is located on flat terrain. Many of South America's cities, however, lie on hilly, even mountainous, sites. Both Rio de Janeiro and São Paulo have been affected in their development by steep slopes. In the former capital, some of these are high and rocky, remote from the main roads, and unserved by city utilities; slums have developed there, because, in "Rio," elevation is less desirable than proximity to the cooling bay and ocean. In São Paulo, on the other hand, the higher elevations soon attracted high-class residential development; this city lies more than 50 kilometers (30 miles) inland; here it is height, not seaward location, that affords coolness. Buenos Aires exhibits another departure from the ideal model pattern: the urban area is elongated parallel to the Plata estuary, focused on the central port area, which stretches for several miles along the banks of this river mouth. A third empirical "distortion" of the ideal structural arrangement is evident in the morphology of metropolitan Santiago, where topography has induced a strong sectoral character, an influence also visible in the Rio de Janeiro urban complex. Such variations notwithstanding, the ideal model's spatial elements remain recognizable in these individual applications, and the abstraction continues to be a useful guide to the understanding of South America's intraurban geography.

As we saw in the preceding chapter, when the Spanish colonizers laid out their New World cities and towns, they created a central square or *plaza* dominated by a church and flanked by imposing government buildings (Fig. 4–3). Lima's *Plaza de Armas*, Montevideo's *Plaza de la Constitución*, and Buenos Aires's *Plaza de Mayo* are good examples of the genre; a true classic is Bogotá's *Plaza Bolivar*, which is still one of Colombia's most important public gathering places (Fig. 5–12). Early in the South American city's development, the plaza formed the hub and focus of the city, surrounded by shopping streets and arcades. But, eventually, the city outgrew its old center and new commercial districts formed, elsewhere in the CBD and along the radial Spine (Fig. 4–15), leaving the plaza largely to serve as a ceremonial link with the past.

Cities, as we have also noted in previous chapters, perform particular functions. Some are capital cities, others are university towns, still others are mining centers. Several geographers have tried to establish a classification of cities based on their leading functions. Perhaps the best known of these efforts is a classification of U.S. cities by Chauncy Harris, first proposed in 1943. He suggested that a sound

**Figure 5–12**
The *plaza* still dominates the old center of the South American city: Bogotá's *Plaza Bolivar.*

way to classify urban centers would be through the measurement of the activities of the labor force. Which are the urban economic sectors that employ most, or a significant portion, of the labor force? If in industry, a city would be a manufacturing center; if in selling, it would be a retail center; if in moving goods, a transportation center. Harris also recognized cities where wholesaling prevails; diversified cities, where there is a mix of economic activities; mining towns; university cities; resort and retirement places; and political centers, including national and subnational capitals.

South America's cities, like several major U.S. cities, in certain instances contain so diversified an economic base that no single activity dominates. Multifunctional places such as Buenos Aires, Rio de Janeiro, and Caracas serve as industrial, wholesale, retail, and educational centers. On the other hand, Valparaiso (Chile), Callao (Peru), La Guaira (Venezuela), and Santos (Brazil) are all transportation centers, serving as ports for larger nearby urban areas located inland. Again, several large mining towns exist, including Cerro de Pasco in Peru, Potosí and Oruro in Bolivia, and Cerro Bolivar in Venezuela. Manufacturing cities are led prominently by São Paulo and Belo Horizonte in Brazil, and also include Guayaquil in Ecuador (an important port as well) and Medellín in Colombia. Only one primarily political center exists in South America—Brasilia—but Sucre in Bolivia and Quito in Ecuador owe in large part what prominence they have to their political roles in the national state.

# The Republics: Regional Geography

We turn now to the countries of South America and their principal regional-geographical properties. In general terms, it is possible to group South America's countries into regional units, because several have qualities in common. Thus, the northernmost Caribbean countries, Venezuela and Colombia, form a regional unit (that might include as well adjacent Guyana, Suriname, and French Guiana). On the basis of their Indian cultural heritage, Andean physiography, and modern populations, the republics of Ecuador, Peru, and Bolivia constitute a regional entity, to which Paraguay may be added on grounds to be discussed later. In the south, Argentina, Uruguay, and Chile have a common regional identity as the realm's mid-latitude, most strongly European states. And Brazil, by itself, constitutes a geographic region in South America, because it contains nearly half the land and about half the people of the continent; in fact, we will treat that burgeoning country in a separate vignette following the end of this chapter.

## The North: Caribbean South America

As another look at Fig. 5–8 confirms, the countries of the northern coast have something in common other than their coastal location: each has an area of "tropical plantation," signifying early European plantation development, the arrival of black laborers, and the absorption of this African element into the population matrix. Not only did myriad black workers arrive: many thousands of Asians (from India) came to South America's northern shores as contract laborers. The pattern is familiar—in the absence of large local labor sources, the colonists turned to slavery and indentured workers to serve their lucrative plantations. Between Spanish Venezuela and Colombia, on one hand, and the non-Hispanic "three Guianas," on the other, there is this difference: in the former, the population center of gravity soon moved into the interior and the plantation phase was followed by a totally different economy, whereas, in the Guianas, coastal settlement and the plantation economy still predominate.

Venezuela and Colombia have what the Guianas lack (Fig. 5–13): their territories and populations are much larger, their natural environments more varied, their economic opportunities greater. Each has a share of the Andes Mountains (Colombia's being larger), and each produces oil from an adjacent joint reserve, which ranks among the world's major deposits (here Venezuela is the primary beneficiary). Much of what is important in Venezuela is concentrated in the northern and western parts of the country, where the Venezuelan Highlands form the eastern spur of the Andes system. Most of Venezuela's 19 million people are concentrated in these uplands, which include the capital (Caracas), its early rival (Valencia), the commercial and industrial center of Barquisimeto, and San Cristóbal near the Colombian border. The Venezuelan

Figure 5–13

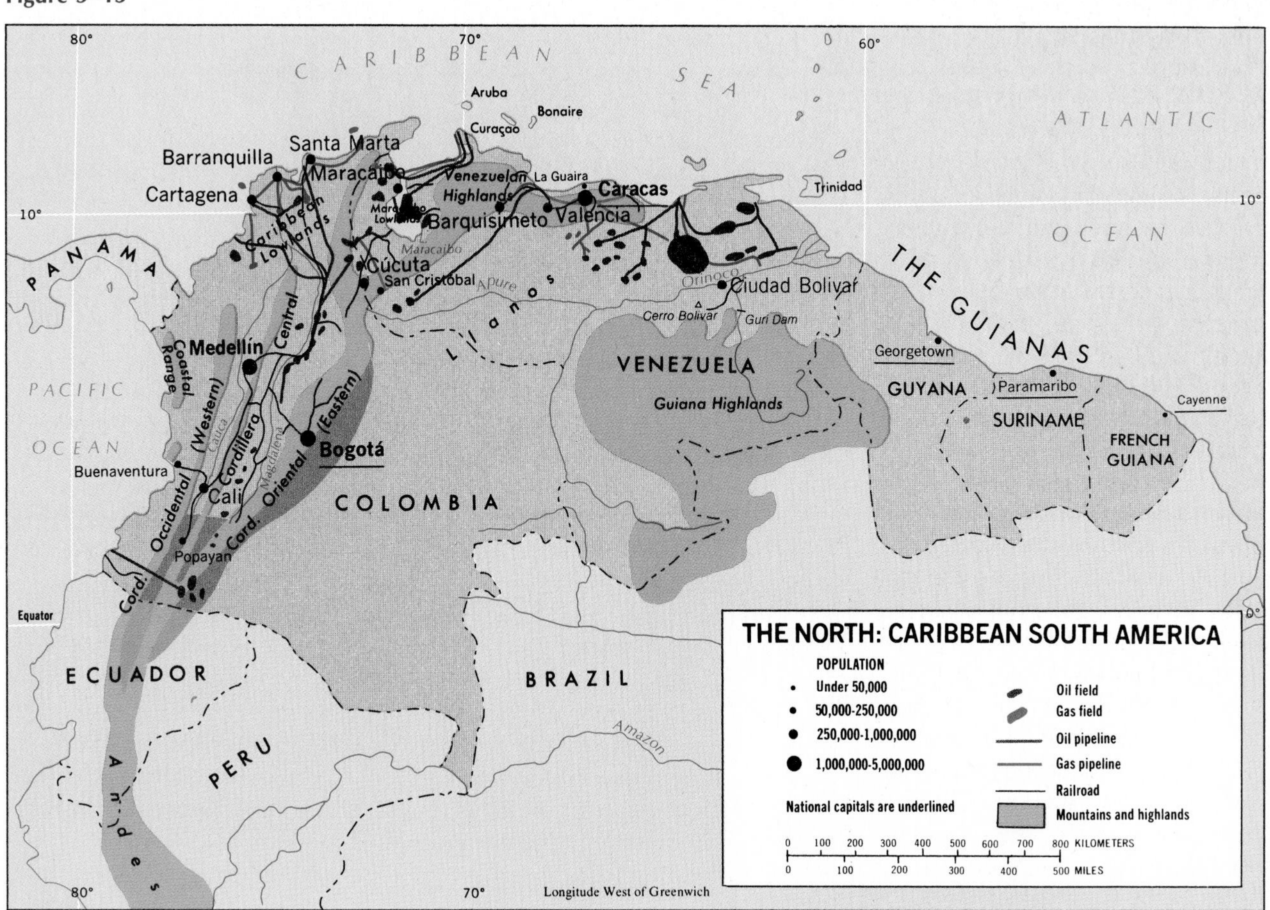

Highlands are flanked by the Maracaibo Lowlands and Lake Maracaibo to the northwest, and by a region of savanna country called the *llanos* in the Orinoco Basin to the southeast. The Maracaibo Lowland, once a disease-infested, sparsely peopled coastland, is today one of the world's leading oil-producing areas; much of the oil is drawn from reserves that lie beneath the shallow waters of the lake itself. Actually, Lake Maracaibo is a misnomer, for the "lake" is open to the ocean and is in fact a gulf with a very narrow entry. Venezuela's second city, Maracaibo, with over 1 million people, is the focus of the booming oil industry that has transformed the Venezuelan economy; in the 1970s, over 90 percent of the country's annual exports by value were crude oil and petroleum products. Large refineries on the offshore Dutch islands of Curaçao and Aruba have for many years refined the Venezuelan crude oil prior to its transportation to U.S. and European markets, but the capacity of local Venezuelan refineries is steadily increasing.

The llanos on the southern side of the Venezuelan Highlands, and the Guiana Highlands in the country's far southeast, are two of those areas that contribute to South America's image as "underpopulated" and "awaiting development." Although the llanos do not yet share in Venezuela's oil boom (reserves have been discovered here), the agricultural potential of these savannas—and of the *templada* areas of the Guiana Highlands—has hardly begun to be realized. There are good opportunities for a major pastoral industry in the llanos and for commercial agriculture in the Guiana Highlands. Of course, the llanos and Guiana Highlands are Venezuela's interior regions, and transportation remains an impediment. The discovery of rich iron ores on the northern flanks of the Guiana Highlands

# THE GUIANAS

Three small countries lie on the north coast of South America, immediately to the east of Venezuela and adjacent to northernmost Brazil (Fig. 5–13). In "Latin" South America, these territories are anomalies of a sort: their colonial heritage is British, Dutch, and French. Formerly, they were known as British, Dutch, and French Guiana, and called the three *Guianas*. But two of them are independent now: Guyana, Venezuela's neighbor, and Suriname, the Dutch-influenced country in the middle. The easternmost territory, French Guiana, still continues under colonial rule.

None of the three countries has a population over 1 million. Guyana is the largest with 840,000, Suriname has about 375,000 inhabitants, and French Guiana a mere 82,000. Culturally and spatially, patterns here are Caribbean: Asian (Indian) and black people are in the majority, whites a small minority. In Guyana, Asians make up just over half the population, blacks about one-third, and others (mixed and European peoples) small minorities. In Suriname, the population picture is even more complicated, for the colonists brought not only Asian Indians (now about 40 percent of the population) and blacks (nearly 33 percent), but also Indonesians (16 percent) to the territory in servitude; the remaining 10 percent of the Surinamese population consists of a separate category of black communities, peopled by descendants of African slaves who escaped from the coastal plantations and fled into the forests of the interior. French Guiana is the most European of the three countries—three-quarters of its small population is French; the majority is of mixed African, Asian, and European ancestry—the *creoles*.

French Guiana constitutes a French *département* and is represented in Paris by a senator, but this is the least developed of the three Guianas. The capital, Cayenne, has a mere 40,000 inhabitants; there is a small fishing (shrimping) industry, and some lumber is exported. But food must be brought in from overseas. By contrast, Suriname has progressed more rapidly, although it has been governed by an erratic socialist regime since the 1980 revolution. The plantation economy has given way to production on smaller farms where rice, the country's staple crop, is grown in adequate quantities to make Suriname self-sufficient. Citrus fruits are exported, and a wide variety of fruits and vegetables for the home market grow on farms in the coastal zone where once sugar plantations prevailed. But Suriname's big income earner is bauxite, mined in a zone across the middle of the country. The search for other minerals, including oil offshore, has had to be postponed until the country's political situation stabilizes.

British-influenced Guyana became independent in 1966 amid much internal conflict that, basically, pitted people of African origins against people of Asian ancestries. Problems of this kind have continued to afflict this country, where the great majority of the people live in small villages in the coastal zone. Plantation products still produce more than half the country's annual revenues, although the contribution of bauxite is rising. In the early 1980s, leadership problems contributed to a persistent economic crisis.

Unsettled boundary problems involve all three countries. Venezuela has laid claim to all of mineral-rich western Guyana—the Essequibo region, which purportedly was "stolen" by the British in 1899—a territorial dispute under continuing negotiation today. Guyana, in turn, claims a corner of southwestern Suriname. And part of the border between Suriname and French Guiana is also under dispute.

(chiefly near Cerro Bolivar) has begun to integrate one part of this region with the rest of Venezuela. A railroad was constructed to the Orinoco River; from there the ores are shipped directly to the steel plants of the U.S. northeast seaboard. San Feliz de Guayana (formerly Ciudad Guayana), hardly three decades old, has over 500,000 inhabitants; neighboring Ciudad Bolivar now contains 180,000 residents. The nearby Guri Dam has been put into service and is supplying much of Venezuela with electricity. Two new paved roads link this eastern frontier with Caracas.

Colombia, too, has a vast area of llanos, covering about 60 percent of the country; in Colombia as in Venezuela, this is comparatively empty land far from the national core and much less productive than it could be. Eastern Colombia consists of the headwaters of major tributaries of the Orinoco and Amazon Basins, and it lies partly under savanna and partly under rainforest conditions. In recent years, the Colombian government has begun in earnest to promote settlement east of the Andes. But it will be a long time before any part of eastern Colombia matches the Andean part of the country or even the Caribbean Lowlands of the north. In these regions, the vast majority of all Colombians live, and here lie the major cities and productive areas.

Western Colombia is dominated by mountains, but there is some regularity to this terrain. In broad terms, there are four parallel mountain ranges, generally aligned north-south, and separated by wide valleys. The westernmost of these four ranges is a coastal belt, less continuous and lower than the other three. These latter are the real Andean mountain chains of Colombia, for, in this country, the Andes separate into three ranges: the Eastern, Central, and Western Cordilleras. The valleys between the Andean Cordilleras open onto the Caribbean Lowland, where the two important Colombian ports of Barranquilla and Cartagena are located.

Colombia's population of nearly 29 million consists of several separate clusters (there are different ways of identifying them, but, in any case, there are more than a dozen), some of them in the Caribbean Lowlands, others in the inter-Cordilleran valleys, and still others in the intermontane basins within the Cordilleras themselves. This distribution existed before the Spanish colonizers arrived, with the Chibcha civilization (Fig. 5–4) concentrated in the intermontane basins of the Eastern Cordillera. The capital city, Bogotá (corrupted from the Chibchan *Bacatá*), was founded in one of the major basins in this Andean range, at an elevation of 2650 meters (8700 feet). For centuries, the Magdalena Valley between the Cordillera Oriental and the Cordillera Central was one link in a cross-continental communication route that began in Argentina and ended at the port of Cartagena, and Bogotá benefited by its position adjacent to this route. Today, the Magdalena Valley is still Colombia's major transport route, but Bogotá's connections with much of Colombia still remain quite tenuous; thus, to a considerable degree, the population clusters of this country, like those in Venezuela, exist in isolation from one another.

Colombia's physiographic variety is matched by its demographic diversity. In the south, it has a major cluster of Indian inhabitants, and the country begins to resemble its southern neighbors. In the north, vestiges remain of the plantation period and the African population it brought. Bogotá is a great "Latin" cultural headquarters, whose influence extends beyond the country's borders. In the Cauca Valley (between the Cordillera Central and Cordillera Occidental), Cali is the urban commercial focus for a hacienda district where sugar, cacao, and some tobacco are grown. Farther to the north, the Cauca River flows through a region comprising the departments (provinces) of Antioquia and Caldas, whose urban focus is the textile manufacturing city of Medellín, but whose greater importance to the Colombian economy entails the production of coffee. With its extensive *templada* areas along the Andean slopes, Colombia is one of the world's largest producers of coffee (Fig. 5–14). In Antioquia-Caldas, it is grown on small farms by a remarkably unmixed European population cluster; elsewhere, it is produced on the large estates so common in this geographic realm.

Coffee and oil are Colombia's two leading exports, with coffee accounting for about 60 percent by value and oil about 15 percent. Colombia's oilfields are extensions of Venezuela's reserves, and the oil is piped to its Caribbean ports. This country has also been identified as a leading source of illicit drugs that are smuggled into the United States, and the Caribbean Lowlands are believed to be the staging area for the northward shipment of narcotics manufactured from locally grown cocaine and marijuana. The importance of Colombia's window on the Caribbean can be read from the map; two major and several mi-

**Figure 5–14**
Coffee under cultivation in Colombia's wide *templada* zone.

nor ports handle goods brought by land and water from near the southern boundaries of the country, which, in total, far overshadow the volume of goods transferred at Colombia's lone west coast port of Buenaventura on the Pacific.

Venezuela and Colombia both exhibit a pronounced clustering of population, share a relatively empty interior, and depend on a small number of products for the bulk of their export revenues. The majority of the people of Colombia and Venezuela subsist agriculturally, and labor under the social and economic inequalities common to most of Iberian America.

## The Andean West: Indian South America

The second regional grouping of South American states encompasses Peru, Ecuador, Bolivia, and Paraguay (Fig. 5–15), a contiguous group of countries that includes South America's only two landlocked republics. The map of culture regions (Fig. 5–8) indicates the common Indo-subsistence sphere extending along the Andes Mountains. These are the countries of South America that have large Indian components in their populations. Nearly half the people of Peru (population: 20.3 million), are of Indian stock, and in Ecuador and Bolivia, too, the figure is near 50 percent. In Paraguay it is over 90 percent, but all these percentages are only approximate, since it is often impossible to distinguish between Indian and "mixed" people of strong Indian character. But there are other similarities among these four countries. Their incomes are low, they are comparatively unproductive, and, unhappily, they exemplify the grinding poverty of the landless peonage—a problem that looms large in the future of Ibero-America. These, too, are Hispanic South America's least urbanized countries; the capitals of three of the four states have under 1 million inhabitants, and only Lima, the capital of Peru, ranks with Bogotá and Santiago as a major-scale urban center.

In terms of territory as well as population, Peru is the largest of the four republics. Its 1.3 million square kilometers (nearly half a million square miles) divide both physiographically and culturally into three subregions: (1) the desert coast, the European-mestizo region; (2) the Andes Mountains or Sierra, the Indian region; and (3) the eastern slopes and the Montaña, the sparsely populated Indian-mestizo interior (Fig. 5–15). It is symptomatic of the cultural division still prevailing in Peru that the capital, Lima, is located not in a populous basin of the Andes but in the coastal zone. Here the Spanish avoided the greatest of the Indian empires and chose a site some 13 kilometers (8 miles) inland from a suitable anchorage, which became the modern port of Callao. From an economic point of view, the Spanish choice of a coastal headquarters proved to be sound, for the coastal region has become commercially the most productive part of the country. A thriving fishing industry based on the cool, productive waters of the Humboldt (Peru) Current offshore contributes a quarter of all exports by value. Irrigated agriculture in some 40 oases distributed all along the coast produces cotton, sugar, rice, vegetables, fruits, and some wheat; the cotton and sugar are important export products, and the other crops are grown mostly for the domestic market.

The Andean region occupies about one-third of the country, and here are concentrated the majority of the country's Indian peoples, most of them Quechua-speaking. But despite the fact that this territory is one-third of Peru and contains nearly one-half of its people, the political influence of this region is slight, and its economic contribution (except for the mines) is meager. In the valleys and intermontane basins, the Indian people are clustered either in isolated villages around which they practice a precarious subsistence agriculture or in the more favorably located and fertile areas where they are tenants—peons—on white- or mestizo-owned haciendas. Most of the Indian people never receive an adequate daily caloric intake or balanced diet of any sort; the wheat produced around Huancayo, for instance, is sent to Lima's European market and would be too expensive for the Indians themselves to buy. Potatoes (which can be grown at altitudes up to 4250 meters/14,000 feet), barley, and corn are among the subsistence crops, and in the high altiplanos the Indians graze their llamas, alpacas, cattle, and sheep. The major products that are derived from the Sierra are copper, silver, lead, and several other metallic minerals. These are mined in a number of areas, of which the one centered on Cerro de Pasco is the leader.

Of Peru's three subregions, the east—the interior slopes of the Andes and the Amazon-drained,

Figure 5–15

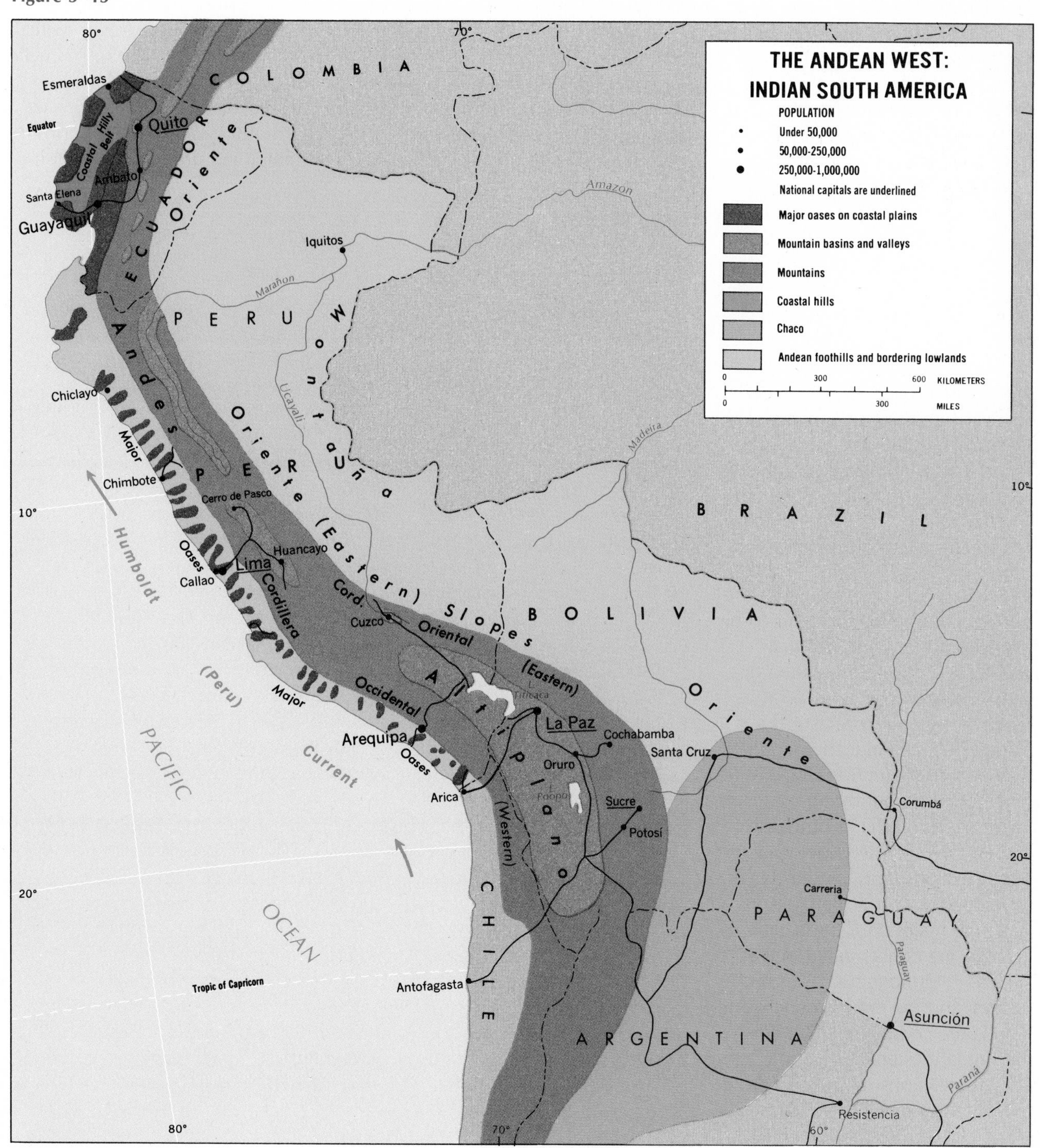

rainforest-covered Montaña—is the most isolated. A look at the map of permanent (as opposed to seasonal) routes, including railroads, shows how disconnected Peru's regions still are. However marvelous an engineering feat, the railroad that connects Lima and the coast to Cerro de Pasco and Huancayo in the Andes does not even begin to link the country's east and west. The focus of the eastern region, in fact, is a town that looks east rather than west, namely Iquitos, which can be reached by oceangoing vessels sailing up the Amazon River. Iquitos grew rapidly during the Amazon wild rubber boom earlier in this century and then declined, but now it is growing again, reflecting Peruvian plans to begin development of the east. In 1977, oil began flowing from wells at San José de Saramuro, through an 850-kilometer (525-mile) pipeline across the Andes to the coast at the port of Bayóvar (see *Box*). Meanwhile, traders plying the navigable rivers above Iquitos continue to collect such products as chicle, nuts, rubber, herbs, special cabinet woods, and small quantities of coffee and cotton.

Ecuador—smallest of the four republics—on the map appears to be just a corner of Peru. But that would be a misrepresentation because Ecuador possesses a full range of regional contrasts: it has a coastal belt, an Andean zone that may be narrow (under 250 kilometers/150 miles) but by no means of lower elevation than elsewhere, and an *Oriente*—an eastern region that is just as empty and just as undeveloped as that of Peru. As in Peru, the majority of the people of Ecuador are concentrated in the Andean intermontane basins and valleys, and the most productive region is the coastal belt. But here the similarities end. Ecuador's coastal area consists of a belt of hills interrupted by lowland areas, of which the most important one lies in the south between the hills and the Andes, drained by the Guayas River and its tributaries. The largest city and commercial center of the country (but not the capital), Guayaquil, forms the focus for this subregion. Ecuador's lowland west, however, is not desert country—it is a fertile tropical lowland not bedeviled by excessive rainfall (see Fig. I–9).

Ecuador's west-coast lowland is also far less Europeanized than Peru's Pacific-facing plain, because the white element in the total population of 9.4 million is a mere 10 percent; a sizable portion of the latter is engaged in administration and hacienda ownership in the interior, where most of the 50 percent of the Ecuadorians who are Indian also reside. Of the remaining national population, over 10 percent is black and mulatto; the rest are mestizo, many with a stronger Indian ancestry. The products of this coastal region, too, differ from those of Peru. Ecuador is the world's top banana exporter, with that commodity accounting for half of all exports by value; small farms owned by black and mulatto Ecuadorians and located in the north, in the hinterland of Esmeraldas, con-

## OIL FROM THE AMAZON

Peru's limited success in the search for oil in the Western Amazon region lying along its eastern flanks is an excellent illustration of the effect of relative location in economic geography. This eastern (*Oriente*) region became the country's leading source of oil and natural gas only in the 1970s; before then, the Talara area in the far northwest had been the chief source (Peru was an oil exporter until the early 1960s when these northwestern fields began their decline).

Although yet modest, present known petroleum reserves in the Amazonian region amount to about 75 percent of the country's overall total. The major discoveries have been made between Iquitos and the Ecuadorian border, and the river town of Iquitos has become the country's fastest-growing urban center. In 1965, the population was just over 50,000, but by 1985 it was approaching 225,000. In effect, Iquitos is Peru's *Atlantic* Ocean port, and it proved simpler to bring equipment and supplies to that city up the Amazon River (3700 kilometers [2300 miles] of waterway) than over the Andes Mountains. Iquitos's small oil refinery serves local needs and sends its surplus petroleum to Brazil on river barges; trucking the oil through the *selva* (tropical rainforest) and across the mountains would be impractical.

But Peru's core area needs the Amazonian oil, thus, an 850-kilometer (525-mile) pipeline was laid from San José de Saramuro through the forest, across the Andes (buried below the ground to minimize earthquake danger), and over the narrow coastal zone to the new and expanding refinery at Bayóvar in the northwest, not far from the old Talara fields. Since Peru needed Japanese capital to help build this pipeline, it was forced to commit part of the oil to Japan; in any case, transshipment is required to move the refined oil from Bayóvar to Lima and its hinterland. Thus, Brazil and Japan consume some of Peru's badly needed domestic energy supplies, in large part because of their situation vis-à-vis Peru's regional-geographical framework.

tribute to this total, as do farms on the eastern and northern margins of the Guayas Lowland. Cacao (12 percent) is another lowland crop, and coffee (22 percent) is grown on the coastal hillsides as well as in the Andean *templada* areas; cotton and rice are also cultivated, and cattle can be raised. In recent years, the production of petroleum in the jungles of the eastern (*Oriente*) region has reached substantial proportions. A trans-Andean pipeline was opened in the 1970s to connect the vast, rich interior oilfields to the port of Esmeraldas; as a result, Ecuador joined the Organization of Petroleum Exporting Countries (OPEC) and has now become Latin America's third largest exporter of crude oil.

Ecuador is not a poor country, and the coastal region, especially in recent years, has seen vigorous development. But the Andean interior, where the white and mestizo administrators and hacienda owners are outnumbered by Indians by about three or four to one, is a different story—or rather, a story similar to that of other Andean regions. Quito, the capital city, lies in one of the several highland basins in which the Andean population is clustered. Its functions remain primarily administrative; there is not enough productivity in the Andean region to stimulate commercial and industrial development. The Ecuadorian Andes do differ from those of Peru, in that they apparently lack major mineral deposits. Despite the completion of a railroad linking Quito to Guayaquil on the coast, the interior of Ecuador remains isolated and, economically, comparatively inert.

From Ecuador southward through Peru, the Andes broaden until, in Bolivia, they reach a width of some 720 kilometers (450 miles). In both the Cordillera Oriental and the Cordillera Occidental, peak elevations in excess of 6000 meters (20,000 feet) are recorded; between these two great ranges lies the Altiplano proper. On the boundary between Peru and Bolivia, freshwater Lake Titicaca—the highest large lake on earth—lies at over 3700 meters (12,507 feet). Here, in its west, lies the heart of modern Bolivia; here, too, lay one of the centers of Inca civilization—and, indeed, of pre-Inca cultures. Bolivia's capital, La Paz, is also on the Altiplano, at 3570 meters (11,700 feet) one of the highest cities in the world.

It is Lake Titicaca that helps make the Altiplano livable, for this large body of water ameliorates the coldness in its vicinity wherein the snow line lies just above the plateau surface (Fig. 5–16). On the surrounding cultivable land, grains, which have been raised for centuries, are grown in the Titicaca Basin to the extraordinary elevation of 3850 meters (12,800 feet); to this day the Titicaca area, in Peru as well as Bolivia, is a major cluster of subsistence-farming Indians. Modern Bolivia is the product of the European impact, an influence that has passed by some of the Indian population clusters. Of course, the Bolivian Indians no more escaped the loss of their land than did their Peruvian or Ecuadorian counterparts, especially east of the Altiplano. What made the richest Europeans in Bolivia wealthy, however, was not land, but minerals. The town of Potosí in the Eastern Cordillera became a legend for the immense deposits of silver nearby; copper, zinc, and several alloys were also discovered there. Most recently, Bolivia's tin deposits, among the richest in the world, have yielded some two-thirds of the country's annual export income. Oil and natural gas, however, are contributing a growing share. Bolivia exports gas to Argentina and Brazil; in return, Brazil is committed to assisting the development of an economic growth pole (see p. 320) in the area of Santa Cruz.

Bolivia has had a turbulent history. Apart from internal struggles for power, the country lost its seacoast in a disastrous conflict with Chile, lost its northern territory of Acre to Brazil in a dispute involving the rubber boom in the Amazon Basin, and then lost 140,000 square kilometers (55,000 square miles) of Gran Chaco territory to Paraguay in the war of 1932–1935. By far the most critical was the loss of its outlet to the sea (see *Box*); although Bolivia has rail connections to the Chilean ports of Arica and Antofagasta, it is permanently disadvantaged by its landlocked situation. Since the Cordillera Occidental and the Altiplano form the country's inhospitable western margins, one might suppose that Bolivia would look eastward and that its Oriente might be somewhat better developed than that of Peru or Ecuador—but such is not the case. The densest settlement clusters occur in the valleys and basins of the Eastern Cordillera, where the mestizo sector is also stronger than elsewhere in the country. Cochabamba, Bolivia's second city, lies in a basin that forms the country's largest concentration of population; Sucre, the legal capital, is located in another. Here, of course, lie the chief agricultural districts of the country, between the barren Altiplano to the west and the tropical savannas to the east.

Paraguay is the only non-Andean country in this region, but it is no less Indian. Of 3.7 million people, about 95 percent are Indian or mestizo, with so pervasive an Indian influence that any white ancestry is almost totally submerged. Although Spanish is the country's official language, Guaraní is more commonly spoken. By any measure, Paraguay is the poorest of the four countries of Indian South America, although it does have opportunities for pastoral and agricultural industries that have thus far gone unrealized. One of the rea-

**Figure 5–16**
Altiplano countryside near Bolivia's Lake Titicaca, close to the snow line.

sons for this must be isolation—the country's landlocked position. Paraguay's exports, in their modest quantities, must be exported via Buenos Aires, a long river haul from the Paraguayan capital of Asunción. Meat (dried and canned), timber (sold to Uruguay and Argentina), oilseeds, quebracho extract (for tanning leather), cotton, and some tobacco reach foreign markets. Grazing is the most important commercial activity, but the cattle generally do not compare well to those of Argentina.

## The South: Mid-Latitude South America

South America's three southern countries—Argentina, Chile, and Uruguay—are grouped in one region. By far the largest in terms of both population and territory is Argentina, whose 2.8 million square kilometers (1.1 million square miles) and 31 million people rank second only to Brazil in this geographic realm. Argentina exhibits a great deal of physical-environmental variety within its boundaries; the vast majority of population—three-quarters of it—is concentrated in the physiographic subregion known as the Pampa. Figure I–14 indicates the degree of clustering of Argentina's inhabitants on the land and in the cities of the Pampa (the word means "plain"); it also shows the relative emptiness of the other six subregions—the scrub-forest Chaco in the northwest, the mountainous Andes in the west (along whose crestline the boundary with Chile runs), the arid plateaus of Patagonia south of the Rio Colorado, and the undulating transitional terrain of intermediate Cuyo, Entre Rios, and the North (Fig. 5–17).

The Argentine Pampa is the product of the last hundred years. During the second half of the nineteenth century, when the great grasslands of the world were being opened up (including those of the United States, Russia, Australia, and South Africa), the economy of the long-dormant Pampa began to emerge. The food needs of industrializing Europe grew by leaps and bounds, and the advances of the industrial revolution—railroads, more efficient ocean transport, refrigerated ships, and agricultural machinery—helped make large-scale commercial meat and grain production in the Pampa not only feasible but very profitable. Large haciendas were laid out and farmed by tenant workers who would prepare the virgin soil and plant it with wheat and alfalfa, harvesting the wheat and leaving the alfalfa as pasture for livestock. Railroads radiated ever farther outward from Buenos Aires, and brought the entire Pampa into production. Today, Argentina has South America's densest railroad network, and once-dormant Buenos Aires is now one of the world's largest urban centers (10.8 million). Yet, the Pampa itself has hardly begun to fulfill its productive potential, which might double with more efficient and intensive agricultural practices.

Over the decades, several specialized agricultural areas appeared on the Pampa. As we would expect, a zone of vegetable and fruit production has become established near the huge Buenos Aires conurbation located alongside the estuary of the Rio de la Plata. In the southeast is the predominantly pastoral district, where beef cattle and sheep (for both mutton and wool) are raised. To the west, northwest, and southwest, wheat becomes the important commercial grain crop, but half the land still remains devoted to grazing. Among the exports, meat usually leads by value, followed by cereals, wool, and such lesser products as vegetable oils and oilseeds, hides and skins, and quebracho-tree extract.

Argentina's wealth and vigor are reflected in its fast-growing cities, which epitomize the European-commercial cultural character of the realm's southernmost countries (Fig. 5–18). Depending on the criteria used, 84 percent of the Argentinian population may be classified as urbanized, which is an exceptionally high figure for South America. More than one-third of all Argentinians live in the conurbation of Greater Buenos Aires alone, a metropolitan complex anchored by a capital city that is also a classic example of the primate city (Fig. 5–19); here, too, are most of the industries, many of them managed by Italians, Spaniards, and other relatively recent immigrants. The Córdoba area (1.2 million) is also a focus of industrial growth. Much of the manufacturing in the major cities is associated with the processing of Pampa products, and the production of consumer goods for the domestic market. One of every six wage earners in the country is engaged in manufacturing—another indication of Argentina's advanced economic standing.

Argentina's population shows a high degree of clustering and a decidedly peripheral distribution. The Pampa region covers only a little over 20 percent of Argentina's area, but with three-fourths of the people residing here the rest of the country

## WAR OF THE PACIFIC

Bolivia today is a landlocked country, but its access to the Pacific Ocean may soon be restored—after a century of confinement.

When Bolivia became an independent country in 1825, long stretches of South America's boundaries were only vaguely defined. Bolivia, at that time, had a sphere of influence along what is today the North Chilean coast. In 1866, the boundary between Chile and Bolivia in this area was defined by treaty (the first step in boundary creation) as lying along the 24th parallel south latitude. After the boundary was delimited (the second step) and put on the map, Chile and Bolivia entered into a complicated agreement whereby Chile would receive part of the revenues derived from the sale of resources found in the Bolivian coastal zone.

What at first seemed just a stretch of the coastal Atacama Desert—one of the driest spots on earth—turned out to have value far beyond its ports (Antofagasta and Arica served as Bolivian outlets). Guano deposits and nitrate fields attracted Chilean investment, but the Bolivians did not have the capital to exploit their own raw materials. As the Bolivian government saw the Chileans encroach, it sought help from Peru. In 1873, Bolivia and Peru signed a treaty of alliance, and the Bolivians then tried to impose higher taxes on Chilean enterprises operating in their Atacama region. Chile responded in 1879 by attacking both Bolivia and Peru; by the middle of 1880 the Chileans had captured Bolivia's coastal zone, eventually winning the "War of the Pacific" in 1884.

The loss of its maritime outlets has had a severely negative impact on Bolivia, and successive Bolivian governments have tried for many years to restore a Bolivian corridor to the sea. In 1976, it appeared that years of quiet negotiation with Chile (interrupted by the political upheavals in Santiago) were about to bear fruit. Chile's government announced its intention to grant Bolivia a zone of land along the Peruvian-Chilean boundary, as yet undefined, to permit Bolivia to reestablish its own direct connections with the outside world. But political instability in Bolivia (the country had three different presidents in 1978 alone) and Peruvian military threats again slowed the process; in the early 1980s, the long-awaited Bolivian Corridor had still not materialized. If the Chilean commitment does bear fruit, Bolivia will finally be able to demarcate (the final step in boundary creation) its border in an area it lost more than a century ago.

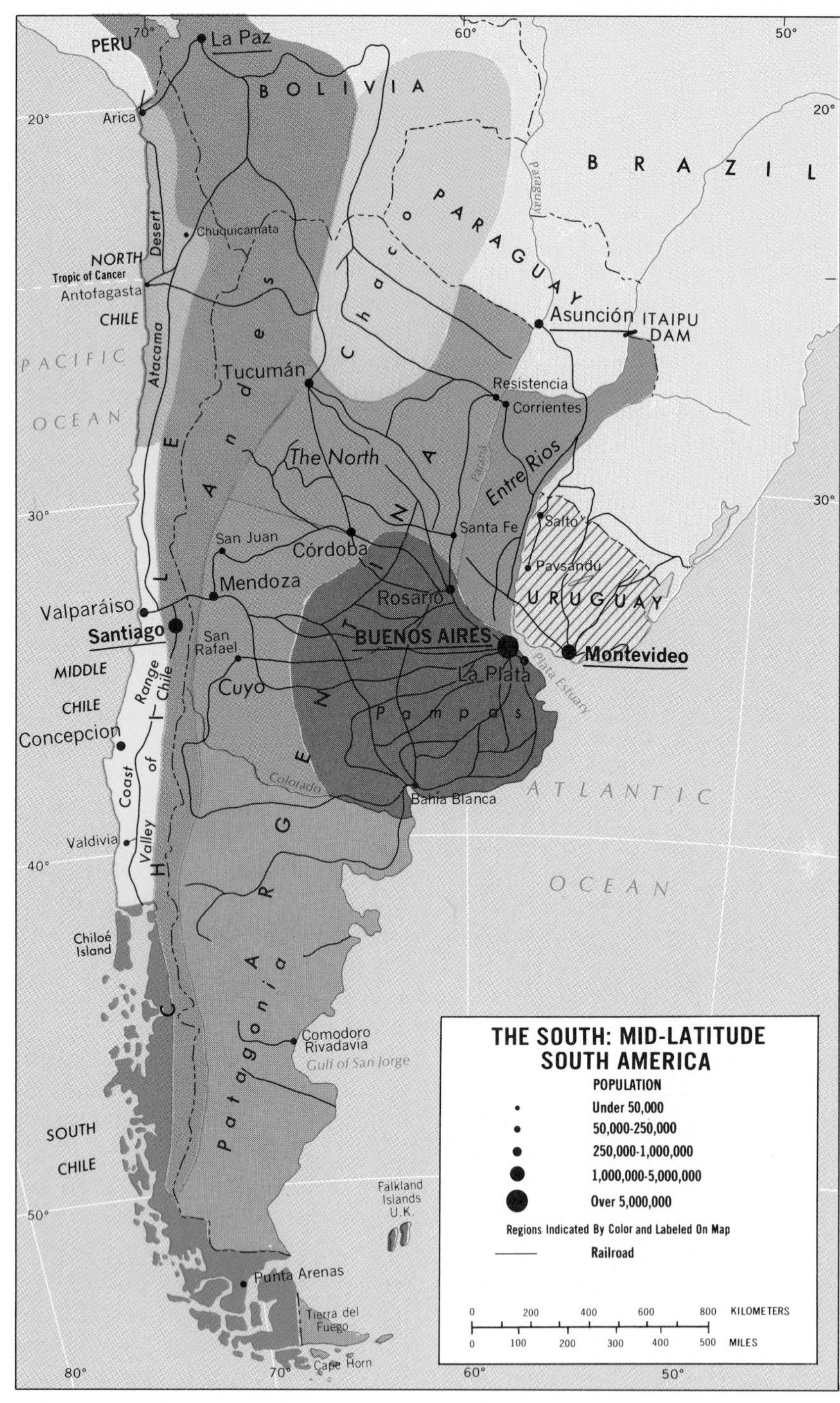


**Figure 5–17**

cannot be densely populated. Outside the Pampa, pastoralism is an almost universal pursuit (except, of course, in the high Andes), but the quality of the cattle is much lower than in the Pampa. In semiarid Patagonia, sheep are raised. Some of the more distant areas are of actual and potential significance to the country: an oilfield is in production near Comodoro Rivadavia on the San Jorge Gulf and in the far northwest, in the Chaco, Argentina may share with Paraguay significant oil reserves that are yet to be brought into production. Yerba maté, a local tea, is produced in the North and Entre Rios subregions (Paraguay has a small output of this product as well), and the quebracho extract that both countries export comes from the valley of the Paraguay-Paraná River.

In addition, the streams that flow eastward off the Andes provide opportunities for irrigation. Tucumán (460,000), Argentina's major sugar-producing district, developed in response to a unique set of physical circumstances and a rapidly growing market in the Pampa cities, which the railroad made less remote; at Mendoza (725,000) and San Juan to the north and San Rafael to the south, vineyards and fruit orchards reflect the economy of the Cuyo subregion. But despite these sizable near-Andean outposts, effective Argentina still remains the area within a radius of 560 kilometers (350 miles) from Buenos Aires.

The early 1980s was a particularly stressful period for Argentina, involving deep economic crisis, a humiliating military defeat by the British in the disastrous 1982 war for the offshore Falkland Islands (see *Box*), and the unraveling of an inept and brutal military junta that had ruled since the overthrow of the second Peronist government in 1976. By the opening of the 1980s, the junta had mismanaged the national economy to the point where hyperinflation exceeded 100 percent annually, unemployment surpassed 20 percent, and the foreign debt soared beyond $40 billion; at the same time, thousands of non-rightist political activists were "disappearing" as a dismal human-rights record accumulated. Although the 1982 invasion to reclaim the Falklands briefly rallied the nation, crushing defeat wiped out the remaining popularity of the junta (which even talked about sparking another war—with Chile—over some obscure islands off Tierra del Fuego, a matter now under amicable negotiation). By 1983, the military regime was forced to call free elections. Surprisingly, the rightist Peronista Party was soundly defeated (its first national loss) by a new center-left coalition; as the new government took office at the end of 1983, a new optimism raised hopes that the country's recent problems could be resolved—a spirit pervading southeastern South America as democratic rule also may return to Uruguay and Brazil in the mid-1980s.

Uruguay, unlike Argentina or Chile, is compact, small, and fairly densely populated. This buffer state of old has become a fairly prosperous agricultural country, in effect a smaller-scale Pampa (though possessing less favorable soils and topography); Figs. I–5 and I–10 show the similarity of physical conditions on the two sides of the Plata estuary. Montevideo, the coastal capital with 1.5 million residents, contains over 50 percent of the country's population; from here, railroads and roads radiate out into the productive agricultural interior. In the immediate vicinity of Montevideo lies Uruguay's major farming area; here vegetables and fruits are produced for the city as well as wheat and fodder crops. Just about all of the rest of the country is used for grazing sheep and cattle; wool constitutes half of the annual exports by value, meat about a quarter. Of course, Uruguay is a small country, and its 176,000 square kilometers (68,000 square miles), less even than Guyana, do not leave much room for population clustering. But it is, nevertheless, a special quality of the land area of Uruguay that this republic is rather evenly peopled right up to the boundaries with Brazil and Argentina. Of all the countries in South America, Uruguay is the most truly European, notably lacking the racial minorities that mark even Argentina and Chile but with a sizable non-Spanish European component in its population.

**Figure 5–18**
Street scene in Buenos Aires, reflecting Argentina's European-commercial culture.

For 4000 kilometers (2500 miles) between the crestline of the Andes and the coastline of the Pacific lies a narrow strip of land that is the Republic of Chile. On the average just 150-kilometers/93-miles wide (and only rarely over 300 kilometers [186 miles]), Chile is the textbook example of what political geographers call an "elongated" state, one whose shape tends to contribute to external political, internal administrative, and general economic problems. In the case of Chile, the Andes Mountains do form a barrier to encroachment from the east, and the sea constitutes an avenue of north-south communication; history has shown the country to be well able to cope with its northern rivals, Bolivia and Peru.

As Figs. I–5 and I–10 as well as Fig. 5–8 indicate, Chile is a three-subregion country. About 90 percent of Chile's 11.8 million people are concentrated in what is called Middle Chile, where Santiago (Fig. 5–21), the capital and largest city, and Valparaiso, the second city and chief port, are located. North of Middle Chile lies the Atacama Desert, wider and colder than the coastal desert of Peru. South of Middle Chile, the coast is broken by a plethora of fjords and islands, the topography is mountainous, and the climate—wet and cool near

**Figure 5–19**
The heart of the Buenos Aires conurbation, home to 11 million Argentinians.

the shore—soon turns drier and colder against the Andean interior. South of the latitude of the island of Chiloé no permanent land transport routes of any kind exist, and hardly any settlement.

These three subregions are clearly apparent in the map of culture spheres (Fig. 5–8), which displays a mestizo north, a European-commercial zone in Middle Chile, and an undifferentiated south. Moreover, a small Indo-subsistence zone in northern Chile's Andes is shared with Argentina and Bolivia. The Indian element in the two-thirds of the Chilean population that is mestizo largely originated from the million or so Indians who lived in Middle Chile.

Despite the absence of a large landless Indian class, Chile nonetheless had—and continues to have, although in decreasing numbers—tenant farmers on the haciendas and estates. Although they are not peons in the strict sense of the word, in that there is no debt bondage in Chile, these people (the *rotos*) are little if any better off than their counterparts in other South American countries. However, there were always a few means of escape from this system. The army provided one of these avenues; after service in the Atacama Campaign in the late nineteenth century, several thousands of these people were settled on land in the open areas of southern Middle Chile. Another escape route was to leave the land and head for the cities to seek work; this migration, of course, still goes on. In recent decades, the land situation has become a major political issue in Chile, and gradually some of the large landholdings are being converted into smaller farms.

Some intraregional differences exist between northern and southern Middle Chile, the country's core area. Northern Middle Chile, the land of the hacienda and of

## FALKLANDS OR MALVINAS?

Off the southern tip of Argentina, about 500 kilometers (300 miles) from the mainland, lies a small archipelago that, in 1982, suddenly became the scene of a bitter war between the Argentines and the islands' owners, the British. This conflict over the Malvinas—as the Falklands are called in Argentina—was a reminder of the potential for strife in the aftermath of the colonial era. On the world politico-geographical map, there are numerous locales capable of causing similar crises: Gibraltar, Guantánamo (Cuba), Hong Kong, the Antarctic Peninsula, and many less familiar places. In this age of superpowers, the consequences might be far more serious.

Judging from the map (Fig. 5–17), the Falklands hardly seem likely to occasion a full-scale war. The archipelago consists of two main islands (East and West Falkland) and about 100 smaller ones, with a total area of just over 12,000 square kilometers (4600 square miles)—which is smaller than Connecticut. The countryside reflects the high latitude: behind the rugged coastline lies a treeless, rolling landscape that looks like the tundra of the northern hemisphere. The weather is cold and damp, with much wind and cloud. When the war broke out in 1982, there were only some 1800 permanent residents on the islands, most of them in some way dependent on the main industry, sheep farming.

The Falklands' small population is mainly of British ancestry. The British first took control of the Falklands in the 1760s, but soon they were ousted by the Spanish. When Spanish power in southern South America weakened, and the new republic of Argentina emerged, the Spanish abandoned the area, leaving it to the Argentines. In 1820, the Argentines claimed the Malvinas and followed this up by establishing a base on East Falkland. In 1833, a British naval force arrived

Figure 5–20
The Falklands, scene of the bitter short-lived 1982 military conflict.

and expelled the Argentines; British seapower was at its zenith. In Argentina, that action has never been forgotten and British rule has never been acknowledged.

Negotiations over the future of the islands had been continuing without progress when, on April 2, 1982, an Argentine force invaded the Falklands, overpowered the small British garrison, and hoisted the Argentinian flag over Stanley, the capital. In the United Kingdom, 12,000 kilometers (7500 miles) away, this act had an impact similar to the Japanese attack on Hawaii in the United States of 1941. The British government launched a counterattack by sea and air and, in the war that cost about 200 British and more than 700 Argentinian lives, defeated the invaders by mid-June (Fig. 5–20).

The British victory did not do anything to solve the fundamental politico-geographical problem however. Argentina continues to insist that the British takeover of the Malvinas in 1833 remains an unresolved act of war and annexation. As the legitimate successors of the Spanish rulers over what is today Argentina, the Argentines say that the Malvinas, part of that legacy, are now legally theirs. The islands' nearness to Argentina and their potential importance in maritime boundary making (see Chapter 10) also figure in Argentina's claim. The British respond that if all forcibly acquired territories in the world were given back to their "original" owners, there would be global chaos. They further point to the English character of the Falklands' population, six generations of close ties with Britain, and the expressed desires of the settlers to remain under the British flag. And the United Kingdom, too, has claims and objectives in nearby Antarctica.

Negotiations obviously are needed, but the chances for a peaceful resolution of the lingering dispute are not good. The British, after their victory, began to strengthen the defensive capabilities of the islands, at enormous cost. A commission was charged with the task of finding ways to bolster the local economy. But much less was done to seek a negotiated settlement to a conflict in which both sides have strong arguments.

Mediterranean climate with its dry summer season, is an area of usually irrigated crops that includes wheat, corn, vegetables, grapes, and other Mediterranean products; livestock raising and fodder crops also take up much of the productive land. Agricultural methods are not particularly efficient: the area could undoubtedly yield a far greater volume of food crops than it does. Southern Middle Chile, into which immigrants from both the north and from Europe (especially Germany) have pushed, is a better-watered area where raising cattle predominates but where wheat and other food crops, including potatoes, are also widely cultivated.

Few of Chile's agricultural products reach external markets however. Some specialized items such as wine and raisins are exported, but Chile remains a net importer of food. Thus, despite the fact that 9 out of 10 Chileans live in what has been defined here as Middle Chile, it is the northern Atacama region that provides most of the country's foreign revenues. At first, the mining of nitrates—the Atacama includes the world's largest exploitable deposits—provided the country's economic mainstay, but the industry declined after the discovery of methods of synthetic production early in this century. Subsequently, copper became the chief export; it is found in several places, but the main concentration lies on the eastern margin of the Atacama near the town of Chuquicamata, not far from the port of Antofagasta. At present, copper in various forms—some of it as pure bars, some as a concentrate refined at Chuquicamata, and some as raw ore—constitutes about 70 percent of the value of Chile's exports. Of course, the country is particularly vulnerable to fluctuations in the price of this commodity on the world market.

Copper has been mined for over

**Figure 5–21**
The urban landscape of Santiago, capital of Chile, set against the nearby Andean ranges.

a century, and American investment in the mining industries of Chile has been heavy. The Chilean government, anxious for a greater share of the profits, expropriated some mining properties in the early 1970s. The resulting disagreement over compensation created a political storm that illuminated some of the problems of direct foreign investment in an economy. Large-scale investment by giant multinational firms throughout the Americas and the developing world is a sensitive political issue today, one that could prove troublesome as Third World countries attempt to gain greater control over their economic futures.

## References and Further Readings

Aguilar, L. *Latin America 1984* (Washington: Stryker-Post Publications, 1984).

Augelli J. "The Controversial Image of Latin America: A Geographer's View," *Journal of Geography*, 62 (1963), 103–112.

Bailey, H. & Nasatir, A. *Latin America: The Development of Its Civilization* (Englewood Cliffs, N.J.: Prentice-Hall, 1973).

Blakemore, H. & Smith, C., eds. *Latin America: Geographical Perspectives* (London and New York: Methuen, 2 rev. ed., 1983).

Blouet, B. & Blouet, O., eds. *Latin America: An Introductory Survey* (New York: John Wiley & Sons, 1982).

Bromley, R. & Bromley, R. *South American Development: A Geographical Introduction* (New York: Cambridge University Press, 1982).

Butland, G. *Latin America: A Regional Geography* (New York: John Wiley & Sons, 3 rev. ed., 1972).

Caviedes, C. *The Southern Cone: Realities of the Authoritarian State* (Totowa, N.J.: Rowman & Allanheld, 1984).

Cole, J. *Latin America: An Economic and Social Geography* (Totowa, N.J.: Rowman & Littlefield, 3 rev. ed., 1976).

Denevan, W. *The Native Population of the Americas in 1492* (Madison: University of Wisconsin Press, 1976).

"Falklands: 150 Years British," *Geographical Magazine*, January 1983, 30–39.

Gilbert, A., ed. *Latin American Development: A Geographical Perspective* (Baltimore: Penguin, 1974).

Griffin, E. & Ford, L. "Cities of Latin America," in Brunn, S. & Williams, J., eds. *Cities of the World: World Regional Urban Development* (New York: Harper & Row, 1983), pp. 198–240.

Grigg, D. *The Agricultural Systems of the World: An Evolutionary Approach* (London: Cambridge University Press, 1974).
Harris, C. D. "A Functional Classification of Cities in the United States," *Geographical Review*, 33 (1943), 86–99.
Harris, D., ed. *Human Ecology in Savanna Environments* (New York and London: Academic Press, 1980).
Hauser, P., ed. *Urbanization in Latin America* (New York: Columbia University Press, 1961).
Hennessy, A. *The Frontier in Latin American History* (London: Edward Arnold, 1978).
James, P. E. *Latin America* (New York: Odyssey Press, 5 rev. ed., 1975).
Jones, C. *South America* (New York: Holt, 1930).
Jumper, S., et al. *Economic Growth and Disparities: A World View* (Englewood Cliffs, N.J.: Prentice-Hall, 1980).
Maos, J. *The Spatial Organization of New Land Settlement in Latin America* (Boulder, Colo.: Westview Press, 1984).
Morris, A. *Latin America: Economic Development and Regional Differentiation* (Totowa, N.J.: Barnes & Noble, 1981).
Morris, A. *South America* (Totowa, N.J.: Barnes & Noble, 2 rev. ed., 1982).
Odell, P. & Preston, D. *Economies and Societies in Latin America* (Chichester, Eng.: John Wiley & Sons, 2 rev. ed., 1978).
Platt, R. *Latin America: Countrysides and United Regions* (New York: McGraw-Hill, 1942).
Prescott, J. *The Geography of Frontiers and Boundaries* (Chicago: Aldine, 1965).
Preston, D. *Environment, Society and Rural Change in Latin America: The Past, Present and Future in the Countryside* (New York: Halsted Press/John Wiley & Sons, 1980).
Sanchez-Albornoz, N. *The Population of Latin America: A History* (Berkeley: University of California Press, 1974).
Steward, J. & Faron, C., eds. *Native Peoples of South America* (New York: McGraw-Hill, 1959).
Taylor, A., ed. *Focus on South America* (New York: Praeger/American Geographical Society, 1973).
Theroux, P. *The Old Patagonian Express: By Train Through the Americas* (Boston: Houghton Mifflin, 1979).
Wagley, C. *The Latin American Tradition: Essays on the Unity and Diversity of Latin American Culture* (New York: Columbia University Press, 1968).
Webb, K. *Latin America* (Englewood Cliffs, N.J.: Prentice-Hall, 1972).
Weil, C., ed. *Medical Geographic Research in Latin America* (Elmsford, N.Y.: Pergamon, 1982).
Wheeler, J. & Muller, P. *Economic Geography* (New York: John Wiley & Sons, 2 rev. ed., 1986).
Wilbanks, T. *Location and Well-Being: An Introduction to Economic Geography* (New York: Harper & Row, 1980).
Wilkie, J., ed. *Statistical Abstract of Latin America* (Los Angeles: UCLA, Latin American Center, annual).

# EMERGING BRAZIL

IDEAS AND CONCEPTS

**Growth-Pole Concept**
**Takeoff Stage of Development**
**Multinational Economic Influences**

Brazil, South America's long-dormant giant, is awakening. The Brazilian economy in recent years has grown at rates that exceed those of underdeveloped countries (UDCs). The population is increasing at such a pace that, in the final 15 years of this century, it will expand by 50 million. Symbolized by its modern forward capital—Brasilia (Fig. B–1)—the vast interior is finally being penetrated by roads and opened up. The cities are mushrooming, and, despite being staggered by short-term economic problems in the early 1980s, Brazil shows signs of taking off in the foreseeable future.

With its 8.5 million square kilometers (3.3 million square miles) of equatorial and mid-latitude South America, Brazil is a giant in world as well as regional terms: it is the only country on earth to contain both the Equator and a Tropic—the Tropic of Capricorn (23½° south latitude). Territorially, it is exceeded in size only by the Soviet Union, Canada, China, and the United States. In terms of population, Brazil now ranks among the world's half-dozen largest countries (1985 estimated population—137.5 million). Economically, its trend has been one of steadily rising in the international ranks. Brazil, therefore, seems likely to become a world force in the twenty-first century.

Brazil is so large that it has common boundaries with all the other South American countries except Chile and Ecuador (Fig. 5–7). Its environments range from the tropical rainforest of the Amazon River Basin (nearly all of which lies within Brazil's borders) to the temperate conditions of the Argentine Pampa (Fig. I–10). Its resources are known to include enormous iron ore reserves, extensive bauxite and manganese deposits, large coalfields, and numerous alloys; however, Brazil's sprawling territory is still incompletely explored, and additional discoveries will undoubtedly be made. For example, oil and natural gas have long been produced from fields in the state of Bahia on the central east coast, but never in sufficient quantities to satisfy Brazil's demand. In the mid-1970s, a major offshore oil deposit was discovered not far from the coastal city of Campos, less than 200 kilometers (125 miles) northeast of Rio de Janiero. Other major energy developments include massive hydroelectric plants and the world's largest dam—Itaipu Dam (Fig. 5–1)—which is discussed below; the Brazilians have also successfully substituted sugarcane-based alcohol ("gasohol") for gasoline, to such an extent that nearly half of its cars utilize this fuel in the mid-1980s.

# Regions

Brazil is a very large country, but its landscapes are not as spectacularly diverse as those of several much smaller South American republics. Brazil has no Andes Mountains, and the countryside consists mainly of plateau surfaces and low hills. Even the lower-lying Amazon Basin is not entirely a plain: between the tributaries of the great

**Figure B–1**
Brasilia: inland capital of Brazil symbolizing the country's modern era and (hoped-for) westward orientation.

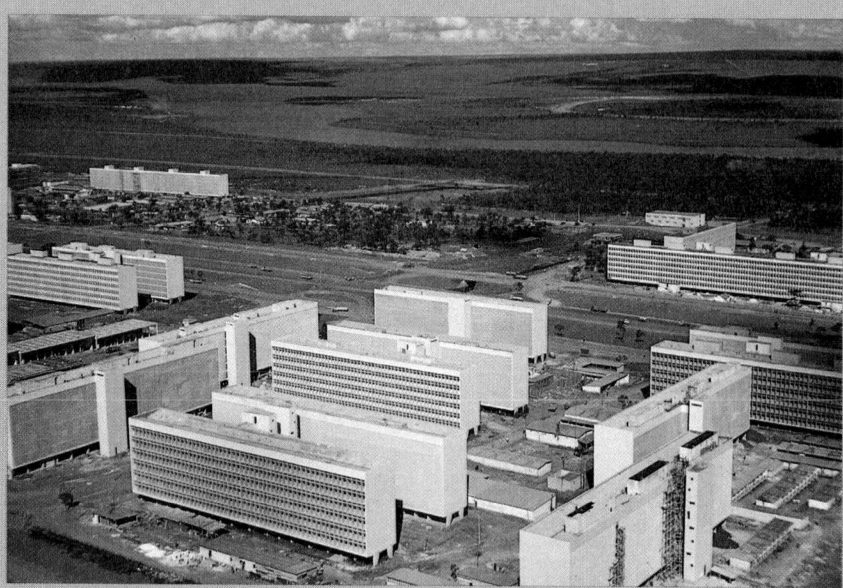

river lie low but extensive *mesas* sustained by layers of sedimentary rocks. To the southeast, the surface of the great Brazilian plateau rises slowly eastward, but the highest areas fail to reach 3000 meters (under 10,000 feet). Along the coastline, there is a steep escarpment leading from plateau surface to sea level that leaves almost no living space along its base. Thus, although Brazil has some 7500 kilometers (4600 miles) of Atlantic coastline, there is relatively little coastal plainland, and cities such as Rio de Janeiro are squeezed between mountain slopes and the oceanfront. Under these physiographic circumstances, Brazil is most fortunate to possess several very good natural harbors.

Brazil is a federal republic consisting of 22 states, 4 territories, and the federal district of the capital, Brasilia (Fig. B–2). As in the United States, and for similar reasons, the smallest states lie in the

Figure B–2

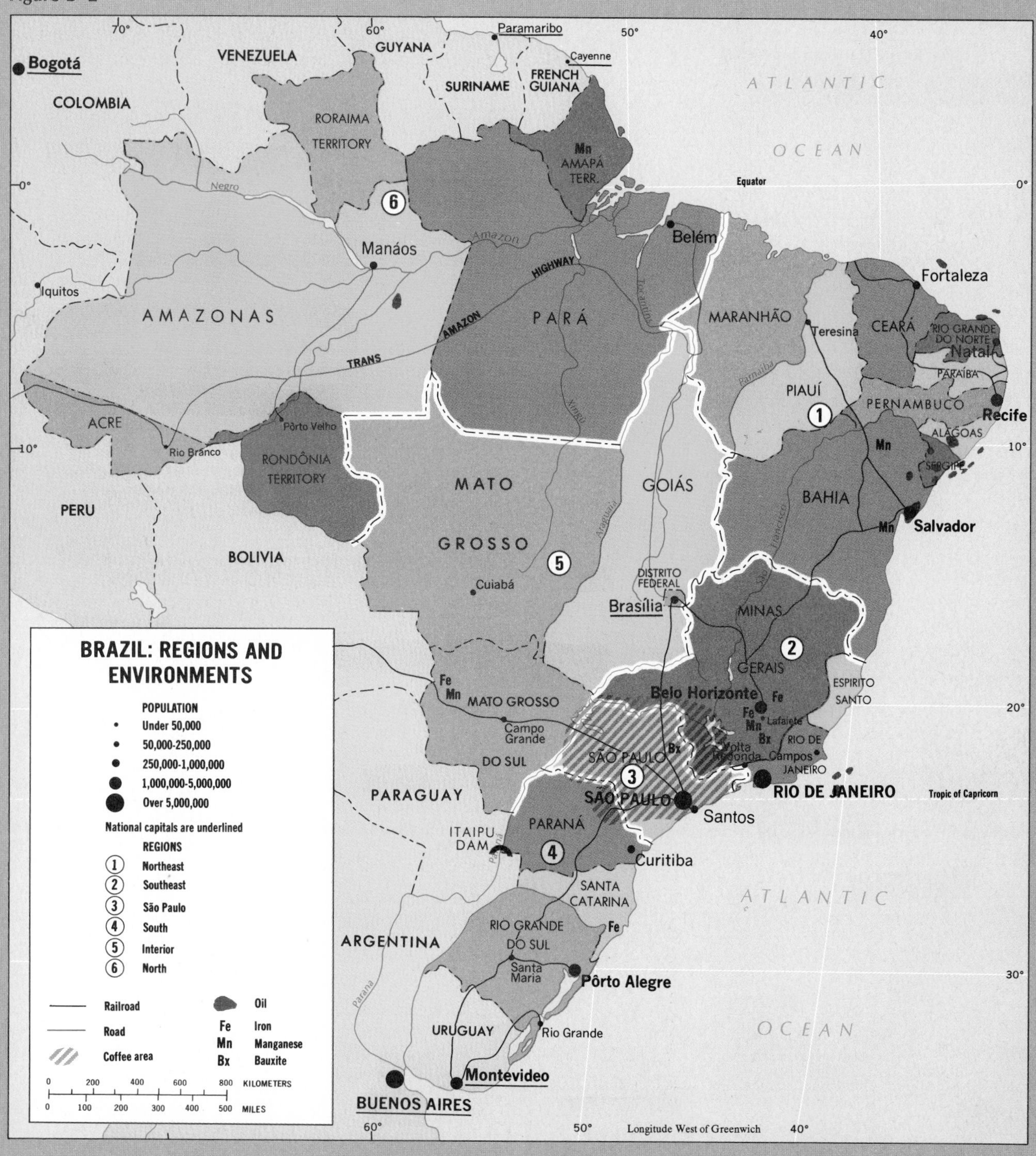

northeast and the larger ones farther west. The state of Amazonas is the biggest, with over 1.5 million square kilometers (more than 600,000 square miles, which is twice the size of Texas)—but Amazonas's huge area contains barely 1 million people. At the other extreme, Rio de Janeiro State (44,000 square kilometers/17,000 square miles) on the southeast coast has a population of almost 15 million. The state with the largest population by far is São Paulo, now approaching 30 million and growing rapidly.

Although Brazil is about as large as the 48 contiguous United States of America, it does not possess the clear physiographic regionalism familiar to us. Apart from the Great Escarpment marking the eastern edge of the Brazilian Plateau, there are no mountain barriers, no well-defined coastal plains, no naturally demarcated desert regions, and no large lakes. Thus, the six regions outlined below have no absolute or even generally accepted boundaries. On Fig. B–2, the regional boundaries have been drawn to coincide approximately with the borders of states, making identifications easier.

## TEN MAJOR GEOGRAPHIC QUALITIES OF BRAZIL

1. Territorially, Brazil ranks among the world's five largest countries. Its area covers nearly half of the continent, and Brazil has common boundaries with every South American country except Chile and Ecuador.
2. Brazil is large, but its physiographic diversity does not match its size. The great Andes lie outside Brazil; landscapes consist mainly of plateaus, low hills, and the undulating Amazon Basin.
3. Brazil's population is about as large as that of all other South American countries combined.
4. In recent decades, Brazil has had one of the world's highest rates of population growth.
5. Brazil's regional development has proceeded most rapidly along its eastern margins and has been slowest in the Amazonian interior.
6. Brazilian governments have strived to focus national attention on the opportunities of the interior. Brasilia, the new forward capital, is a manifestation of that effort.
7. Brazil exhibits a strong national culture, wherein a single language and the domination of one religious faith constitute strong centripetal forces.
8. Although still ranked as an underdeveloped country, Brazil generates economic indicators that point to "takeoff" conditions. Rapid urbanization and growing industrial strength also mark the country.
9. Although it is the giant of South America, Brazil's relationships with its neighbors remain comparatively distant.
10. Brazil is a federation in which the eastern states of its modern core area dominate national affairs. Strong central authority and military involvement in government have recently prevailed.

## 1. The Northeast

The Northeast was Brazil's source area, its culture hearth. The plantation economy took root here at an early date; to this area, there came not only the Portuguese planters, but also the largest number of African slaves to work in the sugar fields. But the ample, dependable rainfall that occurs along the coast soon gives way to lower and more variable patterns in the interior (see Fig. I–10), and today most of the Northeast is poverty stricken, hunger afflicted, overpopulated, and subject to devastating droughts. Extended dry periods occur so regularly that this region has become known as the Polygon of Drought; one of the worst of these episodes suffered by its 21 million residents occurred in the early 1980s, during which almost no rainfall was received in over four years. Moreover, human and animal overpopulation have combined to deplete the natural vegetation, and thereby hasten the encroachment of aridity. Sugar still remains the chief crop along the moister coast, and livestock herding prevails in the drier inland backcountry (*sertão*), with beef cattle in the better grazing zones and goats elsewhere. The comparatively small areas of successful commercial agriculture—cotton in Rio Grande do Norte, sisal in Paraíba, and sugar cane in Pernambuco—stand in sharp contrast to patches of shifting subsistence agriculture located nearby.

The Northeast today is Brazil's great contradiction. In the cities of Recife and Salvador (Fig. B–3), the architecture still bears the imprint of an earlier age of wealth, but thousands of peasants without

**Figure B–3**
The history-filled cityscape of Salvador in Brazil's Northeast culture hearth.

hope—driven from the land by deteriorating conditions—arrive constantly in the surrounding *favelas* or shantytowns. And few of the generalizations about expanding, emerging Brazil yet apply here, in South America's largest most poverty-stricken corner. The Brazilian government, aware of the Northeast's plight, has directed investments to the region to help diversify its economic geography, including the multifaceted "Super Northeast" program of the mid-1980s to alleviate the impact of the latest disastrous drought. The most massive effort so far has been the huge petrochemical complex (one of the world's largest) built near Salvador. This project, 20 years in the making, became fully operational in 1979, creating thousands of jobs, attracting further (including foreign) investment, and boosting the Northeast's industrial base in general. But even an enormous project such as this has had only limited impact on the overall economic geography of the impoverished Northeast, where subsistence remains the rule rather than the exception.

## 2. The Southeast

In the state of Bahia, a transition occurs toward the south. The Great Escarpment becomes more prominent, the plateau higher, and the terrain more varied. Annual rainfall increases, and it is seasonally more dependable. The Southeast has been modern Brazil's core area, the scene of successful as well as abortive gold rushes, with its major cities and biggest population clusters. Gold first drew many thousands of settlers, and other mineral finds also contributed to the influx (Rio de Janeiro itself served as the endpoint of the "Gold Trail"), but ultimately the region's agricultural possibilities ensured its stabilization and growth. The mining towns needed food, and prices for foodstuffs were high; farming was stimulated and many farmers came (with their slave workers) to Minas Gerais State to till the soil. Eventually, a pastoral industry came to predominate, with large herds of beef cattle grazing on planted pastures.

The postwar era brought another mineral age to the region, based not on gold or diamonds but on the iron ores around Lafaiete and the manganese and limestone carried to the steel-making complex at

Volta Redonda (Fig. B–4). Iron mining has now become one of Brazil's leading economic activities: in 1960, the country produced only 9 million gross tons (vs. 89 million in the United States that year), but by 1981 the output was 100 million tons (against only 74 million in the United States). Moreover, industrial diversification in the Southeast region has proceeded apace. Belo Horizonte is the rapidly growing metallurgical center of Brazil, with a 1985 population estimated at 3.8 million, ten times what it was in 1950. Rio de Janeiro, of course, continues in second place nationally (after São Paulo) among the manufacturing cities, and is now home to 12.7 million.

## 3. São Paulo

The state of São Paulo, long a part of the Southeast region, today deserves its own regional identity, because it is the focus of ongoing Brazilian development. Largely from its vast coffee plantations (*fazendas*), this state is the leading foreign-exchange earner of the republic, and it leads Brazil in the production of numerous other crops as well. With the help of Japanese technicians (and with the prospect of sale on the Japanese market), Brazilian farmers have enormously increased their soybean production, surpassing China in the 1970s and taking second place among world producers. Matching this prodigious agricultural output is the state's industrial strength. The city of São Paulo (Fig. 5–11) is now the country's leading manufacturing center; the state does not have a mineral base to compare to that of Minas Gerais, but it has, nevertheless, become the leading industrial region not only of Brazil, but of all Latin America. The revenues derived from the coffee plantations provided the necessary capital; hydroelectric power from the slopes of the Great Escarpment produced the needed energy, as will Itaipu Dam in the future (see p. 317); immigration not only from Portugal, but also from Italy, Japan, and elsewhere contributed the labor force. São Paulo lay juxtaposed between raw material-producing Minas Gerais and the states of the South. Communications were improved, notably with the outport of Santos, but also with the interior hinterland; as the capacity of the domestic market grew, the advantages of central location and agglomeration secured São Paulo's primacy. São Paulo today is truly the pulse of Brazil, with a population fast approaching 17 million.

**Figure B–4**
Volta Redonda's iron and steel complex in the heart of southeastern Brazil's industrial workshop zone.

## 4. The South

Three states make up the southernmost Brazilian region: Paraná, Santa Catarina, and Rio Grande do Sul. The contribution of recent-immigrant Europeans to the agricultural development of southern Brazil, as in Uruguay and Argentina, has been considerable. Many came not to the coffee areas of São Paulo, but to the available lands farther south. Here they occupied fairly discrete areas; for example, Portuguese rice farmers clustered in the valleys of the major rivers of Rio Grande do Sul, and the state today produces between one-quarter and one-fifth of Brazil's annual rice crop. The Germans, on the other hand, occupied the somewhat higher areas to the north and in Santa Catarina, where they were able to carry on the type of mixed farming with which they were familiar: corn, rye, potatoes, and hogs as well as dairying. The Italians selected the highest slopes and established thriving vineyards. The markets for this produce, of course, are the growing cities to the north. Paraná, on the other hand, exports its coffee harvest to overseas markets.

The South never was a boom area, as were the Northeast and Southeast at various times in Brazilian history. It does, however, have a stable, progressive, modern agricultural economy; farming methods here are the most advanced in Brazil. The diversity of its European heritage is still reflected in the regional towns, where German and other European languages are preserved. Some 20 million people live in the three southern states, and the region's importance is increasing. Coal from Santa Catarina and Rio Grande do Sul, shipped north to the steel plants of Minas Gerais, was a crucial element in Brazil's industrial emergence. Local industry is growing as well, especially in Porto Alegre and Turbarão (where Latin America's largest steelmaking facility opened in 1983).

The Brazilian South, today, is also the scene of one of the world's greatest construction projects—Itaipu Dam—which is becoming so classic a symbol of the newest development thrust into the South American interior that this facility was chosen as the subject of the opening photograph (Fig. 5–1) of Chapter 5. Located astride the Paraná River on the Brazil-Paraguay border in the southwest corner of Paraná State (Fig. B–2), Itaipu's dimensions are truly awesome: at (180 meters) 600 feet in height and (8 kilometers) 5 miles in length, it is six times bigger than Egypt's Aswan High Dam, and its peak electricity output equals more than 50 percent of Brazil's total electrical consumption in 1982 (the equivalent of a half-million barrels of oil a day). Begun in response to the oil crisis of the early 1970s—Brazil was and is heavily dependent on foreign imports—75 percent of the funding and 85 percent of the labor inputs were supplied domestically. The first of the dam's 18 massive turbines was opened in early 1984, with the rest scheduled to come on line at regular intervals until Itaipu's full capacity is achieved in 1989. Much of this energy will be transmitted to the São Paulo area (800 kilometers [500 miles] to the east) to fuel new rounds of industrialization. The Itaipu vicinity is likely to become a growth pole and enjoy considerable benefits as well; in anticipation, hundreds of thousands of Brazilians have moved into this area and nearby Paraguay since the late 1970s.

## 5. The Interior

Interior Brazil is often referred to as the Central-West—*Centro-Oeste*. This is the region Brazil's developers hope to make a part of the country's productive heartland, and the new capital of Brasilia was deliberately positioned on its margins (Fig. B–2). But such economic integration will take considerable time. This is a vast upland area, largely a plateau at an elevation over 1000 meters (3300 feet) covered with savanna vegetation (see Fig. I–12). Like Minas Gerais, the interior was the scene of gold rushes and some discoveries; unlike its eastern neighbor, it did not present agricultural alternatives when the mineral search petered out. Today, the three large states of the Interior (Goiás, Mato Grosso, and Mato Grosso do Sul) have a combined population of somewhat more than 8 million, so that the average density is under 10 per square kilometer (25 per square mile). Nevertheless, this represents a noteworthy increase, for in 1960 the Centro-Oeste contained just 2.9 million people. Thus, this Brazilian region's population has nearly tripled in 25 years—a pace well ahead of the already prodigious national rate of growth.

Still, the Centro-Oeste faces the problems common to tropical savanna regions everywhere: soils are not especially fertile and the vegetation is susceptible to damage by overgrazing. In the absence of major mineral finds (the region still awaits systematic exploration), pastoralism remains the chief economic activity. There is always the fear that what happened to the Northeast could happen here and that Brazil's determination to open the interior might create serious environmental problems. But at present that day seems far off. You can fly for hours over this area and see only an occasional small settlement and a few widely spaced roads (communications are best in the south). What the Interior needs is investment—in the clearing and opening of the alluvial soils in the region's river valleys, in experimental farms, in the provision of electric power, in mineral exploration. The first great step was the relocation of the national capital. It was an enormously expensive venture—but it was a mere beginning in view of what the Interior really requires.

## 6. The North

The largest Brazilian region is also the most remote from the core of Brazilian settlement: the three states (and three federal territories) of the Amazon Basin. This was the scene of the great rubber boom at the turn of the century when the wild-growing rubber trees in the tropical rainforest (*selva*) produced huge profits and the Amazon city of Manáos enjoyed a brief period of wealth and splendor. But the rubber boom ended in 1910 when plantations elsewhere—notably in Southeast Asia—began to produce rubber more cheaply, efficiently, and accessibly. For most of the following seven decades, the North was a stagnant hinterland—but all that is beginning to change in the 1980s.

A harbinger of things to come was the Jari project of American millionaire Daniel Ludwig in the 1970s. His huge pioneering development scheme involved the clearing of thousands of square kilometers of rainforest in order to establish extensive tree and rice plantations; his centerpiece was an entire paper-pulp factory and hydroelectric power plant that was built in Japan and floated on barges to a site 500 kilometers (300 miles) inland from the Amazon's mouth (Fig. B–5). Although this particular project failed financially, it was purchased from Ludwig by a consortium of public and private Brazilian interests, and it continues to operate. Similar government-business cooperation has underwritten a number of other large-scale projects in the North since 1980, the largest (ten times the size of Jari) being the Great Carajás scheme in southern Pará State that emphasizes cattle raising, forestry, farming, and iron mining. As dam building, highway and railroad construction, land development, and mineral exploitation transform large tracts of the North, environmentalists are beginning to wonder if such repeated efforts could endanger the survival of the rainforest itself.

This concern notwithstanding, the North's opportunities have barely been touched. In 1985, the region's total population (in an area half the size of Europe) surpassed 7 million, with a majority residing in the eastern state of Pará—the vast Amazon Basin portion of the region contains less than 2 million, but it shows signs of a new in-migration (especially in its westernmost reaches). As more is known about the once-obscure North, some myths are being dispelled. The North is not all swampy rainforest as its popular image sometimes suggests. True, there are low-lying areas at the mouth of the great Amazon River and near the interior streams, but the areas between the rivers are well above the flood-waters; from the air, the view is one of undulating, even hilly countryside. The forest cover is dense, because the crowns of the trees interlace, but there is little undergrowth; tree trunks are large and straight. The rubber boom of the late 1800s was not followed by a plantation, lumbering, or agricultural economy, because of the shortage of workers. But all that is changing today as one of the world's last frontiers begins to yield—the wise development of its rich Northern resource base could help catapult Brazil into the ranks of the leading countries in the next century.

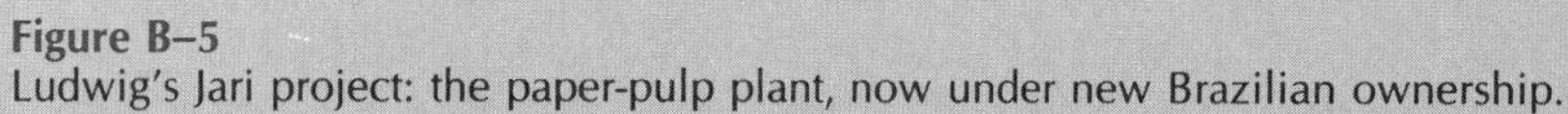

**Figure B–5**
Ludwig's Jari project: the paper-pulp plant, now under new Brazilian ownership.

# Population Patterns

Brazil's population of 137.5 million (1985) is as diverse as that of the United States, and Brazilian society has been a melting pot, perhaps to an even greater degree than the United States itself. In a pattern that is familiar in the Americas, the original Indian inhabitants of the country were decimated following the European invasion; estimates of the number of Indians who survive today in small communities deep in the Amazonian interior vary, but their total is not likely to exceed 250,000—less than 10 percent of the number thought to have been there when the whites arrived. Africans came in great numbers, too; today, there are nearly 16 million blacks in Brazil. Significantly, however, there was also much mixing, and about 40 million Brazilians (almost 30 percent) have combined European, African, and minor Indian ancestries. The majority, over 75 million, is European.

Until Brazil became independent in 1822, the Portuguese were virtually the only Europeans to settle in this country. But after independence other European settlers were encouraged to come, and many Italians, Germans, and Eastern Europeans arrived to work on the coffee plantations, farm in the south, or try their luck in business. Immigration reached a peak during the decade of the 1890s, when nearly 1.5 million newcomers reached Brazilian shores. The complexion of the population was further diversified by the later arrival of Lebanese and Syrians, many of whom opened small shops, and of Japanese, who formed settlements in southern São Paulo, northern Paraná, and around the Amazon's mouth.

Brazilian society, to a far greater degree than is true in North America, has overcome problems of dualism (see Chapter 3). Blacks still are the least advantaged among Brazil's population groups (except for the Indians), and, as in the United States, it is proving difficult to overcome the obstacles of history. But the Brazilian ethnic mix is so all-pervasive that hardly any group is unaffected, and official statistics about "blacks" and "Europeans" are rather meaningless. What Brazil does have is a true national culture, expressed in an overwhelming adherence to the Catholic faith (it is the world's largest Roman Catholic country), in the universal use of a modified form of Portuguese as the common language, and in a set of life-styles in which vivid colors, distinctive music, and a growing national pride are fundamental ingredients.

This is not to suggest that Brazilian population patterns do not exhibit regional variations. The black component of the population in the Northeast, for example, remains far stronger than it is in the Southeast. After their initial concentration in the Northeast, black workers were taken southward to Bahia and Minas Gerais as the economic core moved in that direction. Today, the black population remains strongest in these areas and in Rio de Janeiro; it is weak in the states of the South, more recently settled and more completely taken over by Europeans. Nor has Brazil escaped the problems of pluralism faced in the United States. Black leaders and others in the country have in recent years begun to express publicly their objections against racial discrimination. Yet, although Brazil may not be the multiracial society it is sometimes portrayed to be, it also does not have the history of overt and legally sanctioned racism that has characterized certain other plural societies.

Another population problem that besets Brazil—one that could jeopardize the overall development potential—is its persistently high birth rate: in the early 1980s, the country grew by 2.3 percent annually, a higher relative increase than in India and China. Long ignored by the Brazilian authorities, population-limitation policies are finally being espoused. In 1983, the nation inaugurated its first official family-planning project, despite opposition from the Church. The new ideal is 2 children per family; that goal may require some time to achieve—the average number of children born to Brazilian women in 1980 was 4.0 (which did represent a sizable decline from the figure of 5.7 reported in 1970). Brazil belatedly appears to be getting its overpopulation problem under control: if no more progress occurs, the country can expect another 50 million citizens to be on hand by 2000.

# Development Problems

Until the energy crisis of the early 1970s, Brazil's development had been impressive. The economy was expanding at about 9 percent per year. Since the mid-1960s, the value of Brazilian exports had been doubling every two years. Investments were being made to defeat illiteracy, improve health conditions, assist agriculture, and ameliorate urban living problems. The programs begun after the military regime ended democratic government in 1964 seemed to be succeeding. But the energy crisis severely affected Brazil and caused a major setback in its economic growth. Plans were made to exploit the country's large oil shale deposits for production, and exploration for oil was intensified—leading to the offshore discovery near Campos. But production from the new sources is still in the future. In the meantime, Brazil has seen its forward drive slowed—a trend shared with many developing countries severely affected by the recent energy crisis.

A far more serious development problem than the energy crisis emerged in the early years of the 1980s—like Mexico, Argentina, and several other countries, Brazil amassed a staggering debt in foreign loans that was necessary to sustain the pace of national economic growth. By 1984, this debt exceeded $90 billion and even threatened to bankrupt the Brazilian treasury. To avoid such a disaster, the International Monetary Fund agreed to supply emergency funds, so that Brazil could begin to repay its loans, but only if a severe domestic austerity program was initiated at once—a truly ironic situation because Brazil had strongly resisted foreign financial involvements before 1980 in order to preserve full control over its developing economy. What went wrong? Basically, the answer rests with the worldwide economic recession of the early 1980s: falling commodity prices on the international market suddenly and sharply reduced Brazilian income and caused a sizable decline in exports. The government, deeply committed to numerous development programs and unwilling to trim them back, began to borrow abroad (at ever higher interest rates) to make up the difference. As runaway deficits mounted up at home from continued massive government spending, inflation soared from 80 to nearly 150 percent annually between 1980 and 1983. By late 1983, social unrest among the poor began to boil over (exacerbated by the persistent drought in the Northeast), and Rio de Janeiro experienced a serious food riot. The new austerity program, however, would come down hardest on the burgeoning middle class, and could also provoke serious economic disruptions by labor unions in several key industries. The Brazilian "economic miracle" of the 1970s is, thus, in danger of being overshadowed by the perilous new forces of the 1980s.

Agriculture presents Brazil with great opportunities and serious problems. Apart from the revenues derived from coffee sales overseas, Brazil has also been selling more sugar to the United States, the Soviet Union, and China, and the volume of soybeans harvested is rising as well. The range of crops Brazilian farmers can grow is considerable: corn, rice, and wheat (among the staples), beans and vegetables, cotton, tobacco, and grapes. But much land that could be cultivated is not, agricultural methods still need improvement, and land reform in the *fazendas* has come slowly. Millions of peasants in Brazil practice shifting cultivation when environmental conditions do not really demand it, so that returns from the soil are often minimal. It has been estimated that the productivity per farm worker in the United States is at least 50 times higher than in Brazil; mechanization has barely begun in some Brazilian regions. However, despite the rush to the cities, the labor force engaged in agriculture has been growing, reflecting Brazil's still underdeveloped condition.

In this context, we should keep in mind just how far Brazil has to go from its present takeoff point. Its total area is nearly twice that of Europe, but its total annual production of goods and services in 1981 was only about 60 percent higher than that of the Netherlands alone (with a population of about 14.5 million). Norway by itself produces the same amount of hydroelectric power (for a population of only 4.1 million); Venezuela produces more oil in a few weeks than Brazil does in a year; and Brazil's annual coal production—according to John Cole in his book, *Latin America*—"is equal to that of two or three average-sized coal mines in Britain." Brazil's awakening is sometimes compared to the first period of Japan's modernization, but such comparisons are clearly premature.

## Growth Poles

When the military regime took control of Brazil in 1964, it embarked on a development program that is sometimes referred to as the *Brazilian Model.* Essentially, this program involved the government in shared participation with private enterprise, so that public interests, private concerns, and foreign investors would not operate at mutual disadvantage—and at the disadvantage of Brazil as a whole. Agriculture and industry were supported and promoted, but under certain priorities and guidelines.

Among the problems always facing national development planners is the remote region where opportunities lie, but where the investments needed to exploit them are very high. If an area is already developing and attracting immigrants, should money be spent to improve already-existing facilities or should a new, more distant location be stimulated? In such decisions, the *growth-pole* concept becomes relevant. The term is almost self-explanatory: a growth pole is a location where a set of industries, given a start, will expand and set off ripples of development in the surrounding area. Certain conditions must exist of course: there would be little point in selecting Manáos as a growth pole in the middle of the Amazon Basin where there is no real prospect for development in its hinterland and where there are practically no people and no agricultural or industrial activities to stimulate. But growth-pole theory certainly helped to shape the decision to build Brasilia, for in underpopulated Goiás a new market of 700,000 people constituted a major stimulus for all kinds of activity. We will encounter this concept again as we study the UDCs of East Africa and South Asia.

## Multinational Economic Influences

As we noted earlier, the Brazilian Model involves government intervention in the economy, including various forms of control over foreign investors. This was a significant dimension of Brazilian development policy, for large corporations played a major role in Brazil's rapid economic surge between the late 1960s and 1980. In fact, the power of multinational corporations in underdeveloped countries has recently become a matter for concern among the leaders of these countries, because these global corporations, with their enormous financial resources, can influence the economics as well as the politics of entire states.

Brazilian leaders had long welcomed foreign investment, but, as Brazil's economic progress accelerated in the 1960s, they perceived the risks involved in foreign control over Brazilian firms. Multinational corporations can introduce and spread technological advances, they can provide capital and increase employment—but, in the process, they gain control over the industrial and agricultural export sectors of the economy, and, by their efficiency, they can throttle local competition and damage business oriented toward local markets. Aware of such impacts, the government of Brazil for a time prohibited foreign commercial banks from entering the Brazilian economy, and imposed regulations requiring multinational corporations active in Brazil to keep more of their huge profits inside the country. But the economic misfortunes of the 1980s forced the government to abandon its strictures against foreign borrowing, which subsequently accelerated so rapidly that by early 1984 Brazil had become Latin America's biggest debtor, to the tune of nearly $100 billion.

Brazil's economic progress has been achieved at a cost to its society. The last elected government in the early 1960s faced food shortages, serious unemployment, high inflation, and achieved little in the way of economic progress. The generals who took power in 1964 provided the stability Brazil needed, and they brought improvement to the economic situation. But dissent was suppressed, sometimes ruthlessly, and Brazil's reputation as an open, good-natured, fermenting society was marred by reports of torture, intimidation, and other human rights violations. Although the effect of the new policies can be seen in the modernizing cities and near the highways, the lot of the poor in Brazil has hardly begun to change. The economic crisis of the early 1980s further complicates the country's prospects for resuming its development push. The military junta is scheduled to withdraw in favor of a freely elected civilian government in early 1985. Whether this transition will be successful and can restore Brazil's forward progress to improve the lives of its citizens is the crucial question in South America today.

## References and Further Readings

*Area Handbook for Brazil* (Washington: U.S. Government Printing Office, 1971).

Barnet, R. & Muller, R. *Global Reach: The Power of the Multinational Corporations* (New York: Simon & Schuster, 1974).

Burns, E. *A History of Brazil* (New York: Columbia University Press, 2 rev. ed., 1980).

Cole, J. *Latin America: An Economic and Social Geography* (Totowa, N.J.: Rowman & Littlefield, 3 rev. ed., 1976). Quotation taken from p. 214.

de Castro, J. *Death in the Northeast* (New York: Vintage Books, 1969).

Dickenson, J. *Brazil* (London and New York: Longman, 1983).

Dickenson, J. *Brazil: Studies in Industrial Geography* (Folkestone, Eng.: Dawson, 1978).

Epstein, D. *Brasilia: Plan and Reality* (Berkeley: University of California Press, 1973).

Goodland, R. & Irwin, S. *Amazon Jungle: Green Hell to Red Desert?* (Amsterdam: Elsevier, 1975).

Haller, A. "A Socioeconomic Regionalization of Brazil," *Geographical Review*, 72 (1982), 450–464.

Henshall, J. & Momsen, R. *A Geography of Brazilian Development* (London: Bell & Sons, 1974).

Hodder, B. *Economic Development in the Tropics* (London and New York: Methuen, 1980).

Katzman, M. *Cities and Frontiers in Brazil: Regional Dimensions of Economic Development* (Cambridge: Harvard University Press, 1977).

McDowell, E. "The Amazon: A New Search for El Dorado," *New York Times Magazine*, November 22, 1981, 170–179.

McGrath, P. "Paraguayan Powerhouse (Itaipu Dam)," *Geographical Magazine*, April 1983, 192–197.

Merrick, T. & Graham, D. *Population and Economic Development in Brazil: 1800 to the Present* (Baltimore: Johns Hopkins University Press, 1979).

Moran, E. *Developing the Amazon* (Bloomington: Indiana University Press, 1981).

Morse, R. *From Community to Metropolis: A Biography of São Paulo* (New York: Farrar, Straus & Giroux, 1974).

Robock, S. *Brazil: A Study in Development Progress* (Lexington, Mass.: D. C. Heath, 1975).

Roett, R., ed. *Brazil in the Seventies* (Washington: American Enterprise Institute, 1976).

Smith, N. *Rainforest Corridors: The Transamazon Colonization Scheme* (Berkeley: University of California Press, 1982).

Sternberg, H. *The Amazon River of Brazil* (Wiesbaden, Ger.: Geographische Zeitschrift, Monograph No. 40, 1975).

Wagley, C. *An Introduction to Brazil* (New York: Columbia University Press, 1963).

Wagley, C., ed. *Man in the Amazon* (Gainesville, Fla.: University of Florida Press, 1974).

Webb, K. *The Changing Face of Northeast Brazil* (New York: Columbia University Press, 1974).

Weil, C. "Amazon Update: Developments Since 1970," *Focus*, March–April 1983.

# CHAPTER 6

# NORTH AFRICA AND SOUTHWEST ASIA

IDEAS AND CONCEPTS

**Cultural Geography**
**Culture Hearth (2)**
**Spatial Diffusion Models**
**Boundary Morphology**
**Energy Resources**
**Remote Sensing**
**Urban Dominance**
**Irrigation Methods**
**Nomadism**
**Political Development Cycles**
**Ecological Trilogy**
**Von Thünen's Isolated State (2)**
**Irredentism (2)**

From Morocco on the shores of the Atlantic Ocean to Afghanistan in Asia and from Turkey between the Black and Mediterranean Seas to the Somali Republic in the "Horn" of Africa lies a vast realm of enormous historical and cultural complexity. It lies at the crossroads of Europe, Asia, and Africa and is part of all three; throughout history, its influences have radiated to these continents and to practically every other part of the world as well. This is one of humanity's source areas. On the Mesopotamian Plain between the Tigris and Euphrates and on the banks of the Nile arose civilizations that must have been among the earliest. In the soils, plants were domesticated that are now grown from the Americas to Australia. Along its paths walked the prophets whose religious teachings are still followed by hundreds of millions of people. And in the late twentieth century, the heart of this realm is beset by some of the most bitter and dangerous conflicts on earth—ones that may yet provoke a direct confrontation between the superpowers.

It is tempting to characterize this geographic realm in a few words, to stress one or more of its dominant features. It is, for example, often called the "Dry World," containing as it does the vast Sahara and Arabian Deserts. But most of the people in the region live where there is water—in the Nile Delta, along the coastal strip (or *tell*) of Tunisia, Algeria, and Morocco, along the eastern and northeastern shores of the Mediterranean Sea, in the Tigris-Euphrates Basin, in the desert oases, and along the mountain slopes south of the Caspian Sea. True, we know this region as one where water is almost always at a premium, where peasants often struggle to make soil and moisture yield a small harvest, where nomadic peoples and their animals circulate across dust-blown flatlands, where oases are islands of sedentary farming and trade in a

**Figure 6–1**
Embattled and fragmented Jerusalem: holy city of Christianity, Islam, and Judaism, and a personification of this realm's bitter divisions.

# SYSTEMATIC ESSAY

## Cultural Geography

The essence of **cultural geography** is captured by Joseph Spencer in his 1978 survey of this subdiscipline:

> *[It] focuses on the ways in which human achievement produces many different living systems among the varied peoples of the whole earth. Interest lies in the spatial distribution and the functioning patterns of all culture systems that become operative on earth. . . . Cultural geographers typically take the long-term view of human development, seeing contemporary issues as time-period manifestations of the slow development of cultural systems.*

In the introduction (pp. 1–24) to their 1962 edited volume, *Readings in Cultural Geography*, Philip Wagner and Marvin Mikesell describe the field as the application of the idea of culture to geographic problems. They define its core as consisting of five interrelated themes: (1) culture, (2) culture area, (3) cultural landscape, (4) culture history, and (5) cultural ecology. This quintet provides the framework for the following overview.

The concept of *culture* was discussed in the Introduction (pp. 4–7) and, in subsequent chapters, has proven to be one of the most important keys to the systematic understanding of differences and similarities among human societies. In terms of structure, each culture may be viewed as an interconnected whole that can be dissected into a number of interacting components. The basic building block is the *culture trait*, a single element of normal practice in a culture; the collective individual traits of each culture form a larger discrete combination referred to as a *culture complex*; these, in turn, are aggregated into a national-scale *culture system*, a set of culture complexes united by common characteristics and strong historical-cultural bonds. Language is a leading cultural component, the medium wherein people communicate by spoken and written words, facial expressions and numerous other interpersonal gestures. As the accepted belief system, religion is another major component, one that can demonstrate the close integration of cultural traits; depending on the society, religious traditions can markedly affect several aspects of daily life. In the North African-Southwest Asian realm, religion can influence clothing styles (women must be cloaked and veiled in public in many Islamic countries), shopping days (many businesses are closed on the Moslem and Jewish sabbaths, respectively Fridays and Saturdays), dietary practices (fasting on religious holidays, food avoidances), and even governmental administration (Iran's fundamentalist Shi'ite mullahs are in firm control of the levers of power). Among the many other components of culture—most possessing clear spatial dimensions—are legal and political institutions, settlement patterns, architecture, family structures, adornment, art, literature, theater, and music.

The study of *culture areas* entails the identification and mapping of appropriate phenomena to recognize and delimit territories occupied by human communities that share a particular culture. By aggregating individual traits into culture complexes and systems, hierarchical geographic classifications can be developed. In the Introduction (p. 4), the three-tier hierarchy of culture realm/region/subregion was outlined; in fact, a culture system, on the map, constitutes a culture region, and an assemblage of culture systems creates a culture realm. Cultural geography's concern with distributions of cultural phenomena is especially sensitive to changes over time, as innovations originate in source areas and subsequently disperse across great distances. In addition to transmitting agents and channels, cultural barriers are studied as well. For instance, the spatial expansion of the Islamic religion occurred far more rapidly than the spreading of the Arabic language, because recipient cultures were relatively amenable to religious change but resisted attempts to modify their linguistic traditions.

*Cultural landscape* interpretation focuses on the physical imprint of material culture on the land surface that forms the geographic content of a culture region. In addition to the artifacts of material culture, many geographers now also analyze the emerging landscapes associated with popular culture. This represents a broadening of the field to include vernacular culture, culture in the context of its sustainers—the people who maintain and nurture it. *Popular culture* is the always changing "mass" culture of urbanized and industrialized society that is less tradition-bound, more open, individualistic, and class-structured than strongly traditional *folk culture*, the durable way of life found in the comparatively isolated rural areas. The North African-Southwest Asian realm, like most underdeveloped regions, is dominated by the latter (the overthrow of the Shah's regime in Iran was an object lesson in the limitations to bucking this *status quo*). Among the topics recently researched by vernacular-culture geographers are music styles, sources of professional athletes, front-yard ornaments, fast-food restaurants, and even urban graffiti.

*Culture history* is the explicit concern with the cultural-geographical past. In this chapter, we will formally treat the time-space interface through an examination of *spatial diffusion* in Model Box 4. Here we briefly consider the topic of food taboos—a practice that affects tens of millions in the realms of the three upcoming chapters, and an outstanding example of current cultural behavior that evolved over the long course of human development in the Old World.

The processes whereby food is produced, distributed, and consumed form a fundamental part of every culture. The way land is allocated to individuals or families (or bought and sold), the manner in which it is used for food production, the functions of livestock, and the consumption of food from crops and animals are all aspects of culture. Food consumption itself is often related in important ways to religious influence and dogma. Adherents of Islam and Judaism avoid pork; Hindu believers do not consume beef or other kinds of meat (Fig. 6–2). Various other forms of partial or total abstinence occur among human cultures, including periodic fasts. Such proscriptions, like the religions that generated them, tend to be old and persistent and change only very slowly.

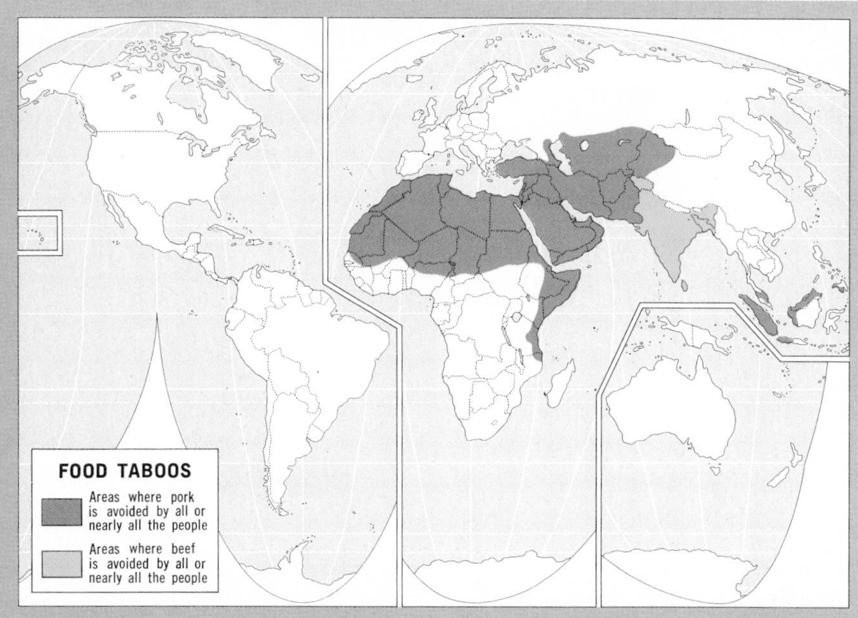


Figure 6–2

*Cultural ecology* involves the multiple relationships between human cultures and their physical environments. However, as we saw in Chapter 4, this tradition of geography in the past attracted proponents of environmental determinism and similar viewpoints, and the idea of a mutual and balanced human-habitat interaction did not gain currency until the 1930s. Environmentalism in American geography was countered in the 1920s from two directions. From Western Europe came the short-lived doctrine of *possibilism*, which argued against climatic and physiographic determinism by claiming that people, through their cultures, are free to choose from a number of environmental "possibilities." The second and far more persuasive attack came from within, through the scholarship of Carl Sauer (this century's leading cultural geographer). His position, the one all but unanimously adopted today, regards humans and the environment as coequal partners in an interacting unity, with humankind the agency of the systematic modification of the physical landscape.

sea of aridity. But it also is the land of the Nile, the lifeline of Egypt, and the great Gezira irrigation scheme, the mainstay of Sudan, and the crop-covered *tell* of Algeria.

North Africa-Southwest Asia is also often referred to as the "Arab World." However, once again, this implies a uniformity that does not actually exist. In the first place, the name Arab is given loosely to the peoples of this area who speak Arabic and related languages, but ethnologists normally restrict it to certain occupants of the Arabian Peninsula—the Arab "source." Anyway, the Turks are not Arabs, and neither are the Iranians nor the Israelis. Moreover, although it is true that Arabic is spoken over a wide region that extends from Mauritania in the west across North Africa to the Arabian Peninsula, Syria, and Iraq in the east, there are many areas within the realm where it is not used by most of the people. In Turkey, for example, Turkic is the major language, and it has Altaic rather than Semitic or Hamitic roots. In Iran, the Iranian language belongs to the Indo-European family. In Ethiopia, Amharic is spoken by the ruling plateau people; although it is more closely related to Arabic than is Iranian, it, nonetheless, remains a distinct language as well. The same is true of Hebrew, spoken in Israel. Other "Arab World" languages that have separate identities are spoken by the Tuareg people of the Sahara Desert, the Berbers of Northwest Africa, and the population residing in the transition zone between North Africa and Black Africa to the south.

Another name given to this realm is the "World of Islam." The prophet Mohammed (Muhammad) was born in Arabia in A.D. 571, and in the centuries that followed his death in 632, Islam spread into Africa, Asia, and Europe. This was the age of Arab conquest and expansion, during which their armies penetrated southern Europe, caravans crossed the deserts, and ships plied the coasts of Asia and Africa. Along these routes, they carried the Moslem (Islamic) faith, converting the ruling class of the states of the West African savanna, threatening the Christian stronghold in the highlands of Ethiopia, penetrating the deserts of inner Asia, pushing into India and even Indonesia. Islam was the religion of the marketplace, the bazaar, the caravan. Where necessary it was imposed by the sword, and its protagonists aimed directly at the political leadership of the communities they entered. Today, the Islamic religion with its 600 million followers extends well beyond the limits of the region discussed here: it is the major religion in northern Nigeria, in Pakistan, and in Indonesia; it has strength in parts of the Soviet Union and survives even in Eastern Europe, notably in Albania and Yugoslavia (Fig. 6–3). On the other hand, the "World of Islam" is not entirely Moslem either. In Israel, Judaism is the prevailing faith; in Lebanon, perhaps as much as half the population adheres to an old form of Christianity; in Ethiopia, the Amharic-speaking ruling class, having managed centuries ago to stave off the Islamic onslaught, also practices an ancient Christian religion, and Coptic Christian churches still exist in Egypt. Thus, the connotation "World of Islam" for North Africa and Southwest Asia is far from satisfactory—the religion prevails far beyond these areas, and, within the realm, there are countries in which Islam is not the faith of the majority.

Finally, this realm is frequently called the "Middle East." This must sound quite odd to someone, say, in India, who might think of a Middle West rather than a Middle East! The name, of course, reflects the biases of its source: the "Western" world, which saw a "Middle" East in Egypt, Arabia, and Iran, and a "Far" East in China and Japan. Still, the term has taken hold, and it can be seen and heard in general use by journalists as well as members of the United Nations. In view of the complexity of the realm, its transitional margins, and its far-flung areal components, at least the name *Middle East* has the merit of being imprecise. It does not make a single-factor region of North Africa and Southwest Asia, as do the terms *Dry World*, *Arab World*, and *World of Islam*.

# TEN MAJOR GEOGRAPHIC QUALITIES OF NORTH AFRICA AND SOUTHWEST ASIA

1. This realm contains several of the world's great ancient culture hearths and some of its most durable civilizations.
2. North Africa and Southwest Asia is the source region of several world religions, including Judaism, Christianity, and Islam.
3. This realm is predominantly, but not exclusively, Islamic (Moslem); that faith pervades cultures from Morocco to Afghanistan.
4. North Africa and Southwest Asia is the "Arab World," but significant population components in the realm are not of Arab ancestry.
5. Population in North Africa and Southwest Asia is widely dispersed in discontinuous clusters.
6. Natural environments in this realm are dominated by drought and unreliable precipitation; population concentrates where water supply is adequate to marginal.
7. The realm contains a pivotal area in the "Middle East," where Arabian, North African, and Asian regions intersect.
8. North Africa and Southwest Asia is a realm of intense discord and conflict, reflected by frequent territorial disputes and boundary frictions.
9. Both the United States and the Soviet Union have vital interests and key client-states in this realm, thereby magnifying internal stresses as potential global problems in the relationship between the two superpowers.
10. Enormous reserves of petroleum lie beneath the realm, but this wealth has affected the living standards of only a minority of the total population.

# A Greatness Past

In our discussions of North and Middle America, reference was made to the *culture hearth* concept—a source area or innovation center from which cultural traditions are transmitted. In such regions of comparative success, growth clusters of population developed, and natural increase was supplemented by the immigration of those attracted from afar. New ways were found to exploit local resources, and power was established over resources located farther away. Farming techniques improved, and so did yields. Settlements could expand, and began to acquire urban characteristics. The circulation of goods and ideas intensified. Traditions emerged in various spheres of life, and these traditions, along with inventions and innovations, radiated outward into the area beyond. Among the ideas that took hold and developed were political ideas, especially the theory and practice of the political organization necessary to cope with society's growing complexity.

Such developments occurred on several continents (Fig. 6–4). The Middle East was a pre-eminent source area, where culture hearths lay in the Tigris-Euphrates Basin (Mesopotamia), the Nile Valley of Egypt, and the Indus Valley of what is now Pakistan. The main hearth of Chinese culture was centered on the confluence of the Wei and Huang He Rivers. South Asia's Ganges Delta was another key cultural spawning ground. In Middle America, as we noted, the Yucatan Peninsula and later the Mexican Plateau were significant culture hearths. In South America, the Central Andes Mountains witnessed the rise of a major civilization. And, as Fig. 6–4 shows, in West Africa's savannalands, a culture hearth also emerged.

The Middle East as it now exists as a geographic realm, thus, included several of the earliest (if not the first) hearths of human culture. Mesopotamia, the "land between the rivers," anchored one end of the Fertile Crescent, which stretched to the southeast Mediterranean Coast, one of the places where people first learned to domesticate plants and gather harvests in an organized way. Mesopotamia became a crossroads for a whole network of routes of trade and movement across Southwest Asia, and its accessibility enhanced its adoption of numerous new ideas and inventions. Many innovations originated in Mesopotamia itself, to be spread to other regions of development. In addition to the organized and planned cultivation of grain crops such as wheat and barley, the Mesopotamians knew how to make tools and implements out of bronze, had learned to use draft animals to pull vehicles (including plows designed to prepare the fields), and employed the wheel, a revolutionary invention, to build carts, wagons, and chariots.

The ancient Mesopotamians also built some of the world's earliest cities. This development was made possible by their accomplishments in many spheres, especially agriculture, whose surpluses could be stored and distributed to the city dwellers. Essential to such a system of allocation, of course, was a body of decision makers and organizers, an elite group who controlled the lives of others. Such an urban-based elite could afford itself the luxury of leisure and could devote time to religion and philosophy. Out of such pursuits came the concept of writing and record keeping, an essential step in the rise of urbanization. Writing made possible the codification of laws and the confirmation of traditions; it was a crucial element in the development of systematic administration in urbanizing Mesopotamia and in the evolution of its religious-political ideology. The rulers in the cities were both priests and kings, and the harvest the peasants brought to be stored in the urban granaries was a tribute as well as a tax.

Mesopotamia's cities emerged about 6000 years ago, and some may have had as many as 10,000 residents (certain archaeologists' estimates go even higher). Today, these urban places are extinct, but careful excavations tell the story of their significant and sometimes glorious past. Mesopotamia's cities had their temples and shrines, priests and kings; there were also wealthy merchants, expert craftspeople, and respected teachers and philosophers. But Mesopotamia was no unified political state. Each city had a hinterland over which its power and influence prevailed; although occasional alliances existed among cities, there was more often competition and conflict in a *feudal* society marked by hostility and strife in political relationships. Still, Mesopotamia made progress in political life as well as other areas, and regional unification eventually was achieved during the fourth millennium B.C. (between 5000 and 6000 years ago) in at least two early states named Sumer and Elam.

Another area that witnessed very early cultural and political development was Egypt. Possibly, Egypt's evolution started even earlier than Mesopotamia's, but Egypt certainly possessed urban centers 5000 years ago, and there are archaeologists who believe that even older remains of urban places may lie buried beneath the silt of the Nile Delta. Actually, the focus of the ancient culture hearth of Egypt lay above (south of) the delta and below (north of) the Nile's first cataract, and this segment of the Nile Valley lies surrounded by rather inhospitable country (see Fig. I–1 facing p.1). The region was open to the Mediterranean; otherwise it was quite inaccessible by overland contact. In contrast to Mesopotamia, which was something of a marchland, the Nile Valley was a natural fortress of sorts. There, the ancient Egyptians converted the security of their isolation into progress. The Nile waterway was the area's highway of trade and association, its true lifeline. It also sustained agriculture by irrigation, and the cyclic regime of the Nile's water level was a great deal more predictable than that of the Tigris or Euphrates. By the time ancient Egypt finally began to fall victim to outside invaders, from about 1700 B.C. onward, a full-scale, urban civilization had emerged, whose permanence and continuity are reflected in the massive stone monuments its artist-engineers designed and created. The political practices and philosophies of Pharaoic Egypt were dispersed far and wide, especially into Africa. Egypt survived as a political entity longer, perhaps, than any state (China being its only rival); in the process, the state changed from a theocratic to a militaristic one, eventually to fall to colonial status, but now to rise again as a modern nation. Ancient Egypt's armies were well disciplined and effective, and the cities were skillfully fortified. Where Egypt's armies ranged, peoples were subjugated and exploited for the state's benefit; Egypt far outlasted its Mesopotamian contemporaries.

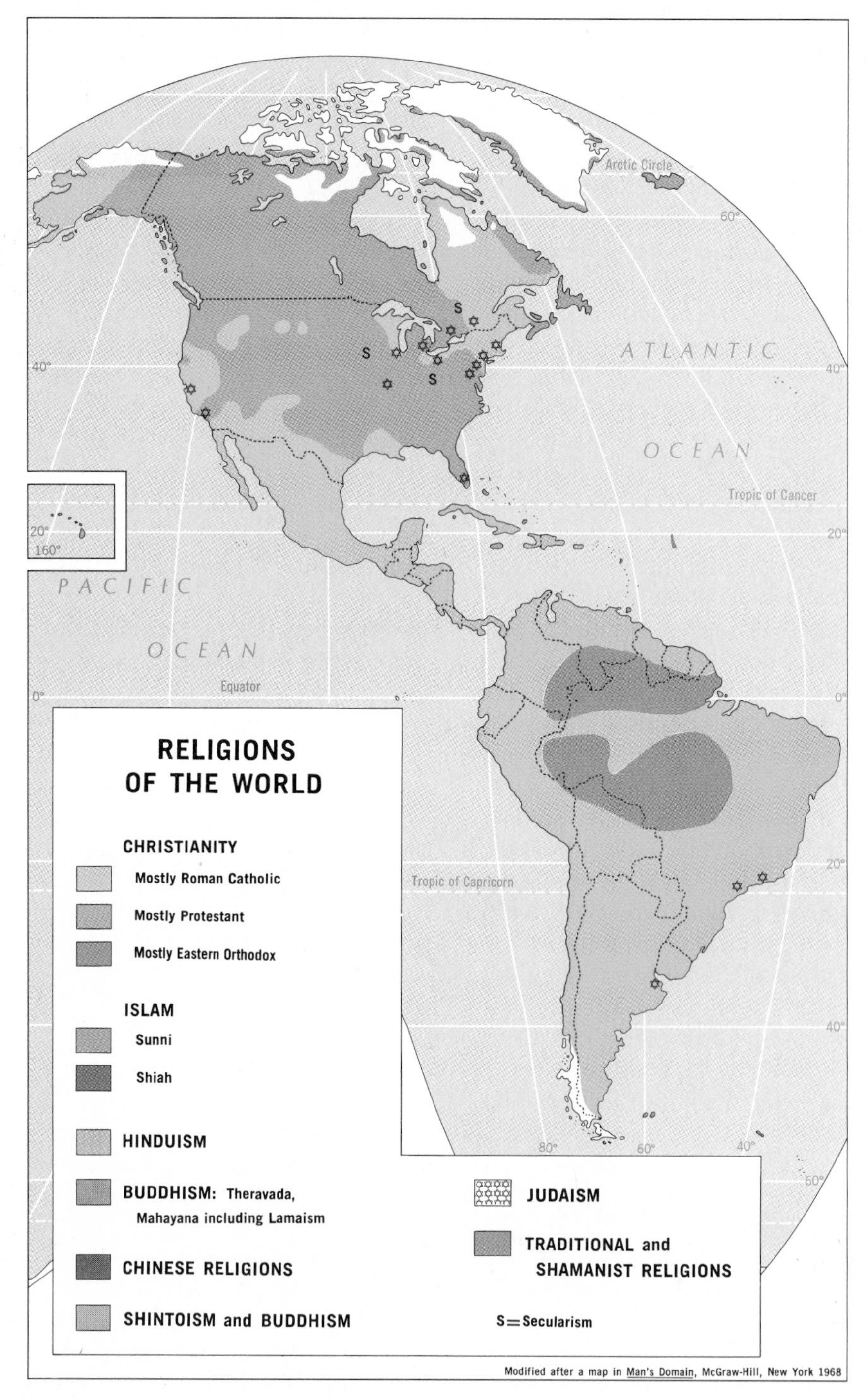


Figure 6–3

Egypt's culture hearth lay on the western fringe of the great Middle Eastern source area. As Fig. 6–4 indicates, Mesopotamia was at the heart of this region; to the east and separated by desert from the land of Babylon lay a third culture hearth—the civilization of the Indus Valley, in what is today Pakistan. By modern criteria, this eastern region lies outside the realm under discussion, but in ancient times it had effective ties with lands to the west. It is believed that Mesopotamian innovations reached the Indus cities of Harappa and Mohenjo-Daro. When the civilization that was centered on those cities reached its greatest extent, it penetrated the upper Ganges Valley of present-day India and extended as far south as the Kathiawar Peninsula.

In those distant times, as today, the key to life in the Middle East was water; where it is available in sufficient quantity—as in the Nile Delta—the "Dry World" turns green (Fig. 6–5). The Mesopotamians and Egyptians achieved breakthroughs in the control of

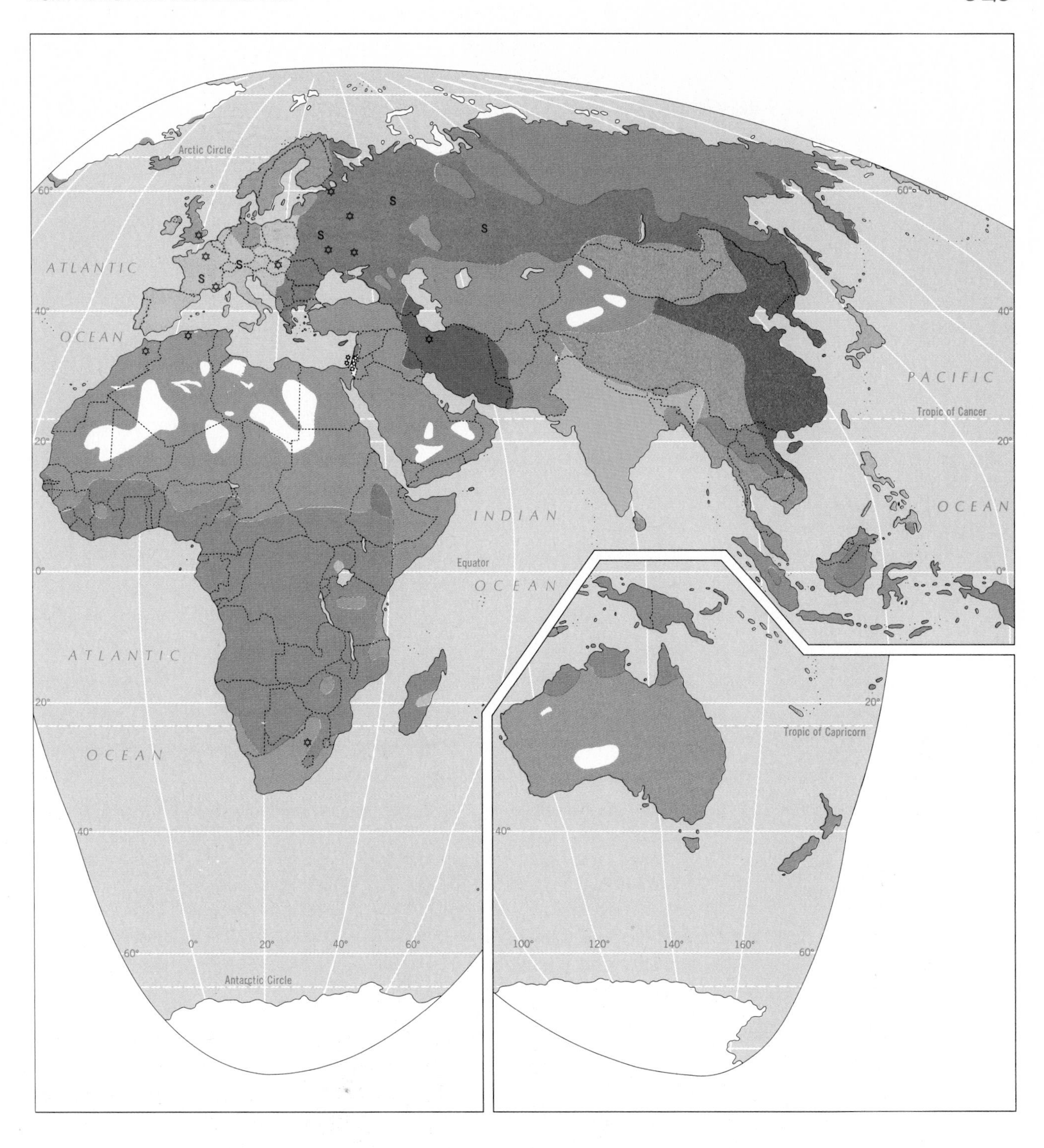


river water for irrigation and in the domestication of several staple crops. Today, people throughout the world continue to benefit from these contributions—they include cereal crops (such as wheat, rye, and barley), vegetables and fruits (peas, beans, grapes, olives, apples, and peaches), and domesticated animals (horses, pigs, sheep, and goats). And the range of indirect contributions is quite beyond measure. The ancient Mesopotamians advanced not only irrigation and agriculture, but also calendrics, mathematics, astronomy, government administration, engineering, metallurgy, and a host of other fields. As time went on many of their innovations were adopted and then modified by other cultures in the Old World and, eventually, even in the New World. Europe was the greatest beneficiary of the

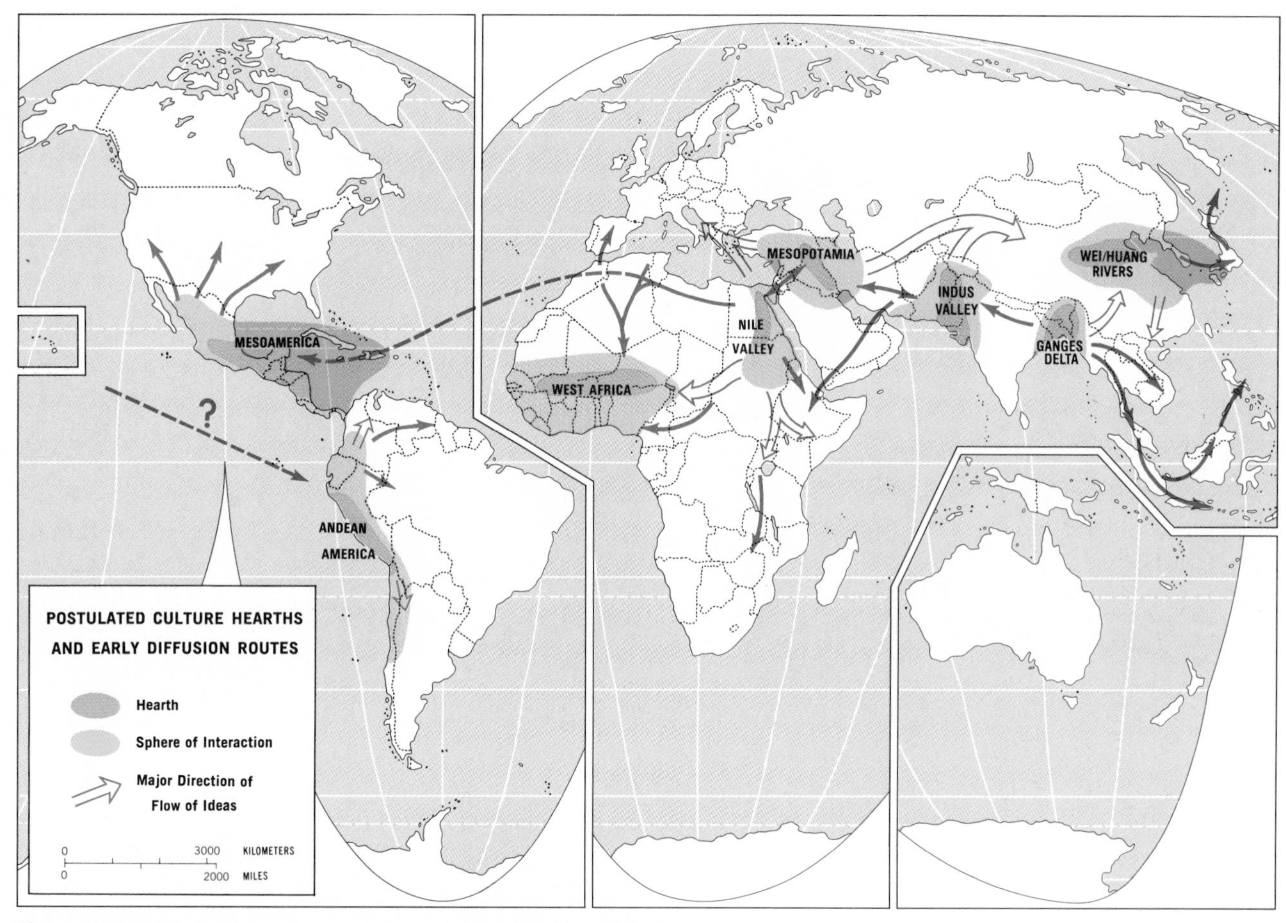


Figure 6–4

legacies of Mesopotamia and Egypt, whose achievements constituted the very foundations of "Western" civilization.

# Decline and Rebirth

Today the great cities of this realm's culture hearths are archaeological curiosities. In some instances, new cities have been built on the sites of the old, but the great ancient cultural traditions of the Middle East went into a deep decline after many centuries of continuity. It cannot escape our attention that a large number of the ruins of these ancient urban centers are located in what is now desert. Assuming that they were not built in the middle of desert areas, it is tempting to conclude that climatic change (associated with shifting environmental zones in the wake of the last Pleistocene retreat) destroyed the old civilizations. Indeed, some geographers have suggested that the momentous innovations in agricultural planning and irrigation technology may have been made in response to changing environmental conditions, as the river-basin communities tried to survive. The scenario is not difficult to imagine: as outlying areas began to fall dry and farmlands were destroyed, people congregated in the already crowded river valleys—and every effort was made to increase the productivity of lands that could still be watered. Eventually overpopulation, destruction of the watershed, and perhaps reduced rainfall in the rivers' headwater areas combined to deal the final blow. Towns were abandoned to the encroaching desert; irrigation canals filled with drifting sand; remaining farmlands dried up. Those who could, migrated to areas reputed to still be productive. Others stayed, their numbers dwindling, increasingly reduced to subsistence.

As old societies disintegrated, new power emerged elsewhere. First the Persians, then the Greeks, and later the Romans imposed their imperial designs on the tenuous lands and disconnected peoples of the Middle East. Roman technicians converted North Africa's farmlands into irrigated plantations whose products went by the boatload to Roman Mediterranean shores. Thousands of people were carried in bondage to the cities of the new conquerors. Egypt was a colony; the Arab settlements on the Arabian

**Figure 6–5**
The fertile fields of the Nile Delta—also shown from orbiting satellite in Figure 6–17.

Peninsula lay more remote and somewhat more secure in their isolation.

In one of these relatively remote places on the Arabian Peninsula an event occurred that was to change the course of history and affect the destinies of people all over the world. In Mecca, a town 72 kilometers (45 miles) from the Red Sea coast and positioned in the Jabal Mountains, a man named Mohammed in about A.D. 613 began to receive the truth from Allah (God) in a series of revelations. Mohammed (A.D. 571–632) already was in his early 40s when these revelations began, and he had barely 20 years to live. But in those two decades commenced the transformation of the Arab world. Convinced, after some initial self-doubt, that he was indeed chosen to be a prophet, Mohammed committed his life to the fulfillment of the divine commands he had received. The Arab world was in social and cultural disarray; in the feudal chaos, Mohammed soon attracted enemies who feared his new personal power. He fled Mecca for the safer haven of Medina (but Mecca's place as a holy center was soon assured), and, from there, he continued his work.

The precepts of Islam constituted, in many ways, a revision and embellishment of Judaic and Christian beliefs and traditions. There is but one god, who occasionally reveals himself to prophets (Islam acknowledges that Jesus was such a prophet). What is earthly and worldly is profane; only Allah is pure. Allah's will is absolute; he is omnipotent and omniscient. All humans live in a world created for their use, but only to await a final judgment day.

Islam brought to the Arab world not only a unifying religious faith, but also a whole new set of values, a new way of life, a new individual and collective dignity. Apart from dictating observance of the "five pillars" of Islam—repeated expressions of the basic creed, frequent prayer, a month of daytime fasting, almsgiving, and at least one pilgrimage to Mecca—the faith prescribed and proscribed in other spheres of life as well. Alcohol, smoking, and gambling were forbidden. Polygamy was tolerated, although the virtues of monogamy were acknowledged. Mosques appeared in Arab settlements, not only for the Friday prayer, but also as social gathering places to bring communities closer together. Mecca became the spiritual center for a divided, far-flung people for whom a joint focus was something new.

The stimulus provided by Mohammed, spiritual as well as political, was such that the Arab world was mobilized almost overnight.

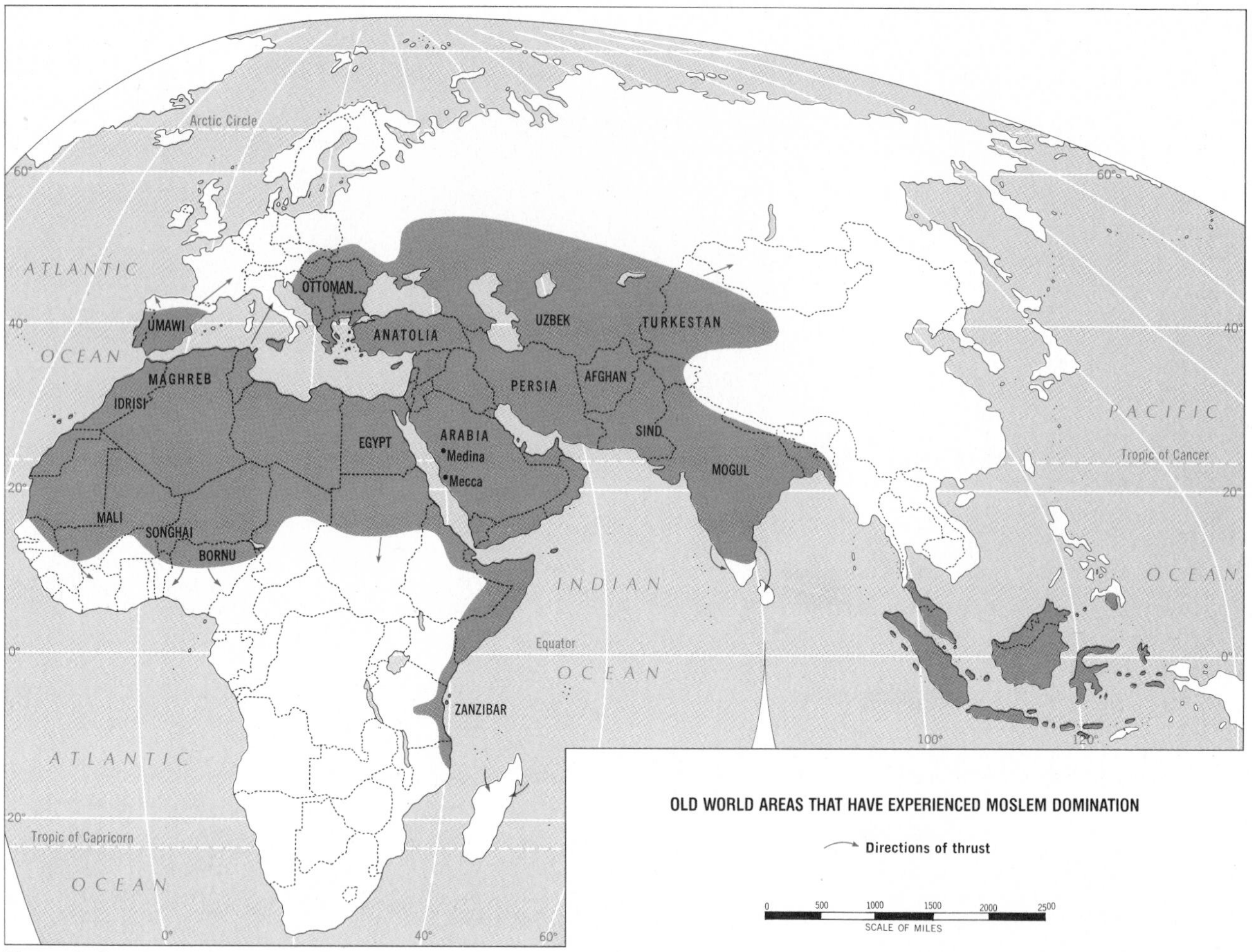


Figure 6–6

The prophet died in 632, but his faith and fame continued to spread like wildfire. Arab armies formed, they invaded and conquered, and Islam was carried throughout North Africa. By the early ninth century, the Moslem world included emirates or kingdoms extending from Egypt to Morocco, a caliphate occupying most of Spain and Portugal, and a unified region encompassing Arabia, the Middle East, Iran, and much of Pakistan (Fig. 6–6). Moslem influences had attacked France and Italy and penetrated what is today Soviet Central Asia as far as the Aral Sea. Ultimately, the Arab empire extended from Morocco to India and from Turkey to the southern margins of the Sahara Desert (where it is still the faith of millions in this part of Black Africa—Fig. 6–7). The original capital was at Medina in Arabia; in response to these strategic successes, it was moved, first to Damascus and then to Baghdad. In the fields of architecture, mathematics, and science, the Arabs far overshadowed their European contemporaries, and they established institutions of higher learning in many cities, including Baghdad, Cairo, and Toledo in Spain. The faith had spawned a culture; it is still at the heart of that culture today.

The spread of the Islamic faith throughout the realm (and beyond its borders as well) occurred in waves radiating from Medina and the holy city of Mecca. Islam went by camel caravan and by victorious army, it was carried by pilgrim and boatsman, scholar and sultan. Its dissemination through so vast and discontinuous a realm is an example of the process of *diffusion*, which is elaborated in Model Box 4.

The worldwide diffusion of Islam is still going on. In the United States, it is emerging in the religious movement commonly called the Black Muslims but officially known as the Nation of Islam. There are Islamic communities in South Africa, in the Philippines, and other far-flung places. This is not the first time that the Middle East generated an idea that affected much of the world. Agricultural methods, metallurgical techniques, architectural styles, and countless other innovations made in this realm have, in the course of his-

**Figure 6–7**
Friday at the mosque in Kano, Nigeria: Islam's impact has spread far beyond its Arabian core.

tory, been adopted by societies across the globe.

# Boundaries and Barriers

If Islam constituted such a strong unifying force and if Middle Eastern armies could penetrate Europe, Asia, and Africa south of the Sahara—why is the Middle East today a realm of boundaries and barriers, tension and conflict, hostility and disunity?

In part, the answer lies in the rise of another religion, one that emerged even earlier than Islam—Christianity. Until the challenge of Islam, the Christian faith (which began as a movement within another Middle Eastern religion, Judaism) had won acceptance in the Roman Empire, could afford the luxury of an east-west ideological split, and diffused to the tribes of Europe north of Rome's boundaries. Then, the Arabs rode on Mohammed's teachings into Iberia and threatened the Roman core of Italy itself. Europeans mobilized to meet the chal-

MODEL BOX 4

# SPATIAL DIFFUSION PRINCIPLES

**Spatial diffusion** may be defined as the geographic spreading or dissemination of ideas, innovations, and phenomena. The study of diffusion is of interest to regional geographers for several reasons. First, like human movement and migration, these are spatial processes in which information flow plays a large role. The diffusion of Islam to West Africa's savanna states, the approximate dates of its arrival, the particular segment of African society that was converted—all these dimensions help us reconstruct trans-Saharan cultural contact between Black West Africa and the Arab North. In the case of Islam, it is not difficult to identify the origins of the innovation, but, in other instances, an understanding of the nature of the process can help us trace back toward a postulated source. Second, a knowledge of diffusion processes can have a practical side. The spread of a disease, as we will see, involves a diffusion process too; if we know how and at what rate it is communicated, it may be possible to slow it down. On the other hand, if a government wants to disseminate something new (fluoridation of drinking water for instance), it is helpful to be able to forecast its adoption pattern by identifying potential areas of resistance as well as places of quick implementation.

The study of spatial diffusion, because of its implicit time dimension, is rooted in historical geography. A particular concern with the regional spreading of culture traits and landscape artifacts led Carl Sauer (1889–1975) to devote considerable attention to the origins and dispersals of numerous cultural phenomena—his 1952 investigation of the diffusion of agriculture is regarded as a classic piece of scholarship in the field. The very same year also saw the publication of the doctoral dissertation of the Swedish geographer Torsten Hägerstrand (1916– ); entitled *The Propagation of Innovation Waves*, it laid the theoretical foundations for the contemporary spatial analysis of diffusion processes that still comprises one of human geography's most active research frontiers. Hägerstrand's model was based on an elaborate set of assumptions and concluded that four stages mark the diffusion mechanism: a *primary* stage (when the innovation appears at its source and begins to be adopted there), a *diffusion* stage (when the innovation radiates rapidly outward, generating new centers of dispersal in more distant areas), a *condensing* stage (in which remaining areas are penetrated), and a *saturation* stage (marking the slowdown and end of the diffusion process). Collectively, these stages describe the progress of an *innovation-wave*, which pulses outward from the hearth and "washes" over increasingly distant areas until it peters out; source areas may also propagate many such waves over time before the adoption process is complete. The remainder of this Model Box treats additional spatial diffusion principles, building on the work of Hägerstrand and his intellectual heirs.

There are two major types of spatial diffusion. The first is *relocation diffusion*, whereby whatever is being diffused enters an area with its carrier agent. The most common form of relocation diffusion involves the spreading of innovations by a migrating population. An example would be the first appearance of the log cabin during colonial times in the northeastern United States—this innovation was transplanted from Sweden by seventeenth-century emigrants to the Philadelphia area and was subsequently adopted by thousands of westward-moving Americans passing through Pennsylvania and New Jersey. The second type—by far the most significant—is *expansion diffusion*. Here we are dealing with a fixed population through which innovations diffuse from a source area in accordance with Hägerstrand's four-stage model. The historical spreading of Islam is a classic example: this religious faith remained strong at its core and progressively disseminated its message over time and across space until its converts encompassed the huge Old World area mapped in red in Fig. 6–6.

Expansion diffusion also assumes two major forms, and they are simultaneously diagrammed (after Keith Chapman) in Fig. 6–8, which displays the four-year adoption pattern of an innovation in a hypothetical agricultural region containing farms of three different size categories. Ignoring these size differences for a moment, we observe an obvious distance-controlled diffusion pattern in Years 2–4 with respect to the source farm of Year 1. This process is called *contagious diffusion*—because of its analogy to the spatial spreading of a disease from an individual to the next closest individual; since its expansion is shaped by proximity to the source location, it is usually confined to local areas. Now taking farm size variation into account, we can also detect a second and rather different pattern of expansion diffusion in Years 2–4. Quite clearly, the innovation is also spreading in a manner that favors larger-sized farms. This "trickling down" from large

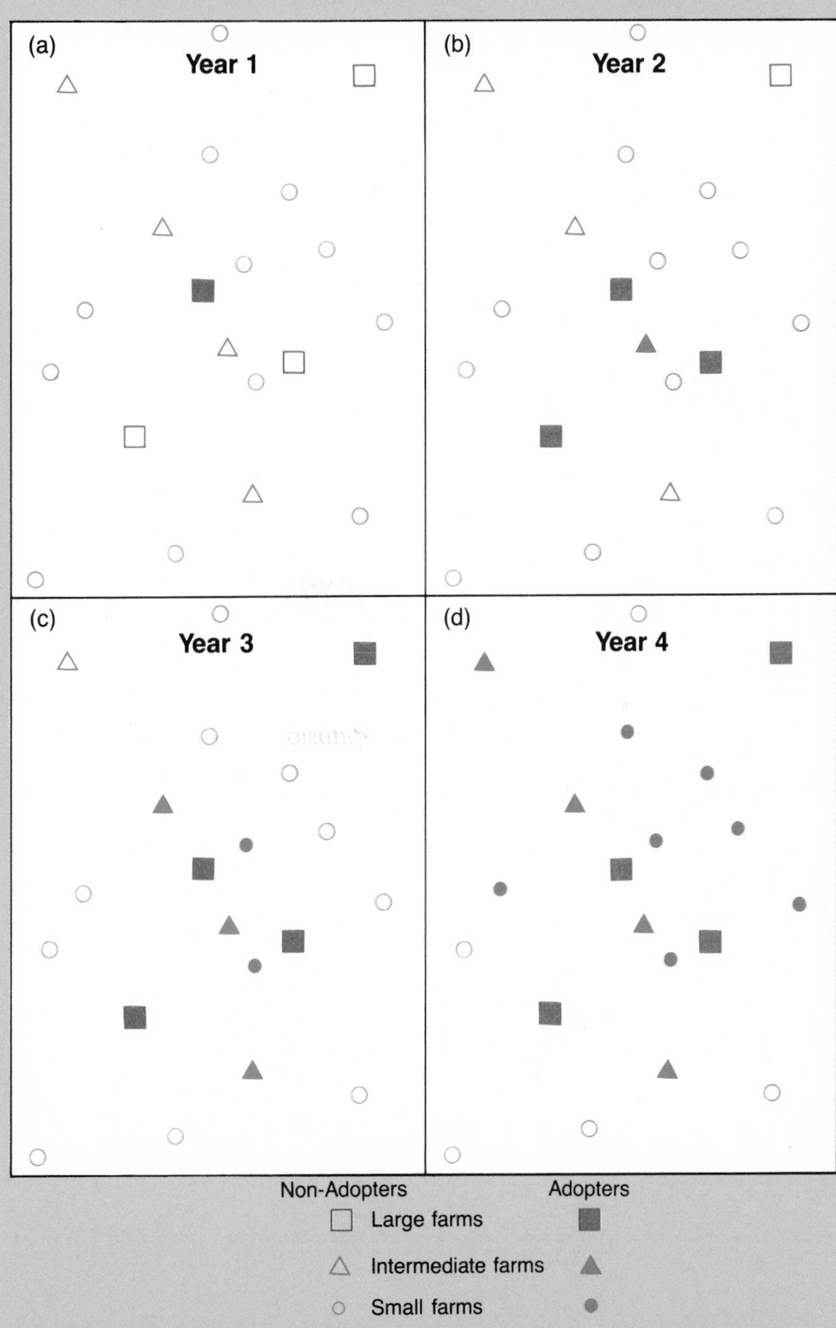


Figure 6–8

to intermediate to small farmsteads is called *hierarchical diffusion,* which is dominated by the leapfrogging of innovations over a wider area than the local dispersion pattern associated with contagious diffusion. In large geographic areas of national and continental scale, the urban hierarchy is the communication channel within which innovations spread—starting in the primate city and diffusing downward through cities, towns, and villages.

Some empirical examples demonstrate the utility of these conceptualizations of horizontal (contagious) and vertical (hierarchical) diffusion dynamics. Figure 6–9, using line and area symbols, presents maps made by John Hunter and Jonathan Young that show the advance of the flu epidemic that struck England and Wales in 1957. Beginning in north-central England, influenza steadily spread outward from this source area until it engulfed the entire region. The diffusion pattern is predominantly a contagious one, although some leapfrogging into South Wales and Southeastern England can be noted; this is not so much due to hierarchical effects as it is to coastal location—because this was part of a worldwide flu outbreak, passengers and crews arriving at certain seaports introduced the disease to those cities in advance of the epidemic that was already approaching them via overland routes. In Fig. 6–10, we observe a classic hierarchical diffusion pattern mapped by Yehoshua Cohen in his study of regional shopping-center diffusion in the United States. He demonstrates vividly that this innovation coursed downward through the national urban system in the 1950s and 1960s, progressing in orderly fashion from larger to smaller metropolitan areas; however, as with all models, perfect accordance between theory and reality rarely occurs—note carefully that the biggest city (New York) did not participate in the earliest years, whereas such smaller centers as Sacramento, Buffalo, and Elyria, Ohio (a suburb of Cleveland), did. The historical significance of the urban hierarchy is displayed in the two maps of Fig. 6–11, which are based on

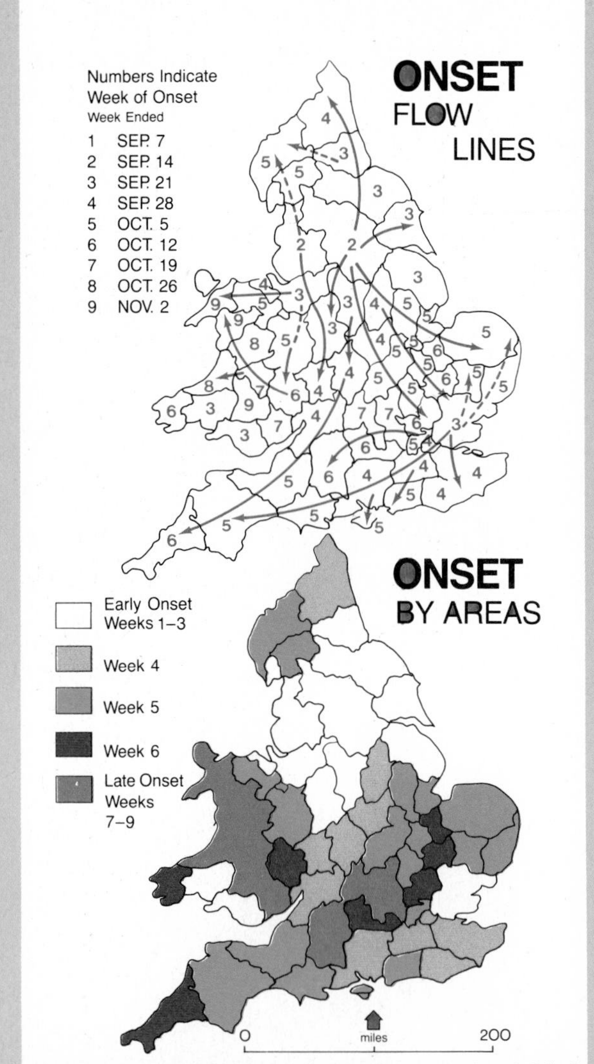


Figure 6–9 Reproduced by permission from the *Annals of the AAG*, Vol. 61, 1971, p. 645, fig. 7.

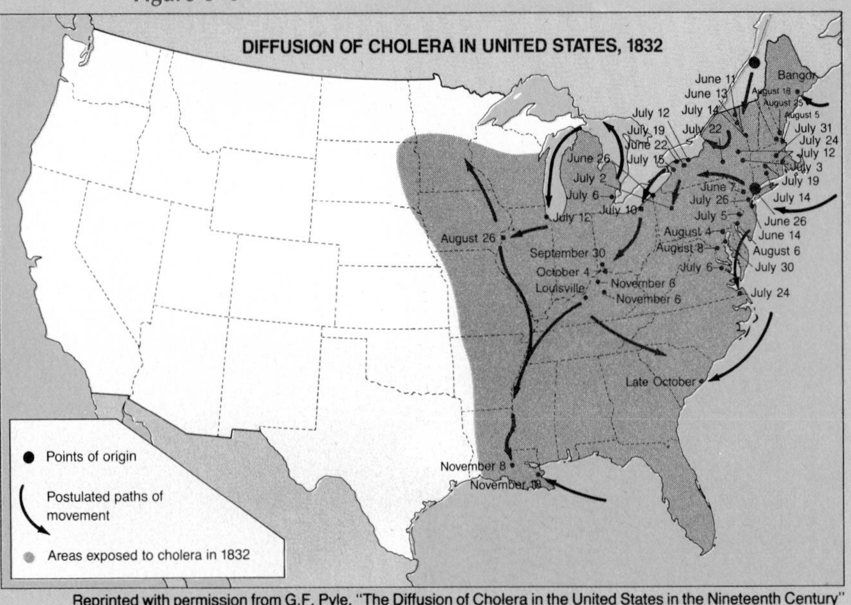


Figure 6–11 Reprinted with permission from G.F. Pyle, "The Diffusion of Cholera in the United States in the Nineteenth Century" *Geographical Analysis*, 1 January 1969, pp. 59-75. Copyright © 1969 by the Ohio State University Press.

**DIFFUSION OF THE PLANNED REGIONAL SHOPPING CENTER IN THE UNITED STATES, 1949-1968**

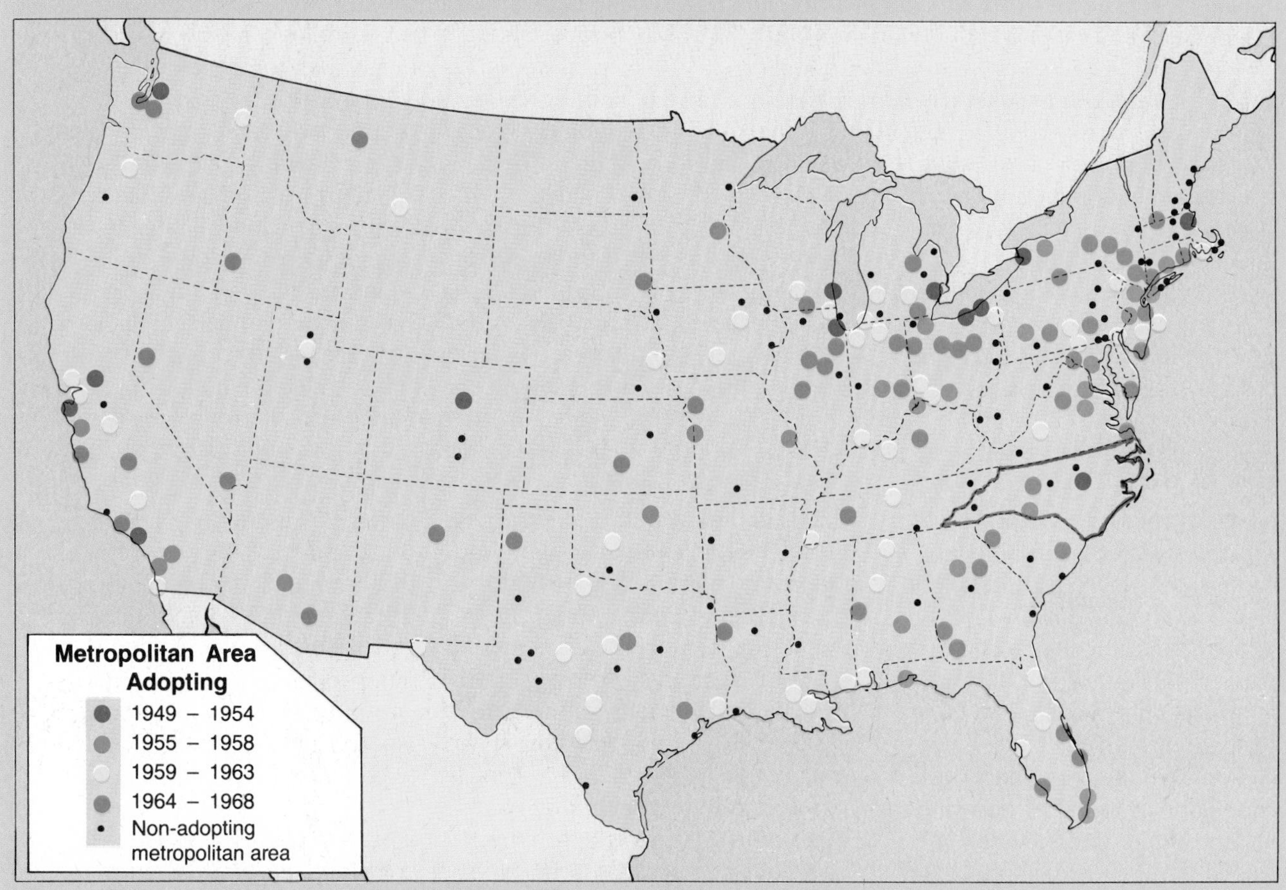


Figure 6–10

*Source:* Adapted from Cohen (1972).

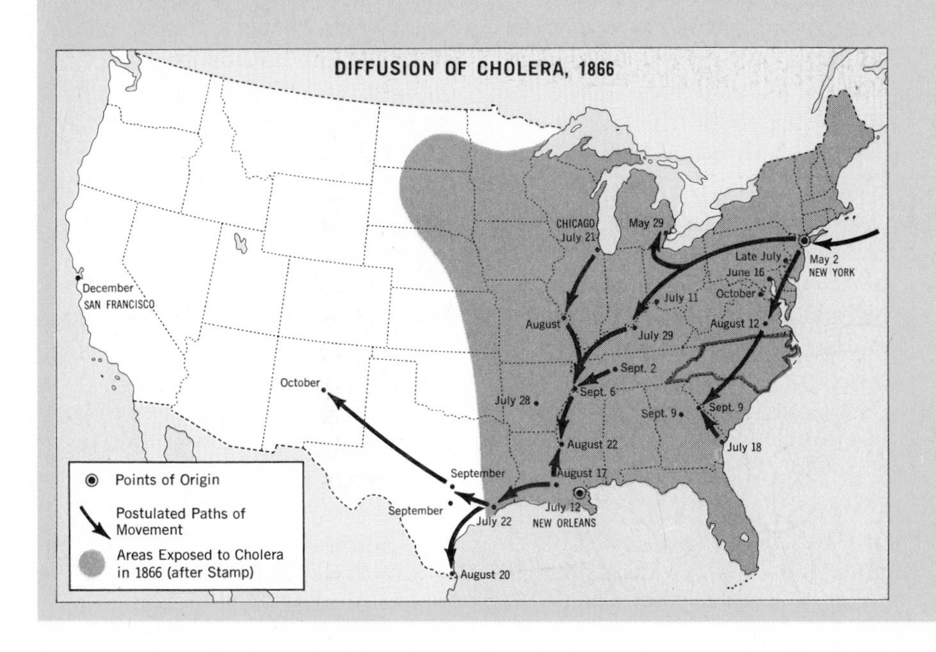

Gerald Pyle's research of nineteenth century cholera epidemics. In the 1832 map, the disease diffuses contagiously from New York State into the interior, its rate and direction of spread controlled by distance along such main water routes as the Erie Canal and the Great Lakes (the railroad had not yet arrived). The 1866 map shows a radically different picture: from a similar source area, cholera now advances far more swiftly along a maturing long-distance rail network that connects the nation's largest cities; therefore, distant Detroit receives the epidemic before nearby Baltimore, and the disease even arrives in the lower Ohio Valley before it reaches Virginia. These maps, thus, provide "before" and "after" snapshots vis-à-vis the emergence of the U.S. national urban system, which became firmly established around the middle of the nineteenth century.

These carrying processes of diffusion notwithstanding, the spreading of innovations is also affected by barriers of various kinds. In extreme cases, these may be absorbing barriers in which the impulse is completely halted; more commonly, spreading innovations encounter permeable barriers that slow down or "filter" the diffusion wave. A physical barrier is a good example: a mountain range or unbridged river will act in this way because communication across such an obstacle is more difficult. Cultural barriers function similarly, often impeding the adoption of innovations from other cultures; where internal linguistic or religious differences exist, diffusion inside a country may also be shaped by cultural barriers. Among the other kinds of diffusion barriers that occur are psychological barriers—in every population, no matter how culturally homogeneous, one finds avant-garde groups that readily adopt new things as well as laggards who only accept changes after everyone else has.

lenge; as soon as Arab power showed signs of weakening (as early as the eleventh century), the first of many crusades by Christians to the eastern Mediterranean took place. The aftermath of that contest left parts of the Middle East converted to Christianity; still, today, about half of the population of Lebanon adheres to a Christian faith. Eight centuries later, Judaism regained territorial expression as the state of Israel. The city of Jerusalem is a holy place for Jews, Moslems, and Christians alike (Fig. 6–1); fragmented and embattled, it personifies the region's divisions.

But the realm is vast, and adherents to faiths other than Islam form only small minorities. Most of the boundaries on the map of the Middle East today (Fig. 6–12) were established after the last of the great Islamic empires collapsed. The Moslems' last hurrah in Europe came when the Ottoman Empire, centered on the Turkish city of Constantinople (now Istanbul), extended its power over the Balkans and beyond. Greece, Bulgaria, Albania, Yugoslavia, Hungary, Romania, and the Black Sea area fell to the sultans, whose empire also extended far to the west along Mediterranean shores and eastward into Persia. As the twentieth century opened, the Ottoman Empire was already decaying, and when the Turks chose the (losing) German-Austrian side in World War I, its fate was sealed.

From the fragments of the Ottoman Empire, the political divisions of the Middle East emerged. Turkey survived as a discrete entity, but other areas were assigned to the victorious European powers. Syria and Lebanon went to France, adding to the colonies the French had already acquired in Algeria, Tunisia, and Morocco. Palestine wound up in the hands of the British who also administered Iraq, Transjordan, Sudan, and Egypt. The Italians established a sphere of influence in Libya and along the lower Red Sea coast in Eritrea; they also controlled part of Somalia in Africa's Horn. A combination of European imperialism and the final breakdown of Moslem power, thus, produced a mosaic of political boundaries to suit colonial powers, not Arab interests.

As Fig. I–14 (p. 27) underscores, this realm's population of 300 million is clustered, fragmented, and strung out in narrow coastal belts and confined, crowded valleys and oases. It is useful to compare Fig. 6–12 and Fig. I–14, for it is clear that thousands of kilometers of political boundaries in North Africa and Southwest Asia lie in virtually uninhabited terrain. Geographers classify boundaries on various grounds, one of which is morphological: Does the boundary coincide with a cultural break or transition in the landscape? Does it coincide with a physical feature such as a river or the crest of a mountain range? Is the boundary visibly straight? As our maps show, North Africa contains a framework of straight-line boundaries, long segments of which lie in the nearly empty Sahara. These are *geometric* boundaries (Egypt in Africa is enclosed by such borders), drawn to coincide with lines of latitude or longitude (or at an angle, from point to point), mostly through territory over which, so it seemed, no one would ever quarrel. However, that was before some of the region's oil reserves were discovered! Other boundaries, such as the Sudan-Ethiopia border along the foot of the Ethiopian Highlands and the Israel-Jordan boundary along the Jordan River, coincide with physical features and are *physiographic* boundaries. Still, another group of boundaries recognize changes in the cultural landscape. Essentially, the borders between Israel and its Arab neighbors are cultural dividing lines, notwithstanding the large Arab minority within Israel's territory. The boundary between Arab Iraq and Persian Iran may be seen on language maps as well as political maps and is of the *anthropogeographic* type. Boundaries will be classified further in the Systematic Essay that opens Chapter 10.

The North African-Southwest Asian realm may also need some boundaries it does not have. Various proposals to create a state for Palestinians displaced by the creation of Israel have been made; partition has long been discussed as a solution for strife-torn Lebanon; and Moslem Eritreans and Christian Ethiopians are locked in a conflict that may be solved only by the resurrection of an old colonial boundary.

# Arabian Oil Bonanza

The political fragmentation of this realm into two dozen countries has major ramifications in relation to the distribution of its leading export commodity: oil. Fig. 6–13 displays recent estimates of the world's fossil fuel reserves. A majority of this realm's countries possess substantial oil and natural gas reserves, notably the states and emirates bordering the Persian Gulf as well as Libya and Algeria in North Africa. As new discoveries are made, estimates change, but, in combination, the states of North Africa and Southwest Asia still probably contain over 50 percent of the world's petroleum reserves. Oil has become the realm's most valuable export commodity, providing Saudi Arabia and even such small countries as Kuwait with huge revenues and giving the region a new form of world influence. As the bar graph on Fig. 6–13 indicates, the great consumers of energy are not the countries of the Middle East but the nations of the developed world. Production in the United States and several European countries continues to lag behind consumption, and Western dependence on Middle East oil was underscored by the two supply disruptions that caused the energy crises of 1973–1974 and 1979.

If the oil exporting states of the North African-Southwest Asian realm were united ideologically, the advantages accruing from their petroleum wealth would be far greater, and the position of the energy-dependent Western states would be far worse. But the area's traditional disunity has once again worked to its disadvantage. Of the 13 member-states of the Organization of Petroleum Exporting Countries (*OPEC*), 8 lie in this realm, but their efforts to agree on petroleum policy and to develop an effective cartel have been only partly successful. The global recession of the early 1980s helped create an oil glut, forcing OPEC to end its upward spiral of price increases through at least the middle years of this decade.

What has been the impact of the realm's "black gold" on the countries and societies with the good fortune to possess substantial oil reserves? In small states such as Kuwait (population: 1.9 million), a true transformation has taken (and continues to take) place; in the 1970s, this country possessed the world's highest per capita income. In the 1980s, that position has been taken over by the United Arab Emirates at the other end of the Persian Gulf. Formerly known as British-administered "Trucial Oman" and "The Seven Sheikdoms," this tradition-bound country is today a union of seven emirates on the Persian Gulf coast of the Arabian Peninsula: Abu Dhabi, Dubai, Ajman, Sharjah, Umm al-Qaiwan, Ras al-Khaimah, and Fujairah. By far the largest in area of the United Arab Emirates is Abu Dhabi, with nearly 68,000 square kilometers (26,000 square miles) and a 1985 population of about 500,000. The seven emirates' total area is only 84,000 square kilometers (32,000 square miles); total population (permanent and transient) is about 1.7 million. Oil income is enormous, in 1983 averaging $28,110 per person (a figure vying with that of neighboring Qatar for highest on earth).

In countries with larger populations, the picture is quite different. In Iran, with a population of 45 million, the deposed Shah's government poured huge sums of its enormous petroleum-derived income into a major military complex. Sizable investments also allegedly disappeared into the personal coffers of the Shah himself and into the accounts of those in favor. The funds that remained were, nevertheless, sufficient to support a number of major development programs, including an export-oriented group of industries, a series of reforms in agriculture, and an attack on illiteracy. Modernization during this period made its mark on the capital, Tehran, and on other cities (Abadan has long been the oil refinery center and export point). But, in the countryside, there was much less evidence of progress, especially in the east and the northwest. By some counts, Iran in the last year of the Shah's rule had about 60,000 villages. Of these, only some 3000 had running water provided; in the remainder, the old system—wells and pumps—still prevailed. Because Iran has been in near chaos since the 1979 revolution, it is hard to assess the changes that have occurred since the Ayatollah Khomeini came to power (but it is hardly likely that the lot of

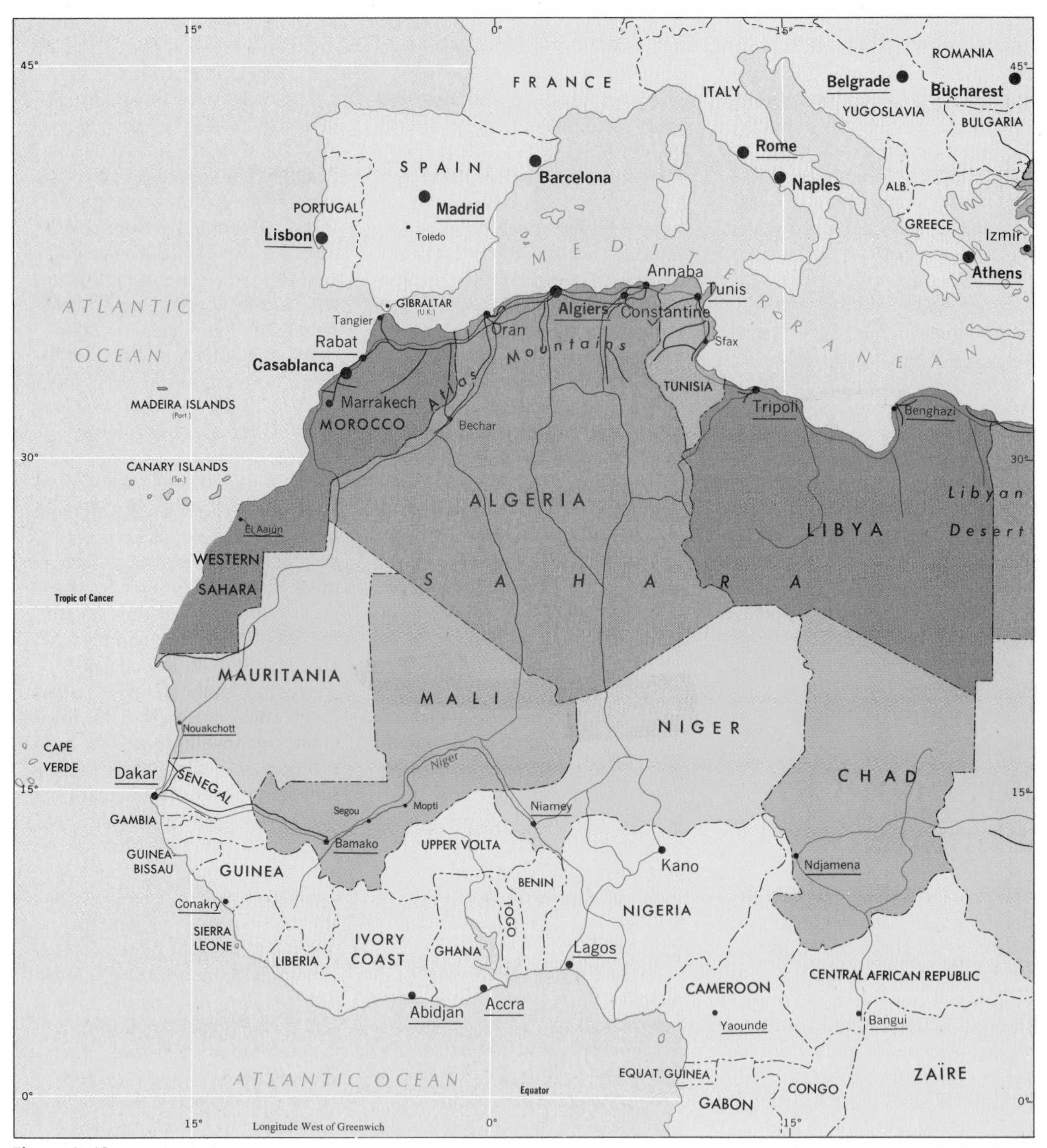


Figure 6–12

the average Iranian has improved).

Neither has Iran, with its 15 million inhabitants, found its petroleum income to be a shortcut to general national welfare. To modernize tradition-bound agriculture is an enormously costly proposition, and Iraq has lacked what Iran's authoritarian government did achieve until 1979: political continuity and stability. Internal political struggles, corruption and factionalism, together with urban-oriented investment policies that persuaded hundreds of thousands of villagers to leave the countryside and come to Baghdad, deprived Iraq of much of the development its oil revenues should have brought. Moreover, a costly stalemated war between Iraq and Iran, which began in late 1980, has also retarded progress in both countries.

Figure 6–14 shows a system of oil and gas pipelines and export locations that very much resembles the exploitative railroads in a min-

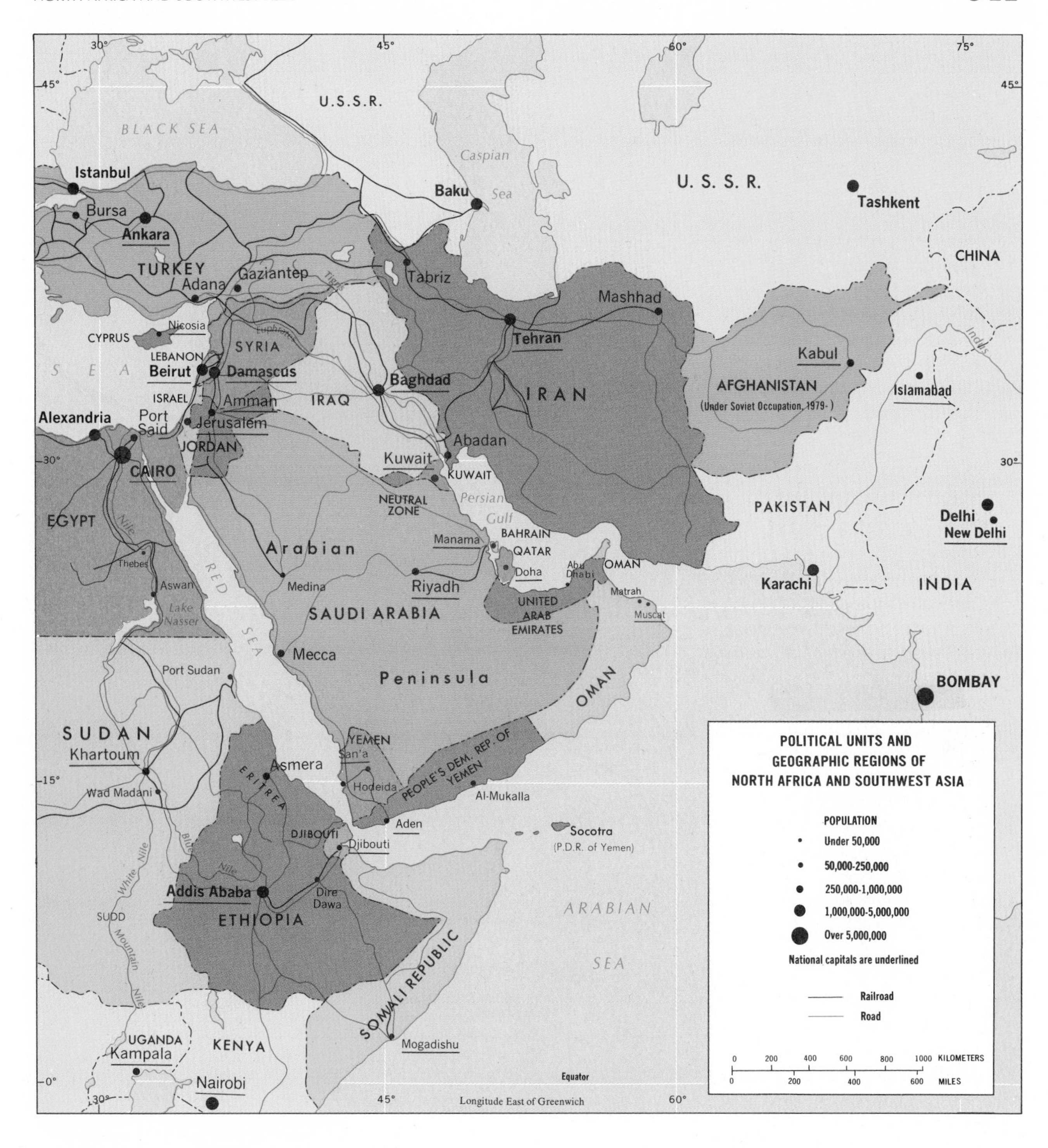


eral-rich colony, a pattern that spells disadvantage for the exporter whether colony or independent country. In North Africa, Libya again reflects the advantages of scale: its mere 3.9 million people have felt the impact of their oil bonanza far more strongly than neighboring Algeria's 22 million, and the Libyan government has used its oil revenues to gain a measure of political influence in the realm far beyond its modest dimensions.

Oil-derived wealth has transformed parts of the Arab world, and it has brought the twentieth century into sometimes rude contact with local traditions. High-rise buildings and luxury automobiles vie for space with mud-walled huts and donkey carts. In the desert, refineries shimmer in sun-drenched expanses, which until recently were disturbed only by a passing camel caravan. In Fig. 6–15, the Petromin Refinery in Saudi Arabia produces petroleum destined for Japan, while Japanese technicians help build the country's petrochemical industry. It is also evident that the impact of the realm's oil wealth is felt most strongly in the less populous oil-rich countries than in the larger states. Furthermore, petroleum revenues have had the effect of intensifying urban dominance and regional disparities, even where, as in Saudi Arabia, a conservative government implements national-development programs funded by oil revenues and designed to limit the abrasive effects of modernization.

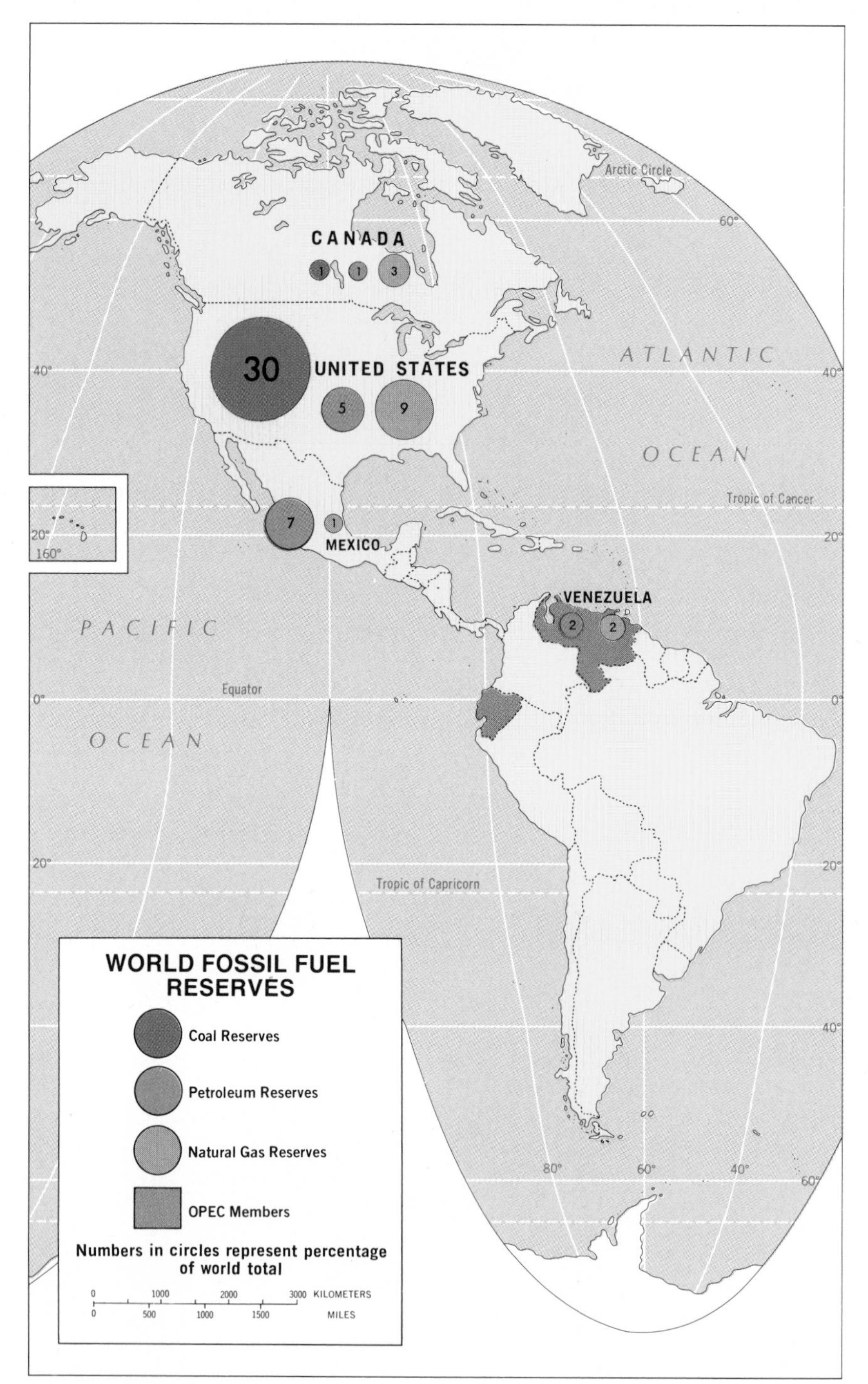


Figure 6–13

## Supranationalism in the Middle East

Their many divisions and quarrels notwithstanding, countries of North Africa and Southwest Asia have forged some effective international organizations. One of these is the League of Arab States founded in 1945 "to strengthen relationships between members and to promote Arab aspirations." The Arab League, today, counts 21 members: Algeria, Bahrain, Djibouti, Iraq, Jordan, Kuwait, Lebanon, Libya, Mauritania, Morocco, Oman, Palestine Liberation Organization (PLO), Qatar, Saudi Arabia, Somalia, Sudan, Syria, Tunisia, United Arab Emirates, North Yemen (Arab Republic), and South Yemen (People's Democratic Republic).

Member states have sometimes been in conflict. Since 1976, Morocco and Algeria have disputed (and warred) over the future of former Spanish Sahara (following partition between Morocco and Mauritania, all of Spanish Sahara became a Moroccan area); and Syria and Iraq have argued over the sharing of waters of the Euphrates River. The League has been helpful in at least the temporary settlement of several problems; in the late 1970s, a combined Arab League armed force entered Lebanon to help end civil war there for a time (which soon broke out again following the 1982 Israeli invasion). In 1979, the League moved its headquarters from Cairo to Tunis, and members voted to suspend Egypt, because of its peace agreement with Israel.

In another international organization, the OPEC cartel, states from this region have taken a leading role. In 1984, OPEC's membership included: Algeria, Ecuador, Gabon, Indonesia, Iran, Iraq, Kuwait, Libya, Nigeria, Qatar, Saudi Arabia, United Arab Emirates, and Venezuela.

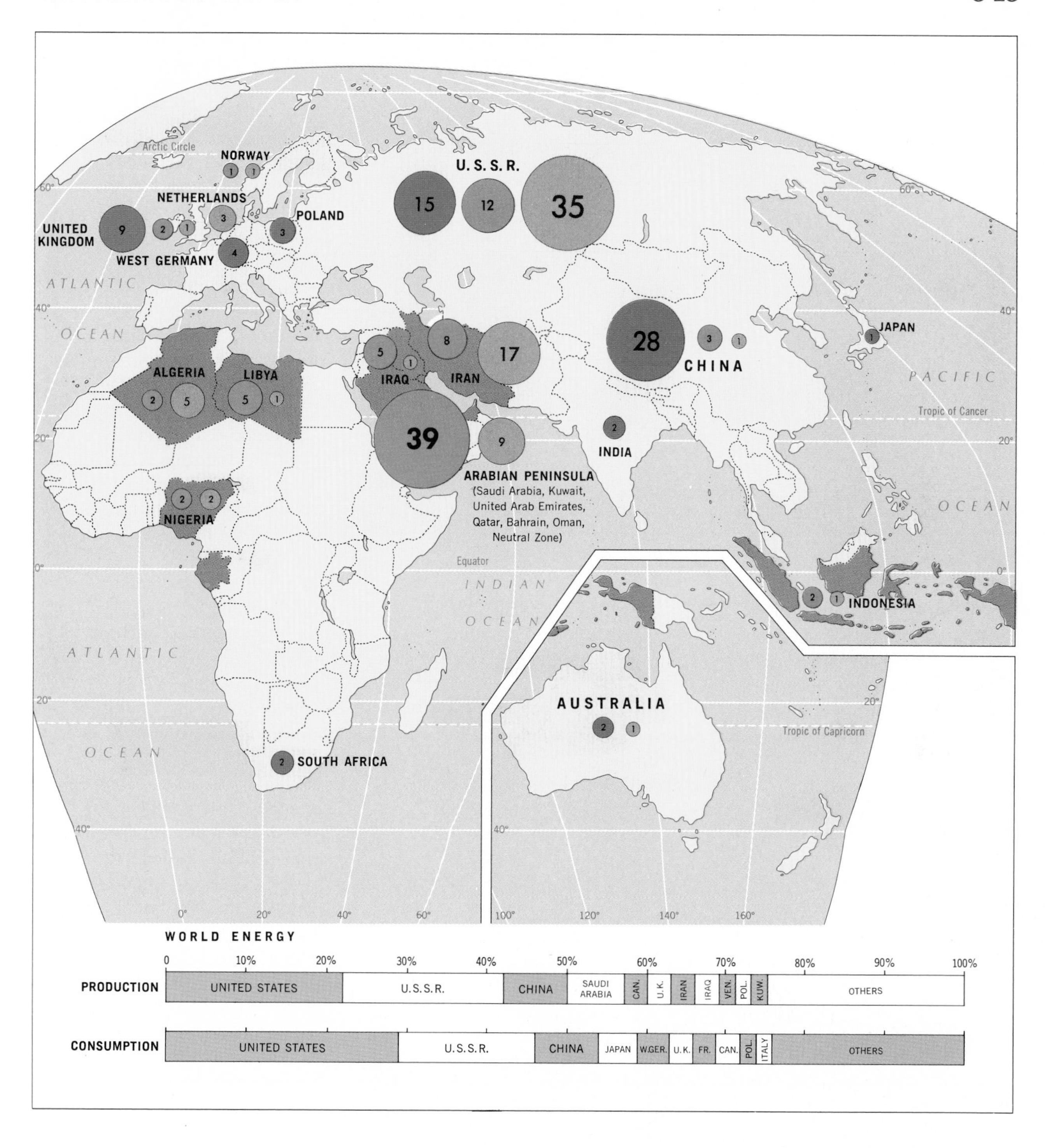


# Regions and States

This vast, sprawling realm is not easily divided into geographic regions. Not only are population clusters widely scattered, but cultural transitions—internal as well as external—make it difficult to discern a regional framework. If we constructed a series of maps based on environmental aridity, religion, ethnic units, and languages, there would be few areas of agreement. Certainly, the regionalization that follows is debatable, but it does highlight the major geographic components of the realm:

1. *Egypt and the Nile Basin.* This region constitutes, in many ways, the heart of the realm as a whole. Egypt (together with Turkey) is one of

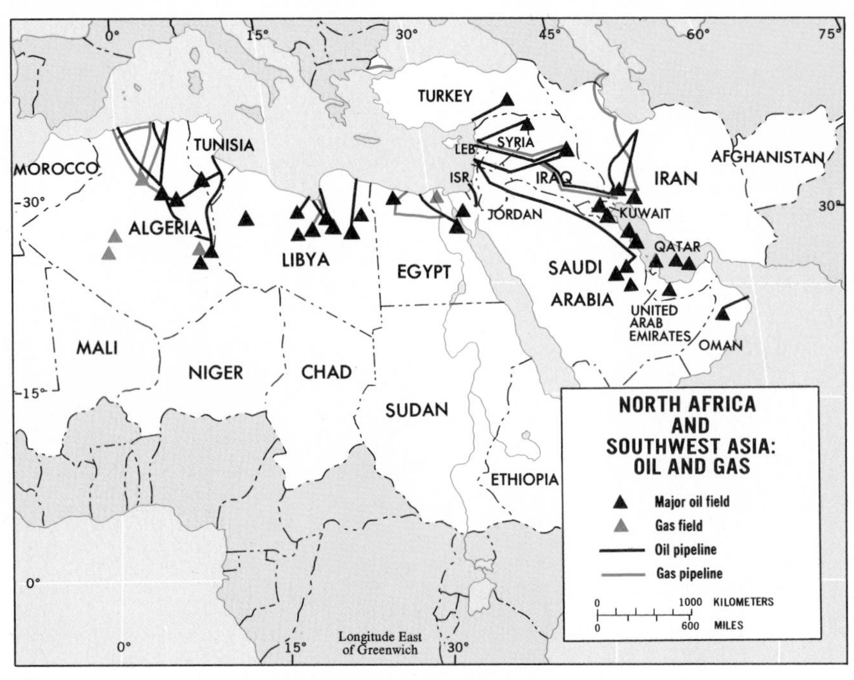


**Figure 6–14**

the realm's two most populous countries. It is the historic focus of this part of the world, a major political force. It shares with its southern neighbor, Sudan, the waters of the Nile River.

2. *The Maghreb and the West.* Western North Africa (the Maghreb) and the areas that border it also form a region, consisting of Algeria, Tunisia, and Morocco at the center and Libya, Chad, Niger, Mali, and Mauritania along the periphery. The last four of these countries lie astride or adjacent to the African transition zone where the Arab-Islamic realm merges into Black Africa.
3. *The Middle East.* This region includes Israel, Jordan, Lebanon, Syria, and Iraq. In effect, it is the zone of countries that extend from the Mediterranean Sea to the Persian Gulf.
4. *The Arabian Peninsula.* Dominated by the large territory of Saudi Arabia, the Arabian Peninsula also includes the United Arab Emirates, Kuwait, and North and South Yemen. Here lies the source and focus of Islam, the holy city of Mecca; here, too, lie many of the realm's great oil deposits.
5. *The North.* Across the northern tier of the realm lie four countries, each of which, in one way or another, constitutes an exception to the prevailing norms. Turkey, Cyprus, Iran, and Afghanistan are not Arab countries. Most Iranians adhere to a minority Islamic faith, the Shiah; these Shi'ite Moslems do not share certain important values with their Sunni contemporaries, who form the majority sect of Islam.
6. *The African Transition Zone.* From Mauritania in West Africa clear across to Ethiopia and Somalia in the east, the realm dominated by Islamic culture interdigitates with the northern region of Black Africa. No sharp dividing lines can be drawn; this is a wide zone of transition south of the Maghreb, Libya, and Egypt. As a result, this is the least well defined of the six regions of the North African-Southwest Asian realm.

**Figure 6–15**
Petromin Refinery, located on Saudi Arabia's Persian Gulf shore.

## Egypt and Its Neighbors

Egypt occupies Africa's northeastern corner, and geometric boundaries separate its 1 million square kilometers (387,000 square miles) from Libya to the west and Sudan to the south. Alone among states on the African continent, Egypt extends into Asia through its foothold on the Sinai Peninsula—a hold that has, in modern times, proved rather tenuous. Egypt lost its entire Sinai territory during the Six-Day War with Israel in 1967; then, it regained a toehold across the Suez Canal in the ensuing conflict of 1973. The years that followed saw intense diplomatic efforts to persuade Israel to yield more of the occupied Sinai to Egypt; in 1979, the terms of the Camp David peace agreement between the two states included a staged return (now completed) of Egyptian lands in the Sinai area.

Egypt now faces an adversary capable of striking into its very heartland in a matter of minutes, a totally new condition for a country protected for centuries (indeed, millennia) by desert and distance from its enemies. Herodotus described Egypt as the gift of the Nile, but Egypt also was a product of natural protection. The middle and lower Nile lie enclosed by inhospitable country, open to the Mediterranean Sea but otherwise rather inaccessible to overland contact. To the south, the upstream Nile is interrupted by a series of cataracts that begin near the present boundary of Egypt and the Sudan. To the east, the Sinai Peninsula never afforded an easy crossing, and the southwestern arm of the ancient Fertile Crescent lay some distance beyond. To the west, there is the endless Sahara Desert. The ancient Egyptians had a natural fortress; in their comparative isolation, they converted security into progress. Internally, the Nile was then what it remains to this day—the country's lifeline and highway of trade. Externally, the Egypt of the pharaohs had to trade, for there was little wood and metal, but most of this trade was left to Phoenicians and Greeks. Egypt's cultural landscape still carries the record of antiquity's accomplishments in the great pyramids and stone sculptures that bear witness to the rise of a culture hearth 5000 years ago.

**The Nile** Perhaps 95 percent of Egypt's 48 million people live within a dozen miles of the Nile River or one of its distributaries. The great river is the product of headwaters that rise in two different parts of Africa: Ethiopia, where the Blue Nile originates, and the lakes region of East Africa, source of the White Nile (the two streams join at Khartoum in Sudan). Before dams were constructed across the river, the Nile's dual origins assured a fairly regular natural flow of water, making the annual floods predictable both in terms of timing and intensity. The Nile normally is at its lowest level in April and May, but rises during July, August, and September to its flood stage at Cairo in October, when it may be more than 7 meters (23 feet) above its low stage. After a rapid fall during November and December, the river declines gradually until its May minimum.

In ancient times, this regime made possible the invention of irrigation, for the mud left behind by the Nile floods was manifestly cultivable and fertile. Eventually, a system of cultivation known as *basin irrigation* developed, whereby fields along the low banks of the Nile were partitioned off by earth ridges into a large number of artificial basins. The mud-rich river waters would pour into these basins during flood time; then, the exits would be closed, so that the water would stand still, depositing its fertile load of alluvium. Then, after six to eight weeks, the exit sluices were opened and the water drained away, leaving the rejuvenated soil ready for sowing. This method, although revolutionary in ancient times, had disadvantages—the susceptible lands must lie near or below the flood level, only one crop could be grown annually, and if the floods were less intense (as they were in some years), some basins remained unirrigated because floodwaters did not reach them. Still, traditional basin irrigation prevailed all along the Egyptian Nile for thousands of years; not until the late nineteenth century did more modern methods develop. Even as late as 1970, as much as one-tenth of all irrigation in Egypt was basin irrigation, most of it practiced in Upper Egypt (the region nearest the Sudan border).

The construction of dams, begun during the nineteenth century, made possible the *perennial irrigation* of Egypt's farmlands. By building a series of artificial barriers (with locks for navigation) across the river, engineers were able to control the floods, raise the Nile's water level, and free the farmers from their dependence on the sharp seasonal fluctuations of the river's natural regime. Not only did the country's cultivable area expand substantially, but it also became possible to grow more than one crop per year. In the early 1980s, all farmland in Egypt came under perennial irrigation, a transformation that was completed within a single century.

The greatest of all Nile control projects, the Aswan High Dam, was begun in 1958 and completed in 1971. The Dam is located some 1000 kilometers (600 miles) upstream from Cairo, in a comparatively narrow, granite-sided section of the valley. The dam wall, 110 meters (364 feet) high, creates Lake Nasser, one of the largest artificial lakes in the world (Fig. 6–16). This

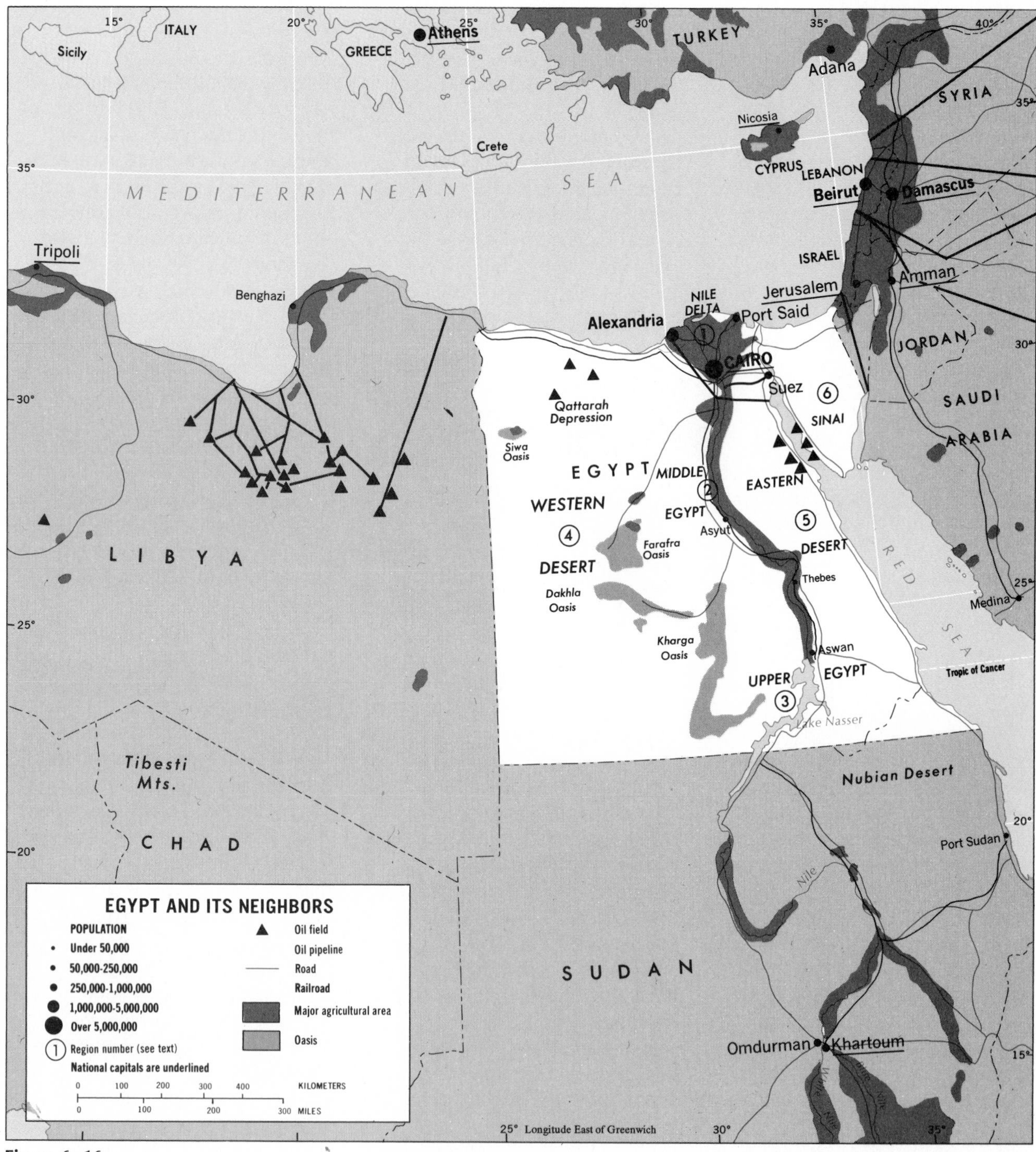


Figure 6–16

reservoir inundates well over 500 kilometers (300 miles) of the Nile's valley not only in Upper Egypt, but in Sudan as well, and the cooperation of this southern neighbor was required since some 50,000 Sudanese had to be resettled. The impact of the Dam on Egypt's cultivable area was momentous: before construction, the Nile's waters could irrigate some 2.53 million hectares (6.25 million acres) of farmland. To this, the Aswan High Dam has added 550,000 hectares (1,360,000 acres); moreover, nearly 400,000 hectares (1 million acres) of farmland in Middle and Upper Egypt that were still under basin irrigation could now be converted to perennial irrigation, resulting in increased crop yields. And the dam also supplies Egypt with about 50 percent of its energy requirements.

Egypt has often been described as one elongated oasis, and any map of its human geography confirms the appropriateness of that description in a country where all of the people are concentrated on 4 percent of the land area. In Upper and Middle Egypt, the strip of abundant, intense green, 5-to-25-kilometers (3-to-5-miles) wide, lies in stark contrast to the barren, harsh, dry desert immediately adjacent, a reminder of what Egypt would be without its lifeline of water. But north of Cairo, the great river fans out across its wide delta—160 kilometers (100 miles) in length and 250 kilometers (155 miles) in width—along the Mediterranean coast between Alexandria and Port Said. In ancient times, the river's waters reached the sea via several channels, and the delta was flood-prone, inhospitable country; ancient Egypt was Middle and Upper Egypt, the Egypt of the Nile Valley. But, today, the delta waters are diverted through two controlled channels, the Rosetta and Damietta distributaries. The delta contains twice as much cultivable land than Middle and Upper Egypt combined; it also contains some of Africa's most fertile soils and is able to support almost half of Egypt's population. But the increasing use of Nile waters upstream and the diminished flow of the river (and its vital supply of silt) now pose a threat to all the progress that has been made in the delta. There is a danger that brackish water will invade the channels from the sea, seep into the soil, and, once again, reduce large areas to an inhospitable and unproductive state. The consequences of such a sequence of events would, of course, be an environmental disaster of great magnitude.

**A pivotal location** Egypt has changed substantially in modern times; it is a more populous, more highly urbanized counry today than ever before. But Egypt's farmers, the *fellaheen*, still struggle to make their living off the land, as did the peasants in the Egypt of 5000 years ago. In a society that at one time was the source of countless innovations and the testing ground for many others, the tools of the farmer often are as old as any still in use—the hand-hoe, the wooden buffalo-drawn plow, and the sickle. Water is still drawn from wells by the wheel and bucket, and the hand-turned Archimedean screw (to move water over riverbank ridges) can still be seen in service. The peasant still lives in a small mud dwelling, which would look very familiar to the farmer of many centuries ago; neither would that distant ancestor be surprised at the poverty, recurrent disease, high infant mortality rates, and the lack of tangible change in the countryside. Despite the Nile dams and irrigation projects, Egypt's available farmland per capita has steadily declined during the past two centuries as rapid population growth keeps nullifying gains in crop productivity.

Yet, of all the countries in the region loosely termed the Middle East, Egypt is not only the largest in terms of population, but also, by many measures, the most influential. Its rival, Turkey, has in recent decades oriented its attentions and energies principally toward Europe; in Africa, Egypt has no Islamic competitors to match its status. Indeed, Egypt is continental Africa's second most populous state, after Nigeria. Thus, Egypt alone is spatially, culturally, and ideologically at the heart of the Arab world. What factors have combined to place Egypt in this position?

Location is clearly a primary element in Egypt's eminence. The country's position at the southeastern corner of the Mediterranean Sea presented an early impetus as Phoenicians and Cretans linked the Nile Valley with the east-coast Levant and other parts of the Mediterranean Basin. Egypt lay protected but immediately opposite the Arabian Peninsula; in the centuries of Arab power and empire-building, it sustained the full thrust of the new wave. The Arab victors founded Cairo in A.D. 969 and made it Egypt's capital (Alexandria, founded in 332 B.C., had functioned as the capital for centuries); they selected a fortuitous site at the place where the Nile Valley opens onto the delta.

Egypt, too, lies astride the land bridge between Africa and Southwest Asia (and between the Mediterranean and the Red Seas). In modern times, this became the major asset to the country as the Suez Canal, completed in 1869, became the vital, bottleneck link in the shortest route between Europe and South and East Asia. Built by foreign interests and with foreign capital, the Suez Canal also brought a stronger foreign presence to Egypt. Until 1922, Egypt was a British protectorate; even when Egyptian demands for sovereignty led to the later establishment of a kingdom, British influence remained paramount and the Suez Canal a foreign operation.

The 1952 uprising against King Farouk and the proclamation of an Egyptian Republic in the following year presaged the showdown over the Suez Canal that was now all but inevitable. In 1956, President Nasser announced the nationalization of the canal and the expropriation of the company that had controlled it; three months later, Israel, Britain, and France undertook a miscalculated and strategically disastrous invasion of the Sinai Peninsula. Pressured and pushed into retreat, the invaders soon abandoned the canal area, and Egyptian stature in the Arab world (and in the Third World generally) was immeasurably enhanced.

The pivotal location of Egypt was expressed in other ways as well. To the west, Algeria was waging a war of liberation against French colonialism, and no Arab country was in a better position to support this campaign than Egypt. To the south, the relationship between Sudan and Egypt had long been strained, but, in 1953, the new Egyptian government offered Sudan the option of retaining a political tie with Cairo or full sovereignty; Sudan became fully independent in 1956. And to the east, beyond the Suez Canal and Sinai, lay Israel, its southern flank exposed to the Arab world's strongest partner.

**Subregions** As Fig. 6–16 reveals, Egypt contains six subregions: (1) the Nile Delta, or Lower Egypt; (2) Middle Egypt, consisting of the Nile Valley from Cairo to Thebes; (3) Upper Egypt, the Nile Valley from Thebes to the Sudanese border, including Lake Nasser; (4) the Western Desert, including several large oases; (5) the Eastern Desert and Red Sea coast; and (6) the Sinai Peninsula. The great majority of the Egyptian people live in Lower and Middle Egypt, which are highlighted here.

The Nile Delta (see *Box*, p. 348–349) covers an area of just under 25,000 square kilometers (10,000 square miles). For thousands of years, only the land nearest the Nile Valley was farmed, for the main part of the delta was flood-prone, sandy, lagoon-infested, and excessively salty; seven Nile distributaries found their way to the Mediterranean Sea. Today, however, the Nile channels are controlled, as modern engineering converted the region into one of perennial irrigation, fertile farmland, and multiple cropping of rice, the staple, and the cultivation of cotton, the chief cash crop. The completion of the Aswan Dam now makes possible the reclamation and eventual cultivation of an additional 400,000 hectares (1 million acres) of delta land; by 1990, it is likely that fully half of Egypt's population will reside in this region.

The urban focus of Lower Egypt is Alexandria, Egypt's leading seaport, containing 3.2 million residents. Industrial growth accelerated during the period between the two World Wars (today, Alexandria is Egypt's leading industrial center), and during the royal period the city served as second capital. Today, Alexandria has a canal connection to the Nile as well as fine road and rail connections to Egypt's heartland. The city has also become a resort, and, of course, it has shared in the contemporary Third World migration toward burgeoning urban areas.

Middle Egypt begins where the Nile Delta ends—at Cairo, the capital and interface between valley and delta. This primate city's location is quintessentially nodal: its delta connections are good and a railroad extends along the entire length of Middle Egypt and beyond to Aswan. A string of towns lies along this route, and centrally positioned Asyut is Middle Egypt's subregional focus.

## REMOTE SENSING

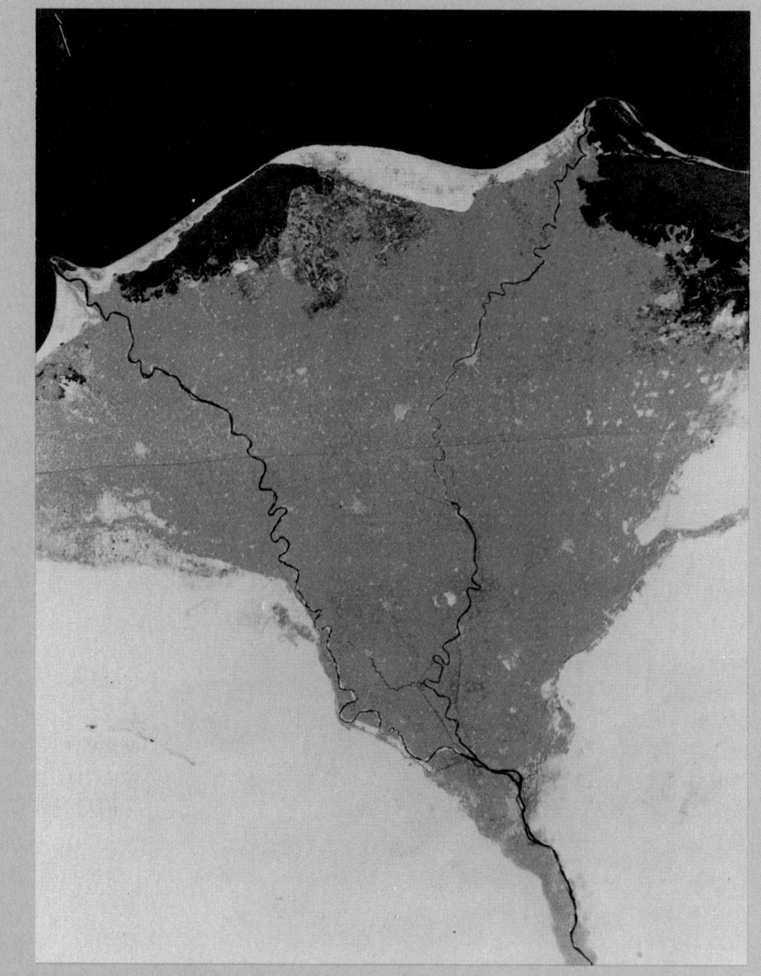

Figure 6–17
A satellite view of the Nile Delta, with each color indicating a specific land-use.

Geographers have for centuries been improving their methods of observing and interpreting the earth's surface. The achievement of aerial photography within the past century has enabled analysts to study all kinds of visible-light patterns that are reflected by ground objects. In the last two decades, new technologies have greatly and rapidly expanded the utilization of this tool for monitoring the environment and human spatial organization. The term now used to refer to these techniques is **remote sensing**, which Arthur and Alan Strahler (in their book *Modern Physical Geography*) define as the gathering of geographic information from great distances and over broad areas, usually through instruments mounted on high-flying aircraft or space vehicles, like the Landsat satellites.

Highly sophisticated instruments called remote sensors, which are far more sensitive and versatile than conventional cameras, collect the necessary data. They are designed to detect both nonvisible and visible light across the entire electromagnetic spectrum as emitted from the surface, and many of today's remote sensing installations use multispectral scanners (including radar) that are capable of automatically producing color imagery for detailed interpretation by analysts. Since each surface feature emits its own *spectral signature* (unique waveband pattern), it is now possible to identify everything from a blacktop parking lot to a cluster of desert shrubbery down to the size of a small city block. Hundreds of spatial phenomena around the globe are being monitored by remote sensing today, among them soil and vegetation change, crop health (diseased fields have different temperatures than normal ones and can easily be detected), air pollution, land-use change, population shifts, and heat losses by buildings.

A pertinent example of remotely sensed imagery is Fig. 6–17, which shows the Nile Delta in the mid-1970s from the Landsat satellite 900 kilometers (560 miles) above the earth. The well-watered delta (red-orange) is easily distinguished from the surrounding desert (light tan); within the delta region, population centers are clearly visible in solid gray—the Cairo metropolis is at the southern apex of the delta—as is the variation of crop types shown by the different shadings of red-orange. Since comprehensive geographic investigations also include information on the landscape of the study area—so-called *ground truth*—the satellite imagery should be compared to the view of the delta's surface provided in Fig. 6–5.

The metropolitan region of Cairo in 1985 had about 8.8 million residents, and the city's rapid growth rate of recent decades continues unabated. Cairo itself is one of the 20 largest urban centers in the world, and it shares with other large cities of the underdeveloped world the problems of crowding, sanitation, health, and education of its huge numbers. Cairo is a city of stunning contrasts. Along the Nile waterfront, elegant hotel-skyscrapers rise above surroundings that are frequently Parisian and carefully manicured in appearance. But beyond this facade lies the maze of depressing ghettoes, narrow alleys, overcrowded slums (Fig. 6–18), and a low skyline dominated by a mass of minarets, the towers of mosques pointing skyward. Hundreds of architectural achievements, many of them mosques, shrines, and tombs, are scattered throughout Cairo, but, seemingly everywhere, one encounters the mud huts and hovels of the very poor. Still, Cairo is a truly cosmopolitan center, the cultural capital of the Middle East and Arab world, with a great university, magnificent museums, renowned zoological and botanical gardens, a symphony orchestra, national theater, and opera. Although Cairo has always been primarily a center of government and administration, it is also a river port and industrial city (textiles, food processing, and iron and steel production). Countless thousands of small handicraft industries exist in the traditional neighborhoods of the city, and the grand *bazaar* throbs daily with the trade in small items.

From the waterfront skyscrapers of Cairo to the monuments of Thebes, ancient capital of the pharaohs, the Nile River gives life to the attenuated oasis that is Middle Egypt. In places, the walls of the valley approach the river so closely that the farmlands are interrupted, but, along most of the river's course in Middle Egypt, the strip of cultivation is between 8-to-16-kilometers (5-to-10 miles) wide. But everywhere, the sharply outlined contrast between luxuriant green and barren desert is intense and startling (see Fig. I–1). Clover (*berseem*, a fodder crop), cotton, corn, wheat, rice, millet, sugarcane, and lentils are among the crops that thrive on fields now under perennial irrigation.

Despite the expansion of its irrigated farmlands, Egypt, nonetheless, must import food to balance local diets. Although the birth rate has declined somewhat since the 1960s, development gains are still being offset by faster population growth: land reform, increased yields, and expanded cultivation are simply not enough to overcome Egypt's greatest obstacle to economic advancement. Egyptian planners have tried to stimulate industry to help the country escape from its

**Figure 6–18**
A street scene in one of inner Cairo's less prosperous neighborhoods.

economic treadmill; the most ambitious project is the steel plant at Hulwân (Helwân) near Cairo, based on the iron deposits found about 50 kilometers (30 miles) west of Aswan, manganese from the Eastern Desert, and local limestone. A search for mineral deposits has yielded some petroleum reserves in the Eastern Desert, in Sinai, and near the Libyan border. But income per person in the early 1980s was still under $500 annually, and the late President Sadat's policy of accommodation with Israel and alignment with the U.S. (continued by his successor) entails certain political risks. Egypt has, at least, had years of political stability—the smooth transition following the 1981 assassination of Anwar Sadat proved that; if order were to break down, all hope that a development takeoff might occur would vanish.

## North Africa: The Maghreb

West of Egypt lies Libya, beyond is the *Maghreb*, the western region of the Arab realm. The countries of Northwest Africa are collectively called the Maghreb, based on the Arab term meaning *western isle;* this suggests the physiographic basis of the region, with the great Atlas Mountains rising like a vast island from the waters of the Mediterranean Sea and the sandy flatlands of the Sahara. Taken together, the four states of Morocco, Algeria, Tunisia, and Libya have a population only slightly larger than Egypt's. In 1985, Morocco's population was estimated to be 24 million; Algeria, the largest Maghreb state territorially, had 22 million inhabitants; Tunisia, the smallest country in North Africa, had just over 7 million people; and Libya's population was about 4 million.

Whereas Egypt is the gift of the Nile, the Atlas Mountains form the nucleus of the settled Maghreb. The high ranges wrest from the rising air the orographic rainfall that sustains life in the valleys intervening—valleys that contain good soils and sometimes rich farmlands. From the area of Algiers eastward along the coast into Tunisia, annual rainfall averages in excess of 75 centimeters (30 inches), a figure more than three times as high as is recorded at Alexandria in Egypt's delta. Even 240 kilometers (150 miles) into the interior, the slopes of the Atlas still receive over 25 centimeters (10 inches) of rainfall—the effect of the topography can be read on the map of precipitation (see Fig. I–9). Where the Atlas terminates, desert conditions immediately begin.

The Atlas Mountains are structurally an extension of the Alpine system that forms the orogenic backbone of Europe, of which Switzerland's Alps and Italy's Appennines are also parts. In Northwest Africa, these mountains trend southwest-northeast and commence in Morocco as the High Atlas, with elevations in excess of nearly 4000 meters (13,000 feet). Eastward, two major ranges appear that dominate the landscapes of Algeria proper: the Tell Atlas to the north, facing the Mediterranean Sea, and the Saharan Atlas to the south, overlooking the Desert. Between these two mountain chains, each consisting of several parallel ranges and foothills, lies a series of intermontane basins (analogous to the Andean altiplanos but at lower elevations), markedly drier than the northward-facing slopes of the Tell Atlas. In these valleys, the rain shadow effect of the Tell Atlas is reflected not only in the steppelike natural vegetation, but also in land-use patterns: pastoralism replaces cultivation and stands of esparto grass and bush cover the countryside.

The countries of the Maghreb (Fig. 6–19) sometimes are referred to as the Barbary states, in recognition of the region's oldest inhabitants, the Berbers. The Berbers' livelihoods (nomadic pastoralism, hunting, some farming) changed as foreign invaders entered their territory, first the Phoenicians and then the Romans. The Romans built towns and roads, laid out farm fields and irrigation canals, and introduced new methods of cultivation. Then came the Arabs, conquerors of a different sort. They demanded the Berbers' allegiance and their conversion to Islam, radically changed the political system, and they organized an Arab-Berber alliance (the *Moors*) that pushed across the Straits of Gibraltar into Iberia and colonized a large part of southern Europe. After the Moors' power declined, the Ottoman Turks established a sphere of influence along North Africa's coasts. But the most pervasive foreign intervention came during the nineteenth and twentieth centuries when the European colonial powers—chiefly France, but also Spain—established control and (in the now-familiar sequence of events) delimited the region's political boundaries.

During the colonial era, nearly 1.4 million Europeans came to set-

Figure 6–19

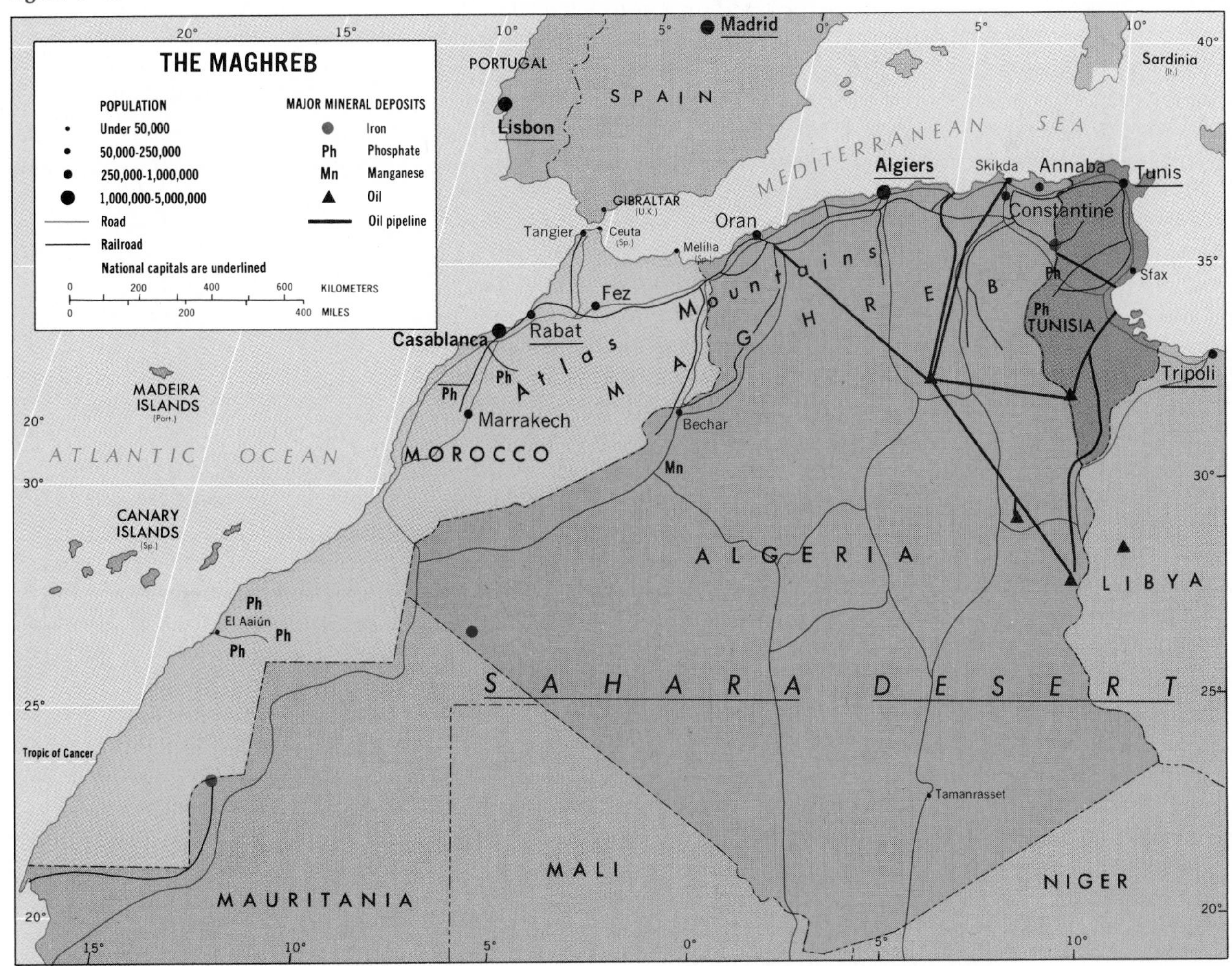

tle in North Africa—most of them French, and a large majority to Algeria—and these immigrants soon dominated commercial life. They stimulated the renewed growth of the region's towns, and Casablanca (2.7 million), Algiers (1.7 million), and Tunis (1.2 million) rose to become the foci of the colonized territories. Although the Europeans dominated trade and commerce and integrated the North African countries with France and the European Mediterranean world, they did not confine themselves to the cities and towns. They recognized the agricultural possibilities of the favored parts of the *tell* and established thriving farms. Agriculture here, naturally, is of the Mediterranean variety: Algeria soon became known for its vineyards and its wines, its citrus groves, and its dates; Tunisia has long been the world's leading exporter of olive oil; Moroccan oranges went to many European markets. And staples such as wheat and barley were also among the exports.

Despite the proximity of the Maghreb to Mediterranean Europe and the tight integration of the region's territories within the French political framework, nationalism emerged as a powerful force in the region. Morocco and Tunisia secured independence mainly through negotiation; in Algeria, a costly revolutionary war began in 1954 that lasted until 1962. It was not difficult for the nationalists to recruit followers for their campaign; the justification for it was etched in the very landscape of the country, in the splendid shining residences of the landlords and the miserable huts of the peasants. But the revolution's success brought new troubles. Hundreds of thousands of French people left Algeria and the country's agricultural economy fell to pieces, making an orderly transition impossible. Productive farms went to ruin, exports declined badly and needed income was lost.

But there are some bright spots in the Maghreb. Algeria has one resource to compensate for its losses in agriculture—oil, the same oil that led the French to resist Algerian independence throughout the 1950s. Petroleum has now become the leader among Algeria's export commodities; it is extracted from the Sahara Desert and piped to Algerian as well as Tunisian ports for transhipment to Europe (Fig. 6–14). Libya's production from its Saharan oilfields has risen spectacularly to place that country among the world's leading exporters. And, quite unlike most of the remainder of the Middle East, the countries of the Maghreb have a varied set of mineral resources. Chief among these is phosphate of lime, used to manufacture fertilizer. Morocco is the world's leading exporter of this commodity; it also occurs in Algeria, and in substantial quantities in Tunisia, where it ranks as the most valuable export as well. Both Morocco and Algeria have Atlas-related iron ores, exported to the United Kingdom from their favorably located sources. Manganese, lead, and zinc are also mined in all three countries, but sufficient coal is not locally available. Despite the new sources of energy from oil and natural gas, this not inconsiderable range of minerals has stimulated little domestic industrial development.

The oil boom has not yet substantially improved life for the majority of the people, who remain undernourished, illiterate (over 60 percent), poor (about $1200 income per person annually), unemployed (over 30 percent), and far removed from the foci of change and progress. Tens of thousands of Algerians followed the European emigrants and went to France's cities in search of jobs; the great majority of those who stayed at home remained trapped in the *bidonvilles*, the poverty-stricken shantytowns that surround the Maghreb's cities.

The three countries of the Maghreb and Libya reflect the ideological division that so strongly marks the entire realm. Morocco is a kingdom, a relatively conservative force in Arab-world politics. Algeria is nonaligned, anti-imperialist, and socialist in its external and internal political orientations and a vigorous supporter of Arab causes. Tunisia, poorest of North Africa's states in some ways, became a model of stability under the leadership of its long-time president and national hero, Habib Bourguiba; this is one country where one-party democracy seems to work without the fanfare that is needed to obscure its failures elsewhere. But adjacent Libya, with its oil riches and tiny population, has become North Africa's radical state, whose unpredictable leader, Muammar Qaddafi, can afford to supply military hardware to extremist allies while engaging in political subversion in Arab countries not committed to Libyan causes. The political geography of the Arab world's western flank—which also features a continuing conflict between Morocco and Algeria over Western Sahara (former Spanish Sahara, now attached to Morocco)—is complex indeed.

## The Middle East

The Middle East, as a region within the Southwest Asian-North African realm, consists chiefly of the pivotal area positioned between Turkey and Iran to the north and east, Saudi Arabia and Egypt to the south and southwest, and the Mediterranean Sea to the west (Fig. 6–12). Five countries lie in this region: Iraq, largest in terms of population (more than 15 million in 1985) as well as territory; Syria, next in both categories; and Jordan, Lebanon, and Israel.

**Israel** Israel lies at the very heart of the Arab world. Its neighbors are Lebanon and Syria to the north, Jordan to the east, and Egypt to the southwest—all of them in some measure resentful of the creation of the Jewish state in their midst. Since 1948, when Israel was created as a homeland for the Jewish people on recommendation of a U.N. commission, the Arab-Israeli conflict has overshadowed all else in the Middle East.

Indirectly, Israel was the product of the collapse of the Ottoman Empire. Britain gained control over the Mandate of Palestine, and it became British policy to support the aspirations of European Jews for a homeland in the Middle East, embodied in the concept of Zionism. In 1946, the British granted independence to the territory lying east of the Jordan River, and "Transjordan" (now the state of Jordan) came into being. Shortly afterward, the territory west of the Jordan River was partitioned by the United Nations, and the Jewish people got the bulk of it—including, of course, land that had been occupied since time immemorial by Arab and other Semitic peoples. The original U.N. plan proposed to allot 55 percent of all Palestine to the Jewish sector, although only 7 percent of the land was actually owned by Jews (about one-third of the total population of Palestine was Jewish), but this partition plan was never implemented as it was intended. As soon as the Jewish people declared the independent state of Israel in 1948, the new country was attacked by its Arab neighbors, who rejected the scheme. In the ensuing battle, Israel not only held its own but gained some territory in the southern Negev Desert (Fig. 6–20). At the end of the first war (in 1949), the Jewish population controlled 80 percent of what had been Palestine. Of course, this success was not won by overnight organization: at the time of Israeli independence, there already were 750,000 Jews in Palestine. Indeed,

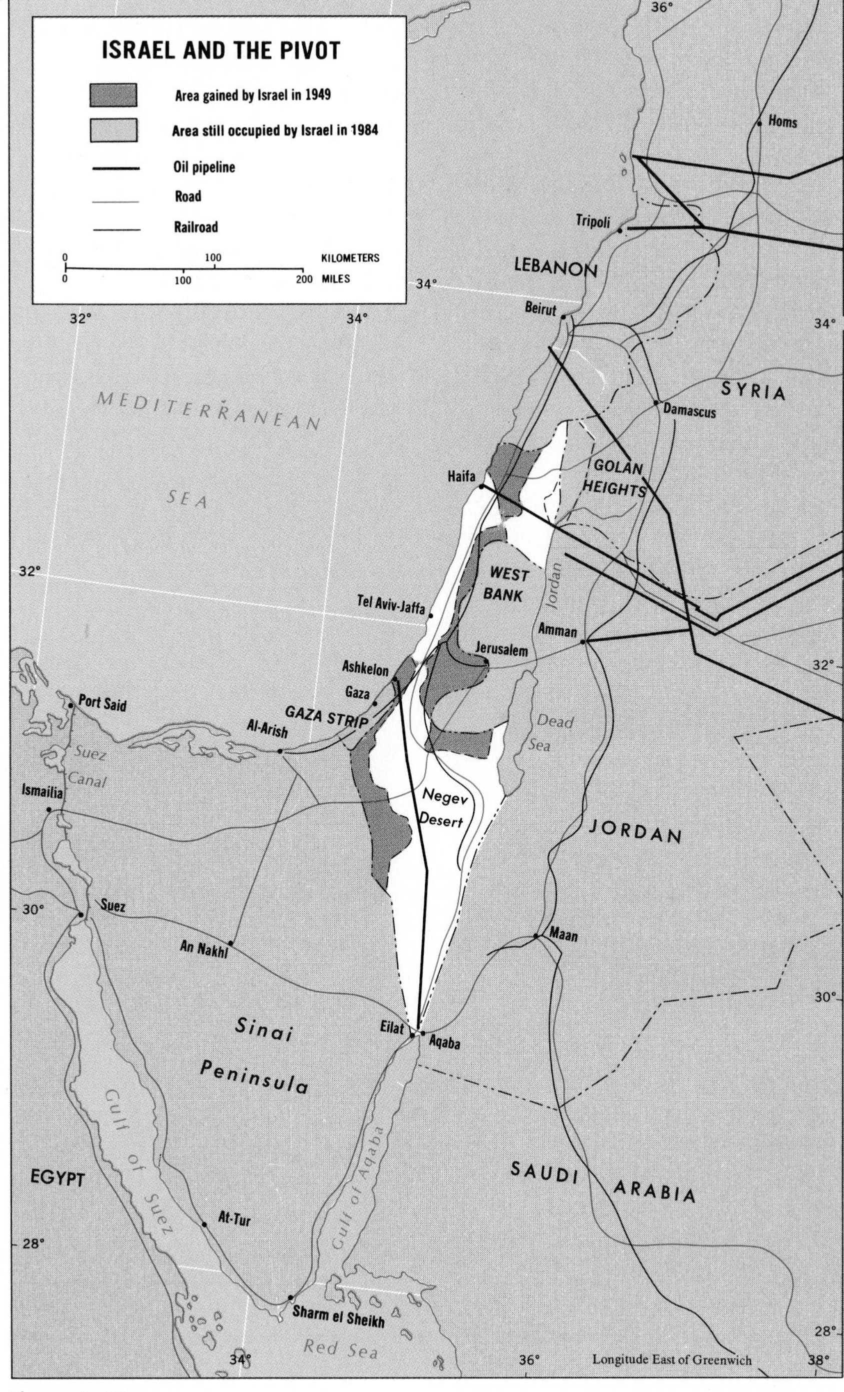

Figure 6–20

the world Zionist movement had been assisting Jews in their return to Palestine since the late nineteenth century.

For more than a third of a century now, a state of latent—and at times actual—war has existed between Israel and the Arab world. In 1967, a week-long conflict resulted in the Israeli occupation of Syrian, Jordanian, and Egyptian territory, including the Sinai Peninsula to the edge of the Suez Canal—a facility the Egyptians had not allowed Israeli vessels to use. In 1973, a brief war led to Israel's withdrawal from the Suez Canal to truce lines in the Sinai, but it also extended Israel's control over Syria's Golan Heights, creating a new source of friction. Since then, peace has been sustained largely because of Egypt's willingness to negotiate a staged Israeli withdrawal from the Sinai (completed in 1982); but the level of hostility has remained dangerously intense, as evidenced by Israel's 1982 invasion of southern Lebanon to halt terrorist attacks on its northern flank. In addition, the area is a power vacuum, subject to Soviet and U.S. moves and pressures.

So many issues divide the Jews and the Arabs that a permanent solution or even a peace imposed by the superpowers does not seem to be attainable. One prominent issue has been the refugee problem: when Israel was created, over 750,000 Palestinian Arabs were forced to leave the area to seek new lives in Jordan, Egypt, Lebanon, or Syria (see *Box*). Another problem has been the city of Jerusalem, a holy place for Jews, Moslems, and Christians. In the original U.N. blueprint for Palestine, the city was to have become internationalized; however, Jerusalem was then divided between the Jewish state and Jordan, a source of friction with neither side satisfied. But the 1967 War saw the holy city fall into Israeli hands as the entire West Bank portion of Jordan was captured; Israel has since begun to settle these lands west of the Jordan River—occupying perhaps two-thirds of the West Bank's land in the mid-1980s—scattering its new settlements throughout the area to make any future territorial concession very difficult. A further problem has been the use of the waters of the Jordan River. In 1964, Israel unilaterally diverted over half of the Jordan's flow for its own use. The river's headwaters lie in Lebanon and Syria, and these two countries have joined with Jordan in plans to divert the water completely away from its present course. But the river is crucial to Israel's economy, and the Israelis have threatened to destroy any diversion works or dams built to cut the lower Jordan's water level.

A major irritant in the whole matter has been Israel's rapid rise to strength and prosperity amid the poverty so common to the Middle East. There is nothing particularly productive about most of Israel's land, and neither is it very large (20,000 square kilometers/8000 square miles—smaller than Massachusetts). But Israel has been transformed by the energies of its settlers and, importantly, by heavy investments and contributions made by Jews and Jewish organizations elsewhere in the world, especially the United States. It is often said that in Israel desert has been converted into farmland, and it is partly so—the irrigated acreage has been enlarged to many times its

# THE PALESTINIAN DILEMMA

Ever since the 1948 creation of the state of Israel in what was the British Mandate of Palestine, hundreds of thousands of Arabs who called Palestine their homeland have lived as refugees in neighboring countries. Many have been assimilated in the societies of Israel's neighbors, but a still larger number reside in makeshift refugee camps (Fig. 6–21). The Palestinians call themselves a nation without a state (much as the Jews were before Israel was founded), and they demand that their grievances be heard. Until Palestinian hopes for a national territory in the Middle East are considered by the powers that have influence in the region, a full settlement between Israel and the Arab countries may remain unattainable. Estimated Palestinian populations in the Middle East are:

| | |
|---|---|
| Jordan | 1,250,000 |
| Occupied Jordan (West Bank) | 20,000 |
| Egypt and Gaza | 550,000 |
| Lebanon | 350,000 |
| Syria | 230,000 |
| Kuwait | 300,000 |
| Saudi Arabia | 150,000 |
| Other Arab States | 100,000 |
| Israel | 550,000 |
| TOTAL | 3,500,000 |

In recent years, militant Palestinian organizations have emerged in Lebanon and Syria. The PLO has generally represented Palestinian views, but has been challenged (even militarily) by more radical groups in the 1980s. Palestinian forces have fought Israeli troops, Lebanese factions, and among themselves in Lebanon, and Palestinian terrorists continue to attack targets in Israel and around the world.

Various territorial solutions have been proposed. One suggests that a Palestinian state be created from the West Bank—the Israeli-

1948 proportions, and water is carried even into the Negev Desert itself.

With nearly 4.1 million people, however, Israel's future obviously does not lie in agriculture—for export or self-sufficiency. Already, the country must import a large volume of wheat. Despite the general intensity of agriculture and the dairying and vegetable gardens that have grown up around the large cities and towns, the gap between local supply and demand is widening. No effort is spared to maintain as high a degree of self-sufficiency as is possible. Despite the efficient *kibbutz* form of collectivized farm settlement (Fig. 6–22) and other innovations designed to combine communal living, productivity, and defensibility, Israel depends heavily on external trade and support.

Without an appreciable resource base, industrialization also presents quite a challenge to Israel. Evaporation of the Dead Sea waters has left deposits of potash, magnesium, and salt, and there is rock phosphate in the Negev. But there is very little fuel available within Israel; no coal deposits are known, although some oil has been found in the Negev. To circumvent the Suez Canal, an oil pipeline leads from the port of Eilat on the Gulf of Aqaba to the refinery at Haifa; a second, larger pipeline has been constructed to link Eilat and Ashkelon. Oil formerly came from Iran, but, since the Iranian revolution, Israel has been forced to turn elsewhere for

occupied part of Jordan west of the Jordan River—plus the Gaza Strip. Another proposes a Palestinian state located between Israel and Egypt in the Sinai. Still another proposal would partition Lebanon into various segments, including one for the Palestinians. But the Palestinian argument is having difficulty even being heard against the international strife that continues to engulf the Middle East, which is compounded by growing internal divisions between the PLO and its competitors. The territorial objectives of the Palestinians seem further than ever from realization.

**Figure 6–21**
The Baqaa Palestinian refugee camp in Jordan.

**Figure 6–22**
Kibbutz Dagania—the nation's oldest—reflects superior Israeli resourcefulness and productivity.

its petroleum. One source is Egypt; when Israel ceded a section of the occupied Sinai Peninsula where oilfields had been found and developed, Egypt committed itself to sell oil to Israel in return. The small steel plant at Tel Aviv uses imported coal as well as iron from foreign sources and was built for strategic reasons; it is not an economic proposition. Thus, the only industry for which Israel has any domestic raw materials at all is the chemical industry, and it, in response, has seen considerable growth. But for the rest, Israel must depend mostly on the skills of its labor force; diamond cutting, for example, is one of the leading industries. Many technicians and skilled craftspersons have been among the hundreds of thousands of Jewish immigrants who have come to Israel since its creation, and the best course, naturally, is to make maximum use of these people.

In effect, then, Israel is a Western-type developed country in the Middle East. It is highly urbanized, with just over 90 percent of its population living in towns and cities. Israel's core includes the two major cities, Tel Aviv-Jaffa (1.4 million) and Haifa (440,000) and the coastal area between them; in total, this core area incorporates over three-quarters of the country's population, although the proportion is less if the territories conquered in the 1967 War are added (including Jerusalem's Arab sector of 120,000 population).

Despite the persistence of antagonisms, time may be slowly eroding some barriers between Israel and the Arabs. There are growing behind-the-scene contacts between the Jewish state and many of its neighbors, tourist exchanges with Egypt and Lebanon, and Israel's Arab residents (one-sixth of the population) are for the first time being received cordially by the Saudis on pilgrimages to Mecca. In Jerusalem, now Israel's largest central city (450,000), tensions and physical barriers among Jews, Arabs, and Christians are beginning to dissipate. As some external problems ease, however, divisions within Israel loom ever more threateningly. Today, the country's 3.5 million Jews are deeply split by a social gap based on ethnicity, income level, and cultural heritage. On one hand, there are the majority Sephardic Jews of Asian-North African origin who largely occupy the lower echelons of Israeli society; on the other, there are the minority Ashkenazim Jews of European and American background who dominate the professions and higher-income groups. Constant frictions affect the relationships between these groups, reflecting a clash between the Islamic and Western cultural values of their source areas, and residential and educational segregation mark the social geography of the two populations. Thus, the younger generations of the Sephardim and Ashkenazim are not encouraged to integrate and commingle, perpetuating a schism that could one day threaten national unity.

**Neighbors and Adversaries** Not only does Israel have the misfortune to be located in the heart of the Arab world rather than in one of the region's peripheral or transition zones, it also has an inordinate number of neighbors for so small a country with so much coastline—five (counting Saudi Arabia). Of these five, three lie along Israel's northern and eastern boundary: Lebanon, Syria, and Jordan.

Lebanon, Israel's coastal neighbor on the Mediterranean Sea, is one of the exceptions to the rule that the Middle East is the world of Islam: perhaps 40 percent or even more of the population of 2.6 million adheres to the Christian rather than Moslem faith. Smaller than Israel (in fact, only a little over half of Israel's territorial size), Lebanon has a long history of trade and commerce, beginning with the Phoenicians of old—who were based here. Like Israel, Lebanon must import much of its staple food, wheat. The coastal belt below the mountains, although intensively cultivated, normally cannot produce enough grain to feed the entire population.

But normality has not prevailed in Lebanon for several years. The country fell apart in 1975 when a civil war broke out between Moslems and Christians. It was a conflict with many causes. Lebanon for several decades had functioned with a political system that divided power between the two dominant communities, but the basis for that system had become outdated. In the 1930s, Moslems and Christians in Lebanon were at approximate parity, but, in the 50 years since, Moslems have increased their numbers at a much faster rate than the more urbanized and generally wealthier Christians. The Moslems' displeasure at an outdated political arrangement (developed during the French occupation) was expressed during several outbreaks of rebellion prior to full civil war. Troops of the United States entered Lebanon in 1958 to help suppress anti-Western opposition and recently returned again (1982–1984) to try to help establish peace. In addition, Lebanon had become a base for over 350,000 Palestinian refugees; these people, many of them living in squalid camps, were never satisfied with Lebanon's moderate posture toward Israel. When the first fighting between Moslems and Christians broke out in the northern coastal city of Tripoli, the Palestinians joined the conflict on the Moslem side. In the process, Lebanon was wrecked. Beirut, the capital (2.3 million), often described as the Paris of the Middle East and a city of great architectural beauty, was almost completely destroyed as Christian, Moslem, Palestinian, and, eventually, Syrian forces fought for control (Fig. 6–23). As the Moslems' strength increased, the Christians concentrated in an area along the coast between Beirut and Tripoli; a multinational peacekeeping force of U.S., French, British, and Italian troops entered Lebanon in 1982 seeking to end the civil war and prompt the combatants to work out a lasting political solution (but withdrew without success in 1984).

**Figure 6–23**
Beirut in late 1983: everyday life struggles to go on amid widening destruction.

Syria's role as a force to be reckoned with in any solution of the Lebanon crisis, has enhanced that country's status since its loss of the Golan Heights to Israel. For a long time, Syria was politically unstable, a poor second to Egypt in their abortive 1958–1961 union known as the United Arab Republic. But the government of Hafez Assad, after taking power in 1970, began to reap the benefits of continuity. In 1976, Syrian armed forces intervened and brought a measure of peace to embattled Lebanon; by backing various client factions, Syria has so entrenched itself since that it has become the key to any peace agreement.

Syria, like Lebanon and Israel, has a Mediterranean coast where unirrigated agriculture is possible. Behind this densely populated coastal belt, Syria has a much larger interior than its neighbors, but the areas of productive capacity are quite dispersed. Damascus, the capital (1.8 million), was built on an oasis and is considered to be the world's oldest continuously inhabited city; although in the dry rain shadow of the coastal mountains, it is surrounded by a district of irrigated agriculture. It lies in the southwestern corner of the country (Fig. 6–20), in close proximity to Israel. In the far north, near the Turkish boundary, lies another old caravan-route center, Aleppo (1.2 million), now on the little-used railroad from Damascus to Turkey and at the northern end of Syria's important cotton-growing area; here the Orontes River is the chief source of irrigation water. Syria's wheat belt, stretching east along the northern border, also centers on Aleppo. In the eastern part of the country, the Euphrates Valley and the far northeast (the Jezira) are being developed for large-scale mechanized wheat and cotton farming with the aid of pump-irrigation systems. Production of these crops is rising rapidly, and more than half the Syrian harvest now comes from this area. Recent discoveries of oil also adds to the potential importance of this long-neglected part of the country.

Southward and southeastward, Syria turns into desert, and the familiar sheep, camel, and goat herders move endlessly across the countryside (see *Box*). There may be as many as half million of them, a noticeable proportion of Syria's 10.4 million people. In contrast to Israel and, to a lesser degree, to Lebanon as well, Syria is very much a country of farmers and peasants; only about 50 percent of the people live in cities and towns of any size. But, again, Syria produces adequate harvests of wheat and barley and normally does not need staple imports: in fact, it exports wheat from its northern wheat zone. It exports barley as well, but its biggest source of external revenue remains cotton. It is, in addition, a country where opportunities for the expansion of agriculture still exist; their realization will improve the cohesion of the state and bring its separate subregions into a tighter framework.

None of this can be said for Jordan (3.6 million), the desert kingdom that lies east of Israel and south of Syria. It, too, was a product of the Ottoman collapse, but it suffered heavily when Israel was created—perhaps more so than any other Arab state. In the first place, Jordan's trade used to flow through Haifa, now an Israeli port, so Jordan has to depend on unreliable Beirut or the tedious route via Aqaba. Second, Jordan's final independence in 1946 was achieved with a total population of perhaps 400,000, including nomads, peasants, villagers, and a few urban dwellers; it was a poor country. Then, with the partition of Palestine and the creation of Israel, Jordan received more than a half million Arab refugees; it soon also found itself responsible for another half million Palestinians who, although living on the western side of the Jordan River, were incorporated into the state. Thus, refugees outnumbered residents by more than 2 to 1, and internal political problems were added to external ones—not to mention the economic difficulties. Jordan has survived with U.S., British, and other aid, but its problems have hardly lessened. Many Jordanian residents have little commitment to the country, do not consider themselves its citizens, and give little support to the hard-pressed monarchy. Extremist groups constantly threaten to drag the country into another war with Israel; the 1967 War was disastrous for Jordan, which lost its sector of Jerusalem (the kingdom's second largest city). Where hope for progress might lie—for example, in the development of the Jordan Valley—political conflicts intrude. The capital city, Amman (850,000), reflects the limitations and poverty of the country. Without oil, without much farmland, without unity or strength, and overwhelmed with refugees, Jordan presents one of the bleakest pictures in the Middle East.

Iraq, by comparison, is well endowed, containing the lower valleys of both the historic Euphrates and Tigris Rivers. Because its agricultural potential is far greater than what is now used, Iraq is a rarity in the Middle East: it can be described as underpopulated, in that it could feed a far larger number of people than it presently does. This is a leftover from the decline that Mesopotamia went through during the Middle Ages, but steps are being taken

# PEOPLE ON THE MOVE

Countless thousands of people in Southwest Asia and North Africa are on the move; movement is a permanent part of their lives. They travel with their camels, goats, and other livestock along routes that are almost as old as human history in this realm. Most of the time, they do so in a regular seasonal pattern, visiting the same pastures year after year, stopping at the same oases, pitching their elaborate tents near the same stream. It is a form of cyclical migration—*nomadism*.

Nomadic movement, then, is not simply an aimless wandering across boundless dry plains. Nomadic peoples know their domain intimately; they know when the rains have regenerated the pastures they will make their temporary locale, and they know when it is necessary to move on. Some nomadic peoples remain in the same location for several months every year, and their portable settlements take on characteristics of permanence—until, on a given morning, an incredible burst of activity accompanies the breaking of camp. Amid ear-piercing bleating of camels, the clanking of livestock's bells, and the shouting of orders, the whole community is loaded on the backs of the animals and the journey resumes (Fig. 6–24). The leaders of the groups will know where they are going; they have been making this circuit all their lives.

Among those in the community there will be a division of labor, and some are skilled at crafts and make leather and metal objects for sale or trade when contact is made with the next permanent settlement. A part of the herd of livestock traveling with the caravan may belong to a townsperson, who will pay for their care. Nomads are not aimless wanderers—nor are they free from the tentacles of the cities.

**Figure 6–24**
A camel caravan forms, and the purposeful journey of these nomads resumes.

to improve conditions and raise standards of living. Iraq has been a major beneficiary of the oil reserves of the Middle East, and it needs the income from petroleum to make the necessary investments in industry and agriculture; oil accounts for about 90 percent of the country's export revenues. Unfortunately, most of Iraq's oil income has been diverted to the fighting of a bitter war with neighboring Iran throughout the first half of the 1980s. This has also made oil exporting via the Persian Gulf hazardous and piping crude to Mediterranean ports an impossibility, because intervening Syria (Iran's ally) has shut off the pipeline; a new pipeline is being laid through Saudi Arabia, but it will not be operational until after 1985.

How can a country that exports food crops and has a huge income from oil have so low a standard of living? There are many answers to this question. For over five decades since its independence (in 1932, after a decade as a British mandate), Iraq has suffered from administrative inefficiency, corruption in government, misuse of the national income, inequities in landholding and tenure systems—a set of problems that practically define the notion of underdevelopment. Apart from western Iraq, where nomads herding camels, sheep, and goats traverse the desert of Jordan-Syria-Iraq, most of the people live in small villages strung along the riverine lowland from the banks of the Shatt-al-Arab (the joint lowest course of the Tigris and Euphrates) to the land of the Kurds near the Turkish border. The Kurds, who number about 1.5 million out of Iraq's 15.5 million people, have at times been opposed to the Baghdad government, and there still is a serious minority problem of strong regional character. But the general impression of rural Iraq reminds one of rural Egypt, although Egypt is ahead in terms of its irrigation-works technology; the peasants face similar problems of poverty, malnutrition, and disease.

## The Arabian Peninsula

South of Jordan and Iraq lies the Arabian Peninsula, environmentally dominated by desert conditions and politically dominated by the Kingdom of Saudi Arabia (Fig. 6–25). With its huge area of 2,150,000 square kilometers (830,000 square miles), Saudi Arabia is the realm's third biggest nation-state; only Sudan and Algeria are somewhat larger. On the peninsula, Saudi Arabia's neighbors (moving clockwise from the Persian Gulf) are: Kuwait, Bahrain, Qatar, the United Arab Emirates, the Sultanate of Oman, Yemen (Aden)—also known as South Yemen or the People's Democratic Republic of Yemen, and Yemen (San'a)—also known as the Yemen Arab Republic. Together, these countries on the eastern fringes of the peninsula contain 12 million inhabitants; the largest

Figure 6–25

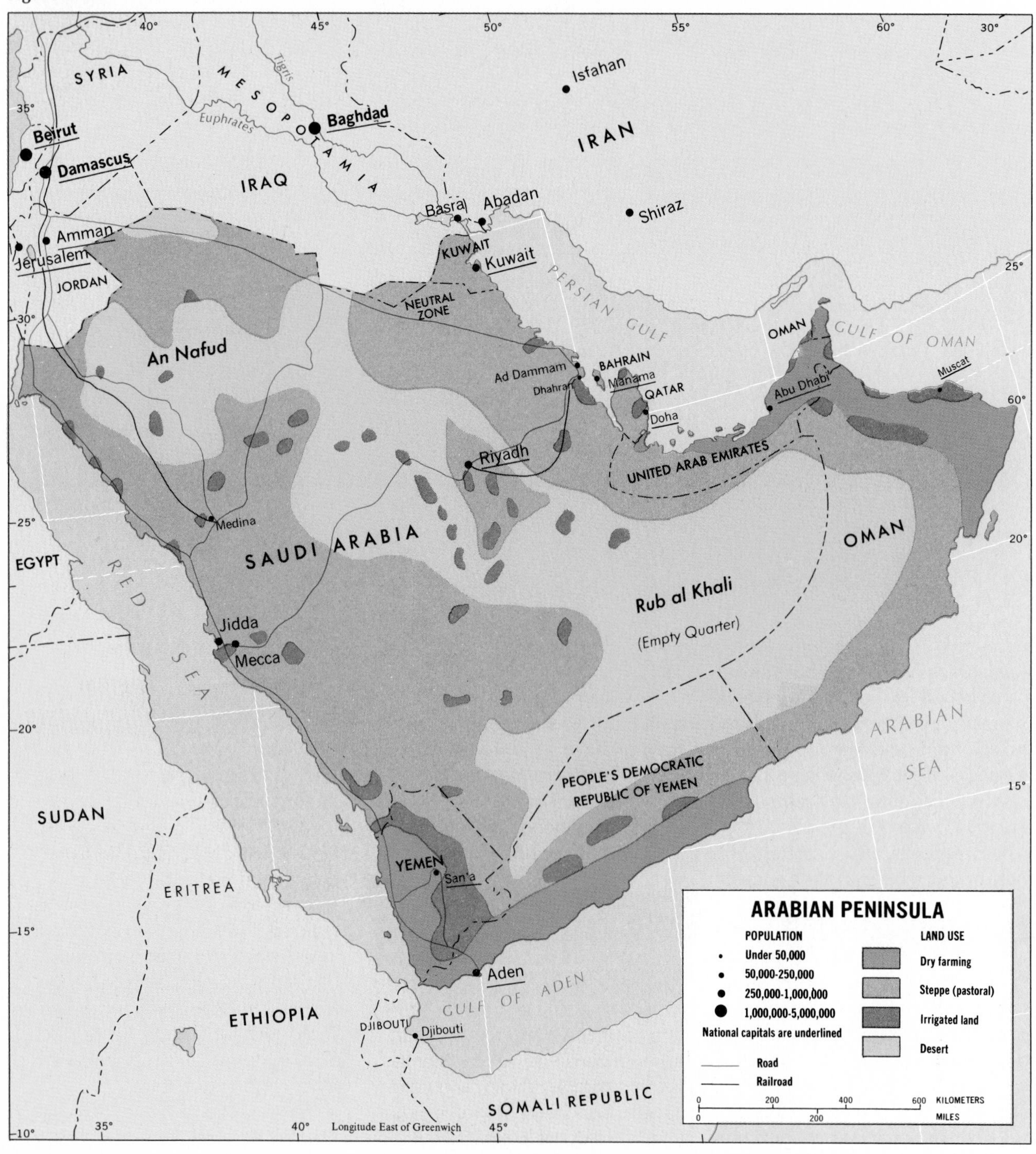

by far is the Yemen Arab Republic, with 6.1 million. Boundaries are still inadequately defined here, one of the world's last frontiers.

Saudi Arabia itself has only about 11.2 million inhabitants in its vast territory, but the kingdom's importance is reflected by Fig. 6–13: the Arabian Peninsula contains the world's largest concentration of known petroleum reserves. Saudi Arabia occupies most of this area; by some estimates, the country may possess one-quarter of all the world's remaining oil. The reserves lie in the eastern part of the country, on the Persian Gulf and in the Rub al Khali (Empty Quarter), a southward continuation of the Iraq-Kuwait-Neutral Zone oilfield.

The national state that is Saudi Arabia was only consolidated as recently as the 1920s through the energies and organizational abilities of King Ibn Saud. At the time, it was a mere shadow of its former greatness as the source of Islam and the heart of the Arab world. Apart from some permanent settlements along the coasts and in scattered oases, there was little to stabilize the country; most of it is desert, with annual rainfall almost everywhere under 10 centimeters (4 inches). The land surface rises generally from east to west, so that the Red Sea is fringed by mountains that reach as high as 3000 meters (nearly 10,000 feet). Here the rainfall is slightly higher, and there are some farms (coffee is a cash crop). The mountains also contain known deposits of gold, silver, and other metals; at one time, these were economic mainstays. Exploration has begun again, and the Saudis hope to diversify their exports by adding minerals from the west to the oil from the east.

As Fig. 6–25 reveals, human activity in Saudi Arabia tends to be concentrated in a wide belt across the "waist" of the peninsula, from the boom town of Dhahran on the Persian Gulf through the national capital of Riyadh in the interior to the Mecca-Medina area near the Red Sea. A fully effective internal transport network has not yet developed, but such an infrastructure is under construction and should be in place by the 1990s. In the more remote interior, however, Bedouin nomads still traverse their ancient caravan routes across the vast deserts. For several decades, Saudi Arabia's aristocratic royal families were virtually the sole beneficiaries of their country's incredible wealth, and there was hardly any change in the lives of villagers and nomads.

But, in recent years, the country's rulers have begun to institute reforms in housing, medical care, and education, and they have spent more than $300 billion since 1970 on a national-development program. A huge petrochemical industry is being developed for future diversification of the economy. The centerpiece is the new city of Jubail (Fig. 6–26), located on the Gulf coast 90 kilometers (55 miles) north of Dhahran. Begun in 1977, this huge construction project will turn a sleepy fishing port of 8000 into an industrial city of over 350,000 by the year 2000. Based on a complex of oil refineries, petrochemical and metal-manufacturing plants, Jubail is to be ready for occupancy by its first 60,000 residents during the mid-1980s. Together with Yanbu, a smaller new city on the Red Sea coast 300 kilometers (185 miles) north of Jidda, Jubail represents the country's plans to develop a diversified industrial base to sustain high living standards after the oil boom ends sometime in the next century. Still, Saudi Arabia remains very much tradition bound, and in most of the villages—even in Medina and Mecca—the modernizing impacts of the oil bonanza seem far away in place as well as time.

**Cycle theory** The new era in the Arab world is one of rejuvenation and reconstruction: rejuvenation in the ouster of colonial regimes and foreign influences and reconstruction with the billions of dollars derived from oil. Historical geographers, noting the rise, decline, and resurrection of nation-states, have attempted to explain the sequence of events in various

**Figure 6–26**
Jubail under construction, future heart of Saudi Arabia's petrochemical industry.

ways. Environmentalists believe that it all has to do with changing climates and, consequently, changing human energies. Ratzel believed (p. 74) that the state functions like a biological organism to grow and die. Another geographer, Samuel van Valkenburg, proposed a "cycle in the political development of nations, recognizing four stages, namely youth, adolescence, maturity and old age . . . after completion [a] cycle may renew itself, possibly with a change in political extension, while the cycle can also be interrupted at any time and brought back to a former stage. The time element (the length of a stage) differs greatly from nation to nation."

In this cycle theory of state evolution, the stage of youth might be called the *organizing* stage, and Tunisia in this realm would be a classic example. Adolescence may follow as an *expanding* stage, since the dynamic qualities of the state now emerge, and Iran during the Shah's administration showed such properties; a greater Persian sphere of influence was not beyond Tehran's dreams and is still nurtured today by fundamentalist Shi'ite leaders who seek to export the Iranian revolution. Maturity, the *stabilizing* stage, leads the state to concern itself, once again, with internal problems (as Egypt is now doing); external relationships involve moderation and internal cooperation. Old age may witness a kind of national senility, but it may also be a *disintegrating* period. Until the cycle was rejuvenated in Saudi Arabia after World War I, there was little left of the regional association forged during the rise of Moslem power.

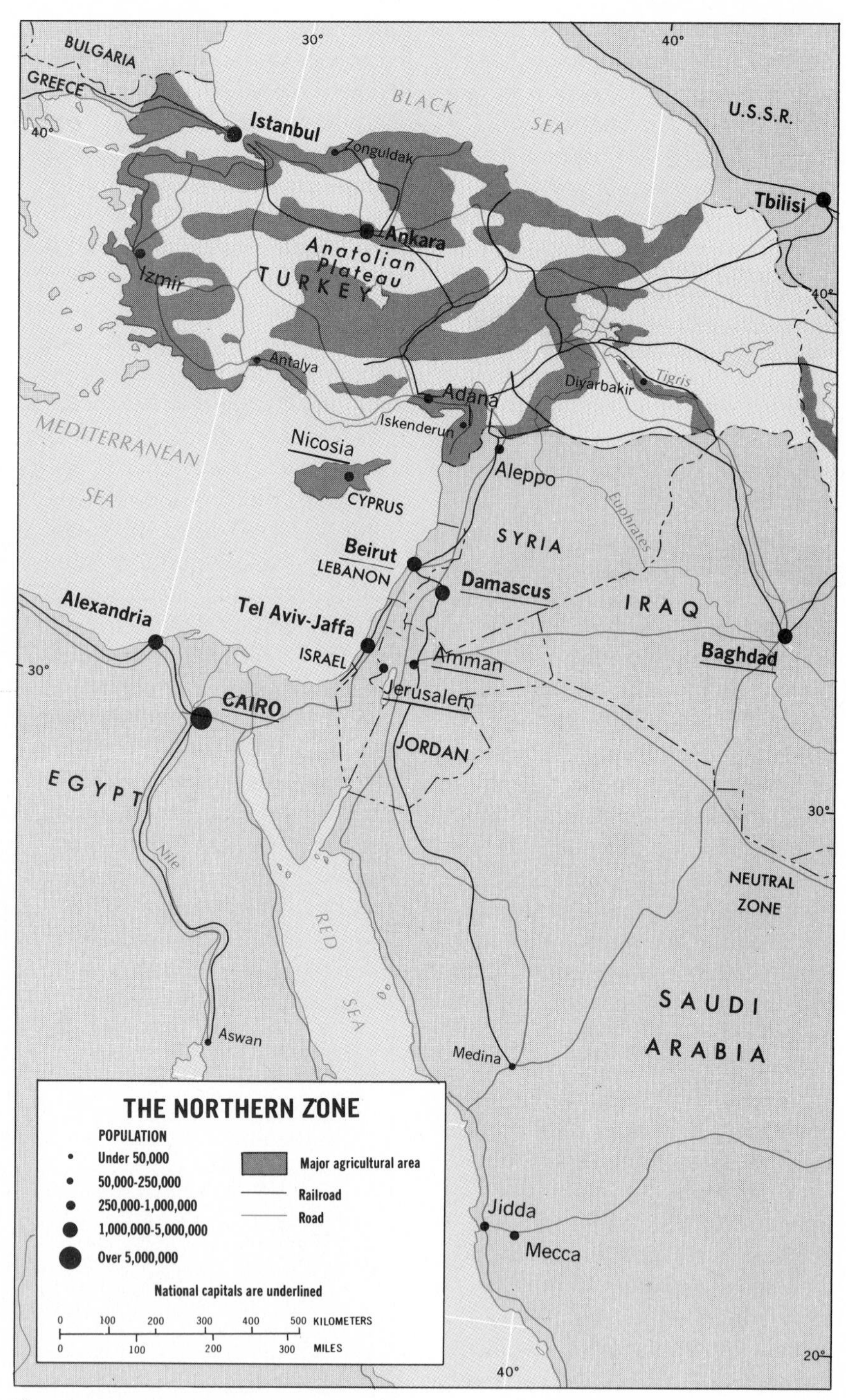


Figure 6–27

## The Non-Arab North

Across the northern tier of the Islamic realm lie three geographically distinct countries: Turkey, Iran, and landlocked Afghanistan (Fig. 6–27). Turkey, although a Moslem country, has been strongly oriented toward Europe in recent decades (it is a member of the North Atlantic Treaty Organization [NATO]) and has remained aloof from many of the issues that excite Arab countries. Iran has long kept its ideological distance as well; when the Suez Canal was closed to Israeli shipping by the Arab countries, it was Iranian oil that flowed, on orders of Shah Reza Pahlavi, through Israel's Aqaba pipelines. And Afghanistan, although a Moslem country as well, lies remote and peripheral, vulnera-

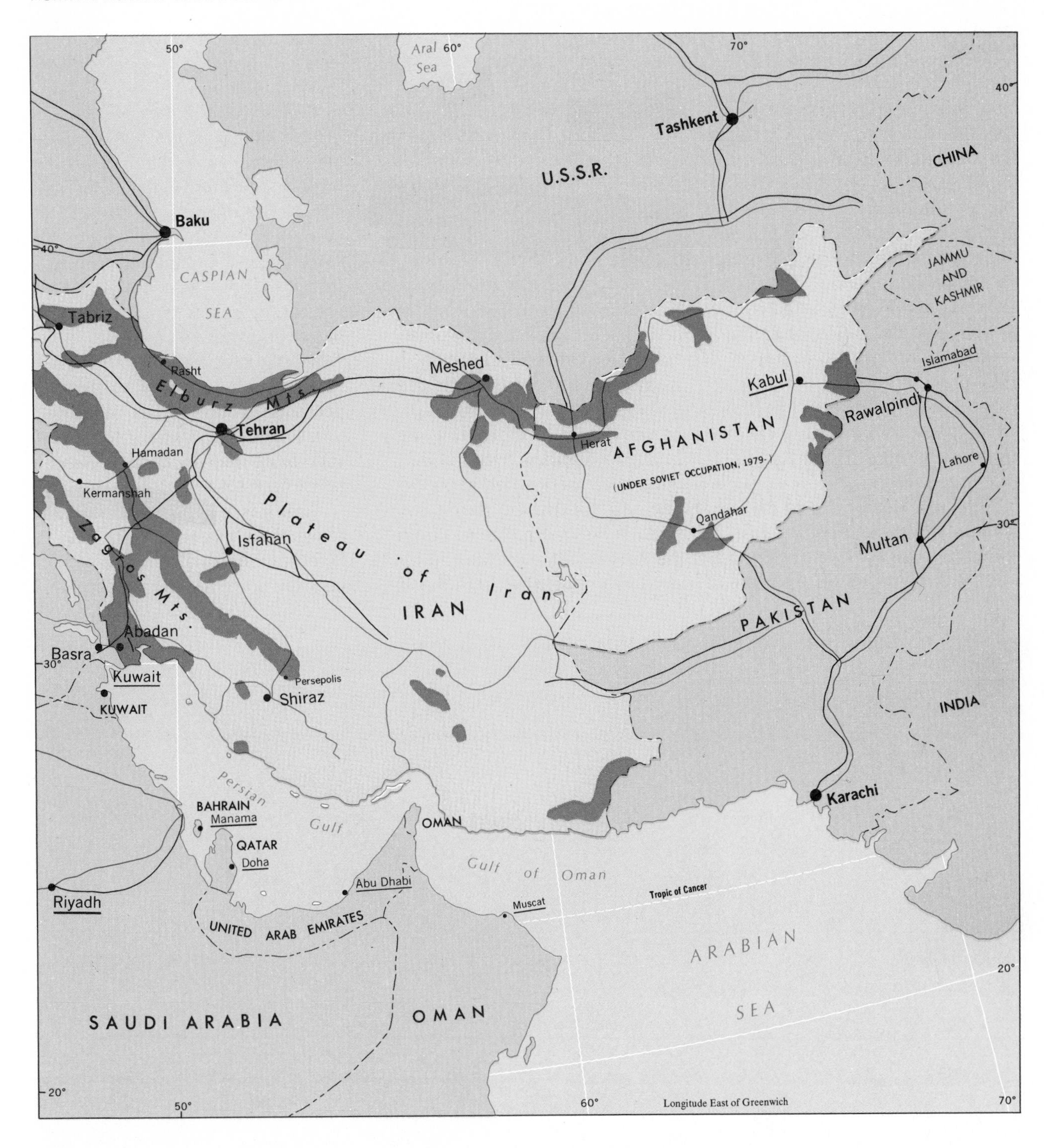


ble to foreign invasion, in its mountainous Asian interior.

Turkey, when it entered the twentieth century, was at the center of a decaying, corrupt, and reactionary empire whose sphere of influence had extended across much of the Middle East and Eastern Europe. At that time, conditions were such that it would undoubtedly have had to be considered a part of the Middle East, the world of its religious-cultural imprint, Islam. But the Turks are not Arabs (their source area lies in the Sino-Soviet border area in Central Asia), and, in the early 1920s, a revolution occurred that thrust into national prominence a leader who has become known as the father of modern Turkey—Mustafa Kemal Ataturk.

The ancient capital of Turkey was Constantinople (now Istanbul), but the struggle for Turkey's survival had been waged from the heart of the country, the Anatolian

Plateau, and it was here that Ataturk wanted to place the new seat of government. Thus, Ankara had many advantages: it would remind the Turks that they were, as Ataturk always said, Anatolians; it lay nearer the center of the country than Istanbul and could act as a stronger unifier; and its strategic position was far better than that of the ancient capital. Istanbul (1985 population: 6.8 million), located on the western side of the Bosporus (middle distance in Fig. 6–28), lies on the threshold of Europe, with the minarets and mosques of this largest and most varied Turkish city rising above a somewhat Eastern European townscape.

Although Ataturk moved the capital eastward and inward, his orientation was westward and outward. To implement his plans for Turkey's modernization, he initiated reforms in almost every sphere of life in the country. Islam, the state religion, lost its official status. The state took over most of the religious schools that had controlled education. The Roman alphabet replaced the Arabic. Moslem law was replaced by a modified Western code. Symbols of old—wearing beards, wearing the fez—were prohibited. Monogamy was made law, and the emancipation of women was begun; in Moslem society, women have generally been denied access to education, freedom of movement, or social contact, and, in a few countries, they still must cover their faces when in public. The new government took pains to stress Turkey's separateness from the Arab world; ever since, it has remained quite aloof from the affairs that involve the other Islamic states.

What Ataturk wanted for Turkey, of course, could not be achieved in a short time. Turkey is largely an agricultural country, and, when three-quarters of its inhabitants are subsistence or near-subsistence farmers, most of whom live in small villages in rather isolated rural areas, some opposition to change, especially rapid change, has to be expected. So it was in Turkey; in fact, the government had to yield to the devoutly Moslem peasants on some issues. Still, Ataturk's directives continue to be pursued more than six decades after he came to power.

Turkey is a mountainous country of generally moderate relief. The highest mountains lie in the east, near the Soviet border, and in Kur-

**Figure 6–28**
Istanbul, located astride the European-Southwest Asian interface, exhibits features of both realms.

distan. Here elevations reach 2000 to 3000 meters (6600 to 10,000 feet), but westward the altitudes decline. The Plateau of Anatolia in the heart of the country is dry enough to be classified as a steppe (see Fig. I–10, the area marked *BSk*). Here the people live in small villages, grow subsistence cereals, and raise livestock. Sheep and goats bring some revenues, from the wool and hides—and from the mohair of the Angora goat, used in the manufacture of valuable textiles and rugs. But the best farmlands lie in the coastal areas, small as they are. There is little coastal lowland along the Mediterranean, where the areas around Antalya and Adana (northeast of Iskenderun) are the only places where the mountains retreat somewhat from the sea; offshore, Cyprus is another important agricultural concentration, despite the long-standing conflict between Turks and Greeks in that island country (see *Box*). The northern Black Sea coast is also quite narrow, but it is comparatively moist: it gets winter as well as summer rainfall (though with a winter maximum) totaling 75 to 260 centimeters (30 to 105 inches) yearly—uncommonly high for the Dry World. Thus, as Fig. I–14 reflects, these coastal areas, especially the Aegean-Marmara lowlands, are the most densely populated zones of Turkey and the most productive ones.

In the farming areas where the export crops are grown, the agricultural complex is the familiar Mediterranean type, with wheat and barley the staples (they occupy over three-quarters of all the cultivated land), and tobacco, cotton, hazelnuts, grapes, olives, and figs for the external market. Cotton, famous Turkish tobacco, and hazelnuts (the Black Sea coast is the world's chief source of these) form the three top exports by value in a country 90 percent of whose external revenues are derived from the land. As elsewhere in this part of the world, production could be increased but, unlike Iraq, the problems involved in irrigation expansion are serious: there is little level land and the streams are deeply incised into narrow valleys in this mountainous country.

Turkey has the largest population among the countries of the North African-Southwest Asian realm, with 51.3 million inhabitants. As Fig. 6–27 indicates, this country has the best-developed network of surface communications in the entire realm; it also is the most industrialized, notwithstanding its chiefly agricultural export economy. The production of cotton has stimulated a textile industry, and a small steel industry has been established based on a coalfield near

## THE PROBLEM OF CYPRUS

In the northeastern corner of the Mediterranean, much farther from Greece than from Turkey or Syria, lies Cyprus. Since ancient times, this island has been dominantly Greek in its population, but, in 1571, it was conquered by the Turks, under whose control it remained until 1878. Following the decay of the Ottoman state, the British soon took control of the island. By the time Britain was prepared to offer independence to much of its empire, it had a problem in Cyprus, for the Greek majority among the 600,000 population preferred *enosis*—union with Greece. It is not difficult to understand that the 20 percent of the Cypriot population that was Turkish wanted no such union; in fact, the answer the Turks gave was partition—the division of the island into a Turkish and a Greek sector. By 1955, the dispute had reached the stage of violence: differences between Greeks and Turks are deep, bitter, and intense. It was really impossible to find a solution to a problem in which the residents of an island country think of themselves as Greeks or Turks first rather than Cypriots; yet, in 1960, Cyprus was made an independent country with a complicated constitution designed to permit majority rule but also to guarantee minority rights.

The fragile order finally broke down in 1974. As a civil war engulfed the island, Turkish armed forces intervened, and a major redistribution of population occurred. The northern one-third of Cyprus became the stronghold of more than 100,000 Turkish Cypriots; only a few thousand Greeks remained there. The rest of the island—south of the "Green Line" that was demarcated across the country to divide the two ethnic communities—became the domain of the Greek majority, now including nearly 200,000 in refugee camps. As in Lebanon, partition became a serious prospect. In the early 1980s, the fundamental differences between Cypriot Greeks and Turks were still unresolved, and a form of *de facto* apartheid existed on the island; in 1983, to again bring the stalemated conflict to the attention of the world, the Turkish community seceded by declaring itself the new independent Turkish Republic of Northern Cyprus. However, a similar ploy failed in the 1970s, and it appeared that this latest unilateral action would meet a similar fate. Until a permanent solution is arrived at, Cyprus (1985 population: 670,000), a pawn of foreign powers for centuries, remains a casualty of history.

Zonguldak, not far from the Black Sea, and an iron ore deposit several hundred kilometers away in east-central Turkey. In the southeast, Turkey has found some oil, and it may share in the zone of oilfields of which the famed Kirkuk reserve in nearby Iraq is a part. With its complex geology, Turkey has proved to have a variety of mineral deposits, which gives much scope for future development. Spatially, this development seems destined, at present, to reinforce the country's westward orientation. The Mediterranean coastal zone is already the economic focus of the country; of what remains, the western half is again the more developed.

Iran constitutes another exception to the rule that this geographic realm is the Arab world. Iranians are not Arabs; nor are they adherents to the majority (Sunni) Moslem faith. Most Iranians are members of the Shi'ite group, who represent about one-tenth of all Islam, mainly in Iran and neighboring Iraq. Persia (as Iran was formerly called) was a kingdom as long as 2500 years ago; its royal succession faltered repeatedly during the twentieth century. The most recent period of monarchy ended in 1979, when an Islamic revolution that had long been intensifying achieved the overthrow of the Shah. Nevertheless, the year 1971 had witnessed the 2500th anniversary of Persia's first monarchy amid celebrations and scenes of royal splendor—and without indications of the upheaval that lay ahead.

Iran is a country of mountains and deserts. The heart of the country is a plateau, the Plateau of Iran, that lies surrounded by even higher mountains, including the Zagros Mountains in the west, the Elburz Mountains along the southern shores of the Caspian Sea, and the mountains of the Khurasan region to the northeast. The Iranian Plateau, therefore, is actually a high-elevation basin marked by salt flats and expanses of stone and sand. On the hillsides, where the topography wrests some moisture from the air, lie some fertile soils. Elsewhere, only oases break the arid monotony—oases that for countless centuries have been stopping places on the area's caravan routes. With so little usable land (such as the moist narrow ribbon along the Caspian Sea coast and the small zone east of the Shatt-al-Arab in the southwest) and with most of the people depending directly on agricultural or pastoral subsistence, it is noteworthy that Iran's population is as large as 45.2 million.

In ancient times, Persepolis in southern Iran was the focus for a powerful kingdom. Then, as now, people clustered in and around the oases or depended on *qanats* (underground tunnels from the mountains) for their water supply, as Persepolis did. The focus of modern Iran, Tehran (6.9 million), lies far to the north on the slopes of the Elburz Mountains. Tehran, which also continues to depend to some degree on the *qanat* system for its water, rose from a caravan station to become the capital of a modernizing state.

Iran, indeed, modernized during the Shah's regime, although the villages and nomadic communities are reminders that change does not come quickly or easily to a country as large and tradition bound as Iran. Thus, the modernization process operated to intensify local and regional contrasts, contributing to the success of the revolution that was to come, an upheaval that had its roots in the Islamic traditions of the great majority of the people. Iran's national economy is based on the sale of its oil on overseas markets; unlike some other countries in this realm, Iran has been selling oil for 75 years. But never had its annual income been greater than in the 1970s; at the direction of the Shah, substantial investments were made in industrial diversification projects, power-generating equipment, and the modernization of agriculture—including the official abolition of peasant serfdom and the redistribution of some lands. Yet, the real impact of the petroleum age still remained confined to the south between the Zagros Mountains and the Persian Gulf, where the oil reserves, refineries, and port facilities are located—and in Tehran, with its skyscrapers and wide tree-lined boulevards.

Like Ataturk and his successors in Turkey, Shah Mohammed Reza Pahlavi, who ruled the country from 1941 until his ouster in 1979, sought to bring major reforms to Iran. Apart from the social changes brought by his "White Revolution," the Shah enfranchised women in 1963, introduced profit sharing in some enterprises, and created a kind of domestic peace corps (aimed principally at improving medical services and literacy in rural areas). But his reign was sustained by the intervention of foreign interests, and his policies ran counter to the Islamic traditions of the great majority of his people. Moslem leaders opposed him; a network of secret police (SAVAK) protected the almost unchecked privileges of the Shah and his elite. Among exiled religious leaders was the Ayatollah Khomeini; in time, he became the symbol of the Islamic revolution that exploded at the end of the 1970s (Fig. 6–29).

Among the circumstances that brought Khomeini to power in Iran were the searing material inequities that Pahlavi's modernization effort had created. The contrasts between the advantaged elite in Tehran and the poverty of the city's slum dwellers, the "haves" in places like Isfahan and Shiraz and the "have-nots" who remained mired in a web of debt and dependency—all this produced wide opposition to the Shah's policies. Ruthless political repression, reports of torture,

**Figure 6–29**
Five years after the revolution, hysterical fervor continues to be the hallmark of Khomeini's Iran.

and a lack of avenues for the expression of opposition also contributed to the breakdown of the order the Shah's authoritarian rule had wrought.

The wealth generated by petroleum could not transform Iran in ways that might have staved off the revolution. Modernization remained but a veneer: in the villages away from Tehran's polluted air, the holy men continued to dominate the life of ordinary Iranians. As elsewhere in the Moslem world, urbanites, villagers, and nomads remained enmeshed in a web of production and profiteering, serfdom, and indebtedness that has always characterized traditional society here. This ecological trilogy (as Paul English called it in a 1967 article in the *Journal of Geography*) is not unique to Iran; it is characteristic of much of this entire realm. It ties the people who live in cities and towns to villagers and nomads; it is the urbanites who dominate, because they have the money and own the land and the livestock. Thus, as English wrote:

> *Society is divided into three mutually dependent types of communities—the city, the village, and the tribe—each with a distinctive lifemode, each operating in a different setting, each contributing to the support of the other two sectors and thereby to the maintenance of total society. . . . the [wealthy and powerful] city dwellers are principally engaged in collecting raw materials from the hinterland—wool for carpets and shawls, vegetables and grain to feed the urban population, and nuts, dried fruit, hides, and spices for export. In return, the urbanites supply peasants and nomads with basic economic necessities such as sugar, tea, cloth, and metal goods as well as cultural imperatives such as religious leadership, entertainment, and a variety of services. This concept of urban dominance is basic to the idea of an interdependent urban trilogy.*

Again, Iran's twentieth-century oil age did little to change these relationships. The peasant tilling the fields near his village still worked on land owned by someone else. The nomadic herdsman moving his flock along centuries-old routes in search of grazing grounds was also not freed from the tentacles of the city; most of the animals belonged to someone living in a distant town—far away, but in control of his existence nevertheless. In modernizing Iran, the new and the old were never brought into a harmony that would have averted the turbulence of the 1979 revolution, which continues almost unabated into the 1980s.

The easternmost country in the realm's northern tier—and the only landlocked state—is Afghanistan (Fig. 6–27). Its territory is compact in shape, except for a narrow but lengthy land extension, the Wakhan Corridor, that has the effect of adding a significant neighbor, China, to its other bordering states: the U.S.S.R., Iran, and Pakistan.

Afghanistan was for centuries a battleground for outsiders, including Turks and Indians, Tatars and Persians, but the country became a recognized political unit because two other competitors—the Russians and the British—agreed to guarantee its integrity. As a buffer state, Afghanistan's high mountains, deep valleys, remote provinces, and inaccessible frontiers were ideal. The nineteenth-century boundary created a country of nearly 650,000 square kilometers (250,000 square miles) that today incorporates 14.8 million people, about half of them Pathans (also called Pushtuns or Pakhtuns) and the remainder Tajiks, Uzbeks, Turkmens, Hazaras, and smaller groups. Thus, Afghanistan was hardly a nation-state, although virtually its entire population adheres to the Sunni form of Islam.

Although Afghanistan does not possess oil reserves (some natural gas has been found in the north) and remains a country of herders, farmers, and nomads, it does have one coveted asset it cannot sell: its relative location. Southern Afghanistan lies only 400 kilometers (250 miles) from the Arabian Sea, not far from the mouth of the Persian Gulf. Along Afghanistan's eastern flank lies a province of Pakistan that has displayed separatist tendencies (see Chapter 8). Afghanistan's western border with Iran does not reflect a clear break in the cultural landscape; Dari, a Persian language, has official status inside Afghanistan. And along Afghanistan's northern border lies the Soviet Union which, late in 1979, became the latest of the country's foreign invaders.

Although foreigners may attempt to erect puppet governments in the capital, Kabul (700,000), the fact remains that Afghanistan is one of the realm's least developed countries. Urbanization is still under 20 percent, nonmilitary communications are minimal, agriculture and pastoral subsistence remain the dominant livelihoods, and there is little national integration. The post-1979 Soviet intervention supported an unpopular Marxist regime whose modernizing intentions were viewed with suspicion by millions of traditional and conservative Moslem Afghans. After five years, however, the Soviets are well entrenched and have the mujahedin rebels under control. Given the proximity of Afghanistan to the strategic Persian Gulf, and their success in preventing the establishment of an anti-Communist Islamic country on the borders of the U.S.S.R.'s Central Asian Moslem republics, the Russians seem certain to maintain their new hegemony here for a long time to come.

Geographically, Afghanistan not only lies in an historic buffer zone, it is also a transitional region. It differs culturally from Iran as well as Pakistan; its economic geography is unlike either. On various grounds, Afghanistan may be viewed as part of each of the three world realms here juxtaposed—perhaps even four, given its new involuntary political orientation to the north.

## VON THÜNEN IN THE UNDERDEVELOPED WORLD

The Von Thünen Model—treated on pages 72–74 and diagrammed in Fig. 1–12A—can also be applied to many farming areas in the developing realms. Unlike the case of the United States (pages 210–211), where contemporary Thünian effects are observed at the continental scale, the much more localized movement systems of the Third World provide a transportation setting much like that of Von Thünen's time in early-nineteenth-century Europe. Therefore, the agricultural lands surrounding large cities would be the likeliest places to find evidence of Thünian spatial organization, and Ronald Horvath discovered just such a pattern around Addis Ababa. The original Isolated State model showed a forest zone in the second land-use ring, from which the preindustrial city drew wood for both construction and fuel. In his 1969 study, Horvath found striking evidence of such an inner wood-producing ring in the form of the continuous belt of eucalyptus forest that surrounded the Ethiopian capital (Fig. 6–30). The configuration of this forest nicely conforms to Von Thünen's own empirical application of his model—instead of a circular ring, a girdling wedge-shaped zone develops that reflects the greater accessibility to the city along radial road corridors. Horvath also captured some interesting dynamics: the rapid outward expansion of the eucalyptus zone between 1957 and 1964 represents the improvement of surface transport in the Addis Ababa area, permitting the importing of wood from a greater distance (and the freeing up of land nearer the city to raise more vegetables—a leading activity of the innermost ring in the ideal Thünian scheme). Additional research has documented the existence of Thünian production patterns in the hinterlands of a number of other Third World cities, and this variation of the model may well apply to the vicinity of any major preindustrial urban area.

## The African Transition Zone

All across Africa south of the Sahara, from Mauritania and Senegal in the west to Ethiopia and Somalia in the east, Islam's culture yields to the life-styles of the Black African realm. However, it is a transition zone quite different from the northern tier. Turkey may be Europe-oriented, but it is a Moslem country; so is individualistic, non-Arab Iran. But in Africa, the transition exhibits

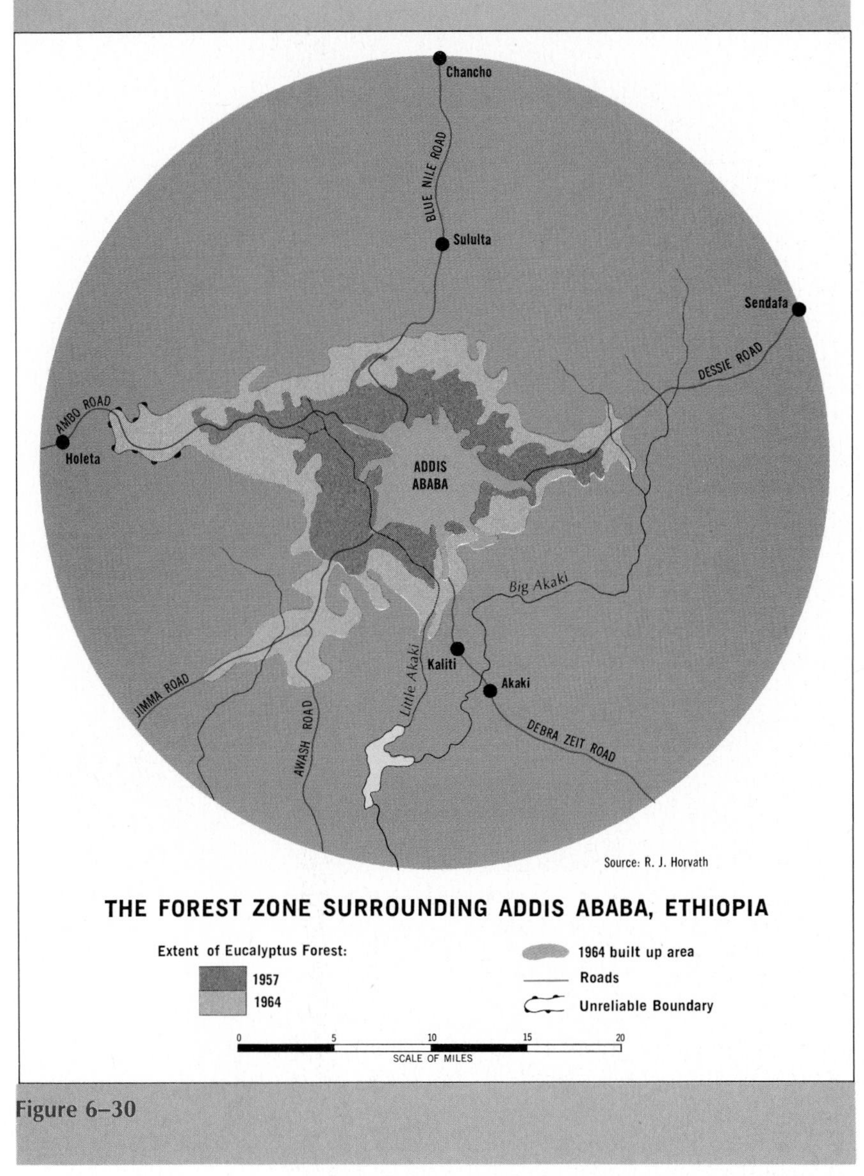


**THE FOREST ZONE SURROUNDING ADDIS ABABA, ETHIOPIA**

Figure 6–30

Reproduced by permission from the *Annals of the AAG*, vol. 59, 1969, p. 314, fig. 3.

local variations in several countries. Northern Nigeria is Moslem, the south is not. Northern Sudan is Moslem (including the core area between the White and the Blue Nile and the capital of Khartoum), but the south is converted to Christianity. Ethiopia's heartland, centered on its capital of Addis Ababa, is the land of the Amharic people, who profess the Ethiopian Orthodox (Coptic Christian) faith—but its coastal province of Eritrea and its eastern region centered on Harar are Moslem.

Life in the Transition Zone has never been easy. The heart of this region is the notorious *Sahel* (Arabic for "border"), where a multiyear drought in the 1970s—exacerbated by widespread overgrazing practices—starved perhaps 300,000 people and 5 million head of cattle. The Sahel itself is a 330-kilometer-wide (200-mile-wide) zone that stretches entirely across Africa along the southern edge of the Sahara Desert. This steppe environment is clearly discernible in Fig. I–10 as an east-west band of *BSh* climate; annual rainfall is highly variable here—ranging from 10 centimeters (4 inches) to 60 centimeters (24 inches)—making livestock raising an always risky proposition that courts disaster if too many animals upset the delicate ecological balance. Sedentary agriculture is therefore rare (Fig. 5–3), limited to those few Transition Zone places where highlands, such as in Ethiopia (see *Box*), and exotic river valleys such as the Sudan Nile offer sufficient water supplies.

The Transition Zone is not always peaceful. In 1984, Ethiopia's northern province of Eritrea was still fighting a long war for independence from the increasingly Marxist Addis Ababa government. Southern Sudanese throughout the 1960s and 1970s claimed brutal subjugation by their Moslem rulers in the north. In order to better unify that country, the 350-kilometer-long (220-mile-long) Jonglei Canal is being constructed as a bypass around the White Nile's huge central Sudan swamp known as the Sudd (Arabic for "barrier"); with its gigantic excavating machine advancing at the rate of 9 kilometers (5.5 miles) a month (Fig. 6–31), the scheduled 1985 opening of the waterway should markedly reduce the evaporation loss of Nile water in the Sudd and thereby release a larger supply for northern irrigation projects—to the chagrin of the southern Sudanese, who fear further exploitation and the diversion of their own (abundant) water supplies. Yet another trouble spot is Chad (5.2 million), where a civil war has pitted Islamic northerners against non-Moslem southerners. When expansionist Libya sent in ground troops and air power to help the northern rebel forces advance to the south in 1983, the French (supported by state-of-the-art U.S. reconnaissance aircraft) moved their troops and planes in to reclaim their old sphere of influence; the rebels decided against further

**Figure 6–31**
Central Sudan's Jonglei Canal, which will soon divert much of the Nile's water around the swamps of the Sudd.

engagement, creating at least a temporary stalemate that may buy enough time to work out a peaceful political solution.

All this regionalism within the African Transition Zone and many of its countries came about, of course, because the European colonial invasion took place after the diffusion of Islam had penetrated deep into West and East Africa, and the Europeans colonized Moslems and non-Moslems alike. The resulting problems are far from solved. The Somali Republic, where Islam is the state religion, lies adjacent to Ethiopia's Moslem east, and periodically Somalia aspires to a Greater Somali Union. This has led to a strong irredentist movement in the Ogaden (eastern Ethiopia), thereby reinforcing long-standing local political instability, because the Somalia-Ethiopia boundary has been in dispute since its superimposition during the colonial heyday of the late nineteenth century. By the early 1970s, sporadic violence had escalated into guerilla warfare; although the Somalis at first held their own, Ethiopia launched a major counterattack in 1978 that has since given them the upper hand. As a result, tens of thousands of Somalis have been driven out of the Ogaden and are forced to reside in refugee camps in Western Somalia; more than 1.5 million now populate these miserably overcrowded camps (Fig. 6–32), which are beset with every Third World problem of disease and malnutrition. Nonetheless, Pan-Somaliism is still kept alive by the government, which also nurtures irredentism in Kenya's bordering northeastern territories. Both Ethiopia (32.7 million) and Somalia (6.5 million) also aspire to control the young republic of Djibouti (290,000), a former French dependency that became independent in 1977 and contains the port city of Djibouti—Ethiopia's principal outlet to the sea. Whether Moslem and non-Moslem coexistence can succeed in Nigeria or Sudan is still an unanswered question. That it has failed in Chad and Ethiopia is beyond doubt.

**Figure 6–32**
Ogaden refugees in Somalia, barely surviving the misery of life in one of the squalid refugee camps.

## References and Further Readings


Abu-Lughod, J. *Cairo: 1001 Years of "The City Victorious"* (Princeton, N.J.: Princeton University Press, 1971).

al Fārūqi, I. & Sopher, D., eds. *Historical Atlas of the Religions of the World* (New York: Macmillan, 1974).

Barbour, K. *The Growth, Location, and Structure of Industry in Egypt* (New York: Praeger, 1972).

Beaumont, P. et al. *The Middle East: A Geographical Study* (London: John Wiley & Sons, 1976).

Blake, G. & Lawless, R., eds. *The Changing Middle Eastern City* (Totowa, N.J.: Barnes & Noble, 1980).

Brett, M., ed. *Northern Africa: Islam and Modernization* (London: Cass, 1973).

Brown, L. *Innovation Diffusion: A New Perspective* (London and New York: Methuen, 1981).

Chapman, K. *People, Pattern and Process: An Introduction to Human Geography* (New York: John Wiley/Halsted, 1979). Diagram adapted from p. 141.

Clarke, J. & Bowen-Jones, H., eds. *Change and Development in the Middle East* (London and New York: Methuen, 1981).

Cohen, Y. *Diffusion of an Innovation in an Urban System* (Chicago: University of Chicago, Department of Geography, Research Paper No. 140, 1972). Map adapted from p. 40.

Cook, E. *Man, Energy, Society* (San Francisco: W. H. Freeman & Co., 1976).

Coon, C. *Caravan: The Story of the Middle East* (New York: Holt, Rinehart & Winston, 1958).

Cottrell, A., ed. *The Persian Gulf States: A General Survey* (Baltimore: Johns Hopkins University Press, 1980).

Cressey, G. *Crossroads: Land and Life in Southwest Asia* (Philadelphia: J. B. Lippincott, 1960).

Devlin, J. *Syria: Modern State in an Ancient Land* (Boulder, Colo.: Westview Press, 1983).

Dohrs, F. & Sommers, L., eds. *Cultural Geography: Selected Readings* (New York: Thomas Y. Crowell, 1967).

English, P. *City and Village in Iran: Settlement and Economy in the Kirman Basin* (Madison: University of Wisconsin Press, 1966).

English, P. "Urbanites, Peasants and Nomads: The Middle Eastern Ecological Trilogy," *Journal of Geography*, 66 (1967), 54–59. Quotation taken from pp. 54–55.

Fahim, H. *Dams, People and Development: The Aswan High Dam Case* (Elmsford, N.Y.: Pergamon, 1981).

"A Fever Bordering on Hysteria: After Five Years, Khomeini Still Seems in Full Control of Iran's Revolution," *Time*, March 12, 1984, 36–39.

Fisher, W. B. *The Middle East: A Physical, Social and Regional Geography* (London and New York: Methuen, 7 rev. ed., 1978).

Friedlander, D. & Goldscheider, C. "Israel's Population: The Challenge of

Pluralism," *Population Bulletin*, 39 (April 1984), 1–39.
Gilbert, M. *Atlas of the Arab-Israeli Conflict* (New York: Macmillan, 1974).
Gordon, D. *The Republic of Lebanon: Nation in Jeopardy* (Boulder, Colo.: Westview Press, 1983).
Gould, P. R. *Spatial Diffusion* (Washington: Association of American Geographers, Commission on College Geography, Resource Paper No. 4, 1969).
Hägerstrand, T. *Innovation Diffusion as a Spatial Process* (Chicago: University of Chicago Press, trans. A. Pred, 1967).
Halpern, B. *The Idea of the Jewish State* (Cambridge: Harvard University Press, 1961).
Holz, R., ed. *The Surveillant Science: Remote Sensing of the Environment* (New York: John Wiley & Sons, 2 rev. ed., 1985).
Horvath, R. "Von Thünen's Isolated State and the Area Around Addis Ababa, Ethiopia," *Annals of the Association of American Geographers*, 59 (1969), 308–323.
Hunter, J. & Young, J. "Diffusion of Influenza in England and Wales," *Annals of the Association of American Geographers*, 61 (1971), 637–653. Maps adapted from p. 645.
Isaac, E. *Geography of Domestication* (Englewood Cliffs, N.J.: Prentice-Hall, 1970).
Karan, P. & Bladen, W. "Arabic Cities," *Focus*, January–February 1983.
Karmon, Y. *Israel: A Regional Geography* (New York: John Wiley & Sons, 1971).
Longrigg, S. *The Middle East: A Social Geography* (Chicago: Aldine, 2 rev. ed., 1970).
Mikesell, M. "Tradition and Innovation in Cultural Geography," *Annals of the Association of American Geographers*, 68 (1978), 1–16.
Murphey, R. "The Decline of North Africa Since the Roman Occupation: Climatic or Human?," *Annals of the Association of American Geographers*, 41 (1951), 116–132.
Orni, E. & Efrat, E. *Geography of Israel* (Jerusalem: Israel Universities Press, 3 rev. ed., 1976).
"Paradise Lost: Lebanon," *ARAMCO World Magazine*, September–October 1982.
Pyle, G. "The Diffusion of Cholera in the United States in the Nineteenth Century," *Geographical Analysis*, 1 (1969), 59–75. Maps adapted from pp. 63, 72.
Rogers, E. *Diffusion of Innovations* (New York: Free Press, 3 rev. ed., 1983).
Ruedisili, L. & Firebaugh, M., eds. *Perspectives on Energy* (New York: Oxford University Press, 3 rev. ed., 1982).
Sauer, C. O. *Agricultural Origins and Dispersals* (New York: American Geographical Society, 1952).
Shipler, D. "Jews from Asia and Europe: A Bitterness Gnaws at Israel," *New York Times*, April 6, 1983, pp. 1, 6.
Simoons, F. *Eat Not This Flesh: Food Avoidances in the Old World* (Madison: University of Wisconsin Press, 1961).
Spencer, J. "The Growth of Cultural Geography," *American Behavioral Scientist*, 22 (1978), 79–92. Quotation taken from p. 79.
Strahler, A. N. & Strahler, A. H. *Modern Physical Geography* (New York: John Wiley & Sons, 2 rev. ed., 1983). Definition on p. 476.
Van Dyk, J. *In Afghanistan: An American Odyssey* (New York: Putnam Publishing Group, 1983).
van Valkenburg, S. *Elements of Political Geography* (New York: Prentice-Hall, 1939).
von Grunebaum, G., ed. *Islam: Essays on the Nature and Growth of a Cultural Tradition* (London: Routledge & Kegan Paul, 1955).
Wagner, P. & Mikesell, M., eds. *Readings in Cultural Geography* (Chicago: University of Chicago Press, 1962).
Zartman, I., ed. *Man, State, and Society in the Contemporary Maghreb* (London: Pall Mall, 1973).

# CHAPTER 7

# AFRICAN WORLDS*

IDEAS AND CONCEPTS

**Medical Geography**
**Geography of Language**
**Continental Drift**
**Complementarity (2)**
**Transport Development Model**
**Colonialism**
**Négritude**
**Periodic Markets**
**Sequent Occupance**
**Unified Field Theory**
**Separate Development**

This chapter focuses on Africa south of the Sahara, the geographic realm of Black Africa. Although census data are not very reliable, its total population was approximately 417 million in 1985, about 8.5 percent of the world's total. But the realm is divided into nearly 50 countries, territories, and island entities, so that it contains almost one-third of the political units in the world today. Black Africa's most populous state is Nigeria, with 91.2 million (and perhaps 20 million more people uncounted by the last census). Next come Zaïre (33.1 million), Ethiopia (32.7 million), and South Africa (32.5 million); at the lower end of the continuum, Africa has several countries containing fewer than 1 million, including Equatorial Guinea and Gambia.

As with other world realms, it is difficult to identify one or two factors that can be used to define the boundaries of the Black African realm as well. None of the three variables that are often applied in attempts at boundary definition—religion, language, and race—are entirely satisfactory. Black Africa meets the Islamic realm in a wide transition zone that appears on Fig. I–19 (p. 39) as a line. We have already discussed the penetration of Islam into Black Africa; the outer limit of the farthest Islamic diffusion wave would hardly constitute a satisfactory boundary for the African geographic realm—it would bisect Nigeria and several other West African countries.

Another criterion lies in the languages of Africa and, in West Africa at least, the linguistic map (Fig. 7–2) comes close to marking the northern limit of the African realm. However, in the Horn on the other side of the continent, the close relationships between Amharic and Arabic make the boundary less meaningful than the religious one. As shown in Fig. 7–2, most of Africa's languages (as many as 1000 different languages are spoken on the

**Figure 7–1**
An image of Black Africa, a realm of farmers: market day in the Ivory Coast's capital of Abidjan.

* Revised by Stephen S. Birdsall

# SYSTEMATIC ESSAY

## Medical Geography

**Medical geography** is that branch of the discipline most specifically concerned with people's health. Practitioners of medical geography study the spatial aspects of health and illness: Where are diseases found? How do they spread? Are there specific kinds of environments in which certain illnesses are located? How are disease and environment related? How do the changes that societies undergo affect the health of their populations? Where do people go when seeking health care? How does the location of a health-care facility affect the opportunity to obtain care? These and many other related questions are of interest to medical geographers.

Insights into disease occurrence or disease origin can sometimes be gained simply by mapping the distribution of that disease. In a famous early case, Dr. John Snow was able to relate the deadly effects of cholera in 1854 to a contaminated water source by mapping the locations of deaths in a portion of London (Fig. I–23B, p. 50); the pattern clearly showed that many deaths occurred within a few blocks of a pump used as a major source of drinking water. Although we now know much more about this disease and its causes, the medical community in the mid-nineteenth century was still arguing whether such invisible things as "germs" actually existed. It took a map to provide evidence that, in this case at least, something about an apparently healthful water supply was fatal to many who drank from it.

This example illustrates a major concern among medical geographers with *disease ecology*. Disease ecology studies the manner and consequences of interaction between the environment and the causes of morbidity and mortality. Morbidity is the condition of illness, whereas mortality is the occurrence of death. By "environment" we mean here the varied aspects of nature and society that bear on people's lives; geographers have a long tradition of dealing with the full array of physical and human environmental factors. The complex interactions of a disease *agent* (the pathogen itself), a possible disease *vector* (the intermediate transmitter of the pathogen), the physical and social environments, and even the cultural behavior patterns of individuals potentially at risk are studied in a holistic manner by medical geographers.

Geographers map the distributions of disease (as did Dr. Snow) in order to identify the spatial context in which the illness occurs and to monitor changes in these patterns. For example, even when a primary agent is known and can be treated, as in the case of rubella (German measles), outbreaks continue to occur. Medical geographers studied the location of recurrent appearances of measles and suggested that one of the secondary causes might well be the lack of information by the population at risk about the availability of vaccines. More difficult is the study of *degenerative* diseases (such as heart disease, cancer, and stroke), because of the multiplicity of factors that lie behind these illnesses. The contributions of medial geographers—with their ecological, holistic, and descriptive approach—are especially useful for suggesting hypotheses to other medical researchers who deal directly with the causes of these diseases and their effects on individuals.

Analysis of the diffusion of illness—its spread across areas—has also been given a great deal of attention by medical geographers. The notion of *contagious* diffusion is drawn directly from our experience with diseases. In Africa, for example, the gradual spread of cerebrospinal meningitis from its early appearance in Ethiopia in 1927 across the Sahel to Senegal in 1941 was associated with the main east-west transport routes traversing the continent. Just as important, however, was the more rapid spread that occurred during the dry seasons; this was the season that contact at water sources was at its highest.

In a separate study of the diffusion of cholera in Africa during the early 1970s, R. Stock identified at least four distinct route patterns in the complex sequence of cholera diffusion that affected much of West and East Africa. Each of the four—coastal, riverine, urban hierarchical, and radial contact diffusion—was identified by a careful mapping of the direction and timing of the disease's spread. Work such as Stock's is especially useful for those attempting to prevent disease diffusion; policies designed to interrupt one type of disease spread may be unsuccessful if applied in areas where other route patterns are more important.

Not all studies of disease ecology by medical geographers suggest natural environmental causes for disease occurrence. When studying cancer of the esophagus in Malawi and Zambia during the late 1960s and early 1970s, Neil McGlashan noted a higher rate among people living in the eastern part of the region. After mapping a number of physical and cultural features and considering a variety of explanations, he argued that the distribution of this form of cancer was matched by the distribution of a particular form of locally distilled spirit. Therefore, the argument went, the cancer was related to either the distillation process or the materials used in this particular drink. The actual cause of the cancer remains unknown, but McGlashan's early association of several mapped patterns provided a focus for subsequent medical research.

There are times, too, when studies focus on secondary patterns of activity as the source of the illness. During the early 1970s, when most gasoline produced in the United States contained a lead additive to boost octane, a number of researchers examined the possible relationship between this fuel additive and patterns of lead poisoning among children. One study, for example, found a much higher level of lead in children's blood if those youngsters lived within 100 meters (330 feet) of a main road than if they lived at greater distances. Another study found that lead levels in the blood of youthful residents of a northern U.S. city varied seasonally; levels were significantly higher during the summer months when children spent more time outside their houses.

If studies of disease ecology are one major area of concern for medical geographers, a second, equivalent area involves the delivery of health care. At a very basic level, equitable health-care delivery is made difficult by the uneven spatial distribution of those who seek care and those who provide it. If people live close to a hospital or clinic, they are more likely to use those facilities when illness strikes—or so it would be sensible to expect. But how far will an individual travel to obtain care? The answer, studies show, depends on the individual's physical characteristics (age, sex), the social context (marital status, proximity to family or neighbors), economic resources (ability to pay for travel or service), attitude toward "illness" (When is a doctor necessary?), barriers to treatment (racial or religious discrimination), and much more. An answer to the question—Where does one go to obtain health care?—is not obvious. A great deal of research has been done in attempts to identify the importance of each variable in terms of its effect on the health care that is sought or obtained.

In some countries, such as the United States, doctors exercise considerable personal choice in selecting where they will practice medicine. Some researchers have pointed to the heavy concentration of doctors in urban areas—more specifically in the more affluent sections of urban areas—as an example of inequitability in health care; thus, patients least able to afford it must bear high travel costs or do without the care. In other countries, the government may regulate the distribution of health care to some degree, but many parts of the world have too few medical professionals for the population's need. In the late 1970s, there were more than 50,000 people per physician in Zaïre, and over 100,000 people per physician in Ethiopia. These figures compare with about 560 people per physician in the United States.

Related to the study of health-care delivery, medical geographers are also involved at the planning stage. As discussed in earlier chapters, one of the major research efforts by geographers since the early 1960s has been to develop methods of determining where activities *should* be located under ideal conditions. This has been used in medical geography by those seeking to identify the optimum locations for hospitals, clinics, medical offices, and even sites for emergency medical-service centers. These so-called "location-allocation" models have been developed primarily for the developed world; elsewhere, questions about the number of practitioners and the available alternatives to "Western" medicine may be more important than their spatial distribution.

Different types of questions are pressing in low-income countries. How does economic development affect the health of a population? Although the ability to pay for health care may improve on the average, new diseases may also be introduced or spread by development projects. Irrigation schemes, for example, have been found to provide new habitats for disease vectors and spread the incidence of malaria and schistosomiasis. The labor migration encouraged by urban and industrial growth has been shown to spread contagious diseases such as hepatitis when workers return home after contracting the illness, or they may bring it to the urban center when they arrive looking for work.

Clearly, there are no easy or obvious paths to the resolution of health-care questions, however vital that resolution may be. Medical geography makes its contribution to what must be a truly multidisciplinary approach through the identification and analysis of spatial patterns, locational associations, and the full environmental contexts within which morbidity and mortality occur.

continent) belong to the Niger-Congo family. In the classification of languages, we use terms employed in biology, so that languages thought to have a shared origin are grouped into a *family*; when their relationship is closer, they belong together in a *subfamily*. The Niger-Congo family of languages, therefore, contains five subfamilies with the Bantu subfamily the most extensive of these.

# TEN MAJOR GEOGRAPHIC QUALITIES OF AFRICA

1. The physical geography of Africa is dominated by the continent's plateau character, variable rainfall, soils of low fertility, and persistent environmental problems in farming.
2. The majority of Africa's peoples remain dependent on farming for their livelihood. Urbanization is accelerating, but most countries' populations remain below 40 percent urban.
3. The people of Africa continue to face a high incidence of disease including malaria, sleeping sickness, and river blindness.
4. Most of Africa's political boundaries were drawn at the beginning of the colonial period without regard for the human and physical geography of the areas they divided. This has caused numerous problems.
5. Considerable economic development has occurred in many scattered areas of Africa, but much of the region's population continues to have little access to the goods and services of a world economy.
6. Patterns of raw material exploitation and export routes set up during the colonial period still prevail in most of Black Africa. Interregional connections are poor.
7. The region is rich in raw materials vital to industrialized economies. Africa has increasingly been drawn into the competition and conflict between the world's major powers. The continent contains almost half of the world's refugee population.
8. Africa's population growth rate is by far the highest of any continent's, in spite of a difficult agricultural environment, numerous hazards and diseases, and periodic food shortages in certain areas. Some of the best land is used to produce such cash crops as coffee, tea, cocoa, and cotton for sale overseas.
9. Even though postindependence dislocations, civil wars, and massive losses of life have occurred in some areas of Africa, other sections have shown relative stability, cohesion, and economic growth.
10. South Africa, Africa's only true temperate-zone country, is also the sole remaining state in which minority rule is exercised by peoples of European descent. This region continues to experience conditions leading to fundamental change.

Scholars studying African languages are not yet in agreement over the regional distribution of African language clusters. It is agreed that the Khoisan family (6), including the Bushman languages, represents the oldest surviving African languages, spoken over a far larger area of the continent before the Niger-Congo diffusion occurred. It is also evident that Madagascar's languages belong to a non-African, Malay-Polynesian family (12), revealing the Southeast Asian origins of a large sector of that island's population. Afrikaans (1A) is an Indo-European language, a derivative of Dutch spoken by a majority of the 4.5 million whites living in South Africa, the realm's southernmost state.

The concept of race remains under constant debate among anthropologists and other scholars. Attempts to define race on purely genetic bases have not proven satisfactory. Race, like culture, is a complex and multifaceted concept. In addition, race is frequently misused in everyday thinking as a substitute for preconceptions about social potential. Categories of individual physical characteristics are erroneously assumed to represent visible indicators of behavior. In spite of a great deal of evidence to the contrary, this socialized approach to the concept of race also continues to create a great deal of disagreement. In spite of all these difficulties, it is useful to treat human races as broad categories of genetically defined features and to compare Fig. 7–3 and Fig. I–19. The comparison reveals the inclusion of both the European and North Africa/Southwest Asia realms within the area of European (Caucasian) racial distribution, the wide dispersal of Asian peoples in the Soviet realm, and the clear delimitation of the transition from African to non-African regions. The boundary across West Africa and the eastern Horn comes very close to marking the northern limit of the African geographic realm as well.

The realm of Black Africa, then, is not easily defined by any one cultural feature and is most generally referred to as Africa South of the Sahara. Even this physical definition of the realm is hardly satisfactory, for people have settled on the desert margins and have engaged for many centuries in trade and conflict with those living to the north and northeast. Adding to the difficulties of definition, there is no clear geographic focus to this

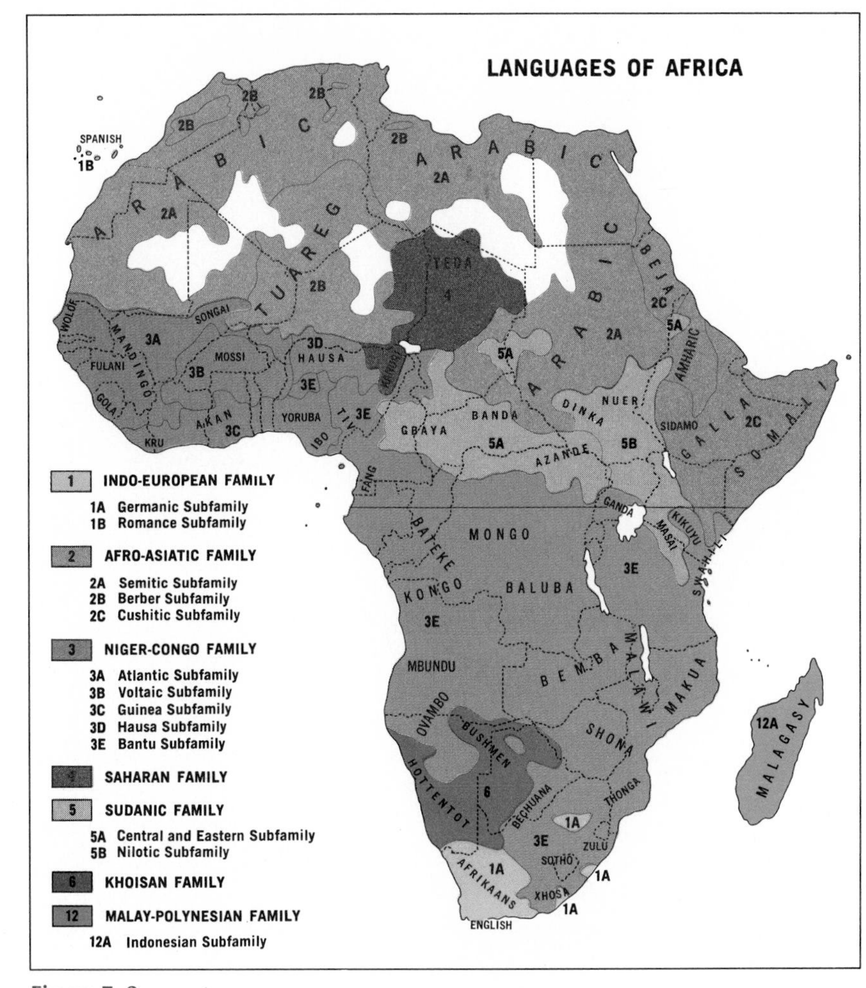


Figure 7–2

realm. There is no equivalent to the Fertile Crescent here, or to the well-known cores of Europe or North Africa. The broad sweep of Africa's physical environment does suggest a coherent pattern or two, but the continent's size and its absolute location provide little support for focused human patterns at the continental scale. At the same time, however, there is much in its physical environment that provides unity to the realm, just as there is a kind of uniformity to the heavy imprint of European colonialism in Black Africa. Both need to be considered before examining the regional variations within the realm.

# The Environmental Base

Africa is an unusual continent. Even the world map suggests this—no other landmass is so squarely positioned astride the equator, reaching almost as far to the north as to the south of this imaginary "middle line" around the earth. This location has much to do with Africa's general climatic patterns and, thus, with Africa's vegetation, soils, agricultural potential, and population distribution. No other continental landmass is without a Pacific Ocean coastline. Moreover, Africa possesses a central location vis-à-vis the world's landmasses, with the Americas to the west, Eurasia to the north and east, Australia to the southeast, and Antarctica to the south; this means that Africa enjoys good global accessibility, with a minimum aggregate distance to all the other continents. Africa is also a very large continent, second in size only to Eurasia. If Cape Mendocino, the westernmost location in California, were superimposed on the city of Dakar in Senegal, the westernmost margin of Africa, the most easterly cape along the coast of Maine would still lie west of the Nile River where it crosses into Egypt—and Africa extends eastward beyond the Nile for another 2000 kilometers (1300 miles). The continent is even longer in north-south extent than its 7200-kilometer (4500-mile) east-west breadth, reaching across 7700 kilometers (4800 miles) between the northernmost point in Tunisia and Cape Agulhas at the southern tip of South Africa.

## Physiography

Africa's physiography is also unusual when compared to other continents. In our discussions of other continental regions, for example, mountain ranges figure prominently. It is hard to visualize a South America without the Andes, North America without the Rockies, Europe without the Alps, Asia without the Himalayas. Yet Africa, covering one-fifth the land surface of the earth, has nothing comparable to these ranges. The Atlas Mountains of the Maghreb occupy a mere corner in the far northwest of the vast African bulk and, in any case, represent an extension of the Alps. The Cape Ranges in the extreme south exhibit local, not regional dimensions. And the high mountains of Ethiopia, the great volcanoes of East Africa such as Kilimanjaro (Fig. 7–4), and the sometimes snowcapped Drakensberg of South Africa are hardly comparable to the Andes and the Himalayas. There are no elongated parallel ranges here, no altiplanos, no continental-scale mountain "chain."

This discovery ought to stimulate us to scrutinize other aspects of Africa's physiography (Fig. 7–5). In East Africa lies a set of great lakes. With the single exception of Lake Victoria, these lakes are remarkably elongated, from Lake Malawi in the south to Lake Turkana in the north. What causes this elongation and the persistent north-south trend that can be observed in these lakes? The lakes occupy portions of deep trenches cutting through the East African plateau, trenches that can be seen to extend well beyond the lakes themselves. Northeast of Lake Turkana such a trench cuts the Ethiopian massif into two sections, and the Red Sea itself looks much like a northward continuation of it. On both sides of Lake Victoria, smaller lakes lie in similar trenches, of which the western one runs into Lake Tanganyika and the eastern one extends completely across Kenya, Tanzania, and Malawi. The technical term for these trenches is *rift valleys*; as the name implies, they are formed when huge parallel cracks or *faults* appear in the earth's crust, and the in-between strips of land sink or are pushed down to form great linear valleys (Fig. 7–6). Altogether, these rift valleys stretch more than 9600 kilometers (6000 miles) from the north end of the Red Sea to Swaziland in Southern Africa. In general, the rifts from Lake Turkana southward are between 30 and 90 kilometers (20 and 60 miles) wide, and the walls, sometimes sheer and sometimes steplike, are well defined.

Next, our attention is drawn to Africa's unusual river systems (Fig. 7–5). Africa has several great rivers, with the Nile and Congo (or Zaïre)

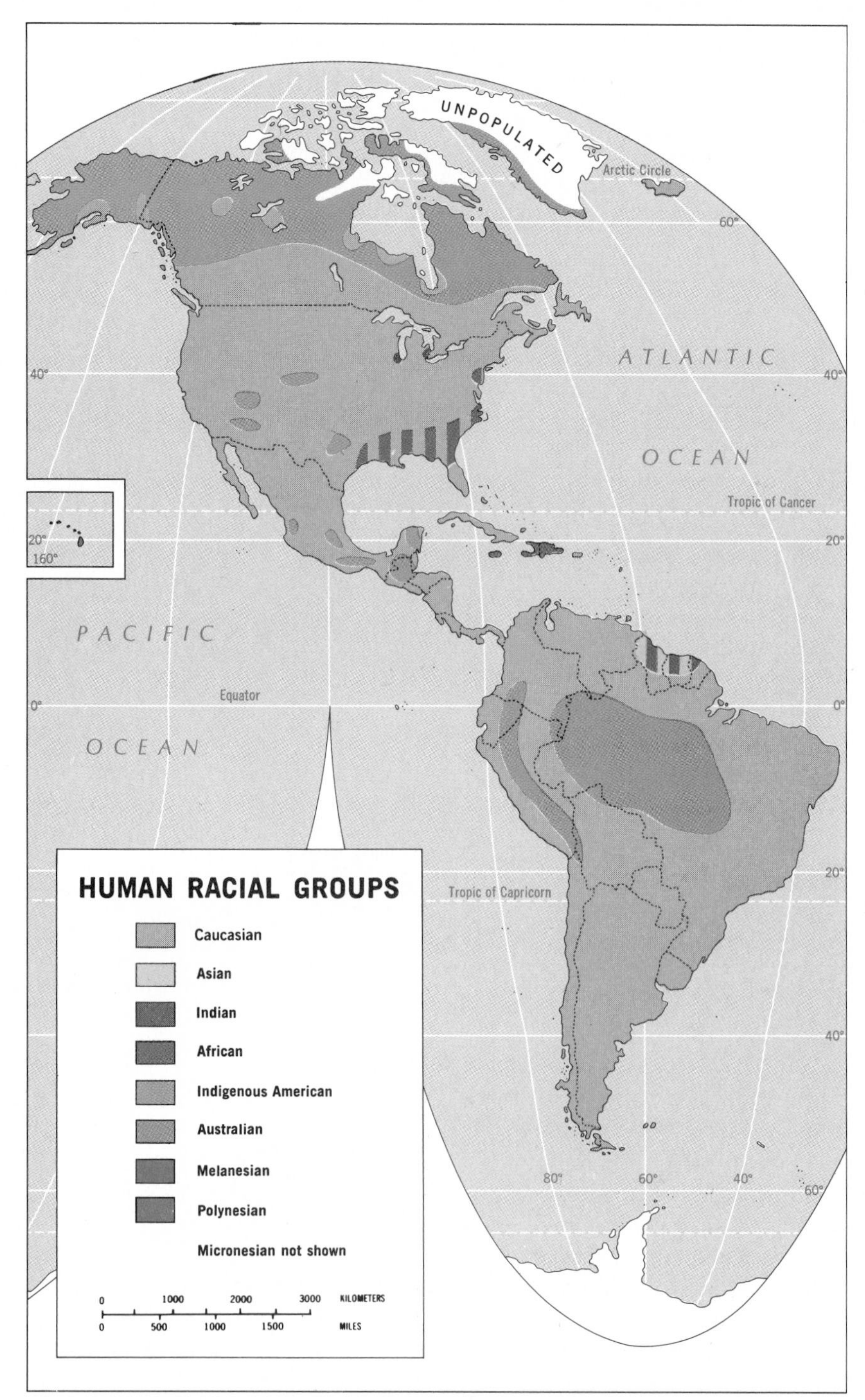


Figure 7–3

ranking among the most noteworthy in the world. The *Niger* rises in the far west of Africa, on the slopes of the Futa Jallon Highlands, but first flows inland toward the Sahara Desert. Then, after forming an interior delta, it suddenly elbows southward, leaves the desert area, plunges over falls as it cuts through the plateau area of Nigeria, and creates another large delta at its mouth. The *Congo (Zaïre)* River begins as the Lualaba River on the Zaïre-Zambia boundary; for some distance, it actually flows northeast before turning north, then west and

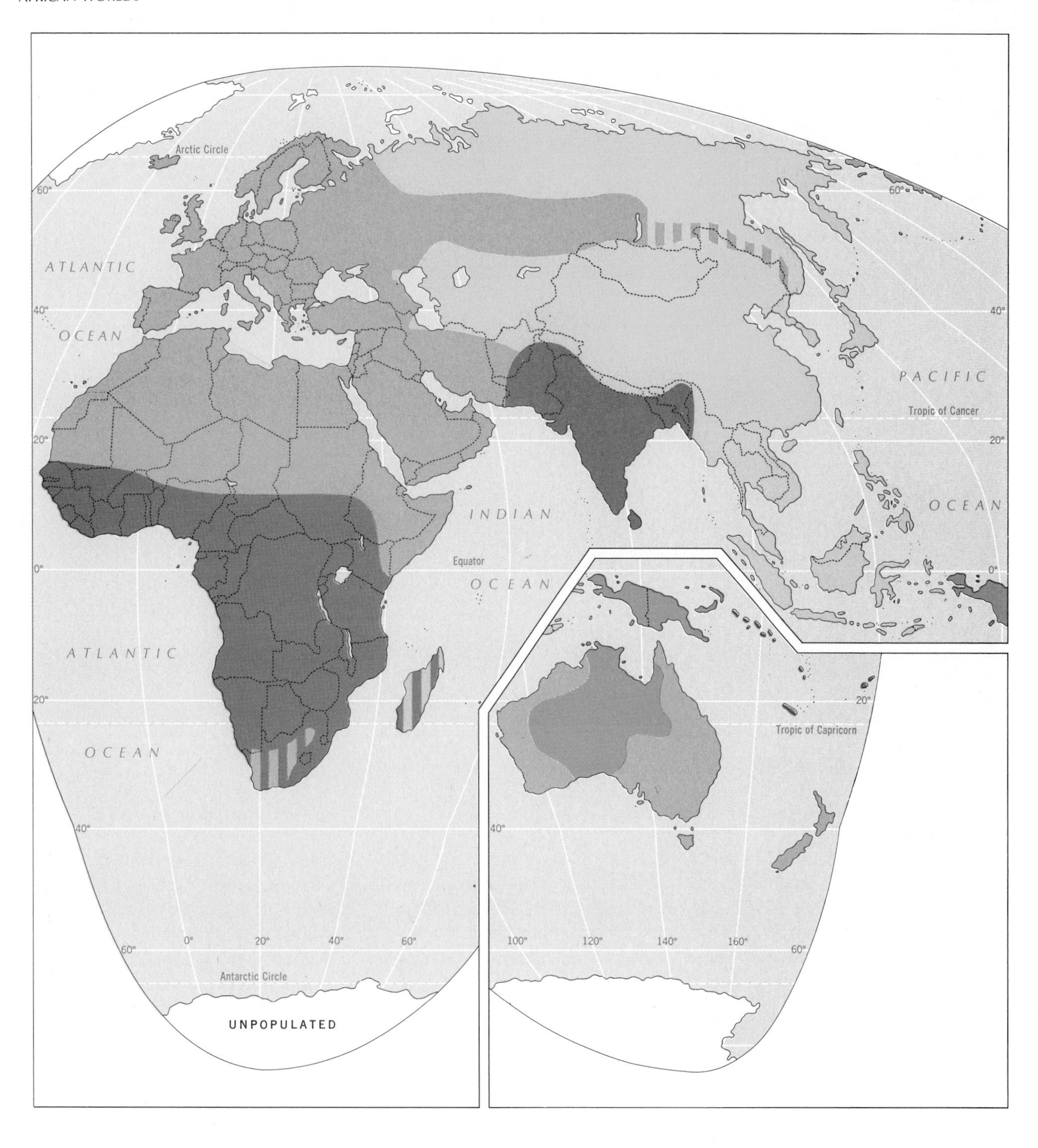


southwest, finally cutting through the Crystal Mountains to reach the ocean. Note that the upper courses of these first two rivers appear quite unrelated to the continent's coasts where they eventually exit. In the case of the *Zambezi* River, whose headwaters lie in Angola and northwestern Zambia, the situation is the same; the river first flows south, toward the inland delta known as the Okovango Swamp, then it turns northeast and southeast, eventually to reach its delta immediately south of Lake Malawi (and the southern terminus of the eastern rift-valley system). Finally, there is the famed erratic course of the *Nile* River, which braids into numerous channels in the Sudd area of southern Sudan, and in its middle course actually reverses direction and flows southward before resuming its flow

**Figure 7–4**
The twin volcanic peaks of Kilimanjaro: Kibo, the taller cone, carries permanent ice and snow.

toward the Mediterranean delta in Egypt. With so many peculiarities among Africa's river courses, could it be that all have been affected by the same event at some time in the continent's history? Perhaps—but first let us look further at the map.

All continents have low-lying areas; witness the interior lowlands of North America, or the coastal and river lowlands of Eurasia and Australia. But as the map shows, coastal lowlands are few and of small extent in Africa. In fact, it is reasonable to call Africa a *plateau continent*; except for coastal Moçambique and Somalia, and along the northern and western coasts, nearly all of the continent lies above 300 meters (1000 feet) in elevation and fully half of it is over 800 meters (2500 feet) high. Even the Congo Basin, Equatorial Africa's huge tropical lowland, lies well over 1000 feet above sea level, in contrast to the much lower-lying Amazon Basin across the Atlantic. Although Africa is mostly plateau, this does not mean that the surface is completely flat and unbroken. In the first place, the rivers have been eroding the surface for millions of years and have made some pretty good dents in it. For example, Victoria Falls on the Zambezi (Fig. 7–7) is 1600 meters (1 mile) wide and over 100 meters (330 feet) high. Volcanoes and other types of mountains, some of them erosional "leftovers," stand well above the landscape in many areas (Fig. 7–4); the Sahara Desert is no exception, where the Ahaggar and Tibesti Mountains both reach about 3000 meters (10,000 feet) elevation. In several places, the plateau has sagged down under the weight of accumulating sediments. In the Congo (Zaïre) Basin, for example, rivers have transported sand and sediment downstream for tens of millions of years and, for some reason, dropped their erosional loads into what was once a giant lake the size of an interior sea. Today the lake is gone, but the thick sediments that press this portion of the African surface into a giant basin are proof that it was there. And this was not the only inland sea. To the south, the Kalahari Basin was filling with sediments that, today, comprise that desert's sand; and to the north, in the Sahara, three similar basins lie in Sudan, in Chad, and in what is Mali today.

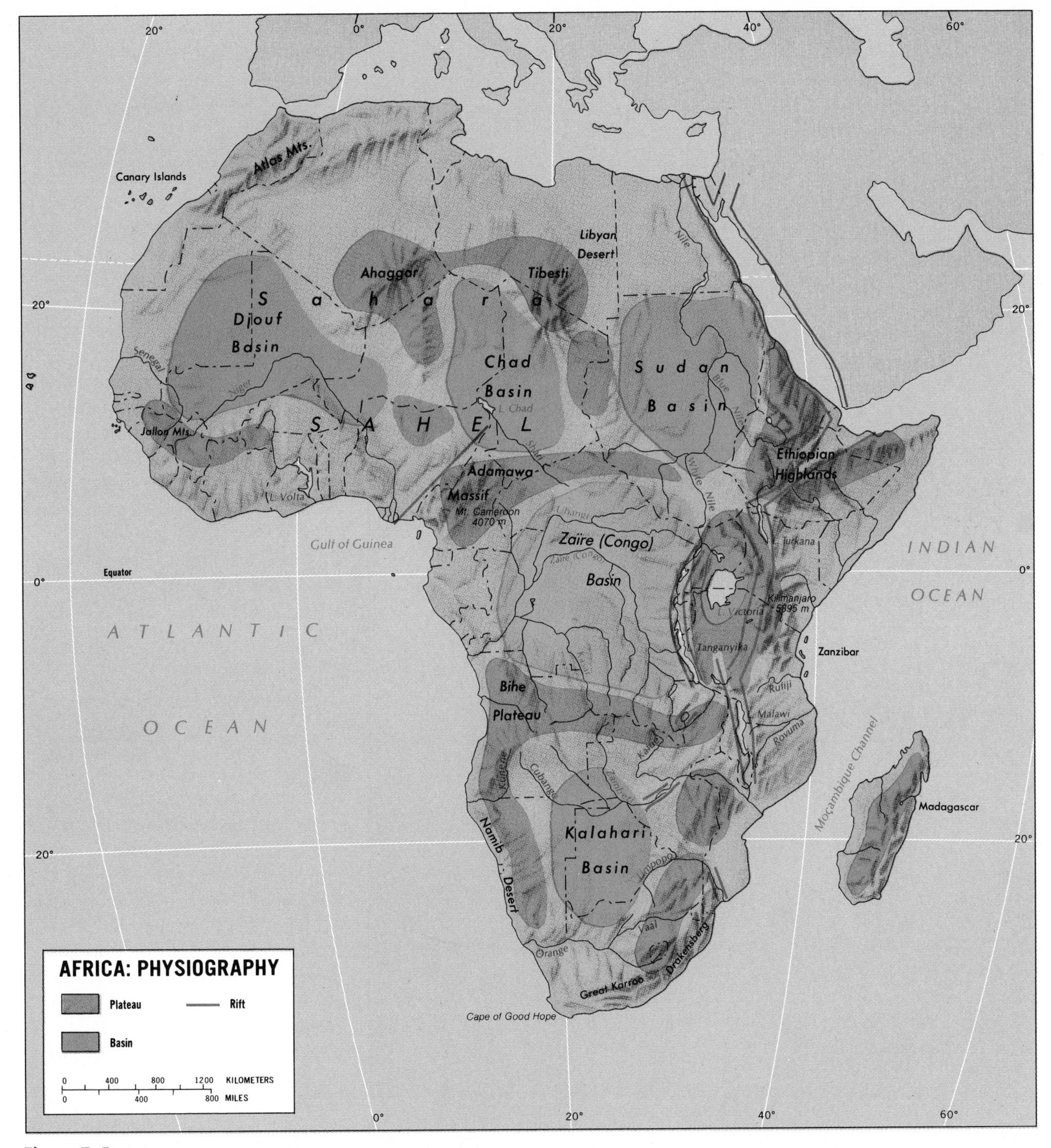


Figure 7–5

The margins of Africa's plateau, too, are of significance. Much of the continent, because it is a plateau, is surrounded by an escarpment. In Southern Africa, where this feature is especially pronounced, the *Great Escarpment* (as it is called there) marks the plateau's edge along many hundreds of kilometers, where the land drops precipitously from more than 1500 meters in elevation to a narrow, hilly coastal belt. From Zaïre to Swaziland, and intermittently on or near most of the African coastline, a scarp bounds the interior upland. Such escarpments are found in other parts of the world too: Brazil at the eastern margins of the Brazilian Highlands, India at the western edge of its Deccan Plateau. But Africa, even for its size, has a disproportionately large share of this topographic phenomenon.

**Figure 7–6**
Rift valleys fracture the East African landscape: the edge of the rift adjacent to Lake Manyara in Tanzania.

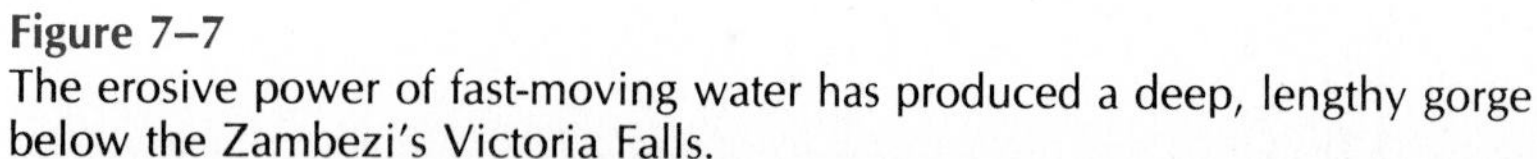

**Figure 7–7**
The erosive power of fast-moving water has produced a deep, lengthy gorge below the Zambezi's Victoria Falls.

And to this remarkable list of peculiar African physiographic features we may add one other characteristic: the continent's western coastline, with a configuration that matches the east coast of South America. In fact, this jigsaw-puzzle-like fit is so remarkable that it was recognized almost as soon as the first maps of Africa and South America were drawn (Sir Francis Bacon drew attention to the match-up as early as the 1620s). In combination, Africa's world location with respect to the other continents, its distinctive physiography, and its coastal configurations suggest a double-barreled hypothesis: (1) the landmasses at one time were part of a giant single continent and have since drifted apart, and (2) Africa's unique physiography stems from its position at the heart of this ancient supercontinent.

## Continental Drift

The hypothesis of **continental drift** is now widely accepted. As we saw in the introductory chapter in the discussion of plate tectonics (pp. 9,12), the idea is not new. However, it is worth recording at this point the extent to which such geographical considerations as direction, distance, shape, and relative location can contribute to questions that lead to such a hypothesis. But what can continental drift as a hypothesis do to help explain the general physical characteristics of Africa? First, let us consider the basic idea: Africa, along with South America, Antarctica, Australia, Madagascar, and even southern India at one time formed a supercontinent called Gondwana (Fig. 7–8), which was the southern portion of the even larger landmass known as Pangaea (discussed on p. 9). The significant point of this map is that Africa occupied the *central* position in Gondwana and was, thus, surrounded by the other landmasses. After a long period of unity, the supercontinent began to break apart more than 100 million years ago. The various fragments have moved radially away from Africa and are continuing to do so; Africa itself has moved least of all from its earlier location near the South Pole.

In the hypothetical reassembly of Gondwana, then, Africa was the core-shield; it occupied the heart of the supercontinent, and, as such, it did not have the coasts it has today. The rivers that arose in the interior failed to reach the sea: the upper Niger flowed into Lake Djouf, the Shari River into Lake Chad, the Upper Nile into Lake Sudan, the Lualaba into Lake Congo, and the upper Zambezi into the Okovango Delta on the shores of Lake Kalahari. Other rivers also drained toward these great basins, which filled with sediments. Eventually, the great Gondwana landmass, for reasons still in doubt, began to break up. South America began to drift westward and the African west coast formed. Antarctica moved away to the south, Australia to the east, India to the northeast, and Africa itself drifted slightly northward (Fig. 7–8 and Fig. I–6).

Now the coasts were created all around the continent, where huge escarpments marked the great continental fractures (probably much like today's rift valleys as in Fig. 7–6), and rivers began to attack the Great Escarpment. Before long, these rivers, by headward erosion, reached the long-isolated lake basins in Africa's interior. With the release of huge volumes of lake water, these rivers cut deep gorges and formed fast-retreating waterfalls (Fig. 7–7). It is no accident that Africa, not exactly a well-watered continent today, still accounts for nearly half of the hydroelectric power potential in the world (Fig. 7–9). According to the drift explanation, then, Africa's major rivers have a double history: the upper courses are predrift and filled the basins; their lower courses are

**Figure 7–8**

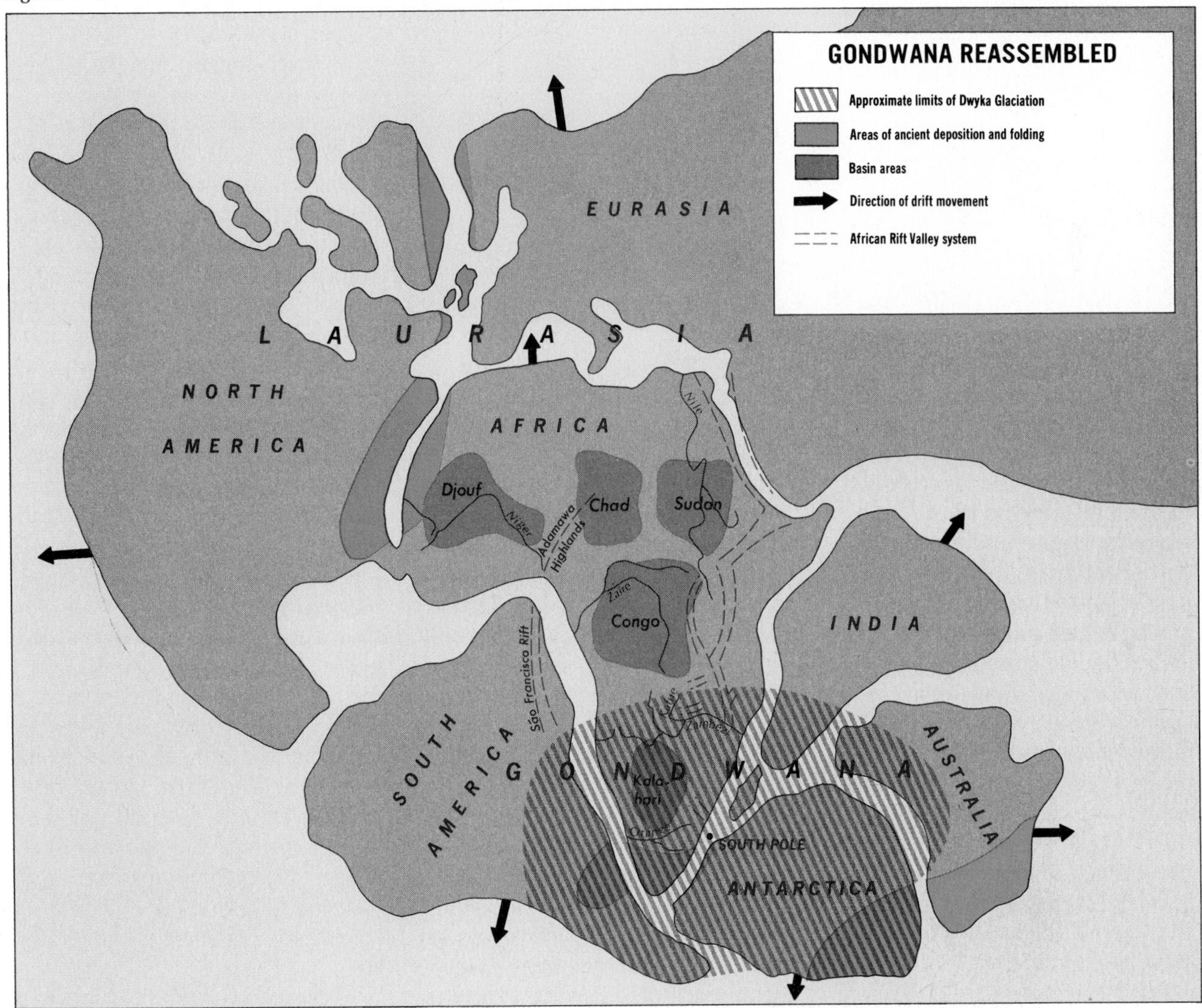

**Figure 7–9**
Zaïre's Inga Dam, one of Africa's largest hydroelectric projects, located at the rapids where the Zaïre (Congo) River crosses the Great Escarpment.

younger and resulted from the release of the pent-up lakes. It is tempting to carry the explanation further and to argue that, since the landmasses moved away from Africa both east and westward, Africa was under great tensional stress and began to yield along the longitudinal rift valleys we can observe in the landscape today. In fact, it is probably not so simple, although the rifts are certainly related in some way to the drift process. For the absence of mountains, then, a deceptively easy answer appears: all the other landmasses moved many hundreds, even thousands, of miles; in the process, their leading edges were crumpled up into giant folded and crushed mountain arcs. This would explain the location of the American and Australian mountains on the side away from Africa. But Africa itself, which moved only slightly to the north, shows such evidence only along its northwestern margin.

Consider how the drift hypothesis ties the African—and, indeed, the whole Gondwanaland—physiography together. It relates the rivers and their peculiarities to the Great Escarpment, the Escarpment to the rift valleys, the rift valleys to the absence of folded mountains, the absence of folded mountains to Africa's limited lateral movement, and the factor of lateral movement to the distribution of major world mountain chains. It provides an explanation for the correspondence between Africa's west coast and South America's east. It has permitted geological and paleontological predictions to be made. And it suggests, on a truly grand scale, that the imbalance inherent in the concentration of continents in the Land Hemisphere and the virtual emptiness of the Pacific-centered Oceanic Hemisphere, is being corrected by a slow redistribution of the landmasses toward the Pacific Basin.

## Climate and Vegetation

Africa's climatic and natural vegetation distributions are almost symmetrical about the equator (Figs. I–10, I–12). With the equator virtually bisecting the continent's north-south extent, the atmospheric conditions that affect the African surface tend to be similar in both halves. The hot rainy climate of the Congo (Zaïre) Basin merges gradually both north and south into climates with distinct winter dry seasons. Outside the Basin, therefore, trees are less majestic and then less frequent as one moves away from the equator. Eventually, the humid equatorial region is left entirely behind and annual rainfall becomes less abundant and less reliable. The alternation of trees with grass vegetation in the tropical savanna and the full grasslands of the tropical steppe, both of which are green only after the rains, give way to the arid conditions of the Sahara and Kalahari Deserts. Even the dry-summer subtropical (Mediterranean) climate of the northern extreme is matched by a small equivalent zone at the continent's southern tip.

The exceptions to this symmetry

are almost all related to the continent's asymmetric outline and the topographic irregularities on its plateau surface. Rainfall is received in sufficient quantities along parts of the western bulge to maintain forested conditions. The Horn of Africa south of the Arabian Peninsula is drier than would be expected solely from its latitude, and the highlands of Ethiopia are wetter. In fact, since the entire plateau is higher in the east and south than in the west and north, the neat climatic pattern of the central span is disrupted altogether along the eastern side of the continent. The implications of this general climatic scheme—almost entirely tropical even though diverse—are many. Among the most serious are those that bear on environmental hazards and disease and those related to the livelihoods pursued by most Africans.

## Environmental Hazards and Diseases

Africa is not a densely populated realm. The total population of approximately 417 million includes several large concentrations (Fig. I–14) in Nigeria, in the region surrounding Lake Victoria, and in southern Africa, but much of Africa south of the Sahara remains sparsely peopled. African environments are difficult, as we have discussed. In the 1970s, parts of Ethiopia, West Africa (in the Sahel), Tanzania, and Moçambique suffered from regional or local famines. Diets are not well balanced, and life expectancies are nowhere lower (Fig. 7–10). Millions of Africans are unwell their entire lives, always fighting illnesses that can afflict them almost from the day they are born. Children do not ingest sufficient proteins, and such deficiency disorders as kwashiorkor and marasmus are common, especially in places where people depend on starchy root crops for their staples.

Diseases strike populations in different ways. When a sudden outbreak occurs, leading to a high percentage of afflictions in a population (and perhaps a large number of deaths), the outbreak is an *epidemic*. An epidemic is a regional phenomenon, or even a local one. When an outbreak that begins locally spreads regionally and then worldwide as various forms of influenza have done in recent times, it is described as having *pandemic* proportions. But a disease can inhabit a human population in still another way. Some illnesses do not strike in a violent attack but exist in a sort of equilibrium with the population. Many people will have it, and the disease saps energies and shortens lifetimes. These, nevertheless, are the survivors of the first attack during childhood or later, who are now able to withstand the parasites. Such *endemic* diseases affect many millions of Africans (examples of endemic diseases in the United States are syphilis and mononucleosis).

Undoubtedly, malaria, endemic to most of Africa and a killer of small children, is the worst. In recent years, malaria has again been on the increase. The malarial mosquito is the vector (carrier) of the parasite, and the mosquito prevails in almost all of inhabited Africa. Africans who survive childhood are likely to suffer from malaria to some degree, with a debilitating effect. Sickle-cell anemia, a blood condition that weakens and frequently is fatal to those who have it, appears related to the endemic presence of malaria in Africa.

African sleeping sickness is transmitted by the tsetse fly (Fig. 7–11) and now affects most of tropical Africa. This fly infects not only people, but also their livestock. Its impact on Africa's population has been incalculable. The disease appears to have originated in a West African source area about A.D. 1400; since then, it has inhibited the development of livestock herds where meat would have provided a crucial balance to seriously protein-deficient diets. It channeled the migrations of cattle herders through fly-free corridors in East Africa (Fig. 7–11), destroying herds that moved into infested zones. Most of all, it ravaged the human population, depriving it not only of potential livelihoods, but also of its health.

Still, another serious and widespread disease is yellow fever, also a mosquito-transmitted malady. Yellow fever is endemic in the wetter tropical zones of Africa, but it sometimes appears in other areas in epidemic form, as in 1965 in Senegal when 20,000 cases were reported (there were undoubtedly thousands more). This is another disease that strikes children, who, if they survive, acquire a certain level of immunity.

To this depressing list must be added schistosomiasis, also called bilharzia, which is transmitted by snails. The parasites enter via body openings when people swim or wash in slow-moving or standing water infested by the snails. Many development projects in Africa involving irrigation have inadvertently introduced schistosomiasis into populations previously not exposed to the parasites. Internal bleeding, loss of energy, and pain result, although schistosomiasis is not by itself fatal. It is endemic today to more than 200 million people worldwide, most of them residing in Africa.

These are the major and more or less continentwide diseases, and there are many others of regional and local distribution. The dreaded river blindness, caused by a parasitic worm transmitted by a small fly, is endemic in the savanna belt south of the Sahara from Senegal east to Kenya; in northern Ghana alone, it blinds a large percentage of the adult villagers (Fig. 7–12). And animals and plants as well are attacked by Africa's ravages. Besides sleeping sickness, livestock herds are also afflicted by rinderpest. Crops and pastures are stripped periodically by swarms of locusts, which number an average of 60 million insects each, travel thousands of kilometers, and devour the vegetation of entire countrysides. Add to these problems those of widespread poor soils and inadequate precipitation, and Black Africa's population of 417 million appears a great deal more impressive.

# The Predominance of Agriculture

A great many Africans live today as their predecessors lived—by subsistence farming, herding, or both. The oldest means of survival, hunting and gathering, still sustains the Bushmen of the Kalahari and the Pygmies of Zaïre, but, for many centuries, the great majority of Africans have been farmers. Not that the African environment is particularly easy for the cultivator who tries to grow food crops or raise cattle; tropical soils are notoriously unproductive, and much of Africa suffers from excessive drought.

It seems strange that on a continent located astride the equator, most of it in the tropics and flanked by two oceans, water supply can be a problem. Consider Fig. I–9 (p. 17); as we have seen, only small parts of West Africa and interior Zaïre exhibit really high annual precipitation totals, and the rainfall drops off very quickly both northward and southward. In West Africa, only a relatively narrow coastal zone is well watered; East Africa is actually quite dry, with steppe conditions (see Fig. I–12) penetrating far into Kenya (Fig. 7–13); in Southern Africa, the rainfall is a modest 50 to 100 centimeters (20 to 40 inches) only in the east and south, whereas, westward, the Kalahari and Namib Deserts take over. And what the map does not tell us is that the hot tropical sun, by evaporating a good part of the rainwater that does reach the ground, reduces the moisture available for plant growth even further (see *Box*, p. 14). Moreover, much of the rainfall of Africa outside the wettest areas is concentrated in one or perhaps two seasons, and the intervening periods may be bone-dry;

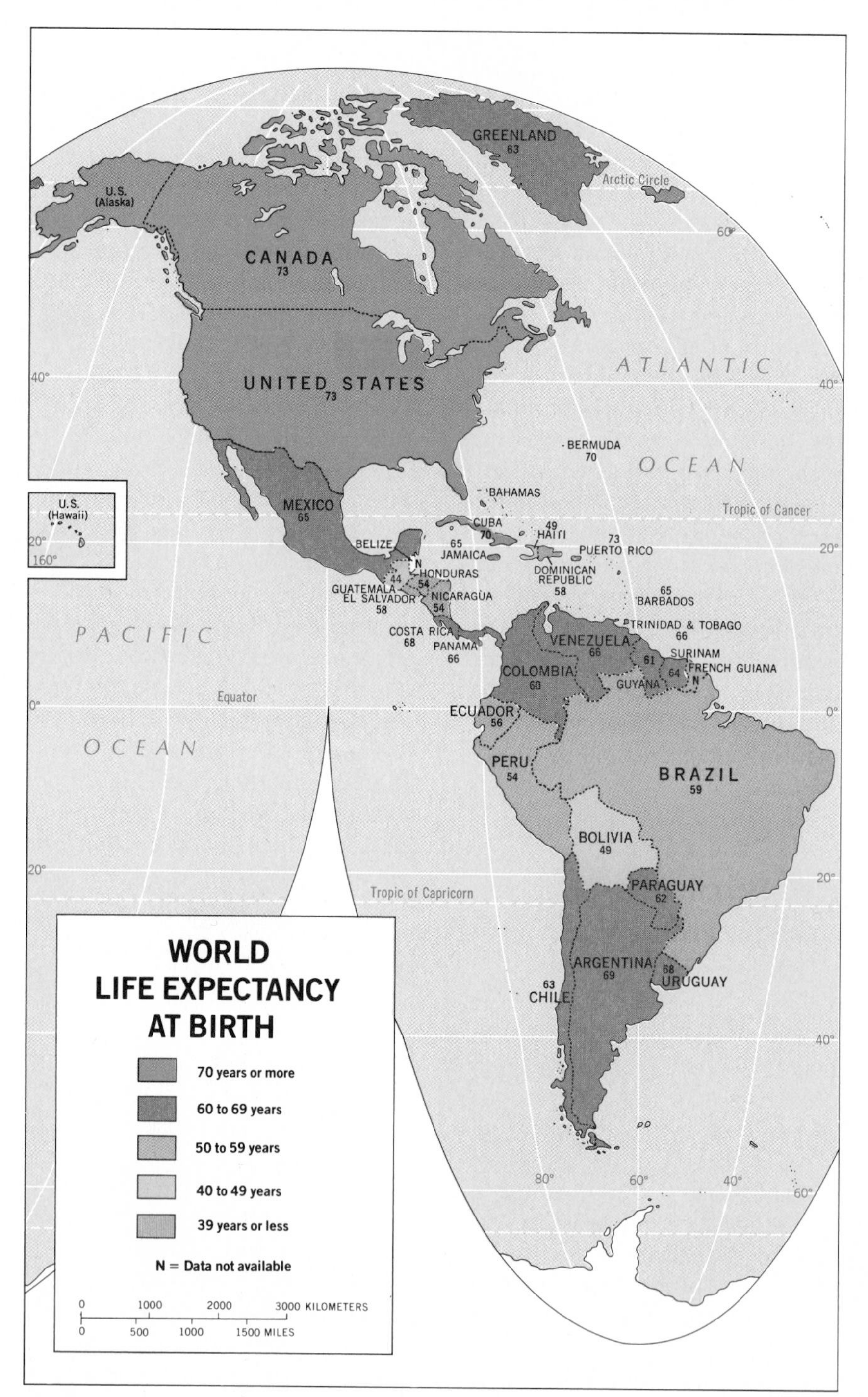


**Figure 7–10**

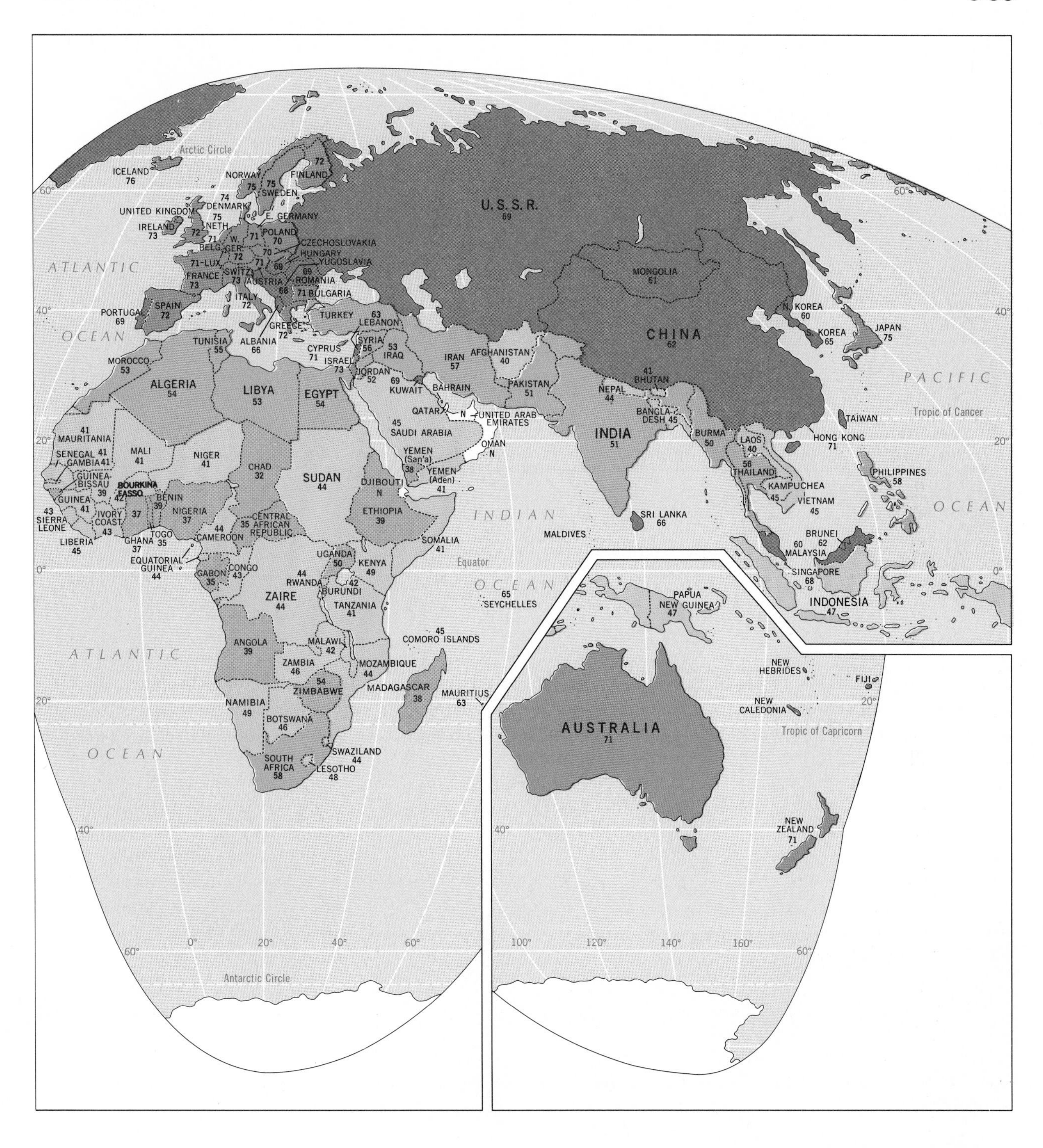


before long, the herds of livestock have used up the water reserve from the short wet season, and their owners must drive them in search of water somewhere else. This is not just a problem faced by the Masai pastoralists of Kenya and Tanzania, it also confronts the ranchers of European cultural background in South Africa. The Masai drive cattle in a never-ending search for pasture and water; the South African ranchers may have to ship their cattle out by rail before it is too late.

The vast majority of African families still depend on subsistence farming for their living (Fig. 7–14), although more and more now include some cash cropping in their efforts. Only a very few still survive on hunting and gathering alone; even the Pygmies today trade for vegetables with their neighbors. Along the coast—especially the west coast—and the major rivers, some communities depend primar-

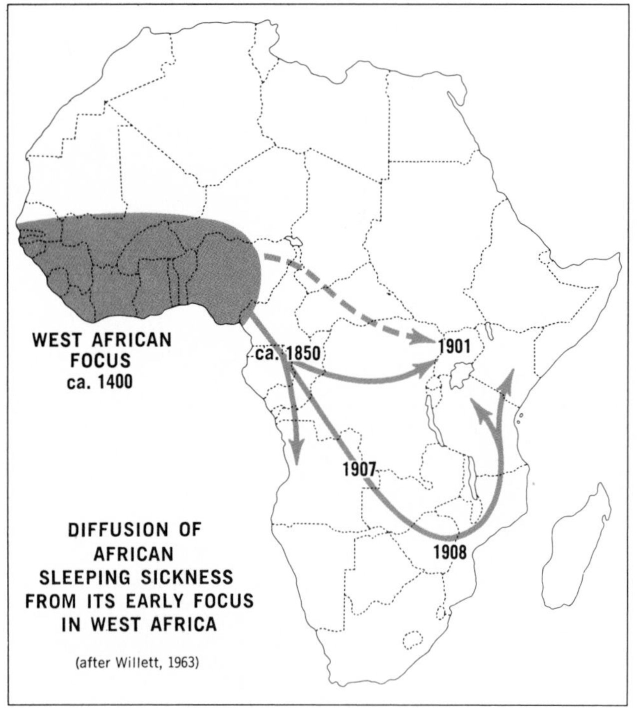


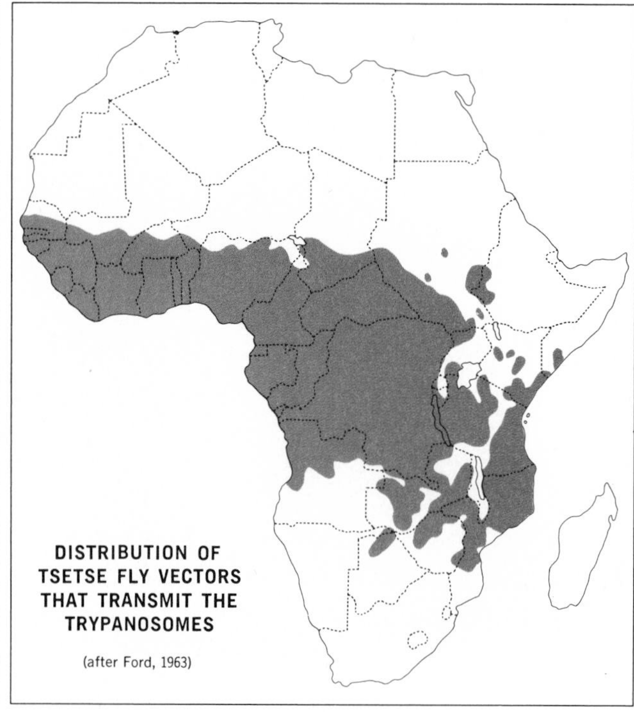


Figure 7–11

ily on fishing. Otherwise, the principal mode of life is farming—with grain crops dominant in the drier areas and root crops in the more humid zones. In the methods of farming, the sharing of the work between men and women, the value and prestige attached to herd animals, and other cultural aspects, subsistence farming provides the opportunity for insight into the Africa of the past. Moreover, the subsistence form of livelihood was changed only very indirectly by the colonial impact; tens of thousands of villages all across Africa were never fully brought into the economic orbit of the European invaders, and life in these settlements went on more or less unchanged.

Figure 7–12
A young child acts as the "eyes" of an elderly victim of river blindness.

Africa's herders more often than not mix farming with their pastoral pursuits, and very few of them are actually "pure" herders. In Africa south of the Sahara, there are two belts of herding activity: one extending laterally along the West African savanna-steppe and connecting with the East African area (where the famous Masai drive their herds), the other centering on the plateau of South Africa. Especially in East and South Africa, cattle are less important as a source of food than they are as a measure of the wealth and prestige of their owners in the community; hence, African cattle owners in these areas have always been more interested in the size of their herds than in the quality of the animals. The sta-

**Figure 7–13**
East Africa can be surprisingly dry: the drought-devastated vegetation of a part of the savanna.

ple food in these areas entails the grain crops of corn, millet, and sorghum (together with some concentrations of rice), which overlap with the herding areas. Probably a majority of Africa's cattle owners are sedentary farmers, although some people—such as the Masai—engage in a more or less systematic cycle of movement, following the rains and seeking pastures for their livestock. In West Africa, the pastoralists who will sell their cattle for meat face the problem of considerable distances to the major coastal markets. If the animals are taken by truck across the several hundred kilometers from the savanna-steppe margins to the coast, the cost is high, but the cattle do not lose much weight between farm and market; on the other hand, if the animals are driven to the markets on the hoof, the cost of moving them is low, but their weight loss is substantial. And the cattle herder, too, faces environmental problems, not the least of which is the dreaded tsetse fly.

Cattle, of course, are not the only livestock in Africa. Chickens are nearly ubiquitous, and there are millions of goats everywhere—in the forest, in the savanna, in the steppe, and even in the desert; the goats always seem to survive, and no African village would be complete without them. Where conditions are favorable goats multiply swiftly, and then they denude the countryside and promote soil erosion; in Swaziland, for example, goats constitute both an asset to their individual owners and a serious liability to the state.

A gradual change in the involvement of Africans in the cash economy has taken place during the last quarter century. Excluded during the colonial period from most activities that could produce income for themselves and their families, many Africans have chosen since independence to make changes that promise to bring in money. Farmers have introduced cash crops onto their small plots, at times replacing the subsistence crops altogether. Itinerant traders, moving from market to market, carry small packets of manufactured goods into the most remote villages. Young people and the not-so-young have, by the millions, migrated to major urban centers and other established areas of employment in search of work. Although these changes, in general terms, have accelerated during the last generation, the activities are hardly new to the realm, because Africa's past is as rich with the successes and failures of individual human effort as any continent's.

# Africa's Past

Africa is the cradle of humanity. Research in Kenya, Ethiopia, and Tanzania has steadily pushed back the date of the earliest origin of human prototypes by hundreds of thousands, even millions, of years. It is, therefore, something of an irony that comparatively little is known about Black Africa from 5000 to 500 years ago—that is, prior to the onset of European colonialism. This is only partly due to the colonial period itself, during which African history was neglected, many African traditions and artifacts were destroyed, and many misconceptions about African cultures and institutions arose and became entrenched. It is also a result of the absence of a written history over most of Africa south of the Sahara until the sixteenth century—and over a large part of it until much later than that. The best records are those of the savanna belt immediately south of the Sahara Desert, where contact with North African peoples was greatest and where Islam achieved a major penetration.

The absence of a written record does not mean, as some scholars suggested, that Africa does not have a history as such prior to the coming of Islam and Christianity.

**Figure 7–14**
Subsistence farming of staple crops still dominates African agriculture: shelling corn (the realm's leading staple) in a village in Zambia.

Nor does it mean that there were no rules of social behavior, no codes of law, no organized economies. Modern historians, encouraged by the intense interest shown by Africans generally, are now trying to reconstruct the African past, not only where this can be done from the meager written record, but also from folklore, poetry, art objects, buildings, and other such sources. But much has been lost forever. Almost nothing is known of the farming peoples who built well-laid terraces on the hillsides of northeastern Nigeria and East Africa or of the communities that laid irrigation canals and constructed stone-lined wells in Kenya; and very little is known about the people who, perhaps a thousand years ago, built the great walls of Zimbabwe. Porcelain and coins from China, beads from India, and other goods from distant sources have been found at Zimbabwe and other points in East and Southern Africa, but the trade routes within Africa itself—let alone the products that circulated on them and the people who handled them—still remain the subject of guesswork.

## The Pre-European Prelude

Africa on the eve of the colonial period was in many ways a continent in transition. For several centuries, the habitat in and near one of the continent's most culturally and economically productive areas—West Africa—had been changing. For 2000 years, probably more, Africa had been innovating as well as adopting ideas. In West Africa, cities were developing on an impressive scale; in Central and Southern Africa, peoples were moving, readjusting, sometimes struggling with each other for territorial supremacy. The Romans had penetrated to the southern Sudan, North African peoples were trading with West Africans, and Arab *dhows* were plying the eastern coasts, bringing Asian goods in exchange for gold, copper, and a comparatively small number of slaves.

Consider the environmental situation in West Africa as it relates to the past. As Figs. I–9 to I–13 (pp. 17–25) indicate, the environmental regions in this part of the continent have a strong east-west orientation. The isohyets run parallel to the coast (Fig. I–9); the climatic regions, now positioned somewhat differently from where they were two millennia ago, still trend strongly east-west (Fig. I–10). Soil regions are similarly aligned (Fig. I–13), and the vegetation map, although generalized, also reflects this situation (Fig. I–12), with a coastal forest belt yielding to savanna (tall grass with scattered trees in the south, shorter grass in the north), which, in turn, gives way to steppe and desert.

Essentially, then, the situation in West Africa was such that over a north-south span of not too many hundreds of kilometers there was an enormous contrast in environments, economic opportunities, modes of life, and products. The people of the tropical forest produced and needed goods that were quite different from the products and requirements of the peoples of the dry, distant north. To give an example, salt is a prized commodity in the forest, where the humidity precludes its formation, but salt is in plentiful supply in the desert and steppe. Hence, the desert peoples could sell salt to the forest peoples, but what could be offered in exchange? Ivory and spices could be sent north; there were elephants in the forest, as were certain plants that yield valuable condiments. Thus, there was a degree of complementarity between the peoples of the forests and the peoples of the drylands. And the peoples of the savanna—those who were located in between—were beneficiaries of this situation, for they found themselves in a position to channel and handle the trade, and that activity is always economically profitable. The markets in which these goods were exchanged prospered and grew; so, there arose a number of true cities in the savanna belt of West Africa. One of these old cities, now an epitome of isolation, was once a thriving center of commerce and learning and one of the leading urban places in the world—Timbuktu. Others, predecessors as well as successors of Timbuktu, have declined, some of them into oblivion. Still other savanna cities continue to have considerable importance, such as Kano in the northern part of Nigeria.

States of impressive strength and truly amazing durability arose in the West African culture hearth. The oldest state about which anything at all is known is Ghana. Ancient Ghana was located to the northwest of the coastal country that has taken its name in recent years. It covered parts of present-day Mali and Mauritania, along with some adjacent territory. It lay astride the upper Niger River and included gold-rich streams flowing off the Futa Jallon Highlands, where the Niger has its origins. For a thousand years, perhaps longer, old Ghana managed to weld various groups of people into a stable state. The country had a large capital city, complete with markets, suburbs for foreign merchants, religious shrines, and, some miles from the city center, a fortified royal retreat. There were systems of tax collection for the citizens and extraction of tribute from subjugated peoples on the periphery of the territory; tolls were levied on goods entering the Ghanaian domain; and an army maintained control. Moslems from the northern drylands invaded Ghana in about 1062 when the state may already have been in decline. Even so, Ghana continued to show its strength: the capital was protected for no less than 14 years. However, the invaders had ruined the farmlands, and the trade links with the north were destroyed. Ghana could not survive, and it finally broke apart into a number of smaller units.

In the centuries that followed, the focus of politico-territorial organization in the West African culture hearth shifted almost continuously eastward—first to ancient Ghana's successor state of Mali, which was centered on Timbuktu and the middle Niger River Valley, and then to the state of Songhai, whose focus was Gao, also a city on the Niger and one that still exists today. One possible explanation for this eastward movement may lie in the increasing influence of Islam; Ghana had been a pagan state, but Mali and its successors were Moslem and sent huge, rich pilgrimages to Mecca along the savanna corridor south of the desert. Indeed, hundreds of thousands of citizens of the modern Republic of Sudan trace their ancestry to the lands now within northern Nigeria, their ancestors having settled there while journeying to or from Mecca.

In any event, the West African savanna region was the scene of momentous cultural, political, and economic developments for many centuries—but it was not alone in its progress in Africa. In what is today southwestern Nigeria, a number of urban farming communities became established, the farmers being concentrated in these walled and fortified places for reasons of protection and defense; surrounding each "city of farmers" were intensely cultivated lands that could sustain thousands of people clustered in the towns. In the arts, too, Nigeria produced some great achievements, and the bronzes of Benin are true masterworks (Fig. 7–15). In the region of the Congo (Zaïre) River mouth, a large state named Kongo existed for centuries. In East Africa, trade on a large scale with China, India, Indonesia, and the Arab world brought crops, customs, and merchandise from these distant parts of the world. In Ethiopia and Uganda, populous kingdoms emerged. Nonetheless, much of what Africa was in those earlier centuries has yet to be reconstructed. But with all this external contact, it was clearly not isolated.

## The Colonial Transformation

The period of European involvement in Black Africa began in the fifteenth century. This period was to interrupt the path of indigenous

**Figure 7–15**
Artistic expression has long characterized West African cultures: a Benin bronze head from Nigeria, perhaps 5 centuries old.

African development and irreversibly alter the entire cultural, economic, political, and social makeup of the continent. It started quietly enough, with Portuguese ships groping their way along the west coast and rounding the Cape of Good Hope not long before the turn of the sixteenth century. Their goal was to find a route to the spices and riches of the Orient. Soon, other European countries sent their vessels to African waters, and a string of coastal stations and forts sprang up. In West Africa, the nearest part of the continent to European spheres in Middle and South America, the initial impact was strongest. At their coastal control points, the Europeans traded with African middlemen for the slaves that were wanted on American plantations, for the gold that had been flowing northward across the desert, and for ivory and spices. Suddenly, the centers of activity lay not with the cities of the savanna, but in the foreign stations on the coast. As the interior declined, the coastal peoples thrived. Small forest states rose to power and gained unprecedented wealth, transferring and selling slaves captured in the interior to the white men on the coast. Dahomey (now called Benin) and Benin (now part of neighboring Nigeria) were states built on the slave trade; when the practice of slavery eventually came under attack in Europe, abolition was vigorously opposed in both continents by those who had inherited the power and riches it had brought.

Although it is true that slavery was not new to West Africa, the kind of slave trading introduced by the Europeans certainly was. In the savanna states, African families who had slaves usually treated them comparatively well, permitting marriage, affording adequate quarters, and absorbing them into the family. The number of slaves held in this way was small, and probably the largest number of persons in slavery in precolonial Africa were in the service of kings and chiefs. In East Africa, however, the Arabs had introduced—long before the Europeans—the sort of slave trading that was first brought to West Africa by the white man: African middlemen from the coast raided the interior for slaves and marched them in chains to the Arab *dhows* that plied the Indian Ocean. There, packed by the hundreds in specially built vessels, they were carried off to Arabia, Persia, and India (Fig. 7–16). It is sad but true that Europeans, Arabs, and Africans combined to ravage the Black Continent, forcing perhaps as many as 30 million persons away from their homelands in bondage; destroying families, whole villages, and cultures; and bringing those affected a degree of human misery for which there is no measure.

The European presence on the West African coast brought about a complete reorientation of trade routes, for it initiated the decline of the interior savanna states and strengthened coastal forest states. And it ravaged the population of the interior through its insatiable demand for slaves. But it did not lead to any major European thrust toward the interior, nor did it produce colonies overnight. The African middlemen were well organized and strong, and they managed to maintain a standoff with their European competitors, not just for a few decades but for centuries. Although the European interests made their initial appearance in the fifteenth century, West Africa was not carved up among them until nearly 400 years later, in many areas not until after the beginning of the twentieth century.

Figure 7–16

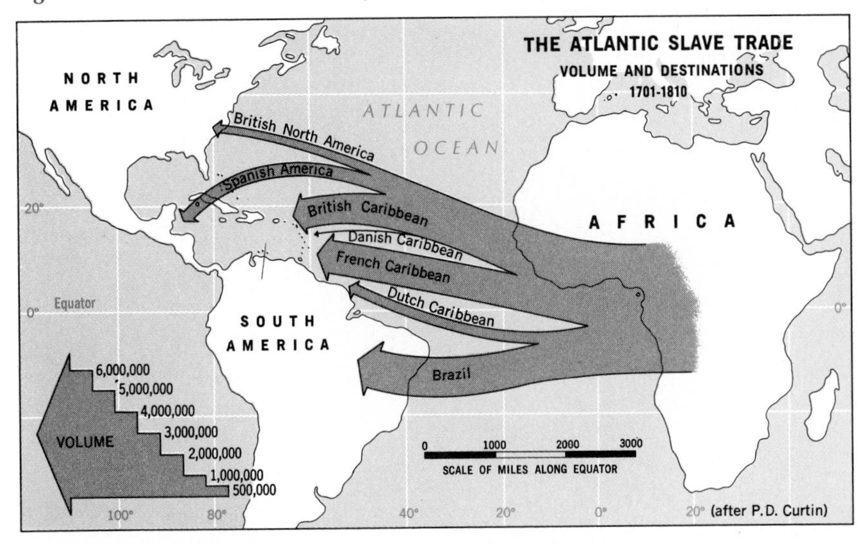


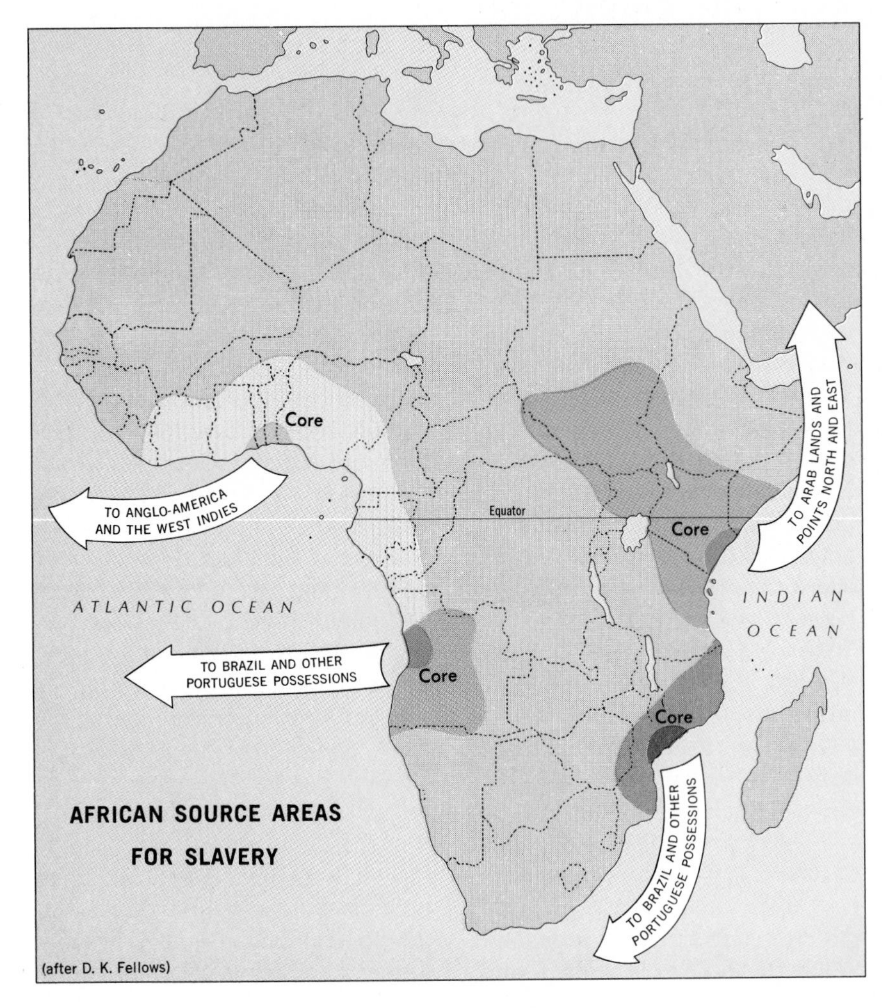


As fate would have it, European interest was to grow strongest—and ultimately most successful—where African organization was weakest. In the middle of the seventeenth century, the Dutch chose the shores of South Africa's Table Bay, where Cape Town lies today (Fig. 7–17), as the site for permanent settlement. Their initial purpose was not colonization, but rather to establish a resupply station on the months-long voyage to and from Southeast Asia and their East Indies colonies; the southern tip of Africa was the obvious halfway house. There was no intent to colonize there; Southern Africa was not known as a productive area, and the more worthwhile East lay in the spheres of the Portuguese and the Arabs. Probably, the Hollanders would have elected to build their station at the foot of Table Bay whatever the indigenous population of the interior, but they happened to choose a location about as far away from the centers of Bantu settlement as they could have found. Only the Bushmen and their rivals, the Hottentots, occupied Cape Town's hinterland. When conflicts developed between Amsterdam and Cape Town and some of the settlement's residents decided to move into the hinterland, they initially faced only the harassment of small groups of these two peoples rather than the massive resistance likely to have been offered by the well-organized Bantu states. To be sure, a confrontation eventually did develop between the advancing Europeans and the similarly mobile Bantu Africans, but it began decades after Cape Town was founded and hundreds of miles from it. Unlike some of the West African way-stations, Cape Town was never threatened by African power, and it became a European gateway into Southern Africa.

Elsewhere in Black Africa, the European presence remained confined almost entirely to the coastal

**Figure 7–17**
Table Mountain dominates the site of Cape Town near Africa's southern tip, where the first Dutch settlers landed in 1652.

trading stations, whose economic influence was very strong. No real frontiers of penetration developed; individual travelers, missionaries, explorers, and traders went into the interior, but nowhere else in Africa south of the Sahara was there an invasion of white settlers comparable to Southern Africa's.

**Penetration** After more than four centuries of contact, Europe finally laid claim to all of Africa during the second half of the nineteenth century. Parts of the continent had been "explored," but now representatives of European governments sought to expand or create African spheres of influence for their homelands; Cecil Rhodes for Britain, Karl Peters for Germany, Pierre de Brazza for France, and Henry Stanley for the king of Belgium—these were some of the leading figures who helped shape the colonial map of the continent. In some areas, such as along the lower Congo (Zaïre) River and in the vicinity of Lake Victoria, the competition among the European powers was especially intense. Spheres of influence began to crowd each other; sometimes they even overlapped. So, in late 1884, a conference was convened in Berlin to sort things out. At this conference, the groundwork was laid for the now-familiar political boundaries of Africa (see *Box*).

As the twentieth century opened, Europe's colonial powers were busily organizing and exploiting their African dependencies. The British, having defeated the Dutch-heritage Boers in Southern Africa, came very close to achieving their Cape-to-Cairo axis: only German East Africa interrupted a vast empire that stretched from Egypt and the Sudan through Uganda, Kenya, Nyasaland, and the Rhodesias to South Africa (Fig. 7–18). The French took charge of a vast realm that reached from Algiers in the north and Dakar in the west to the Congo River in Equatorial Africa. King Leopold II of Belgium held personal control over the Congo. Germany had colonies scattered in all sections of the continent, except the north. The Portuguese controlled two huge territories, Angola and Moçambique, along the flanks of Southern Africa, and a small entity in West Africa known as Portuguese Guinea. Italy's possessions in tropical Africa were confined to the Horn, and even Spain got into the act with a small dependency consisting of the island of Fernando Póo and the mainland area called Rio Muni (Equatorial Guinea). The only places where the Europeans did not overwhelm African desires to remain independent were Ethio-

## COLONIZATION AND LIBERATION

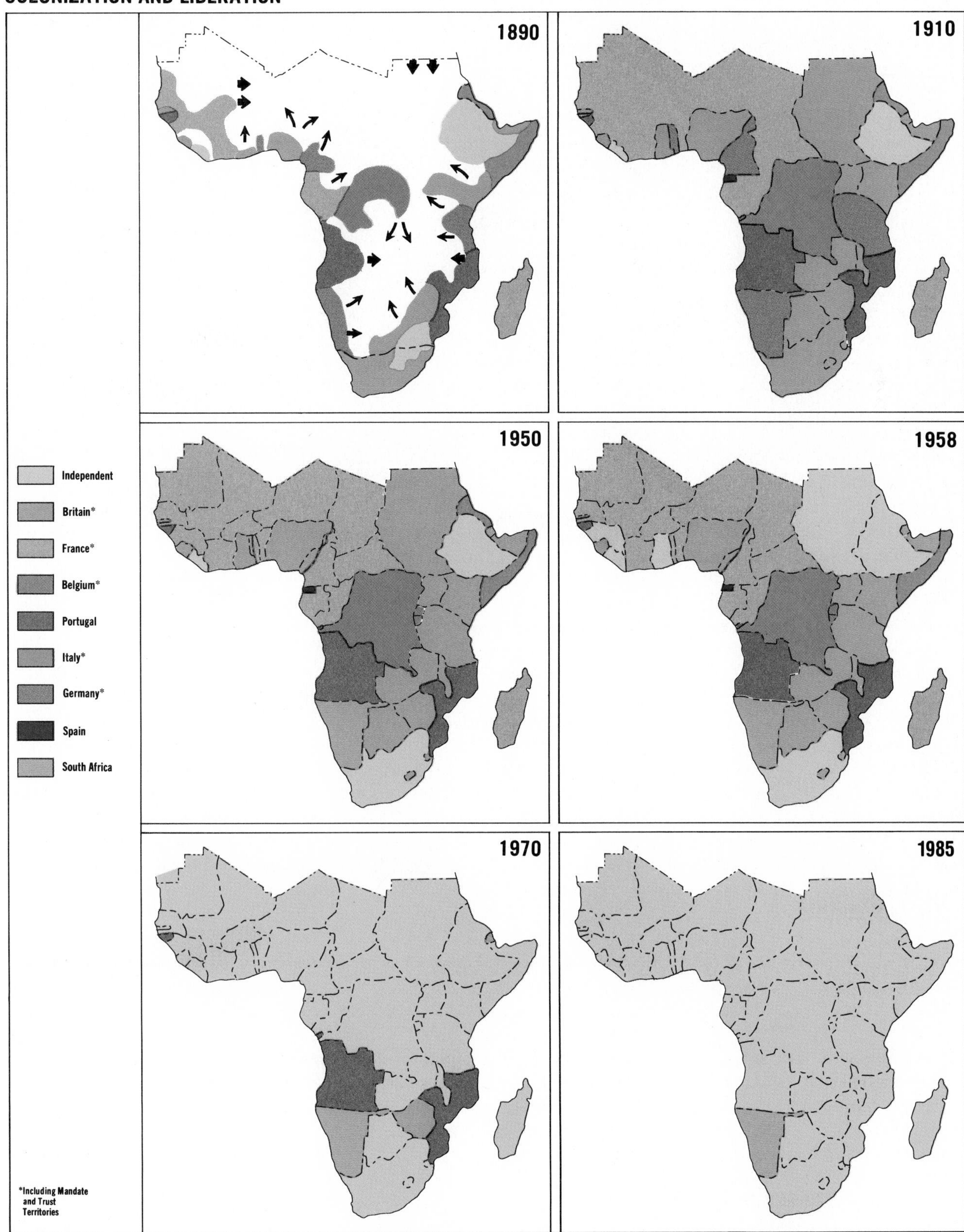


Figure 7–18

## THE BERLIN CONFERENCE

In November 1884, the imperial chancellor and architect of the German Empire, Otto von Bismarck, convened a conference of 14 powerful states (including the United States) to settle the political partitioning of Africa. Bismarck not only wanted to expand German spheres of influence in Africa, he sought to play Germany's colonial rivals off against one another, to the Germans' advantage. The major colonial contestants in Africa were the British, who held beachheads along the West, South, and East African coasts; the French, whose main sphere of activity was in the area of the Senegal River and north of the Congo; the Portuguese, who now desired to extend their coastal stations in Angola and Moçambique deep into the interior; King Leopold II of Belgium, who was amassing a personal domain in the Congo (Zaïre); and Germany itself, active in areas where the designs of other colonial powers might be obstructed, as in Togo (between British holdings), Cameroon (a wedge into French spheres), South West Africa (taken from under British noses in a swift strategic move), and East Africa (where the effect was to break the British design for a Cape-to-Cairo axis).

When the conference convened in Berlin, more than 80 percent of Africa was still under traditional African rule. Nevertheless, the colonial powers' representatives drew their boundary lines across the entire map. The lines were drawn through known as well as unknown regions, pieces of territory were haggled over, boundaries were erased and redrawn, and sections of African real estate were exchanged in response to urgings from European governments. In the process, African peoples were divided, unified regions were ripped apart, hostile societies were thrown together, hinterlands were disrupted, and migration routes were closed off. All this was not felt in the beginning, of course, but these were some of the effects when the colonial powers began to consolidate their holdings and the boundaries on paper soon became barriers on the African landscape (Fig. 7–18). The Berlin Conference was Africa's undoing in more ways than one. Not only did the colonial powers superimpose their domains on the African continent; when independence returned to Africa after 1950, the realm had by then acquired a legacy of political fragmentation that could neither be eliminated nor made to operate satisfactorily. The African politico-geographical map is, therefore, a permanent liability, resulting from three months of ignorant, greedy acquisitiveness during the period of Europe's insatiable search for minerals and markets.

pia, which fought some heroic battles against Italian forces, and Liberia, where Afro-Americans retained control.

The two world wars also had some effect on this colonial map of Africa. In World War I, Germany's defeat resulted in the complete loss of its colonial possessions. The territories in Africa were placed under the administration of other colonial powers by the League of Nations' mandate system. In World War II, fascist Italy launched a briefly successful campaign against Ethiopia, but the ancient empire was restored to independence when the allied forces won the war. Otherwise, the situation in colonial Africa in the late 1940s—after a half-century of colonial control—remained quite similar to the one that arose out of the Berlin Conference.

**Policies and Repercussions** Geographers—especially political geographers—are interested in the ways in which philosophies and policies of the colonial powers were reflected in the spatial organization of the African dependencies. These colonial policies can be expressed in just a few words. For example, Britain's administration in many parts of its vast empire was referred to as "indirect rule," since indigenous power structures were sometimes left intact and local rulers were made representatives of the crown. Belgian colonial policy was called "paternalism," in that it treated Africans as though they were children who needed to be tutored in Western ways. Although the Belgians made no real efforts to make their African subjects culturally Belgian, the French very much wanted to create an "Overseas France" in their African dependencies. French colonialism has been identified as a process of assimilation—the acculturation of Africans to French ways of life—and France made a stronger cultural imprint in the various parts of its huge colonial empire. Portuguese colonial policy had objectives similar to the French, and the African dependencies of Portugal were officially regarded as "Overseas Provinces" of the state. If you sought a one-word definition of Portuguese colonial policy, however, the term *exploitation* would emerge most strongly: few colonies made a greater contribution (in proportion to their known productive capacities) to the economies of their colonial masters than did Moçambique and Angola.

Colonial policies have geographical expressions as well, and the spatial organization fostered by the colonial powers has become the infrastructure of contemporary independent Africa. As the map shows (Fig. 7–18), Britain possessed the most widely distributed colonial empire in Africa. British colonial policy tended to adjust to individual situations: in *colonies*, white-settler minorities had substantial autonomy, as in Kenya and Southern Rhodesia (now Zimbabwe); in *protectorates*, the rights of African peoples were guarded more effectively; in *mandate* (later *trust*) *territories*, the British undertook to uphold League of Nations' (later U.N.) administrative rules; and in one case, the British shared administration with another government, in Sudan's *condominium*. Britain's colonial map of Africa was a patchwork of these different systems, and the independent countries that emerged reflect the differences. Nigeria, which had been a colony in its south and an indirectly ruled protectorate in its north, became a federal state containing major internal differences. Kenya, the former colony, became a highly centralized unitary state after the Africans wrested control of the productive core area from the whites.

In contrast to the British, the French placed a cloak of uniformity over their colonial territories in Black Africa. Contiguous and vast, although not very populous, France's colonial empire extended from Senegal eastward to Chad and southward to the former French Congo (now the Congo Republic). This huge area was divided into two units, French West Africa (focused on Dakar) and French Equatorial Africa (whose headquarters was in Brazzaville). After World War I, France was granted a mandate over the former German colony of Kamerun (Cameroon) in Equatorial Africa and over a part of Togo in West Africa; its only other dependencies were French Somaliland (Djibouti), the gateway to Ethiopia on the Horn, and the island of Madagascar. France itself, as we know, is the classic example of the centralized, unitary state, whose capital is the cultural, political, and economic focus of the nation, overshadowing all else. The French brought their concept of centralization to Africa as well. In France, all roads lead to Paris; in Africa all roads were to lead to France, to French culture, to French institutions. For the purposes of assimilation and acculturation, French West Africa, half the size of the entire United States of America (although with a population of only some 30 million in 1960), was divided into administrative units, each centered on the largest town, all oriented toward the governor's headquarters at Dakar. As the map shows, great lengths of these boundaries were straight lines delimited across the West African landscape; history tells us to what extent they were drawn, not on the basis of African realities, but for France's administrative convenience. The present-day state of Upper Volta, for example, existed as an entity prior to 1932 when, because of administrative problems, it was divided among Ivory Coast, Soudan (now Mali), and Dahomey (now Benin). Then, in 1947, the territory was suddenly recreated. Little did the boundary makers expect that what they were doing would one day affect the national life of an independent country.

During its period of stewardship, France created French-speaking, acculturated African elites, trained at French universities and often experienced in French politics through direct representation in Paris. From these elites, headquartered in the colonies' capital cities, the governments of the newly independent states were forged. During the period of transition and its aftermath, the Africanization of Francophone (French-speaking) Africa was bolstered by African artists, especially writers. Since the 1930s there had been a movement among black writers in the French-speaking Caribbean, a movement that involved some of the most prominent local authors, to emphasize the contradictions between French colonial policies of "assimilation" and the realities of discrimination and inequality. In Paris these writers met their African counterparts, and African and French Caribbean intellectuals labored to bring new vitality to their cultures. Thus, what had begun as a regional literary movement, *négritude*, evolved into a whole cultural philosophy. Through their poems, novels, and plays the Francophone-African writers inspired their readers, urging them to renew the strength of their traditions. *Négritude* became a focus for the revival of African self-knowledge and self-esteem. Old concepts of people's closeness to nature, their special contacts with ancestors, their disdain for natural wealth were contrasted against the arrogance and acquisitiveness of European culture. Not surprisingly, *négritude* also became a political philosophy. One of West Africa's most respected authors, Léopold Senghor, became the first president of Senegal.

The maturing of Francophone West African society, and its considerable political stability, saw many of the objectives and aspirations of the *négritude* philosophy achieved. Thus, the focus of African artistic expression in this region shifted and diversified, and *négritude* ceased to be dominated by its proselytizing spirit. Today, manifestations of *négritude* are more prominent in Nigerian and Ghanaian writing than in French-speaking West Africa, where the movement's noble goals were essentially attained. It is, thus, no longer the regional phenomenon it was in the 1950s and early 1960s.

One reason for the considerable stability of West African political systems and economies has been their continuing connection with France. The great majority of their modern institutions, including their political machinery, is based on French models. France has made heavy investments in its former colonial realm. Much of this investment can be observed in the capital cities, which have become true primate cities. Dakar (925,000) and Abidjan (950,000) rank among Africa's largest cities, disproportionately so in context of the national populations of Senegal (6.8 million) and Ivory Coast (9.5 million). Much investment also was directed toward the improvement of surface communications, again especially between the capital cities and the various regions. Core-area development was stimulated in anticipation of a "ripple effect" into more distant areas. France continues, through aid projects, loans, educational and military assistance, and other programs to maintain its presence in the region. In 1983, it involved itself in the Libya-instigated civil war in Chad, and Paris stands ready to continue such interventions on behalf of its former dependencies.

Belgian administration in the Congo (now Zaïre) provides another insight into the results, in terms of spatial organization, of a particular set of policies of colonial government. Unlike the French, the Belgians made no effort to acculturate their African subjects. The policies of "paternalism" actually consisted of rule in the Congo by three sometimes competing interest groups: the Belgian government, the managements of huge mining corporations, and the Roman Catholic Church. Each of these groups had major regional spheres of activity in this vast country.

As the map shows (Fig. 7–28), Zaïre has a corridor to the ocean along the Zaïre River between Angola to the south and the Congo Republic (former French Congo) to the north. The capital, long known as Leopoldville (now Kinshasa), lies at the eastern end of this corridor; not far from the ocean lies the country's major port, Matadi. It was in this corridor that the administrative and transport core area of Zaïre developed; and this was the place from which the decisions made in Brussels were promulgated by a governor-general. The economic core of the Congo (and now of Zaïre) lay in the country's southeast, in the province then known as Katanga, which was administered from the copper-mining headquarters of Lubumbashi (formerly Elizabethville). This is a portion of the northernmost extension of Southern Africa's great mining belt, the balance of which continues into Zambia as that country's Copperbelt. From hundreds of miles around, African workers streamed toward the mines (Fig. 7–19), a pattern of regional labor migration that has persisted for decades and is still going strong. Between the former Belgian Congo's administrative and economic core areas lay the Congo (Zaïre) Basin, once a profitable source of wild rubber and ivory. The colony's six administrative subdivisions were laid out in such a way that each incorporated part of the area's highland rim and part of the forested basin. As the 1960 date of independence approached, it seemed briefly that each of these Congolese provinces, centered on its administrative capital, might break away and become an independent African country (as each French-African dependency did in West and Equatorial Africa). But, in

Figure 7–19

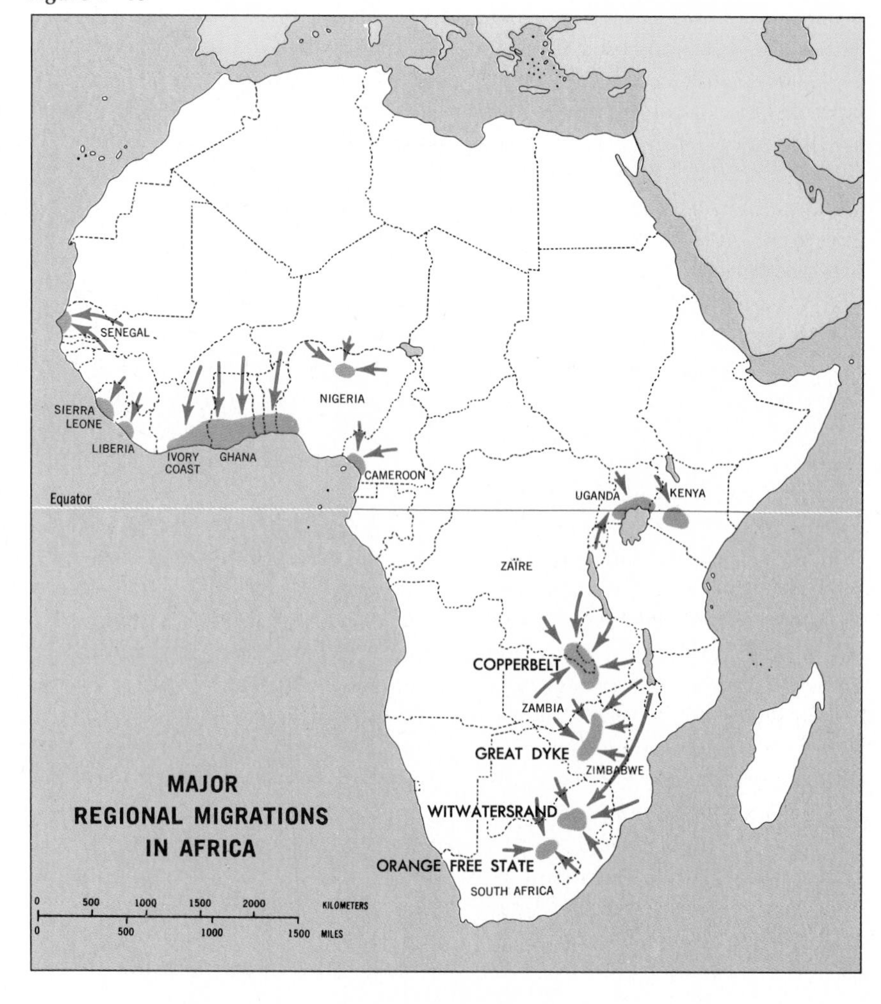

the end, the country held together, and Leopoldville became Kinshasa, capital of Zaïre, Black Africa's largest state in terms of territory.

Portugal's rule in Angola and Moçambique was designed to exploit four assets: (1) labor supply for the interior mines (especially in Moçambique), (2) transit functions and port facilities (Moçambique from South Africa and Southern Rhodesia—Angola from the Copperbelt and Katanga), (3) agricultural production (especially cotton from northern Moçambique and coffee from Angola), and (4) minerals (mainly from diamond- and oil-rich Angola). In this effort, the Portuguese created a system of rigid control that involved forced labor and the compulsory farming of certain crops. In Moçambique, particularly, the country was split into a large number of small districts that were tightly controlled. Movement and communication, even within a single African ethnic area, were kept to a minimum. Accordingly, Portuguese colonial rule was often described as the most harsh of all the European systems; for a long time, it seemed unlikely that an independence movement could be mounted. But when independence came to Angola's and Moçambique's neighbors (Zaïre and Tanzania), Portugal's days were numbered. As elsewhere in recently colonial Africa, however, the imprint of colonialism in former Portuguese Africa remains strong, a pervasive element in the regional geography of the continent.

Besides the geographic consequences that flowed from the various forms of colonial policy, the current (and future) map of the realm was affected in many other ways as well. In spite of the differences among individual colonial policies, all were paternalistic, assimilative, and exploitive to some degree. However difficult it was to unify territories within colonial spheres, it has proven virtually impossible to resolve the problems of redefining truly African regional interests without regard to the colonial heritage. With few exceptions, the primary development of capital cities within each colonial territory has been strengthened even further since independence. And much of the transport network at independence reflected the colonial approach to development; railways (if they existed) and major roads (however defined) almost always facilitated movement of goods between the interior and coastal outlets but offered little basis for interior circulation (see *Model Box 5*). There is much in Africa that is new and promising, but the geographic patterns established during the colonial period are more often a hindrance to development than an aid.

# Africa's Regions Today

On the face of it, Africa would seem so massive, so compact, and so unbroken that any attempt to justify a contemporary regional breakdown is doomed to failure. No deeply penetrating bays or seas create peninsular fragments as in Europe. No major islands (other than Madagascar) provide the broad regional contrasts we see in Middle America. Nor does Africa really taper southward to the peninsular proportions of South America. And Africa is not cut by an Andean or a Himalayan mountain barrier. Does any clear regional division exist? Indeed it does. As we noted earlier, the Sahara Desert, extending from west to east across the entire northern part of the continent, constitutes a broad transition zone between Arab Africa and Black Africa. Whereas busy trade routes have crossed it for centuries and millions of people who live south of the Sahara share the religious outlook of the Arab world, Black Africa retains a cultural identity quite distinct from that of the Arab realm.

Within Africa south of the Sahara, regional identities exist as well, although the regional boundaries involved are not easily defined and have at times been the subject of debate (Fig. 7–21). Three or four such regions are in common use: *West Africa,* which includes the countries of the west coast and Sahara margins from Senegal and Mauritania to Nigeria and Niger; *East Africa,* which normally refers to the three states of Kenya, Uganda, and Tanzania; and *South Africa* (or *Southern* Africa to get away from the political connotation), of which South Africa, Zimbabwe, and Zambia are parts. Less clear has been the use of such regions as *Equatorial* Africa, by which is generally meant Zaïre and the countries that lie between it and Nigeria. Although the appellation "Equatorial" presumably has to do with the location of these countries astride or near the equator, the similarly positioned countries of East Africa are practically never referred to in this way. Chad, too, offers problems of regional assignment; it does not fit very neatly into either the Equatorial or the West African (or the North African) regional definitions. *Equatorial* has come to mean hot, tropical, and low lying rather than equatorially-located Africa; thus Gabon is a part of Equatorial Africa, whereas Uganda is not. The regional divisions shown in Fig. 7–21 should therefore be viewed in general terms and as a matter of orientation; these boundaries, as will be clear from the following discussion, are not beyond dispute.

## West Africa

West Africa occupies most of Africa's "bulge," extending south from the margins of the Sahara

MODEL BOX 5

# COLONIAL SEQUENCE OF TRANSPORT DEVELOPMENT

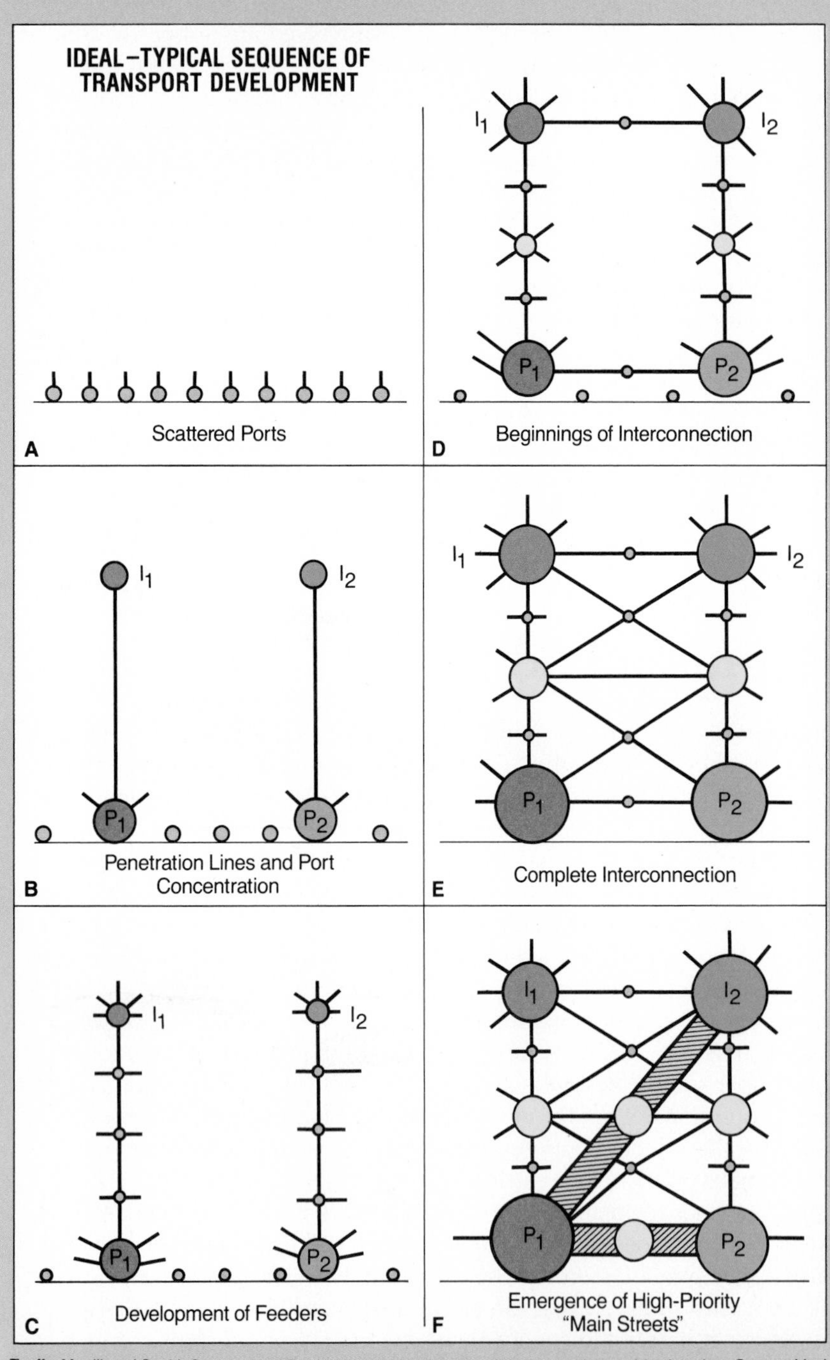


Taaffe, Morrill, and Gould, *Geographical Review*, Vol. 53, 1963, p. 504. With the permission of the American Geographical Society.

Figure 7–20

When a region experiences economic growth, how does its transportation network change? It was this question that led three American geographers—Edward Taaffe, Richard Morrill, and Peter Gould—to propose a descriptive model of network development. Ostensibly a model of an "ideal-typical" sequence of network-growth stages, their model actually represents a sequence of changes that would be initiated by the efforts of a colonial power as it exerted increasing control over a region. This model was supported by information collected in two West African countries, Ghana (formerly Gold Coast) and Nigeria.

The generalized sequence of network expansion is reflected in a series of six stages, with each a progressive response to the forces operating in a region as it undergoes economic development (Fig. 7–20). At the beginning of the sequence (*A*), the only apparent overland routes are those that reach short distances into the interior from each of a number of small fishing ports. Significant here is the unstated assumption of an interior area filled with population and resources awaiting contact with world markets—a classic presumption that lay behind many colonial powers' efforts at exploitation. The second stage (*B*) occurs when several of the ports begin to grow as they gain greater access to the resources of interior hinterlands. It is not important to the model which seaports begin this process, because there are many reasons why one or two might grow, whereas others do not: a slightly better harbor for overseas shipping, a somewhat richer hinterland, the rise of an ambitious local trading organization, a higher birth rate, and so on. In the classical manner of town-hinterland relationships, the response to better transport access is growth in the size of the most favorably connected ports.

By the third stage (*C*), major transport lines are extended inland from the favored ports ($P_1$ and $P_2$) and produce various ripples of development. The ports respond to their heightened access to the interior in several ways: by bringing in more materials, by growing in population, and, by implication, with an increase in economic diversity. Additional lines of access to the interior are constructed to tap other parts of the hinterland, eventually cutting off potential competition from smaller nearby ports. Meanwhile, a growing settlement becomes established at the terminus of each transport line ($I_1$ and $I_2$) as a collection, supply, and service center for the goods, people, and vehicles moving to or from the coast. The parallel with colonial penetration and its desire for exploitation is clear by this stage: What else could have provided an impetus for so great a penetration?

The interplay between each settlement's growth, the changing access to the various hinterlands, and the transport network that ties the entire set of places together, produces the three stages exhibited in the remainder of the sequence. There are four major relationships shaping this continuing pattern of change: (1) places that are connected by transport routes will grow in size more than those that remain isolated from the network, (2) places located at transport junctions are more likely to grow than places that lie along a single transport line, (3) larger settlements are more likely to demand and receive improved connections than smaller settlements, and (4) places that are close to much larger settlements will grow more slowly than more distant places not in the "shadow" of a larger place. As these conditions operate through time, the two parts of the region (initially penetrated by two separate transport lines) find it is advantageous to exchange goods, services, and people, and they become interconnected (*D* and *E*). The alternative faced by the two largest places—the initial coastal settlements ($P_1$ and $P_2$)—and many of the smaller ones is a very long overland shipment or two transfers of goods between land and water transport. This alternative becomes expensive, then intolerable, as demands for exchange increase between what have now become two major cities. The result is a coastal highway or rail line that, in turn, closes the circuit of regional transportation connections. Finally, in the last stage (*F*), the variability of places and people is given some attention in the model as one of the two large coastal centers ($P_1$) and one of the two major interior centers ($I_2$) begin to exert dominance over the others; this creates transport demands for construction of higher-access arteries superimposed on the underlying, more basic network.

As with most models, this "ideal-typical" sequence attempts to clarify the relationships among forces operating in a complex manner. It does this by making assumptions about what is important and what is not and by simplifying the outcomes of spatial processes. Here it was assumed that (1) population growth and economic growth occur together, (2) transportation improvements will follow from the demands generated by this growth, (3) proximity is more important than distance (though distance is not unimportant), and (4) the distributions of population and resources need not be considered in order to understand the way in which these kinds of transport systems develop. It is possible, of course, to question the validity of any or all of these assumptions. Also, since no explicit recognition is given to the obvious connection with (and dependence of) the model on a specifically colonial experience, it seems to have been assumed that the postcolonial impact on the region does not affect the pattern or sequence of transport development—a rather tenuous assumption because these circulatory patterns have continued to evolve since 1960. Since nearly a quarter-century has passed since this model was proposed, a revision and extension would do much to elucidate the changing interaction patterns of the spatial economy that have evolved since independence.

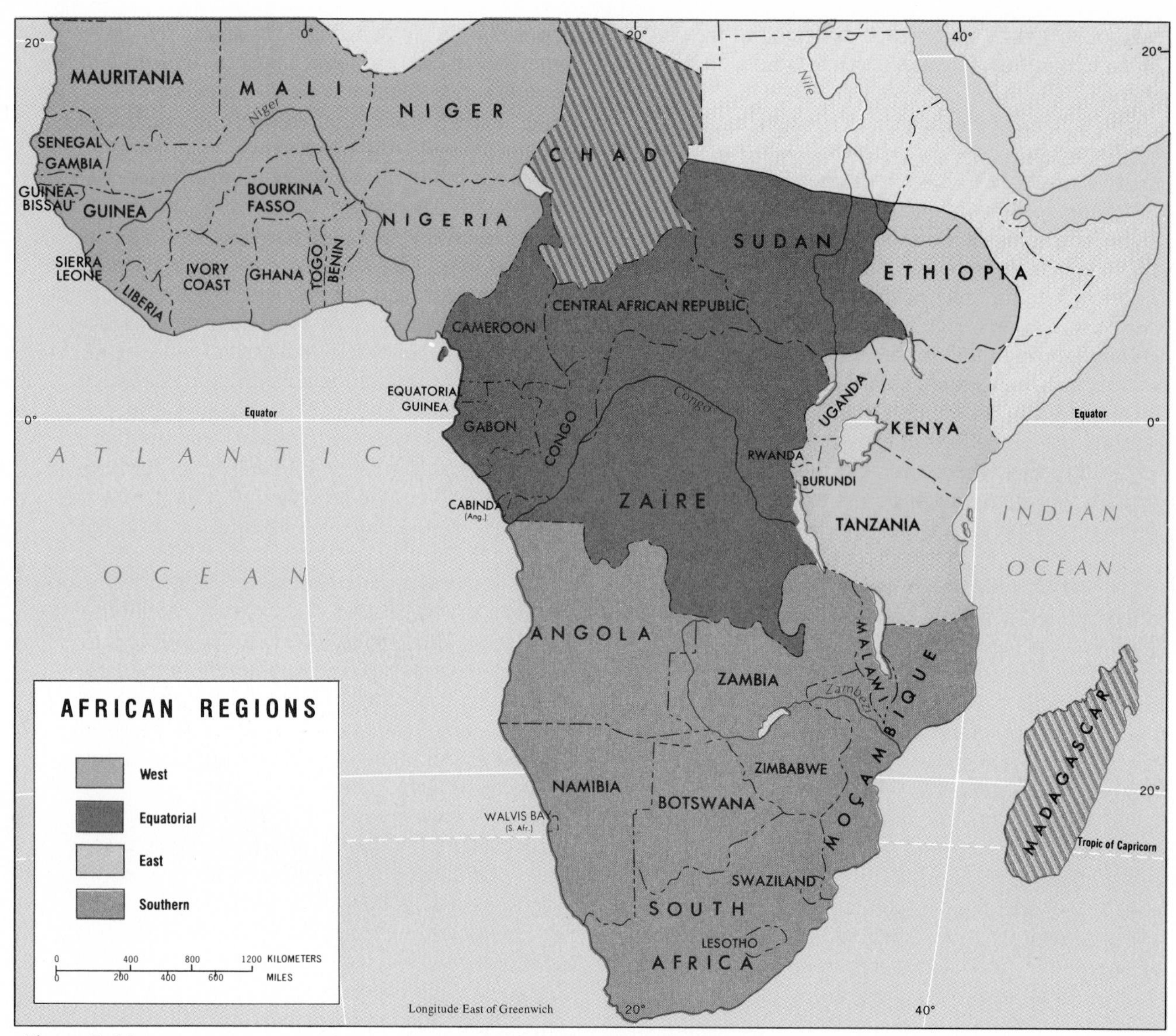


Figure 7–21

Desert to the coast, and from Lake Chad west to Senegal (Fig. 7–22). Politically, the broadest definition of this region includes all those states that lie to the south of Morocco, Algeria, and Libya, and west of Chad (itself sometimes included) and Cameroon. Within West Africa, a rough division is sometimes made between the very large, mostly steppe and desert states that extend across the southern Sahara (Chad could also be included here), and the smaller better-watered coastal states. Upper Volta, small but dry and landlocked, does not fit into either group.

Apart from once-Portuguese Guinea-Bissau and long-independent Liberia, West Africa comprises only former British and French dependencies (four British and nine French). The British-influenced countries (Nigeria, Ghana, Sierra Leone, and Gambia) lie separated from one another, whereas Francophone West Africa is contiguous. As Fig. 7–22 shows, political boundaries extend from the coast into the interior, so that from Mauritania to Nigeria the West African habitat is parceled out among parallel, coast-oriented states. Across these boundaries, especially across those between former British and former French territories, moves very little trade. For example, in terms of value, Nigeria's trade with Britain is about one hundred times as great as its trade with nearby Ghana. The countries of West Africa are not interdependent economically, and their incomes are to a large extent derived from the sale of their products on the international market. But the African countries do not control the prices their goods can command on the world markets; when these prices fall, they face serious problems.

Given these crosscurrents of subdivision within West Africa, what are the justifications for the concept of a single West African region?

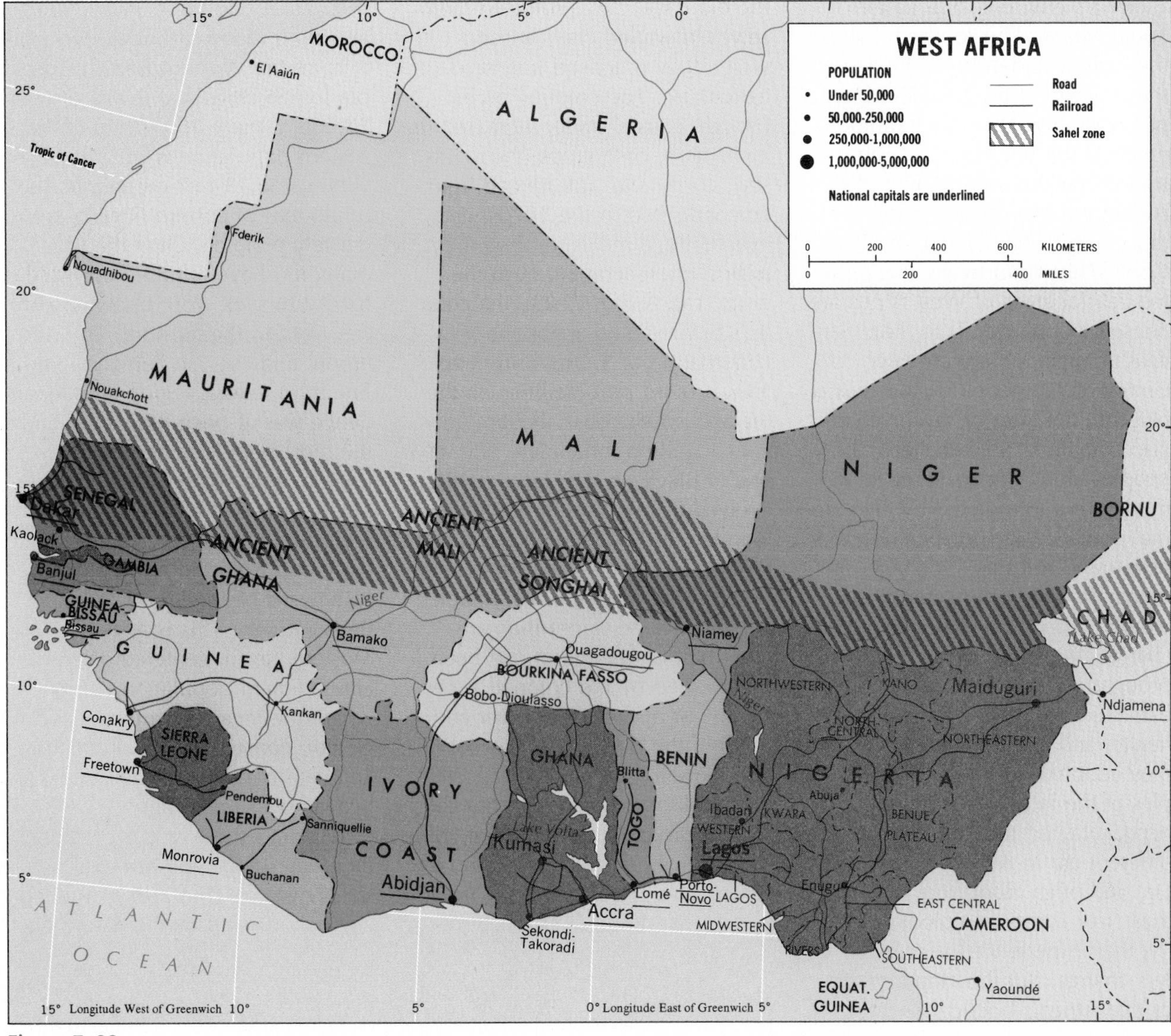


Figure 7–22

First, there is the remarkable cultural and historical momentum of this part of Africa. The colonial interlude failed to extinguish West African vitality, expressed not only by the old states and empires of the savanna and the cities of the forest, but also by the vigor and entrepreneurship, the achievements in sculpture, music, and dance of peoples from Senegal to Nigeria's Iboland. Second, West Africa contains a set of parallel east-west ecological belts, clearly reflected by Figs. I–9 to I–13, whose role in the development of the region is pervasive. As a map of transport routes in West Africa indicates, overland connections within each of these belts, from country to country, are quite poor; no coastal or interior railroad was ever built to connect the coastal tier of countries. On the other hand, spatial interaction is far better across these belts, and some north-south economic exchange does take place, notably in the coastal consumption of meat from cattle raised in the northern savannas. Third, West Africa received an early and crucial impact from European colonialism, which—with its maritime commerce and the slave trade—transformed the region from one end to the other. This impact was felt all the way to the heart of the Sahara, and it set the stage for the reorientation of the whole area—out of which emerged the present patchwork of states.

The effects of the slave trade notwithstanding, West Africa today is Black Africa's most populous region (Fig. I–14). In these terms, Nigeria, with 92 million (but perhaps as many as 110 million) people, is Africa's largest state, and Ghana, with 14.3 million, also ranks high. As Fig. I–14 shows, West Africa also claims regional identity, in that it constitutes one of Africa's five major population clusters (the others are northern Morocco and Algeria, the Nile Valley and Delta, Lake

Victoria's environs, and eastern South Africa). The southern half of the region, understandably, carries the majority of the people. Mauritania, Mali, and Niger include too much of the unproductive Sahel's steppe and the Sahara Desert to sustain populations comparable to those of Nigeria, Ghana, or Ivory Coast. This is not to say that only the coastal areas of West Africa are densely populated. True, from Senegal to Nigeria there are large concentrations of people in the coastal belt, but the interior savannalands also contain sizable clusters. The peoples along the coast reflect the modern era introduced by the colonial powers: they prospered in their new-found roles as middlemen in the coastward trade. Later, they were in a position to undergo the changes the colonial period brought: in education, religion, urbanization, agriculture, politics, health, and many other endeavors they adopted new ways. The peoples of the interior, on the other hand, retained their ties with a very different era in African history; distant and often aloof from the main theater of European colonial activity, they experienced very much less change. But the map reminds us that Africa's boundaries were not drawn to accommodate such differences. Both Nigeria and Ghana possess population clusters representing the coastal as well as the interior peoples, and, in both countries, the wide gap between north and south has produced political problems.

When, in 1960, Nigeria achieved full independence, it was endowed with a federal political structure that consisted of three regions based on the three major population clusters within its borders—two in the south and one in the north. Around the Yoruba core in the southwest lay the Western Region. The Yoruba are a people with a long history of urbanization, but they are also good farmers; in the old days, they protected themselves in walled cities around which they practiced intensive agriculture. The colonial period brought coastal trade, increased urbanization, cash crops (the mainstay, cocoa, was introduced from Fernando Póo in the 1870s) and, eventually, a measure of security against encroachment from the north. Lagos (Fig. 7–23), the country's first federal capital and now a conurbation of 3.4 million people, grew around port facilities on the region's south coast. Ibadan, now with 1 million inhabitants and also one of Black Africa's largest cities, evolved from a Yoruba settlement founded in the late eighteenth century. At independence, the Western Region counted some 10 million inhabitants and, more than any other part of Nigeria, it had been transformed by the colonial experience. East of the Niger River and south of the Benue, the Ibo population formed the core of the Eastern Region. Iboland, although coastal, lay less directly in the path of colonial change, and its history and traditions, too, differed sharply from those of western Nigeria. Little urbanization had taken place here, and even today, although over one-third of the Western Region's people live in cities and towns of over 20,000, a mere 10 percent of the Eastern population is urbanized. With some 13 million people, the rural areas of eastern Nigeria are densely peopled. Over the years, many Ibo have left their crowded habitat to seek work elsewhere, in the west, in the far north, in Cameroon, and even in Fernando Póo. The third federal region at independence was at once the largest and the most populous: the Northern Region, with some 30 million people. It extended across the full width of the country from east to west, and from the northern border southward beyond the Niger and Benue Rivers. This is Nigeria's Moslem North, centered on the Hausa-Fulani population cluster, where the legacy of a feudal social system, conservative traditionalism, and resistance to change hung heavily over the country.

Nigeria's three original regions, then, lay separated—not only by sheer distance, but also by tradition and history, and by the nature of

**Figure 7–23**
Lagos, which has evolved from its historic core around the seaport in the foreground.

Figure 7–24
The swampy Niger Delta contains Nigeria's considerable oil reserves, now the basis of its export economy.

colonial rule. In Nigeria, even physiography and biogeography conspire to divide south from north: across the heart of the country (and across much of West Africa in about the same relative location) stretches the so-called Middle Belt—poor, unproductive, disease- and tsetse-ridden country that forms a relatively empty barrier between the northern and southern regions. Not surprisingly, Nigeria's three-region federal system failed. Interregional rivalries and interethnic suspicions led to civil war between 1967 and 1971. The original three regions were subdivided, rearranged, and subdivided, once again, in an attempt to devise a system that would prevent such disasters in the future.

Nigeria today is a cornerstone of the new Africa, a country on the move whose economic windfall has been the discovery of substantial oil reserves in the area of the Niger Delta (Fig. 7–24). By the early 1980s, over 90 percent of Nigeria's export revenues were derived from the sale of petroleum and petroleum products, and cities from Port Harcourt to Lagos reflected the oil boom of the south. This heavy dependence on petroleum, however, made the country's ambitious development plans hostage to world oil prices. Dropping oil prices in the first half of the 1980s threw the Nigerian economy into disarray in spite of tremendous growth during the previous decade. As unemployment rose, the government in 1982 expelled virtually all non-Nigerian workers and their families, forcing millions to return to their countries of origin, where local economies were no more able to absorb them. Several other changes are also reshaping Nigeria's human geography. Urbanization, barely 20 percent 15 years ago (but 25 percent today), is still accelerating, surface communications are being improved, and national integration is proceeding. A new capital, Abuja, is being built in a more central location (Fig. 7–22). Nigeria's regional disparities still exist, of course, and the former Eastern Region's war of independence (the tragic Biafra affair of the late 1960s) may not have been the last civil conflict to disrupt the country. But Nigeria has, nonetheless, made major strides toward national integration and laying the foundations for sustained economic development.

**Periodic markets** The great majority of the people of West Africa are not involved in the production of exports for world markets, but subsist on what they can grow and raise—and trade. Their local transactions take place at small markets in villages (Fig. 7–25). These village markets are not open every day, but operate at regular intervals. In this way, several villages in a region get their turn to attract the day's trade and exchange, and each benefits from its participation in the wider network of interactions. People come to these *periodic markets* on foot, by bicycle, on the backs of their animals, or by whatever other means are available. Periodic markets are not exclusively a West African phenomenon. They also occur in interior Southeast Asia, in China, and in Middle and South America as well as in other parts of Africa. The intervals between market days vary. In much of West Africa, village markets tend to be held on every fourth day, although some areas have two-day and eight-day cycles; in Ghana's northern region, markets are held in a particular village every third day.

Periodic markets, then, form a sort of interlocking network of exchange places that serves areas where there are no roads; as each market in the network gets its turn, it will be near enough to one portion of the area so that the people who live in the vicinity can walk to it, carrying what they wish to sell or trade. In this way, small amounts of produce filter through the market chain to a larger regional market, where shipments are collected for interregional or perhaps even international trade. What is traded, of course, depends on where the market is located. A visit to a market in West Africa's forest zone will produce impressions that are very dif-

**Figure 7–25**
A periodic market in a village in Mali, not far from the ancient city of Timbuktu.

ferent from a similar visit in the savanna zone. In the savanna, sorghum, millet, and shea-butter (an edible oil drawn from the shea-nut) predominate, and you will see some Islamic influences here; in the south, such products as yams, cassava, corn, and palm oil change hands. In the southern forest zone, too, one is more likely to see some imported manufactured goods passing through the market chain, especially near the relatively prosperous cocoa, coffee, rubber, and palm oil areas. But, in general, the quantities of trade are very small—a bowl of sorghum, a bundle of firewood—and their value is low: these markets serve people who mostly live near the subsistence level.

## East Africa

Despite the fact that they are neighbors, the three onetime British territories of East Africa—Kenya, Uganda, and Tanzania (formerly Tanganyika and its offshore islands of Zanzibar and Pemba)—and the former Belgian wards of Rwanda and Burundi display numerous differences. These differences have arisen less out of any contrasts in colonial rule than out of the nature of the East African habitat, the course of African as well as European settlement, and the location of the areas of productive capacity. This, of course, is highland, plateau Africa, mainly savanna country that turns into steppe toward the drier northeast. Great volcanic mountains rise above a plateau that is cut by the giant rift valleys (Figs. 7–4, 7–6). The pivotal physical feature is Lake Victoria, where the three major countries' boundaries come together (Fig. 7–26) and on whose shores lie the primary core area of Uganda and secondary cores of both Kenya and Tanzania. With limited mineral resources (the chief ones are diamonds in Tanzania, not far south of Lake Victoria, and copper in Uganda, to the west of the lake), most people depend on the land—and on the water that allows crops to grow and livestock to graze. In much of East Africa, rainfall amounts are marginal or insufficient. The heart of Tanzania is dry, tsetse- and malaria-ridden, and still the locale of occasional food shortages. Eastern and northern Kenya consist largely of dry steppe country, land that is subject to frequently recurring drought. As Fig. I–9 shows, the wettest areas lie spread around Lake Victoria and Uganda receives more rainfall than its neighbors.

Tanzania is the largest of the East African countries. Its area exceeds that of the four other countries combined, although its population of 21.9 million is only slightly larger than Kenya's (20.2 million). Uganda, with 14.7 million, ranks next; however, the small countries of Burundi and Rwanda together exceed 10 million, so that East Africa contains some densely populated areas. Although all these states lie in the same region, they display some strong internal differences. Tanzania, for example, has been described as a country without a primary core area, because its areas of productive capacity (and its population, Fig. I–14) lie spread about, mostly near the country's margins on the east coast, near the shores of Lake Victoria in the northwest, near Lake Tanganyika in the west, and near Lake Malawi in the south. Kenya, on the other hand, has several good-quality agricultural areas and a strongly concentrated core area centered on its capital of Nairobi (population: 1.9 million). Again, Tanzania is a country of many ethnic groups, none with a numerical or locational advantage to enable domination of the state, but Kenya comes closest to being dominated by the Kikuyu, the country's largest national group, whose traditional domain includes much of the productive farmland of the core. In terms of political and politico-geographical philosophy, Tanzania has gone the route of African socialism, whereas Kenya has become a capitalist state. Although

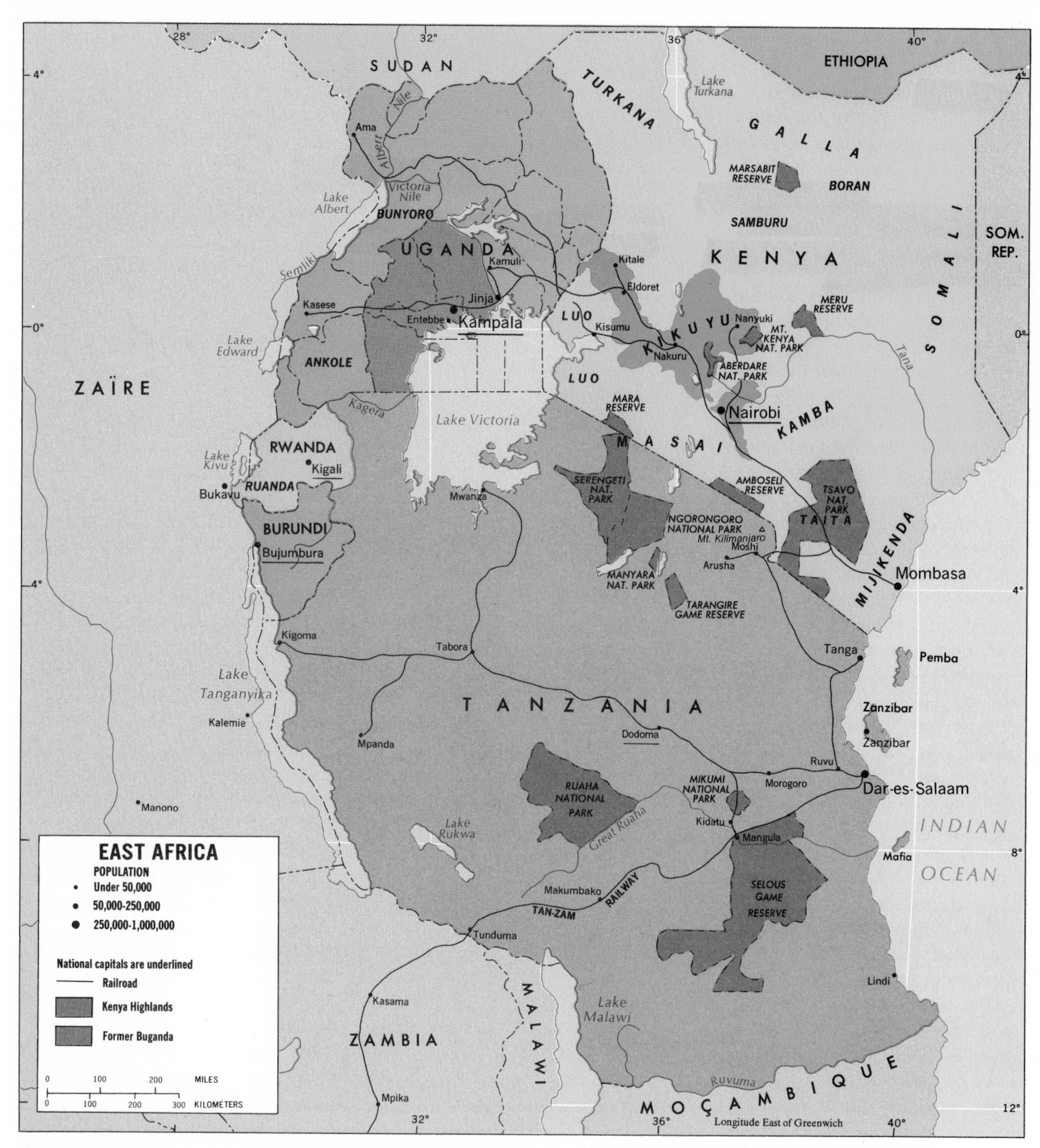


Figure 7–26

latent problems always remain, both Kenya and Tanzania have come to terms with their non-African minorities, but in different ways. Kenya had some 70,000 white settlers, over 200,000 Asians, and perhaps 30,000 Arabs; alternatively, Tanzania never had such large foreign minorities and the white population never reached beyond 20,000. In African-socialist Tanzania, Asians and other minorities have been integrated within a new economic and social order; but, in Kenya, many of the old conditions still prevail—e.g., Asians are still prominent in commerce. However, neither Tanzania nor Kenya took the drastic step ordered by the Amin regime in Uganda (1971–1979), where all 75,000 Asians were ordered to leave the country within a three-month period in 1972. It was a classical modern case of forced migration (see p. 124), and emigrants left behind some $400 million in assets as well as a commercial system that soon fell apart.

Tanzania has been viewed as an important example for the rest of Africa. Notwithstanding its limited resources and fragmented population, the country achieved political stability in the face of pressures generated by its transition to self-help African socialism. Tanzania has some commercial agriculture (sisal plantations along the north coast, coffee farms on the slopes of Mount Kilimanjaro near the Kenya border, cotton south of Lake Victoria, and tea in the southwest near Lake Malawi), but the great majority of Tanzanians are subsistence farmers. The government of president Julius Nyerere undertook a major reorganization of agriculture, creating new villages, forming cooperatives, and supporting improved farming methods. There was opposition, of course, but the new order prevailed. All this occurred while Tanzania was a haven for insurgents fighting the Portuguese in Moçambique, a difficult merger with Zanzibar was accomplished, and the capital was moved from coastal Dar es Salaam (1.8 million inhabitants) to Dodoma in the interior. At the same time, the Chinese were in Tanzania building the Tan-Zam or TAZARA Railway. Even when Tanzania faced the impact of the same drought that caused such devastation in the West African Sahel (which was followed by another drought during the early 1980s), things did not fall apart. However, the pressures of a fickle environment and periodic food shortages remain great.

Kenya's comparative prosperity is reflected by the tall skyscrapers of Nairobi (Fig. 7–27) and the productive farms of the nearby highlands. But Kenya's development is concentrated, not spatially dispersed like Tanzania's, so that the evidence of postindependence development is strong in the core area, but very limited in the sparsely peopled interior. As Fig. 7–26 shows, both Tanzania and Kenya have a single-line railroad that traverses the whole country from the major port in the east (Mombasa in the case of Kenya) to

# SEQUENT OCCUPANCE

The African realm affords numerous opportunities to illustrate the geographic concept of *sequent occupance* (a notion first introduced on p. 7). This concept involves the study of an area that has been inhabited—and transformed—by a succession of residents, each of whom has left a lasting cultural imprint. A place and its resources are perceived differently by peoples of different technological and other cultural traditions. These contrasting perceptions are reflected in the contents of their cultural landscapes. The cultural landscape today, therefore, is a collage of these contributions, and the challenge is to reconstruct the contributions made by each successive community.

The idea of sequent occupance is applicable in rural as well as urban areas. The ancient Bushmen used the hillsides and valleys of Swaziland to hunt and to gather roots, berries, and other edibles. Then, the cattle-herding Bantu found the same slopes to be good for grazing, and, in the valleys, they planted corn and other food crops. Next came the Europeans who laid out sugar plantations in the lowlands, but, after using the higher slopes for grazing, they planted extensive forests and lumbering became the major upland industry.

The Tanzanian coastal city of Dar es Salaam provides an interesting urban example. Its Indian Ocean site was first chosen for settlement by Arabs from Zanzibar to serve as a mainland retreat. Next it was selected by the German colonizers as a capital for their East African domain, and it was given a German layout and architectural imprint. When the Germans were ousted following their defeat in World War I, a British administration took over in "Dar," and the city entered still another period of transformation; a large Asian population soon left it with a zone of three- and four-story apartment houses that seem transplanted directly from India. Then, in the early 1960s, Dar es Salaam became the capital of newly independent Tanzania, under African control for the first time. Thus, Dar es Salaam in less than one century experienced four quite distinct stages of cultural dominance, and each stage of the sequence remains etched into the cultural landscape. Indeed, a fifth stage has now begun, since the national-government functions were recently moved from Dar es Salaam to the new interior capital of Dodoma.

Because of Africa's cultural complexity and historical staging, numerous possibilities of this kind exist. The Dutch, British, Union, and Republican periods can be observed in the architecture and spatial structure of Cape Town and other urban places in South Africa's Cape Province. Mombasa in Kenya carries imprints from early Persian, Portuguese, Arab, British, and several African communities. African traditional farmlands taken over by Europeans and fenced and farmed are now being reorganized by new African proprietors, and contrasting attitudes toward soil and slope are clearly imprinted in the landscape.

**Figure 7–27**
The modern center of Nairobi is the focus of East Africa's largest metropolis, which surpasses the 2 million mark in 1986.

the far west, and feeder lines come from north and south to meet this central transport route. In Kenya, that central railroad and its branches in the highlands really represent the essence of the country, but in Tanzania the railroad lies in the "empty heart" of the country, and the branches lead to the productive and populated peripheral zones. The new Tan-Zam Railway provides Tanzania with a link to neighboring landlocked Zambia as well as access to a poorly developed domestic area with relatively good agricultural potential in the southern highlands.

Kenya does not possess major known mineral deposits; besides its coffee and tea exports, it receives substantial revenues from a tourist industry that grew rapidly during the 1960s and 1970s, until East African political difficulties that began in the late 1970s reduced the flow of visitors. In the mid-1970s, the total number of foreign visitors to Kenya was approaching half a million, and the industry became the largest single earner of foreign exchange. Kenya's famous wildlife reserves lie mostly in the drier and more remote parts of the country, but population pressure is a growing problem, even in those areas, and poses still another threat to the future of tourism.

If Tanzania is an example for Africa, neighboring Kenya is as well, but of another kind. Unlike Tanzania, Kenya chose a capitalist route in its quest for accelerated development. In the process, Kenya in many respects forged ahead of its larger neighbor, exceeding it in GNP, per capita income, and in a host of other economic indicators. Certainly, this was accompanied by a less equitable division of wealth (in Nairobi, a class of people owning expensive automobiles are referred to as *Wa-Benzi*), but entrepreneurship and private initiative had fuller reign in Kenya. In 1978, Kenya passed a major test when its first president, Jomo Kenyatta, died and, in an orderly succession, the vice president, Daniel arap Moi, assumed the state's highest office; a further test was passed when a *coup d'état* was attempted in 1982 but failed and stability was restored.

Landlocked Uganda depends on coastal Kenya for its exit to the ocean, but relations between Uganda and Kenya have not been good in recent years. For several years after independence, Uganda seemed set on a course similar to that of many other African countries, but a military coup in 1971 brought to power the unpredictable General Idi Amin. The totalitarian Amin regime soon claimed a large part of western Kenya as Ugandan territory, and Amin also spoke of a Ugandan corridor to the sea through northern Tanzania. For many years, before and after the independence of these three countries, there was much talk about a greater East African international community. The pugnacious claims by Amin were the final blow to this idea, already threatened by the growing differences between Tanzania and Kenya.

Uganda contained the most important African political entity in the region when the British entered the scene during the second half of the nineteenth century. This was the Kingdom of Buganda, which faced the north shore of Lake Victoria, had an impressive capital in Kampala, and was stable—as well as ideally suited for indirect rule over a large hinterland. The British established their headquarters at Entebbe on the lake (thus adding to the status of the kingdom), and proceeded to solidify their hold over their Uganda Protectorate, using Buganda representatives among other peoples. In the territory that came to be Uganda, the Buganda were always in the minority, but when the British departed they endowed the country with a complicated federal constitution that was designed to maintain Buganda primacy by making them the most powerful among nominal "equals." But the system failed, and in the new Uganda movement became restricted and modernization and interaction reversed.

Under the chaotic and brutal Amin regime during the 1970s, upheavals brought ruin to once-prosperous Uganda. Submerged animosities resurfaced, and tumultuous waves of retribution and revenge flooded the countryside. After the ouster of the Asians and the collapse of the economy, Uganda largely reverted to a state of fragmentation. Markets closed, and subsistence living quickly reappeared. Wildlife in the national parks was hunted out. Thousands of people escaped to neighboring Kenya, but many more were killed; some estimates put the death-by-violence toll as high as 300,000. Ultimately, Amin began a conflict with Tanzania over an area along Uganda's border near Lake Victoria; the Tanzanian army eventually invaded Uganda and, in 1979, drove the paranoid despot into exile. Milton Obote, the country's first prime minister, soon returned from his exile, but the troubles continued. The economy remained a shambles, the population weakened and badly divided. For Uganda, the 1970s constituted a decade of almost incalculable setbacks from which the country had not recovered by the mid-1980s and may never fully recover.

# UNIFIED FIELD THEORY

A *unified field theory* was proposed by the political geographer Stephen Jones as a description of the means by which political territories become defined. More correctly a model than a theory, Jones argued that whenever we see a politically organized area, we are looking at the result of a number of distinctly different "hubs" of activity that have succeeded each other over a period of time. The first of these is an *idea*—a political idea or an idea with political implications. If the idea is to lead to creation of a political territory, the idea must be related to a *decision* that will provide the basis for territorial organization. Implementation of the decision almost always involves some *movement*—of people, goods, ideas—that represents the diffusion of the idea. The movement takes place in a *field* of activity, quite often the area eventually occupied by the political territory when it is made legitimate. In the end, there will be a politically organized *area*, the territory defined politically but legitimized as well by the various "hubs" of activity that occurred earlier. His five-step model may be diagrammed in this manner:

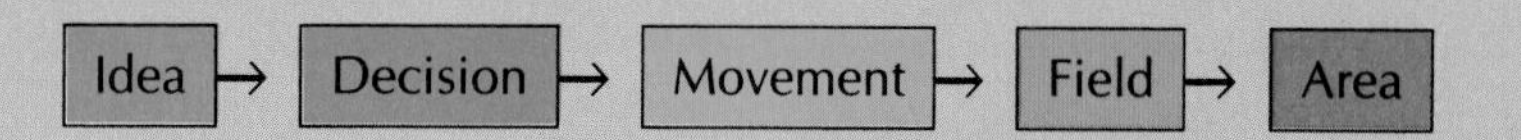


This sequence became known as the *idea-area chain*, but you should not assume that it is simply a progression in which one step always gives rise to the next. Rather, the several elements ("hubs") in the model interact with each other all the time, backward as well as forward.

In Uganda, the fundamental political idea involved the welding of the Buganda Kingdom to various smaller kingdoms nearby (and more distant peoples as well) under the general philosophy of indirect rule. The specific decision was to use Buganda people to administer the resulting protectorate on behalf of the British crown. The movement phase witnessed Buganda dispersal into the non-Bugandan areas of Uganda, but it also meant the diffusion of new ideas, goods, and practices. Roads, police stations, water traffic on rivers and lakes, post offices, and other kinds of movement-related innovations were brought to Uganda's interior. The field phase of the model produced the actual boundaries of Uganda, the political territory within which these changes had taken effect. Finally, there emerged the politically organized area: Uganda as it was until the British departed in 1962.

## Equatorial Africa

As the term is used, *Equatorial Africa* lies to the west of highland East Africa, and consists of Zaïre, Congo (formerly French Congo; capital, Brazzaville), Gabon, the Central African Republic, Cameroon, and Equatorial Guinea (Fig. 7–28). Chad, located to the north of the Central African Republic, was an administrative part of French Equatorial Africa and is often still included in this region, but it lies between Niger and Sudan in the West African-related savanna belt, and only by the most tenuous of arguments can it be considered a part of Equatorial Africa.

The giant of Equatorial Africa is Zaïre, the former Belgian Congo. With nearly 2.4 million square kilometers (905,000 square miles) and 33.1 million inhabitants, Zaïre is one of Black Africa's largest politico-geographical units. It contains the bulk of Equatorial Africa's human and natural resources, its largest cities, its best-developed communications, and, undoubtedly, its greatest opportunities and potential. The regional geography of Zaïre is most distinctively tied to its physical geography: most of the country's margins occupy the plateau rim surrounding the Zaïre (Congo) Basin, but the interior and northeast quadrant fall within the basin. Virtually all of Zaïre's considerable mineral wealth lies within the ba-

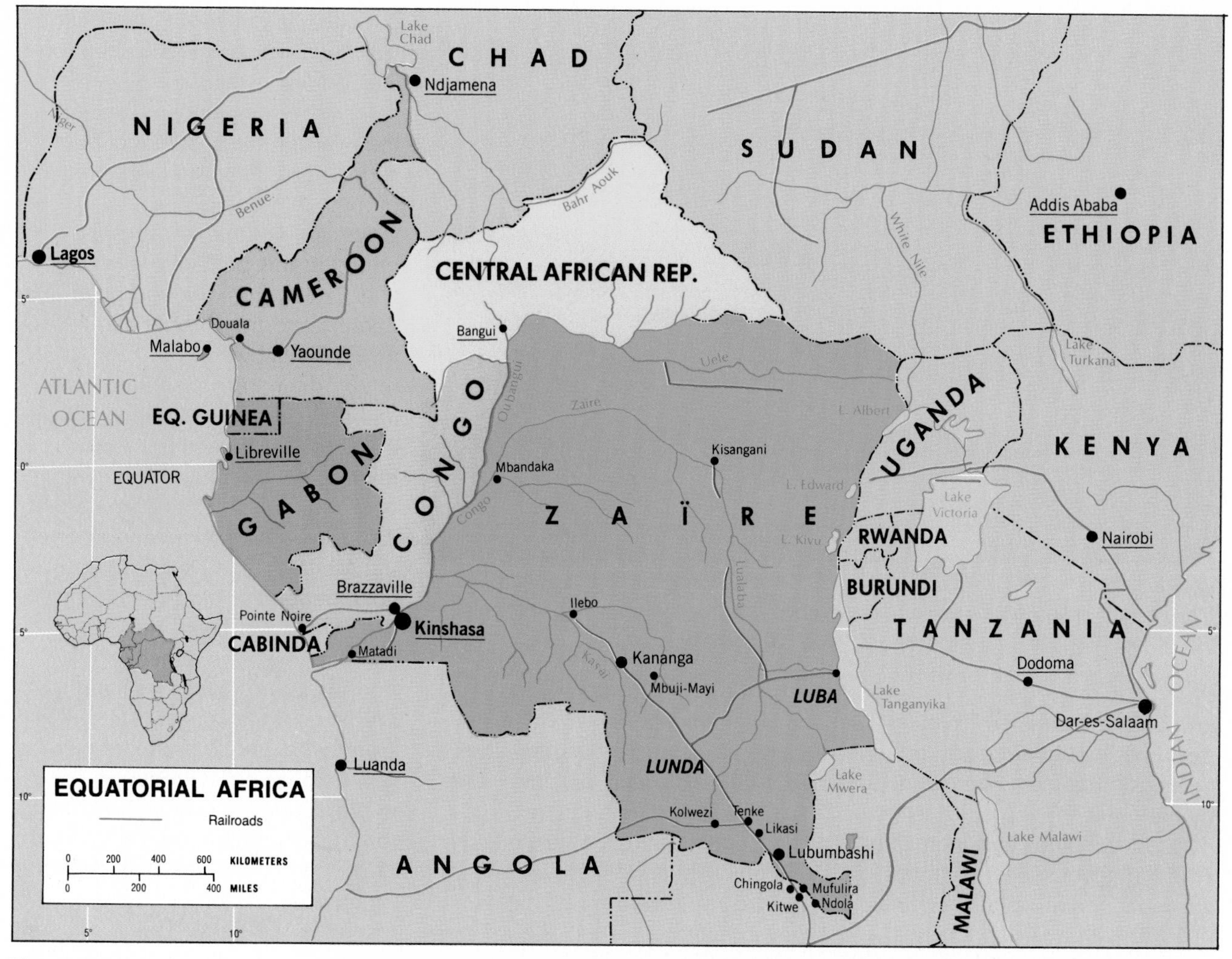


Figure 7–28

sin's rim, and this distribution exaggerates the politico-geographical difficulties described earlier (p. 400). The apparent transport focus provided by Zaïre's abundant river system has been neutralized by the rapids found in many key locations, a lack of attention by the colonial Belgians, and the great distances involved. It, therefore, took nearly two decades after independence to connect the mines of the southeast with the main port of Matadi via a feasible, wholly domestic transport system. In spite of Zaïre's development potential, it faces a number of continuing tests imposed by its geography. The country's size and diversity represent both strengths and weaknesses, and its brief history since independence in 1960 demonstrates the opposing tendencies toward consolidation and fragmentation. Zaïre will, no doubt, continue to be pulled in several directions at once, but the longer the country remains whole and continues toward a more integrated development than that left by the Belgians, the more likely Zaïre's potential will be realized.

The fragments of former French Equatorial Africa add comparatively little to the region's contents. France never managed to create an interconnected whole among these onetime equatorial possessions, and there is no unity in this part of the region today. The total population of the five former French dependencies is only 20 million, and the rate of natural increase remains a bit below the continent's average of 3 percent per year. Independence has not made the economic picture much brighter, although as small as this group of states is, some variation does exist within the region (Fig. 7–29). It is the coastal countries of Gabon and Cameroon that have fared better economically since independence, whereas the other three states lagged; the landlocked Congo (Brazzaville) and Central African Republic both have limited known resources and a heavy transport burden in getting what they do produce to the coast. Gabon, small, compact, and thinly settled, has modest oil reserves, substantial metallic mineral deposits, and abundant forest resources. Cameroon benefited from

**Figure 7–29**
One of Equatorial Africa's few non-subsistence farming landscapes: the banana-plantation zone around Buea near the Cameroon coast.

unusually constructive colonial treatment by the Germans and four decades of relatively benign neglect by the French; so, despite several colonial traditions and more than 200 indigenous ethnic groups, it began its period of independence in 1960 with a fair commercial agricultural base (tea, bananas, coffee, oil palm) and fewer of the curses of the colonial period than many African states. Nonetheless, Equatorial Africa remains a region of subsistence and raw-material exports; only in the growth areas of Zaïre and a few tiny pockets elsewhere can the beginnings of a new age be detected.

## Southern Africa

As a geographic region, Southern Africa consists not only of the Republic of South Africa and its immediate neighbors, but contains Angola, Zambia, and Malawi as well (Fig. 7–30). Not long ago, the latter two countries (then named Northern Rhodesia and Nyasaland) were locked into an ill-fated federation dominated by white interests in Southern Rhodesia (which became Zimbabwe on independence in 1980). For many decades, Zambia, whose economic core area is the Copperbelt (lying between the Lubumbashi and Ndola areas), has looked southward for its electric-power supply, its fuel, and for its outlets to the sea; the Tan-Zam Railway was built, in large measure, to redirect this trade, but Zambia's involvement with Southern African countries set a pattern that will probably emerge again after the postindependence political and economic turmoil is resolved. Malawi, too, has had to look south for its outlets; like Zambia, Malawi is a landlocked country, and its core area lies in the southern half of its elongated territory. The region's largest state in area is the former Portuguese colony of Angola, whose Soviet-backed Marxist regime does not deal with neighboring countries and is engaged in repelling a continuing guerilla war along its southern border.

Southern Africa is the continent's richest region in material terms. A great zone of mineral deposits extends from Zambia's Copperbelt, through Zimbabwe's Great Dyke and South Africa's Bushveld Basin and Witwatersrand, to the gold fields of the Orange Free State in the heart of the Republic of South Africa—and attracts a large number of migrant laborers (Fig. 7–19). The range and volume of minerals mined in this belt are enormous (Fig. 7–31), from the copper of Zambia and the chrome and asbestos of Zimbabwe to the gold, chromium, diamonds, platinum, coal, and iron ore of South Africa. However, not all of Southern Africa's mineral deposits lie in this central backbone. There is coal in western Zimbabwe at Wankie, and in central Moçambique near Tete. In Angola, petroleum from fields along the north coast heads the export list; diamonds are mined in the northeast, and manganese and iron on the central plateau. Namibia (Southwest Africa) produces copper, lead, and zinc from a major mining complex centered on the town of Tsumeb in the north, and diamond deposits are worked along the beaches facing the Atlantic Ocean in the south. This is a mere summary of Southern Africa's mineral wealth, and it is matched by the variety of crops cultivated in the region. Vineyards drape the valleys of the Cape Ranges in South Africa; apple orchards, citrus groves, banana plantations, and fields of sugarcane, pineapples, and cotton reflect South Africa's diverse natural environments. In Zimbabwe, tobacco has long been the leading commercial crop, but cotton also grows on the plateau and tea thrives along the eastern escarpment slopes. Angola is a major coffee producer, and Moçambique has huge cashew and coconut plantations. The staple crop for the majority of Southern Africa's farmers is corn (maize), but wheat and other cereals are also grown. Even the pastoral industry is marked by variety: not only are dairying and beef-cattle herds large, but South Africa also is one of the world's leading wool exporters.

But this mass of diverse resources has not been sufficient to improve the lives of all of Southern Africa's peoples—nor do the region's 10 countries share equally in these assets. South Africa is by far the richest country in Southern Africa, indeed the wealthiest in all of Black Africa. On the other hand, Lesotho, the mountainous country that is completely surrounded by South Africa, is poor by any standards. Botswana—landlocked and bounded by South Africa, Namibia, and Zimbabwe—is mainly desert and steppe country; in 1985, this Texas-sized entity still had only 1.1 million inhabitants. Moçambique, heavily exploited during more than four centuries of Portuguese colonial rule, remains an example of a seriously underdeveloped state. Less than 2 million of its 14 million residents were urbanized by the mid-1980s, and the popula-

**Figure 7–30**

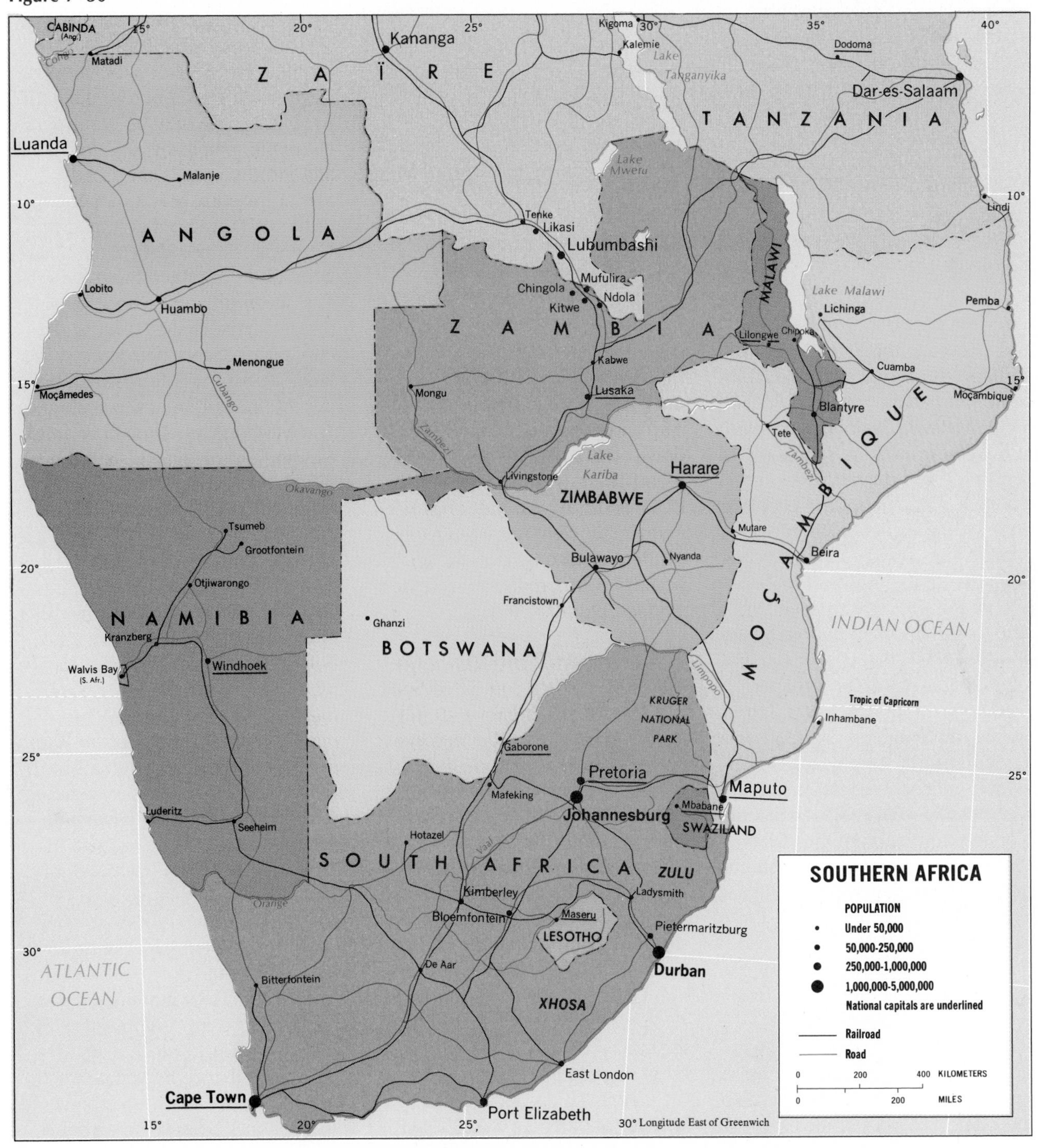

**Figure 7–31**
Open-pit copper mining at Chingola in the heart of Zambia's ore-rich Copperbelt.

tion's life expectancy at birth was only 46 years; the literacy rate was perhaps 20 percent, and annual income per person remained under $300. Malawi, Zambia, and Angola also suffer from these symptoms, albeit to a lesser degree. The overwhelming majority of the people in these countries live a life of often difficult subsistence, tilling patches of land near their villages that may or may not produce an adequate crop to sustain them during the approaching dry season, and eating meals that are ill balanced at best.

Even in South Africa and Zimbabwe, skyscrapered cities, air-polluting industrial complexes, and modern mechanized farms still adjoin areas of rural Africa almost unchanged from earlier times, where all those manifestations of modernization seem far away and of little relevance to daily village life. On world maps showing developed and underdeveloped countries, South Africa often appears as one of the DCs—but there are two South Africas. The cities and their factories, the mines, and the commercial farms generate the incomes and other measures of development that place South Africa in the DC category. But it is white South Africa that is described in these figures, with only a minority of native African peoples who have been drawn into their sphere of development. There is substantial wealth in black South Africa, but it has been neither sufficient nor adequately distributed among its multiracial population to produce a truly developed plural society.

**Emergent Zimbabwe** At the heart of black-controlled Southern Africa lies Zimbabwe (9 million), a state of considerable economic potential, but one that did not come under majority rule until 1980. The protracted struggle for control here in Britain's former dependency of Southern Rhodesia illustrates the regional-geographical problems of decolonization in Africa. The country's core area is well defined, lying astride the mineral-rich Great Dyke and extending, approximately, from the environs of the capital, Harare (formerly Salisbury) (population: 1.3 million) to the vicinity of the second city, Bulawayo (population: 575,000). This is the stronghold of the former colony's white population, although white-owned farms and plantations lie scattered far beyond its limits.

The drive for independence by Zimbabwe's Africans was not simply a white-black confrontation however. If there are strong animosities between white and black, there are also serious divisions among the African peoples of Zimbabwe, and these divisions have regional expression. To begin with, Zimbabwe's African population is fragmented into about 40 different peoples. These may be grouped culturally, ethnically, and locationally into two major units: the Shona-speaking Mashona peoples of the east and northeast, and the Nguni-speaking Ndebele (or Matabele) of the southwest and west. The seven principal Shona-speaking peoples have created a loose federal alliance, and their territory extends across the Zimbabwean border into neighboring Moçambique. The Mashona are Zimbabwe's oldest modern inhabitants, and the famous relict city of Zimbabwe—from which the country takes its name—is an ancient cultural center of the nation. Shona society is quite egalitarian and comparatively passive. The Mashona outnumber all other peoples in Zimbabwe combined, so that they have the capacity to control the country's government. In contrast, the Ndebele peoples are recent arrivals in Rhodesia/Zimbabwe. When the British arrived to colonize the area during the second half of the nineteenth century, the Ndebele had just invaded the west and southwest of Zimbabwe, and they were in the process of colonizing Shona communities. The European intrusion stopped this process, but Ndebele society retains its stratified character, with a "pure" Nguni-speaking elite as the upper class, a somewhat assimilated and acculturated middle sector, and an overpowered and colonized lower class consisting of Shona (and other) groups in the process of transformation under Ndebele cultural dominance.

Like other African countries dominated by one or two national groups, Zimbabwe has a number of smaller minority peoples, but the Mashona and Ndebele constitute the bulk of the population, now at 9 million. Even as Ndebele and Shona leaders were caught up in the struggle for independence in the 1960s and 1970s, they clearly differed on ideological grounds and saw Zimbabwe's postindependence future in different ways. The intervention of the British government in the late 1970s, which finally brought all parties to the conference table, ended the escalating war of independence; the subsequent transfer of sovereignty to the elected representatives of the African majority in 1980 was a major diplomatic achievement. But political events within independent Zimbabwe have shown that old divisions strongly persist, and it is likely that the new country will also have to confront new pressures that arise from its location adjacent to the continent's last bastion of white rule and supremacy.

### South Africa's prospects

South Africa contains the greatest material wealth in the entire African realm, rich in both natural resources and in the diversity of its cultures. The Republic's 32.5 million people (1985) include over 23 million Africans representing several nations; 4.5 million whites, of whom the majority are Afrikaners, whereas most of the remainder have British ancestries; over 2.5 million "Coloured" people of mixed African/white ancestry, mainly concentrated in the southernmost Cape Province; and about 1 million Asians, mostly of Indian descent and largely concentrated in the port city of Durban and surrounding rural Natal.

South Africa's modern history has set it apart from the rest of Black Africa. The old Boer republics were founded before the colonial scramble elsewhere really gathered momentum, and the diamonds of Kimberley attracted thousands of whites long before individual European travelers first saw much of tropical Africa. While African laborers from Lesotho made their way in the 1860s and 1870s to the mines of Kimberley to seek work and wages, Asians from India were arriving by the thousands on the shores of Natal Province, under contract to work on the sugar plantations there. Together, black, white, Asian, and Eurafrican contributed to the emergence of Africa's most developed country. Whites discovered the mineral resources of South Africa and provided the capital for their exploitation, but blacks did the work in the mines at wages low enough so that their extraction was an economic proposition. Whites laid out the farms and plantations that produce so wide a range of crops, but Africans and Asians constituted the essential labor force to make them successful. Whites built the factories of Johannesburg, but blacks form the majority of the wage earners that work in them. Whites built or bought the fishing boats that annually bring in a catch large enough to rank South Africa among the world's top 10 fishing nations, but the nets are manned mostly by Coloureds.

With such interracial cooperation in common economic pursuits, one might expect that South Africa's plural society over the years would have become more and more integrated. But in fact this has decidedly not been the case. South Africa stands today as a world symbol of racial injustice, an outpost of white-minority rule on a black continent. It largely acquired this reputation after World War II, although the foundations were laid much earlier. While much of the rest of the world reflected on the racial injustices involved in that war and the decolonization period of the 1950s transformed the international political scene, South Africa became known for its official policy of racial subjugation and separation as summarized by one word—**apartheid**. With this policy, South Africa moved in a direction that

**Figure 7–32**
An entrance to Durban's beach: subtlety is not observed when racial segregation is frequently expressed in South Africa's landscape.

was more or less opposite to those ideals that were championed—if not necessarily practiced—in other parts of the free world.

Within South Africa and in the world at large, the concept of *apartheid* has often been incorrectly defined. *Apartheid* involved not only the separation of the black peoples of South Africa from the white, but also the Coloured from the white and the Asian from the African. Behind it lay a vision of South Africa as a community of racial states, in which each racial sector would have its own "homeland," although residents of one such "homeland" could travel to another to work. What gave South Africa its international reputation was not so much this scheme as the *ad hoc*, "petty" *apartheid* that was designed to set the wheels of ultimate segregation in motion. The

Figure 7–33

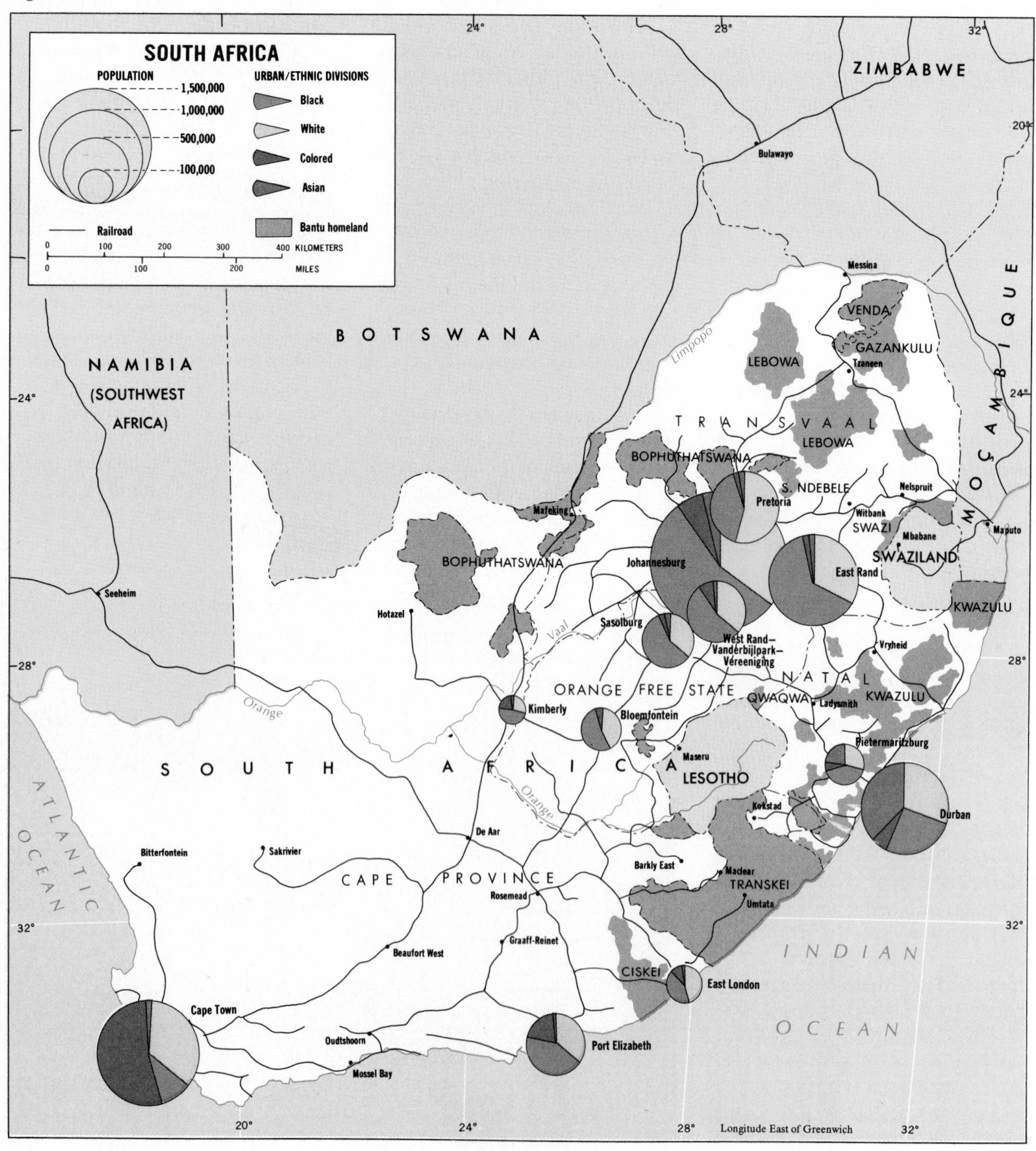

number of "Europeans Only" signs on such public facilities as park benches, bus stops, restrooms, and the like, rapidly multiplied (Fig. 7–32). Discriminatory pass (access) laws appeared, and police powers were increased to enforce them. Persons who were not white began to disappear from the white universities; persons of liberal political views lost their freedom; and African political organizations were banned, their leaders exiled or jailed.

Nevertheless, South Africa, with the view that the end justifies the means, is proceeding in its program of *separate development*—now the official name for the *apartheid* scheme. The first of the African Bantustans, or "homeland states," the Transkei, has been functioning for some years; it was formally declared independent in 1976. Bophuthatswana followed in 1978, Venda in 1979, and Ciskei in 1982, with a total of 10 homelands scheduled for "independence" by the late 1980s (Fig. 7–33). Initially, there were plans for a massive relocation of industry from the core areas of the country to the "borderlands," the boundaries between the Bantustans and the white areas. If industries could be located there, it was argued, then skills and capital could come from white South Africa and labor from the black states without any need for racial mixing. This part of the program has proved unworkable on any large scale, but other aspects of it are being vigorously pursued.

But the majority of the people of South Africa are black, and the separate development program's first objective has been the allocation of "homelands" for these people. Taking into consideration the spatial characteristics of the country, it is not surprising that a program that might seem totally unworkable in most other countries is viewed in South Africa as the ultimate and inevitable solution to racial problems in plural societies. When the Boers penetrated the Highveld, they left behind them the Xhosa and other peoples of Southern Natal and the Eastern Cape, who were hemmed in by the Great Escarpment on one side and the ocean on the other. On the plateau itself, the Boers found relatively little effective opposition, and, before long, virtually the whole of the Highveld was white country with the Africans confined in reservations around the drier Kalahari margins, the rugged Northern and Eastern Transvaal, and below the Great Escarpment in Natal and the Eastern Cape. Thus, the horseshoe arrangement of reserves existed before the current separate development program came into being.

What made the whole separate development program viable was the agricultural, mining, and manufacturing development of the Highveld. From Swaziland, Zululand, Lesotho, the Transkei, and every other African territory, black laborers have come to the farms, factories, and mines of the Highveld. In fact, they came by the hundreds of thousands; for example, the population of Johannesburg today is 2.1 million, nearly two-thirds of it African. All too often, nonwhite families would find themselves in miserable slums without adequate shelter, food, medical care, schools, or other necessary facilities (Fig. 7–34). Building programs could not keep pace with the influx; yet, the government could not prohibit the immigration of workers for they were always needed. This situation was high on the list of problems the separate development program was designed to solve. If an African man from an autonomous "homeland"—say, the Transkei—wanted to work in a white-owned mine or factory, he would apply for a permit to enter white South Africa on a temporary visa, would live in a dormitory-type residence in the city near the mine or factory, and would send his wife and children part of his income for their subsistence. Thus, the municipal government would no longer be faced with the task of making available the facilities necessary for the man's family, and the economy would not lose its labor supply.

Naturally, such a program could not be implemented overnight. The homelands must be established,

**Figure 7–34**
The juxtaposition of poverty and prosperity, here in the cityscape of Johannesburg, is all too common a sight throughout South Africa.

given the trappings of semiautonomous states, governments installed, and arrangements made for the accommodation of repatriated families; and, of course, people who may have been living in the big cities for generations must be classified according to their "race." In this process, once again, the South African government paid little heed to the sensitivities of many of its subjects. People were classified and —regardless of their race—were made to carry identity cards to prove their racial heritage. In many cases, of course, there was little doubt: the spoken language quickly tells whether a man or woman is a Zulu or a Xhosa. But race in South Africa has long been correlated with status. If someone can prove he or she is white rather than Coloured, chances for a decent life are immeasurably improved and even if individuals can prove they are Coloured, not black, they may be substantially better off than otherwise. Some of the saddest tales ever to come from South Africa tell of people desperately trying to disavow their own ancestries, hoping to climb the social ladder one rung, to give themselves and their children a better chance in life. Separate development is ultimately designed to solve all this. Instead of struggling for advancement in the white society, black or Coloured people would do so within their own societies, without racial barriers, job restrictions, and other racially imposed obstacles. Thus, separate development is touted as *parallel* development; eventually, all the states of South Africa are supposed to have every facility, including universities, hospitals, transport services, and so forth.

The critical flaw in the design appears to lie in the allocation of the country's areas of productive capacity to the homelands of South Africa's 1.2 million square kilometers (nearly 475,000 square miles) of territory. Only about one-sixth is presently being considered for the African homelands; yet, blacks constitute well over two-thirds of the total population. Four-fifths of the territory of the country will be kept as "white" South Africa—controlled by the one-sixth of the population that is of European ancestry. Obviously, these figures by themselves mean very little: in many of the countries of the world, most of the productive lands lie on less than one-sixth of the total area. But a look at maps of the distribution of resources, cities, transport lines, and the other assets of South Africa quickly shows that the black African homelands, far from having a real share of South Africa's prosperity, are among the poorest and most isolated parts of the country. No major mineralized areas lie in these homelands; from 1965 to 1975, only about 3 percent of South Africa's income from its mineral resources came from the Bantu areas. As far as agricultural resources are concerned, there is good farmland in the eastern homelands—in Natal and the Eastern Cape—but these areas, especially the Transkei and Zululand, also face a serious overpopulation problem. Those reserves that lie in the Northern Cape and on the Western and Northern Transvaal fringe suffer from drought and terrain problems. In terms of industrialization, the homelands have hardly begun to change, and even secondary industries are practically absent there; they possess few resources, have never had much capital, and lie far from markets and the sources of power supply. South Africa's railway network and the country's major roads skirt the homelands or cross them only to connect cities of white South Africa; internal communications remain inferior in the homelands. And none of South Africa's major metropolitan complexes lies in a present or potential Bantu homeland.

Such realities have begun to stimulate rethinking of the separate development program among South Africa's decision makers. The permanence of black residents in the large urban areas has been acknowledged through changing legislation in favor of black homeowners: property rights can now be acquired by black city residents. As a further sign of change, a division developed within the ruling (white) National Party between those who would maintain the *status quo* and those who support flexibility, however modest. When the issue between these party segments came to a vote, the so-called "enlightened" won the day. This encouraged the government to press ahead with a program of limited power sharing, initially with the Asians and Coloureds, who would have representation in a newly structured parliament. That plan was approved in a national referendum (in which whites only voted) late in 1983. It illustrates South Africa's difficult contradictions: while fundamental separatist policy continues, the old *apartheid* doctrine is being dismantled.

Will South Africa ultimately reverse its course toward spatial segregation by race? There may not be time to allow a peaceful transition. Black opposition within the Republic is growing, punctuated by terrorist bombings and acts of sabotage. As internal polarization grows, external pressure also intensifies. Meanwhile, 8 out of 10 farm laborers in "white" South Africa are black Africans, as are nearly 9 out of 10 miners and 9 of 20 factory workers; in the ultimate sense, the country's social geography itself is a denial of any separate development. Geography and history may have conspired to divide South Africa's peoples, but economic realities have thrown them irretrievably together.

## References and Further Readings

"Africa's Woes," *Time*, January 16, 1984, 24–41.

Barbour, K. "Africa and the Development of Geography," *Geographical Journal*, 148 (1982), 317–326.

Bernard, F. & Thom, D. "Population Pressure and Human Carrying Capacity in Selected Locations of Machakos and Kitui Districts (Kenya)," *Journal of Developing Areas*, 15 (1981), 381–406.

Best, A. & de Blij, H. J. *African Survey* (New York: John Wiley & Sons, 1977).

Beyer, J. "Africa," in Klee, G., ed. *World Systems of Traditional Resource Management* (New York: Halsted Press/V. H. Winston, 1980), pp. 5–37.

Christopher, A. *Colonial Africa: An Historical Geography* (Totowa, N.J.: Rowman & Allanheld, 1984).

Christopher, A. *South Africa* (London: Longman, 1982).

Church, R. *West Africa: A Study of the Environment and Man's Use of It* (London: Longman, 8 rev. ed., 1980).

Clarke, J. & Kosinski, L., eds. *Redistribution of Population in Africa* (Exeter, N.H.: Heinemann Educational Books, 1982).

Craton, M. *Sinews of Empire: A Short History of British Slavery* (Garden City, N.Y.: Anchor Books/Doubleday, 1974).

Crush, J. "The Southern African Regional Formation: A Geographical Perspective," *Tijdschrift Voor Economische en Sociale Geografie*, 73 (1982), 200–212.

Curtin, P. *The Atlantic Slave Trade* (Madison: University of Wisconsin Press, 1969).

de Blij, H. J. *Dar es Salaam: A Study in Urban Geography* (Evanston, Ill.: Northwestern University Press, 1963).

Dostert, P. *Africa 1984* (Washington: Stryker-Post Publications, 1984).

Fage, J. "Slaves and Society in Western Africa, c.1445-c.1700," *Journal of African History*, 21 (1980), 289–310.

Fage, J. & Oliver, R., eds. *Papers in African Prehistory* (London: Cambridge University Press, 1970).

Fair, T. *South Africa: Spatial Frameworks for Development* (Cape Town: Juta, 1982).

Fincham, R. "Economic Dependence and the Development of Industry in Zambia," *Journal of Modern African Studies*, 18 (1980), 297–313.

Floyd, B. & Tandap, L. "Intensification of Agriculture in the République Unie du Cameroun," *Geography*, 65 (1980), 324–327.

Franke, R. & Chasin, B. *Seeds of Famine: Ecological Destruction and the Development Dilemma in the West African Sahel* (Montclair, N.J.: Allanheld, Osmun, 1980).

Gould, P. R. "Tanzania 1920–63: The Spatial Impress of the Modernization Process," *World Politics*, 22 (1970), 149–170.

Gourou, P. *The Tropical World: Its Social and Economic Conditions and Its Future Status* (London and New York: Longman, 5 rev. ed., trans. S. Beaver, 1980).

Graubard, S. et al. "Black Africa: A Generation After Independence," *Daedalus*, 111 (1982), 1–273.

Griffiths, I., ed. *An Atlas of African Affairs* (London and New York: Methuen, 1984).

Grove, A. *Africa* (Oxford: Oxford University Press, 3 rev. ed., 1978).

Hibbert, C. *Africa Explored: Europeans in the Dark Continent, 1769–1889* (London: Allen Lane, 1982).

Hunter, J. "River Blindness in Nangodi, Northern Ghana: A Hypothesis of Cyclical Advance and Retreat," *Geographical Review*, 56 (1966), 398–416.

Kloos, H. & Thomson, K. "Schistosomiasis in Africa: An Ecological Perspective," *Journal of Tropical Geography*, 48 (1979), 31–46.

Knight, C. & Newman, J., eds. *Contemporary Africa: Geography and Change* (Englewood Cliffs, N.J.: Prentice-Hall, 1976).

Learmonth, A. *Patterns of Disease and Hunger* (London: David and Charles, 1978).

Lele, U. "Rural Africa: Modernization, Equity, and Long-Term Development," *Science*, 211 (1981), 547–553.

Levi, J. & Havinden, M. *Economics of African Agriculture* (Harlow, Eng.: Longman, 1982).

Martin, E. & de Blij, H. J., eds. *African Perspectives: An Exchange of Essays on the Economic Geography of Nine African States* (London and New York: Methuen, 1981).

May, J. *Studies in Disease Ecology* (New York: Hafner, 1961).

McGlashan, N. *Medical Geography: Techniques and Field Studies* (London: Methuen, 1972).

Mosley, P. "Agricultural Development and Government Policy in Settler Economies: The Case of Kenya and Southern Rhodesia, 1900–1960," *Economic History Review*, 35 (1982), 390–408.

Murdock, G. *Africa: Its Peoples and Their Culture History* (New York: McGraw-Hill, 1959).

Nwafor, J. "The Relocation of Nigeria's Federal Capital: A Device for Greater Territorial Integration and National Unity," *GeoJournal*, 4 (1980), 359–366.

O'Connor, A. *The African City* (New York: Holmes & Meier, 1983).

O'Connor, A. *The Geography of Tropical African Development: A Study of Spatial Patterns of Economic Change Since Independence* (Elmsford, N.Y.: Pergamon, 2 rev. ed., 1978).

Oguntoyinbo, J. "Climatic Variability and Food Crop Production in West Africa," *GeoJournal*, 5 (1981), 139–149.

Pirie, G. "The Decivilizing Rails: Railways and Underdevelopment in Southern Africa," *Tijdschrift Voor Economische en Sociale Geografie*, 73 (1982), 221–228.

Prothero, R. *Migrants and Malaria in Africa* (London: Longmans, Green, 1965).

Pyle, G. *Applied Medical Geography* (New York: Halsted Press/V. H. Winston, 1979).

Richards, P. "The Environmental Factor in African Studies," *Progress in Human Geography*, 4 (1980), 589–600.

Rodenwaldt, E. & Jusatz, H., eds. *World Atlas of Epidemic Diseases* (Hamburg: Falk, 3 vols., 1952–1961).

Smith, R., guest ed. "Spatial Structure and Process in Tropical West Africa," *Economic Geography*, 48 (1972), 229–355.

Stock, R. *Cholera in Africa* (London: International African Institute, 1976).

Taaffe, E. et al. "Transport Expansion in Underdeveloped Countries: A Comparative Analysis," *Geographical Review*, 53 (1963), 503–529. Diagram adapted from p. 504.

Udo, R. *The Human Geography of Tropical Africa* (Exeter, N.H.: Heinemann Educational Books, 1982).

Western, J. *Outcast Cape Town* (Minneapolis: University of Minnesota Press, 1981).

SHIV FABRICS
SHIV EMBROIDERY
RAJMAL PAHARCHAND
EXPORTERS, IMPORTERS, AGENTS
apsara
SAREE
VINOD
BMF 5141

# CHAPTER 8

# INDIA AND THE INDIAN PERIMETER

IDEAS AND CONCEPTS

**Geomorphology**
**Boundary Superimposition**
**Green Revolution**
**Demographic Transition Model**
**Population Growth Dynamics**
**Physiologic and Arithmetic Density**
**Social Stratification**
**Centripetal and Centrifugal Forces**
**Irredentism (3)**

Southern Asia consists of three protruding landmasses. In the southwest lies the huge rectangle of Arabia. In the southeast lie the slim peninsulas and elongated islands of Malaysia and Indonesia. And located between these flanking protrusions is the familiar triangle of India, a subcontinent in itself. Bounded by the immense Himalaya Ranges to the north, by the mountains of eastern Assam to the east, and by the rugged arid topography of Iran and Afghanistan to the west, the Indian subcontinent is a clearly discernible physical region (Fig. 8–2).

This chapter focuses on India's huge human population in a world context. India alone, with 763 million people in 1985, has a larger population than South America, Black Africa, and Australia combined. Neighboring Bangladesh (103 million) and Pakistan (100 million) also rank among the world's 10 most populous countries. With just under 1 billion inhabitants, this realm constitutes one of the great human concentrations on earth. It was also the scene of one of the world's oldest known civilizations and later became the cornerstone of the British colonial empire.

**Figure 8–1**
Tradition and modernity are frequently juxtaposed in India's urban landscape: a sacred cow brings traffic to a halt near Bombay's Marine Drive.

The giant of South Asia, India, lies surrounded by the smaller countries. Bangladesh and Pakistan flank India on the east and west. To the north lie the disputed territory of Kashmir and the landlocked states of Nepal and Bhutan; Sikkim, formerly an Indian protectorate, was in 1975 incorporated into India as that federal country's 22nd state. And off the southern tip of the Indian triangle lies the island state of Sri Lanka (formerly Ceylon). The landlocked countries of the north lie in the historic buffer zone between the former British sphere of influence in South Asia and its Eurasian competitors; India, Pakistan, Bangladesh, and Sri Lanka all formed part of what was once known as British India.

British colonial rule threw a false cloak of unity over a region that possesses enormous cultural vari-

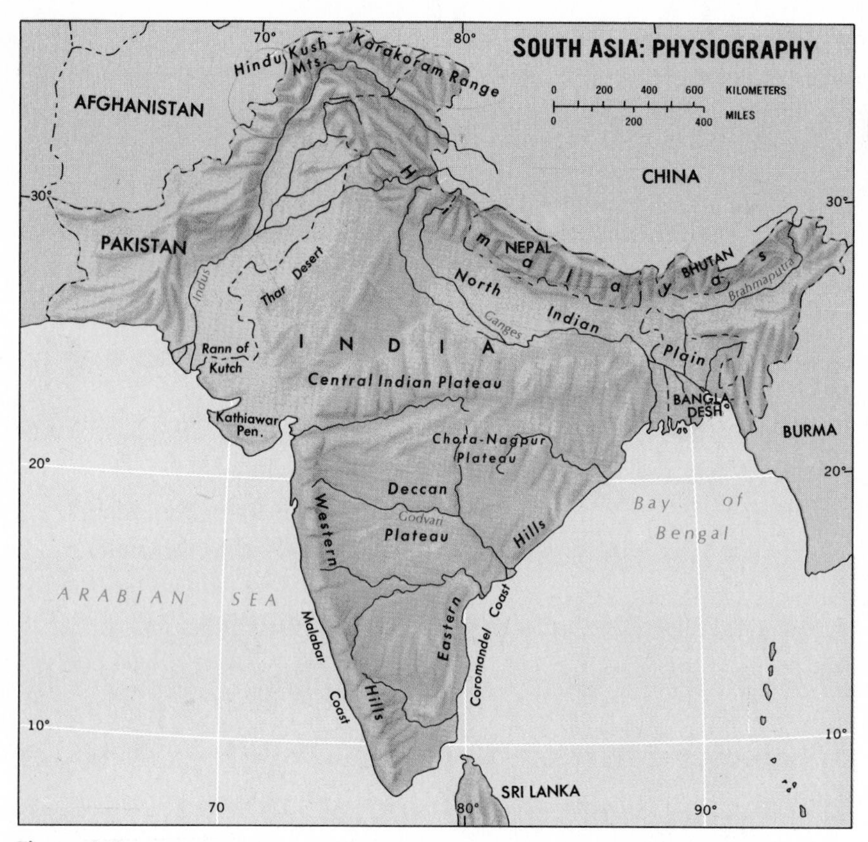


Figure 8–2

ety; even before the British departure in 1947, plans had been developed for the separation of a Moslem state, Pakistan, from Hindu-dominated India. Islamic Pakistan was divided into a discontinuous state that consisted of two territories, West Pakistan (now Pakistan) and East Pakistan (now Bangladesh); Sri Lanka became an independent state in 1948. In 1971, however, following a bitter war, Bangladesh (with rival India's support) freed itself from its ties with Pakistan. Nevertheless, it is something of a miracle that a region as large and populous as South Asia has generated only four postcolonial states. Within India itself, regional differences remain strong, but the country has held together for nearly 40 years. In Pakistan, persistent separatist movements affect Pakhtunistan in the northwest and Baluchistan in the southwest, but these pressures have largely been accommodated and the integrity of the state is sustained. In Sri Lanka, the Buddhist Sinhalese majority has clashed from time to time with the minority Hindu Tamils, but Sri Lanka as yet is no Cyprus.

# Physiographic Regions

As the systematic essay pointed out, the Indian subcontinent is a land of immense physiographic variety. Nonetheless, it is possible to recognize three rather clearly defined regions: the northern mountains, the southern peninsular plateau, and between these two, the belt of river lowlands. In this division, the northern mountains consist of the Baluchistan, Kashmir, Himalayan and Assam uplands. The southern tableland is constituted mainly by the huge Deccan Plateau. And the intervening tier of river lowlands extends from Sind (Pakistan's lower Indus Valley) through the Punjab and Ganges Plain and on through the great delta into Assam's Brahmaputra Valley (Fig. 8–2).

But this set of regions only introduces the varied Indian landscape. Depending on the degree of detail employed, it would be easy to subdivide each of these three into more specific physiographic subregions. Thus, in the mountain wall that separates India and Pakistan from the Asian interior, we may recognize the desert ranges of Baluchistan and the Afghanistan border, the towering ranges of the Hindu Kush, the Karakoram, and the Himalayas, and the jungle-clad mountains of the eastern Assam-Burma margin. Moreover, these northern mountains do not simply rise out of the river valleys below: there is a persistent belt of transitional foothills between the lofty ranges and the low-lying river basins.

The belt of alluvial lowlands that extends from the Indus to the Brahmaputra also is anything but uniform. This is sometimes called the North Indian Plain, and some geographers consider it to include the western Kathiawar Peninsula and part of the Thar Desert as well as the zone of river lowlands. Even without these extensions, this is a region full of contrasts, well reflected, incidentally, by Figs. I–9 to I–13. The Indus, which has its source in Tibet and which penetrates the Himalayas to the west in its course to the Arabian Sea, receives its major tributaries from the Punjab ("land of five rivers").

The physiographic region of the Punjab—located between the Indus and North Indian Plains—extends into Pakistan as well as India. The region of the lower Indus is characterized by its minimal precipitation, its desert soils (Fig. I–13), and its irrigation-based cluster of settlement. This is the heart of Pakistan, and

# TEN MAJOR GEOGRAPHIC QUALITIES OF SOUTH ASIA

1. South Asia is a well-defined physiographic region, extending from the southern slopes of the Himalayas to the island of Sri Lanka.
2. Two river systems (the Ganges-Brahmaputra and the Indus) form crucial lifelines for hundreds of millions of people in this realm. The annual wet monsoon is a critical environmental element.
3. India lies at the heart of the world's second largest population cluster.
4. No part of the world faces demographic problems with dimensions and urgency comparable to South Asia.
5. All the states of South Asia suffer from underdevelopment. Food shortages exist, nutritional imbalance prevails, and famines occur.
6. Agriculture in South Asia, in general, is comparatively inefficient and not as productive as it is in other parts of Asia.
7. The great majority of South Asia's peoples live in villages and subsist directly from the land.
8. Strong cultural regionalism marks South Asia. The Hindu religion dominates life in India; Pakistan is an Islamic state; Buddhism thrives in Sri Lanka.
9. The South Asian realm's politico-geographical framework results from the European colonial period, but important modifications took place after the European withdrawal.
10. India constitutes the world's largest and most complex federal state.

the farmers grow wheat for food and cotton for sale. Some 70 percent of all cultivated land in Pakistan is irrigated, most of it by an elaborate system of canals; wheat occupies more than one-third of the cropped land.

Hindustan, the region extending from the vicinity of the historic city of Delhi to the Ganges Delta at the head of the Bay of Bengal, is wet country; the precipitation exceeds 250 centimeters (100 inches) in sizable areas and 100 centimeters (40 inches) almost everywhere (Fig. I–9). Deep alluvial soils cover much of the region (Fig. I–13), and the combination of rainfall, river flow, good soils, and a long growing season make this India's most productive area. As Fig. I–14 verifies, this is also India's largest and most concentrated population core; not only do rural densities in places approach 400 people per square kilometer (1035 people per square mile), but also, lying between Delhi and Calcutta, is a series of the country's leading urban centers, connected by the densest portion of its rail- and road-transport network. In the moister east, rice is the chief food crop; jute is the main commercial product, notably in the delta, in Bangladesh. To the west, toward the Punjab and around the drier margins of the Hindustan region, wheat and such drought-resistant cereals as millet and sorghum are cultivated.

The easternmost extension of the North Indian Plain comprises the Brahmaputra Valley in Assam. This valley is much narrower than the Gangetic Plain, very moist, suffers from frequent flooding, and the river has only limited use for navigation. Up on the higher slopes, tea plantations have been developed, and in the lower elevations rice is grown. Assam is just barely connected with the main body of India through a corridor of only a few miles between the southeast corner of Nepal and the northern boundary of Bangladesh, and it remains one of South Asia's frontier areas.

Turning now to the peninsular part of India, referred to previously as plateau India, we find, once again, that there is more variety than first appearances suggest. There are physiographic bases for dividing the plateau into a northern area (the Central Indian Plateau would be an appropriate term) and a southern sector, which is already well known as the Deccan Plateau. The dividing line can be drawn on the basis of roughness of terrain (along the Vindhyas) or rock type (much of the Deccan is lava covered), and the Tāpi and Godāvari Rivers—in the west and east, respectively—form a clear lowland corridor between the two regions. The Deccan (meaning "south") has been tilted to the east, so that the major drainage is eastward; the plateau is marked along much of its margin by a mountainous, steplike descent formerly called the *Ghats* but now, simply, *Hills*. Parallel to the surrounding coasts are the Eastern and Western Hills; they meet near the southern tip of the subcontinent on the Cardamon Upland, recognized by some as a physiographic subregion.

Peninsular India also possesses a coastal lowland zone of varying width. Along the southwestern litto-

# SYSTEMATIC ESSAY

## Geomorphology

The geographic study of the configuration of the surface of the earth is called **geomorphology**. This is not the only term used in connection with this work. The term *physiography* also appears in this connection, and geographers call regions where a particular landscape prevails (such as the U.S. Great Plains or the Rocky Mountains) *physiographic provinces*. But physiography has a wider connotation than geomorphology: physiographic regions are created not only by landscape, but also by climate and weather, soils and vegetation. Geomorphology, on the other hand, concentrates on landscapes and landforms alone.

As in other fields of geography, the study of the earth's surface features can be approached at various levels of detail. Some geomorphologists focus on one particular *landform* and try to learn what forces and processes shape it, testing hypotheses wherever such a landform occurs. Thus, a landform is a single feature; a volcanic mountain of a special shape, a segment of a river valley, a deposit laid down by a glacier, a dune, and a sinkhole all are landforms. Other physical geographers take a wider view and study whole *landscapes* in an effort to discover what led to their formation and present appearance. A landscape, thus, is an assemblage of landforms. Some knowledge of landform formation may help in the analysis of landscapes, but it is not the whole story.

Obviously, geomorphology is a complicated field. To understand why landscapes appear the way they do, it is necessary to know the forces that bend and break the earth's crust *and* the agents of weathering and erosion that constantly change the exposed surface. The internal, crustal movements are caused by *tectonic* forces, and anyone who has experienced even a mild earthquake will know their power. Whole continents are slowly moved by these forces; in recent decades, it has become clear to geologists that *continental drift* is only a manifestation of a greater global process whereby crust is "made" along midoceanic ridges and "recycled" where the pieces (*plates*) collide and are pushed under. The continental landmasses are dragged along as the plates move, their contact zones buckling and breaking (forming mountain belts from Alaska to Argentina and across all Eurasia) and their trailing edges warping and bending. As all this happens, the exposed surface is attacked by *weathering* (rock disintegration by continuous temperature change and by chemical action in the moist atmosphere) and by *erosion* (the removal of weathered rock and its further breakdown by streams, glaciers, wind, and waves). But this is not all. The material loosened by weathering and erosion is carried away and then deposited elsewhere, so that although there is wearing down (*degradation*, to use the technical term) in one area, there is accumulation (*aggradation*) in another. A river system such as the Mississippi and its tributaries, thus, erodes or degrades the landscapes of its interior and aggrades the landscape in its lower valley and delta. To keep track of all the forces—tectonic, erosional, depositional—involves much scientific detective work.

Tectonic geomorphic processes work at different rates, accentuating surface relief at any given time. Each process is associated with its own assemblage of landforms and, therefore, leaves a distinct imprint on the landscape. A good example is volcanism, shown in Fig. 4–17, which can produce conical moun-

Figure 8–3
The Mount Everest massif on the Nepal-China border: the peak itself (at left) is the earth's highest point at 8848 meters/29,028 feet above sea level.

tains built by the violent ejection and sudden cooling of molten rock on the surface; the Mt. St. Helens explosion of 1980, however, demonstrated that such building up can also be interrupted by episodes of destruction (Fig. I–7). Volcanoes, like all landforms, are composed of crustal materials that are sculpted by tectonic and erosional forces. The type of rock associated with volcanism is known as *igneous,* which is formed by the cooling and solidification of molten material (called magma if it hardens underground, lava if it solidifies on the surface). Other igneous landforms besides volcanoes are plateaus formed from horizontal lava flows (such as India's Deccan or the Columbia Plateau of the U.S. Pacific Northwest) and huge granitic magma reservoirs called batholiths, which may even be uplifted later to form mountain ranges (such as California's Sierra Nevada). The second major type of rock, and the most widespread, is known as *sedimentary,* because it is formed by the deposition of loose eroded material that is then slowly compacted (lithified) into solid rock by the heat and pressure generated from the increasing weight of newer sediments deposited above. South Asia's Himalayas—capped by Mount Everest (Fig. 8–3)—are composed of geologically-young sedimentary materials that were laid down in a shallow sea (tiny marine fossils are contained in the rocks collected atop the highest peaks); the enormous tectonic forces responsible for this rapid and extreme uplifting were unleashed as the Indian subcontinent, cast adrift as the Gondwana landmass broke apart (Fig. 7–8), smashed into Asia and caused an accordionlike crumpling of the crust, thrusting up the Himalayas in a process that still continues as the Australian plate grinds against the Eurasian plate (Fig. I–6). Such transformations of the surface help create the final type of rock, appropriately called *metamorphic,* because it is altered from a preexisting igneous or sedimentary rock by the reintroduction of great heat and pressure (e.g., sedimentary shale is metamorphosed into slate). Metamorphism is often associated with horizontal deformation of the crust, which can make sedimentary rock formations plastic, producing warping or *folding* of parallel rock strata. When these forces are too great, however, fracturing or *faulting* occurs, producing formations like East Africa's rift valleys (Fig. 7–6); besides vertical faulting, horizontal fracturing can also take place (California's San Andreas Fault is a famous example), a dislocation that can trigger sudden and devastating seismic (earthquake) activity.

Gradational processes wear down the land surface through weathering and erosion, reshaping landscapes and creating new ones where significant quantities of eroded materials are deposited. The progressive actions of erosional agents produce an orderly sequence of landforms. For example, the recently uplifted Himalayas will eventually be reduced by erosion to a low, rounded mountain chain (much like the U.S. Appalachians), bordered on the south by vast depositional plains and plateaus composed of the eroded Himalayan materials transported by running water in *streams*—the most widespread and effective erosional agent (even in deserts!). The steepness of the slope determines, in part, the work of the river, dragging rock fragments along, carrying others in suspension, and even dissolving some on the way. Eventually, the stream's velocity slows down as it approaches its mouth, and aggradation becomes the dominant process. Valleys are filled (sometimes even choked) by sediment, and deltas extend from the river mouth into the sea. Where rock strata are soluble, for example in limestone areas, the water not only erodes but also dissolves, producing a unique landscape known as *karst* (after a Yugoslavian area where it prevails) pocked by sinkholes and caves. Water in a different form fashions shorelines as *waves* perform gradational functions. Where the coastline coincides with one of those crunching, mountain-creating plate contacts, as on the west coast of the United States, the deep-water waves attack with great power and sculpt an even more spectacular high-relief landscape. Where the coastal landscape is plainlike, as in the southeastern United States, the waves deposit sediment and prove that they, too, can be aggradational agents, forming wide beaches and offshore sandbars. Landscapes and landforms associated with *glaciation* usually have complex histories. Where the glaciers formed huge icesheets during the most recent (Pleistocene) glaciation, they scraped the topography beneath into a nearly flat plain, carrying off the rubble, grinding it down, and depositing it in thick layers far to the south, thus burying much of the U.S. Midwest. Where the deepening cold caused mountain glaciers to form, as in the Rocky Mountains and European Alps, the glaciers descended into valleys that were formerly fashioned by rivers, deepening and widening them and creating an unmistakable, peak- and ridge-studded landscape with waterfalls and elongated lakes. The erosive power of *wind* is at its greatest in the drier climates, where it picks up and hurls loose surface particles against obstacles, wearing them away by abrasion; although thick sand deposits are shaped by the wind into various dune formations in certain deserts, the majority of the world's desert surfaces exhibit rocky (not sandy) landscapes. Wind also creates depositional features, the most important of which is *loess,* a highly fertile windblown silt that can accumulate to depths of over 30 meters (100 feet) as we will see in Chapter 9.

Geomorphology, a vital inquiry in and of itself, must also be linked to the other elements of physical geography, because all together form a dynamically interacting total environmental system. Climate is the ultimate generator of gradational agents; soils are intimately related to both geomorphic processes and climatic influences and to vegetation and hydrography in varying degrees as well. Finally, human influences must also be considered in the late twentieth century as physical features can increasingly be transformed by earthmoving machinery, and as agricultural land-use, urban development, water diversion, surface mining, and the like dominate an ever-expanding artificial landscape.

ral lies the famous Malabar Coast, and, north of it, the Konkan; along the eastern shore is the Coromandel Coast. These physiographic regions lie wedged between the interior plateau and the Indian Ocean; the Malabar Coast is the more clearly defined, since the Western Hills are more prominent and higher in elevation than the Eastern Hills. Thickly forested and steep, the Malabar escarpment dominates the west coast and limits its width to an average of less than 80 kilometers (50 miles); the Coromandel coastal plain is wider and its interior margins are less pronounced. Although not very large in terms of total area, these two regions have had and continue to have great importance in the Indian state. Figure I–9 indicates how well watered the Malabar Coast is (the triggering onshore airflow, or *wet monsoon*, was discussed in the climatology essay on page 141). This rainfall supply and the balmy tropical temperatures have combined with the coast's fertile soils to create one of India's most productive areas. On the lowland plain, rice is grown; on the adjacent slopes stand spices and tea. Of course, this combination of favorable circumstances for intensive agriculture led to the emergence of southern India's major population concentration (Fig. I–14). Along these coasts, the Europeans made their contact with India, beginning with the Greeks and Romans. Later these regions became spheres of British influence, and it was during this period that two of India's greatest cities, Bombay and Madras, began their growth.

# The Human Sequence

The Indian subcontinent is a land of great river basins. Between the mountains of the north and the uplands of the peninsula in the south lie the broad valleys of the Ganges, the Brahmaputra, and the Indus. In one of these—the Indus—lies evidence of the realm's oldest urban civilization, contemporary to, and interacting with, ancient Mesopotamia (see *Box*). Unfortunately, much of the earliest record of this civilization lies buried beneath the present water table in the Indus Valley, but those archaeological sites that have yielded evidence indicate that here was a quite sophisticated urban culture, with large and well-organized cities, in which houses were built of fired brick and consisted of two and sometimes even more levels. There were drainage systems, public baths, and brick-lined wells. As in Mesopotamia and the Nile Valley, considerable advances were made in the technology of irrigation, and the civilization was based on the productivity of the Indus lowlands' irrigated soils. In the pottery and other artistic expressions of this literate society lies evidence of contact—and, thus, trade—with the lowland civilizations of Southwest Asia (Fig. 8–4).

But the Indus Valley did not escape the invasion of the Aryans any more than did Europe or the Mediterranean Basin. After about 2000 B.C., people began to move into the Indus region from western Asia, through what is today Iran and through passes in the mountains to the north. Culturally, these people were not as advanced as the Indus Valley inhabitants, and they brought destruction to the cities of the Indus. But they also adopted many of the innovations of the Indus civilization and pushed their frontier of settlement out beyond the Indus Valley into the Ganges area and southward into the peninsula. They absorbed the tribes they found there, through conquest and enslavement, and the language they had brought to India, Sanskrit, began to differentiate into the linguistic complex that is India's today. Their village culture spread over a much larger part of the subcontinent than the urban civilization of the Indus had done.

In the centuries during and following the Aryan invasion, Indian culture went through a period of growth and development. From a formless collection of isolated tribes and their villages, regional organization began to emerge. Towns developed; local rulers became something more as surrounding areas fell under their control. Arts and crafts blossomed once again, and trade with Southwest Asia was renewed. Hinduism emerged from the religious beliefs and practices brought to India by the Aryans; tribal priests took on the roles of religious phi-

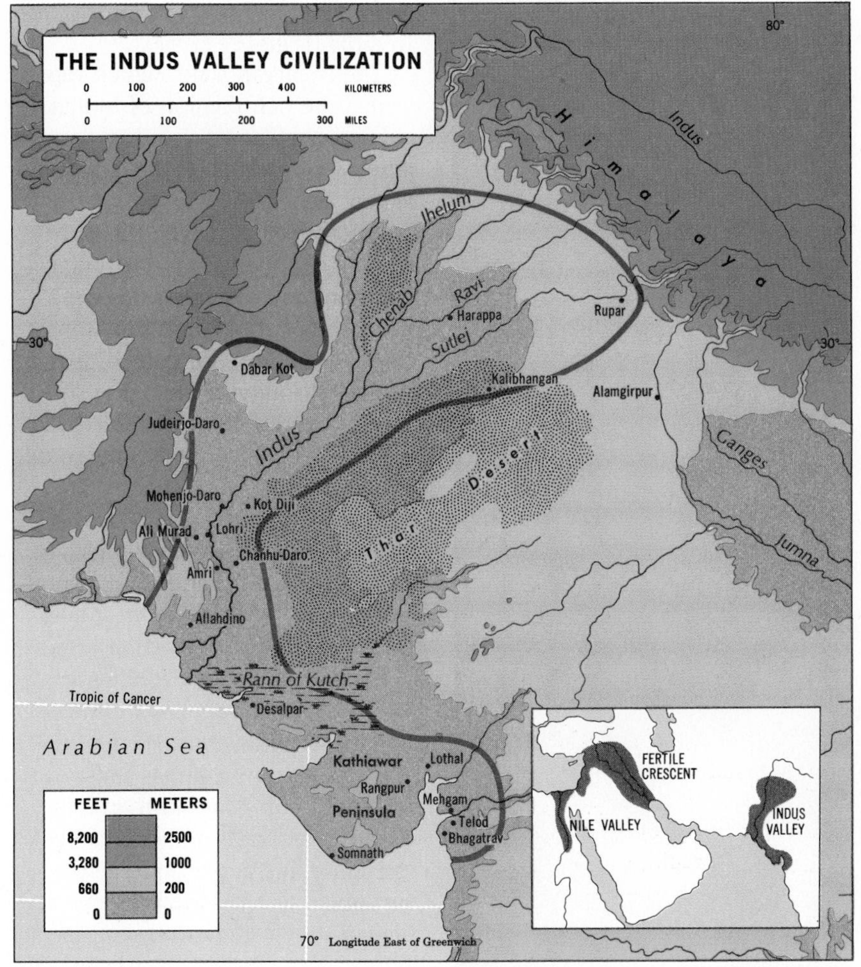


Figure 8–4

losophers, shaping a whole new way of life. Social stratification evolved, with the ruling Brahmans, administered by powerful priests, at the head of a complex bureaucracy—a *caste system*—in which soldiers, artists, merchants, and all others had their place. Aggressive and expansion-minded kingdoms arose, always competing with each other for greater power and control. It was in one of these kingdoms, in northeastern India, that Prince Siddhartha, or Buddha, was born in the sixth century B.C. Buddha voluntarily gave up his princely position to seek salvation and enlightenment through religious meditation. His teachings demanded a rejection of earthly desires and a reverence for all life. But although Buddha had a substantial following during his lifetime, the real impact of Buddhism was to come during the third century B.C. after an interval marked by the end of Persian domination in the northwest and a brief but intense intervention by Alexander the Great (326 B.C.). Alexander's Greeks pushed all the way to the Hindu Indian heartland of the Ganges Valley, where the river today remains a sacred symbol that attracts the masses to bathe in it at the holy city of Varanasi (Fig. 8–5).

Thus, Northern India was the theater of culture infusion and innovation. The south lay removed, protected by distance from much of this change. Here, a very different culture came into being. The darker skins of the southerners today still reflect their direct ties with ancient forebears such as the Negritos and black Australians who lived here, even before the Indus civilization arose far to the northwest. Their languages, too, are distinctive and not related to those of the Indo-Aryan region. Both the peoples and the languages of southern India are known collectively as *Dravidian*. The four major Dravidian languages—Telugu, Tamil, Kanarese (Kannada), and Malayalam—all have long literary histories, and Telugu and Tamil are the languages of nearly one-fifth of India's 763 million inhabitants (Fig. 8–6).

## "INDIA"

Our use of the name *India* for the heart of the South Asian realm derives from the Sanskrit word *sindhu*, used to identify the ancient civilization in the Indus Valley. This word became *sinthos* in Greek descriptions of the area, and then *sindus* in Latin. Corrupted to *indus*, which means "river," it was first applied to the region that now forms the heart of Pakistan; subsequently it was again modified to *india* to refer generally to the land of river basins and clustered peoples from the Indus in the west to the Brahmaputra in the east.

## Mauryan Era and Islam

The Maurya Empire first incorporated most of the subcontinent, emerging with the decline of Hellenistic influence shortly before 300 B.C. Its heartland lay in the middle Ganges Valley; quite rapidly it extended its power over India as far

**Figure 8–5**
Bathing in the holy Ganges River is a Hindu tradition that draws huge crowds to Varanasi (Banaras) each year.

west as the Punjab and the Indus Valley, as far east as Bengal, and as far south as modern Bangalore. The state was led by a series of capable rulers, among whom the greatest was undoubtedly Aśoka, who ruled for nearly 40 years during the middle period of the third century B.C. and who was a convert to Buddhism. In accordance with Buddha's teachings, Aśoka diverted the state's activities from conquest and expansion to the attainment of internal stability and peace; had he not done so, it is likely that all of the subcontinent would have fallen to the Maurya's rule. A vigorous proponent of Buddhism, Aśoka sent missionaries to the outside world to carry Buddha's teachings to distant peoples; in so doing, he contributed to the further spread of Indian culture. Thus, Buddhism became permanently established as the dominant religion in Ceylon (now Sri Lanka), and it achieved temporary footholds even in the eastern Mediterranean lands—although it eventually declined to minor status in India itself.

The Mauryan state, which represented the culmination of Indian cultural achievements up to that time, was not soon to be repeated. When the state collapsed late in the second century A.D., the old forces—regional-cultural disunity, recurrent and disruptive invasions, and failing central authority—again came to the fore. Of course, the cultural disunity of India was a reality even during the era of the Mauryans, who could not submerge it. Although India did not see anything like the Aryan invasions again, there were almost constant infusions of larger and smaller population groups from the west and northwest. Persians, Afghans, and Turks entered the subcontinent, mostly along the obvious route: across the Indus, through the Punjab, and into the Ganges Valley. Thus, beginning late in the tenth century, the wave of Islam came spreading like a giant tide across the subcontinent from Persia and Afghanistan to the northwest. In the Indus region, there was a major influx of Moslems and a nearly total conversion to Islam by the local population; in the Punjab, perhaps two-thirds of the population was converted. Islam crossed the bottleneck in which Delhi is situated and diffused southeastward into the Indian heartland of the Ganges (Ganga) Valley—Hindustan. Whereas the Moslem impact in Hindustan was far less significant, about one-eighth of the local population did become adherents to the new faith. In the Ganges Delta region, where up to three-quarters of the people became Moslems, Islam seems to have spread at the expense of Buddhism especially. Southward into the peninsula, however, the force of Islam was spent quite quickly, and the far south was never under Moslem control.

Islam was an alien faith to India, as it was to southern Europe, and it

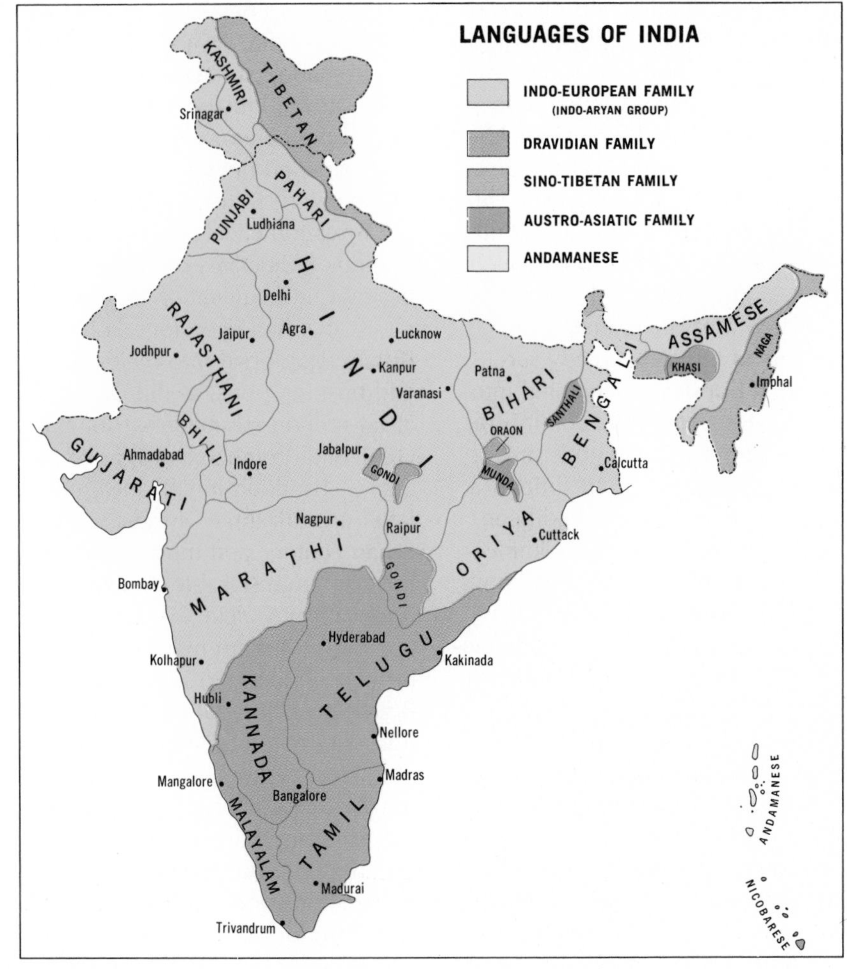


Figure 8–6

brought great changes to existing ways of life. It was superimposed by political control; by early in the fourteenth century, a sultanate centered at Delhi controlled all but the extreme southern, eastern, and northern margins of the subcontinent. There were constant struggles for control, as challengers to existing authority came from the Afghan empire to the west and from Turkestan and inner Asia to the northwest. Out of one of these challenges arose the largest unified Indian state since Aśoka's time, the Islam-dominated Mogul Empire, which in about 1690 under the rule of Aurangzeb comprised almost the entire subcontinent from Baluchistan to the Ganges Delta and from the Himalayan foothills to Madras. But Islam in India was neither the monopoly of the invaders from outside nor the exclusive religion of the rulers and new aristocracy. Islam provided a welcome alternative to Hindus who had the misfortune of being of low caste; it was also an alternative for Buddhists and others who faced absorption into the prevailing Hindu system. In the majority of cases, the Moslems of the Indian subcontinent are racially indistinguishable from their non-Moslem neighbors; there is no correlation here between race and religion. Most of the subcontinent's Moslems today are descendants of converts rather than descendants of any invading Moslem ruling elite.

## European Intrusion

Into this turbulent complexion of religious, political, and linguistic disunity still another element intruded—the European powers in search of raw materials, markets, and political influence. The Europeans profited from the Hindu-Moslem contest; they were able to exploit local rivalries, jealousies, and animosities. British merchants gained control over the trade with Europe in spices, cotton, and silk goods, and, in time, they ousted their competitors, the French, Dutch, and Portuguese. The British East India Company's ships also took over the intra-Asian sea trade between India and Southeast Asia, long in the hands of Arab, Indonesian, Chinese, and Indian merchants. In effect, the East India Company became India's colonial administration.

As time went on, the East India Company faced problems it could not solve. Its commercial activities remained profitable, but it became entangled in an ever-growing effort to maintain political control over an expanding Indian domain. It proved an ineffective governing agent at a time when the increasing Westernization of India brought to the fore new and intense frictions. Christian missionaries were challenging Hindu beliefs, and many Hindus thought that the British were out to destroy the caste system. Changes came also in public education, and the role and status of women began to improve. Aristocracies saw their positions threatened as landowners had their estates expropriated. Finally, in 1857, the three-month Sepoy Rebellion occurred and changed the entire situation. It took a major military effort to put down, and, from that time, the East India Company ceased to function as the government of India. Administration was turned over to the British government, the Company was abolished, and India formally became a British colony—a status it held for the next 90 years—until 1947.

## Colonial Transformation

Four centuries of European intervention in India greatly changed the subcontinent's cultural, economic, and political directions. Certainly, the British made positive contributions to Indian life, but colonialism also brought serious negative consequences. In this respect, there are important differences between the Indian case and that of Black Africa, for when the Europeans came to India, they found a considerable amount of industry, especially in metal goods and textiles, and an active trade with both Southwest and Southeast Asia in which Indian merchants played a leading role. The British intercepted this trade, changing the whole pattern of Indian commerce. India now ceased to be South Asia's manufacturing area: soon, the country was exporting raw materials and importing manufactured goods—from Europe of course. India's handicraft industries declined; after the first stimulus, the export trade in agricultural raw materials also suffered as other parts of the world were colonized and linked in trade to Europe. Thus, the majority of India's people, who were farmers then as now, suffered an economic setback as a result of the manipulations of colonialism. Therefore, although in total *volume* of trade the colonial period brought considerable increases, the *composition* of the trade India now supported was by no means a way to a better life for its people.

Neither did the British manage to accomplish what the Mauryans and the Moguls had tried to do: unify the subcontinent and minimize its internal cultural and political divisions. When the crown took over from the East India Company in 1857, nearly 2 million square kilometers (about 750,000 square miles) of Indian territory were still outside the British sphere of influence; slowly, the British extended their control over this huge unconsolidated area, including several pockets of territory already surrounded but never integrated into the previous corporate administration. Also, the British government found itself obligated to support a long list of treaties that had been made by the Company's administrators with numerous Indian kings, princes, regional governors, and feudal rulers. These treaties guaranteed various degrees of autonomy for literally hundreds of political entities in India, ranging in size from a few hectares to Hyderabad's more than 200,000 square kilometers (80,000 square miles). The British crown saw no alternative but to honor these guarantees; so, India was carved up into an administrative framework under which, during the late nineteenth century, there were over 600 "sovereign" territories in the subcontinent. These "Native States" had British advisors; the large British provinces such as Punjab, Bengal, and Assam had British governors or commissioners who reported to the Viceroy of India who, in turn, reported to Parliament and the monarch in London. In all, this was a near-chaotic amalgam of modern colonial control and traditional feudalism, reflecting and in some ways deepening the regional and local disunities of the Indian subcontinent. Although certain parts of India quickly adopted and promoted the positive contributions of the colonial era, other areas rejected and repelled them, thereby adding yet another element of division to an increasingly complicated human spatial mosaic.

Indeed, colonialism did produce assets for India. As a glance at a world map of surface communications shows, India was bequeathed one of the better railroad and highway transport networks of the colonial domain. British engineers laid out irrigation canals through which millions of hectares of land were brought into cultivation. Settlements that had been founded by Britain developed into major cities and bustling ports, as did Calcutta (10.4 million inhabitants), Bombay (10.1 million inhabitants), and Madras (7.1 million inhabitants); the latter are still three of India's largest urban centers, and their cityscapes bear the unmistakable imprint of colonialism (Calcutta is shown in Fig. 8–7). Modern industrialization, too, was brought to India by the British on a limited scale. In education, an effort was made to combine English and Indian traditions; the Westernization of India's elite was supported through the education of numerous Indians in Britain. Modern practices of medicine were also introduced. Moreover, the British administration tried to eliminate features of Indian culture that were deemed undesirable by any standards—such as the burning alive of widows on the funeral pyres of their husbands, female infanticide, child marriage, and the caste system. Obviously the task was far too great to be achieved successfully in less than four generations of rule, but India itself has continued the efforts initiated by the British where necessary.

## Partition

Even before the British government decided to yield to Indian demands for independence, it was clear that British India would not survive the coming of self-rule in one piece. As early as the 1930s, the idea of a separate Pakistan was being promoted by Moslem acti-

**Figure 8–7**
The unmistakable imprint of British colonialism survives in the urban landscape of central Calcutta.

vists, who circulated pamphlets arguing that India's Moslems were a distinct nation from the Hindus and that a separate state consisting of the Punjab, Kashmir, Sind, Baluchistan, and a section of Afghanistan should be created. The first formal demand for such a partitioning was made in 1940, and the idea had the almost universal support of the region's Moslems as subsequent elections proved. As the colony moved toward independence, a crisis developed: India's majority Congress Party would not consider partition, and the minority Moslems refused to participate in any unitary government. But partition was not simply a matter of cutting off the Moslem areas from the main body of the country; true, Moslems were in the majority of what is now Pakistan and Bangladesh, but other Moslem clusters were scattered throughout the subcontinent (Fig. 8–8). The boundaries between India and Pakistan would have to be drawn right through transitional areas, and people by the millions would be dislocated.

In the Punjab, for example, a large number of Sikhs—whose leaders were fiercely anti-Moslem—faced incorporation into Pakistan. Even before independence day, August 15, 1947, Sikh leaders had been talking of revolt, and there were some riots; but no one could have anticipated the horrible killings and mass migrations that followed independence and official partition. Just how many people felt compelled to participate in these migrations will never be known, but the most common estimate is 15 million, representing a mass of human suffering that is indeed incomprehensible. Even that huge number of refugees hardly began to purify either India of Moslems or (former East and West) Pakistan of Hindus. After the initial mass exchanges, there were still tens of millions of Moslems in India, and East Pakistan (now Bangladesh) remained about 20 percent Hindu. The remarkable photograph shown in Fig. 8–9, which fully captures the atmosphere of these exchanges, was taken in mid-1947 and shows thousands of Moslem refugees jamming the railroad cars of a train leaving the New Delhi area for the Islamic republic of Pakistan. Facing the difficult alternatives, many people decided to stay where they

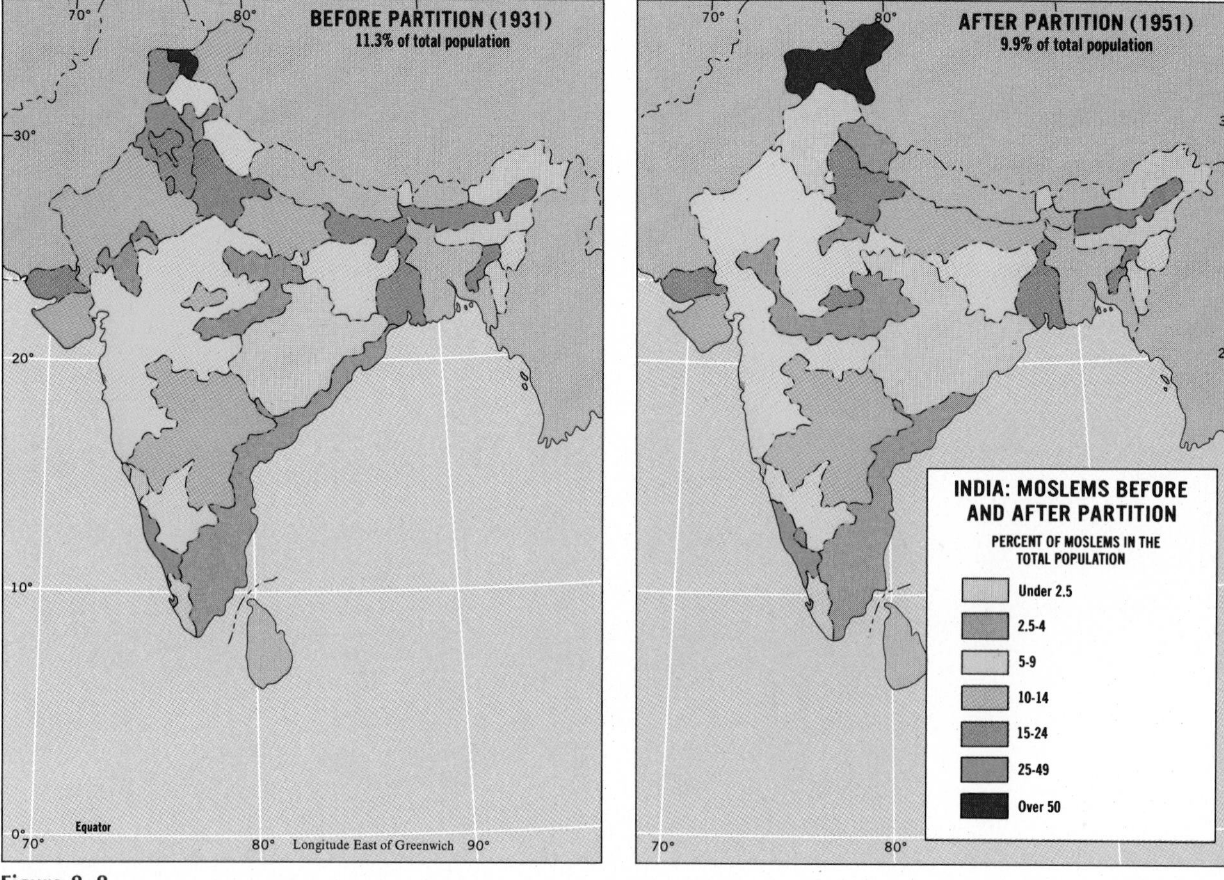


Figure 8–8

were and make the best of the situation. For most, it turned out to be a wise choice.

The actual process of partition was done quite quickly and, of necessity, rather arbitrarily, supervised by a joint commission whose chairman was a neutral British representative. Using data from the 1941 census of India, Pakistan's boundaries were defined in such a manner that the Moslem state would incorporate all contiguous civil divisions and territories in which Moslems formed a majority. The commission had to make decisions it knew in advance would be highly unpopular, although no one foresaw the proportions of the Sikh-initiated violence. What took place, then, was the *superimposition* of a political boundary on a cultural landscape in which such a boundary had not previously existed or functioned; this is one of the four genetic boundary-types that will be discussed in the essay on political geography that opens Chapter 10.

Figure 8–9
Flight was a widespread response to the 1947 partition: thousands of Moslem refugees jammed aboard an Indian train departing for newly created Pakistan.

# Federal India

India today is the world's most populous federal state. Including recently incorporated Sikkim, the Indian federation consists of 22 states and 8 union territories (Fig. 8–10). A comparison between Fig. 8–10 and the map of languages in India (Fig. 8–6) indicates a considerable degree of coincidence between linguistic and state boundaries. The present framework is not the one with which India was born as a sovereign state in 1947, and it is likely to be modified again in the future. Soon after the British withdrew, the several hundred "princely states," whose rights had been protected during the colonial period, were absorbed into the states of the federation, although their rulers' privileged "princely orders" were not terminated until

**Figure 8–10**

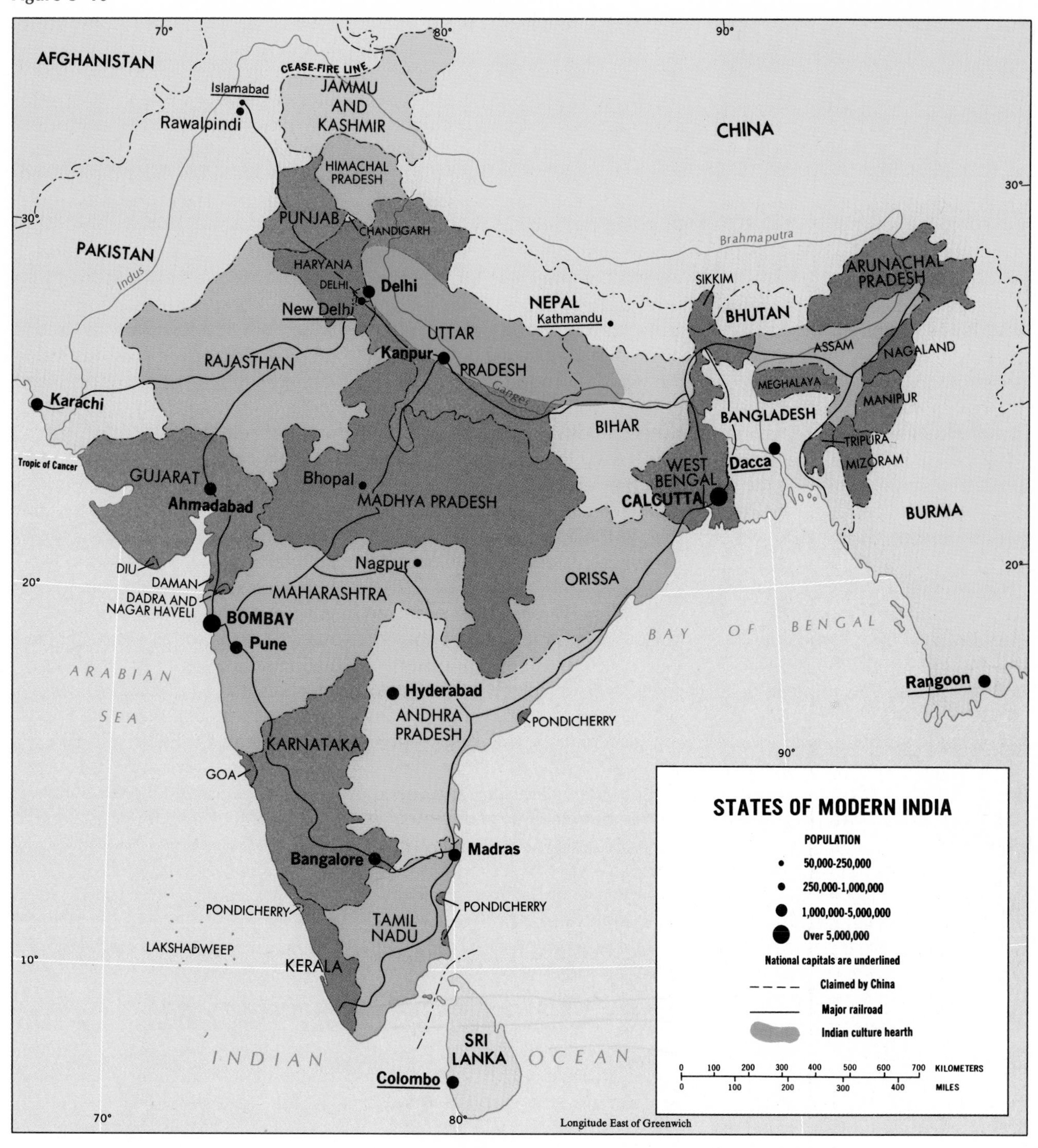

1972. Next, the Indian government reorganized the country on the basis of its regional languages, and Hindi, spoken by more than one-third of all Indians, was designated the official language. Hindi was one of 14 languages given national status by the Indian constitution—10 of them spoken in the northern and central part of the country, and 4 in the Dravidian south. English, it was anticipated, would remain a *lingua franca* when Hindi could not serve as a medium of communication at government and administrative levels; this it has proven to be, and English has also become the chief language of the business world of emerging urban India, increasingly regarded as the key to good jobs, financial success, and personal advancement.

The politico-geographical framework based on the major regional languages, however, proved to be unsatisfactory to many communities in India. In the first place, many more languages are in use than the 14 that were officially recognized—even after Sindhi (spoken in the Rann of Kutch and Kathiawar Peninsula adjacent to Pakistan) was added to the list. As early as 1953, the government yielded to demands for the creation of a Telugu-speaking state from Tamil-dominated Madras; thus, the state of Andhra Pradesh was formed. In 1960, the state of Bombay was fragmented into two linguistic states, Gujarat and Maharashtra (Fig. 8–10). In the distant eastern borderlands, the Naga peoples, numbering less than half a million, put up a struggle against federal authority and local Assamese administration; after Indian armed forces were sent into the area, Nagaland was established as a separate Indian state. In the west, the Sikhs demanded the breakup of the original state of Punjab into a Sikh-dominated west (now Punjab) and a Hindu east (now Haryana). That action temporarily defused Sikh demands for more control over their homeland; by 1984, the persistent refusal of New Delhi to grant further concessions prompted violent protests and attacks by the Indian army on Sikh extremists. Pressure for greater regional autonomy continues in several other parts of India (especially Assam and Tamil Nadu), and it is remarkable that the federal state has been able to survive such **centrifugal forces** (see *Box*).

India's unity is all the more impressive in view of its enormous population and the problems inherent in its demographic condition. The country's most populous states (Uttar Pradesh, approaching 120 million; Bihar, nearly 75 million; Maharashtra, over 65 million) have populations larger than most of the world's countries. No country contains greater cultural diversity than India, and variety in India comes on a scale unmatched anywhere in the world. Even today, after four decades of religious partition between Hindus and Moslems in South Asia, India remains 11 percent Moslem. That might seem to be a rather small minority, except that, in India, an 11 percent minority represents over 84 million people.

A still-pervasive aspect of Indian society is its stratification into *castes*. Under Hinduism, castes are fixed layers in society whose ranks are based on ancestries, family ties, and occupations. The caste system may have evolved from an early social differentiation into priests and warriors, merchants and farmers; it may also have a racial foundation (the Sanskrit term for caste is color). Over the centuries its complexity grew until India came to possess several thousand castes, some with a few hundred members, others containing millions. In village as well as city, communities were, thus, segregated according to caste, ranging from the highest (priests, princes) to the lowest (the untouchables). A person was born into a caste based on his or her actions in a previous existence; hence, it would not be appropriate to counter such ordained caste assignment by permitting movement (or even contact) from a lower caste to a higher one. Persons of a particular caste could only perform certain jobs, only wear certain clothes, only worship in prescribed ways at particular places. They or their children could not eat with, play with, or even walk with people of a higher social status. The untouchables occupying the lowest tier were the most debased, wretched members of this rigidly structured social system. Although the British ended the worst excesses associated with the caste system and modern Indians, including Gandhi and Nehru, have worked to modify it, centuries of class consciousness are not wiped out in a few decades. In traditional India, caste provided stability and continuity. In modernizing India, it constitutes an often painful and difficult legacy.

What **centripetal forces** (see *Box*) have helped keep India unified? Without a question, the dominant binding force in India is the cultural strength of Hinduism, its holy writings (read in all languages), its holy rivers, and its many other influences over Indian life. Hinduism is a way of life as much as it is a faith, and its diffusion over virtually the entire country (Moslem, Christian, and Sikh minorities notwithstanding) brings with it a national coherence that constitutes a powerful antidote to regional divisiveness. Moreover, communications in much of densely populated India are better than they are in many other Third World countries, and the constant circulation of people, ideas, and goods helps bind the nation together. Before independence, opposition to British rule was a shared philosophy throughout the country, but India remained divided and separated internally. Independence

## CENTRIFUGAL AND CENTRIPETAL FORCES

Political geographers use the terms *centrifugal* and *centripetal* to identify forces within a state that tend, respectively, to pull that political system apart and to bind it together.

*Centrifugal* forces are disunifying or divisive. They can cause deteriorating internal relationships. Religious conflict, racial strife, linguistic polarization, and contrasting regional outlooks are among major centrifugal forces. During the 1960s and early 1970s, the issue of the Vietnam War became a major centrifugal force in the United States, one whose aftermath will be felt for years to come. Newly independent countries find tribalism a leading centrifugal force, sometimes strong enough to threaten the very survival of the whole state system, as the Biafra conflict of the late 1960s did in Nigeria.

*Centripetal* forces tend to bind the state together, to unify and strengthen it. A real or perceived external threat can be a powerful centripetal force, but more important and lasting is a sense of commitment to the governmental system, a recognition that it constitutes the best option. This commitment is sometimes focused on the strong charismatic qualities of one individual—a leader who personifies the state, who captures the population's imagination. (The origin of the word *charisma* lies in the Greek expression meaning "divine gift.") At times, such charismatic qualities can submerge nearly everything else; in India, Gandhi and Nehru had personal qualities that went far beyond political leadership, and their binding effect on the young nation extended far beyond their lifetimes.

The degree of strength and cohesion of the state depends on a surplus of centripetal forces over divisive centrifugal forces. It is difficult to measure such intangible qualities, but some attempts have been made in this direction—for example, by determining attitudes among minorities and by evaluating the strength of regionalism as expressed in political campaigns and voter preferences. When the centrifugal forces gain the upper hand and cannot be checked (even by external imposition) the state can break up—as once-united East and West Pakistan did in the early 1970s.

brought with it a first taste of national planning, national political activity, and national debates over priorities. India also produced some strong leaders whose compelling personalities constituted a binding force. Gandhi and Nehru did much to forge the India of today, and Nehru's daughter, Indira Gandhi, twice took decisive control (in 1966 and 1980) following episodes of weak national leadership. Finally, federal India has proved capable of accommodating far-reaching change, thereby avoiding the alternative of secession. Although political leaders in some states have on occasion openly endorsed the prospect of secession, India's flexibility on the language issue, its ability to tolerate individuality in its states, and its capacity to modify and remodify the federal map all offer evidence that change *is* possible and that political, economic, educational, and other objectives can, in fact, be achieved within India's complex federal framework.

# Indian Development

If India has faced problems in its great effort to achieve political stability and national cohesion, these are more than matched by the difficulties that lie in the way of economic growth and development. The large-scale factories and power-driven machinery of the colonial powers wiped out a good part of India's indigenous industrial base. Indian trade routes were taken over. European innovations in health and medicine sent the population growth rate soaring, without introducing solutions for the many problems this spawned. Surface communications were improved and food distribution systems became more efficient, but local and regional famines occurred nonetheless. Since 1970, India has been frequently struck by droughts and crop failures, as were Africa's Sahel, Northeast Brazil, and other parts of the underdeveloped world, and food shortages were widespread.

## Agriculture

India's underdevelopment is nowhere more apparent than in its agriculture. Traditional farming methods continue to prevail, and yields per hectare and per worker remain low for virtually every crop grown in the region. The scene in Fig. 8–11, set in a rural area near Madras, is depressingly typical. Water is brought to the surface not by a pump but by oxen pulling a bucket from a well; the animals walk endlessly back and forth along the

**Figure 8–11**
Agriculture in India suffers from numerous problems as inefficient traditional methods continue to prevail.

short path in the center of the photograph, but they are capable of hauling only a comparative trickle of water to the surface (as shown in the ditch at the far left).

As the total population grows, the amount of cultivated land per person declines; in 1980, the latest year for which data were available, the *physiologic density* was estimated at 405 per square kilometer (1049 per square mile). This is nowhere near as high as it is in neighboring Bangladesh—where it is almost 2½ times greater (963 and 2494, respectively)—or in Egypt, but India's farming is so inefficient that this is a deceptive comparison. More than two-thirds of India's huge working population depends directly on the land for its livelihood, but the great majority of Indian farmers are poor, unable to improve their soils, equipment, or yields. Those areas in which India has made substantial progress toward the modernization of its agriculture (as in the Punjab's wheat zone) still remain islands in a sea of agrarian stagnation.

This stagnation has persisted in large measure because India failed, after independence, to implement a much-needed nationwide land reform program. In the early 1980s, more than one-quarter of India's entire cultivated area was still owned by only 4 percent of the country's farming families; about half of all rural families either owned as little as a half hectare (1¼ acres) or less—or no land at all. Independent India inherited inequities from the British colonial period, but the individual states of the union would have had to cooperate in any national land reform program. The large landowners had much political influence, and so the program never got off the ground, although several states in recent years have placed upper limits on the holdings families may possess, and some redistribution of land has taken place. In addition, much of India's farmland is badly fragmented as a result of local rules of inheritance, thereby inhibiting cooperative farming, mechanization, shared irrigation, and other opportunities for progress. Land consolidation efforts have had only limited success, except in the states of Punjab, Haryana, and Uttar Pradesh, where modernization has gone farthest. Certainly, official agricultural development policy, at the federal as well as state level, has also contributed to India's agricultural malaise. Unclear priorities, poor coordination, inadequate information dissemination, and other failures have been reflected in the country's disappointing output.

It is useful to compare Fig. 8–12, showing the distribution of crop regions and water-supply systems in India, with Fig. I–9 (p. 17) which maps mean annual precipitation in India and the world. In the comparatively dry northwest, notably in the Punjab and neighboring areas of the Upper Ganges, wheat is the leading cereal crop; in this zone India has made major gains in annual production through the introduction of high-yielding varieties developed experimentally in Mexico. This innovation was part of the so-called Green Revolution of the 1960s, when strains of wheat and rice (the latter in the Philippines) were developed that were so much

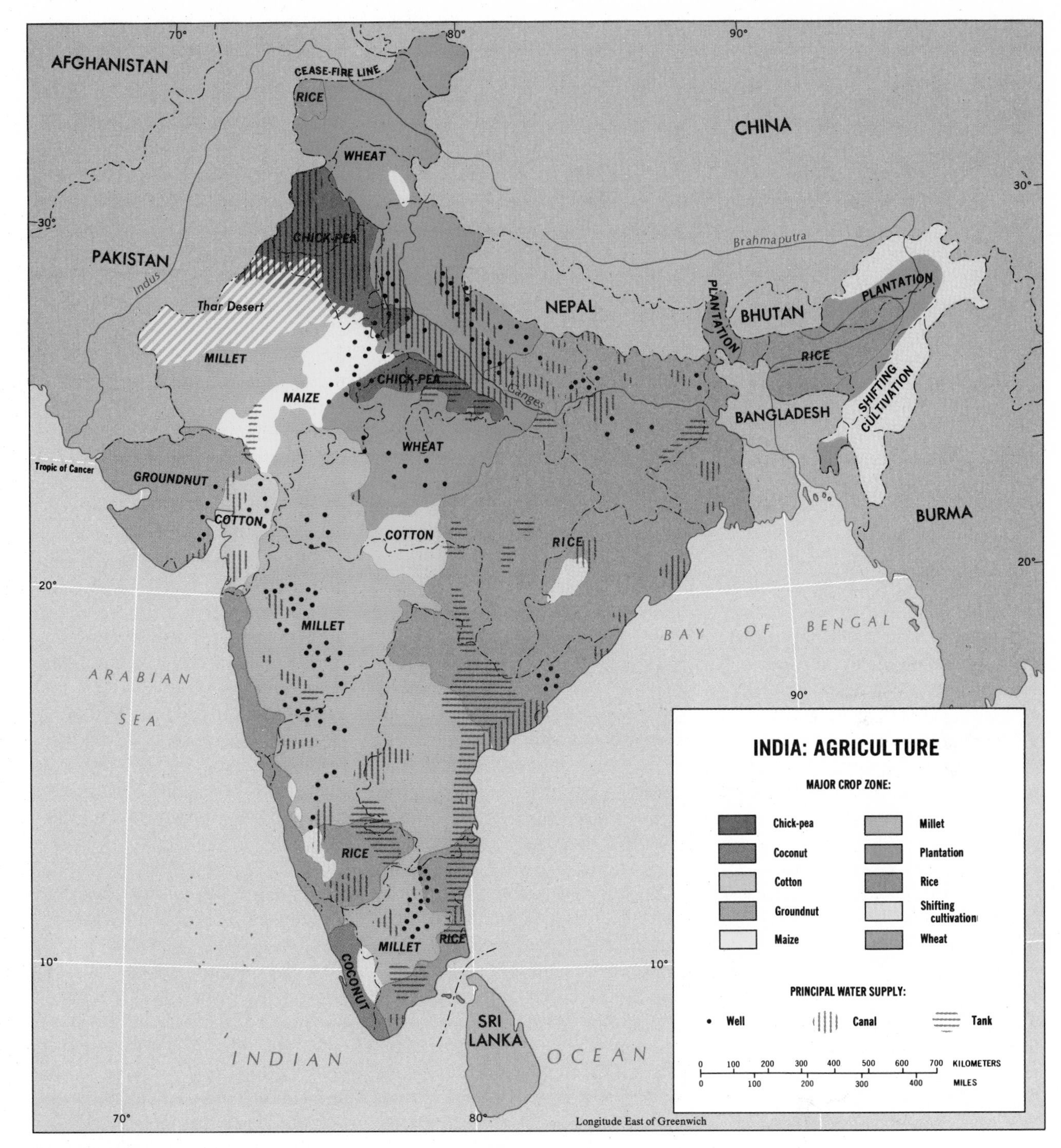


**Figure 8–12**

more productive than other varieties that they were dubbed "miracle" crops. Introducing these new seeds also led to the expansion of cultivated areas, the development of new irrigation systems, and the more intensive use of fertilizer (a mixed blessing, for fertilizers are expensive and the "miracle" crops are more heavily dependent on fertilizers).

Toward the moister east and in the summer-monsoon-drenched south, rice takes over as the dominant staple. Over one-fourth of all of India's farmland lies under rice cultivation, most of it in the states of Assam, West Bengal, Bihar, Orissa, and eastern Uttar Pradesh; this area has over 100 centimeters (40 inches) of rainfall, and irrigation supplements precipitation where necessary. India does have the largest acreage of rice among all the world's countries, but despite the introduction of "miracle rice" in parts of India, the average yield per hectare still is below 1000 kilograms (900 pounds per acre). The world population map (Fig. I–14) reveals a high degree of coincidence between India's rice areas and its most densely peopled zones, but rice yields per unit area in India are among the lowest on earth.

As the map of agricultural zones indicates, India's varied environments support the cultivation of millet, barley, corn (maize), and other cereals. But subsistence remains a way of life for countless millions of Indian villagers, who cannot afford to buy fertilizers, cannot cultivate the new strains of wheat or rice, and cannot escape the cycle of abject poverty. At least 100 million Indians do not even have a plot of land and must live as tenants, always uncertain of their fate. This is the background against which optimistic predictions of better food conditions in the India of the future must be regarded. True, wheat production increased over 50 percent in five years during the Green Revolution, and rice production has approximately doubled since 1950. But the gap between population and food supply has not closed. When the worldwide droughts of the mid-1970s struck India, there was famine and widespread malnutrition. In 1974, people in the state of Gujarat rioted when food supplies ran out, and the government was overthrown; the federal administration was forced to take control. People by the hundreds of thousands left the land and headed for the cities and towns, and the all-too-familiar specter of hunger faced India once again.

## Industrialization

Notwithstanding the problems faced by the farmers, agriculture must be the basis for development in India. Agriculture employs the vast majority of the workers (about 70 percent), generates most of the government's tax revenues, contributes its chief exports by value (cotton textiles, jute manufactures, tea, and tobacco all rank high), and produces most of the money the country can spend in other sectors of the economy. Add to this the compelling need to grow ever-more food crops, and India's heavy investment in agriculture is understandable.

In 1947, India inherited the mere rudiments of an industrial framework. After a century of British control over the economy, only 2 percent of India's workers were involved in industry, and manufacturing and mining combined produced perhaps 6 percent of the national income. Textile and food-processing industries dominated; although India's first iron-making plant was opened in 1911 and the first steel mill began production in 1921, the initial major stimulus for heavy industrialization came after the outbreak of World War II. Manufacturing was concentrated in the major cities: Calcutta led, Bombay was next, and Madras ranked third.

The geography of manufacturing today still reflects these beginnings, and industrialization and urbanization in India have proceeded slowly, even after independence (Fig. 8–13). Calcutta now forms the center of India's eastern industrial region, the Bihar-Bengal area, where jute manufactures dominate, but engineering and chemical and cotton industries also exist. In the nearby Chota Nagpur district, coal mining and iron and steel manufacturing have developed, and Bhilai is a growing nucleus of heavy industry (Fig. 8–14). On the opposite side of the subcontinent, two industrial areas combine to form the western manufacturing region: one centered on Bombay, the other focused on Ahmadabad. This Maharashtra-Gujarat area specializes in cotton and chemicals, with some engineering and food processing. Cotton textiles have long been an industrial mainstay in India, and this was one of the few industries to derive some benefit from the nineteenth-century economic order brought by the British. With the local cotton harvest, the availability of cheap yarn, abundant and inexpensive labor, and the power supply from the Western Hills' hydroelectric stations, the industry thrived and today outranks Britain itself in the volume of its exports. Finally, the southern industrial region chiefly consists of a set of linear, city-linking corridors centered on Madras, specializing in textile production and light engineering activities.

When India achieved independence, its government immediately set out to develop its own industries, to lessen dependency on imported manufactures, and to cease being an exporter of raw materials to the developed world. In the process, emboldened by the early results of the Green Revolution, Indian planners actually overspent on their industrial programs—but the problem of rapid population growth bedeviled industry as it did agriculture. Unemployment was to be reduced as industrialization progressed; instead, unemployment rose. Per capita incomes would rise as high-value products rolled off new assembly lines; but incomes rose very little—today, they are still below $200 per year. India has limited coking-coal deposits in the Chota Nagpur and large lower-grade coalfields, but the country does rank among the world's 10 largest coal producers. In the absence of known major petroleum reserves (some oil comes from Assam, Gujarat, and Punjab and offshore from Bombay), it must spend heavily on energy every year. Major investments have been made in hydroelectric plants, especially multipurpose dams that provide

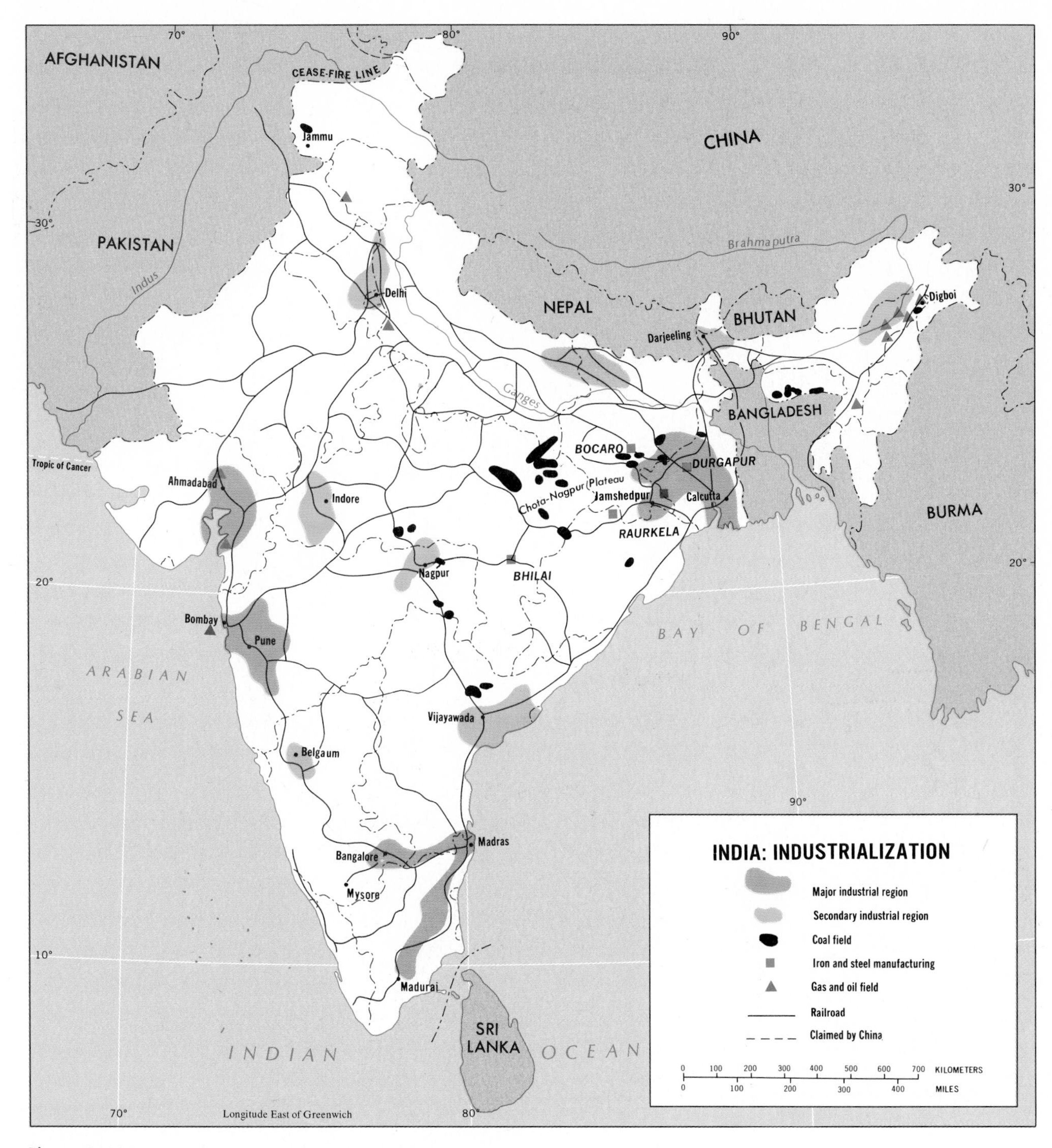


Figure 8–13

electricity, make irrigation possible, and facilitate flood control. India's iron ores in Bihar State (northwest of Calcutta) and Karnataka State (in the heart of the Deccan) may be the largest in the world. Jamshedpur, a leading steelmaking and metal-fabrication center located at the southwestern edge of the Calcutta industrial region, has emerged as India's Pittsburgh; yet, India still exports iron ore as a raw material to developed countries. In the Third World, entrenched patterns are difficult to break.

## Urbanization

In 1985, only 24 percent of India's population resided in urban areas. India may be known in part for its teeming, crowded cities and its homeless, sidewalk-sleeping ur-

**Figure 8–14**
Steel mill at Bhilai, in Madhya Pradesh State, exhibiting the air-quality problems that accompany Third World industrialization.

colonies of newcomers, increasingly defined by language and custom. Thus, even by Third World standards, Indian cities are places of great social contrasts (Fig. 8–16). However, as crowding intensifies, social stresses multiply: a spate of Hindu-Moslem riots in the Bombay area in 1984, in which hundreds died, was largely attributable to the actions of rootless urban youths who were unable to find employment.

India's modern urbanization also has its roots in the colonial period, when the British selected Calcutta, Bombay, and Madras as regional trading centers and as coastal focal points for their colony's export and import traffic. All were British military outposts by the late seventeenth century. Madras, where a

banites (Fig. 8–15), but urbanization has proceeded very slowly. The largest city in a country of 760+million, Calcutta, in 1985, had 10.4 million inhabitants; Bombay, the second city, had 10.1 million, and Delhi's conurbation fewer than 7 million. Once again, we must take note of India's massive proportions, and realize that 24 percent of its population involves 183 million people. The already enormous dimensions of India's urban crisis portended by this figure are now expanding, because there are definite signs that Indian urbanization is accelerating in the mid 1980s. The latest findings show an unprecedented upsurge in cityward migration, and urban India is today growing more than twice as rapidly as the country's overall population. Among the reasons for this shift that are cited by Indian planners are a dramatic loosening of ties between poor peasants and their villages, and the widespread establishment in the cities of villagemen or "caste brothers" who are finally able to assist their relatives and friends to make similar moves to the burgeoning urban residential

**Figure 8–15**
Bombay's street dwellers: Indian cities are home to thousands who have no housing at all.

**Figure 8–16**
Strong social contrasts mark India's urban scene, here beside the Cooum River in Madras.

fort was built in 1640, lay in an area where the British faced few challenges; Bombay (1664) had the advantage of being located closest of all Indian ports to Britain, and Calcutta (1690) was positioned on the margin of India's largest population cluster and had the most productive hinterland. Calcutta lies some 130 kilometers (80 miles) from the coast on the Hooghly River, and a myriad of Gangetic delta channels connect it to its hinterland, a natural transport network that made the city an ideal colonial headquarters. But the British had to contend with Indian rebelliousness in the region; in 1912, they moved the colonial capital from Calcutta to the safer interior city of New Delhi, built adjacent to the old Mogul headquarters of Delhi (today these urban areas have coalesced).

Indian urbanization reveals several regional patterns. In the northern heartland, the west (the wheat-growing area) is more urbanized than the east (where rice forms the main staple crop). This undoubtedly relates to the differences between wheat farming and the labor-intensive, small-plot cultivation and multiple cropping of rice. In the west, urbanization may be as much as 40 percent (high by Indian standards); in the east only about 10 percent of the population resides in urban centers. Furthermore, India's larger cities (over 100,000) are concentrated in three regions: (1) the northern plains from Punjab to the Ganges Delta, (2) the Bombay-Ahmadabad area, and (3) the southern end of the peninsula, which includes Madras and Bangalore. The only interior cities with populations over 1 million not located within one of these zones are centrally positioned Nagpur and the capital of Andhra Pradesh, Hyderabad (both in the 1.5-to-3.0-million range).

In contrast to many other former colonial and now underdeveloped countries, India has a relatively well-developed network of railroads with over 100,000 kilometers (63,000 miles) of track. Once again, the British colonizers built the first railroads (the first train ran 34 kilometers [21 miles] from Bombay to Thana in 1853), and, as usual, they bequeathed India with a liability as well as an asset. The railroads were laid out, of course, to facilitate exploitation of interior hinterlands and to improve India's governability; this effective transport system permitted the British to move their capital from coastal Calcutta to the deep interior at Delhi. But the railroad system never developed as a unified whole. Different British colonial companies constructed their own systems, and no fewer than four separate railway gauges came into use. The two widest gauges now constitute about 90 percent of the whole system, but many transhipments are still necessary, and the country must buy different sets of equipment. India's planners have begun to standardize the system when parts of the track must be replaced, but the expense is enormous and progress remains slow. Yet, India's railway network connects most parts of the country, and, in terms of overall length, it ranks as the world's fourth largest.

# Crisis of Numbers

The population of the South Asian realm, which will surpass 1 billion in 1986, constitutes one-fifth of all humanity. This reality is alarming enough, especially because so many millions of these people are unable to secure adequate food and there is still death by starvation. But even more frightening is the rate at which the population continues to grow. In the early 1980s, 20 million people were being added to the South Asian population every year—over 15 million in India alone, and nearly 3 million each in Pakistan and Bangladesh. Here lies the region's greatest obstacle to progress—whatever the gains in crop production and industrial output, the demands of an ever-larger population overcome them.

The situation in India reflects an equally disturbing global pattern. Not only is the world's population growing: the *rate* at which it is expanding increases all the time. This is dramatically illustrated when we take a backward look. At the time of the birth of Christ, world population was probably about 250 million; it took until about A.D. 1650 to reach 500 million. Then, however, the population grew to 1 billion by 1820, and in 1930 it was 2 billion. In 1975 it passed the 4 billion mark, and reached 4.80 billion in mid-1984; if current growth rates continue, the world's population will be 8 billion shortly after 2010.

We can look at these figures in another way. It took nearly 17 centuries for 250 million to become 500 million, but then that 500 million grew to 1 billion in just 170 years. The billion doubled to 2 billion in 110 years, and 2 billion became 4 billion only 45 years later. The world population's *doubling time* at present growth rates is only 38 years. This steady decrease in the doubling time reflects an accelerating population growth rate, or what mathematicians call an *exponential* increase—growth along an upward curve rather than along a straight line. Geographically, however, it is important to note that the world's population is not increasing at the same rate in all regions. Today's overall growth rate, producing a doubling time of 38 years, is 1.8 percent. But, as Fig. 8–17 shows, the population of some Middle and South American countries is growing more rapidly than that—in excess of 3 percent (which produces a doubling time of just 23 years). In Africa too, as we saw, a number of countries exhibit very high birth and growth rates.

Annual population growth is calculated as the number of excess births over deaths recorded during a given year, usually expressed as a percentage, as shown on Fig. 8–17. In Bangladesh, for example, the annual growth rate is given as 2.7. The birth rate in Bangladesh during 1980 was 44 per thousand, and the death rate 17 per thousand. Thus, in that year for every thousand persons in Bangladesh there were 27 excess births over deaths, or 2.7 percent. In the United States, the population growth rate in 1980 was 0.8 percent, with 16 births per thousand population and 8 deaths. To compare what such figures mean in terms of doubling time, for Brazil's population it was just 30 years, whereas for the United States it was 87. Some populations are growing even faster than Brazil's: Mexico, with a population now surpassing 80 million, has an annual growth rate of 2.7 percent and a current doubling time of only 26 years. Therefore, in just two and a half decades, Mexico will contain twice as many people as today.

## The Dilemma in India

At the beginning of the twentieth century, India had under 250 million inhabitants. This population fluctuated before 1900 with the vagaries of climate, disease, and war, but historical demographers believe that this average size had remained almost stationary for centuries prior to the European intervention—and did not change significantly until the impact of Europe's industrial revolution was diffused to the colonies. Then, the gap between birth rates and death rates immediately began to widen. Birth rates remained high, but death rates declined as medical services proliferated, food distribution systems improved, agricultural production expanded, and costly wars were suppressed. The last time death rates (from famine and outbreaks of disease) approached the level of the birth rate was in 1910, when births were near 50 per thousand and deaths in the high 40s. Since then, the birth rate has declined to 36 (it first went below 40 as recently as 1970), but the death rate in India is now only about 15. The population doubled to about 500 million by the mid-1960s and now, two decades later, another 260-plus million have been added. These are staggering figures that make one wonder about this planet's ultimate capacity to support human populations (see *Box*).

In terms of population growth—as in culture, economy, and urbanization—there is not just one India but several distinct and different Indias. The growth of population is most rapid in Assam and in West Bengal, both adjacent to Bangladesh (Fig. 8–18). In Nagaland and Manipur, it actually exceeds 3.5 percent, and it is over 2.5 percent (as it is in Bangladesh) throughout much of India's great eastern population cluster in the lower Ganges-Brahmaputra region. However, in Uttar Pradesh (the central and upper Gangetic lowland), it has been below 2.0 percent. In the southern peninsula, the growth rate trends near the national average; in the west, it is somewhat higher. All this mirrors India's regional food shortages; hunger and starvation can afflict the crowded east, whereas the food supply is adequate in Punjab.

India obviously must reduce its birth rate, but in a country as vast and heavily rural as this, such a national task confronts virtually insurmountable obstacles. Indian governments have endorsed family planning as part of official development programs, but it has proved difficult to disseminate the practice throughout tradition-bound Indian society. Not all the states of the union cooperated equally in the effort; when, in the mid-1970s, a program of compulsory sterilization of persons with three children or more was instituted in Maharashtra state, there were riots. After more than 3.7 million people were sterilized (bringing the 1977 total in India to about 22.5 million), the government was forced to declare that sterilization would no longer be compulsory. By the mid-1980s, however, family planning was reemerging from years of disrepute, and birth control appeared to be slowly gaining acceptance; about one-third of India's married couples in 1984 were believed to be practicing contraception (vs. 10 percent in 1970), a proportion that population planners say must surpass 60 percent by 2000 if India is to get off the demographic treadmill.

## Demographic Cycles

Some people predict that India will overcome its population crisis

## PREDICTING THE NIGHTMARE

In 1650, the world's human population was approximately 500 million; not until 1820 did the world accommodate 1 billion inhabitants. Even at that time, some scientists realized that the world was headed for trouble. The most prominent among these was Thomas Malthus, an English economist who sounded the alarm as early as 1798 in an article entitled *An Essay on the Principle of Population as It Affects the Future Improvement of Society*. In this essay, Malthus reported that population in England was increasing faster than the means of subsistence; he described population growth as geometric and the growth of food production as arithmetic. Inevitably, he argued, population growth would be checked by hunger. Malthus's essay caused a storm of criticism and debate, and between 1803 and 1826 he revised it several times. He never wavered from his basic position, however, and continued to predict that the gap between the population's requirements and the soil's productive capacity would ultimately lead to hunger, famines, and the cessation of population growth.

We know now that Malthus was wrong about several things. The era of colonization and migration vastly altered the whole pattern of world food production and consumption, and worldwide distribution systems made possible the transportation of food surpluses from one region of the world to another. Malthus could not have foreseen this; nor was he correct about the arithmetic growth of agricultural output. Expanded acreages, improved seed strains, and better farming techniques have produced geometric increases in world food production, but Malthus was correct in his prediction that the gap between need and production would widen. It has—and there are areas in the world in the 1980s where population growth rates are being checked in the way he predicted, as famines claim the lives of hundreds of thousands in Africa and Asia.

Even today there are scholars and specialists who essentially adhere to Malthus's position, modified, of course, by contemporary knowledge and experience. These people are sometimes referred to as prophets of doom by those who take a more optimistic view of the future. They are also called neo-Malthusians, tying their concerns to those of that farsighted Englishman who first warned that there are limits to this earth's capacity to sustain our human numbers.

because other countries, too, went through periods of explosive growth only to reach a demographic near-equilibrium. Figure 8–17 indicates that the low-growth countries are the developed countries, many of them with an annual population increase of less than 1 percent. In the nineteenth century, many of those countries also experienced explosive population expansion—but, when they became industrialized, urbanized, and modernized, their rapid growth declined. As we saw in the Systematic Essay in Chapter 1 (pp. 57–58), it is possible to discern a population cycle of three major stages that comprises the Demographic Transition Model (Fig. 1–4). First, birth as well as death rates are high, and the population fluctuates within a certain low-growth range. Next, as industrial, medical, and technological innovations are introduced, the death rate declines but the birth rate stays high until it begins to decline much later (this is the period of "population explosion"). Finally, birth rates plunge sharply, rejoining the level of the (now low) death rate, and the overall growth rate is reduced below 1 percent. This is essentially what occurred in Western Europe and, some scholars insist, underdeveloped countries will also experience such a sequence.

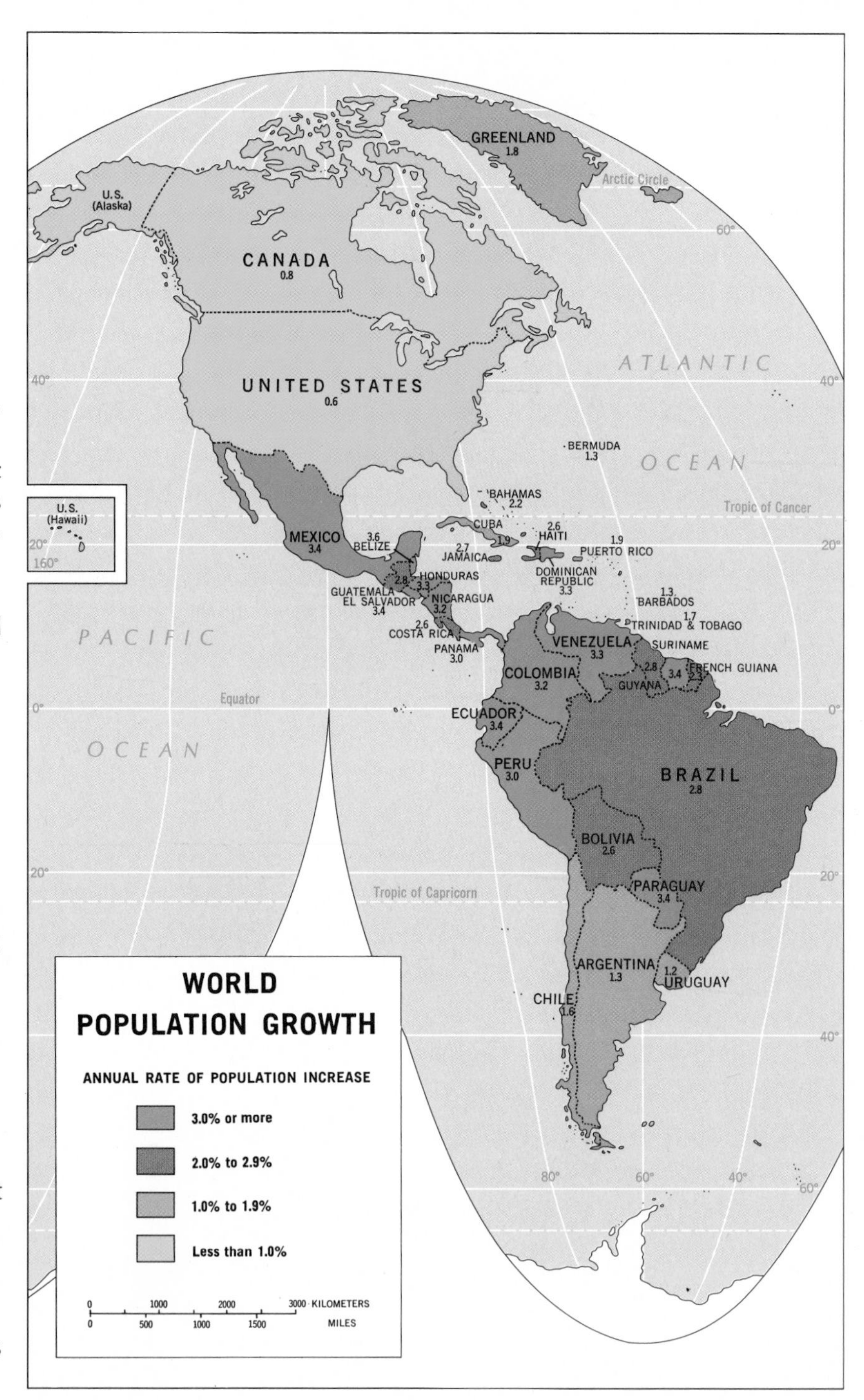


Figure 8–17

But the European model may not fit countries of the late-twentieth-century underdeveloped world. To begin with, the population of a country such as the United Kingdom was under 10 million when its rapid-growth stage began. Two doublings would still only produce 40 million from this base; although England went through its explosive period, there was mass emigration to alleviate the population pressure further. Present population totals in countries such as Pakistan and Bangladesh (let alone India) are on a very different scale. In two doubling periods, Pakistan would go from 100 to 400 million, Bangladesh from 103 to 412 million, and India from 763 million to 3.05 billion. Moreover, many underdeveloped countries do not at present appear to have the raw materials to sustain the kind of development that Europe witnessed. As we noted earlier, all the oil in the Middle East has not generated genuine development, even though it has produced money with which the products of development can be bought. Therefore, the Indian dilemma is far from solution.

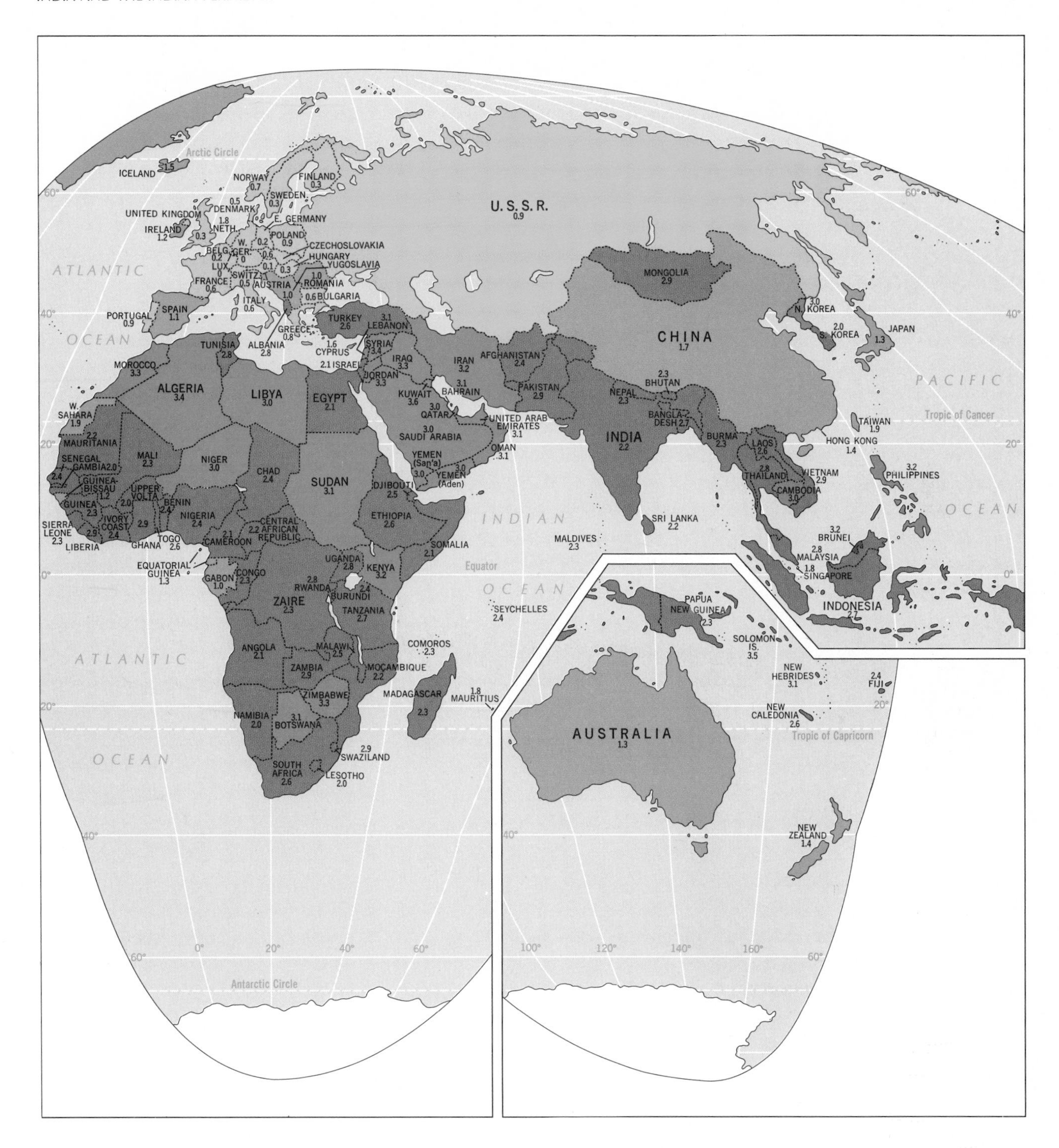


# Bangladesh

The state of Bangladesh was born in 1971 following a war of independence against Pakistan—of which it had been a part (as East Pakistan) since 1947. With its economy shattered by many months of conflict, Bangladesh began its sovereign existence desperately impoverished, ill fed, and overcrowded. Bangladesh is a comparatively small country (144,000 square kilometers/55,000 square miles), somewhat smaller than the state of Florida. Its population, now edging past 100 million, constituted 55 percent of formerly united Pakistan. Its national territory, nearly surrounded by India (a short stretch of boundary adjoins Burma on the southeast), is essentially the deltaic plain of the Ganges-Brahmaputra

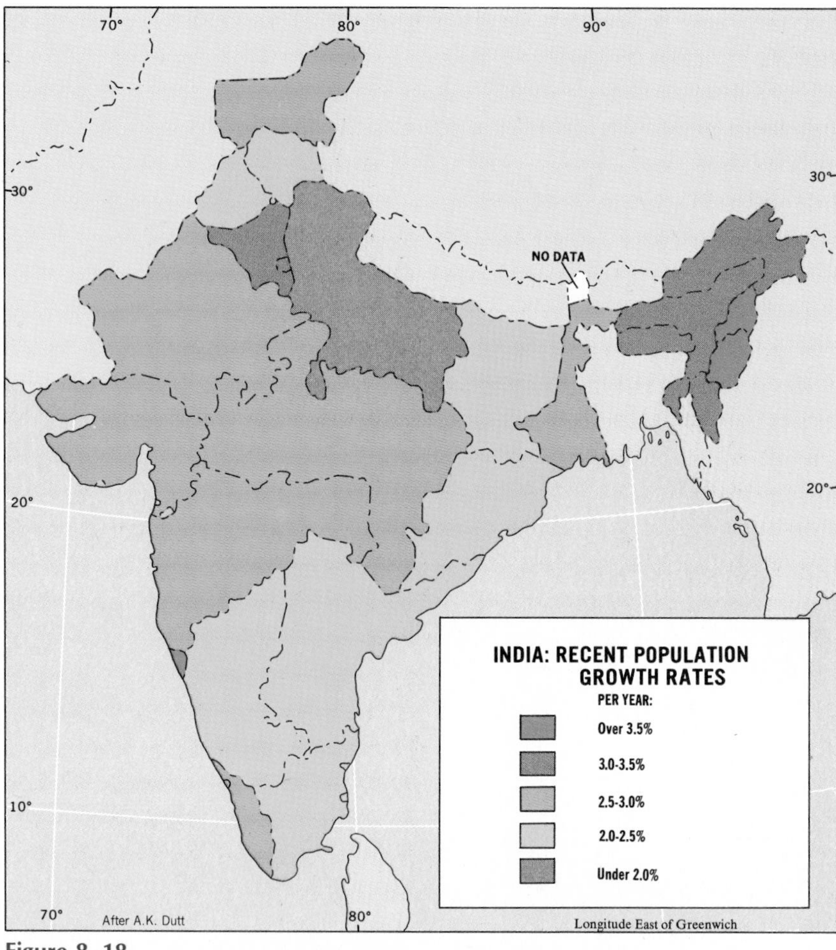


Figure 8–18

system, which empties into the Bay of Bengal through numerous channels (Fig. 8–19). Only in the extreme east and southeast, in the hinterland of Chittagong, does the flat terrain of the floodplains rise into hills and mountains.

The land of Bangladesh is extremely fertile, and practically every cultivable foot of soil is under crops—rice for subsistence, jute and tea for cash. The jute industry made a major contribution to the economy of Pakistan prior to "the" Bangladesh secession, producing (with other products) well over half of the country's annual export revenues. It was always a bone of contention that Bangladesh's share of the Pakistani budget was only about 40 percent, so that Bangladesh served as an exploited colony to West Pakistan in the eyes of many of its people.

The staple food for Bangladesh's enormous population is rice, grown in fields whose fertility is renewed by the silt swept down by the river systems' annual floods. In most places, three harvests of rice per year are possible, and the country normally was able to produce some 80 percent of its people's food requirements. Then the dislocation brought on by the war of secession threatened widespread starvation as harvests rotted and crops were abandoned. Large-scale emergency imports saved much of the situation, but Bangladesh never recovered. The country remains the epitome of Malthusian forecasts, in a constant state of hunger and local starvation, dislocation and misery.

Bangladesh's land is fertile because it is low lying and subject to river flooding (Fig. 8–20). This condition has negative aspects as well, because, to the south, the country lies open to the Bay of Bengal where destructive tropical storms are born and sometimes make landfall. With much of southern Bangladesh less than 4 meters (13 feet) above sea level, the penetration of cyclones—as these hurricane- or typhoon-type storms are called there—can do incalculable damage. In early 1971, a devastating tropical cyclone hit Bangladesh, and the rising waters and high winds exacted a toll of perhaps 600,000 persons. It was the second greatest natural disaster of the twentieth century (after the 1976 earthquake that killed upwards of 700,000 in Tangshan, China), but it is unlikely to be the last such assault on the land and people of Bangladesh. Crowding onto low-lying farmlands, inadequate escape routes and mechanisms, and insufficient warning time continue to exist.

The people of Bangladesh have in recent years faced disaster of several kinds. The indifference of the Pakistani government during the aftermath of the 1971 cyclone was one of the leading factors in the outbreak of open revolt against established authority. The war of secession brought unspeakable horror to villages and towns; millions left their homes and streamed across the border into neighboring India. Hunger and rampant disease soon followed. It is impossible to be certain of the consequences, but estimates of the total loss of life (not counting those permanently disabled and ruined) run as high as 3 million. Destructive flooding again besieged the country in the mid-1970s, forcing it to rely on emergency grain shipments from the United States. The ever-present threat of mass starvation, however, has lessened a bit in the 1980s.

Although rice yields per unit

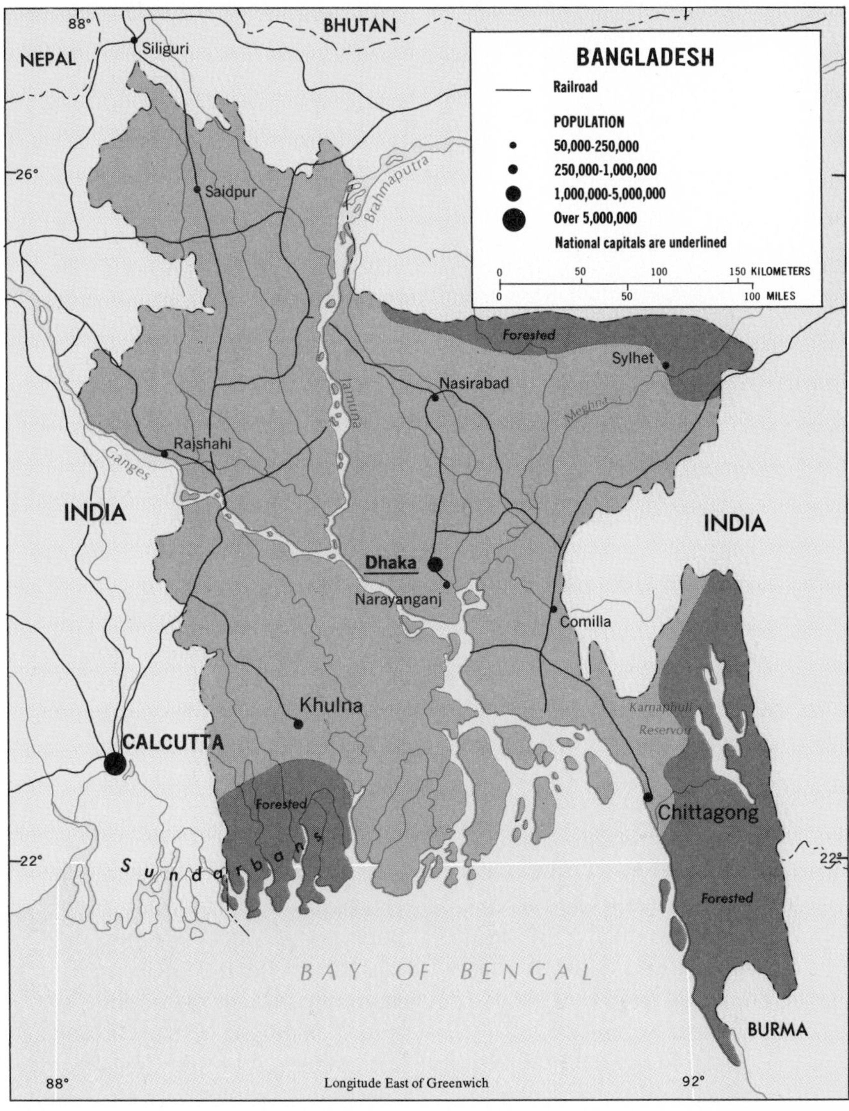


Figure 8–19

area have been low in Bangladesh, the country has made recent gains with higher-yielding varieties and could achieve self-sufficiency in food by 1987—though at a rather low caloric level. The problem, of course, is once again a population that keeps growing too rapidly, thereby limiting the impact of these advances in rice production to sustaining a nutritional level that is barely adequate. Thus, Bangladeshi energies are almost totally devoted to fighting the daily battle against chronic malnutrition, and the country's resources of natural gas, coal, timber, and several minerals goes unexploited. As in India, effective birth control (involving today less than 15 percent of all married couples) lags far behind the 60-percent level needed to turn the demographic corner, but the outlook for improvement in the near future is less encouraging. Bangladesh's level of poverty is such that its 1981 per capita GNP was $144—only Ethiopia and Chad (together containing one-third of the Bangladeshi population) ranked lower among all the world's countries—and no nation anywhere suffers more from the effects of underdevelopment. After centuries of British colonialism, Bangladesh reaped a harvest of Pakistani neglect, and the country's economy is still mired in general disorder. The jute industry has suffered severely, and the already-inadequate communications system is in disarray. Governmental instability, corruption in administration, and terrorism have further bleakened Bangladesh's first decade and a half of independence.

Dhaka, the centrally positioned capital (4.3 million), and the port of Chittagong (1.8 million) are urban islands in a country where only 13 percent of the people live in towns and cities. A further measure of Bangladeshi economic misfortunes can be gained from a look at the transport network. There is still no road bridge over the Ganges anywhere in the country, and only one railroad bridge; there is neither a road nor a railroad crossing over the Brahmaputra anywhere in Bangladesh. Road travel from Dhaka to the eastern town of Comilla involves two ferry transfers across the same river (distributaries of the Meghna); there is hardly any means of surface travel from Dhaka to the western half of the country, except by boat. Thousands of boats ply Bangladesh's ubiquitous waterways, which still form a more effective interconnection system within the country than do the roads. In Bangladesh, survival is still the leading industry; all else is luxury.

# Pakistan

The Islamic Republic of Pakistan (Fig. 8–21) is also an underdeveloped country—but it differs in almost every respect from its former federal partner. Indeed, so strong are the contrasts between former East Pakistan and the state that now carries the name that the generation-long survival of two-part Pakistan was something of a miracle. Islam was the unifier (across a thousand miles of Hindu Indian territory), but there were differences

**Figure 8–20**
Periodic floods affect the farms of the low-lying alluvial plain and double delta of Bangladesh.

even in the paths whereby the Moslem faith reached west and east. In dry-world West Pakistan, it arrived over land, carried eastward by invaders from the west and northwest. In monsoon East Pakistan it came by sea, brought to the double delta by Arab traders. In fact, it is difficult to find any additional area of correspondence or similarity between former East and West Pakistan. The two wings of the country lay in different culture regions, with all that this implied. Pakistan lies on the fringes of Southwest Asia; Bangladesh adjoins Southeast Asia. In Pakistan, the problem has always been drought, and the need is for irrigation. In Bangladesh, the problem is flooding, and the need is for dikes. Pakistanis build their houses of adobe and matting, eat wheat and mutton, and grow cotton for sale; the people of Bangladesh build with grass and thatch, eat rice and fish, and grow jute.

## Livelihoods

Contrary to the predictions made at the time Pakistan became a separate state, it has managed to survive economically and has even made a good deal of progress. This occurred, despite the country's poverty in known mineral resources; apart from natural gas in Baluchistan (which may be related to oil reserves) and chromite, Pakistan has only minor iron deposits which are being used in a small plant at Multan. So, the country must make up in agricultural output what it lacks in minerals; since independence, it has made considerable strides. Favored by the high prices generated during the Korean War, the economy got an early boost from jute and cotton sales. Then the mid-1950s brought a period of stagnation, made worse by continuing political conflicts with India. But during the 1960s and 1970s a considerable expansion of irrigated agriculture, land reform, the improvement of marketing techniques, and the success of the textile industry all contributed to noteworthy progress.

Certainly, there was room for such progress. Pakistan shares with India the familiar low yields per hectare for most crops, per capita annual incomes are quite low, and manufacturing is only now emerging as a significant contributor to the economy. The low yields for vital crops such as wheat and rice have long been due to the inability of the peasants to buy fertilizers, to the poor quality of seeds, and to inadequate methods of irrigation. In many areas, there has not been enough irrigation water; in other places, there is excessive salinity in the soil. In Sind, where large estates farmed by tenant farmers existed before the British began their irrigation programs in the Punjab to the northeast, yields are kept down by outdated irrigation systems and by the paucity of incentives for landless peasants.

Pakistan's virtually new textile industry, based on the country's substantial cotton production, has developed rapidly. It now satisfies all of the home market, and textiles have quickly risen to become the top contributor to exports. Other industries have also been stimulated; increasingly, Pakistan is able to produce at home what it formerly had to import from foreign sources. A chemical industry is emerging, and an automobile assembly plant has opened at Karachi, where a small steel-producing plant now operates as well. Thus, Pakistan is trying to reduce its dependency on foreign aid, investing heavily in the agricultural sector, and encouraging local industries as much as possible. But there is a long road to travel, and a difficult one. There are few natural resources, and export products are subject to both sudden price fluctuations on world markets and to the increasing pressure of competition.

Pakistan still faces a host of unsolved problems. Literacy is low. Family planning is government-approved, but its dissemination and acceptance are just beginning. Many millions of people still live in poverty, subsist on an ill-balanced diet, have a short life expectancy, and experience little tangible prog-

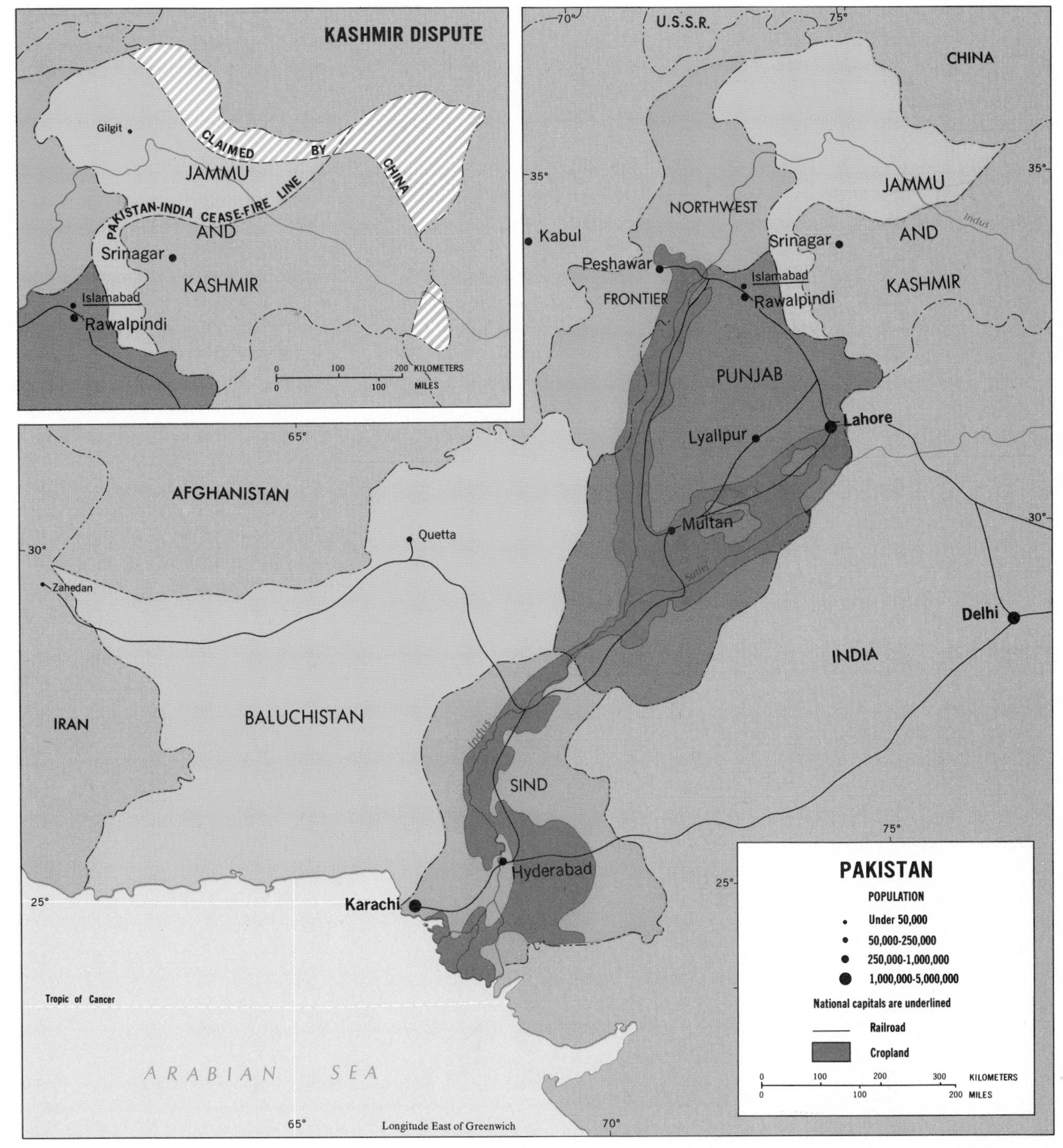


Figure 8–21

ress during their lifetimes. Of course, there are successful farmers in the Punjab, in Sind, and elsewhere. But they are still far outnumbered by those for whom life remains a daily struggle for survival.

## The Cities

It is understandable that the young state of Pakistan chose Karachi as its first capital. Clearly, it was desirable to place the capital city in the Moslem stronghold of the country, West Pakistan. But the outstanding center of Islamic culture, Lahore, lay too exposed to the sensitive nearby boundary with India.

Lahore, with 3.8 million residents, grew rapidly as a result of the Punjab's partitioning in 1947.

In 1950, its population was only about one-third as large as today. Founded in the first or second century A.D., Lahore became established as a great Moslem center during the Mogul period. As a place of royal residence, the city was adorned with numerous magnificent buildings, including a great fort, several palaces, and mosques, which remain to this day monuments of history with their marvelous stonework and excellent tile and marble embellishments (Fig. 8–22). The site of magnificent gardens and an old university, Lahore was also the focus of a large area in prepartition times, when its connections extended south to Karachi and the sea, north to Peshawar and east to Delhi and Hindustan. Its Indian hinterland was cut off by the partitioning, but Lahore has retained its importance as the center of one of Pakistan's major population clusters, as a place of diverse industries (textiles, leather goods, and gold and silver lacework), as the unchallenged historic headquarters of Islam in this region, and as an educational center.

Karachi (6.5 million) grew even faster than Lahore. After independence it was favored not only as the first capital of Pakistan, but also as West Pakistan's only large seaport. Its overseas trade rose markedly, new industries were established, and new regional and international connections developed between Karachi and former East Pakistan as well as other parts of the world. A flood of Moslem refugees came to the city, presenting serious problems—there simply was not enough housing or food available to cope with the half-million immigrants who had arrived by 1950. Nevertheless, Karachi survived, and, in its growth, has reflected the new Pakistan. As a result of the expansion of cultivated acreages in the Pakistani Punjab and in nearby Sind, Karachi's trade volume has increased greatly, especially in wheat and cotton. Imports of oil and other critical commodities required construction of additional port facilities.

But Karachi never became the cultural or emotional focus of the nation. It lies symbolically isolated, along a desert coast, almost like an island; the core of Pakistan still lies far inland. In 1959, after just over a decade as the federal capital of

**Figure 8–22**
Lahore is the cultural focus of Islamic Pakistan: Badshai Mosque, with a half million worshipers in attendance.

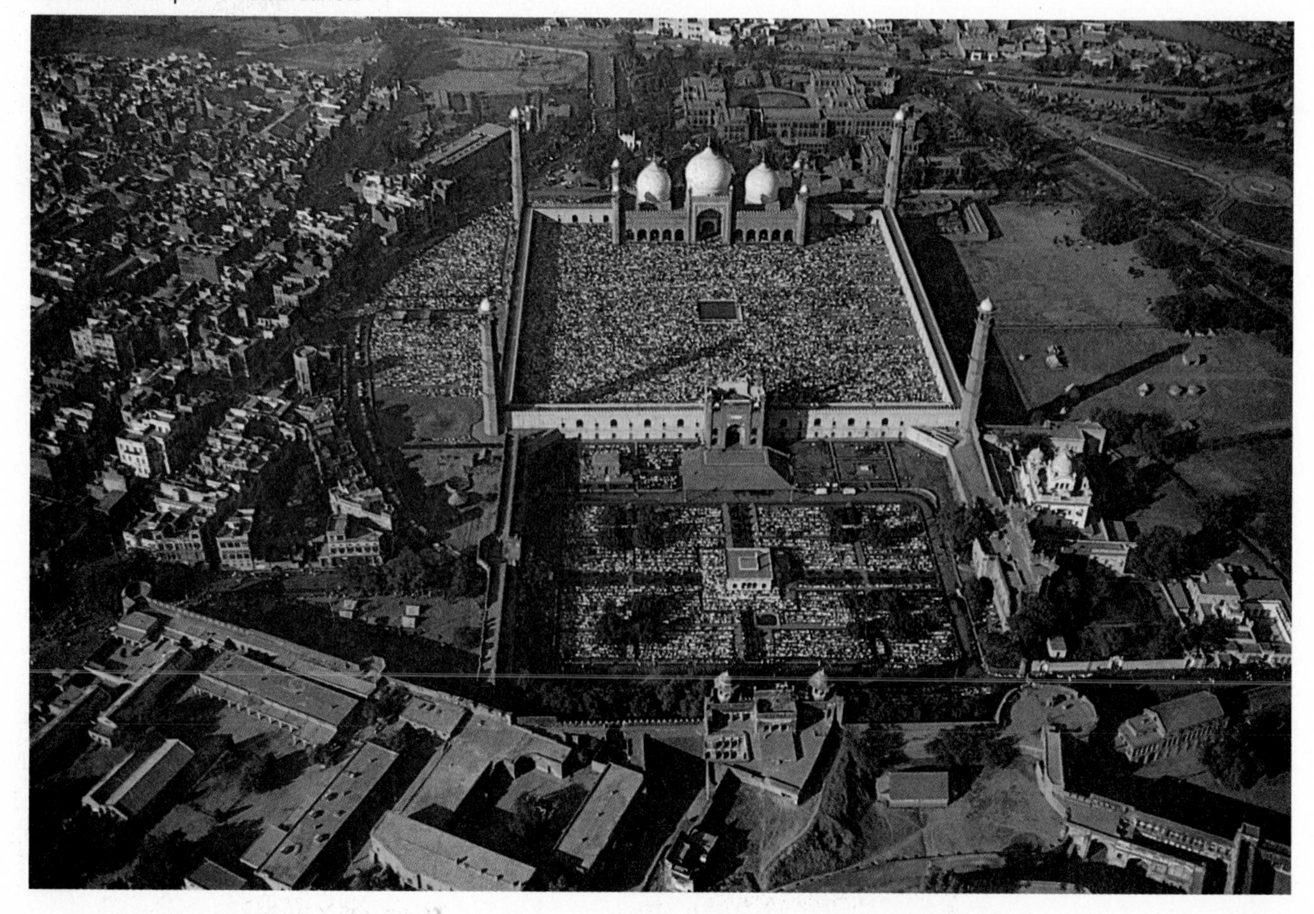

Pakistan, Karachi surrendered its political functions to an interior city, Rawalpindi. This was a temporary measure: an entirely new government headquarters was completed in the 1970s a short distance from Rawalpindi, at Islamabad. This new planned city, as Fig. 8–21 shows, lies near the boundary of Kashmir. It confirms not only the internal positioning of Pakistan's cultural and economic heartland, but also the state's determination to emphasize its presence in the contested north. It is evidence of a sense of security that Pakistan did not yet possess when it chose Karachi over exposed Lahore. In this context, Islamabad is a classic example of the principle of the *forward* capital.

## Irredentism

For many years, Pakistan has been beset by problems in its marginal areas. The boundary with India in Kashmir (Fig. 8–21 inset) is a truce line that appears to be stabilizing (see *Box*); but in Baluchistan in the west and along the boundary with Afghanistan in the northwest, Pakistan has faced more recent difficulties. In Baluchistan, a still-remote area where the Iran-Pakistan border has little practical meaning, government efforts to exercise its authority were met by opposition from local chiefs. Baluchistan is a region of rugged mountains, severe desiccation, scattered oases, and ancient nomadic routes. It resembles neighboring Iran and is quite unlike the settled farmlands of the Indus Valley. Baluchistan's peoples were not ready to be controlled by the Pakistani government, and, in their rebelliousness, they were supported by the government of Afghanistan. Afghanistan's majority population, the Pakhtuns (also called Pashtuns, Pathans, or Pushtuns), constitute about 50 per-

# THE PROBLEM OF KASHMIR

Kashmir is a territory of high mountains surrounded by Pakistan, India, China, and along a few kilometers in the far north, by Afghanistan (Fig. 8–21). Although known simply as Kashmir, the area actually consists of several political divisions, including the state properly referred to as Jammu and Kashmir (one of the 562 Indian states at the time of independence) and the administrative areas of Gilgit in the northwest and Ladakh (including Baltistan) in the east. Ladakh gained world attention as a result of Chinese incursions there in the 1960s, but the main conflict between India and Pakistan over the final disposition of this territory has focused on the southwest, where Jammu and Kashmir are located.

When partition took place in 1947, the existing states of British India were asked to decide whether they would go with India or Pakistan. In most of the states, this issue was settled by the local authority, but, in Kashmir, there was an unusual situation. There were about 5 million inhabitants in the territory at that time, nearly half of them concentrated in the so-called Vale of Kashmir (where the capital, Srinagar, is located). Another 45 percent of the people were concentrated in Jammu, which leads down the foothill slopes to the edge of the Punjab. The small remainder of the population were scattered through the mountains, including Pakhtuns in Gilgit and other parts of the northwest. Of these population groups, the people of the mountain-encircled Vale of Kashmir are almost all Moslems, whereas the majority of Jammu's population is Hindu. But the important feature of the state of Jammu and Kashmir was that its rulers were Hindu, not Moslem, although the overall population was more than 75 percent Moslem. Thus, the rulers were faced with a difficult decision in 1947—to go with Pakistan and thereby exclude themselves from Hindu India, or to go with India and thereby incur the wrath of the majority of the people. Hence, the Maharajah of Kashmir sought to remain outside both Pakistan and India and to retain the status of an autonomous separate unit. This decision was followed, after the partitioning of India and Pakistan, by a Moslem uprising against Hindu rule in Kashmir. The maharajah asked for the help of India, and Pakistan's forces came to the aid of the Moslems. After more than a year's fighting and through the intervention of the United Nations, a cease-fire line was established that left Srinagar, the Vale, and most of Jammu and Kashmir in Indian hands, including nearly four-fifths of the territory's population. In due course, this line began to appear on maps as the final boundary settlement, and Indian governments have proposed that it be so recognized.

Why should two countries, whose interests would be served by peaceful cooperation, allow a distant mountainland to trouble their relationship to the point of war? There is no single answer to this question, but there are several areas of concern for both sides. In the first place, Pakistan is wary of any situation whereby India would control vital irrigation waters needed in Pakistan—as the map shows, Kashmir is traversed by the Indus River, the country's lifeline. Moreover, other tributary streams of the Indus originate in Kashmir, and it was in the Punjab that Pakistan learned the lessons of dealing with India for water supplies. Second, the situation in Kashmir is analogous to the one that led to the partition of the whole subcontinent: Moslems are under Hindu domination. The majority of Kashmir's people are Moslems, so Pakistan argues that free choice would deliver Kashmir to the Islamic Republic; a free plebiscite is what the Pakistanis have sought and the Indians have thwarted. Furthermore, Kashmir's connections with Pakistan prior to partition were much stronger than those between Kashmir and India—although India has invested heavily in improving its links to Jammu and Kashmir since the military stalemate. Finally, Pakistan argues that it needs Kashmir for strategic reasons, in part to cope with the Pakhtun secessionist movement that extends into this area. As the mid-1980s arrived, it appeared likely that the cease-fire line would, indeed, become a stable boundary between India and Pakistan. The incorporatioan of the state of Jammu and Kashmir into the Indian federal union was accomplished as far back as 1975 when India was able to reach an agreement with Sheikh Muhammad Abdullah, the state's chief minister and leader of the majority.

cent of that country's 15 million people and are concentrated in the region centered on the capital (Kabul), positioned opposite the Khyber Pass less than 200 kilometers (120 miles) from the Pakistan border. Kabul's government has also encouraged Pakhtuns living in Pakistan's northwest to demand their own state of Pakhtunistan (Pathanistan). Pakistan's moderate response has been to hasten the integration of its northwestern areas into the national state through improved communications, education, and other facilities, but Afghan irredentism continues. Islamabad, therefore, is situated between pressure areas in Kashmir *and* Afghanistan. The Soviet invasion and subsequent entrenchment in Afghanistan, and its resultant problem of refugees (perhaps 1 million are now in Pakistan) have further aggravated internal regional difficulties; thus, Pakistan's territorial problems did not end with the secession of Bangladesh.

# Sri Lanka

Sri Lanka (formerly Ceylon), the compact, pear-shaped island located off the southern tip of the peninsula of India, is the fourth independent state to have emerged from the British sphere of influence in South Asia (Fig. 8–23). Sovereign since 1948, Sri Lanka has had to cope with political as well as economic problems, some of them quite similar to those facing India and Pakistan and others quite different. There were good reasons to create a separate independence for Sri Lanka. This is neither a Hindu nor a Moslem country; the majority—some 75 percent—of its 16 million people are Buddhists. And, unlike India or Pakistan, Sri Lanka is plantation country (a legacy of the European period), with export agriculture still the mainstay of the external economy.

The majority of Sri Lanka's people are not Dravidian, but are of Aryan origin with a historical link to ancient northern India. After the fifth century B.C., their ancestors began to migrate to Ceylon, a relocation that took several centuries to complete and brought to this southern island the advanced culture of the northwestern portion of the subcontinent. Part of that culture was the Buddhist religion; another component was the knowledge of irrigation techniques. Today, the descendants of these early invaders, the Sinhalese, speak a language belonging to the Indo-Aryan language family of northern India.

The darker-skinned Dravidians from southern India never came in sufficient numbers to challenge the Sinhalese. They introduced the Hindu way of life, brought the Tamil language to Sri Lanka, and eventually came to constitute a substantial minority (12 percent) of the country's population. Their number was strengthened substantially during the second half of the nineteenth century, when the British brought many thousands of Tamils from the mainland to Ceylon to work on the plantations that were being laid out. Sri Lanka has sought the repatriation of this element in its population, and an agreement to that effect was even signed with India. In 1978, however, Tamil was granted the status of a national language of Sri Lanka.

Sri Lanka is not a large island (65,000 square kilometers/25,000 square miles), but it has considerable topographic diversity. The upland core lies in the south, where elevations reach 2500 meters (over 8000 feet) and sizable areas exceed 1500 meters (5000 feet). Steep, overgrown slopes lead down to a surrounding lowland, most of which lies below 300 meters (1000 feet); northern Sri Lanka is entirely low-lying. Rivers, the sources of ricefield irrigation waters, flow radially from the interior highland across this lowland rim.

Since the decline of the Sinhalese empire, focused on centrally-located Anuradhapura, the moist upland southwest has been the ma-

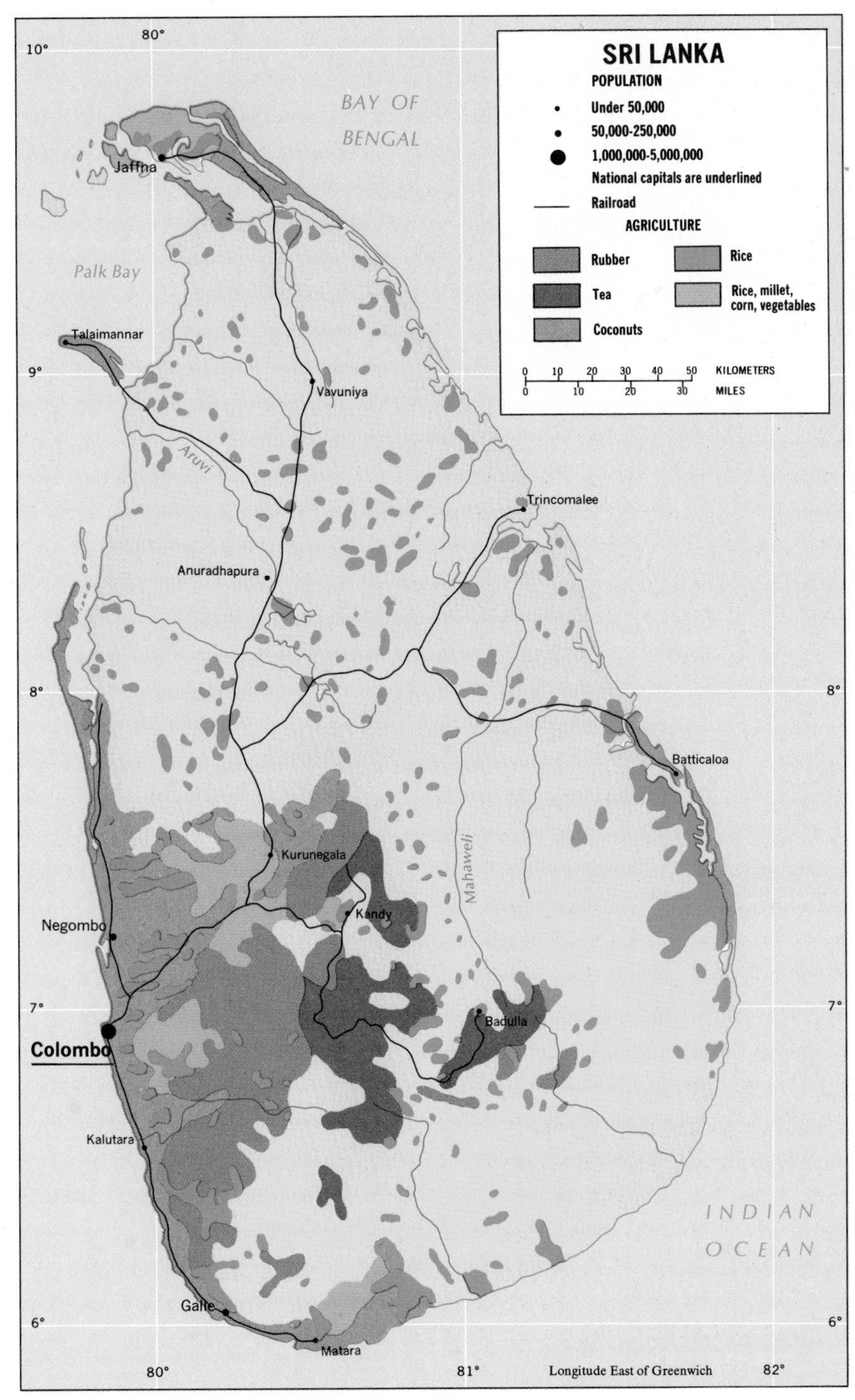


**Figure 8–23**

jor area of productive capacity (the plantations are concentrated here) and the population core. Three plantation crops have been successful: coconuts in the hot lowlands, rubber up to about 600 meters (2000 feet), and tea in the highlands above (Fig. 8–24), the product for which Sri Lanka is most famous. Usually, tea continues to account for two-thirds of Sri Lanka's annual exports by value. Whereas Sri Lanka's plantation agriculture is very productive and quite efficient, the same cannot be said for the island's ricelands. As recently as the early 1960s, it was necessary to import half the rice consumed in Sri Lanka, a situation that was detrimental to the general economic situation. Accordingly, the government made it a priority to reconstruct plainland irrigation systems, repopulate the lowlands (until its recent eradication, malaria was an obstacle to settlement there), and intensify rice cultivation. The result has been a substantial increase in rice production, but Sri Lanka has still not attained self-sufficiency. Although lower today than in the 1970s, population growth remains a worrisome 1.8 percent annually, and Sri Lanka spends one-fifth of its import expenditures on grains.

In a country so heavily agricultural, it is not surprising to find very little industrial development except for factories processing plantation and farm products. Sri Lanka appears to have very little in the way of mineral resources; graphite is the most valuable mineral export, although gemstones (sapphires, rubies) once figured importantly in overseas commerce. The industries that have developed, other than those processing agricultural products, depend on Sri Lanka's relatively small local market. Predictably, these include cement, shoes, textile, paper, china, glassware, and the like. The majority of these establishments are located in Colombo (740,000), the country's capital, largest city, and leading port.

Although an overwhelmingly agricultural country, Sri Lanka is still in a position to close the gap between the demand for staple foods and the local supply capacity. As elsewhere in South Asia, its chief objectives must be to increase the productivity of its soil while decreasing the growth rate of its population. Unfortunately, Sri Lanka also shares in this realm's catalogue of ongoing conflicts. In 1983, extremist Tamils rioted to demand the creation of a separate homeland in the island's northern lowland; this triggered a violent response by rov-

**Figure 8–24**
Sri Lanka's best-known product, tea, under cultivation in the plantation zone of the island's south-central highlands.

ing Sinhalese bands, who murdered and brutally destroyed the property of hundreds of Hindu Tamils. Although the government did manage to calm things down (this festering hatred could easily explode into bloodshed again), it admits that this episode has set back overall economic development by at least five years.

## References and Further Readings

Ahmed, N. *An Economic Geography of East Pakistan* (London: Oxford University Press, 1968).

Berry, B. J. L. et al. *Essays on Commodity Flows and Spatial Structure of the Indian Economy* (Chicago: University of Chicago, Department of Geography, Research Paper No. 111, 1966).

Bhardwaj, S., ed. *Hindu Places of Pilgrimage in India: A Study in Cultural Geography* (Berkeley: University of California Press, 1983).

Breese, G. *Urban and Regional Planning for the Delhi-New Delhi Area* (Princeton: Princeton University Press, 1974).

Brush, J. "Spatial Patterns of Population in Indian Cities," in Dwyer, D., ed., *Cities in the Third World* (London: Macmillan, 1974), pp. 105–132.

Butzer, K. *Geomorphology From the Earth* (New York: Harper & Row, 1976).

Chakravati, A. "Green Revolution in India," *Annals of the Association of American Geographers* 63 (1973), 319–330.

Dikshit, R. *The Political Geography of Federalism: An Inquiry into Origins and Stability* (New York: John Wiley & Sons, 1975).

Dumont, L. *Homo Hierarchus: The Caste System and Its Implications* (Chicago: University of Chicago Press, 1970).

Dutt, A. "Cities of South Asia," in Brunn, S. & Williams, J., eds., *Cities of the World: World Regional Urban Development* (New York: Harper & Row, 1983), pp. 324–368.

Dutt, A., ed. *Contemporary Perspectives on the Medical Geography of*

*South and Southeast Asia* (Elmsford, N.Y.: Pergamon, 1981).

Dutt, A. et al. *India: Resources, Potentialities, and Planning* (Dubuque, Iowa: Kendall/Hunt, 1972).

Farmer, B. *An Introduction to South Asia* (London and New York: Methuen, 1984).

Feldman, H. *Pakistan: An Introduction* (New York: Oxford University Press, 1968).

Fox, R., ed. *Urban India: Society, Space and Image* (Durham, N.C.: Duke University Program in Comparative Studies on Southern Asia, 1970).

Greenland, D. & de Blij, H. J. *The Earth in Profile: A Physical Geography* (New York: Canfield Press/Harper & Row, 1977).

Gwatkin, D. & Brandel, S. "Life Expectancy and Population Growth in the Third World," *Scientific American*, May 1982, 57–65.

Hall, A. *The Emergence of Modern India* (New York: Columbia University Press, 1981).

Hutton, J. *Caste in India: Its Nature, Function and Origins* (New York: Oxford University Press, 4 rev. ed., 1963).

Johnson, B. *Bangladesh* (Totowa, N.J.: Barnes & Noble, 2 rev. ed., 1982).

Johnson, B. *India: Resources and Development* (Totowa, N.J.: Barnes & Noble, 1979).

Johnson, S., ed. *The Population Problem* (New York: John Wiley & Sons, 1973).

Karan, P. *Bhutan* (Lexington: University Press of Kentucky, 1967).

Karan, P. *Nepal: A Cultural and Physical Geography* (Lexington: University Press of Kentucky, 1960).

Kaufman, M. "Across India, A Rural Tide Is Engulfing the Cities," *New York Times*, February 19, 1982, p. 2.

Khan, A. "Rural-Urban Migration and Urbanization in Bangladesh," *Geographical Review*, 72 (1982), 379–394.

Long, R. *The Land and People of Pakistan* (Philadelphia: J. B. Lippincott, 1968).

McLane, J. *India: A Culture Area in Perspective* (Boston: Allyn & Bacon, 1970).

Murphey, R. "*The City in the Swamp*, Aspects of the Site and Early Growth of Calcutta," *Geographical Journal*, 30 (1964), 241–256.

Murton, B. "South Asia," in Klee, G., ed. *World Systems Traditional Resource Management* (New York: Halsted Press/V.H. Winston 1980), pp. 67–99.

Myrdal, G. *Asian Drama: An Inquiry into the Poverty of Nations* (New York: Pantheon, 3 vols., 1968).

Myrop, R. *Area Handbook for Ceylon* (Washington: U.S. Government Printing Office, 1971).

Newman, J. & Matzke, G. *Population: Patterns, Dynamics, and Prospects* (Englewood Cliffs N.J.: Prentice-Hall, 1984).

Noble A. & Dutt, A., eds. *India: Cultural Patterns and Processes* (Boulder, Colo.: Westview Press, 1982).

Pitty, A. *Introduction to Geomorphology* (New York: Barnes & Noble, 1971).

Schwartzberg, J., ed. *An Historical Atlas of South Asia* (Chicago: University of Chicago Press, 1978).

Singhal, D. *Pakistan* (Englewood Cliffs, N.J.: Prentice-Hall, 1972).

Sopher, D., ed. *An Exploration of India: Geographical Perspectives on Society and Culture* (Ithaca, N.Y.: Cornell University Press, 1980).

Sopher, D. *Geography of Religions* (Englewood Cliffs, N.J.: Prentice-Hall, 1967).

Spate, O. & Learmonth, A. *India and Pakistan: A General and Regional Geography* (London: Methuen, 2 vols., 1971).

Spencer, J. & Thomas, W. *Asia, East by South: A Cultural Geography* (New York: John Wiley & Sons, 2 rev. ed., 1971).

Stanford, Q. *The World's Population: Problems of Growth* (New York: Oxford University Press, 1972).

Stevens, W. "Bangladesh Threatens to Burst at the Seams," *New York Times*, April 6, 1983, p. 4.

Stevens, W. "English Gains as India's Chief Tongue," *New York Times*, May 19, 1983, p. 4.

Stevens, W. "Rioting Reduces Sri Lanka's Hope to Ruins," *New York Times*, August 4, 1983, pp. 1, 6.

Taylor, A., ed. *Focus on South Asia* (New York: Praeger/American Geographical Society, 1974).

Thornbury, W. *Principles of Geomorphology* (New York: John Wiley & Sons, 2 rev. ed., 1969).

Whyte, R. *Land, Livestock and Human Nutrition in India* (New York: Praeger, 1968).

民共和国万岁

# CHAPTER 9

# CHINA OF THE FOUR MODERNIZATIONS

IDEAS AND CONCEPTS

**Geography of Development**
**Hegemony**
**Colonialism (2)**
**Extraterritoriality**
**Regional Complementarity (3)**
**Activity Space**
**Collectivized Agriculture**
**Central Place Theory (2)**
**Confucianism**

More than one-fifth of all humanity exists within the borders of a single country—China. China's size in land area is nearly the same as the United States's, but its population is more than four times as large. In 1980, the Chinese government announced what scholars had long suspected: that China's population had surpassed 1 billion (reaching about 1.1 billion in 1985).

China's one thousand million people are heirs to what may well be the world's oldest continuous national culture and civilization. The present capital, Beijing (Peking), lies not far from the region where the nucleus of ancient China emerged and where cities were built almost 4000 years ago. During its lengthy history, the Chinese state alternately grew and contracted as its fortunes changed, sometimes gaining territory only to lose it again to adversaries. The state was united only to fragment into feudal rival entities. But, in the heartland, there always was a China—hearth of culture, source of innovations, and focus of power. Over the centuries, indeed over millennia, the Chinese created for themselves a world apart, a society with strong traditions, values, and philosophies that could reject and repel influences from the outside. It was an isolated society, confident of its strength and superiority. Even when its strength was finally broken and its superiority disproved, that confidence remained. China is a world unto itself.

This is not to suggest that the Chinese themselves did not deliberately contribute to the isolation of their country from foreign contacts and influences. Indeed, a certain degree of isolation is itself an integral part of Chinese tradition, and spatially the Chinese were certainly in a position to sustain this posture. To China's north lie the great mountain ranges of eastern Siberia and the barren country of Mongolia. To the northwest, beyond Xinjiang (Sinkiang), the mountains open—but onto the vast Kirghiz

**Figure 9–1**
The Gate of Heavenly Peace on Beijing's Tiananmen Square, where Mao Zedong proclaimed the birth of the People's Republic.

SYSTEMATIC ESSAY

# Geography of Development

Spatial perspectives on the multidisciplinary field of **development studies** are an important concern in regional economic geography. The brief introduction to this subject in the opening chapter (pp. 34–38) differentiated between developed countries (DCs) and underdeveloped countries (UDCs). It showed that the widening development gap that separates DCs and UDCs is deeply rooted in the colonial era, which saw the rapidly advancing Western countries gain a decisive head start economically while their little-changing colonies remained the suppliers of raw materials and the consumers of manufactured goods. The inequalities of wealth, advantage, and human well-being that resulted from this schism, of course, constitute a continuing crisis of great magnitude, because today 70 percent of the world's population resides in the UDCs of the Third World realms.

Underdevelopment, as we also saw in the Introduction, is characterized by a number of specific symptoms. *Demographically*, UDCs exhibit rapidly increasing populations, relatively short life expectancies, and frequently overcrowded living conditions. *Health and nutrition standards* are modest at best, with widespread hunger and disease, high rates of infant mortality and dietary deficiencies, poor sanitation, and primitive health care services. Among the symptomatic *social ills* are rigid, elite-dominated class structures that retard upward mobility, inferior educational systems, and high illiteracy rates. *Economic problems* are pervasive, including far-reaching poverty and unemployment, inferior circulation systems, internal regional imbalances, and labor forces in which women and children perform grueling physical work. *Rural areas* are often dominated by low-yield subsistence agriculture, and inefficient land-tenure practices and farming techniques; *urban centers* are marked by massive overcrowding, a chronic lack of sufficient jobs and human-services provision, and abysmal housing conditions in both city slums and peripheral shantytowns.

The geography of development is also a theme pursued further in our discussion of the underdeveloped realms (Chapters 4–10). As we saw, the search for meaningful generalizations has led to the building of several spatial models; most approach development as a *center-periphery* process wherein the spatial-organization changes introduced by modernization systematically spread outward from coastal urban agglomerations (or other foreign contact points) toward increasingly remote rural areas. The Taaffe-Morrill-Gould model of colonial-era transport network development (pp. 402–403) typifies this approach, and is one cornerstone of a body of research that focuses on spatial structures and impacts that result from the diffusion of modernization within Third World countries. Many planning strategies are based on similar conceptualizations—for instance, the Growth Pole model discussed in the vignette on Brazil (pp. 320–321) presumes that the beneficial spreading effects of development will radiate across the hinterland of a promising young economic center if sufficient capital investments to spur industrial growth are made there. Among the more comprehensive global models of the development process is the five-stage growth theory proposed by the economist Walt Rostow. This model created a furor when it appeared in the early 1960s, because it claimed that all countries follow a similar path through a number of interrelated growth stages; although controversy continues to surround the Rostow model, after two decades, it has still not been displaced in the social science literature and is still frequently discussed in the writings of human geographers.

The Rostow model is based on the progressive absorption of technology into a national economy, and every country in the world may be measured and assigned to a position within one of the five stages of growth. The initial stage is the *traditional society*, which exhibits limited local economic activity (mainly subsistence farming), low levels of technology, a rigid social structure, and a pervasive resistance to change. Those societies able to overcome such obstacles enter the transitional phase of Stage 2, *preconditions for takeoff.* Led by an avant-garde elite, national unity is achieved, birth rates are lowered, and the rigid class structure gives way to a more open society. Meanwhile, the agriculture-dominated traditional economy begins to diversify as goods- and services-production increases, transportation and communications facilities are improved, and the once-hostile physical environment is increasingly managed to maximize its usefulness for human endeavors. If sustained, self-perpetuating growth becomes the norm, the third stage—*takeoff*—is reached. This is the industrial revolution phase of development, involving not only technological and mass-production breakthroughs, but also the slowing of population growth, rising rates of investment, the steady expansion of manufacturing and urbanization, and a growth-oriented social and political framework. If successful, a country enters the fourth stage—*the drive to maturity*. This phase entails several decades of persistent progress marked by the nationwide diffusion of new technologies, growing involvements in advantageous international trade, and increasingly specialized industrial development backed by high investment rates and an ever more skilled labor force. Finally, a select few of the most advanced countries reach the fifth stage of *high mass consumption*, characterized by high per capita incomes, the dominance of consumer goods and services production, and a labor force in which tertiary and quaternary workers predominate. Clearly, the achievement of a *postindustrial* society represents the culmination of this overall growth process, perhaps even comprising a new sixth stage as such countries as the United States and Japan now amass unprecedented combinations of advanced economic activity and ultrasophisticated technology.

On the world map, the traditional-society Stage 1 UDCs are most heavily concentrated in Equatorial Africa, the Arabian Peninsula, Southeast Asia, mainland Middle America, and interior South America. The changing UDCs of the preconditions-for-takeoff Stage 2 are found in East and West Africa, North Africa and the Middle East, South Asia, and northwestern South America. Countries at the takeoff Stage 3—which arguably comprise a middle tier of "emerging" states on the UDC/DC continuum—are Mexico and Cuba in Middle America; Argentina, Brazil, Chile, Uruguay, and Venezuela in South America; South Africa; Egypt and Turkey in Southwest Asia; China and both Koreas in East Asia; Spain, Portugal, and Greece in Mediterranean Europe; and Yugoslavia, Hungary, Romania, and Bulgaria in Eastern Europe. The Soviet Union is the leading example of a Stage 4 drive-to-maturity DC in the 1980s; other candidates are many of the Western and Northern European countries that have overcome the economic setbacks of two world wars in this century. The only five countries that have fully entered the high-mass-consumption Stage 5 are the United States and Japan (also the only two postindustrial societies), Canada, Australia, and West Germany; other European states such as the Netherlands, Switzerland, and the three Scandinavian nations are at the threshold of Stage 5, but ongoing industrial decline may well block the United Kingdom, France, Italy, and Belgium from this achievement in the foreseeable future.

The diagnosis of a country's current level of development and its prospects for the future is a complicated task, one that far transcends a superficial pulse taking by various quantitative measures (see *Box*, p. 37). A good example, particularly appropriate for introducing Chapter 9 (in which the topic is treated at some length), is the appraisal of China's potential provided by Clifton Pannell and Laurence Ma in their 1983 book, *China: The Geography of Development and Modernization*. Despite impressive recent growth, a number of problems are recognized that will require amelioration if ongoing progress is to be sustained. *Environmental difficulties* include chronic water shortages and the limitations to arable land—which covers only about 10 percent of China but is already utilized at near-capacity levels. The *demographic dilemma* is formidable—trying further to reduce the steady increase of the world's largest population cluster. *Geographic dimensions* continue to be a hindrance, especially the overcoming of huge distances with a less than adequate transport and communication system; but the infrastructure must be improved if the country's natural and human resources are to be more effectively marshaled, its regional inequities equalized, and its spatial economy integrated into a well-organized whole. *Economic problems* center on low productivity and the widespread persistence of poverty; the "Four Modernizations" program (elaborated in this chapter) now seems to offer considerable promise, but its success will depend on sustained agricultural progress and the stabilizing of industrialization from a sporadic to a continuous developmental thrust. Perhaps the most important need of all is the elimination of *political instability*, following the recent upheavals of the Cultural Revolution and the Gang of Four episode. The new regime seems well on its way toward providing the needed atmosphere; with its novel encouragement of individual incentives and rewards, less centralized planning, and foreign trade and technological infusions, the outlook for truly meaningful development is the rosiest since the 1949 Communist Revolution.

The goals of national development are succinctly stated by Anthony de Souza and Philip Porter in their 1974 survey of modernization trends in the Third World: (1) a balanced healthful diet, (2) adequate medical care, (3) environmental sanitation and disease control, (4) labor opportunities commensurate with individual talents, (5) sufficient educational opportunities, (6) individual freedom of conscience and freedom from fear, (7) decent housing, (8) economic activities in harmony with the natural environment, and (9) a social and political milieu promoting equality. But the realities of the contemporary world do not offer much hope for improving the quality of life in the UDCs by achieving these objectives anytime soon. The DCs have been thrust into a world system of exchange and capital flow over which they have no control, leaving Mexico, Brazil, and many other UDCs in the middle 1980s prone to such economic disasters as the rapid accumulation of stupendous foreign debts during a time of global recession and skyrocketing interest rates. Thus, as present conditions prevail, the UDCs are likely to fall further behind, not because they resist "progress" but because the character of world development has been that the DCs, already way ahead, will make the gap still bigger. And so the DCs, if they are serious about the role they say the UDCs are to play in preserving a stable and secure world, must respond to the demands of the UDCs that the discriminatory system of international commerce be changed, that the exports of the UDCs be given a greater chance and greater returns.

Steppe. To the west and southwest lie the Tian Shan and Pamirs, the forbidding Plateau of Xizang (Tibet), and the Himalayan wall. And to the south there are the mountain slopes and tropical forests of Southeast Asia—the same that mark India's eastern flank with Burma (Fig. 9–2). But more telling even than these physical barriers to contact and interaction is the factor of distance. China has always been far from the modern source areas of industrial innovation and technological change. True, China itself was such an area, but China's contributions to the outside world were very limited, and European capacities and ideas did not change China as they transformed other societies. China did interact to some extent with Japan and with areas of Southeast Asia. But compare this to the Arabs, who ranged far and

Figure 9–2

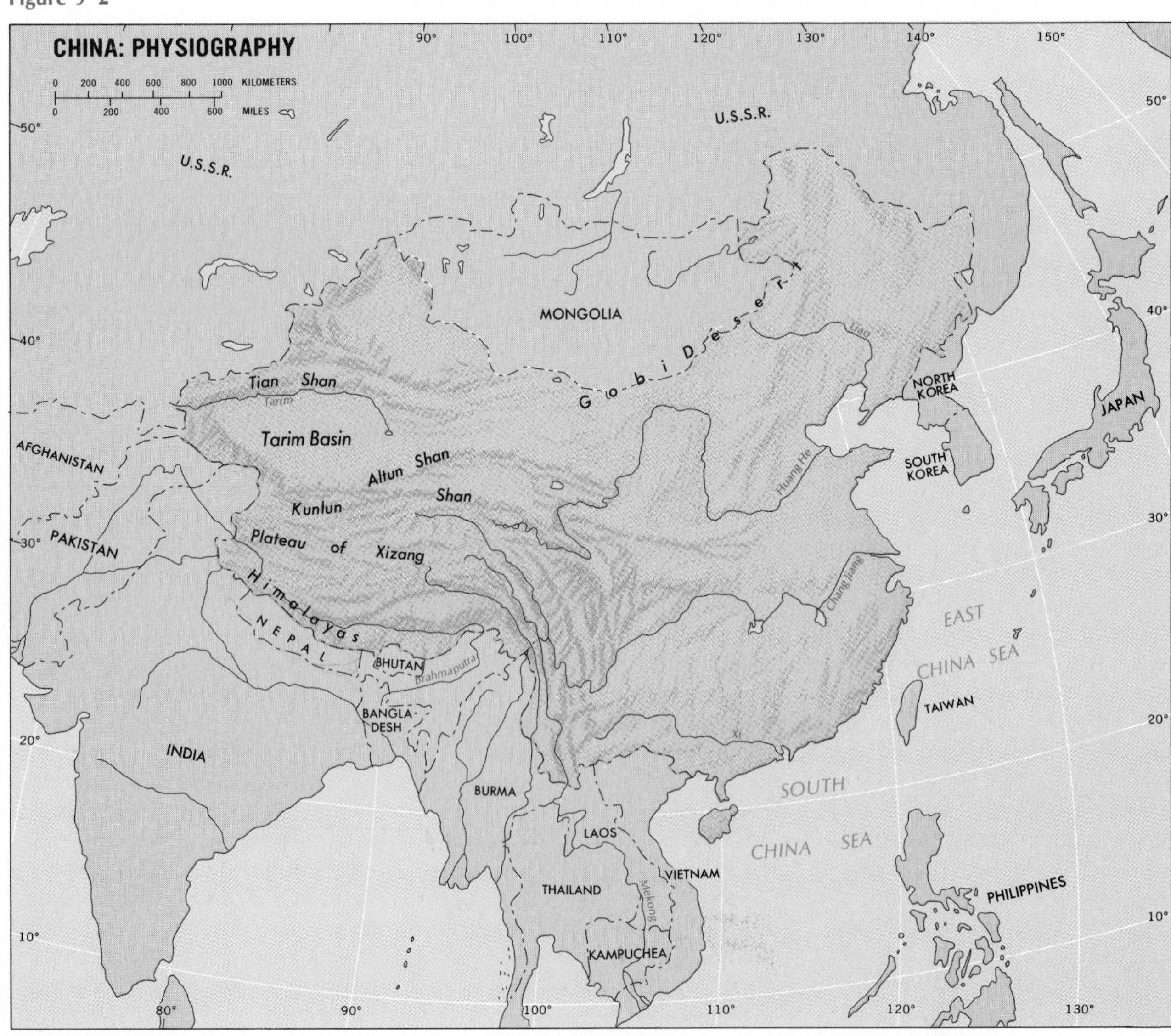

# TEN MAJOR GEOGRAPHIC QUALITIES OF CHINA

1. China's population represents over one-fifth of all humanity; territorially, China ranks among the world's three largest states.
2. China is one of the world's two oldest continuous civilizations.
3. The Chinese state and national culture evolved from a core area that emerged in the north, near the present capital of Beijing (Peking), and China's culture hearth has remained there ever since.
4. China's civilization developed over a long period in considerable isolation, protected by physiographic barriers and by sheer distance from other source areas.
5. Foreign intervention had disastrous impacts on Chinese society, from European colonialism to Japanese imperialism; intensified regionalism and territorial losses are only two of many resulting afflictions.
6. China occupies the eastern flank of Eurasia; it sphere of influence was reduced by Russian expansionism in East Asia.
7. China's enormous population is strongly concentrated in the country's eastern regions; western zones remain comparatively empty and open, and are also more arid and less productive.
8. China's Communist-designed transformation after World War II involved unprecedented regimentation and the imposition of effective central authority, with results that are perhaps permanently imprinted on the cultural landscape.
9. China's recent modernizing drive notwithstanding, the country remains a dominantly rural society with limited urbanization and industrialization.
10. Rural China is a land of enduring traditions. Neither the Communist Revolution nor the modernization drive has truly changed the villagers' way of life. Many old values persist, and Confucian teachings are still remembered.

wide and who brought their religion and political influence to areas from Southern Europe to Indonesia, India to East Africa. When Europe became the center of change, China lay farther away by sea than almost any other part of the world, farther even than the coveted Indies.

Today, modern communications notwithstanding, China is still distant from almost anywhere else on earth. Going by rail from the heartland of China's Eurasian neighbor, the Soviet Union, is a long and tedious journey of several days. Direct overland communications with India are virtually nonexistent. Communications with Southeast Asia are still tenuous. And China's long Pacific coastline adjoins the world's largest ocean; Japan is China's only nearby, industrialized trading partner.

# China in the World Today

For all its isolation and remoteness, China has moved to the center stage of world attention in recent years. It was Napoleon who long ago remarked that China was asleep and whoever would awaken the Chinese giant would be sorry. Today, China is wide awake. It was stung by Japanese aggression in the 1930s and 1940s; after the end of World War II, the growing Communist tide took power following a bitter civil war in which the United States supported the losing side. European colonialism, Japanese imperialism, and a communist ideology combined to stir China into action. The foreigners were ousted, China's old order was rejected and destroyed, and since 1949 the Communist regime has been engaged in a massive effort to remake China in a new image of unity and power.

This effort has taken China from a position of backwardness and weakness to one of considerable strength. China is not yet a superpower equal to the United States or the Soviet Union. But China is on the move and, by almost any combination of measures, now ranks in third place. What is most remarkable about all this, of course, is that after barely four decades of Communist control, China has been able to place itself in a position of serious potential contention for world power, something that hardly seemed possible at the end of World War II. Chinese assistance is going to countries in Asia and America; Chinese technicians have built railroads in Africa. At a time when the two greatest powers on earth are attempting to achieve a world in which they can coexist, China looms as a new threat to this joint monopoly over ultimate power.

China's potential for world influence was underscored by its initial explosion of an atomic device in the 1960s, and its subsequent development of a nuclear arsenal as well as an expanding capacity to deliver these weapons over a long range. After the Communist takeover in 1949, the Chinese were assisted by the Soviets in their industrial and military development, but an ideological quarrel ended Sino-Soviet cooperation and the Russian technicians were sent home. The Chinese accused the Soviets of "revisionism"—the dilution of Communist ideology with doses of capitalism. So it was that China, awakened to communism by the revolutionary example of Russia, took on the mantle of the "purest" of Communist systems and rejected the modern Soviet version. When anti-Soviet feeling in China reached its peak in the late 1960s and Soviet and Chinese military forces exchanged shots across the Amur and Ussuri Rivers, it was not difficult to hear the echo of Napoleon's famous words.

Momentous changes came to China during the 1970s. Following China's entry into the United Nations in 1971 (replacing the delegation from the nationalist holdout island of Taiwan), a U.S. president visited Beijing (Peking) in 1972. Diplomatic relations with Japan were established, and several trade agreements followed; China was an important oil supplier to Japan during the energy crises of the 1970s. In 1976, both Premier Zhou Enlai (Chou En-lai) and the architect of modern communist China, Mao Zedong (Mao Tse-tung), died, and a power struggle ensued after Mao's death. This struggle involved "moderate" and "radical" factions of the Communist party and had been building even before Mao's death. The basic issues centered on the course of economic development and the future of education in China. The pragmatic moderates wanted to speed economic growth

# THE *PINYIN* SYSTEM

If a Western traveler in China were to stop and ask for directions to Peking, the response would quite probably be a shake of the head. The Chinese call their capital Beijing, not Peking. For centuries the romanization of Chinese has led to countless errors of this kind, in part because Chinese dialects are also inconsistent and the Chinese in northern China may pronounce a place name somewhat differently from those living in the south.

In 1958, the Chinese government adopted the *pinyin* system of standard Chinese—not to teach foreigners how to spell and pronounce Chinese names and words, but to establish a standard form of the Chinese language throughout China. The *pinyin* system is based on the pronunciation of Chinese characters in Northern Mandarin, the Chinese spoken in the region of the capital and the north in general.

During the 1970s, China's contact with the outside world expanded, but the *pinyin* system was not in use elsewhere. Of course the Chinese preferred the new standard, which represented not only a linguistic modernization but also something of a psychological break with the past. Since 1979, all press reports from China have only contained the *pinyin* spelling of Chinese names. In the United States and in other parts of the world, newspapers and magazines began to adopt the system. Atlases, wallmaps, and other geographic materials (including textbooks) were revised. In some instances, the change was resisted, and the old spellings persist. But *pinyin*, undoubtedly, will become the world standard for romanized Chinese.

Strange as the familiar Chinese place names seem when transcribed into *pinyin*, the new system is not difficult to learn or understand. Most of the sounds are clear to someone who speaks English, except the *q* and the x. The *q* sounds like the *ch* in *cheese*; the *x* is pronounced *sh* as in *sheer*. Thus, the Chinese spell China *Xinhua* (which actually means "new China"), to more closely reflect how the name should sound.

Some familiar place names, spelled the old and the *pinyin* way, are listed below:

| | |
|---|---|
| Peking | Beijing |
| Shanghai | Shanghai |
| Canton | Guangzhou |
| Tibet | Xizang |
| Sinkiang | Xinjiang |
| Yangtze Kiang | Chang Jiang (River) |
| Nanking | Nanjing |
| Szechwan | Sichuan |
| Hwang Ho | Huang He (River) |
| Tientsin | Tianjin |

And some personal names:

| | |
|---|---|
| Mao Tse-tung | Mao Zedong |
| Chou En-lai | Zhou Enlai |
| Teng Hsiao-ping | Deng Xiaoping |

In this chapter both the old and new *pinyin* spellings are given as needed, so that it will be easy to become familiar with the new names. The maps have been revised to conform to the *pinyin* system. In the text, the *pinyin* version is given first and the old spelling is shown in parentheses.

by offering workers some material rewards; they also proposed that the schools spend less time on ideological teaching and more on practical technical training. The radicals feared that such changes would weaken China's "pure" communism and produce new elite people, setting back three decades of class struggle. Eventually, the pragmatists prevailed; by the opening of the 1980s, a new China had begun to emerge. Thousands of American visitors were admitted, and diplomatic relations returned to normal. In the 1979 municipal elections, non-Communist as well as Communist candidates were allowed to compete for office. And China's planners laid out a new course for the new China: the "Four Modernizations." In industry, defense, science, and agriculture, China opened an era of modernization as rapid as the country could achieve.

On July 28, 1976, two devastating earthquakes struck the northeast of China. Perhaps as many as 700,000 people died as the important industrial city of Tangshan in Hebei (Hopeh) Province was destroyed (Fig. 9–3); also damaged were the port of Tianjin (Tientsin, China's third largest city), and the capital, Beijing (Peking). It was by far the worst disaster of its kind in this and the preceding three centuries, and it dealt a staggering blow to China's economy, which was already burdened by the ideological struggle and associated labor problems. The 1970s in China were turbulent years indeed.

China's expanding role in the world is a major portent for the future. Its government is opposed to what it views as the socialist imperialism of the Soviet Union as well as the capitalist imperialism of the United States, but it has come to consider the Soviet threat the greater. A major reorientation took place when the Chinese joined the United Nations, invited U.S. leaders for discussions, normalized relations with Japan—all this despite the continuing Vietnam War (then still in progress) and the U.S. presence in Taiwan and South Korea. But the map of East Asia helps explain some of China's recent actions. As the widened Vietnam or Indochina War came to a close in the mid 1970s, the Soviet presence in Southeast Asia strengthened even while the Americans withdrew. Although there are large Chinese minorities in Southeast Asian countries and the Chinese had given assistance to Communist insurgencies during the war in Vietnam, much anti-Chinese hostility exists in that geographic realm. The Soviets capitalized on the situation to begin developing a sphere of influence there, a prospect that greatly concerned China. And, of course, China lies an ocean away from California's beaches, but only a stone's throw from Soviet soil. It was Russian imperialism that cost China dearly, not America's. It is Soviet territory that China claims on historic grounds as its own, not American soil. And as China's capacity for war grows, it is Russia that will lie in its reach first; even in this age of intercontinental missiles, an ocean provides a time cushion and, therefore, constitutes an advantage to the technologically superior foe. That was proved by the Cuban missile crisis during the Kennedy administration; rockets poised to strike the United States from Russia were one thing, but rockets aimed at America from Cuba were something else. China is testing rockets and bombs, but it still does not have the capacity to strike effectively—it does not as yet possess a sufficient arsenal. When

**Figure 9–3**
Damage to the capital, Beijing, following the catastrophic 1976 earthquake that leveled Tangshan 150 kilometers/95 miles to the east.

the time comes, will the Soviet Union tolerate missiles poised to strike it from Xinjiang (Sinkiang) any more than the United States did in the case of Cuba? Thus, China's approaches to the United States are a matter of self-interest, consistent with the policies of the pragmatic moderates who are now trying to speed the country's economic development along revisionist lines.

# Evolution of the State

China may have developed in comparative isolation for more than 4000 years, but there is evidence that China's earliest core area, which was positioned around the confluence of the Huang He (Yellow) and the Wei Rivers (Fig. 9–4), received stimuli from other distant, and possibly slightly earlier civilizations—the river cultures of Southwest and South Asia. From Mesopotamia and the Indus, techniques of irrigation and metalworking, innovations in agriculture, and possibly even the practice of writing reached the Yellow River Basin, probably overland along the almost endless routes across interior deserts and steppes. The way these early Chinese grew their rice crops gives evidence that they learned from the Mesopotamians; the water buffalo probably came from the Indian subcontinent. But quite soon the distinctive Chinese element began to appear; by the time the record becomes reliable and continuous, shortly after 1800 B.C., Chinese cultural individuality was already strongly established.

The oldest dynasty of which much is known is the Shang Dynasty (sometimes called Yin), centered in the Huang He-Wei confluence from perhaps 1775 B.C. to about 1125 B.C. Walled cities were built during the Shang period, and the Bronze Age commenced during Shang rule. For more than a thousand years after the beginning of the Shang Dynasty, North China was the center of development in this part of Asia. The Chou Dynasty (about 1120–221 B.C.) sustained and consolidated what had begun during the Shang period.

Eventually, agricultural techniques and population numbers combined to press settlement in the obvious direction—southward, where the best opportunities for further expansion lay. During the brief Chin Dynasty, the lands of the Chang (Yangtze) Jiang (*jiang* means "river") were opened up, and settlement spread even as far southward as the lower Xi Jiang (Hsi Kiang or West River). A pivotal period in the historical geography of China lay ahead: the Han Dynasty (202 B.C.–A.D. 220). The Han rulers brought unity and stability to China, and they enlarged the Chinese sphere of influence to include Korea, the Northeast, Mongolia, Xinjiang (Sinkiang), and, in Southeast Asia, Annam (located in what is today Vietnam). This was done to establish control over the bases of China's constant harassers—the nomads of the surrounding steppes, deserts, and mountains—and to protect (in Xinjiang) the main overland avenue of westward contact between China and the rest of Eurasia.

The Han period was a formative one in the evolution of China. Not only was Chinese military power stronger than ever before, but there were also changes in the systems of land ownership as the old feudal order broke down and private, individual property was recognized, and the silk trade grew into China's first external commerce. To this day, most Chinese, recognizing that much of what is China first came about during this period, still call themselves the "people of Han."

Han China was the Roman Empire of East Asia. Great achievements were made in architecture, art, the sciences, and other spheres. Like the Romans, the Han Chinese had to contend with unintegrated hostile peoples on their empire's margins. In China, these peoples occupied the mountainous country south of the national territory, and the Han rulers built military outposts to keep these frontiers quiet. Again, like the Roman Empire, the China of Han went into decline and disarray, and more than a dozen states arose after A.D. 220 to compete for primacy. Not until the Sui Dynasty (581–618) did consolidation begin again, to be continued during the Tang Dynasty (618–906), another period of national stability and development.

Following the Tang period, when China once again was a great national state, the Song Dynasty (960–1279) carried the culture to unprecedented heights, in spite of continuing problems with marauding nomadic peoples. This time the threat came from the north; in 1127, the Song rulers had to abandon their capital at Kaifeng on the Huang He in favor of Hangzhou (Hangchou) in the southeast. Nevertheless, China during the Song Dynasty was in many ways the world's most advanced state; it had several cities with more than 1 million people, paper money was in use, commerce intensified, literature flourished, schools multiplied in number, the arts thrived as never before, and the philosophies of Confucius, for many centuries China's guide, were modernized, printed, and mass-distributed for the first time. Eventually, Song China fell to conquerors from the outside, the Mongols led by Kublai Khan. The Mongol authority, known as the Yuan Dynasty (1279–1368), made China a part of a vast empire that extended all the way across Asia to Eastern Europe. But it lasted less than a century and had little effect on Chinese culture. In-

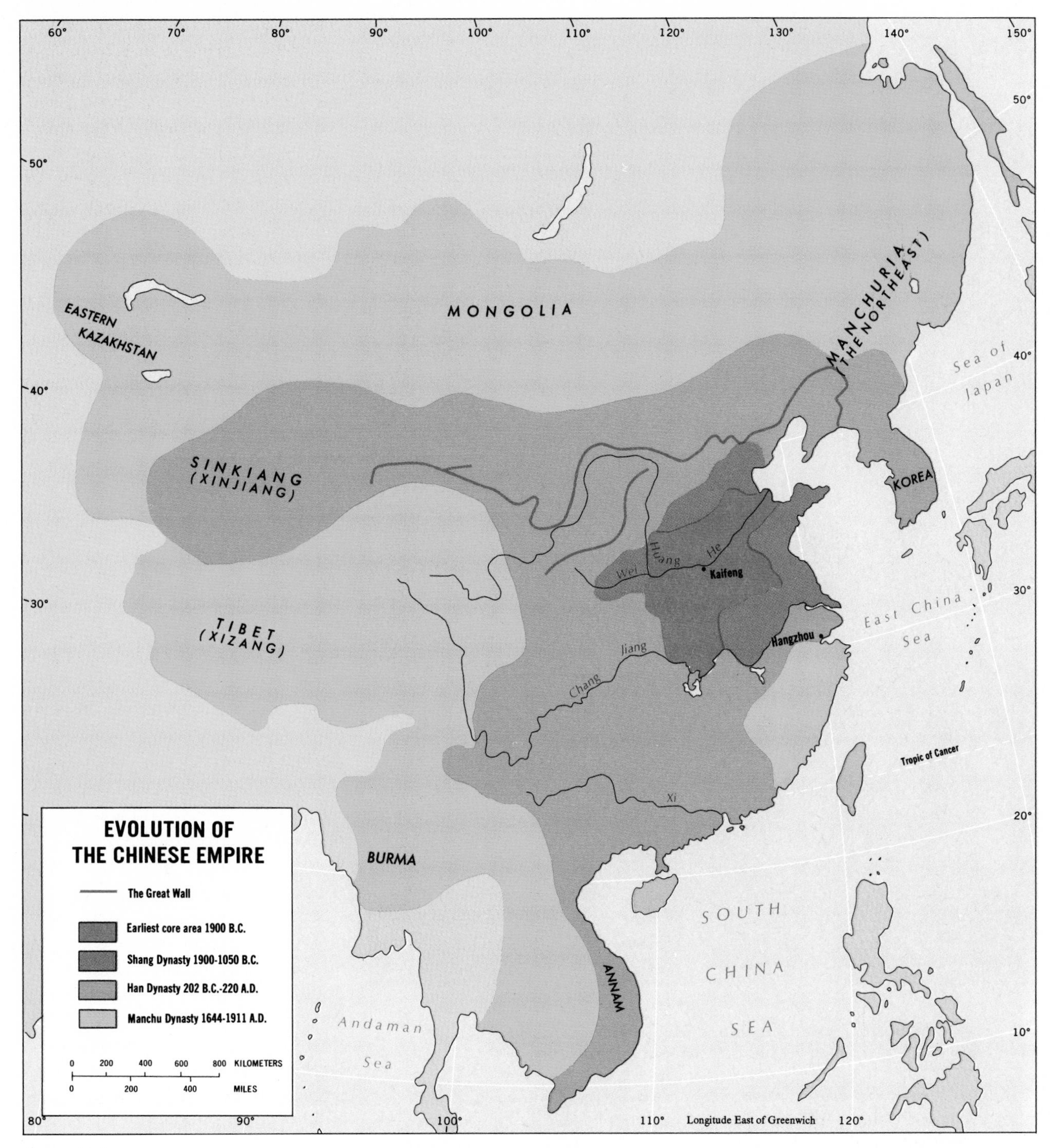


Figure 9–4

stead, it was the Mongols who adopted Chinese civilization.

In 1368, a Chinese local ruler led a rebellion that ousted the Mongols, and this signaled the ascent of a great indigenous dynasty, the Ming Dynasty (1368–1644). Under the Ming rulers, China's greatness was restored, its territory once again consolidated from the Great Wall in the north (Fig. 9–5) to Annam in the south. Finally, in 1644, another northern, foreign nomadic people forced their way to control over China. They came from the far Northeast (see *Box*); unlike the Mongols, they sustained and nurtured Chinese traditions of administration, authority, and national culture. The Manchu (or Qing) Dynasty (1644–1911) extended the Chinese sphere of influence to include Xinjiang and Mongolia, a large part of southeastern

**Figure 9–5**
The Great Wall of China, completed ca. 200 B.C., was an attempt to separate eastern China's farmers from the pastoral herders of the interior.

Siberia, Xizang (Tibet), Burma, Vietnam, eastern Kazakhstan and Korea (Fig. 9–4). But the Manchus also had to contend with the rising power of Europe and the West and the encroachment of Russia. Large portions of the Chinese spheres in Siberia, interior Asia, and Southeast Asia were lost to China's voracious competitors.

It is often said that the year 1949, when Communist Party Chairman Mao Zedong proclaimed the People's Republic of China at Beijing's (Peking's) historic Gate of Heavenly Peace (Fig. 9–1), marked the beginning of a new dynasty that is not so different from the old—Communist doctrine notwithstanding. Certainly, some of China's old traditions continue in the new Communist era, but, in many more ways, the new China is a totally overhauled society. Benevolent or otherwise, the rulers of old China headed a country in which—for all its splendor, strength, and culture—the fate of the landless and the serf was often indescribably miserable, famine and disease could decimate the population of entire regions, local lords could repress the people with impunity, children were sold, and brides were bought. Contact with Europe, however peripheral, brought ruin to Chinese urban society where slums, starvation, and deprivation were commonplace. Mao's long term and apparent omnipotence may remind us of the dynastic rulers' frequent longevity and absolutism, but the new China he left behind is quite different from the old.

# A Century of Convulsion

China long withstood the European advent in East Asia with a self-assured superiority, based on the strength of its culture and the continuity of the state. There was no market for the British East India Company's rough textiles in a country long used to finely fabricated silks and cottons. There was little interest in the toys and trinkets the Europeans produced in the hope of bartering for Chinese tea and pottery. And even when Europe's sailing ships made way for steam-driven vessels and newer and better factory-made textiles were offered in trade for China's tea and silk, China continued to reject the European imports that were still, initially at least, too expensive and too inferior to compete with China's handmade materials. Long after India had fallen into the grip of mercantilism and economic imperialism, China was able to maintain its established order. This was no surprise to the Chinese; after all, they had held a position of undis-

# MANCHURIA: THE NORTHEAST

Three provinces, Liaoning, Jilin (Kirin), and Heilongjiang (Heilungkiang), constitute China's Northeast, a region bounded by the Soviet Union on the north and east, Mongolia in the west and North Korea in the southeast (see Fig. 9–20 p. 490). This was the home of the conquering Manchus and was later the scene of foreign domination. Russians and Japanese struggled for control over the area; ultimately, Japan created a dependency here in the 1930s that it called *Manchukuo*. More recently, the region came to be called *Manchuria*, but this is not an accepted regional appellation in China. Chinese geographers refer to the region encompassed by these three northeastern provinces as simply the *Northeast*; this practice is followed in the present discussion.

puted superiority among the countries of Eastern Asia as long as could be remembered, and they had dealt with invaders from land and from the sea before.

The nineteenth century shattered the self-assured isolationism of China as it proved the superiority of the new Europe. On two fronts, the economic and the political, the European powers destroyed China's invincibility. In the economic sphere, they succeeded in lowering the cost and improving the quality of manufactured goods, especially textiles, and the handicraft industry of China began to collapse in the face of unbeatable competition. In the political sphere, the demands of the British merchants and the growing English presence in China led to conflicts. In the early part of the nineteenth century, the central issue was the importation into China of opium, a dangerous and addictive intoxicant; opium was destroying the very fabric of Chinese culture, weakening the society, and rendering China an easy prey for colonial profiteers. As the Manchu government moved to stamp out the opium trade in 1839, armed hostilities broke out and soon the Chinese sustained their first defeats. Between 1839 and 1842, the Chinese fared badly, and the first "Opium War" signaled the end of Chinese sovereignty. British forces penetrated up the Chang Jiang (Yangtze) and controlled several areas south of that river; Beijing hurriedly sought a peace treaty. As a result, leases and concessions were granted to foreign merchants. Hong Kong Island was ceded to the British, and five ports, including Guangzhou (Canton) and Shanghai, were opened to foreign commerce. No longer did the British have to accept a status that was inferior to the Chinese in order to do business; henceforth, negotiations would be pursued on equal terms. Opium now flooded into China, and its impact on Chinese society became even more devastating. Fifteen years after the first Opium War, the Chinese tried again to stem the disastrous tide—and, again, they were bested by the foreign parasites that had attached themselves to their country. Now the cultivation of the opium poppy in China itself was legalized. Chinese society was disintegrating, and the scourge of this narcotic was not defeated until the revival of Chinese power in the early twentieth century.

But before China could reassert itself, much of what remained of China's sovereignty was steadily eroded away (see *Box*). The Germans obtained a lease on Qingdao (Tsingtao) in 1898, and in the same year the French acquired a sphere of influence at Kwanchowan, now Zhanjiang (Fig. 9–6). The Portuguese took Macao, the Russians obtained a lease on Liaodong (Liaotung) in the Northeast as well as railway concessions there, and even Japan got into the act by annexing the Ryukyu Islands and, more importantly, Formosa (Taiwan) in 1895. After millennia of cultural integrity, economic security, and political continuity, the Chinese world lay open to the aggressions of foreigners whose innovative capacities China had denied to the end. But now ships flying European flags lay in the ports of China's coasts and rivers; the smokestacks of foreign factories rose above the landscapes of its great cities. The Japanese were in Korea, which had nominally been a Chinese vassal; the Russians had entered China's Northeast. The foreign invaders even took to fighting among themselves, as did Japan and Russia in Manchuria (as the foreigners now called the Northeast) in 1904.

## Rise of a New China

In the meantime, organized opposition to the foreign presence in China was emerging, and this century opened with a large-scale revolt against all outside elements. Bands of revolutionaries roamed the cities as well as the countryside, attacking not only the hated foreigners but also Chinese citizens who had adopted Western cultural traits. Known as the Boxer Rebellion (after a loose translation of the Chinese name for these revolutionary groups), the 1900 uprising was

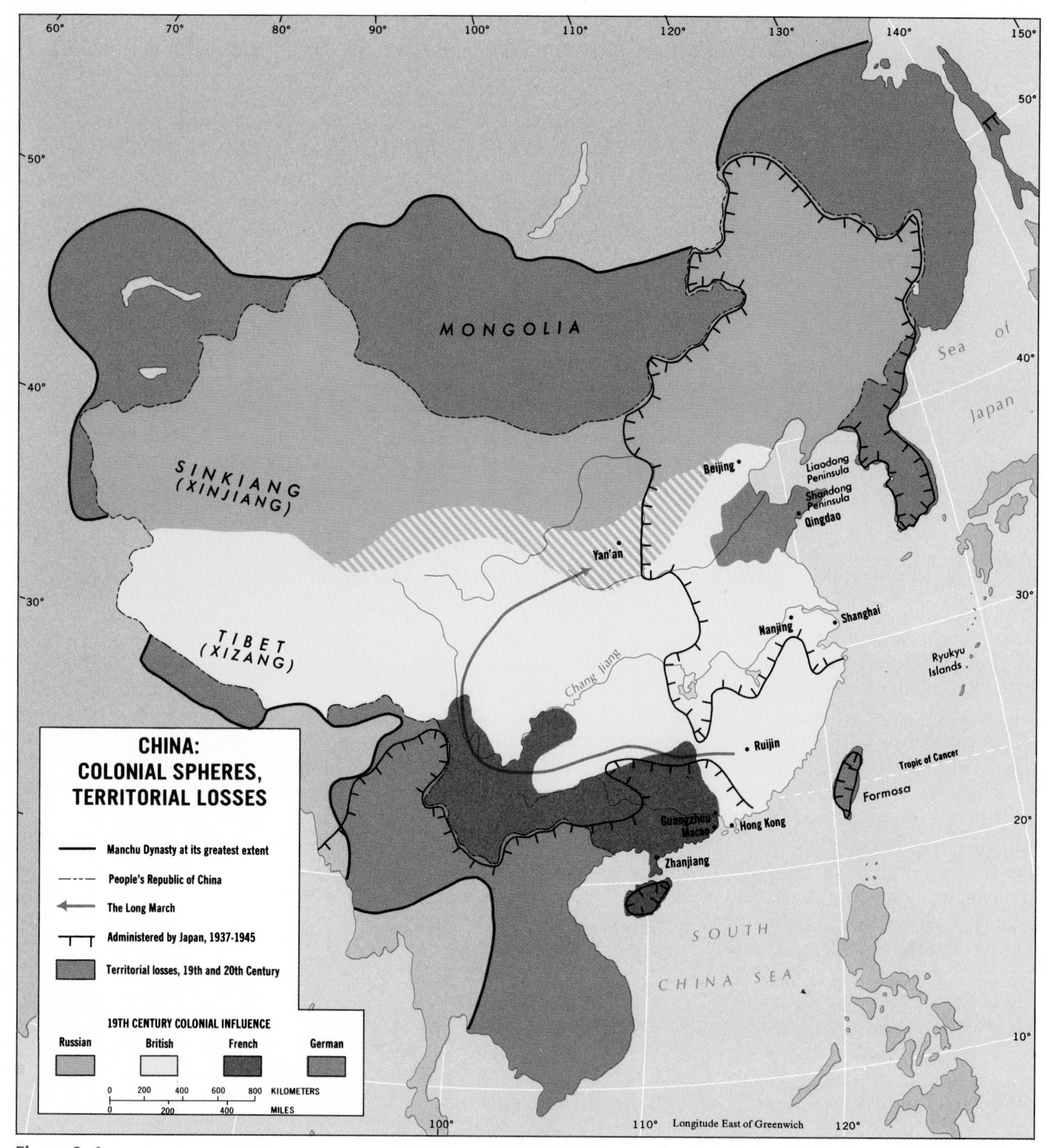

Figure 9–6

put down with much bloodshed. Simultaneously, another revolutionary movement was gaining strength, aimed against the Manchu leadership itself. In 1911, the emperor's garrisons were attacked all over China, and, in a few months, the two-and-a-half-century-old dynasty was overthrown. Indirectly, it too was yet another casualty of the foreign intrusion, and it left China divided and disorganized.

The end of the Manchu era and the proclamation of a republican government in China did little to improve the country's overall position. The Japanese captured Germany's holdings on the Shandong (Shantung) Peninsula, including the city of Qingdao (Tsingtao) during World War I; when the victorious European powers met at Versailles to divide the territorial spoils, they confirmed Japan's rights in the

## EXTRATERRITORIALITY

Figure 9–7
Sha Mian Island in Guangzhou (Canton), a colonial-era enclave where foreigners resided beyond the jurisdiction of the Chinese.

A sign of China's weakening during the second half of the nineteenth century was the application in its cities of a European doctrine of international law, *extraterritoriality*. This principle originated with the French Jurist Ayraut (1536–1601); it denotes a situation in which foreign states or international organizations and their representatives are immune from the jurisdiction of the country in which they are present. This, of course, constitutes an erosion of the sovereignty of the state hosting these foreign elements, especially when the practice goes beyond the customary immunity of embassies and persons in diplomatic service. In China, extraterritoriality reached unprecedented proportions. The best residential suburbs of the large cities, for example, were declared to be "extraterritorial" parts of foreign countries and were made inaccessible to Chinese citizens. Sha Mian Island in the Pearl River in Canton (Guangzhou) was a favorite extraterritorial area of that city (Fig. 9–7). A sign at the only bridge to the island stated "No Dogs or Chinese" and still stands today as a reminder of foreign excesses. In this way, the Chinese found themselves unable to enter their own public parks and many buildings without permission from foreigners. Christian missionaries fanned out into China, their bases fortified with extraterritorial security. To the Chinese, this involved a loss of face that contributed to the bitter opposition against the presence of all foreigners, a resentment that finally exploded in the Boxer Rebellion of 1900.

area. This led to another significant demonstration of Chinese reassertion as nationwide protests and boycotts of Japanese goods were organized in what became known as the May Fourth Movement. One participant in these demonstrations was a young man named Mao Zedong (Mao Tse-tung).

Nevertheless, China after World War I remained a badly divided country. By the early 1920s, there were two governments—one in Beijing (Peking) and another in the southern city of Guangzhou (Canton) where the famous Chinese revolutionary, Sun Yat-sen, was the central figure. Neither government could pretend to control much of China. The Northeast was in complete chaos, petty states were emerging all over the central part of the country, and the Guangzhou (Canton) "parliament" controlled only a part of Guangdong (Kwangtung). Yet, it was just at this time that the power groups that were ultimately to struggle for supremacy in China were formed. While Sun Yat-sen was trying to form a viable Nationalist government in Guangzhou (Canton), the Chinese Communist party was formed by a group of intellectuals in Shanghai. Several of these intellectuals had been leaders in the May Fourth Movement, and, in the early 1920s, they received help from the Communist party of the Soviet Union. Mao Zedong was already a prominent figure in these events.

Initially, there was cooperation between the new Communist party and the Nationalists led by Sun Yat-sen. The Nationalists were stronger and better organized, and they hoped to use the Communists in their antiforeign (especially anti-British) campaigns. By 1927, the foreigners were on the run; the Nationalist forces entered cities and looted and robbed at will while aliens were evacuated or, failing that, sometimes killed. As the Nationalists continued their drive northward and success was clearly in the offing, internal dissension arose. Soon, the Nationalists were as busy purging the Communists as they were pursuing foreigners, and the central figure to emerge in this period was Chiang Kai-shek. Sun Yat-sen died in 1925, and, when the Nationalists established their capital at Nanjing (Nanking) in 1928, Chiang was the country's leader.

## Three-Way Struggle

The post-Manchu period of strife and division in China was quite similar to other times when, following a lengthy period of comparative stability under dynastic rule, the country fragmented into rival factions. In the first years of the Nanjing government's hegemony, the campaign against the Communists intensified and many thousands were killed. Chiang's armies drove them ever deeper into the interior (Mao himself escaped the purges only because he was in a remote rural area at the time); for a while, it seemed that Nanjing's armies would break the back of the Communist movement in China.

**The Long March** A core area of Communist peasant forces survived in the zone where the provinces of Jiangxi (Kiangsi) and Hunan adjoin in southeastern China (Fig. 9–8), and defied Chiang's attempts to destroy them. Their situation grew steadily worse, however, and, in 1933, the Nationalist armies were on the verge of encircling this last eastern Communist stronghold. The Communists decided to avoid inevitable strangulation by leaving. Nearly 100,000 people—armed soldiers, peasants, local leaders—gathered near Ruijin and started to walk westward in 1934. It was a momentous event in modern China, and among the leaders of the column were Mao Zedong and Zhou Enlai. The Nationalists rained attack after attack on the marchers, but they never succeeded in wiping them out completely; as the Communists marched, they were joined by sympathizers. The Long March (see Fig. 9–6), as this drama has come to be called, first took them to Yunnan Province, where they turned north to enter western Sichuan (Szechwan). They traversed Gansu (Kansu) Province and eventually reached their goal, the mountainous interior near Yan'an (Yenan) in Shaanxi (Shensi) Province. The Long March covered nearly 10,000 kilometers (6000 miles) of China's most difficult terrain, and the Nationalists' continuous attacks killed an estimated 75,000 of the original participants. Only about 20,000 survived the epic migration, but among them were Mao and Zhou who were convinced that a new China would emerge from the peasantry of the rural interior to overcome the urban easterners whose armies could not eliminate them.

**The Japanese** While the Nanjing government was pursuing the Communists, foreign interests made use of the situation to further their own objectives in China. The Soviet Union held a sphere of influence in Mongolia, and was on the verge of annexing a piece of Xinjiang (Sinkiang). Japan was dominant in Manchuria, where it had control over ports and railroads. The Nanjing government tried to resist the expansion of Japan's sphere of influence; it failed, and the Japanese even set up a puppet state in the region. They appointed a ruler and called their new dependency Manchukuo. The inevitable war between the Japanese and Chinese broke out in 1937. There were calls for a suspension of the Nationalist-Communist struggle in the face of the common enemy, but, after a brief armistice, the factional conflict arose again while both sides fought the Japanese. China became divided into three regions: the areas taken by the Japanese during their quick offensive of 1937–1938 (when they took much of eastern China, including the principal ports), the Nationalist zone centered on China's wartime capital of Chongqing (Chungking), and the Communist-held areas of the interior west. The Japanese, by engaging Chiang's forces, gave the Communists an opportunity to build their strength and prestige in China's western regions.

**Victory** After Japan was defeated by the U.S.-led Western powers in 1945, the full-scale Nationalist-Communist struggle was quickly resumed. The United States, hopeful for a stable and friendly government in China, sought to mediate the conflict, but did so while recognizing the Nationalist faction as the legitimate government. Its chances of mediation were impaired by this position and also by the military aid given to the Nationalists' forces. By 1948, it was clear that Mao Zedong's armies would defeat the forces of Chiang Kai-shek and that the final victory was only a matter of time. Chiang kept moving his capital—back to Guangzhou (Canton, where Sun Yat-sen had assembled the first Nationalist government), and then to Chongqing (Chungking), where the Nationalists had held out during World War II. But late in 1949, following disastrous defeats in which hundreds of thousands of Nationalist forces were killed, the remnants of Chiang's faction fled to the island of Taiwan, taking control there and proclaiming the Republic of China. Meanwhile, in Beijing, standing in front of the assembled masses at the Gate of Heavenly Peace on Tiananmen Square, Mao Zedong announced the birth of the People's Republic of China.

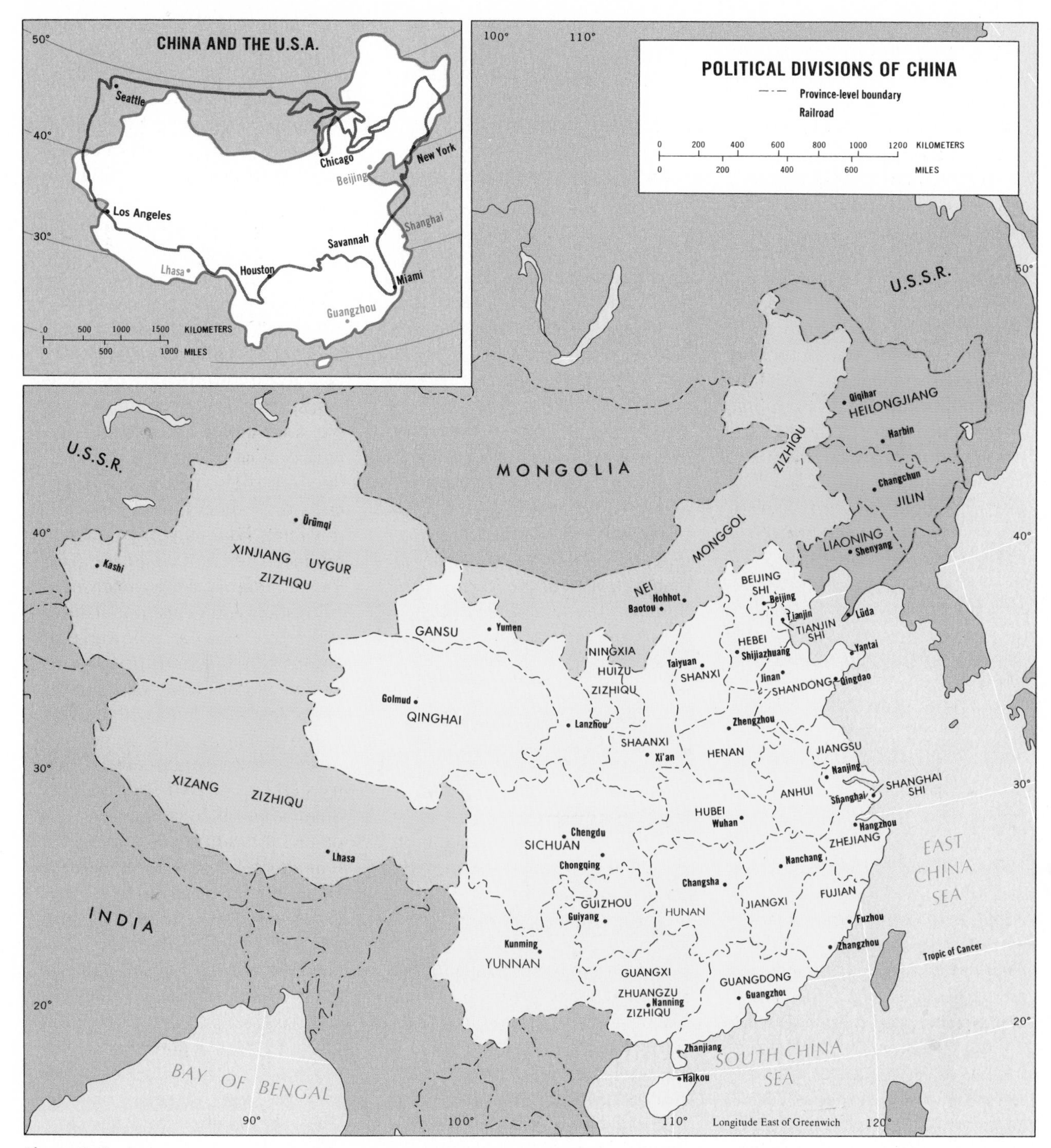


Figure 9–8

# Regions of China

The human drama of modern China's evolution was played out on a physical stage of immense variety.

As we noted earlier, China's territorial proportions are almost identical to those of the continental United States; latitudinally, China lies in the same general range as the contiguous United States (although it extends somewhat farther both to the north and south, as the inset map in Fig. 9–8 shows). There are some other similarities. In China as well as the United States, the core area lies in the eastern part of the national territory; both China and the United States have lengthy east coasts. But on the other hand, there is nothing in China to compare with California: there is no west coast and no subsidiary core areas or large conurbations are emerging in China's far west. Yet, China contains well over four times as many people as the United States, a context in which the world population distribution map (Fig. I–14) should be reviewed.

In the following paragraphs, we consider China's general physiographic layout, and then focus on the country's human regions. In very general terms, we can observe still another similarity to the United States: politically (Fig. 9–8), China has small provinces in the east and Texas-sized units toward the west. And physiographically, China divides into about the same number of regions as the United States, ranging from the snow-capped mountains of Xizang (Tibet) to the warm river lowlands of Guangdong and from the deserts of Xinjiang to the wheatfields of the North China Plain.

## Physiographic Regions

We have already come to know China as a land of river basins, fertile alluvium and loess (Fig. I–13), temperate to continental climates (Fig. I–10), severe dryness in the north and west (Fig. I–9), and great mountain chains (Fig. I–5). To get a picture of the spatial arrangement of things in China, a general frame of reference is useful. As in the case of India, we can identify several major regions and then break them down into more detailed subregions. Accordingly, China's four major regions (Fig. 9–9 inset) are: (1) the river basins and highlands of East China and the Northeast, (2) the plateau steppe of Mongolia, (3) the high plateaus and mountains of Xizang (Tibet), and (4) the desert basins of Xinjiang (Sinkiang). Strictly speaking, of course, these are not physiographic *provinces*—they are broader than that. Similar "regions" in the United States would be, say, the Interior Lowland—which includes several different kinds of landscapes—or the Rocky Mountains, diverse enough to be divided into many distinct areas as well.

**Region 1, River Basins and Highlands of East China and the Northeast,** is a land of river valleys separated by highlands, with the lowlands covered by farms (Fig. 9–10). This region also contains the greater length of the courses of China's three most important rivers: the Huang He (Yellow), Chang Jiang (Yangtze), and Xi (West), marked as *A*, *B*, and *C*, respectively, on the larger map (Fig. 9–9). All three rivers rise on eastern slopes of the Xizang (Tibet)-Yunnan Plateau and flow eastward, the Huang He through the most circuitous and longest course, the Xi through the most direct and shortest. The upper courses of the Huang and Chang Rivers lie in close proximity, but in two distinct physiographic provinces; their lower reaches, however, lie in an area that can with justification be identified as a single region, the Eastern Lowlands. In human terms, the Eastern Lowlands region is China's most important one, including as it does the North China Plain and the cities of Beijing and Tianjin (Tientsin), and the productive lower Chang Valley with Nanjing and China's largest city, Shanghai. In every respect, this is China's core area with its greatest population concentration, highest percentage of urbanization (about 40 percent, as opposed to 28 percent for the country as a whole), enormous agricultural production, growing industrial complexes, and intensive communication networks. China's heartland, as Fig. 9–9 shows, extends into the Northeast, where the lowland of the Liao River and the city of Shenyang form part of it.

The Huang He, in its upper basin, makes an immense clockwise bend and in the process almost encircles one of China's driest areas, the Ordos Desert. Downstream from the Ordos, the river enters the Loess Plateau, and here conditions are quite different. Loess is a windborne deposit whose origin in this area is attributable to nearby deserts (possibly the Ordos) and the Pleistocene glacial epoch, during which the deposits were laid down in a mantle up to 75-meters (250-feet) thick, covering the preexisting landscape. This loess is quite fertile, and, unlike ordinary soil, its fertility does not decrease with depth. It is also easily excavated, and hundreds of thousands of dwellings here lie partly underground. The loess landscape (Fig. 9–11) is dominated by intensive cultivation and dense population, although life in the hollowed-out houses entails serious risk—countless inhabitants have, over the centuries, lost their lives when earthquakes collapsed their homes.

The Huang He has always been marked by violent floods and frequent changes in course. Alternately, it has drained north into the Gulf of Chihli or to the south of the Shandong (Shantung) Peninsula, with numerous distributaries forming and shifting position over time. Flowing as it does through the Loess Plateau, the river brings enormous quantities of silt to the North China Plain, one of the world's most fertile and productive farming regions. There it deposits all this sediment and then proceeds to flow over the new accumulation; for uncounted centuries, local inhabitants have constructed dikes and artificial levees in order to stabilize the river's various channels. All that was needed for a disastrous flood was a season with a particularly high volume of water, enough to overflow and breach the dikes. It has happened dozens of times, and the lives lost directly in these

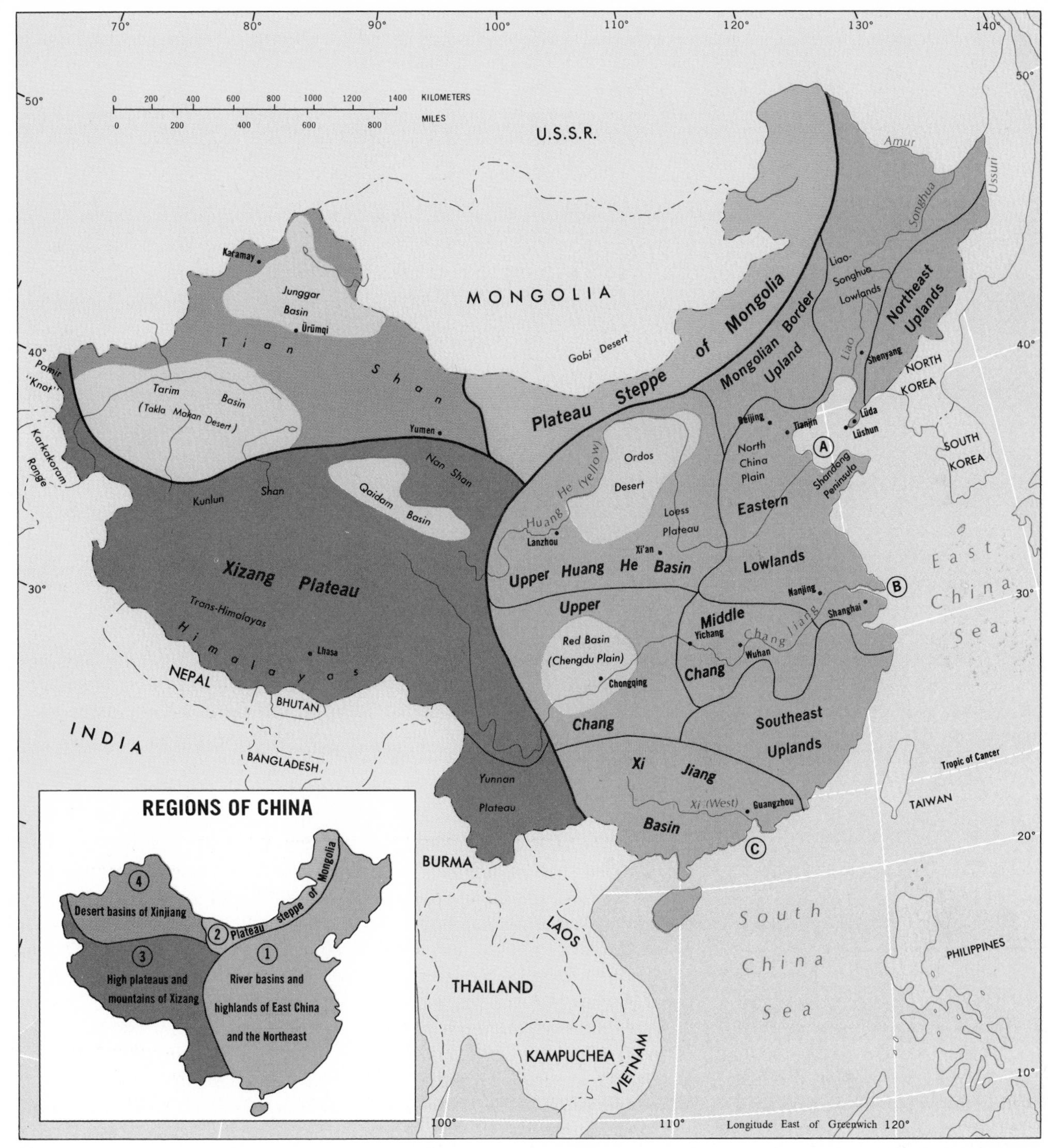


Figure 9–9

floods or subsequently in the inevitable famine must number in the tens of millions.

The course of China's middle river, the Chang (Yangtze) Jiang, is usually divided into three basins, of which the westernmost, the Red Basin, contains one of China's largest population clusters. The 1982 census showed the population of Sichuan (Szechwan) Province to be 99.7 million; about 75 million of these are clustered in the Red Basin, one of the most intensively cultivated areas in the world. Apart from the Chengdu (Chengtu) Plain, where there is relatively level land, the slopes of the basin's hilly country have been transformed by innumerable terraces where rice grows in summer and wheat in winter. Besides such other cereals as corn, major crops grown here are soybeans, sweet potatoes, sugarcane,

**Figure 9–10**
The lowlands of East China are clothed in farm fields, here along the fertile banks of the Li River.

and a wide range of fruits; on the warmer slopes, tea flourishes.

The Chang's middle course begins in the vicinity of Yichang, where the river emerges from a series of gorges that limit navigation as far as Chongqing (Chungking) to small motorized vessels. By the end of the 1980s, the huge Gezhouba Dam will be completed here. In 1983, it was already partially in operation and produced more electricity than all of China in 1949; in the more distant future, the Chinese hope to complete Three Gorges Dam, 40 kilometers (25 miles) upstream from Yichang, which will have a hydroelectric output twice the size of Brazil's massive Itaipu project. Below Gezhouba Dam, the river enters an area of more moderate relief and flows through middle China's lake country—agriculturally one of the most productive parts of the nation. This is the southern part of the Eastern Lowlands, and along or near the Chang lie the large cities of Wuhan (3.4 million), Nanjing (2.4 million), Hangzhou (Hangchow) (1.3 million), and China's leading urban center, Shanghai (15.6 million).

China's southernmost major river is actually called the West (Xi) River, but neither in length nor in terms of the productivity of the region through which it flows is this river comparable to the Huang He or the Chang. In this Xi Jiang Basin, there is much less level land; in the hills and mountains of South China, there remain many millions of local indigenous peoples not yet acculturated to the civilization of the people of Han. Hence, the western part of the Xi Jiang Basin is constituted largely by the Guangxi Zhuangzu Zizhiqu (Kwangsi-Chuang Autonomous Region). To the east lie Guangdong (Kwangtung) Province, the populous Xi Delta, and Guangzhou (Canton), the south's largest and most important city (3.7 million).

Between the deltas of the Xi and Chang Rivers lie the Southeast Uplands, a region of rugged relief which for a long time in Chinese history remained a refuge for southern tribes against the encroachment of the people of Han from the north. With its steep slopes and narrow valleys, this region has very little agricultural potential, although a massive land-development program was begun here in the 1960s.

**Figure 9–11**
The distinctive landscape of the Loess Plateau, situated in the Huang He's central basin.

For many years, this has been one of China's most outward-looking regions, with a considerable emigration (via Guangzhou) to the Philippines and Southeast Asia; substantial overseas trade in the leading commercial product, tea; and a seafaring tradition for which the people of Fujian (Fukien) Province have become famous.

The Northeast in some respects resembles Eastern China, although not in others. The Liao-Songhua Lowland does not compare to the great basins of the Huang, Chang, or Xi; it is essentially an erosional plain rather than a depositional basin. But its rising population densities and growing productivity are in a class with those of the heartland provinces farther to the south. The relief of the Northeast Uplands, which border North Korea, is also quite similar to that of the highlands of the southeast between the Chang and Xi Rivers. What really makes the landscape look different here is the colder, more wintry climate of this, China's northern frontier. The Liao-Songhua Lowland and the Northeast Uplands converge on the silt-plagued Liaodong (Liaotung) Gulf, where Lüda (Dairen) is the port city near the tip of the adjacent Liaodong Peninsula.

**Region 2, the Plateau Steppe of Inner Mongolia,** actually constitutes the southern rim of the basin of the Gobi Desert. Near the physiographic boundary line, elevations in a series of ranges reach an average of 1700 meters (5600 feet); toward the heart of the Gobi, the land lies much lower, as low as 750 meters (2500 feet). As is suggested by Figs. I–9 and I–12, the moistest parts of this region lie on the slopes of those southern higher hills, although even there the vegetation is only poor steppe grass with some scrub; some of the mountains are rocky and barren, and the whole aspect of the area is one of drought. Not surprisingly, this is a difficult environment in many ways: summer temperatures are often searingly hot, whereas winters are bitterly cold, and vicious winds often blow up sand and dust. Altogether, this is unlike the image we have of a populous, productive, rice-dominated China. Despite the government's efforts to extend sedentary agriculture here through river-dam and irrigation projects, the Plateau Steppe as a whole remains one of the country's more sparsely populated areas, with an average of just 12 people per square kilometer (30 per square mile) in a country where 300 people per rural square kilometer is no rarity. Only Xinjiang (Sinkiang) and Xizang (Tibet) are less populous.

**Region 3, the High Plateaus and Mountains of Xizang (Tibet),** comprises perhaps the world's greatest assemblage of lofty, snow-covered mountain ranges and high-elevation plateaus. The southern margin of this region is the great Himalaya Mountain wall, and to the north lie the Kunlun Shan and Nan Shan Mountains. From the Pamir "Knot" in the extreme west, these great mountain chains spread out eastward in a series of vast arcs that eventually converge again and turn southward into the Yunnan Plateau. The *average* elevation of this region, as defined in Fig. 9–9, is probably around 4500 meters (15,000 feet), with the higher mountain ranges standing 3000 meters (10,000 feet) above this level, and the valleys—where most of the people live—going down to about 1500 meters (5000 feet) above sea level. The central plateau of Xizang is desolate and barren, cold, windswept, and treeless; here and there some patches of grass sustain a few animals. In terms of human development, there are two areas of interest.

The first of these is the zone between the Himalayas in the south and the Trans-Himalayas that lie not far to the north. In this area, there are some valleys below 2000 meters (under 7000 feet) where the climate is milder and cultivation is possible; here one finds Xizang's main population cluster. The capital of Lhasa is situated at the intersection of roads leading east-west along the valley and northward into Qinghai and China Proper; looming above this mountain city is the Potala (Fig. 9–12), the magnificent palace of the Dalai Lama, the Tibetan Buddhist leader forced to flee to India after the Communist takeover. The Chinese government, since its confirmation of control in Tibet, has made investments to speed up the snail-paced development that prevailed under the previous administration. The southern valleys contain excellent sites for hydroelectric-power projects, and some of these have been put to use in a few light industries. More importantly, the south of Xizang may be an area of considerable mineral wealth; despite the enormous difficulties of distance and terrain, any such minerals may become vitally important to the developing China of the future.

One other part of the Xizang region is of politico-economic importance—the Qaidam (Tsaidam) Basin, in Qinghai on the edge of Xinjiang (Sinkiang). This basin lies hundreds of meters below the surrounding Kunlun Shan and Nan Shan Mountains, and, as such, it has always contained a cluster of nomadic Xizang (Tibetan) pastoralists—in fact, there were times when these nomads plundered the trade route to the west that skirted the northern Nan Shan. Recently, however, exploration has revealed the presence of oilfields and coal reserves below the surface of the Qaidam Basin, and the development of these resources is under way.

**Region 4, the Desert Basins of Xinjiang (Sinkiang),** comprises two huge mountain-enclosed basins and several smaller basins. The two largest of Xinjiang's basins are separated by the Tian Shan, a mountain range that stretches across the region from the Kirghiz border with the U.S.S.R. in the west to Mongolia in the east. Climatically, these are dry areas; the Tarim Basin is in fact a desert (the Turkestan or Takla Makan Desert), whereas the basin of Junggar (Dzungaria) is steppe country. Both the Junggar and the Tarim Basins are areas of internal drainage, that is, the rivers that rise in the adjacent mountains and flow to the basin floor do not continue to the sea. The rivers coming off these highlands are the chief supply of water flowing onto the region's rough gravels that have been washed from the mountains. They disappear below the surface as they reach these coarse deposits. But, then, they reappear where the gravels thin out; there, oases have long existed. Along the southern margin of the Takla Makan Desert lie a string of these oases, which at one time formed stations on the long trade route to the west.

Since 1949, the Chinese have made a major effort to develop the agricultural potential of the Tarim area. Canals and *qanats* were built, oases enlarged, and the acreage of productive farmland has at least quadrupled by now from what it was 40 years ago. Especially the northern rim of the Tarim Basin, long neglected, has been brought into the sphere of development, and fields of cotton and wheat now attest to the success of the program.

Junggar (Dzungaria), although it contains perhaps just one-quarter of Xinjiang's 14 million inhabitants, has a number of assets of importance to China. First, it has long been the site of strategic east-west routes. Second, the main westward rail link toward the Kazakh S.S.R. and Soviet Russia runs from Xi'an (Sian) in China Proper via Yumen in Gansu and Ürümqi (Urumchi) in Junggar, which is its present terminus. Third, Junggar has proved to contain sizable oilfields, notably around Karamay, not far from the Soviet border. Pipelines have been laid all the way to Yumen and Lanzhou (Lanchou), where refineries have been built. Thus, it is

**Figure 9–12**
Lhasa's Potala, former base of the Dalai Lama and the pre-revolutionary center of Buddhism in Xizang (Tibet).

not altogether surprising that the region's capital of Ürümqi (975,000) lies in the less populous but strategically important northern part of Xinjiang.

## Human Regional Geography

China's human regional geography displays a strong contrast between the populous east (sometimes called China Proper, the "real" China) and the sparsely peopled west. The great cities, the wheatfields and ricelands, the factories and railroads lie in the east. The west and northwest are the China of inner Asia, of mountains and deserts, oases and widely spaced towns.

**Inner Asian China** Interior China—the China of Xinjiang, Xizang, and Inner Mongolia—is still frontier China, where China faces neighbors across unsettled boundaries and where human occupance remains spotty and tenuous. Xinjiang's population of 14 million is now over one-third Chinese as a result of Beijing's determination to integrate this distant province into the national framework; just 30 years ago, the population was only about 5 percent Chinese. The majority of the people in Xinjiang are Moslem Uighurs, Kazakhs, and Kirghiz, with cultural affinities across the borders to the Soviet Asian republics. The Uighurs, most numerous (about 5 million) among these peoples, have for centuries been concentrated in the oases of the Tarim Basin, and the mobile Kazakhs and Kirghiz moved along nomadic routes across the Soviet borders. In recent decades, Chinese from eastern China have come to build and develop towns and roads, to secure the Soviet boundaries, and to exploit the region's resources. In the process, they have accomplished the Sinicization of an area that always was beyond China's national sphere, and, before the end of the century, the majority of Xinjiang's population could be Chinese.

Xizang (Tibet) was pressed into the Chinese fold in the 1950s, first through frontier settlement and economic interference, and, in 1958, by force of arms after Tibetan villagers tried to resist the Chinese presence. Tibetan society had been organized around the fortresslike monasteries of Buddhist monks who paid allegiance to the Dalai Lama (Fig. 9–12). The Chinese wanted to modernize this feudal system, but the Tibetans clung to their traditions. In 1958, they proved no match for the Chinese forces, the Dalai Lama was ousted, and the monasteries were emptied.

The Chinese destroyed much of Tibet's cultural heritage, looting the region of its religious treasures and what remained of the magnificent Tibetan works of art. Their harsh rule and heavy-handed effort to erase memories of the old order took a severe toll of Tibetan society, but after Mao's death in 1976 the Chinese softened their control. Thousands of Tibetan religious treasures were returned to Tibet, and reconstruction of destroyed monasteries was permitted. Since its annexation, Xizang has been administered as a Chinese Autonomous Region; although it is large (1,222,000 square kilometers/ 472,000 square miles) its population remains only about 2 million. The Chinese presence is far weaker here than in Xinjiang, but Xizang may yet come to play a crucial role in China's development because of its considerable mineral potential.

Inner Mongolia lies closer to China's eastern heartland—it also has a history of closer association with Han China than either Xinjiang or Xizang. Here (and in adjacent Mongolia, now an independent country in the Soviet sphere) Chinese farmers long competed with nomadic horse- and camel-riding Mongols for control over an area that consists of vast expanses of steppe and desert as well as riverine ribbons and oases of farmland. Inner Mongolia's population of about 20 million is now overwhelmingly sedentary and Chinese (fewer than 1 million Mongols concentrate in the northern border area) and depends largely on irrigation systems built in the basin of the great bend of the Huang He.

The three regions of Inner Asian China cover a huge territory—some 3.3 million square kilometers (1.3 million square miles), an area larger than all of India—but they contain fewer than 40 million of China's more than 1 billion people. Small wonder that China's east, where the great majority of Chinese are concentrated, is known as the "real" China, China Proper. Even more than in the United States, "the east" is China's historic and contemporary heartland, where the civilization of the People of Han had its roots and where the modern core of the state evolved. China's east consists of China Proper and the Northeast; of these two regions, China Proper still retains its primacy in every way. This is humid China, intensive-agricultural China, the China of villages and cities, workshops, and factories. Above all, it is the China of the masses, the throngs whose regimentation on the land and in the towns has brought the country an era of comparative stability and order.

**China Proper** China Proper is the China south of the Great Wall, the China of the three rivers (Fig. 9–13). Of the three, the middle river—the Chang (Yangtze) Jiang—and its basin are in almost every respect the most important. From the Red Basin of Sichuan in its upper catchment area to its delta in the Eastern Lowlands (Fig. 9–13), the Chang traverses China's most populous and most productive areas. As a transportation route, it is China's most navigable waterway; oceangoing ships can sail over 1000 kilometers (600 miles) up the river to the Wuhan conurbation (Wuhan is short for Wuchang, Hanyang, and Hankow), and boats of up to 1000 tons can even reach Chongqing (Chungking). Several of the Chang's tributaries are also navigable, and—depending on the size of the ships capable of using these various stretches—over 30,000 kilometers (18,500 miles) of water transport routes exist in the Chang Basin. Therefore, the Chang Jiang is one of China's leading transport arteries, and, with its tributaries, attracts the trade of a vast area, including nearly all of middle China and large parts of the north and south. Funneled down the Chang, most of this enormous volume is transhipped at Shanghai (Fig. 9–14), whose metropolitan size of 15.6 million in 1985 reflects the productivity and great population of this hinterland.

Early in China's history, when the Chang's basin was being opened up and rice and wheat cultivation began, a canal was built to link this granary to the northern core of old China. Over 1600 kilometers (1000 miles) in length, this was the longest artificial waterway in the world, but during the nineteenth century it fell into disrepair. Known as the Grand Canal (Fig. 9–13), it was dredged and rebuilt during the period when the Nationalists held control over Eastern China; and after 1949, the Communist regime continued this restoration effort. During the 1960s, the southern section of the canal was opened once again to barge traffic, supplementing the huge fleet of vessels that transports domestic interregional trade along the east coast.

The bulk of China's internal trade in agricultural as well as industrial products is either derived from, or distributed to, the Chang region. International trade also goes principally through Shanghai, whose port normally handles half the country's overseas tonnage. The rest is split among China's other leading ports, including Tianjin (Tientsin) on the northern coast and Guangzhou (Canton) in the south.

Shanghai, just a regional town until the mid-nineteenth century, rose to prominence as a result of its selection as a treaty port by the British; ever since then its unparalleled locational advantages have

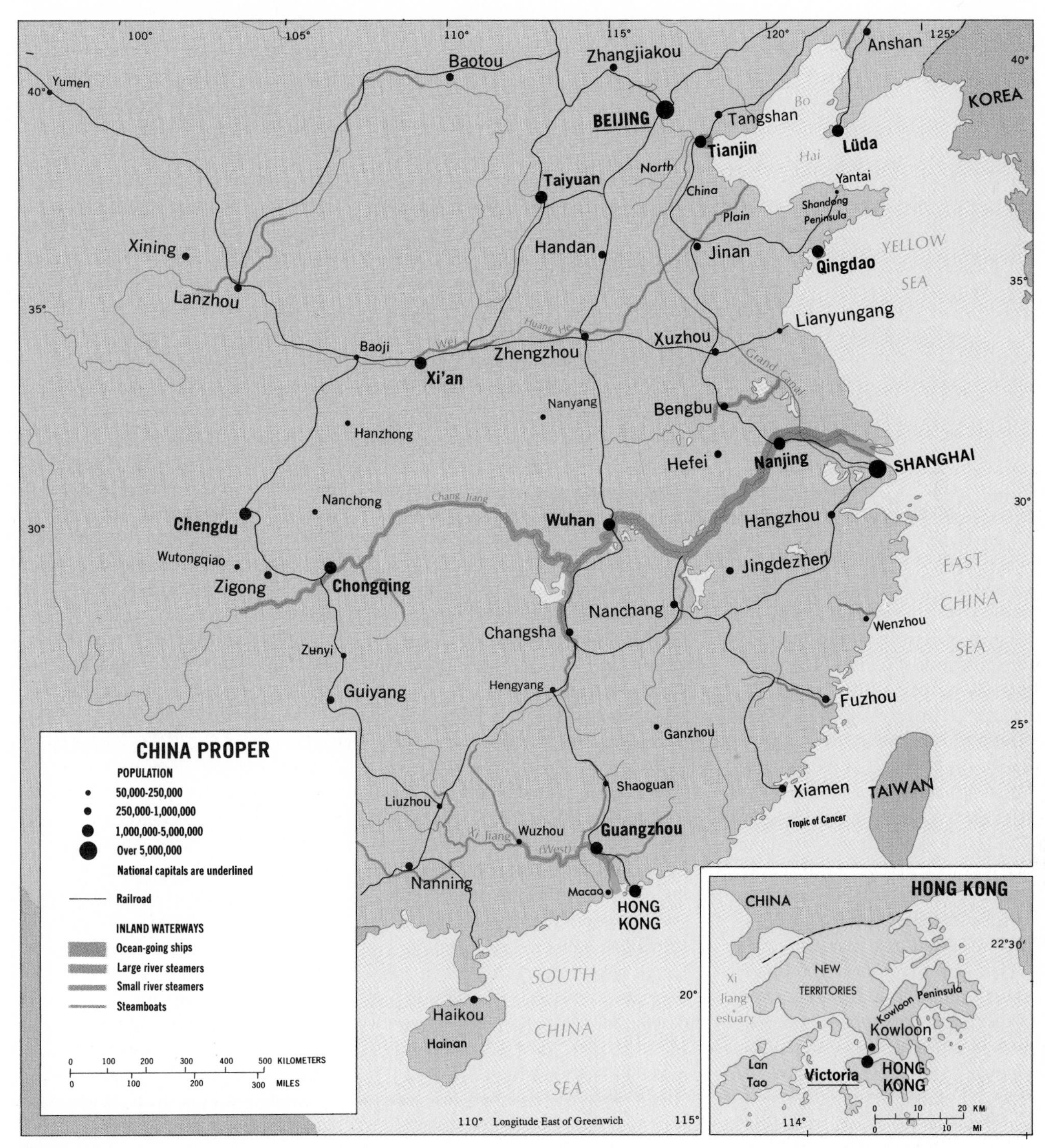


Figure 9–13

sustained its position as China's leading city in almost every respect. This metropolis lies at one corner of the Chang Delta, an area of about 50,000 square kilometers (20,000 square miles) containing more than 50 million people. Some two-thirds of these are farmers who produce food, silk filaments, and cotton for the city's industries. Thus, Shanghai has as its immediate hinterland one of the most densely populated areas on earth, and beyond the Delta lies what must be the most populous region in the world to be served by one major outlet. During the nineteenth century and up to the war with Japan, the principal exports to pass through Shanghai were tea and silk; large quantities of cotton textiles and opium were imported. At that time, Shanghai's prominence was undisputed; it handled two-thirds of all of China's external trade. But its fortunes suffered during and after World War II. First, the Nationalists

**Figure 9–14**
The center of Shanghai, China's largest city and trade focus for the burgeoning Chang Basin.

blockaded the port (1949) and made bombing raids on it; then the Beijing government decided to disperse its industries up-country, thereby reducing their vulnerability to attack. Meanwhile, safer Tianjin (Tientsin) had taken over as the leading port. But Shanghai's situational advantages promised a comeback, and it soon occurred. In the 1960s, the port regained its dominant position, and the industrial complex (textiles, food processing, metals, shipyards, rubber, chemicals) resumed its expansion.

The vast majority of the Chang Basin's 350-odd million people are farmers on cooperatives and collectives that produce the country's staples and cash crops. We have already mentioned the amazing variety of crops grown in Sichuan; here, as in the Chang's lower basin, rice predominates (Fig. 9–15). However, a look at Figs. I–9 and I–10 will suggest why the Sichuan-Chang line is more or less the northern margin for rice cultivation. To the north, temperatures decline and rainfall diminishes as well; even in the irrigated areas, wheat rather than rice is the grain crop. In the Chang Basin, rice rotates with winter wheat and, in effect, this is the transition zone; a line drawn from southern Shaanxi Province east to mid-coastal Jiangsu Province approximates the northern limit of rice cultivation.

As Fig. 9–13 shows, the Eastern Lowlands merge northward into the lower basin of the Huang He and the North China Plain. Here spring wheat is grown in the northern areas, whereas winter wheat and barley are planted in the south; in the spring, other crops follow the winter wheat. Millet, sorghum (kaoliang), soybeans, corn, a variety of vegetables, tobacco, and cotton are cultivated in this northern part of China Proper; on the Shandong Peninsula's higher ground, fruit orchards do well. This part of China was marked by very small parcels of land, and the Communist regime has effected a major reorganization of landholding here. At the same time, an enormous effort has been made to control the flood problem that has bedeviled the Huang He Basin for uncounted centuries, and to expand the land area under irrigation.

The North China Plain is one of the world's most overpopulated agricultural areas, with about 500 people per square kilometer (1300 people per square mile) of cultivated land (Fig. 9–16). Here the ultimate hope of the Beijing government lay less in land redistribution than in raising yields through improved fertilization, expanded irrigation facilities, and the more intensive use of labor. A series of dams on the Huang He now reduce the flood danger, but outside the irrigated areas the ever-present problem of rainfall variability and drought recurs. The North China Plain has not produced any substantial food surplus, even under normal circumstances; when the weather is unfavorable, the situation soon becomes precarious. The specter of famine has receded, but the food situation is still uncertain in this northern part of China Proper.

The layout of villages, farmlands, and market areas here in Northern China affords an opportunity to reexamine a concept introduced in Model Box 3 (pp. 192–193): *central place theory*. You will recall that Walter Christaller developed a location theory for the hierarchy of market centers (central places)—which he verified for southern Germany—and that one of his key assumptions involved a flat uninterrupted plain that could be traversed in all directions with equal ease. The North China Plan comes close to just such a circumstance. Over centuries of human settlement, China's rural population has become distributed among nearly 1 million farm villages, many of them in the North China Plain. These villages each have populations of several hundred, and they are surrounded by their own farm fields. Groups of such villages (averaging about 18) lie in the hinterlands of larger market towns. Each market town lies within walking distance (6 to 7 kilometers or about 4 miles) of these villages. For centuries, the market towns were the crucial places

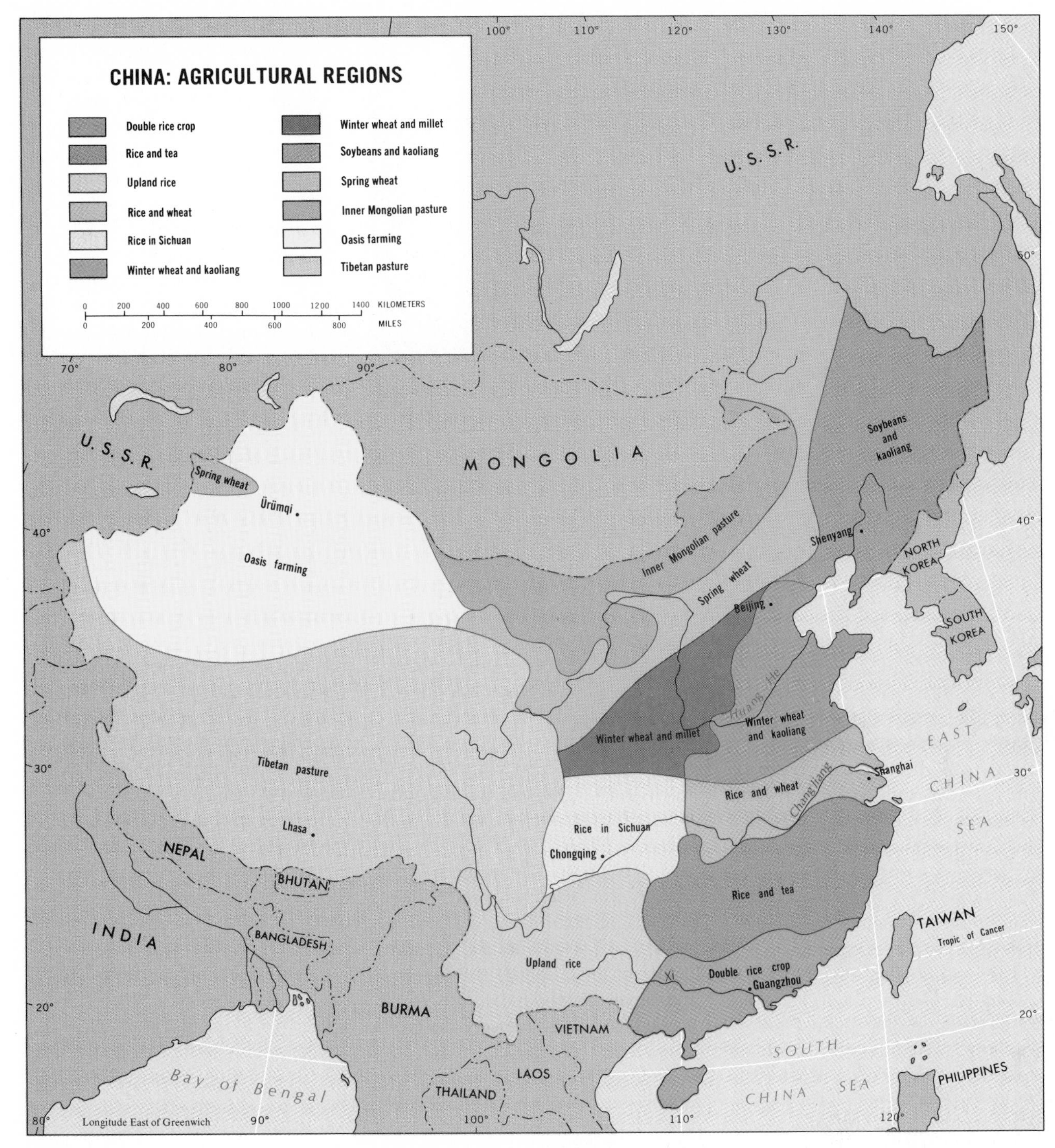


Figure 9–15

where peasant and merchant met and goods and ideas were exchanged. Here the peasant would sell the farm's produce, buy goods and services, and talk to people from other villages. The *activity space* (territorial range) of the Chinese peasant always was very limited, but the market town was part of the network of still larger market centers, including, at the top of the hierarchy, China's largest cities. In the market towns, the peasants were exposed to the Chinese world beyond their own villages, so that these places served as more than markets for the exchange of goods—they also functioned to integrate the Chinese nation.

The Chinese central-place network in certain areas strongly resembled elements of the Christaller model (p. 192), as G. William Skinner described in a 1965 article enti-

**Figure 9–16**
The overcrowded North China Plain, southeast of Beijing, contains one of the highest physiologic densities on earth.

tled "Marketing and Social Structure in Rural China." But even on the North China Plain, (similar to Skinner's study area in the Sichuan region) there are distortions; fo, example, villages tend not to be evenly spaced but cluster on higher ground away from streams in order to reduce the flood danger. Still, the Chinese pattern confirms that Christaller's concepts, based as they were on observations in a totally different geographic realm, have universal application.

The two major cities of the Huang He (Yellow River) Basin are the historic capital, Beijing, and the port city of Tianjin, both positioned near the northern edge of the Plain. In common with many of China's harbor sites, that of the river port of Tianjin is not particularly good; but this city is well situated to serve the northern sector of the Eastern Lowlands, the capital (now with a population of about 13 million), the Loess Plateau to the west, and Inner Mongolia beyond. Like Shanghai, Tianjin had its modern start as a treaty port, but the city's major growth awaited Communist rule. For many decades, it had remained a center for light industry and a flood-prone port; after 1949, a new artificial port was constructed and flood canals were dug. More important, Tianjin was chosen as a site for major industrial development, and large investments were made in the chemical industry (in which Tianjin now leads China); in heavy industries, especially iron and steel production; in heavy machinery manufacturing; and in the textile industry. Today, with a population of 5.4 million, Tianjin is the center of the country's third-largest industrial complex, after the Northeast's Liaoning Province and its old competitor, Shanghai.

Beijing, on the other hand, has chiefly remained the political, educational, and cultural center of China; although industrial development also occurred here after the Communist takeover, its dimensions have not been comparable to Tianjin. The Communist administration did, however, greatly expand the municipal area of Beijing, which is not controlled by the province of Hebei (Hopeh) but is directly under the central government's authority. In one direction, Beijing was enlarged all the way to the Great Wall—50 kilometers (30 miles) to the north—so that the "urban" area includes hundreds of thousands of farmers. Just as Shanghai symbolizes bustling, crowded, industrial, trade-oriented China, so Beijing represents Chinese history, power, and order. Facing Beijing's main thoroughfare are the Forbidden City—the treasure-laden preserve of the rulers and elite of old—and, directly opposite, the starkly contrasting Tiananmen Square flanked by modern buildings, the scene of mass assemblies of millions during the Communist era. The great Gate of Heavenly Peace overlooks both ancient city and modern square, and here Mao Zedong proclaimed the birth of the People's Republic of China on October 1, 1949 (Fig. 9–1). Beijing's relative location is also remarkable. Not only does the city lie close to the northern margin of the North China Plain and not far from the Great Wall and the Ming Tombs, but China's greatest archaeological discovery, Peking Man, also was made near the city. These fossil human remains prove that communal life, organized hunting, and the use of fire existed at the present site of the city as long as 350,000 years ago. Beijing and its environs are indeed a truly historic locale.

The middle Huang Basin—and that of the tributary Wei River—includes the Loess Plateau, an area of winter wheat and millet cultivation whose environmental problems were discussed earlier. Beneath the loess lie highly significant coal deposits (Fig. 9–17); Taiyuan in Shanxi (Shansi) is already the site of a large iron and steel complex, machine-manufacturing plants, and chemical industries; a reassessment of this province's coal resources in the mid-1980s has concluded that they surpass the reserves of such world-class coalfields as the Soviet Donbas and West Germany's Ruhr Valley. As the map indicates, the

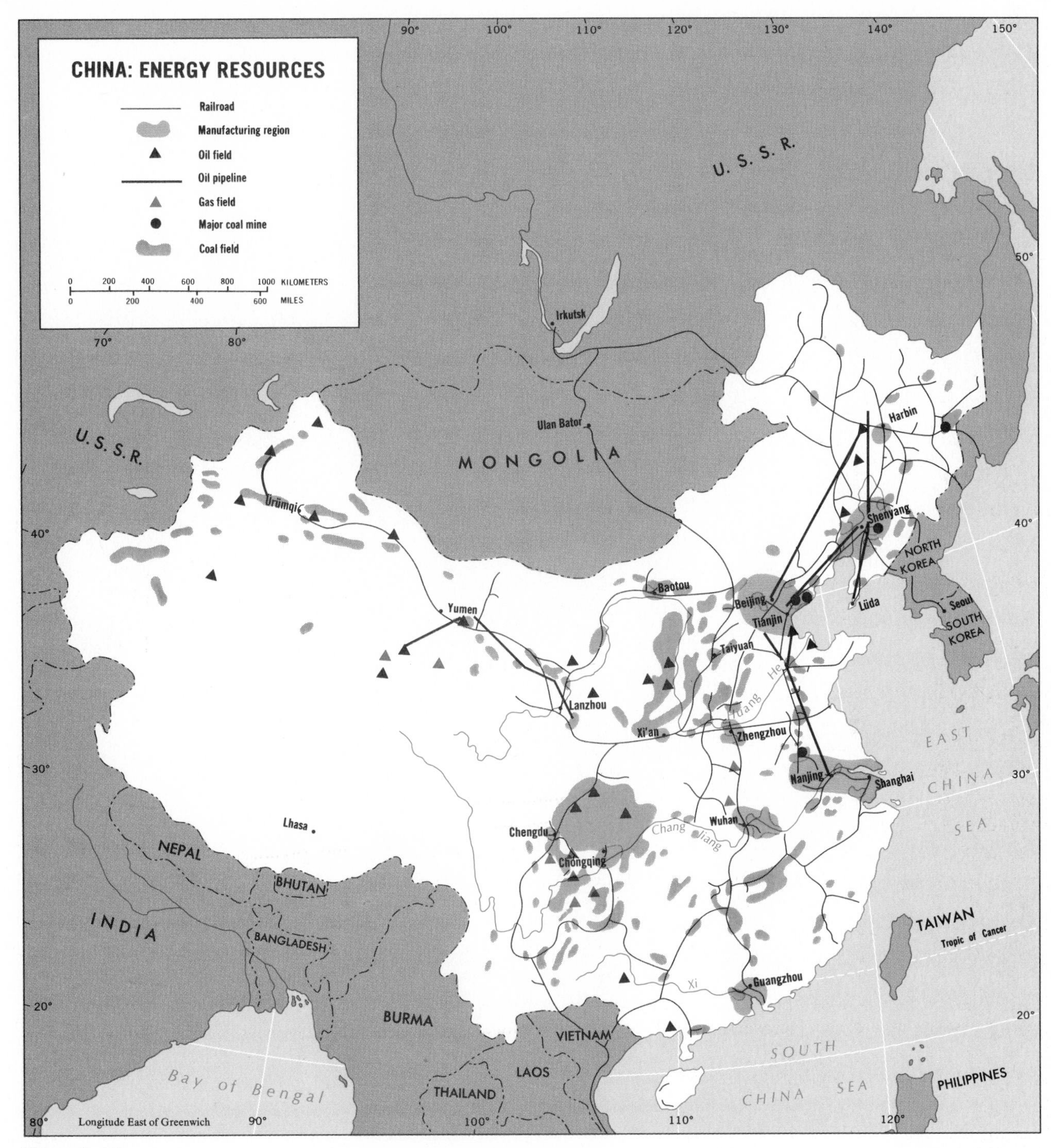


Figure 9–17

central Huang He area is not especially well positioned for the transportation of raw materials, although rail connections do exist to the Northeast's source area.

Southern China provides a strong regional contrast to the north, the differences reflected by the cities as well as the countryside. The south is dominated by the basin of the Xi Jiang—the West River. Whereas northern China is the land of the ox and even the camel, southern China uses the water buffalo. Northern China grows wheat for food and cotton for sale; southern China grows rice and tea. Northern China is rather dry and continental in its climate; southern China is subtropical and humid. And the people of northern China are much more clearly Mongoloid in their appearance. Northern China has long looked inward to

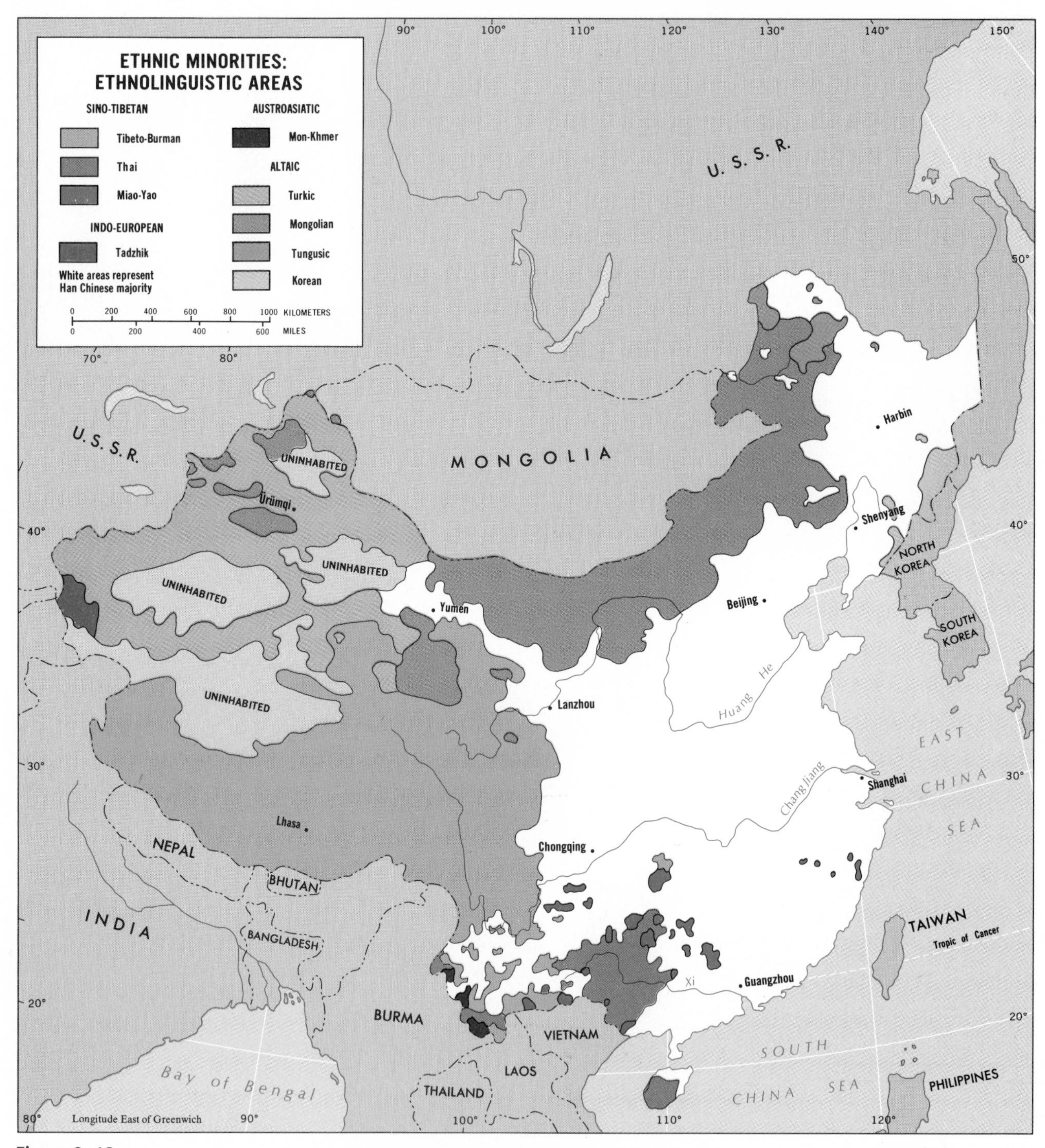


**Figure 9–18**

interior Asia, whereas southern China has oriented itself outwardly to the sea and even to lands beyond. With its very mixed population and multilingual character, southern China carries strong Southeast Asian imprints (Fig. 9–18).

The map (Fig. 9–2) suggests that the Xi Jiang is no Chang, and this impression is soon verified. Not only is the West River much shorter, but for a great part of its course it lies in mountainous or hilly terrain. Above Wuzhou (Wuchou), the valley is less than a half-mile wide and, confined as it is, the river is subject to great fluctuations in level. Except for the delta, whose 8500 square kilometers (3100 square miles) support a population of over 15 million, there is little here to compare with the lower Chang Basin. Cut by a large number of distributaries and by levee-protected flood canals, the delta of the Xi Jiang is southern

China's largest area of flat land and the site of its largest population cluster; here, too, lies Guangzhou (Canton), the south's leading city.

Subtropical and moist, southern China provides a year-round growing season. In the lower areas, rice is double-cropped; one planting takes place in mid- to late winter, with a harvest in late June or shortly thereafter; a second crop is then planted in the same paddy field and harvested in midautumn. It is even possible to raise some vegetables or root crops between the rice plantings! As in Sichuan, whole areas of hillside have been transformed into a multitude of artificial terraces; here, too, rice is grown, although the higher areas permit the harvesting of only one crop. But, again, there is time for a vegetable crop or some other planting before the next rice is sown. Fruits, sugarcane, tea, corn—wherever farming is possible southern China is tremendously productive, and the range of produce is almost endless. If only there were more level land—and fewer people; this is one of China's food-deficient regions, always requiring imports of grain. The problems of China's southern zone also relate in part to the distribution of the country's ethnic minorities (Fig. 9–18). Southern China is quite far from the heart of Han China, and its peoples are unlike the people of Han. Here in the south, China takes on an almost Southeast Asian character. The minority peoples are large and numerous, and they have been somewhat exempt from certain Chinese rules (e.g., those limiting family size). As a result, natural rates of population growth are the highest of all Chinese regions, land pressure is enormous, and environmental deterioration (especially soil erosion) is a serious problem. Small wonder that the south cannot feed itself.

With its 3.7 million people, Guangzhou (Canton) is the urban focus for the whole region. Despite its rather narrow valley, the Xi Jiang is navigable for several hundred miles (Fig. 9–13), so that hinterland connections are effective. Hong Kong (see *Box*), just over 160 kilometers (100 miles) away, overshadows Guangzhou in size as well as trade volume. Guangzhou has faded as an industrial center; there was a time when its factories and production made it a competitor for Shanghai, but, in recent times, the paucity of natural resources in the West River Basin has caused its industrial development to lag. In current development plans, Guangzhou's harbor is undergoing improvement, and port facilities are being expanded (with an eye to the competition of British-owned Hong Kong); but industrial growth is envisaged as relating to the agricultural production of the hinterland—sugar mills, textile factories, fertilizer plants. There is a fuel problem in southern China that has long been countered by coal imports from northern China, but now the hydroelectric potential of the West River Basin, which is good, is being put to use. The entire social atmosphere in Guangzhou and southern China differs from that of Beijing and the north. Compared to orderly Beijing, Guangzhou is lively, open, even boisterous, reminiscent of the Canton of old. Goods from Hong Kong and Southeast Asia can be found on black markets from Guangzhou to Kunming. But in recent years, the south has felt the heavy hand of discipline from Beijing when its individuality went too far, and China's old regional animosities are not buried as yet.

**The Northeast** China's northeastern region, home of the Manchus and embattled frontier, has become a vital part of the new Chinese world. The Manchus were the last to impose a dynastic rule on China, but since its collapse (in 1911) and after the elimination of foreign interests in the Northeast, massive Chinese immigration into the region and vigorous development of its considerable resources have combined to make this an integral and crucial part of the new China.

Northeast China has lengthy, and in places sensitive, foreign boundaries. Korea lies to the southeast, the Soviet Union to the east and north; Chinese Inner Mongolia and Soviet-sphere Mongolia lie to the west (Fig. 9–20). As the map shows, the shortest route from Vladivostok to the Soviet heartland is right across this region via Harbin, China's northernmost large city. This is the rail corridor in which the Soviets long had an interest, finally relinquished by treaty as late as 1950. With the increasing tension between the two countries, the Soviets have come to rely on the route skirting the Chinese border, via Khabarovsk—a trip that is 480 kilometers (300 miles) longer.

Japan also had imperial designs on the Northeast, but the aftermath of Japan's defeat in World War II was the confirmation of Chinese hegemony in the region. The Japanese contributed to the legacy by leaving behind railroads, factories, and a general framework for the region's development. The Soviets in the late 1940s removed machinery, railroad rolling stock, and a large quantity of hoarded gold. China controlled an area of only 1.2 million people, but with enormous economic possibilities. Today, however, the three provinces of Heilongjiang, Jilin, and Liaoning contain nearly 95 million people. The old Manchu culture has been completely submerged.

The physical and economic layout of the Northeast is such that the areas of greatest productive capacity and largest population lie in the south, where they form an extension of the core area of China Proper. The lowland axis formed by the basin of the Liao River and the Songhua Valley extends from the Liaodong Gulf north and then northeastward. Although the growing season here is a great deal shorter than in most of lowland China Proper, it is still long enough in the Liao watershed to permit the cultivation of grains such as spring wheat, kaoliang (sorghum), barley, and corn; and large areas are devoted to soybeans as well. Although the Northeast has drought problems and severe winters, its southern zone contains some excellent fertile soils. Toward the northern margins, extensive stands of oak and other hardwood forests on higher slopes have great value in timber-poor China.

Han Chinese have been moving into the northeastern provinces for centuries, driven by hunger and devastating Huang He floods, dislocated by the ravages of war, and attracted by the region's opportunities of land and space. Since 1949, the settlement and development of the region has been promoted by official policy, and millions more have come to the Northeast. Not all of them came to farm the soil, because this region has become a Chinese industrial heartland. The largest city, Shenyang, has a population of 3.6 million and has emerged as the Chinese Pittsburgh. All this is based on large iron ore and coal deposits, which—like the best farmland—lie in the area of the Liao Basin. Shenyang is located centrally amid these resources: there is coal 160 kilometers (100 miles) to the west and less than 50 kilometers (30 miles) to the east of the city (at Fushun), and iron ore is found about 100 kilometers (60 miles) to the southwest near Anshan. Since the iron ore is not very pure, coal is hauled from Fushun (and also from northern China Proper) to the site of the iron reserves; this is the cheapest way to convert these low-grade ores to finished iron and steel, and Anshan (1.5 million) has, thus, become

# HONG KONG

**Figure 9–19**
Crowded yet thriving Hong Kong, whose future status is the subject of ongoing Sino-British negotiations.

The mouth of China's southernmost major river, the Xi Jiang, opens into a wide estuary below the city of Guangzhou (Canton). At the head of this estuary, on the northeast side, lies the British colony of Hong Kong (Fig. 9–13 inset). Hong Kong consists of three parts: the island of Hong Kong (82 square kilometers/32 square miles), the Kowloon Peninsula on the mainland opposite this island, and, adjacent to the west, the so-called New Territories. The total area of this British dependency is only just over 1000 square kilometers (400 square miles), but the population is very large; in 1985 it had reached 5.5 million, 99 percent of them Chinese. Hong Kong Island and the Kowloon Peninsula were ceded permanently by China to Britain in 1841 and 1860 respectively, but the rest of the New Territories were leased on a 99-year basis in 1898.

With its excellent deep-water harbor, Hong Kong is the major entrepôt of the western Pacific between Shanghai and Singapore. Undoubtedly, the Chinese could have recaptured the territory during the successful campaign of the late 1940s, but both London and Beijing saw advantages in the *status quo*. During its period of isolation and Communist reorganization, Hong Kong provided the People's Republic of China with a convenient place of contact with the Western world without any need for long-distance entanglement. Even when hundreds of thousands of Chinese from Guangdong and Fujian Provinces fled to Hong Kong and the city was a place of rest and recuperation for American forces during the Indochina War, the Chinese government continued to tolerate the British colonial presence. Indeed, the colony depends on China for vital supplies, including fresh water and food. After Japan, China is Hong Kong's chief source of imports.

The colony is extremely crowded. Hong Kong Island, where the capital Victoria is located, is beyond the saturation point, and urban sprawl now extends onto the Kowloon Peninsula where high-rise apartments continue to go up (Fig. 9–19). Until the early 1950s, Hong Kong was a trading colony; then, the Korean War and the U.N. embargo on trade with China cut the dependency's connections with its hinterland, and an economic reorientation was necessary. This came about in an amazingly short time; in just a few years, a huge textile industry and many other light manufacturing industries developed. Today textiles and fabrics make up over 40 percent of the exports by value (their manufacture employs half the labor force), and the volume of electrical equipment and appliances is growing. These consumer goods go to the United States, the United Kingdom, West Germany, and Japan; China still lags far behind these four as an importer of Hong Kong's products. However, China earns about 40 percent of its foreign exchange through Hong Kong, still its major economic gateway to the outside world.

The crucial politico-geographical issue now involving Hong Kong is the approaching end, in 1997, of the U.K. lease to mainland Kowloon and the New Territories (constituting 90 percent of the colony's area). The British government has maintained that the permanent treaties are valid and that only the 99-year lease is subject to negotiation. The Chinese government, citing the duress under which China's colonial cessions were made, has stated that it is not bound by any such treaties, permanent or otherwise, and has announced its intention to reclaim all of Hong Kong—when the time is ripe. Growing uncertainty over the future has caused a slackening of confidence in Hong Kong, where land values have begun to decline, an economic slowdown has commenced, and local conflicts have intensified. China, recognizing that its own interests are affected, softened its stance by promising that Hong Kong, when reunited with the People's Republic, will be permitted to maintain its capitalist economy; only the British imprint must be erased. That has not been enough to overcome all the fears that have now arisen in this productive human beehive, but the Chinese and British continued their discussions in 1984 aimed at reaching a mutually satisfactory solution. Cautious optimism may also be justified by another reason for China's wanting to keep Hong Kong intact—if the People's Republic of China absorbs this colony with minimal disruption of the *status quo*, it could convince Taiwan (a far bigger prize in the Chinese government's eyes) that reunification with the mainland need not be a painful experience.

China's leading iron and steel producing center. But Shenyang remains the Northeast's largest and most diversified industrial city (Fig. 9–21). Machine-fabrication plants and other engineering works in Shenyang supply the entire country with various types of equipment, including drills, lathes, cable, and the like; as the economic capital of northeastern China's most productive farming area, this city is also an agricultural-processing center of national importance.

The southern part of the Northeast, however, is not the only area where coalfields and iron ores exist. From near the Korean border (where the iron ore is of considerably higher quality) to the far northeast, where the coal is of best quality, the region has natural resources in abundance; there is aluminum ore, molybdenum, lead, zinc, and limestone, an important ingredient in the manufacture of steel. Reflecting the northward march of development in the Northeast, the railway crossroads of Harbin has become a city of 2 million, with large machine-manufacturing plants (especially farm equipment such as tractors), a wide range of agricultural processing factories, and such industries as leather products, nylon, and plastics. Not only does Harbin lie at the convergence of five railroads, it also lies at the head of navigation on the Songhua River, connecting the city to the towns of the far northeast.

Communist doctrine in this Northeast frontier zone has always been somewhat relaxed; wages have always been higher than in China Proper in order to attract the skills the industries needed. In terms of planning, the rebuilt industries of Shenyang, Harbin, and Anshan, in some instances, are models of everything the Communist regime would like them to be; schools, apartments, recreational facilities, hospitals, and even old-age homes are all part of the huge industrial complex—and having a job means access for the worker to all of these. But most important, in the northeastern provinces the large-scale and efficient agricultural development stands in contrast to the parceled chaos of old China. Here the farms have been laid out more recently; they have always been larger, more effectively collectivized, and more quickly mechanized. In many ways, the Northeast is the image in which economic planners would like to remake all China.

# The New China

China at mid-century was a country torn by war almost as long as the people could remember. Foreign

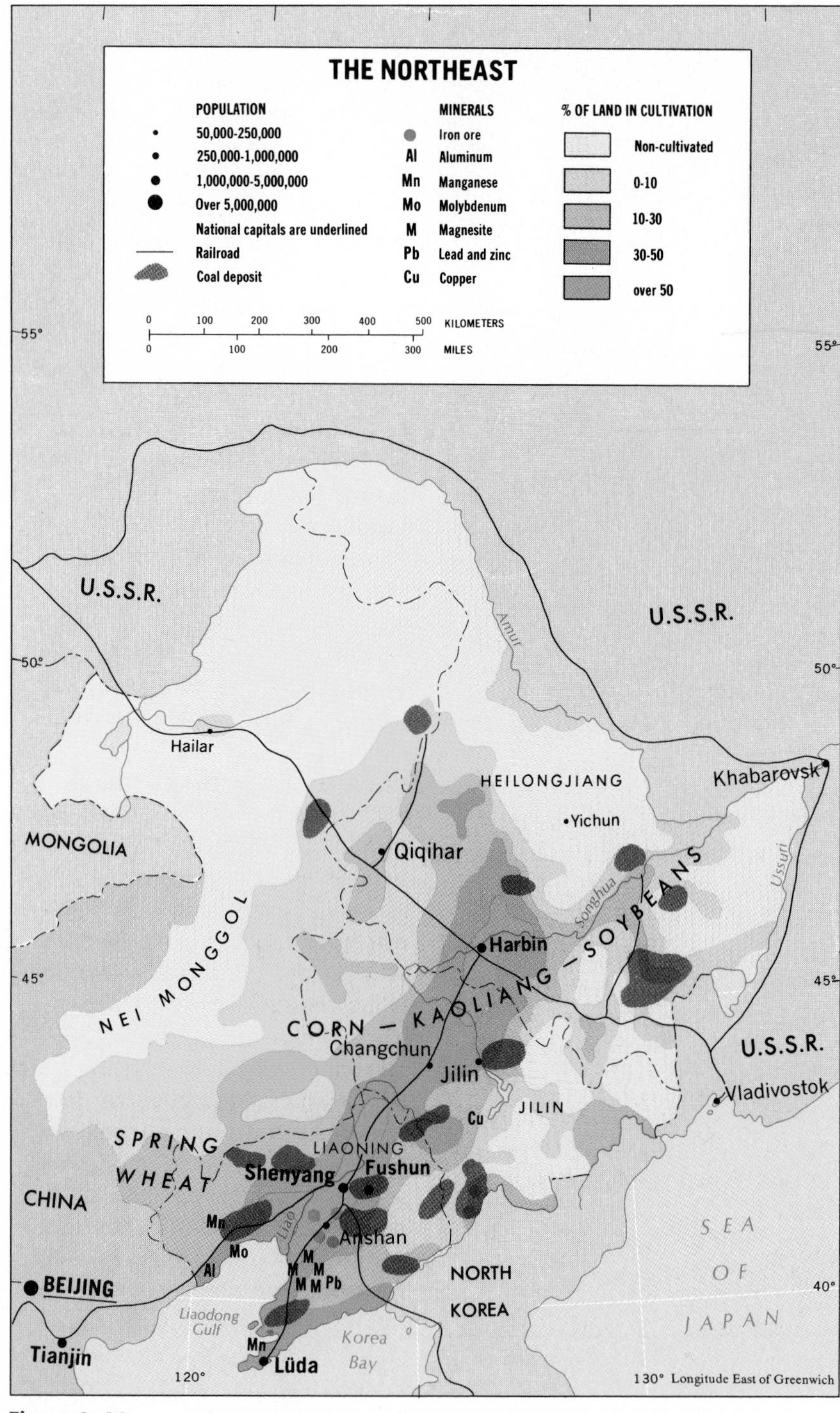


**Figure 9–20**

interference and exploitation and civil strife had been added to the problems China already faced, stemming from its huge and rapidly growing population, the limits to the country's productive capacity, and generally low levels of output. In the absence of any effective central government, the worst aspects of human nature could freely run their course; feudal warlords held parts of the country, village landlords mercilessly exploited their victims, corruption was rife, the cities had the world's most terrible slums where rats thrived by the millions, children were dying of starvation by the thousands, beggars were everywhere. Of course, in so large a country, there were variations. There were areas where things were not quite so bad; in other places, they were even worse. But when the new Communist government took control, it faced an enormous task of reconstruction, one at which the Nationalist regime of the 1920s and 1930s had only begun to nibble.

In 1949, the Communist government immediately initiated a program of reform in China. The central theme of this program, initially, was the reorganization of agriculture, following the doctrines applied earlier in the Soviet Union. But conditions in China were different. In the Soviet Union, there was a distinction—in terms of productivity—between the poor peasants who formed the majority of the inhabitants of thousands of rural villages, and the subsistence farmers; when the Soviet Communist government expropriated the landowning minority, it thereby acquired most of the productive land. In China, on the other hand, no such clear distinction existed. There were poor peasants, rich peasants, and "middle" peasants; almost every peasant family sold some produce on the market. As in Russia, there was a system of tenancy, which the new government was determined to wipe out. But, in China, the reorganization of agriculture was not simply a matter of expropriating the rich. Proportionately fewer peasants lived a pure subsistence existence; the risks that collectivization would reduce production were therefore greater.

Nevertheless, the reform program was initiated almost at once in every village of Han China (the minorities were at first excluded), involving the redistribution of the land of the landlords among all the landless families of the village. This distribution was based on the num-

**Figure 9–21**
Shenyang: the Northeast's leading manufacturing center, but also a classic example of the bleakness and pollution that mark China's industrial landscapes today.

ber of people per family, and was done strictly according to these totals; the landlords themselves were included and received their proportionate share of what had been their own land. Those landlords who had been guilty of the worst excesses were executed; most were absorbed into the new system. Meanwhile, the villagers who were not landlords but who did own the land they worked were allowed to keep their properties; hence, the reform program did not produce a truly egalitarian situation.

## Collectivization

During the early 1950s, the whole pattern of land ownership in China was being radically changed. In response, there *was* an increase in agricultural output, but not as great as the authorities had hoped. True, the taxes the new landowners had to pay on their acquired land were only on the order of one-third of what they previously paid their landlords, so they had some money available to buy fertilizer and improve yields. But in the absence of any prospect of large-scale mechanization of agriculture, the Chinese leaders now began to seek a method whereby still greater productivity could be achieved. The year 1952 produced exceptionally good harvests and encouraged the government to press ahead with a mutual-aid program. This program envisaged the creation of countrywide mutual-aid teams in which all the peasants in the community were encouraged to render assistance to each other during the planting and harvest seasons. In 1955, the first stage of collectivization started whereby all the peasants in a village pooled their land and their labor. Compensation would be in proportion to the size of each share and to the labor one had contributed; the farmers retained the right to withdraw their land from the cooperative if desired. In the beginning, participation was slow: by early 1955, only about 15 percent of China's farmers were in these new Agricultural Producers' Cooperatives (APCs). But then the government pressed for greater participation; by coercion and often quite brutal methods, compliance was raised to nearly 100 percent by 1956. In that year, the cooperatives were being turned into collective farms on which it was not the share of the land but the amount of the farmer's labor that determined returns; as with Soviet collectives, the farmer was able to hold on to a small plot of land for private cultivation. Thus, by the end of 1956, virtually the whole countryside was organized into socialist collectives; the Chinese peasants who had briefly held some land under the Agrarian Reform Law of 1950 lost it again to the collectives onto which they had been pressed.

## Communization

In 1958, the program of collectivization was carried one step further. In a very short time—less than a year—over 120 million peasant households, most of them already organized in collectives, were reorganized into about 26,500 People's Communes numbering about 20,000 people each. This was not just a paper reorganization: it was a massive modification of China's whole socioeconomic structure. This was to be China's Great Leap Forward from socialism to communism. Teams of party organizers traveled throughout the country; opposition was harshly put down while the new system was being imposed. In effect, the new communes were to be the economic, political, and social units of the Chinese Communist state. But most drastic was the way these communes affected the daily lives of the people. The adults, men and women, were organized into a hierarchy of "production teams" with military designations (sections or teams, companies, battalions, brigades). Communal quarters were built, families were disrupted through distant work assignments, children were put in boarding schools, and households were viewed as things of the past. (Later, the Chinese Communist party rescinded this order and allowed families to stay together.) The private lands of the collective system were abolished, and even the private and personal properties of the peasants were ruled communal, although this rule also was soon relaxed. The wage system of the collectives was changed in favor of an arrangement whereby the farmer or worker received free food and clothing plus a small salary—another step toward the Communist ideal.

The impact of the commune system on the face of China was enormous and immediate. Workers by the thousands tackled projects such as irrigation dams and roads; fences and hedges between the lands of former collectives were torn down and the fields consolidated. Villages were leveled, others enlarged. New roads were laid out to better serve the new system. Schools to accommodate the children of parents in the workers' brigades were built. Each commune was given the responsibility to maintain its own budget, to make capital investments, and to pay the state a share of its income to replace the taxation formerly levied on the collectives and individual peasants.

As might be expected, there was opposition to the introduction of the communes. Peasants in some areas destroyed their crops; elsewhere, they were left to be overgrown with weeds. But there was also peasant support for the concept, for most Chinese farmers know that some form of communal organization is necessary in their heavily overcrowded country. Nonetheless, the commune system faltered—not so much because of peasant rebellion but because of two other factors, one human and one environmental. The human factor must be obvious from the preceding description of the communes' introduction: it was all done with too much haste, too little planning (although there had been pilot communes in Henan Province and elsewhere), and too little preparation for what was truly an immense switch from private family living to a communal existence. But more devastating was the environment factor; China had successive bad years from 1959 to 1961, with severe droughts and destructive floods in various parts of the country. In combination with the negative effects of such resistance as the peasants did offer, it was enough to cause a moderation of the communization program and the restoration of such incentives as private peasant plots.

## Cultural Revolution

It was natural that China's revolutionary leaders would attempt to transform agriculture before turning their attention to other programs of change. China's earth had to produce far greater yields if China's people were to be fed; from China's land would have to come the revenues to sustain industrial and technological development; on China's farms there prevailed much of the worst corruption and exploitation in China's pre-Communist society. But Mao Zedong and his associates also wanted to rid Chinese society of its Confucian traditions and prescriptions (see *Box*), and to substitute the principles of communism as a philosophical basis for the new Chinese order. Further, the new leaders intended to give China industrial strength based on domestic resources and generated by heavy investment in this sector.

The industrial growth of China has been spectacular, helped at first by Soviet technicians and loans and sustained by new mineral discoveries resulting from intensified exploration. The energy picture brightened as oil reserves were located in the far Northeast, Sichuan and Xinjiang; China's coal reserves extend from the Northeast to the Chang Basin and Sichuan (Fig. 9–17). China's coal production in recent years has ranked third in the world (734 million tons in 1982 as against 791 million for the Soviet Union and 838 million for the United States); in terms of iron ore and other vital raw materials, China's domestic reserves are as good as those of the United States or the Soviet Union. China's planners have sought to diversify the country's manufacturing base and to disperse industry into the interior (the east has always been China's industrial heartland); to a considerable extent, they have succeeded. Compared to the slow progress of agriculture, China's industrial march has been powerful.

China's First Five-Year Plan (1952–1957) gave an indication of the aspirations of China's new leaders: more than half the state's investments were poured into industry, against only 8 percent into agriculture. Of the enormous allotment to the manufacturing sector, about four-fifths went to heavy industry. True, when the Second Five-Year Plan (1957–1962) faltered as a result of the problems associated with the communization program, these priorities were reconsidered and agriculture got a much larger share. But the first plan gives a good indication of Communist Chinese priorities. A strong military establishment needs a heavy industrial base, and the Beijing government was determined to create it. Recent problems notwithstanding, those priorities still exist.

China's major difficulties in the sociopolitical arena have involved the so-called Cultural Revolution and the reorientation of China's population from Confucian to Communist precepts. Mao Zedong, apparently fearing that the China of the 1960s and 1970s might abandon its revolutionary fervor and become "revisionist" like the modern Soviet Union, initiated a "Cultural Revolution" to rekindle the old enthusiasm and to recoup and conceal losses sustained during the ill-fated Great Leap Forward. The Cultural Revolution centered initially on education, and programs of adult education were begun in the communes. University and college teachers joined workers on assembly lines and peasants in the

# CONFUCIUS

Confucius (*Kongfuzi* in Chinese) was China's most influential philosopher and teacher, whose ideas dominated Chinese life and thought for over 20 centuries. Confucian ideals were considered incompatible with Communist doctrine by the leaders who took control of China in 1949, and the elimination of Confucian principles was one of Mao Zedong's primary objectives. Confucius left his followers a wealth of "sayings," many of which were frequently quoted as a part of daily life in China. Mao's "Red Book" of quotations was part of the campaign to erase the Confucian tradition.

Confucius was born in 551 B.C. and died in 479 B.C. He was one of many philosophers who lived and wrote during China's classical age; during his lifetime he was prominent, although no more so than a number of other philosophers. The Confucian school of thought was one of several to arise during this period; the philosophies of Daoism also emerged at this time. Confucius was appalled at the suffering of the ordinary people in China, the political conflicts, and the harsh rule by feudal lords. In his teaching, he urged that the poor assert themselves and demand the reasons for their treatment at the hands of their rulers (thereby undermining the absolutism of government in China); he also tutored the indigent as well as the privileged, giving the poor an education that had hitherto been denied them, thereby ending the aristocracy's exclusive access to the knowledge that constituted power.

Confucius, therefore, was a revolutionary in his time—but he was no prophet. Indeed, he had an aversion to supernatural mysticism and argued that human virtues and abilities should determine a person's position and responsibilities in society. In those days, it was believed that China's aristocratic rulers had divine ancestors and governed in accordance with the wishes of these godly connections. Confucius proposed that the dynastic rulers give the reins of state to ministers chosen for their competence and merit. This was another Confucian heresy, but, in time, the idea came to be accepted and practiced.

Notwithstanding his earthly philosophies, Confucius took on the mantle of a spiritual leader after his death. His ideas spread to Korea, Japan, and Southeast Asia, and temples were built in his honor all over China. As so often happens, Confucius was a leader whose teachings were far ahead of his time. His thoughts emerged from the mass of philosophical writing of his day—the Zhou period, classical dynastic period of Chinese civilization—to become guiding principles during the formative Han Dynasty. Confucius had written that the state should not exist for the pleasure and power of the aristocratic elite; it should be a cooperative system, and its principal goal should be the well-being and happiness of the people. As time went on, a mass of Confucian writings evolved, much of these Confucius never wrote but they were attributed to him nevertheless. At the heart of this body of literature lies the Confucian Classics, 13 texts that became the focus of education in China for 2000 years. In the fields of government, law, literature, religion, morality, and in every conceivable way the Confucian Classics were the Chinese civilization's guide. The whole national system of education (including the state examinations through which everyone, poor or privileged, could achieve entry into the arena of political power) was based on the Confucian Classics. Confucius was a champion of the family as the foundation of Chinese culture, and the Classics prescribe a respect for parents and the aged that was a hallmark of Chinese society. It has been said that to be Chinese—whether Buddhist, Christian, or Communist—one would have to be a Confucian; hardly any conversation of substance could be held without reference to some Confucian principle.

When the Western powers penetrated China, Confucian philosophy came face to face with practical Western education; for the first time, a segment of China's people (initially small) began to call for reform and modernization, especially in teaching. Confucian principles could guide an isolated China, but they were found wanting in the new age of competition. The Manchus resisted change, and during their brief tenure the Nationalists under Sun Yat-sen tried to combine Confucianism and Western knowledge into a neo-Confucian philosophy. But it was left to the Communists, beginning in 1949, to attempt to substitute an entirely new set of principles to guide Chinese society. Confucianism was attacked on all fronts, the Classics were abandoned, ideological indoctrination pervaded the new education, and even the family was assaulted during the early days of communization. But it is difficult to eradicate two millennia of cultural conditioning in a few decades. The dying spirit of Confucius will haunt physical and mental landscapes in China for years to come.

fields; the idea was to eliminate the old Confucian distinction between mental and manual labor and to prevent the emergence of an educated elite in the urban areas, who would live in comfort off the labors of the peasants. But serious problems arose. In 1966, China's schools were closed on the grounds that the entry system was unfair to the mass of the students and the teachers perpetuated bourgeois principles. In this way, millions of young people found themselves at loose ends—ready to be recruited into the Red Guards, the organization that was to become the heart of the revolution. But just as the Great Leap Forward simply did not mobilize enough of the needed energy in China, so the Cultural Revolution failed to get the necessary commitment from the people. The Red Guards met with opposition, and there was even fighting between them and cadres organized to support and protect local party leaders. Once again, the country was badly dislocated by a wholesale revolutionary program, and it did not stop there. People were encouraged to criticize their superiors if they thought they might be corrupt or incompetent. Further factions were created; even the army was threatened. Violent battles occurred between workers and Red Guards and even between rival pro-Mao groups.

Since 1967, the Communist Party as well as the country has at times been badly divided; suspicions and accusations had been brought to the surface and could not easily be buried again. After Mao's death, these divisions led to a power struggle that nearly plunged China into civil war. But worst of all, China had, once again, been set back by the instability the whole affair had caused—the disruption of education, the diversion of the activities of workers who should have been on their jobs, the loss of managerial personnel accused and forced out by the Maoists, the interruption of the country's transport services as millions of people moved about for no other than political reasons. That China could be able to absorb the collectivization drive, the Great Leap Forward, and the Great Proletarian Cultural Revolution—and could still show substantial progress in agriculture, industry, and several other spheres (e.g., such as the production of nuclear weapons)—is proof of the amazing capacity of this huge country and its people to overcome enormous odds.

## The "Four Modernizations"

Historical geographers of the future will undoubtedly point to the decade following Mao's death in 1976 as a crucial time in the development of the contemporary Chinese state. With the ideological struggle between the radicals and the moderates resolved in favor of the pragmatic moderates, the new Chinese leadership was able to turn its attention to the country's urgent needs in several spheres. Premier Hua Guofeng decided to include Deng Xiaoping, previously dismissed as too "moderate," as the country's deputy premier. Deng's pragmatic approach to China's future soon made its mark on government policies; in 1981–1982, when a new constitution was drawn up, he became the strongest member of the ruling triumvirate. Hua Guofeng was demoted in favor of one of Deng's proteges, Hu Yaobang, and the third member of the powerful threesome, Zhao Ziyang, became the premier. The rise of Zhao Ziyang illustrates China's determination to maintain its "revi-

**Figure 9–22**
Billboards and posters blanket most Chinese public places: this one advises on population growth, a crucial issue in the 1980s.

sionist'' course. As the Communist Party Secretary in the province of Sichuan, he had pioneered agricultural and industrial reforms that allowed peasants and workers to keep and sell their surpluses. His system was so productive that it spread all over China, and Zhao became nationally prominent. Soon, he was appointed premier, symbolizing the new order.

Deng Xiaoping, however, remained the leading force in China's overall development. In a country where exhortations, slogans, and other calls for action—including ubiquitous wall posters and huge billboards (Fig. 9–22)—are commonplace, Deng introduced his plea for ''Four Modernizations.'' (Although Deng is often credited with the concept, it was Zhou Enlai who introduced it as long ago as 1964—only to see it engulfed by the Cultural Revolution.) Briefly, these involved: (1) the rapid modernization and mechanization of agriculture, (2) the immediate upgrading of defense forces, (3) the modernization and expansion of industry, and (4) the development of science, technology, and medicine. Importantly, Deng proposed that Mao Zedong's long-prevailing policy of self-reliance be abandoned through the expansion of foreign trade, the purchase of foreign technology and machinery, and the use of foreign scientists to help China in its modernization drive. Moreover, capitalist-type incentives to spur production would be used, and education would return to practices where success in examinations rather than political attitudes and associations counted most.

It was not difficult to see in this new determination to modernize the reflection of Japan just slightly more than a century ago. Japan, too, broke its long-term isolation and embarked on an intensive modernizing drive—one that in a remarkably short time produced an imperial power whose colonial acquisitions encompassed much of East Asia. China's initial target year is 2000. By that year, its planners hope, agriculture will be substantially mechanized and self-sufficiency, even in years of drought, assured; the gap between China and the Soviet Union in military capacity will no longer exist; industrialization will rank China as one of the world's three leading producers; and the temporary dependence on foreign technology and science will have ended.

These are optimistic predictions, but, in the 1980s, the groundwork was being laid. Japanese technicians were in China planning some of the 120 large industrial complexes to be operational by 1985 (according to the new regime's First Ten-Year Plan, announced in 1978, but now behind schedule). In return, Japan was to receive Chinese oil and coal. American researchers were in China to help in mineral exploration; a U.S. corporation was contracted to develop additional Chinese iron mines. China also signed a major trade agreement with the European Economic Community. The need for foreign exchange was among the factors that led China to open its doors to hundreds of thousands of tourists. As the Maoist doctrines of the previous 30 years faded, the modernization effort accelerated.

China's cultural landscapes reflect the successes as well as the failures of the modernizing drive. In much of rural China, the negative impacts of three turbulent decades are more clearly etched than anything else. The symbols are many: traditional dwellings in disrepair, brick buildings of the communal period rising above the old village walls, fading slogans on their walls. In China's villages, the uncertainties and dislocations of the 1960s and 1970s pervade the environment. The cities present scenes that resemble those of nineteenth-century urban England during the first half-century of the industrial revolution. Pollution-belching smokestacks rise as a forest from city blocks; factories large and small vie for space with apartment buildings, schools, and architectural treasures of the past. There seems to be no zoning plan, nor any protection for the environment. Turn a street corner, and the deafening roar of machines drowns out everything else; walk along the street, and as the roar fades another ear-splitting sound replaces it. Across the street is a school, and one wonders how students can learn under such conditions, day after day. Gaze upward, and the sun is shrouded by an ever-present cloud of pollution so dense that the air is dangerous to breathe (for an idea of conditions see Fig. 9–21). A bridge ahead brings another vivid portrait of the cost of China's industrial march: the water below is little more than an industrial sewer, filled with waste and colored gray by tons of chemicals.

All this must be seen against the reality of a China that, in the middle 1980s, was still far from a developed country. Grain imports were still needed to supplement inadequate production at home. Annual income per person was only around $300. Urbanization hovered around 28 percent. More than 75 percent of the huge labor force worked in agriculture. Such statistics emphasize the magnitude of the effort on which the new China is embarked, and they suggest that more than two decades will be required to achieve its goals. On the other hand, there is no doubt that China has the human and material resources to bring its "four modernizations" within ultimate reach.

# Taiwan

Less than 200 kilometers (120 miles) off the southeast coast of China lies Taiwan, the island where the defeated Chiang Kai-shek and the remainder of his Nationalist faction took refuge in 1949. In all, about 2 million refugees reached Taiwan between the end of World War II and the mainland revolution four years later. Assisted by the United States, Taiwan has survived as "Nationalist China" on the very doorstep of the People's Republic.

Mountainous Taiwan is one island in a huge arc of islands and archipelagoes that stretches along Asia's eastern coasts from the Kuriles north of Japan to Indonesia south of the Philippines. Unlike many of the other islands, Formosa (as the Portuguese called the island when they tried to colonize it) is quite compact and has rather smooth coastlines; very few good natural harbors exist. In common with several of its northern and southern neighbors, the island's topography has a linear, north-south orientation and a high mountainous spine. In Taiwan, this backbone lies in the eastern half of the island (Fig. 9–23), and elevations in places exceed 3000 meters (10,000 feet). Eastward from this forested mountain spine, the land surface drops rapidly to the coast and there is very little space for settlement and agriculture; but westward there is an adjacent belt of lower hills and, facing the Taiwan Strait, a substantial coastal plain. The overwhelming majority of Taiwan's 19.5 million inhabitants live in this western zone, near the northern end of which the capital, Taipei (3.9 million), is positioned.

There have been times when Taiwan was under the control of rulers based on the mainland, but history shows that its present separate status is really nothing new. Known to the Chinese since at least the Sui Dynasty, it was not settled by mainlanders until the 1400s. For a time, there was intermittent Chinese interest in Taiwan, then the Portuguese arrived. During the 1600s, there was conflict among the Europeans, including the Dutch and the Spaniards, but, in 1661, a Chinese general landed with his mainland army and ousted the foreign invaders. This was during the decline of the Ming Dynasty and the rise of the Manchus, and, much like the Nationalists of the 1940s, thousands of Chinese refugees fled

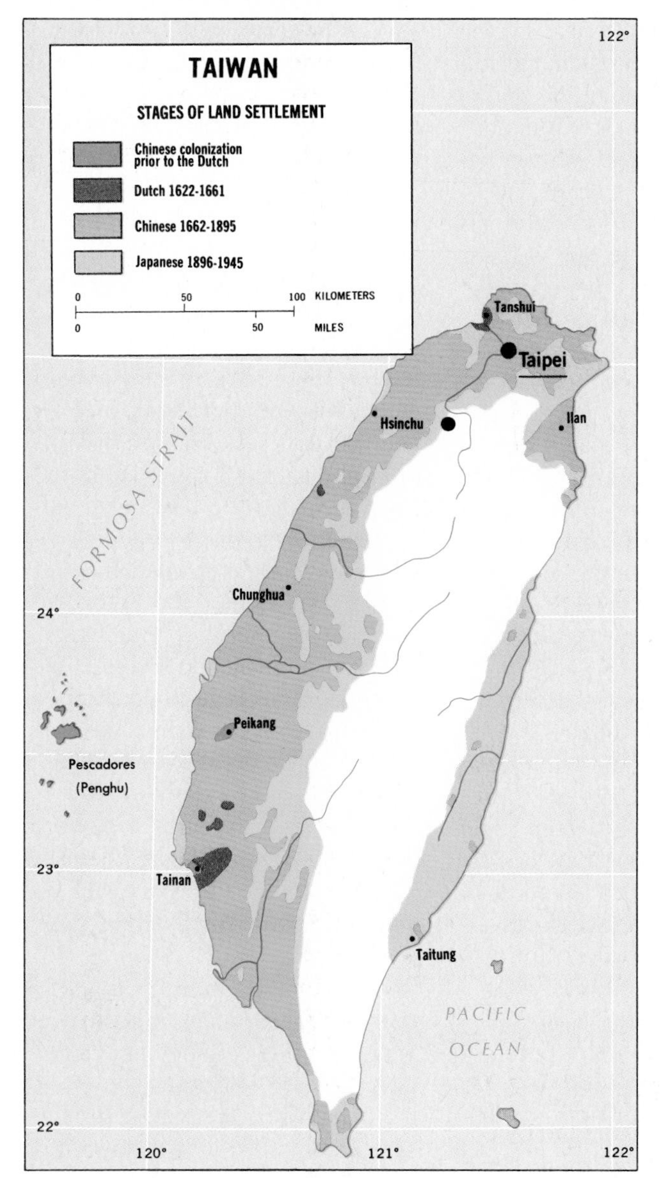


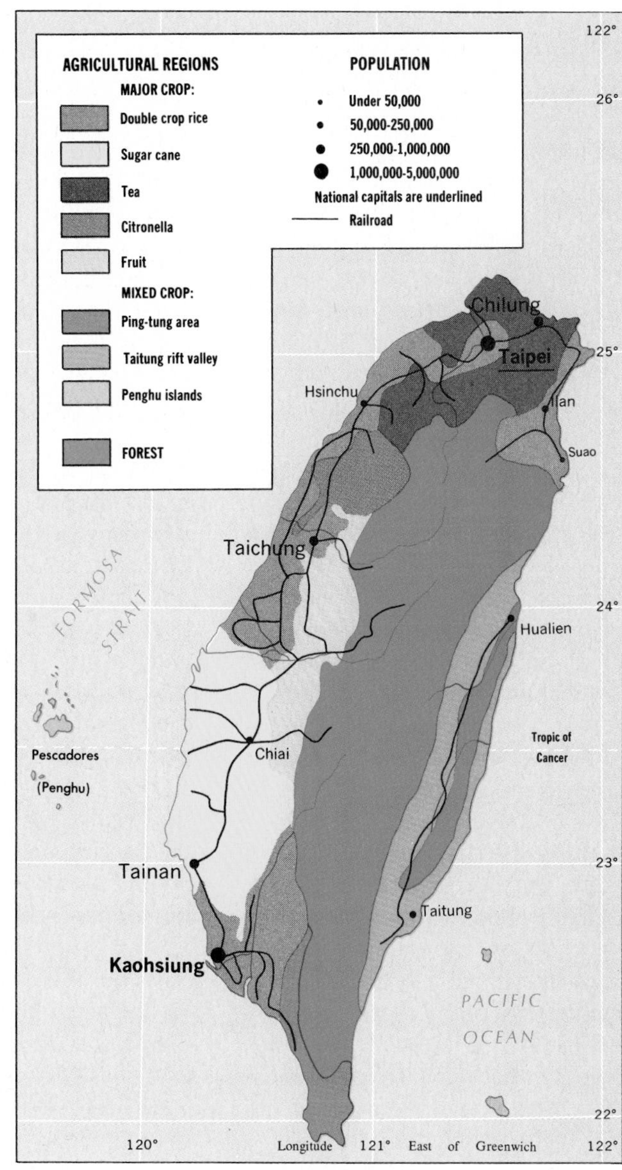


Figure 9–23

to Taiwan rather than face Manchu domination. But before 1700, the island fell to the Manchu victors and became administratively a part of Fujian (Fukien) Province. By then, the indigenous population, of Malayan stock, was already far outnumbered and in the process of retreating into the hills and mountains.

Taiwan did not escape the fate of China itself during the second half of the nineteenth century. In 1895, it fell to Japan as a prize in the war of 1894–1895; for the next half century, it was under foreign rule. The Japanese saw more of the same possibilities in Formosa as they saw in Manchuria: the island could be a source of food and raw materials and, if developed effectively, a market for Japanese products. Thus, Japan engaged in a prodigious development program in Taiwan, involving road and railroad construction, irrigation projects, hydroelectric schemes, mines (mainly for coal), and factories. The whole island was transformed. Farm yields rose rapidly as the area of cultivated land was expanded and better farming methods were introduced; in the sphere of education, the Japanese attacked the illiteracy problem and oriented the entire system toward their homeland.

Japanese rule ended with Japan's collapse in World War II; after 1945, the island was briefly returned to Chinese control, but, before long, it became the last stronghold of the Nationalists and, once again, its mainland connections were severed. A large influx of refugee Nationalists arrived in the late 1940s, and they were fortunate; here was one of the few parts of China where they could have found a well-functioning economy, productive farmlands, the beginnings of industry, good communications—and the capacity to absorb an immigrant population of more than 2 million. True, U.S. assistance made a major difference, but Taiwan itself had much to offer.

From 1945 until 1949, Taiwan was officially part of China, torn as it was by the civil war. This period was one of quite ruthless and generally unpopular government on the island, which had to do with the degree to which the Japanese had transformed its life and culture, and the large residue of sympathy they had left behind. But 1949 brought a new beginning with the installation of Chiang Kai-shek's regime in Taipei and the arrival of the mainland immigrant horde, which constituted about 20 percent of the total population in 1950. In some ways, the problems faced by Chiang's regime were similar to those of the new mainland government. The Japanese had achieved much, but the war had brought destruction; and, as on the nearby continent, there was a need for land reform and increased agricultural yields. While U.S. assistance helped reconstruct Taiwan's transport network and industrial plants, the Taipei government initiated a program whereby the farm tenancy system was attacked. In 1949, most Taiwanese farmers were still tenant farmers who paid rent amounting to as much as 70 percent of the annual crop. By law, the maximum allowable rent was reduced to 37.5 percent; the land of large landowners was bought by the government in exchange for stock holdings in large government-owned corporations, and this land was resold to farmers at low interest rates. Through incentives, seed improvement, expanded irrigation, more fertilizers, and new double-crop rotations, it has been possible to double the per hectare yields of Taiwan's major crops since 1950.

These efforts to raise the volume of the grain harvest are familiar to anyone who has studied the Beijing regime's attempts to do the same on a much larger scale. Taiwan lies astride the Tropic of Cancer (latitude 23½°N), thus, its lowland climate is comparable to that of the Xi Jiang Basin; rice is the leading staple, and about two-thirds of the harvest is from double-cropped land. Wheat and sweet potatoes are also important staples, and sugarcane in the lower areas and tea higher up are grown for cash. But in at least one respect Taiwan faces problems even more serious than China itself. In recent decades, the population's annual growth rate has been very high; in the 1950s and early 1960s, it was 3.4 percent, one of the highest in the world. In the early 1970s, it averaged 2.4 percent, still a high rate of increase, so that annual grain harvests must rise continuously just to keep pace; the latest figures show a decline to about 1.8 percent. The Green Revolution has helped Taiwan overcome its shortage of good farmland, but there are limits to what can be accomplished—in parts of the coastal plain, the rural population density exceeds 1500 per square kilometer (4000 per square mile).

In contrast to overwhelmingly rural China, four-fifths of Taiwan's inhabitants live in urban areas. The Japanese began an industrialization program that temporarily faltered during the Nationalist takeover, but the 1960s and 1970s witnessed the island's industrial resurgence. In the early 1980s, agriculture accounted for under 25 percent of Taiwan's labor force and income per capita was over $2100 per year, statistics no other Asian country could match (except, of course, Taiwan's model, Japan). All this has been accomplished despite the limitations of Taiwan's raw material base. Mineral resources are very limited (although power sources are ample); there is no large iron ore reserve, for example. Along the western flank of the mountain backbone there are coal deposits, and numerous streams on this well-watered isle provide good opportunities for hydroelectric-power development. There even is some petroleum and natural gas. But the textile industry, Taiwan's major foreign-revenue earner, depends on imported raw cotton; the aluminum industry gets its bauxite from Indonesia, and ores and minerals rank high on the list of imports. On the other hand, the Taiwanese in 1976 concluded an agreement with Saudi Arabia whereby Chinese technicians would help develop the Arabian economy in return for large loans (and oil shipments) to support Taiwan's vigorous industrialization program. In recent years, Taiwan's planners have moved toward the establishment of a domestic iron and steel industry, nuclear power plants, shipyards, a larger chemical industry, and an electrified and more comprehensive railroad network.

In the 1980s, Taiwan's urban-commercial economy is also advancing rapidly. Taipei, the capital city, contrasts strongly against mainland Chinese cities of similar size; busy traffic, ample consumer goods, and other aspects of Taiwan's capitalistic economics produce a strikingly different atmosphere and urban landscape (Fig. 9–24). The latest effort is to shift the Taiwanese economy away from labor-intensive manufacturing toward today's high-technology industries, especially microelectronics and computers. To that end, such projects as the science-based Hsinchu Industrial Park (40 miles southwest of Taipei) are being given the highest development priority. In fact, the western Pacific rim is increasingly described as the domain of the "Four New Japans"—Taiwan, South Korea, Hong Kong, and Singapore—as these burgeoning

**Figure 9–24**
Taipei, Taiwan: Chinese capitalism produces a cityscape radically different from that of the communist People's Republic.

beehive countries successfully follow the Japanese example with recent annual size increases in their economies that have ranged up to 9 percent (more than double Japan's yearly 4 percent gain in the early 1980s).

In the international arena, Taiwan's position has become more difficult as a result of China's emergence from its isolation. Taiwan was expelled from the United Nations to make way for China; the United States, Taiwan's vital ally, closed its embassy, and, since 1979, has been under pressure to lower its commitments to Taipei in favor of more productive relationships with Beijing. In 1976, the Canadian government refused to permit Taiwan to participate in the Montreal Olympic Games. But Taiwan remains in effect a U.S. protectorate, a bastion of support for the American presence in East Asia. A military alliance between the U.S. and Taiwan was signed in 1954. At any rate, mainland China has opposed a policy wherein the United States would recognize both Beijing and Taipei, so that a two-China policy is no solution to this persistent problem.

# Korea

The peninsula of Korea reaches from the East Asian mainland toward Japan and, not surprisingly, has experienced a turbulent history. For uncounted centuries it has been a marchland, a pawn in the struggles among the three powerful countries that surround it (Fig. 9–25). Korea has been a dependency of China and a colony of Japan; when it was freed of Japan's oppressive rule in 1945, it was divided for administrative purposes by the victorious allied powers. This division gave North Korea beyond the 38th parallel to the forces of the Soviet Union, and South Korea to U.S. forces. In effect, Korea traded one master for two others. The country was never reunited, as North Korea immediately fell under the Communist ideological sphere and South Korea, with massive American aid, became part of East Asia's non-Communist perimeter, of which Japan and Taiwan are also parts. Once again, it was the will of external powers that prevailed over the desires of the Korean people themselves. In 1950, North Korea sought to reunite the country by force and invaded the South across the 38th parallel, an attack that drew a U.N. military response led by the United States. This was the beginning of a devastating conflict (1950–1953) in which North Korea's forces pushed far to the south, only to be driven back into their

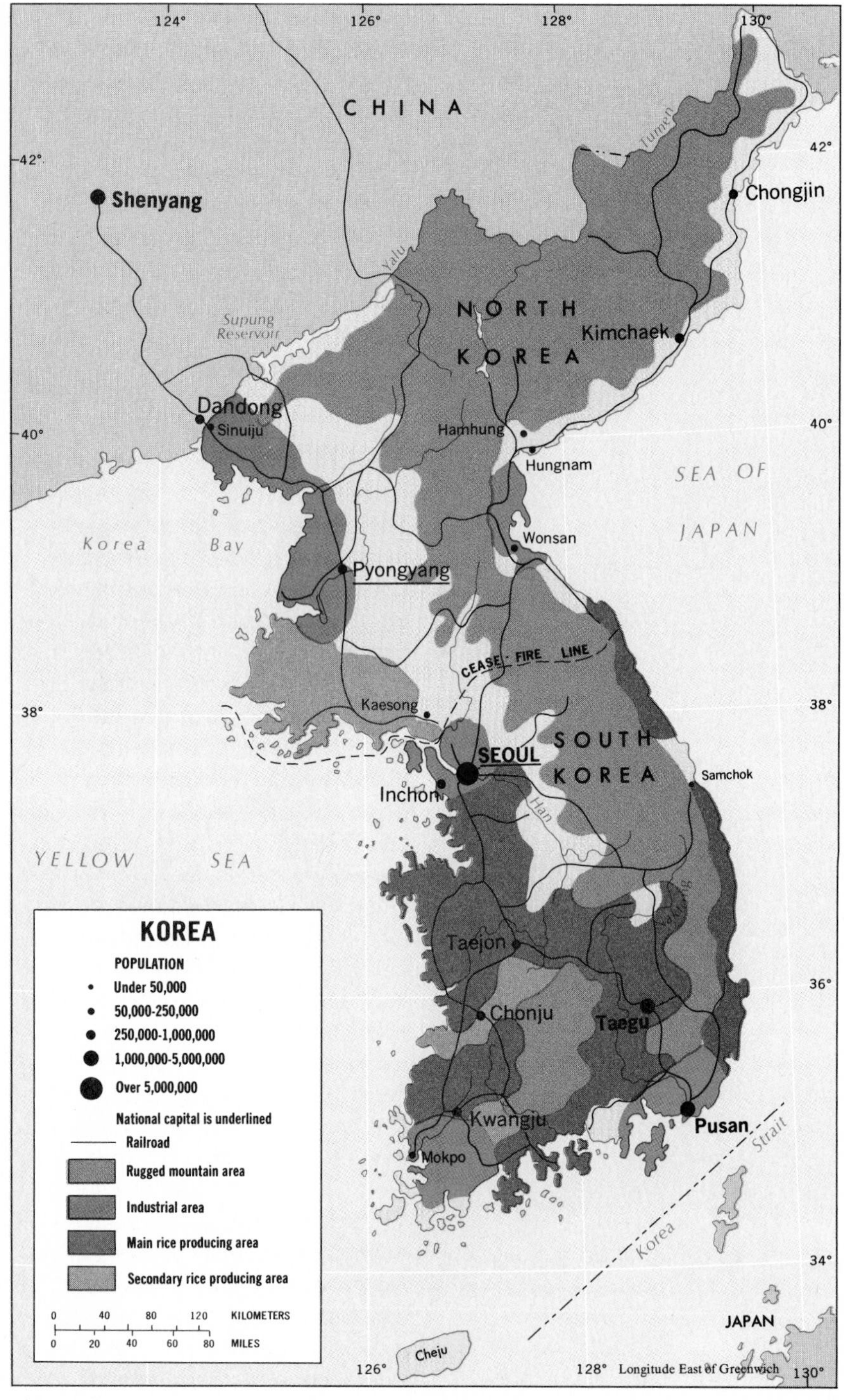


Figure 9–25

own half of Korea almost to the Chinese border, whereupon the Red Chinese army entered the war to drive the U.N. forces southward again. Eventually, a cease-fire was arranged in mid-1953, but not before the people and land of Korea had been ravaged in a way that was unprecedented even in its violent past. The boundary between North and South, demarcated in 1953, remained unchanged 30 years later (Fig. 9–26), an ever-present symbol of divided Korea.

The fragmentation of Korea into two political units happens, to a considerable extent, to coincide with the regional division any physical geographer might suggest as an initial breakdown of the country. As Fig. I–9 shows, South Korea is moister than the North; Fig. I–10 emphasizes the temperate maritime climatic conditions that prevail over most of the South as opposed to the more extreme continental character of the North. Fig. I–13 shows much of the Korean peninsula to lie under mountain soils, but the most extensive belt of more productive soils is found in a zone that lies largely in South Korea. Fig. I–12 proves South Korea to possess a sizable area of broadleaf evergreen trees like that of southern Japan; most of the remainder of the peninsula has a deciduous forest as its natural vegetation. And Fig. I–14 shows that the majority of Korea's 63 million people live in a zone in the western part of the country, an area that widens southward, so that the great majority of the population is in South Korea, roughly in a triangle to the west of a line drawn from Seoul to Pusan.

Thus, a great number of contrasts exist between North and South Korea. The North is continental, the South peninsular; the North is more mountainous than the South; the North can grow only one crop annually and depends on wheat and millet, whereas, in much of the South, multiple cropping is possible and the staple is rice; the North has significantly fewer people than the South (20.1 million against 42.6 million), but the North has a food deficit, whereas the South comes close to feeding itself (it has even had surpluses in the past). But perhaps the most striking contrast lies in the distribution of Korea's raw materials for industry. North Korea has always produced vastly more coal and iron ore than the South, and the overwhelming majority of all other Korean production also comes from the North. Similarly,

**Figure 9–26**
The heavily-guarded demarcated boundary between North and South, symbol of divided Korea.

North Korea has maintained its great lead in hydroelectric-power development, an advantage that was initiated by the Japanese and has been sustained. In recent years, several discoveries of coal and iron have been made in South Korea, but the overall balance relating to the bases for heavy industry remains strongly in favor of North Korea.

Although it is practically impossible to obtain data concerning North Korea's external trade, one thing is certain: the superimposed political boundary between the two Koreas has been virtually airtight, and no trade has passed across it. North Korea's trading connections have been with China and the Soviet Union, those of South Korea with the United States, Japan, and West Germany. Yet, it must be clear from what has just been said that North Korea, the focus of heavy industry on the peninsula, could transact a great deal of business with more heavily agricultural South Korea. The two Koreas are potentially interdependent in so many ways; even within the industrial sector itself there are complementarities. While North Korea specializes in heavy manufacturing, such light industries as cotton textiles and food processing still prevail in South Korea—although heavy industries are now developing (especially around Pusan) since the discovery of iron ores and Samchok's coal (Fig. 9–25). North Korea's chemical industries produce fertilizer; South Korea needs it. South Korea long exported food to what is today North Korea. North Korea has much electric power; the transmission lines that used to carry it to the South were cut soon after the postwar division of the country.

South Korea also has the largest part of the domestic market as well as the biggest cities, the old capital of Seoul (10.2 million) and the southern metropolis of Pusan (3.9 million). Pyongyang, the North Korean headquarters (Fig. 9–27), is just approaching 1.7 million. Seoul is the country's primate city (Fig. 9–28), whose location in the waist of the peninsula, midway between the industrialized northern and agriculturally-productive southern regions of Korea, has been an advantage. Pusan is closest to Japan and grew in phases during this century, first under Japanese stimulus and later as the chief U.S. entry point. Pyongyang was also developed by the Japanese, and it lies at the center of Korea's leading industrial region, which is merging with the mining-industrial area of the northwest.

The Koreans, like the Vietnamese, are one people, with common ways of life, religious beliefs, historic and emotional ties, and with a common language. In the entire twentieth century, the Koreans have barely known what self-determination in their own country would be like, but they have not forgotten their aspirations. After seemingly endless suffering through conflicts and wars, most of them precipitated by the "national" interests of other states, Korea today, ideologically and politically divided as it is, still retains the ingredients of a cohesive national entity. In 1972, a first effort was begun to thaw the hostilities between North and South Korea through the reopening of trade connections, but this attempt failed.

The next attempt at reconciliation will almost surely have to await the passing of the present North Korean regime, led by the aged Kim Il Sung (b. 1911)—a long-time leader installed by Stalin—who has made himself the focus of a fanatical personality cult. The juxtaposition of the cityscapes of Pyongyang and Seoul (Figs. 9–27, 9–28)—only 200 kilometers (125 miles) from each other—underscores the current ideological polarization that keeps the two Korean entities apart. Perhaps in the not-too-distant future, with great power "spheres of influence" in a state of flux, it might become more feasible for a united progressive Korea to function as a cultural and economic whole. The potential advantages for both North and South will certainly continue to exist until such a time.

**Figure 9–27**
Speckles of colored balloons cannot mask Pyongyang's drabness, dominated by Kim Il Sung's Tower of the "Juche Idea" (rear).

**Figure 9–28**
Burgeoning Seoul, only 200 kilometers but a world apart from gloomy communist Pyongyang.

# References and Further Readings


Appel, B. *Why the Chinese Are the Way They Are* (Boston: Little, Brown, 2 rev. ed., 1973).

Bartz, P. *South Korea* (Oxford, Eng.: Clarendon Press, 1972).

Baum, R., ed. *China's Four Modernizations: The New Technological Revolution* (Boulder, Colo.: Westview Press, 1980).

Buchanan, K. *The Transformation of the Chinese Earth* (London: Bell & Sons, 1970).

"Capitalism in the Making," *Time*, April 30, 1984, 25–35.

Chang, S. "Modernization and China's Urban Development," *Annals of the Association of American Geographers*, 71 (1981), 202–219.

Chisholm, M. *Modern World Development* (Totowa, N.J.: Barnes & Noble, 1982).

Chiu, T. & So, C., eds. *A Geography of Hong Kong* (London: Oxford University Press, 1983).

Cole, J. *The Development Gap: A Spatial Analysis of World Poverty and Inequality* (Chichester, Eng.: John Wiley & Sons, 1981).

Cressey, G. *Land of the 500 Million: A Geography of China* (New York: McGraw-Hill, 1955).

de Souza, A. & Porter, P. *The Underdevelopment and Modernization of the Third World* (Washington: Association of American Geographers, Commission on College Geography, Resource Paper No. 28, 1974). List of goals adapted from pp. 3–4.

East, W. et al. *The Changing Map of Asia* (London: Methuen, 1971).

Fairbank, J. et al. *East Asia: Tradition and Transformation* (Boston: Houghton Mifflin, 1973).

Fullard, H., ed. *China in Maps* (Chicago: Denoyer-Geppert, 1968).

Gilbert, A., ed. *Development Planning and Spatial Structure* (New York: John Wiley & Sons, 1976).

Ginsburg, N. *The Pattern of Asia* (Englewood Cliffs, N.J.: Prentice-Hall, 1958).

Hinton, H. *The Far East and Western Pacific, 1984* (Washington: Stryker-Post Publications, 1984).

Ho, S. *Economic Development of Taiwan* (New Haven, Conn.: Yale University Press, 1978).

Hofheinz, R. & Calder, K. *The East Asia Edge* (New York: Basic Books, 1982).

Hoyle, B., ed. *Spatial Aspects of Devel-*

*opment* (London: John Wiley & Sons, 1974).
Hsieh, C. *Atlas of China* (New York: McGraw-Hill, 1973).
"Inside the Hermit Kingdom [North Korea]," *Time*, May 30, 1983, 34.
Johnson, E. *The Organization of Space in Developing Countries* (Cambridge: Harvard University Press, 1970).
Knapp, R., ed. *China's Island Frontier: Studies in the Historical Geography of Taiwan* (Honolulu: University Press of Hawaii, 1980).
Lattimore, O. *Inner Asian Frontiers of China* (New York: American Geographical Society, 1940).
Leung, C. & Ginsburg, N., eds. *China: Urbanization and National Development* (Chicago: University of Chicago, Department of Geography, Research Paper No. 196, 1980).
Leung, C. et al., eds. *Hong Kong: Dilemmas of Growth* (Canberra: Australian National University Press, 1980).
Liang, E. *China: Railways and Agricultural Development, 1875–1935* (Chicago: University of Chicago, Department of Geography, Research Paper No. 203, 1982).
Lohr, S. "Four 'New Japans' Mounting Industrial Challenge," *New York Times*, August 24, 1982, pp. 1, 35.
Ma, L. "Preliminary Results of the 1982 Census in China," *Geographical Review*, 73 (1983), 198–210.
Ma, L. & Hanten, E., eds. *Urban Development in Modern China* (Boulder, Colo.: Westview Press, 1981).
Ma, L. & Noble, A., eds. *Chinese and American Perspectives on the Environment* (London and New York: Methuen, 1981).
Mabogunje, A. *The Development Process: A Spatial Perspective* (London: Hutchinson, 1979).
March, A. *The Idea of China: Myth and Theory in Geographic Thought* (New York: Praeger, 1974).
McCune, S. *Korea's Heritage: A Regional and Social Geography* (Rutland, Vt.: Tuttle, 1956).
McGee, T. G. "Western Geography and the Third World," *American Behavioral Scientist*, 22 (September-October 1978), 93–114.
Murphey, R. *The Fading of the Maoist Vision: City and Country in China's Development* (London and New York: Methuen, 1980).
Murphey, R. *Shanghai: Key to Modern China* (Cambridge: Harvard University Press, 1954).
Myers, R. *The Chinese Economy: Past and Present* (Belmont, Cal.: Wadsworth, 1980).
Pannell, C., ed. *East Asia: Geographical and Historical Approaches to Foreign Area Studies* (Dubuque, Iowa: Kendall/Hunt, 1983).
Pannell, C. & Ma, L. *China: The Geography of Development and Modernization* (New York: Halsted Press/ V. H. Winston, 1983). Problems discussed on pp. 314–320.
Parish, W. & Whyte, M. *Village and Family in Contemporary China* (Chicago: University of Chicago Press, 1978).
Rostow, W. *The Stages of Economic Growth* (London and New York: Cambridge University Press, 2 rev. ed., 1971).
Shabad, T. *China's Changing Map: National and Regional Development, 1949–1971* (New York: Praeger, 1972).
Skinner, G. W. "Marketing and Social Structure in Rural China," *Journal of Asian Studies*, 24 (1964–1965), 3–44, 195–228, 363–400.
Snow, E. *The Other Side of the River: Red China Today* (New York: Random House, 1962).
Spencer, J. & Thomas, W. *Asia, East by South: A Cultural Geography* (New York: John Wiley & Sons, 2 rev. ed., 1971).
Tregear, T. *China: A Geographical Survey* (New York: Halsted Press/John Wiley & Sons, 1980).
Tuan, Y. *China* (Chicago: Aldine, 1969).
Wheatley, P. *The Pivot of the Four Quarters: A Preliminary Enquiry into the Origins and Character of the Ancient Chinese City* (Chicago: Aldine, 1971).
White, T. H. "China: Burnout of a Revolution," *Time*, September 26, 1983, 30–49.
Wiens, H. *Han Chinese Expansion in South China* (Hamden, Conn.: Shoe String Press, 1967).
Wittfogel, K. *Oriental Despotism: A Comparative Study of Total Power* (New Haven, Conn.: Yale University Press, 1957).
Wortman, S. "Agriculture in China," *Scientific American*, 232 (1975), 13–21.

# CHAPTER 10

# SOUTHEAST ASIA: BETWEEN THE GIANTS

IDEAS AND CONCEPTS

**Political Geography**
**Genetic Boundary Classification**
**Third World City Structure (2)**
**Insurgent States**
**Refugee Flows**
**Spatial Morphology**
**Domino Theory**
**Maritime Boundaries**
**World-Lake Concept**
**Territoriality**

Southeast Asia, the largest continent's southeastern corner of peninsulas and islands, is bounded by India on the northwest and by China on the north; its western coasts are washed by the Indian Ocean, and to the east stretches the vast Pacific. From all these directions, Southeast Asia has been penetrated by outside forces. From India came traders, and from China came settlers. From across the Indian Ocean came the Arabs to engage in commerce, and the Europeans in pursuit of empires. And from across the Pacific came the Americans. Southeast Asia has been the scene of countless contests for power and primacy; the competitors have come from near and far. Like Eastern Europe, Southeast Asia is a region of great cultural diversity, and it is also a *shatter belt*.

Southeast Asia is nowhere nearly as well defined a geographic realm as India or China. It does not have a single dominant core area of indigenous development. Culturally, it is diverse in every respect; there are numerous ethnic and linguistic groups, various religions, and different economies. Even spatially, the region's discontinuity is obvious: it consists of a peninsular mainland and hundreds of islands that form the archipelagoes of Indonesia and the Philippines. The mainland zone, smaller than India, is divided into seven political entities (Fig. 10–2). This underscores not only the cultural complexity of this realm, but also the historical geography of European intervention. Except for Thailand, none of the states on the map of Southeast Asia today was independent when World War II came to an end 40 years ago.

**Figure 10–1**
Silk dyeing in Thailand's capital, Bangkok: waterways criss-cross this city, often called "Venice of the East," which is well known for the floating markets of its Klong canals.

## TEN MAJOR GEOGRAPHIC QUALITIES OF SOUTHEAST ASIA

1. The Southeast Asian realm is fragmented into numerous peninsulas and islands.
2. Southeast Asia, like Eastern Europe, exhibits shatter-belt characteristics. Pressures on this realm from external sources have always been strong.
3. Southeast Asia exhibits intense cultural fragmentation, reflected by complex linguistic and religious geographies.
4. The legacies of powerful foreign influences (Asian as well as non-Asian) continue to mark the cultural landscapes of Southeast Asia.
5. Southeast Asia's politico-geographical traditions involve frequent balkanization, instability, and conflict.
6. Population in Southeast Asia tends to be strongly clustered, even in the rural areas.
7. Compared to neighboring regions, mainland Southeast Asia's physiologic population densities remain relatively low.
8. Explosive population growth has prevailed in the island regions of Southeast Asia, mainly in Indonesia, during the middle part of the twentieth century.
9. Intraregional communications in Southeast Asia remain poor. External connections are often more effective than internal ones.
10. The regional boundaries of Southeast Asia are problematic. Transitions occur into the adjoining Indian, Chinese, and Pacific realms.

# Population Patterns

Compared to the population numbers and densities in the habitable regions of India and China, demographic totals for the countries of Southeast Asia seem almost to suggest a sparseness of people (see Appendix). Laos, territorially quite a large country, has a population of less than 4 million and an average population density that is lower than in desert-dominated Iran. Much of interior mainland Southeast Asia has population densities similar to those of savanna Africa, the margins of South America's Amazon Basin, and Soviet Central Asia (Fig. I–14). Even the more densely inhabited coastal areas of Southeast Asia have fewer people and smaller agglomerations than elsewhere in South and East Asia. There is nothing in this realm to compare to the immense human clusters in the Ganges Basin or the Huang He and Chang Jiang Lowlands. The whole pattern is different: Southeast Asia's few dense population clusters are relatively small and lie separated from each other by areas of much sparser human settlement.

Why the difference? When there is such population pressure and such land shortage in adjacent regions, why has Southeast Asia not been flooded by waves of immigrants? Several factors have combined to inhibit large-scale invasions. In the first place, the overland routes into Southeast Asia are not open and unobstructed. In discussing the Indian subcontinent, we noted the barrier effect of the densely forested hills and mountains that lie along the border between India's Assam and northwestern Burma. North of Burma lies forbidding Xizang (Tibet) and northeast of Burma and north of Laos is the high Yunnan Plateau. Transit is easier into northern Vietnam via southeastern China, and, along this avenue, considerable contact and migration have indeed occurred. Neither is contact within Southeast Asia itself helped by the rugged, somewhat parallel ridges that hinder east-west communications between the fertile valleys of the region (Fig. 10–4). Second—and the population map reflects this—Southeast Asia is not an area of limitless agricultural opportunities. Much of the region is covered by dense tropical rainforest, part of an ecological complex whose effect on human settlement we have previously observed in other low-latitude areas of the world. Except in certain locales, the soils of mainland Southeast Asia are excessively leached (diluted of chemical nutrients) by the realm's generally heavy rains. In the areas of monsoonal rainfall regimes and savanna climate (the latter prevailing across much of the interior), there is a dry season, but the limitations imposed by savanna conditions on agriculture are all too familiar. High evapotranspiration rates, long droughts, hard-baked soils, meager fertility, high runoff, and erosional problems add up to anything but a peasant farmer's paradise in these parts of Southeast Asia.

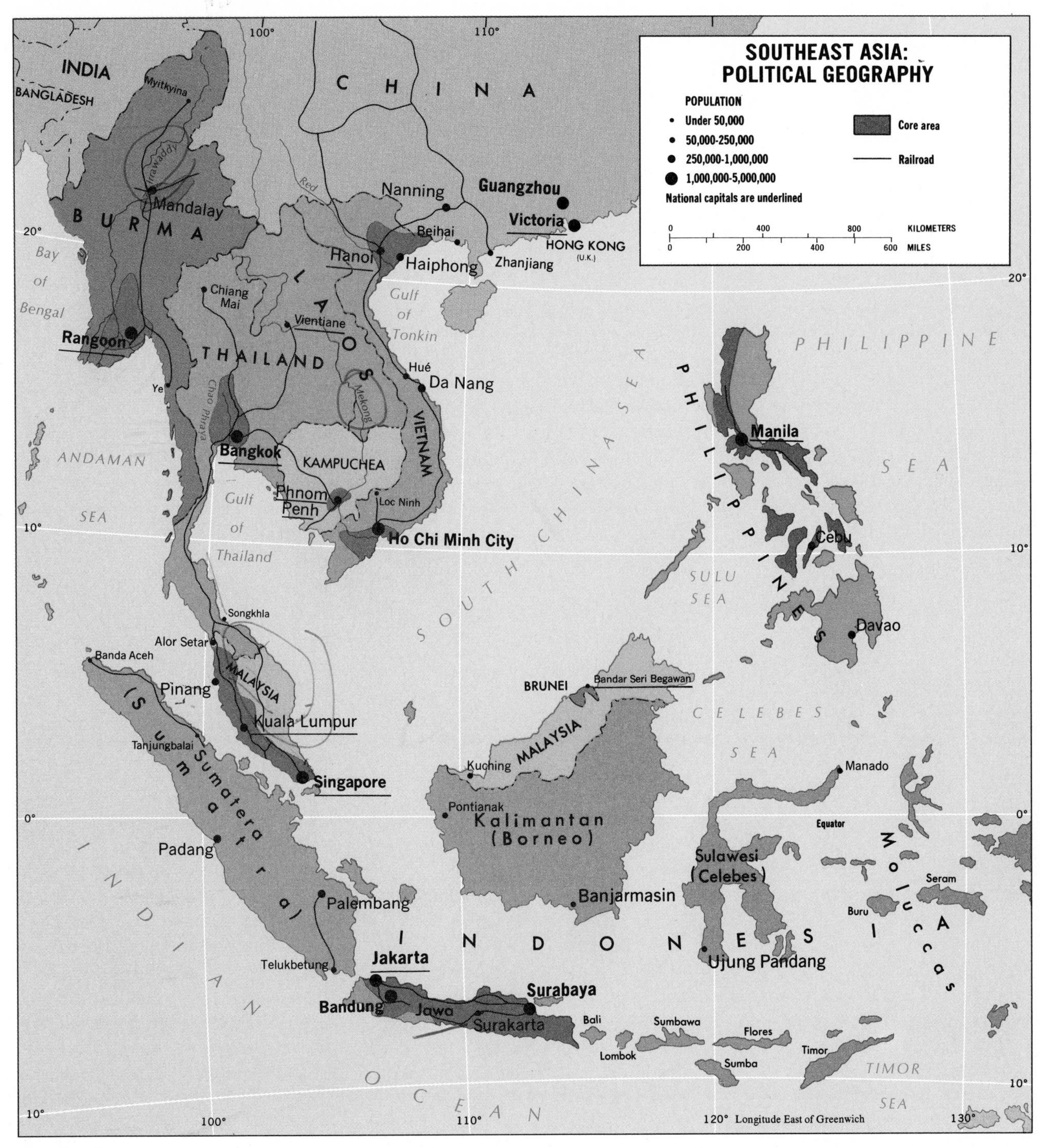


Figure 10–2

## The Clusters

Population in Southeast Asia has become concentrated in three kinds of natural environments. First, there are the valleys and deltas of Southeast Asia's major rivers, where alluvial soils have been formed. Four major rivers stand out. In Burma, the Irrawaddy rises near the border with Xizang (Tibet) and creates a delta as it empties into the Bay of Bengal. In Thailand, the Chao Phraya traverses the length of the country and flows into the Gulf of Thailand. In southern Vietnam, there is the extensive delta of the great Mekong River, which rises in the high mountains on the Qinghai-Xizang border in interior China and traverses the whole Indochinese peninsula. And in northern Vietnam lies the Red River, whose lowland (the Tonkin Plain) is probably the

# SYSTEMATIC ESSAY

## Political Geography

**Political geography** is the study of the spatial expressions of political behavior. Aspects of this subject appear frequently in this book; the world map of states and territories (Fig. I–17) is itself one of these expressions. Political activity is one of the most basic ingredients of culture, and it began with the earliest organized human communities as competition for leadership roles and hunting territories became a part of daily life. Ever since, political behavior by individuals, social groups, and nations has expressed the human desire for power and influence and the pursuit of personal and public goals. The results of these activities are etched on the globe in countless ways, and political geographers study them from many different viewpoints.

One dimension of political geography lies in the analysis of the state, the nation, and the nation-state. The *states* (or, popularly, the *countries*) of the world are the products of millennia of territorial conflict and adjustment, changing economic forces, and altered power relationships. Some modern states are truly *nation-states* in which the great majority of the inhabitants are bound by the history and traditions of a single nation. Other states incorporate two or more national groups and the nation becomes more a legal than an emotional concept. Political geographers try to assess the *centripetal* (binding, uniting) forces in the state against the *centrifugal* (divisive) influences present. Linguistic and religious conflicts within the state can constitute powerful centrifugal forces. Other topics in the political geography of the state are the dissection of the territorial *morphology* (the size and shape of countries and what they mean in national life), the analysis of *core areas* (the heart of the state), the study of the role and effectiveness of *capital cities*, and the development and functions of *boundaries* on land and in the seas.

Political geographers, therefore, study states as political regions: their structure, internal divisions, circulation systems, external relationships. Long ago, *geopolitics* became an international preoccupation as political geographers tried to evaluate and predict the power relationships of states and groups (blocs) of states. Sir Halford Mackinder published an article in 1904 in which he suggested that a certain "Heartland" in Eurasia would become the key to world power (see p. 165); that treatise on *geostrategy* became one of the most hotly debated statements in all of geography, and the debate still goes on.

Political geographers also study political behavior at closer (and less speculative) quarters. The question of human *territoriality* has led geographers and psychologists to cooperate in the search for an understanding of our attachment to space as individuals as well as nations. Various kinds of *decision making* by political leaders as well as voters have been subjected to geographical analysis, and voting behavior is now a major area of interest among political geographers. This leads naturally to questions of political-spatial *perception*, or the way politically active individuals and groups think about their social environments.

Like other geographers, political geographers try to formulate rules or models to help us understand the political world. Sometimes these models seem to represent the real world quite accurately; at other times, they are more abstract. They are useful in practical ways too, for example when a boundary between countries (states) becomes a matter of disagreement and perhaps dispute. Then political geographers go to work trying to determine just how that boundary came about, whether the treaties involved are soundly drawn, and how an international court should view the issue. The terminology of political geography often turns up in legal briefs!

The creation of an international boundary begins with *definition*, a treatylike document signed by neighboring states that describes in words the location of the boundary line that divides them; cartographers then *delimit* the border (draw it on a large-scale map); in certain cases, the boundary may also be *demarcated* or actually placed on the landscape in the form of a fence or some similar obstacle. Several schemes for classifying political boundaries have been proposed. One is the *genetic* classification—devised by Richard Hartshorne a half-century ago—that refers to the stage of development of the cultural landscape at the time a boundary was established. Accordingly, we may identify four types of boundaries: antecedent, subsequent, superimposed, and relict.

*Antecedent boundaries* were defined and delimited before the main elements of the present cultural landscape began to develop. Geometric (straight-line) boundaries drawn through empty areas are examples; much of the territory of the still-empty Sahara Desert was divided in this fashion, as was discussed in Chapter 6. A much more

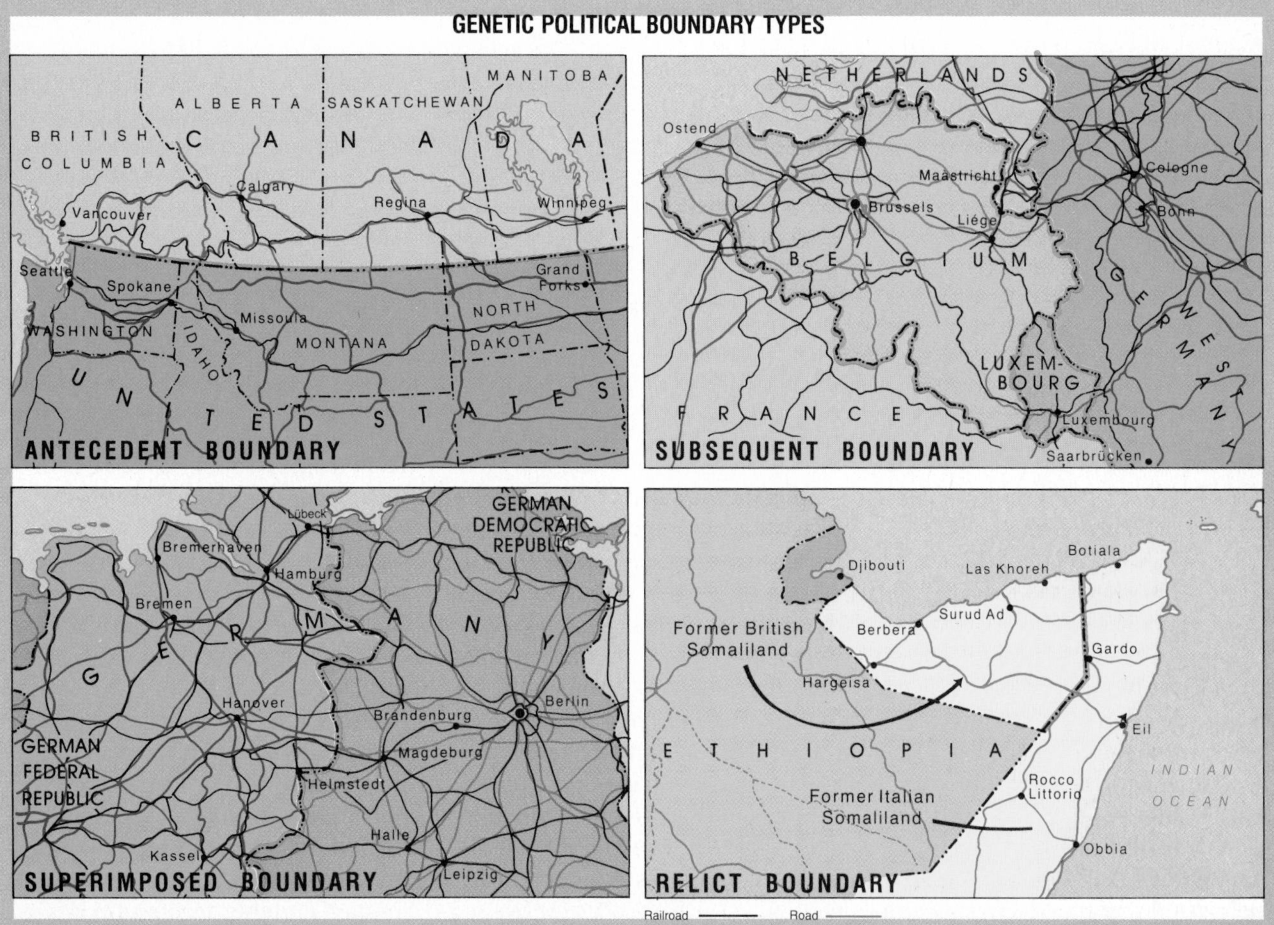


Figure 10–3



interesting case, however, is shown in the upper-left map in Fig. 10–3: the United States-Canada border that follows the 49th parallel between the Pacific and the Great Lakes area. Here, a settlement pattern has evolved since the boundary was established in the early nineteenth century, and human adjustment to the border is clearly reflected in the parallel long-distance transport routes that carefully stay within each country and only rarely interconnect.

*Subsequent boundaries* display a certain degree of conformity to the cultural space through which they lie, adjusting to linguistic, religious, or ethnic breaks and transition zones. They come about as part of the process of spatial organization in the regions they now fragment, developing contemporaneously with the evolution of major elements in the cultural landscape. In the 1947 partitioning of the Indian subcontinent, the boundaries between India and the two Pakistans were subsequent in nature, seeking to follow existing cultural (largely religious) breaks in the settlement pattern. Partition, however, ignored (and quickly reversed) the centuries-long cultural accommodation between Hindus and Moslems in many parts of India and resulted in a frenzied migration of millions across the new borders that, in effect, now divided separate Islamic and Hindu states (Figs. 8–8 and 8–9). The example in the upper-right map in Fig. 10–3 shows other problems in delimiting a subsequent boundary: because of the complex subdivisioning of inherited lands, complicated further by generations of international intermarriages, the modern borders of several Western European countries had to follow tortuous zigzag paths to conform to individual property lines (a local-level convolution only suggested on this small-scale map).

*Superimposed boundaries* are forcibly placed upon an existing, unified cultural landscape. They result from the intervention of a state (or alliance of states) more powerful than the country through which the boundary lies. A group of victorious allies may divide the state they vanquished, as they did to Germany after World War II (lower-left map in Fig. 10–3). When the European colonial powers divided Africa among themselves, the boundaries they defined were superimposed on indigenous African cultures. Over time, some superimposed boundaries take on the properties of subsequent boundaries. The superimposed boundary, if it endures through successive generations, may become a part of life and lose its divisive character, thus acquiring the character of a subsequent boundary. When African countries became independent after nearly a century of colonial rule, few of the European-superimposed boundaries were eliminated, and many of them are now classified as subsequent boundaries.

*Relict boundaries* are those that have ceased to function but whose imprints are still evident on the cultural landscape. The example in the lower-right map in Fig. 10–3 is the former boundary between the British and Italian sectors of Somaliland in the Horn of Africa, running northeast and then north from the far eastern corner of Ethiopia. Each sector was distinguished by a different set of place names—Arabic in the former British zone and Italian in the other zone. When the British withdrew in 1960 and the two sectors were joined to create the new independent state Republic of Somalia, the old boundary was discarded. Yet, the difference in toponymy (place names) has endured on the cultural landscape and can only be understood today by recognizing that a boundary once functioned here.

most densely settled area in Southeast Asia. Each of these four river basins contains one of mainland Southeast Asia's major population clusters—in effect, the core areas of those countries.

Second, Southeast Asia is known for its volcanic mountains—at least in the island archipelagoes. In certain parts of these islands the conditions are present for the formation of deep, dark, and rich volcanic soils, especially in much of Jawa (Java).[1] The population map indicates the significance of this fertility: the island of Jawa is one of the world's most densely populated and intensively cultivated areas. On Jawa's productive land live about 110 million people—approximately two-thirds of the inhabitants of all the islands of Indonesia and more than a quarter of the population of the entire Southeast Asian realm.

Another glance at Fig. I–14 indicates that one additional area remains to be accounted for: the belt of comparatively dense population that extends along the western coast of the Malay Peninsula, apparently unrelated to either alluvial or volcanic soils. This represents the third basis for population agglomeration in Southeast Asia—the plantation economy. Actually, plantations were introduced by the European colonizers throughout most of insular Southeast Asia, but nowhere did they so thoroughly transform the economic geography as in Malaya. Rubber trees were planted on tens of thousands of acres, and Malaya became the world's leading exporter of this product. Undoubtedly, Malaysia would not have developed so populous a core area without its plantation economy.

Where there are no alluvial soils, no volcanic soils, and no profitable plantations, as in the rainforest areas and the steep-sided uplands where there is little level land, far fewer people manage to make a living. Here the practice of shifting subsistence cultivation prevails, sometimes augmented by hunting, fishing, and the gathering of wild nuts, berries, and the like. This practice is similar to what we observed in South America and Equatorial Africa, and it is only capable of sustaining a rather sparse population; land that was once cleared for cultivation must be left alone for years in order to regenerate. Here in the forests and uplands of Southeast Asia, these nonsedentary cultivators still live in considerable isolation from the peoples of the core areas; the forests are dense and difficult to penetrate, distances are great, and steep slopes only add to the obstacles that retard efficient surface communications. Over a major part of their combined length, the political boundaries of Southeast Asia traverse these rather thinly peopled inland areas. But here are the roots of some of the region's political troubles: these unstable interior zones have never been effectively integrated into the states of which they are part. The people who live in the forested hills may not be well disposed toward those who occupy the dominant core areas. Therefore, these frontierlike inner reaches of Southeast Asia—with their protective isolation and distance from the seats of political power—are fertile ground for revolutionary activities if not for agriculture (which can turn illicit too, as in the case of the huge opium-poppy harvests that emanate from the notorious "Golden Triangle" located where the borders of Burma, Laos, and Thailand converge).

# Indochina

The French colonialists called their Southeast Asian possessions *Indochina*, and the name is appropriate to the bulk of the mainland region, suggesting the two leading Asian influences that have affected the realm for the past 2000 years. Overland immigration was mainly southward from southern China, re-

[1] As with Africa, names and spellings have changed with independence. In this chapter, the contemporary spellings will be used, except when reference is made to the colonial period. Thus, Indonesia's four major islands are Jawa, Sumatera, Kalimantan (the Indonesian part of Borneo), and Sulawesi. The Dutch called them Java, Sumatra, Borneo, and Celebes.

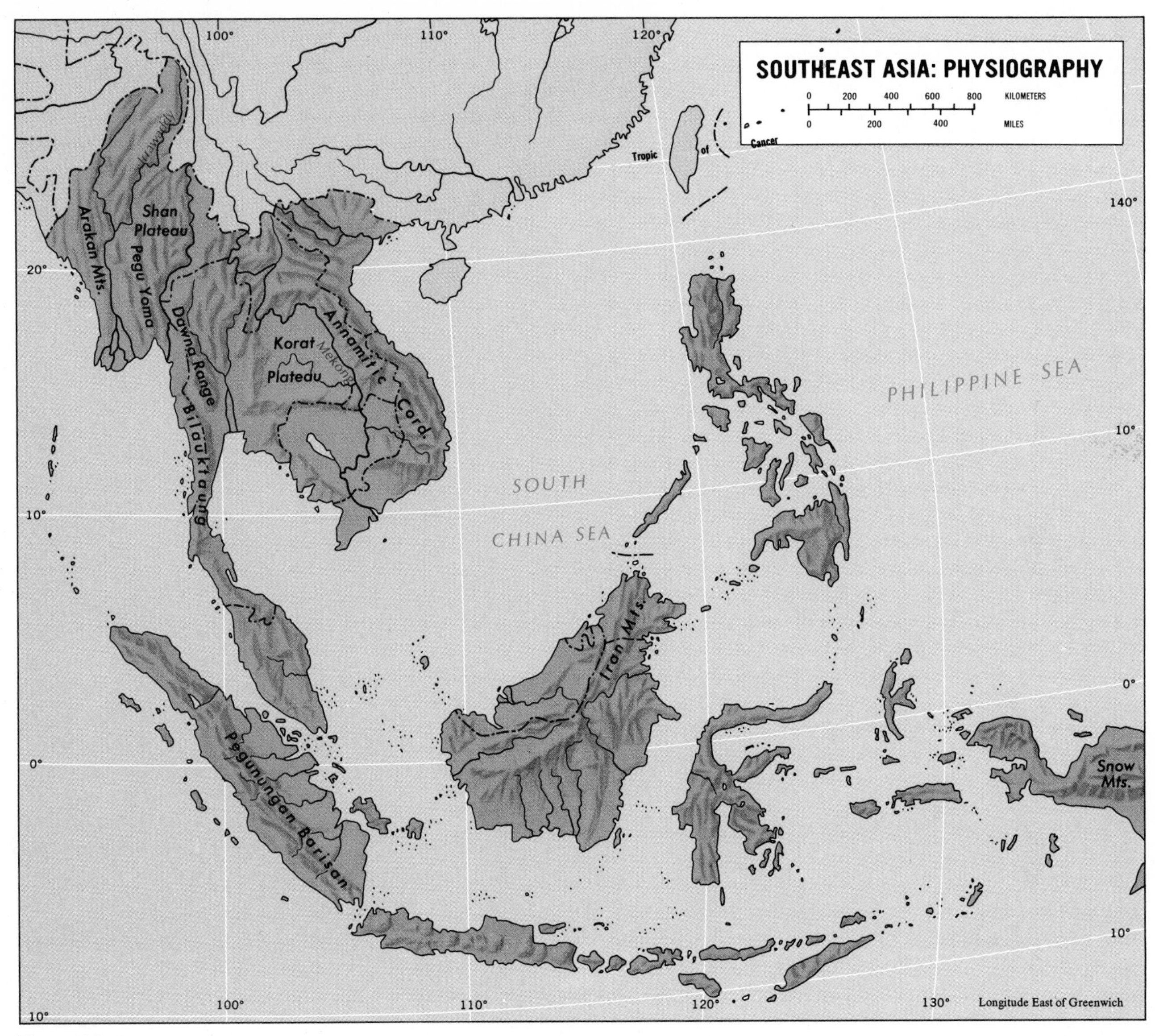


Figure 10–4

sulting from the expansion of the Chinese empire. The Indians came by way of the seas, as Indian trading ships plied the coasts and Indian settlers founded colonies on Southeast Asian shores in the Malay Peninsula, the Mekong Plain, and on Java and Borneo.

With the migrants from the Indian subcontinent came their faiths: first Hinduism and Buddhism, later Islam. The Moslem religion was also promoted by the growing number of Arab traders who appeared on the scene, and Islam became the dominant religion in Indonesia where nearly 90 percent of the population adheres to it today. But, in Burma, Thailand, and Cambodia, Buddhism remained supreme, and, in all three countries over 80 percent of the people are now adherents. In Malaysia, the Malays are Moslems (to be a Malay is to be a Moslem), and almost all Chinese are Buddhists; but Hinduism retained its early strength, and still today over 70 percent of the country's South Asians live in accordance with Hindu principles. Although Southeast Asia has generated its own local cultural expressions, most of what remains in tangible form has resulted from the infusion of foreign elements. For instance, the main temple at Angkor Wat, constructed in Kampuchea (Cambodia) during the 12th century, remains a monument to Indian architecture of that time (Fig. 10–5).

This, then, represents the *Indo* part of Indochina: the Buddhist and Hindu faiths (Ceylonese merchants played a big role in the introduction of the former), Indian architecture, art (especially sculpture), writing and literature, and social

**Figure 10–5**
Southeast Asia's most famous relic, Kampuchea's Angkor Wat temple, which fortunately survived that country's recent war with only minor damage.

structuring. But the Chinese role in Southeast Asia has been even greater. Apart from the fact that upheavals within China itself contributed to the intermittent southward push of Sinicized peoples, Chinese traders, pilgrims, sailors, fishermen, and others sailed from southeast China to the coasts of Malaya and the islands and settled there. Invasions and immigrations continued into modern times, and the relations between recent Chinese arrivals and the earlier settlers of Southeast Asia have often been strained.

The Chinese initially profited from the arrival of the Europeans, who stimulated the growth of agriculture, trade, and industries; here they found opportunities they did not have at home. They established rubber holdings, found jobs on the docks and in the mines, cleared the bush, and transported goods in their sampans. They brought with them skills that proved to be very useful, and as tailors, shoemakers, blacksmiths, and fishermen they did well. The Chinese also proved to be astute in business; soon, they not only dominated the region's retail trade, but also held prominent positions in banking, industry, and shipping. Thus, their importance has always been far out of proportion to their modest numbers in Southeast Asia. The Europeans used them for their own designs but found the Chinese to be stubborn competitors at times—so much so that, eventually, they tried to impose restrictions on Chinese immigration. Previously, the United States, when it took control of the Philippines, also had sought to stop the influx of Chinese into those islands.

When the European colonial powers withdrew and Southeast Asia's independent states emerged, Chinese population sectors ranged from nearly 50 percent of the total in Malaysia (1963) to barely over 1 percent in Burma (Fig. 10–6). The separation of Singapore, where Chinese constitute over 75 percent of the population of 2.6 million (1985), reduced the Malaysian Chinese component to about 35 percent—still a very large minority. In Indonesia, the percentage of Chinese in the total population is not high (an estimated 3 percent), but the Indonesian population is so large that even this small percentage represents a Chinese sector of about 5 million. Indonesia's Chinese were suspected of complicity in a 1965 Communist plot to overthrow the government (a plot that was almost certainly of indigenous origin and not fomented by Communist China), and hundreds of thousands of Chinese were killed. In Thailand, on the other hand, many Chinese have married Thais, and the Chinese minority of about 15 percent has become a cornerstone of Thai society, dominant in trade and commerce. In recent years, however, the Bangkok government has begun to express concern over the number of Chinese immigrants crossing the northern boundary (entering via Laos or Burma). Possibly, the changing situation in the China of the 1980s will stem that tide at its source.

In general, in Southeast Asia, the Chinese communities remained quite aloof and formed their own separate societies in the cities and towns; they kept their culture and language alive by maintaining social clubs, schools, and even residential suburbs that, in practice if not by law, were Chinese in character. There was a time when they were in the middle between the Europeans and Southeast Asians, and when the hostility of the local people was directed toward white people as well as toward the Chinese. But, since the withdrawal of the Europeans, the Chinese have become the main target of this antagonism, which remained strong because of the Chinese involvement in moneylending, banking, and trade monopolies. Moreover, there is the specter of an imagined or real Chinese political imperialism along Southeast Asia's northern flanks.

The *china* in Indochina, thus, represents a wide range of penetrations; the source of most of the old invasions was in southern China, and Chinese territorial consolidation provided the impetus for successive immigrations. The Mongoloid racial strain carried southward from East Asia mixed with the earlier Malay stock to produce a transition from Chineselike people in the northern mainland to dark-skinned, Malay types in the distant Indonesian east. Although Indian cultural influences remained strong, Chinese modes of dress, plastic arts, types of boats, and other cultural attributes were adopted throughout Southeast Asia. During the past century, especially during the last half-century, renewed Chinese immigration brought Chinese skills and energies that propelled these minorities to positions of comparative wealth and power in this realm.

Figure 10–6

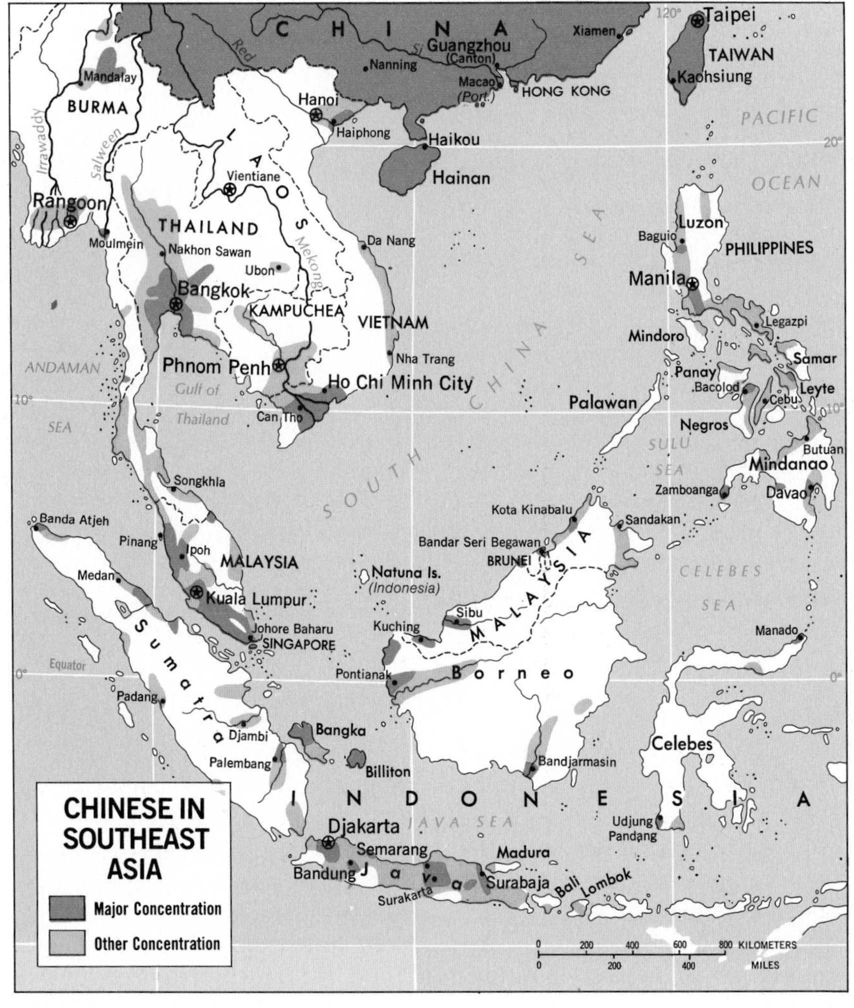


# European Colonial Frameworks

One shortcoming of the term *Indochina* is its failure to reflect the third cultural force that has molded

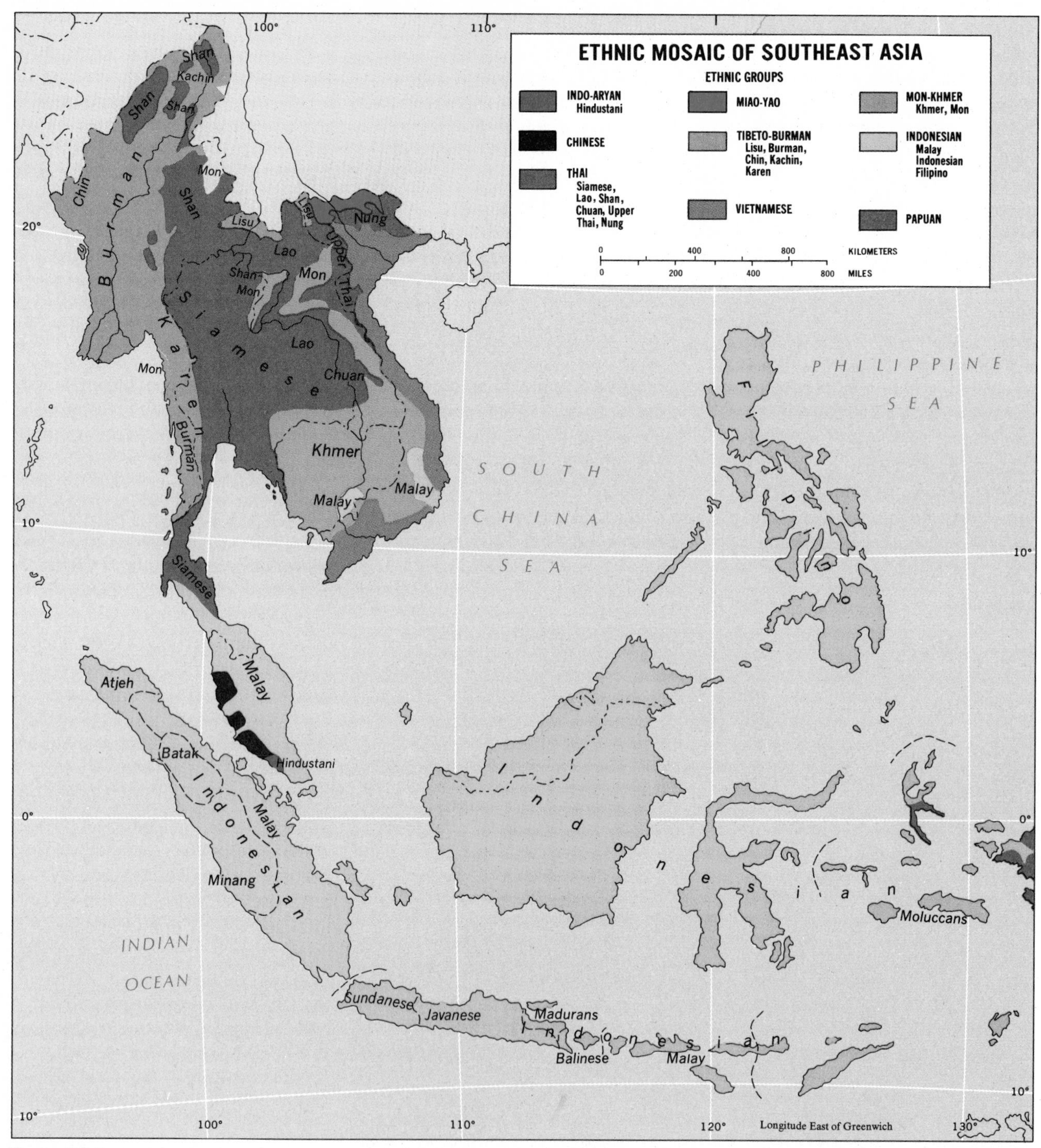


Figure 10–7

modern Southeast Asia: Europe's. As in Africa, the colonial politico-geographical structure that emerged in Southeast Asia during the nineteenth century had the effect of throwing diverse peoples together within larger political units, yet, at the same time dividing groups possessing strong ethnic and cultural affinities. The Thais, Malays, and the Shans of northern Thailand and eastern Burma all experienced the latter at the hands of the British, as did the Khmers and the Annamese in Indochina under the French. The peoples unified by the Dutch in their East Indies colony are numerous and varied; at least 25 distinct languages are spoken in this insular archipelago now comprising the state of Indonesia. Again, as in Africa, the internal divisions within the respective colonial territories had lasting significance, for, eventually, the individual fragments were to emerge as separate states. Thus, the French organized their Indochinese dependency into five units—four protectorates (Cambodia, Laos, Tonkin, and Annam) and the colony of Cochin China in the Mekong Delta. Eventually, Cambodia—now called Kampuchea—and Laos emerged as independent states; but Vietnam (consisting of Tonkin, Annam, and Cochin China) remained fragmented into a North (capital Hanoi) and a South (capital Saigon, now Ho Chi Minh City) until their reunification under the Communist North following the end of the Indochina War in 1976.

The British held two major entities in Southeast Asia—Burma and Malaya—in addition to a number of islands in the South China Sea and the northern sector of Borneo (which contains the realm's newest independent state, Brunei—see *Box*). In Burma, the British created a colony that incorporated many different peoples, but, from 1886 to 1937, the government was based in far-away New Delhi. Burma's lowland core area is the domain of the majority of Burmese, but the surrounding areas are occupied by other peoples (Fig. 10–7). The Karens live to the southeast of the Irrawaddy Delta, the Shans inhabit the eastern plateau that adjoins Thailand, the Kachins live near the Chinese border in the north, and the Chins reside in the highlands along the Indian border in the northwest. Other groups also form part of Burma's complex population of 38.9 million, and this heterogeneity was further intensified during the period of India-connected administration, when more than 1 million Indians entered the country. These Indians came as shopkeepers, moneylenders, and commercial agents, and their presence heightened Burmese resentment against British policies. The national division of Burma was brought sharply into focus during World War II when the lowland Burmese welcomed the Japanese intrusion, whereas the peoples of the surrounding hill country, who had seen less of the British maladministration and who had little sympathy for their Burmese compatriots anyway, generally remained pro-British. Since independence in 1948, at least half of the Indian population has been expelled.

In Malaya, the British developed a complicated system of colonies and protectorates that eventually gave rise to the equally complex, far-flung Malaysian federation. Included were: the former Straits Settlements (Singapore was one of these colonies), the nine protectorates on the Malayan Peninsula (born of sultanates of the Moslem era and developed from tiny rivermouth settlements into states covering entire drainage basins by the British), the British dependencies of Sarawak and Sabah on the island of Borneo, and numerous islands in the Strait of Malacca and the South China Sea. The original federation of Malaysia was created in 1963 by the political unification of recently independent mainland Malaya, Singapore, and the former British dependencies on the Indonesian island of Borneo (Sarawak and Sabah). Singapore, however, left the federation in 1965 to become a sovereign city-state, and the remaining units were restructured into West Malaysia (the peninsula) and East Malaysia (on Borneo). In 1972, still another change was introduced: West Malaysia now became Peninsular Malaysia, and East Malaysia once again was called Sarawak and Sabah. Thus, the term *Malaya* properly refers to the geographic area of the Malayan Peninsula, including Singapore and other nearby islands; the term *Malaysia* identifies the politico-geographical unit of which Kuala Lumpur is the capital city.

Malaysia's ethnic and cultural divisions are etched into its land-

## BRUNEI

Brunei is an anomaly in Southeast Asia—an oil-exporting Islamic Sultanate far from the Middle East. Located on the north coast of Borneo, sandwiched between Malaysian Sarawak and Sabah, the Brunei sultanate is a former British-protected remnant of a much larger Islamic kingdom that once controlled all of Borneo and areas beyond. Brunei achieved full independence in 1984. With a mere 5800 square kilometers (2230 square miles)—slightly larger than Delaware—and only 230,000 people, Brunei is dwarfed by the other political entities of Southeast Asia. But the discovery of oil in 1929 heralded a new age for this remote territory. Today, Brunei is one of the largest oil producers in the British Commonwealth, and new offshore discoveries suggest that production will increase. As a result, the population is growing rapidly by immigration (about 70 percent of the people are Malay, one-fourth Chinese), and the sultanate enjoys one of the highest standards of living in Southeast Asia ($22,000 per capita in 1984). Most of the people live near the oilfields in the western corner of the country, at Kuala Belait and Seria, and in the capital in the east, Bandar Seri Begawan. The evidence of a boom period can be seen in modern apartment houses, shopping centers, and hotels—a sharp contrast to many other towns on Borneo. There are some sharp internal contrasts as well; Brunei's interior still remains an area of agricultural subsistence and rural isolation, virtually untouched by the modernization of the coastal zone.

**Figure 10–8**
The Malaysian capital of Kuala Lumpur, whose landscape is an amalgam of multiple cultural influences.

scapes. The Malays are traditionally a rural people. They originated in this region and displaced the aboriginal peoples, now no longer significant numerically. The Malays, who constitute just under 50 percent of the Malaysian population of 15.7 million, possess a strong cultural unity expressed in a common language, adherence to the Moslem faith, and a sense of territoriality that arises from their Malayan origins and their collective view of Chinese, South Asian, and other foreign intruders. Although they have held control over the government, the Malays often express a fear of the more aggressive, commercially oriented, and urbanized Chinese minority. Malay-Chinese differences were exacerbated during the Second World War when the Japanese (who occupied the area) elevated the Malays into positions of authority but ruthlessly persecuted the Chinese, driving many of them into the forested interior where they founded a Communist-inspired resistance that long continued to destabilize the region. The British returned, then yielded after a system of interracial cooperation had been achieved, but racial tensions continued and, in 1969, resulted in racial clashes in Kuala Lumpur (1985 population: 1.4 million). This capital city's townscape reflects Malaysia's cultural mosaic (Fig. 10–8): minarets and domes of Islamic shrines rise above a city whose major structures (the railroad station, State House) exhibit a Moorish architectural imprint; but trade and commerce are in the hands of the Chinese, around whose shops and businesses the city's life revolves. Such contrasts, often reinforced by the visible remnants of colonialism, are typical of the Southeast Asian urban experience (see *Box*).

The Hollanders took control of the "spice islands" through their Dutch East India Company, and the wealth that was extracted from what is today Indonesia brought the Netherlands its Golden Age. From the mid-seventeenth to the late eighteenth century, Holland could develop its East Indies sphere of influence almost without challenge, for the British and French were preoccupied with the Indian subcontinent. By playing the princes of Indonesia's states off against each other in the search for economic concessions and political influence, by placing the Chinese in positions of responsibility, and by imposing systems of forced labor in areas directly under its control, the Company had a ruinous effect on the Indonesian societies it subjugated. Java (Jawa), the most populous and productive island, became the focus of Dutch administration; from its capital at Batavia (now Jakarta), the company extended its sphere of influence into Sumatra (Sumatera), Borneo (Kalimantan), Celebes (Sulawesi), and the smaller islands of the East Indies. This was not accomplished overnight; the struggle for territorial control was carried on long after the company had yielded its administration to the Netherlands government. Northern Sumatra, in fact, was not subdued until early in this century.

Thus, Dutch colonialism threw a girdle around Indonesia's more than 13,000 islands, creating the realm's largest political unit and one of its most complex states. Broadly, Indonesia's 168 million people can be divided into three groups: the aborigines (now a small minority), the peoples who grow rice on the islands' well-watered slopes, and the coastal communities. The aboriginal peoples have been pushed into the less productive, drier parts of the islands, where they practice shifting cultivation. The rice growers make up over half the total population and form the core of modern Indonesian society; their cultural traditions have ancient roots and are expressed in architecture, dance, music, and literature as well as in social structure, farming, and land-use. The coastal peoples accumulated from immigrations during and after the arrival of Islam, and Moslem communities exist on all the large islands.

This broad differentiation, however, barely begins to reveal Indonesia's cultural complexity. There are dozens of distinct aboriginal cultures; virtually every coastal community has its own roots and traditions. And the rice-growing In-

# THE SOUTHEAST ASIAN CITY

In their survey of current urbanization trends in this realm, Thomas Leinbach and Richard Ulack offered a few salient generalizations about the larger cities of postcolonial Southeast Asia. First, they are all experiencing rapid growth; between 1950 and the mid-1980s, the region's urban population has nearly doubled in relative size (from 15 to 27 percent) and more than quadrupled in absolute numbers (26 to 111 million). Second, despite a waning overall presence, foreigners still play a decisive role in the commercial lives of cities, with the Japanese influence particularly prominent in this decade. Third, the most recent episode of urban agglomeration has been heavily concentrated on the large coastal cities, reinforcing their old colonial-era dominance through renewal of their high-order functions as collection-distribution nodes for interior hinterlands as well as leading ports for external trade and shipping. And fourth, they exhibit similar internal land-use patterns, a spatial structuring worth examining in some detail.

The general intraurban pattern of residential and nonresidential activities is nicely summarized in the model presented by T. G. McGee in his 1967 book, *The Southeast Asian City* (Fig. 10–9). The old colonial port zone, its functions renewed in the postcolonial period, is the city's focus together with the largely commercial district that surrounds it. Although no formal central business district is evident, its elements are present as separate clusters within the land-use

**Figure 10–9**

A GENERALIZED MODEL OF LAND-USE AREAS IN THE LARGE SOUTHEAST ASIAN CITY

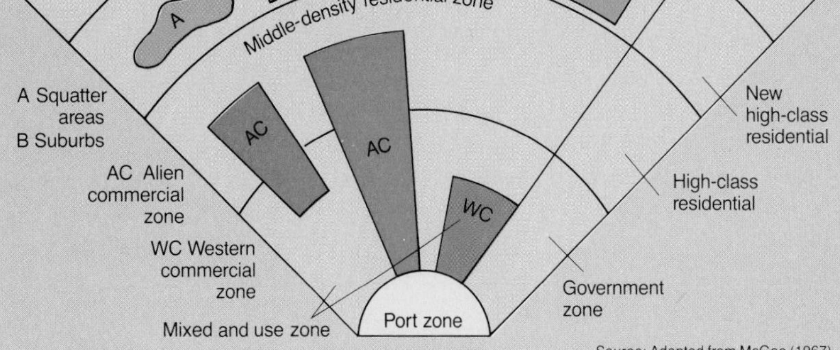

*Source:* Adapted from McGee (1967).

belt beyond the port: the government zone, the western commercial zone (a colonialist remnant, which is practically a CBD by itself), the alien commercial zone—usually dominated by Chinese merchants whose residences are attached to their places of business—and the mixed land-use zone that contains miscellaneous economic activities, including light industry. The other nonresidential areas are the market-gardening zone at the urban periphery and, still further from the city, an industrial park or "estate" of recent vintage. The residential zones in McGee's construct are quite reminiscent of the Griffin-Ford model of the Latin American city (Fig. 4–15, p. 271) which, as we saw, could be extended to the Third World city in general. Among the similarities between the two are: the hybrid sector/ring framework, an elite residential sector that includes new suburbanization, an inner-city zone of comfortable middle-income housing (with new suburban offshoots in the McGee schema), and peripheral concentrations of low-income squatter settlement. The differences are relatively minor and can partly be accounted for by local cultural and historical variations. If the Griffin-Ford model can be viewed as a generalization of the spatial organization of the Third World city, then the McGee model illustrates the departures that occur in coastal cities that were laid out as major colonial ports but continue their development within a now-independent country.

donesians include not only the numerous Javanese, who are Moslems largely in name only and have their own cultural identity, but also the Sundanese (who constitute about 15 percent of Indonesia's population), the Madurese, and others; especially impressive levels of rice production have been achieved by the Balinese, whose meticulously terraced paddy fields (Fig. 10–10) yield some of the realm's largest harvests per unit area. Perhaps the best impression of the cultural mosaic comes from the string of islands that extends east from Jawa to Timur (Timor). The Balinese are mainly Hindu adherents; the population of Lombok is mainly Moslem, with some Hindu immigrants from Bali. Sumbawa is a Moslem community, but the next island, Flores, is mostly Roman Catholic. On Timur Protestant groups predominate, and this island remains marked by its long-time division into a Dutch-controlled and a Por-

**Figure 10–10**
Rice-growing on Bali exhibits such impeccable terracing that the farming landscape often assumes a finely manicured appearance.

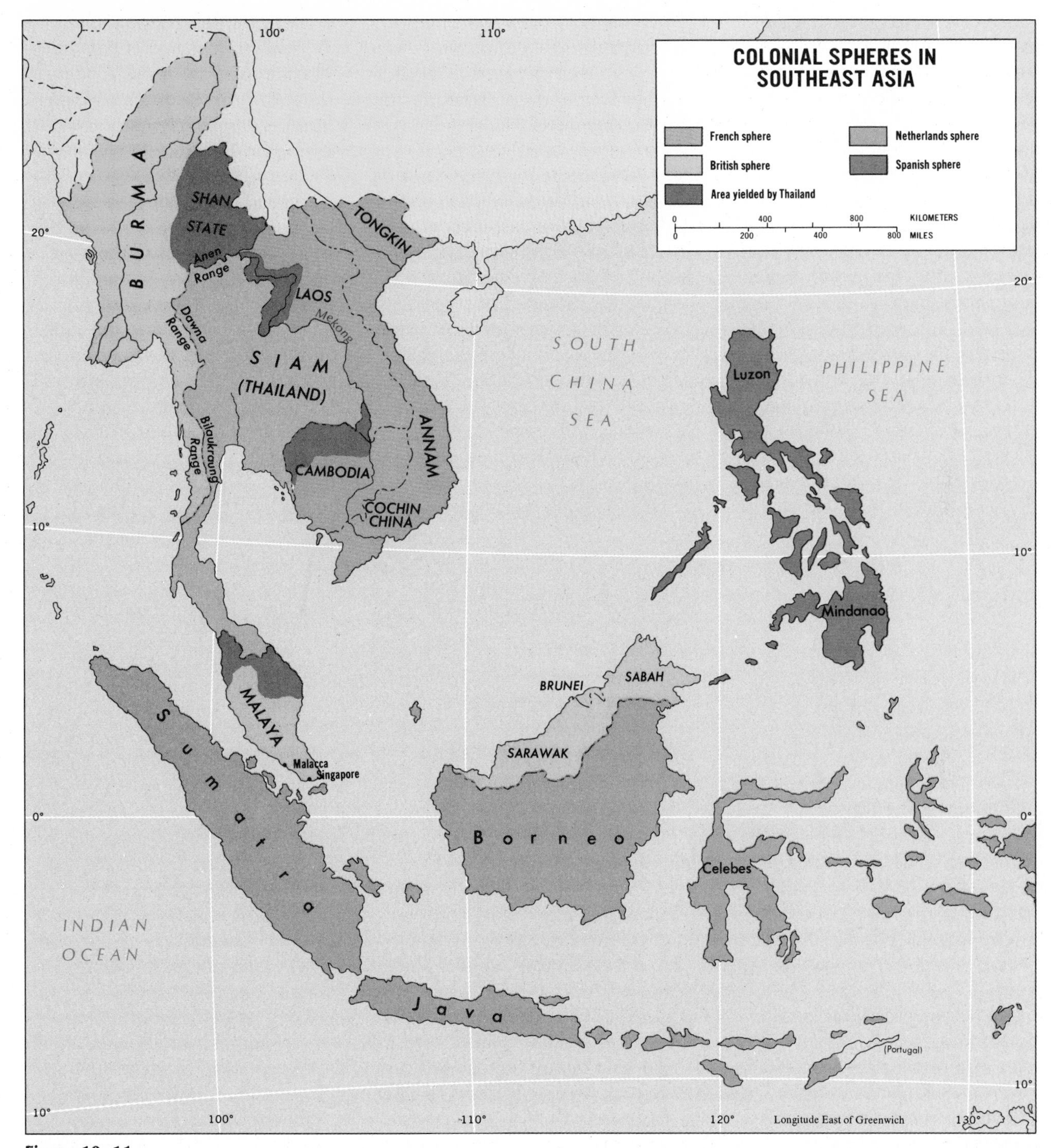


**Figure 10–11**

tuguese-owned sector (Fig. 10–11). An independence movement on former Portuguese Timur was subdued by invading Indonesian forces, a campaign that was followed by severe dislocation and famine.

Dutch colonialism unified all this diversity and exploited Indonesia's considerable productive capacity under the so-called *Culture System (Cultuur Stelsel)*. This involved forced-crop and forced-labor practices, requiring the production of stipulated crops in predetermined areas; Indonesians on the outer islands who had no land to cultivate were forced to make themselves available as laborers for the state. Undeniably, this stimulated production: sugar, coffee, and indigo were the leading exports, and they brought in rising revenues as the Culture System was made to work. Crops such as tea, tobacco, tapioca, and the oil palm (especially in Sumatera) were introduced or their strains were improved. But the price was high: the Javanese resented the principle, and harsh and cruel methods were used to sustain the System—to such an extent that there was an outcry about it in the Netherlands. With the Dutch and Chinese in control of all phases of the trade—based on the products of the involuntary croplands, including collection, treatment, packaging, and dispatch—the local people got little or no experience in this aspect of their island's economy. There was little incentive, other than fear, to cultivate what was required. And lands that had been producing rice and other staples were periodically turned over to these cash crops with (initially) satisfactory arrangements for the adequate replacement of the food supplies thus lost.

The Culture System was officially abandoned in 1870 (although the forced growing of coffee continued until 1917), and a new and more liberal colonial policy evolved. But throughout the twentieth century, the Indonesian drive for self-determination intensified, and, following World War II, independence was proclaimed in 1945. The Dutch fought a losing battle to regain their East Indies, finally yielding in 1949; the eastern territory of Irian Jaya or West Irian (on New Guinea) was awarded to Indonesia as recently as 1969. And formerly Portuguese Timur became Indonesia's 27th province in 1976 after Indonesian armed forces invaded the area.

The persistence of Indonesia as a unified state is another politico-geographical wonder on a par with that of India. Wide waters and high mountains have helped perpetuate cultural distinctions and differences; centrifugal political forces have been powerful and, in the late 1950s and 1960s, nearly pulled the country apart. But Indonesia's unity appears to have strengthened, just as the national motto, *Unity in Diversity*, underscores. With more than 300 discrete ethnic clusters and over 250 individual languages (and just about every religion practiced in the world), Indonesian nationalism has faced enormous odds—overcome to some extent by development based on the country's considerable resource base. This includes sizable petroleum reserves (Fig. 6–13), large rubber plantations (Sumatera shares Malaya's environments), extensive timber resources, major tin deposits offshore from eastern Sumatera, and soils that produce tea, coffee, and other cash crops; copra and palm oil still rank among the exports as well. However, Indonesia's population continues to grow at the annual rate of 2.1 percent, a long-term threat to the country's future—and rice and wheat already figure prominently among annual imports.

# Territorial Morphology

Having viewed territories long under British and Dutch colonial administration, we now consider French and Spanish imprints (and Thailand's singular independence) in another context. It is difficult to look at the political map the Europeans left behind in Southeast Asia (Fig. 10–2) without being struck by the several different spatial forms or shapes that these states display. Indonesia is dispersed over a number of large and small islands, Vietnam extends narrowly along the east coast of Indochina, Burma and Thailand share the narrow northern part of the Malay Peninsula. This spatial aspect of the state—its shape—is a very important variable that can have a decisive effect on its cohesion and political viability. A state that consists of several separate parts located far away from each other obviously has to cope with problems of national cohesion not afflicting a state that has a single contiguous territory. We saw what problems Pakistan's division caused; alternatively, we know the advantages Uruguay derives from its compactness.

Political geographers identify five major shape categories on the world map, and four of them are well illustrated by the countries of Southeast Asia; hence, the realm should be considered in this context. For example, consider Kampuchea (Cambodia), whose territory is quite nearly round in shape: it looks a great deal like Uruguay and, as it happens, it is almost the same size (181,000 square kilometers against Uruguay's 176,000). Kampuchea is a *compact* state, which means that points on its boundary lie at about the same distance from the geometric center, near Kompong Thom on the Sen River. Theoretically, a compact state encloses a maximum of territory within a minimum of boundary, and this may hold advantages where boundaries lie in sensitive areas. In the case of Kampuchea, the common boundary with Vietnam is less than half as long as that of adjacent Laos; the Kampucheans have sought to secure their border areas against intrusions, whereas the Laotian boundary is violated as though it did not exist. Compact countries such as Kampuchea, moreover, do not face the problems confronting some other underdeveloped states of integrating faraway islands, lengthy peninsulas, or other distant territorial extensions into the national framework. Effective political control is most easily established in a compact area, and, for centuries, the Kampucheans came closer to national integration than perhaps any other Southeast Asian state. Kampuchea also had the advantage of remarkable ethnic and cultural homogeneity in so complex a region: over 90 percent of its people are Khmers, with the rest divided about equally between Vietnamese and Chinese origin.

Following the end of the Indochina War in the mid-1970s, a new Communist government sought to reconstruct Kampuchea as a rural society. The country's borders were sealed, thousands of suspected dissidents were eliminated, and the urban population of the capital, Phnom Penh, and other towns was forced to march into the countryside, there to engage in an unfamiliar subsistence life-style. Reports from the few who managed to cross into Thailand stated that hundreds of thousands died, especially the old, the very young, and the sick. In the aftermath of the Indochina War, Kampuchea's compact manageability spelled an internal disaster.

In contrast, the states of Malaysia, Indonesia, and the Philippines are *fragmented* states, meaning exactly what the term implies—states whose national territory consists of two or more individual parts, separated by foreign territory or by international waters. Indonesia and the Philippines represent one of the three subtypes of this category, in that their areas lie entirely on islands. Malaysia is a second subtype, because it lies partly on a continental mainland and partly on islands. The third subtype is not present in Southeast Asia: in that case, the major components of the fragmented state all lie on a continental mainland (such as Alaska and the conterminous 48 states of the United States of America).

Fragmented states must cope with problems of internal circulation and contact, and often with the friction effects of distance. Indonesia's government, based on Jawa, has had to put down secessionist uprisings in Sumatera, Sulawesi, and in the Moluccas (between Sulawesi and West Irian). The Luzon-centered government of the Philippines has also had to combat rebels (on Mindanao, Mindoro, and other islands) who took advantage of the insular character of the national territory. Far-flung Malaysia was forced to yield to the centrifugal forces inherent in its ethnic complexity and its spatial and functional structure, and it expelled the microstate of Singapore in the mid-1960s (see *Box*). Even in the choice of a capital, the fragmented state has difficulty: of all the separate parts of the country, one must be chosen to become the seat of government, the national headquarters. The choice may bring resentment elsewhere in the state, as it did in Pakistan. In Southeast Asia's fragmented states, however, there could not have been much doubt. Jawa contains almost two-thirds of Indonesia's 168 million people, and the choice of Jakarta (1985 population: 9.4 million) as the capital—the Hollanders' Batavia—can hardly be disputed; nevertheless, the peoples of the outer islands have shown resentment against Jawanese domination in Indonesia's affairs. In the Philippines, Luzon is the most populous island and Manila (7.1 million), without doubt, the country's primate city. In Malaysia, sparsely-peopled Sarawak and Sabah can hardly compete with the mainland core area, and naturally Kuala Lumpur, the preindependence headquarters, continued as the new state's capital city.

Still another spatial form is represented by Burma and Thailand. The main territories of these two states, which contain their core areas, are essentially compact—but to the south they share sections of the Malay Peninsula. These peninsular portions are long and narrow, and states with such extensions leading away from the main body of territory are referred to as *prorupt* states. Obviously, Thailand is the best example: its proruption extends nearly 1000 kilometers (600 miles) southward from the vicinity of Bangkok. Where the Thailand-Burma boundary runs along the peninsula, the Thai proruption is in places less than 32 kilometers (20 miles) wide. Naturally, such proruptions can be troublesome, especially when they are as lengthy as this. In the whole state of Thailand, no area lies as far from the core or from Bangkok as the southern extremity of its very tenuous proruption. But at least Thailand's railroad system extends all the way to the Malaysian border; in the case of Burma, not only does the railway terminate more than 480 kilometers (300 miles) short of the end of the proruption, but, in 1982, there was not even a permanent road over its southernmost 240 kilometers (150 miles), the railway reaching only as far as Mergui.

In Burma, the spatial morphology of the state is complicated by a shift in the core area that took place during colonial times. Prior to the colonial period, the focus of the embryonic state was in the so-called dry zone, between the Arakan Mountains and the Shan Plateau. Mandalay in Upper Burma, today containing 740,000 people, was the main urban node. Then the British developed the rice potential of the Irrawaddy Delta, and Rangoon (now with 2.8 million residents), a less important city occupied earlier by the British, became the new capital. The old and new core areas are connected by the Irrawaddy in its function as a water route, but the center of gravity in modern Burma lies in the south.

The fourth category of territorial shape represented in Southeast Asia is the *elongated* or attenuated state. By this is meant that a state is at least six times as long as its average width. Familiar examples on the world map are Chile in South America, Norway and Italy in Europe, and Malawi in Africa. Elongation presents obvious and recurrent problems; even if the core area lies in the middle of the state, as in Chile, the distant areas at either end may not be effectively connected and integrated in the state system. In Norway the core area lies near one end of the country, with the result that the opposite extremity takes on the characteristics of a frontier. In Italy, contrasts between north and south pervade all aspects of life, and they reflect the respective exposures of these two areas to different mainstreams of European-Mediterranean change. If a state is elongated and at the same time possesses more than one core area, strong centrifugal stresses will arise. This has been the classic case in Vietnam, one of the three political entities into which former French Indochina was divided.

French Indochina incorporated three major ethnic groups: the Laotians, the Cambodians, and the Vietnamese. Even before independence, the Laotians and the Cambodians possessed their own political areas (which later were to become the states of Laos and Cambodia [Kampuchea]), and that left a 2000-kilometer (1200-mile) belt of territory to the east, averaging under 240 kilometers (150 miles) in width, facing the South China Sea. This was the domain of the Annamites that extended all the way from the Chinese border to the Mekong Delta. Administratively, the French divided this elongated stretch of land into three units—Cochin China in the south, Annam in the middle, and Tonkin in the north (Fig. 10–11). The capitals of these areas became familiar to us during the Vietnam War: Saigon (now Ho Chi Minh City) was the focus of Cochin China and the headquarters of all Indochina, the ancient city of Hué was the center of Annam, and Hanoi was the capital of Tonkin. Cochin China and Tonkin were both incipient core areas, for they lay astride the populous and productive deltas of the Mekong and Red Rivers, respectively.

In 1940 and 1941, the Japanese

# SINGAPORE

**Figure 10–12**
The ultramodern skyline of the beehive city-state of Singapore, one of the western Pacific's four "New Japans" (see pp. 498–499).

Singapore, in 1965, became Southeast Asia's smallest independent state in terms of territory (just over 600 square kilometers/239 square miles) as well as in population (2.6 million in 1985). Situated at the southern tip of the Malay Peninsula where the Straits of Malacca open into the South China Sea and the waters of Indonesia, the port city had been a part of Britain's Southeast Asian empire and, briefly, a member state in the Malaysian federation. It had grown to become the world's fourth largest port by number of ships served, and it was always a distinct and individual entity—physically (the Johor Strait separates the island from Malaysia's mainland) and culturally (the population is 76 percent Chinese, 14 percent Malay, and 7 percent South Asian). When Singapore seceded from Malaysia in 1965, a major conflict between city and federation was averted.

This city-state (Fig. 10–12) has thrived since independence as Southeast Asia's only developed country, capitalizing on its relative location and its function as an entrepôt between the Malay Peninsula, Southeast Asia, Japan, and the other industrialized nations. Crude oil from the Middle East is unloaded and refined at Singapore, then shipped to Asian destinations. Raw rubber from Malaya and Sumatera is shipped to Japan, the United States, China, and other countries. Timber from Malaysia, rice, and spices are also assembled and forwarded via Singapore; in turn, automobiles, machinery, and equipment are imported for transhipment to Malaysia and other countries in the realm.

Significantly, Singapore's manufacturing sector is expanding rapidly and has now overtaken the entrepôt function in terms of its contribution to the national income. Taking a page from Hong Kong's book and aware of world price fluctuations and their effects on international trade, Singapore's planners have encouraged the diversification of industries. Foreign investment is attracted and given very advantageous terms. In addition to the refineries, Singapore now has shipbuilding and repair yards, food-processing plants, sawmills, and a growing number of small industries serving the local market. High technology is particularly stressed in the 1980s as the city-state relishes its role as one of the four "New Japans" of eastern Asia.

Singapore's population may be 76 percent Chinese, but its Chinese sector is by no means homogeneous. The Chinese immigration came from separate sources, Fujian Province and Guangdong Province among them. The Chinese communities speak different Chinese languages, but only one (Mandarin) is recognized as one of Singapore's official languages (the others are English—the *lingua franca*—Malay and Tamil). Southeast Asia's largest port is truly a cultural crossroads as well as an economic hub.

entered and occupied Indochina, and during their nearly four years of occupation the concept of a united Vietnam emerged. The Japanese did not discourage rising Vietnamese nationalism, especially when it became apparent that the tide of war was turning against them; they would rather have an Annamese government in Vietnam than a French one. So, during the period of Japanese authority, a strong coalition of proindependence movements arose in Indochina—the Viet Minh League. In 1945, when the Japanese surrendered, this organization seized control and proclaimed the Republic of Vietnam a sovereign state under the leadership of Ho Chi Minh. For some time after the Japanese defeat, the Chinese actually occupied Tonkin and northern Annam, but theirs was a sympathetic involvement as well and the Viet Minh grew in strength during that time. Meanwhile, the French had proposed that Vietnam, Laos, and Cambodia should join as associated states in the fourth-republic concept of the French union, but this plan was rejected everywhere in Indochina. As the French reestablished themselves where they had always been strongest—in Saigon and the Mekong Delta of Cochin China—they started negotiations with the Viet Minh. These talks soon broke down; in 1946, a full-scale war broke out between the French and the Viet Minh. France's base lay in Cochin China, and the Viet Minh had their greatest strength in Tonkin 1000 miles away. Vietnam's pronounced elongation favored the Viet Minh and their numerous nationalist, Communist, and revolutionary sympathizers. After an enormously costly eight-year war, the crucial battle was fought in the north (at Dien Bien Phu) not far from the Chinese border, and it was lost by the French.

## The Insurgent State

The sequence of events in Vietnam in the 1960s and 1970s is still fresh in memory. Opposition to the Saigon government in South Vietnam turned into armed insurrection; North Vietnam provided arms and personnel to aid the anti-Saigon insurgents; U.S. advisors and armed forces in support of the Saigon regime at one time exceeded a half million in number. Despite the severe bombing of North Vietnam and its capital, Hanoi, and extensive defoliation efforts in the protective forests of the South, what had begun as a scattered rebellion ended in the defeat of a national government and the ouster of its ally, the United States. Political scientists recognize stages in the sequence of these events; the first is a period of *contention*, when armed rebellions start to occur in remote locales, followed by *equilibrium*, when the insurgents manage to withstand government forces in critical areas of the country, and finally *counteroffensive*, a stage of regular ground and air war. The result may be the defeat of the insurgents, as happened in Malaya in the 1950s, or their success and takeover of the national government, as in Cuba in 1959.

These stages have been translated into territorial terms by Robert McColl, who suggested that they chronicle the emergence of an *insurgent state* within the boundaries of a national state. During the stage of equilibrium, the emerging insurgent state attains a degree of stability. It may be fragmented and consist of several cores, but it has headquarters, boundaries, an increasingly complex administrative structure, schools, hospitals, and

**Figure 10–13**

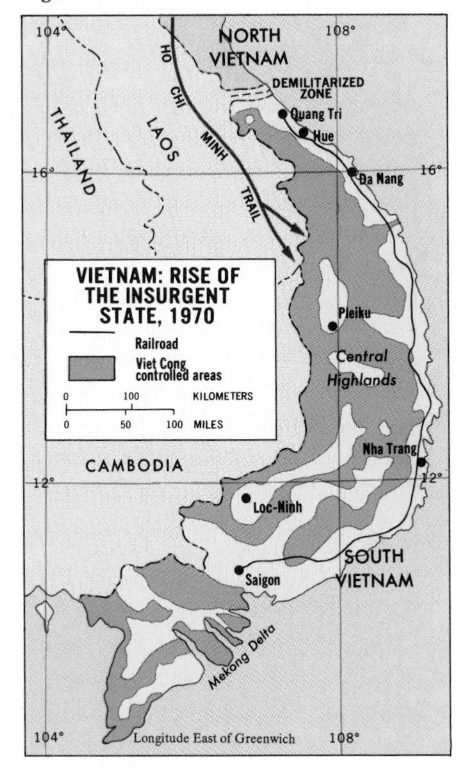

even air traffic with neighboring countries that may sympathize with the insurgents' cause. The map of (then) South Vietnam in 1970 (Fig. 10–13) shows areas still under Saigon's control (in green) actually surrounded by territory held by the insurgent Viet Cong. In 1978, the 10-year-old insurgent state that had existed in Moçambique took control of the government. In Ethiopia, an insurgent state still exists today in the provice of Eritrea; in Pakistan there are signs that an insurgent state might emerge in the northwest. For years, the Philippine government has been combating efforts by Moslem separatists in the southern islands, especially Mindanao, to create an insurgent state (and by Maoists throughout the fragmented country to accomplish the same); in 1976, the Philippine authorities promised to grant greater autonomy to the country's Moslem regions, thereby reducing the level of contention. The insurgent state is a product of current ideological contests, the proliferation and availability of modern weapons, and the effectiveness of modern communications.

The unification of North and South Vietnam in 1976 produced one of the world's most decidedly elongated states and, with a population of 59.6 million (1985), Southeast Asia's second largest. And the country's attenuation continues to dominate its regional geography. A totalitarian central government notwithstanding, the old divisions between north, center, and south still prevail. Saigon may now be Ho Chi Minh City (1985 population: 2.7 million), but it remains a world apart from Hanoi (1.3 million) and the distant north. Since unification, Vietnam has been involved in an intermittent armed conflict with China along its northern boundary, and it still pursues a campaign of invasion and domination in neighboring Kampuchea. In these efforts, Vietnam functions as a client state of the Soviet Union which, together with its Eastern bloc satellites, provides the financial aid (about $2 billion per year in the mid-1980s) to keep the Vietnamese economy from collapsing; in return, the Soviets get to use Vietnam as a vital military base for their warplanes and navy—the only Russian outpost on the 15,000-mile-long Asian coastline between Vladivostok and South Yemen.

One of the by-products of successful insurgency movements is the human upheaval that occurs when these types of states become fully established. Large numbers of those who opposed the rise of the new order, fearing for their lives, feel compelled to flee the country—and thereby join the growing world population of refugees. We have described many such involuntary migrations in earlier chapters, but, since 1975, none have been more tragic than the refugee flows generated by the turbulence in Indochina. These movements are mapped in Fig. 10–14 and are dominated by the exodus of "Boat People" from Vietnam; of the 1 million estimated to have set sail on the wide South China Sea following the Communist takeover in the mid-1970s, more than half perished at the hands of storms, pirates, leaky boats, exposure, and starvation (only about 400,000 survivors were counted in refugee centers and resettlement programs in 1980). The map also shows a second major emigration flow—the

Figure 10–14

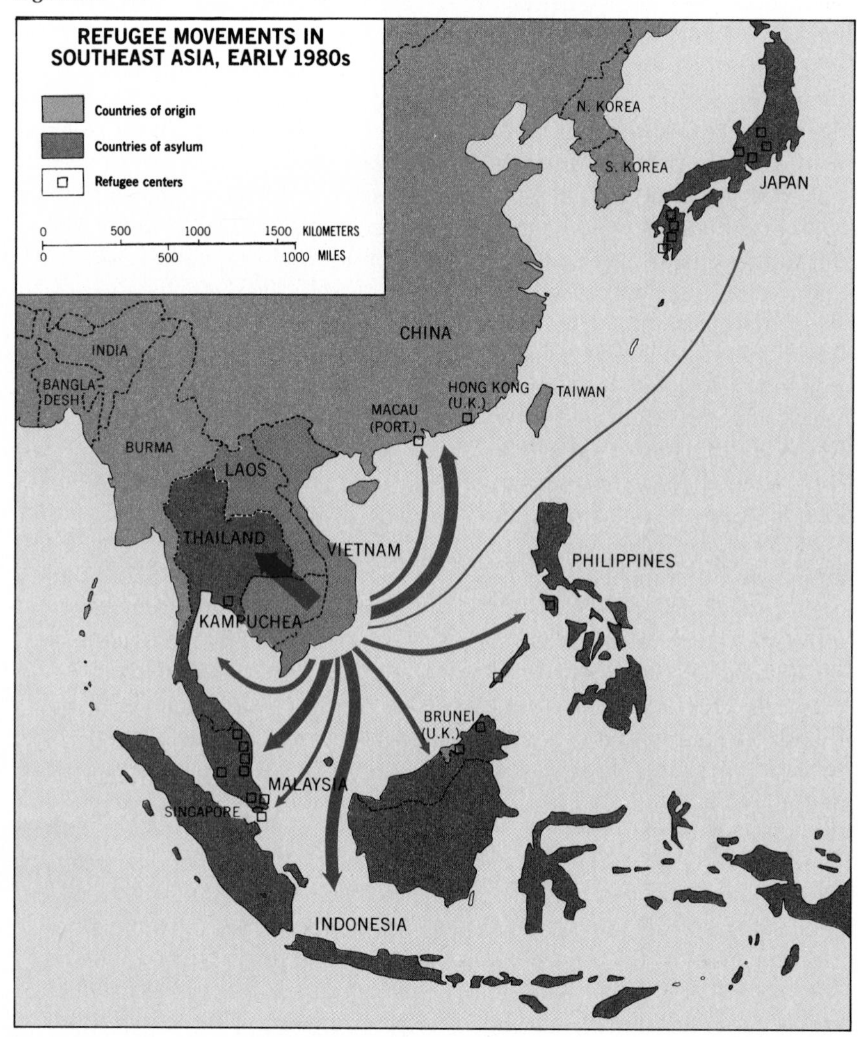

200,000-plus "Land People" who staggered into Thailand across Kampuchea's western border, escaping from the genocide perpetrated by the Pol Pot regime and, later, the internal hostilities surrounding the fall of that ultrarepressive government in 1979. Because the non-Communist countries of Southeast Asia did not permit most of these migrants to settle permanently within their borders (only 44,000 remained in the entire realm in 1983), the refugees were primarily resettled elsewhere in the world. Of the approximately 1 million such surviving refugees who fled the various Communist purges in Vietnam, Kampuchea, and Laos between 1975 and 1981, 500,000 were resettled in the United States, 265,000 in China, 75,000 in both Canada and France, 52,000 in Australia, 30,000 in the United Kingdom (including Hong Kong), 18,000 in West Germany, and 30,000 in Switzerland and other smaller European countries.

## The Domino Theory and Thailand

During the 1960s, as one country after another became engulfed by the Vietnam War (later to be called the Indochina War to signify its wider arena), Southeast Asian specialists began to refer to the "domino theory"—the idea that the fall of South Vietnam would inevitably lead to Communist takeovers in Kampuchea, Laos, Thailand, Burma, Malaysia, and, ultimately, Indonesia and the Philippines. This was no mere speculation, for the effect of adjacent sanctuary on the progress of the Vietnamese insurgency was crucial. The Ho Chi Minh Trail brought supplies from North Vietnam to the South via Laos; Viet Cong insurgents enjoyed safety beyond South Vietnam's borders in Kampuchea. When Laos and Kampuchea were drawn into the conflict and the war took its course, those who supported the domino idea appeared to have been correct.

Further support for the domino effect can be derived from Africa, where Moçambique's FRELIMO insurgents could not mount their ultimately successful effort without staging areas in neighboring Tanzania, and where Rhodesia (Zimbabwe) and Namibia became areas of contention only after adjacent Moçambique and Angola had seen revolutionary movements succeed. Opponents of the idea pointed to Malaysia, where a large-scale Communist insurgency was defeated more than 20 years ago and where, political conflicts notwithstanding, no renewed threat of similar proportions arose during the Indochina War.

Both supporters and opponents of the domino concept point to Thailand, from where American warplanes took off on their bombing missions to North Vietnam and to where those Kampucheans who could, fled when a vicious Communist regime took power in Kampuchea. Thailand's strongly prorupt territory (containing a total population of 52.7 million in 1985) adjoins Laos as well as Kampuchea (Fig. 10–15); its northern areas lie within 150 kilometers (slightly under 100 miles) from China's borders. These northern areas are also among Thailand's most mountainous, remote, and frontierlike, but there was colonial design in the creation of a British-French buffer zone between Thailand and China. Thailand retained its independence during the colonial period (when it was called Siam; in 1939, it was renamed *Prathet Thai*—"Land of the Free"), but it was purposely separated from southern China. Today, Thailand's core area lies in the lower basin of the Chao Phraya, centered on the capital, Bangkok (the metropolitan area, including Thon Buri, has a population of about 6.2 million).[2] This is the focal region for the 75 percent of the population who are Thais (Siamese); in the country's marginal areas, there are peoples with other roots. Lao and Khmer in the east, Shan and Mon in the north, Karen in the west (movement continues across the border to and from Burma), and Malay on the southern peninsula—these are but some of the peripheral minorities (Fig. 10–7). Add to these some significant differences, ethnic as well as cultural, between Thais in the lowlands and the mountains, and the country's regional vulnerability becomes clear.

Already, Communist-inspired insurgents have established an area of contention in the far north, and separatist problems have emerged in the deep peninsular south, where the dominantly Buddhist country's Moslem minority is concentrated. The divisions between left and right, mountain peasant and central government, student and soldier have intensified in recent years. There can be no doubt about it: external ideological pressures and actual guerilla support have begun to destabilize a country that survived the colonial era, but became heavily involved with the United States after the Second World War. As the U.S. withdrew from Vietnam and began to establish new relations with China, Thailand was forced to modify its long-term commitment to the West—a process that has already contributed to governmental instability, political clashes, and regional unrest. Should the insurgent state emerge in Thailand, it will be difficult to argue against a domino effect in Southeast Asia.

[2]The inset map in Fig. 10–15 shows the fifth of the spatial forms of states, the only case not represented in Southeast Asia—the *perforated* state. South Africa is shown, perforated by the state of Lesotho.

## Philippine Fragmentation

After Indonesia and Vietnam, the Philippines, with 55.8 million people, is Southeast Asia's next most populous state (Fig. 10–16). However, few of the generalizations that can be made about the realm would apply without qualification to this country, and the Philippines's location relative to the mainstreams of change in this part of the world has had much to do with this. The islands, inhabited by peoples of Malay ancestry with Indonesian strains, shared with much of the rest of Southeast Asia an early period of Indian cultural influence, strongest in the south and southwest and diminishing northward. Next came a Chinese invasion, felt most strongly in Luzon in the northern part of the archipelago. Islam's arrival was delayed somewhat by the position of the Philippines well to the east of the mainland and to the north of the Indonesian islands. The southern Moslem beachheads were soon overwhelmed by the Spanish invasion of the sixteenth century; today, the Philippines, adjacent to the world's largest Moslem state (Indonesia), is about 85 percent Roman Catholic and only 5 percent Moslem.

Out of the Philippine melting pot, where Mongoloid-Malay, Arab, Chinese, Japanese, Spanish, and American elements have met and mixed, has emerged the distinctive culture of the Filipino. It is not a homogeneous or a unified culture, but in Southeast Asia it is in many ways unique. One example of its absorptive qualities lies in the way the Chinese infusion has been accommodated: the "pure" Chinese minority numbers a mere 1 percent of the total population (far less than normal for Southeast Asian countries); in fact, a much larger portion of the Philippine population carries a decidedly Chinese ethnic imprint. What has happened is that the Chinese have intermarried, producing a sort of Chinese-mestizo element—consituting more than 10 percent of the entire population. In another cultural sphere, the country's ethnic mixture and variety is paralleled by its great linguistic diversity. Nearly 90 Malay languages, major and minor, are spoken by the 56 million people of the Philippines; only about 1 percent still uses Spanish. Visayan is the language most commonly spoken; nearly 50 percent of the population is able to use En-

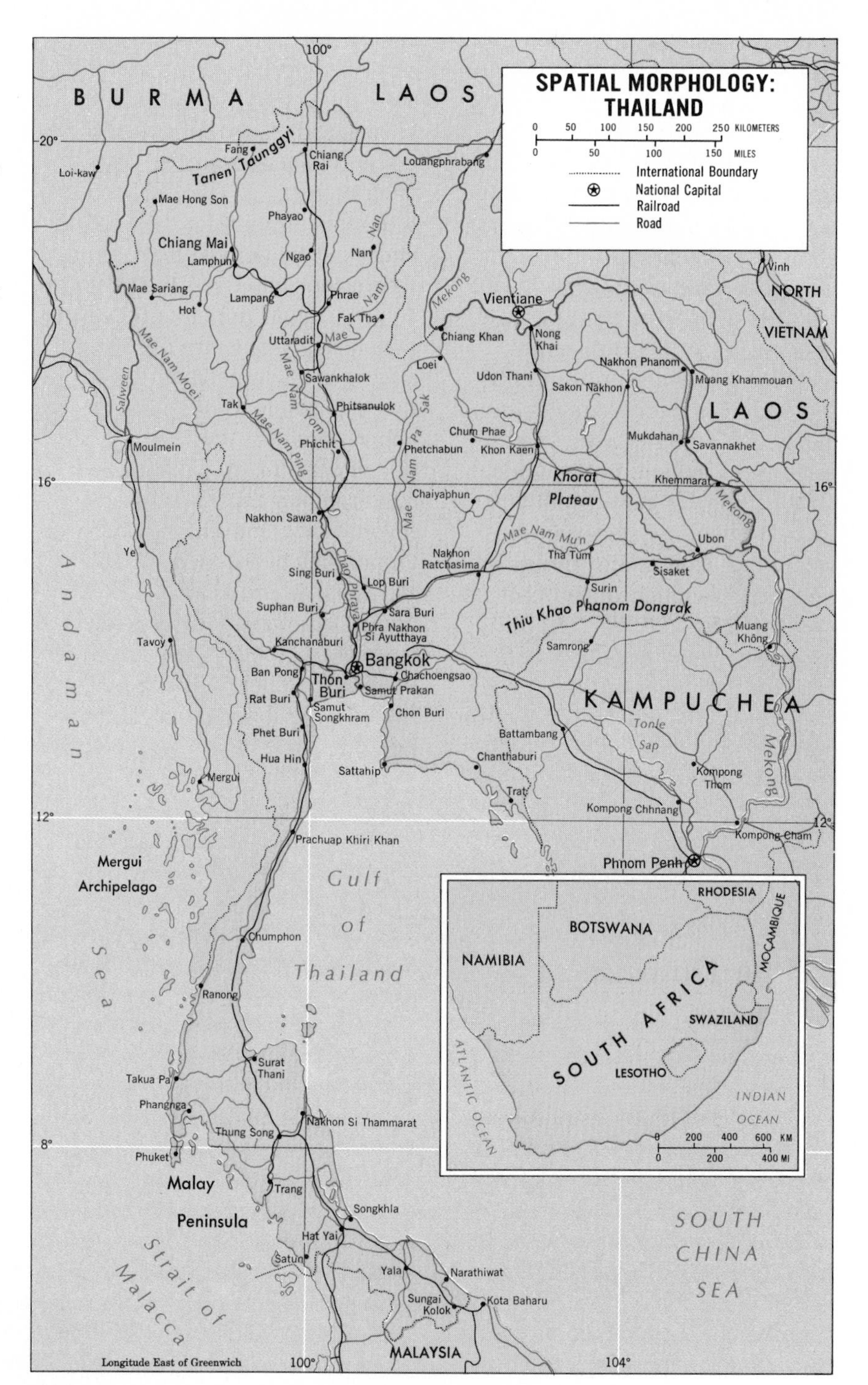


Figure 10–15

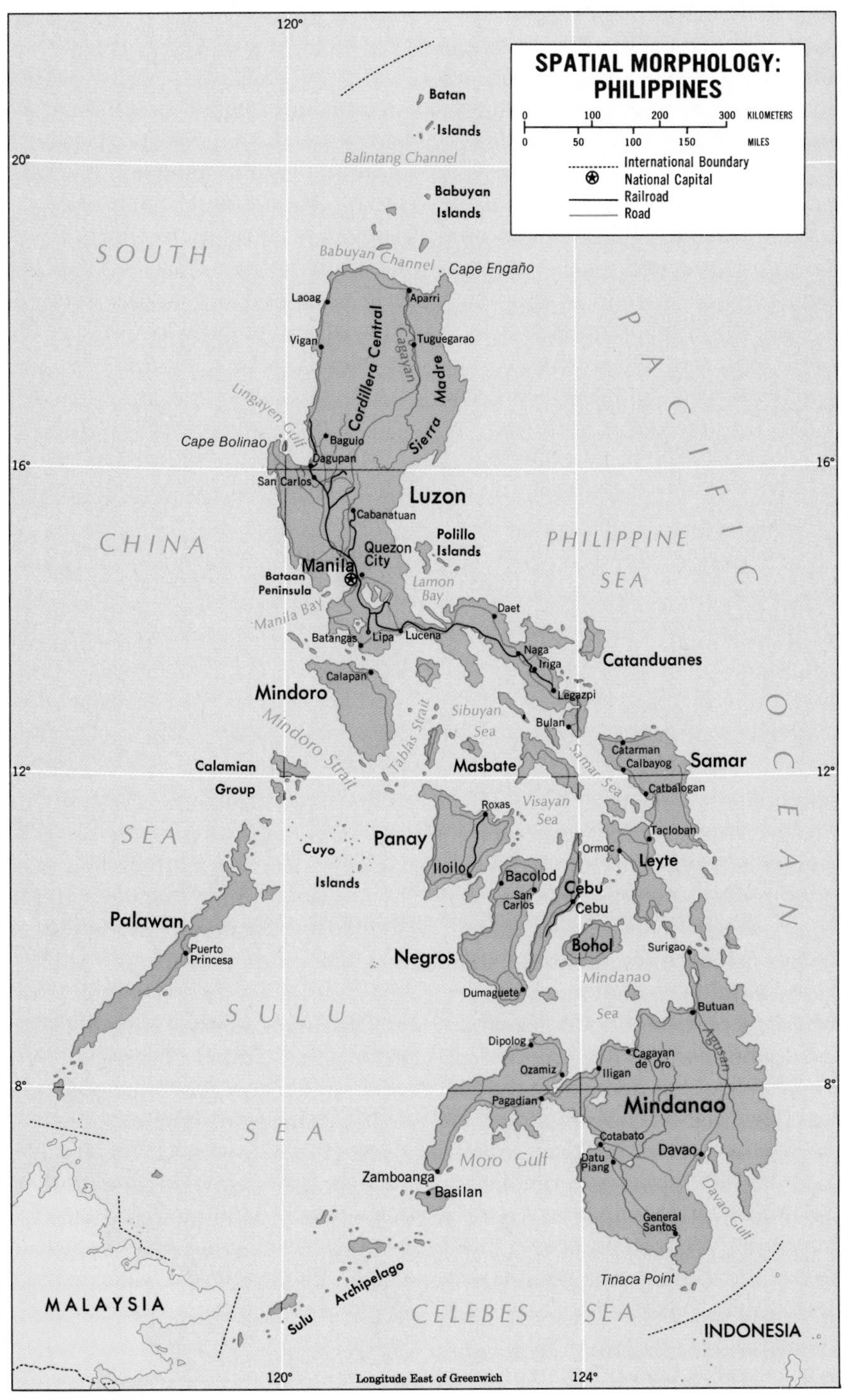


Figure 10–16

glish. At independence, the largest of the Malay languages, Tagalog or Filipino, was adopted as the country's official language, and its general use is strongly promoted through the educational system. English is learned as a subsidiary language and remains the chief *lingua franca*; an English-Tagalog hybrid—"Taglish"—is increasingly heard in the 1980s, remarkably cutting across all levels of society.

The widespread use of English in the Philippines, of course, results from a half-century of American rule and influence, beginning in 1898 when the islands were ceded to the United States by Spain under the terms of the Treaty of Paris, following the Spanish-American War. The United States took over a country in open revolt against its former colonial master (the Philippines had declared its independence from Spain in 1898) and proceeded to destroy the Filipino independence struggle, now directed against the new foreign rulers. It is a measure of the subsequent success of U.S. administration in the Philippines that this was the only dependency in Southeast Asia that during World War II sided against the Japanese in favor of the colonial power (anti-Japanese movements did develop in Malaya and Burma as well). U.S. rule had its good and bad features, but the Americans did initiate reforms that were long overdue, and they were already in the process of negotiating a future independence for the Philippines when the War intervened.

The reforms begun by the United States in the Philippines and continued by the Filipinos themselves after postwar independence in 1946 were designed to eliminate the worst aspects of Spanish rule. Spanish provincial control was facilitated through allocations of good farmland as rewards to loyal representatives of church and state. The quick acceptance of Catholicism and its diffusion throughout the islands helped consolidate this system; exploitation of land and labor (often by force) were the joint objectives of the priests and political rulers alike. When, after three centuries of such exploitation, Spanish colonial policy showed signs of relaxation during the nineteenth century, it was too late. Crops from the Americas were introduced (tobacco became one lucrative product), and the belated effort was made to integrate the Philippine economy with that of Spanish-influenced America. But what the Philippines needed most, the Spaniards could not provide: land reform. As everywhere else in the colonial world, the main issue between the colonizers and the colonized in the Philippines was land, agricultural land. And, as elsewhere, the Spanish colonizers found that what had been easy to give away was almost impossible to retrieve. Long after the Spaniards had lost their Philippine dependency, the Americans, with a much freer hand, found the same still to be true—and even today the Filipino government, after four decades of sovereignty, still faces the same issue.

The Philippines's population, concentrated where the good farmlands lie in the plains, is densest in three general areas: the south-central and northwestern part of Luzon (metropolitan Manila with one-eighth of the national population lies at the southern end of this zone), the southeastern proruption of Luzon, and the Visayan Islands between Luzon and Mindanao. The Philippine archipelago consists of over 7000 mostly mountainous islands, of which Luzon and Mindanao are the two largest (accounting for nearly two-thirds of the total area). In Luzon the farmlands, producing rice and sugar cane, lie on alluvial soils; in extreme southeastern Luzon and in the Visayan Islands, there are good volcanic soils. When world market prices are high, sugar is the most valuable export of the agriculture-dominated Philippines; the forests yield valuable timber, and copra and coconut oil rank next—but most Filipino farmers are busy growing the subsistence crops, rice and corn. As in the other Southeast Asian countries, there is a considerable range of supporting food crops. Along with other countries in the region, the Philippine state does not yet face a major overpopulation problem; however, in parts of Luzon and the Visayan Islands that threshold is being reached, and the national annual growth rate was 2.5 percent in the early 1980s. When it comes to increasing the intensity of farming, few peoples in the world could provide a better example than northern Luzon's Igorots who have transformed their hillsides into impeccably terraced, irrigated paddy fields.

Thus, the Philippine state exhibits both rule and exception in Southeast Asia. A fragmented island country, it must cope with internal revolutionary forces and intermittent encroachment from outside (Indonesia's former leader, Sukarno, envisaging a Greater Indonesia sphere, at times expressed irredentist support for Mindanao's Moslems). Culturally, it is in several ways unique; economically, it shares with the rest of Southeast Asia the liability of imbalanced and painfully slow development. The Philippines's lumber and agricultural exports are supplemented by copper and iron ores, while revenues must be spent on the purchase of machinery, various kinds of equipment, fuels, and even some foods. All this exemplifies the country's persistent underdevelopment, and regional frustrations are in the process of deepening a political crisis that has been building for many years. As in the case of Thailand, the Philippine government is reconsidering its long-term alignment with Western powers in this part of the world while the politico-geographical map of Southeast Asia changes. And internally, the long-entrenched, repressive Marcos government grows increasingly unpopular, as the election results of 1984 revealed; in the face of this autocratic regime's limitations on individual freedoms—and such shocking incidents as the 1983 assassination of Opposition Leader Benigno Aquino—democracy struggles to survive.

# Land and Sea

In a realm of peninsulas and islands such as Southeast Asia, the surrounding and intervening waters are of extraordinary significance. They may afford more effective means of contact than the land itself; as trade and migration routes, they sustain internal as well as external circulation. On the other hand, the waters between the individual islands of a fragmented state also function as a divisive force. There are literally tens of thousands of islands in Southeast Asia; although some are productive and in effective maritime contact with other parts of the region and the world, many of these islands and islets are comparatively isolated. In this respect, of course, Southeast Asia is not unique in the world. In previous chapters, we discussed the peninsular character of Western Europe and the insular nature of Caribbean America. In all these areas, peoples and governments possess a keen awareness of the historic role of the seas. States with coastlines have extended their sovereignty over some of their adjacent waters. This is not a new principle; in Europe, the concept that a state should own some offshore waters is centuries old. Initially, such coastal claims—on a world scale—were quite modest. But, as time went on, some states began to extend their territorial waters farther and farther out, and other countries followed suit.

These widening maritime claims involved not only the ocean itself, but also the seabed below, the floor of the ocean and the rocks beneath. Technological advances in recent decades have made it possible to explore and mine the bottom of the ocean to ever greater depths, and soon only the deepest trenches may lie beyond the grasp of modern machines. Already the continental shelves (submerged zones adjacent to the continents to a depth averaging 200 meters [660 feet]) are intensively exploited, and more than 25 percent of all the oil brought to the surface each year now comes from offshore wells. Minerals are also being mined, even from the deep floors of abyssal plains in the central areas of ocean basins. Manganese nodules, potato-sized concentrations of valuable industrial minerals, are being drawn to the surface through pipes lowered more than 3500 meters (11,500 feet) to the ocean floor; besides containing up to 40 percent manganese (an important ferroalloy), these nodules also yield copper, nickel, and cobalt.

These developments place the underdeveloped countries in a disadvantageous position, for they do not possess the capital and technology to extract the resources con-

Figure 10–17

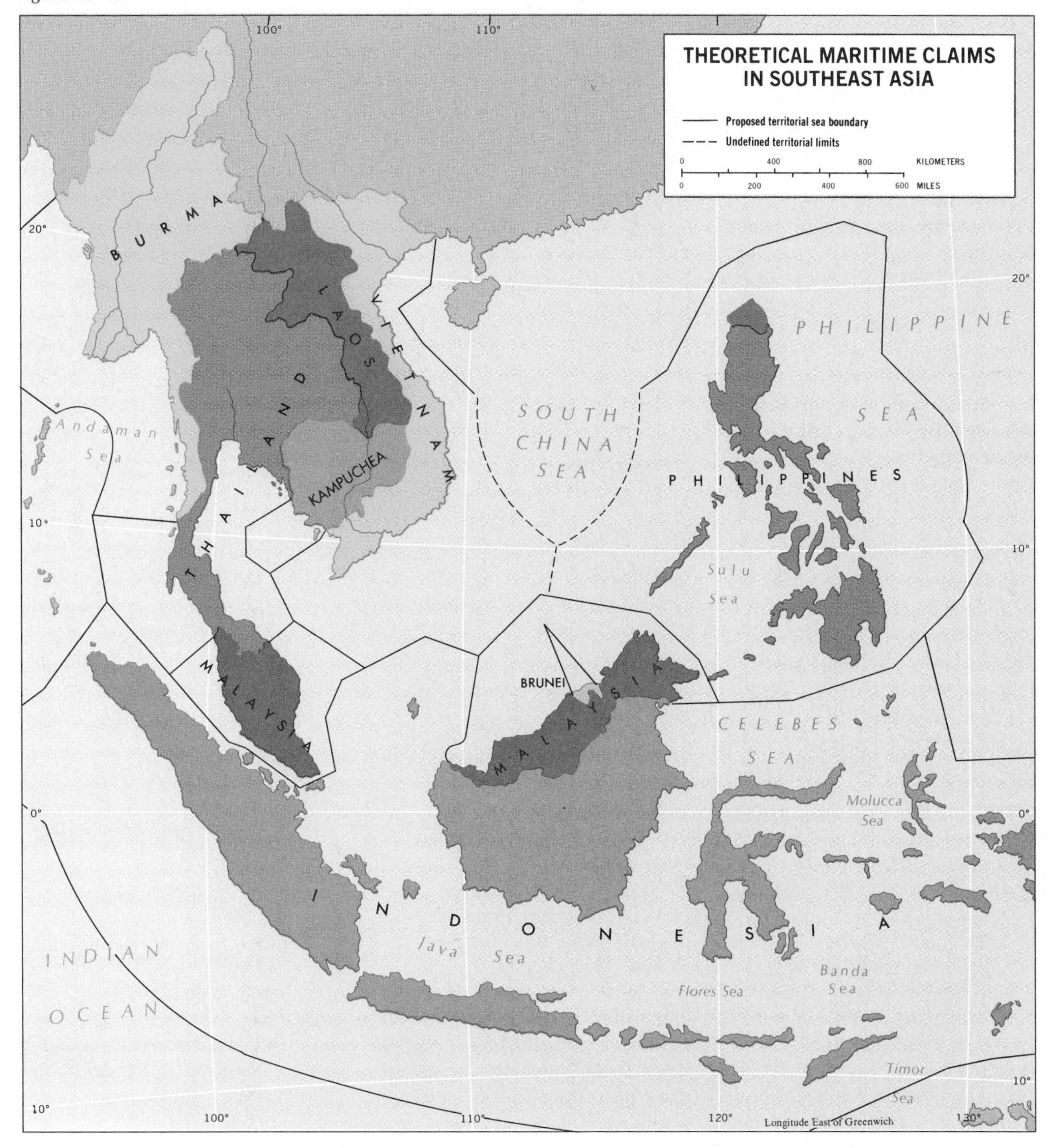

tained within their own offshore zones. They do, however, have collective influence in the Untied Nations, and their response has been to extend their claims to territorial waters far beyond prevailing limits. This process has become known as the "scramble for the oceans," and it is still in progress today. When World War II ended, as many as 40 countries claimed a mere 3 nautical miles of territorial waters, and only 9 countries had territorial waters wider than 6 nautical miles. But by 1980, only 24 countries still adhered to their 3-mile claim, 60 countries claimed a 12-mile territorial sea, and 10 countries demanded sovereignty over 100 to 200 miles of adjacent ocean and seabed.

In Southeast Asia, Indonesia announced in 1957 that it would claim as national territory all waters within 12 nautical miles of its outer islands *and* all waters between all the islands of its far-flung archipelago (Fig. 10–17). This had the effect of making the entire Java Sea, Banda Sea, and most of the Celebes (Sulawesi) Sea territorial waters. A similar claim by Malaysia, following establishment of its federation with Sarawak and Sabah, appropriated much of the southern South China Sea. These claims to territorial waters in Southeast Asia, as elsewhere, also involve the underlying continental shelf—known to contain abundant petroleum reserves but still only partially explored.

In the early 1980s, the scramble for the oceans continued unabated. International agreement was reached in 1977 on a 320-kilometer (200-mile) fishing zone; subsequently, countries within 200 miles of one another began the task of defining and delimiting their maritime boundaries, a difficult undertaking even for allies as close as the United States and Canada (let alone such hostile pairings as the United States and Cuba). The effect

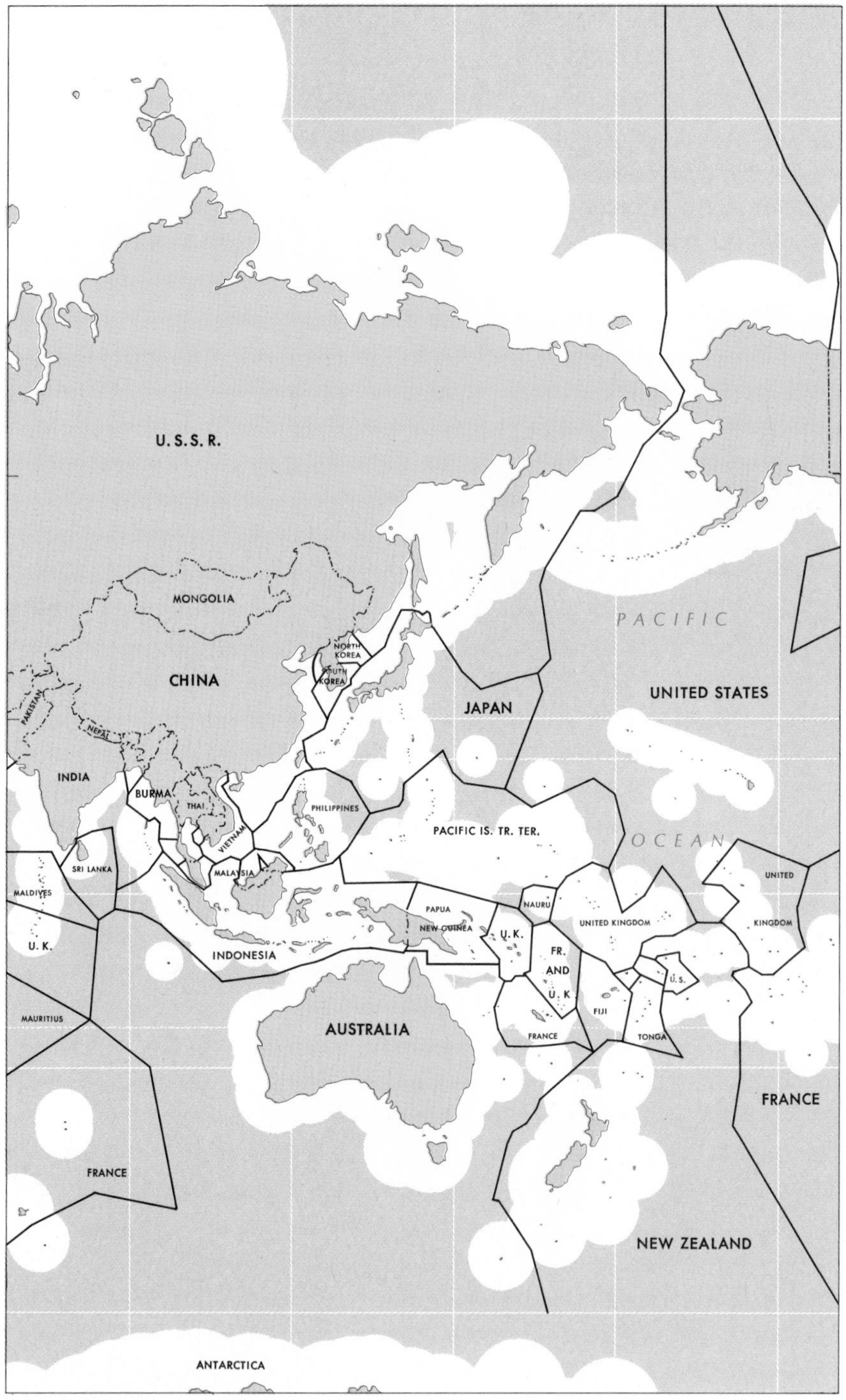


**Figure 10–18**

of a 200-mile zone of national sovereignty on the world's oceans is shown in Fig. 10–18. The open oceans, or "high seas," are much diminished by the 200-mile territorial limit; but the first steps *beyond* this 200-mile principle have already been taken in some areas. In Southeast Asia, the Caribbean region, and the North Sea, the 200-mile zone is irrelevant because the waters between countries are often narrower. In such cases, other techniques for boundary definition are employed, and the result is a map such as Fig. 1–14 (p. 81), allocating geometrically delimited regions of the sea to coastal countries. Al-

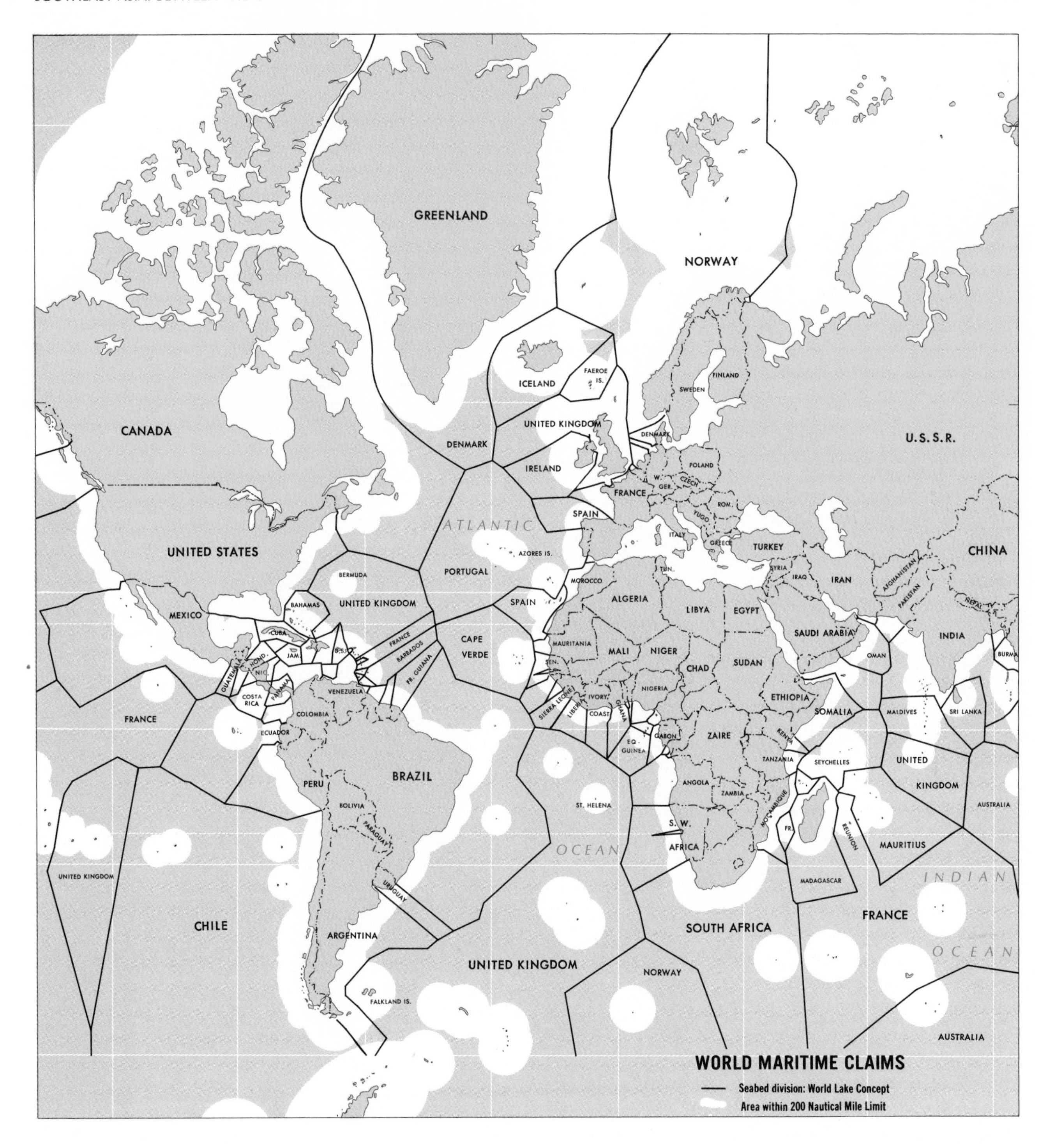


ready this process has been extended to include all the world's oceans, and Fig. 10–18 reveals the global boundary framework if the "world-lake concept" were to be adopted. It seems inconceivable, at present, that this could actually come about; but it seemed equally inconceivable, 30 years ago, that a 200-mile limit would ever receive international sanction. The oceans—together with Antarctica, where other considerations have shaped the scramble for territorial jurisdiction (see *Box*)—constitute this planet's last frontiers.

## References and Further Readings

Aiken, S. et al. *Development and Environment in Peninsular Malaysia* (Singapore: McGraw-Hill International, 1982).

Alexander, L. *World Political Patterns* (Chicago: Rand McNally, 2 rev. ed., 1963).

Atlas of Southeast Asia (New York: St. Martin's Press, 1964).

Bergman, E. *Modern Political Geography* (Dubuque, Iowa: W. C. Brown, 1975).

Broek, J. "Diversity and Unity in Southeast Asia," *Geographical Review*, 34 (1944), 175–195.

Burling, R. *Hill Farms and Padi Fields: Life in Mainland Southeast Asia* (Englewood Cliffs, N.J.: Prentice-Hall, 1965).

Christian, J. *Modern Burma: A Survey of Political and Economic Development* (Berkeley: University of California Press, 1942).

Couper, A., ed. *The Times Atlas of the Oceans* (New York: Van Nostrand Reinhold, 1983).

Couper, A. "Who Owns the Oceans?" *Geographical Magazine*, September 1983, 450–457.

de Blij, H. J. "A Regional Geography of Antarctica and the Southern Ocean," *University of Miami Law Review*, 33 (1978), 299–314.

Dobby, E. *Southeast Asia* (London: London University Press, 7 rev. ed., 1969).

Dutt, A., ed. *Southeast Asia: Realm of Contrasts* (Dubuque, Iowa: Kendall/Hunt, 1974).

East, W. & Moodie, A., eds. *The Changing World: Studies in Political Geography* (Yonkers, N.Y.: World, 1956).

Fisher, C. *Southeast Asia: A Social, Economic and Political Geography* (New York: E. P. Dutton, 2 rev. ed., 1966).

Fryer, D. *Emerging South-East Asia: A Study in Growth and Stagnation* (New York: McGraw-Hill, 2 rev. ed., 1979).

Glassner, M. "The Law of the Sea," *Focus*, March–April 1978.

Glassner, M. & de Blij, H. J. *Systematic Political Geography* (New York: John Wiley & Sons, 3 rev. ed., 1980).

Hill, R., ed. *South-East Asia: A Systematic Geography* (New York: Oxford University Press, 1979).

Huke, R. *Shadows on the Land: An Economic Geography of the Philippines* (Manila: Bookmark, 1963).

Hunter, G. *South-east Asia: Race, Culture, and Nation* (New York: Oxford University Press, 1966).

Jackson, W. & Samuels, M., eds. *Politics and Geographic Relationships* (Englewood Cliffs, N.J.: Prentice-Hall, 2 rev. ed., 1971).

Kasperson, R. & Minghi, J. *The Structure of Political Geography* (Chicago: Aldine, 1969).

# THE SOUTHERN REALM: ANTARCTICA AND SURROUNDINGS

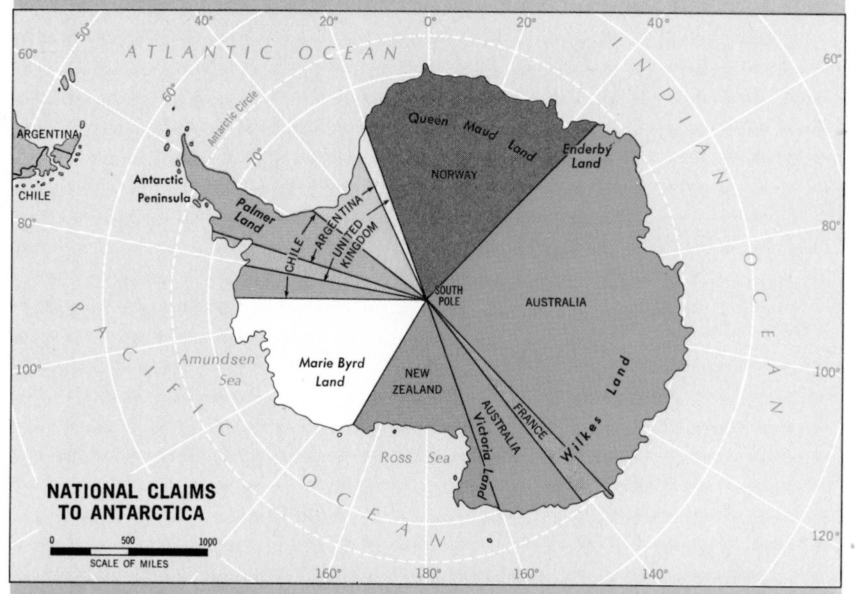


Figure 10–19

The Southern Realm is the least populous and most remote of the world's geographic regions. It consists of the Antarctic continent and the waters surrounding it, the Southern Ocean. The Antarctic landmass is almost totally covered by an enormous and thick icesheet; the Southern Ocean is a giant swirl of frigid water, moving in an easterly direction around Antarctica. This is not an attractive picture, but the Southern Realm did attract pioneers and explorers. Antarctica's coasts were visited by navigators from various countries, by whale and seal hunters, and by explorers who established temporary stations on the margins of the landmass and planted the flags of their nations there. Between 1895 and 1914, the journey to the South Pole became an international obsession; Roald Amundsen, the Norwegian, reached it first in 1911. All this led to the creation of national claims during the ensuing interwar period. The geographic effect was the partitioning of Antarctica into pie-shaped sectors centered on the South Pole (Fig. 10–19). One of the areas of contention among states was the Antarctic Peninsula (facing South America), where British, Argentinian, and Chilean claims overlapped—a situation that is not yet resolved. In 1985, only one Antarctic sector remained unclaimed: Marie Byrd Land (shown in white on Fig. 10–19).

Why should states be interested in territorial claims in so remote and difficult an area? Both land and sea contain resources that, some day, may become crucial: protein in the waters, and fuels and minerals beneath the land surface. Antarctica (14.2 million square kilometers/5.5 million square miles) is twice as large as Australia, and the Southern Ocean is nearly as large as the Atlantic. However distant actual exploitation may be, countries want to keep their stakes in the Southern Realm. But the *claimant* states (those with territorial claims) recognized the need for cooperation and the potential for conflict. In the 1950s, there was a major international program of geophysical research, the so-called *International Geophysical Year.* The spirit of cooperation that made this program possible extended to the political sphere and led, in 1961, to the signing of the Antarctic Treaty, an agreement that ensured continued scientific cooperation, prohibited military activities, and held national claims in abeyance. But this Treaty will expire in 1991, and there is growing concern that it may not be extended. It does not address the question of resource exploitation; in an age of growing national self-interest and increasing resource needs, there is the possibility that international rivalry in the Southern Realm will intensify and produce a confrontation the Treaty cannot prevent. The Southern Realm is truly the globe's last frontier, and the partition of its lands and waters is a process fraught with dangers.


Leinbach, T. & Ulack, R. "Cities of Southeast Asia," in Brunn, S. & Williams, J., eds., *Cities of the World: World Regional Urban Development* (New York: Harper & Row, 1983), pp. 370–407. See especially p. 383.

Leung, C. et al., eds. *Hong Kong: Dilemmas of Growth* (Canberra: Austrialian National University, Research School of Pacific Studies, 1980).

Lovering, J. & Prescott, J. *Last of Lands . . . Antarctica* (Melbourne, Austral.: Melbourne University Press, 1979).

McColl, R. "The Insurgent State: Territorial Bases of Revolution," *Annals of the Association of American Geographers*, 59 (1969), 613–631.

McGee, T. G. *The Southeast Asian City: A Social Geography* (New York: Praeger, 1967). Diagram adapted from p. 128.

Pounds, N. *Political Geography* (New York: McGraw-Hill, 2 rev. ed., 1972).

Prescott, J. *The Geography of Frontiers and Boundaries* (Chicago: Aldine, 1965).

Prescott, J. *Political Geography* (New York: St. Martin's Press, 1972).

Prescott, J. *The Political Geography of the Oceans* (New York: Halsted Press, 1975).

Pryor, R., ed. *Migration and Development in Southeast Asia* (Kuala Lumpur: Oxford University Press, 1978).

Purcell, V. *The Chinese in Southeast Asia* (New York: Oxford University Press, 2 rev. ed., 1965).

Samuels, M. *Contest for the South China Sea* (London and New York: Methuen, 1982).

Soja, E. *The Political Organization of Space* (Washington: Association of American Geographers, Commission on College Geography, Resource Paper No. 8, 1971).

Spencer, J. *Shifting Cultivation in Southeastern Asia* (Berkeley: University of California Press, 1976).

Spencer, J. & Wernstedt, F. *The Philippine Island World: A Physical, Cultural, and Regional Geography* (Berkeley: University of California Press, 1967).

Sugden, D. *Arctic and Antarctic: A Modern Geographical Synthesis* (Totowa, N.J.: Barnes & Noble, 1982).

Taylor, A., ed. *Focus on Southeast Asia* (New York: Praeger/American Geographical Society, 1972).

"When Will the Peace Begin? An Exclusive Look Inside Troubled, Still Divided [Vietnam]," *Time*, April 25, 1983, 82–85.

White, G. F. "The Mekong River Plan," *Scientific American*, April 1963, 49–60.

Wilhelm, D. *Emerging Indonesia* (Totowa, N.J.: Barnes & Noble, 1980).

Williams, L. *The Future of the Overseas Chinese in Southeast Asia* (New York: McGraw-Hill, 1966).

Yeung, Y. & Lo, C., eds. *Changing South-East Asian Cities: Readings on Urbanization* (Singapore: Oxford University Press, 1976).

Zegarelli, P. "Antarctica," *Focus*, September–October, 1978.

# PACIFIC REGIONS

IDEAS AND CONCEPTS

**Maritime Environments**
**High-Low Insular Island Cultures**
**Acculturation (2)**

**Figure P–1**
A representative village and landscape near Papua New Guinea's Mount Hagen, scene of the 1984 visit by Pope John Paul II.

Between the Americas to the east and Asia and Australia to the west lies the vast Pacific Ocean, larger than all the world's land areas combined. In this great ocean lie tens of thousands of islands, some large (New Guinea is by far the largest), others small (many are uninhabited). This fragmented, culturally complex realm, despite the preponderance of water, does possess regional identities. It includes the Hawaiian Islands, Tahiti, Fiji, Tonga, Samoa—fabled names in a world apart.

Indonesia and the Philippines are not part of the Pacific realm; neither are Australia and New Zealand. Before the European invasion, Australia and New Zealand would have been included: Australia as a discrete Pacific region on the basis of its indigenous black population, and New Zealand because its Maori population has Polynesian affinities. But black Australians and Maori New Zealanders have been engulfed by the Europeanization of their countries, and the regional geography of Australia and New Zealand today is dominantly Western, not Pacific. Only on the island of New Guinea do Pacific peoples remain the dominant cultural element.

The Pacific realm is fragmented, scattered, remote—but it does possess regions. New Guinea lies at the western end of a Pacific region that extends eastward to Fiji and includes the Solomon Islands, New Hebrides, and New Caledonia (Fig. P–2). These islands are inhabited by Melanesian peoples who have very dark skins and dark hair (*melas* means black); the region as a whole is called Melanesia. Some cultural geographers include the Papuan peoples of New Guinea in the Melanesian race, but others suggest that the Papuans are more closely related to the aboriginal Australians. In any case, Melanesia is the most populous Pacific region

# TEN MAJOR GEOGRAPHIC QUALITIES OF THE PACIFIC REALM

1. The Pacific realm's total area is the largest of all geographic realms. Its land area, however, is among the smallest.
2. The bulk of the land area of the Pacific realm lies on the island of New Guinea.
3. Papua New Guinea, with an estimated population of 3.5 million, alone contains almost three-fifths of the Pacific realm's population.
4. The Pacific realm consists of three regions: Melanesia (including New Guinea), Micronesia, and Polynesia.
5. The Pacific realm is the most strongly fragmented of all world realms.
6. The Pacific realm's islands and cultures may be divided into volcanic "high-island" cultures and coral "low-island" cultures.
7. The Hawaiian Islands, the 50th state of the United States, lie in the northern sector of Polynesia. As in New Zealand, indigenous culture has been submerged under Westernization.
8. In Polynesia, local culture is nearly everywhere under severe strain in the face of external influences.
9. Indigenous Polynesian culture has a remarkable consistency and uniformity throughout the Polynesian region, its enormous dimensions and dispersal notwithstanding.
10. The Pacific realm is in politico-geographical transition as islands attain independence or revise their political associations.

(4.1 million in 1985). New Guinea alone has a population of about 7 million (although statistics are unreliable), but this island is divided into two halves by a geometric boundary that separates non-Melanesian West Irian, now an Indonesian province, from newly independent Papua New Guinea (P.N.G.). With about 3.5 million inhabitants today, P.N.G. became a sovereign state in 1975 after nearly a century of British and Australian administration. It is one of the world's poorest and least developed countries, with much of the mountainous interior—where the Papuan population is concentrated in tiny villages (Fig. P–1)—hardly touched by the changes that transformed neighboring Australia. The largest town and capital, Port Moresby, has 380,000 residents; only about 38 percent of the people of P.N.G. live in urban areas, but the current rate of urbanization is phenomenal (only 10 percent of the population was urbanized as recently as 1970). Although English is used by the educated minority, about 85 percent of the population remains illiterate and over 700 languages are spoken by the Papuan and Melanesian communities. The Melanesians are concentrated in the northern and eastern coastal areas of the country, and here, as in the other islands of this region, they grow root crops and bananas for subsistence.

North of Melanesia and east of the Philippines, lie the islands that comprise the region known as Micronesia (*micro* means small). In this case, the name refers to the size of the islands, not the physical appearance of the population. The islands of Micronesia are not only small (many of them no larger than one square kilometer), but they are also much lower-lying, on an average, than those of Melanesia. There are some volcanic islands ("high islands," as the people call them), but they are outnumbered by islands composed of coral, the "low islands" that barely lie above sea level. Guam, with 550 square kilometers (210 square miles), is Micronesia's largest island, and no island elevation anywhere in Micronesia reaches 1000 meters (3300 feet). The region is largely a U.S. Trust Territory and includes the Mariana, Marshall, and Caroline Islands and Truk (Fig. P–2). The Micronesians are not nearly as numerous as the Melanesians (totaling only 350,000 in 1985), but they nevertheless comprise a distinct racial group in the Pacific realm. Culturally, it is useful to distinguish their communities as "high-island" cultures based on the better-watered volcanic islands where agriculture is the mainstay, or "low-island" cultures on the sometimes drought-plagued coral islands where fishing is the chief mode of subsistence. These numerous Micronesian communities have developed a large number of locally spoken languages, many of them mutually unintelligible.

But these islands do not exist in isolation. There is a certain complementarity between the islands of farmers and the islands of fishing people, and Micronesians—especially the low islanders—are skilled seafarers. The trade for food and basic needs encourages circulation; sometimes the threat of a devastating typhoon compels the "low-islanders" to seek the safety of higher ground elsewhere. Thus,

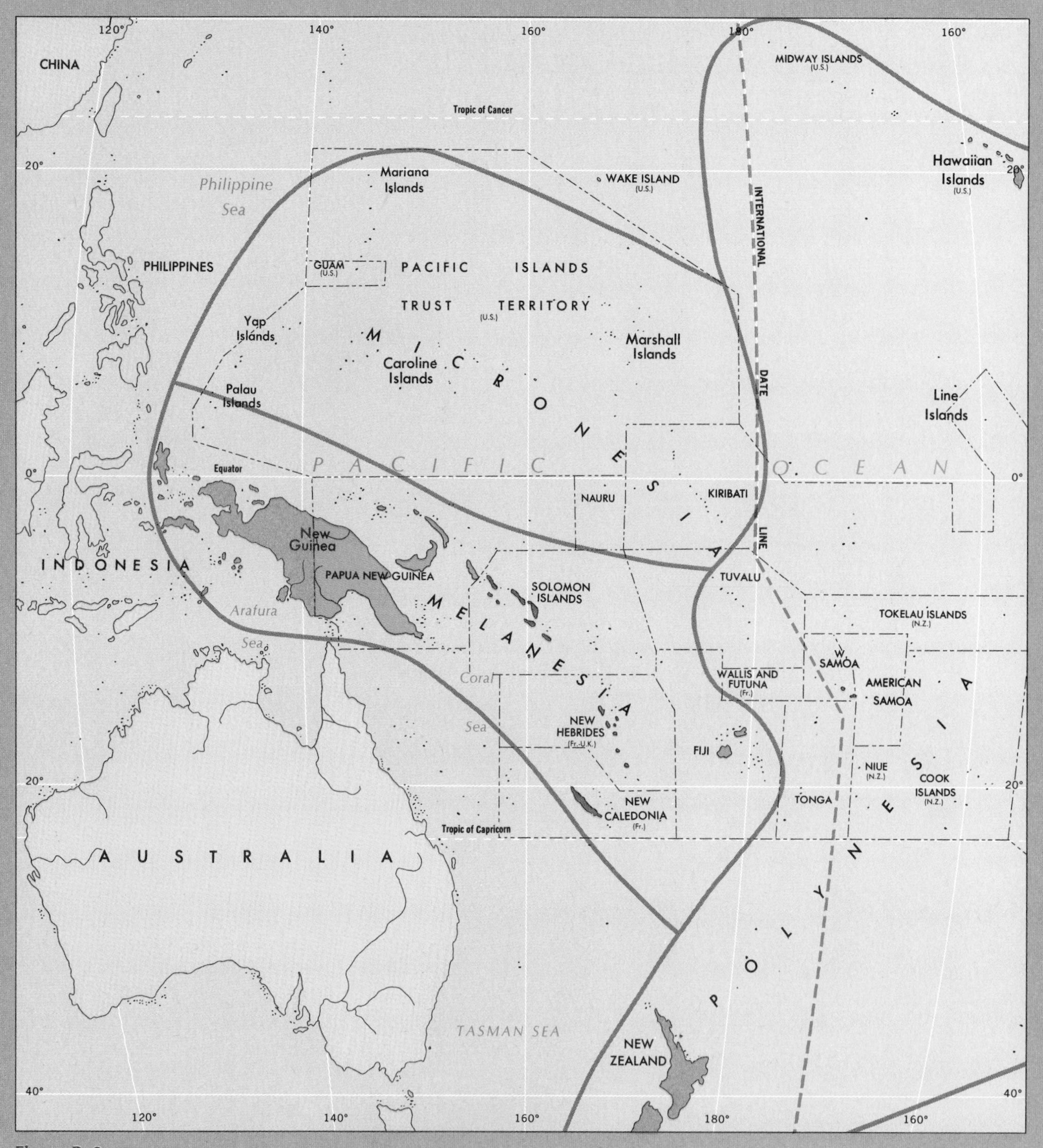


Figure P–2

movement in the Pacific realm has always been by water. Even today, many islanders are expert boaters, often choosing a water route when a road is also available. In Fig. P–3, a canoe loaded with people and goods, sets off from Suva (Fiji) to a small town a few kilometers away; the trip could also be made by land vehicle around the bay, but these Fijians prefer the familiar alternative.

To the east of Micronesia and Melanesia lies the heart of the Pacific, enclosed by a great triangle stretching from the Hawaiian Islands to Chile's Easter Island to New Zealand. This is Polynesia (Fig. P–2), a region of numerous islands (*poly* means many), ranging from volcanic mountains rising above the Pacific's waters (Mauna Kea on Hawaii reaches over 4200 meters or nearly 13,800 feet),

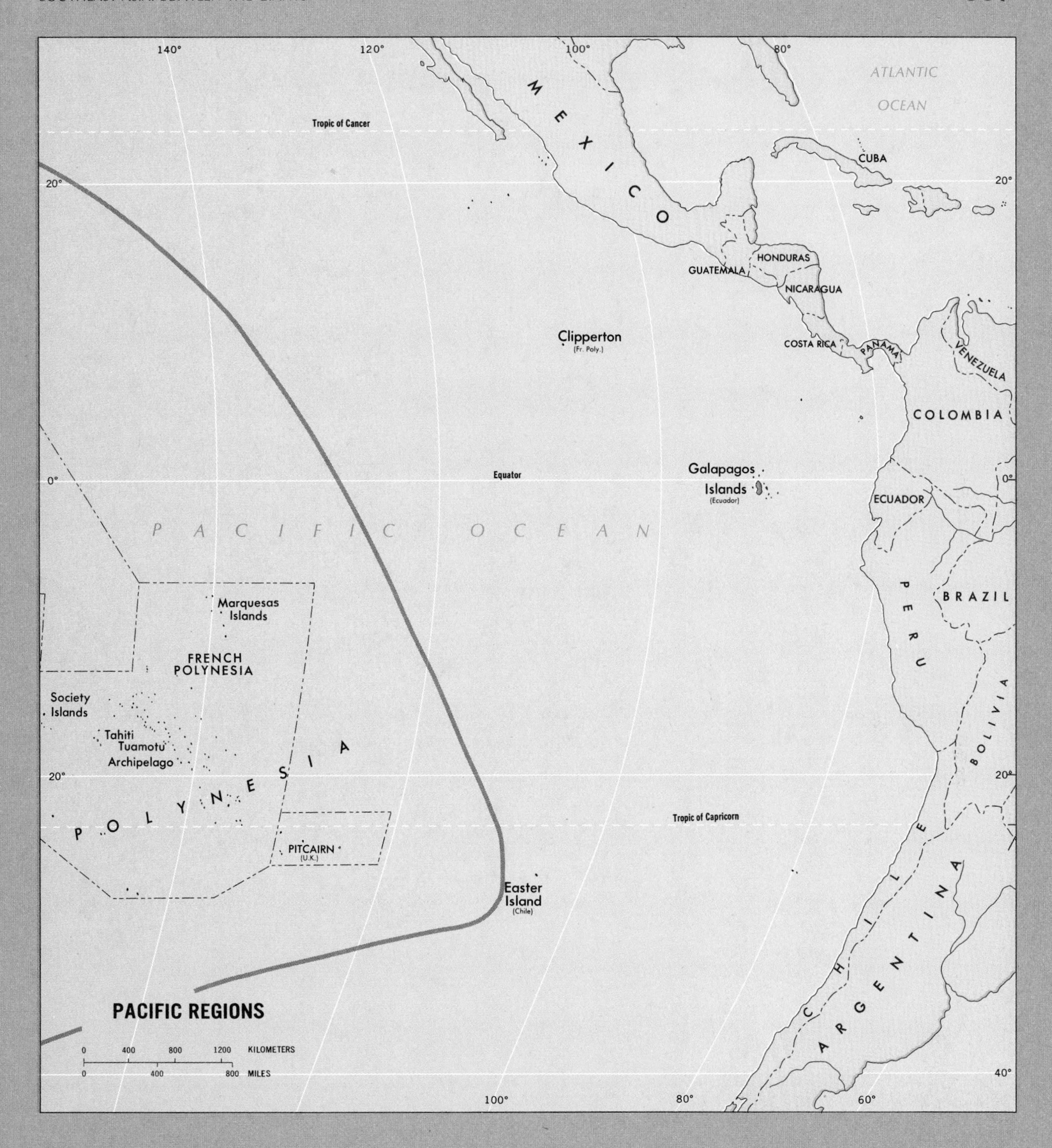


clothed by luxuriant tropical vegetation, and drenched by hundreds of centimeters of rainfall each year, to low coral atolls where a few palm trees form the only vegetation and where drought is a persistent and recurrent problem. The Polynesians have somewhat lighter skin and wavier hair than do the other peoples of the Pacific realm; they are often also described as having an excellent physique. Anthropologists differentiate between these original Polynesians and a second group, the Neo-Hawaiians, who are a blend of Polynesian, European, and Asian ancestries. In the U.S. state of Hawaii—actually an archipelago of more than 130 islands—Polynesian culture has not only been Europeanized but also Orientalized.

Its vastness and the diversity of its natural environments notwith-

Figure P–3
Ubiquitous water transportation marks the Pacific realm's islands, here on Fiji.

standing, Polynesia clearly constitutes a geographic region within the Pacific realm (1.2 million population in 1985). Polynesian culture, spatially fragmented though it is, exhibits a remarkable consistency and uniformity from one island to the next, from one end of this scattered region to the other; this consistency is particularly expressed in vocabularies, technologies, housing, and artforms. The Polynesians are uniquely adapted to their maritime environment, and long before European sailing ships began to arrive in their waters, Polynesian seafarers had learned to navigate their wide expanses of ocean in huge double canoes as much as 45 meters (150 feet) in length. They traveled hundreds of kilometers to favorite fishing zones and engaged in inter-island barter trade, using maps constructed from bamboo sticks and cowrie shells and navigating by the stars. However, modern descriptions of a Pacific Polynesian paradise of emerald seas, lush landscapes, and gentle people distort harsh realities; Polynesian society was forced to accommodate much loss of life at sea when storms claimed their boats, families were ripped apart by accident as well as migration, hunger and starvation afflicted the local communities on smaller islands, and the island communities were often embroiled in violent conflicts and cruel retributions.

The political geography of Polynesia is complex. The Hawaiian Islands in 1959 became the 50th state to join the United States. The state's population surpassed 1 million in 1982, over 80 percent resident on the island of Oahu; there, the superimposition of cultures is exquisitely illustrated by the panorama of Honolulu's skyscrapers, capped by the famous extinct volcano at nearby Diamond Head (Fig. P–4). The Kingdom of Tonga, with a population of 108,000, became an independent country in 1970 after 70 years as a British protectorate. Western Samoa, a German possession before its occupation by New Zealand during World War I, also achieved independence (1962) as one of the world's smallest countries with a mere 2934 square kilometers (1133 square miles) and 163,000 inhabitants. The British-administered Ellice Islands were renamed Tuvalu; along with the Gilbert Islands to the north, they received independence from Britain in 1978. Other islands continued under French control (including the Marquesas Islands and Tahiti), under New Zealand's administration (Rarotonga), and under British, U.S., and Chilean flags. In the process of politico-geographical fragmentation, Polynesian culture has been dealt some severe blows.

Figure P–4
The circle closes: our regional survey returns us to the developed world of familiar Honolulu, capital of the 50th U.S. state.

Land developers, hotel builders, and tourist dollars have set Tahiti on a course along which Hawaii has already traveled far. The Americanization of Eastern Samoa has created a new society quite different from the old. Polynesia has lost much of its ancient cultural consistency; today, the region is a patchwork of new and old—the new often bleak and barren, the old under increasing pressure.

## References and Further Readings

Alley, R., ed. *New Zealand and the Pacific* (Boulder, Colo.: Westview Press, 1984).

*Atlas of Hawaii* (Honolulu: University Press of Hawaii, 2 rev. ed., 1983).

Brookfield, H., ed. *The Pacific in Transition: Geographical Perspectives on Adaptation and Change* (New York: St. Martin's Press, 1974).

Brookfield, H. & Hart, D. *Melanesia: A Geographical Interpretation of an Island World* (New York: Barnes & Noble, 1971).

Chapman, K. "Population Geography and the Island Pacific," *Pacific Viewpoint*, 16 (1975), 177–188.

Freeman, O. *Geography of the Pacific* (New York: John Wiley & Sons, 1951).

Howard, A., ed. *Polynesia: Readings on a Culture Area* (Scranton, Pa.: Chandler, 1971).

Howells, W. *The Pacific Islanders* (New York: Charles Scribner's Sons, 1973).

Levison, M. et al. *The Settlement of Polynesia* (Minneapolis: University of Minnesota Press, 1977).

Morgan, J., ed. *Hawaii* (Boulder, Colo.: Westview Press, 1983).

Spate, O. *The Pacific Since Magellan: Monopolists and Freebooters* (Minneapolis: University of Minnesota Press, 1983).

Turner, W. "90 Years After American Takeover, Some Hawaiians Still Distrust U.S.," *New York Times*, July 9, 1983, p. 7.

Vayda, A., ed. *Peoples and Cultures of the Pacific* (New York: Natural History Press, 1968).

Ward, R., ed. *Man in the Pacific Islands: Essays on Geographical Change in the Pacific* (New York: Oxford University Press, 1972).

Wurm, S. & Hattori, S., eds. *Language Atlas of the Pacific Area* (Canberra: Australian Academy of the Humanities, 1982).

# APPENDIX

# AREA AND DEMOGRAPHIC DATA FOR THE WORLD'S STATES

(Microstates Omitted)

| Country | Area 1000 Km² | Area 1000 Miles² | Population (Millions) 1976 | 1980 | 1984 | 2000 | Life Expectancy at Birth (Newest Est.) | Average Annual Growth Rate (%) 1980–1985 | 1984 Population Density No./Km² | No./Miles² |
|---|---|---|---|---|---|---|---|---|---|---|
| **Europe** | 4,867.6 | 1,879.6 | 475.2 | 483.2 | 489.6 | 512.3 | 73 | 0.3 | 101 | 261 |
| Albania | 28.7 | 11.1 | 2.5 | 2.7 | 2.9 | 3.9 | 68 | 2.1 | 101 | 261 |
| Austria | 83.8 | 32.4 | 7.6 | 7.6 | 7.6 | 7.7 | 72 | 0.1 | 91 | 235 |
| Belgium | 30.5 | 11.8 | 9.8 | 9.8 | 9.9 | 10.0 | 71 | 0.1 | 325 | 839 |
| Bulgaria | 110.9 | 42.8 | 8.8 | 8.9 | 9.0 | 9.5 | 71 | 0.3 | 81 | 210 |
| Czechoslovakia | 127.9 | 49.4 | 14.9 | 15.3 | 15.5 | 16.3 | 71 | 0.3 | 121 | 314 |
| Denmark | 43.1 | 16.6 | 5.1 | 5.1 | 5.1 | 5.1 | 74 | −0.1 | 118 | 307 |
| Finland | 337.0 | 130.2 | 4.7 | 4.8 | 4.9 | 5.2 | 73 | 0.5 | 15 | 38 |
| France | 547.0 | 211.2 | 52.9 | 53.8 | 54.9 | 57.6 | 74 | 0.5 | 100 | 260 |
| East Germany | 108.2 | 41.8 | 16.8 | 16.7 | 16.7 | 16.4 | 72 | −0.3 | 154 | 400 |
| West Germany | 248.6 | 96.0 | 61.5 | 61.6 | 61.4 | 60.2 | 73 | −0.1 | 247 | 640 |
| Greece | 131.9 | 51.0 | 9.2 | 9.6 | 10.0 | 11.4 | 72 | 0.9 | 76 | 196 |
| Hungary | 93.0 | 35.9 | 10.6 | 10.7 | 10.7 | 10.3 | 70 | −0.8 | 115 | 298 |
| Iceland | 103.0 | 39.8 | 0.2 | 0.2 | 0.2 | 0.3 | 77 | 1.1 | 2 | 5 |
| Ireland | 70.3 | 27.1 | 3.2 | 3.4 | 3.6 | 4.2 | 71 | 1.2 | 51 | 133 |
| Italy | 301.2 | 116.3 | 55.7 | 56.2 | 56.4 | 56.8 | 73 | 0.1 | 187 | 485 |
| Luxembourg | 2.6 | 1.0 | 0.4 | 0.4 | 0.4 | 0.4 | 72 | 0.1 | 154 | 400 |
| Netherlands | 41.2 | 15.9 | 13.8 | 14.1 | 14.4 | 15.2 | 76 | 0.5 | 350 | 906 |
| Norway | 324.2 | 125.1 | 4.0 | 4.1 | 4.1 | 4.3 | 76 | 0.4 | 13 | 33 |
| Poland | 312.7 | 120.7 | 34.3 | 35.6 | 36.9 | 41.6 | 71 | 0.9 | 118 | 306 |
| Portugal | 88.8 | 34.3 | 9.7 | 9.9 | 10.0 | 10.9 | 71 | 0.4 | 113 | 292 |
| Romania | 237.5 | 91.7 | 21.4 | 22.2 | 22.8 | 25.3 | 70 | 0.6 | 96 | 249 |
| Spain | 504.8 | 194.9 | 36.0 | 37.5 | 38.4 | 42.7 | 73 | 0.6 | 76 | 197 |
| Sweden | 450.0 | 173.7 | 8.2 | 8.3 | 8.3 | 8.3 | 76 | 0.1 | 18 | 48 |
| Switzerland | 41.3 | 15.9 | 6.3 | 6.4 | 6.5 | 6.5 | 75 | 0.3 | 157 | 409 |
| United Kingdom | 244.0 | 94.2 | 56.0 | 56.0 | 56.0 | 56.8 | 73 | 0.0 | 230 | 594 |
| Yugoslavia | 255.8 | 98.8 | 21.6 | 22.3 | 23.0 | 25.4 | 70 | 0.8 | 90 | 233 |
| **Australia** | 7,950.9 | 3,070.1 | 17.0 | 17.7 | 18.7 | 22.1 | 74 | 1.2 | 2 | 5 |
| Australia | 7,682.7 | 2,966.2 | 13.9 | 14.6 | 15.5 | 18.5 | 75 | 1.4 | 2 | 5 |
| New Zealand | 268.2 | 103.9 | 3.1 | 3.1 | 3.2 | 3.6 | 72 | 0.4 | 12 | 31 |
| **Soviet Union** | | | | | | | | | | |
| Soviet Union | 22,275.0 | 8,600.4 | 256.8 | 265.5 | 274.5 | 309.6 | 69 | 0.8 | 12 | 32 |
| **North America** | 19,464.7 | 7,515.3 | 241.2 | 252.0 | 261.7 | 301.7 | 74 | 1.0 | 13 | 35 |
| Bahamas | 14.0 | 5.4 | 0.2 | 0.2 | 0.2 | 0.3 | 68 | 2.0 | 14 | 37 |
| Canada | 9,922.3 | 3,831.0 | 23.0 | 24.1 | 25.1 | 31.6 | 74 | 1.1 | 3 | 7 |
| United States | 9,528.4 | 3,678.9 | 218.0 | 227.7 | 236.4 | 269.8 | 74 | 0.9 | 25 | 64 |
| **Japan** | | | | | | | | | | |
| Japan | 377.4 | 145.7 | 112.8 | 116.8 | 120.0 | 130.6 | 76 | 0.6 | 318 | 824 |
| **Middle America** | 2,712.0 | 1,047.0 | 109.5 | 120.3 | 132.2 | 190.6 | 68 | 2.1 | 49 | 126 |
| Barbados | 0.5 | 0.2 | 0.2 | 0.2 | 0.3 | 0.3 | 70 | 0.3 | 586 | 1,518 |
| Belize | 23.1 | 8.9 | 0.1 | 0.1 | 0.2 | 0.3 | NA | 1.8 | 9 | 22 |
| Costa Rica | 51.0 | 19.7 | 2.1 | 2.4 | 2.7 | 3.5 | 71 | 2.8 | 53 | 137 |
| Cuba | 114.5 | 44.2 | 9.4 | 9.7 | 10.0 | 11.5 | 70 | 0.8 | 87 | 226 |
| Dominican Republic | 48.7 | 18.7 | 5.2 | 5.8 | 6.4 | 9.1 | 60 | 2.6 | 131 | 342 |
| El Salvador | 21.0 | 8.1 | 4.3 | 4.7 | 4.8 | 7.6 | 63 | 0.1 | 229 | 593 |
| Guatemala | 108.8 | 42.0 | 6.3 | 7.1 | 8.0 | 12.4 | 58 | 2.8 | 74 | 190 |
| Haiti | 27.7 | 10.7 | 5.1 | 5.4 | 5.8 | 8.6 | 51 | 1.9 | 209 | 542 |

| Country | Area 1000 Km² | Area 1000 Miles² | Population (Millions) 1976 | 1980 | 1984 | 2000 | Life Expectancy at Birth (Newest Est.) | Average Annual Growth Rate (%) 1980–1985 | 1984 Population Density No./Km² | No./Miles² |
|---|---|---|---|---|---|---|---|---|---|---|
| Honduras | 112.1 | 43.3 | 3.3 | 3.8 | 4.4 | 6.9 | 56 | 3.6 | 39 | 102 |
| Jamaica | 10.9 | 4.2 | 2.1 | 2.2 | 2.4 | 2.9 | 69 | 1.4 | 220 | 571 |
| Mexico | 1,972.5 | 761.6 | 63.2 | 70.1 | 77.7 | 114.3 | 67 | 2.7 | 39 | 102 |
| Nicaragua | 130.0 | 50.2 | 2.3 | 2.5 | 2.9 | 4.7 | 56 | 3.8 | 22 | 58 |
| Panama | 77.2 | 29.8 | 1.8 | 1.9 | 2.1 | 2.9 | 69 | 2.3 | 27 | 70 |
| Puerto Rico | 8.8 | 3.4 | 3.0 | 3.2 | 3.3 | 4.1 | 74 | 1.0 | 375 | 971 |
| Trinidad and Tobago | 5.2 | 2.0 | 1.1 | 1.2 | 1.2 | 1.5 | 68 | 1.5 | 231 | 600 |
| **South America** | 17,714.5 | 6,839.6 | 221.2 | 242.4 | 265.5 | 368.3 | 64 | 2.4 | 15 | 39 |
| Argentina | 2,766.9 | 1,068.3 | 26.5 | 28.2 | 30.1 | 36.6 | 69 | 1.6 | 11 | 28 |
| Bolivia | 1,098.7 | 424.2 | 4.9 | 5.5 | 6.0 | 9.0 | 49 | 2.6 | 5 | 14 |
| Brazil | 8,512.0 | 3,286.5 | 111.3 | 122.4 | 134.4 | 187.5 | 63 | 2.3 | 16 | 41 |
| Chile | 756.5 | 292.1 | 10.4 | 11.0 | 11.7 | 14.6 | 67 | 1.5 | 15 | 40 |
| Colombia | 1,138.8 | 439.7 | 24.6 | 26.1 | 28.2 | 37.7 | 62 | 2.0 | 25 | 64 |
| Ecuador | 283.6 | 109.5 | 7.1 | 8.0 | 9.1 | 13.9 | 61 | 3.1 | 32 | 83 |
| Guyana | 215.0 | 83.0 | 0.8 | 0.8 | 0.8 | 1.0 | 68 | 0.5 | 4 | 10 |
| Paraguay | 406.6 | 157.0 | 2.9 | 3.2 | 3.6 | 5.4 | 66 | 2.7 | 9 | 23 |
| Peru | 1,285.2 | 496.2 | 15.8 | 17.6 | 19.7 | 30.5 | 58 | 2.8 | 15 | 40 |
| Suriname | 163.2 | 63.0 | 0.4 | 0.4 | 0.4 | 0.6 | 68 | 1.4 | 2 | 6 |
| Uruguay | 176.1 | 68.0 | 2.9 | 2.9 | 2.9 | 3.2 | 69 | 0.3 | 16 | 43 |
| Venezuela | 911.9 | 352.1 | 13.6 | 16.3 | 18.6 | 28.3 | 67 | 3.2 | 20 | 53 |
| **North Africa and Southwest Asia** | 15,361.4 | 5,931.0 | 234.6 | 262.7 | 292.3 | 442.6 | 60 | 2.9 | 19 | 49 |
| Afghanistan | 647.5 | 250.0 | 14.5 | 15.2 | 14.4 | 20.9 | 41 | −0.6 | 22 | 58 |
| Algeria | 2,381.8 | 919.6 | 16.6 | 18.8 | 21.4 | 35.4 | 55 | 3.1 | 9 | 23 |
| Bahrain | 0.8 | 0.3 | 0.3 | 0.3 | 0.4 | 0.7 | 67 | 4.1 | 617 | 1,598 |
| Cyprus | 9.3 | 3.6 | 0.6 | 0.6 | 0.7 | 0.8 | 74 | 1.3 | 75 | 194 |
| Egypt | 1,002.1 | 386.9 | 37.7 | 42.1 | 47.1 | 66.2 | 56 | 2.8 | 47 | 122 |
| Iran | 1,648.0 | 636.3 | 34.4 | 38.8 | 43.8 | 65.8 | 55 | 3.1 | 27 | 69 |
| Iraq | 434.9 | 167.9 | 11.5 | 13.1 | 15.0 | 23.8 | 57 | 3.3 | 34 | 89 |
| Israel | 20.2 | 7.8 | 3.4 | 3.8 | 4.0 | 5.2 | 74 | 1.7 | 198 | 513 |
| Jordan | 97.6 | 37.7 | 2.7 | 3.1 | 3.5 | 6.1 | 61 | 3.1 | 36 | 93 |
| Kuwait | 17.9 | 6.9 | 1.1 | 1.4 | 1.8 | 3.1 | 68 | 6.2 | 101 | 261 |
| Lebanon | 10.4 | 4.0 | 2.7 | 2.7 | 2.6 | 3.6 | 66 | −0.2 | 250 | 650 |
| Libya | 1,759.6 | 679.4 | 2.6 | 3.0 | 3.7 | 6.3 | 57 | 5.1 | 2 | 5 |
| Morocco | 712.5 | 275.1 | 18.8 | 21.1 | 23.7 | 36.8 | 56 | 2.8 | 33 | 86 |
| Oman | 212.4 | 82.0 | 0.8 | 0.9 | 1.0 | 1.6 | 51 | 3.1 | 5 | 12 |
| Qatar | 10.9 | 4.2 | 0.2 | 0.2 | 0.3 | 0.5 | 58 | 3.7 | 28 | 71 |
| Saudi Arabia | 2,149.7 | 830.0 | 7.7 | 9.4 | 10.8 | 18.2 | 57 | 3.4 | 5 | 13 |
| Sudan | 2,505.8 | 967.5 | 16.5 | 18.7 | 21.1 | 32.8 | 47 | 2.9 | 8 | 22 |
| Syria | 185.2 | 71.5 | 7.7 | 8.8 | 10.1 | 17.6 | 62 | 3.4 | 55 | 141 |
| Tunisia | 163.7 | 63.2 | 5.9 | 6.5 | 7.2 | 12.2 | 58 | 2.6 | 44 | 114 |
| Turkey | 779.3 | 300.9 | 41.8 | 46.0 | 50.2 | 70.2 | 63 | 2.2 | 64 | 167 |
| United Arab Emirates | 83.7 | 32.3 | 0.6 | 1.0 | 1.5 | 2.8 | 64 | 10.7 | 18 | 46 |
| Yemen (Aden) | 333.1 | 128.6 | 1.7 | 1.9 | 2.1 | 3.2 | 45 | 2.9 | 6 | 16 |
| Yemen (San'a) | 195.0 | 75.3 | 4.8 | 5.3 | 5.9 | 8.8 | 44 | 2.7 | 30 | 78 |
| **Black Africa** | 21,741.7 | 8,394.4 | 321.3 | 360.3 | 405.6 | 649.1 | 48 | 2.9 | 19 | 48 |
| Angola | 1,246.8 | 481.4 | 6.0 | 7.0 | 7.8 | 11.9 | 42 | 2.7 | 6 | 16 |

| Country | Area 1000 Km² | Area 1000 Miles² | Population (Millions) 1976 | 1980 | 1984 | 2000 | Life Expectancy at Birth (Newest Est.) | Average Annual Growth Rate (%) 1980–1985 | 1984 Population Density No./Km² | No./Miles² |
|---|---|---|---|---|---|---|---|---|---|---|
| Benin | 112.7 | 43.5 | 3.1 | 3.5 | 3.9 | 6.4 | 47 | 3.0 | 35 | 90 |
| Botswana | 600.4 | 231.8 | 0.8 | 0.9 | 1.0 | 1.6 | 55 | 3.6 | 2 | 4 |
| Burundi | 27.7 | 10.7 | 3.8 | 4.2 | 4.7 | 6.9 | 42 | 2.8 | 170 | 439 |
| Cameroon | 475.5 | 183.6 | 7.7 | 8.6 | 9.5 | 13.7 | 47 | 2.6 | 20 | 52 |
| Central African Republic | 622.9 | 240.5 | 2.1 | 2.3 | 2.6 | 3.8 | 42 | 2.8 | 4 | 11 |
| Chad | 1,284.1 | 495.8 | 4.2 | 4.4 | 5.1 | 7.6 | 41 | 3.4 | 4 | 10 |
| Comoro Islands | 2.1 | 0.8 | 0.3 | 0.4 | 0.5 | 0.7 | 46 | 2.8 | 238 | 616 |
| Congo | 341.9 | 132.0 | 1.4 | 1.6 | 1.7 | 2.6 | 47 | 2.9 | 5 | 13 |
| Equatorial Guinea | 28.0 | 10.8 | 0.2 | 0.3 | 0.3 | 0.4 | 47 | 2.4 | 11 | 28 |
| Ethiopia | 1,223.5 | 472.4 | 28.8 | 29.8 | 32.0 | 48.7 | 42 | 1.9 | 26 | 68 |
| Gabon | 267.5 | 103.3 | 0.7 | 0.8 | 1.0 | 1.6 | 44 | 4.4 | 4 | 10 |
| Gambia | 11.4 | 4.4 | 0.5 | 0.6 | 0.7 | 1.0 | 43 | 2.6 | 61 | 159 |
| Ghana | 238.5 | 92.1 | 10.7 | 12.1 | 13.8 | 23.5 | 49 | 3.2 | 58 | 150 |
| Guinea | 245.8 | 94.9 | 4.5 | 5.0 | 5.6 | 8.5 | 44 | 2.7 | 23 | 59 |
| Guinea-Bissau | 36.0 | 13.9 | 0.7 | 0.8 | 0.8 | 1.1 | 41 | 1.8 | 22 | 58 |
| Ivory Coast | 320.6 | 123.8 | 7.0 | 8.1 | 9.2 | 14.4 | 47 | 3.2 | 29 | 74 |
| Kenya | 582.8 | 225.0 | 14.0 | 16.4 | 19.4 | 36.6 | 53 | 4.1 | 33 | 86 |
| Lesotho | 30.3 | 11.7 | 1.2 | 1.3 | 1.5 | 2.1 | 52 | 2.4 | 50 | 128 |
| Liberia | 111.4 | 43.0 | 1.7 | 1.9 | 2.2 | 3.6 | 54 | 3.2 | 20 | 51 |
| Madagascar | 587.2 | 226.7 | 7.8 | 8.7 | 9.6 | 14.7 | 46 | 2.7 | 16 | 42 |
| Malawi | 118.4 | 45.7 | 5.4 | 6.0 | 6.8 | 11.8 | 47 | 3.2 | 57 | 149 |
| Mali | 1,240.1 | 478.8 | 6.3 | 6.9 | 7.6 | 11.9 | 42 | 2.3 | 6 | 16 |
| Mauritania | 1,030.8 | 398.0 | 1.4 | 1.5 | 1.6 | 2.8 | 43 | 1.9 | 2 | 4 |
| Mauritius | 2.1 | 0.8 | 0.9 | 1.0 | 1.0 | 1.2 | 64 | 1.5 | 500 | 1,294 |
| Moçambique | 783.0 | 302.3 | 10.8 | 12.1 | 13.4 | 21.0 | 46 | 2.9 | 17 | 44 |
| Namibia | 824.4 | 318.3 | 0.9 | 1.0 | 1.1 | 1.7 | 52 | 3.0 | 1 | 3 |
| Niger | 1,267.0 | 489.2 | 4.9 | 5.5 | 6.3 | 10.3 | 43 | 3.2 | 5 | 13 |
| Nigeria | 923.9 | 356.7 | 67.8 | 77.1 | 88.1 | 146.9 | 49 | 3.4 | 95 | 247 |
| Rwanda | 26.4 | 10.2 | 4.5 | 5.1 | 5.8 | 9.6 | 48 | 3.3 | 220 | 567 |
| Senegal | 196.8 | 76.0 | 5.1 | 5.8 | 6.5 | 9.8 | 42 | 3.2 | 33 | 86 |
| Sierra Leone | 72.3 | 27.9 | 3.1 | 3.4 | 3.8 | 5.9 | 45 | 2.6 | 53 | 136 |
| Somalia | 637.7 | 246.2 | 3.7 | 5.4 | 6.4 | 9.4 | 44 | 3.9 | 10 | 26 |
| South Africa | 1,220.9 | 471.4 | 26.1 | 28.7 | 31.7 | 46.8 | 60 | 2.4 | 26 | 67 |
| Tanzania | 945.1 | 364.9 | 16.3 | 18.6 | 21.2 | 34.8 | 53 | 3.2 | 22 | 58 |
| Togo | 56.7 | 21.9 | 2.3 | 2.6 | 2.9 | 4.7 | 46 | 3.0 | 51 | 132 |
| Uganda | 235.9 | 91.1 | 11.4 | 12.8 | 14.3 | 23.6 | 52 | 2.8 | 61 | 157 |
| Upper Volta | 274.3 | 105.9 | 5.7 | 6.1 | 6.7 | 10.6 | 41 | 2.4 | 24 | 63 |
| Zaïre | 2,345.5 | 905.6 | 25.7 | 28.6 | 32.2 | 49.7 | 48 | 2.9 | 14 | 36 |
| Zambia | 752.7 | 290.6 | 5.1 | 5.8 | 6.6 | 10.4 | 50 | 3.2 | 9 | 23 |
| Zimbabwe | 390.6 | 150.8 | 6.7 | 7.6 | 8.7 | 14.8 | 52 | 3.4 | 22 | 58 |
| **India and Perimeter** | 4,429.7 | 1,710.3 | 813.8 | 890.2 | 977.4 | 1,312.0 | 50 | 2.4 | 221 | 571 |
| Bangladesh | 144.0 | 55.6 | 77.8 | 88.1 | 99.6 | 149.0 | 47 | 3.1 | 692 | 1,791 |
| Bhutan | 46.9 | 18.1 | 1.2 | 1.3 | 1.4 | 1.9 | 43 | 2.2 | 30 | 77 |
| India | 3,204.1 | 1,237.1 | 630.6 | 685.1 | 746.4 | 973.5 | 51 | 2.1 | 233 | 603 |
| Maldives | 0.3 | 0.1 | 0.1 | 0.2 | 0.2 | 0.3 | 48 | 3.0 | 581 | 1,504 |
| Nepal | 140.9 | 54.4 | 13.6 | 15.0 | 16.6 | 23.2 | 45 | 2.5 | 118 | 305 |
| Pakistan | 828.5 | 319.9 | 76.6 | 85.7 | 97.3 | 142.8 | 50 | 3.0 | 117 | 304 |
| Sri Lanka | 65.0 | 25.1 | 13.9 | 14.8 | 15.9 | 21.3 | 64 | 1.8 | 245 | 633 |

| Country | Area 1000 Km² | Area 1000 Miles² | Population (Millions) 1976 | 1980 | 1984 | 2000 | Life Expectancy at Birth (Newest Est.) | Average Annual Growth Rate (%) 1980–1985 | 1984 Population Density No./Km² | No./Miles² |
|---|---|---|---|---|---|---|---|---|---|---|
| **China and Its Sphere** | 11,380.8 | 4,394.2 | 1,029.4 | 1,090.6 | 1,156.7 | 1,386.3 | 69 | 1.5 | 102 | 263 |
| China | 9,560.9 | 3,691.5 | 957.8 | 1,013.6 | 1,074.0 | 1,280.2 | 70 | 1.4 | 112 | 291 |
| North Korea | 120.4 | 46.5 | 16.3 | 17.9 | 19.6 | 26.8 | 65 | 2.3 | 163 | 422 |
| South Korea | 98.4 | 38.0 | 37.3 | 39.6 | 42.0 | 52.1 | 66 | 1.5 | 427 | 1,105 |
| Mongolia | 1,565.1 | 604.3 | 1.5 | 1.7 | 1.9 | 2.8 | 50 | 2.8 | 1 | 3 |
| Taiwan | 36.0 | 13.9 | 16.5 | 17.8 | 19.2 | 24.4 | 73 | 1.8 | 533 | 1,381 |
| **Southeast Asia** | 4,494.6 | 1,735.4 | 332.6 | 361.9 | 394.6 | 532.8 | 54 | 2.1 | 88 | 227 |
| Brunei | 5.7 | 2.2 | 0.2 | 0.2 | 0.2 | 0.3 | 68 | 4.0 | 35 | 91 |
| Burma | 676.5 | 261.2 | 31.2 | 34.4 | 38.0 | 54.1 | 53 | 2.4 | 56 | 146 |
| Indonesia | 1,919.2 | 741.0 | 139.3 | 151.2 | 164.3 | 218.4 | 50 | 2.1 | 86 | 222 |
| Kampuchea | 181.0 | 69.9 | 6.2 | 5.7 | 6.1 | 8.6 | 40 | 1.9 | 34 | 87 |
| Laos | 236.7 | 91.4 | 3.3 | 3.5 | 3.7 | 5.0 | 45 | 2.0 | 16 | 40 |
| Malaysia | 332.6 | 128.4 | 12.7 | 14.0 | 15.3 | 20.8 | 63 | 2.2 | 46 | 119 |
| Philippines | 299.9 | 115.8 | 44.3 | 49.3 | 54.5 | 76.3 | 63 | 2.5 | 182 | 471 |
| Singapore | 0.6 | 0.2 | 2.3 | 2.4 | 2.5 | 3.0 | 73 | 1.2 | 4,089 | 10,590 |
| Thailand | 513.1 | 198.1 | 43.5 | 47.7 | 51.7 | 68.3 | 62 | 2.0 | 101 | 261 |
| Vietnam | 329.4 | 127.2 | 49.6 | 53.5 | 58.3 | 78.0 | 55 | 2.1 | 177 | 458 |
| **Pacific** | 481.2 | 185.8 | 4.7 | 5.0 | 5.6 | 7.5 | 66 | 2.7 | 9 | 22 |
| Fiji | 18.4 | 7.1 | 0.6 | 0.6 | 0.7 | 0.9 | 68 | 2.0 | 38 | 99 |
| Papua New Guinea | 462.8 | 178.7 | 2.7 | 3.0 | 3.4 | 4.8 | 54 | 2.8 | 7 | 19 |
| **World** | **149,757.2** | **57,821.3** | **4,177.6** | **4,478.4** | **4,802.7** | **6,185.9** | **62** | **1.8** | **32** | **83** |

# GLOSSARY

**Absolute location**
The position or place of a certain item on the surface of the earth as expressed in degrees, minutes, and seconds of **latitude**,* 0° to 90° north or south of the equator, and **longitude**, 0° to 180° east or west of the *prime meridian* passing through Greenwich, England.

**Accessibility**
The degree of ease with which it is possible to reach a certain location from other locations. Accessibility varies from place to place and can be measured.

**Acculturation**
Cultural modification resulting from intercultural borrowing. In cultural geography, the term is used to designate the change that occurs in the culture of indigenous peoples when contact is made with a technologically more advanced society.

**Activity space**
The local area within which a person moves in the course of daily activity. The territorial range of this daily effective environment depends on the mode of transportation to which an individual has access.

**Agglomerated (nucleated) settlement**
A compact, closely-packed settlement (usually a hamlet or larger village) sharply demarcated from adjoining farmlands.

**Agglomeration**
Process involving the clustering of people or activities. Often refers to manufacturing plants and businesses that benefit from close proximity because they share skilled-labor pools and technological and financial amenities.

**Agrarian**
Relating to the allocation and use of land, to rural communities, and to agricultural societies.

**Agriculture**
The purposeful tending of crops and livestock.

**Alluvial**
Refers to the mud, silt, and sand (*alluvium*) deposited by rivers and streams. *Alluvial plains* adjoin many larger rivers; they consist of such renewable deposits that are laid down during floods, creating fertile and productive soils. Alluvial **deltas** mark the mouths of rivers such as the Mississippi and the Nile.

**Altiplano**
High-elevation plateau, basin, or valley between even higher mountain ranges. In the Andes Mountains of South America, altiplanos are at 3000 meters (10,000 feet) and even higher.

**Antecedent**
*Antecedent*, like **subsequent** and **superimposed**, is a term used in human as well as physical geography. Antecedent is something that goes before. In physical geography, a river that is antecedent is one that is older than the landscape through which it flows. In political geography, an antecedent boundary is one that was there before the cultural landscape emerged—it "went before" and stayed put while all around it people moved in to occupy the surrounding area. An example is the 49th parallel bound-

*Words in boldface type are defined elsewhere in this Glossary.

ary, dividing the United States and Canada between the Pacific Ocean and Lake of the Woods in Minnesota.

**Anthracite coal**
Highest carbon-content coal (therefore, of the highest quality), which is formed under conditions of high pressure and temperature that eliminated most impurities. Anthracite burns almost without smoke and produces high heat.

**Apartheid**
Literally, apartness; the Afrikaans term given to South Africa's policies of racial separation. The term no longer has official sanction and has been replaced by *separate development*.

**Aquaculture**
The use of a river segment or an artificial body of water such as a pond for the growing of food products, including fish, shellfish, and even seaweed. Japan is among the world's leaders in aquaculture. When the activity is confined to the raising and harvesting of fish, it is usually referred to as *fish farming*.

**Aquifer**
An underground reservoir of water contained within a porous, water-bearing rock layer. This rock stratum is often overlain by impervious layers to prevent upward seepage of groundwater by capillary action.

**Arable**
Literally, cultivable—that is, land fit for cultivation by one method or another.

**Archipelago**
A set of islands grouped closely together, usually elongated into a chain.

**Area**
A term that refers to a part of the earth's surface with less specificity than **region**. *Urban area* alludes very generally to a place where urban development has taken place, whereas *urban region* requires certain specific criteria upon which a delimitation is based.

**Areal interdependence**
A term related to **functional specialization**. When one area produces certain goods or has certain raw materials or resources, and another area has a different set of resources and produces different goods, their needs may be *complementary* and, by exchanging raw materials and products, they can satisfy each other's requirements. The concepts of areal interdependence and **complementarity** are related; both have to do with exchange opportunities between regions.

**Arithmetic density**
A country's population, expressed as an average per unit area (square kilometer or square mile), without regard for its distribution or the limits of **arable** land—see **physiologic density**.

**Aryan**
From the Sanskrit *Arya*—noble—a name applied to an ancient people who spoke an Indo-European language and who moved into northern India from the northwest. Although properly a language-related term, Aryan has assumed additional meanings, especially racial ones.

**Autocratic**
An autocratic government holds absolute power; rule is often by one person or a small group of persons who control the country by despotic means.

**Atmosphere**
The earth's envelope of gases that rests on the oceans and lands and penetrates the open spaces within soils. This layer of nitrogen (78 percent), oxygen (21 percent), and traces of other gases is densest at the earth's surface and thins with altitude. It is held against the planet by the force of gravity.

**Balkanization**
The fragmentation of a region into smaller, often hostile, political units.

**Bantustan**
Term formerly used to denote one of South Africa's African territories designated for "independence" under the terms of *separate development* (see **apartheid**). The official designation today is Bantu Homeland.

***Barrio***
Slum development in a Middle American city; increasingly applied to inner-city concentrations of Hispanics in such southwestern U.S. cities as Los Angeles.

**Bauxite**
Ore of aluminum, an earthy, reddish-colored material that usually contains some iron as well. Soil-forming processes such as leaching and redeposition of aluminum and iron compounds contribute to bauxite formation, and many bauxite deposits exist at shallow depths in tropical areas of high precipitation.

**Bituminous coal**
Softer coal of lesser quality than **anthracite** (more impurities remain) but of higher grade than **lignite**. Usually found in relatively undisturbed, extensive horizontal layers, often close enough to the surface to permit strip-mining. When heated and converted to coking coal or *coke*, it is used in blast furnaces to make iron and steel.

**Break-of-bulk point**
A location along a transport route where goods must be transferred from one carrier to another. In a port, the cargoes of oceangoing ships are unloaded and put on trains, trucks, or perhaps smaller river boats for inland distribution.

**Buffer zone**
A set of countries separating ideological or political adversaries. In South Asia, Afghanistan, Nepal,

Sikkim, and Bhutan formed a buffer zone between British and Russian-Chinese imperial spheres. Thailand was a *buffer state* between British and French colonial domains in Southeast Asia.

***Caliente***
See ***tierra caliente***.

**Cartel**
An international syndicate formed to promote common interests in some economic sphere through the formulation of joint pricing policies and the limitation of market options for consumers. The Organization of Petroleum Exporting Countries (OPEC) is a classic example.

**Cartography**
The art and science of making maps, including data compilation, layout, and design. Also concerned with the interpretation of mapped patterns.

**Caste System**
The strict social segregation of people—specifically in Hindu society—on the basis of ancestry and occupation.

**Central business district (CBD)**
The heart of a city, the CBD is marked by high land values, a concentration of business and commerce, and the clustering of the tallest buildings.

**Centrality**
The strength of an urban center in its capacity to attract producers and consumers to its facilities: the city's "reach" into the surrounding region. A *central place* possesses a certain measure of centrality and forms the urban focus for a particular region.

**Central Place Theory**
**Location theory** devised by Christaller and his followers; it seeks to explain the spatial distribution of urban places of varying size and function.

**Centrifugal forces**
A term employed to designate forces that tend to divide a country—such as internal religious, linguistic, ethnic, or ideological differences.

**Centripetal forces**
Forces that unite and bind a country together—such as a strong national culture, shared ideological objectives, and a common faith (Islam constitutes a centripetal force in Arab countries).

**Charismatic**
Personal qualities of certain leaders that enable them to capture and hold the popular imagination, to secure the allegiance and even the devotion of the masses. Gandhi, Hitler, and Franklin D. Roosevelt are good examples in this century.

**City-state**
An independent political entity consisting of a single city with (and sometimes without) an immediate **hinterland**. The ancient city-states of Greece have their modern equivalent in Singapore.

**Climate**
A term used to convey a generalization of all the recorded weather observations over time at a certain place or in a given area. It represents an "average" of all the weather that has occurred there. In general, a tropical location such as the Amazon Basin has a much less variable climate than areas located, say, midway between the equator and the pole. In low-lying tropical areas the *weather* changes little, and so the climate is rather like the weather on any given day. But, in the middle latitudes, there may be summer days to rival those in the tropics, winter days so cold that they resemble polar conditions.

**Climax vegetation**
The final, stable vegetation that has developed at the end of a succession under a particular set of environmental conditions. This vegetative community is in dynamic equilibrium with its environment, including other ecosystems such as its animal life.

**Collectivization**
The reorganization of a country's agriculture that involves the expropriation of private holdings and their incorporation into relatively large-scale units, which are farmed and administered cooperatively by those who live there. This system has transformed Soviet agriculture, has been resisted in Communist Poland, and has gone beyond the Soviet model in China's program of communization.

**Common Market**
Name given to a group of 10 European countries (as of 1985) that belong to a **supranational** association to promote their economic interests. Official name is European Economic Community (EEC), sometimes shortened to European Community (EC).

**Compact state**
A politico-geographical term to describe a state that possesses a roughly circular, oval, or rectangular territory in which the distance from the geometric center to any point on the boundary exhibits little variance. Poland, Uruguay, and the U.S. state of Iowa are examples of this shape category.

**Complementarity**
Regional complementarity exists when two regions, through an exchange of raw materials and finished products, can specifically satisfy each other's demands.

**Concentric Zone Model**
A geographical model of the American central city that suggests the existence of five concentric rings (that is, arranged around a common center).

**Condominium**
In political geography, this denotes the shared administration of a territory by two governments. The

North African state of Sudan was long administered jointly by the United Kingdom and Egypt as a condominium.

**Coniferous forest**
As the word suggests, a forest of cone-bearing trees—needleleaf evergreen trees with straight trunks and short branches, including spruce, fir, and pine.

**Contagious diffusion**
The distance-controlled spreading of an idea, innovation, or some other item through a local population by contact from person to person—analogous to the communication of a contagious illness.

**Contiguous**
A word of some importance to geographers that means, literally, to be in contact with, adjoining, or adjacent. Sometimes we hear the continental ("lower 48") United States minus Alaska referred to as contiguous. Alaska is not contiguous to these states, for Canada lies between; neither is Hawaii, separated by thousands of kilometers of ocean.

**Continental shelf**
Beyond the coastlines of the continents the surface beneath the water, in many offshore areas, declines very gently until a depth of about 100 fathoms (somewhat under 200 meters/600 feet). Thus, the edges of the landmasses today are inundated, for only at the 100-fathom line does the surface drop off sharply to the deeper ocean bottom along the *continental slope*. The submerged continental margin is called the continental shelf, and it extends from the shoreline to the continental slope.

**Conurbation**
General term used to identify large metropolitan complexes formed by the coalescence of two or more major urban areas. The Boston-Washington **Megalopolis** along the U.S. northeastern seaboard is an outstanding example.

**Copra**
Meat of the coconut; fruit of the coconut palm.

**Cordillera**
Mountain chain consisting of sets of parallel ranges, especially the Andes in northwestern South America.

**Core area**
In geography, a term with several connotations. *Core* refers to the center, heart, or focus. In physical geography, the core area of a continent identifies the ancient shield zone. In human geography, the core area of a **nation-state** is constituted by the national heartland—the largest population cluster, most productive region, the area with greatest centrality and accessibility, probably containing the capital city as well.

**Corridor**
Specific meaning in politico-geographical context is a land extension that connects an otherwise landlocked state to the ocean. History has seen several such corridors come and go. Poland once had a corridor (it now has a lengthy coastline); Bolivia lost a corridor to the Pacific Ocean between Peru and Chile; Finland possessed a corridor to the Arctic Ocean. *Secondary corridors* may lead to another means of egress, for example, a navigable river. Colombia, although it has coasts on both the Caribbean Sea and the Pacific Ocean, has a secondary corridor to the Amazon River.

**Cultural diffusion**
The spreading and adoption process of a cultural element, from its place of origin over a wider area.

**Cultural ecology**
The interactions between a developing culture and its natural environment.

**Cultural landscape**
The forms and artifacts sequentially placed on the physical landscape by the activities of various human occupants. By this progressive imprinting of the human presence, the physical landscape is modified into the cultural landscape, forming an interacting unity between the two.

**Culture area**
A distinct, culturally discrete spatial unit; a region within which certain cultural norms prevail. Culture *region* might be more appropriate.

**Culture complex**
A related set of culture traits such as prevailing dress codes, cooking and eating utensils.

**Culture hearth**
Heartland, source area, place of origin of a major culture.

**Culture realm**
A cluster of regions in which related culture systems prevail. In North America, the United States and Canada form a culture realm, but Mexico belongs to a different one.

**Culture trait**
A single element of normal practice in a culture—such as the wearing of a turban.

**Cyclic movement**
Movement (e.g., nomadic migration) that has a closed route repeated annually or seasonally.

**Death rate**
The *crude death rate* is expressed as the annual number of deaths per 1000 individuals within a given population.

**Deciduous**
A deciduous tree loses its leaves at the beginning of winter or the start of the dry season.

**Definition**
In political geography, the written legal description of a boundary between two countries or territories—see **delimitation**.

**Delimitation**
In political geography, the translation of the written terms of a boundary treaty (the **definition**) into cartographic representation.

**Delta**
**Alluvial** lowland at the mouth of a river, formed when the river deposits its alluvial load on reaching the sea. Often triangular in shape—hence, the use of the Greek letter whose symbol is Δ.

**Demarcation**
In political geography, the actual marking of a political boundary on the landscape by means of barriers, fences, walls, or other markers.

**Demographic Transition**
Three-stage model, based on Western Europe's experience, of changes in population growth exhibited by countries undergoing modern industrialization. High birth and death rates are followed by plunging death rates, producing a huge net population gain; this is followed by the convergence of birth and death rates at a low overall level.

**Demographic variables**
Births (fertility), deaths (mortality), and migration (population redistribution) are the three basic demographic variables.

**Demography**
Interdisciplinary study of population—especially birth and death rates, growth patterns, longevity, and related characteristics.

**Density of population**
The number of people per unit area. See also **arithmetic** and **physiologic density** measures.

**Desert**
An arid area supporting very sparse vegetation, receiving less than 25 centimeters (10 inches) of precipitation per year. Usually exhibits extremes of heat and cold, because the moderating influence of moisture is absent.

**Desertification**
The encroachment of **desert** conditions on moister zones along the desert margins. Here plant cover and soils are threatened by desiccation, in part through overuse by humans and their domestic animals and, possibly, also because of inexorable shifts in the earth's environmental zones.

**Determinism**
See **environmental determinism**.

**Devolution**
In political geography, the disintegration of the **nation-state** as a result of emerging or reviving regionalism.

**Dhows**
Wooden boats with characteristic triangular sail, plying the seas between Arabian and East African coasts.

**Diffusion**
The spatial spreading or dissemination of a culture element (such as a technological innovation) or some other phenomenon (e.g., a disease outbreak). See also **contagious, expansion, hierarchical,** and **relocation diffusion**.

**Dispersed settlement**
In contrast to **agglomerated** or **nucleated** settlement, dispersed settlement is characterized by the wide spacing of individual homesteads. This lower-density pattern is characteristic of rural North America.

**Distance decay**
The various degenerative effects of distance on human spatial structures. The degree of spatial interaction diminishes as distance increases; therefore, people and activities try to arrange themselves in geographic space to minimize the "friction" effects of overcoming distance, which involves the costs of time as well as travel.

**Diurnal**
Daily.

**Divided capital**
In political geography, a country whose administrative functions are carried on in more than one city is said to have divided capitals.

**Domestication**
The transformation of a wild animal or wild plant into a domesticated animal or a cultivated crop to gain control over food production. A necessary evolutionary step in the development of humankind—the invention of **agriculture**.

**Double cropping**
The planting, cultivation, and harvesting of two crops successively within a single year on the same piece of farmland.

**Doubling time**
The time required for a population to double in size.

**Ecology**
Strictly speaking, this refers to the study of the many interrelationships between all forms of life and the natural environments in which they have evolved and continue to develop. The study of ecosystems focuses on the interactions between specific organisms and their environments. **Cultural ecology** involves the study of the relationships between human cultural manifestations and the natural environment.

**Economies of scale**
The savings that accrue from large-scale production whereby the unit cost of manufacturing decreases as the level of operation enlarges. Supermarkets operate on this principle and are able to charge lower prices than small groceries.

**Elite**
A small but influential, upper-echelon social class whose power and privilege give it control over a country's political, economic, and cultural life.

**Emigrant**
A person migrating away from a country or area; an outmigrant.

**Empirical**
Relating to the real world, as opposed to theoretical abstraction.

**Enclave**
A piece of territory that is surrounded by another political unit of which it is not a part.

**Entrepôt**
A place, usually a port city, where goods are imported, stored, and transhipped. Thus, an entrepôt is a **break-of-bulk point**.

**Environmental determinism**
The view that the natural environment has a controlling influence over various aspects of human life, including cultural development. Also referred to as *environmentalism*.

**Erosion**
A combination of gradational forces that shape the earth's surface landforms. Running water, wind action, and the force of moving ice combine to wear away soil and rock. Human occupation often speeds erosional processes through the destruction of natural vegetation, careless farming practices, and overgrazing by livestock.

**Escarpment**
A cliff or steep slope; frequently marks the edge of a plateau.

**Estuary**
The widening mouth of a river as it reaches the sea. An estuary forms when the margin of the land has subsided somewhat (or as ocean levels rise following glaciation periods) and seawater has invaded the river's lowest portion.

**Evapotranspiration**
The loss of moisture to the atmosphere through the combined processes of evaporation from the soil and transpiration by plants.

**Exclave**
A bounded (non-island) piece of territory that is part of a particular state, but lies separated from it by the territory of another state.

**Expansion diffusion**
The spreading of an innovation or an idea through a fixed population in such a way that the number of those adopting grows continuously larger, resulting in an expanding area of dissemination.

**Extraterritoriality**
Politico-geographical concept suggesting that the property of one state lying within the boundaries of another actually forms an extension of the first state. The principle was used by colonial powers in China to carve their own city neighborhoods out of the Chinese national state.

***Favela***
**Shantytown** on the outskirts or even well within an urban area in Brazil.

***Fazenda***
Coffee plantation in Brazil.

**Federal state**
A political framework wherein a central government represents the various entities within a **nation-state** where they have common interests—defense, foreign affairs, and the like—yet allows these various entities to retain their own identities and to have their own laws, policies, and customs in certain spheres.

**Federation**
The association and cooperation of two or more **nation-states** or territories to promote common interests and objectives.

**Ferroalloy**
A metallic mineral smelted with iron to produce steel of a particular quality. Manganese, for example, provides steel with tensile strength and the ability to withstand abrasives. Other ferroalloys are nickel, chromium, cobalt, molybdenum, and tungsten.

**Fertile Crescent**
Semicircular zone of productive lands extending from near the southeast Mediterranean coast through Lebanon and Syria to the **alluvial** lowlands of Mesopotamia. Once more fertile than today, this is one of the world's great source areas of agricultural innovations.

**Feudalism**
Prevailing politico-geographical system in Europe during the Middle Ages, when land was owned by the nobility and was worked by peasants and serfs. Feudal lords held absolute power over their domains and inhabitants, and Europe was a mosaic of private estates. Feudalism also existed in other parts of the world, and the system persisted into this century in Ethiopia and Iran, among other places.

**Fjord**
Narrow, steep-sided, elongated, and inundated coastal valley deepened by glacier ice that has since melted away, leaving the sea to penetrate. Norway's coasts are marked by fjords, as are parts of the coasts of Canada, Greenland, and southern Chile.

**Floodplain**
Low-lying area adjacent to a mature river, often covered by **alluvial** deposits and subject to the river's floods.

**Forced migration**
Human migration flows in which the movers have no choice but to relocate.

**Formal region**
A type of region marked by a certain degree of homogeneity in one or more phenomena; also called *uniform* region or *homogeneous* region.

**Forward capital**
Capital city positioned in actually or potentially contested territory, usually near an international border; it confirms the state's determination to maintain its presence in the region in contention.

**Fragmented state**
A state whose territory consists of several separated parts, not a **contiguous** whole. The individual parts may be isolated from each other by the land area of other states or by international waters.

**Francophone**
Describes country or region where other languages are also spoken, but where French is the ***lingua franca*** or the language of the **elite**. Quebec is Francophone Canada.

***Fria***
See ***tierra fria***.

**Frontier**
Area of advance penetration, of contention; an area not yet fully integrated into a national state.

**Functional region**
A region marked less by its sameness than its dynamic internal structure; because of its focus on a central node, also called *nodal* or *focal* region.

**Functional specialization**
The production of particular goods or services as a dominant activity in a particular location. Certain cities specialize in producing automobiles, computers, or steel; others serve tourists.

**Geometric boundaries**
Political boundaries **defined** and **delimited** (and occasionally **demarcated**) as straight-lines or arcs.

***Geopolitik*** **(Geopolitics)**
A school of political geography that involved the use of quasi-academic research to encourage a national policy of expansionism and imperialism. Its origins are attributed to a Swede by the name of Kjellen, but its most famous practitioner was a German named Haushofer.

**Ghetto**
An intraurban region marked by a particular ethnic character. Often an inner-city poverty zone (such as the *black ghetto*) in the American city, whose residents are involuntarily segregated from other income and racial groups.

**Green Revolution**
The successful recent development of higher-yield, fast-growing varieties of rice and other cereals in certain Third World countries. This has led to increased production per unit area and a temporary narrowing of the gap between population growth and food needs.

**Gross national product (GNP)**
The total value of all goods and services produced in a country during a certain year. In some **underdeveloped countries (UDCs)**, where a substantial number of people practice subsistence and where the collection of information is difficult, GNP figures may be unreliable.

**Growing season**
The number of days between the last frost in the spring and the first frost of the fall.

**Growth pole**
An urban center with certain attributes that, if augmented by a measure of investment support, will stimulate regional economic development in its **hinterland**.

**Hacienda**
Literally, a large estate in a Spanish-speaking country. Sometimes equated with **plantation**, but there are important differences between these two types of agricultural enterprise (See pp. 260–261).

**Hegemony**
The political dominance of a country (or even a region) by another country; the Soviet Union's grip on Eastern Europe is an outstanding example.

**Hierarchical diffusion**
A form of diffusion in which an idea or innovation spreads by "trickling down" from larger to smaller adoption units. Urban **hierarchies** are usually involved, encouraging the leapfrogging of innovations over wide areas, with geographic distance a less important influence.

**Hierarchy**
An order or gradation of phenomena, with each level or rank subordinate to the one above it and superior to the one below. A national urban hierarchy consists of hamlets, villages, towns, cities, and (usually) the **primate city**.

**High seas**
Areas of the oceans away from land, still beyond national jurisdiction, open and free for all to use.

***Highveld***
A term used in southern Africa to identify the grass-covered, high-elevation plateau that dominates much of the region. ***Veld***, sometimes misspelled veldt, means grassland in Dutch and Afrikaans. The lowest-lying areas in South Africa are called *lowveld*; areas that lie at intermediate elevations are the *middleveld*.

**Hinterland**
Literally, "country behind," a term that applies to a surrounding area served by an urban center. That center is the focus of goods and services production for its hinterland, and is its dominant urban influence as well. In the case of a port city, the hinterland, also includes the inland area whose trade flows through that port.

**Humus**
Dark-colored upper layer of a soil that consists of decomposed and decaying organic matter such as leaves and branches, nutrient-rich and giving the soil a high fertility.

**Iconography**
The identity of a region as expressed through its symbols; its particular cultural landscape and atmosphere.

**Immigrant**
A person migrating into a particular country or area; an inmigrant.

**Imperialism**
The drive toward the creation and expansion of a colonial empire and, once established, its perpetuation.

**Indentured workers**
Contract laborers who sell their services for a stated period of time. Tens of thousands of South Asians came to East Africa as indentured workers at British behest; working conditions often did not match contract stipulations, but indentured workers did not have the option to cancel their agreements.

**Infrastructure**
The foundations of a society: urban centers, communications, farms, factories, mines, and such facilities as schools, hospitals, postal services, and police and armed forces.

**Insular**
Having the qualities and properties of an island. Real islands are not alone in possessing such properties: an **oasis** in the middle of a large desert also has qualities of insularity.

**Insurgent state**
Territorial embodiment of a successful guerilla movement. The antigovernment insurgents establishing a territorial base in which they exercise full control; thus, a state within a state.

**Intermontane**
Literally, "between mountains." The location can bestow certain qualities of natural protection or isolation to a community.

**Internal migration**
Migration flow within a **nation-state**—such as ongoing westward and southward movements in the United States and eastward movement in the Soviet Union.

**International migration**
Migration flow involving movement across international boundaries.

**Intervening opportunity**
The presence of a nearer opportunity that diminishes the attractiveness of sites farther away.

**Irredentism**
A policy of cultural extension and potential political expansion aimed at a national group living in a neighboring country.

**Irrigation**
The artificial watering of croplands. *Basin* irrigation is an ancient method that involves the use of floodwaters that are trapped in basins on the floodplain and released in stages to augment rainfall. *Perennial* irrigation requires the construction of dams and irrigation canals for year-round water supply.

**Isolation**
The condition of being cut off or far removed from mainstreams of thought and action. It also denotes a lack of receptivity to outside influences, caused at least partially by inaccessibility.

**Isoline**
Line connecting points of equal value. An *isotherm* connects points of equal temperature, an *isobar* points of equal atmospheric pressure, and an *isohyet* points of equal rainfall totals.

**Isthmus**
A land bridge, a comparatively narrow link between larger bodies of land. Central America forms such a link between North and South America.

**Juxtaposition**
Contrasting places in close proximity to one another.

**Land alienation**
One society or culture group taking land from another. In Africa, white Europeans took land from black Africans and put it to new uses, fencing it off and restricting settlement. Land alienation also occurred in the Americas and Asia.

**Land reform**
The spatial reorganization of agriculture through the allocation of farmland (often expropriated from landlords) to peasants and tenants who never owned land; also, the consolidation of excessively fragmented farmland into more productive, perhaps cooperatively-run farm units.

**Landlocked**
An interior country or state that is surrounded by land. Without coasts, a landlocked state is at a disadvantage in a number of ways—in terms of access to international trade routes and in the scramble for possession of areas of the **continental shelf**.

**Latitude**
Latitude lines are **parallels** that are aligned east-west across the globe, from 0° latitude at the equator to 90° north and south latitude at the poles. Areas of low latitude, therefore, lie near the equator in the tropics; high latitudes are those in the polar Arctic and Antarctic regions.

**Lignite**
Also called brown coal, this is a low-grade variety of coal somewhat higher in fuel content than *peat* but not nearly as good as the next higher grade, **bituminous coal**. Lignite cannot be used for most industrial processes, but it is important as a residential fuel in certain parts of the world.

***Lingua franca***
The term derives from "Frankish language," and applied to a tongue spoken in ancient Mediterranean ports that consisted of a mixture of Italian, French, Greek, Spanish, and even some Arabic. Today, it refers to a "common language," a second language that can be spoken and understood by many peoples, although they speak other languages at home.

**Littoral**
Coastal; along the shore.

***Llanos***
Name given to the savannalike

grasslands of the Orinoco River's wide basin in parts of Colombia, Venezuela, and Guyana. Elsewhere, these grassy lowlands have different names—for example, *chaco* in northern Argentina.

**Location theory**
A logical attempt to explain the locational pattern of an economic activity, and the manner in which its producing areas are interrelated. **Central Place Theory** is the leading example.

**Loess**
Deposit of very fine silt or dust that is laid down after having been windborne over distances as great as several hundred kilometers. Loess is characterized by its fertility under irrigation and its ability to stand in steep vertical walls when eroded by a river or (as in China) excavated for cave-type human dwellings.

**Longitude**
Angular distance (0° to 180°) east or west as measured from the prime **meridian** (0°) that passes through the Greenwich Observatory in London, England. For much of its length, the 180th meridian functions as the International Date Line.

**Maghreb**
Westernmost segment of the North African–Southwest Asian realm, consisting of Morocco, Algeria, and Tunisia.

**Malthusian**
Designates the early nineteenth-century viewpoint of Thomas Malthus who argued that population growth was outrunning the earth's capacity to produce sufficient food. Neo-Malthusian refers to those who subscribe to such positions in modern contexts.

**Marchland**
A **frontier** or area of uncertain boundaries that is subject to various national claims and an unstable political history. The term refers to the movement of various national armies across such zones.

**Megalopolis**
Term used to designate large coalescing supercities that are forming in diverse parts of the world. Once used specifically to refer to the Boston-Washington multimetropolitan corridor in the northeastern United States, the term is now used generically with a lowercase *m* as a synonym for **conurbation**.

**Mental map**
The structured spatial information an individual acquires in his or her perception of the surrounding environment and more distant places. A *designative* "mental map" involves the objective recording of geographical information received (e.g., Florida is a narrow peninsular state at the southeastern corner of the United States). An *appraisive* "mental map" entails the subjective processing of spatial information according to a person's biases (e.g., Florida is a nice place to live because it has a balmy climate, opportunities for water recreation, and a rapidly growing postindustrial economy).

**Mercantilism**
Protectionist policy of European states during the sixteenth to the eighteenth centuries that promoted a state's economic position in the contest with other countries. The acquisition of gold and silver and the maintenance of a favorable trade balance (more exports than imports) were central to the policy.

**Meridian**
Line of **longitude**, aligned north-south across the globe, that together with **parallels** of **latitude** forms the global grid system. All meridians converge at both poles and are at their maximum distances from each other at the equator.

**Mestizo**
The root of this word is the Latin for *mixed*; it means a person of mixed white and American Indian ancestry.

**Metropolitan area**
See **urban (metropolitan) area**.

**Mexica Empire**
The name the ancient Aztecs gave to the domain over which they held **hegemony** on the north-central mainland of Middle America.

**Migration**
A change in residence intended to be permanent. See also **forced, internal**, **international**, and **voluntary migration**.

**Migratory movement**
Human relocation movement from a source to a destination without a return journey, as opposed to **cyclic movement**.

**Model**
An idealized representation of reality created to demonstrate certain of its properties. A **spatial** model focuses on a geographical dimension of the real world.

**Monotheism**
The belief in, and worship of, a single god.

**Mulatto**
A person of mixed African (black) and European (white) ancestry.

**Multinationals**
Internationally, active corporations that strongly influence the economic and political affairs of many countries.

**Multiple-Nuclei Model**
The Harris-Ullman model that showed the mid-twentieth-century American central city to consist of several zones arranged around nuclear growth points.

**Nation**
Legally a term encompassing all the citizens of a state, it has also taken on other connotations. Most definitions tend to refer to bonds of language, ethnicity, religion, and other shared cultural attributes; such

homogeneity prevails within very few states.

**Nation-state**
A country whose population possesses a substantial degree of cultural homogeneity and unity.

**Natural increase rate**
Population growth measured as the excess of live births over deaths per 1000 individuals per year. Natural increase of a population does not reflect either **emigrant** or **immigrant** movements.

**Natural resource**
As used in this book, any valued element (or means to an end) of the environment, including minerals, water, vegetation, and soil.

**Nautical mile**
By international agreement, the nautical mile—the standard measure at sea—is 6076.12 feet in length, equivalent to approximately 1.15 statute miles (1.85 kilometers).

**Négritude**
Philosophy promoting black cultural pride, with roots in Caribbean America, French universities, and **Francophone** West Africa.

**Network (transport)**
The entire physical system of transportation connections and nodes through which movement can occur.

**Nomadism**
**Cyclic movement** among a definite set of places. Nomadic people are mostly **pastoralists**.

**Norden**
A regional appellation for northern Europe's three Scandinavian countries (Denmark, Norway, and Sweden), Finland, and Iceland.

**Nucleated settlement**
A clustered pattern of population distribution.

**Nucleation**
Clustering, agglomeration.

**Oasis**
An area, small or large, where the supply of water transforms the surrounding desert into a green cropland; the most important focus of human activity for miles around. It may encompass a large, densely populated zone along a major river where irrigation projects stabilize the water supply—as along the Nile River in Egypt, which is really an elongated set of oases.

**Occidental**
See **Oriental**.

**Oligarchy**
Political system involving rule by a small minority, an often corrupt **elite**.

**Organic theory**
Concept that suggests that the state is in some ways analogous to a biological organism, with a life cycle that can be sustained through cultural and territorial expansion.

**Oriental**
The root of the word *oriental* is from the Latin for *rise*; thus, it has to do with the direction in which one sees the sun "rise"—the east; *oriental* therefore means Eastern. *Occidental* originates from the Latin for *fall*, or the "setting" of the sun in the west; *occidental* means Western.

**Orographic precipitation**
Mountain-induced precipitation, especially where airmasses are forced over topographic barriers. Areas beyond such a mountain range experience the **rain shadow effect**.

**Pacific ring of fire**
Zone of crustal instability along tectonic **plate** boundaries, marked by earthquakes and volcanic activity, that rings the Pacific Ocean basin.

**Pangaea**
A vast landmass consisting of most of the areas of the present continents, including the Americas, Eurasia, Africa, Australia, and Antarctica, that existed until near the end of the Mesozoic era when plate divergence and continental drift broke it apart. The "northern" segment of Pangaea is called *Laurasia*, the "southern" part *Gondwana*.

**Parallel**
An east-west line of **latitude** that is intersected at right angles by a **meridian**, a line of **longitude**.

***Paramós***
Highest zone in the Latin American altitudinal zonation, above the tree line in the Andes; sometimes called *tierra helada* (frozen land).

**Pastoralism**
A form of economic pursuit; the practice of raising livestock, although many peoples described as herders actually have a mixed economy—they may also fish, hunt, or even grow a few crops. But pastoral peoples' lives do revolve around their animals.

**Peasants**
In a stratified society, peasants are the lowest class of people who depend on agriculture for a living, but they often own no land at all and must survive as tenants or day workers.

**Peninsula**
A comparatively narrow, fingerlike stretch of land extending from the main body into the sea. Italy constitutes a peninsula.

**Peon**
Term used in Latin America to identify people who often live in serfdom to a wealthy landowner, landless **peasants** in continuous indebtedness.

**Per capita**
*Capita* means individual. Income, production, or some other measure is often given per individual.

**Periodic market**
Village market that opens every third or fourth day or at some other regular interval. Part of a regional network of similar markets in a preindustrial, rural setting where goods

are brought to market on foot (or perhaps by bicycle) and barter remains a major mode of exchange.

**Permafrost**
Permanently frozen water in the soil and bedrock, as much as 300 meters (1000 feet) in depth, producing the effect of completely frozen ground. Can thaw near surface during brief warm season.

**Permanent capital**
Capital city that has historically been the headquarters of the state, such as Rome and Athens.

**Physiographic-political boundaries**
International political boundaries that coincide with prominent physical breaks in the natural landscape—such as rivers or the crest-ridges of mountain ranges.

**Physiographic region (province)**
A region within which there prevails substantial physical-landscape homogeneity, expressed by a certain degree of uniformity in relief, climate, vegetation, and soils.

**Physiologic density**
The number of people per unit area of **arable** land.

**Pidgin**
A language that consists of words borrowed and adapted from other languages; originally developed from commerce among peoples speaking different languages.

**Pilgrimage**
A journey to a place of great religious significance by an individual or by a group of people. When Islam entered West Africa, pilgrimages to Mecca attracted tens of thousands of Moslems annually. Many took up permanent residence along the way and the human map of savanna Africa was transformed.

**Plantation**
A large estate owned by an individual, family, or corporation and organized to produce a cash crop. Almost all plantations were established within the tropics; since 1950, many have been divided into smaller holdings or reorganized as cooperatives.

**Plate tectonics**
Bonded portions of the earth's mantle and crust, averaging 100 kilometers (60 miles) in thickness. More than a dozen such plates exist, most of continental proportions, and they are in motion. Where they meet one slides under the other, crumpling the surface crust and producing significant volcanic and seismic (earthquake) activity. A major mountain-building force.

**Pleistocene Epoch**
Recent period of geologic time that spans the rise of humanity (beginning about 3 million years ago), marked by repeated advances of continental icesheets and milder *interglacials*. Although the last 10,000 years are known as the *Recent* Epoch, Pleistocene-like conditions seem to be continuing and the present is likely to be another Pleistocene interglacial; the glaciers will return.

**Population explosion**
The rapid growth of the world's human population during the past century, attended by ever-shorter **doubling times** and accelerating *rates* of increase.

**Population structure**
Graphic representation of a population by sex and age, as in Fig. 1–3 (p. 58).

**Postindustrial economy**
Emerging economy in the United States and a handful of other highly advanced countries as industry gives way to a high-technology productive complex dominated by services, information-related and managerial activities.

**Primary economic activity**
Activities engaged in the direct extraction of **natural resources** from the environment—such as mining, fishing, lumbering, and especially **agriculture**.

**Primate city**
A country's largest city—ranking atop the urban hierarchy—most expressive of the national culture and usually (but not always) the capital city as well.

**Process**
Causal force that shapes spatial pattern or structure as it unfolds over time.

**Proletariat**
Lower-income working class in a community or society; people who own no capital or means of production and who live by selling their labor.

**Prorupt**
A type of state territorial morphology or shape that exhibits a narrow, elongated land extension.

**Protectorate**
In Britain's system of colonial administration, the protectorate was a designation that involved the guarantee of certain rights (such as the restriction of European settlement and **land alienation**) to peoples who had been placed under the control of the Crown.

**Push-pull concept**
The idea that **migration** flows are simultaneously stimulated by conditions in the source area that tend to drive people away and by the perceived attractiveness of the destination.

**Quaternary economic activity**
Activities engaged in the collection, processing, and manipulation of *information*.

**Rain shadow effect**
The relative dryness in areas downwind of or beyond mountain ranges caused by **orographic precipitation**, whereby moist airmasses are forced to deposit most of their water content in the highlands.

**Region**
A commonly used term and a geographic concept of central impor-

tance. An **area** on the earth's surface marked by certain properties.

**Relative location**
The position or **situation** of a place relative to the position of other places. Distance, **accessibility**, and connectivity affect relative location.

**Relict boundary**
An international political boundary that has ceased to function but the imprint of which can still be detected on the cultural landscape.

**Relief**
Vertical difference between the highest and lowest elevations within a particular area.

**Relocation diffusion**
Sequential **diffusion** process in which the items being diffused are transmitted by their carrier agents as they evacuate the old areas and move (relocate) to new areas. A disease can move from population cluster to population cluster in this way, running its course in one area before fully invading the next. The most common form of relocation diffusion involves the spreading of innovations by a migrating population.

**Rural density**
A measure that indicates the number of persons per unit area living in the rural areas of a country, outside of the urban concentrations; provides an index of rural population pressure.

**Sahel**
Semiarid zone across most of Africa between the southern margins of the arid Sahara and the moister forest and savanna zone to the south. Chronic drought, **desertification**, and overgrazing have contributed to severe famines in this area since 1970.

**Satellite state**
A state whose sovereignty has been compromised by the **hegemony** (dominating influence) of a larger power. Such a situation can arise when the ruling political party in such a country represents the ideology of the larger power, or when the economic influence of the larger power is so great that it is in virtual control of production in the satellite state.

**Scale**
Representation of a real-world phenomenon at a certain level of reduction or generalization. In **cartography**, the ratio of map distance to ground distance; shown on map as bar graph, representative fraction and/or verbal statement.

**Secondary economic activity**
Activities that process raw materials and transform them into finished products; the *manufacturing* sector.

**Sector Model**
A structural model of the American central city that suggests land-use areas conform to a pie-shaped pattern focused on the downtown core.

**Sedentary**
Permanently attached to a particular area, a population fixed in its location; the opposite of **nomadic**.

**Sequent occupance**
The notion that successive societies leave their cultural imprints on a place, each contributing to the cumulative cultural landscape.

**Shantytown**
Unplanned slum development on the margins of Third World cities, dominated by crude dwellings and shelters mostly made of scrap wood, iron, and even pieces of cardboard. Brazil's ***favelas***, Mexico's ***barrios***, North Africa's *bidonvilles* and India's *bustees* are examples.

**Sharecropping**
Relationship between a large landowner and farmers on the land in which the farmers pay rent for the land they farm by giving the landlord a share of the annual harvest.

**Shatter belt**
**Region** caught between stronger external cultural-political forces, under persistent stress and often fragmented by aggressive rivals. Eastern Europe and Southeast Asia are classic examples.

**Shifting agriculture**
Cultivation of crops in forest clearings, soon to be abandoned in favor of more recently cleared nearby forest land. Also known as **slash-and-burn agriculture**, and by various regional names.

**Sinicization**
Giving a Chinese cultural imprint; Chinese **acculturation**.

**Site**
The internal locational attributes of a place, including its local spatial organization and physical setting.

**Situation**
The external locational attributes of a place; its **relative location** or position with reference to other nonlocal places.

**Slash-and-burn agriculture**
The system of "patch" or **shifting agriculture** that involves the cutting down and burning of the vegetation in a part of the rainforest, and the temporary cultivation of crops in the clearing.

**Spatial**
Pertaining to space on the earth's surface; synonym for *geographic(al)*.

**Spatial model**
See **model**.

**Squatter settlement**
See **shantytown**.

**Stratification**
Layering. In a stratified society, the population is divided into a **hierarchy** of social classes. In industrialized society, the **proletariat** is at the lower end; **elites** that possess capital and control the means of production are at the upper level. In traditional India, the "untouch-

ables" form the lowest **caste**, while the still-wealthy remnants of the princely class are at the top.

**Subsequent boundary**
An international political boundary that developed contemporaneously with the evolution of the major spatial elements of the cultural landscape in an area.

**Subsistence**
Existing on the minimum necessities to sustain life; spending most of one's time in pursuit of survival.

**Superimposed boundary**
An international border placed by powerful outsiders on a developed human landscape; usually ignores preexisting cultural-spatial patterns—such as the boundaries that now divide Germany and Korea.

**Supranational**
A venture involving three or more national states—political, economic, or cultural cooperation to promote shared objectives. The European **Common Market** is such an organization, as is OPEC.

**System**
Any group of objects or institutions and their mutual interactions; geography treats systems that are expressed spatially—such as **regions**.

**Takeoff**
Economic concept to identify a stage in a country's development when conditions are set for a domestic industrial revolution, as happened in the United Kingdom in the late eighteenth century and in Japan after the Meiji Restoration; it is happening in China today.

**Tell**
The lower slopes and narrow coastal plains along the Atlas Mountains in northwest Africa where the majority of the **Maghreb** region's population is clustered.

***Templada***
See ***tierra templada***.

**Terracing**
The transformation of a hillside or mountain slope into a steplike sequence of horizontal fields for intensive cultivation (as in Fig. 10–10).

**Territoriality**
A community's sense of property and attachment toward its territory as expressed by its determination to keep it inviolable and strongly defended.

**Territorial sea**
Zone of water adjacent to a country's coast, held to be part of the national territory and treated as a segment of the sovereign state.

**Tertiary economic activity**
Activities that engage in *services*—such as transportation, banking, retailing, finance, education, and routine office-based jobs.

**Theocracy**
State whose government is under the control of a ruler who is deemed to be divinely guided or under the control of a group of religious leaders, as in Khomeini's Iran. The opposite of the theocratic state is the *secular* state.

***Tierra caliente***
The lowest of three vertical zones into which the topography of Middle and South America is divided according to elevation. The *caliente* is the hot humid coastal plain and adjacent slopes up to between 750 and 900 meters (2500 and 3000 feet) above sea level. The natural vegetation is the dense and luxuriant tropical rainforest; the crops are bananas, sugar, cacao, and rice in the lower areas and coffee, tobacco, and corn along the somewhat higher slopes.

***Tierra fria***
The highest-lying zone below the snow line in Middle and South America, from about 2100 meters (7000 feet) up to 3000 meters (10,000 feet). Coniferous trees stand here; upward they change into scrub and grassland. There are also important pastures within the *fria*, and wheat can be cultivated. Several major population clusters in the region lie at these altitudes. Above the *fria* lie the ***paramós***, the highest and coldest zone.

***Tierra templada***
The intermediate altitudinal zone in Middle and South America, between 750 to 900 meters (2500 to 3000 feet) and 1800 to 2100 meters (6000 to 7000 feet). This is the "temperate" zone, with moderate temperatures compared to the ***tierra caliente***. Crops include tobacco, coffee, corn, and some wheat.

**Totalitarian**
A government whose leaders rule by absolute control, tolerating no differences of political opinion.

**Transculturation**
Cultural borrowing that occurs when different cultures of approximately equal complexity and technological level come into close contact. In **acculturation**, by contrast, an indigenous society's culture is modified by contact with a technologically more developed society.

**Transferability**
The capacity to move a good from one place to another at a bearable cost; the ease with which a commodity may be transported.

**Transhumance**
Seasonal movement of people and their livestock in search of pastures. This movement may be vertical—that is, into highlands during the summer and back to lowlands in winter—or horizontal, in pursuit of seasonal rainfall.

**Underdeveloped countries (UDCs)**
Countries that, by various measures, suffer seriously from negative economic and social conditions, including low **per capita** incomes, poor nutrition, inadequate health, and related circumstances.

**Unitary state**
A **nation-state** that has a centralized government and administration that exercises power equally over all parts of the state.

**Urbanization**
A term with several connotations. The proportion of a country's population living in urban places is its level of urbanization. The **process** of urbanization involves the movement to, and clustering of, people in towns and cities, a major force in every geographic realm today. Another kind of urbanization occurs when a sprawling city absorbs rural countryside and transforms it into suburbs; in the case of the Third World city, this also generates peripheral **squatter settlements**.

**Urban (metropolitan) area**
The entire built-up, nonrural area and its population, including the most recently constructed suburban appendages. Gives a better picture of the dimensions and population of such an area than the delimited municipality (central city) that forms its heart.

***Veld***
Open grassland on the South African plateau, becoming mixed with scrub at lower elevations where it is called *bushveld.* As in Latin America, there is an altitudinal zonation into **highveld**, *middleveld* and *lowveld.*

**Voluntary migration**
Population movement in which people relocate in response to perceived opportunity, not because they are forced to move.

**Von Thünen Model**
Explains the location of agricultural activities in a commercial, profit-making economy. A process of spatial competition allocates various farming activities into concentric rings around a central market place, with profit-earning capability the determining force in how far a crop locates from the market. The original (1826) *Isolated State* model now applies to the continental scale (Fig. 1–12B); in certain Third World areas, however, transportation conditions still engender an application reminiscent of the original (Fig. 6–30).

**Water table**
When precipitation falls on the soil, some of the water is drawn downward through the pores in soil and rock under the force of gravity. Below the surface it reaches a level where it can go no farther; there it joins water that already saturates the rock completely. This water that "stands" underground is *groundwater,* and the upper level of the zone of saturation is the *water table.*

# PHOTO CREDITS

| Figure | Photographer | Source | Page |
|---|---|---|---|
| **I–1** | René Burri | *Magnum* | xvi |
| **I–2** | Craig Aurness | *West Light* | 3 |
| **I–4** | Harm J. de Blij | | 8 |
| **I–7** | Steve Raymer | | 12 |
| **I–18** | Brian Seed | | 36 |
| **I–20** | Tom Mc Hugh | *Photo Researchers* | 41 |
| **I–21** | Mehmet Biber | *Photo Researchers* | 43 |
| **1–1** | Sepp Seitz | *Woodfin Camp* | 54 |
| **1–7** | Sepp Seitz | *Woodfin Camp* | 62 |
| **1–9** | Brian Brake | *Photo Researchers* | 65 |
| **1–15** | Allen Green | *Photo Researchers* | 83 |
| **1–17** | Anthony Howarth | *Woodfin Camp* | 84 |
| **1–20** | Mopy | *Rapho/Photo* | 92 |
| **1–21** | Alan D. Harkrader | *Black Star* | 93 |
| **1–22** | Tom Mc Hugh | *Photo Researchers* | 95 |
| **1–23** | Craig Aurness | *Woodfin Camp* | 98 |
| **1–26** | George Holton | *Photo Researchers* | 102 |
| **1–28** | Owen Frank | *Stock, Boston* | 109 |
| **1–30** | Chris Niedenthal | *Black Star* | 112 |
| **A–1** | G. R. Roberts | *Photo Researchers* | 122 |
| **A–4** | David Hiser | *The Image Bank* | 127 |
| **A–7** | David Moore | *Black Star* | 133 |
| **A–7** | David Moore | *Black Star* | 133 |
| **A–9** | Harm J. de Blij | | 136 |
| **2–1** | | *AP Photo Color* | 138 |
| **2–4** | George Holton | *Photo Researchers* | 147 |
| **2–6** | | *Sovfoto* | 149 |
| **2–9** | Henry N. Michael | | 152 |
| **2–11** | George Holton | *Photo Researchers* | 154 |
| **2–13** | George Holton | *Photo Researchers* | 156 |
| **2–16** | Ilkka Ranta | *Woodfin Camp* | 160 |
| **2–19** | | *Tass from Sovfoto* | 170 |
| **3–1** | Adam Woolfitt | *Susan Griggs Agency* | 172 |
| **3–3** | Ted Speigel | *Black Star* | 176 |
| **3–6** | Georg Gerster | *Photo Researchers* | 184 |
| **3–8** | Ted Speigel | *Black Star* | 186 |
| **3–15** | Dan Mc Coy | *Black Star* | 199 |
| **3–16** | | *The Irvine Company* | 201 (top) |
| **3–18** | C. Vergara | *Photo Researchers* | 201 (bottom) |
| **3–23** | William W. Bacon, III | *Photo Researchers* | 209 |
| **3–24** | Earl Roberge | *Photo Researchers* | 210 |
| **3–26** | Eunice Harris | *Photo Researchers* | 213 |

| Figure | Photographer | Source | Page |
|---|---|---|---|
| 3–29 | Patrick Grace | *Photo Researchers* | 218 |
| 3–30 | Helen Marcus | *Photo Researchers* | 219 |
| 3–32 | Allen Green | *Photo Researchers* | 221 |
| 3–33 | Robert Phillips | *The Image Bank* | 222 |
| J–1 | Graig Davis | *Black Star* | 228 |
| J–5 | Harm J. de Blij | | 236 |
| J–6 | Harm J. de Blij | | 237 |
| J–7 | René Burri | *Magnum* | 238 |
| J–9 | Eiji Miyazawa | *Black Star* | 241 |
| 4–1 | Albert Moldvay | *Woodfin Camp* | 246 |
| 4–5 | Gilles Peress | *Magnum* | 254 |
| 4–6 | Albert Moldvay | *Woodfin Camp* | 257 |
| 4–10 | Harm J. de Blij | | 263 |
| 4–11 | Jason Laure | *Woodfin Camp* | 265 (top) |
| 4–12 | Harm J. de Blij | | 265 (bottom) |
| 4–13 | Harm J. de Blij | | 266 |
| 4–17 | Nicholas De Vore III | *Bruce Coleman, Inc.* | 273 |
| 4–18 | John Lopinot | *Black Star* | 275 |
| 4–19 | Cindy Karp | *Black Star* | 276 |
| 5–1 | O. Louis Mazzatenta | | 278 |
| 5–5 | Eric Simmons | *Stock, Boston* | 286 |
| 5–9 | Loren Mc Intyre | *Woodfin Camp* | 292 |
| 5–10 | Owen Franken | *Stock, Boston* | 293 |
| 5–11 | Dick Davis | *Photo Researchers* | 294 |
| 5–12 | Loren Mc Intyre | *Woodfin Camp* | 296 |
| 5–14 | Carl Frank | *Photo Researchers* | 300 |
| 5–16 | Victor Englebert | *Black Star* | 304 |
| 5–18 | Larry Dale Gordon | *The Image Bank* | 307 |
| 5–19 | Georg Gerster | *Photo Researchers* | 308 |
| 5–20 | J. C. Criton | *Sygma* | 309 |
| 5–21 | Rick Merron | *Magnum* | 310 |
| B–1 | René Burri | *Magnum* | 312 |
| B–3 | Jack Fields | *Photo Researchers* | 315 |
| B–4 | Ellis Herwig | *Stock, Boston* | 316 |
| B–5 | Martin Rogers | *Woodfin Camp* | 318 |
| 6–1 | Jodi Cobb | | 322 |
| 6–5 | Robert Azzi | *Woodfin Camp* | 331 |
| 6–7 | Marc & Evelyn Bernheim | *Woodfin Camp* | 333 |
| 6–15 | Alain Nogues | *Sygma* | 344 |
| 6–17 | | *NASA* | 348 |
| 6–18 | Thomas Nebbia | *Woodfin Camp* | 350 |
| 6–21 | F. Sautereau | *Rapho/Photo* | 355 |
| 6–22 | William Katz | *Photo Researchers* | 356 |
| 6–23 | René Burri | *Magnum* | 357 |
| 6–24 | George Holton | *Ocelot* | 359 |
| 6–26 | Robert Azzi | *Woodfin Camp* | 361 |
| 6–28 | Wolfgang Steinmetz | *The Image Bank* | 364 |
| 6–29 | David Burnett | *Woodfin Camp* | 367 |
| 6–31 | Robert Caputo | | 370 |
| 6–32 | Mike Yamashita | | 371 |
| 7–1 | Marc & Evelyn Bernheim | *Woodfin Camp* | 374 |
| 7–4 | Harm J. de Blij | | 382 |
| 7–6 | Harm J. de Blij | | 384 (top) |
| 7–7 | Pete Turner | *The Image Bank* | 384 (bottom) |
| 7–9 | Georg Gerster | *Photo Researchers* | 386 |
| 7–12 | Ray Witlin | | 390 |
| 7–13 | Harm J. de Blij | | 391 |
| 7–14 | Marc & Evelyn Bernheim | *Woodfin Camp* | 392 |
| 7–15 | | *Woodfin Camp* | 394 |
| 7–17 | Richard & Mary Magruder | *The Image Bank* | 396 |
| 7–23 | Georg Gerster | *Photo Researchers* | 406 |
| 7–24 | Georg Gerster | *Photo Researchers* | 407 |
| 7–25 | George Holton | *Photo Researchers* | 408 |
| 7–27 | Stephen S. Birdsall | | 411 |
| 7–29 | Tomas D.W. Friedman | *Photo Researchers* | 414 |
| 7–31 | Tomas D.W. Friedman | *Photo Researchers* | 416 |
| 7–32 | Harm J. de Blij | | 417 |
| 7–34 | Donald N. Rallis | | 419 |
| 8–1 | Van Bucher | *Photo Researchers* | 422 |
| 8–3 | Nicholas de Vore III | *Bruce Coleman, Inc.* | 426 |
| 8–5 | Jehangir Gazdar | *Woodfin Camp* | 430 |
| 8–7 | Cary Wolinsky | *Stock, Boston* | 433 |
| 8–9 | | *Wide World* | 434 |
| 8–11 | Jacques Jangoux | *Peter Arnold* | 438 |
| 8–14 | Brian Brake | *Rapho/Photo* | 442 (top) |
| 8–15 | Raghubir Singh | | 442 (bottom) |
| 8–16 | Paolo Koch | *Photo Researchers* | 443 |

| Figure | Photographer | Source | Page |
|---|---|---|---|
| **8–20** | Albert Moldvay | *Photo Researchers* | 450 |
| **8–22** | Frank Keating | *Photo Researchers* | 452 |
| **8–24** | Peter Ward | *Bruce Coleman, Inc.* | 456 |
| **9–1** | Harm J. de Blij | | 458 |
| **9–3** | | *UPI* | 465 |
| **9–5** | Paolo Koch | *Photo Researchers* | 468 |
| **9–7** | Harm J. de Blij | | 471 |
| **9–10** | Terry Madison | *The Image Bank* | 476 |
| **9–11** | Harm J. de Blij | | 477 |
| **9–12** | Reinhold Messner | | 479 |
| **9–14** | Marvin E. Newman | *The Image Bank* | 482 |
| **9–16** | Harm J. de Blij | | 484 |
| **9–19** | Robin Moyer | *Black Star* | 488 |
| **9–21** | Larry Mulvehill | *Photo Researchers* | 491 |
| **9–22** | Harm J. de Blij | | 495 |
| **9–24** | P. J. Griffiths | *Magnum* | 499 |
| **9–26** | David Burnett | *Stock, Boston* | 501 |
| **9–27** | Hiroji Kubota | *Magnum* | 502 (top) |
| **9–28** | Dennis Stock | *Magnum* | 502 (bottom) |
| **10–1** | C.A. Peterson | *Rapho/Photo* | 504 |
| **10–5** | George Holton | *Photo Researchers* | 512 |
| **10–8** | Jackie Foryst | *Bruce Coleman, Inc.* | 516 |
| **10–10** | Hans Hoefer | *Woodfin Camp* | 518 |
| **10–12** | Allen Green | *Photo Researchers* | 522 |
| **P–1** | Jack Fields | *Photo Researchers* | 534 |
| **P–3** | Harm J. de Blij | | 538 (top) |
| **P–4** | Harm J. de Blij | | 538 (bottom) |

# MAP INDEX


| Figure | Map Title | Page |
|---|---|---|
| **I–3** | Effect of Scale | 5 |
| **I–5** | World Landscapes | 10–11 |
| **I–6** | World Tectonic Plates | 12 |
| **I–8** | Extent of Glaciation During the Pleistocene | 13 |
| **I–9** | Mean Annual Precipitation of the World | 16–17 |
| **I–10** | World Climates | 18–19 |
| **I–11** | Hypothetical Continent Model | 20 |
| **I–12** | World Vegetation | 22–23 |
| **I–13** | World Soil Distribution | 24–25 |
| **I–14** | World Population Distribution | 26–27 |
| **I–15** | Cartogram of the World's National Populations | 28–29 |
| **I–17** | Political Units, 1985 | 34–35 |
| **I–19** | World Geographic Realms | 38–39 |
| **I–23A** | The Triangular Trade | 49 |
| **I–23B** | Dr. John Snow's Map of London Cholera Deaths | 50 |
| **1–2** | Europe's Population Distribution | 57 |
| **1–5** | Relative Location: Europe in the Land Hemisphere | 59 |
| **1–6** | European Landform Regions | 61 |
| **1–8** | The Roman Empire | 64 |
| **1–10** | Formative Europe Ca. 1200 A.D. | 67 |
| **1–11** | Europe: Core and Regions | 70 |
| **1–12B** | Von Thünen Rings in Europe | 73 |
| **1–13** | Languages of Europe | 76 |
| **1–14** | The British Isles | 81 |
| **1–16** | London Region | 84 |
| **1–18** | Western Europe | 89 |
| **1–19** | Paris Region | 91 |
| **1–24** | Northern Europe | 99 |
| **1–25** | Acid Rain in Europe | 101 |
| **1–27** | Mediterranean Europe | 104–105 |
| **1–29** | Eastern Europe | 110 |
| **1–31** | European Supranationalism | 120 |
| **A–2** | Human Migrations in Modern Times | 124 |
| **A–3** | Australia: Physiographic Regions | 126 |
| **A–5** | Australia: Political Divisions, Capitals, and Communications | 128 |
| **A–6** | Australia: Agriculture and Mineral Resources | 129 |
| **A–8** | New Zealand | 135 |
| **2–2** | Climates of the Soviet Union | 141 |
| **2–3** | Growth of Russian Empire | 146 |
| **2–5** | U.S.S.R.: Physiography and Physiographic Regions | 148 |
| **2–7** | Political Units of the U.S.S.R. and Major Urban Areas | 150–151 |
| **2–8** | Peoples of the Soviet Union | 152 |
| **2–10** | Percent Russians in Soviet Republics | 153 |
| **2–12** | Soviet Agriculture | 155 |
| **2–14** | Soviet Railway Network | 158 |
| **2–15** | U.S.S.R. Manufacturing Regions | 159 |
| **2–17** | Eurasian Heartlands | 165 |
| **2–18** | The Sensitive East: Soviet–Chinese Borderlands | 167 |
| **3–4** | North America | 178 |
| **3–5** | North America: Physiography | 181 |
| **3–7** | Geographic Patterns of Cancer | 186 |
| **3–9** | Population Distribution: 1980 | 187 |
| **3–10** | North America: Expansion of Settlement | 190 |

| Figure | Map Title | Page |
|---|---|---|
| 3–12 | Continental Core Region | 195 |
| 3–13 | North America: Projected Megalopolitan Growth | 196 |
| 3–19 | Minority Population Distribution | 203 |
| 3–20 | Perceptual Regions of North America | 205 |
| 3–21 | Future Residential Preferences of University of Miami Students, 1981 | 206 |
| 3–22 | North America: Major Deposits of Fossil Fuels | 208 |
| 3–25 | Agricultural Regions of the United States | 211 |
| 3–27 | Change in Personal Income Per Capita, by State, 1975–1982 | 215 |
| 3–28 | Regions of the North American Realm | 216 |
| 3–31 | Distribution of Corn and Wheat Production | 220 |
| 3–34 | Garreau's Nine Nations of North America | 225 |
| J–2 | Japan: Land and Livelihoods | 229 |
| J–3 | Japanese Colonial Empire | 232 |
| J–4 | Manufacturing Regions of Japan | 236 |
| J–8 | Japan's Trade With the World | 239 |
| 4–2 | Regions of Middle America | 248–249 |
| 4–4 | Mesoamerica: Historical Geography | 253 |
| 4–7 | Indian Languages of Middle America | 258 |
| 4–8 | Caribbean Region: Colonial Spheres | 259 |
| 4–9 | Middle America: Mainland and Rimland | 261 |
| 4–14 | Mexico | 267 |
| 4–16 | Central America | 272 |
| 5–2 | South America: Physiography | 281 |
| 5–3 | World Agriculture | 282–283 |
| 5–4 | Indian Regions of South America | 285 |
| 5–6 | South America: Colonial Realms | 287 |
| 5–7 | South America: Political Units and Modern Regions | 288 |
| 5–8 | South America: Culture Spheres | 291 |
| 5–13 | The North: Carribean South America | 297 |
| 5–15 | The Andean West: Indian South America | 301 |
| 5–17 | The South: Mid–Latitude South America | 306 |
| B–2 | Brazil: Regions and Environments | 313 |
| 6–2 | Food Taboos | 325 |
| 6–3 | Religions of the World | 328–329 |
| 6–4 | Postulated Culture Hearths and Early Diffusion Routes | 330 |
| 6–6 | Old World Areas That Have Experienced Moslem Rule | 332 |
| 6–9 | Diffusion of Influenza in England and Wales, 1957 | 336 |
| 6–10 | Diffusion of the Planned Regional Shopping Center in the United States, 1949–1968 | 337 |
| 6–11 | Diffusion of Cholera in the U.S.: 1832 and 1866 | 337 |
| 6–12 | Political Units and Geographic Regions of North Africa and Southwest Asia | 340–341 |
| 6–13 | World Fossil Fuel Reserves | 342–343 |
| 6–14 | North Africa and Southwest Asia: Oil and Gas | 344 |
| 6–16 | Egypt and Its Neighbors | 346 |
| 6–19 | The Maghreb | 351 |
| 6–20 | Israel and the Pivot | 353 |
| 6–25 | Arabian Peninsula | 360 |
| 6–27 | The Northern Zone | 362–363 |
| 6–30 | The Forest Zone Surrounding Addis Ababa, Ethiopia | 369 |
| 7–2 | Languages of Africa | 379 |
| 7–3 | Human Racial Groups | 380–381 |
| 7–5 | Africa: Physiography | 383 |
| 7–8 | Gondwana Reassembled | 385 |
| 7–9 | World Life Expectancy at Birth | 388–389 |
| 7–11 | Diffusion of African Sleeping Sickness from Its Early Focus in West Africa | 390 |
| 7–11 | Distribution of Tsetse Fly Vectors that Transmit the Trypanosomes | 390 |
| 7–16 | The Atlantic Slave Trade | 395 |
| 7–16 | African Source Areas for Slavery | 395 |
| 7–18 | Colonization and Liberation | 397 |
| 7–19 | Major Regional Migrations in Africa | 400 |
| 7–21 | African Regions | 404 |
| 7–22 | West Africa | 405 |
| 7–26 | East Africa | 409 |
| 7–28 | Equatorial Africa | 413 |
| 7–30 | Southern Africa | 415 |
| 7–33 | South Africa | 418 |
| 8–2 | South Asia: Physiography | 424 |
| 8–4 | The Indus Valley Civilization | 429 |
| 8–6 | Languages of India | 431 |
| 8–8 | Moslem India Before and After Partition | 434 |
| 8–10 | States of Modern India | 435 |
| 8–12 | India: Agriculture | 439 |
| 8–13 | India: Industrialization | 441 |
| 8–17 | World Population Growth | 446–447 |
| 8–18 | India: Recent Population Growth Rates | 448 |
| 8–19 | Bangladesh | 449 |
| 8–21 | Pakistan | 451 |
| 8–23 | Sri Lanka | 455 |
| 9–2 | China: Physiography | 462 |
| 9–4 | Evolution of the Chinese Empire | 467 |
| 9–6 | China: Colonial Spheres, Territorial Losses | 470 |
| 9–8 | Political Divisions of China | 473 |

| Figure | Map Title | Page |
|---|---|---|
| **9–9** | Regions of China | 475 |
| **9–13** | China Proper | 481 |
| **9–15** | China: Agricultural Regions | 483 |
| **9–17** | China: Energy Resources | 485 |
| **9–18** | Ethnic Minorities: Ethnolinguistic Areas | 486 |
| **9–19** | The Northeast | 490 |
| **9–23** | Taiwan: Stages of Land Development | 497 |
| **9–25** | Korea | 500 |
| **10–2** | Southeast Asia: Political Geography | 507 |
| **10–3** | Genetic Political–Boundary Types | 509 |
| **10–4** | Southeast Asia: Physiography | 511 |
| **10–6** | Chinese in Southeast Asia | 513 |
| **10–7** | Ethnic Mosaic of Southeast Asia | 514 |
| **10–11** | Colonial Spheres in Southeast Asia | 519 |
| **10–13** | Vietnam: Rise of the Insurgent State | 523 |
| **10–14** | Refugee Movements in Southeast Asia, 1980 | 524 |
| **10–15** | Spatial Morphology: Thailand | 526 |
| **10–16** | Spatial Morphology: Philippines | 527 |
| **10–17** | Theoretical Maritime Claims in Southeast Asia | 529 |
| **10–18** | World Maritime Claims | 530–531 |
| **10–19** | National Claims to Antarctica | 532 |
| **P–2** | Pacific Regions | 536–537 |

# INDEX


Aarhus, 104
Abadan, 339
Abidjan, 400
Aborigines, 131, 517
Absolute location, 2
Abu Dhabi, 339
Abuja, 407
Acapulco, 258
Acculturation, 153
  Middle America and, 257–258
  Pacific regions and, 535–538
Achaean League, 63
Acid rain, 101–102, 185
Activity space (territorial range), China and, 483
Adams, J., 196–197
Adana, 365
Addis Ababa, 73–74, 368, 369
Adelaide, 125, 126, 127
Aden, *see* Yemen
Adriatic Sea, 59, 62, 108
Aegean Sea, 59, 107, 116
Afghanistan, 96, 168, 169, 344, 362–363, 367–368, 433, 453, 454
Africa, 5, 7, 8, 9, 13, 15, 18, 21, 24–25, 30, 31, 266, 287–288, 305, 368–370. *See also* Black Africa; North Africa and Southwest Asia; Southern Africa; West Africa
African Plate, 9
African sleeping sickness, 387, 390, 391
Afrikaans, 378
Aggradation, 426, 427
Agrarian revolution, 68, 71
Agribusinesses, 210
Agricultural Producers' Cooperatives (APCs), China and, 491
Agriculture, 33–34
  agrarian revolution, 68, 71
  commercial, 33, 284
  collectivization, 113, 115, 117, 154, 490–492
  definition, 33
  Green Revolution, 438–440
  hacienda, 260–261, 268
  *latifundia*, 107
  plantation, 33–34, 260–261, 274, 283, 290, 510
  rotation system, 127–128
  slash-and burn, 33
  soils, 14, 15, 18–19, 21–22, 24–25, 183
  subsistence, 283–284, 388–389, 390
  systems of, 280–282, 282–284
  transhumance, 97
  underdeveloped countries, 36–37
  vegetation patterns, 16–18, 22–23
  Von Thünen's spatial structure of, 72–74
  *see also* Irrigation
Ahaggar Mountains, 382
Ahmadabad, 440, 443
Air pollution, 185
Air pressure, 140
Alaska, 209, 222, 223
Albania, 110, 111, 112, 115, 116, 326, 338
Alberta, 179, 209
Aleppo, 357
Alexander the Great, 429
Alexandria, 347, 348
Alfisols, 21, 24–25
Algeria, 43, 323, 325, 338, 339, 341, 344, 347, 350, 352, 401, 405
Algiers, 252, 350, 352, 396
Allah, 331
Allemanni, 66
Alpine Mountains, 60, 61
Alps, 59, 60, 96, 108, 427
Altai Mountains, 168
Altiplanos, 281, 303
Altitude, 140
Altitudinal zonation, 269
Amazonas, 314
Amazon Basin, 2, 16, 284, 289, 291, 303, 312, 317, 318, 382
Amazon River, 14, 280, 291, 302, 312
America, 9, 124, 145. *See also* Latin America; North America; United States of America
American Manufacturing Belt, 194, 195, 198, 201, 202, 206, 211, 212, 214, 215–217
American Virgin Islands, 262
Amharic, 325, 326, 369
Amin, I., 409, 411, 412
Amman, 358
Amsterdam, 75, 78, 94, 96, 295
Amundsen, R., 532
Amur River, 145, 164, 167
Analogue area analysis, 157
Andes Mountains, 16, 269, 279, 280, 281, 289, 291, 297, 299, 300, 301, 302, 303, 304, 307
Andhra Pradesh, 436, 443
Andropov, Y., 169
Angara River, 163
Angkor Wat, 511
Angles, 80, 84
Anglo-Saxons, 66
Angola, 381, 398, 400–401, 414, 416, 525
Ankara, 364
Annam, 514, 515, 522, 523
Anshan, 488, 489
Antalya, 365
Antarctica, 385, 532–533
Antarctic Peninsula, 532
Antarctic Treaty, 533
Antecedent boundaries, 508–509
Anthropogeographic boundaries, 339
Antigua, 263
Antioquia, 299
Antofagasta, 303, 305, 309
Antwerp, 90, 94, 95, 96
Apartheid, 417–420
Appalachia, 208–209
Appalachian Highlands, 180
Appalachian Mountains, 427
Appennine Mountains, 60, 106, 108
Appraisive mental maps, 204–205
Aqaba, 358, 362
Aquaculture, 240

Aqueducts, 64, 65
Aquifers, 185
Aquitaine, 92
Arabia, 332
Arabian Peninsula, 325, 330–331, 344, 360
Arabian Sea, 424
Arab League, *see* League of Arab States
Arab world, 325–326. *See also* North Africa and Southwest Asia
Arakan Mountains, 521
Aral Sea, 147, 170
Arawaks, 260
Archipelago, 45
  of Southeast Asia, 132, 496, 505, 510, 528
Area, of regions, 2
Areal functional organization, 235
Areal functional specialization, 65
Areal representation, 4
Area symbols, on maps, 51
Argentina, 15, 34, 284, 286, 289, 290, 291, 294, 297, 303, 304–307, 308–309
Arica, 303
Aridisols, 21, 24–25
Aristotle, 63
Arithmetic density, population and, 56
Arizona, 184, 221, 223
Armenia, 151–152, 153
Arnhem Land, 133
Aruba, 262, 298
Aryans, 428, 454
Ashkelon, 355
Asia, 13, 18, 21, 24–25, 26, 30, 31, 33, 34. *See also* China; North Africa and Southwest Asia; South Asia; Southeast Asia
Aśoka, 430
Assad, H., 357
Assam, 423, 424, 425, 436, 439, 440, 445
Asunción, 303
Aswan High Dam, 345–346, 348
Asyut, 348
Atacama Desert, 289, 291, 305, 307, 309
Ataturk, Mustafa Kemal, 363, 364
Athens, 63, 78, 106, 107
Atlanta, 220, 221
Atlas Mountains, 60, 350–351
Attenuated state, 521–522
Auckland, 135, 136
Augelli, J.P., 38, 260–261, 290, 291
Aurangzeb, 431
Austin, 222
Australia, 9, 13, 18, 21, 24–25, 34, 40, 122–137, 288
  agriculture, 123, 127–129
  Canberra, 133–134
  economy, 125, 130–131
  federalism, 132–134
  gold and, 129, 130, 131
  migration and transfer and, 123–125
  mineral resources, 129–130
  New Zealand and, 134–136
  outback, 125
  population, 30, 122, 123, 126, 130, 131–132
  urbanization, 125–127, 133–134
Australian Capital Territory, 133
Australian Desert, 133
Austria, 66, 88, 96–97, 116
Austro-Hungarian Empire, 111, 114, 115
*Autobahns*, 79
Ayraut, 471
Azerbaydzhan S.S.R., 162
Aztec Empire, 252, 254–256, 258, 284, 285

Bacon, Sir F., 384
Baghdad, 332, 340, 359
Bahia, 315, 319
Bahrain, 360
Baku, 162
Bali, 518
Balkanization, 111, 203. *See also* Eastern Europe
Balkan Mountains, 117
Balkan Peninsula, 4, 66, 111, 114–118
Baltic Sea, 59, 88, 103, 113, 139, 144
Baltimore, 216
Baltistan, 453
Baluchistan, 424, 433, 450
Bandar Seri Begawan, 515
Banda Sea, 530
Bangalore, 430, 433
Bangkok, 521, 525
Bangladesh, 22, 26, 34, 44, 423, 425, 438, 444, 445, 446, 447–449, 450
Bantu, 395, 410, 420
Bantu language, 377
Bantustans, 419
Barbados, 262, 263
Barbary states, 351
Barcelona, 106, 107
Barents Sea, 103
Bar graph, scale and, 4, 5
Barquisimeto, 297
Barranquilla, 299
Basin irrigation, 345
Basque provinces, 86, 106
Bass Strait, 130
Batavia, 517, 521. *See also* Djakarta
Batholiths, 427
Bathurst, 130
Baton Rouge, 184
Bay of Bengal, 425, 448
Bay of Biscay, 107
Baykal-Amur Mainline (BAM) Railroad, 148, 169
Bay of Campeche, 269
Bayóvar, 302
Bedouins, 361
Beijing (Peking), 25, 166, 459, 464, 465, 471, 474, 484
Beirut, 357, 358
Belfast, 86
Belgian Congo, *see* Zaïre
Belgium, 26, 68, 77, 88, 94, 95–96, 398, 400
Belgrade, 64, 114, 116
Belize, 252, 253, 258, 262, 271, 272, 274, 277
Bell, D., 214
Belmopan, 272
Belo Horizonte, 296, 316
Belorussia, 151
Benelux, 77, 79, 88, 94–96, 119. *See also* Belgium; Luxembourg; Netherlands
Benelux Agreement, 118
Bengal, 430, 440
Benin, 393, 394, 399
Benue River, 406
Berbers, 326, 351
Bergen, 98, 100
Berlin, 77, 78, 88, 93–94, 295
Berlin Conference (1884), 396, 398
Bern, 97
Berry, B., 201–202
Bhutan, 168, 423
Biafra, 86, 407, 437
Bihar, 436, 439, 440, 441
Bilbao, 106
Bilharzia, *see* Schistosomiasis
Biomes, 17–18, 22–23
Birdsall, S.S., 157
Birmingham:
  England, 220
  U.S.A., 82, 84
Birmington, 81
Bismarck, O. von, 88, 398

Black Africa, 34, 374–421
  agriculture, 44, 388–391
  climate and vegetation, 386–387
  colonialism in, 395–401, 402–403
  continental drift, 384–386
  environmental hazards and diseases in, 387–388
  as geographic realm, 39, 40, 43–44, 48
  history, 391–401
  physiography, 379–384
  population, 375, 387
  regions, 401–420
Black Belt, 69
Black Muslims, 332
Black Sea, 114, 117, 139, 143, 151, 158, 365
Blue Nile, 345, 369
Bluff, 136
Boca Raton, 213
Boers, 396, 417, 419
Bogotá, 292, 294, 295, 299, 300
Bohemia, 114
Bohemian Basin, 113, 114
Bolivar, S., 289
Bolivia, 34, 285, 289, 291, 295, 296–297, 300, 303, 305, 307
Bolshevik Revolution, 142, 144, 149, 153–154
Bolton, 81
Bombay, 432, 436, 440, 442, 443
Bonaire, 262
Bophuthatswana, 419
Borchert, J., 194, 198
Bordeaux, 89
Borneo, 515 *See also* Kalimantan
Bosnia-Hercegovina, 115
Bosporus Sea, 364
Boston, 194, 198, 205, 218
Botswana, 415
Boundaries, 166, 338–339, 508–510
  anthropogeographic, 339
  creation of, 305
  geometric, 338, 508–509
  linear political, 32–33
  morphology, 338–339
  of nation states, 4, 32
  physiographic, 338–339
  regional, 4
  relicit, 510
  superimposition of, 434, 509–510
Bourguiba, H., 352
Boxer Rebellion, 469–470, 471
Bradford, 82, 85
Brahmaputra, 424, 445
Brahmaputra Valley, 425, 428
Brasilia, 294, 295, 296, 312, 313, 317, 320
Bratislava, 114
Bratsk, 163
Brazil, 4, 13, 18, 32, 284, 286, 288, 289, 290, 291, 292, 293–294, 296, 297, 303, 307, 312–321
Brazilian Model, 320, 321
Brazza, P., 396
Brazzaville, 399, 412, 413
Break-of-bulk point, 90
Bremen, 90
Brest, 89
Brezhnev, L., 169
Brisbane, 125, 126, 127, 130
Britain, *see* England
British Columbia, 15, 179, 224
British Honduras, *see* Belize
British Isles, 26–27, 80–88. *See also* England; Ireland; Northern Ireland; Scotland; Wales
British West Indies, 260, 263
Brno, 114
Broek, J.O.M., 203
Broken Hill, 130
Brown, R., 250–251
Brunei, 515
Brunhes, J., 11
Brunn, S., 198
Brussels, 94, 96
Bucharest, 117
Budapest, 114, 115
Buddha, 429, 430
Buddhism, 44, 45, 328, 329, 429, 430, 454, 478, 479–480, 511
Budjovice, 114
Buenos Aires, 292, 294, 295, 296, 303, 305, 306
Buffer state, 168
Buganda, 411, 412. *See also* Uganda
Bulawayo, 416
Bulgaria, 110, 112, 113, 115, 116, 117, 118, 338
Bulli, 130
Bureya River Valley, 164
Burgenland, 116
Burgess, E.W., 174
Burma, 507, 510, 511, 512, 514, 515, 521
Burundi, 408
Bushman language, 378
Bushmen, 7, 388, 395, 410

Cabral, P., 287
Cairo, 332, 347, 348–349
Calais, 88
Calcutta, 425, 432, 440, 442, 443–444
Caldas, 299
Cali, 299
California, 15, 184, 185, 189–190, 202, 204, 205, 207, 209, 211, 212, 214–215, 223, 224
Callao, 296, 300
Cambodia, 511, 515. *See also* Kampuchea
Cameroon, 398, 399, 404, 406, 412, 413–414
Campos, 312
Canada, 2–3, 4–5, 13, 21, 30, 34, 41, 86, 173, 177, 183, 185, 186, 190, 508–509
  regions, 215–219, 222–224
  *see also* North America
Canadian Shield, 180
Canberra, 127, 133–134
Cantabrian Mountains, 105, 106, 107
Canterbury Plain, 135
Cape of Good Hope, 394
Cape Province, 410, 417
Cape Ranges, 414
Cape Town, 295, 395, 410
Cape Verde Islands, 287
Cape York, 125
Capital cities, 293–295, 508
Caracas, 293, 295, 296, 297, 299
Cardamon Upland, 425
Cardiff, 85
Caribbean America, 247–248, 250–251, 259–260, 262–266
Caribbean Lowlands, 299
Caribbean Sea, 248
Carinthia, 116
Caroline Islands, 535
Carpathian Mountains, 60, 113, 116, 117
Cartagena, 65, 299
Cartogram, 24–25, 28–29
Cartography, 46
Casablanca, 352
Cascade Mountains, 182
Caspian Sea, 170, 323, 366
Caste system, 44, 429, 436
Catalonia, 106, 107
Catherine the Great, 144
Caucasian race, 75
Caucasus Mountains, 144, 148, 155
Cauca Valley, 299
Cayenne, 298
Celebes, 517. *See also* Sulawesi

Celtic language, 86
Cenozoic era, 9
Center-pivot irrigation, 183
Central African Republic, 412, 413
Central America, 4, 249, 253, 271–275
Central Asiatic Ranges, 148
Central business district (CBD), 7, 79, 175, 199–200
of Latin America city, 270–271
Central Indian Plateau, 425
Central place theory, 174, 192–193
Northern China and, 482–484
Central Uplands, 60, 61
Centrifugal forces, 508, 521
India and, 436, 437
Centripetal forces, 508
India and, 436-437
Cerro Bolivar, 296, 298
Cerro de Pasco, 296, 302
Ceylon, 430. *See also* Sri Lanka
Chaco, 304, 306
Chad, 344, 369, 370, 399, 400, 401, 404, 412
Chang Basin, 482
Chang Jiang (Yangtze) River, 25, 30, 466, 474, 475–476, 480
Chang Jiang Valley, 474
Chao Phraya, 507, 525
Chaparral, 184
Chapel Hill, 214
Chapman, K., 334, 335
Charlemagne, 66
Charleroi, 94
Chengdu (Chengtu) Plain, 475
Chen Pao (Damansky) Island, 167
Cherbourg, 89
Chernenko, K., 169
Chiang Kai-shek, 472, 496, 498
Chiatura, 161
Chibcha Indians, 295, 299
Chicago, 194, 216
Chile, 13, 15, 34, 284, 289, 290, 291, 294, 296, 297, 304, 305, 306, 307–310, 521
China, 13, 15, 68, 116, 145, 166–167, 240, 459–503, 515
agriculture, 25–26
communists in, 463–465, 468, 471, 472, 482, 489–496
current world position, 463–466, 494–496
dynastic history, 466–468
"Four Modernization" program, 461, 494–496
as geographic realm, 44–45, 48
nineteenth century, 468–470
*pinyin* system, 464–465
population, 22, 24, 25, 32
regions, 468, 473–489
Southeast Asia and, 511–513
twentieth century, 469–473
Chin Dynasty, 466
Ching Dynasty, 467–468, 469, 470
Chinook, 182
Chittagong, 448, 449
Cholera, 51, 337, 338, 376
Chongqing (Chungking), 472, 476, 480
Chota Nagpur, 440
Chou Dynasty, 466
Christaller, W., 74, 174, 192–193, 235, 482, 483, 484
Christchurch, 135, 136
Christianity, 66, 75, 80, 326, 328, 329, 333, 338, 357
Chuquicamata, 309
Ciskei, 419
City, 7, 30, 64
capital, 293–295, 508
functional classification of, 295–296
grid-plan for, 251–252
models, 174–175
nongridiron, 252
*see also* Primate city; Urbanization
City-state, 63, 64
Ciudad Bolivar, 299
Civil War, 189
Claimant states, 533
Cleveland, 205, 216
Climate:
Hypothetical Continent model for, 20–21
rain shadow effect and, 182–183
regionalization of, 14–16, 18–19
Climatology, 140–141
Climax vegetation, 16, 17
Cluj, 117
Clyde River, 69
Coahuila, 269
Coastal landscape, 427
Cochabamba, 303
Cochin China, 515, 522, 523
Cocos plate, 9
Cohen, Y., 336, 337
Cold polar climates, 15, 18–19
Cole, J., 320
Collective agriculture, 113, 115, 117, 154, 490–492
Cologne, 64
Colombia, 274, 284, 286, 289, 291, 294, 295, 296, 297, 299–300
Colombo, 455
Colonia Agrippensis (Cologne), 64
Colonialism, 513–520
in Black Africa, 395–401, 402–403
in China, 469, 470–471
in India, 432
in West Africa, 405
Colorado, 205, 214–215, 222, 223
Colorado Plateau, 180
Colorado River, 185
Columbia Plateau, 180–181, 427
Columbia River, 181, 185, 224
Comilla, 449
Commercial agriculture, 33, 284
Common Market, *see* European Economic Community
Communes (China), 492. *See also* Collective agriculture
Communication:
in Europe, 78
facilities for per person, 37
Communism, *see* China; Soviet Union
Comodoro Rivadavia, 306
Compact state, 520–521
Complementarity, 75, 393
Comprehensive Soil Classification System (CSCS), 19
Computer graphics, 46
Concentric zones, 174, 175
Condominium, as colonialism, 399
Confucianism, 492, 493–494
Confucius, 466
Congo, 14, 396, 398, 400, 412. *See also* Zaïre
Congo Republic, 399, 400
Congo (Zaïre) Basin, 382, 386, 400
Congo (Zaïre) River, 380–381, 393, 396, 400
Coniferous forests, 15
Connecticut, 214, 217
Constanta, 117
Constantinople, 66, 338. *See also* Istanbul
Contagious diffusion, 334–335, 336,376
Contiguous empire, 145
Continent, 9
Continental core area, in North America, 194, 195
Continental drift, 9, 384–386, 426
Continentality, 140, 141, 182
Continental shelf, 528, 530
Conurbation, 83, 85, 194–195
of American South, 220

of Australia, 134
of Netherlands, 94
Convection, 182
Cook, J., 123
Copenhagen, 103–104
Copperbelt, 400, 414
Coptic church, 326, 369
Cordillera (Andes), 299, 303
Cordoba, 305
Core area, 158, 508
Corn Belt, 2, 3, 209, 211
Cornwall, 84
Cortés, H., 255–256, 284
Cossacks, 143
Costa Rica, 253, 260, 261, 271, 272, 273, 274, 277
Council of Europe, 118, 119
Council for Mutual Economic Assistance, 120
Counter-urbanization, 31
Creoles, 298
Cretans, 347
Croatia, 115, 116
Crust, of earth, 9
Crystal Mountains, 381
Cuba, 248, 262, 263, 264, 266, 276, 308, 523
Cultural ecology, 325
Cultural geography, 46, 201–206, 324–325
Cultural landscape, 5, 7–8, 256–257, 324
Cultural processes, 7
Cultural region, 2–3
Culture, 4–8, 324
Culture areas, 324
Culture complex, 324
Culture hearth, 189, 327, 330
Brazil, 314
Middle America, 252
Middle East, 327–329
Culture history, 325
Culture realm, 4
Culture regions, 4
Culture spheres, of South America, 290–292, 308
Culture system, 324, 519–520
Culture trait, 324
Culture world, 4
Curaçao, 248, 262, 265, 298
Curtin, P., 395
Cuyo, 304, 306
Cuzco, 280, 281, 284, 286, 295
Cycle theory, of state evolution, 361–362
Cyclone, 182, 448
Cyprus, 32, 86, 344, 365
Czechoslovakia, 110, 111, 112, 113, 118, 143

Dahomey, 394, 399. *See also* Benin
Dakar, 396, 399, 400
Dakota, 206
Dalai Lama, 478, 480
Dallas, 198, 205, 213, 222
Damascus, 332, 357
Danube River, 62, 96, 97, 114, 115
Danube Valley, 96, 113
Dar-es-Salaam, 410
Dari, 368
Dark Ages, 66
Dartmoor, 84
Darwin, 127
Dead Sea, 355
Deccan, 427
Deccan Plateau, 383, 424, 425
Degradation, 426
Delhi, 425, 431, 442, 443, 444
Demographic Transition Model, 57–58, 446
Demography, 56
Demonstration effect, tourists and, 265
Deng Xiaoping, 494, 495
Denmark, 98–99, 100, 103–104
Density, population, 56
Denver, 185, 198, 200, 205, 214, 219, 222
Dependent territories, 32
Derby, 81
Desert biome, 17–18
Desertification, 17
Designative mental mapping, 204
de Souza, A., 461
Detroit, 194, 252
Developed countries, 34, 35, 36, 37, 38, 460, 461, 462
Development:
geography of, 460–462
takeoff stage of, 320
Devolution, 86
Dhahran, 361
Dhaka, 449
Dialects, 202
Diamond Head, 538
Díaz, P., 268
Dien Bien Phu, 523
Diffusion, *see* Spatial diffusion
Dinaric Ranges, 60
Disadvantaged countries, *see* Underdeveloped countries
Disease:
Black Africa and, 387–388
medical geography and, 185, 186, 376–377
Disease ecology, 376–377
Dispersion, population and, 56
Djakarta, 517, 521
Djawa, 30. *See also* Jawa
Djibouti, 370, 399
Dnepropetrovsk, 161
Dnieper River, 143
Dobruja, 112
Dodoma, 410
Dominica, 260, 262
Dominican Republic, 248, 260, 262, 263, 264, 265, 266
Domino theory, Thailand and, 525
*Donbas*, 161
Donets Basin, 161
Donetsk, 162
Don River, 161
Dortmund-Ems Canal, 90
"Double maximum," 15
Drakensberg, 380
Dravidian languages, 429, 436
Dravidians, 454
Dresden, 93
Dry climates, 15, 18–19
Dry monsoon, 141
Dry World, 322. *See also* North Africa and Southwest Asia
Dualism, Brazilian society and, 319
Dubai, 339
Dublin, 87–88
Dunedin, 136
Durango, 269
Durban, 417
Durban Beach, 417
Durham, 198, 214
Dust bowl, 182
Dutch East India Company, 516–517
Dvina River, 158, 170

Earth, 8–11
Earthquakes, 9
Earth science, 46
East Africa, 9, 376, 380, 388, 392, 393, 394, 396, 401, 408–412
East Anglia, 84
Easter Island, 536
Eastern Cape, 419
Eastern Desert, 350
Eastern Europe, 2, 4, 63, 66, 78, 109–118, 120
Eastern Samoa, 539

East Germany, 26, 88, 92–94, 112–113. *See also* Germany
East India Company, 431, 432
East Indies, 287, 517, 520
East Malaysia, 515
East Pakistan, 424, 433, 437, 450. *See also* Bangladesh
East Prussia, 113
Ebro Valley, 108
Eburacum (York), 64
Ecological trilogy, 367
Economic geography, 33–34, 46, 282–284, 460
 of Mexico, 268–270
 of North America, 206–214
Economies of scale, 212
Ecosystems, 101
Ecuador, 284, 285, 289, 291, 294, 295, 296, 300, 302–303
Ecumene, 56
Edinburgh, 85
Edo, *see* Tokyo
Egypt, 2, 22, 326, 332, 338, 343–344, 345–350, 354, 356
 ancient, 327–329, 396
Eilat, 355
Eire, *see* Republic of Ireland
*Ejidos*, 261, 268, 269
Ekofisk, 82, 85, 103
Elam, 327
Elba, 65
Elbe River, 62, 90, 113, 114
Elburz Mountains, 366
Elizabethville, 400
Ellice Islands, 538
Elongated state, 521–522
 Chile as, 307
El Salvador, 253, 272, 273, 274, 276, 277
Emden, 90
Endemic diseases, 387
Energy consumption per person, economic development measured by, 37
England, 80, 338, 399
 agriculture, 68, 78, 82
 European Economic Community and, 119
 Falkland Islands, 308–309
 highland and lowland, 80–83
 Hong Kong and, 488–489
 India and, 432
 industrial revolution in, 69, 81–82, 85
 industry in, 68
 population, 446
 regions of, 83–86
 urbanization, 78, 79
English, P.W., 367
English Channel, 59
Entebbe, 411
Entisols, 21, 24–25
*Entrepôt*, Copenhagen as, 103
Entre Rios, 304
Environmental determinism, 253
Environmental engineering, 46
Epidemic, 387
Epochs, 9
Equatorial Africa, 2, 382, 396, 399, 400, 401, 412
Equatorial areas, 14
Equatorial Guinea, 375, 396, 412
Equatorial zone, 13
Eras, geologic, 9–11
Eratosthenes, 46
Erie Canal, 194
Eritrea, 338, 369, 524
Erosion, 426, 427
Erzgebirge (Ore) Mountains, 113
Escarpments, 383, 385, 386
Eskimos, 178, 179
Esmeraldas, 302, 303
Essequibo, 299
Estonia, 111, 151
Ethiopia, 22, 30, 34, 86, 96, 325, 326, 338, 344, 368, 369, 370, 375, 376, 377, 387, 391, 393, 396, 398, 524
Ethnicity, North America and, 202–203, 204
Euphrates River, 358
Euphrates Valley, 357
Eurasia, 9, 13, 21, 24–25
Europe, 21, 34, 66, 75, 76, 77, 78
 agriculture, 62, 68, 71, 78
 ancient civilizations in, 62–66
 economy, 69, 70, 71, 77
 geographic features, 33, 40, 48, 60, 62, 63, 70,74–75, 79–80, 88
 industry, 71, 77
 migration from, 123–125
 nation-states in, 66, 68, 74, 75
 population, 24–25, 26–27, 30, 31, 56–57
 revolutions in, 66–70
 spatial interaction in, 75–77
 supranationalism in, 118–120
 urbanization, 71, 74, 78–79
 *see also* British Isles; Eastern Europe; Mediterranean Europe; Northern Europe; Western Europe
European Atomic Energy Commission (Euratom), 118, 120
European Coal and Steel Community (ECSC), 118, 119
European Economic Community (EEC), 77, 88, 96, 119–120
European Free Trade Association (EFTA), 118, 119–120
European Monetary Agreement (EMA), 118
European Parliament, 119
European race, 75
European Space Research Organization (ESRO), 120
Evapotranspiration, 14, 87
Evolution, 9
Exmoor, 84
Expansion diffusion, 334–336
Exploitation, colonialism and, 398
External migrations, 58
Extractive sector, of economy, 207
Exurbia, 200

Falkland Islands, 306, 308–309
Family, linguistic, 375, 377
Farming, *see* Agriculture
Faults(ing), 380, 427
Federal state, Australia as, 132–134
*Fellaheen*, 347
Fellows, D.K., 395
Fenneman, N., 47–48
Fernando Póo, 396, 406
Fertile Crescent, 327
Fiji, 45, 534, 535–536
Finland, 63, 98, 100, 102, 104
Five-stage growth theory, 460–461
Flanders, 68, 84
Florence, 108
Flores, 518
Florida, 183–184, 202, 205, 211, 213, 214, 220, 221, 224
Folk culture, 324–325
Food taboos, 325
Forbidden City, 484
Forced migration, 124–125, 409
Ford, L., 270–271
Ford, T.R., 390
Forest biome, 17–18
Formal regions, 3
Formosa, 469. *See also* Taiwan
Fort-de-France, 265
Ft. Lauderdale, 198
Fort Worth, 222
Forward capital, Brasilia as, 295
Fossil fuels, 208

Fragmented state, 521
Philippines as, 525–528
United States as, 177
France, 61, 62, 68, 69–70, 74, 77, 78, 79, 86, 90, 91–92, 332, 338, 398
in Equatorial Africa, 413
French Revolution, 69–70
Germany compared with, 88–93
population, 26–27
in West Africa, 398, 399–400
Franks, 66
French Canada, 218–219, 224
French Congo, 400. *See also* Congo Republic
French Equatorial Africa, 399
French Guiana, 287, 296, 298, 299
French Revolution, 69–70
French Somaliland (Djibouti), 399
French West Africa, 399
Frontiers, 166–168
of United States, 189–191
Fronts, 182
Fujairah, 339
Fujian (Fukien) Province, 477, 497
Fukuoka, 237–238
Functional regions, 3
Functional specialization, England and, 82
Fushun, 488
Fuson, R.H., 3–4
Futa Jallon Highlands, 380, 393

Gabon, 401, 412, 413
Gaelic, 88
Galactic metropolis, 198, 200
Galicia, 107
Gambia, 375, 404
Gandhi, M., 436, 437
Ganges Delta, 327, 425, 430, 443, 445
Ganges Plain, 424
Ganges River, 22, 26, 30
Ganges Valley, 425, 428, 429
Gangetic lowland, 445
Gant, 115
Gao, 393
Garonne River, 61, 62, 90
Garonne Valley, 92
Garreau, J., 177, 224–225
Geiger, R., 14, 16, 18–19, 21
Geneva, 97
Genoa, 106, 109
Gentrification, 200
Geographic realms, 38–45
Geography, 45–48
Geometric (straight-line) boundaries, 338, 508–509
Geomorphology, 46, 426–428
Geopolitics, 75, 508
*Geopolitik*, 75
Georgia (U.S.A.), 212
Georgian soviet republic, 151–152
German Democratic Republic, *see* East Germany
German Federal Republic, *see* West Germany
Germanic people, 80
Germany, 74, 75, 77, 93–94, 295, 396, 398
East, 26, 88, 92–94, 112–113
France compared with, 88–93
geographic characteristics, 62, 89–90
industry in, 68, 77, 93
West, 26, 30, 32, 78, 79, 82, 92–93, 94
Gezhouba Dam, 476
Ghana, 387, 393, 399, 403, 404, 405, 406
Gibraltar, 308
Gilbert Islands, 538
Gilgit, 453
Giurgiu, 117
Glaciation, 9–10, 12, 427
Glasgow, 85
Gloucestershire, 84
Gobi Desert, 168, 477
Godāvari River, 425
Goiás, 317, 320
Golan Heights, 354, 357
Gold Coast, *see* Ghana
Golden Triangle, 510
Gondwana, 384–385, 386, 427
Gor'kiy, 160
*Gorods*, 143
Göteborg, 101, 104
Gottmann, J., 7
Gould, P., 402, 403, 460
Gran Chaco, 303
Grand Canyon, 180
Grassland biome, 17–18, 386
Great Artesian Basin, 130
Great Britain, *see* England; Scotland; Wales
Great Dividing Range, 125, 133
Great Dyke, 416
Greater Antilles, 260, 262, 264
Greater Buenos Aires, 305
Greater Vancouver, 179
Great Escarpment:
African, 383, 385, 386, 419
Brazilian, 314, 315, 316
Great Lakes, 184, 225
Great Plains, 9, 180, 183, 185, 189, 209, 211, 219
Great Salt Lake, 181
Greece, 105, 106–107, 112, 338
ancient, 62–63, 104
Green Belt, London and, 83
Greenbelt, 79
Green Revolution, 438–440
Taiwan and, 498
Greenwich Observatory, 50
Grenada, 262
Grid, world political, 4
Griffin, E., 270–271
Griffin, W.B., 133
Ground truth, 349
Growth-pole concept, 317, 320, 460
Guadalajara, 267, 268
Guadalquivir River, 106
Guadarrama Range, 108
Guadeloupe, 260, 262, 266
Guam, 264, 535
Guangdong (Kwangtung), 471, 476
Guangxi Zhuangzu Zizhiqi (Kwangsi-Chuang Autonomous Region), 476
Guangzhou (Canton), 469, 471, 472, 476, 487, 480
Guantánamo, 308
Guarani, 303
Guatemala, 252, 253, 254, 257, 260, 261, 271, 272, 273, 274, 276
Guatemala City, 272
Guayaquil, 294, 296, 302
Guayas River, 302
Guiana Highlands, 298
Guianas, 284, 297, 298–299
Guinea-Bissau, 404
Gujarat, 436, 440
Gulf of Aqaba, 355
Gulf of Bothnia, 59
Gulf of Finland, 61, 139
Gulf of Genoa, 109
Guyana, 290, 298–299

Haciendas, 260–261, 268
Hadrian's Wall, 32–33, 36
Hägerstrand, T., 244, 251, 334
Haggett, P., 20
Hague, The, 94, 295
Haifa, 355, 356, 358
Haiti, 34, 132, 248, 260, 262, 263, 264, 265, 266
Hamburg, 90, 113
Hampshire, 84

Han Dynasty, 466, 487, 488
Hangzhou (Hangchow), 466, 476
Hankow, 480
Hanoi, 515, 522, 523, 524
Hanyang, 480
Harar, 369
Harare, 416
Harbin, 489
Harris, C.D., 174, 238, 295–296
Harris, M., 6
Hartshorn, T., 31
Hartshorne, R., 508
Haryana, 436, 438
Hausa-Fulani, 406
Havana, 264
Hawaiian Islands, 534, 536, 537, 538, 539
Hazaras, 368
Heartland theory, 165–166
Hebei (Hopeh) Province, 465, 484
Hebrew, 325
Heilongjiang (Heilungkiang), 469
Held, C., 73, 74
Helsinki, 102
Hepatitis, 377
Herodotus, 345
Herskovits, M.J., 6
Heyerdahl, T., 250
Hierarchical diffusion, 335–336
High Atlas Mountains, 351
High-island cultures, 535
Highs, climate and, 140
High-technology, 213, 214
Highveld, 419
Himalaya Mountains, 26, 423, 424, 427, 478
Hinduism, 44, 45, 325, 329, 428–429, 431, 436, 509, 511
Hindu Kush, 424
Hindustan, 425, 430
Hinterland, 3
Hiroshima, 237
Hispanics, 202, 204, 221
Hispaniola, 260, 262, 264. *See also* Haiti
Historical geography, 250–252, 334
Historical inertia, 212
Histosols, 21, 24–25
Hobart, 127
Ho Chi Minh, 523
Ho Chi Minh City, 515, 522, 524
Ho Chi Minh Trail, 525
Hoebel, E.A., 5–6
Hokkaido, 228, 233, 235–237
Holland, *see* Netherlands
Holocene epoch, 9, 10
Holy Roman Empire, 66
Homogeneity, of regions, 2
Honduras, 252, 253, 261, 272, 274, 277
Hong Kong, 308, 487, 488–499
Hong Kong Island, 469, 488, 489
Honolulu, 538
Honshu, 237, 238
Hooghly River, 443
Hooson, D.M., 165, 166
Horizontal fracturing, 427
Horn of Africa, 387
Horvath, R.J., 73–74, 368
Hottentots, 395
Houston, 198, 214, 220, 221, 222
Hoyt, H., 174
Hua Guofeng, 494, 496
Huancayo, 301, 302
Huang He (Yellow) River, 22, 25, 30, 44, 327, 466, 474–475, 482, 488
Hudson/James Bay drainage systems, 4, 5
Hué, 522
Huguenots, 67
Hull, 85
Hulwân (Helwân), 350
Human ecology, 46
Humboldt Current, 300
Humid cold climates, 15, 18–19
Humid equatorial climates, 14–15, 18–19
Humid temperate climates, 15, 18–19
Humus, 14
Hunan, 472
Hungarian Basin, 114, 117
Hungary, 63, 66, 68, 110, 111–112, 113, 114–115, 118, 143, 328
Hunter, J., 336
Hunting and gathering, 388, 389
Huntington, E., 253–254
Hu Yaobang, 494
Hyderabad, 432, 443
Hydrography:
  North America and, 184–185
  water, 11–14
Hydrologic cycle, 12
Hydrosphere, 12
Hypothetical Continent model, for climate, 20–21

Ibadan, 406
Iberian Peninsula, *see* Portugal; Spain
Iberian Plateau, 106
Ibn Saud, King, 361
Ibo, 406
Iboland, 405, 406
Ice Age, 9–10, 12
Iceland, 32, 98, 100, 103, 104
Icesheets, 9–11, 12
Idaho, 202, 223
Idea-area chain, 412
Igneous rock, 427
Igorots, 528
Île de la Cité, 90
Illinois, 184, 205, 219
Immigration, Australia and, 131–132. *See also* Migration
Inca Empire, 280–281, 284–285, 286, 291, 303
Inceptisols, 21, 22, 24–25
India, 18, 32, 44, 48, 69, 332, 385, 423–424, 509
  caste system in, 44, 429, 436
  federal, 435–446, 448
  history, 428–434
  physiographic regions, 424–425, 428
  population, 24, 26
  *see also* Bangladesh; Pakistan; Sri Lanka
Indirect rule, 398
Indochina, 2, 45, 510–513. *See also* Southeast Asia
Indochina War, 515, 525. *See also* Vietnam War
Indonesia, 32, 34, 45, 326, 505, 510, 514, 516–520, 521, 529–530
Indus River, 424
Industrial location, 71
  Japan and, 234
Industrial revolution, 36, 68–69, 71, 81–82, 85
Indus Valley, 26, 327, 428–429
Influenza, 336
Inland Sea, 237
Inner Mongolia, 479, 480
Insurgent State, 523–525
Intercontinental migration, 167
Interglacials, 10
Intermontane topography, 181
Internal migrations, 58, 167
International Date Line, 50
International Geophysical Year, 533
Interurban geography, 174
Intervening opportunity, 75, 124
Intraurban geography, 174
Intraurban growth, stages of, 197
Involuntary migrations, 58
Ionian Sea, 69

Iowa, 3, 26, 184
Iquitos, 298, 302
Iran, 43, 86, 325, 332, 339–341, 344, 355, 362, 366–367, 368
Iraq, 338, 339, 352, 358–359
Ireland, 75, 80
 Northern, 75, 80, 86–87, 88
 Republic of, 80, 87–88
Irian Jaya, 520
Irkutsk, 162, 163
Iron Curtain, 75
Irrawaddy Delta, 515, 521
Irrawaddy River, 507
Irredentism, 112
 in Italy, 112
 in Northern Africa, 370
 in Pakistan, 453, 454
Irrigation:
 basin, 345, 346
 center-pivot, 183
 perennial, 345, 346
Irtysh River, 147, 170
Isfahan, 366
Iskenderun, 365
Islam, 7, 44, 45, 86, 326, 328, 329, 331–332, 333, 334, 338, 344, 351, 357, 364, 366, 393, 430–431, 433–434, 436, 449–450, 509
Islamabad, 295, 453, 454
Islamic realm, *see* North Africa and Southwest Asia
Islamic Republic of Pakistan, *see* Pakistan
Isohyet, 183, 392
*Isolated State*, 71, 72–74, 368
Isolation, South America and, 288–289, 291–292
Israel, 32, 34, 325, 326, 344, 345, 348, 352–356
Istanbul, 338, 364
Isthmus, 249
Itaipu Dam, 312, 316, 317
*Italia Irrendenta*, 112
Italy, 26–27, 68, 75–76, 88, 104, 105, 106, 108–109, 112, 338, 521–522
Ivanovo, 160
Ivory Coast, 399, 400, 406

Jabal Mountains, 331
Jaffa, 356
Jamaica, 248, 260, 262, 263, 264, 265, 266, 274, 290
Jammu, 453
Jamshedpur, 440, 441
Janelle, D.G., 180
Japan, 15, 34, 42, 130, 131, 228–241, 294, 469, 472, 497
Jari project, 317
Java Sea, 530
Jawa (Java), 30, 510, 517, 521
Jefferson, M., 78, 193, 293
Jefferson, T., 191, 202
Jerusalem, 338, 354, 356, 358
Jet streams, 140
Jezira, 357
Jiangsu, 482
Jiangxi (Kiangsi), 472
Jidda, 361
Jilin (Kirin), 469
João, Dom, 289
Johannesburg, 417, 419
Johor Strait, 522
Jones, S.B., 412
Jordan, 344, 352, 353, 354, 356, 358
Jordan, D.K., 6
Jordan River, 338, 354, 358
Jordan Valley, 358
Jubail, 361
Judaism, 326, 328, 329, 338
Julian March, 112
Junggar (Dzungaria), 478–479
Jura Mountains, 92
Jutland Peninsula, 103–104

Kabul, 368, 454
Kachins, 515
Kaifeng, 466
Kalahari Basin, 382
Kalahari Desert, 13, 386, 388, 419
Kalgoorlie, 130
Kalimantan, 517
Kambalda, 130
Kamchatka Peninsula, 147, 164
Kamerun, 399. *See also* Cameroon
Kampala, 411
Kampuchea (Cambodia) 132, 511, 520–521, 524, 525
Kanarese (Kannada), 429
Kano, 393
Kansas, 209
Kansas City, 30, 219
Kanto Plain, 235
Karachi, 295, 450, 451, 452–453
Karaganda-Tselinograd region, 162–163
Karakoram, 424
Karamay, 478
Karens, 515
Karl-Marx-Stadt, 93
Karst, 427
Kashmir, 423, 424, 433, 453–454
Katanga, 400
Kathiawar Peninsula, 424
Kattegat Straits, 100
Kazakhs, 479
Kazakhstan, 163, 170
Kazan, 161
Kentucky, 209
Kenya, 370, 387, 388, 389, 391, 392, 396, 399, 401, 408–409, 410–411, 412
Kenyatta, J., 411
Kerch, 161
Khabarovsk, 164, 169
Kharkov, 161
Khartoum, 369
Khmers, 514, 520
Khoisan language family, 378
Khomeini, A., 339, 366
Khyber Pass, 454
Kibbutz, 355
Kiel, 90
Kiev, 158, 161
Kikuyu, 408
Kilimanjaro, 380
Kimberley, 417
Kim Il Sung, 501
Kingdom of Tonga, 538
Kingston, 264
Kinki District, 237
Kinki Plain, 235
Kinshasa, 400
Kirghiz Steppe, 147, 166, 459, 462, 479
Kirkuk reserve, 366
Kiruna, 101
Kitakyushu, 237
Kluckhohn, C., 6
Kobe, 235, 237
Kodiak Island, 145
*Kolkhoz*, 156, 157
Kompong Thom, 520
Komsomolsk, 164, 169
Kongo, 393
Köppen, W., 14, 16, 18–19, 21, 140
Köppen classification, 140
Korea, 231, 469, 499–501
Korean War, 489, 499–500
Kosovo, 115
Kowloon, 488, 489
Krasnoyarsk, 163
Krivoy Rog, 161
Kroeber, A.L., 6
Kuala Belait, 515
Kuala Lumpur, 515–516, 521
Kublai Khan, 466

Kunlun Shan Mountains, 478
Kurdistan, 364–365
Kurds, 86, 359
Kure, 237
Kurile Islands, 241, 496
Kursk Magnetic Anomaly, 161
Kuwait, 339, 344
Kuybyshev, 161
*Kuzbas*, 163, 164
Kuznetsk Basin, 163
Kwanchowan, 469
Kwashiorkor, 387
Kyoto, 235, 237, 294
Kyushu, 228, 233, 235, 237, 238

Labor migration, 400
Ladakh, 453
Lafaiete, 315
Lagos, 406
La Guaira, 296
Lahore, 451–452
Lake Baykal, 147, 148
Lake Chad, 385, 401
Lake Congo, 385
Lake District, 85
Lake Djouf, 385
Lake Ilmen, 143
Lake Kalahari, 385
Lake Malawi, 380, 381, 408–410
Lake Managua, 273
Lake Maracaibo, 297–298
Lake Nasser, 345–346
Lake Nicaragua, 273
Lake Tanganyika, 380, 408
Lake Titicaca, 303
Lake Turkana, 380
Lake Victoria, 380, 387, 396, 405, 408, 410, 411
Lancashire, 85
Land bridge, 248
Landform, 426, 427
Landlocked location, 96
Landscape, 426
  coastal, 427
  evolution, 250
Land/water heating differential, 140
Language, 304, 375, 377
Laos, 295, 506, 510, 515, 522, 523, 525
La Paz, 295, 303
Larimore, A.E., 6
La Rochelle, 89
Laterites, *see* Oxisols
*Latifundias*, 107
Latin America, 4, 31
  cities of, 270–271, 295–296
  *see also* Central America; Middle America, South America
Latitude, 50, 140
Latium, 108
Latosols, *see* Oxisols
Latvia, 111, 151
Lava, 427
*Laws of the Indies*, 251
Leaching, of soil, 14, 21
League of Arab States, 343
League of Nations, 398, 399
Lebanon, 32, 338, 344, 352, 354, 356–357
Leeds, 82, 85
Legend, of maps, 50
Le Havre, 89
Leinbach, T., 517
Leipzig, 93
Lelystad, 79
Leman Bank, 84
Lenin, V.I., 144, 149–150
Leningrad, 139, 144, 160, 169
Leopoldville (Kinshasa), 400
Lesotho, 96, 415, 417, 419, 525
Lesseps, F. de, 274
Lesser Antilles, 260, 262, 263, 266
Lewis, P., 200
Lhasa, 478
Liaodong (Liaotung), 469
Liaodong (Liaotung) Gulf, 477, 487
Liaoning, 469
Liao River, 474, 487
Liao-Songhua Lowland, 477
Liberia, 398, 404
Libya, 295, 339, 341, 344, 350, 352, 369, 400, 401
Liechtenstein, 32, 118
Liège, 94
Lille, 90, 91
Lima, 279, 286, 295, 300, 302
Limburg, 95
Linear political boundary, 32–33
Linear scale, 4, 5
Line symbols, on maps, 51
*Lingua franca*, 65
Linton, R., 6
Lisbon, 75, 106
Lithgow, 130
Lithosphere, 9
Lithuania, 111, 151
Liverpool, 82, 85
*Llanos*, 298
Location:
  industrial, 234
  of region, 2
  relative, 234
  theory, 192
Locusts, 388
Lódz, 113
Loess, 183, 427–428, 474
Loess Plateau, 474, 484
Loire River, 61, 90
Loire Valley, 92
Lombardy, 108, 109
London, 69, 75, 78, 79, 82, 83–84, 85–86
Longitude, 50
Long March, 472
Long rains, 15
Lorraine, 90
Los Angeles, 30, 41, 185, 200, 202, 223
Louis XIV (France), 67
Louisiana, 184, 189, 214–215
Low Countries, *see* Benelux
Low-island cultures, 535
Lows, climate and, 140
Lualaba River, 385
Lübeck, 90
Lubumbashi, 400
Lüda (Dairen), 477
Ludwig, D., 317–318
Lugdunum (Lyons), 64
Luxembourg, 77, 88
Luzon, 521, 526, 528
Lydolph, P.E., 38
Lyons, 64, 90, 91–92, 109

Ma, L., 461
Maas (Meuse) River, 94, 95
Macao, 469
McColl, R., 523
McCullough, D., 274
McCurdy, A.W., 6
Macedonia, 112, 115, 118
McGee, T.G., 517–518
McGlashan, N., 376–377
Machu Picchu, 285
Mackinder, H.J., 165–166, 508
Madagascar, 399, 401
Madras, 432, 436, 437, 440, 442–443
Madrid, 106, 107, 108
Madurese, 518
Magdalena Valley, 299
Maghreb, 43, 344, 350–352
Magma, 427
Magyars, 115, 118. *See also* Hungary
Maharashtra, 436, 440, 445
Maine, 217

Mainland, Middle America, 290
Makeyevka, 161
Malaria, 377, 387
Malawi, 376, 414, 416, 521
Malaya, 510, 511, 515
Malayalam, 429
Malay Peninsula, 510
Malaysia, 510, 511, 512–513, 515, 521, 530
Mali, 344, 393, 399, 405–406
Malmö, 101
Malthus, T., 445
Malvinas, *see* Falkland Islands
Managua, 272
Manáos, 317
Manchester, 81, 82, 85
Manchu Dynasty, 467–468, 469, 470, 487, 497
Manchukuo, 469, 472
Manchuria, 145, 164, 231, 237, 469. *See also* Northeast China
Mandalay, 521
Mandate territories, 399
Manhattan, 217
Manila, 521, 528
Manipur, 445
Manitoba, 4, 5, 179
Mantle, 9
Manufactured metals per person, economic development measured by, 37
Manufacturing Belt, *see* American Manufacturing Belt
Manufacturing sector, of economy, 207
Maoris, 134, 136
Mao Zedong (Mao Tse-tung), 464, 468, 471, 472, 480, 484, 492, 493, 495
Map projections, 50
Maps, 49–51
Maracaibo, 298
Maracaibo Lowlands, 297, 298
Marasmus, 387
Mariana Islands, 535
Marie Byrd Land, 522
Maritime boundaries, 528–531
Maritime environment, 534–538
Maritime Provinces, 189, 217–218
Maritsa River, 117
Markets, periodic, 407–408
Marne River, 90
Marquesas Islands, 538
Marseilles, 88, 89, 90
Marshall Islands, 535
Marshall Plan, 75, 77, 118, 119
Martinique, 260, 262, 265, 266
Marx, K., 149
Maryland, 189, 217
Masai, 389, 390, 391
Mashona, 416, 417
Massachusetts, 217
Massif Central, 92
Matabele, 416
Matadi, 400
Mato Grosso, 317
Mato Grosso do Sul, 317
Mauna Kea, 536
Mauritania, 32, 325, 344, 401, 414, 405–406
Mauryans, 430
Mayan civilization, 252, 254, 271
Mayer, H., 176
May Fourth Movement, 471
*Mebyon Kernow*, 86
Mecca, 331, 332, 344, 356, 361, 393
Medellín, 296, 299
Medical geography, 185, 186, 376–377
Medina, 331, 332, 361
Mediterranean Basin, 106
Mediterranean climate, 386
Mediterranean Europe, 4, 15, 98, 104–109
Mediterranean Sea, 55, 59, 62, 323
Megalopolis, 30, 83, 194–195, 196, 217
Meghna River, 449
Meiji Restoration, 231, 234
Meinig, D., 251
Mekong Delta, 30, 515, 522, 523
Mekong River, 507, 522
Melanesia, 45, 534
Melbourne, 125, 127, 130, 131, 133, 134
Melting pot, 179, 195, 201
    Brazil as, 319
Memphis, 221
Mendoza, 306
Mental maps, 203–206
Mercantilism, 66
Meseta Plateau, 60, 106
Mesoamerica, 252. *See also* Middle America
Mesopotamia, 327, 329
Mesozoic era, 9
Mestizos, 266, 267, 268, 291
Metamorphic rock, 427
Metropolis:
    European, 79
    galactic, 198, 200
    multicentered, 199
Metropolitan complexes, 31–32
Metropolitan evolution, stages of, 196
Meuse (Maas) River, 90
Mexican Plateau, 249
Mexico, 247, 248, 252, 254–255, 257, 258, 260, 262, 267–271, 272, 444
Mexico City, 267, 268, 269, 272
Miami, 185, 198, 220, 221
Michigan, 205
Micronesia, 45
Microstates, 32
Middle America, 2, 180, 246–277
    Caribbean America, 247–248, 250–251, 259–260, 292–266
    Central America, 4, 249, 253, 271–275
    as culture realm, 4
    definition, 253
    as geographic realm, 39, 42, 48
    Mexico, 247, 248, 252, 254–255, 257, 258, 260, 262, 267–271, 272
    Spanish colonial town in, 251–252
Middle East, 22, 32, 209, 240, 326, 327, 332, 344, 352–359
    boundaries, 333, 338–339
    culture hearth, 327–329
    oil in, 352
    supranationalism in, 342
    *see also* North Africa and Southwest Asia
Middlesbrough, 85
Midlands, 69
Mid-latitude climates, *see* Humid temperate climates
Migration, 58
    to Australia, 123, 124–125
    forced, 124–125, 409
    intercontinental, 167
    internal, 167
    involuntary, 524
    labor, 400
    refugees and, 525
    regional, 167
    to United States, 186–188
    voluntary, 124
Mikesell, M., 324
Milan, 108, 109
Miller, C.S., 2
Milwaukee, 194, 219
Minarets, 7
Minas Gerais, 315, 316, 317, 319
Mindanao, 521, 524, 528
Mindoro, 521
Ming Dynasty, 467, 496
Minicities, 199
Minneapolis, 30, 219

Minnesota, 223
Mississippi, 214–215
Mississippi Delta, 3
Mississippi River, 184
Mittelland Canal, 90
Mobility, 188
Moçambique, 382, 387, 396, 398, 400–401, 410, 414, 415–416, 524, 525
Models, 20–21
Modernization:
- China and, 461, 465, 494–496
- development and, 460–462
- Iran and, 366, 367
- Japan and, 231–233

Mohammed (Muhammad), 326, 331–332, 333. *See also* Islam
Moi, D. arap, 411
Moldavia, 151
Mollisols, 21, 24–25
Moluccas, 521
Mombasa, 410
Monaco, 32
Monclova, 269
Mongolia, 164, 168, 472
- Inner, 479, 480

Monsoon, 15, 141
Montana, 223, 225
Montaña, 301
Mont Blanc, 60
Montego Bay, 265
Montenegro, 115
Montevideo, 294, 295, 307
Montreal, 4, 5, 179, 218
Moors, 66, 351
Moravia, 113, 114
Moravian Gate, 113
Morocco, 32, 86, 323, 332, 338, 344, 350, 351, 352, 401, 405
Morphology:
- boundary, 338–339
- of a city, 249

Morrill, R., 46–47, 402, 403, 460
Moscow, 142–143, 144, 149, 155, 158–160
Moscow Basin, 147
Moslem, *see* Islam
Mountain barriers, climates and, 140
Mountain ranges, 9
Mount Ainslie, 133
Mt. Arenal, 272
Mount Everest, 427
Mount Kilimanjaro, 410
Mt. St. Helens, 9, 12, 427
Mughal Empire, 431
Mulatto, 266
Muller, P., 211
Multan, 450
Multicentered metropolis, 199
Multiple nuclei structure, 174–175
Murray River, 129
Murray River Basin, 127, 129
Murrumbidgee Valley, 129
Mysore, 441

Naga, 436
Nagaland, 436, 445
Nagasaki, 237-238
Nagoya, 235
Nagoya Plain, 237
Nagpur, 443
Nairobi, 408, 410, 411
Namib Desert, 388
Namibia, 414, 525
Nanjing (Nanking), 472, 474, 476
Nan Shan Mountains, 478
Nantes, 89
Naples, 108
Napoleon, 69-70, 88, 91, 463
Nasser, G., 347
Natal, 417, 419, 420
Nation, definition, 68
Nationalism, 352
- supranationalism, 342

Nationalist China, *see* Taiwan
National product per person, economic development measured by, 37
Nation-states, 66, 68, 74, 75, 508
Nazca Plate, 9
Ndebele, 416–417
Nebraska, 211
Needleleaf forests, 21
Negev Desert, 353, 355
Negritos, 429
*Négritude*, 399
Nehru, J., 436, 437
Neocolonialism, 36
Neo-Hawaiians, 45, 537
Neo-Malthusians, 445
Nepal, 168, 423, 425
Netherlands, 26, 56, 61, 68, 77, 78, 88, 94–95, 96, 295
- in Indonesia, 516–520

Netherlands Antilles, 260
Nevada, 181, 206, 223
Newark, 252
New Brunswick, 217, 218, 225
New Caledonia, 534
Newcastle, 69, 81, 85, 130
New Delhi, 433, 443
New England, 189, 194, 202, 205, 214, 217–218, 224
Newfoundland, 217
New Granada, 286, 289, 295
New Guinea, 45, 132, 534
New Hampshire, 205, 217, 218
New Hebrides, 534
New Jersey, 189, 205–206
New Mexico, 221
New Orleans, 184, 220, 221
New South Wales, 125, 126, 127, 129, 130, 132, 133. *See also* Sydney
New Territories, 488, 489
New Towns Movement, 79
New York City, 2, 194, 204, 205, 217
New York State, 189
New Zealand, 14, 34, 134–136, 536
Nicaragua, 253, 258, 261, 271, 272–273, 274, 276, 277
Niger, 344, 401, 405–406, 412
Niger-Congo language family, 377, 378
Niger Delta, 407
Niger River, 380, 385, 393
Nigeria, 30, 32, 86, 326, 347, 369, 375, 387, 392, 393, 394, 399, 401, 403, 404, 405, 406, 437
Niger River Valley, 393, 406
Nikopol, 161
Nile Basin, 343–344
Nile Delta, 327, 329, 348, 349
Nile River, 22, 325, 345–347, 380, 381–382
Nile Valley, 2, 30, 327, 328, 405
Nine Nations hypothesis, 177, 225
Nobi Plain, 237
Nomadism, 7, 351, 358, 361
Nordic Europe, *see* Northern Europe
Norilsk, 147
Normans, 66, 80
North Africa/Southwest Asia realm, 34, 43, 48, 323–372
- boundaries, 333, 338–339
- Christianity in, 333, 338
- Islam in, 331–333
- oil in, 339–343, 344
- past civilizations in, 327
- regions and states of, 343–370
- supranationalism in, 342

North America, 4, 5, 13, 15, 173–227
- cultural geography, 201–206
- as culture realm, 4
- economic activity, 206–214
- as geographic realm, 40, 41, 48
- physical geography, 180–185
- population, 24–25, 31, 185–201

nine nations of, 224–225
regions, 215–225
*see also* Canada; Middle America; United States of America
Northampton, 82
North Atlantic Ocean, 100
North Atlantic Treaty Organization (NATO), 96, 120, 362
North Carolina, 205, 212, 214, 220, 221
North China Plain, 474, 482, 484
Northeast China (Manchuria), 469, 470, 477, 487–489, 490
Northern Europe, 97–104
Northern Ireland, 75, 80, 86–87, 88
Northern Rhodesia, 414. *See also* Zambia
North European Lowland, 60–62, 69
North Indian (Gangetic) Plain, 424, 425
North Island, 134, 135, 136
North Korea, 499–500, 500–501
North Pole, 50
North Sea, 55, 59, 82, 84, 85, 86, 88, 94, 100, 103
North Vietnam, 32, 523, 524
Northwest Territory, 189
North Yemen, 344
Norway, 82, 98, 100, 102, 103, 104
Nottingham, 81, 82, 84
Nova Carthago, 65
Nova Scotia, 189, 217
Novgorod, 143
Novokuznetsk, 163
Novosibirsk, 147, 163
Nyasaland, 396, 414. *See also* Malawi
Nyerere, J., 410

Oahu, 538
Obote, M., 412
Ob River, 147, 163
Occupational structure of the labor force, economic development measured by, 37
Oceans, 12
currents, 140
Ocho Rios, 265
Oder-Neisse system, 90
Oder River, 62
Odessa, 117, 144, 161
Ogaden, 370
Ohio, 205
Oil, 339–343
in Middle East, 352
in North Africa and Southwest Asia, 339
in North Sea, 55, 59, 82, 84, 100, 103
Organization of Petroleum Exporting Countries and, 38, 303, 339, 342
in Romania, 117
Oise River, 90
Oka River, 158
Oklahoma, 209, 214
Okovango Delta, 385
Okovango Swamp, 381
Omaha, 219
Oman, 360
Omsk, 147
Ontario, 179, 189, 194
Opium War, 469
Ordos Desert, 474
Oregon, 15, 189, 224
Organic theory, of state evolution, 74
Organization for Economic Cooperation and Development (OECD), 120
Organization for European Economic Cooperation (OEEC), 77, 118, 119
Organization of Petroleum Exporting Countries (OPEC), 38, 303, 339, 342
Orientation, of maps, 49–51
Orinoco Basin, 298
Orinoco River, 299
Orissa, 439
Orlando, 221
Orographic (mountain) rainfall, 182
Orontes River, 357
Oruro, 296
Osaka, 235, 237
Oslo, 98, 100, 103
Ostrava, 113
*Ostrogs*, 143, 159
Otago, 135
Otanmakiin, 102
Ottoman Empire, 66, 115, 338, 351, 353
Outokumpu, 102
Oxisols, 21, 24–25

Pacific Coast Mountain System, 181
Pacific Ocean, 45, 534
Pacific Plate, 9
Pacific regions, 534–539
"Pacific Ring of Fire," 9
Pahlavi, Shah Reza, 362, 366
Pakhtunistan, 434, 454
Pakhtuns, 368, 453, 454
Pakistan, 22, 26, 44, 86, 295, 326, 423, 424–425, 432–434, 444, 446, 449–454, 509, 524
Paleozoic era, 9
Palestine, 338, 353, 354–355, 358
Palestinian Liberation Organization (PLO), 354, 355
Pamirs, 462
Pampa, 284, 290, 304–305, 306
Panama, 248, 253, 258, 261, 271, 272, 273, 274–275, 277
Panama Canal, 249, 263, 273, 274–275
Panama City, 272
Pandemic, 387
Pangaea, 9
Pannell, C., 461
Papua New Guinea (P.N.G.), 534–535
Papuans, 534, 535
Pará, 318
Paraguay, 289, 291, 297, 300, 303–304, 306
Paraiba, 314
*Paramós*, 269
Paraná, 316, 319
Paraná River, 317
Paris, 69, 71, 75, 78, 79, 88, 90–91
Paris Basin, 90, 92
Pastoralism, 87
Patagonia, 304, 306
Paternalism, 398, 400
Paterson, J.H., 2
Pathanistan, 86
Pattern, population and, 56
*Paulistas*, 287
Pearl Harbor, 231
Pecs, 115
Pedologist, 19
Pedro, Dom, 289
Peking, *see* Beijing
Peking Man, 484
Peloponnesus Peninsula, 106
Pemba, 408
Pennine Mountains, 80, 81, 82, 85
Pennsylvania, 189
*Peones*, 268
People's Republic of China, *see* China
Perception of the past, 250
Perennial irrigation, 345, 346
Perforated state, 525
Periodic markets, 407–408
Perm, 161
Pernambuco, 289, 314
Perry, Commodore, 231
Persepolis, 366
Persia, 338, 366
Persian Gulf, 361, 366, 368
Perth, 125, 126, 127, 134

Peru, 279, 280, 285, 286, 289, 290, 291, 295, 296, 300–302, 307
Peters, K., 396
Petersburg, 144
Peter the Great, 144
Petrograd, 149, 160
Philadelphia, 194, 205, 216
Philbrick, A.K., 235
Philippines, 45, 86, 231, 264, 438, 505, 512, 521, 524, 525–528
Philippines Plate, 9
Phillips, P., 198
Phnom Penh, 521
Phoenicians, 345, 347, 351, 356
Phoenix, 198, 216
Physical geography, 46, 80
Physiographic boundaries, 338–339
Physiographic provinces, 180, 426
Physiography, 426
Physiologic density, 56, 238, 438
Pilsen, 114
*Pinyin* system, 464–465
Piraeus, 107
Pittsburgh, 194, 216
Pivot area, 165
Pizarro, F., 286
Place-name geography, 202
Placering, gold mining and, 258
Plantation, 33–34, 260–261, 274, 283, 290, 510
Plateau continent, 382
Plateau of Iran, 366
Plateau of Xizang, 462
Plate tectonics, 9, 10–12, 384, 426–427
Plato, 63
Pleistocene epoch, 9–11, 16, 24, 62, 427
Ploesti, 117
Plural society, 178–179
  Caribbean America as, 266
  South America as, 290
*Podzol*, 21
Point symbols, on maps, 50–51
Poland, 26, 62, 68, 110, 111, 112, 113, 118, 143
Polar Easterlies, 140
Political development cycles, 361–362
Political geography, 32–33, 34–35, 47, 74, 508–510
Political revolution, 69
Pollution, North America and, 185, 186
Pol Pot, 524
Polynesia, 45, 536–539
Popular culture, 324
Population:
  in cities, 31
  composition, 56, 58
  Demographic Transition Model, 446
  distribution, 56
  doubling time of, 444
  growth, 56–57, 444, 446–447
  indigenous, 131
  Malthus on, 445
  mobility, 188
  movements, 58
  space and, 22–23
  underdeveloped countries, 36
  world, 10, 22–30, 56–58, 444, 446–447
  *see also* Migration
Population geography, 56–58
Po River, 62, 106
Po Valley, 105, 108
Port Antonio, 265
Port-au-Prince, 264, 265
Porter, P., 461
Port Moresby, 535
Porto, 106
Porto Alegre, 317
Portugal, 104, 106, 107, 287, 288, 289, 332, 394, 398, 400–401
Portuguese Guinea, 396
Possibilism, 325
Postindustrial economy, 198
Postindustrial revolution, 173, 213–215, 240
Potala, 478
Potosí, 296, 303
*Povolzhye*, 160–161
Prague, 114
Precambrian era, 9
Precipitation, 15
  acid rain, 101–102
  global patterns, 12
  rain shadow effect, 182–183
Pressure cells, 140
Pretoria, 295
Prevailing Westerlies, 140
Primary activity, of economy, 207
Primate city, 78, 193, 194
  Buenos Aires as, 305
  Cairo as, 348
  Prague as, 114
  Seoul as, 501
  in South America, 293
  Warsaw as, 113
Prime Meridian, 50
Prince Edward Island, 217
Productivity per worker, economic development measured by, 37
Protestants, 86–87
Proxemics, 6
Prudhoe Bay, 145
Puerto Plata, 265
Puerto Rico, 260, 262, 263, 264, 266
Pull factors:
  migration and, 124
  urbanization and, 292
Punjab, 424, 425, 429, 433, 436, 438, 440, 443, 445, 451
Pusan, 501
Push factors:
  migration and, 124
  urbanization and, 292
Puteoli, 65
Pygmies, 7, 388
Pyle, G., 338
Pyongyang, 501
Pyrenees, 60, 104, 106, 108

Qaddafi, M., 352
Qaidam (Tsaidam) Basin, 478
Qanat system, 366
Qatar, 32, 360
Qingdao (Tsingtao), 469, 470
Qinghai, 478
Quaternary activity, of economy, 207
Quebec, 3, 4, 5, 41, 86, 179, 189, 217, 218–219, 224, 225
Quechua, 285, 300
Queensland, 127, 129, 130, 132
Quiché, 271
Quinary activity, of economy, 207
Quito, 284, 294, 295, 296, 303

*Ra*, 250
Race, 378, 380–381
Rainfall, *see* Precipitation
Rainforest, 14, 15, 16
Rain shadow effect, 140, 182–183
Raleigh, 198, 214
Randstad, 94
Randstad-Holland, 94
Rangoon, 521
Rarotonga, 538
Ras al-Khaimah, 339
Rates, economic development measured by, 37
Ratzel, F., 74–75, 362
Ravenstein, E.G., 124
Rawalpindi, 453
Recent epoch, 9, 10

Recife, 289, 314–315
Red Basin, 475–476, 480
Red River, 507, 522
Red Sea, 361
Refugee flows, in Indochina, 524–525. *See also* Migration
Regional concepts, 1–4, 5
Regional geography, 48
Regionalism, 45–47
Regional migrations, 167
Relative location, 2
  Japan and, 234
Relict boundaries, 510
Religions, 324
  of the world, 326, 329
  *see also specific religions*
Relocation diffusion, 334
Remoteness, Canberra having, 133–134
Remote sensing, 46
Renaissance, 66
Representative fraction, on scale, 4, 5
Republic of China, *see* Taiwan
Republic of Ireland, 80, 87–88
Republic of South Africa, *see* South Africa
Revolutionary War, 189
Reykjavik, 103
Rhine-Maas (Meuse) river system, 62
Rhine River, 89, 90, 94, 95
Rhode Island, 217
Rhodes, C., 396
Rhodesia, 396, 525. *See also* Zimbabwe
Rhône River, 90
Rhône-Saône system, 62
Rhône-Saône Valley, 88, 92
Rift valleys, 380, 386, 427
Rimland, 166
  Middle America, 290
Rinderpest, 388
Ring-city, 94
Rio Colorado, 304
Rio de Janeiro, 289, 292, 294, 296, 312, 313, 314, 315, 316, 319, 320
Rio de la Plata, 286, 305
Rio Grande do Norte, 314
Rio Grande do Sul, 316, 317
Rio Muni, 396
River blindness, 387
Riyadh, 361
Robinson, J.L., 2
Rock formations, 427
Rockhampton, 130
Rocky Mountains, 2, 3, 180, 181, 209, 223, 427
Romania, 63, 110, 112, 113, 115, 116–117, 118, 338
Rome (ancient), 7, 8, 32–33, 36, 46, 63–66, 80, 108, 330, 351, 392
Rostov, 161, 162
Rostow, W., 460–461
Rotation system, of farming, 127–128
Rotterdam, 89–90, 94, 95, 96
Rouen, 89
Rub al Khali, 361
Ruhr, 89, 90, 92
Russia, 21, 40–41, 145, 151. *See also* Soviet Union
Russification, 153
Russo-Japanese War of 1905, 145, 149
Rwanda, 408
Ryukyu Islands, 469

Saar, 93
Saba, 262
Sabah, 515, 521, 530
Sabinas, 269
Sadat, A., 350
Sahara Desert, 13, 14, 15, 325, 332, 345, 352, 369, 380, 382, 387, 401, 406, 508
Saharan Atlas Mountains, 351
Sahel, 4, 14, 15, 369, 376, 387, 410
Saigon, 252, 515, 522, 523, 524. *See also* Ho Chi Minh City
St. Eustatius, 262
St. Kitts, 263
St. Lawrence River, 4, 5, 179, 184, 218
St. Lawrence Seaway, 184
St. Louis, 30, 185, 194, 219, 224
St. Lucia, 262, 263
St. Maarten, 262
St. Nazaire, 89
St. Paul, 30
St. Petersburg, 144, 145, 149, 160
St. Vincent, 262
Sakhalin Island, 145, 164, 166, 228
Salford, 81
Salisbury, *see* Harare
Salvador, 314–315
Samchok, 501
San Andreas fault, 427
San Antonio, 198, 222
San Cristóbal, 297
San Diego, 198, 223
Sandinista revolution, 274, 276, 277
San Feliz de Guayana, 299
San Francisco, 30, 223
San Francisco Bay Area, 2, 3, 198, 199, 213
San Joaquin Valley, 223
San Jorge Gulf, 306
San José, 223, 272, 273
San José de Saramuro, 302
San Juan, 264
San Martin, J. de, 289
San Rafael, 306
San Salvador, 272
Sanskrit, 428
Santa Catarina, 316, 317
Santa Clara Valley, 198
Santa Cruz, 303
Santiago, 294, 295, 300, 305, 307
Santo Domingo, 264
Santos, 296
São Paulo, 287, 292, 293, 295, 296, 314, 316, 317, 319
Sarawak, 515, 530
Sardinia, 108
Sargent, C., 251
Saskatchewan, 179
Saudi Arabia, 339, 341, 342, 344, 356, 359, 360–361
Sauer, C.O., 7, 8, 325, 334
Savannas, 15, 17–18, 21, 24–25, 393
Saxons, 80
Saxony, 92, 93
Scale, 4, 5
Scandinavia, 88, 98. *See also* Northern Europe
Schelde River, 95
Scheveningen, 94
Schistosomiasis, 377, 387
Schuman, R., 119
Scotland, 69, 79, 80, 82, 85–86, 87
Scrub-and-bush association, 16
Sea of Japan, 139
Seattle, 223, 224
Secondary activity, of economy, 207
Sedimentary rock, 427
Segovia, 64
Seine River, 61, 89, 90
Semiarid steppe, 15
Semipermanent highs and lows, 140
Senegal, 387, 399, 400, 401, 405, 406
Senegal River, 398
Senghor, L., 399
Sen River, 520
Seoul, 501
Separate development, South Africa and, 419–420
Sepoy Rebellion, 431
Sequent occupance, 7, 8, 252, 410
Serbia, 115, 515
Services sector, of economy, 207

Settlement geography, 251
Seventh Approximation, 19
Shaanxi, 482
Shamanist religions, 328, 329
Sha Mian Island, 471
Shandong (Shantung) Peninsula, 470, 474, 482
Shang Dynasty, 466
Shanghai, 25, 469, 471, 474, 476, 487, 480–482
Shan Plateau, 521
Shans, 515
Shanxi (Shansi), 484
Shari River, 385
Sharjah, 339
Shatt-al-Arab, 359, 366
Shatter belt, 114
Sheffield, 81, 84
Shenyang, 474, 488, 489
Shiah, 344
Shi'ite Moslems, 324, 366
Shikoku, 228
Shintoism, 230, 328, 329
Shiraz, 366
Shona, 416, 417
Short rains, 15
Siam, 525
Siberia, 141, 143, 147, 169, 170, 241
Sichuan (Szechwan) Province, 475, 480, 482, 492
Sicily, 105, 106, 108
Sickle cell anemia, 387
Sierra Leone, 404
Sierra Nevada, 182, 427
Sikhs, 433, 434, 436
Sikkim, 168, 423, 435
Silesia, 93, 113
Silicon Valley, 198, 213, 223
Sinai Peninsula, 345, 347, 353, 356
Sind, 424, 450, 451
Sindhi, 436
Singapore, 34, 499, 512, 515, 521
Singidunum (Belgrade), 64
Sinhalese, 454, 455, 456
Siret River, 116
Site, of city, 90
Situation, of city, 90
Six Day War, 345
Skagerrak-Kattegat Straits, 59
Skinner, G.W., 483
Slash-and-burn agriculture, 33
Slavery, Black Africa and, 394–395
Slavs, 143
Slovakia, 113, 114
Slovenia, 115, 116
Slow-Growth Epoch, 198
Smog, 185
Snow, J., 51, 376
Snow climates, *see* Humid cold climates
Sofia, 117
Soil, 14, 15, 18–19, 21–22, 24–25, 183
Solomon Islands, 534
Somalia, 338, 344, 370, 382
Sommers, L., 78–79
Somoza regime, 277, 534
Songhai, 393
Songhua River, 489
Songhua Valley, 487
Soudan, 399. *See also* Mali
South Africa, 15, 30, 34, 295, 375, 378, 380, 396, 401, 405, 410, 414–415, 416, 417–420
South America, 4, 21, 24–25, 30, 34, 247, 279–321, 386
  Africans in, 287–288
  civilizations in, 280–281, 284–288
  culture areas of, 290–292
  as geographic realm, 39, 42–43, 48
  Iberian invaders in, 285–287
  Incas in, 280–281, 284–285
  independence in, 298–290
  regions, 296–310
  urbanization in, 292–296
  *see also* Central America; Middle America
South Asia, 24–25, 26, 424–434. *See also* Bangladesh; India; Pakistan; Sri Lanka
South China Sea, 515, 522, 530
Southeast Asia, 30, 45, 48, 505–531
  city of, 517–518
  colonialism in, 513–520
  Indochina, 510–513
  population patterns in, 506–507, 510
  seas of, 528–531
  territorial morphology, 520–528
Southern Africa, 387, 388, 392, 395, 400, 401, 414–420
Southern Alps, 134, 136
Southern Ocean, 21, 532, 533
Southern realm, 532–533
Southern Rhodesia, 399, 414, 416. *See also* Zimbabwe
Southern Sudan, 369
South Island, 134, 135, 136
South Korea, 34, 499, 500, 501
South Pole, 50, 532
South Vietnam, 32, 523, 524
Southwest Asia, *See* North Africa and Southwest Asia
South Yemen, 344
Soviet Union, 21, 32, 63, 112, 120, 139–170, 240–241, 276, 368, 461
  centrally planned economy, 153–157
  eastern boundaries, 166–167
  Eastern Europe and, 63, 111, 112–113, 115, 116, 117–118
  as geographic realm, 40–41, 48
  heartlands, 164–166
  physiographic regions, 146–148
  political framework, 149–153
  population, 26–27, 31, 167–169
  regions, 157–164
  as world superpower, 142–145
*Sovkhoz*, 154, 157
Spain, 60, 68, 86, 104, 105, 106, 107–108, 262, 332
  colonial towns of, 251–252
  in Middle America, 251–252, 255–257
  in South America, 285–287, 289
Spanish Sahara, 32
Sparta, 63
Spatial diffusion, 252, 325, 332, 334–338
Spatial economy, 207–214, 282
Spatial extent, of region, 2
Spatial interaction, 75–77
Spatial models, 16
Spatial organization, 46
  of Afrian dependencies, 398–399
Spatial processes, 250
Spatial relationshps, maps and, 51
Spatial systems, regions as, 3
Spencer, J.E., 38, 324
Spodosols, 21, 24–25
Spradley, J.P., 6
Spykman, N.J., 165, 166
Squatters, 37
Sri Lanka, 26, 44, 423, 424, 454–456
Srinagar, 453
Stalin, J., 152
Stalingrad, 161
Stanley, H., 396
Starkey, O.P., 2
State:
  insurgent, 523–525
  nation-state, 66, 68, 74, 75, 508
  spatial aspects of, 520–523
Statfjord, 103

Steppe, 16, 17, 388, 393
Stockholm, 78, 101, 104
Stonehenge structures, 80
Strahler, A.H., 349
Strahler, A.N., 349
Strait of Malacca, 515
Strasbourg, 119
Streams, 427
Subfamily, of language, 377
Subregions, 4
Subsequent boundaries, 509
Subsistence agriculture, 283–284, 388, 389, 390
Suburbia, 196–197
Sucre, 295, 296, 303
Sudan, 32, 44, 325, 338, 344, 346, 348, 369, 370, 393, 399, 412, 518
Sudan Lake, 385
Sudan Nile, 369
Sudeten Mountains, 113
Suez Canal, 345, 347, 353–354, 362
Sui Dynasty, 466, 496
Sukarno, 528
Sulawesi, 521
Sulawesi Sea, 530
Sumatera, 517, 520, 521
Sumatra, 517. *See also* Sumatera
Sumbawa, 518
Sumer, 327
Sunbelt, 2, 186, 188, 194, 198, 199, 206, 212, 214
Sung Dynastyg, 466
Sunni Moslems, 344, 366
Sun Yat-sen, 471, 472, 494
Superimposed boundaries, 434, 509–510
Supranationalism, in Middle East, 342
Suriname, 290, 296, 298, 299
Swansea, 85
Swartz, M.J., 6
Swaziland, 383, 391, 410, 419, 525
Sweden, 30, 32, 56, 58, 98–99, 100–101, 102, 104
Switzerland, 32, 69, 79, 88, 96–97
Sydney, 125, 126–127, 130, 133, 134
Syria, 325, 338, 344, 352, 354, 356, 357–358
System, region as, 3
Systematic geography, 47, 48

Taaffe, E., 402, 403, 460
Table Bay, 395
Taboos, food, 326
Tadzhik, 166
Tagalog, 526
Tahiti, 534, 538–539
Taipei, 496, 498
Taiwan, 34, 231, 472, 489, 496–499
Taiyuan, 484
Tajiks, 368
Takeoff-stage development, 320
Takla Makan Desert, 478
Tamil language, 429, 436, 454
Tamil Nadu, 436
Tamils, 456
Tampere, 102
Tampico, 269
Tanganyika, *see* Tanzania
Tang Dynasty, 466
Tangshan, 465
Tan-Zam Railway, 410, 411, 414
Tanzania, 9, 387, 389, 391, 401, 408–410, 411, 412, 525
Tapiola, 79
Tāpi River, 425
Tarascans, 255
Tarim Basin, 478, 479
Tasman, A., 123
Tasmania, 125, 127, 130
Tatars, 143
Tatra Mountains, 113
TAZARA Railway, 410, 411, 414
Tectonic forces, 9, 10–12, 384, 426–427
Teeside, 103
Tegucigalpa, 272
Tehran, 339, 366, 367
Tel Aviv, 356
Tell, 323, 325, 352
Tell Atlas Mountains, 351
Telugu language, 429, 436
Temperature, 14, 15. *See also* Climate
Tenochtitlan, 255, 256
Teplice-Sanov, 114
Territoriality, 6, 508, 528–531
Territorial morphology, 508
Tertiary activity, of economy, 207
Tete, 414
Texas, 184, 191, 202, 209, 212, 214–215, 221, 222
Thailand, 505, 507, 510, 511, 513, 520, 521, 525
Thames Basin, 80
Thames River, 80
Thar Desert, 424
Third World, 34, 272. *See also* Underdeveloped countries
Thoman, R.S., 38
Thon Buri, 525
Thüringen, 68
Tianjin (Tientsin), 465, 474, 480, 482, 484
Tibesti Mountains, 382
Tibet, 96, 462. *See also* Xizang
Tien Shan Mountains, 462, 478
*Tierra caliente*, 269
Tierra del Fuego, 279, 306
*Tierra fria*, 269
*Tierra helada*, 269
*Tierra templada*, 269
Tigris-Euphrates Basin, 323, 327
Tigris River, 358
Tikhvin, 160
Timbuktu, 393
Time-geography perspective, 251
Time-space convergence, 180, 210
Timur (Timor), 32, 518, 519, 520
Tirane, 116
Tisdale, H., 31
Tisza River, 114
Tito, J.B., 86, 115, 116
Tobago, 262
Togo, 398, 399
Tokyo, 234, 235–237, 240, 294
Toledo, 332
Toltecs, 254
Tomsk, 163
Tonkin, 515, 523
Tonkin Plain, 507–508
Toponymy, 202
Tornadoes, 183
Toronto, 179, 218
Tourism, in Caribbean America, 265–266
Townscape, 7
Township-and-Range land-division system, 191
Tradewinds, 140
Trans-Alaska Pipeline, 209, 222
Transculturation, in Mexico, 268
Transferability, 75
Trans-Himalayan Mountains, 478
Transhumance, 97
Transition Zone, African, 368–370
Transjordan, 338, 353
Transkei, 419, 420
Transportation facilities per person, economic development measured by, 37
Transport development model, 402–403, 460
Trans-Siberian Railroad, 145, 147, 148, 163, 164
Transvaal, 419, 420
Transylvania, 111–112, 113, 116, 118

Transylvanian Mountains, 116
Treaty of Berlin, 117
Triangular trade, 49, 51
Trieste, 112
Trinidad, 262, 263, 266
Tripoli, 357
Trondheim, 98, 100
Tropic of Cancer, 15, 18
Tropic of Capricorn, 15, 18
Trotsky, L., 144
Trudeau, P.E., 180, 185
True desert, 15
Truk, 535
Trust territories, 399
Tsetse fly, African sleeping sickness and, 387, 390, 391
Tsukuba Science City, 240
Tsumeb, 414
Tuareg, 325
Tucamán, 306
Tula, 160, 254
Tumbes, 286
Tundra, 15, 18, 21, 22–23
Tunis, 352
Tunisia, 323, 338, 344, 350, 352
Turbarão, 317
Turin, 108, 109
Turkestan Desert, 478
Turkey, 13, 325, 338, 344, 347, 362, 363–366, 368
Turkic, 325
Turkmens, 368
Turku, 102
Tuvalu, 538
Tylor, E.B., 6
Tyrrhenian Sea, 59

Uganda, 393, 396, 401, 408, 409, 411–412
Uighurs, 479
Ukraine, 15, 113, 143, 151, 160, 161–162
Ulack, R., 517
Ulan Bator, 168
Ullman, E.L., 75, 174
Ulster, *see* Northern Ireland
Ultisols, 21
Umm al-Qaiwan, 339
Underdeveloped countries, 34–38, 40, 321, 460–461, 462, 528–529
 isolated states and, 368–369
Unified field theory, 412
Union of Soviet Socialist Republics (U.S.S.R.), *see* Soviet Union
Unitary state, 132
United Arab Emirates, 339, 344, 360
United Arab Republic, 357
United Kingdom, 30, 57–58, 68, 77, 80. *See also* England; Northern Ireland; Scotland; Wales
United Mexican States, *see* Mexico
United Nations, 32, 529
 China in, 464, 465
 Taiwan and, 499
United States of America, 4, 5, 15, 34, 41, 173, 183, 240, 259, 276, 499
 agriculture, 3, 7, 183–184, 210–211, 219, 220
 border, 508–509
 Canada compared with, 177–180
 cultural geography of, 201–206
 hydrography, 13, 184–185
 manufacturing in, 194, 195, 198, 201, 202, 207, 211–213, 215–217
 pollution, 102, 185, 186
 population, 30, 167, 168, 185–200
 postindustrial revolution in, 173, 213–215, 240
 regions, 215–224
 resources, 207–210
 soils of, 21, 24–25
 urbanization, 191–200, 206–207
 *see also* North America
Upper Ganges, 438
Upper Nile River, 385
Upper Volta, 399, 404
Ural Mountains, 63, 139, 146, 147, 162, 163, 169
Urayasu, 241
Urban economic geography, 175
Urban evolution, 174
Urban geography, 46, 174–176
Urban hierarchy, 193
Urbanization, 30–32
 Australia, 125–127, 133–134
 Europe, 71, 74, 78, 79, 90–92
 North America, 191–207
 South America, 292–296
 urban geography, 174–176
 *see also* City
Urban realms model, 199
Urban social geography, 175
Uruguay, 34, 284, 286, 287, 289, 290, 294, 297, 304, 307
Ürümqi, 478–479
Ussuri River, 164, 167
Utah, 181, 202, 206, 223
Uttar Pradesh, 436, 438, 439, 445
Uzbek, 162, 368

Vale of Kashmir, 453
Valencia, 106, 108, 297
Valley of Mexico, 255, 258
Vallingby, 79
Valparaiso, 296, 307
Vance, J., 174, 199
Vancouver, 190, 223, 224
Varanasi (Banaras), 429
Varangians, 143
Varna, 117
Vatican City, 108
Vegetation, 14, 16–18, 22–23
Venda, 419
Venezuela, 284, 286, 289, 291, 293, 295, 296, 297–299, 300
Venice, 108
Veracruz, 258, 268, 269
Verkhoyansk, 141
Vermont, 205, 217
Vertisols, 21
Vespucci, A., 287
Victoria:
 Australia, 127, 129, 130, 132, 133, 134
 Hong Kong, 489
Victoria Falls, 382
Vienna, 78, 96, 97
Viet Minh League, 523
Vietnam, 22, 132, 230, 507, 515, 520, 522–525
Vietnam War, 2, 30, 132, 437, 465, 522, 523, 525
Vikings, 80
Villahermosa, 269
Vindhyas, 425
Virginia, 189, 205, 214, 217, 220
Virgin Islands, 265
Visayan Island, 528
Visayan language, 526
Vistula River, 62, 113
Vladivostok, 139, 145, 164, 169
Vltava River, 114
Vojvodina, 115
Volcanism, 426–427
Volcanoes, 9, 12, 427
Volga River, 143, 158, 159, 160, 161, 170
Volgograd, 161
Volkhov River, 143
Vologda, 170
Volta Redonda, 316
Voluntary migrations, 58, 124
Von Thünen, J.H., 71, 72–74, 78, 210–

211, 282, 368
Vosges Mountains, 92

Wadden Sea, 94
Wagner, P., 324
Wakhan Corridor, 367
Wakool Valley, 129
Wales, 69, 80, 82, 85–86, 87
Wankie, 414
Warsaw, 78, 113
Washington, D.C., 79, 194, 205, 217, 220
Washington State, 15, 223, 224
Water, 11–14
North American hydrography and, 184–185
pollution, 185
Waterloo, 70
Watt, J., 69
Weathering, 426, 427
Weber, A., 71
Wegener, A., 9
Wei River, 327, 466, 484
Wellington, 136
Weser River, 62, 90
West, R.C., 261
West Africa, 4, 30, 375, 376, 385, 387, 388, 391, 392–393, 394–395, 396, 399, 401, 404–408
West Bengal, 439, 445
Westerlies, 140
Western Europe, 13, 14, 15, 62–66, 68, 71, 75–77, 88–97
Western European Union (WEU), 120
Western Sahara, 352
Western Samoa, 538
Western Uplands, 60, 61
West Germany, 26, 30,32, 78, 79, 82, 92–93, 94
West Irian, 520, 534
West Malaysia, 515
West Pakistan, 424, 437, 450. *See also* Pakistan
Wet monsoon, 141
Wheeler, J., 211
White Nile, 345, 369
Whittlesey, D., 7, 8
Willemstad, 265
Willett, 390
William the Conqueror, 80
Williams, J., 32
Wiltshire, 84
Wind, erosive power of, 427
Wind belts, 140
Windisch, 116
Winnipeg, 219
Wismar, 90
Woodward, C.V., 221
World of Islam, 326. *See also* North Africa and Southwest Asia
World Island, 165
World-lake concept, 531
Wuhan, 476, 480
Wuzhou (Wuchou), 486
Wyoming, 205, 209, 214, 222, 223, 225

Xhosa, 419
Xi Delta, 476
Xinjiang (Sinkiang), 166, 459, 472, 474, 478–479, 492
Xi (West) River, 466, 474, 476, 485, 486, 487
Xizang (Tibet), 474, 478, 479–480
Xizang (Tibet)-Yunnan Plateau, 474

Yablonec, 114
Yakutia, 169
Yanbu, 361
Yaroslavl, 160
Yawata, 237
Yeates, M., 95
Yellow fever, 387
Yellow Sea, 474
Yemen, 360–361
Yenisey River, 163
Yichang, 476
Yin Dynasty, 466
Yokohama, 234, 235–237
Yonne River, 90
York, 64
York Peninsula, 130
Yorkshire, 85
Yoruba, 406
Young, J., 336
Yuan Dynasty, 466–467
Yucatan, 260
Yucatan Peninsula, 252
Yugoslavia, 68, 86, 110, 111, 112, 113–114, 115–116, 326, 338
Yunnan Plateau, 478

Zagros Mountains, 366
Zaïre, 14, 32, 375, 377, 383, 388, 400, 401, 412, 414. *See also* Congo
Zaïre River, *see* Congo (Zaïre) River
Zambezi River, 381, 382, 385
Zambia, 376, 381, 400, 401, 411, 414, 416
Zanzibar, 408, 410
Zeeland, 94, 95
Zelinsky, W., 204, 205
Zero Population Growth, 58
Zhanjiang, 469
Zhao Ziyang, 494–495
Zhou Enlai (Chou En-lai), 464, 472, 495
Zimbabwe, 392, 399, 401, 416–417
Zionism, 353
Zonguldak, 366
Zuider Zee, 94
Zululand, 419, 420
Zurich, 97